AAPG Treatise of Petroleum Geology
Reprint Series

The American Association of Petroleum Geologists
gratefully acknowledges and appreciates the leadership and support
of the AAPG Foundation in the development of the
Treatise of Petroleum Geology.

RESERVOIRS II

SANDSTONES

COMPILED BY
EDWARD A. BEAUMONT
AND
NORMAN H. FOSTER

TREATISE OF PETROLEUM GEOLOGY
REPRINT SERIES, NO. 4

PUBLISHED BY
THE AMERICAN ASSOCIATION OF PETROLEUM GEOLOGISTS
TULSA, OKLAHOMA 74101, U.S.A.

Library of Congress Cataloging-in-Publication Data

Reservoirs.

(Treatise of petroleum geology reprint series; no. 4)
Bibliography: v. 2,p.
Contents: 1. Properties. 2. Sandstones
1. Petroleum—Geology. 2. Gas, Natural—Geology.
I. Foster, Norman H. II. Beaumont, E.A. (Edward A.) III.
American Association of Petroleum Geologists.
TN870.5.R43 1987 553.2'82 87-19599
ISBN 0-89181-403-5 (v.2)

TABLE OF CONTENTS

RESERVOIRS II
SANDSTONES

SANDSTONES

SANDSTONE DEPOSITIONAL ENVIRONMENTS

SANDSTONE DIAGENESIS

Table of Contents

Reservoirs I

Properties

General

Pore Geometry, Permeability and Fluid Interaction

Table of Contents

Reservoirs III
Carbonates

Carbonates

Carbonate Depositional Environments

Carbonate Diagenesis

INTRODUCTION

The reprint volumes (I) Reservoirs: Properties, (II) Reservoirs: Sandstones, and (III) Reservoirs: Carbonates, belong to a series that is part of the *Treatise of Petroleum Geology*. The *Treatise of Petroleum Geology* was born during a discussion we had at the Annual AAPG Meeting in San Antonio in 1984. When our discussions ended, we had decided to write a state-of-the-art textbook in petroleum geology, directed not at the student, but at the practicing petroleum geologist. The project to put together one textbook gradually evolved into a series of three different publications: the Reprint Series, the Atlas of Oil and Gas Fields, and the Handbook of Petroleum Geology; collectively these publications are known as the *Treatise of Petroleum Geology*. Using input from the Advisory Board of the Treatise, we designed this entire effort so that the set of publications will represent the state-of-the-art in petroleum exploration knowledge and application. The Reprint Series collects the most up-to-date, previously published literature; the Atlas is a collection of detailed field studies to illustrate all the various ways oil and gas are trapped; and the Handbook is a professional explorationist's guide to the latest knowledge in the various areas of petroleum geology and related fields.

Papers in the various volumes of the Reprint Series are meant to complement the chapters of the Handbook. Papers were selected mainly on the basis of their usefulness today in petroleum exploration and development. Many "classic papers" that led to our present state of knowledge are not included because of space limitations.

We divided the general topic of reservoirs into three separate volumes: (I) properties, (II) sandstones, and (III) carbonates. We grouped papers that discuss properties common to all reservoirs in the volume entitled *Reservoirs: Properties*. The volumes, *Reservoirs: Carbonates* and *Reservoirs: Sandstones* contain papers dealing specifically with those two general types of reservoirs. Originally, the subject of reservoirs was intended to be one volume, but the number of papers suggested for it became so large that it had to be divided into three separate volumes. Therefore, if you examine the content of the papers in all three volumes you will notice that there is much overlap in content.

Volume I, *Reservoirs: Properties*, comprises five parts: (1) General (papers dealing with general reservoir properties), (2) Pore geometry, permeability, and fluid interactions, (3) Pressures, (4) Fractures, and (5) Production geology. Papers in this volume generally describe properties or production techniques that are common to all reservoirs.

Volumes II and III, *Reservoirs: Sandstones* and *Reservoirs: Carbonates*, are each divided into three sections: (1) a section of general papers, (2) depositional environments, and (3) diagenesis.

Edward A. Beaumont
Tulsa, Oklahoma

Norman H. Foster
Denver, Colorado

Treatise of Petroleum Geology
Advisory Board

SANDSTONE DIAGENESIS

Reprinted by permission of the Geological Society from *Sedimentology: Recent Developments and Applied Aspects*, P.J. Brenchley, editor, pp. 249-310.

Role of clastic sedimentology in the exploration and production of oil and gas in the North Sea

H. D. Johnson & D. J. Stewart

SUMMARY: The nature and quality of clastic reservoir rocks is the result of a complex interplay of tectonic setting, provenance, depositional environment and diagenesis. In particular, it is now well established that the reconstruction of depositional environments in clastic successions provides the optimum framework for describing and predicting reservoir development and reservoir quality distribution on both regional (exploration) and field (production) scales.

This review, in the context of the 21st anniversary of the British Sedimentological Research Group (BSRG), expands the above-mentioned theme in relation to the development of depositional models of each of the major oil- and/or gas-bearing clastic reservoirs in the U.K. sector of the North Sea, which has been the subject of a surge in geological interest following the first gas discoveries approximately 21 years ago.

Each major reservoir interval is used to illustrate specific themes, with emphasis placed on how the application of depositional models has contributed to a better understanding of the exploration for, and the production from, the following reservoirs: (i) Permian aeolian and fluvial sands (Rotliegendes Sandstones, southern North Sea), (ii) Jurassic fluvial, deltaic and marginal marine sands (Lower Jurassic Statfjord Formation and Middle Jurassic Brent Group, northern North Sea), (iii) Jurassic shallow marine and deeper marine (mass flow) sands and conglomerates (Upper Jurassic Fulmar/Piper-type sands and Brae/Magnus-type deposits; central and northern North Sea), and (iv) Tertiary deep marine, turbidite fan deposits (Palaeocene and Eocene sands, central and northern North Sea).

Sedimentology has established itself over the past 20–25 years as an integral part of the multidisciplinary studies that are involved in the exploration and production of oil and gas. It is now well known, for example, that the reconstruction of depositional environments in clastic sequences provides the optimum framework for describing and predicting reservoir development and reservoir quality distribution on both regional-(exploration) and field-(production) scales. Nowhere has this involvement been more apparent than in the North Sea, where the phenomenal success of oil and gas exploration over the past 20 years has transformed the area into one of the most well-known hydrocarbon basins in the world, containing around 50 commercially proven oil and gas fields (Fig. 1). The current geological understanding of the North Sea area has been achieved through a broad range of multidisciplinary and multiorganizational studies involving geophysics, structural geology, stratigraphy, geochemistry, sedimentology and the many branches of petroleum engineering (e.g. production/development geology, petrophysics, reservoir engineering and production technology).

Sedimentology is only one part of the North Sea story but, in the context of the British Sedimentological Research Group's 21st anniversary, an extremely relevant subject for this meeting for several reasons. Firstly, the birth of the BSRG 21 years ago occurred very shortly after the initial realization by the oil industry that the North Sea could be a hydrocarbon-bearing basin, following the discovery of the giant Groningen gas field in the northern part of the Netherlands in 1959. Secondly, over the past 21 years, but particularly during the last 10 years, many members of the BSRG have been involved in North Sea studies in a wide range of different capacities, both directly and indirectly. Finally, the North Sea remains a vitally important and exceptionally interesting geological region which will be the subject of considerable sedimentological attention for at least the next 20–30 years.

The principal aim of this review is to highlight the contribution of sedimentological studies in the exploration of oil and gas specifically in the North Sea. This is achieved by reviewing the sedimentological characteristics of the major clastic reservoir intervals encountered in the U.K. sector within the context of their respective depositional models. The following reservoirs are considered:

(i) Permian aeolian and associated fluvial sands;
(ii) Lower–Middle Jurassic fluvial and deltaic sands;
(iii) Upper Jurassic shallow and deeper marine sands; and
(iv) Tertiary deep marine sands.

The objective of this review is not to give a detailed regional synthesis of these various rock

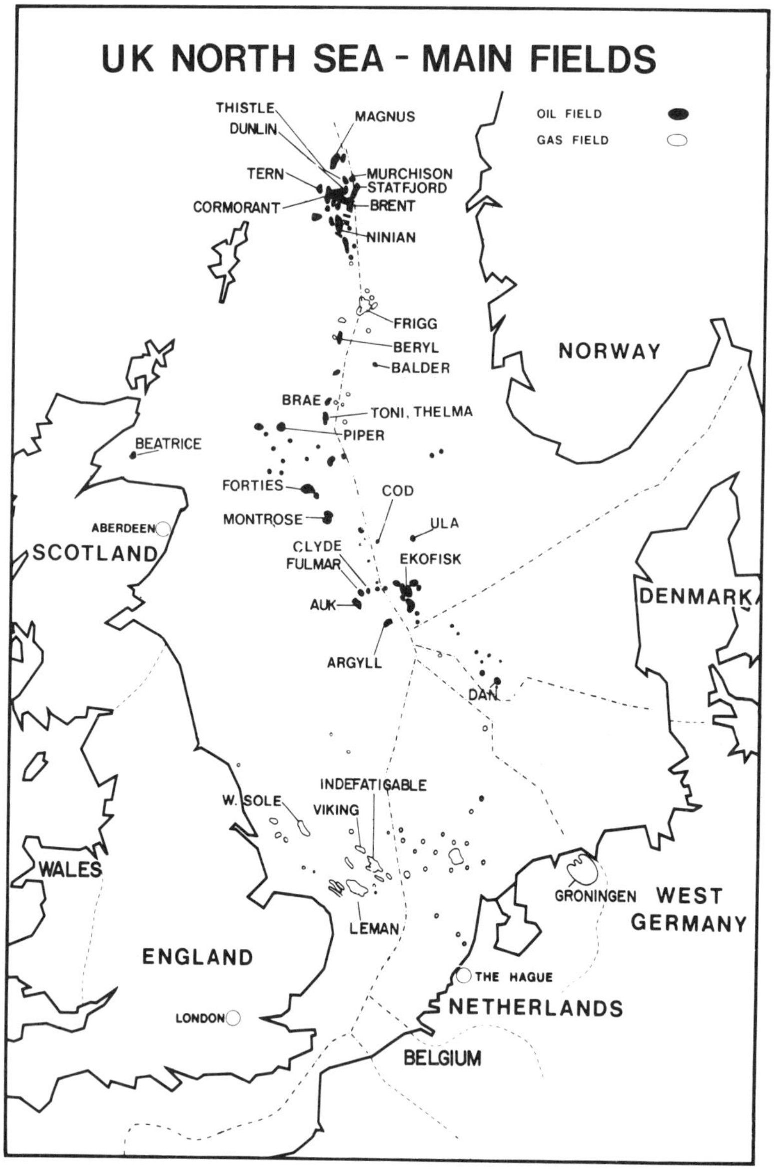

FIG. 1. Geological sketch map of the British Isles indicating the distribution of the main oil and gas fields.

units, since this has been amply covered elsewhere. Instead, emphasis is placed on how facies models and the interpretation of clastic depositional environments have influenced exploration and production activities.

The examples used partly comprise published material and also in-house reports prepared by Shell on behalf of the Shell/Esso North Sea partnership.

Sedimentology in hydrocarbon exploration and production

The nature and quality of clastic reservoirs are determined by a complex interplay of processes involving source area, depositional environment, tectonic setting and diagenesis.

Essentially the composition, climate and geomorphology of the source area determine the primary detrital composition of sediment deposited in the receiving basin which, in turn, will affect any subsequent diagenetic trends (Nagtegaal 1978). Depositional environment exerts the most basic control on clastic facies distribution, notably by determining the size, shape and trend of sand bodies, and it therefore controls primary reservoir quality (porosity-permeability) distribution (e.g. LeBlanc 1972; Shelton 1973; Pryor 1973). Reservoir development (thickness distribution) reflects the relative

EXPLORATION	1. BASIN ANALYSIS (Initial exploration stage)
	2. PROSPECTS EVALUATION (Immature exploration stage)
	3. PROSPECT EVALUATION (Mature exploration stage)
APPRAISAL / EARLY DEVELOPMENT	4. APPRAISAL STAGE
	5. EARLY DEVELOPMENT STAGE
DEVELOPMENT	6. MATURE DEVELOPMENT STAGE
	7. LATE DEVELOPMENT / ENHANCED RECOVERY STAGE

FIG. 2. Tabulation illustrating the various stages of exploration and production in which sedimentological studies are applied.

roles of tectonically induced subsidence, base-level changes and sedimentation rates. Tectonics exerts a fundamental control on all clastic sedimentation by determining the size and location of depositional (sub-) basins, rates of uplift and subsidence, directions of sediment transport, and the thickness and orientation of facies belts. Subsurface reservoir quality is finally determined by diagenesis, initially involving variable burial, compaction and cementation histories, and subsequently including any uplift, dissolution and differential diagenesis associated with hydrocarbon emplacement.

Unravelling the significance and interrelationships of these processes is fundamental to understanding the nature and quality of clastic reservoirs and to predicting their development and distribution in the subsurface. Sedimentological studies contribute to all stages of exploration and production (development) activities, ranging from the initial basin analysis stage, through the prospect evaluation of increasingly mature plays, and continuing into field appraisal and throughout the various stages of field development (Fig. 2). The typical role and application of sedimentological studies to the exploration and production of oil and gas are outlined in Fig. 3. In practice there are two distinct areas of interest, exploration (regional scale) and production (field scale), with each having its own specific objectives. The basic sedimentological data are common to both but the aims and applications of the depositional models are different (Fig. 3). Exploration studies are primarily concerned with locating optimum areas of reservoir development with suitable source rock, seal and trapping potential to allow hydrocarbon accumulation. In production, sedimentological studies are essentially required to provide an accurate reservoir description to ensure optimum recovery of hydrocarbon reserves.

This review expands on all these aforementioned themes in the context of North Sea clastic reservoirs.

Permian, rotliegend sandstones, southern North Sea

In the context of both exploration and production, the southern North Sea was the first major European offshore hydrocarbon province to be developed. Interest in the southern North Sea as a potential hydrocarbon province was triggered by the discovery of the huge Groningen gas field in the northern Netherlands in 1959, which spurred the drive to acquire seismic data over the adjacent offshore areas in the early 1960s. This exploration effort has culminated in the discovery of several similar hydrocarbon accumulations with reserves totalling some 646 billion m^3 (22.8 trillion ft^3) of proven recoverable gas in the U.K. sector (Department of Energy 1983, p.5). Exploration and production in this offshore area were particularly attractive because of the shallow water depths ideal for the use of small jack-up rigs. Nearly 20 years later this province is experiencing a further resurgence owing to a combination of higher gas prices and lower oii prices which have renewed interest in some of the previously marginal commercial prospects.

Geological framework

The main gas accumulations are found in Permian sandstones of the Rotliegendes Group underlying a tight seal of Zechstein dolomites and evaporites. The Rotliegendes Group has a wide distribution in three intracratonic basins (Glennie 1981), the largest being the South Permian Basin which contains the southern North Sea gas fields (Fig. 4).

Although the Rotliegendes Group has a fairly wide distribution, it was quickly realized that it was composed of markedly different lithologies with varying geographical distributions. In particular, two main types of reservoir sandstone were recognized: (1) large scale cross-bedded sandstones with well developed bimodal foreset

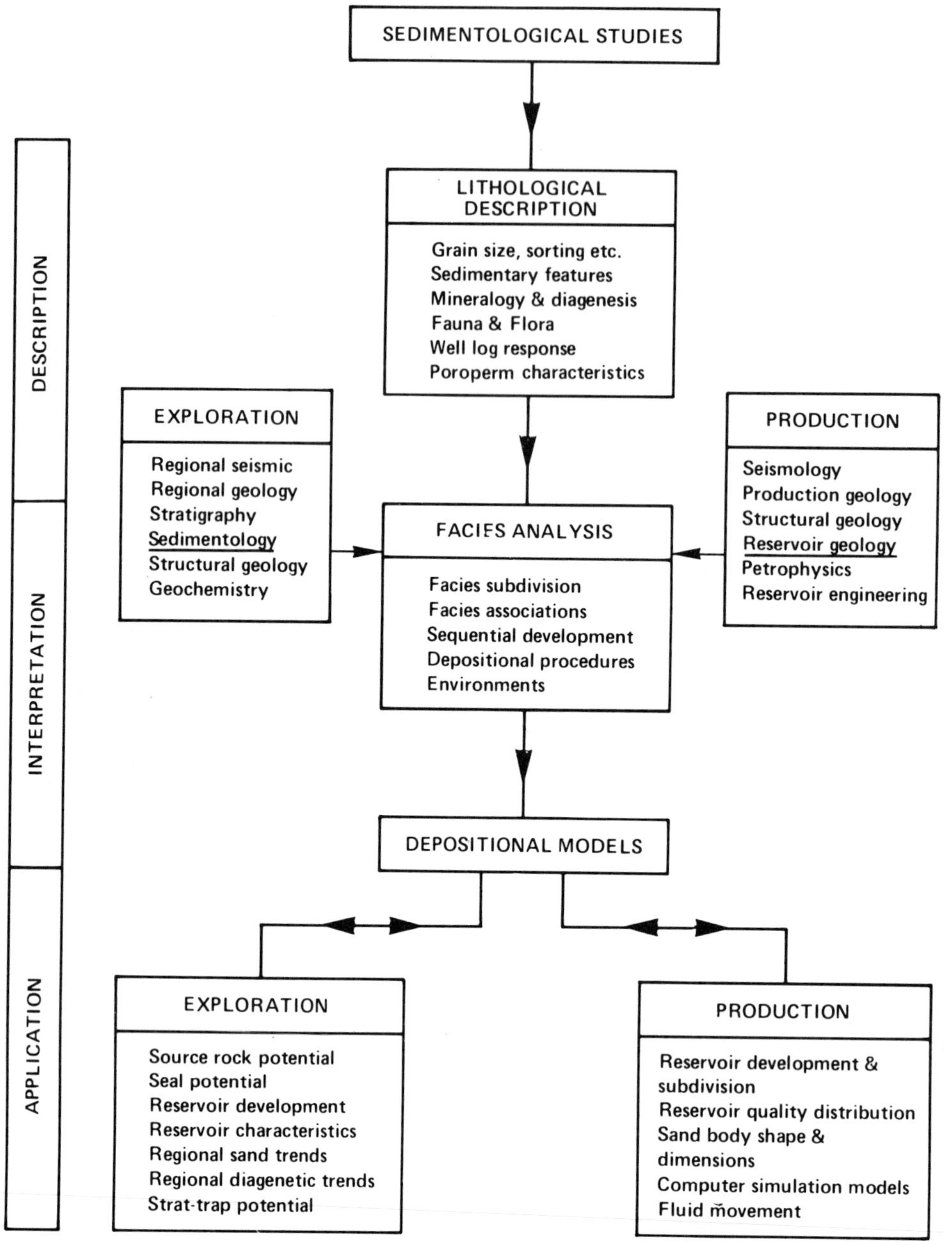

FIG. 3. Tabulation illustrating the nature and application of sedimentological studies in exploration and production.

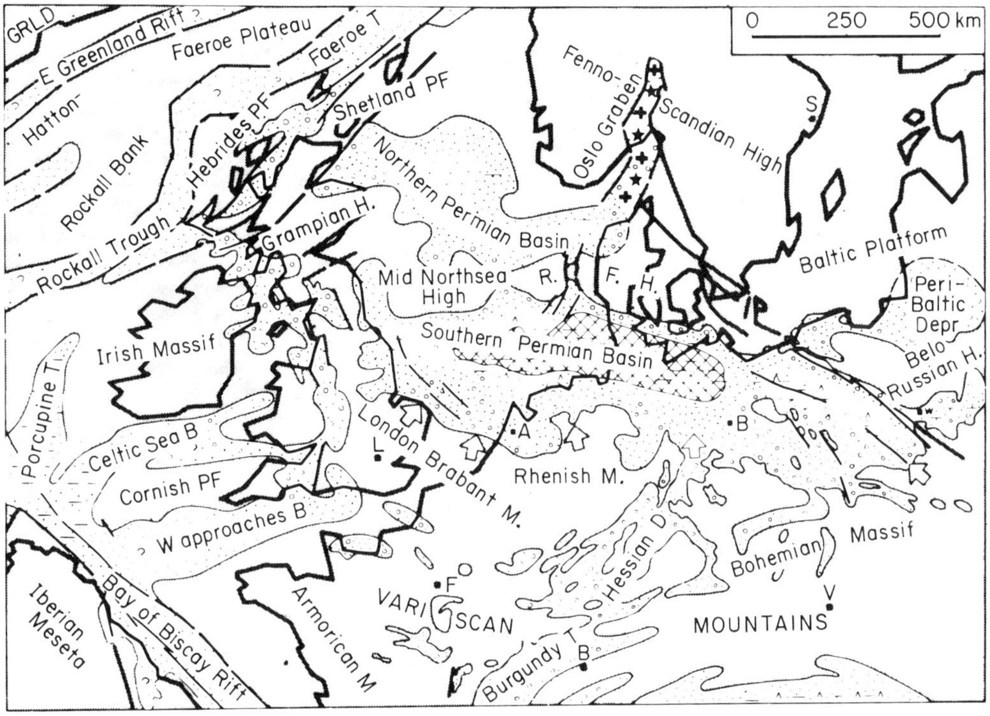

FIG. 4. Early Permian, Rotliegendes palaeogeography (from Ziegler 1981).

laminae believed to be aeolian in origin, and (2) poorly sorted, pebbly sandstones deposited in fluviatile wadis. The association of these sandstones with fine-grained evaporitic deposits and the frequent reddening of the sandstones indicated deposition in an arid, desert basin.

Whilst considerable knowledge was available in the early 1960s on the geomorphology of desert landscapes and on the external morphology of aeolian sand dunes, there was little detailed knowledge of the internal structure of desert deposits. However, this information was necessary to identify facies associations in cores and from other well-log data. In addition, a basin-scale depositional model was required to aid the prediction of sand thickness trends and the geographical distribution of facies belts for exploration purposes.

Facies-related studies

It was against this background that renewed studies of desert environments were initiated, notably by Glennie (1972), to assist the broader interpretation of ancient desert deposits and, more specifically, to assist the search for hydrocarbons in the South Permian Basin.

Glennie's work on arid environments of the present-day Persian Gulf adjacent to the Oman Mountains of the United Arab Emirates (Glennie 1972) illustrates how studies of modern sedimentary environments can aid the search for hydrocarbons in comparable ancient deposits. The depositional model for the Rotliegendes Group consists of facies belts of fluviatile, aeolian and evaporite/lacustrine deposits, roughly paralleling the London–Brabant massif/Variscan highlands which lay to the south of the South Permian Basin (Figs 5 and 6). The facies belts pass northward from wadi/alluvial fan facies, best developed in proximity to the mountain chain, through dune and inter-dune aeolian sands further to the north-east, and finally into sabkha desert lake facies at the basin centre. In building up a basin-scale depositional model, regional facies mapping has proved to be an essential part of exploration studies, largely because the depositional en-

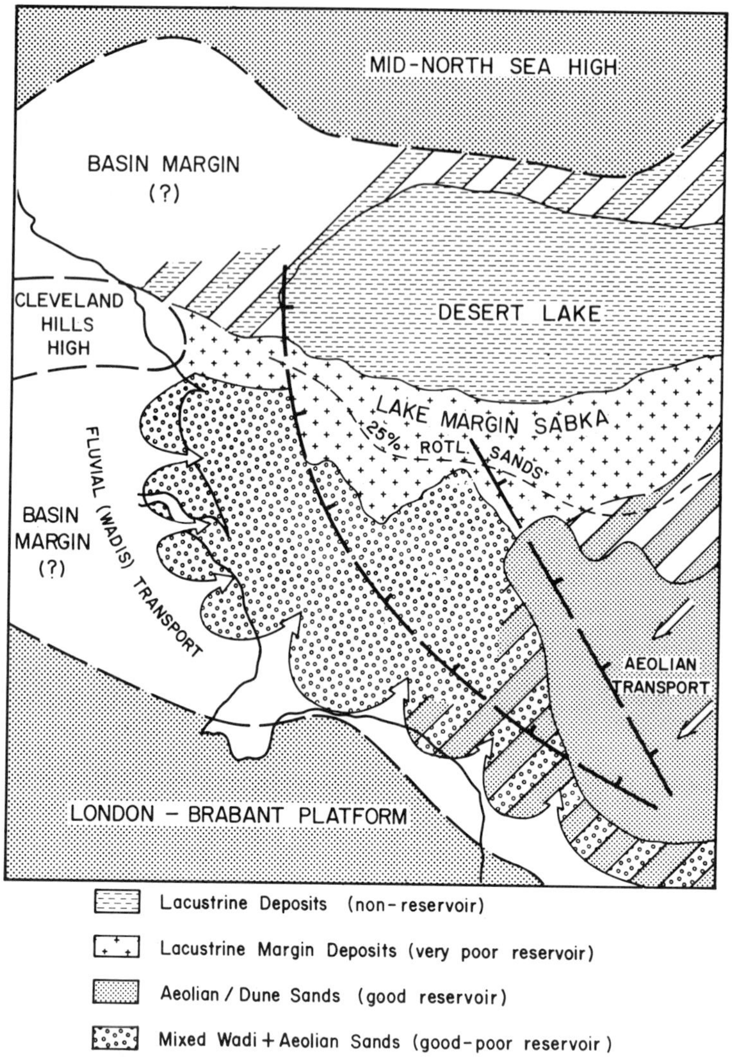

FIG. 5. Rotliegendes facies distribution (based on Glennie *et al.* 1978).

vironment of the sandstones has a fundamental influence on reservoir quality. Hence, detailed sand quality studies have attempted to distinguish the high quality aeolian sandstones from the poorer quality fluviatile sandstones. This type of study is most clearly illustrated in the work of Marie (1975; see also Fig. 7), who used a variety of mappable parameters to identify areas of optimum reservoir development based on the previously outlined facies model. The early stages of basin analysis typically involve the construction of thickness and/or percentage maps of sandstone and claystone, based initially on widely spaced well data. This is followed up by more detailed core studies, which in the case of the Rotliegendes have enabled the two main sandstone facies associations (fluviatile and aeolian) to be identified and mapped. The combination of these maps provide an excellent quantitative framework for evaluating reservoir quality distribution, and for recognizing the relative importance of additional factors such as tectonics and diagenesis.

Facies analysis of the Rotliegendes Group of the U.K. sector shows that it consists of a lower interval of mainly variable alternations of fluviatile/wadi sands and gravels, and an upper interval of aeolian dune and interdune sands (Fig. 8).

From the above it is apparent that core studies are essential in order to build up depositional models. However, many exploration and pro-

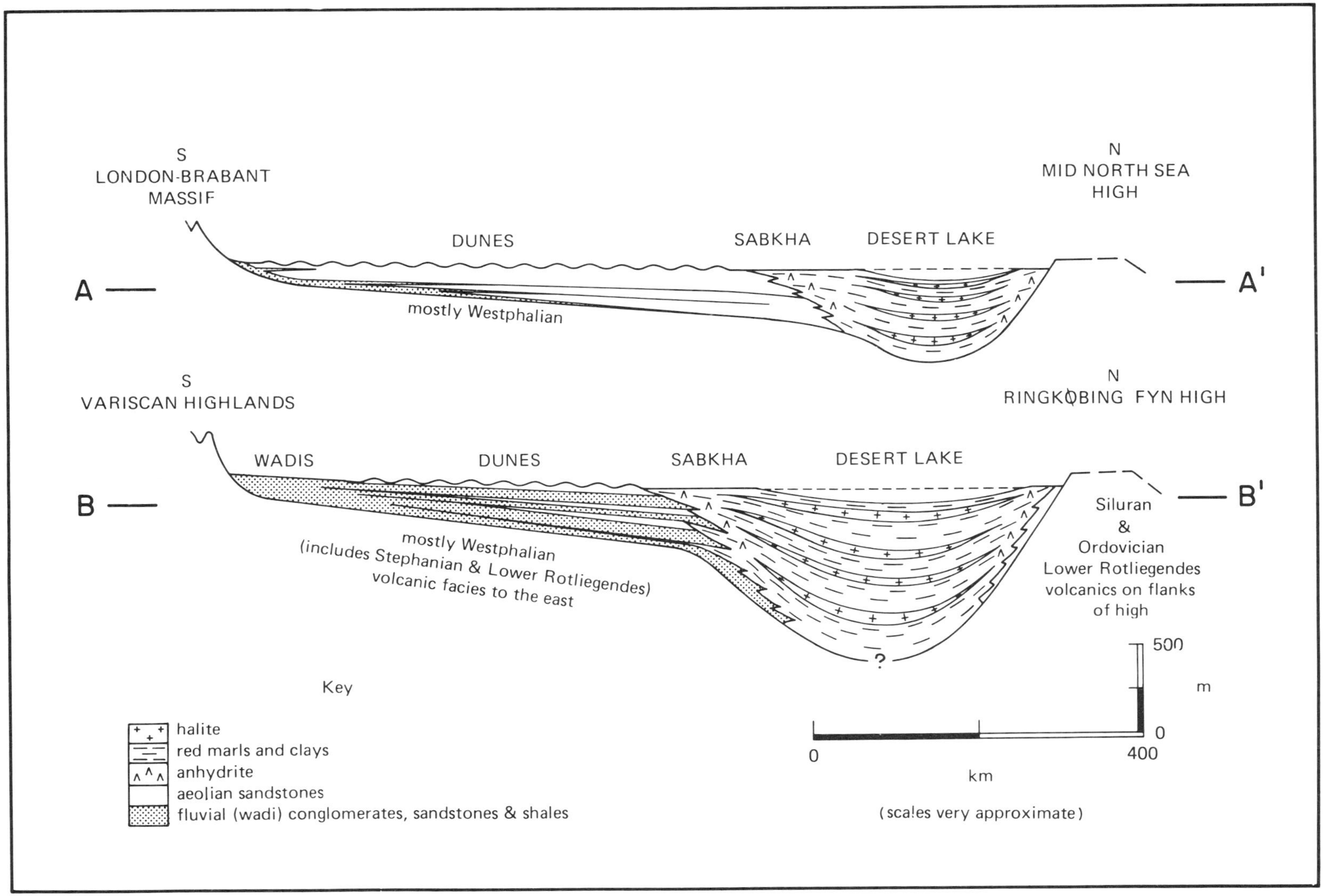

FIG. 6. Schematic cross-sections through the early Permian deposits of the southern North Sea (from Glennie 1972).

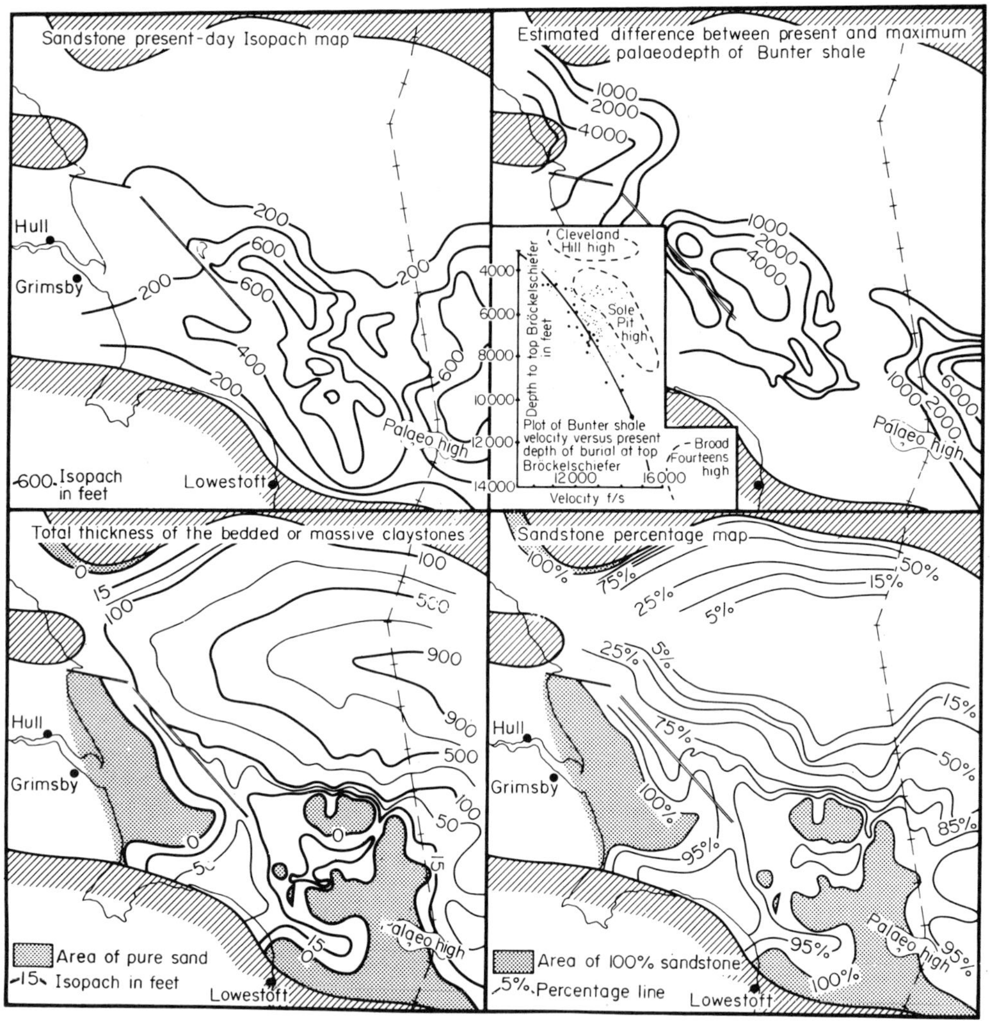

FIG. 7. Various mapping parameters used to reconstruct the depositional framework and geological burial history of the Rotliegendes Group (from Marie 1975).

duction wells are not extensively cored owing to the higher cost. In this context, it proved possible to use wireline log data not only to identify the two main facies associations for work on an exploration scale, but also to identify smaller facies variations for reservoir correlation on a field scale (Van Veen 1975). For example, the large-scale cross-sets of the aeolian dunes and the interposed finer grained, more argillaceous interdune or bottom set (including desert lag and adhesion ripple sands) deposits can be detected accurately using a combination of gamma ray, bulk density (FDC) and microlaterolog records (Fig. 9). This technique depends on the higher clay and cement contents of the dune bottom set beds. The clay causes a peak on the gamma ray log and the cement is detected on the FDC log. The combination of clay and cement causes a similar but large peak on the microlaterolog (resistivity).

The additional use of dipmeter logs (Fig. 10) enables the measurement of dune thickness, by plotting the gradual upwards increase in foreset inclination, and the deduction of wind directions by measuring the direction of foreset inclination. This information can help predict field-scale variations in the quality and thickness of smaller-scale reservoir units necessary for field development studies. Further development information can be gained by studies of bedding characteristics of the dune sequences in cores, with the ultimate aim of identifying the size, geometry and orientation of individual cross-bed sets. The prediction of the size and orientation of dune cross-bedding is important for the construction of reservoir geological

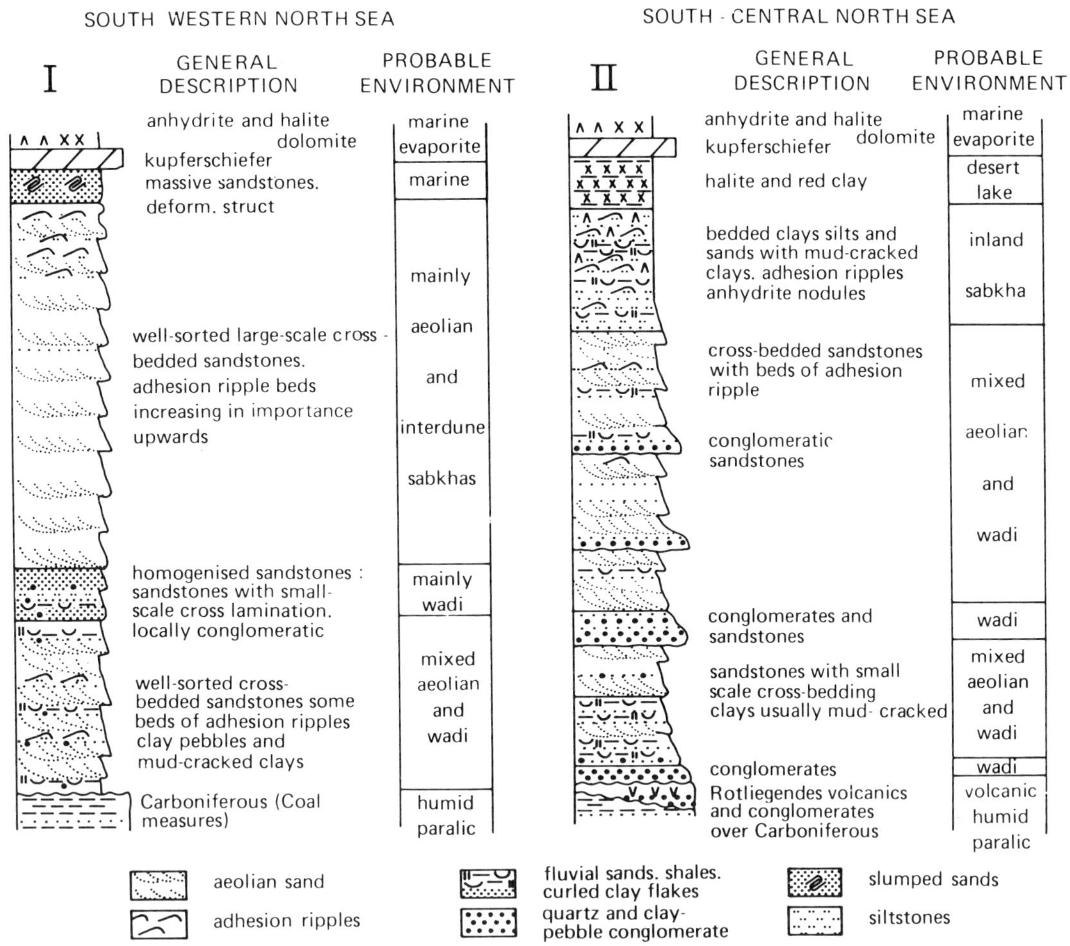

FIG. 8. Generalized sedimentology and vertical lithological profiles through the Rotliegendes of the southern North Sea (from Glennie 1972).

models because permeabilities are significantly greater parallel to the dune bedding as compared to the perpendicular direction (van Veen 1975; Weber 1984). It becomes vital, therefore, to know the internal geometry of the dune deposits when designing the field development programme, since this allows optimum location of production wells with regard to maximum well productivity.

Studies of dune bedding in the Rotliegendes have been carried out by Glennie (1982), who, with the aid of core and dip-meter data, has inferred the presence of two main types of dune: (1) barchan or transverse dunes with the bulk of the bedding dipping downwind (unimodal), and (2) longitudinal seif dunes with bimodal foreset directions at right angles to the main wind direction (Fig. 11). This type of analysis may enhance the interpretation of palaeocurrent data in terms of transport directions and dune bedding permeability characteristics.

Diagenetic studies

The relatively simple picture of reservoir quality based directly on depositional facies distributions is complicated by the intricate burial history of the Rotliegendes Group in the southern North Sea. This has resulted in a diagenetic overprint which has a important detrimental effect on porosity/permeability characteristics of all facies associations. It was Stalder (1973) who, with the aid of the scanning electron microscope, demonstrated that the poorer reservoir quality resulted from the presence of authigenic clay minerals and that the type and morphology of these clay minerals have an important effect on porosity/permeability relationships.

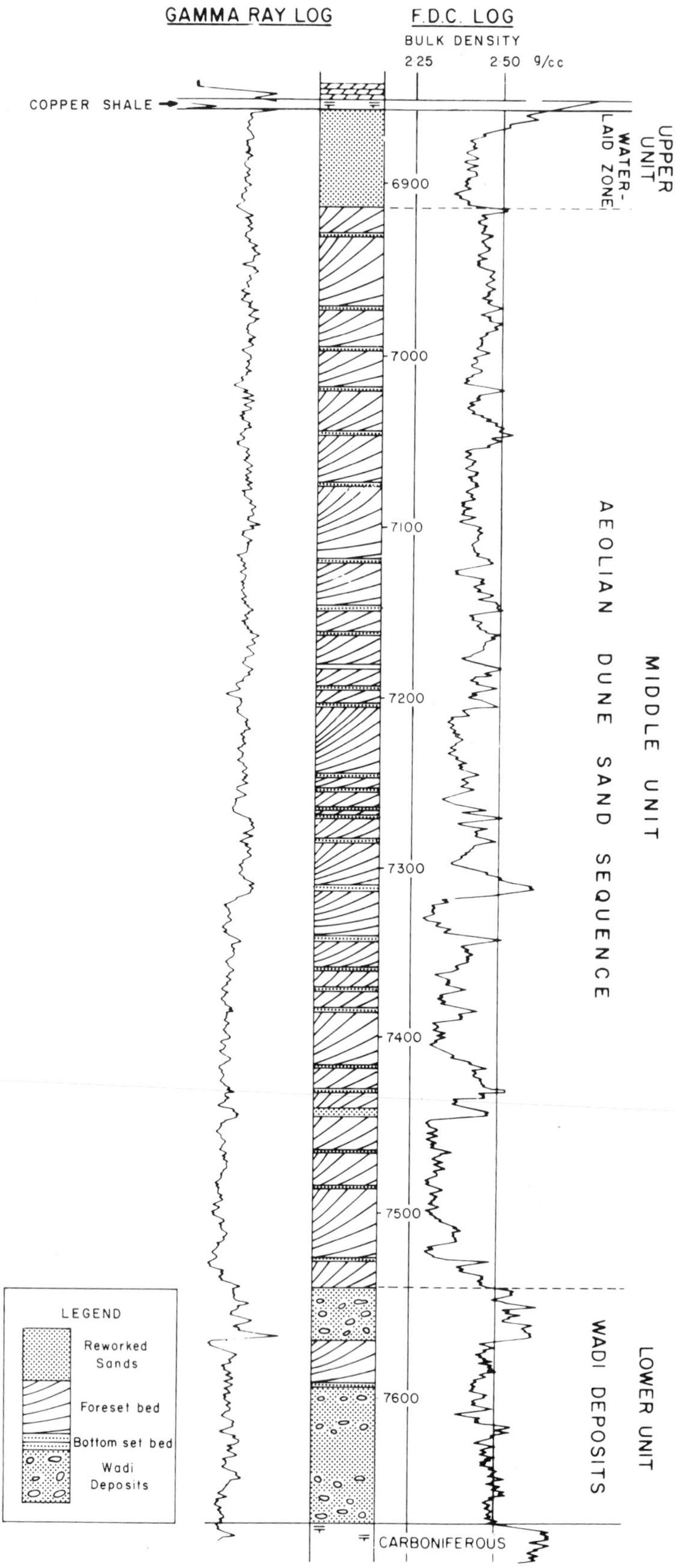
GAMMA RAY LOG
F.D.C. LOG
BULK DENSITY
2.25
2.50 g/cc
COPPER SHALE
6900
7000
7100
7200
7300
7400
7500
7600
CARBONIFEROUS
UPPER UNIT WATER-LAID ZONE
MIDDLE UNIT
AEOLIAN DUNE SAND SEQUENCE
LOWER UNIT
WADI DEPOSITS
LEGEND
Reworked Sands
Foreset bed
Bottom set bed
Wadi Deposits

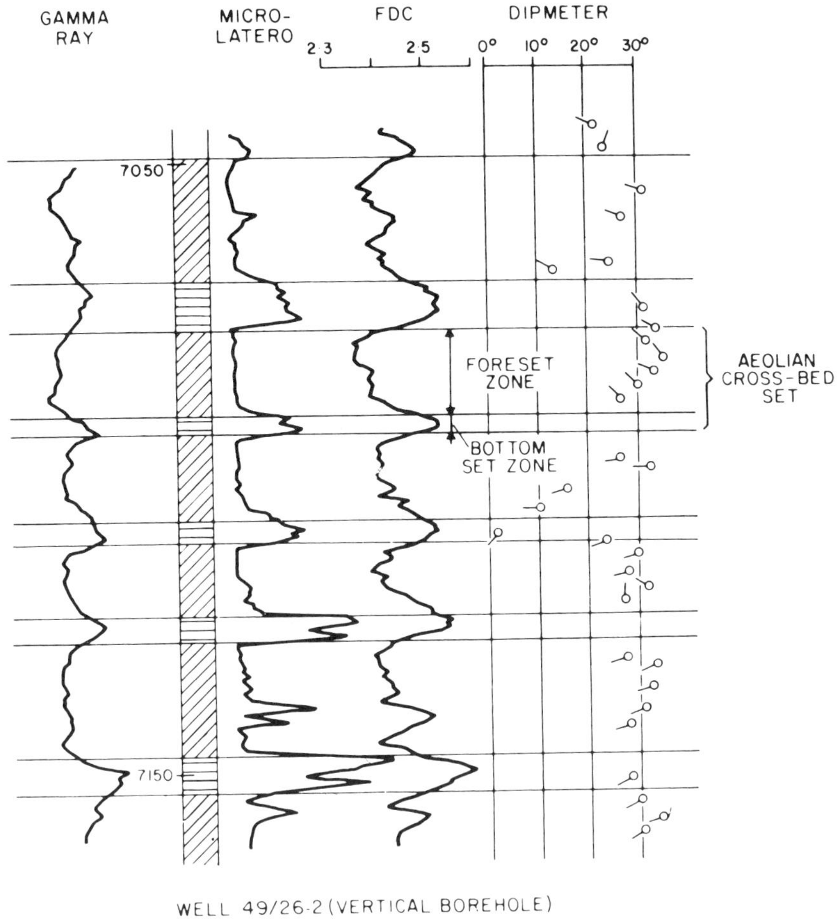

FIG. 10. Dipmeter response through dune-bedded Rotliegendes sandstones (from van Veen 1975).

It is now well known that illite, in particular, is a common cause of lower permeabilities in Rotliegendes reservoirs. The explanation for this is the fibrous bridging habit of the illite crystals (Fig. 12), which reduces the size of pore throats and increases the tortuosity in the interstitial pore spaces, whilst having little effect on porosity (Stalder 1973). Conversely, the more blocky kaolinite, also a common authigenic clay mineral in the Rotliegendes has a lesser effect on permeability owing to its pore-filling, rather than pore-bridging, habit (Fig. 13). It is, therefore, essential to be able to predict the distribution of the various authigenic clays, notably those areas most susceptible to kaolinite or illite development.

Regional mapping of authigenic clays initially showed that poor reservoir quality, illitic sandstones often occur at present-day shallower depths than the kaolinitic sandstones (Seeman 1979). At first sight this appears to suggest that kaolinite is formed at the expense of illite with increasing depths of burial. However, by using a combination of geological well data and sonic velocities (Glennie *et al.* 1978), it can be deduced that the illitic sandstones were in fact, buried more deeply than the kaolinitic sandstones and, in most cases, in excess of 11,000 ft (Fig. 14). In demonstrating this relationship, sonic velocities in the Bunter Shale overlying the gas reservoirs are important since the illitic sandstones are associated with shales of higher sonic velocities than those associated with the kaolinitic sandstones (Glennie & Boegner 1981), suggesting that the former have undergone greater compaction and, therefore, greater burial. The present-day inverse relationship is the result of later uplift of a previously deeply buried sequence (Glennie &

FIG. 9. Example of gamma ray and FDC (density) log responses to various facies types in the Rotliegendes Group of the Leman Field (from Van Veen 1975).

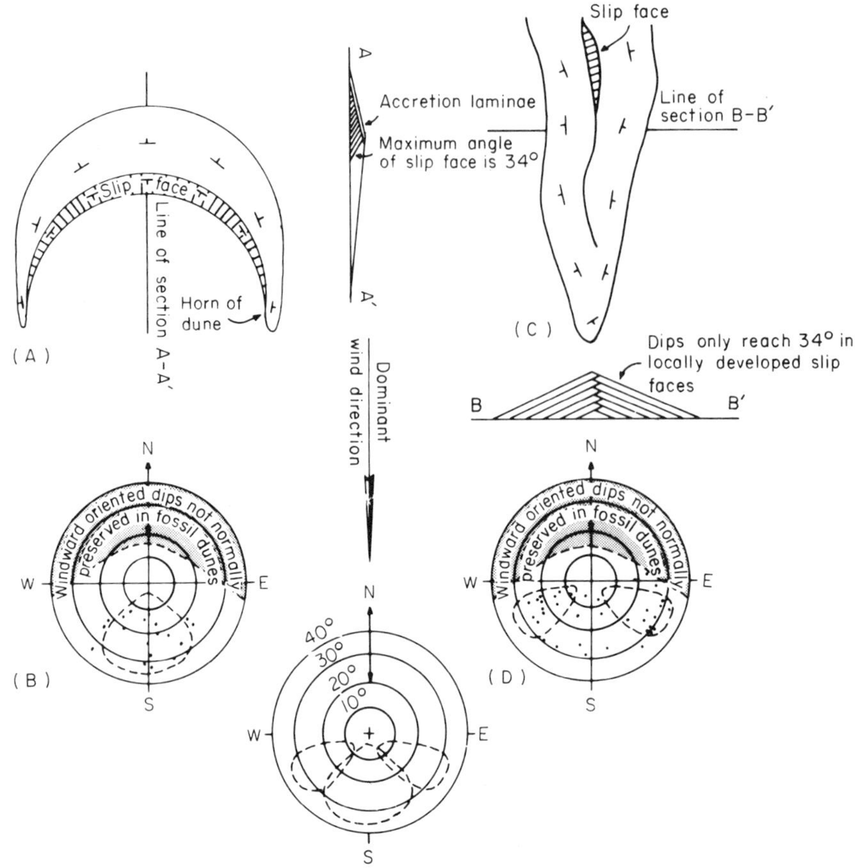
Slip face
Accretion laminae
Maximum angle of slip face is 34°
Line of section B-B'
Line of section A-A'
Horn of dune
(A)
(C)
Dips only reach 34° in locally developed slip faces
B
B'
Dominant wind direction
N
Windward oriented dips not normally preserved in fossil dunes
W
E
S
(B)
(D)
40°
30°
20°
10°

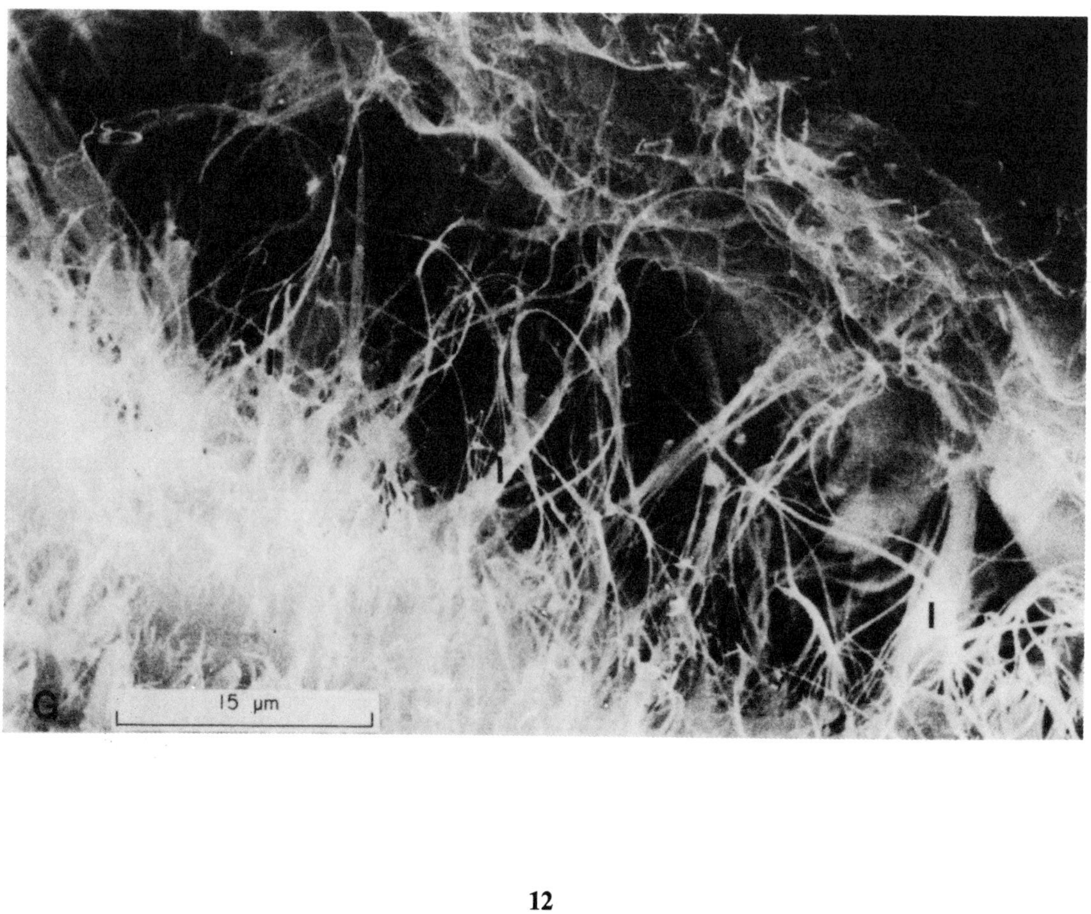
15 μm

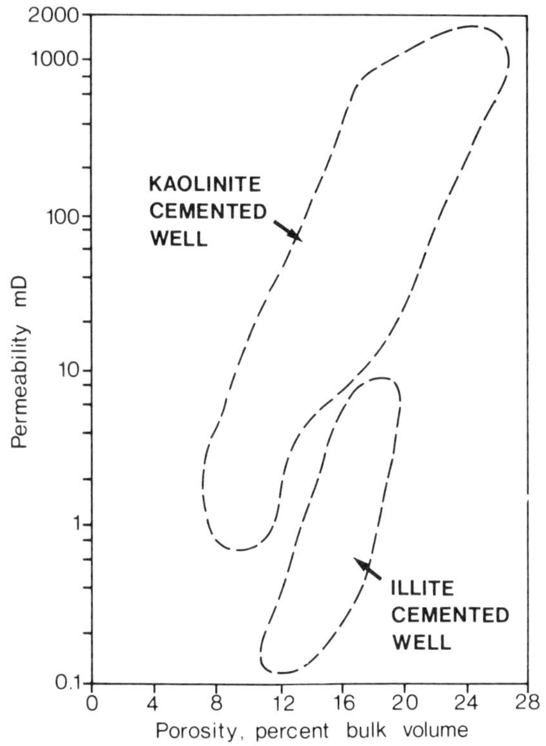

FIG. 13. Different effects of kaolinite and illite on porosity/permeability characteristics in the Rotliegendes sandstones (from Stalder 1973).

Boegner 1981).

Further studies have culminated in the recognition of a complex diagenetic sequence (Fig. 14) ranging from early environment-related features to late, post-uplift features (Glennie *et al.* 1978). These studies also show that depth is not alone in controlling diagenesis. It has been suggested, for example, that feldspar may be a source of kaolinite and, therefore, variations in detrital feldspar may control the abundance of kaolinite. In addition, the degree of feldspar dissolution at a location may be related to the thickness of underlying organic-rich Carboniferous strata available to generate CO_2 gases (Rossel 1982). The situation is further complicated by the possibility of cements such as carbonate or anhydrite being introduced from the overlying Zechstein strata. Nevertheless, porosity distribution at depth is still largely a function of variations in sorting, as reflected in the dune, adhesion and wadi sands (Glennie *et al.* 1978).

Lower and Middle Jurassic Sands, northern North Sea

The Lower Jurassic Statfjord Sands, and particularly the Middle Jurassic Brent Sands, represent the main reservoirs of the prolific North Viking Graben oil province in the northern North Sea. In terms of the U.K. Sector of the North Sea, the discovery of major hydrocarbon reserves in these reservoirs and their subsequent development represented the second *major* phase of hydrocarbon exploration and production, following on from the successes of the southern North Sea gas province (although oil was first discovered in Palaeocene sands of the Forties and Montrose fields). However, the discovery of the Brent Sands in 1971 in well 211/29-1 (Brent field), and the Statfjord Sands in later wells, revealed a more complex and regionally less predictable reservoir sequence compared to the Rotliegend Sandstones (e.g. Bowen 1975).

One early investigation by C. Kruit (Shell E & P Laboratory, Rijswijk, personal communication) recognized the large-scale regressive/transgressive character of the Brent Sands (Fig. 15). Based on limited data (fourteen wells and 2500 ft core) the Brent Sands were interpreted as the product of a wave-dominated delta to interdeltaic system which prograded northwards along the axis of an active rift system (Fig. 16). This interpretation remains essentially correct and it illustrates the value of a conceptual depositional model, particularly during the early stages of basin analysis, in providing a framework for interpreting stratigraphic relationships and predicting regional trends in reservoir distribution.

Increased subsurface control, provided by substantially more wells (with more abundant cores and well logs) and better seismic coverage, has provided a wealth of new sedimentological data. These data, when integrated with the increased knowledge of the regional tectonic evolution and improved stratigraphic control, have resulted in the development of a more comprehensive depositional model. The

FIG. 11 (facing, top). Comparison of bedding architecture in ideal barchan and transverse dunes (a) and seif dunes (c) (adapted from Glennie 1982).

FIG. 12 (facing, bottom). Scanning electron photomicrograph 'hairy' illite crystals.

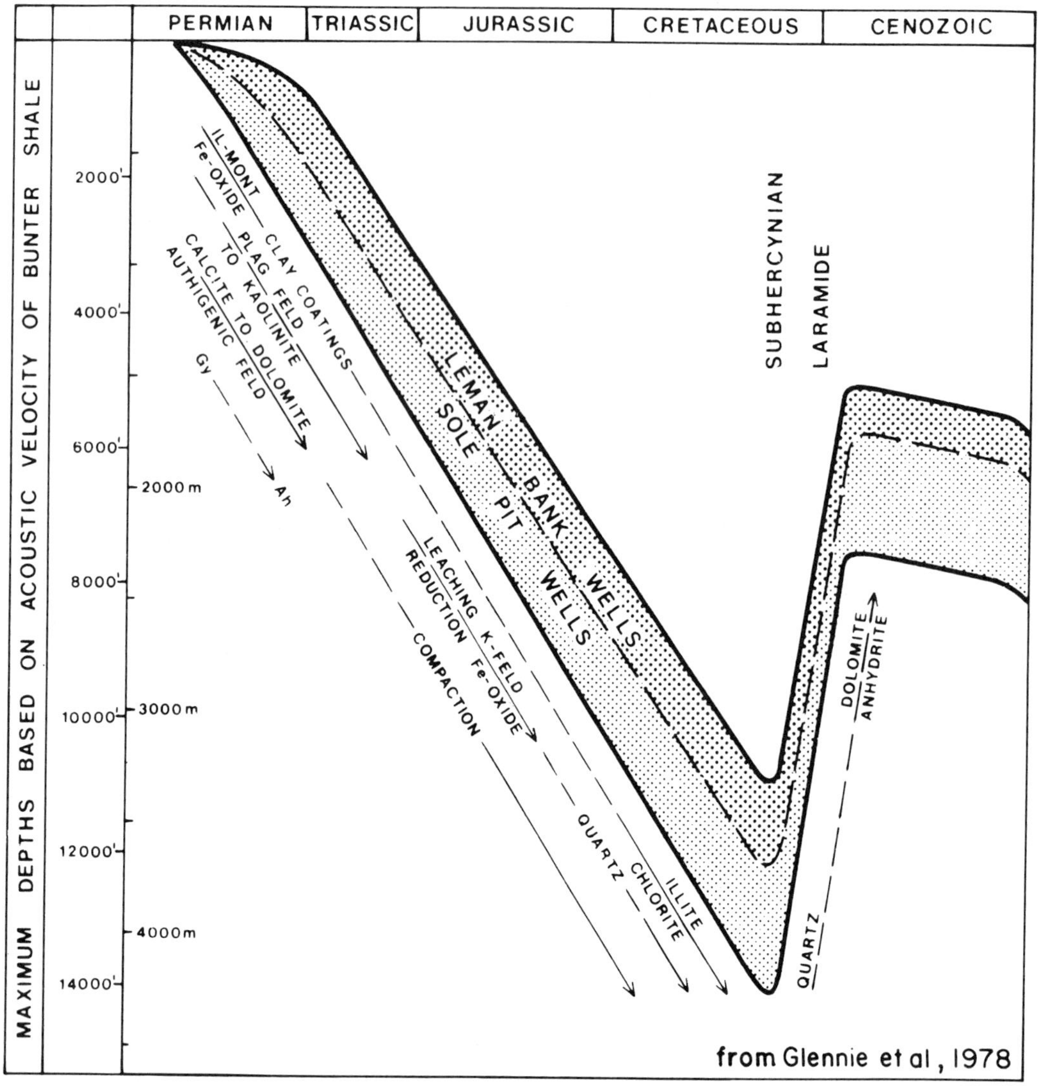
PERMIAN
TRIASSIC
JURASSIC
CRETACEOUS
CENOZOIC
MAXIMUM DEPTHS BASED ON ACOUSTIC VELOCITY OF BUNTER SHALE
2000'
4000'
6000'
8000'
10000'
12000'
14000'
2000 m
3000 m
4000 m
IL-MONT CLAY COATINGS
Fe-OXIDE PLAG FELD TO KAOLINITE
CALCITE TO DOLOMITE
AUTHIGENIC FELD
Gy
Ah
LEMAN BANK WELLS
SOLE PIT WELLS
SUBHERCYNIAN
LARAMIDE
LEACHING K-FELD
REDUCTION Fe-OXIDE
COMPACTION
QUARTZ
CHLORITE
ILLITE
QUARTZ
DOLOMITE
ANHYDRITE
from Glennie et al, 1978

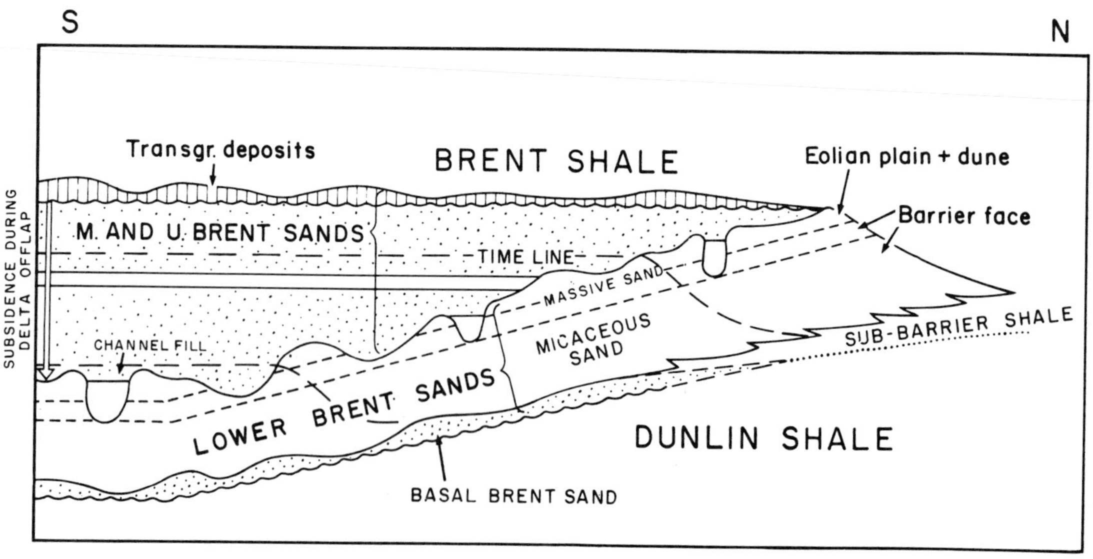
S
N
Transgr. deposits
BRENT SHALE
Eolian plain + dune
Barrier face
SUBSIDENCE DURING DELTA OFFLAP
M. AND U. BRENT SANDS
TIME LINE
MASSIVE SAND
MICACEOUS SAND
SUB-BARRIER SHALE
CHANNEL FILL
LOWER BRENT SANDS
DUNLIN SHALE
BASAL BRENT SAND

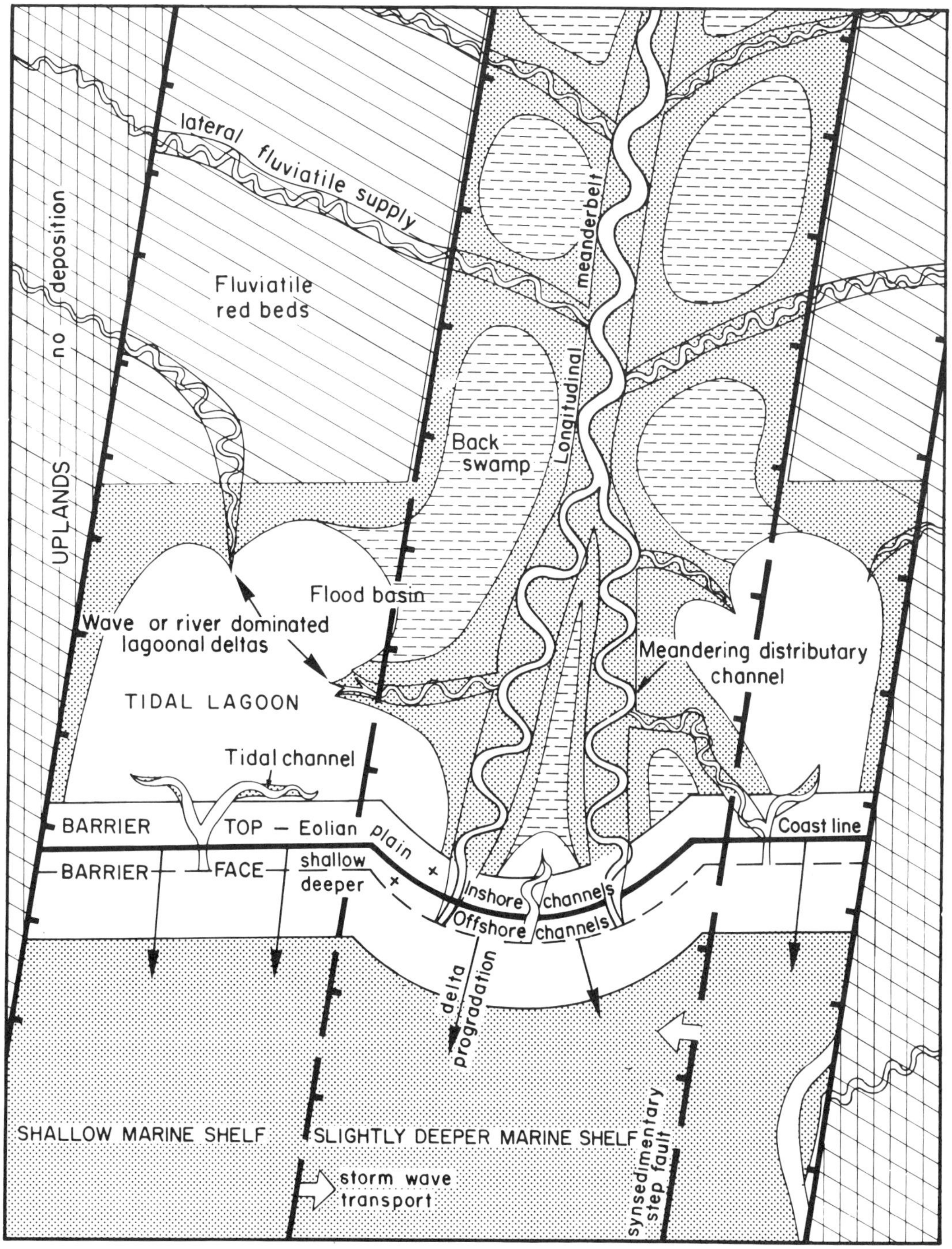

FIG. 14 (facing, top). Rotliegende burial history and sequence of diagenetic events in Sole Pit and Leman Bank areas (from Glennie *et al.* 1978).

FIG. 15 (facing, bottom). Basic depositional framework of the Middle Jurassic, Brent Sand Viking Graben, northern North Sea.

FIG. 16 (above). Model of axial fill of a shallow marine rift valley by a wave-dominated delta.

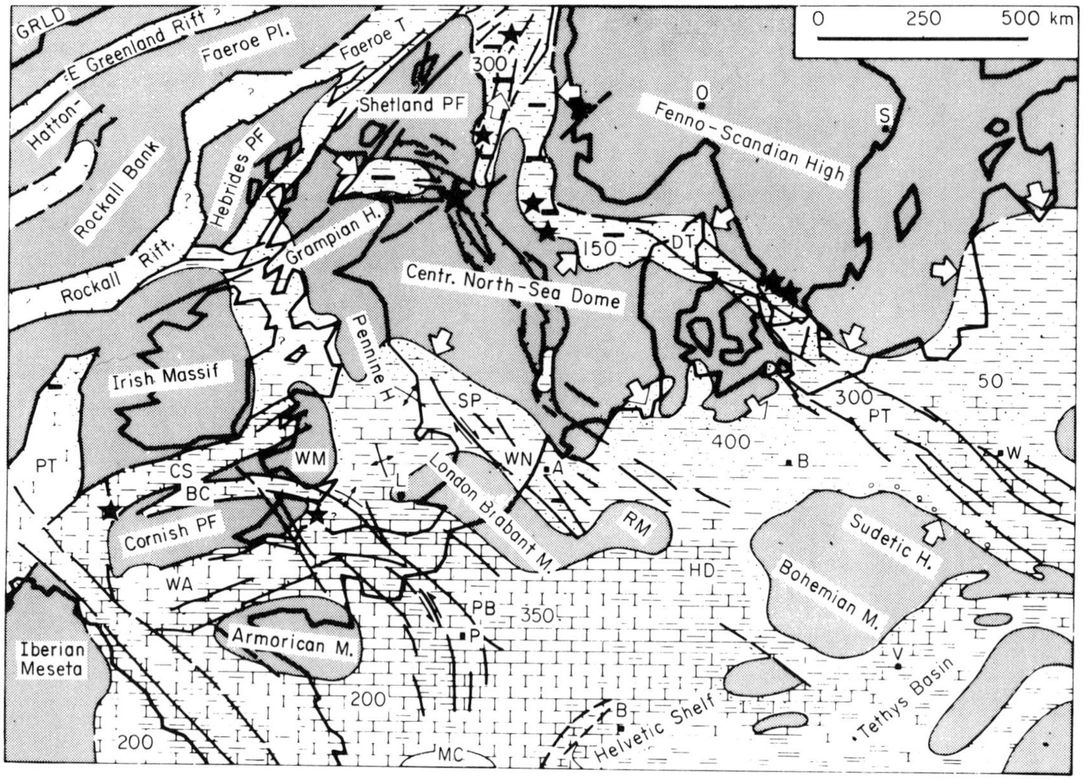

FIG. 17. Bajocian-Bathonian palaeogeography (from Ziegler 1981).

regional framework, for example, indicates that the Brent Sands were deposited following the widespread mid-Cimmerian tectonic phase and a eustatic lowering of sea-level, when a large rift dome was uplifted in the central North Sea which became a major sediment source area (Fig. 17; Ziegler 1982). Clastic sediments were shed radially off this dome and sediment transport paths developed through the continuously subsiding Viking Graben (e.g. Eynon 1981), as predicted in the earlier conceptual depositional model.

The increased well control has also provided positive definition of facies trends, which confirm the south to north progradation of the coastal/deltaic clastic wedge. Detailed facies distribution patterns in part of the Ness Formation, for example, indicate that fluvial deposits gradually merge northwards into lagoonal deposits, which are themselves replaced by coastal barrier and offshore marine sediments (Fig. 18). Furthermore, the convergence of progadational and transgressive shallow marine sands at the northern pinch-out of the clastic wedge, demonstrates that the time-equivalent coastal and delta plain complex (Ness Formation) was located to the south behind a contemporaneous coastal barrier complex (Rannoch–Etive Formations).

The important interplay between sedimentological and graben tectonic processes in influencing the thickness and lateral distribution of the Brent Sands, together with the main facies associations, is illustrated in a series of S–N-trending time-slices (Fig. 19). The sedimentological relationships of the main types of sediment body (facies associations) have been outlined in a later conceptual depositional model (Fig. 20; Budding & Inglin 1981).

The model of a wave-dominated delta to coastal (interdeltaic) coastal barrier complex enables comparison with several modern wave-dominated and fluvial-wave interaction deltas and their associated delta and marginal coastal plain environments (Galloway 1975; Coleman *et al.* 1981; Elliott 1978; cf. Fig. 21).

Reservoir characteristics of the Brent Sands and their relation to exploration and production

The reservoir properties of the Brent Sands are, for the most part, controlled by the depositional environments within which the various facies associations were deposited (Simpson & Whitley 1981; Proctor 1981). Diagenesis is

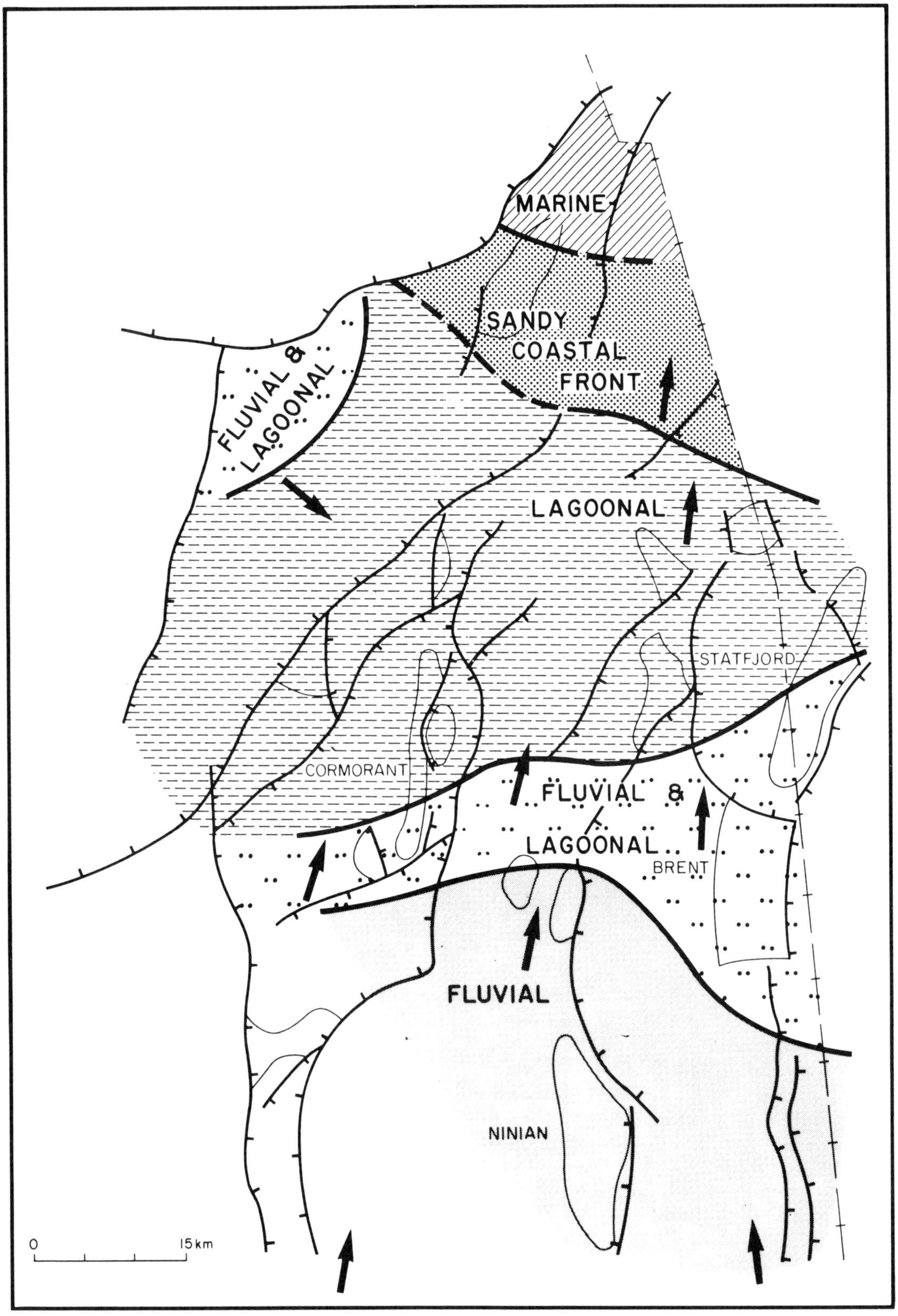

FIG. 18. Lateral facies relationships in late Brent Sands times in the North Viking Graben.

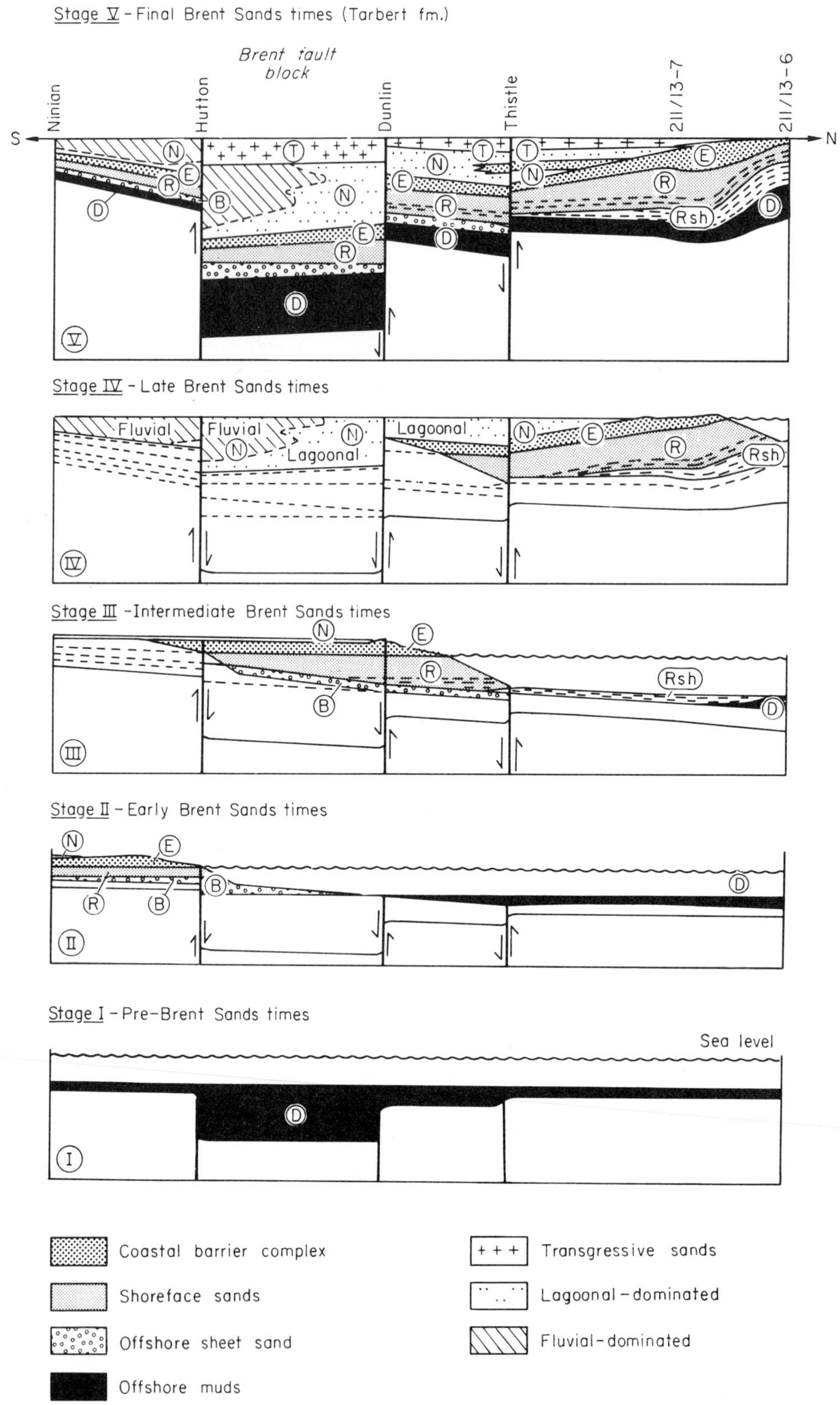

FIG. 19. Conceptual model of the evolution of the Brent Sands depositional system.

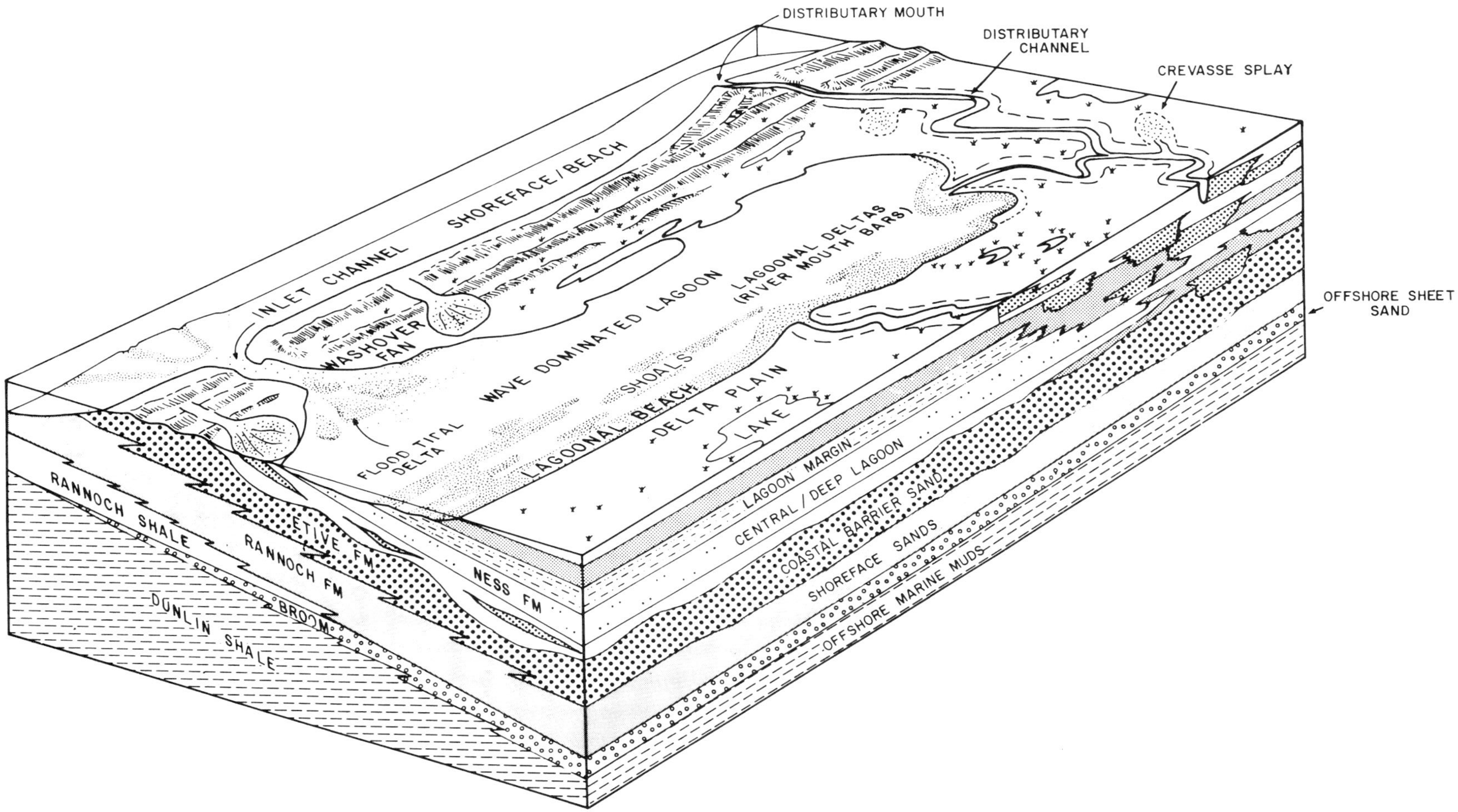

FIG. 20. Conceptual model of deposition in the Brent delta (from Budding & Inglin 1981).

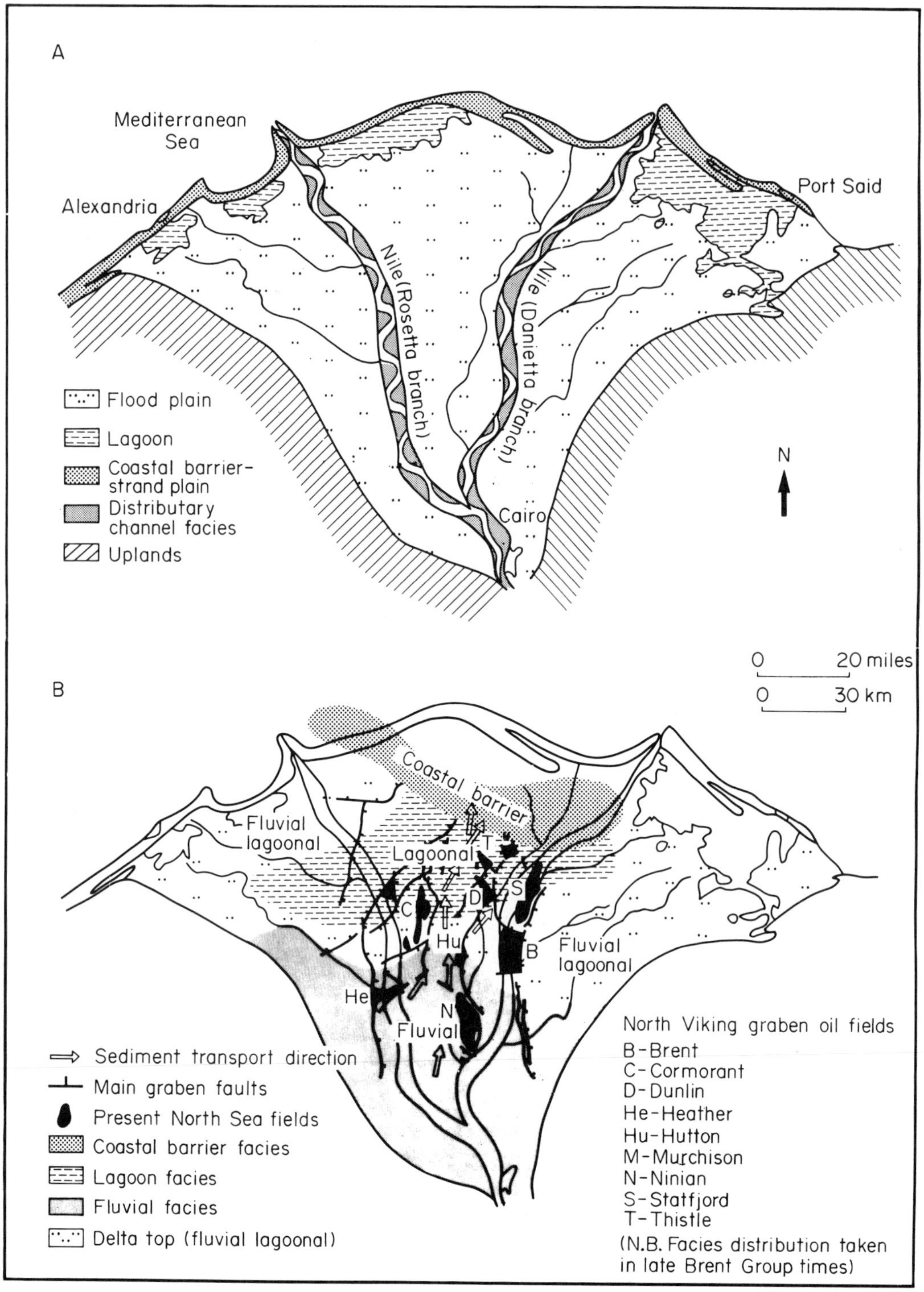

FIG. 21. (a) A possible modern analogue of the Brent Group provided by the Nile delta. (b) Time-slice in late Brent Group times illustrating the observed facies distributions, which are superimposed on the Nile delta to allow comparison of scales.

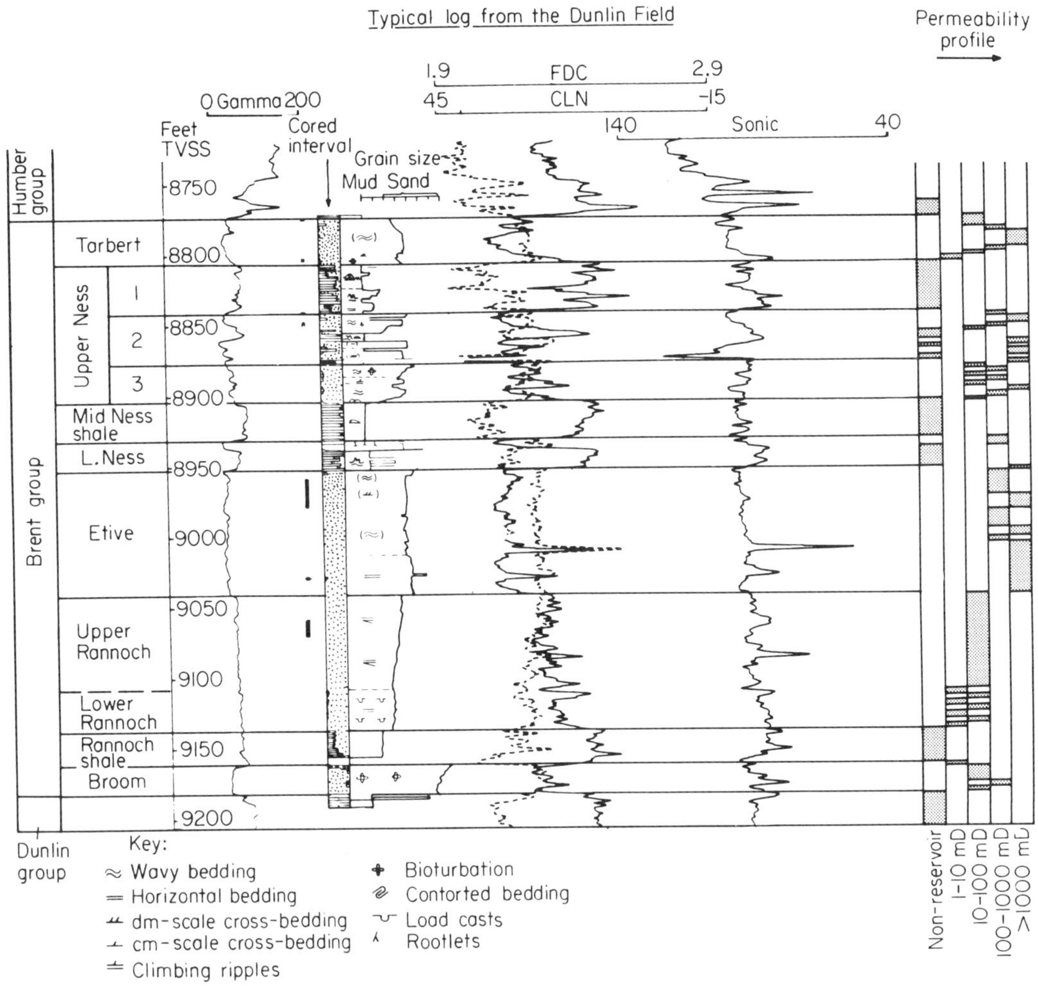

FIG. 22. Typical sedimentary sequences and corresponding wireline log and permeability profiles for Brent Group reservoir sands.

widespread but, in most cases, of secondary importance since it tends to enhance primary variations in reservoir quality rather than overprint them. It has, therefore, been necessary to have a clear understanding of the type and nature of the different facies associations, their reservoir properties, and their vertical and lateral relationships in order to interpret and predict reservoir development and reservoir quality distribution on both regional and field scales.

From a reservoir quality/development point of view, the Brent Sands comprise three main elements: (1) a sand-dominated coastal barrier-shoreface complex (Rannoch and Etive Formations), (2) a variable, alternating sequence of sands and shales deposited in coastal and delta plain environments (Ness Formation), and (3) a relatively thin, transgressive sand sheet (Tarbert Formation).

The *coastal barrier-shoreface complex* forms as essentially tabular-shaped sand sheet characterized by an upward transition from offshore shales and fine grained, micaceous shoreface sands (Rannoch Formation), into coarse grained, well-sorted and non-micaceous sands deposited in upper shoreface/foreshore, distributary channel and beach-ridge environments (Etive Formation). Permeability profiles closely follow the grain size and sorting profiles with the abrupt jump at the Rannoch–Etive boundaries (e.g. Fig. 22) commonly associated with distributary and barrier inlet channels cutting into finer grained shoreface sands.

The *coastal and delta plain complex* comprises a wide range of coal-bearing and rootletted

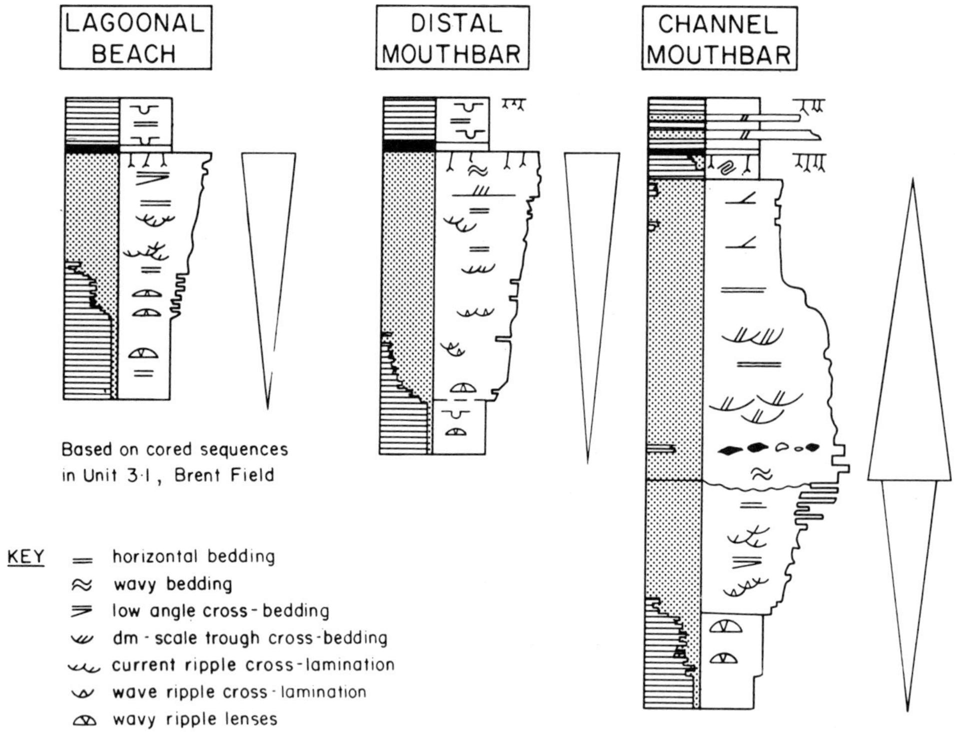

FIG. 23. Examples of lithological sequences through the three main sand body types of the Ness Formation.

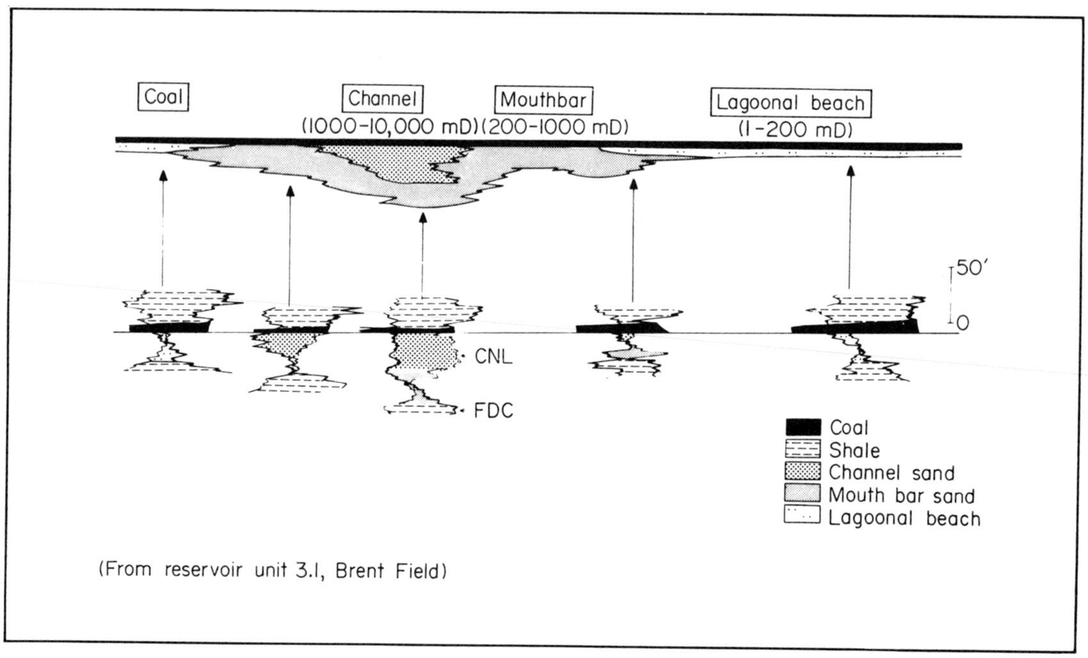

FIG. 24. Examples of FDC/CNL log response through the three main sand body types in the Ness Formation.

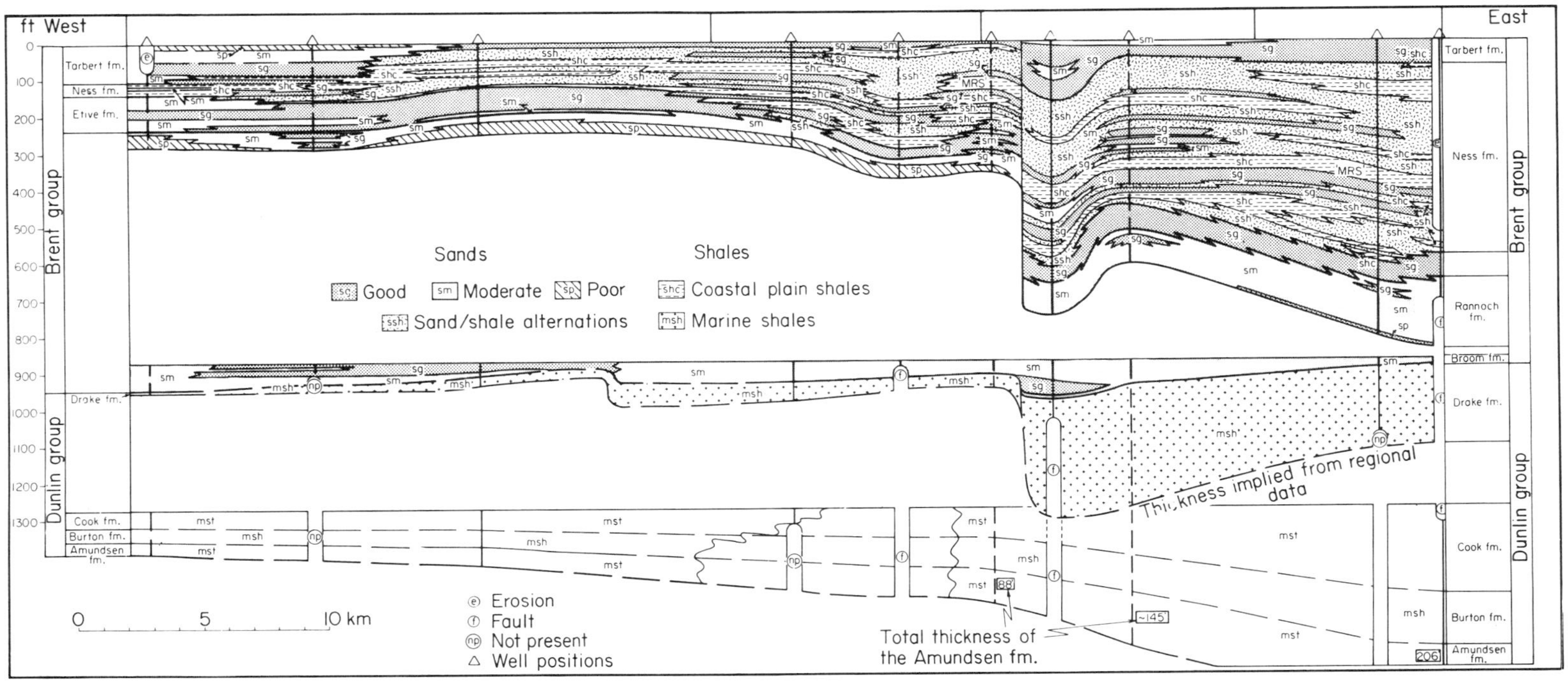

FIG. 25. East to west panel highlighting the influence of syndepositional graben faults on reservoir development in the Brent Group.

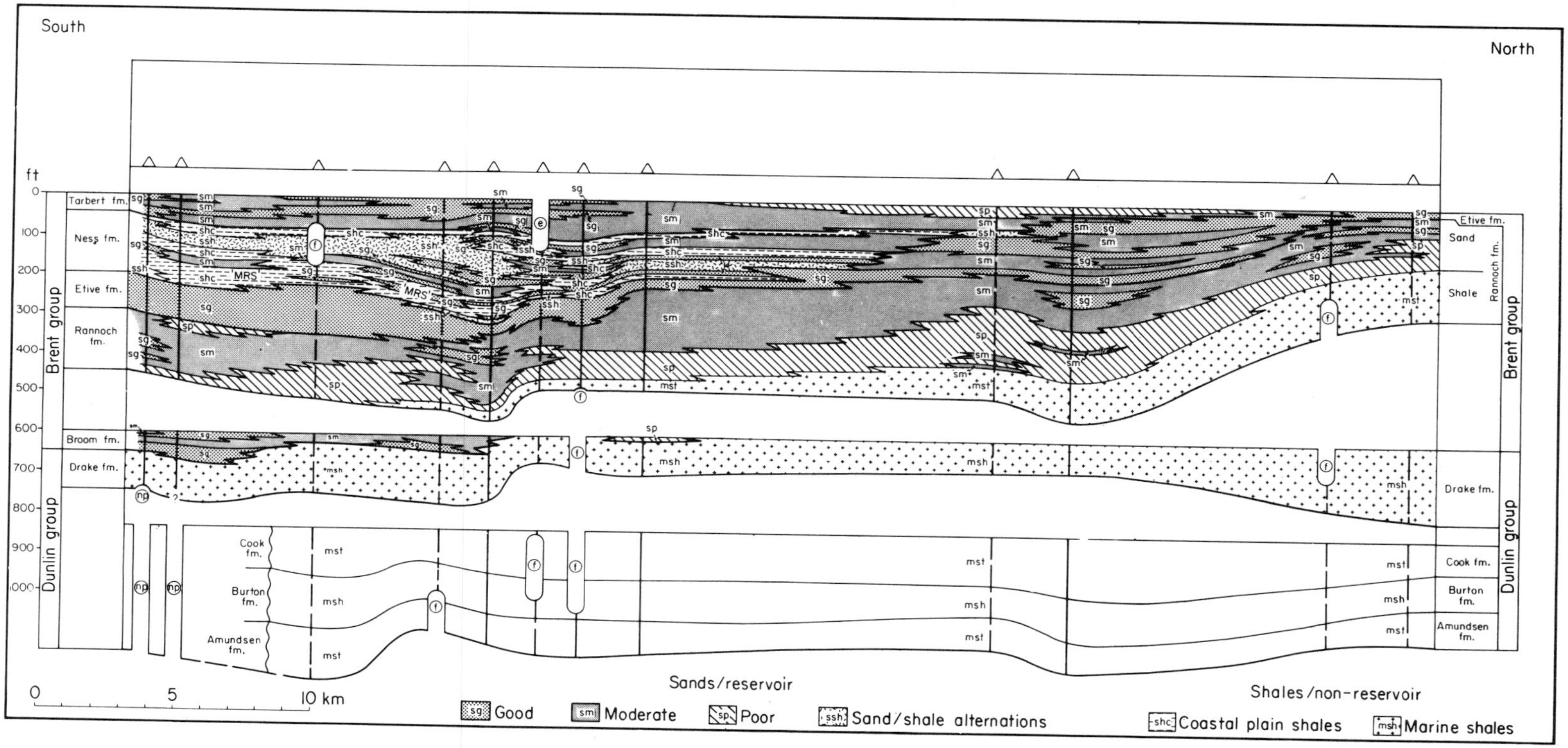

FIG. 26. North to south panel illustrating the northward termination of coastal barrier and concomitant wedge-out of lagoonal shales into barrier and back barrier sands.

sands and shales. Detailed studies of cores and well logs have identified the following facies associations: (1) fluvio-distributary channels, (2) crevasse/sheet splays, (3) river mouth bars/lagoonal deltas, (4) lagoonal beaches, (5) washover fans, (6) floodplain shales, and (7) lagoonal/lacustrine shales (e.g. Fig. 23). Although complex, these deposits occur in vertically and laterally predictable relationships (Figs 23 and 24). Distinguishing between the different types of sand bodies requires careful core/well log calibration and interpretation, which can then enable similar deposits to be distinguished by well logs alone (e.g. Fig. 24). This also represents an essential step in predicting the shape, orientation and porosity/permeability distribution within individual or multistorey sand bodies.

The *transgressive sand sheet* consists of an irregular blanket of reworked, shallow marine sands which were deposited during the final retreat of the Brent shoreline/delta system. Reservoir quality tends to be variable, but in general, porosity and permeabilities are good. In detail this interval appears to comprise several elements of transgressive/onlapping coastal barrier deposits (e.g. shoreface, inlet channels, washover fans and offshore sands).

At the common depth of many North Viking Graben oil fields of around 8000–10,000 ft, and with gross thicknesses of approximately 300–900 ft, the internal characteristics of the Brent Sands have not been able to be resolved by seismic characteristics. This has only been achieved by using the facies characteristics and depositional model, as outlined earlier on the basis of cores and well logs, to provide a framework for predicting reservoir development and reservoir quality distribution on both regional and field scales

Two arbitrary examples of E–W- and N–S-trending correlation panels illustrate typical regional-scale patterns of reservoir distribution. The E–W panel (Fig. 25) highlights the importance of syndepositional graben faults in influencing reservoir development, while the N–S panel (Fig. 26) illustrates the northward termination of the coastal barrier complex and the northward wedge-out of lagoonal shales into barrier and back barrier sands. This type of information enables prediction of anticipated reservoir development and porosity characteristics ahead of exploratory drilling once a basin has reached a relatively mature stage, as in the North Viking Graben. At the same time it provides a framework for more detailed studies associated with appraisal and development drilling.

One application of sedimentological studies to field development is provided by the Brent field in which the Brent Sands are approximately 900 ft thick. The lagoonal/lacustrine deposits noted earlier form four important laterally extensive shale layers which divide the reservoir into five separate drainage units (Fig. 27). Different sedimentological and reservoir characteristics divide the Brent Sands into 15 distinct units which can be correlated over the whole field. Within this framework it is possible to identify the different types of sand body and shale layers and, thereby, reconstruct in some detail their vertical and lateral distribution (Fig. 28). This enables the accurate mapping of a whole suite of reservoir quality parameters such as net sand, porosity, permeability and fluid saturations, amongst others.

A final example of how a sedimentological model can help predict reservoir development trends is provided by the Cormorant field. Here Budding & Inglin (1981) demonstrated how large thickness changes in the Rannoch, Etive and Lower Ness Formations can be interpreted in terms of varying rates of shoreline progradation (Fig. 29). Rapid rates of progradation lead to relatively thin sequences of coastal barrier deposits whereas slow rates of progradation allow accumulation of much thicker sequences. Thickness patterns combined with a knowledge of both regional- and field-scale facies relationships and depositional models were, therefore, able to locate areas of optimum reservoir development, particularly of the high-quality Etive Formation, ahead of extensive development drilling. From a sedimentological point of view the results of this Cormorant field study also gave independent support to the regional model, particularly in terms of shoreline orientation and the direction of progradation (to the NNE).

Geological reservoir modelling of the Statfjord Formation

The Lower Jurassic Statfjord Formation is discussed here in the context of the development of a geological model of the Brent field, which was constructed for a reservoir engineering computer simulation study utilizing a large capacity CRAY computer.

Although economically less important than the Brent Group, the Statfjord Formation contains substantial hydrocarbon accumulations in both the Brent and Statfjord fields (Bowen 1975; Chauvin & Valachi 1980). Regionally, the Statfjord Formation comprises a lower fluviatile succession, which is restricted to the central part

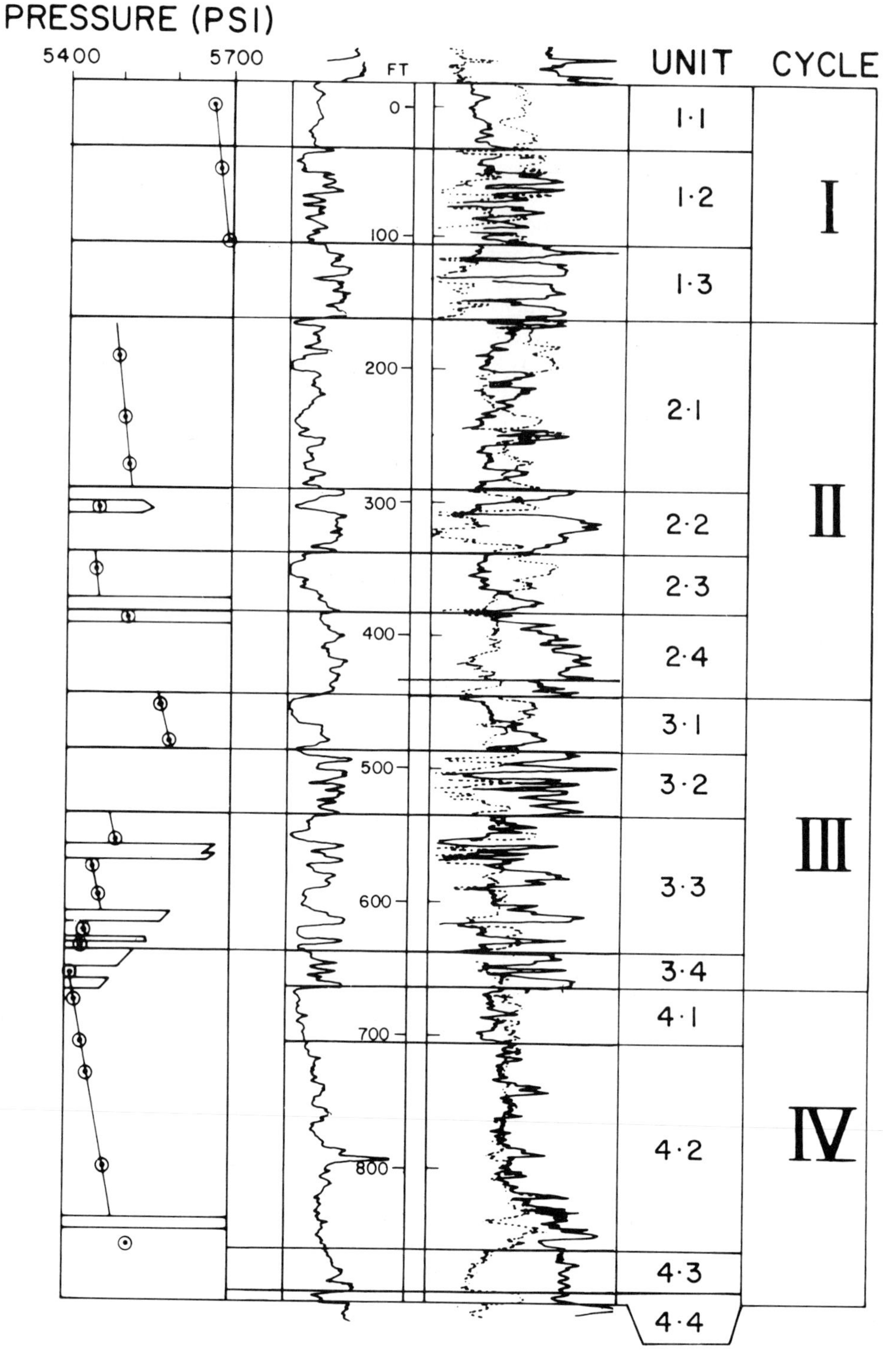

FIG. 27. Subdivision of the Brent Sands in the Brent field illustrating the subdivision of the reservoir into four drainage units and 15 distinct correlation units.

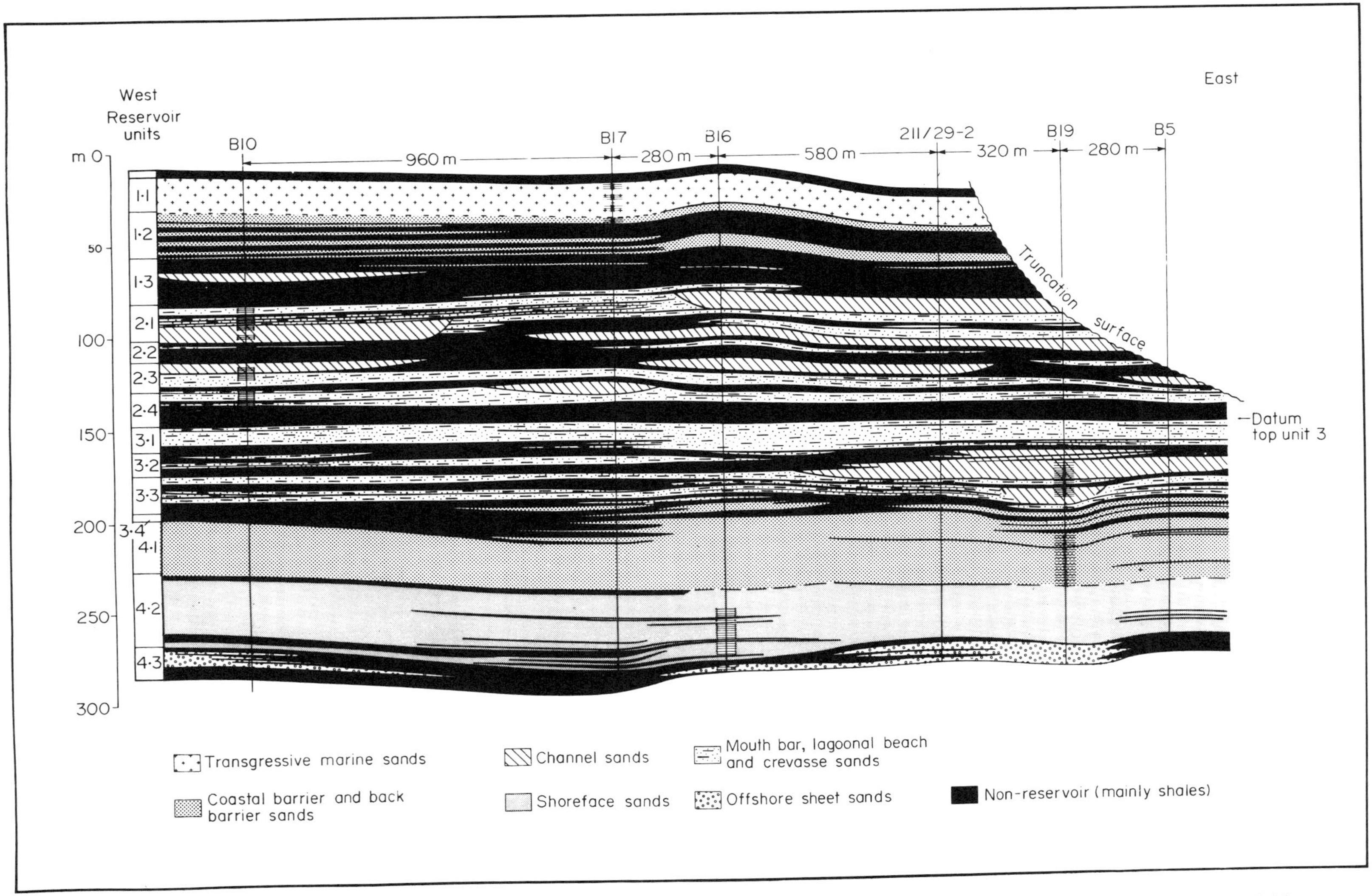

FIG. 28. East to west panel illustrating detailed reconstruction of vertical and lateral distributions of sand and shale bodies in the Brent field.

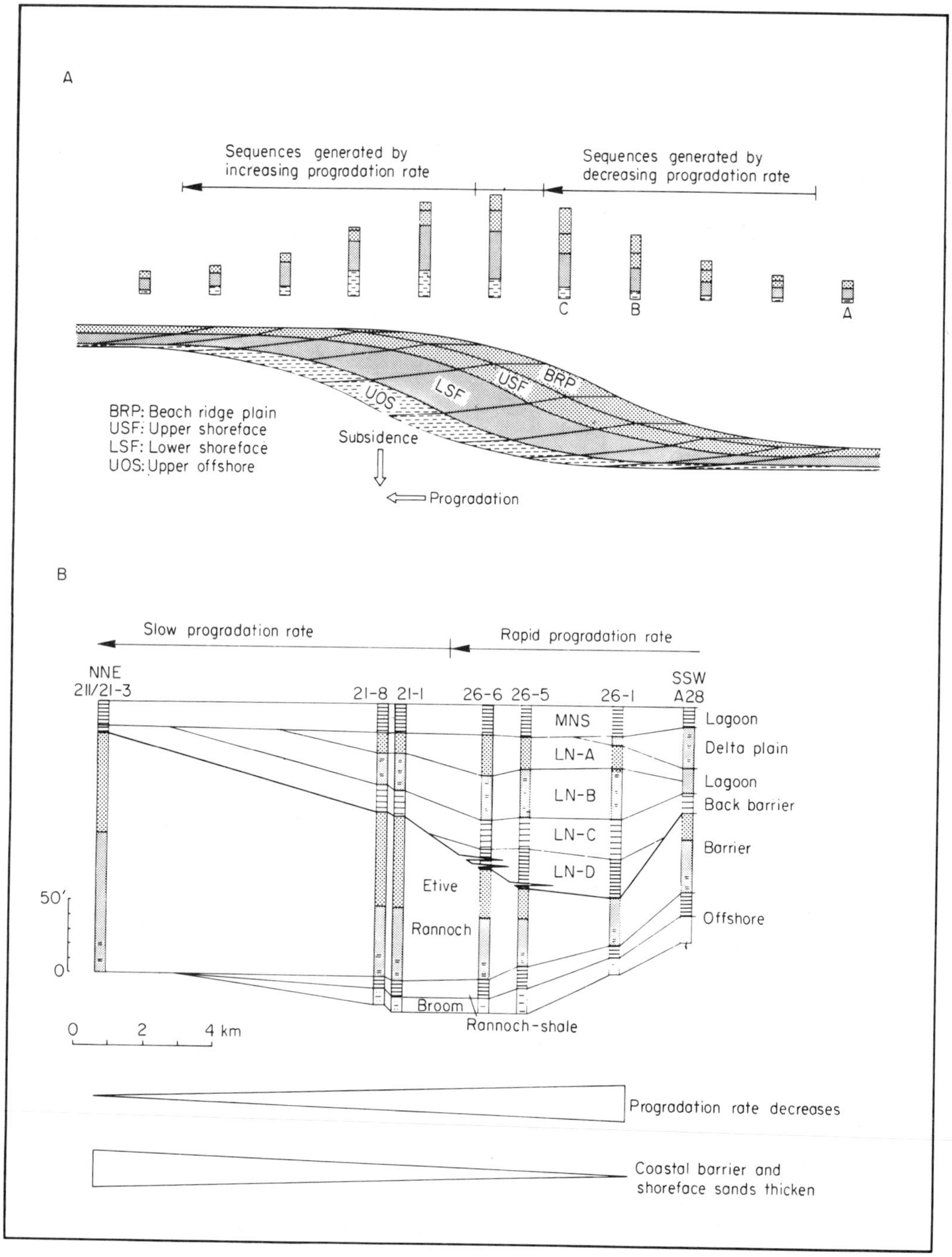

FIG. 29. (A, B) Thickness variations of coastal barrier sediments as a function of progradation rate (from Budding & Inglin 1981). (A) model for thickness variations, (B) example from South Cormorant.

of the graben east of the Hutton–Dunlin fault zone, and an upper and laterally more extensive shallow marine sandstone sequence (Fig. 30). The large-scale depositional history of the Statfjord Formation reflects two main phases of sedimentation: (1) *fluviátile sedimentation* was probably associated with a possibly eustatically induced lowering of the erosional base level, and a gradual increase in surface water run-off accompanying a change from a semi-arid (early Triassic) to a humid/sub-tropical (early Jurassic) climate, and (2) *shallow marine sedimentation* occurred in response to the early Sinemurian (*c.* 189 Ma) eustatic sea-level rise (Vail & Todd 1981) which ultimately led to the deposition of offshore/shelf muds throughout

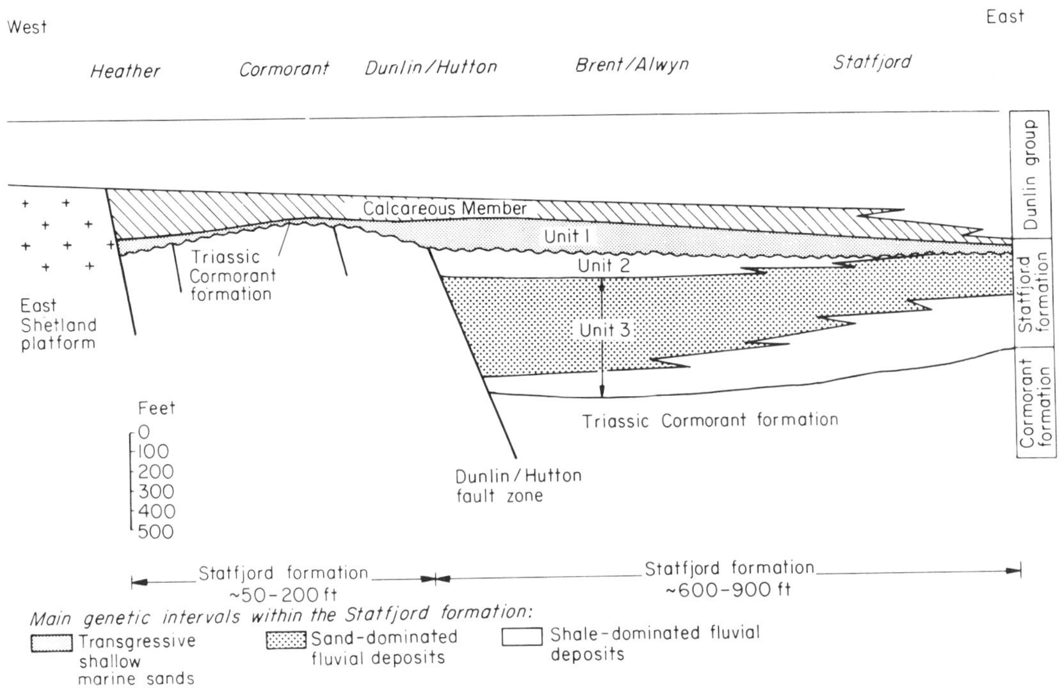

FIG. 30. Cross-section through the Statfjord Formation in the North Viking Graben.

the North Sea area.

In the Brent field, the Statfjord Formation is represented by a large-scale offlapping-onlapping couplet. The offlapping sequence occurs in the lower part of the reservoir (Cormorant Formation → Unit 3.2 → Unit 3.1) and comprises an upward increase in sand content, pebble size and frequency, channel sand body density and channel sand body interconnectedness. The onlapping/transgressive sequence begins with an abrupt change from the fluvial channel sand-dominated Unit 3.1 into the floodplain shale-dominated Unit 2 (Fig. 31), and this is followed by a succession of increasingly lower energy/offshore shallow marine deposits (Unit 1 → Calcareous Member → Amundsen Formation), with the high-quality sands of Unit 1 forming the important, uppermost reservoir unit in the Brent field. A conceptual reservoir geological model of the Brent field area, illustrated in Fig. 32, schematically outlines the inferred distribution of sand bodies and shale layers.

Of particular importance to the development of the Statfjord Formation in the Brent field was the potential for fluid communication between the various reservoir units, in particular between Units 1 and 3.1 across the shaley Unit 2. This stems from an investigation into the possibility of producing the reservoir by injecting gas under miscible conditions into Unit 1 and injecting water into Unit 3 (Fig. 33); such a scheme requires Unit 2 to behave as a seal. In order to test the significance and sealing potential of Unit 2 a detailed reservoir geological model was constructed from the B-platform area of the field where vertical fluid communication is a possibility for the following reasons: (1) the basal shale layer is locally cut through by channel sands, (2) sand/shale ratios are abnormally high for Unit 2, ranging between 0.4 and 0.65 in this area, (3) multistorey channel sands are locally well-developed, and (4) theoretical studies indicate substantial increases in channel interconnectedness when sand/shale ratios exceed 0.5 (Allen 1978).

The geological model was constructed within the framework of both the B-platform area in particular (Fig. 34) and the field as a whole, and to the limitations of the computer. The model comprised 30 layers with each layer being 25 ft thick, which approximates to the thickness of individual channel sand bodies. Areally the model comprised 10 × 11 grid blocks, measuring 5000 × 5500 ft (1.5 × 1.7 km), and included three wells within the model area plus seven wells in the immediate vicinity. The model was constructed in a series of panels depicting both observed and inferred distributions of sand and shale layers (Fig. 35).

In order to model the geological heterogeneity of this type of reservoir realistically it was

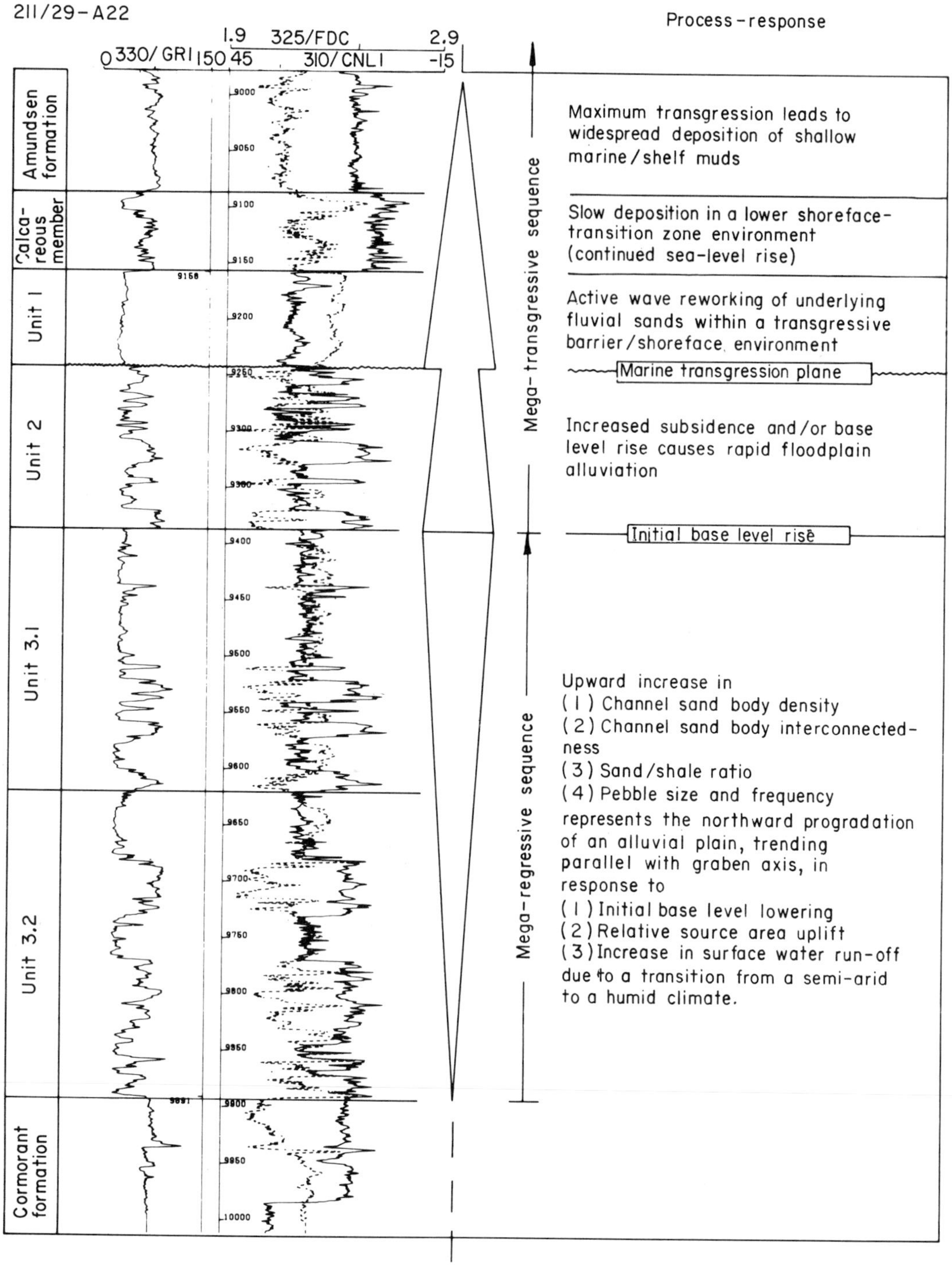

FIG. 31 (above). Interpretation of the large-scale depositional history of the Statfjord Formation in the Brent Field.

FIG. 32 (facing). Conceptual reservoir geological model through the Statfjord Formation in the Brent Field.

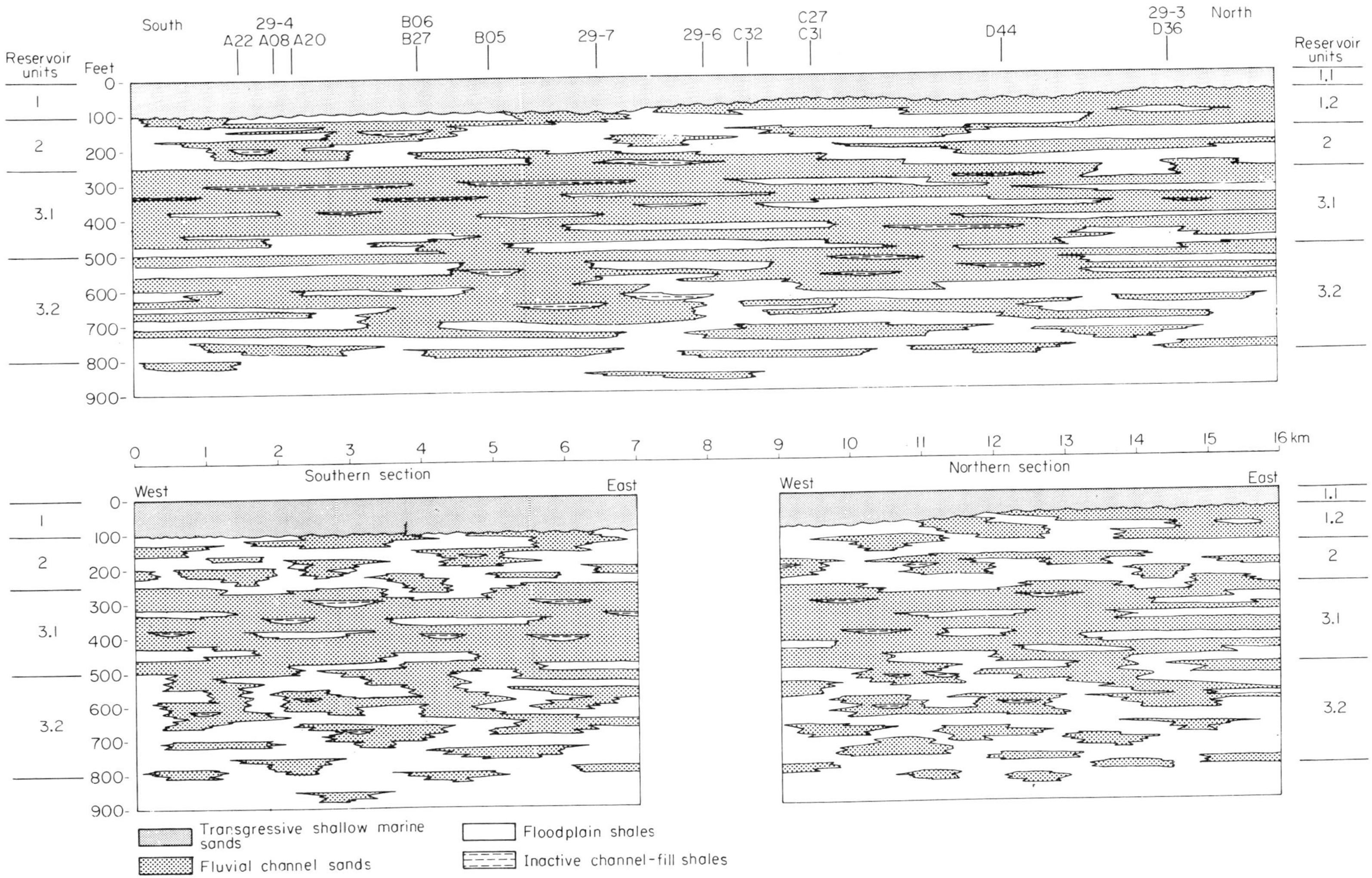
South
North
29-4
A22 A08 A20
B06
B27
B05
29-7
29-6
C32
C27
C31
D44
29-3
D36
Reservoir units
Feet
0
100
200
300
400
500
600
700
800
900
1
2
3.1
3.2
1.1
1.2
0 1 2 3 4 5 6 7 8 9 10 11 12 13 14 15 16 km
Southern section
Northern section
West
East
Transgressive shallow marine sands
Fluvial channel sands
Floodplain shales
Inactive channel-fill shales

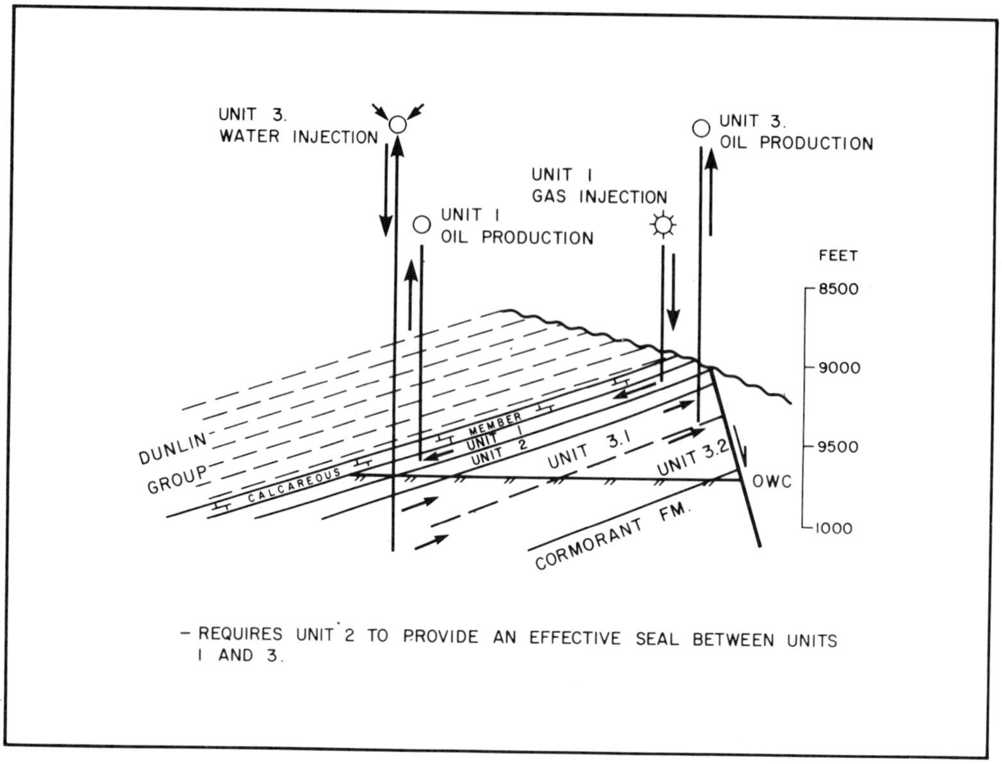

FIG. 33. Statfjord reservoir, Brent field, indicating a possible recovery scheme involving combined miscible gas/water injection.

necessary to utilize a CRAY computer, which has the capacity for handling complex three-dimensional models. Indeed, the increasing use of the new breed of supercomputers in reservoir simulation studies is an important step forward because it allows the use of more realistic geological models than has been previously possible. In all cases, however, the type of model, particularly its complexity, should be developed in the context of the objectives of the simulation exercise and it requires close liaison between the geologist and reservoir engineer. In the case cited above, it is possible to test the sensitivity of the system to different geological conditions (e.g. varying degrees of channel interconnectedness) which can given important insight into the behaviour of fluid movements in complex reservoirs.

Upper Jurassic sandstones, northern and central North Sea

Regionally the Lower and Middle Jurassic reservoirs of the Northern Viking Graben present a relatively simple 'layer cake' geological framework, and the oil accumulates in rather simple, truncated structural traps (tilted fault blocks). In contrast, Upper Jurassic oil accumulations occur in reservoirs which are often of more restricted lateral extent, owing to rapid lateral changes in facies type and reservoir thickness. They therefore confront both the explorationist and production geologist with a somewhat different set of problems.

Geological framework

Four main facies associations have been recognized within the Upper Jurassic, Humber Group (Oxfordian to Portlandian) of the Central and Northern North Sea:

(i) shallow marine sand association (reservoir);
(ii) shallow marine shale association, characterized by low levels of radioactivity (seal);
(iii) gravity/mass flow sands and conglomerates (reservoir); and
(iv) basinal shales (highly radioactive; gamma ray readings approximately over 100 API), deposited in anoxic conditions (seal/source rock).

During the Upper Jurassic much of the northern and central North Sea was the site of argillaceous sedimentation, including the widespread development of organic shales, thought

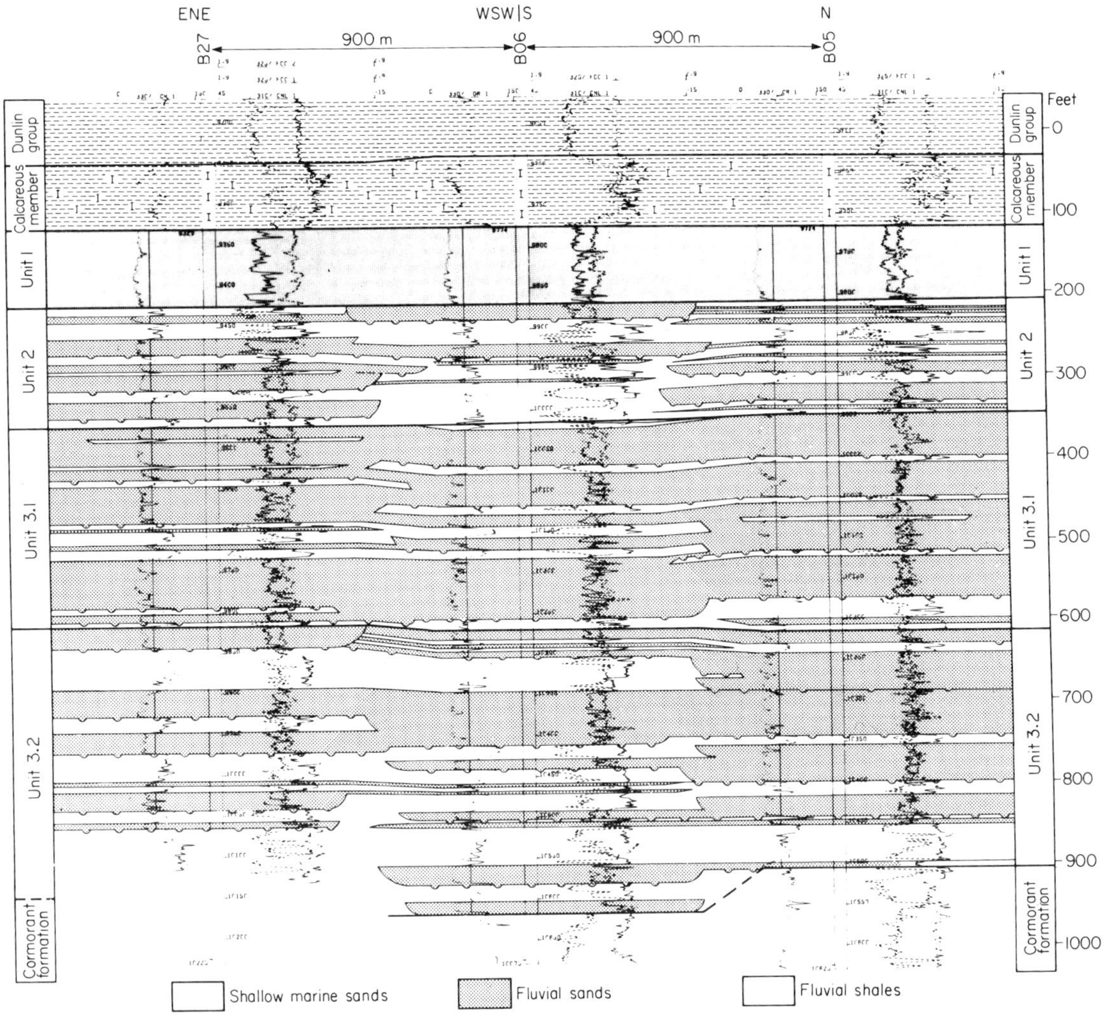

FIG. 34. Well log correlations in the Statfjord Formation in the southern part of the Brent field.

to be the main source rock for North Sea oil (Fig. 36). However, tectonic factors, in particular the continued movement of major faults and the associated development of half grabens, have caused the local accumulation of clastic deposits. In the Central Graben salt movements have had additional influences on sedimentation. Furthermore, the Humber Group forms an overall transgressive sequence, reflecting a relative rise in sea level which may be eustatic in nature (Vail *et al.* & 1977; Hallam 1978).

Two main styles of clastic reservoir are recognized in the Humber Group; shallow marine sand and mass flow sand/conglomerate accumulations (Fig. 37).

The mass flow sand/conglomerate sequences (e.g. Brae, Toni, Thelma, Magnus) are enclosed in, or interfinger with, the organic-rich Kimmeridge Clay, which frequently acts as source rock and seal. Shallow marine sand reservoirs (example Piper, Fulmar, Ula) are also generally capped by Kimmeridge Clay. Syndepositional tectonics appear to have played an important role during the accumulation of sand bodies of both facies types. Geographically, the Upper Jurassic shallow marine sandstones are restricted mainly to the Moray

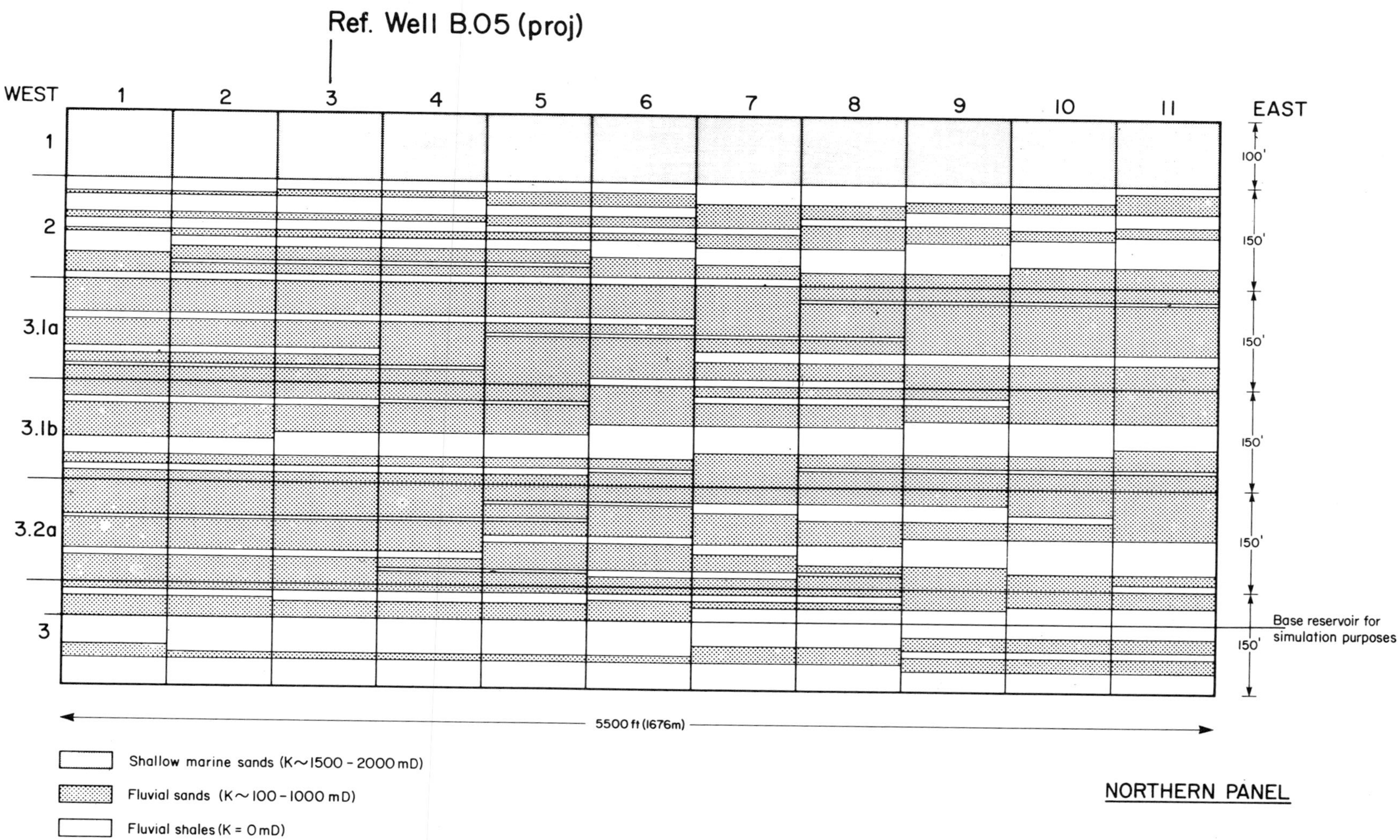

NORTHERN PANEL

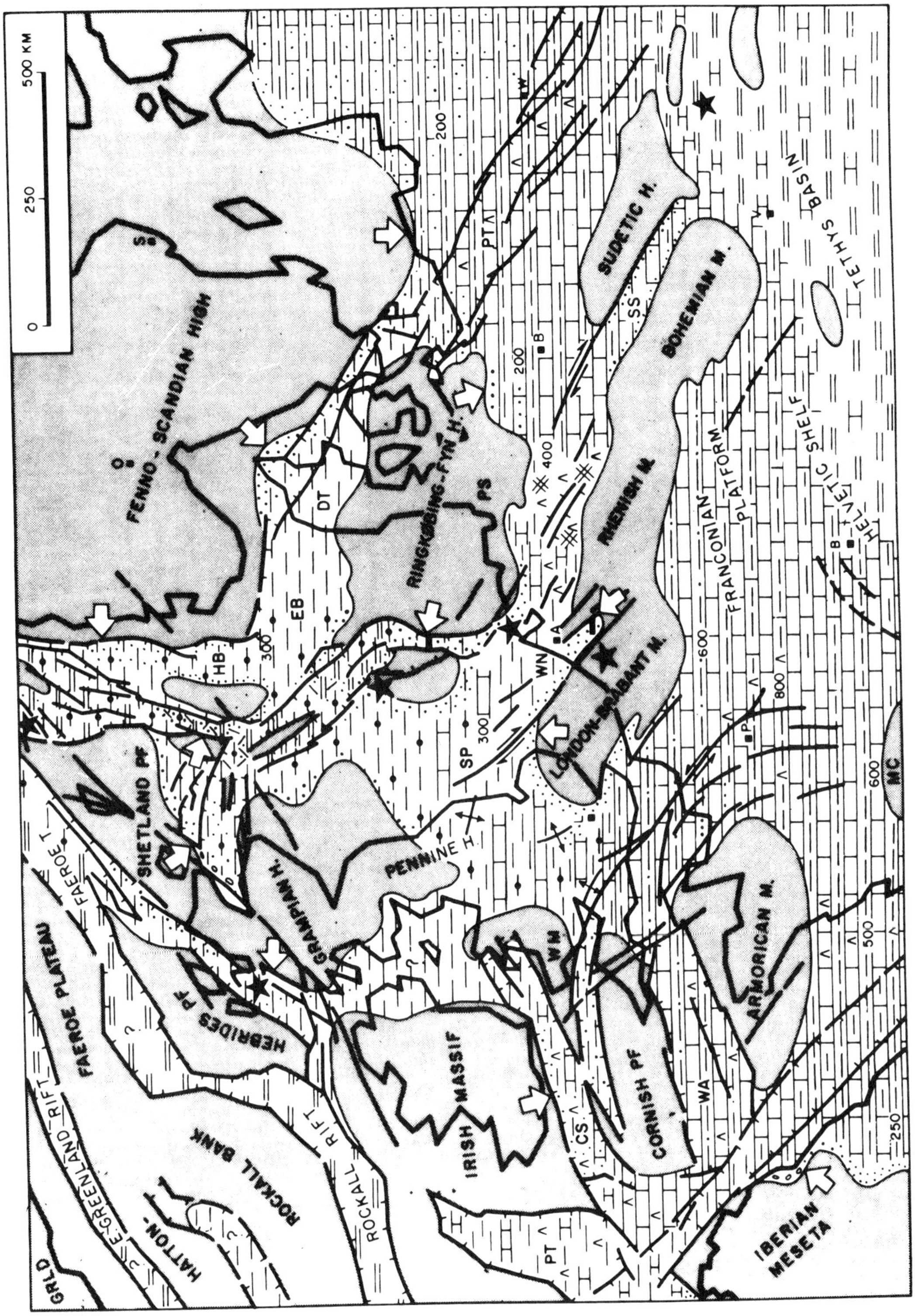

FIG. 35 (facing). An example of a panel used as input for a reservoir simulation model of the Statfjord Formation in the Brent field using a CRAY computer.

FIG. 36 (above). Oxfordian–Portlandian palaeogeography (from Ziegler 1981).

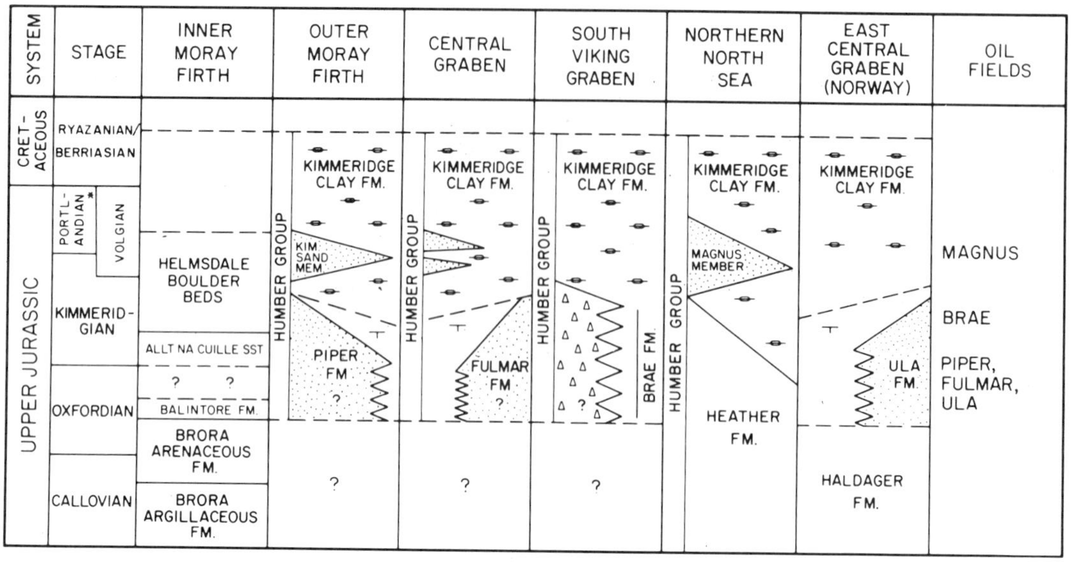

FIG. 37. Stratigraphic framework of the Upper Jurassic of the central and northern North Sea. Age equivalents are approximate.

Firth and central North Sea. Mass flow sands/conglomerates occur over a larger area, but are most common in the South Viking Graben and Moray Firth.

Mass flow sand/conglomerate reservoirs

Reservoirs comprising this reservoir type are formed by two end members:

(i) conglomerate-dominated facies associations; and
(ii) sand-dominated facies associations.

In neither case do we attach any depth connotations to the deposition of these mass flow sands, but in our opinion they were deposited by gravity-driven processes.

Conglomerate-dominated facies associations

The western edge of the South Viking Graben and the Helmsdale region of the Inner Moray Firth are characterized by prominent fault zones with an overall present-day throw on a scale of thousands of feet. In the case of the South Viking Graben the respective fault system delimits the margin of the stable East Shetland Platform from the rapidly subsiding Viking Graben (Fig. 38); Jurassic movement along this fault dramatically affected the type and thickness of sediments deposited in the graben. The sediments were accumulated on the downthrown side of the fault as a thick syntectonic clastic wedge (Fig. 39). This wedge comprises thick scree-like deposits of angular cobbles and boulders (mainly of Devonian sandstones), amalgamated sandstone units and interbedded sandstones and shales. A simplified facies breakdown is outlined on Fig. 40.

The gross facies types, recognized in cores, can also be recognized in uncored wells with fair success by using a combination of gamma ray and FDC/CNL logs, as demonstrated on Fig. 40.

The best known accumulation of this type is the Brae field. Facies changes in Brae are abrupt (Fig. 39) and the conglomeratic facies lies close to the fault scarp, wedging out in 2–5 km into thin sandstones, laminated siltstones and sandy siltstones (Harms *et al.* 1981).

Harms *et al.* (1981) considered two interpretations for these deposits (but favoured the former):

(i) fan delta systems, and
(ii) submarine fan systems.

In their preferred model the breccia and conglomerate is considered to be subaerial and the finer grained deposits submarine.

Other authors (Stow *et al.* 1982), including ourselves, favour the submarine fan interpretation. In this respect, much emphasis is placed on outcrop studies of the Helmsdale Boulder beds (Neves & Selley 1975), where it can be demonstrated that ammonite- and belemnite-bearing marine shales arc intercalated with the boulder beds. The model for these strata has been greatly influenced by the outcrop studies of closely comparable deposits by Surlyk (1978) on the Lower Cretaceous of East Greenland (Fig. 41), and thus serves to emphasize the importance of relevant outcrop data. It cannot be proven whether the apices of the fans were submarine or subaerial. Furthermore, sediment

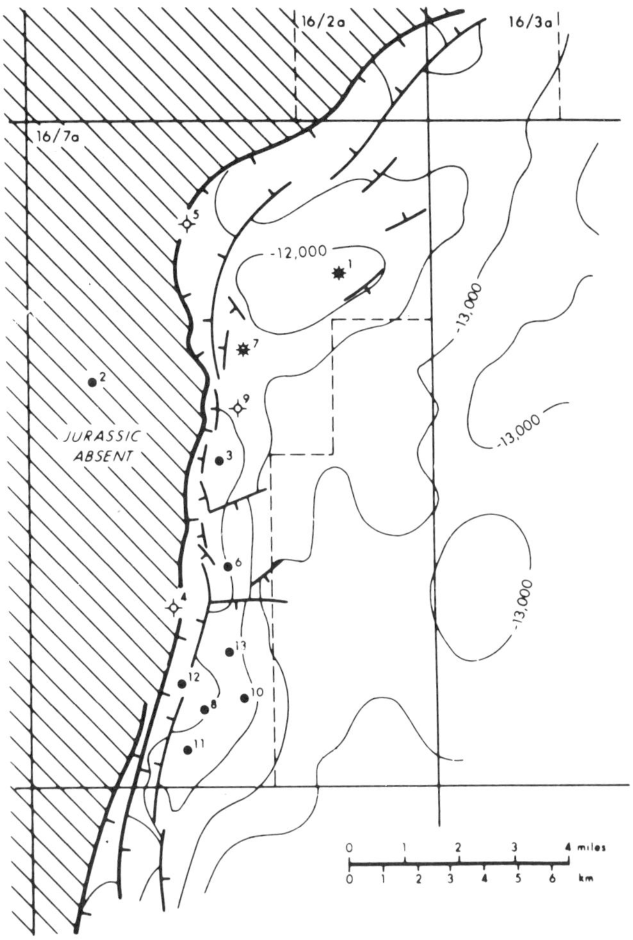

FIG. 38. Structure at top Jurassic level in Block 16/7a with faults cutting through the uppermost Jurassic. The closures approximate to coarse clastic fan deposits which wedge out basinwards (from Harms *et al.* 1981).

may have been transported to the fault edge by a combination of shallow marine, fluvial and mass emplacement processes. Whichever model is used the resultant reservoir distribution model will be very similar.

The main features can be summarized as follows:

(i) The sizes of individual fans are small, the radius of the main reservoir section (i.e. conglomerates and pebbly sands) being 4–5 km. These pass basinward into thinly bedded (mainly non-commercial) sands and silts and then into shales.

(ii) The fans may coalesce laterally to form continuous facies belts parallel to the fault zone but, judging from the location of dry wells in Brae, interfan areas are composed either of tight siltstones or similarly tight breccias.

(iii) The hydrocarbon column in the Brae Field is in the order of 1500 ft. To account for this extreme column length a stratigraphic trapping element probably involving a seal against the East Shetland boundary fault is required. This stratigraphic trapping element is probably provided by the basinward passage into siltstones and shales (Fig. 39). Since the structural closure is only a fraction of the oil column (Fig. 38) it is extremely difficult to locate such traps and once located it is difficult to estimate their reserves. The latter problem is further hampered because it is difficult to calculate net oil sand from wireline logs in some of the more heterogeneous facies due to a lack of resolution in very thinly bedded sequences. In these cases cores become very important for such estimates.

(iv) In the Viking Graben, Brae-type reservoirs are found between 11,000 and 13,000 ft depth, which results in rather low

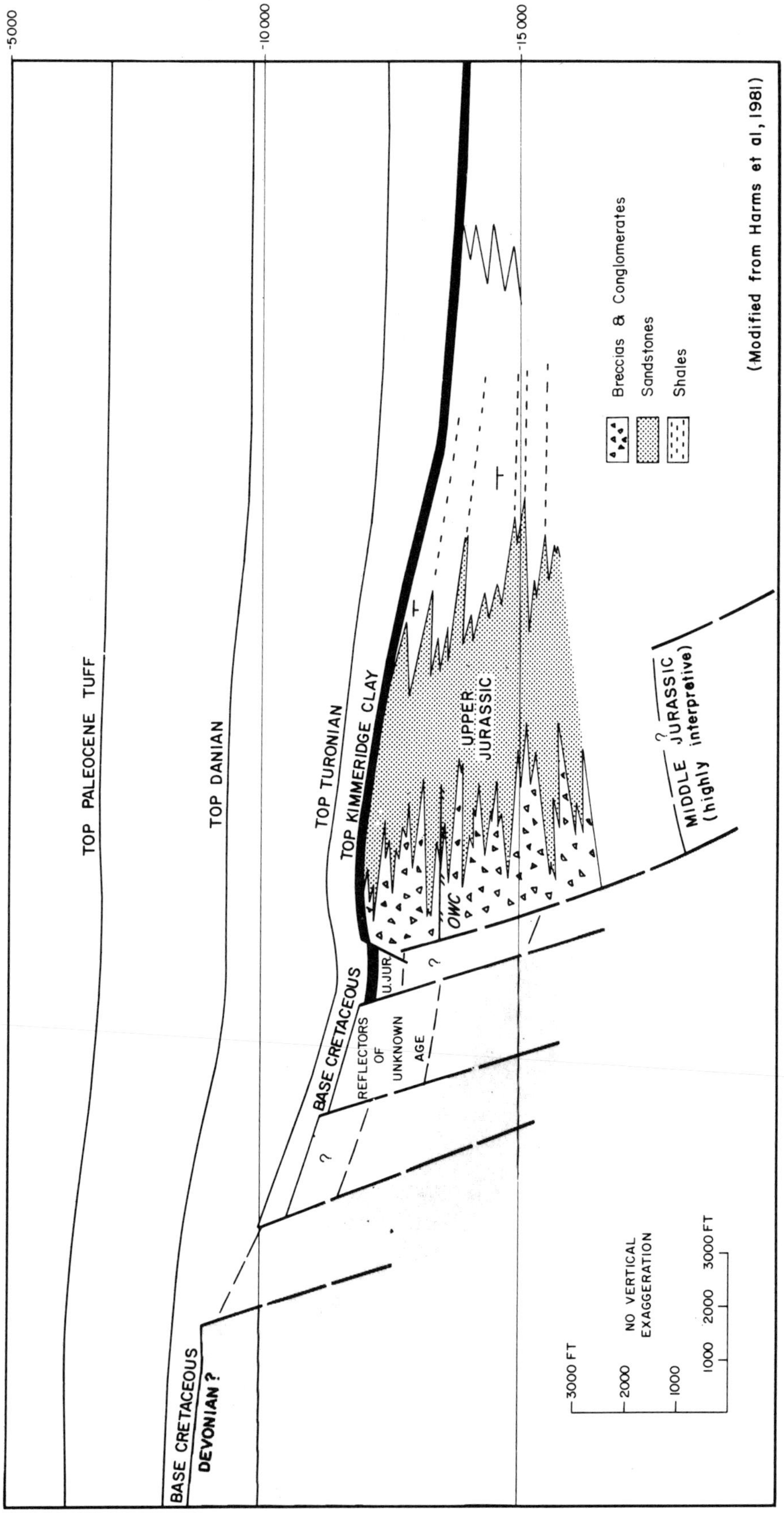
-5000
-10 000
-15 000
TOP PALEOCENE TUFF
TOP DANIAN
TOP TURONIAN
TOP KIMMERIDGE CLAY
BASE CRETACEOUS
BASE CRETACEOUS
DEVONIAN ?
REFLECTORS OF UNKNOWN AGE
U.JUR.
OWC
UPPER JURASSIC
MIDDLE ? JURASSIC (highly interpretive)
Breccias & Conglomerates
Sandstones
Shales
NO VERTICAL EXAGGERATION
3000 FT
2000
1000
1000 2000 3000 FT
(Modified from Harms et al, 1981)

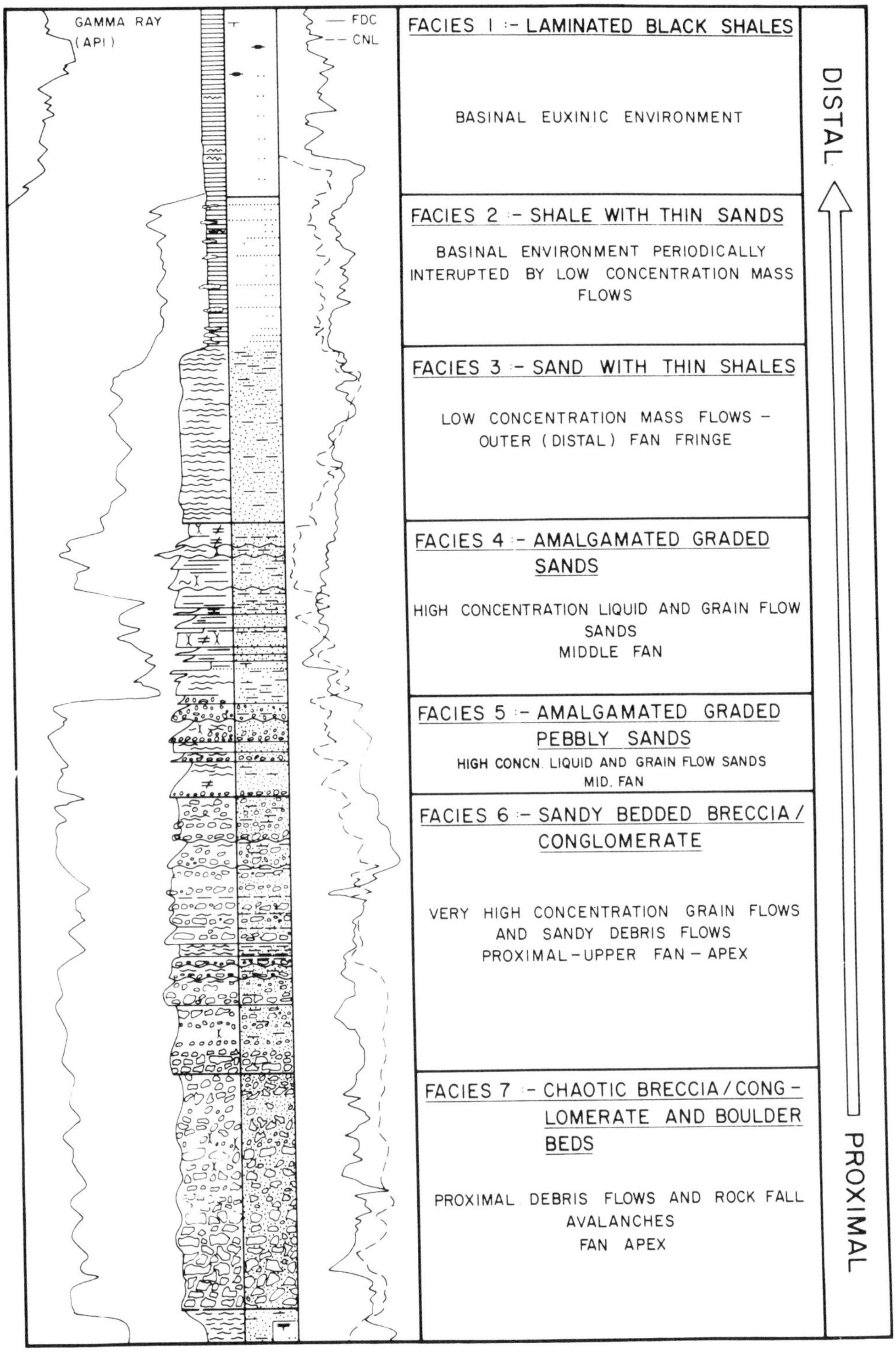

FIG. 39 (facing). East-west cross-section through the southern part of the Brae area (from Harms *et al.* 1981).

FIG. 40 (above). Schematic vertical profiles with inferred well log response through a typical Brae-type sequence.

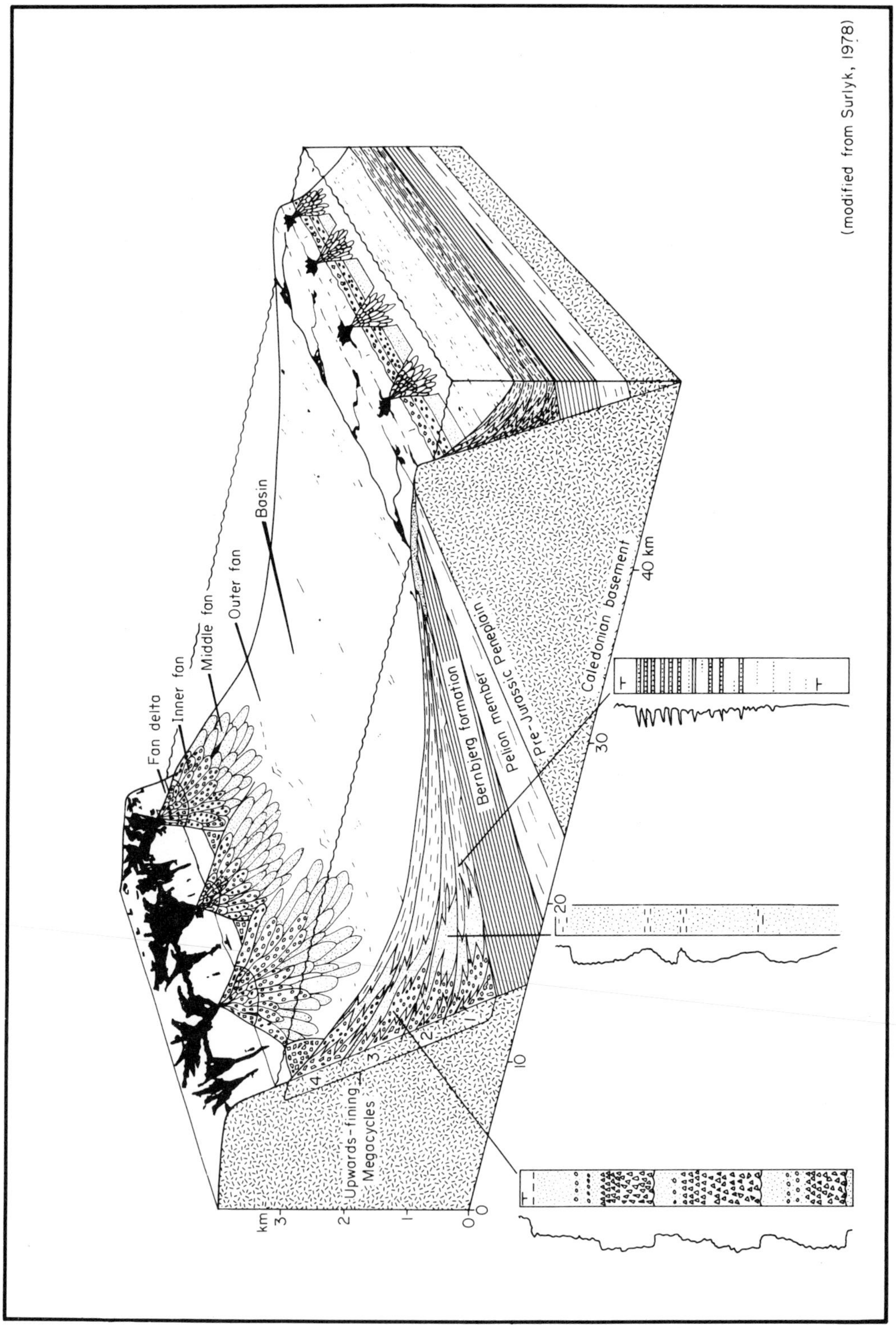

(modified from Surlyk, 1978)

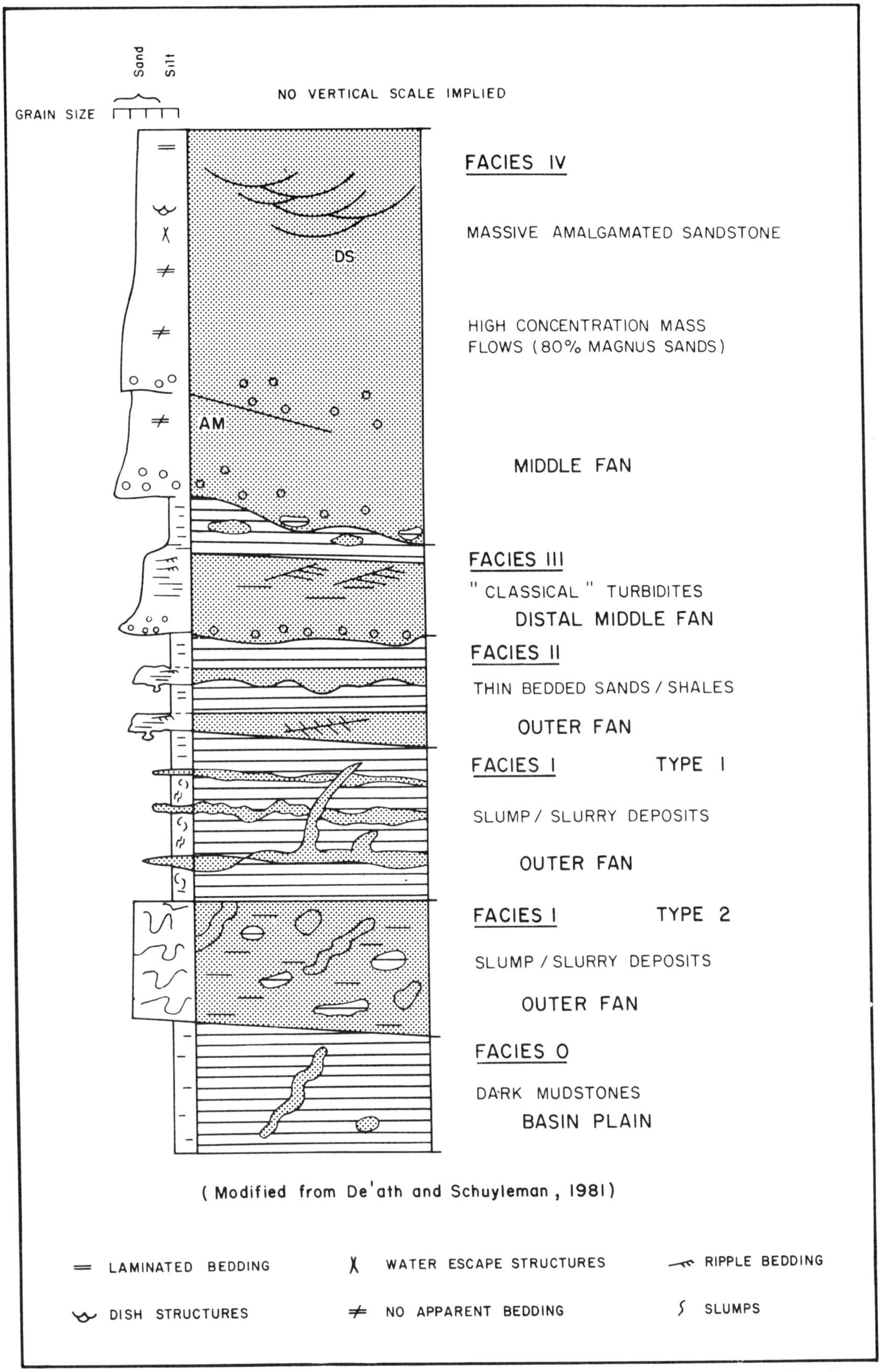

FIG. 41 (facing). Facies relationships in Jurassic submarine fan deposits based on outcrop studies of the Wollaston Forland Group of East Greenland (modified form Surlyk 1978). Vertical profiles depict schematic gamma ray profiles.

FIG. 42 (above). Facies scheme for sandy fan Magnus-type deposits (modified from De'Ath & Schuyleman 1981).

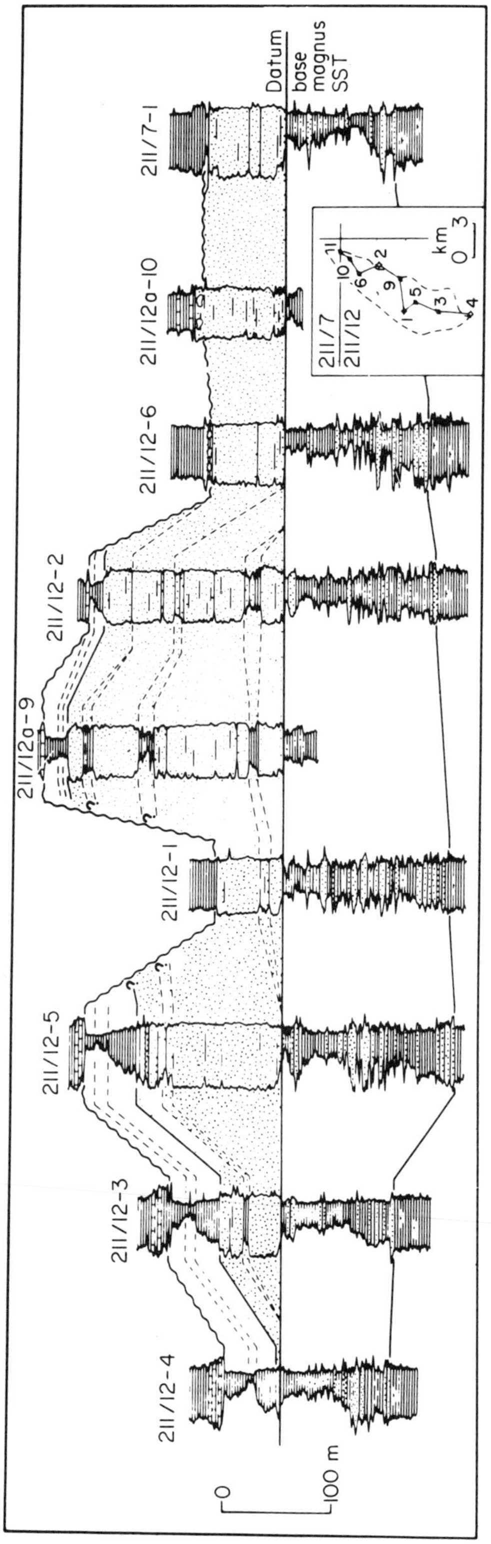

FIG. 43. Cross-section through the Magnus field illustrating typical sand distribution patterns (from De 'Ath & Schuyleman 1981).

average porosities and permeabilities. However, the reservoir rocks interdigitate with basinal shale source rocks, which are sufficiently mature to generate oil. Consequently there is a high probability that oil will migrate into the reservoir.

(v) A further point to consider is the possibility of enhancement in reservoir quality at these depths by the leaching of detrital feldspar (Harms *et al.* 1981).

It is apparent from the above that conglomerate-dominated reservoirs of this type are complex and difficult to develop. This is clearly demonstrated in the drastic reduction in reserve estimates, from 800 to 300 mb, that has taken place between initial discovery and the subsequent drilling of a further twelve exploratory and appraisal wells in the first three years of appraisal of the Brae field (Harms *et al.* 1981).

Sand-dominated facies association

Upper Jurassic sandy mass flows occur throughout the Viking Graben, Central Graben and Moray Firth, but the only field of this reservoir type under development is BP's Magnus field in the North Viking Graben (De'Ath & Schuyleman 1981). Sand-dominated submarine fans appear to be slope-related rather than fault-related, although it cannot be ruled out that small-throw faults have influenced deposition. Like the conglomerate-dominated fans, sand-dominated fans have a patchy distribution, making them difficult exploration targets. They contrast with conglomerate facies associations in tending to be of greater aerial extent, whilst individual fans are isolated rather than coalescent.

The Magnus Sands comprise five basic facies (Fig. 42) as originally described by De 'Ath & Schuyleman (1981). However, the majority of the reservoir comprises massive, amalgamated sandstone units, which represents their facies IV. These sandstones have been interpreted as the Middle Fan deposits of an eastward prograding submarine fan complex (De 'Ath & Schuyleman 1981). This interpretation is based on sedimentological studies which have shown an eastward fining in grain size, an increase in detrital clay content and decrease in bed thickness as the outer fan is approached. In addition, the sands wedge out into shales to the south (Fig. 43).

The main reservoir characteristics of sandy mass flow facies associations are outlined below:

(i) Fans of this type appear to extend over considerable areas. The Magnus fan covers an area of approximately 12 ×

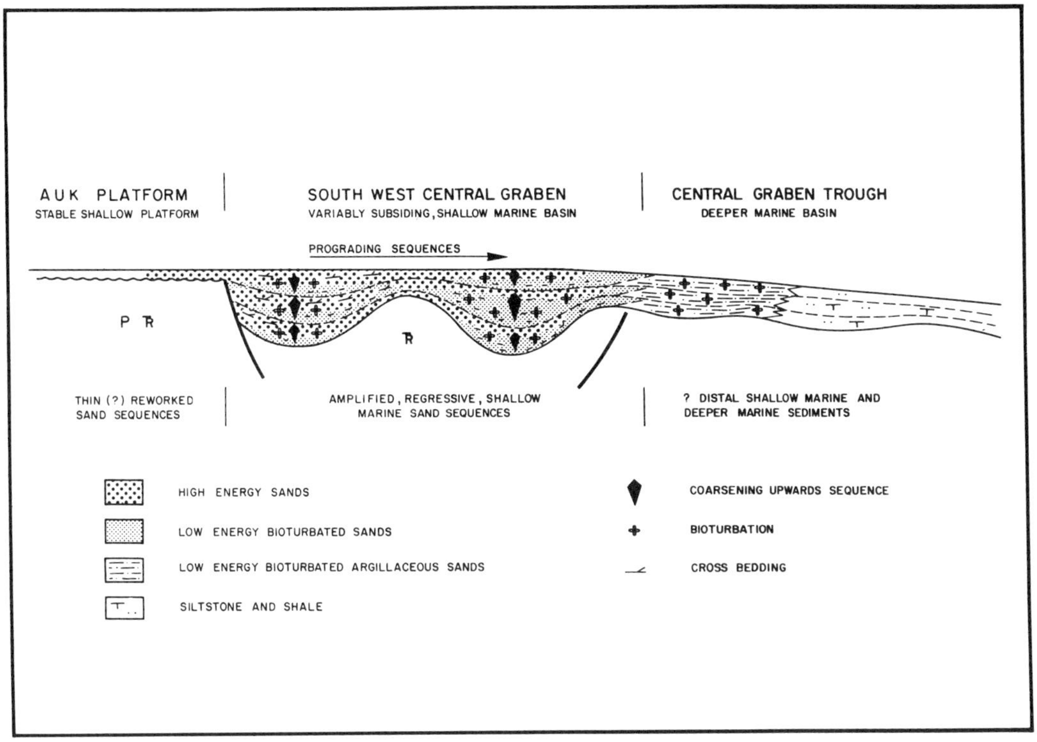

FIG. 44. Schematic cross-section through the Fulmar Sands in the Southwest Central Graben area.

12 km, but thin turbiditic sands could penetrate into the basin to considerable distances. However, these latter sands are unlikely to be of commercial significance.

(ii) Although porosities and permeabilities of these sands are fairly good, average permeabilities are, on the whole, lower than in shallow marine sands at corresponding depths.

(iii) Lateral facies changes are common. However, owing to the sandy nature of most facies the reservoir appears to be homogeneous enough to act as a single producing unit (De 'Ath & Schuyleman 1981). Appraisal of the Magnus field has therefore, been more straightforward than that of the Brae field, with fewer and more predictable appraisal wells required (eight appraisal wells).

(iv) The massive amalgamated sandstones that form the main reservoir of the sandy mass flow facies association can be recognized on wireline logs by their blocky, cylindrical shapes (cf. conglomerate-dominated mass flow association and shallow marine associations).

(v) These reservoirs possess good stratigraphic trap possibilities since they are generally encased in organic shales. These shales may act both as the source for oil and the reservoir seal. The close proximity of the source rock and reservoir rock is an obvious advantage, especially, as in Magnus, when the source rock is mature to generate oil.

Shallow marine sandstone reservoirs

In the U.K. sector of the North Sea, Upper Jurassic, shallow marine reservoirs are mainly located in the Moray Firth and Central Graben; they also occur in adjacent areas of the Norwegian sector. Reservoirs of this type are represented by the Piper (Moray Firth), Fulmar (Central Graben) and Ula (Norwegian sector near U.K./Norway boundary) fields. In comparison to mass flow sand/conglomerate reservoirs, the shallow marine reservoirs show greater lateral continuity, but as a consequence there is a reduced potential for stratigraphic trapping. Nevertheless, these shallow marine sands were deposited over the irregular mid-Kimmerian erosional surface which has contributed to their variable distribution.

For the purpose of this discussion, the Upper Jurassic shallow marine sandstone reservoirs have been divided into two types: one showing strong tectonic influence on sedimentation and the other moderate tectonic influence on

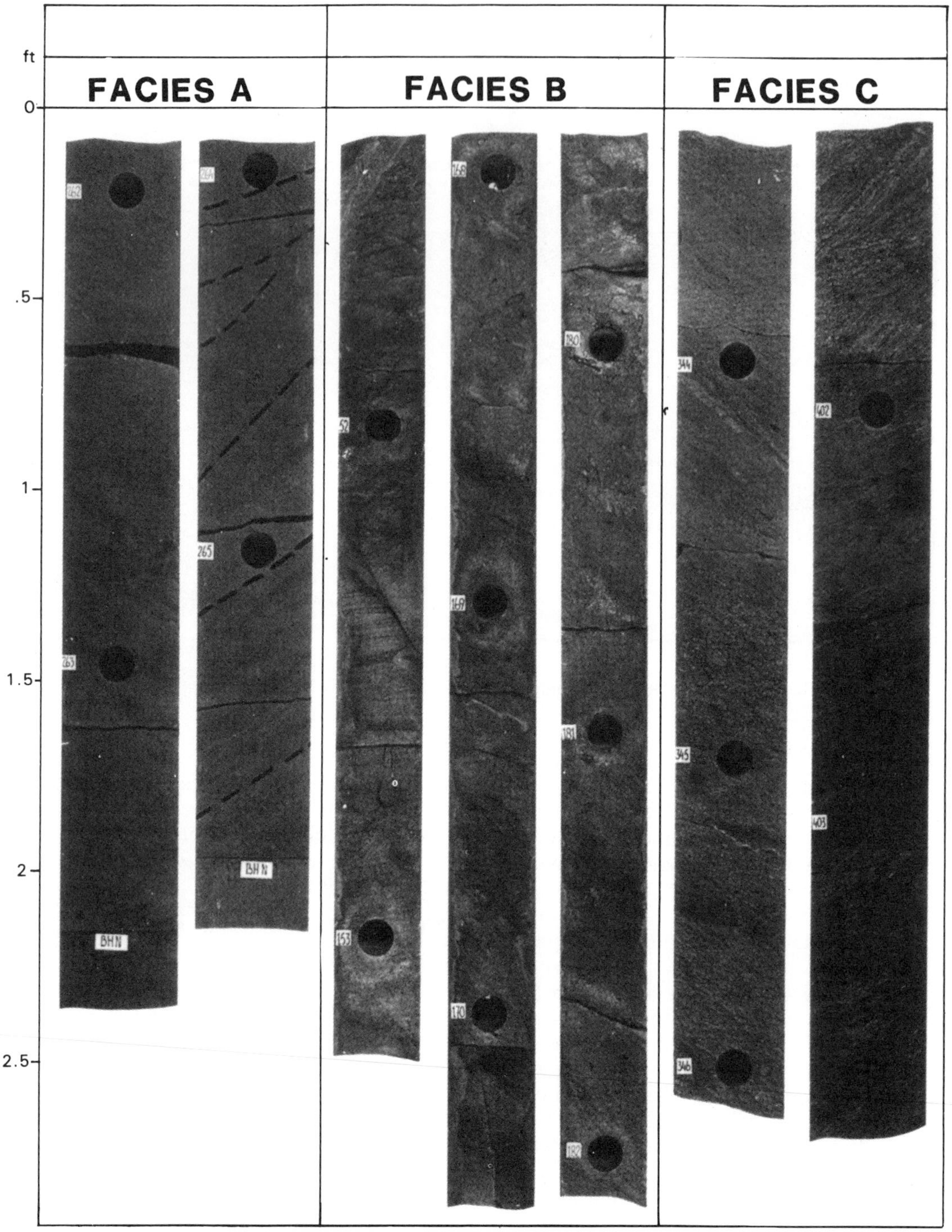

FIG. 45. Core photographs illustrating a typical Fulmar-type regressive sequence.

sedimentation.

Shallow marine sandstone reservoirs belonging to the first category are typified by the Fulmar Sandstones. The main Fulmar reservoir sandstones were deposited in a strongly subsiding basin with an abrupt faulted western margin abutting against the Auk Platform (Fig. 44). Facies identification in the Fulmar Field has required extensive core coverage and the use of X-radiography to aid interpretation, since in normal light much of the core is structureless (Fig. 45). Consequently slabs of

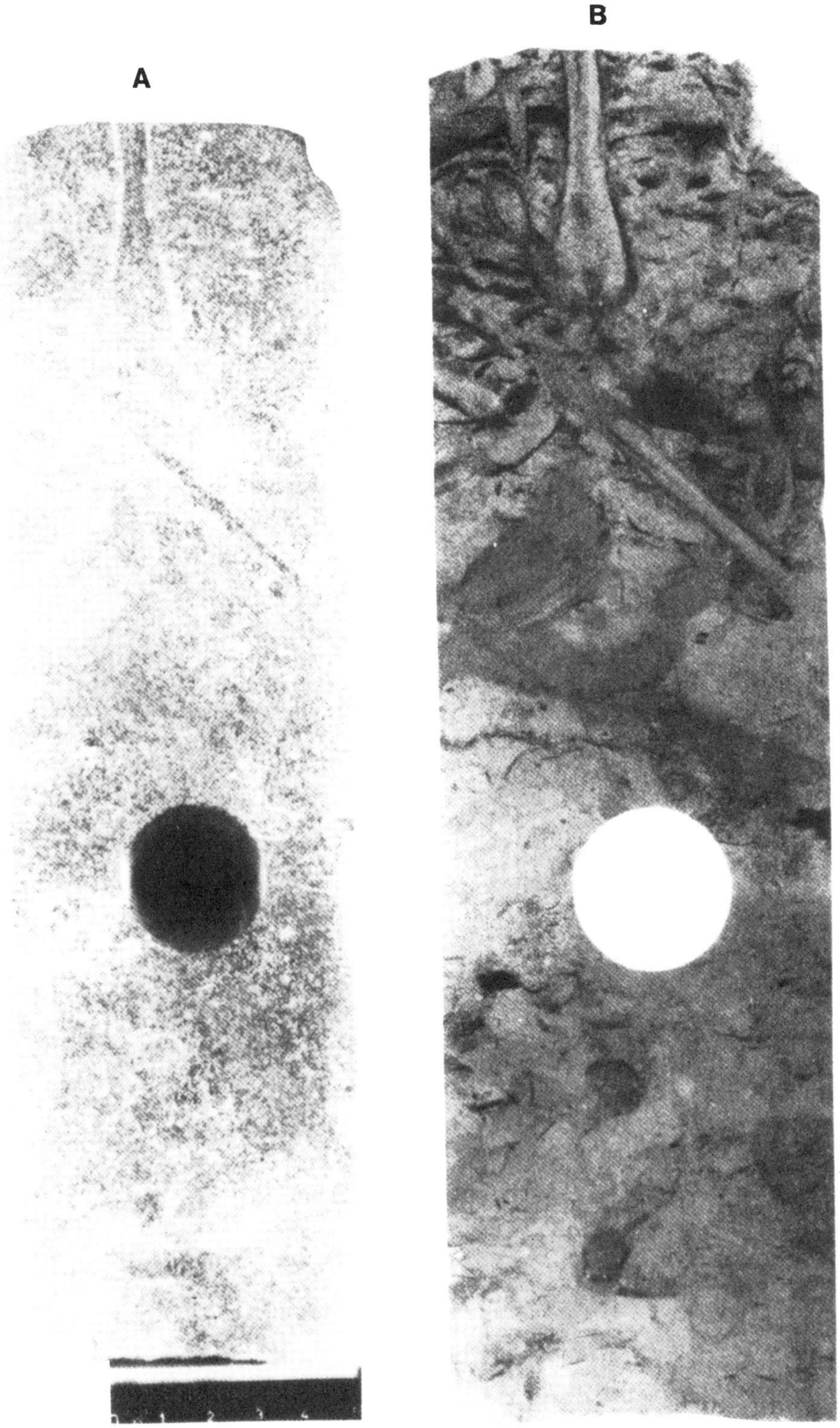

FIG. 46. Normal photograph and X-radiography of the 'massive' Fulmar Sands.

cores subjected to X-radiography reveal extensive bioturbation and burrowing (Fig. 46) and occasional cross-bedding etc., suggesting an overall shallow marine depositional environment. The sandstones form thick packages, up to 1000 ft thick, with the most characteristic feature being 'amplified' coarsening-upwards sequences up to 600 ft thick which display an upward passage from argillaceous fine grained sandstones to clean, medium grained sandstones (i.e. facies A–C, Fig. 47). In the area of the Fulmar field the highest energy sandstones (i.e. facies A) are more common near the Auk platform, with finer grained, more argillaceous and more bioturbated sandstones (facies B and C) encountered in more basinward (?deeper water) locations (Fig. 44). The 'amplified' coarsening upwards sequences can be accounted for by strong subsidence along the Auk boundary fault and/or localized salt withdrawal from the underlying Zechstein salt.

A delicate balance between subsidence and sedimentation resulted in the substantial accumulation of good quality reservoir sands. In general the problems of field appraisal and development are less acute for Fulmar-type reservoirs than for the Brae- and Magnus-type sands. However, extensive core coverage and a good biostratigraphic framework is required. Correlation between wells can be most effectively carried out using core-derived depositional facies sequences, since the use of log markers, which often includes diagenetic features, can be misleading in these relatively homogeneous sandstones.

In parallel with Brae-type reservoirs, facies belts are strongly controlled by the tectonic grain. Consequently, the good reservoir quality

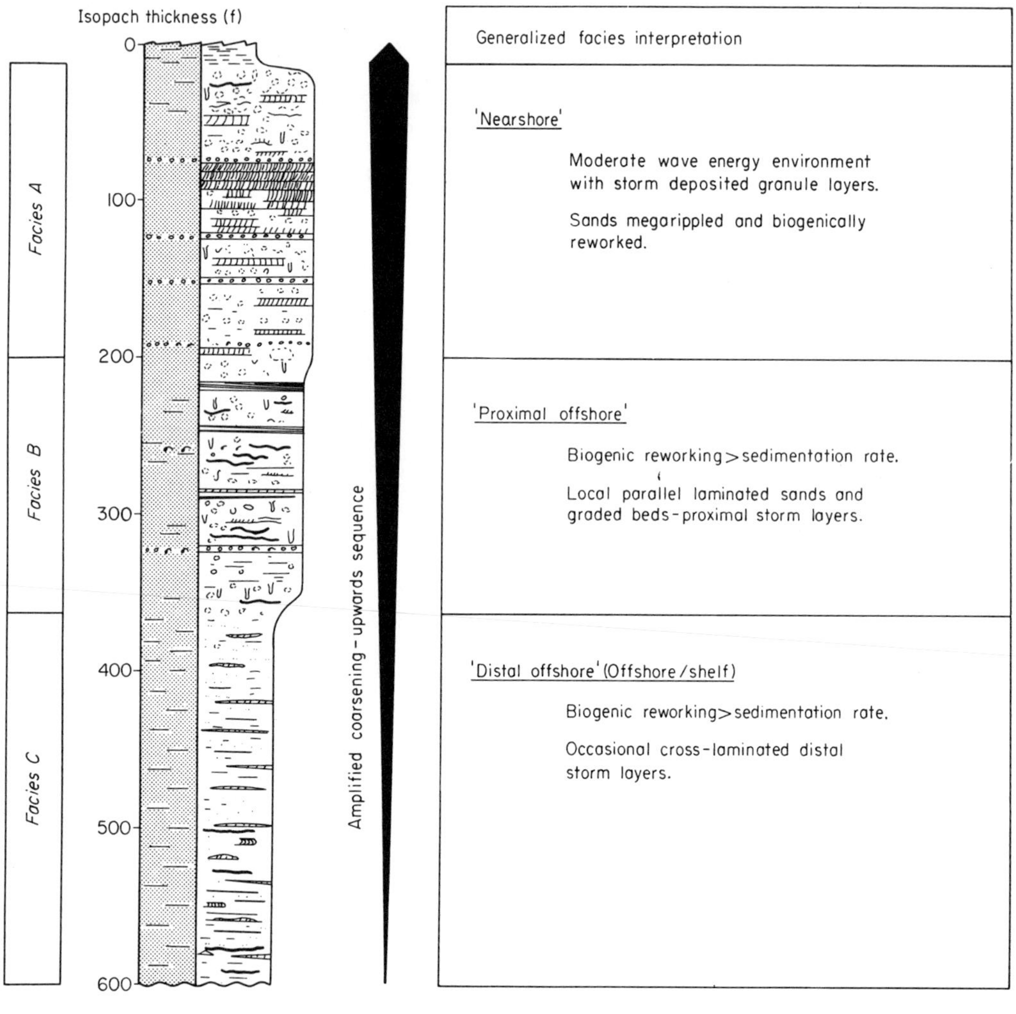

FIG. 47. Schematic vertical profile through Upper Jurassic shallow marine sands (based on the Fulmar Sands in the Fulmar field).

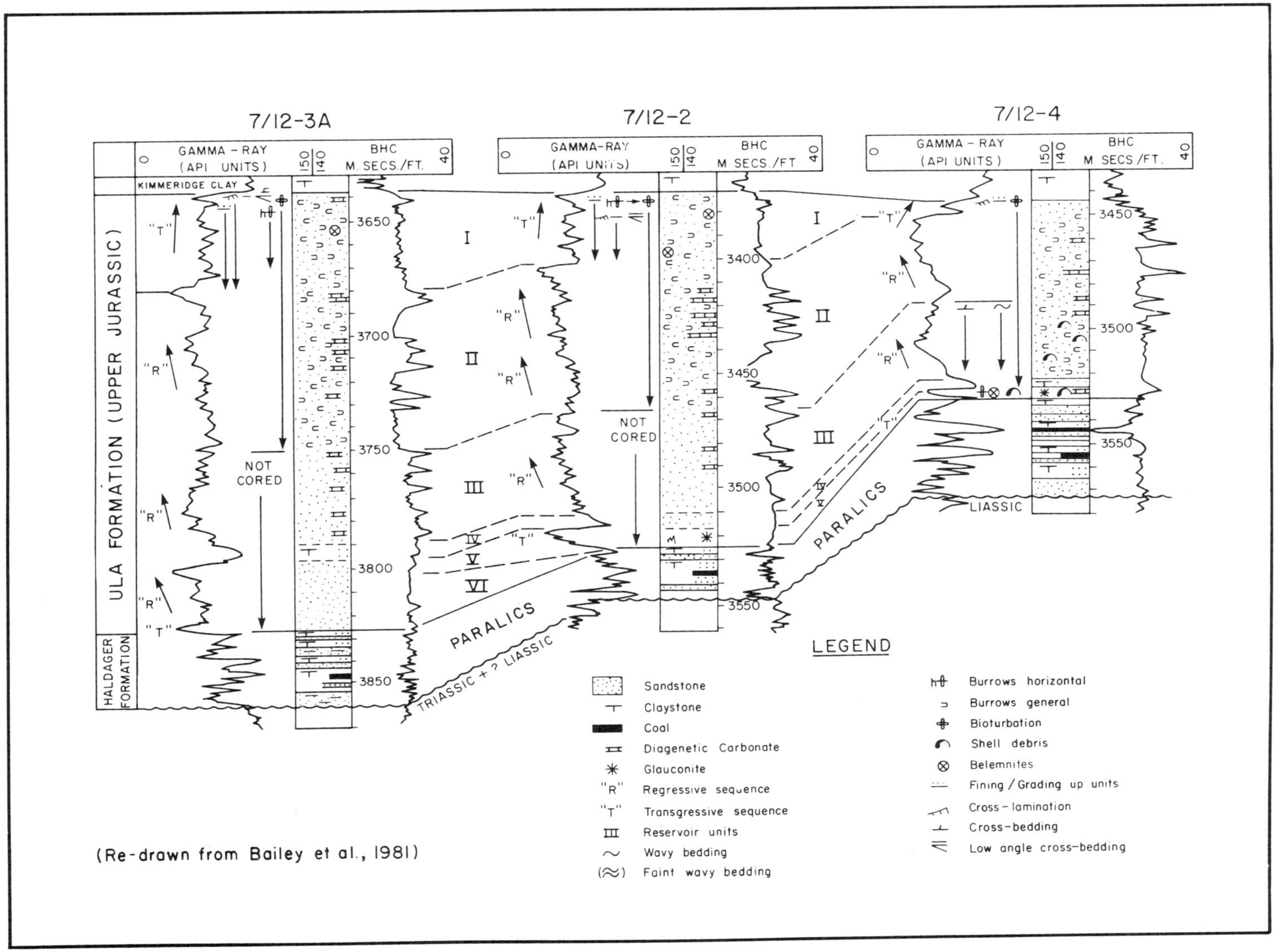

FIG. 48. Sedimentology and reservoir correlation of the Ula Field.

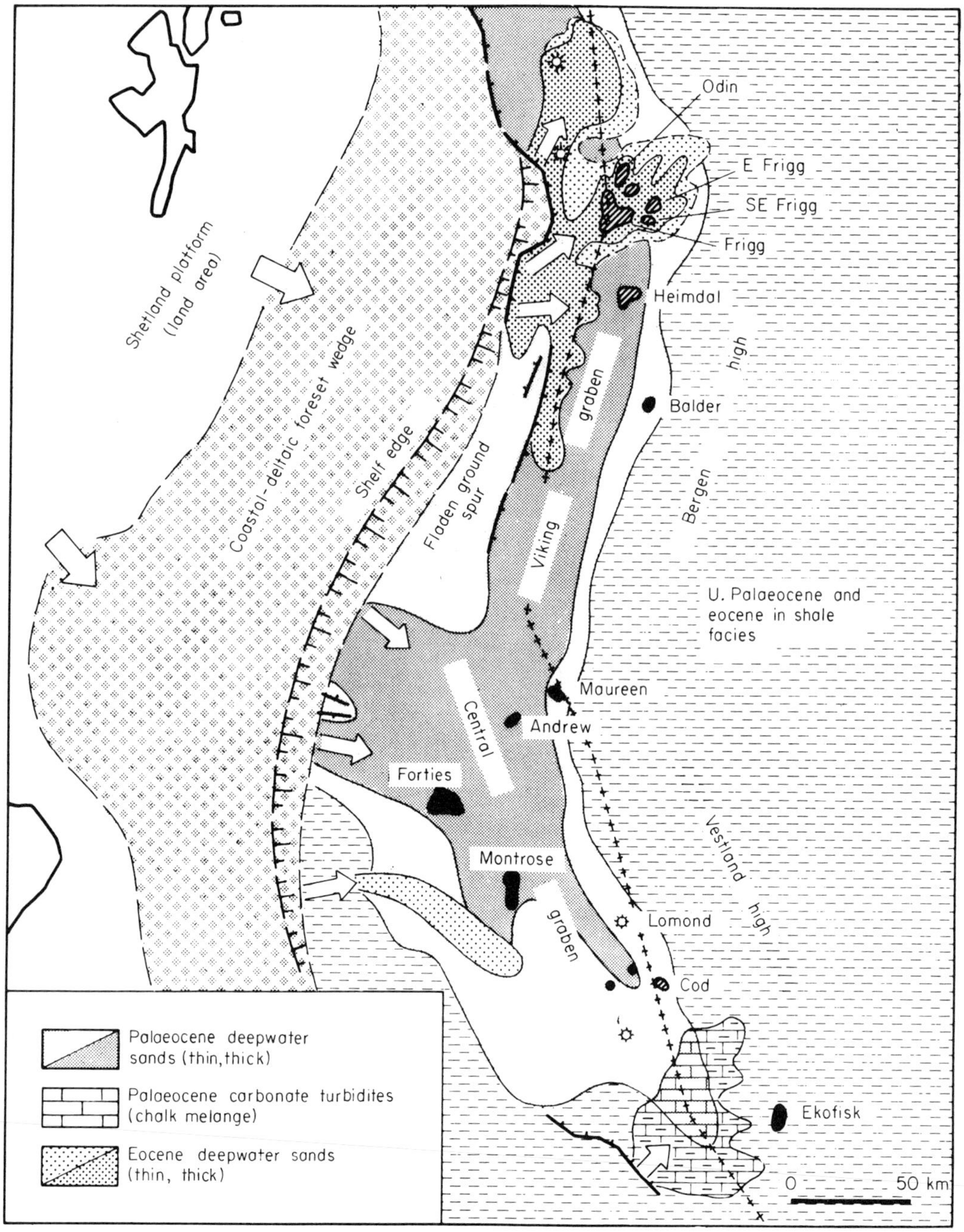

FIG. 49. Regional facies distribution of Lower Tertiary sediments in the central and northern North Sea (from Ziegler 1982).

facies belts are generally elongate and narrow, and parallel major graben faults. However, syndepositional salt diapirism has, in some places, played an equally important role in controlling reservoir development. The above factors indicate why good quality reservoir sands can be difficult to locate on a regional scale.

Reservoirs of the second, moderate tectonic control, category differ from the Fulmar example in only a few respects. The best examples of shallow marine sandstones deposited under these conditions are found in the Ula field (Bailey *et al.* 1981) and Piper field (Williams *et al.* 1975; Maher 1980, 1981). The reservoirs of both these fields, like the Fulmar field, are composed of a stacked sequence of predominantly bioturbated coarsening upwards units. However, in contrast to Fulmar, the coarsening upwards sequences of the Ula and Piper reservoirs are much thinner (90–170 ft). Furthermore, particularly in the Piper field, the coarsening upwards units show greater variations in reservoir quality, ranging from non-reservoir offshore shales at the base and passing upwards into high-energy, cross-bedded sandstones (with Darcy permeabilities) at the top. In both Ula and Piper there is evidence that the reservoir sands are thicker in the vicinity of faults, but faulting has had less impact that it has on Fulmar. Therefore, in the Ula and Piper Fields depositional facies are laterally more homogeneous than in the Fulmar Field (Fig. 48). Owing to this lateral homogeneity of facies and to the thinner and more sharply defined coarsening-upwards units in both Ula and Piper, these reservoirs have the potential to exhibit strong layering of reservoir properties. This could lead to extensive field wide permeability barriers, which could in some cases hamper attempts to maintain reservoir pressure by water injection.

The shallow marine sandstones of both of the above categories can be recognized on wireline logs by funnel-shaped gamma ray profiles and increasing upwards positive separation of FDC/CNL logs. This contrasts with the more blocky (mid-fan) or serrate (thin bedded mass flows/basinal shales) profiles of mass flow deposits discussed earlier in this section. However, extensive core coverage is essential in order to identify positively the various types of Upper Jurassic sandstone reservoirs.

Tertiary sands, northern and central North Sea

The Tertiary deposits of the northern and central North Sea areas include several important oil and gas accumulations, including the Montrose field (Fowler 1975) which was the first commercial oil discovery in the U.K. sector of the North Sea (1969). Tertiary sands also form the reservoir of the giant Forties oil field (Thomas *et al.* 1974), and the giant Frigg gas field (Heritier *et al.* 1979). Other important commercially proven hydrocarbon accumulations, occurring in both the U.K. and Norwegian sectors, include Andrew, Maureen, Lomond, Cod, Odin, Heimdal and Balder (Fig. 49).

The reservoir sands of all these fields were deposited in a variety of submarine fan complexes of Palaeocene and Eocene age. A major similarity between many of these Tertiary sand plays and the previously discussed Upper Jurassic fan bodies is the importance of, and considerable potential for, stratigraphic trapping. Furthermore, all these fan bodies display substantial variability in sand distribution, particularly on a field scale. One major difference between these Tertiary sand plays and other, generally older North Sea sand plays, is the potential for applying seismostratigraphy to both exploration and production studies, which will be discussed more fully later. In addition, this section also highlights the application of a submarine fan facies model to the development of a reservoir geological model of the Forties field.

The regional geological framework of the Cenozoic period has been covered amply elsewhere (e.g. Ziegler 1981, 1982; Rochow 1981; Knox *et al.* 1981; Morton 1979) and is not discussed specifically here.

Seismostratigraphic studies

The regional geological picture of the Cenozoic of the North Sea has been compiled from a combination of sedimentology, palaeontology and seismostratigraphy. The latter has proved particularly valuable in both exploration and production studies for several reasons. Particularly important is the fact that the Cenozoic represents a period of relatively rapid sedimentation in which up to 3.5 km of sands and shales were deposited, mainly in the form of discrete clastic wedges which built out from the basin margins (Fig. 49). This, combined with only minor tectonism and moderate burial/compaction, has resulted in the partial preservation of the primary, large-scale depositional geometry and internal structure of individual clastic bodies, which can, in some instances, be identified on seismic records.

The Palaeocene sands of the central North Sea have proved particularly receptive to seismostratigraphic studies. This is largely due to their relatively shallow burial (c. 3000–9000 ft), and the commensurate high resolution of reflection seismic data. Furthermore, the top of the Danian chalks and the early Eocene volcanic ash marker provide regionally correlative seismic reflectors of considerable lithostratigraphic value (Rochow 1981). The first published seismostratigraphic study of the Palaeocene interval was by Parker (1975), who identified on seismic records the wedge-shaped geometry of the unit and its internal organisation which comprised large-scale topset, foreset and bottomset reflectors (Fig. 50). Sedimentological and palaeontological data established that this geometry and seismic character represents a sequence of prograding coastal/deltaic deposits

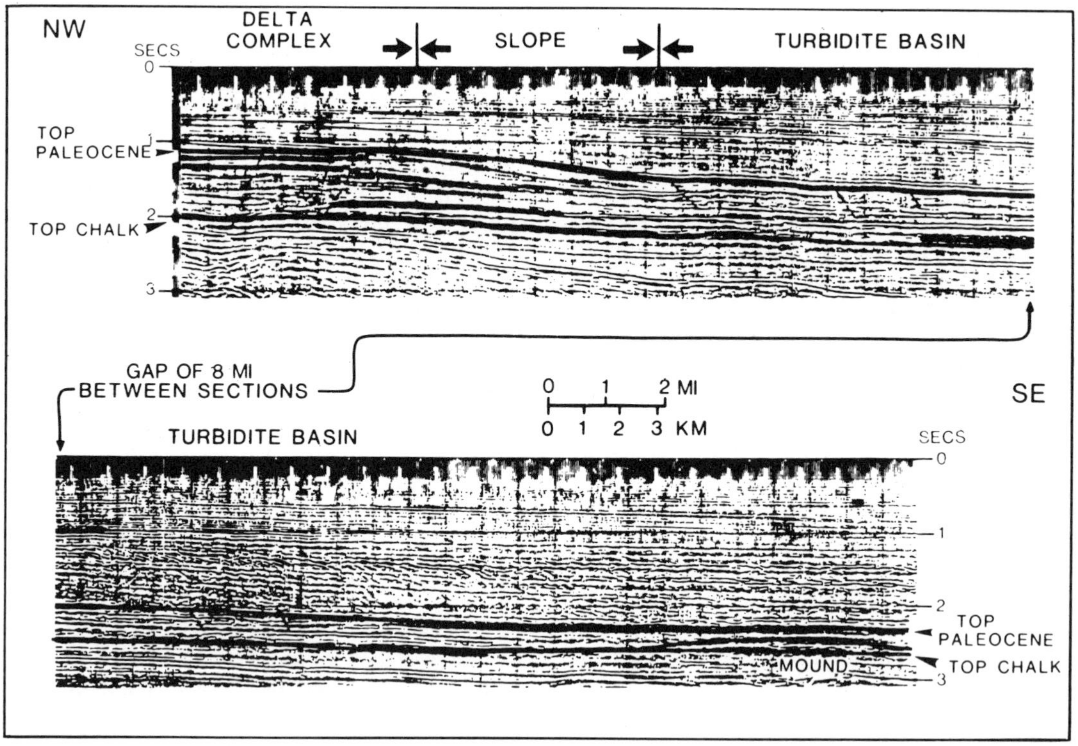

FIG. 50. Seismic sections illustrating seismic reflection patterns of Palaeocene seismic facies in the central North Sea (Parker 1975).

(topset) which pass laterally and downdip into a partly channelized slope complex (foreset), and beyond into a base of slope, deep water turbidite fan system (bottom set; Fig. 51).

The North Sea hydrocarbon productive Palaeocene sands display some of the general seismic attributes considered characteristic of turbidite sequences associated with prograding deltas (Fig. 52; Berg 1982). In the context of recent seismostratigraphic terminology (Vail *et al.* 1977), the slope or pro-delta facies are represented in dip sections by prograding clinoforms, which in this case are of a very low angle and continue far out into the basin. However, the characteristic mounding ('hummocky' to chaotic internal seismic reflection pattern) of turbidite fan systems (Sangree & Widmier 1977) is only faintly developed in Parker's Palaeocene example (observed by Berg 1982) but elsewhere they are more striking and contain hydrocarbons which are partially stratigraphically trapped (e.g. Frigg).

The Frigg field is a particularly striking example of a turbidite fan reservoir complex in which the primary fan geometry is clearly visible from the top structure map. This map displays a radial body with discrete 'fingers' which correspond to sand-rich channel/fan lobe deposits (Fig. 53). The mounded or 'hummocky' character of internal seismic reflectors can be shown from detailed well log correlations to reflect the large number of channel/fan lobe sand bodies encased in interchannel and distal fan lobe shales (Figs 53 and 54). Within the framework of the regional depositional model, as outlined earlier, the recognition on seismic of a fan-shaped body with internal mounded reflectors can be confidently predicted as representing a submarine fan complex.

The Palaeocene turbidite fan reservoirs of the Balder and Cod fields display complex patterns of reservoir distribution which presented considerable problems during appraisal and early development stages (Skjold 1981; Kessler *et al.* 1981). In both examples the complex sand body distribution patterns have resulted in large variations in hydrocarbon distribution, local stratigraphic trapping and even substantial variations in hydrocarbon-water contacts. These complexities are clearly not unusual and represent an integral feature of submarine fan reservoirs. Early recognition of this type of sand play from seismic studies should help in the planning of appraisal and development drilling campaigns. In practice, this needs supplementing by a thorough facies analysis of each reservoir complex as outlined in the following section.

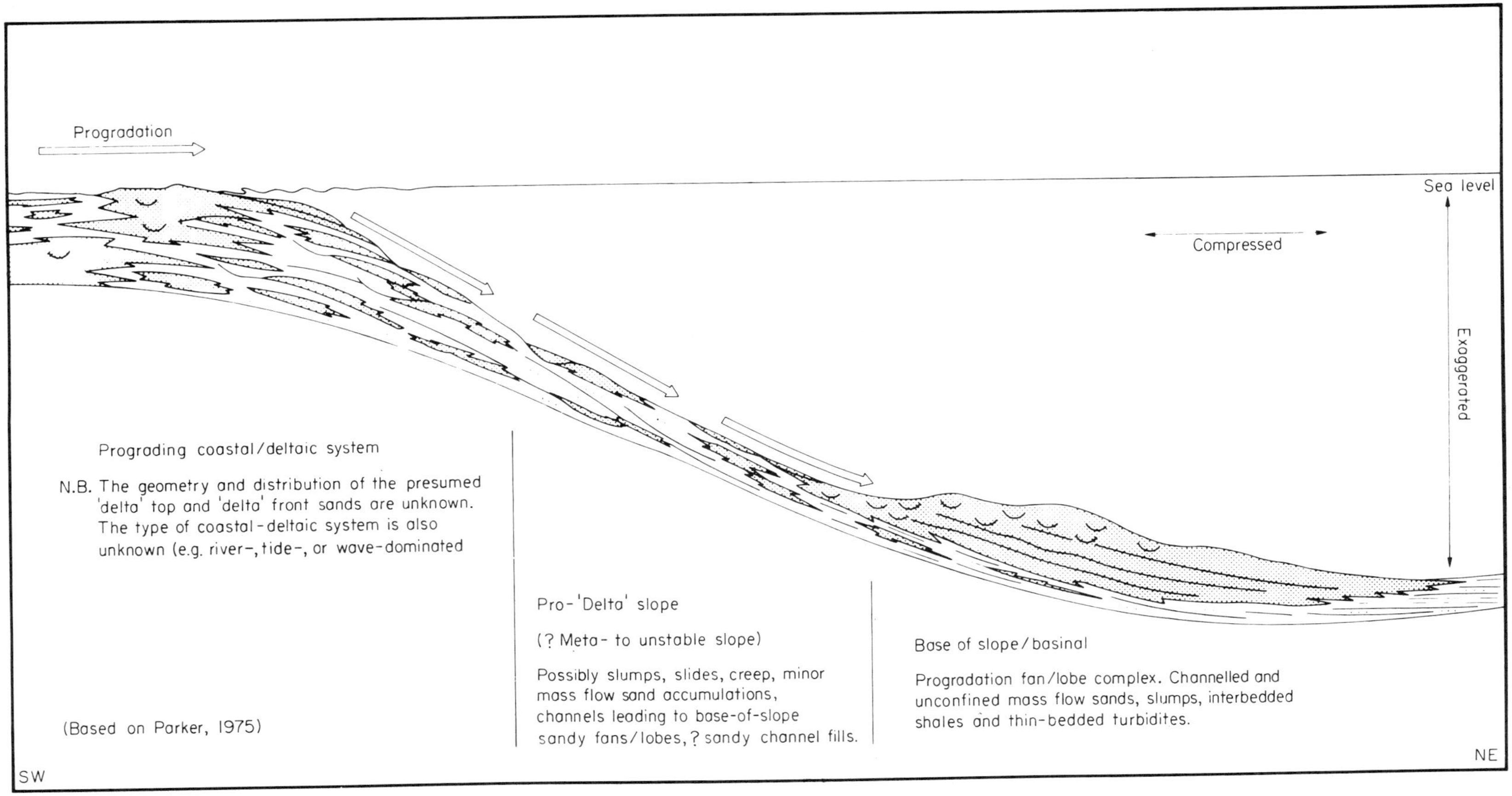

FIG. 51. Facies model for Tertiary clastic deposits illustrating the main depositional elements (Parker 1975).

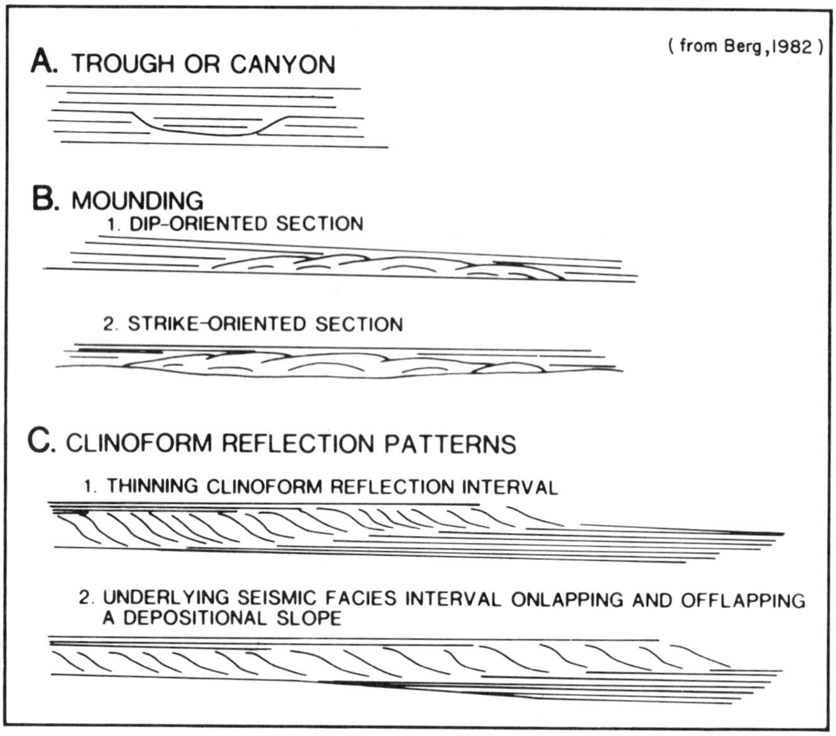

FIG. 52. Seismic indicators of turbidite sequences (from Berg 1982).

Deep sea fan sedimentation and reservoir modelling in the Forties field

The Tertiary sand reservoirs of the Forties field are illustrated here as an example of how the development of a general facies model, in this case the deep sea turbidite fan model, has proved beneficial to the understanding and prediction of subsurface reservoir distribution.

During the appraisal drilling of the Forties field, in the early 1970s, it was recognized on the basis of five wells that reservoir distribution was highly variable. Although the average sand content was generally high, it appeared that optimum sand development occurred in the north and north-western parts of the field, and that these good sands passed laterally into thin sands and shales to the SW (Fig. 55). In the context of the deep sea turbidite facies models available at that time, this sand distribution pattern could have represented a transition from proximal to distal turbidites, moving in a basinward direction towards the SE (Fig. 55). In such a model the sands could have been expected to have tabular, sheet-like geometries, which over the field would have allowed lateral and, possibly, vertical fluid communication. It was therefore anticipated that development wells would be located on a regular drainage pattern, and that the irregular sand distribution patterns noted earlier would not influence the location of such wells nor substantially affect reservoir performance (Walmsley 1975, and 'Discussion' by R. C. Selley, p. 485).

By the mid-1970s, however, submarine fan facies models were being widely applied to ancient turbidite sequences (e.g. Walker & Mutti 1973). Furthermore, and as indicated earlier, seismostratigraphic studies demonstrated that the reservoir sands of the Forties Formation accumulated as deep water submarine fans at the base of actively prograding delta slopes (Parker 1975, predicted water depths of 3000–8000 ft). This regional depositional model of submarine fan sedimentation provided an important framework for the interpretation of current development well drilling results and of reservoir performance data in the Forties field. In addition, it was noted that 'the very variable sand characteristics and geometry discovered did cause numerous changes to the early development drilling and production schedules' (Hillier *et al.* 1978, p. 327). In particular, it was recognized that the main sand bodies were extensively channelized (Fig. 56), with one major channel sand complex (the Charlie Sand on the western flank of the field) isolated from the main reservoir unit (the multistorey Main

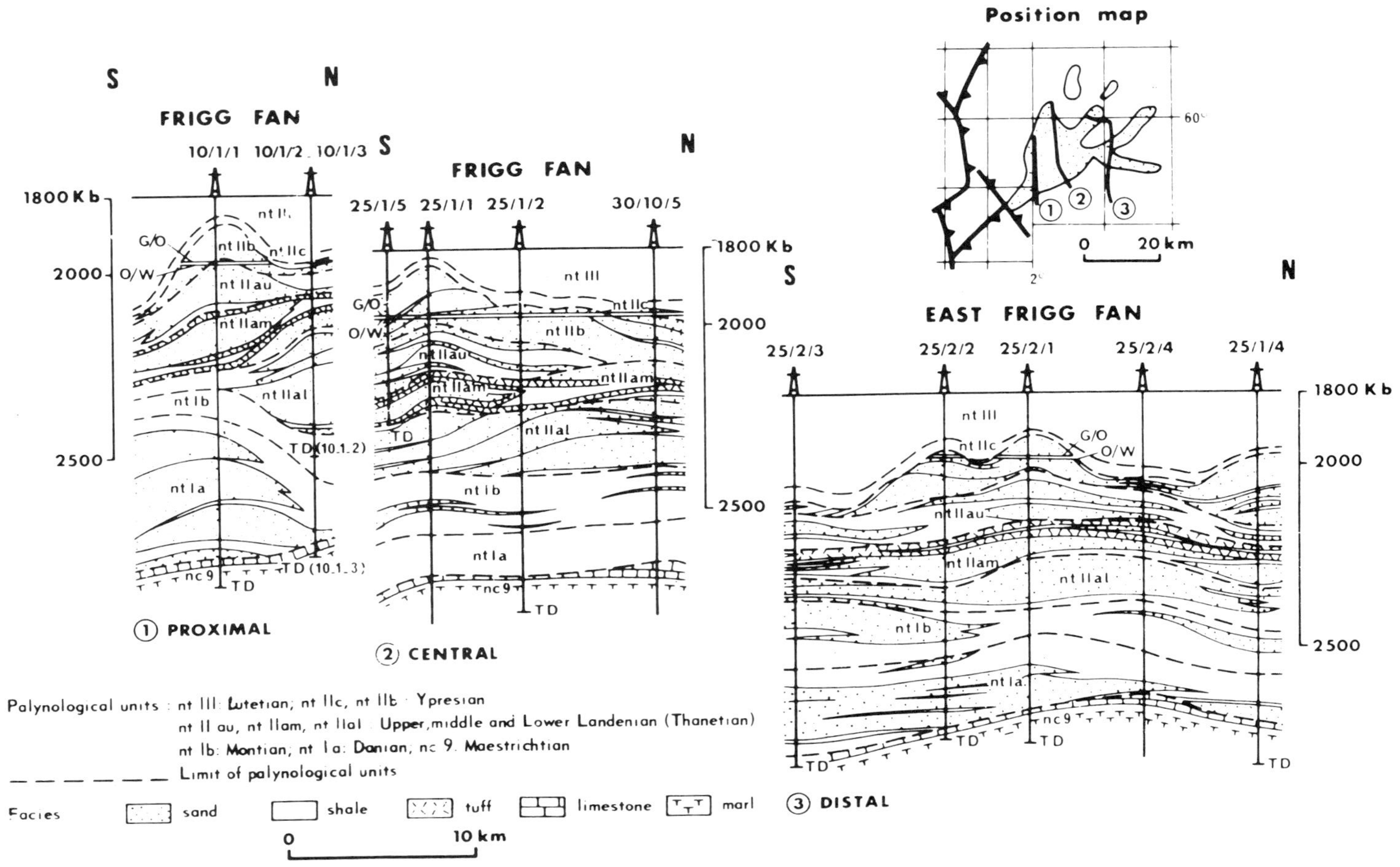

FIG. 53. Top structure map cross-sections of the Frigg Field showing the complex mounded or hummocky character of reservoir sands. (from Heritier *et al.* 1979).

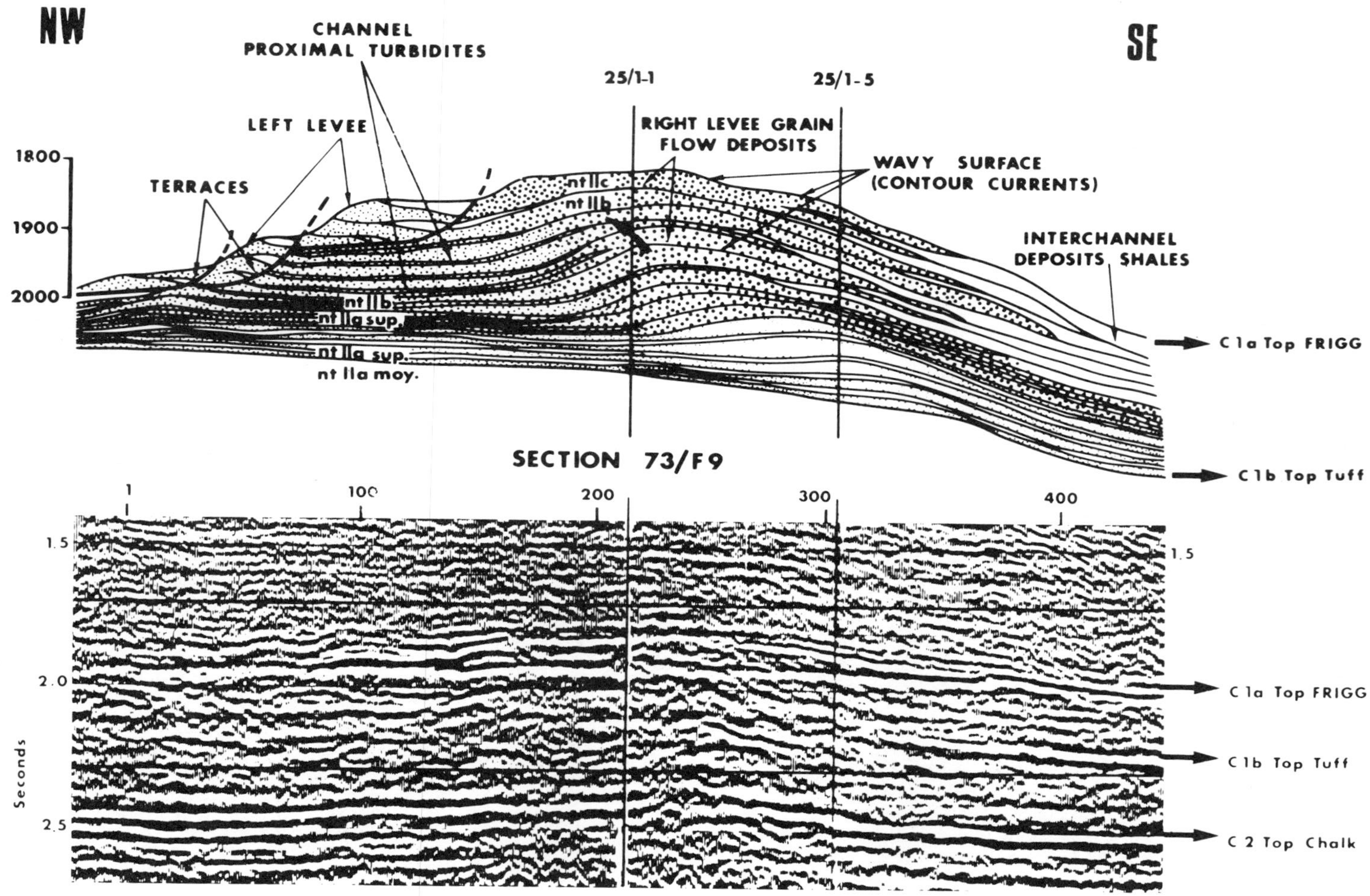

FIG. 54. Seismic section and sedimentological interpretation of a typical Frigg mound (from Heritier *et al.* 1979).

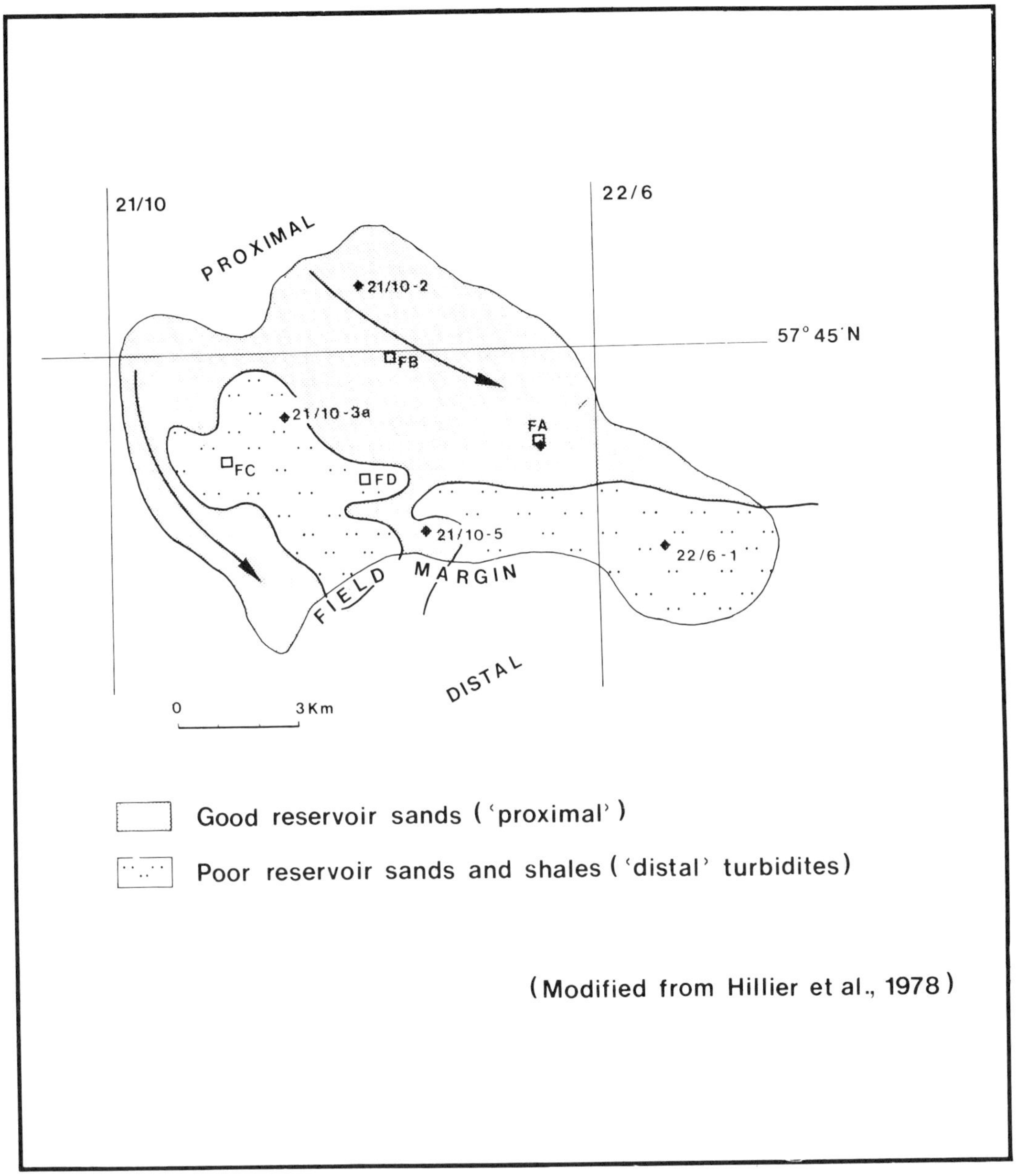

FIG. 55. Pre-development facies distribution model in the Forties field (from Hillier *et al.* 1978).

Sands). Clearly, the accurate prediction of the distribution of channel and interchannel facies will significantly affect all aspects of reservoir management. A prerequisite to reservoir prediction is the correct identification of the different sand and shale body types as seen in cores and well logs.

A facies analysis of the Forties Sands in the Forties field by Carman & Young (1981) outlines the various sedimentological elements of this complex reservoir which comprises four main lithofacies: facies A = mainly medium to thickly bedded sands displaying well-developed Bouma sequences and analogous to the proximal 'classic turbidite' group of Walker (1967, 1978); facies B = mainly massive, thickly bedded sands with subordinate pebbly sands, interpreted as the deposits of fluidized flows (= massive sandstones of Walker 1978); facies C = laminated grey mudstone with 1–5 mm thick, graded siltstone-mudstone couplets and subordinate fine grained sandstone beds probably deposited by relatively distal turbidity currents; facies D = green mudstones with abundant microfauna and microflora, which are interpreted as hemipelagic deposits.

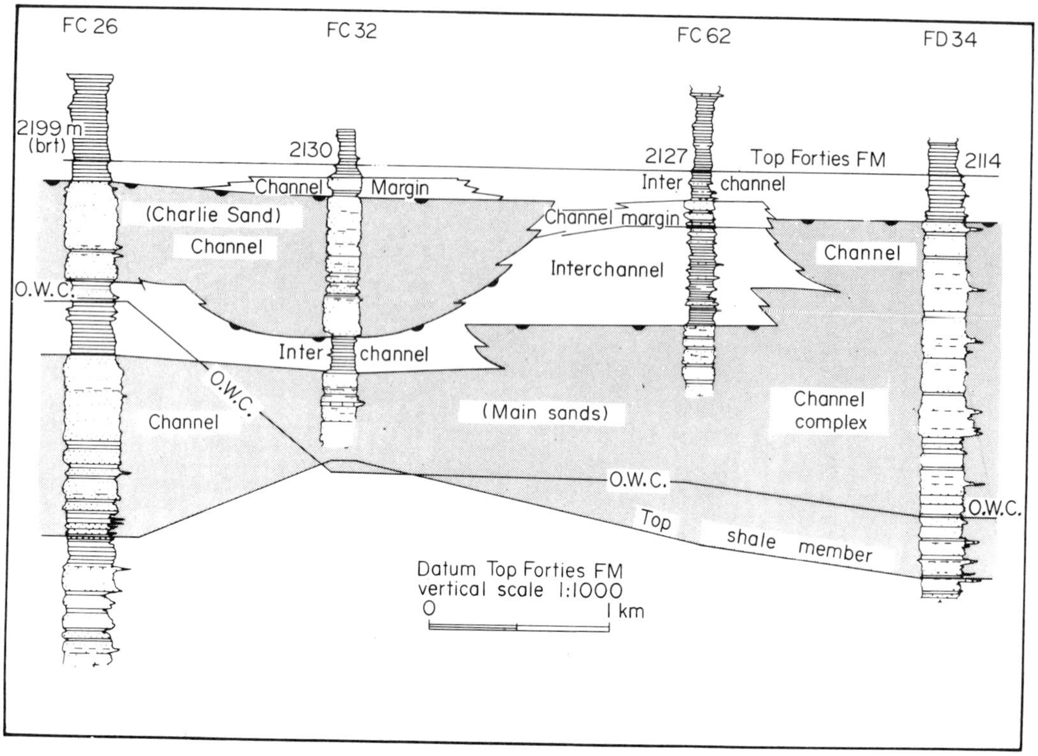

FIG. 56. Cross-section through the Forties field (from Hillier *et al.* 1978).

These four lithofacies are arranged in several distinctive facies associations and sequences which represent deposition in five environments belonging to the middle and lower regions of a submarine fan system (Fig. 57). Particularly important is the fact that all five facies associations have distinctive well log patterns which ensures correct facies interpretation in uncored wells. The present facies distribution model of the Forties field, based on the above facies analysis, identifies a series of mainly channelized sand bodies trending in a NW–SE direction (Fig. 58), which resulted from deposition in a middle fan environment. This model is substantially different in many important respects to the initial sheet-like distribution model outlined earlier. As seen in the Shell/Esso portion of the field (Block 22/6a), there are marked contrasts in reservoir development between the high-quality, multistorey channel sands and the low-quality interchannel sands and shales (Fig. 59). This type of detailed facies mapping provides a three-dimensional model capable of accurately monitoring reservoir performance, and which should also ensure the optimum location of infill wells and provide an essential framework for the application of any enhanced oil recovery techniques.

Concluding remarks

The 21 year history of the British Sedimentological Research Group has been paralleled by the exploration and production of the North Sea's oil and gas resources, and both have experienced a continual expansion in activity during this period.

Sedimentological studies have already made a significant contribution in assisting in the location and development of North Sea oil and gas accumulations, principally by increasing our understanding of the origin, nature and distribution of all the major clastic reservoirs. In particular, the reconstruction of depositional environments in clastic sequences has been applied extensively to provide the essential framework for describing and understanding the qualitative and quantitative reservoir distribution on both regional (exploration) and field (production) scales.

The continuing search for hydrocarbons in the relatively mature sub-basins within the North Sea will require the development of further imaginative models, particularly since many future discoveries are likely to include traps with an increasing stratigraphic component. The continued development of predictive deposition-

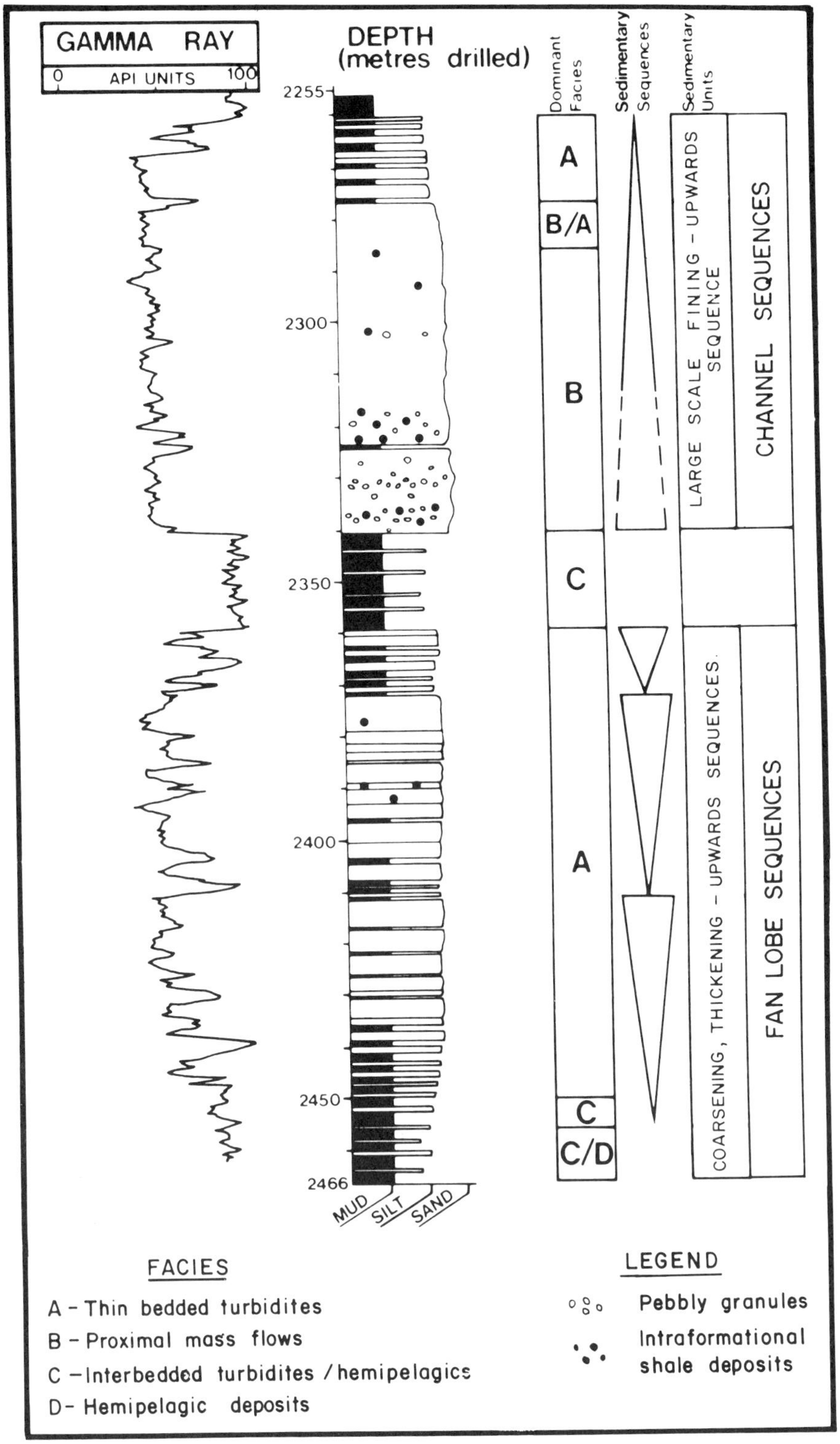

FIG. 57. Vertical facies profile through the Forties Sands (from Carman & Young 1981).

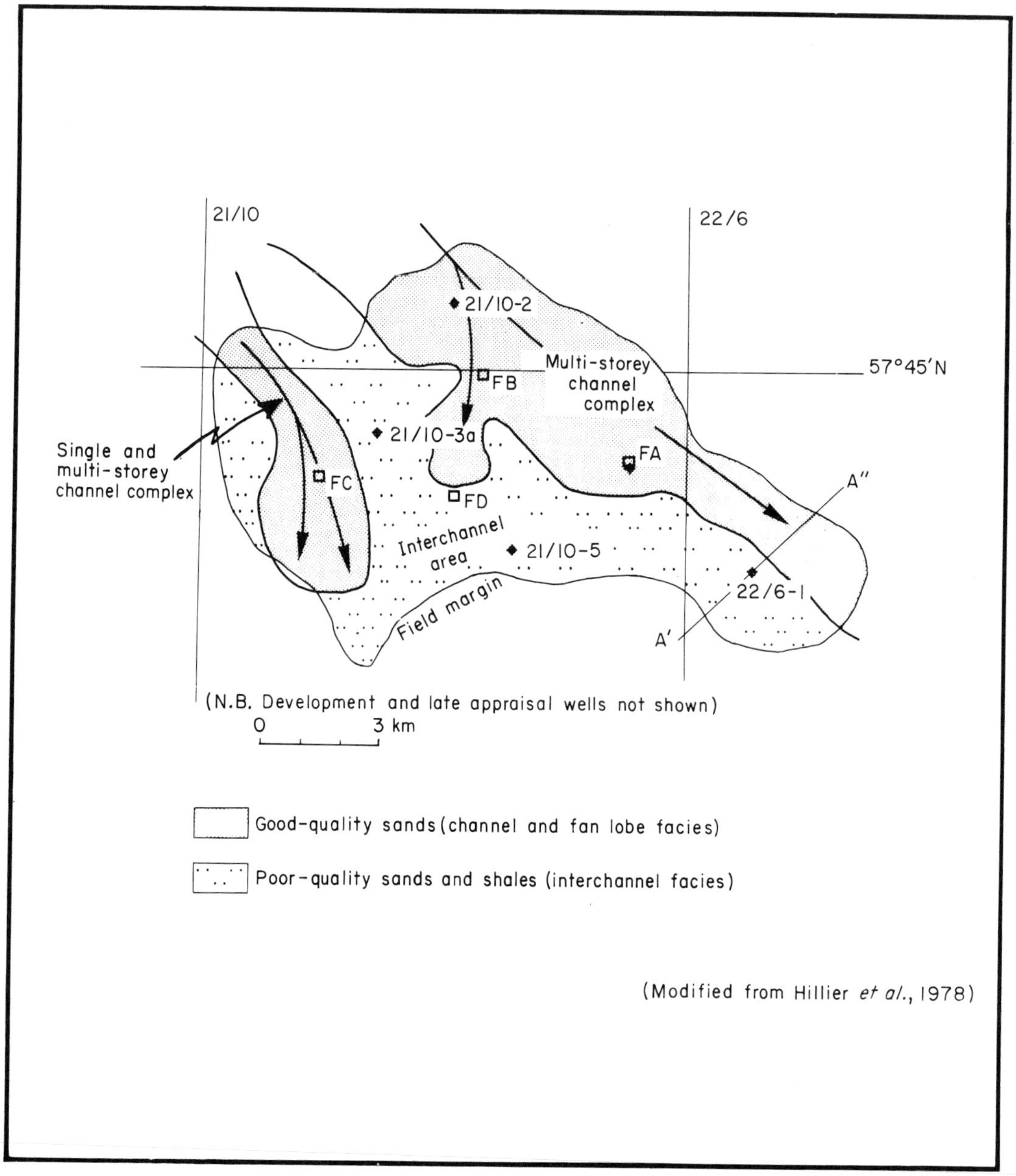

FIG. 58. Post-development facies distribution model in the Forties field (modified from Hillier *et al.* 1978).

al models is required, therefore, to help locate these stratigraphic traps. Furthermore, it remains essential to be able to predict the reservoir potential of undrilled structures in terms of reservoir characteristics (lithology, mineralogy, diagenesis etc.), thickness and quality (porosity-permeability), and to assess the likely extent of such reservoirs.

The appraisal and development of oil and gas accumulations require an accurate geological description of reservoir characteristics, particularly in terms of thickness and quality (porosity-permeability) distribution (e.g. Harris & Hewitt 1977). In clastic reservoirs this is best achieved through a thorough understanding of depositional environments, particularly the size, shape, trend and quality of sand bodies and the thickness and continuity of shale layers (e.g. LeBlanc 1977) and the tectonic framework of the basin and/or hydrocarbon accumulation. The resulting reservoir geological model provides the optimum framework for reservoir subdivision and correlation, reservoir quality mapping (e.g. net sand, porosity, permeability,

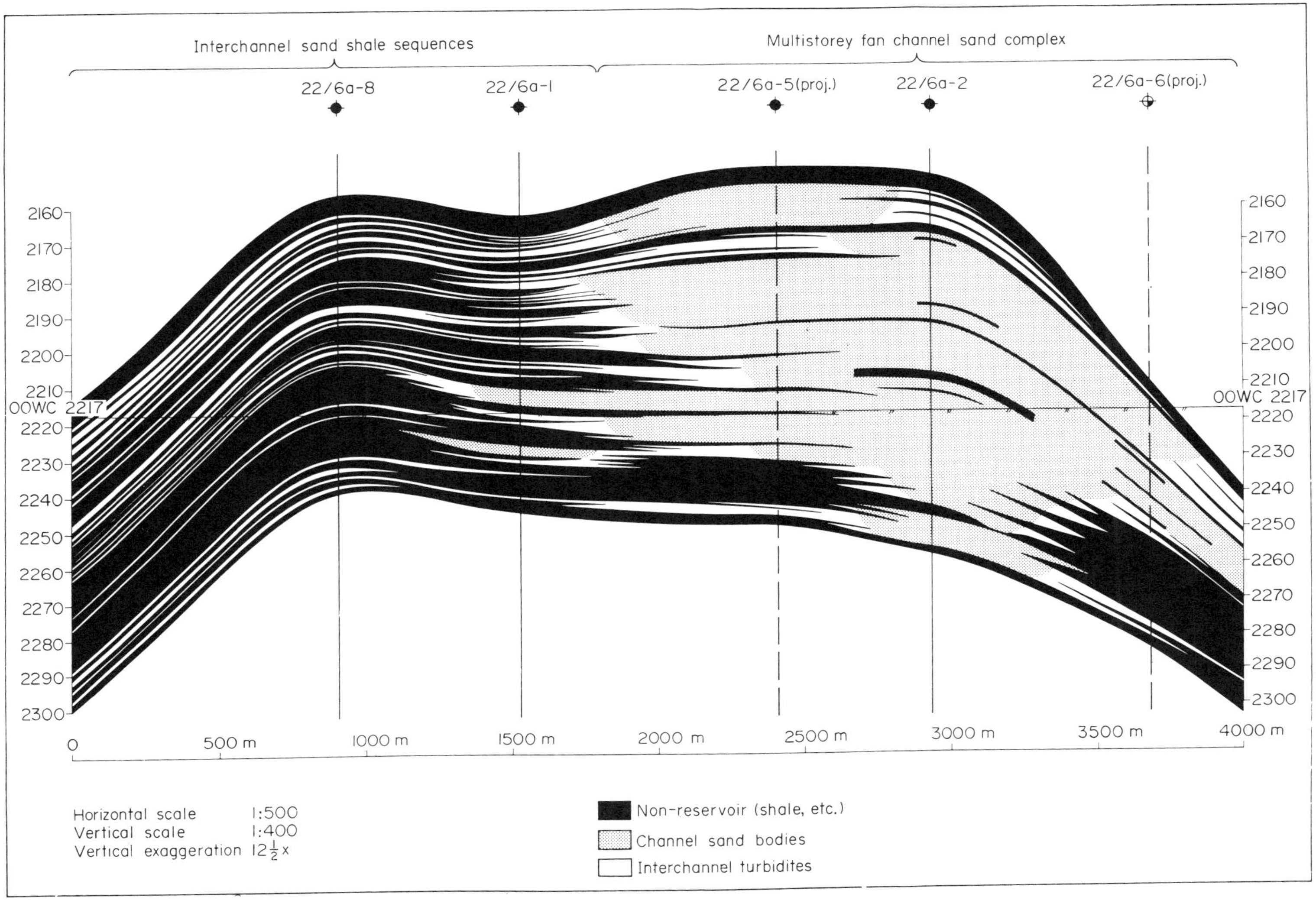

FIG. 59. Facies distribution in the south-eastern part of the Forties field, Block 22/6a.

saturation maps etc; e.g. Sneider *et al.* 1976) estimating reserves, developing reservoir simulation models (e.g. Harris 1975; Weber *et al.* 1978) and monitoring and predicting reservoir performance (e.g. fluid movements; cf. Van Veen 1977; Hartman & Paynter 1979). The relatively long productive life of a field (e.g. 20–30 years) usually requires a continual update of reservoir geological models in order to ensure optimum reservoir management and, thereby, to maximize ultimate recoveries. The quality of the reservoir geological model can prove a particularly significant factor during the latest stages of development, particularly when decisions are taken on the positions of infill wells and the possible application of enhanced oil recovery techniques. Indeed, the success of these latter projects can hinge on the accuracy of the geological description of the reservoir (e.g. Basan *et al.* 1978; Cook 1977; Kossack & Bilhartz 1976; Webb 1978).

In conclusion, it is apparent that sedimentology will continue to play an important role in the multidisciplinary studies being employed in the exploration and production of hydrocarbons from the North Sea.

ACKNOWLEDGMENTS: The authors are indebted to Shell U.K. Ltd and Esso Petroleum Co. Ltd for permission to publish this paper. We have drawn freely from the large amount of data available in existing publications, in-house reports and the personal knowledge and experience of our colleagues. The development of exploration and production sedimentological models requires a multidisciplinary approach and, as a result, the ideas and concepts outlined in this paper have crystallized through discussions between many specialists in various fields (e.g. stratigraphy, geophysics, geochemistry, reservoir engineering, petrophysics etc). Ideas have also evolved over several 'generations' of specialist activities and, consequently, it is not possible to single out all the individuals who originated the different contributions to this text, but we are indebted to all our colleagues. Specifically, however, the authors wish to acknowledge their predecessors in Shell Expro, M. C. Budding (Reservoir Geology/H. D. J.) and A. T. Buller (Sedimentology/D. J. S.), who made significant contributions in their specialist activities, particularly in relation to their separate studies on the Brent Sands. Comments by Drs P. J. C. Nagtegaal and P. Ziegler improved the final manuscript.

Special thanks are also due to Catherine Senior, Yvonne Jensen and Martin Lewis for valuable help with preparing the illustrations for our presentation of this paper at the 21st anniversary meeting of the BSRG in December 1982 at Liverpool University. Finally, we are indebted to Mrs Diane Symmons for all her work on the published figures.

References

ALLEN, J. R. L. 1978. Studies in fluviatile sedimentation: an exploratory quantitative model for the architecture of avulsion-controlled alluvial sites. *Sedim. Geol.* **21**, 129–47.

BAILEY, C. C., PRICE, I. & SPENCER, A. M. 1981. The Ula oil field, block 7/12, Norway. *In*: *The Sedimentation of the North Sea Reservoir rocks.* Norsk Petroleumsforening (NPF), Geilo.

BASAN, P. B., MCCALEB, J. A. & BUXTON, T. S. 1978. Important geological factors affecting the Sloss field micellar pilot project. *Soc. Petrol. Engrs, Pap. no. 7047, 5th Symp. on Improved Oil Recovery*, Tulsa, Oklahoma.

BERG, O. R. 1982. Seismic detection and evaluation of delta and turbidite sequences: their application to exploration for the subtle trap. *Bull. Am. Ass. Petrol. Geol.* **66**, 1271–88.

BOWEN, J. M. 1975. The Brent Oil-field. *In*: WOODLAND, A. W. (ed.) *Petroleum and the Continental Shelf of Northwest Europe*, pp. 353–60. Applied Science Publishers, London.

BUDDING, M. & INGLIN, H. 1981. A reservoir geological model of the Brent Sands in Southern Cormorant. *In*: ILLING, L. V. & HOBSON, G. D. (eds) *Petroleum Geology of the Continental Shelf of North-West Europe*, 326–34. Heyden, London.

CARMAN, C. J. & YOUNG, R. 1981. Reservoir geology of the Forties oil field. *In*: ILLING, L. V. & HOBSON, G. D. (eds) *Petroleum Geology of the Continental Shelf of North-West Europe*, 371–9. Heyden, London.

CHAUVIN, A. & VALACHI, L. 1980. Sedimentology of the Brent and Statfjord Formations of Statfjord field. *In*: *The Sedimentation of the North Sea Reservoir Rocks.* Norsk Petroleumsforening (NPF) Geilo.

COLEMAN, J. M., ROBERTS, H. H., MURRAY, S. P. & SALAMA, M. 1981. Morphology and dynamic sedimentology of the eastern Nile delta shelf. *Mar. Geol.* **42**, 301–26.

COOK, D. L. 1977. Influence of silt zones on steam drive performance—Upper Conglomerate zone, Yorba Linda field, California. *J. Petrol. Technol.* **29**, 1397–404.

DE 'ATH, N. G. & SCHUYLEMAN, S. F. 1981. The geology of the Magnus Oilfield. *In*: ILLING, L. V. & HOBSON, G. P. (eds) *Petroleum Geology of the Continental Shelf of North-West Europe*, 342–51. Heyden, London.

DEEGAN, C. E. & SCULL, B. J. 1977. A proposed standard lithostratigraphic nomenclature for the Central and Northern North Sea. *Rep. Inst. Geol. Sci.* 77/25.

Department of Energy 1983. *Development of the Oil and Gas Resources of the United Kingdom 1983.* HMSO, London. 65 pp.

ELLIOTT, T. 1978. Deltas. *In*: READING, H. G. (ed.) *Sedimentary Environments and Facies*, 97–142. Blackwell Scientific Publications, Oxford.

EYNON, G. 1981. Basin development and sedimentation in the Middle Jurassic of the North Sea. *In*: ILLING, L. V. & HOBSON, G. D. (eds) *Petroleum Geology of the Continental Shelf of North-West Europe*, 196–204. Heyden, London.

FOWLER, C. 1975. The geology of the Montrose Field. *In*: WOODLAND, A. W. (ed.) *Petroleum and the Continental Shelf of North West Europe*, pp. 467–76. Applied Science Publishers, London.

GALLOWAY, W. E. 1975. Process framework for describing the morphologic and stratigraphic evolution of the deltaic depositional systems. *In*: BROUSSARD, M. L. (ed.) *Deltas, Models for Exploration*, pp. 87–98. Houston Geological Society.

GLENNIE, K. W. 1972. Permian Rotliegendes of North-West Europe interpreted in the light of modern desert sedimentation studies. *Bull. Am. Ass. Petrol. Geol.* **56**, 1048–71.

—— 1981. Early Permain Rotliegendes. *In*: *Introduction to the Petroleum Geology of the North Sea*. Joint Association of Petroleum Exploration Courses (U.K.). *Course Notes No. 1*. Burlington House, London.

—— 1982. Early Permian (Rotliegendes) palaeowinds of the North Sea. *Sedim. Geol.* **34**, 245–65.

—— & BOEGNER, P. L. E. 1981. Sole Pit inversion tectonics. *In*: ILLING, L. V. & HOBSON, G. D. (eds) *Petroleum geology of the Continental Shelf of North-West Europe*, 110–20. Heyden, London.

——, MUDD, G. C. & NAGTEGAAL, P. J. C. 1978. Depositional environment and diagenesis of Permian Rotliegendes sandstones in the Leman Bank and Sole Pit areas of the U.K. Southern North Sea. *J. geol. Soc. London*, **135**, 25–34.

HALLAM, A. 1978. Eustatic cycles in the Jurassic. *Palaeogeogr. Palaeoclim. Palaeoecol.* **12**, 1–32.

HARMS, J. C., TACKENBERG, P., POLLOCK, R. E. & PICKLES, E. 1981. The Brae Field area. *In*: ILLING, L. V. & HOBSON, G. D. (eds) *Petroleum Geology of the Continental Shelf of North-West Europe*, 352–7, Heyden, London.

HARRIS, D. G. 1975. The role of geology in reservoir simulation studies. *J. Petrol. Technol.* May, 625–32.

—— & HEWITT, C. H. 1977. Synergism in reservoir management—the geologic perspective. *J. Petrol. Technol.* July, 761–70.

HARTMAN, J. A. & PAYNTER, D. D. 1979. Drainage anomalies in Gulf Coast Tertiary Sandstones. *J. Petrol. Technol.* October, 1313–22.

HERITIER, F. E., LOSSELL, P. & WATHNE, E. 1979. Frigg Field: large submarine fan trap in lower Eocene rocks of North Sea Viking Graben. *Bull. Am. Ass. Petrol. Geol.* **63**, 1999–2020.

HILLIER, G. R. K., COBB, R. M. & DIMMOCK, P. A. 1978. Reservoir development planning for the Forties Field. *Proceedings of the European Offshore Conference*, Vol. II, pp. 325–35. Soc. Petrol. Eng., Dallas.

KESSLER, L. G., ZANG, R. D., ENGLEHORN, J. A. & EAGER, J. D. 1981. Stratigraphy and sedimentology of a Paleocene submarine fan complex, Cod Field, Norwegian North Sea. *In*: *The Sedimentation of North Sea Reservoir Rocks*. Norsk Petroleums forening (NPF), Geilo.

KNOX, R. W. O'B, MORTON, A. C. & HARLAND, R. 1981. Stratigraphical relationships of Palaeocene sands in the UK sector of the Central North Sea. *In*: ILLING, L. V. & HOBSON, G. D. (eds) *Petroleum Geology of the Continental Shelf of North-West Europe*, 267–81. Heyden, London.

KOSSACK, C. A. & BILHARTZ, H. L. 1976. The sensitivity of micellar flooding to reservoir heterogeneities. *Soc. Petrol. Engrs, Pap. no. 5808, 'Improved Oil Recovery Symposium', March, Tulsa, Oklahoma.*

LEBLANC, R. J. 1972. Geometry of sandstone reservoir bodies. *Mem. Am. Ass. Petrol. Geol.* **18**, 133–90.

—— 1977. Distribution and continuity of sandstone reservoirs—Parts 1 and 2. *J. Petrol. Technol.* July, 776–804.

MAHER, C. E. 1980. Development geology of the Piper Field. *Oceanol. Int.* **80**, 13–24.

—— 1981. The Piper Oilfield. *In*: ILLING, L. V. & HOBSON, G. D. (eds) *Petroleum Geology of the Continental Shelf of North-West Europe*, 358–70. Heyden, London.

MARIE, J. P. P. 1975. Rotliegendes stratigraphy and diagenesis. *In*: WOODLAND, A. W. (ed.) *Petroleum and the Continental Shelf of North-West Europe*, 205–12. Applied Science Publishers, London.

MORTON, A. C. 1979. The provenance and distribution of the Palaeocene sands of the Central North Sea. *J. Petrol. Geol.* **2**, 11–21.

NAGTEGAAL, P. J. C. 1978. Sandstone-framework instability as a function of burial diagenesis. *J. geol. Soc. London*, **135**, 101–5.

NEVES, R. & SELLEY, R. C. 1975. A review of the Jurassic rocks of the North-east Scotland. *In*: FINSTEAD, K. S. & SELLEY, R. C. (eds) *Jurassic Northern North Sea Symposium, Stavanger*, JNNSS/5, 1–29 Norwegian Petroleum Society.

PARKER, J. R. 1975. Lower Tertiary sand development in the Central North Sea. *In*: WOODLAND, A. W. (ed.) *Petroleum and the Continental Shelf of North west Europe*, 447–51, Applied Science Publishers, London.

PROCTOR, C. V. 1981. Distribution of Middle Jurassic facies in the East Shetlands basin and their control of reservoir capability. *In*: *The Sedimentation of the North Sea Reservoir Rocks*. Norsk Petroleumsforening (NPF) Geilo.

PRYOR, W. A. 1973. Permeability-porosity patterns and variations in some Holocene sand bodies. *Bull. Am. Ass. Petrol Geol.* **57**, 162–89.

ROCHOW, K. A. 1981. Seismic stratigraphy of the North Sea "Palaeocene" deposits *In*: ILLING, L. V. & HOBSON, G. D. (eds) *Petroleum Geology of the Continental Shelf of North-West Europe*, 255–66. Heyden, London.

ROSSEL, N. C. 1982. Clay mineral diagenesis in Rotliegend aeolian sandstones of the southern North Sea. *Clay Miner.* **17**, 69–77.

SANGREE, J. B. & WIDMIER, J. M. 1977. Seismic stratigraphy and global changes of sea level, part 9: seismic interpretation of clastic depositional facies. *In*: PAYTON, C. E. (ed.) *Seismic Stratigraphy—applications to hydrocarbon exploration. Mem. Am. Ass. Petrol. Geol.* **26**, 165–84.

SEEMAN, U. 1979. Diagenetically formed interstitial clay minerals as a factor in Rotliegende sandstone reservoir quality in the Dutch sector of the North Sea. *J. Petrol. Geol.* **1**, 55–62.

SHELTON, J. W. 1973. Models of sand and sandstone deposits: a methodology for determing sand genesis and trend. *Bull. Oklahoma geol. Surv.* **118**, 122 pp.

SIMPSON, R. D. H. & WHITLEY, P. K. J. 1981. Geological input to reservoir simulation of the Brent Formation. *In*: ILLING, L. V. & HOBSON, G. D. (eds) *Petroleum Geology of the Continental Shelf of North-West Europe*, 310–09. Heyden, London.

SKJOLD, L. J. 1981. Paleocene sands of the Balder Field. *In*: *The Sedimentation of the North Sea Reservoir Rocks.* Norsk Petroleums forening (NPF), Geilo.

SNEIDER, R. M., RICHARDSON, F. H., PAYNTER, D. D., EDDY, R. E. & WYANT, I. A. 1976. Predicting reservoir-rock geometry and continuity in Pennsylvanian reservoirs, Elk City field, Oklahoma. *Soc. Petrol. Engrs, Pap. no 6138, 51st Ann Tech. Conf.* New Orleans.

STALDER, P. J. 1973. Influence of crystallographic habit and aggregate structure of authigenic clay minerals on sandstone permeability. *Geologie Mijnb* **52**, 217–9.

STOW, D. A. V., BISHOP, C. D. & MILLS, S. I. 1982. Sedimentology of the Brae oilfield, North Sea: fan models and controls. *J. Petrol. Geol.* **5**, 129–48.

SURLYK, F. 1978. Submarine fan sedimentation along fault scarps on tilted fault blocks (Jurassic/Cretaceous boundary, East Greenland) *Bull. Grønlands Geologiske Undersøgelse*, **128**, 108 pp.

THOMAS, A. N., WALMSLEY, P. J. & JENKINS, D. A. L. 1974. Forties Field North Sea. *Bull. Am. Ass. Petrol. Geol.* **58**, 396–405.

VAIL, P. R., MITCHUM, R. M. & TODD, R. G. 1977. Eustatic model for the North Sea during the Mesozoic. *In*: *Proc. Mesozoic Northern North Sea Symposium.* MNNSS/12, 1–35. Norwegian Petroleum Society.

—— & TODD, R. G. 1981. Northern North Sea Jurassic unconformities, chronostratigraphy and sea level changes from seismic stratigraphy. *In*: ILLING, L. V. & HOBSON, G. D. *Petroleum Geology of the Continental Shelf of North-West Europe* 216–35. Heyden, London.

VEEN, F. R. VAN 1975. Geology of the Leman field. *In*: WOODLAND, A. W. (ed.) *Petroleum and the Continental Shelf of North-West Europe*, 223–32. Applied Science Publishers, London.

—— 1977. Prediction of permeability trends for water injection in a channel-type reservoir, Lake Maracaibo, Venezuala. *Soc. Petrol. Engrs, Pap. no. 6703, 52nd Ann. Techn. Conf.* Denver, Colorado.

WALKER, R. G. 1967. Turbidite sedimentary structures and their relationship to proximal and distal depositional environments. *J. sedim. Petrol.* **37**, 25–43.

—— 1978. Deep-water sandstone facies and ancient submarine fans: models for exploration for stratigraphic traps. *Bull. Am. Ass. Petrol. Geol.* **62**, 932–66.

WALKER, R. G. & MUTTI, E. 1973. Turbidite facies and facies associations. *In*: MIDDLETON, G. V. & BOUMA, A. H. (eds) *Turbidites and Deep-Water Sedimentation*, 119–57 Society of Economic Paleontologists and Mineralogists (Pacific Section), Anaheim.

WALMSLEY, P. J. 1975. The Forties field. *In*: WOODLAND, A. W. (ed.) *Petroleum and the Continental Shelf of North-West Europe*, 477–85. Applied Science Publishers, London.

WEBB, M. G. 1978. Reservoir description, Kf Sandstone, Redwash field, Utah. *Soc. Petrol. Engrs, Pap. no. 7046, 5th Symp., Improved Methods for Oil Recovery*, Tulsa, Oklahoma.

WEBER, K. J. 1984. Computation of initial well productivities in aeolian sandstone on the basis of a geological model, Leman gas field, U.K. *In*: TILLMAN, R. W. & WEBER, K. J. (eds) *Reservoir Sedimentology. Spec. Publ. Soc. econ. Paleont. Mineral.* (in press).

WEBER, K. J., KLOOTWIJK, P. H., KONIECZEK, J. & VLUGT, W. R. VAN DER 1978. Simulation of water injection in a barrier-bar type, oil rim reservoir in Nigeria. *J. Petrol. Technol.* November, 1555–65.

WILLIAMS, J. J., CONNER, D. C. & PETERSON, K. E. 1975. Piper oilfield, North sea: fault block structure with Upper Jurassic beach/bar reservoir sands. *Bull. Am. Ass. Petrol. Geol.* **59**, 1581–601.

ZIEGLER, P. 1981. Evolution of sedimentary basins in Northwest Europe. *In*: ILLING, L. V. & HOBSON, G. D. (eds) *Petroleum Geology of the Continental Shelf of North-West Europe*, 3–39, Heyden, London.

—— 1982. *Geological Atlas of Western and Central Europe.* Shell Internationale Petroleum Maatschappij. B. V., 130 pp.

H. D. JOHNSON & D. J. STEWART, Shell U.K. Exploration and Production, London.

Methods for Detection and Characterization of Reservoir Rock, Deep Basin Gas Area, Western Canada

Robert M. Sneider,* SPE, Sneider and Meckel Assocs. Inc.
Howard R. King, Canadian Hunter Exploration Ltd.
H. Earl Hawkes, Canadian Hunter Exploration Ltd.
Thomas B. Davis, Canadian Hunter Exploration Ltd.

Summary

Major gas reserves have been discovered in the past 5 years in Cretaceous sandstones and conglomerates at depths of 3,000 to 10,000 ft (915 to 3045 m) within Alberta and British Columbia, Canada. Discovery of these new reserves resulted from a joint geological/petrophysical/reservoir engineering effort that utilized rock/fluid data from cuttings, cores, well logs, and drillstem and production tests. A key element in the exploration search was the rapid detection and characterization of reservoir-rock properties from well cuttings, especially of the low-permeability rocks (1 md to a few microdarcies). The methods used to establish and to characterize the reservoir potential of thick, multiple sandstone and conglomerate intervals are illustrated.

Rock studies of > 10,000 ft (3045 m) of conventional cores integrated with petrophysical studies of well logs and core analyses, which in turn were compared with drillstem tests (DTS's) and production tests in more than 200 wells, provide the basis for establishing reservoir-rock potential. Porous rocks are subdivided into three categories.

Type I rocks are capable of gas production without natural and/or artificial fracturing.

Type II rocks are capable of gas production when interbedded with Type I rocks or with natural and/or artificial fracturing.

Type III rocks are too tight to produce at commercial rates even with natural or artificial fracturing.

Type I rocks are subdivided into four classes, with four air-permeability ranges: "A" (> 100 md), "B" (10 to 100 md), "C" (1 to 10 md), and "D" (±0.5 to 1 md).

Criteria to identify a rock's reservoir potential from well cuttings or conventional and sidewall cores are based on examination of dry, freshly broken fragments with a binocular microscope at 20× magnification. A rock's reservoir potential is established on the basis of estimates of (1) size, volume, and distribution of visible pores, (2) particle size and distribution, and (3) type and amount of cements and pore-filling material and degree of consolidation (the way fragments break). Geologists and engineers can make rapid, accurate estimates of reservoir-rock potential of unknown porous intervals with the help of a number of visual aids. These aids include (1) plastic trays of cuttings-size rock chips crushed from conventional cores of known rock/pore types, porosities, and permeabilities; (2) color photographs of freshly broken surfaces of these rock/pore types taken at 20× magnification; and (3) thin-section microphotographs and scanning electron microscope (SEM) photographs (at 100 to 5,000× magnification) of the rock chips and their pore casts.

Methods and procedures described in this paper continue to be used in western Canada as well as the U.S. for delineation of exploration opportunities, identification of bypassed pays in old wells, interpretation of well logs, and evaluation of intervals for completion.

*Now with Robert M. Sneider Exploration Inc.

0149-2136/83/0091-0072$00.25

TABLE 1—PORE-TYPE CLASSIFICATION AND CHARACTERISTICS, DEEP BASIN, CANADA

Characteristics of Dry, Freshly Broken Rock Surfaces at 20X Magnification	Remarks
Type I	
Visible ϕ—very abundant to common. Pinpoint ϕ—very abundant to common. Pore interconnection is visible on many pores. Needle probe can easily dislodge some grains from rock surface and reveal pore.	Reservoir quality rock for gas without natural and/or artificial fracturing (if thick enough). k* (md) IA: >100 IB: 10 to 100 IC: 1 to 10 ID: ±0.5 to 1
Type II	
Visible ϕ—scattered. Pinpoint ϕ—abundant to common. Needle probe can only occasionally dislodge a grain from rock surface.	Capable of gas production if interlayered with Type I rock, or has natural, open fractures and/or is artificially fractured and is thick enough. k is > ±0.07 to 0.5 to 1.0 md (depending on particle size, sorting, and clay-mineral content).
Type III	
Visible ϕ—none to very isolated. Pinpoint ϕ—None to few, scattered pores. Usually very well consolidated and/or pore filled with clays or other pore-filling material.	Usually too tight to produce at commercial rate with natural or artificial fractures or when interlayered with Type I rock.

*ϕ = porosity; k = permeability.

Introduction

Integrated geological/petrophysical/reservoir engineering studies have resulted in the discovery of significant volumes of natural gas in Cretaceous sandstones and conglomerates in the Alberta Basin, Canada.[1-2] Although the exact size of the reserves must await definition by continued exploration and development, the more than 400 wells drilled in the past 5 years in the 26,000-sq-mile (67 000-km^2) area called the "Deep Basin" indicate that ultimate gas reserves will be many times larger than the several trillion cubic feet of gas accepted now as proved reserves.

The Cretaceous gas-bearing reservoirs have varied and complex lithology and pore systems. Most of the rocks are cemented and compacted and contain varying amounts of the clay minerals kaolinite, illite, chlorite, and smectite. Because of these clay minerals, some reservoir rock is highly sensitive to drilling and completion fluids. In addition, the pore system of most reservoir rocks is complex because of altered primary macropores, micropores in clays or chert, and secondary pores developed through leaching. Commercial production is from rocks that range in porosity from 5 to 15% and in permeability from a few microdarcies to more than 10 darcies. The reservoirs are usually stimulated artificially to obtain or to improve gas production.

Before discovery of the Deep Basin gas reserves in 1975, approximately 95 wells had been drilled completely or partially through the objective Cretaceous section. In the search for hydrocarbons, we studied almost all the cores, cuttings, well logs, and DST's and production tests of these older wells and integrated the rock/fluid data from these studies into regional stratigraphic studies. We found that a thorough understanding of the basic reservoir-rock parameters is a must for correct analysis of the well logs and fluid-flow tests as well as for identifying reservoir rock. A key element in our exploration search is the rapid detection and characterization of reservoir-rock properties from well cuttings. The concepts and methods used by Canadian Hunter Exploration Ltd. to detail and to characterize the reservoir potential of thick, multiple sandstone and conglomerates are presented in this paper.

General Considerations of Reservoir Rocks

The initial porosity, permeability, and pore geometry (pore size, shape, and orientation) of sands are controlled in large part by grain-size distribution. Other textural properties such as grain shape, grain orientation, and packing have minor influence on initial porosity and permeability.[3] Work by Beard and Weyl[4] and Pryor[5] has shown that initial porosity is dependent almost entirely on sorting and is independent of grain size, and that initial permeability is a function of both grain size and sorting. Studies of subsurface reservoir rocks show that cementation, compaction, leaching, and fracturing alter the initial pore space, sometimes drastically. In most subsurface sandstones (e.g., Elk City field, OK[6]),

TABLE 2—CONSOLIDATION CLASSIFICATION FOR SILICATE-RICH CLASTIC ROCKS

Descriptive Term	Sample Description
Unconsolidated	Sample disaggregates into individual particles before or after hydrocarbons are removed.
Slightly consolidated	Sample easily disaggregates or crumbles into individual particles when rubbed between fingers.
Moderately consolidated	Sample disaggregates only after rubbed vigorously between fingers.
Moderately well consolidated	Sample will not disaggregate when rubbed vigorously between fingers. Forceps or steel probe will disaggregate the sample into individual particles and smaller pieces containing several particles.
Well consolidated	Sample disaggregates with great difficulty, into smaller pieces containing several particles when forceps or steel probe is used.
Very well consolidated	Sample will not disaggregate with forceps or probe. A hammer disaggregates the sample into small pieces; pieces break across particles.

we have found that the grain size and sorting still have a significant control on the porosity, permeability, and pore geometry of the reservoir rocks even though the rocks are cemented and compacted. In rocks with low porosity (< 12 to 15%) and low permeability (< 10 to 20 md), detailed petrographic studies by other workers have shown the importance of clays and other fines[7-13] and secondary porosity[14] in determining porosity, permeability, pore geometry, fluid saturation, and productivity. The tighter rocks have pore geometries that approach the complexities that are typical of carbonate reservoir rocks.[15-17]

In our studies of the geological and petrophysical properties of sandstones and conglomerates in rocks of all ages, at all depths, in most of the hydrocarbon-bearing basins in North America, we find that the concepts outlined by Archie[18-19] and Teodorovich[20] for carbonate rocks are extremely valuable in interrelating rock and pore properties and physical-chemical properties of clastics. Studies of reservoir rocks by the senior author the past 25 years have shown that most petrophysical properties of sandstones and conglomerates can be related to (1) grain size and sorting, (2) degree of rock consolidation, (3) volume percent of clays, cements, and other pore-filling materials, and (4) sizes of pores and pore interconnections.

We have found that there are a finite number of rock types and corresponding pore geometries (pore types) when considering the pore-space parameters that control fluid flow. We recognize the complexities of size, shape orientation, and continuity of pores in sandstones; however, our experience shows that we can usually make a good estimate of the important petrophysical properties of clastic rocks from the parameters listed here.

Pore-Type Classification for Clastic Rocks

The simple pore-type classification (Table 1) we use is designed primarily for the wellsite geologist or engineer who looks at rocks in cores and cuttings with a binocular microscope. We divide rocks into three classes (I, II, and III) on the basis of the type, amount, and distribution of visible and small (pinpoint) pores visible on *freshly broken, dry rock surfaces* with a binocular microscope at *20× magnification* and on the way the rock breaks. In deciding on the pore type(s), we pay particular attention to estimating particle size and sorting, the way the rock breaks apart (consolidation), which is an indication of the degree of a rock's cementation and compaction (see Table 2), the amount (volume) of cements and pore-filling materials and an estimate of visible pores, and the size and distribution of pore throats. We make the estimates of grain size and sorting using standard size-sorting comparators and we estimate volume percentages using the Terry-Chilinger charts made for volume estimates. Consolidation is an arbitrary, but practical scheme (Table 2). Our observations are recorded on a data sheet similar to that in Fig. 1.

Figs. 2 and 3 show SEM photographs of typical Pore Types I, II, and III. We find that in many reservoirs (see Fig. 4) the pore types grade into one another or that more than one pore type is present in a sample. We record the relative percentage of each pore type in the samples described.

As demonstrated in the next section, we can usually make reasonably good guesses of permeability of most rocks and therefore have subdivided Type I rocks into four subgroups according to air-permeability ranges (1) IA—> 100 md, (2) IB—10 to 100 md, (3) IC—1 to 10 md, and (4) ID—±0.5 to 1 md.

We cannot overemphasize that all our work involves observing freshly broken, dry rock surfaces (cores, cuttings, outcrop samples) at 20× magnification. Most of the key rock-pore properties we use are not observable at 10 to 12×, the usual magnification of many binocular microscopes. In practice, the observations we make at 20× are usually confirmed with a "quick view" of the rock at 40 to 60× magnification.

Methods and aids used to recognize and classify pore types and to estimate permeability are discussed in the next section.

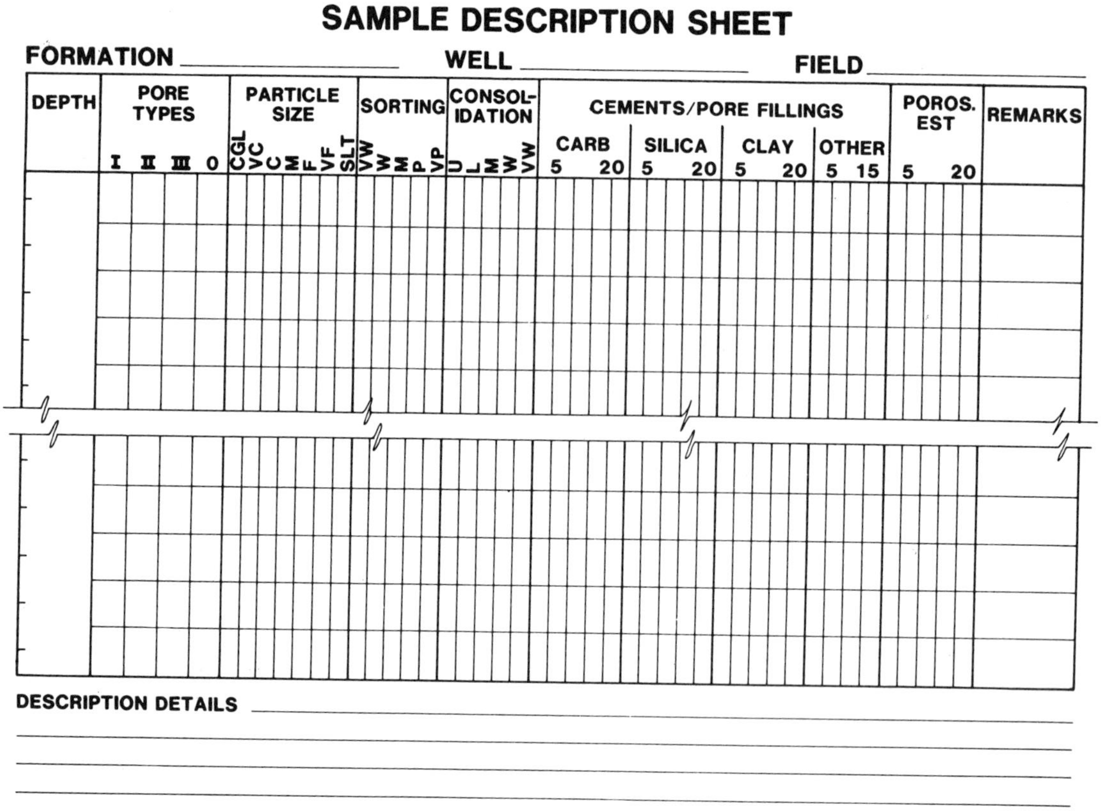

Fig. 1—Sample description sheet for cuttings and cores.

Aids To Recognize Pore Type and To Estimate Permeability

The key to our ability to recognize and classify pore types and estimate permeability is detail studies of a variety of different reservoir rocks made with thin-sections, the SEM, and X-ray diffraction. These rocks are from conventional core. We compare the rock-pore description derived from these studies with detailed petrophysical studies and with our binocular microscope description (at 20 to 60×). Fig. 5 shows some of the tests made on whole and plug-size core samples and the steps taken in preparing and calibrating the pore-type comparators.

The primary observations made with the binocular microscope (at 20×) on the core samples are grain size, sorting, consolidation, and volume and distribution of cements, pore-filling material, visible pores, and pore throats. We compare these with observations from thin-sections, photomicrographs of thin-sections, SEM photographs (at magnifications up to 5,000×) of the rock itself, and plastic replicas (pore casts) of the pore space. Next, we compare all these rock-pore observations with the various petrophysical analyses, especially porosity and permeability (measured unstressed and under stress). The calibration of rock-pore and petrophysical parameters on the same samples is of fundamental importance in learning how to estimate pore types and permeability from chips or plugs of rock.

We have established a reference collection of rock-pore petrophysical properties for typical reservoir rocks for the Alberta Deep formations. In new basins or formations where we have no core—that is, where there are only cuttings, we examine the cutting, decide what rock and pore type may be present, and pick reference rocks from basins where we think we have similar rock-pore types. For example, when we first studied the Cretaceous Falher sandstones and conglomerates in the area that now contains the giant Elmworth field, we used the Pennsylvanian sandstones and conglomerates from the Anadarko basin, OK,[6] for calibration until cores were cut in the Falher. Fig. 4 shows a crossplot of porosity vs. permeability (corrected for Klinkenberg effect) for these sandstones and conglomerates. In this giant Oklahoma oil field, we found that porosity, permeability, and pore geometry are related to grain size and sorting, cementation and compaction, consolidation and the amount of pore-filling clays.

To set up our rock-pore type calibration for the 15 Cretaceous reservoir units in the Alberta basin, we examined more than 10,000 ft (3045 m) of conventional core in more than 200 wells. Over half the core was specifically taken by Hunter to obtain new petrophysical parameters, as well as to provide stratigraphic and depositional environmental information to define plays and prospects and for development drilling. To determine the characteristics of the pore types of intervals that

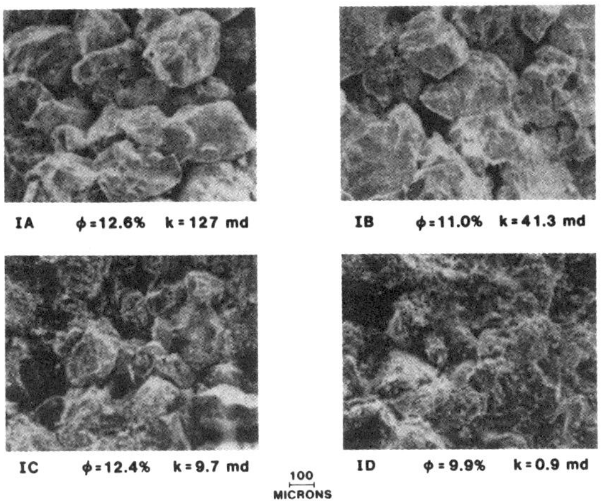

Fig. 2—SEM photographs of Pore Types IA, IB, IC, and ID.

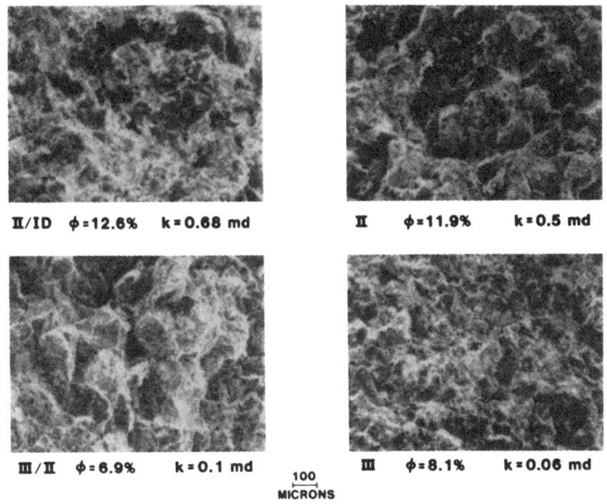

Fig. 3—SEM photographs of Pore Types ID, II, III, and III/II.

will or will not produce, the critical step is to compare (1) the rock properties observed with a binocular microscope (at 20×) aided with petrographic and SEM studies with (2) petrophysical properties established from well logs and cores, which in turn are compared with (3) drillstem and production tests of the intervals of interest. It is essential to know why a rock (pore) type will or will not produce and what stimulation techniques might be useful to obtain or to improve productivity.

Useful Visual Aids

With some practice and the use of aids described in this section, geologists and engineers can make rapid, accurate estimates of reservoir-rock potential of "unknown" intervals from the binocular examination of cuttings.

Permeability-plug-size core samples are cut for a variety of petrophysical tests from whole cores of known porosity and permeability (see Fig. 5). The end pieces of these plugs are used for thin-section, SEM, and X-ray analysis to study texture, mineralogy, and pore structure. Representative permeability plugs of typical rock-pore types are then selected and crushed and sieved to obtain "simulated cuttings." These cuttings are examined, and samples that represent the desired pore types are handpicked and mounted into comparator trays (Fig. 6). In addition, for most comparators we have a calibration set (see Fig. 7) so that the geologist and engineer can examine simulated cuttings as well as a polished and freshly broken surface of the original rock. By combining observations from these calibration sets or comparators with photographs of the rock-pore type taken of (1) freshly broken rock surface at 20×, (2) thin-section at 32× (also at 125×), (3) the rock surface with the SEM at 100 to 5,000×, and (4) replicas of the pores (pore casts), one can learn to estimate pore types. By comparing unknown cuttings with similar pore types of known porosity and permeability in appropriate comparators, one can estimate the permeability of the unknowns. With reasonable practice, one can become very proficient in estimating both porosity and permeability of rock chips. Fig. 8 is a sample page for

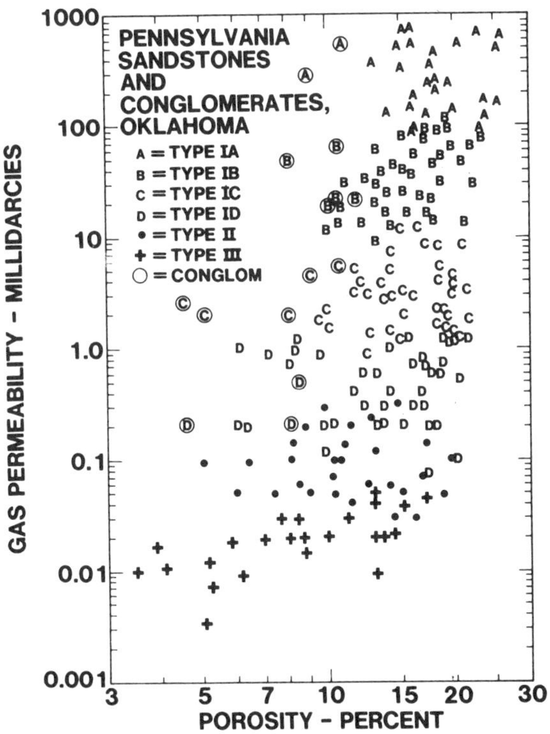

Fig. 4—Porosity/permeability crossplot showing pore types of Pennsylvanian sandstones and conglomerates, Elk City field, OK.

WHOLE CORE ANALYSES	CONVENTIONAL PLUG ANALYSES	PETROGRAPHIC ANALYSES	COMPARATOR PREPARATION
POROSITY (ϕ), PERMEABILITY (k), SATURATION (S), MODULUS TESTS	(PLUG CUT FROM WHOLE CORE) ϕ, k (STRESSED AND UNSTRESSED), CAPILLARY PRESSURE, FORMATION FACTOR, CATION EXCHANGE, RELATIVE PERM, ACOUSTIC TESTS, WATER SENSITIVITY	THIN SECTIONS, X-RAY AND RDX, SCANNING ELECTRON MICROSCOPE	PLUGS SELECTED & CRUSHED, PARTICLES SIEVED, PORE-TYPE SAMPLES SELECTED, SAMPLES MOUNTED

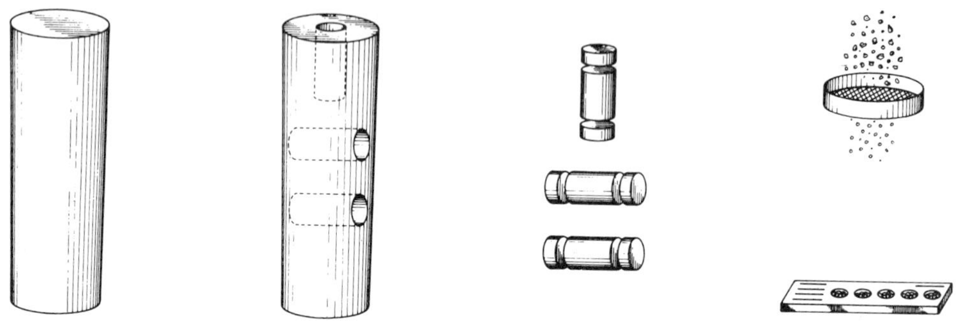

Fig. 5—Laboratory tests, measurements, and procedures for preparing and calibrating pore-type comparators.

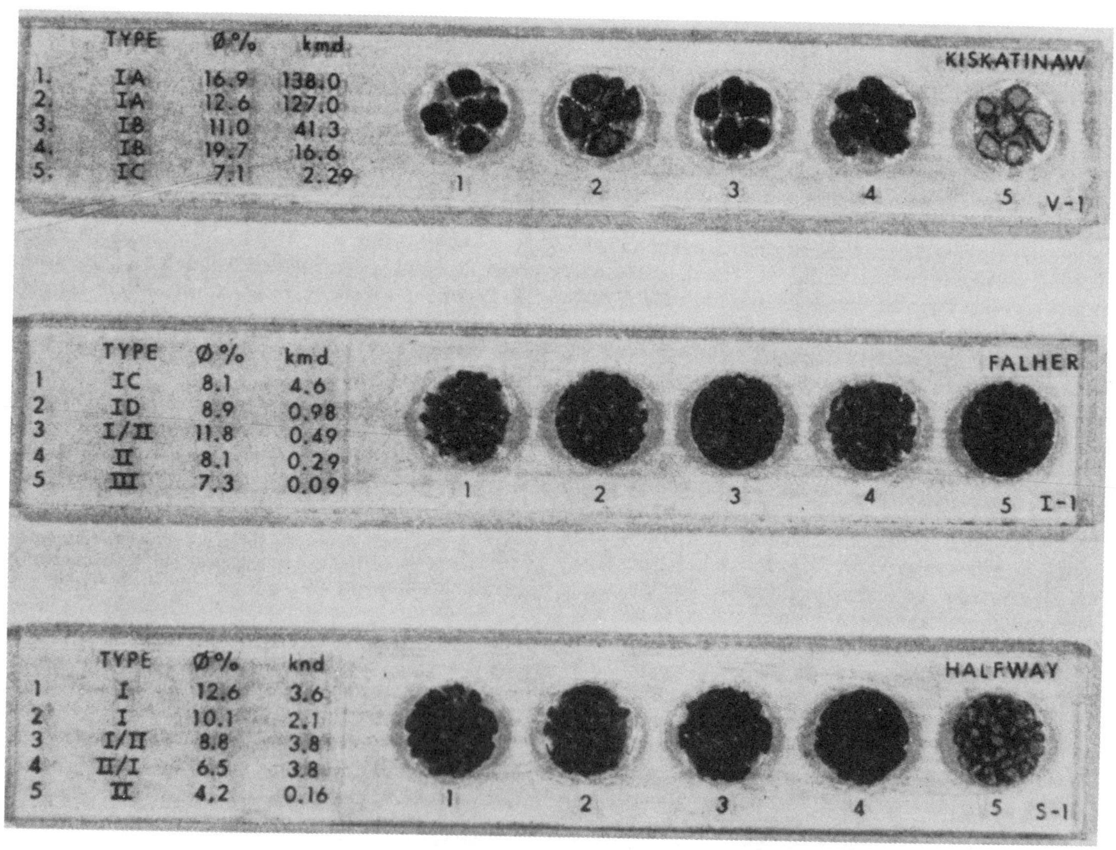

Fig. 6—Pore-type comparators for three Alberta basin producing formations.

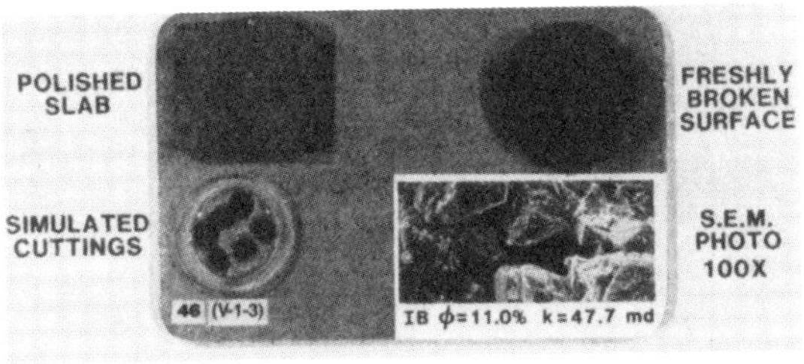

Fig. 7—Pore-rock-type calibration sample, Kiskatinaw formation, Alberta.

the rock-pore type comparator set shown in Fig. 7 and Pore Type IB of the topmost comparator in Fig. 6. As many workers[7,12,13,16,21,22] have found previously, we have found the SEM examination of rock surfaces and plastic replicas of the pores (pore casts) invaluable in understanding pore geometry and the type and distribution of clays and other pore-filling material.

Special Pore Type Problems

When working with cuttings, we find that three rock types present difficult evaluation problems. These are conglomerates, very shaly or clay-mineral rich sandstones, and rocks with abundant microporosity within clay-mineral aggregates or microporous chert.

Fragments of gravel-size particles develop as the drillbit cuts through a conglomerate. The cuttings that result rarely show any pores. Careful attention must be paid to those fragments that show part of the original particle surface. Small crystals of cement on these surfaces commonly are the only clue that an open pore was there before fragmentation into a chip. Grain surfaces that contain parts of large equant carbonate or quartz crystals generally indicate cemented rock in many reservoirs. Our experience suggests it is very difficult to pore-type most conglomerate rock accurately. In cores of the Falher conglomerates, we have intervals with porosity between 5 and 9%, with permeabilities of several hundred millidarcies to several darcies. Since porous and permeable conglomerates generally damage easily, especially by being plugged with whole mud, evaluation of a conglomerate interval by drillstem testing (DST) can be inconclusive or indicate an apparent "tight" rock. We recommend testing conglomerate intervals whenever they are found in cuttings, unless there are strong negative indications from the cuttings, logs, or flow tests. We have found many examples where low-porosity conglomerates (<7 to 8%) flow substantial amounts of hydrocarbons because of excellent permeability.

Evaluation of shaly or clay-rich rocks is a problem encountered by all who work with them.[7-13,23] Our calibration rock-pore sets of these rocks require special handling. The permeabilities used in the comparators must be measured under stress and corrected for Klinkenberg effects[24] to obtain a reasonable value of permeability. Water-sensitivity tests of core samples are very helpful to evaluate the effects of swelling clays. With the aid of SEM studies, especially of pore casts (Fig. 8) we believe we can evaluate the pore type in these difficult rocks

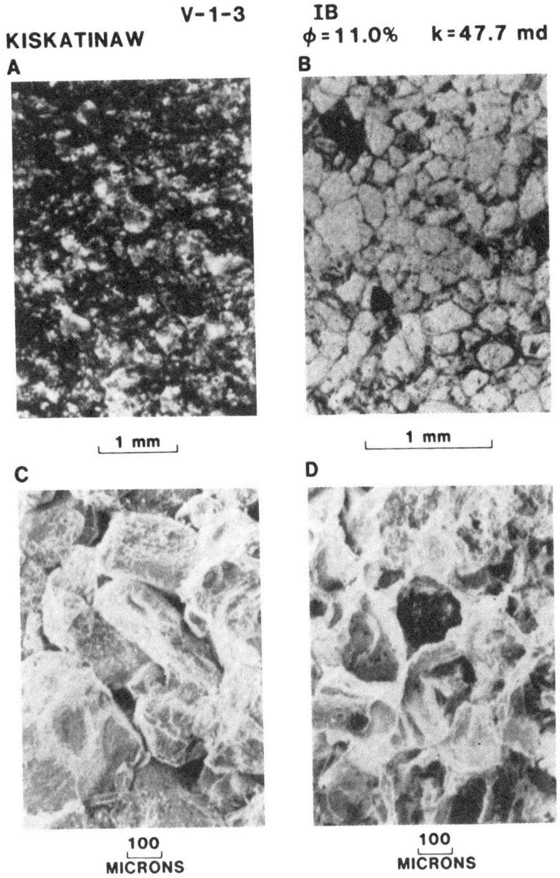

Fig. 8—Photographs illustrating pore structure of a Kiskatinaw IB pore type: (A) freshly broken surface at 20X, (B) thin-section, (C) SEM view, and (D) SEM view of pore cast. A and B are color photographs.

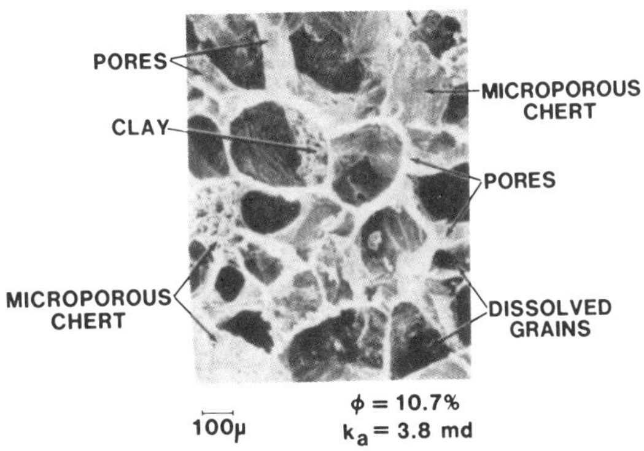

Fig. 9—An SEM photograph of pore cast showing micropores in a Kiskatinaw sandstone, Elmworth field, Alberta.

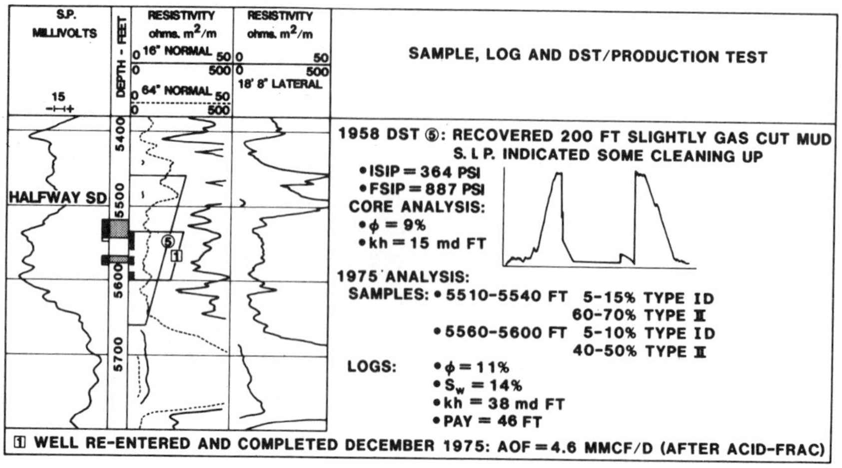

Fig. 10—Logs, sample data, and DST's and production tests, Example 1, Alberta basin, Canada.

with careful study. Putting fresh water on cuttings and seeing how the clays respond helps us to evaluate the influence of clays on the pore system and thus arrive at some basis for assigning a pore type. Most clay-rich, fine-grained sandstones usually are Type II rocks.

Microporosity of cherts and clay aggregates (Fig. 9) is another evaluation problem.[7-9] We believe that using 20× magnification enables us to "see" a lot of particles that have secondary porosity. Using binocular microscopes that can view rocks with 40 to 60× magnification, we believe we can make reasonable estimates of the volume of secondary pores. Study of pore casts (Fig. 9) helps us to recognize and evaluate microporosity better.

In cuttings samples composed mostly of loose grains, one is faced with the decision of whether or not the loose grains represent (1) unconsolidated or loosely consolidated sands or sandstones (Type I), (2) rocks that have abundant clays that fall apart when exposed to drilling fluids, or (3) rocks that "explode" apart when brought to or near the surface. We know of no easy way to tell consistently why a sample contains loose grains. Examination of grain surfaces sometimes is a clue to the origin of loose "grain" cuttings.

Applications

Three examples illustrate the use of the concepts and methods discussed here.

Example 1

Well c-69-J, 94-B-16 is a dry hole examined in our regional hydrocarbon search in 1975. This well (Fig. 10), originally drilled in 1958, was drillstem tested through the Halfway zone that was also partially cored. The DST recovered only 200 ft (61 m) of gas-cut mud, and the buildup and the shut-in pressures indicated little permeability or extreme formation damage. Analysis of the conventional core of the zone showed relatively low porosity and permeability. However, sample examination of the cuttings of the entire zone showed some ID

Fig. 11—Logs, sample data, and DST's and production tests, Example 2, Alberta basin, Canada.

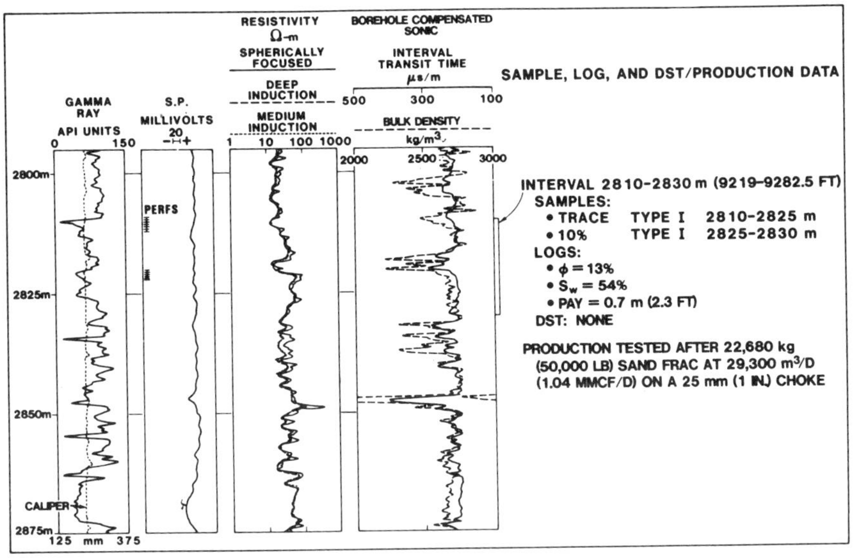

Fig. 12—Logs, sample data, and DST's and production tests, Example 3, Alberta basin, Canada.

pore types throughout. Our log analysis, which was calibrated with the sample and core data, suggested that the halfway zone contained the commercial pay zone of about 46 ft (14 m), with some conventional reservoir rock. The 1958 well was re-entered in 1975, and the zone was acid fractured with 1,500 gal (5.7 m^3) of 15% NCA acid and 60,000 lbm (27 216 kg) sand. The well was completed with an absolute open flow of 4.6 MMcf/D (130,000 m^3/d) gas and has already produced several billion cubic feet of gas at an average daily rate of 1.34 MMcf/D (40,000 m^3/d). The presence of some Type ID rock in the cuttings was an essential clue to indicating the presence of conventional reservoir rock. This well discovered the Town field and gave us a calibration point to continue our search for new Halfway opportunities.

Example 2

The Cretaceous Notikewan marine sand is generally rich in pore-filling clays. Most of the rock is Types II and III, and the few attempts to establish gas production in this formation have been unsuccessful. Examination at the wellsite of cuttings of a recent well (Fig. 11) showed loose grains and some Type ID rock in an interval corresponding to a drilling break. The wellsite geologist drillstem-tested the interval containing the drilling break; the results are shown in Fig. 11. The "gas show" from the DST, together with encouraging pore types in samples and some log-calculated pay, suggested a completion attempt. The interval of indicated pay was fractured with 50,000 lbm (22 680 kg) sand with KCl water, CO_2, and methanol. The zone flowed 3.5 MMcf/D (100 000 m^3/d) gas on a 16-day deliverability test. Other Notikewan zones have been flagged for testing on the basis of a re-examination of cuttings in other wells and the encouraging results of this well.

Example 3

In the well shown in Fig. 12, log analysis indicated a possible thin [2.3-ft (0.7-m)] pay zone in an interval 66 ft (20 m) thick that has a small amount of Type ID rock. The formation in the interval 9,219 to 9,282.5 ft (2810 to 2830 m) is usually a well-consolidated, shaly marine sand. The coincidence of a log show, a minor drilling break, and some Type ID rock suggested the interval in this exploratory well should be production tested. No DST was run during drilling.

Intervals 9,216 to 9,226 ft (2809 to 2812 m) and 9,252 to 9,259 ft (2820 to 2822 m) were perforated and fractured with 50,000 lbm (22 680 kg) sand with KCl water, methanol, and CO_2. After a long cleanup period, the well flowed 1.04 MMcf/D (22 680 kg) on a 1-in. (22-mm) choke. A re-examination of the samples and logs of the same formation in other wells in the area have shown that several better completion intervals are possible. One well has been completed already for 1.86 MMcf/D (52 270 m^3/d) at 2960 kPa (429 psi) on a ¼-in. (7.59-mm) choke after fracture treatment.

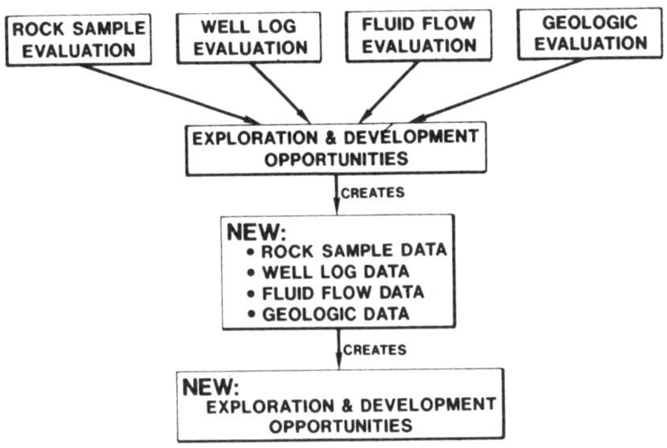

Fig. 13—Value of integrated geological/petrophysical/petroleum engineering studies.

Conclusions

We have found it very useful to classify sandstones and conglomerates into pore types for the purpose of establishing reservoir-rock potential. The system we use, together with the variety of visual aids illustrated in this paper, makes possible rapid detection and characterization of reservoir rocks in drill cutting at the wellsite with a binocular microscope equipped with at least 20× magnification. Comparators are useful to estimate permeability.

Integration of rock-pore-type data with laboratory petrophysical data, log-derived data, and geology has resulted in the discovery of major new reserves. We believe that the concepts and methods we use make for more efficient and effective delineation of exploration opportunities, identification of bypassed pay zones, well-log interpretation, and identification and evaluation of completion intervals. Fig. 13 expresses our view of the value of continued integration of geologic and petroleum engineering data in exploration and development. We have found that continued incorporation of petroleum-engineering data and ideas into exploration is an essential ingredient in successful ventures.

Acknowledgments

We thank J.A. Masters and J.K. Gray of Canadian Hunter Exploration Ltd. for permission to present and publish this paper. We also thank our colleagues D.A. Provias, N.D. Sullivan, and R.E. Wyman for their assistance. Our laboratory work in Canada could not have been successful without the assistance of A.H. Shepard and his staff at the Alberta Energy Resource Conservation Board who made thousands of feet of core available for study. J.W. Neasham and R.E. Tompkins of Reservoirs Inc. did the SEM, pore cast, X-ray diffraction, and petrographic analyses.

Special acknowledgment is given to the late G.E. Archie who introduced the senior author to the study of pore structure of rocks and its influence on hydrocarbon entrapment, production, and evaluation.

Nomenclature

kh = permeability-thickness product, md-ft
S_w = water saturation, %
ϕ = porosity

References

1. Masters, J.A.: "Deep Basin Gas Trap, Western Canada," *Bull.*, AAPG (Feb. 1979) **63**, 152-81.
2. Sneider, R.M. *et al.*: "Petroleum Engineering—Geology Synergy: The Key to Finding Large Reserves in Mature Basins," paper presented to SPE Geological Engineering Session at the 1978 SPE Annual Technical Conference and Exhibition, Houston, Oct. 1-3.
3. Krumbein, W.C. and Pettijohn, F.J.: *Manual of Sedimentary Petrography*, Appleton-Century-Crofts, New York City (1938).
4. Beard, D.C. and Weyl, P.K.: "Influence of Texture on Porosity and Permeability on Unconsolidated Sand," *Bull.*, AAPG (Feb. 1973) **57**, 349-69.
5. Pryor, W.A.: "Permeability-Porosity Patterns and Variations in Some Holocene Sand Bodies," *Bull.*, AAPG (1973) **57**, 162-89.
6. Sneider, R.M. *et al.*: "Predicting Rock Geometry and Continuity in Pennsylvanian Reservoirs, Elk City Field, Oklahoma," *J. Pet. Tech.* (July 1977) 851-66.
7. Kieke, E.M. and Hartmann, D.J.: "Detecting Microporosity To Improve Formation Evaluation," *J. Pet. Tech.* (Oct. 1974) 1080-86.
8. Neasham, J.W.: "The Morphology of Dispersed Clay in Sandstone Reservoirs and its Effect on Sandstone Shaliness, Pore Space and Fluid Flow Properties," paper SPE 6858 presented at the 1977 SPE Annual Technical Conference and Exhibition, Denver, Oct. 9-12.
9. Pittman, E.D. and Thomas, J.B.: "Some Applications of Scanning Electron Microscopy to the Study of Reservoir Rock," *J. Pet. Tech.* (Nov. 1979) 1375-80.
10. Keighin, C.W.: "Influence of Diagenetic Reactions on Reservoir Properties of the Nelsen, Farrer, and Tuscher Formations, Uinta Basin, Utah," paper SPE 7919 presented at the 1979 SPE Symposium on Low Permeability Gas Reservoirs, Denver, May 20-22.
11. Keighin, C.W. and Sampath, K.: "Evaluation of Pore Geometry of Some Low-Permeability Sandstones—Uinta Basin," *J. Pet. Tech.* (Jan. 1982) 65-70.
12. Wilson, M.D. and Pittman, E.D.: "Authigenic Clays in Sandstones: Recognition and Influence in Reservoir Properties and Paleoenvironmental Analysis," *J. Sed. Petrol.* (March 1977) **47**, 3-31.
13. Neasham, J.W.: "Applications of Scanning Electron Microscopy to the Characterization of Hydrocarbon-Bearing Rocks," *Scanning Electron Microscopy* (March 1977) **I**, 101-08.
14. Schmidt, V., McDonald, D.A., and Platt, R.L.: "Pore Geometry and Reservoir Aspects of Secondary Porosity in Sandstones," *Bull.*, Cdn. Pet. Geol. (May 1977) **25**, 271-90.
15. Choquette, P.W. and Pray, L.C.: "Geologic Nomenclature and Classification of Porosity in Sedimentary Carbonates," *Bull.*, AAPG (Feb. 1970) **54**, 207-50.
16. Pittman, E.D.: "Microporosity in Carbonate Rocks," *Bull.*, AAPG (1971) **55**, 1873-81.
17. Swanson, B.F.: "Visualizing Pores and Nonwetting Phase in Porous Rock," *J. Pet. Tech.* (Jan. 1979) 10-18.
18. Archie, G.E.: "Introduction to Petrophysics of Reservoir Rocks," *Bull.*, AAPG (May 1950) **34**, 943-61.
19. Archie, G.E.: "Classification of Carbonate Reservoir Rocks and Petrophysical Considerations," *Bull.*, AAPG (Feb. 1952) **36**, 278-98.
20. Teodorovich, G.I.: "Structure of the Pore Space of Carbonate Oil Reservoir Rocks and Their Permeability as Illustrated by Paleozoic Reservoirs of Bashkiriya," *Doklady Akad. Nauk SSSR* (1943) **39**, 231-34.
21. Wadlaw, N.C.: "Pore Geometry of Carbonate Rocks as Revealed by Pore Casts and Capillary Pressure," *Bull.*, AAPG (Feb. 1976) **60**, 245-57.
22. Pittman, E.D. and Duschatko, R.W.: "Use of Pore Casts and Scanning Electron Microscope to Study Pore Geometry," *J. Sed. Petrol.* (1970) **40**, 1153-57.
23. Simon, D.E., McDaniel, B.W., and Coon, R.M.: "Evaluation of Fluid pH Effects on Low Permeability Sandstones," paper SPE 6010 presented at the 1976 SPE Technical Conference and Exhibition, New Orleans, Oct. 3-6.
24. Klinkenberg, L.J.: "The Permeability of Porous Media to Liquids and Gases," *Drill. and Prod. Prac.*, API (1941) 200-13.

SI Metric Conversion Factors

cu ft	× 2.831 685	E−02	= m^3
ft	× 3.048*	E−01	= m
in.	× 2.54*	E+01	= mm
lbm	× 4.535 924	E−01	= kg
micron	× 1.0*	E−06	= m
psi	× 6.894 757	E+00	= kPa
sq ft	× 9.290 304*	E−02	= m^2

*Conversion factor is exact.

JPT

Original manuscript received in Society of Petroleum Engineers office July 22, 1981. Paper accepted for publication March 24, 1982. Revised manuscript received June 6, 1983. Paper (SPE 10072) first presented at the 1981 SPE Annual Technical Conference and Exhibition held in San Antonio, Oct. 4-7.

Reprinted by permission of the Tulsa Geological Society.
Published in the Tulsa Geological Society Special Publication
No. 1 (1979), pp. 115-162.

DELTAIC DEPOSITS IN THE PENNSYLVANIAN UPPER MORROW FORMATION OF THE ANADARKO BASIN

by

Donald C. Swanson

Esso Production Research Company

Houston, Texas 77001

ABSTRACT

Rocks of late Morrowan age represent a variety of deltaic environmental facies. The important reservoirs are both point-bar and stream-mouth-bar sandstones and conglomerates. Significant associated deposits are prodelta claystones, backswamp-marsh claystones, and meander-channel-fill claystones.

Point-bar deposits are the most important reservoirs, and many well-documented examples show their characteristic sizes, shapes, trends, and reservoir characteristics. Detailed data from these areas show relationships which would give reliable clues to the existence of nearby point-bar deposits when dry holes are drilled.

Study of the upper Morrow gives insight into processes of deltaic sedimentation which should provide help in local exploration, production operations, and regional exploration in similar rocks encountered throughout the world. Environmental information from only a few points in the most virgin area should lead to the visualization of paleogeography and sedimentologic processes which could be a guide to an exploration program. Where much data are available, environmental facies analysis can be invaluable in production operations such as field development or waterflood projects.

Many different fields of geology are utilized in a complete environmental facies analysis. Sedimentology, paleontology, stratigraphy, petrography, geomorphology, and structural geology contributed and were coordinated to give the final Morrow interpretations. These various fields of geology, when combined through environmental facies analysis, should provide the petroleum geologist with a tool which will enhance his ability to predict subsurface conditions and will improve his knowledge of reservoir deposits.

INTRODUCTION

Cores of deposits in the upper Morrow Formation from 45 wells located in nine separate areas of the Anadarko basin were examined during this study (Fig. 1). Core chips and sample cuttings were examined in many other wells during the study. More than 1,000 electric logs and several hundred sample logs were used, as were many maps and cross-sections. This article is based on data available in 1969. Although many wells have been drilled since then, occasional checking has revealed that no major changes in interpretation or concepts have occurred and the conclusions are valid.

The sandstone and conglomerate reservoirs in the upper Morrow Formation have been the most attractive targets in the deep "play" of the past 25 years in the Anadarko basin (Fig. 2).

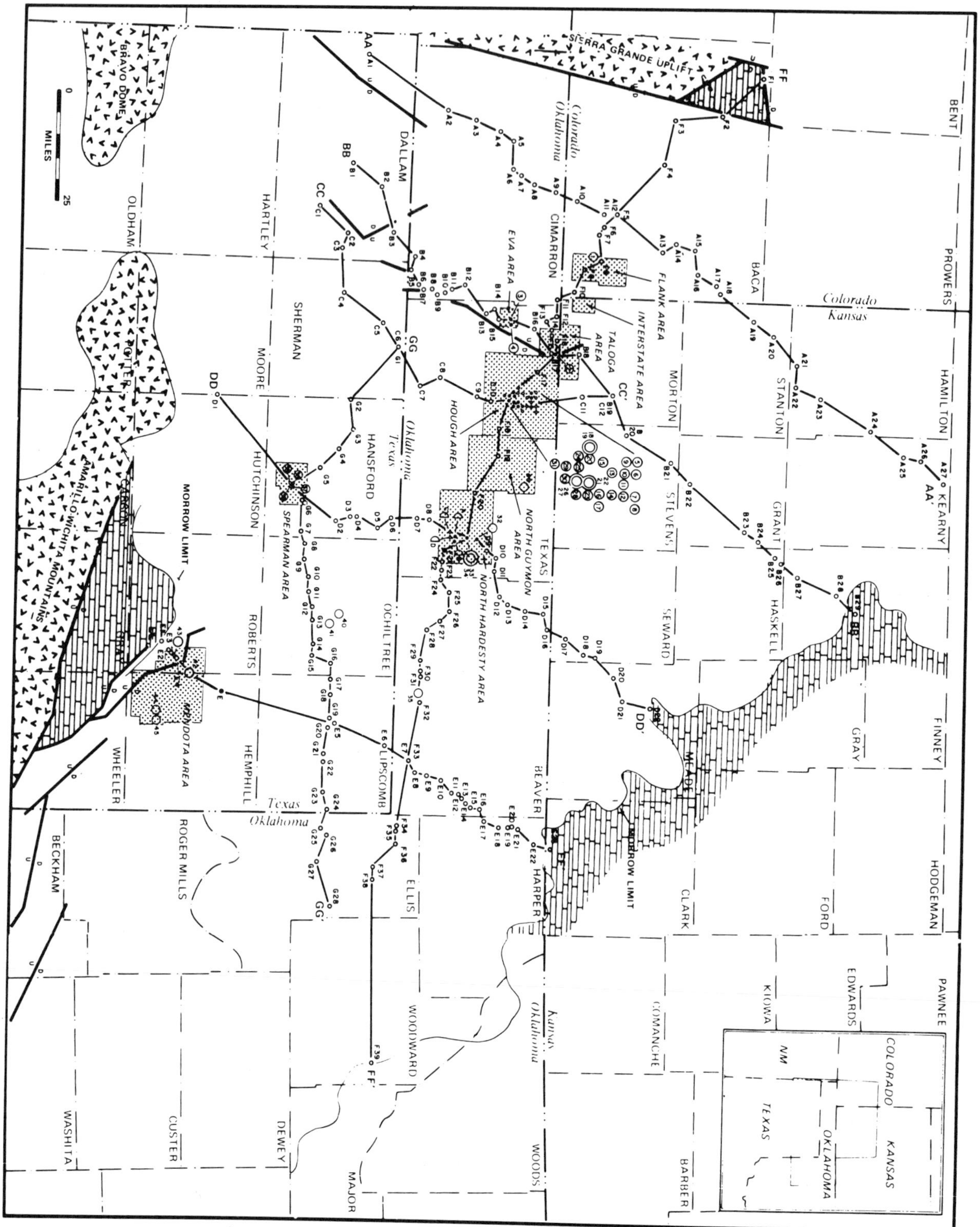

FIG. 1—Index map showing location of cored wells, location of regional sections, and areas of intensive study.

ERODED SEDIMENTARY ROCKS

ERODED IGNEOUS ROCKS

33 CORES SLABBED, EXAMINED, AND LOGGED

13 14 CORES SLABBED, EXAMINED, AND LOGGED – TWO IN THE SAME SECTION

A CORE CHIPS EXAMINED AND LOGGED

E SAMPLE CUTTINGS EXAMINED AND LOGGED

D12 ELECTRIC LOG USED PLUS SAMPLE LOG IF AVAILABLE

BOUNDARY OF DETAILED STUDY AREAS OF INTENSIVE INTEREST

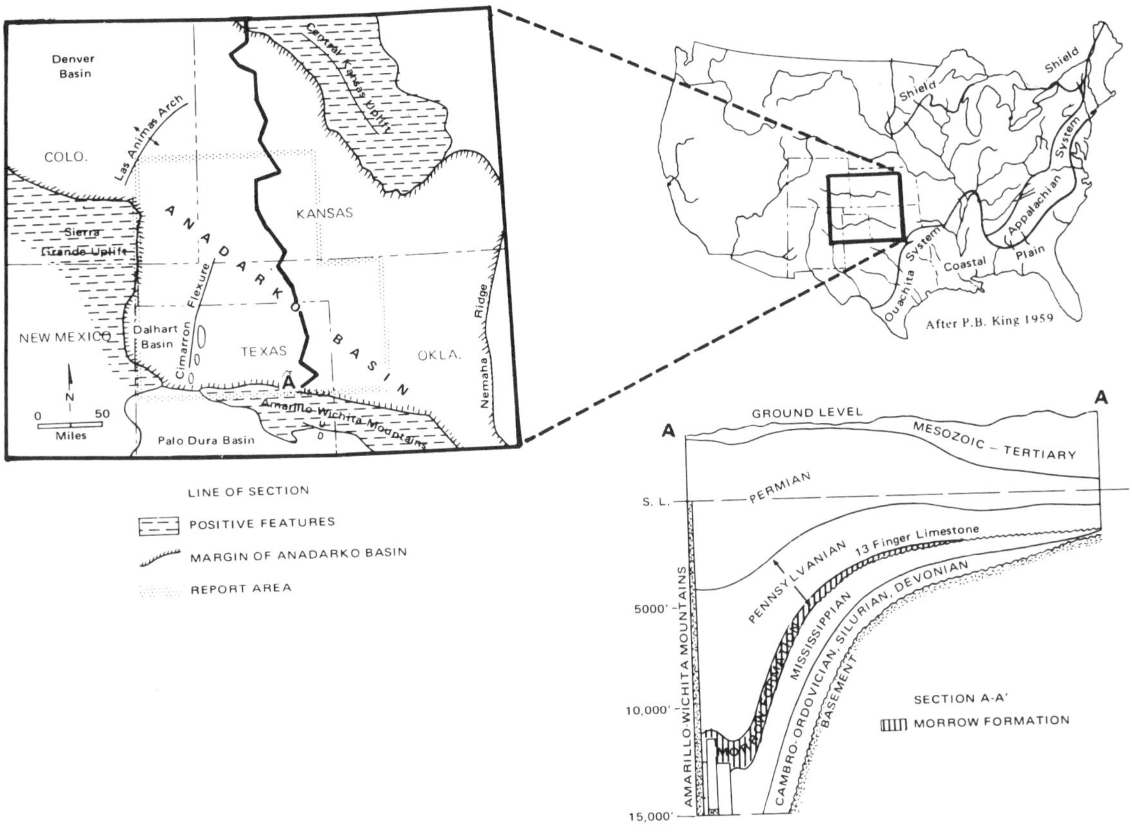

FIG. 2—Index map and cross-section of the Anadarko basin.

Although production is now widespread through the area, the first discovery in the upper Morrow Formation was not made until 1951 in Morton County, Kansas, and Ochiltree County, Texas (DeLong, 1952; Maravich, 1952; Payne, 1952). In 1952, additional discoveries were made in Meade and Seward Counties, Kansas, and in Ochiltree County, Texas (Berg, 1953; Maravich, 1953; Tarver, 1953). The Interstate field of Morton County, Kansas, and the Camrick area of Texas County, Oklahoma, were discovered in 1953 (Maravich, 1954; Totten, 1954; Williams, 1954), after which additional discoveries and development drilling expanded rapidly. In most, a "wet" gas is produced with considerable distillate; however, there are isolated pools of oil production throughout the basin. Good reserves, a favorable production history, and relatively shallow depth (4,500 to 10,000 ft) continue to make these rocks economically attractive.

GEOLOGIC SETTING

Figure 3 is a structure contour map drawn on the base of the Morrow Formation (in most places this datum is the top of Mississippian age rocks) showing various tectonic elements of the basin. A structural cross-section across the Anadarko basin (Fig. 2b) shows the basin to be a typical asymmetrical geosyncline with a pronounced southern mobile rim, a basin deep, a hinge line, and a shelf. It is essentially a Permian-Pennsylvanian basin, with maximum mountain building, basin subsidence, and clastic deposition occurring during that time.

During the Early Morrowan, seas transgressed an eroded Mississippian surface and deposited a series of "normal" shallow-marine claystones, sandstones, limestones and sandy limestones. This onlapping sequence of marine rocks is locally known as the lower Morrow Formation. Above this are the upper Morrow deposits (Fig. 4). Dark clayey limestones and dolomites lie conformably and uncomformably on the upper Morrow. These carbonates are of uncertain Atokan age. Because of their thin, persistent nature and ease of correlation, they provide excellent local time-stratigraphic datums.

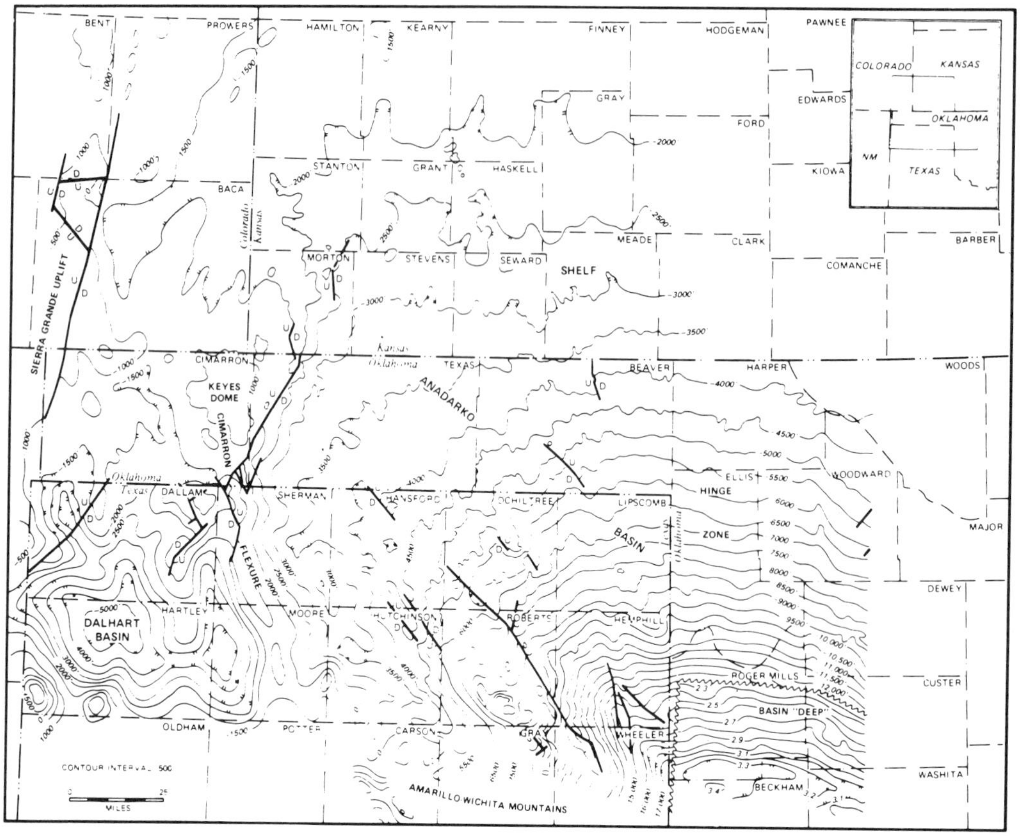

FIG. 3—Structure map of the base of the Morrow Formation (usually the top of the Mississippian) in the Anadarko basin.

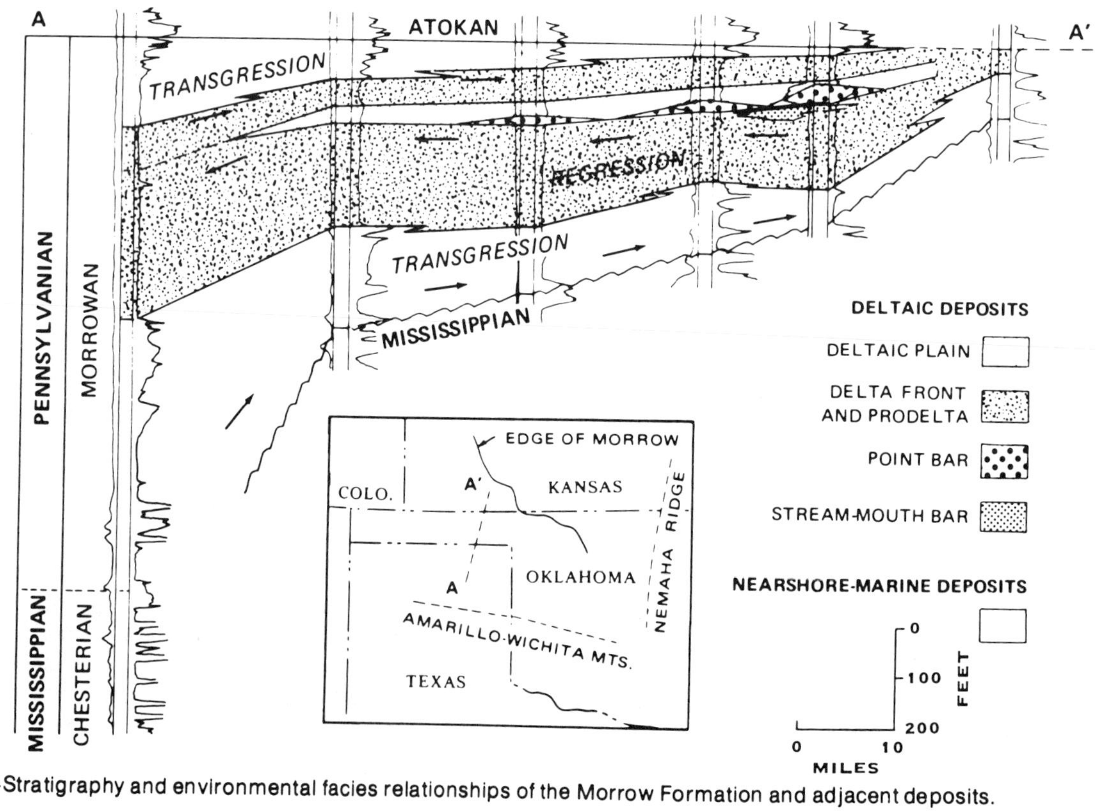

FIG. 4—Stratigraphy and environmental facies relationships of the Morrow Formation and adjacent deposits.

The total Morrow isopach map (Fig. 5) shows that the Morrow Formation is an asymmetrical wedge of sediments that trends northwest-southeast and reflects the major structural elements of the Anadarko basin. These sediments terminate on the south by onlapping the pre-Pennsylvanian sediments and basement in the Amarillo-Wichita Mountains. On the northeast, the Morrow becomes thin and finally disppears on the shelf along a line generally paralleling the central Kansas uplift. The rocks preserved along this edge may be of only early Morrowan age, remnants of Morrowan sediments which once extended further northeast but which were removed by post-Morrowan uplift and erosion. On the northwest, the Morrow extends into southeastern Colorado, over the Las Animas arch, and into the Denver basin. Southeast, the Morrow thickens into the central Anadarko basin. The exact extension of Morrowan rocks into the Dalhart basin and in a general southwestern direction is not known. Although the Morrow reaches thicknesses of greater than 3,000 ft in the basin deep, more than 85% of it is less than 700 ft thick.

The lower Morrow interval varies greatly in thickness over relatively short distances and tends to fill erosional lows on the underlying erosion surface. The top of the lower Morrow is marked in many places by the occurrence of sandy skeletal limestone that may be as thick as 200 ft. This lithologic unit is not a regional stratigraphic marker, but it does separate two distinct environmental facies. The overlying upper Morrow rocks, dated as Morrowan on the basis of spores, pollen, plant remains, and fusulinids, consist primarily of claystone and silty claystone with many discontinuous coarse-grained deposits. Coal and thin, dark limestone are minor constituents. The upper Morrow section is capped by the Atokan Thirteen Finger Limestone and Dolomite. This Atokan-Morrowan boundary is crossed by continuous sedimentation in

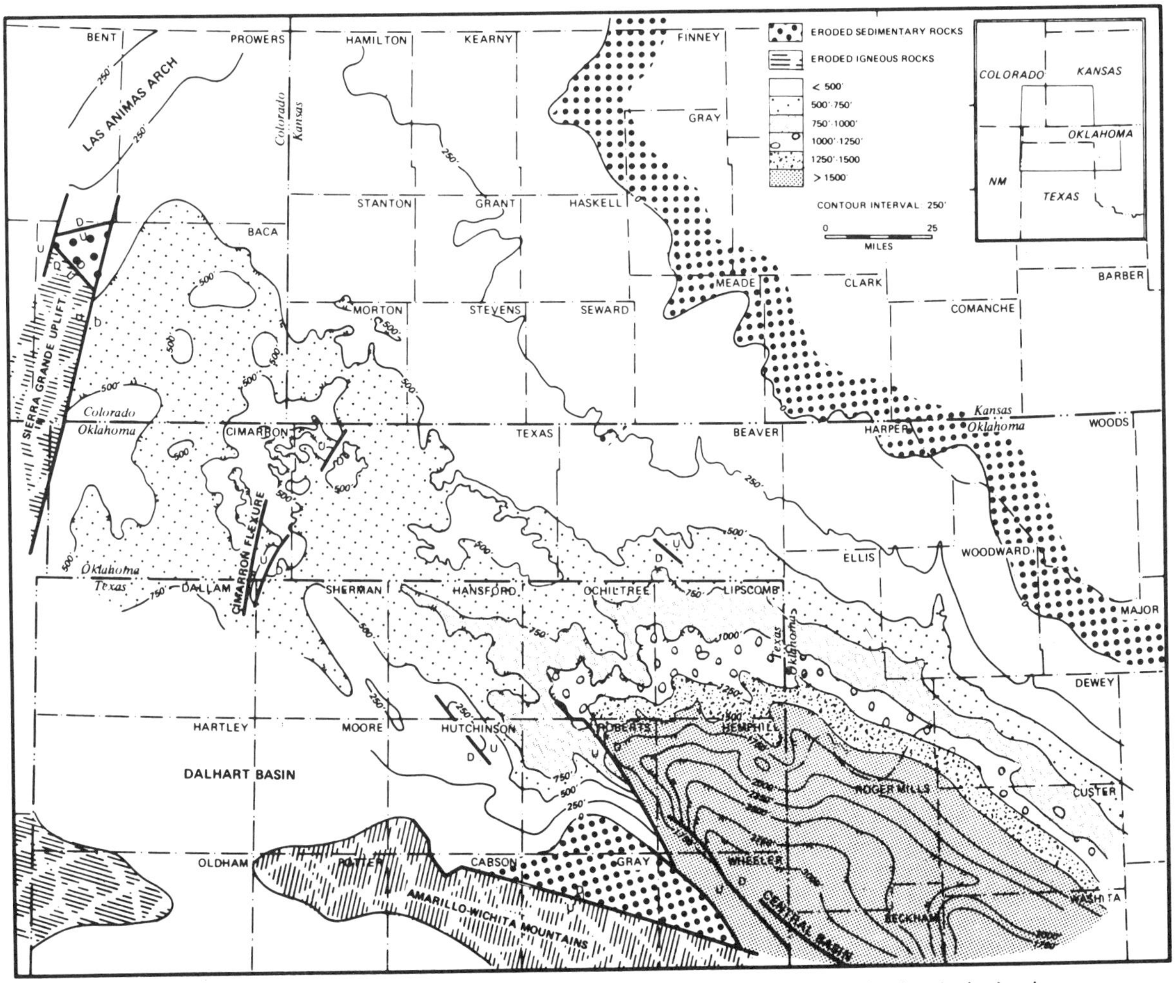

FIG. 5—Isopach map of the total Morrowan Formation (upper and lower Morrowan) in the Anadarko basin.

many areas; however, on the northeastern flank of the basin, and in local areas of structural movement, some of the Morrow has been removed, and this contact is an angular unconformity. Figure 6 is an electric log of a typical Morrowan section.

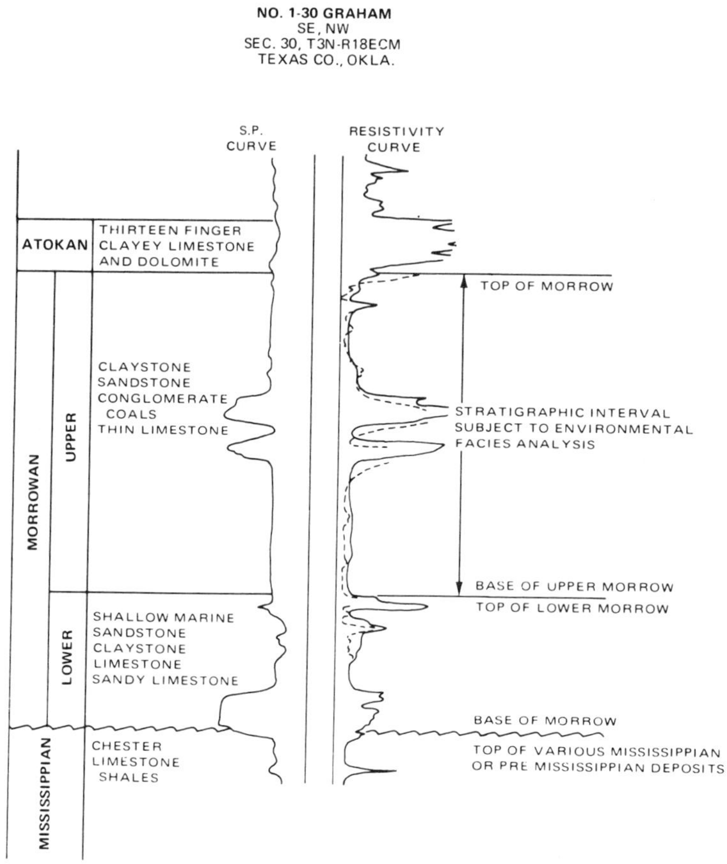

FIG. 6—Typical electric log of the Morrow Formation and adjacent rocks.

ENVIRONMENTAL FACIES ANALYSIS OF THE UPPER MORROW RESERVOIRS

Vertical Facies Relationships

Seven primary environmental facies were identified in the upper Morrow cores. They include (1) point-bar cross-bedded to massive sandstone and conglomerate; (2) stream-mouth-bar or "fan" and bar-finger cross-bedded and laminated sandstone and conglomerate; (3) distributary channel-fill sandstone and conglomerate; (4) meander-channel-fill organic claystone and silty claystone; (5) backswamp-marsh organic claystone and silty claystone; (6) delta-front intercalated sandstone and claystone; and (7) prodelta claystone.

Point-Bar Sandstone and Conglomerate — The majority of the wells studied contains deposits interpreted as point-bar sandstones and conglomerates (Fig. 7). The point-bar sandstones and conglomerates contain mostly quartz grains with some feldspar and rock fragments. Matrix consists primarily of kaolinite, but illite and montmorillonite are also present. Clay content varies from 20% in the upper part to less than 5% in the lower part. Textural sizes range from silt and clay to boulders. The vertical variation is remarkably constant in all the point bars. Except for minor fluctuations, the grain size of the point bars increases downward. The upper part of the point bar has textures of clay, silt, and fine sand. The middle part has fine to coarse textures, and the lower part consists of very coarse sandstone to boulders. A clay-pebble conglomerate occurs in the base of many deposits, and isolated claystone pebbles can

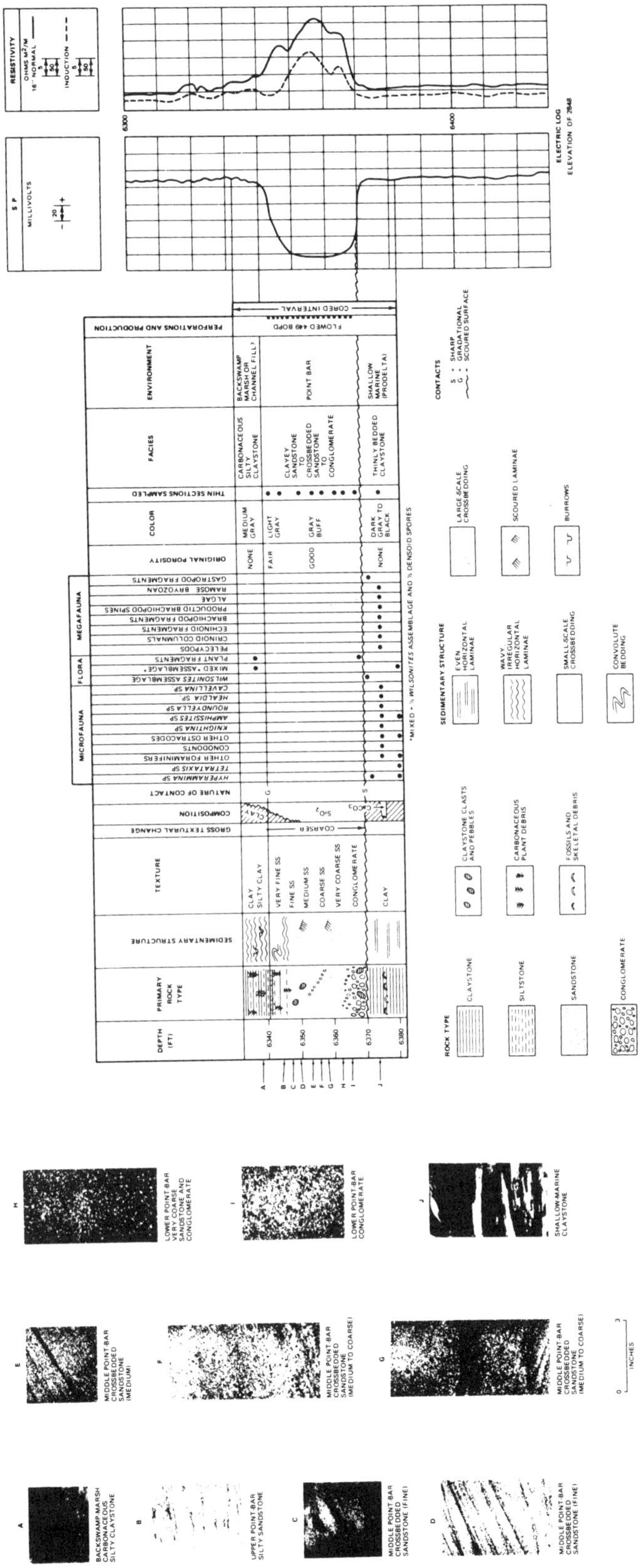

FIG. 7—Detailed log of cores from upper Morrow point-bar deposits, Humble No. 1 Pauls, Hardesty area, Texas County, Oklahoma (Fig. 1, No. 33).

be found throughout the cores (Figs. 8b, 8c, 8d, 9a, 9c). The upper part of the point bar is characterized by small-scale cross-bedding and wavy irregular ripple bedding (Figs. 8a, 10b). Convolute bedding is common, and an occasional burrow can be found in the upper point-bar rocks. Middle point-bar deposits have large-scale festoon cross-bedding (Fig. 9b), whereas most of the lower deposits are characterized by massive or large-scale cross-bedding (Fig. 9c).

The basal contact of the deposits is always sharp and represents scoured surfaces. The lower point bar grades upward successivley into the middle and upper point bar; the upper point bar in turn grades into the overlying backswamp-marsh deposits. Delta-front deposits or prodelta claystones underlie the point-bar deposits. In many places, a stream depositing a point bar cuts into stream-mouth sediments and deposits the point-bar sandstone and conglomerate directly on and cutting into stream-mouth-bar sands and conglomerates. No fauna was found in any of the point-bar deposits; however, considerable amounts of carbonaceous plant remains are present. This material commonly occurs on bedding planes in the upper point bar and as large remains of woody material near the base of the deposits.

Steam-Mouth-Bar Sandstone and Conglomerate — In the Morrow Formation, steam-mouth-bar sandstones and conglomerates consist of mostly quartz grains with minor feldspar and rock fragments (in two wells chert fragments were the main grain type). The matrix is clay, which increases in amount toward the lower part of the deposit. Grains range in size from pebbles to silt and clay, with the textural variations always decreasing in size downward. Most cores observed consist of generally very coarse sand to boulders in the top grading to fine sand and often intercalated sand and clay at the base (Fig. 11). Examples of upper Morrow stream-mouth-bar deposits are shown in Figures 12 and 13.

Although the upper part of the deposits may have large-scale festoon cross-bedding, and the lower part displays wavy irregular bedding, the outstanding structural feature of these deposits is the textural laminations they display. In gross aspect, the texture is from coarse to fine in a downward direction; however, in most parts of the deposit, there are small textural laminations of different sizes (Figs. 11a, 11b), commonly in the form of thin, graded units. the laminations and graded units probably result from different stages of flow in the stream feeding the bar and from turbid flow down the bar slope. The upper parts often have truncated beds and clay pebbles, whereas the lower parts have convolute bedding, burrows, load casts, and slump features. The base of the deposits has a gradational and often intercalated contact with the underlying rocks. The upper contact is usually sharp. Underlying sediments are commonly prodelta claystones. The overlying deposits vary from backswamp marsh to delta front and are in many places point-bar deposits which scoured into the stream-mouth bar. No fauna was recovered in these sediments; however, bedding planes in the upper parts contained carbonaceous plant material. Some spores and pollen were recovered from shaly zones in and surrounding the deposits.

Distributary-Channel-Fill Sandstone and Conglomerate — Two wells in Hansford County, Texas, contain sands which in vertical sequence resemble point-bar deposits. Additional cores and cross-sections, however, show this facies to be a distributary-channel-fill sandstone. The grains of this deposit are mostly quartz with minor feldspar and rock fragments. The matrix is clay and silt increasing in amount upward through the deposit. The grain texture varies from fine to very coarse sand with a generally coarser downward trend. Some convolute bedding is present, and laminae dipping up to 65° were observed. This facies is encountered in wells which penetrated deposits involved with late Morrowan structural movements, and it is difficult to reconstruct original bedding. No identifiable flora was encountered; however, much carbonaceous debris is found in the clay laminae of some bedding planes. The basal contact of this deposit is sharp; no cores were available through the upper contact. The underlying material is probably prodelta claystone.

The definition of this facies as a straight distributary channel is primarily based on lateral environmental facies control afforded by nearby cores in which laterally equivalent facies were identified as backswamp-marsh claystone. The possibility exists that the material identified as backswamp-marsh material is actually meander-channel-fill claystone, and the sand facies encountered is merely the remnant of preexisting point-bar sediments which have been cut away.

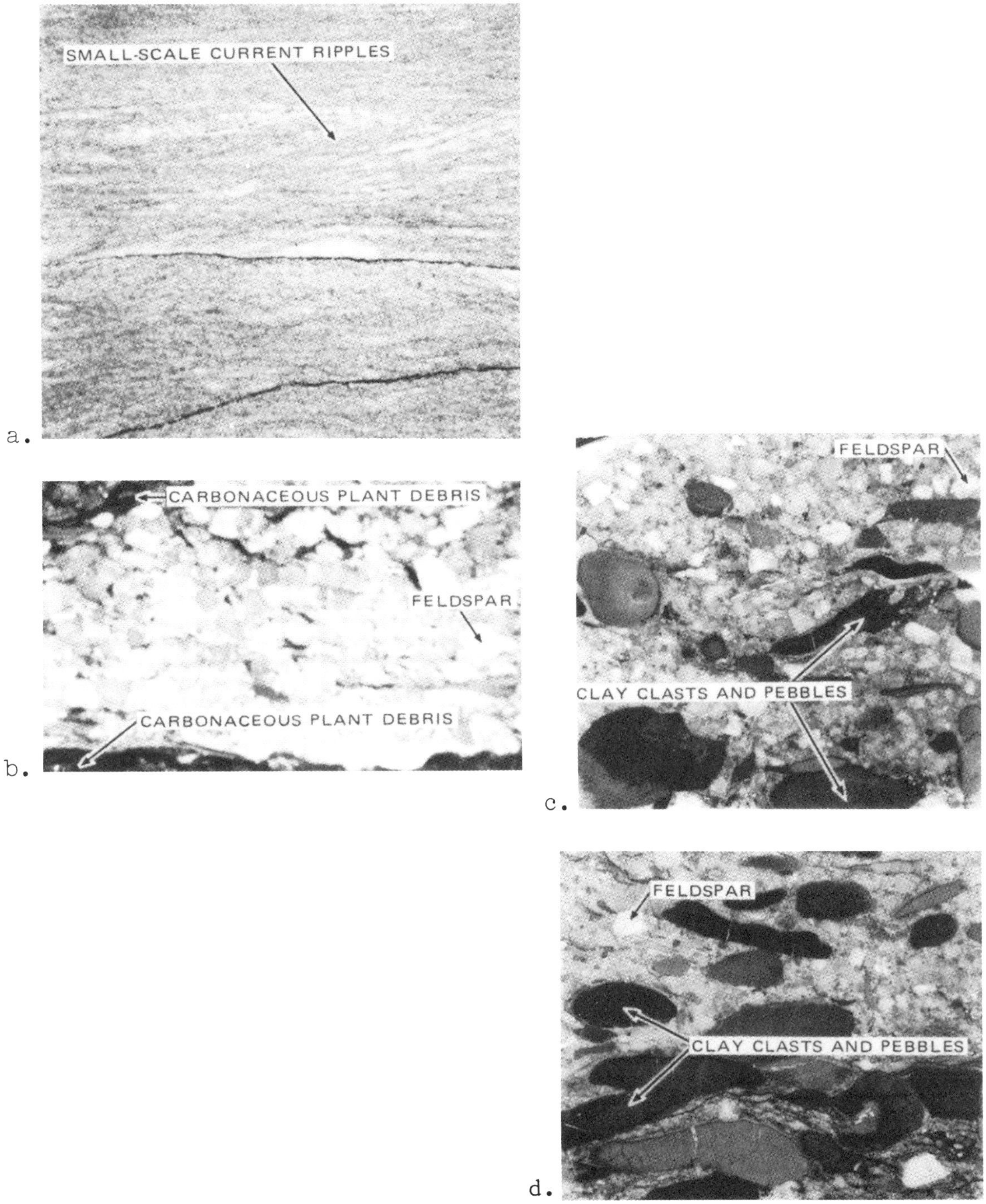

FIG. 8—Characteristics of point-bar deposits in Texas County, Oklahoma (a) Upper point-bar, silty, clayey fine sandstone in the Edwin Cox No. 5-A Buck, sec. 13, T.6N, R.11ECM, Texas County, Oklahoma, from a depth of 4,823 ft x1. (Fig. 1, no. 2). (b) Basal point-bar pebble conglomerate in the Humble No. 1 Schroeder, sec. 2, T.3N, R.18ECM, Texas County, Oklahoma, from a depth of 6,358 ft x1. (Fig. 1, No. 32). (c, d) Basal point-bar clay pebble conglomerates from the Humble No. 1 Jolliffee Trust, sec. 3, T.2N, R.18ECM, x0.75.

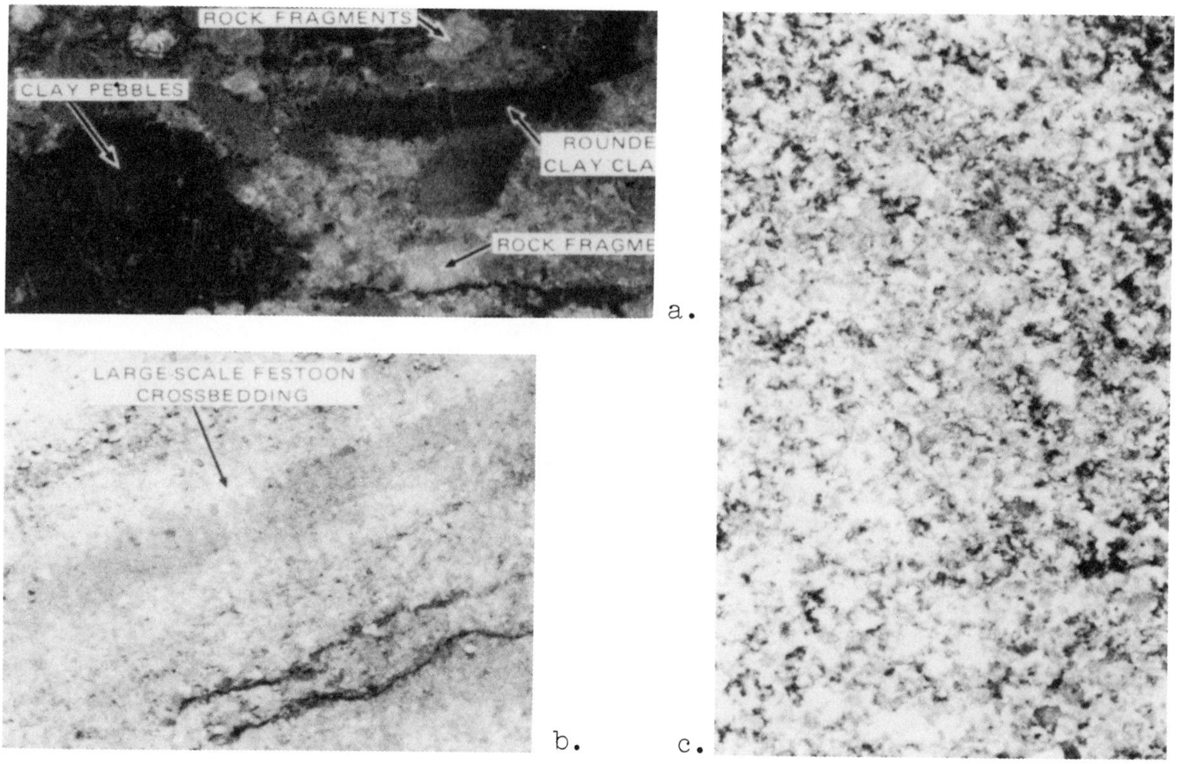

FIG. 9—Representative characteristics of delta-plain deposits from the North Hardesty area Texas County, Oklahoma. (a) Clay-pebble conglomerate from the base of meander-channel-fill deposits in the Humble No. 1 Jolliffee Estate, sec. 2, T.2N, R.18ECM, from a depth of 6,382 ft x1. (b) Middle point-bar sandstone from the Humble No. 1 Ora Craft, sec. 14, T.3N, R.18ECM, from a depth of 6,357 ft x1. (c) Lower point-bar conglomerate from the Humber Oil No. 1 Ora Craft, sec. 14, T.3N, R.18ECM; from a depth of 6,363 ft x1.

Channel-fill deposits often are silt and claystone. However, when a distributary is "cut off" slowly enough that the decreasing flow remains strong, there is a significant amount of sand deposited. Even then there is usually much silt and clay in the deposits (Fisk, 1955). This facies in the upper Morrow did not display these characteristics, but because of its shape and lateral facies relationship, it is interpreted as being distributary-channel-fill sandstone.

Meander-Channel-Fill Claystone — Several wells cored rock identified as meander-channel-fill claystone and silty claystone (Figs. 12, 14). This material is intimately associated with point-bar deposits and is composed of mostly claystone with minor silty claystone and sandstone. Textural variation, although slight, is from finer at the top to coarser at the base. In some cases, thin sand laminae or gall conglomerates were found at or near the base. Bedding tends to be indistinct to wavy irregular with some small-scale cross-bedding. This facies usually has a sharp contact with underlying beds, which are either prodelta claystone or delta-front deposits. Upward, the channel-fill claystone grades imperceptibly into backswamp-marsh facies. Although no fauna was found, a variety of well-preserved carbonaceous plant fragments is present. They are preserved as large woody pieces or as delicate fern imprints.

The meander-loop channel represents generally a very low energy environment associated with a rapid reduction of current activity. In many places, the only coarse material is the original bed load of the meander loop, and sometimes this is not even found. Positive distinction of a very fine-grained channel-fill claystone from a backswamp-marsh facies would be very difficult if lateral facies were not known. In the upper Morrow, meander-channel-fill deposits tend to have a greater silt and fine-sandstone content and to contain more carbonaceous plant remains — sometimes even developing a thin lignite.

Backswamp-Marsh Claystone — Many wells cored rocks which represent backswamp-marsh deposits. (Figs. 7, 13, 14). These rocks are essentially dark-gray claystone with minor

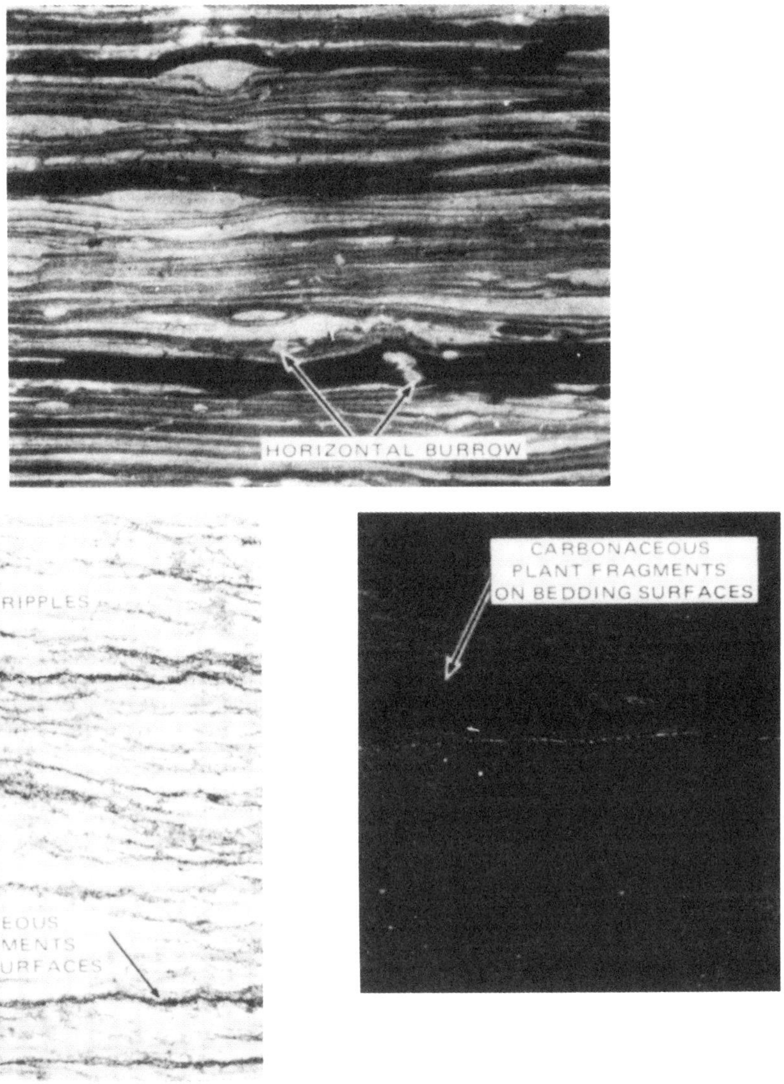

FIG. 10—Typical textures and structures in different deltaic facies from wells in the Oklahoma Panhandle. (a) delta-front intercalated very fine sandstone and claystone from the Humble No. 1 Smith, sec. 4, T.1N, R.24ECM, Beaver County, Oklahoma; from a depth of 7,918 ft x1. (Fig. 1, no. 35). (b) Upper point-bar silty clayey fine sandstone from the Humble No. 1 Curtis Paul, sec. 14, T.3N, R.18ECM, Texas County, Oklahoma, from a depth of 6,326 ft (Fig. 1, no. 33). (c) Backswamp-marsh silty claystone from the Humble No. 1 Curtis Pauls, sec. 14, T.3N, R.18ECM, Texas County, Oklahoma, from a depth of 6,326 ft x1 (Fig. 1, no. 33).

siltstone (Fig. 10c) and are commonly evenly and horizontally bedded or have mild wavy irregular bedding. This facies contains much carbonaceous plant debris and some rootlets and burrows (Fig. 15). It grades into underlying deposits which in most places are upper point bar, delta front or meander-channel fill. Fauna is sparse; only one specimen of *Lingula* sp. was encountered (Fig. 12). This facies has a prolific flora, however. Some of the macroflora identified are *Lepidodendron* sp., *Neuopteris* sp., *Condaites* sp., *Cordicarpus* sp., *Pecopteris* sp., *Sphenopteris* sp., and *Sphenophyllum* sp, indicating a typical coastal swamp (L. R. Wilson, personal communication). Many spore and pollen also occur in this facies.

Stream-Mouth Bar-Finger Deposit — When a distributary stays in a relatively straight course, it cuts into or through stream-mouth-bar deposits. If the distributary is cut off slowly, the channel fills with coarse material and the channel-fill and stream-mouth-bar material becomes incorporated into a single bar-finger deposit (Fisk, 1961).

An example of a core through such a deposit is shown in Figure 16. The upper part of the deposit represents distributary-channel fill. In this example, the load was extremely coarse.

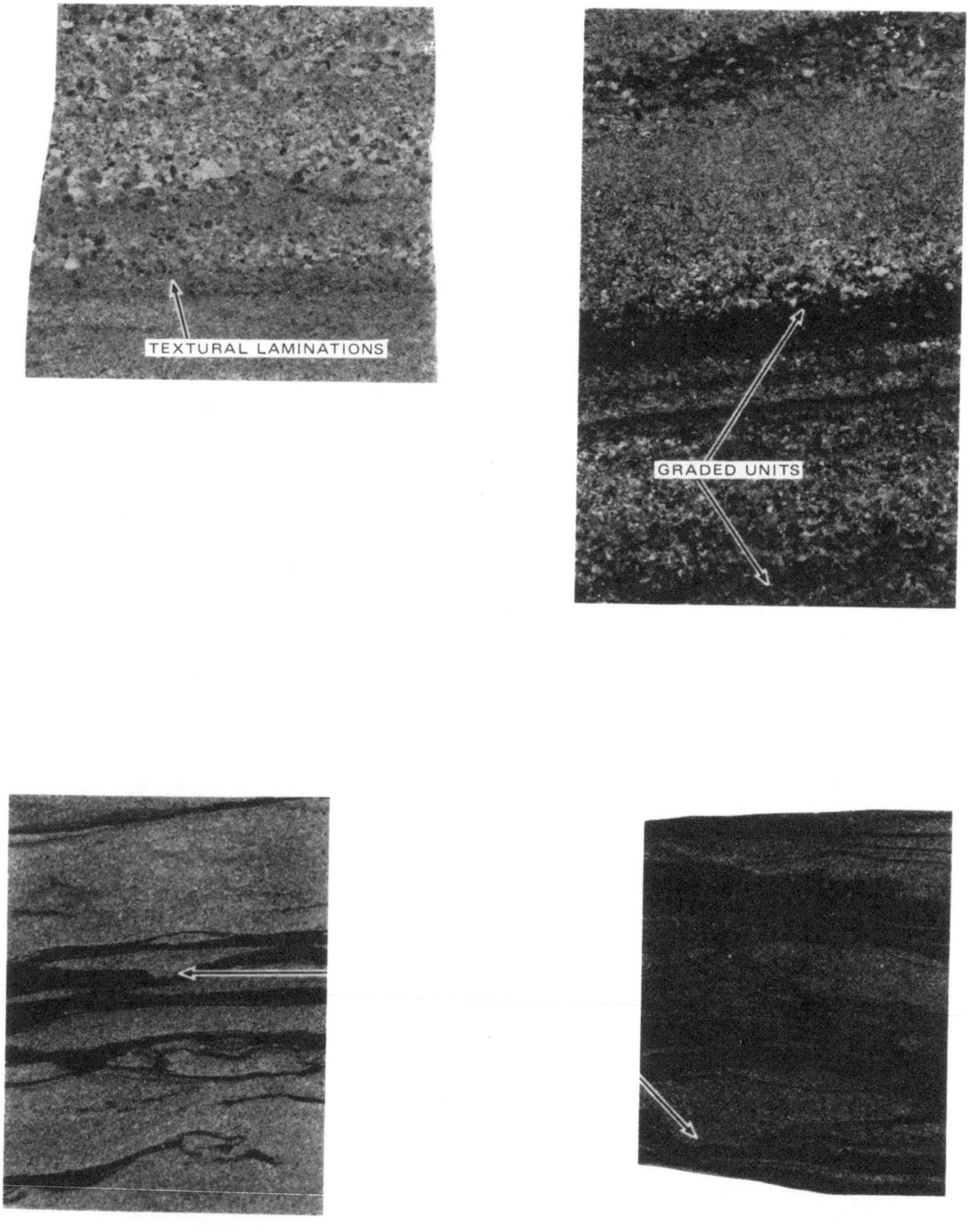

FIG. 11—Representative textures and structures of stream-mouth-bar deposits in the Mobil No. 1 Cora Evans, sec. 3, T.4N, R.13ECM, Texas County, Oklahoma (Fig. 1 no. 29). (a, b) Laminated and graded middle stream-mouth-bar sandstone and conglomerate; from a depth of 6,128-6130 ft x0.8. (c, d) Intercalated lower stream-mouth-bar fine sandstone and prodelta claystone showing load feature; from a depth of 6,134-6,140 ft x0.8.

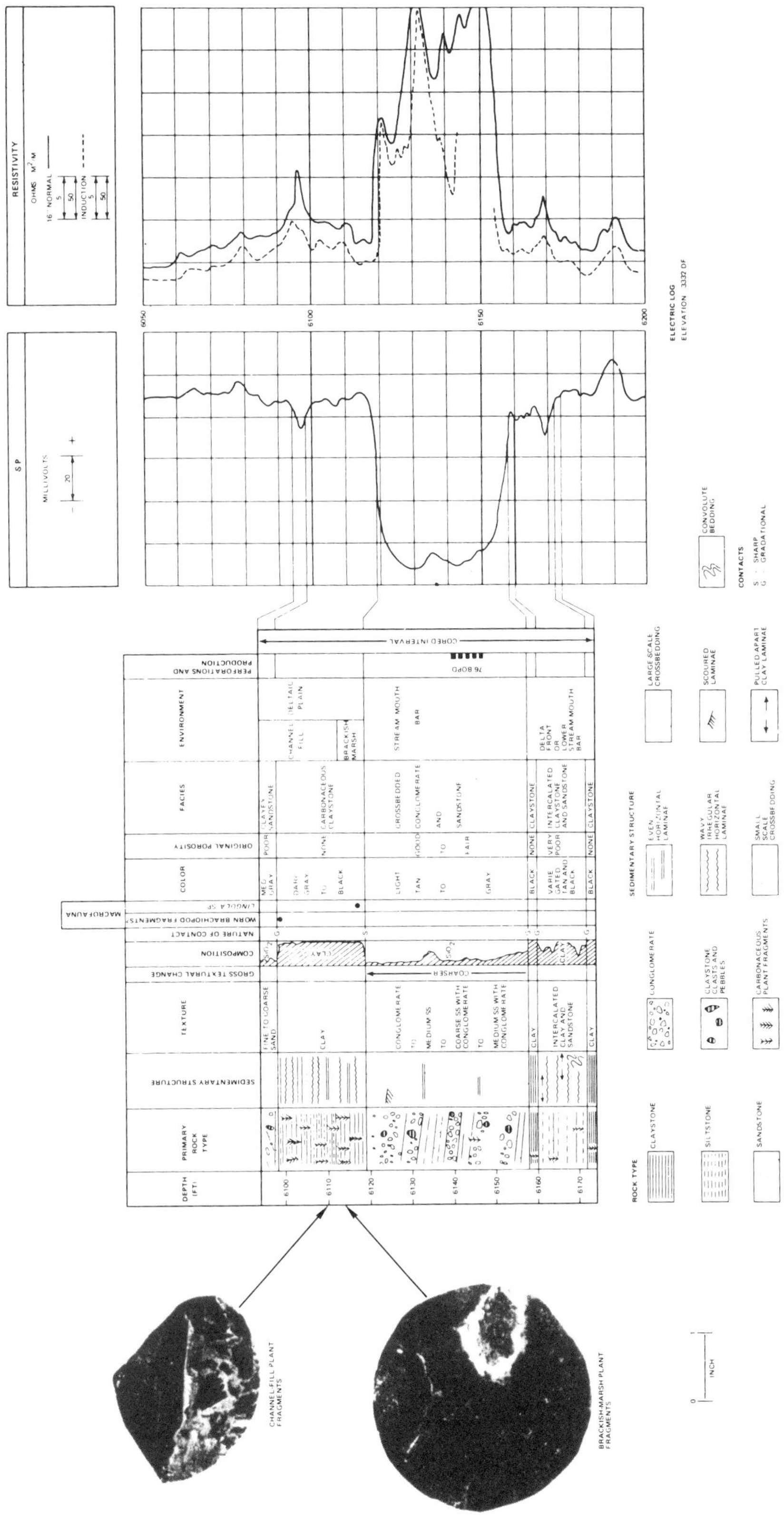

FIG. 12—Detailed log of cores from upper Morrow stream-mouth-bar deposits, Mobil No. 1 Sparks Estate, Hough area, Texas County, Oklahoma (Fig. 1, No. 24).

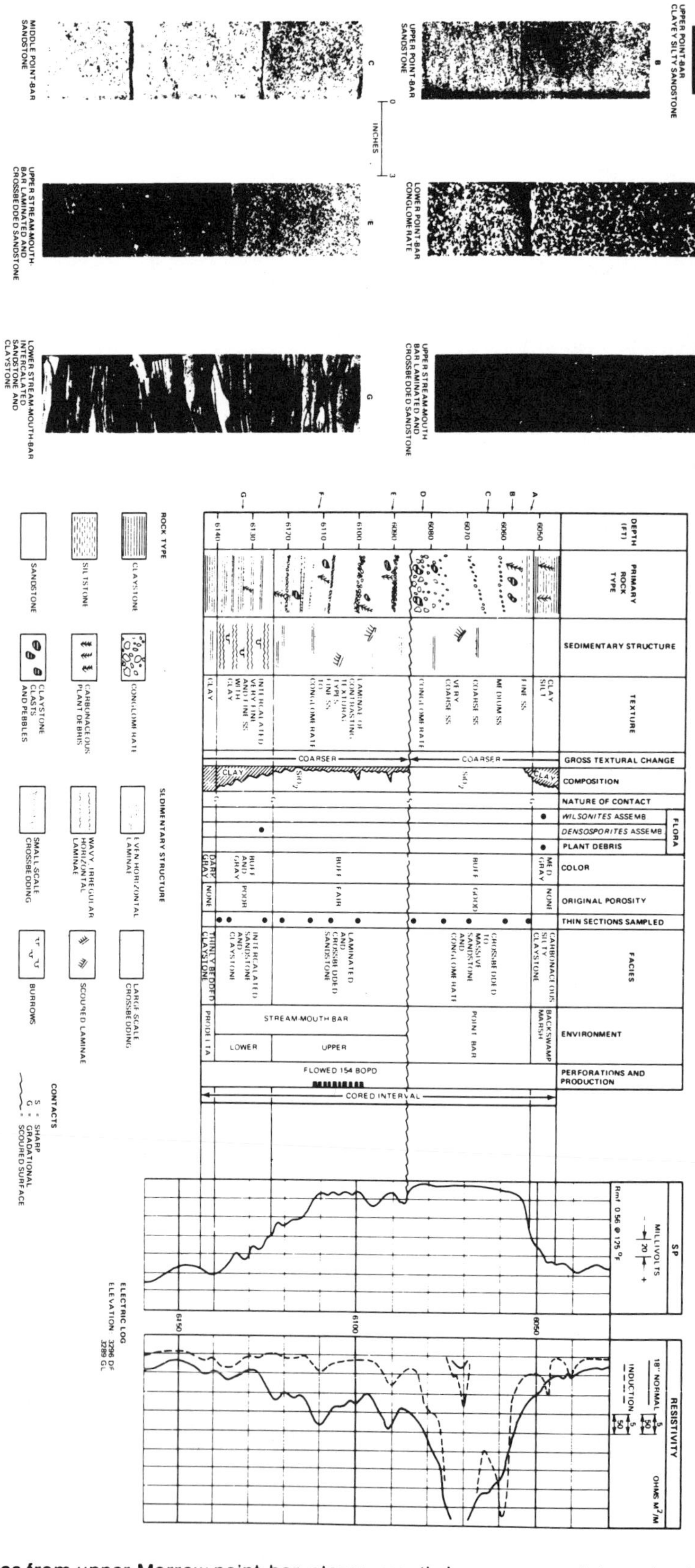

FIG. 13—Detailed log of cores from upper Morrow point-bar, steam-mouth-bar compound deposits, Mobil No. 1 Cora Evans, Hough area, Texas County, Oklahoma (Fig. 1, No. 29).

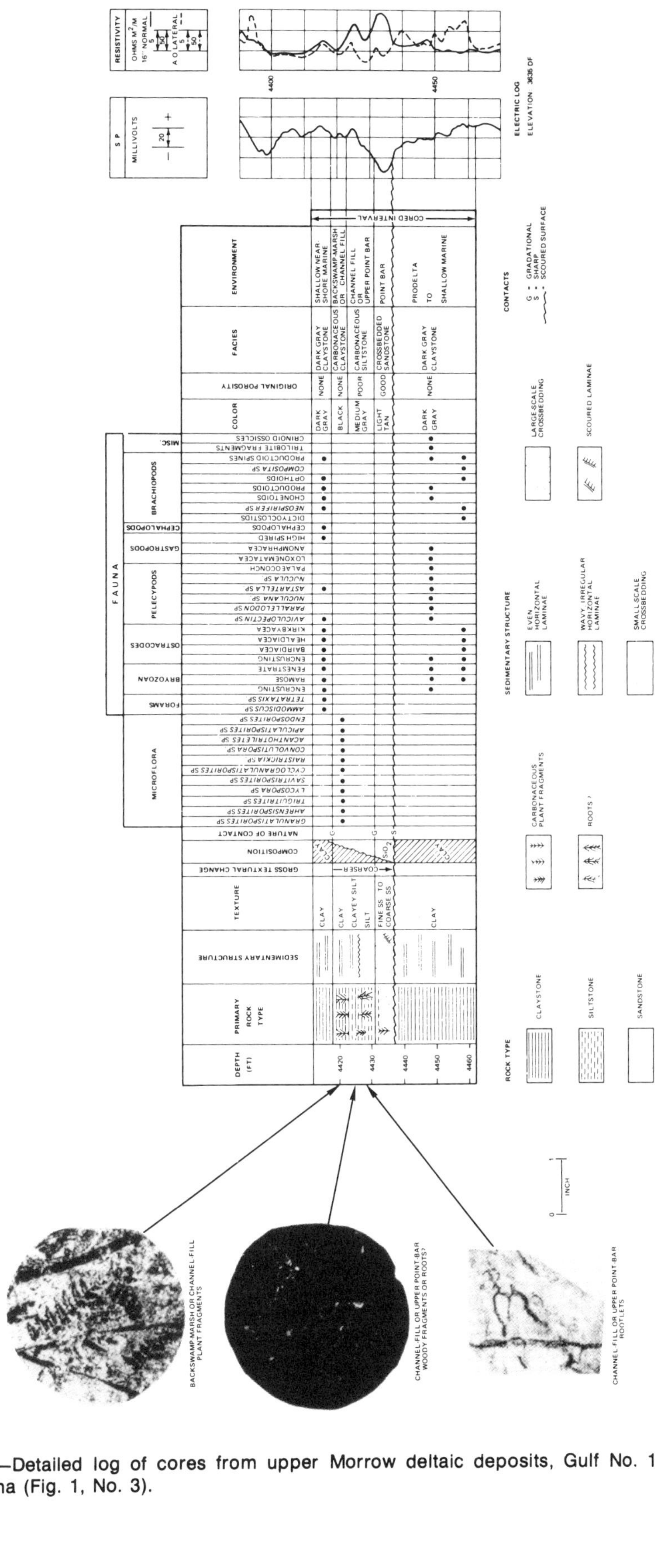

FIG. 14—Detailed log of cores from upper Morrow deltaic deposits, Gulf No. 1 Kelly, Eva area, Texas County, Oklahoma (Fig. 1, No. 3).

FIG. 15—Typical flora of backswamp-marsh deposits; Humble No. 1 Curtis Pauls, sec. 14, T.3N, R.18ECM, Texas County, Oklahoma (Fig. 1, no. 33). Silty claystone containing large carbonaceous fragments of woody material; from a depth of 6,328-6,330 ft x1.

Note the sharp, scoured contact between the channel-fill conglomerate and the underlying lower stream-mouth-bar sandstone. There are several zones containing the large chert pebbles which are so prevalent in the channel-fill deposit. This is probably due to coarse material being carried down the outer slope of the stream-mouth bar during flood stage or by gravity processes. Had the upper stream-mouth bar been preserved in this core, it would have had material in it nearly as coarse as that found in the distributary-channel fill.

Undifferentiated Delta-Front Deposits — A typical delta-front deposit consists of intercalated fine sandstone and claystone (Fig. 10a). This type material was deposited in shallow interdistributary areas away from actively building stream-mouth deposits. It is the result of essentially clay deposition occasionally disrupted by weak currents and by the influx of fine sand.

Prodelta Claystone — This facies is widespread and was encountered in many cores. It consists almost entirely of claystone with minor silt. The facies grades upward into a variety of delta-front deposits and downward into offshore shallow-marine claystone. It is evenly and horizontally laminated and is relatively free of large carbonaceous debris. The facies contains few zones that are obviously fossiliferous; careful processing, however, yielded a prolific fauna. Paleontology is the main criterion for distinguishing between prodelta and offshore shallow-marine environments. Often it was necessary to understand the lateral facies relationships before prodelta and normal offshore shallow-marine environments could be separated. Even then it was sometimes impossible to distinguish between the two.

Other Facies. — No definite bay claystone, natural-levee siltstone or mud-flat claystone were identified in cores examined; these environmental facies, however, are probably present and associated with the facies identified. No rocks interpreted to be beach, barrier-island or shoreface deposits were found in any of the cores analyzed. Apparently, these environments are not important in the area studied. Coarse material was either trapped on point bars or dropped in the stream-mouth bar, where it was not distributed to any great extent by wave or current action.

Point-Bar, Stream-Mouth-Bar Compound Deposits — A vertical environmental facies sequence repeated in many areas of the upper Morrow deltaic deposits is shown in Figure 13. The illustration shows the development of the point-bar sandstone (conglomerate) on top of a typical stream-mouth-bar deposit. In two examples, the deposits are separated by thin backswamp-marsh and delta-front deposits. In many places, the point bar actually cuts into the stream-mouth bar.

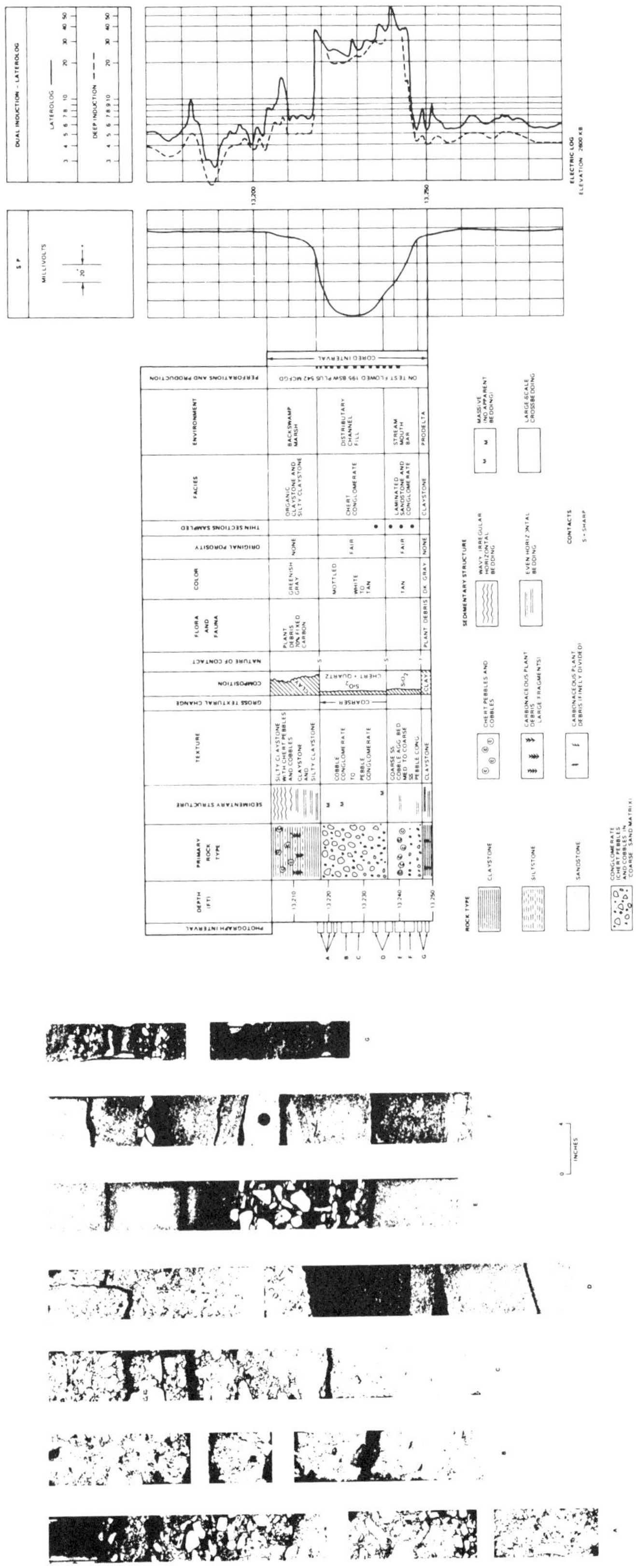

FIG. 16—Detailed log of cores from upper Morrow deltaic deposits, Shell No. 1-21 Hobart, Mendota area, Hemphill County, Texas (Fig. 1, no. 45).

Development and Characteristics of Important Reservoir Facies

Within the upper Morrow, point-bar and stream-mouth-bar facies form the most important reservoirs, with point-bar deposits of the deltaic plain the most widespread and economically the most attractive.

Point-Bar Deposits — Deposits associated with meandering streams are important constituents of deltaic deposits and are in a good position to be preserved, since they are surrounded by clay and covered by backswamp-marsh clays. Meandering streams and their point bars may reach the very edge of the deltaic plain. Figure 17 contains an isometric diagram identifying the subenvironments and deposits associated with a meandering stream or distributary.

The lower point bar includes the bed load and contains the coarsest material moving along the stream as well as clay pebbles from the caved bank. The middle and upper point-bar deposits become finer upward. Typically, the middle point bar has large-scale festoon cross-bedding, and the upper point bar has small-scale cross-beds or wavy irregular bedding. The contact betwen materials of different texture tends to be horizontal. In the upper Morrow deposits, the "time" lines representing depositional surfaces cross the deposits at angles up to 3° to 5°, a good example of a crossing "rock unit" — time-stratigraphic relationship.

Accretion in the point-bar environment is essentially lateral. As the ovoid deposit develops behind the extending meander loop, marsh or swamp deposits begin to cover the point bar. Marsh and swamp environments are also present behind the natural levee on the outer bank of the meander loop. When the meander loop becomes overextended, the stream cuts off the extended loop, usually during a breakthrough at flood stage. The abandoned meander is quickly isolated from stream flow, and the resulting oxbow lake slowly fills with organic-rich clays and silts, forming a "clay plug."

The average maximum thickness of point-bar deposits approaches the maximum depth that the stream scours during flood stage. This depth is a function of the discharge of the stream, since the size, shape, and pattern of stream channels are closely related to flow. Discharge is also the prime factor governing the channel width and meander length. Hence, for a given set of conditions of bank material, gradient, and similar factors, there is a close, predictable relationship that exists between discharge, stream width and depth, radius of curvature of the meander loop, and meander length. This conclusion is supported by a study of 50 rivers of various sizes in which it was demonstrated that the average meander length was 7 to 10 times greater than the channel width (Leopold, Wolman, and Miller, 1964). Therefore, because the meander-loop size determines the extent of the accompanying point-bar deposit, it is apparent that even if only one data-point core is identified as being a point-bar sandstone, this single bit of information together with empirical graphs and mathematical relationships (Leopold, Wolman, and Miller, 1964) can be used to infer the possible size of the point-bar deposit encountered.

Stream deposits, in section, should not always be shown as being flat on top and convex downward by losing section at their base. The actual vertical profile, facies, and stratigraphic relationships of a Recent point-bar sandstone are shown in the cross-section on Figure 18. The deposits have essentially flat bases, except for some relatively minor undulations resulting from either traces of the thalweg or changes in stream discharge. These minor undulations may have several tens of feet of differential elevation, which would be impressive if encountered at an outcrop. Overall, however, their existence does not alter the fact that point-bar deposits have relatively flat scoured bases, and the deposits thin by losing section from the top. The section in Figure 18 shows this thinning into the channel as well as the angular relationship between the depositional shape (bedding planes, that is, "time" lines) and the rock-unit boundaries of lower, middle, and upper point bars.

Point-bar deposits, if developed in a simple and ideal form, are circular to oval in shape, with the size of the ovoid deposits being a function of stream width. Figure 19 shows a series of well-developed point-bar deposits at various points on the Brazos River of Texas. Since the courses of meandering streams and distributaries fluctuate, they tend to destroy old point-

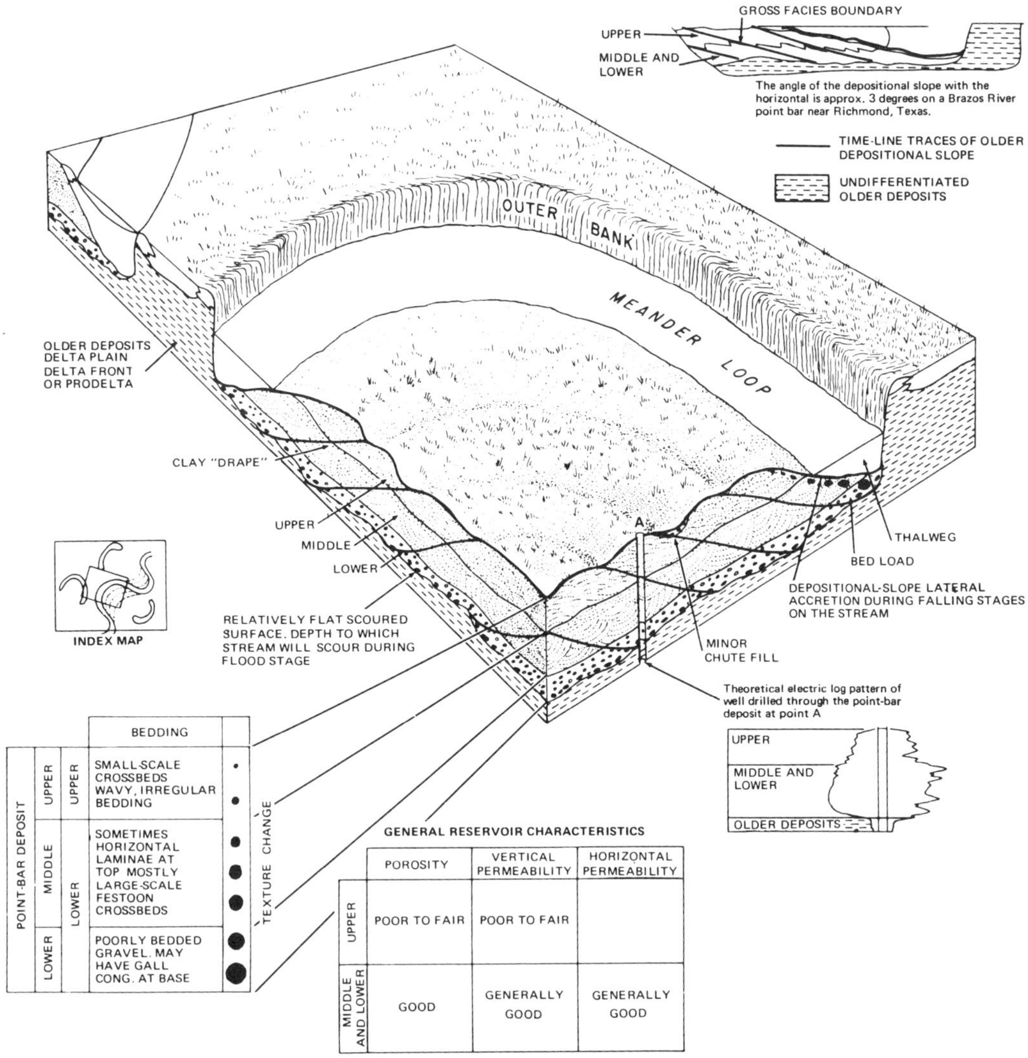

FIG. 17—Morphology and environmental facies relationships in a point-bar deposit.

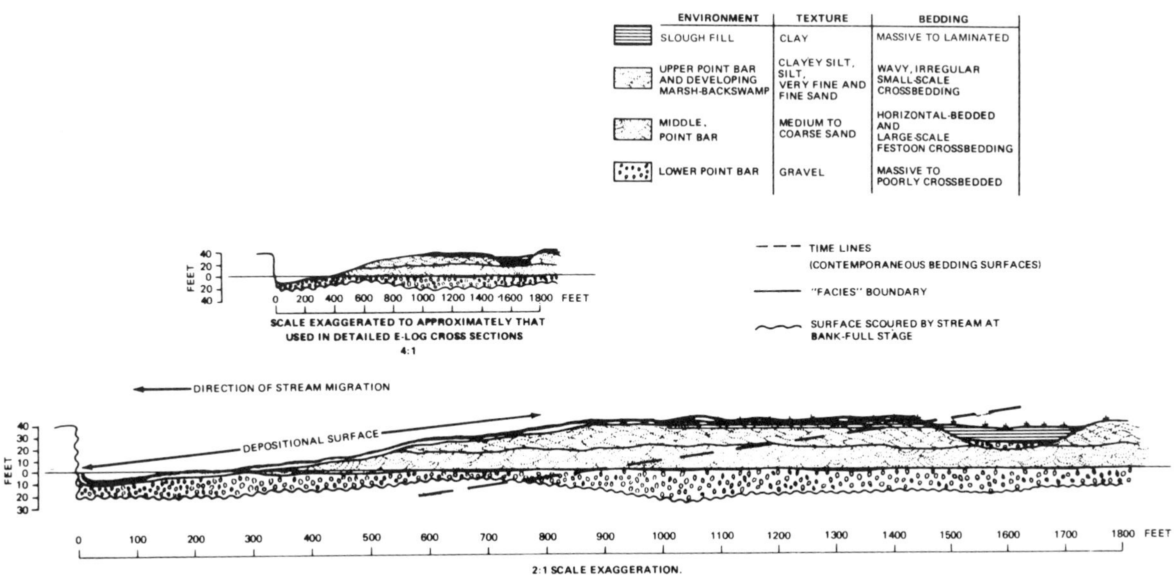

Fig. 18—Cross-sections through a Recent point bar of the Brazos River, Texas.

bar deposits and build new ones. This condition results in a complicated pattern of ovals, half circles, figure eights, and oval deposits within other deposits.

Since the channel course of meander loops fills with organic-rich clay, point-bar deposits contain sinuous and narrow clay bodies which tend to isolate the various individual sand bodies. This separation of point-bar reservoirs often leads to isolated production-pressure areas and to unexpected dry holes. Recognition of channel-fill claystone can be helpful in improving the development of an area of point-bar sandstone reservoirs. Furthermore, as a meandering stream shifts course, if subsidence is rapid enough, the stream will not completely destroy older point-bar deposits but will only "top" them.

Stream-Mouth-Bar (Fan) Deposits — When a stream or distributary enters a standing body of water, the coarse load drops rapidly and builds a mass called the stream- or distributary-mouth bar. Such bars may result from a straight or meandering stream and may be deposited into relatively shallow or deep water in a stable or subsiding basin. All these factors influence the development of stream-mouth-bar deposits. If the supply of sediment continues, sooner or later the stream-mouth bar will build basinward (prograde) and develop a mass of coarse sediment of various shapes and sizes. Figure 20 illustrates the characteristics of the stream-mouth bar.

Ideally, there is a continuous textural variation along the depositional slope of the bar. The coarsest material is deposited near the stream mouth, and as the velocity drops, finer and finer material is deposited outward and down the depositional slope, where, along the outer edge, very fine sand and siltstone interfinger with the prodelta claystone. This is a prograding sequence, with the top of the deposit being at or near sea level. A vertical section through such a sequence reflects the facies variation found laterally along the depositional slope. If the bar is building into relatively deep water, or if the prodelta clays have not built a sufficiently high platform, the angle between the depositional slope and sea level may be quite high, with the result that gravity-flow processes can affect the front of the bar slope. Such flow may also occur as the flood stage of the stream builds an unstable mass of coarse material near the stream mouth which later breaks away and moves down the bar slope.

FIG. 19—Recent point bars on the Brazos River near Richmond, Texas.

The chances of preservation of extensive stream-mouth-bar deposits vary from poor to excellent. A persistent meandering stream following behind a stream-mouth bar building out in relatively shallow water will tend to destroy the stream-mouth-bar complex. Only remnants will be left where either the stream course missed the stream-mouth bar or the bar subsided fast enough to be preserved.

Stream-mouth bars tend to have an arcuate fan shape in plan view and a roughly triangular shape with a flat base in cross-section (Fig. 20). However, a relatively straight distributary building into relatively deep water will develop bar fingers (Fisk, 1961). These bar fingers represent a coarse mass composed of stream-mouth-bar material and relatively straight distributary-channel-fill sand. As the distributary slowly cuts off and shifts to a new location, the channel backfills with sand, resulting in one long thin "finger" of coarse material. The resulting geometry will be roughly the same in cross-section, but in plan view the bar-finger sands will be linear, not arcuate and fan shaped. These long thin sand bodies are heavy and tend to sink into the underlying prodelta clays. Since no meandering stream follows them, and if subsidence is rapid enough, these finger deposits have an excellent chance of preservation.

Whether or not a stream-mouth deposit is preserved depends upon the interrelation of many parameters such as depth of water, rate of subsidence, rate of sedimentation, and nature of the stream. Such complexities result in a variety of sizes and shapes of preserved stream-mouth deposits. Sea level is the critical reference point in the relationship between stream-mouth-bar and deltaic-plain deposits. The stream-mouth bar builds up to sea level, and the distributaries and meandering streams cut down below sea level. Unless subsidence is rapid, both of these environments compete for the same space, and since the deltaic-plain sediments develop last, they are most often preserved.

Both stream-mouth-bar and point-bar deposits are probably more prevalent in the geologic column than previously suspected. They provide attractive objectives for exploration because they are in many places coarse grained and are deposited in association with organic-rich source rocks. They occur in a subsiding basin of deposition where the entire deltaic mass is subject to burial and preservation.

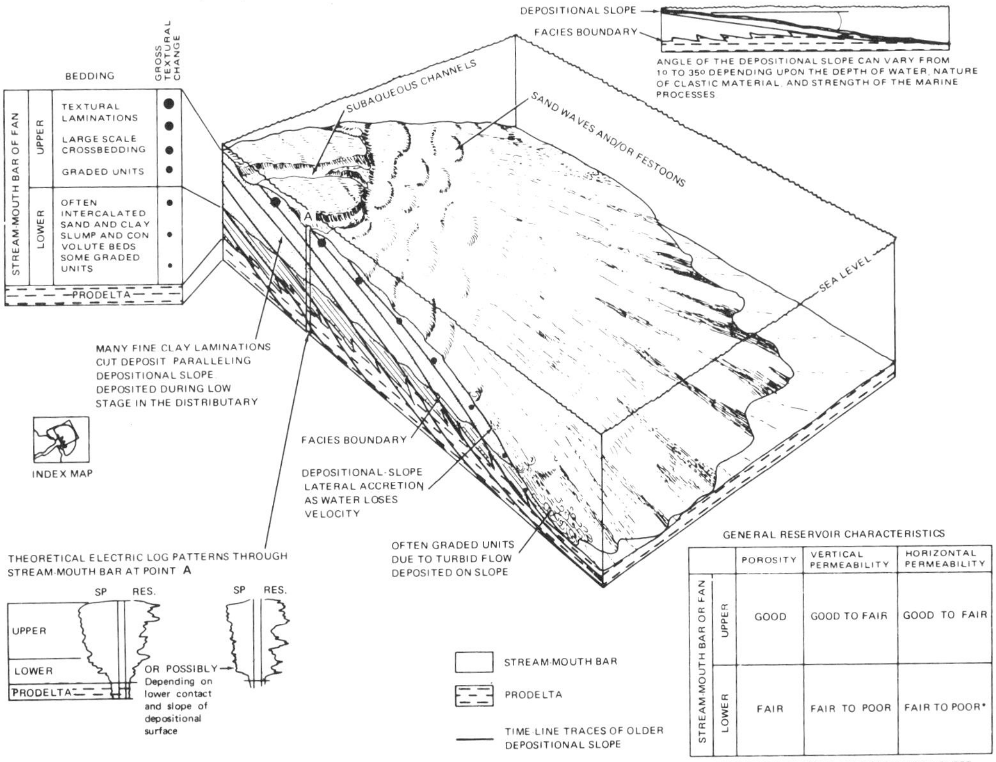

FIG. 20—Morphology and environmental facies relationships in a stream-mouth bar or fan deposit.

Lateral Extension of Facies Relationships

After the vertical sequence of environmental facies was tentatively identified in cores at the basic data points, detailed cross-sections based on sample logs and electric logs were drawn through the cored wells, in order to determine the lateral environmental facies relationships.

Relationship Between Electric Log Pattern and Facies — In the graphic logs of Figures 8, 12, 13, it is shown that each environmental facies has a characteristic electric log pattern. These patterns can be used effectively to expand the environmental facies interpretations in areas where electric logs, but few cores, are available. Of particular interest is that in the upper Morrow Formation, the spontaneous potential (SP) curve, which measures the shaliness of a formation, accurately reflects the overall textural change through point-bar and stream-mouth-bar deposits and, when combined with the resistivity curves, indicates whether contacts between facies are sharp or gradational. The resistivity curves and particularly the conductivity curves of induction logs also commonly indicate the presence of backswamp-marsh and meander-channel-fill organic claystones. These environmental facies are slightly more resistive than prodelta and offshore-marine claystone because of their higher silt content and because of the presence of large quantities of carbonaceous plant remains. Figure 21 shows a selection of electric log patterns from a number of upper Morrow deposits interpreted to be either point bar or stream-mouth bar.

To relate environmental facies on the detailed cross-section, a reliable upper time-stratigraphic datum and good correlation points within the upper Morrow Formation had to be established. The thin, persistent limestones and dolomites of the "Thirteenth Finger Limestone" were used as the datum in local areas, and subtle but unique electric log patterns in the upper Morrow "Shale" provided many additional and supporting correlation lines. On the basis of the resulting correlation, the late Morrowan deltaic events were designated as follows: Stage C — Primarily deltaic-plain deposits (youngest); Stage B — Primarily delta-front and prodelta deposits; and Stage A — Primarily delta-front and prodelta deposits (oldest).

Throughout the western Anadarko basin, the deposits within each individual stage are roughly time-equivalent.

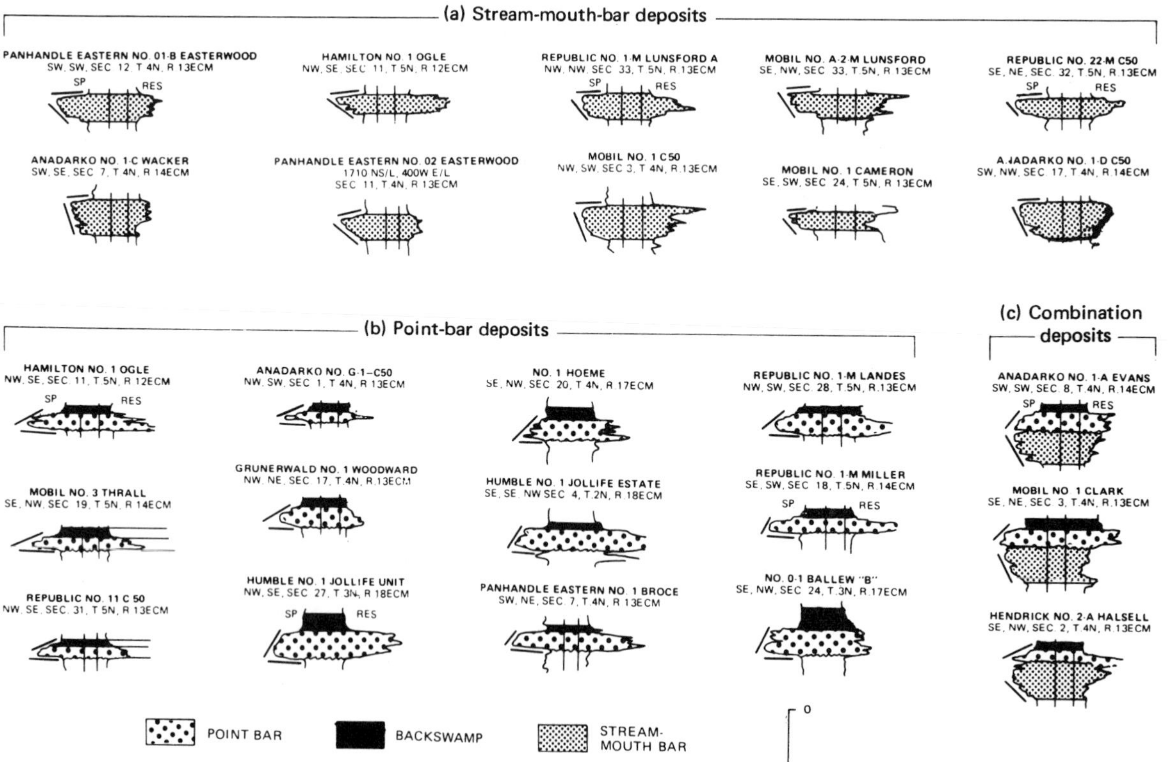

FIG. 21—Typical electric log patterns of upper Morrow deltaic deposits from wells in Oklahoma.

Lateral Facies Relationships — Figure 22 shows cross sections with good control from the Hough area of Texas County, Oklahoma. Sections A-A' and B-B' are good examples of typical sections through point-bar deposits. In B-B', the point bar has a flat base and terminates at the end by being cut out by another meander loop. The other end loses section from the top and is thinning into another channel (fill). Two logs in section A-A' are characteristic of an interpreted meander-channel-fill deposit. The stream-mouth bars in sections C-C' and D-D' have relatively flat tops and lie below the deltaic-plain deposits. The finer texture of the stream-mouth bar downward is expressed in the SP curves of the wells. It should be noted in these sections, as well as in others, that the base of the point-bar deposits tends to parallel the top of the backswamp-marsh (deltaic-plain) claystones.

Sections C-C' and D-D' illustrate stratigraphic relationships in complex deltaic deposits and show a typical point-bar deposit associated with meander-channel-fill and backswamp-marsh claystone. Directly beneath these point-bar facies are deposits which display the characteristics of the stream-mouth-bar environment. Wells which penetrated the greatest thickness of the lower stream-mouth deposits (section C-C') cut facies interpreted to be straight-distributary-channel fill (a typical bar-finger complex). The jagged character of the SP cruve indicates that the thicker parts of this material contain many laminations, and from the lateral facies relations, it appears that this thicker sandstone either cut into or sank into the underlying deposits. Good examples of meander-channel fill and "cutting out" of point-bar deposits are shown in sections A-A', B-B', and D-D'. Note the similarity between the resistivity curves of these channel-fill deposits.

Some of the best documented deltaic-plain deposits were found in the North Hardesty area of Texas County, Oklahoma. Channel-loop cutting, channel-fill characteristics, backswamp-marsh relationships, and typical variations of point-bar deposits are shown in these sections (Fig. 23). The point bars have their characteristic flat base and, in this area, divide into slightly older and younger deposits. Apparently in this meander belt, subsidence was sufficient to give the relationship that is illustrated in sections E-E' and F-F'.

Section G-G' cuts across an older stage B stream-mouth-bar-finger complex. The deltaic-plain deposit in section G-G' is a good example of cored meander-channel fill. At the base is a 6-ft section of clay-pebble conglomerate, which probably represents the bed load of the meander loop when it was cut off and abandoned. Cores from the equivalent interval in other nearby wells contain similar material. Apparently, this clay-pebble conglomerate continues all along the scoured base of the meander-channel-fill deposits in this area and serves as a clue that point-bar deposits are nearby.

Practical Uses of Cross-Sections — During development drilling and local exploration, knowledge of environmental facies can be very useful. Identification of a deltaic-plain environmental facies in the Hough area of Texas County, Oklahoma, led to prediction of the existence of an unknown point-bar reservoir which was later confirmed by drilling.

A major oil company, while developing an upper Morrow oil field, drilled wells A and B (Fig. 24). Well B contained a 40-ft section of point-bar sandstone, whereas well A encountered practically no sand but did penetrate rocks which, from their electric log characteristics, were interpreted to be meander-channel-fill deposits. Wells A and B were only 1/2 mi apart. It was reasoned that if well A encountered meander-channel-fill deposits resulting from a meander loop cutting into the point-bar deposit in well B, then it would follow that in this meander loop and behind the meander-channel-fill deposits another point-bar deposit must exist. The nature of similar deposits in the area indicated that adequate thicknesses of the expected point-bar deposit would be located at least 1/2 mi beyond well A. Wells containing no sand were directly west of well A, indicating that the undrilled deposit would lie in a northwestern direction. The company first drilled three more wells (C, D, and E) in the meander-loop-channel fill at a total cost of more than $200,000. Environmental facies analysis should have prevented this expenditure, although one was completed in another horizon. Another oil company finally drilled well F in the SE-1/4, SE-1/4 of Section 28, and the unknown but predicted point-bar deposit was discovered. Subsequent drilling of well G confirmed the interpretation.

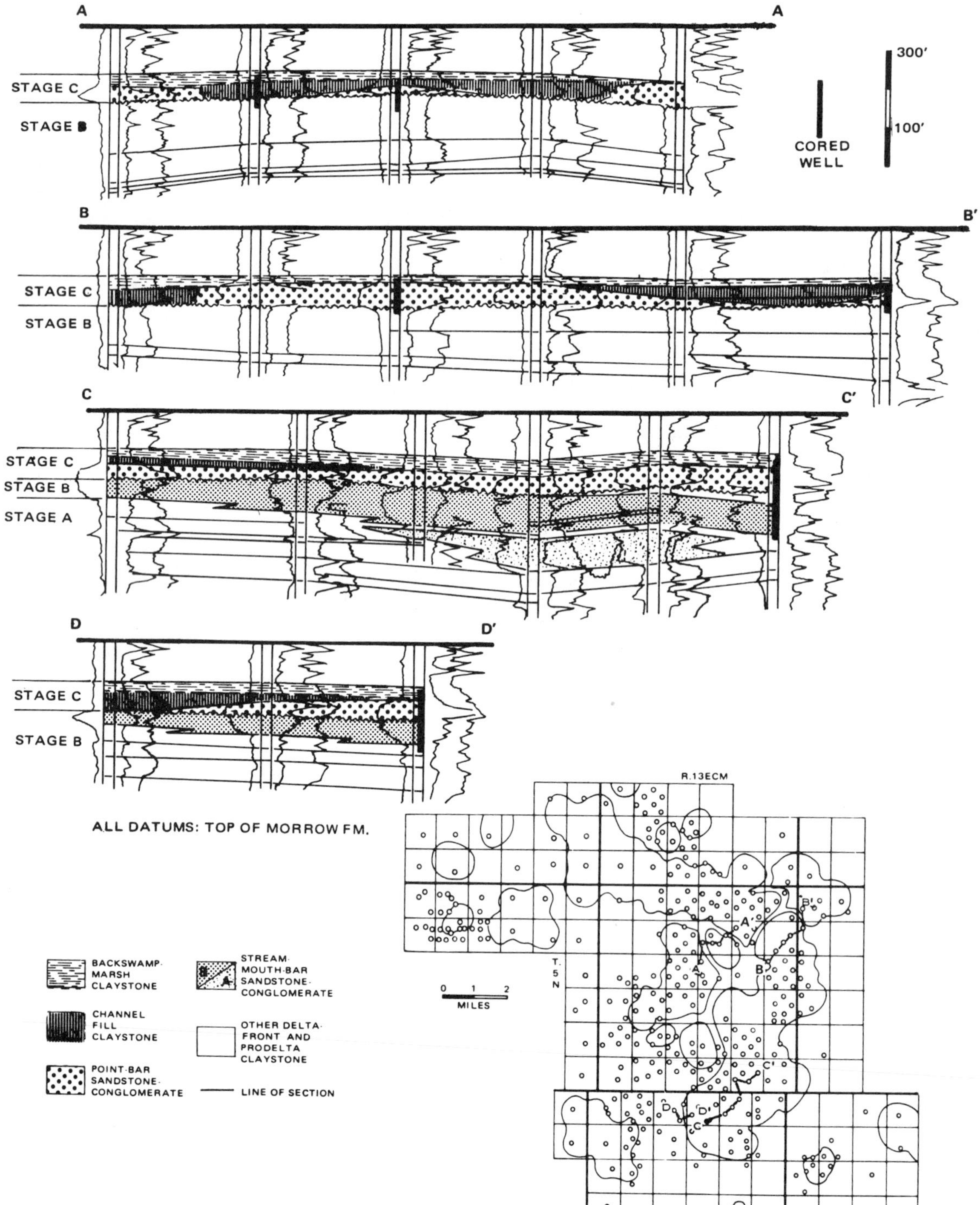

FIG. 22—Lateral extension along cross-sections of upper Morrow deltaic deposits in the Hough area, Texas County, Oklahoma.

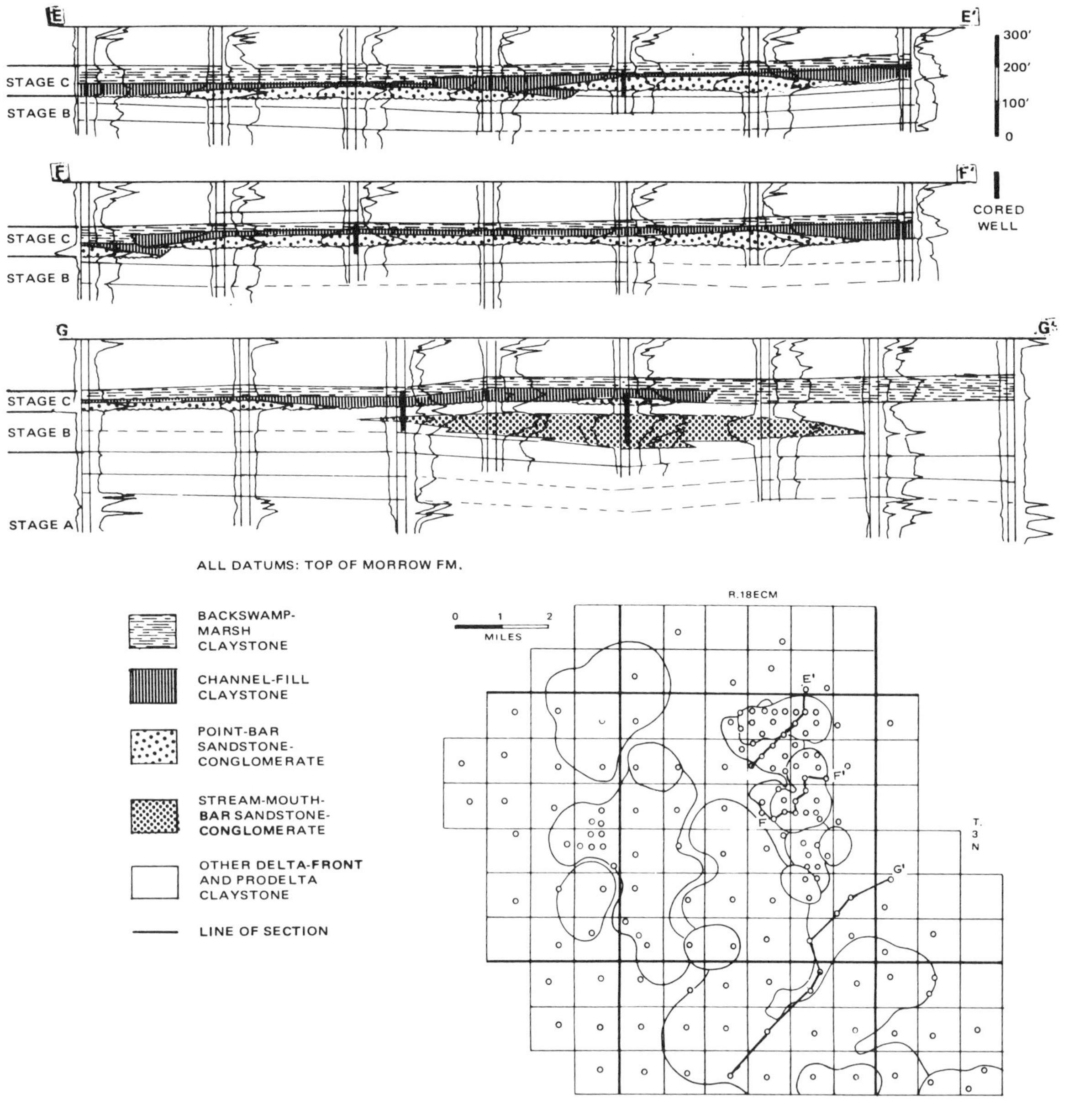

FIG. 23—Lateral extension along cross-sections of upper Morrow deltaic deposits in the North Hardesty area, Texas County, Oklahoma.

Three-Dimensional Environmental Facies Control

In order to develop the three-dimensional aspect of the deltaic deposits, a detailed network of stratigraphic cross-sections was constructed in nine areas. The datum selected for these sections was the base of the "Thirteen Finger Limestones," which are thin argillaceous limestones of unique log character and probably time-synchronous. The environmental facies, as identified by cores, electric log patterns, and sample cuttings, were then plotted on the cross-sections, and their apparent relationships were reconstructed.

Electric log patterns of wells not appearing in the cross-sections were compared and correlated with nearby sections. The various environmental facies were identified and put into the proper stratigraphic position. Isopach maps of important reservoir deposits were then drawn, using all available data. Knowledge of the environmental facies of the deposits guided the con-

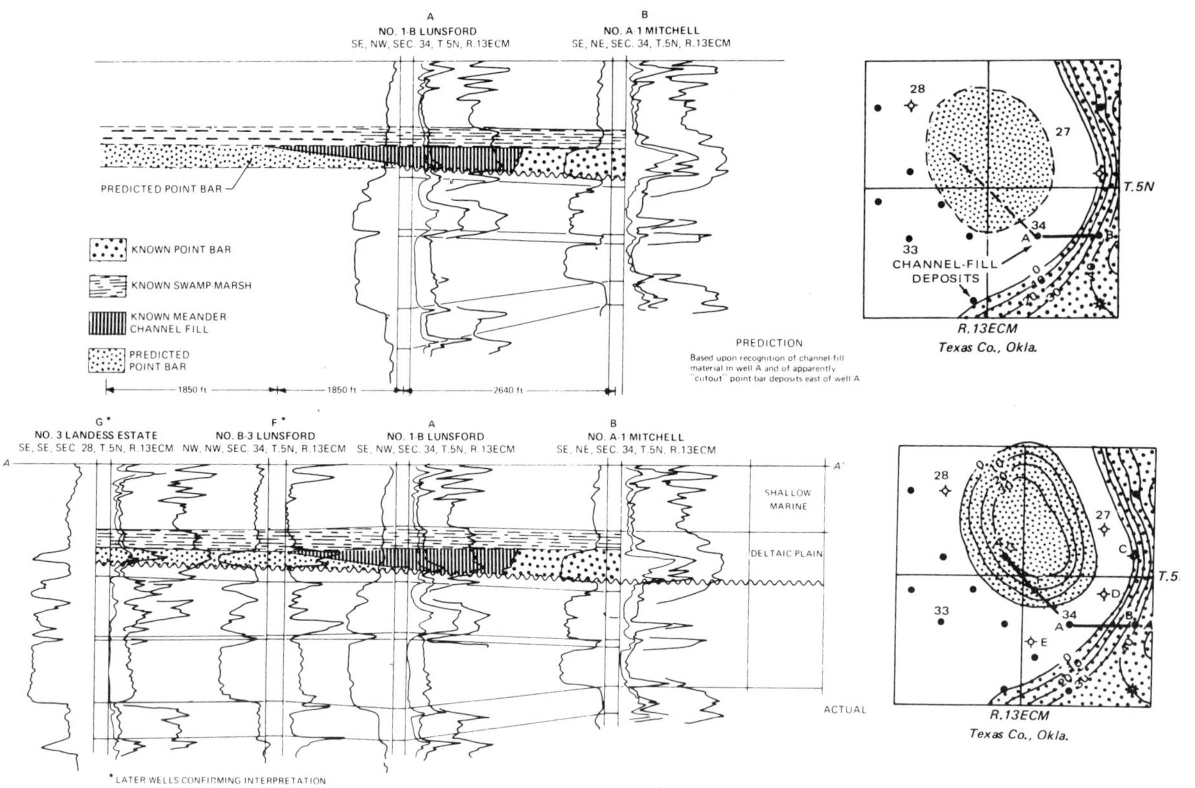

FIG. 24—Prediction of point-bar and channel-fill deposits based on environmental facies analysis.

touring so that wherever possible the contours are consistent with known information regarding the particular environmental facies. For example, in the contouring of point bars, knowledge of their ovoid nature and their relationship with meander-channel fill and "cutouts" was utilized. Speculation as to the size of the stream (function of depth) and the resulting diameter of its meander loops controlled the size and general shape of the ovoid deposits. In some locations, oil wells drilled on 80-acre spacing gave close control on the size and shape of the deposits. For purposes of illustrating the use of this procedure, its application to three areas in the study is presented in Figures 25-30.

Hough Area — This area is located in Texas County, Oklahoma (Fig. 1), and is one in which there has been a high incidence of discovery of oil and gas pools in upper Morrow deltaic deposits. Much of the area has been drilled on close spacing (80-acre); hence, many cores were available for making a detailed study of deltaic deposits.

Figures 25-27 show cross-sections which were drawn to reconstruct the complete facies framework of the Hough area. Sections 1-1' and 2-2' trend generally north-south across the deposits, whereas section 3-3' intersects them and trends northwest-southeast. The upper boundary and datum for these sections is the base of the Thirteen Finger Limestone marker. The base of the sections is the top of the lower Morrow sandy limestone. Although electric log curves are not shown on these sections, the actual logs were used in the "work" sections.

The sections show that this was an area of persistent deposition of coarse deltaic deposits. Several periods of active deltaic sedimentation occurred, and the sediments were preserved. The stages A, B, and C, earlier described in this report, represent the temporal sequence of delta building. Stage A actually contains two unrelated sandstone deposits of different ages, but they are considered together. Little is known about the oldest stage A deposit except that it is probably associated with deltaic sedimentation. The main stage A reservoir deposit is a long, thick, narrow sandstone unit that trends northwest-southeast. It is indicated that this reservoir body is a bar-finger facies complex which resulted from the building of a stream-mouth bar out into relatively deep and quiet water. The distributary was probably straight; as it was slowly cut

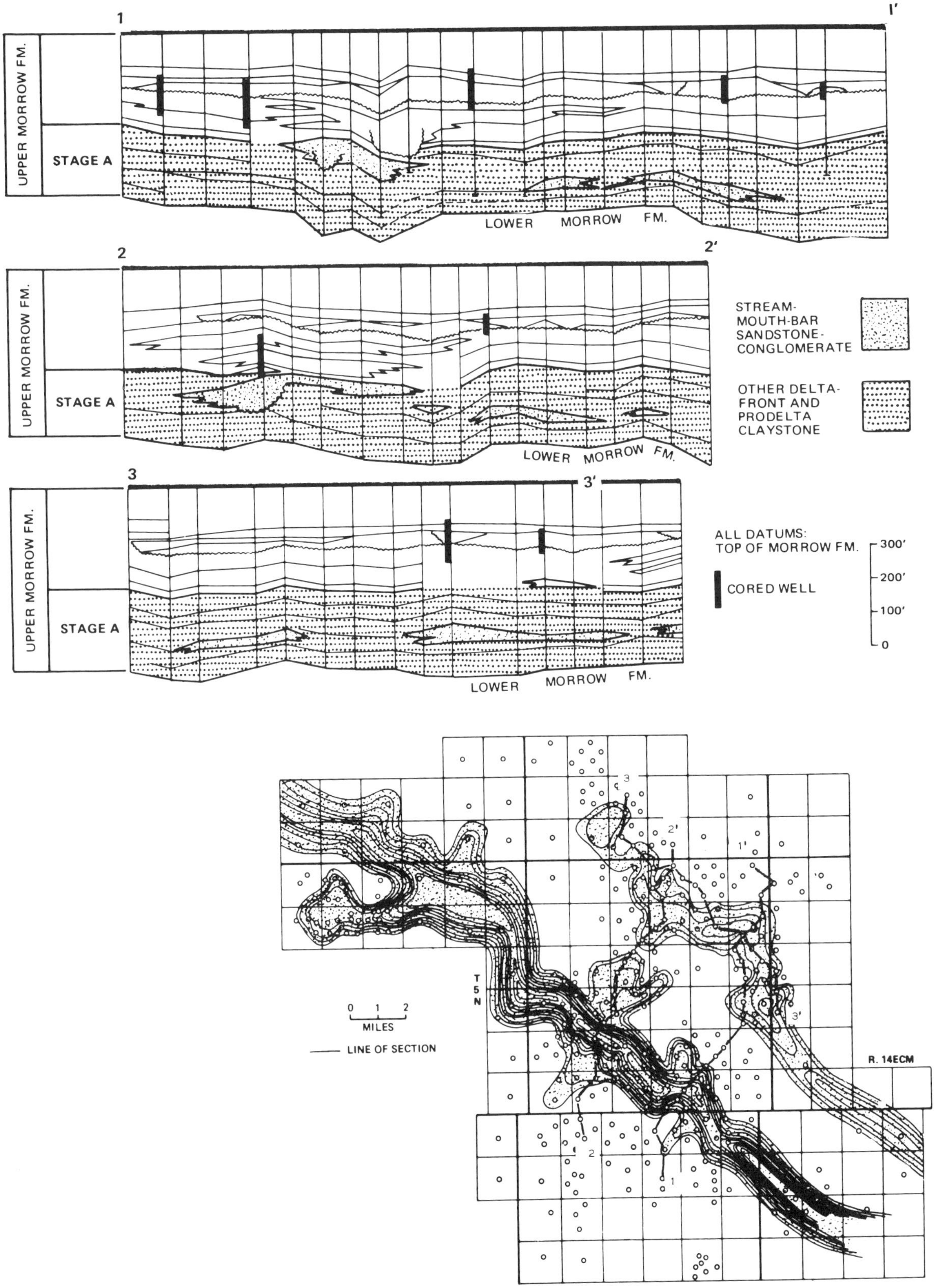

FIG. 25—Isolith map and cross-sections showing the three-dimensional framework of stage A upper Morrow stream-mouth deposits in the Hough area, Texas County, Oklahoma.

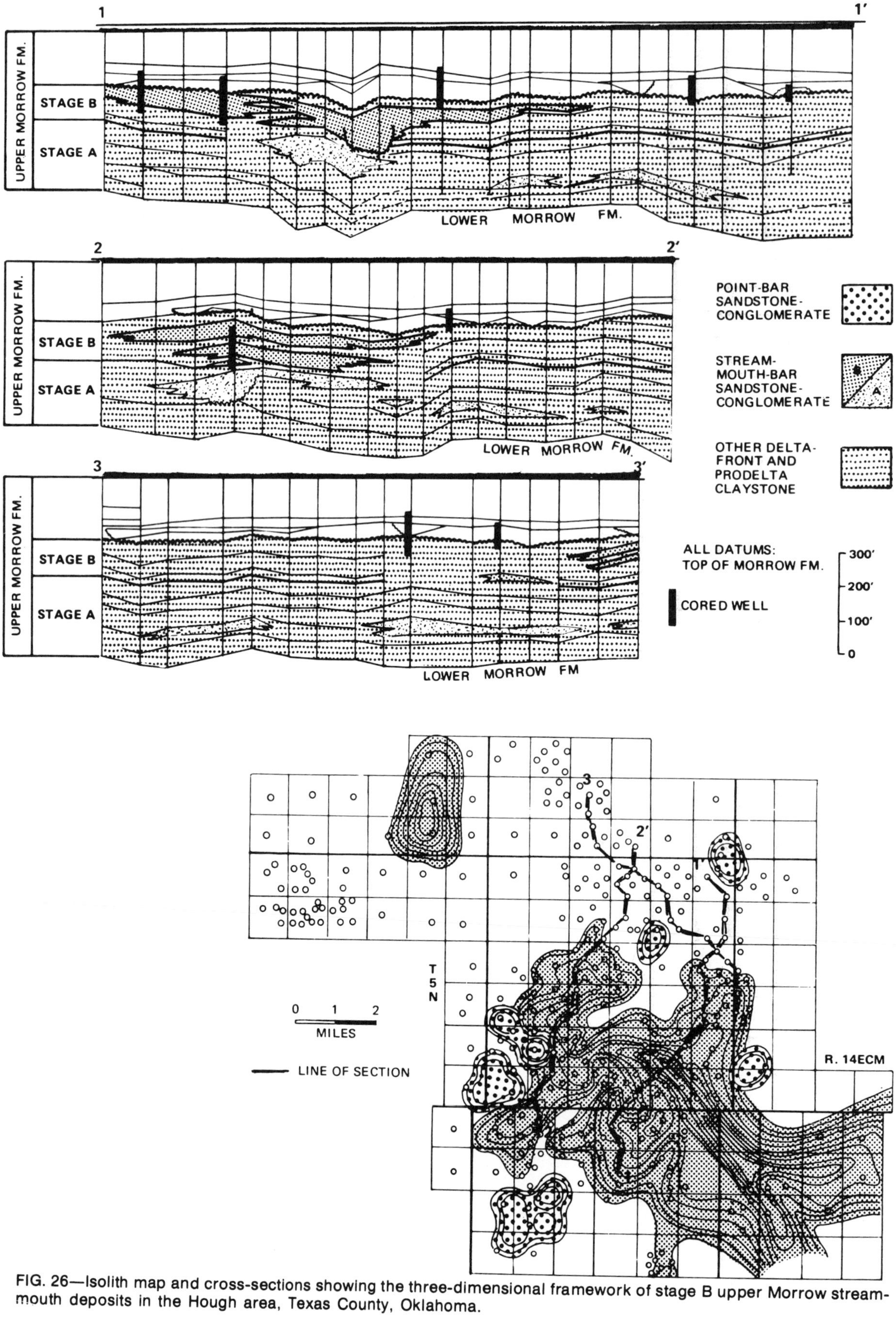

FIG. 26—Isolith map and cross-sections showing the three-dimensional framework of stage B upper Morrow stream-mouth deposits in the Hough area, Texas County, Oklahoma.

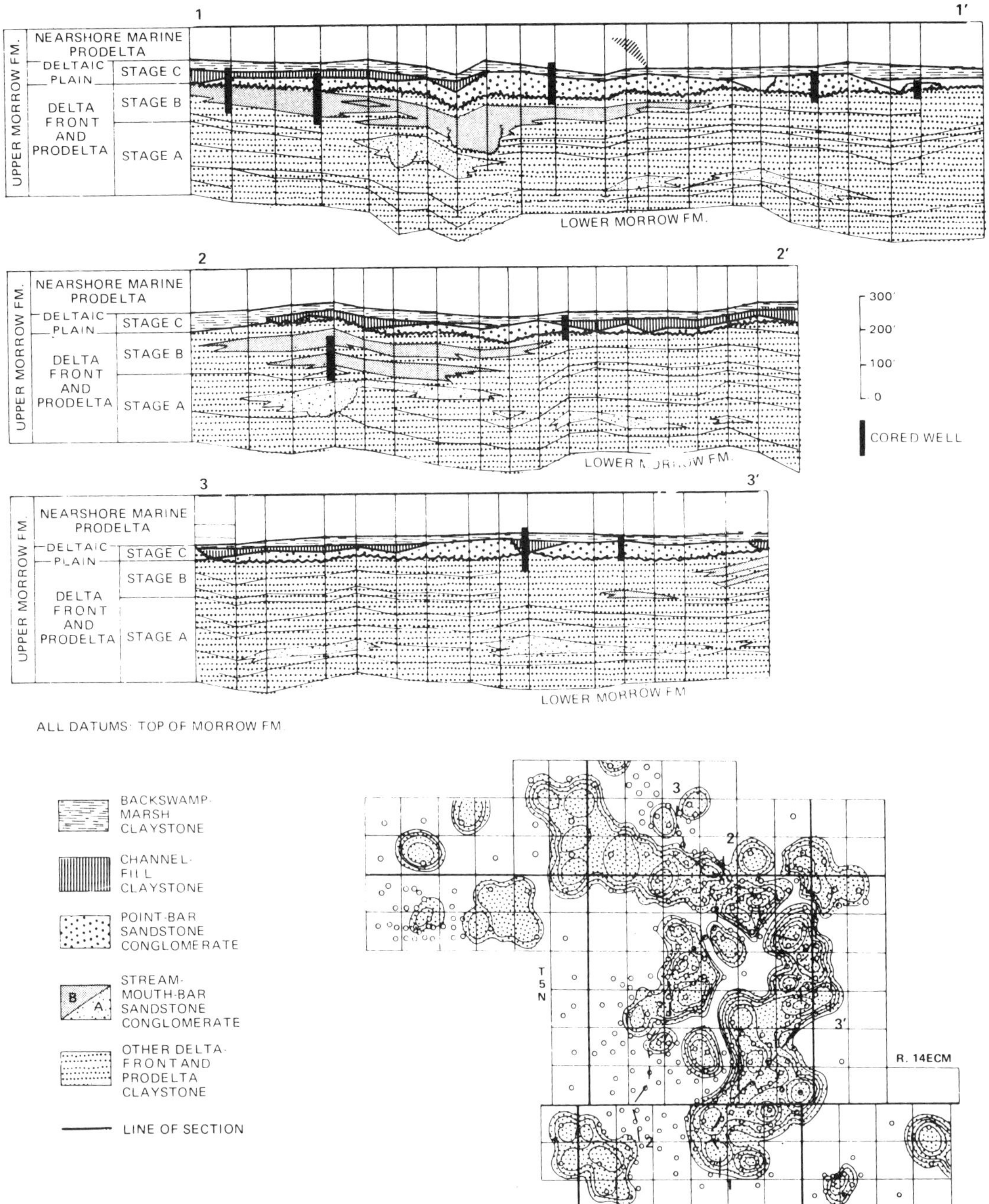

FIG. 27—Isolith map and cross-sections showing the three-dimensional framework of stage C upper Morrow point-bar deposits in the Hough area, Texas County, Oklahoma.

off, it backfilled with sand and formed a typical bar-finger mass having an average width of 1-1/4 mi, a maximum thickness of 110 ft, and a maximum length of about 16 mi (Fig. 25). Apparently, the distributary building this deposit carried coarse material southeastward which was rapidly "dumped" into relatively deep, quiet water, where little redistribution occurred. This situation probably accounts for the narrowness of the bar finger indicated in cross-sections 1-1' and 2-2'.

The most widespread and best documented stream-mouth-bar deposits are those of stage B. The shape and distribution of this environmental facies are shown on sections 1-1', 2-2', and the related maps in Figure 26. The sections show the general flat top of the deposit, the interfingering of the lower part with prodelta claystone, and the downward thickening where the distributary-channel fill is located. Some isolated stage B deposits are identified as point-bar deposits lying on the stream-mouth bar; however, the majority of the stage B coarse material occurs in the stream-mouth complex and forms an important producing oil reservoir in the area. The distributary responsible for this deposit probably built southeastward into relatively shallower water, and some of the coarse material was swept back from the stream mouth and deposited in a band from 3 to 5 mi wide. The distributary-fill central portion reaches thicknesses as great as 100 ft. The trend of this stream-mouth deposit is northwest-southeast and is terminated updip toward the northwest. This deposit represents the last of the delta-front sedimentation before the final stage C deltaic plain which followed.

The stage C deltaic-plain sediments are the most widespread and most easily recognized of all the various upper Morrow deposits. They also contain the deposits which are the major Pennsylvanian reservoirs in the Anadarko basin.

All the sections in Figure 27 show the different aspects of point-bar deposits and their relationship with other environmental facies such as meander-channel-fill and backswamp-marsh claystones. Note the relatively flat base of the point-bar deposits and how this base is roughly parallel to the top of the deltaic-plain backswamp-marsh facies. The sections also indicate a slight structural movement after stage C deposition.

The geometry of this important oil- and gas-producing facies is shown in Figure 27. Note that the deposits tend to form loops which closely approximate the theoretical diameter resulting from a stream that would deposit point bars of the observed thickness. Also note that the individual point bars occur in a meander belt which generally trends northwest-southeast. Many wells have penetrated what is probably meander-channel-fill material and would indicate the nearby presence of additional point bars.

North Hardesty Area — The North Hardesty area is located in the eastern portion of Texas County, Oklahoma (Fig. 1), and is 25 mi southeast of the Hough area. This is also an area of prolific upper Morrow oil and gas production.

Figure 28 shows the detailed cross-sections across the area, the locations of which are shown on Figure 29. The upper boundary and datum for the sections is the base of the "Thirteen Finger Limestone"; the lower boundary is the top of the lower Morrow sandy limestone. The North Hardesty area has coarse deltaic sedimentation only in stages B and C. Cross-sections 5-5' and 7-7' illustrate the cross-sectional shapes of the stage B stream-mouth-bar deposits. Figure 29 shows their thickness and distribution in plan view. The most easterly of the two deposits is a typical stream-mouth-bar, distributary-fill complex trending northwest-southeast. It may actually represent the terminal end (southeast) of this particular deltaic extension. This deposit disappears to the northwest, where it has probably been destroyed by overlying deltaic-plain sediments.

Lying above and occasionally cutting into the stage B deposits is the widespread stage C deltaic plain with its point-bar, meander-channel-fill, and backswamp-marsh deposits. The point-bar deposits are the main oil- and gas-producing reservoirs in the area. The cross-sections illustrate two periods of point-bar deposition of slightly different age. This relationship is evident in most of the cross-sections, which also show the typical flat bottoms and variations of point bars as well as their relationship with the other deltaic-plain sediments.

Figure 29 also shows the thickness and distribution of the stage C point-bar deposits. The contours are compatible with the points of penetrated meander-channel-fill claystone and have been drawn as a series of loops which appear to fit the control best and yet not violate the

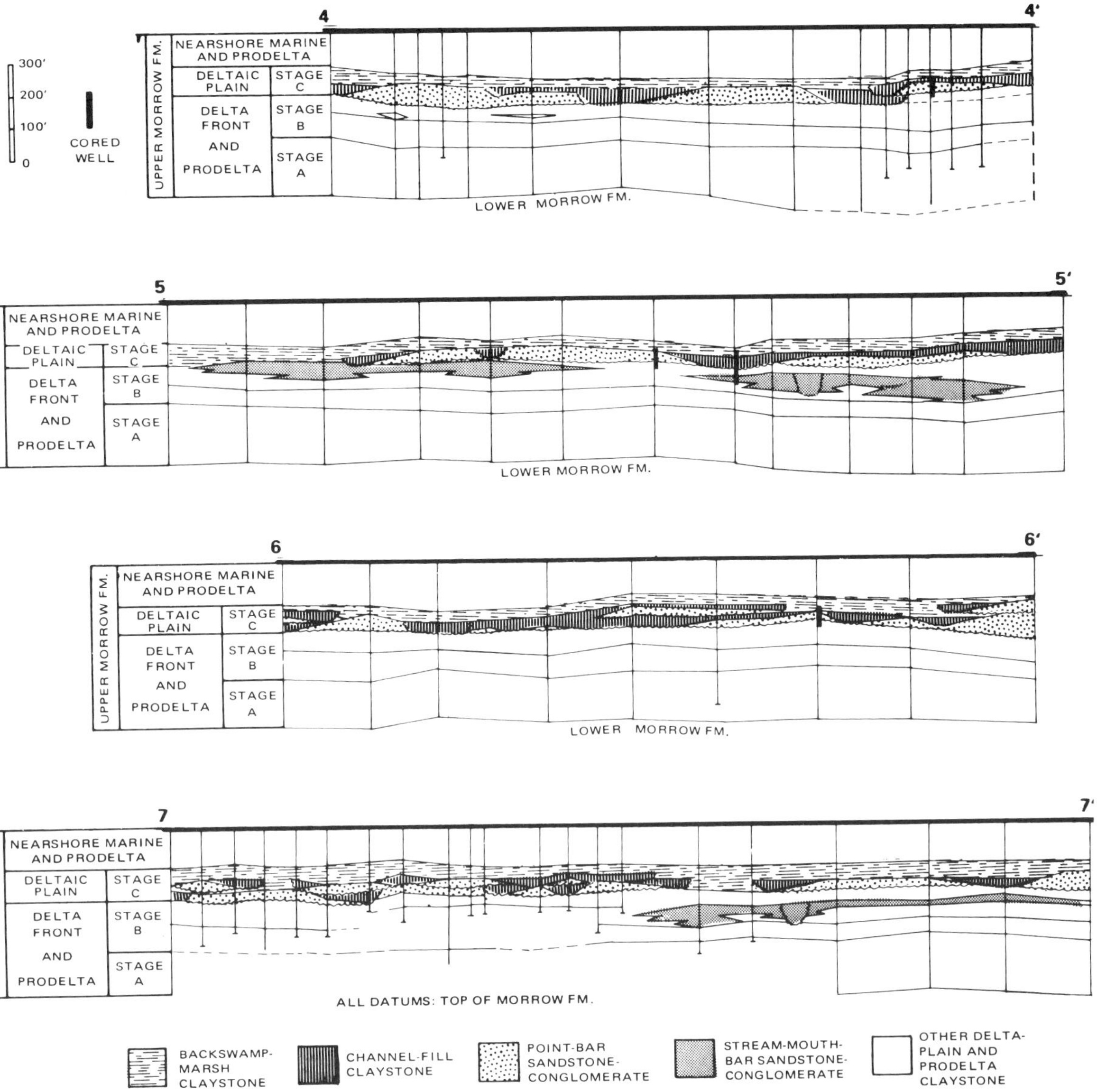

FIG. 28—Cross-sections showing the three-dimensional framework of stage B and stage C upper Morrow deltaic deposits in the Hardesty area, Texas County, Oklahoma.

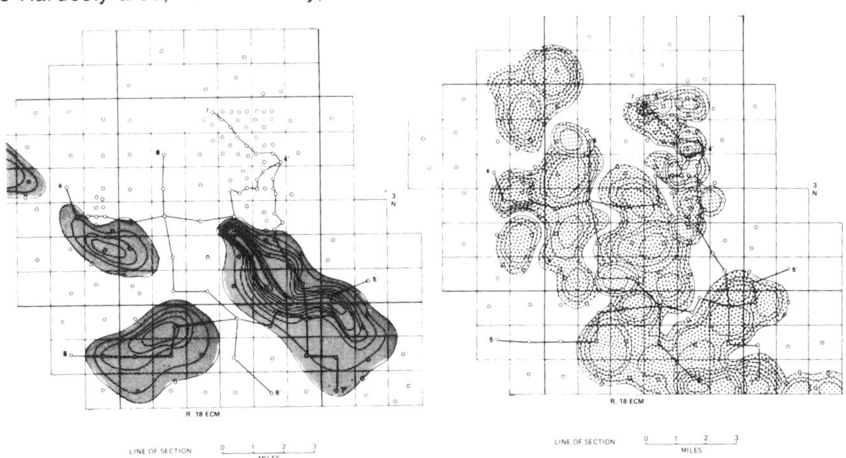

FIG. 29—Isolith maps showing the three-dimensional framework of stage B and stage C upper Morrow stream-mouth and point-bar deposits, Hardesty area, Texas County, Oklahoma.

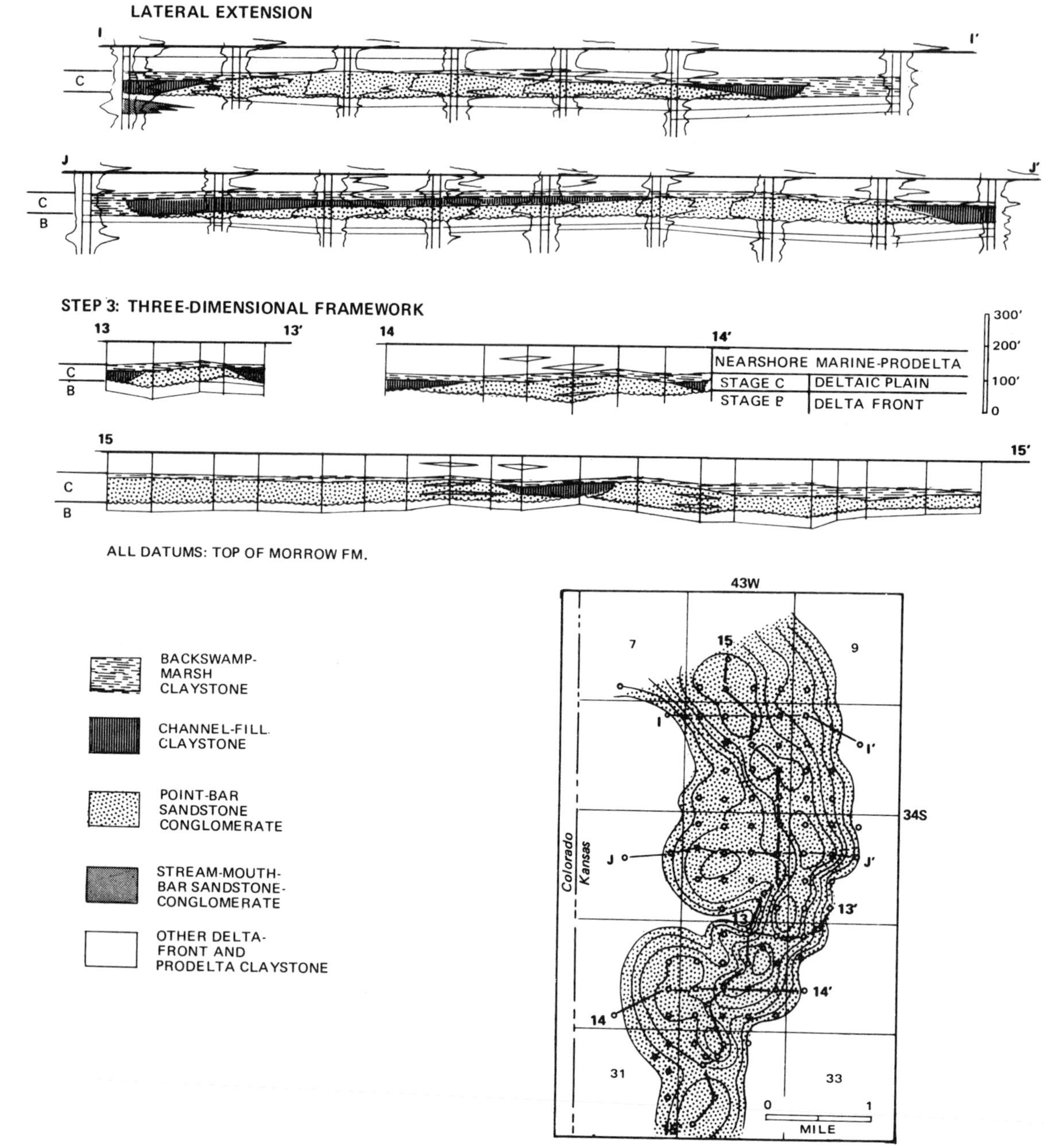

FIG. 30—Lateral extension along cross-sections and cross-sections and isolith maps showing the three-dimensional framework of stages B and C upper Morrow deltaic deposits (Interstate area, Morton County, Kansas).

theoretical size that the hypothetical stream would yield. This point-bar complex occupies a meander belt about 6 mi wide which trends generally northwest-southeast. Figure 30, from the interstate area located in Morton County, Kansas, also shows well-developed point-bar deposits. Environmental facies analysis was also successfully applied to the North Guymon area, Texas County, Oklahoma; Taloga area, Morton County, Kansas, and Texas County, Oklahoma; Eva area, Texas County, Oklahoma; Flank area, Baca County, Colorado; Spearman area, Hansford County, Texas; and Mendota area, Roberts and Hemphill Counties, Texas.

Characteristics and Production Behavior, Upper Morrow Reservoirs

After careful analysis of the data from all nine areas of intensive study, some generalizations can be made about the geometry and important facies relationships of significant reservoir facies in the upper Morrow deltaic deposits.

Point Bars — These deposits (Fig. 31) are ovoids and semiovoids which tend to average from 3/4 to 1 mi in diameter. Maximum thickness is 60 ft, with the average thickness around 35 ft. Point-bar deposits have relatively flat bottoms, with variations in thickness resulting from change in the upper surface. The deposits may be complete ovoids or only partial ovoids where one has been cut out at the expense of another developing system. Point bars either "toe

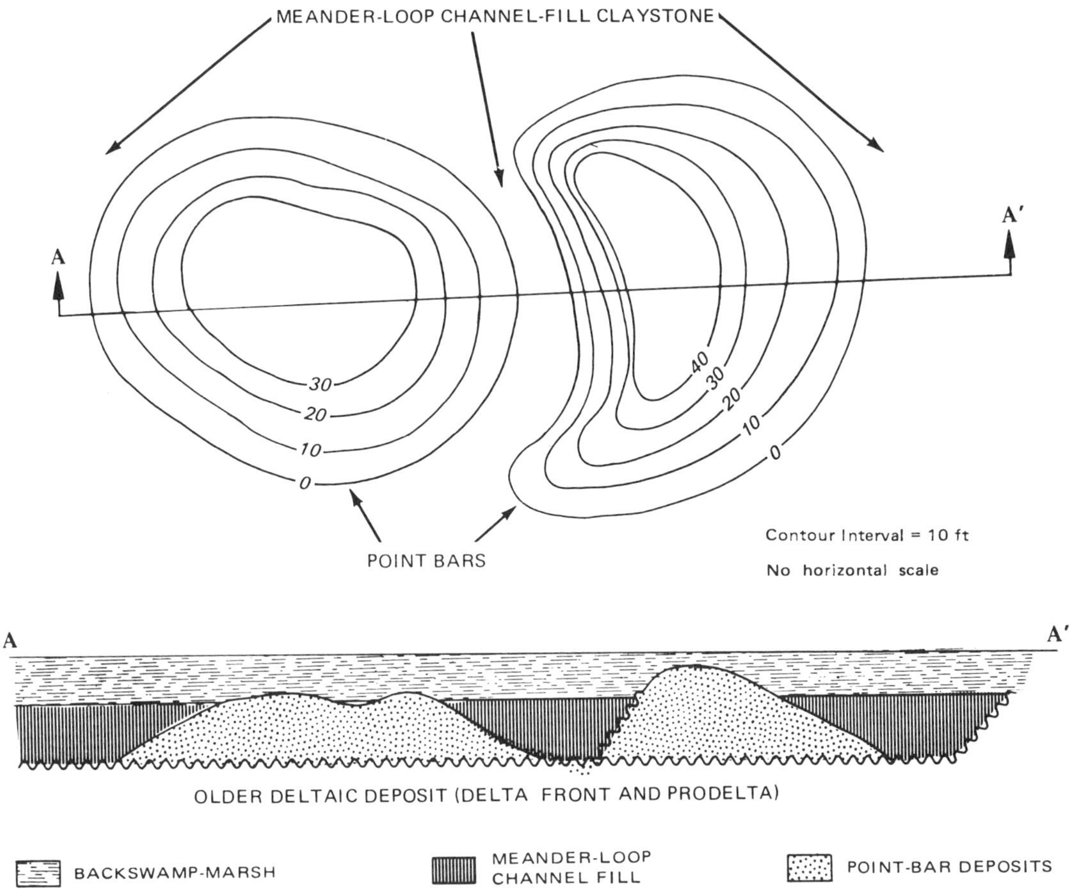

FIG. 31—Generalized model of upper Morrow point-bar deposits.

out" in and are covered by the fill in the loop that is forming them, or they end abruptly against the claystone fill in the loop that is destroying them. Note the relative spacing of the contours (Fig. 31a).

Upper Morrow point bars lie above and cut into delta-front and prodelta sediments. In many places, they lie directly on stream-mouth-b sandstones. In all examples studied, the laterally enclosing facies is a meander-channel-fill organic claystone. The overlying facies is backswamp-marsh organic claystone.

Stream-Mouth Bars and Bar Fingers — Two varieties of stream-mouth-bar deposits (Fig. 32) occur in the upper Morrow: (1) a thick, narrow bar-finger type and (2) a thinner, wider variety. The thick, narrow deposits have an average width of about 1-1/4 mi and thicknesses as great as 120 ft. The thinner deposits have widths as great as 4 to 5 mi and an average thickness of 30 ft, except where the central portion may have been cut and filled by an overlying distributary. There the deposits are up to 80 ft thick. All the stream-mouth-bar deposits studied are linear, except where erosion has left only remnants. The stream-mouth bars and bar fingers tend to have flat tops, except at the very edge, where the initial depositional slope is preserved (note relative contour spacing, Figs. 32a, 32b). Apparently, the thicker parts of the deposit pushed down into the underlying deposits, or they represent distributary-channel deposits that fill cuts in the underlying rocks. These deposits are enclosed by delta-front or deltaic-plain sediments.

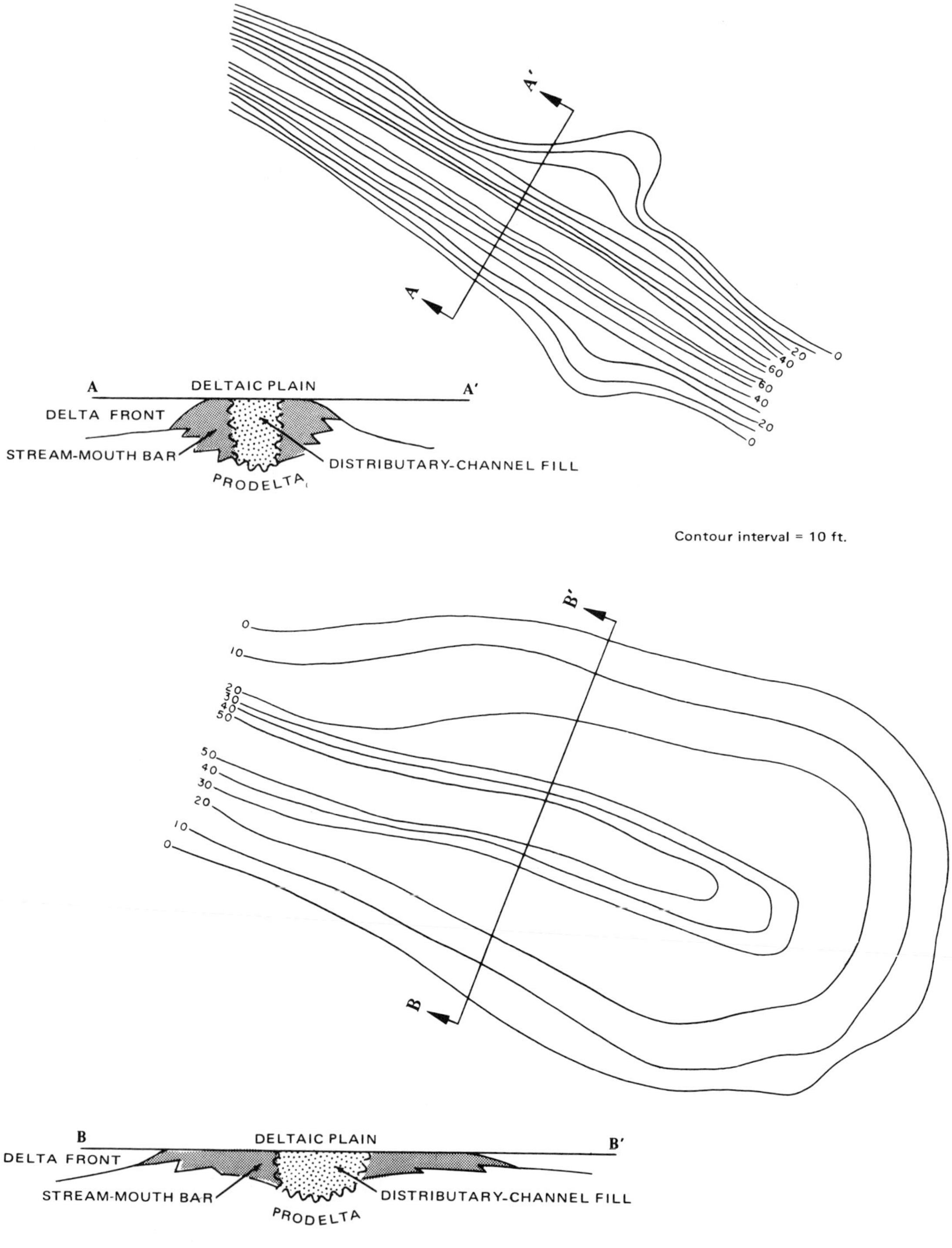

FIG. 32—Generalized models of upper Morrow stream-mouth-bar-finger deposits.

Distributary-Channel Fills — One example of a minor straight-distributary-channel fill (Fig. 33) was found in the Spearman area of Hansford County, Texas. The deposit is 1/2 mi wide and as thick as 60 ft. It trends in a northwest-southeast direction for several miles. The exact shape is not known, but it is probably convex downward, with a flat top. The deposit cuts into the deltaic-plain sediments which enclosed it laterally. In places, it also cuts into and rests upon prodelta claystones. Overlying sediments are backswamp-marsh organic claystones.

Production Behavior — The lower parts of point bars appear to be the best reservoirs, with porosities between 15 and 22%. The texture is coarse, and there is a general absence of intergranular clay and clay laminations. Middle point-bar sandstone, although cross-bedded and containing more clay is also fairly well-sorted sandstone and should hold and transport hydrocarbons better than upper point-bar deposits. The latter are fine grained and contain much matrix clay and clay and silt laminations.

In stream-mouth-bar deposits, the upper portion is the most effective reservoir, although it may contain many laminations. Here, the porosities vary between 12 and 18%. The lower parts of stream-mouth bars are finer, contain more matrix clay, and commonly are intercalated with fine-grained sandstone and claystone. In both point bars and stream-mouth bars, the bedding planes represent original depositional slope. They lie at an angle to the horizontal development of the deposits and the lithic units (Figs. 17, 18, 20).

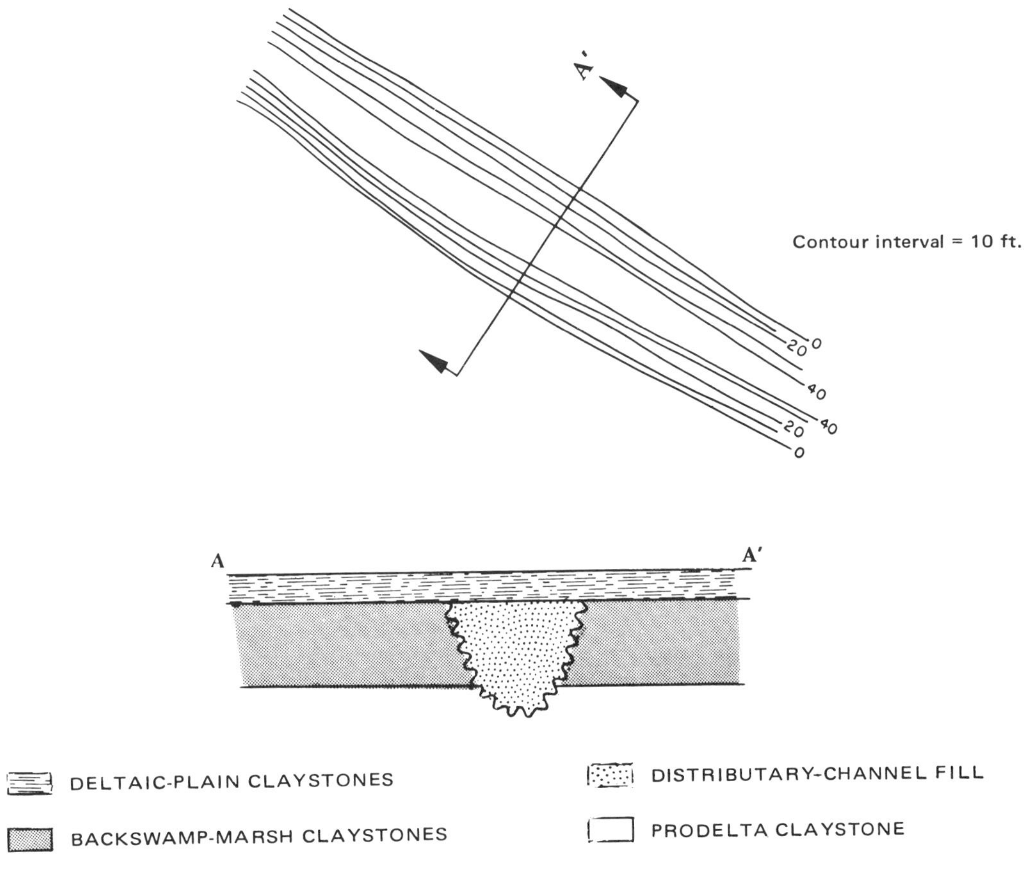

FIG. 33—Generalized model of upper Morrow distributary-channel-fill deposits.

If a significant amount of clay is deposited along these depositional slopes, such as in a clay drape on a point bar, the effective behavior of the deposit as one reservoir will be reduced. Zeito (1965) showed these clay "breaks" to be widespread and continuous in "channel" and deltaic sandstone. If present in a reservoir, they will tend to inhibit or stop vertical and horizontal migration. Figure 34 shows the effective blocking action caused by such clay and silty laminae in a point-bar sandstone from the Cherokee Formation of eastern Kansas. Note how the silty clay laminae contain no hydrocarbons and block movement through them. The best way to assure migration and drainage of deltaic reservoirs is to perforate the lower point bar and the upper stream-mouth bar, which are least likely to contain clay laminae and tend to connect and allow communication throughout the deposit.

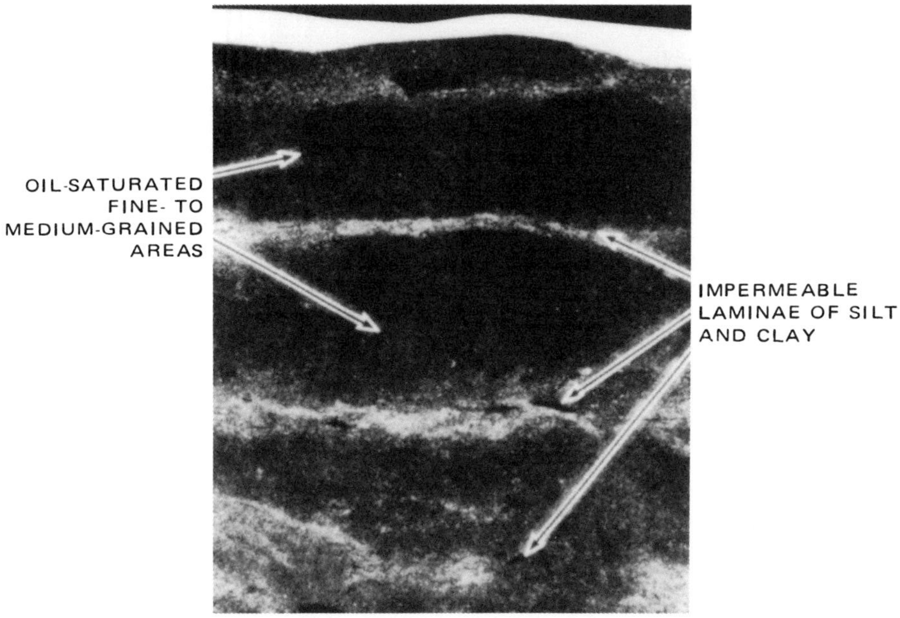

FIG. 34—Representative point-bar sandstones from the Cherokee Formation of eastern Kansas. (a) Point-bar sandstone illustrating how impermeable laminae of silt and clay can inhibit movement of oil in these sandstones, x1.

The validity of these relationships was tested in the Hough area, where upper Morrow deposits in 84 wells are selectively perforated either in the upper, middle or lower point-bar facies or in the upper or lower stream-mouth-bar facies (Fig. 35). The quantity labeled "production factor" was used in an attempt to get a production figure which equalized the effect of choke size and the number of feet perforated; hence, a higher production factor indicates better reservoir conditions. The average, median, and frequency distribution of the "production factor" all show that the best reservoir conditions are found in the lower point-bar deposits; the next best are found in middle point-bar and upper stream-mouth-bar deposits; and the worst are found in upper point-bar and lower stream-mouth-bar deposits. The extremely laminated nature of upper stream-mouth-bar facies probably keeps it from being as good as lower or even middle point-bar facies.

If an operator has a choice, and water is not a problem, these results indicate that the middle and lower parts of point bars should always be perforated. Opening just the top few feet of a point bar is a poor production practice and will not drain the reservoir. Conversely, opening the lower part of a point bar gives a permeable "sheet" that will tend to drain the entire deposit. This phenomenon has been observed in the past during development drilling in the upper Morrow, where operators frequently encountered only thin upper Morrow "Sand" (8 to 10 ft thick). They were surprised to find that, when perforated, this deposit "made a good well" and that the production "held up." This condition is easily explained, however, by the fact that the rocks perforated were the lower porous ones on the edge of a point-bar deposit (Figs. 17, 18). Perforating this thin edge opened up the entire deposit of the well.

Upper Morrowan Deltaic Processes — As a result of information available from the nine areas of intensive study, the hypothesis has been developed that upper Morrow rocks were deposited in a prograding deltaic environment. The isometric diagram in Figure 36, together with appropriate cross-sections, shows the postulated development of deltaic deposits resulting from meandering streams and distributaries entering relatively shallow water. This

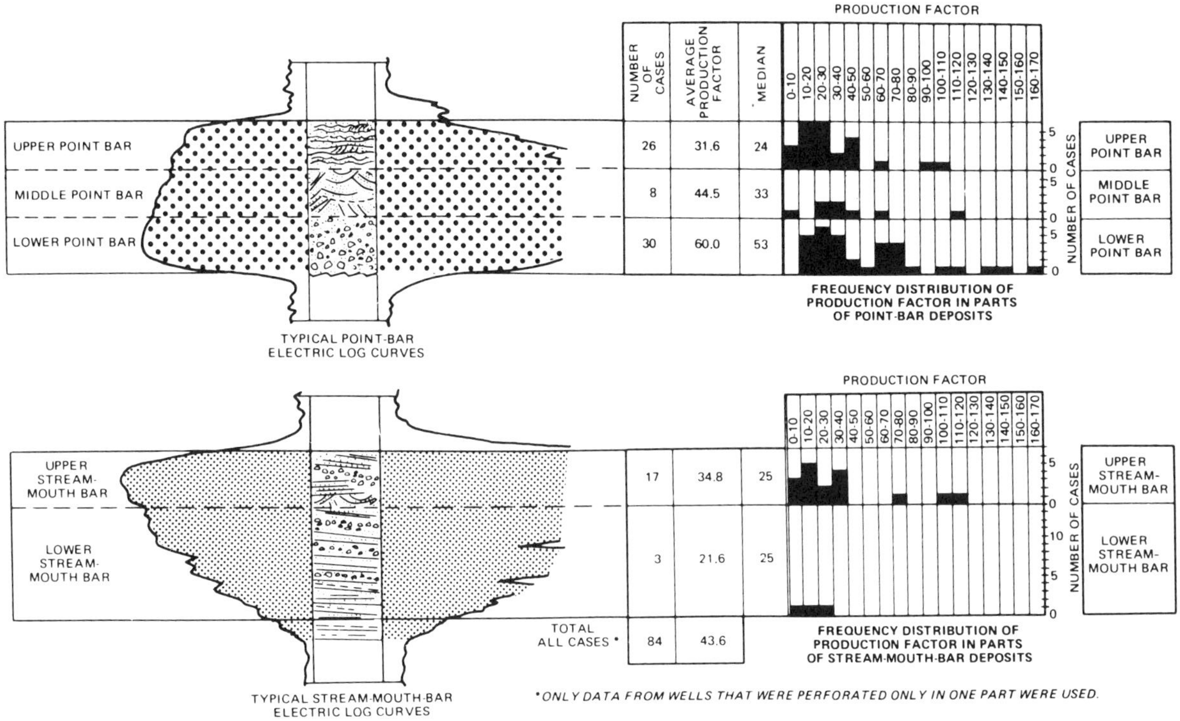

FIG. 35—Statistical analysis of the production behavior of various deltaic deposits in the Hough area, Texas County, Oklahoma.

model approximates conditions which existed during stage B, and particularly stage C, upper Morrow deposition.

Because the water was shallow, the prodelta clays rapidly built up a platform on which the delta-front sediments were deposited. The stream-mouth bar was the main reservoir facies developed in the delta front; however, shoreface sandstone could have been present in front of spit beaches which developed away from the stream mouth. The deltaic plain advanced closely behind the stream mouth. Since the basin was not subsiding fast enough to lower the stream-mouth-bar deposits into a protected position, meandering streams on the deltaic plain destroyed the delta-front deposits. Only relict parts of lower stream-mouth bars were preserved. These were parts that sank deep enough in the prodelta mud to be below the depth of scour of the meandering streams (Fig. 36). The destruction of the delta-front deposits occurred because stream-mouth bars built up to sea level from a depth of 30 to 50 ft, and deltaic-plain meandering streams and distributaries cut 30 to 60 ft below sea level. If the stream-mouth-bar deposits had sunk far enough into prodelta clay, or if the area had subsided very rapidly, the stream-mouth bars would have had a better chance for preservation.

After the meandering stream or distributary built a deltaic lobe out into the basin, it became surrounded by interdistributary delta-front areas (bays, etc.). The distributary system eventually became inefficient as the lobe became overextended and the stream broke through into the interdistributary bay area, where a new deltaic lobe began to build. Because the whole area was subsiding, the older deltaic lobe sank below the water and was eventually covered by fringe clay. An abandoned relict stream-mouth bar could have been buried out in front of the older lobe.

Marine curents did not rework much of the older lobe because it consisted mostly of fine-grained, organic-rich, deltaic-plain sediments which were very resistant to erosion. The point-bar sand deposits were especially protected from marine processes by the backswamp-marsh and meander-channel-fill clays.

Figure 37 is an isometric drawing with appropriate cross-sections illustrating another type

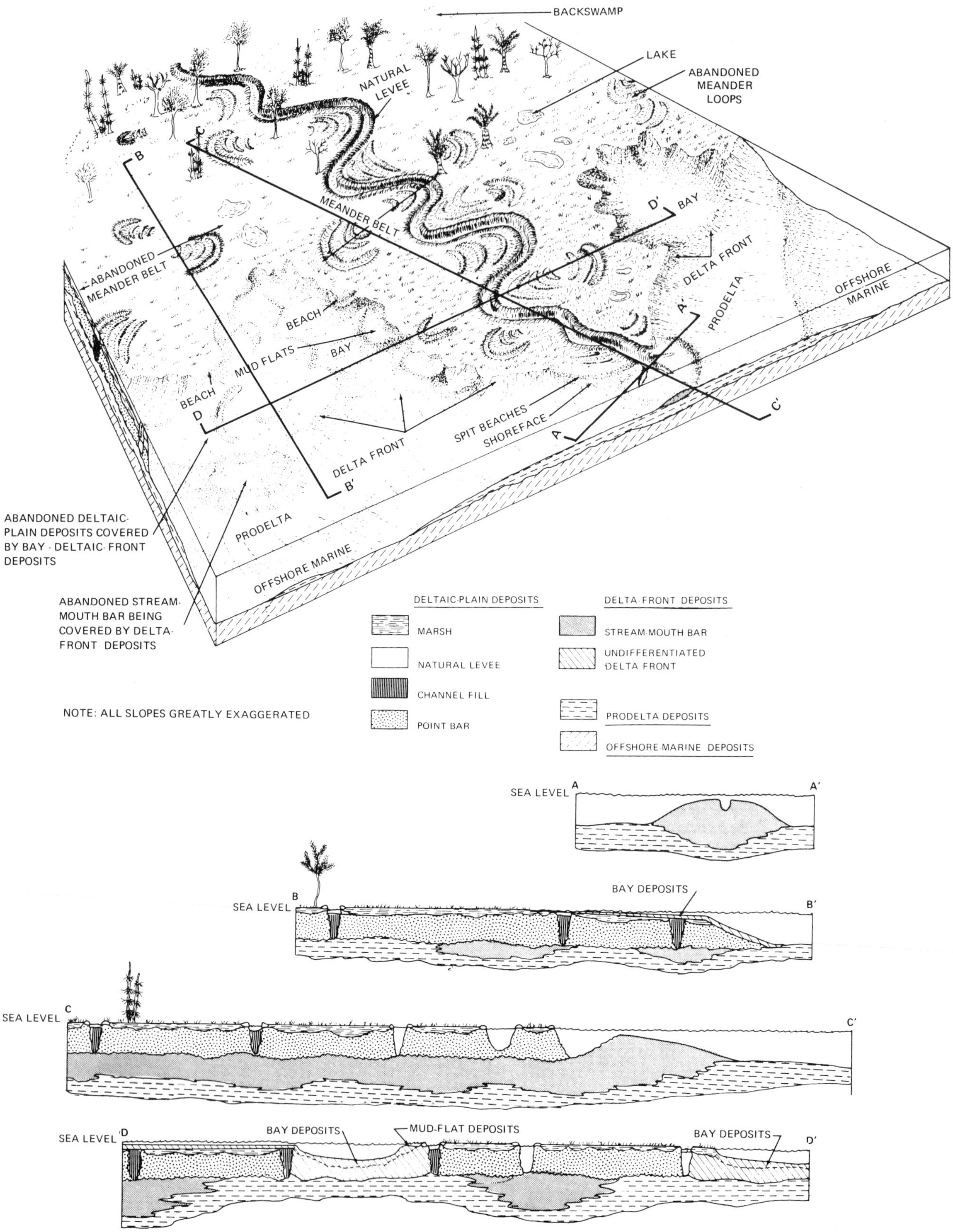

FIG. 36—Isometric diagram and cross-sections showing the theoretical relationships between upper Morrow deltaic facies deposited in relatively shallow water.

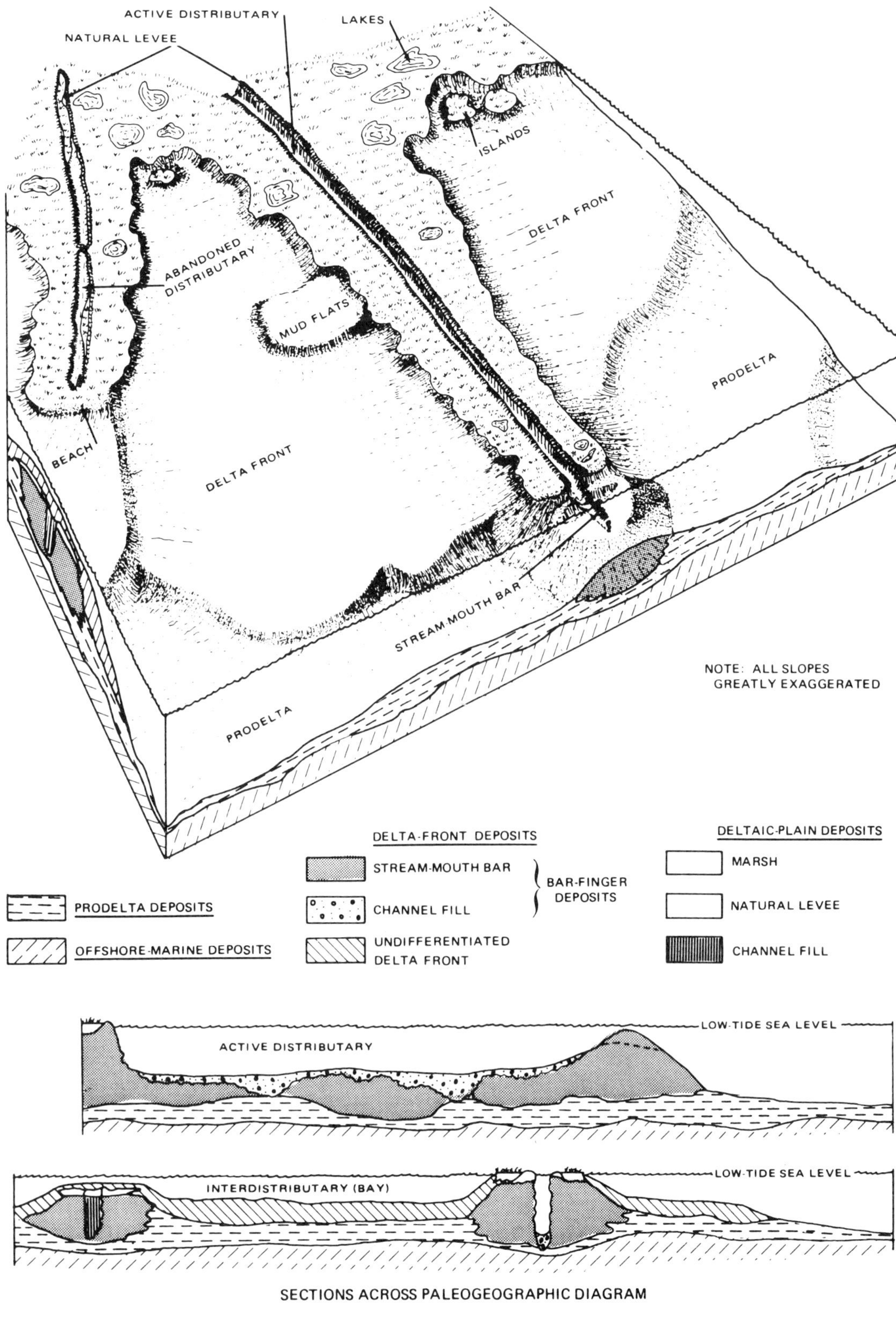

FIG. 37—Isometric diagram and cross-sections showing the theoretical development of upper Morrow straight distributaries building stream-mouth bars or fans in relatively deep water.

of deltaic development that occurs in the upper Morrow Formation. This model shows a system of straight distributaries which deposited their load into relatively deep water. These conditions were possibly present during stage A and stage B deposition. The processes involved were typical of those described by Fisk (1961) as responsible for the development of bar-finger deposits.

Studies show that the upper Morrow distributaries carried a bimodal load of coarse material and clay. As this material was introduced into relatively deep quiet water, the coarse load dropped immediately and was incorporated into a rapidly prograding stream-mouth bar or "fan." The depositional slope on this fan may have been relatively steep. As the deposit advanced, the straight scoured channel of the distributary followed it, cutting deeply into or completely through the stream-mouth-bar deposits. As the overextended point bar was slowly cut off upstream, the lowered velocity of the distributary resulted in backfilling of the channel with sand. This condition could have advanced far back up the bar finger, resulting in the coarse-grained deposit known as the bar-finger complex. A section through such a bar-finger complex is shown in cross-section C-C' in Figure 22 and in sections 1-1' and 2-2' in Figure 25. After the bar finger was isolated, it continued to subside and quickly became subaqueous where it was covered by delta-front clays and protected from marine processes.

The complete sequence of events illustrating the stage-by-stage development of the vertical sequence of environmental facies as they exist in the Hough area of Texas County, Oklahoma, is shown in Figure 38. Stage Aa deposition consisted of bar fingers building basinward and being cut off and buried as the local sites of sedimentation shifted. A major change of the stream and distributary system occurred on the deltaic plain in stage Ab. During this stage, delta building occurred elsewhere, and prodelta and offshore-marine claystones continued to bury the bar fingers under more clay. In stage B, basin subsidence slowed as active sedimentation returned to the model area, where once again deltaic lobes prograded with the advance of a broad stream-mouth bar. A deltaic plain containing meandering streams quickly followed as individual lobes were cut off and local shifting of the depositional site occurred. Subsidence was still great enough to lower the abandoned stream-mouth-bar deposits and preserve most of them. Stage C was the final period of the deltaic sedimentation. The "front" of the delta had advanced far basinward, and the deltaic plain covered the entire complex of buried deltaic deposits. The face of the stage C isometric drawing closely approximates an actual cross-section in the Hough area.

Reservoir Prediction Using Environmental Facies Analysis

The Hough area was selected for testing the predictability of maps based on analysis of the upper Morrow Formation because (1) the environmental facies identifications were reliable, (2) the isopach maps of the reservoirs were contoured utilizing knowledge of the environmental facies, and (3) the area was being developed so rapidly that the accuracy of the maps could be determined in a relatively short time.

Figure 39 shows the results of analyzing an isopach map of stage C point-bar deposits. Knowledge of the environmental facies was used to draw this map during September, 1964, prior to the drilling of 28 wells. The locations of these wells are listed in Figure 39 and are shown on the map. In 24 wells, the original map predicted the thickness of the reservoir close enough to be considered an operational success. In 4 wells, the map missed far enough so that these wells would be considered an operational failure. This is a successful prediction of 86%. If operators had known the origin of the deposits and mapped them accordingly, they could have saved dry-hole costs as well as successfully predicted the existence of unknown point-bar deposits. The looping, oval character of these reservoirs would have led to more successful step-out programs. Compare the original map in Figure 39 with the final interpertation of these deposits shown in Figure 27.

Regional Analysis

Regional Cross Sections and Maps — The initial step in regional analysis of the upper Morrow Formation was the construction of small-scale regional cross-sections (Fig. 40) forming a

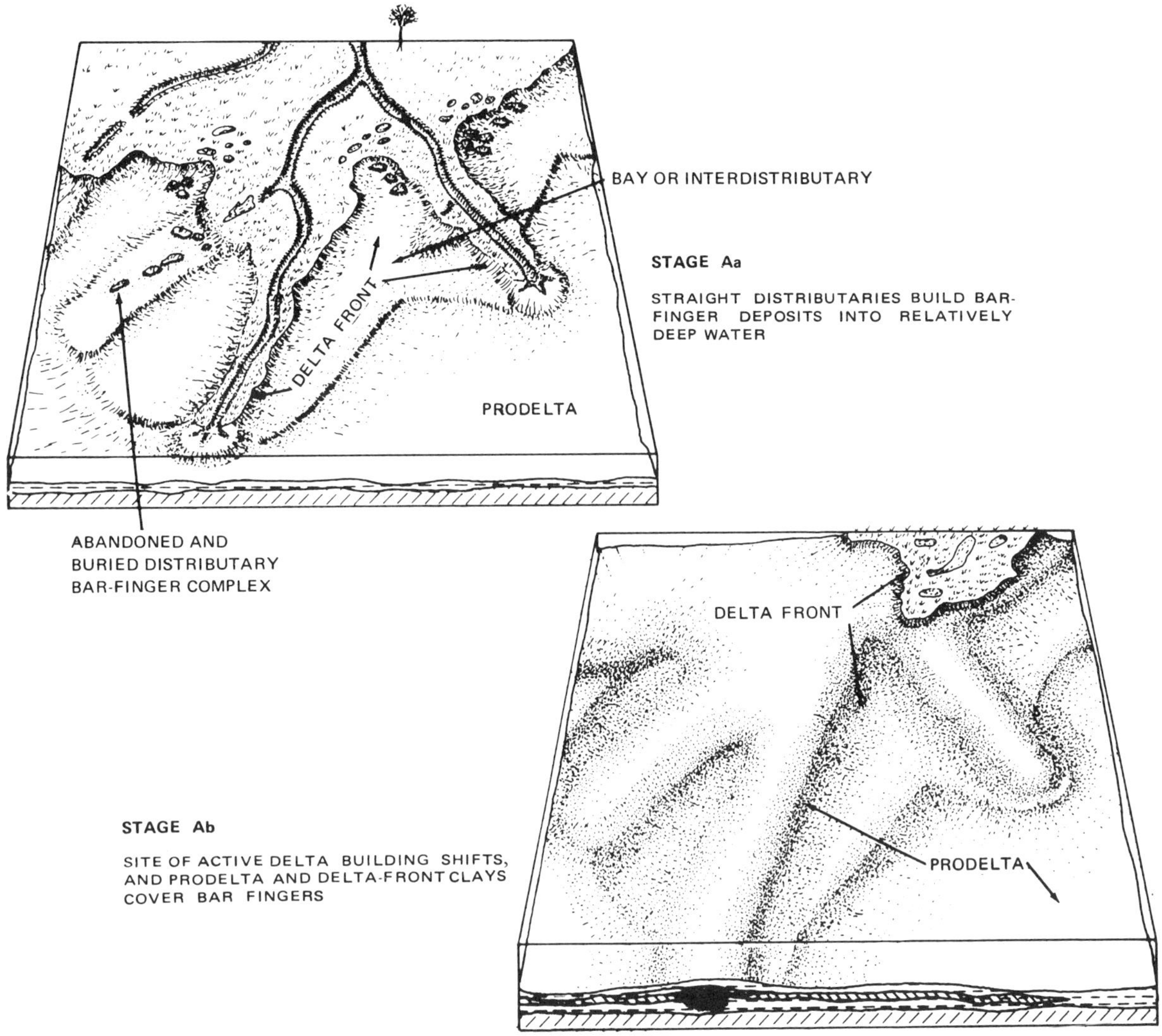

FIG. 38—Isometric diagrams showing the theoretical development of events which resulted in the deltaic deposits found in the Hough area, Texas County, Oklahoma.

network across the basin. These lines of section crossed the areas of intensive study (Fig. 1) and, where possible, included data points for which either cores or sample cuttings were available. Although electric logs were used for correlation, they are not shown in the sections. The best possible correlation markers were used in the regional cross-sections; however, whether or not these lines are absolute timestratigraphic markers is not critical. An actual 50- or 75-ft "bust" in correlation can be tolerated. The important thing in these sections is the spatial relationship of the various gross facies and the location of the specific known environmental facies. The sample cuttings and logs helped in correctly recognizing and locating important facies groups.

In the seven regional cross-sections constructed for the upper Morrow study (Fig. 40), the upper boundary and datum, as in the local area cross-sections, is the base of the "Thirteen Finger Limestone". Although these carbonates afford good time-stratigraphic markers in local areas, they do not represent time lines on a regional basis. The stratigraphic "error" and minor correlation errors, however, do not significantly affect the facies relationship shown on the sec-

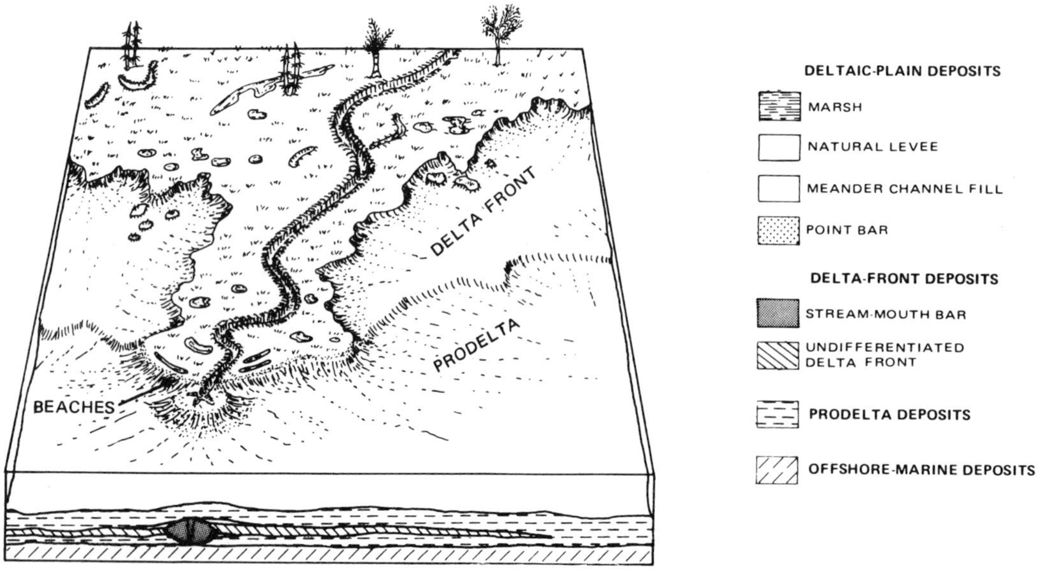

STAGE B

ACTIVE DISTRIBUTARY RETURNS AND BUILDS A BROADER SYSTEM OF STREAM-MOUTH-BAR DEPOSITS INTO RELATIVELY SHALLOWER WATER

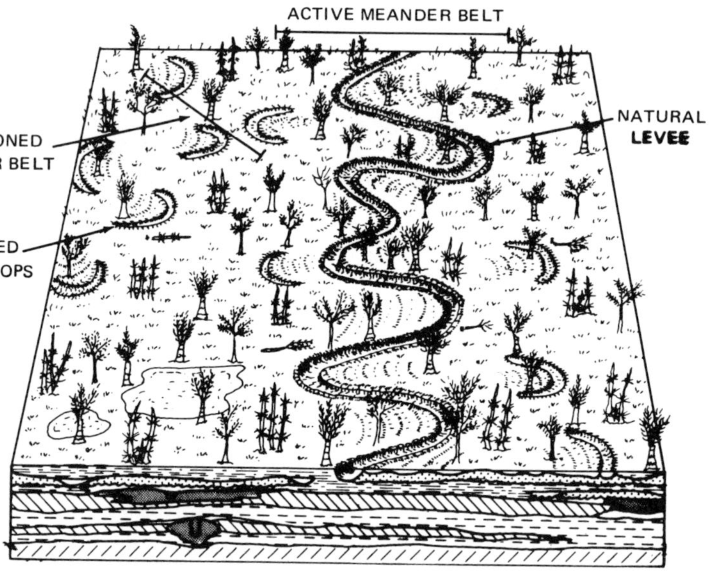

STAGE C

WHOLE AREA IS FINALLY COVERED BY DELTAIC-PLAIN DEPOSITS

tions. The lower boundary is the base of the total Morrow Formation (in most places, the top of the Mississippian). The primary division shown on these sections is the separation between the upper Morrow deltaic clastic sediments and the lower Morrow marine sediments. This contact is generally a conformable one, but in many places it represents an interfingering facies relationship.

Where these sections cross known environmental facies, the facies and their depositional stage are shown. Evidence of late Morrowan structural movement can also be seen. This is particularly evident in sections CC-CC' and FF-FF'.

Identification of probable lower Morrow marine sediments on the shelf end (right) of the sections leads to the conclusion that there was some uplift and erosion of Morrowan deposits in this area; however, these marine sandstones, limestones, and clays in many places are probably facies equivalents of "late Morrowan" deltaic-derived claystones which were deposited "farther out" in the Morrow basin. Sections BB-BB', DD-DD', and FF-FF' show this relationship.

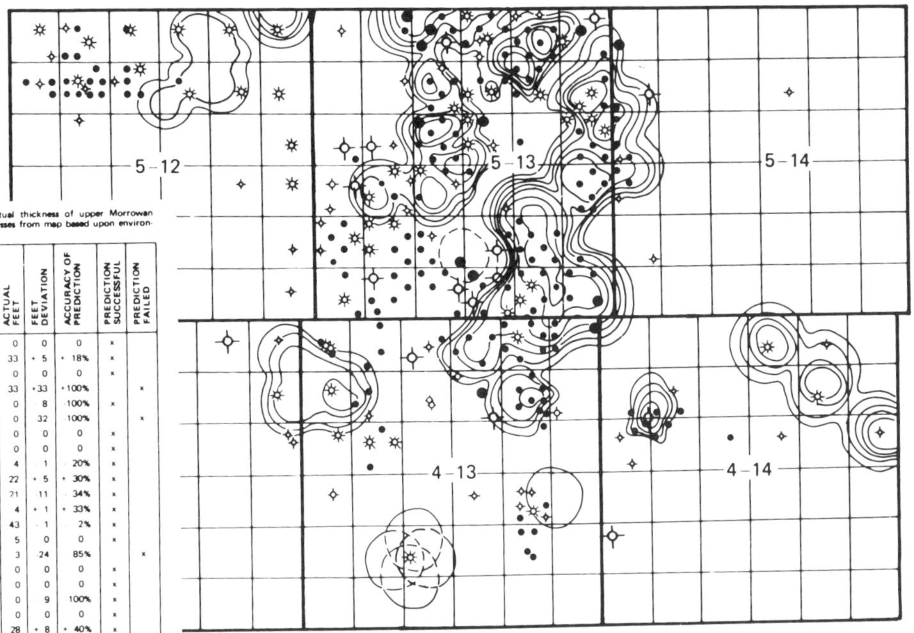

Comparison of predicted thickness and actual thickness of upper Morrowan point-bar sand reservoirs. Predicted thicknesses from map based upon environmental facies analysis

LOCATION	PREDICTED FEET	ACTUAL FEET	FEET DEVIATION	ACCURACY OF PREDICTION	PREDICTION SUCCESSFUL	PREDICTION FAILED
SW, NE sec 2, T 4N, R 12ECM	0	0	0	0	x	
NE, NE sec 1, T 4N, R 13ECM	28	33	+ 5	+ 18%	x	
NW, SW sec 4, T 4N, R 13ECM	0	0	0	0	x	
SW, NE sec 10, T 4N, R 13ECM	0	33	+33	+100%		x
SE, SE sec 10, T 4N, R 13ECM	8	0	- 8	-100%	x	
SE, SE sec 7, T 4N, R 14ECM	32	0	-32	-100%		x
NE, NE sec 30, T 4N, R 14ECM	0	0	0	0	x	
NW, NE sec 1, T 5N, R 13ECM	0	0	0	0	x	
SE, NW sec 1, T 5N, R 13ECM	5	4	- 1	- 20%	x	
NW, SW sec 4, T 5N, R 13ECM	17	22	+ 5	+ 30%	x	
SE, NW sec 4, T 5N, R 13ECM	32	21	-11	- 34%	x	
NW, SE sec 4, T 5N, R 13ECM	3	4	+ 1	+ 33%	x	
NE, NE sec 5, T 5N, R 13ECM	44	43	- 1	- 2%	x	
NE, NW sec 15, T 5N, R 13ECM	5	5	0	0	x	
NW, NW sec 16, T 5N, R 13ECM	27	3	-24	- 85%		x
NW, SW sec 17, T 5N, R 13ECM	0	0	0	0	x	
NW, SE sec 18, T 5N, R 13ECM	0	0	0	0	x	
SE, NE sec 19, T 5N, R 13ECM	9	0	- 9	-100%	x	
NW, SE sec 27, T 5N, R 13ECM	0	0	0	0	x	
SE, SE sec 28, T 5N, R 13ECM	20	28	+ 8	+ 40%	x	
NW, NW sec 32, T 5N, R 13ECM	0	0	0	0	x	
SE, NE sec 33, T 5N, R 13ECM	0	0	0	0	x	
NW, NW sec 34, T 5N, R 13ECM	12	18	+ 6	+ 50%	x	
NW, NE sec 34, T 5N, R 13ECM	0	0	0	0	x	
NW, SW sec 34, T 5N, R 13ECM	0	0	0	0	x	
NW, SE sec 36, T 5N, R 13ECM	40	19	-21	- 50%	x	
NW, SW sec 6, T 5N, R 14ECM	11	13	+ 2	+ 18%	x	
NW, SE sec 7, T 5N, R 14ECM	10	0	-10	-100%		x
AVERAGE FEET DEVIATION OF PREDICTION + = MORE THAN PREDICTED – = LESS THAN PREDICTED			6.3			
PERCENT SUCCESSFUL PREDICTION					86%	

FIG. 39—Map and table showing the successful prediction, based upon environmental facies analysis, of the distribution of stage C point-bar reservoirs in the Hough area, Texas County, Oklahoma.

Careful analysis of the regional sections leads to conclusions concerning the sequence of development of the Morrow Formation. Initially, lower Morrow "marine" deposits were formed in a series of onlapping shoreline deposits. The first influx of prograding delta-derived claystones probably began early in this initial phase (section DD-DD'). In one local area, the upper Morrow clastics rest directly on the Mississippian and are probably a lateral time equivalent of the adjacent lower Morrow marine deposits. After deltaic sedimentation became widespread, local areas of marine sedimentation persisted almost to the end of Morrowan time (sections AA-AA', CC-CC', and FF-FF').

In addition to developing the regional framework, regional facies cross-sections serve as the key factor in formulating a mapping program to (1) show the trend and development of the deltaic reservoir deposits, (2) relate them to the general deltaic development, (3) locate the major source areas of the sediments, and (4) project the existence of reservoir deposits into virgin areas where they would be exploration objectives.

The first map to be made for extending the study regionally was a map showing the areal extent of the major deltaic wedge containing the reservoir facies. This was accomplished in two ways — by an isopach contour map of the thickness of the upper Morrow wedge (Fig. 41) and by a contour map of the ratio of upper Morrow deltaic sediments to lower Morrow "marine" sediments (Fig. 42). The regional sections and areas of intensive study are located on each map.

The isopach map (Fig. 41) shows that the deltaic sediments are in a long, relatively narrow trough trending northwest-southeast, with the thickest portion being centrally located. The deposits thicken southeast into the central Anadarko basin of Oklahoma. The sediments

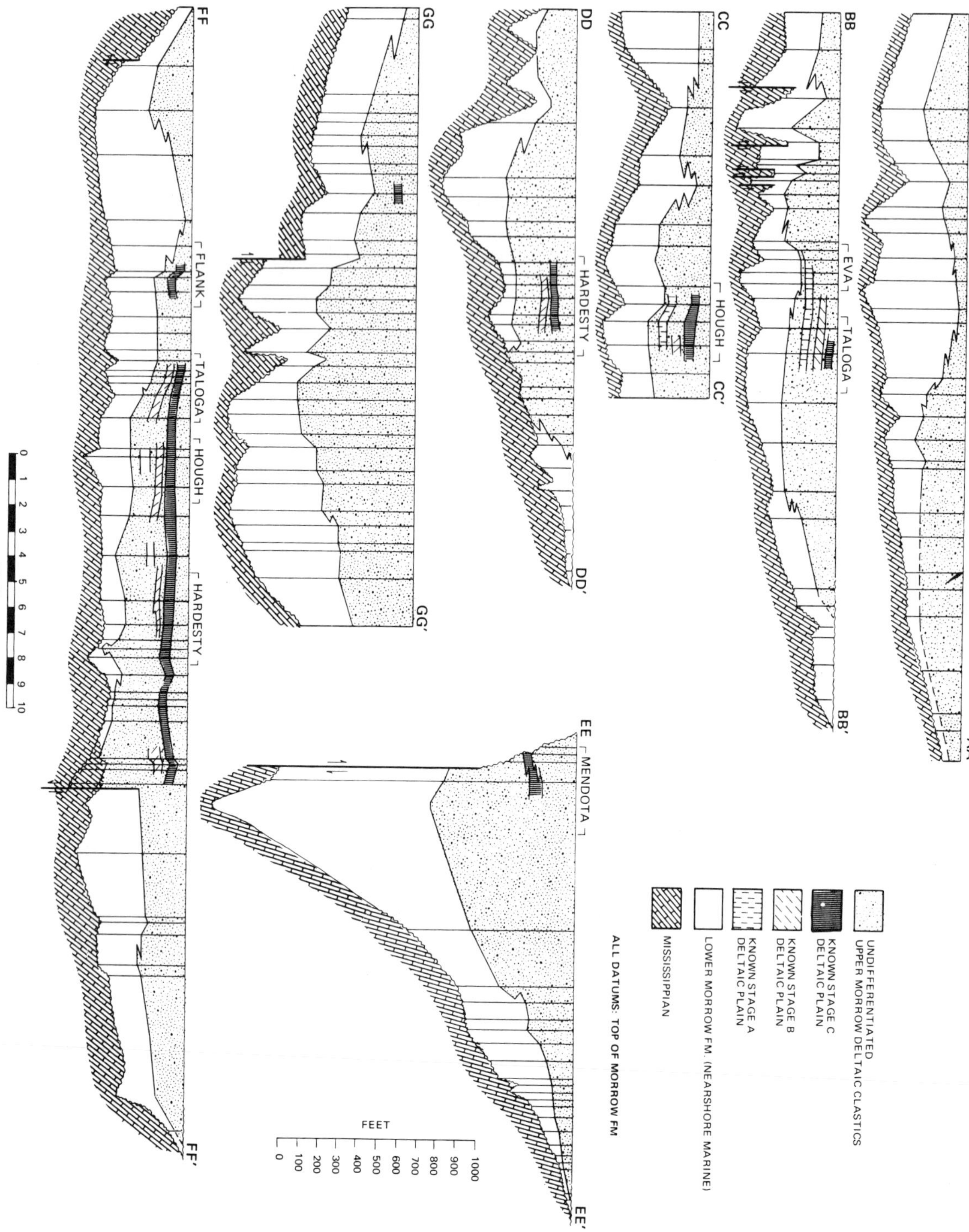

FIG. 40—Regional cross-sections showing the relationship between Morrowan environmental facies in the Anadarko basin. (See Figure 1 for location of cross-sections.)

generally terminate on the Sierra Grande uplift in the west and are removed by slight erosion on the northeast flank. The exact nature and thickness of these deposits southeast in the Dalhart basin are not known. An anomalously thin area exists in south-central Baca County, Colorado. Whether this is due to tectonic activity or to a facies change is not evident from the isopach map.

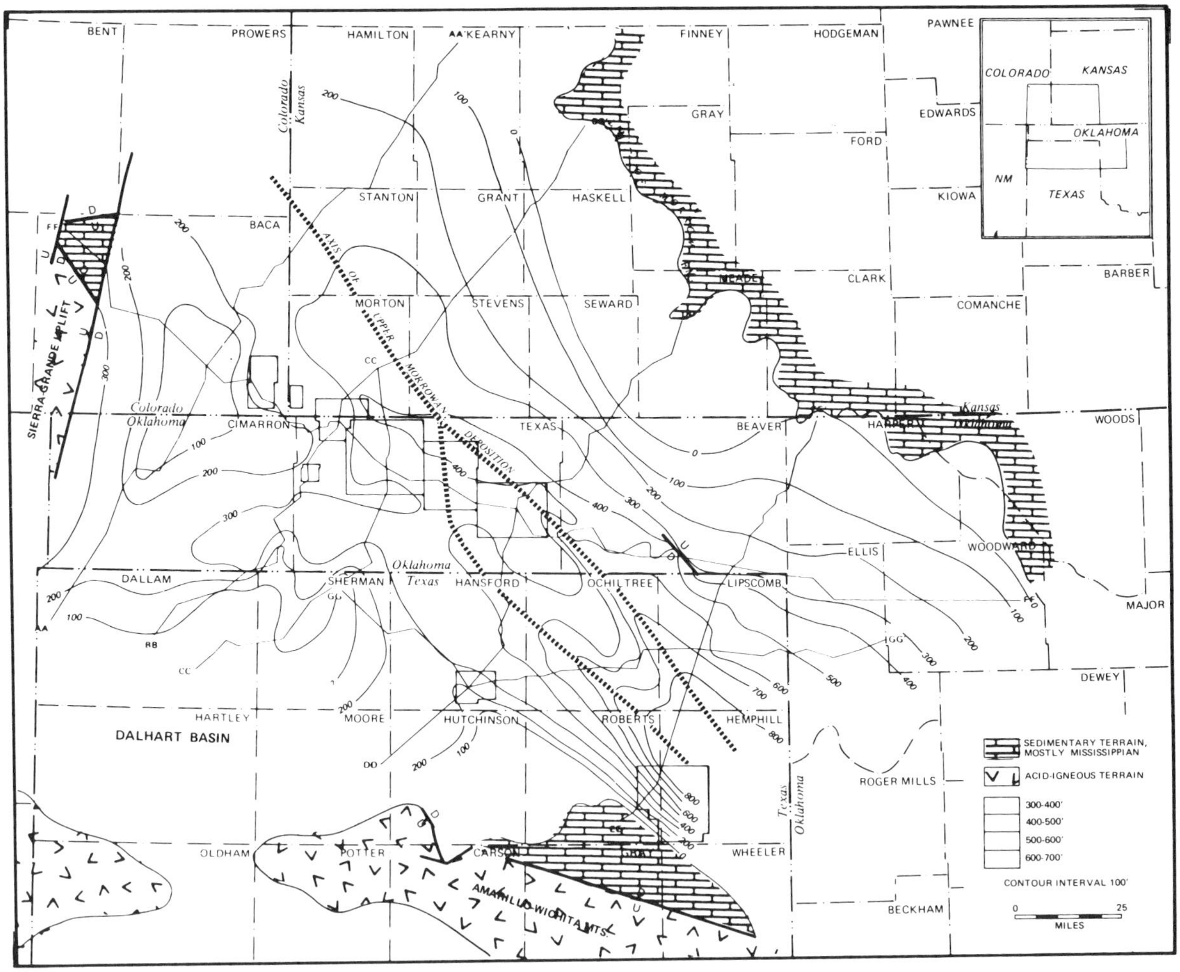

FIG. 41—Isopach of upper Morrow deltaic clastics.

The best map for isolating and showing key facies relationships in the Morrow Formation is the ratio map (Fig. 43). The following examples illustrate the differences in information provided by the two maps. First, the isopach map shows a thin area in Baca County, Colorado; the ratio map, however, indicates that this is an anomalous facies relationship. Second, in the eastern part of Texas County, Oklahoma, and the western part of Beaver County, Oklahoma, another anomalous area occurs where "upper Morrow" deltaic sediments persist throughout the whole Morrow interval (section DD-DD', Fig. 40). This facies condition does not manifest itself on the isopach map but stands out on the ratio map. Third, the ratio map isolates the main deltaic wedge from a minor basin next to the Sierra Grande uplift and from another upper Morrow deltaic wedge in the Dalhart basin. It limits the main deltaic wedge toward the southeast, whereas the isopach map shows progressive thickening in this direction. The high ratio numbers in the Amarillo-Wichita Mountains show only that the thin Morrow Formation present is part of the upper Morrow deltaic sediments. Overall, the ratio map better limits and delineates the actual deltaic wedge of clastics.

Sediment Source and Reservoir Trend Analysis — With the regional framework of the sedimentary wedge and its reservoir facies identified, the source and trend of the reservoir facies can be analyzed. The two concepts that govern this part of the analysis are that (1) the long axis of meander belts containing point-bar deposits and stream-mouth-bar-finger complexes trends in the direction of general delta development, and (2) if deltaic-plain or front deposits are found at location B and the source of the clastic material is known to be A, then the chance of point-bar and stream-mouth-bar deposits existing between A and B is great. In determining as conclusively as possible the source and trend of these reservoir bodies, two

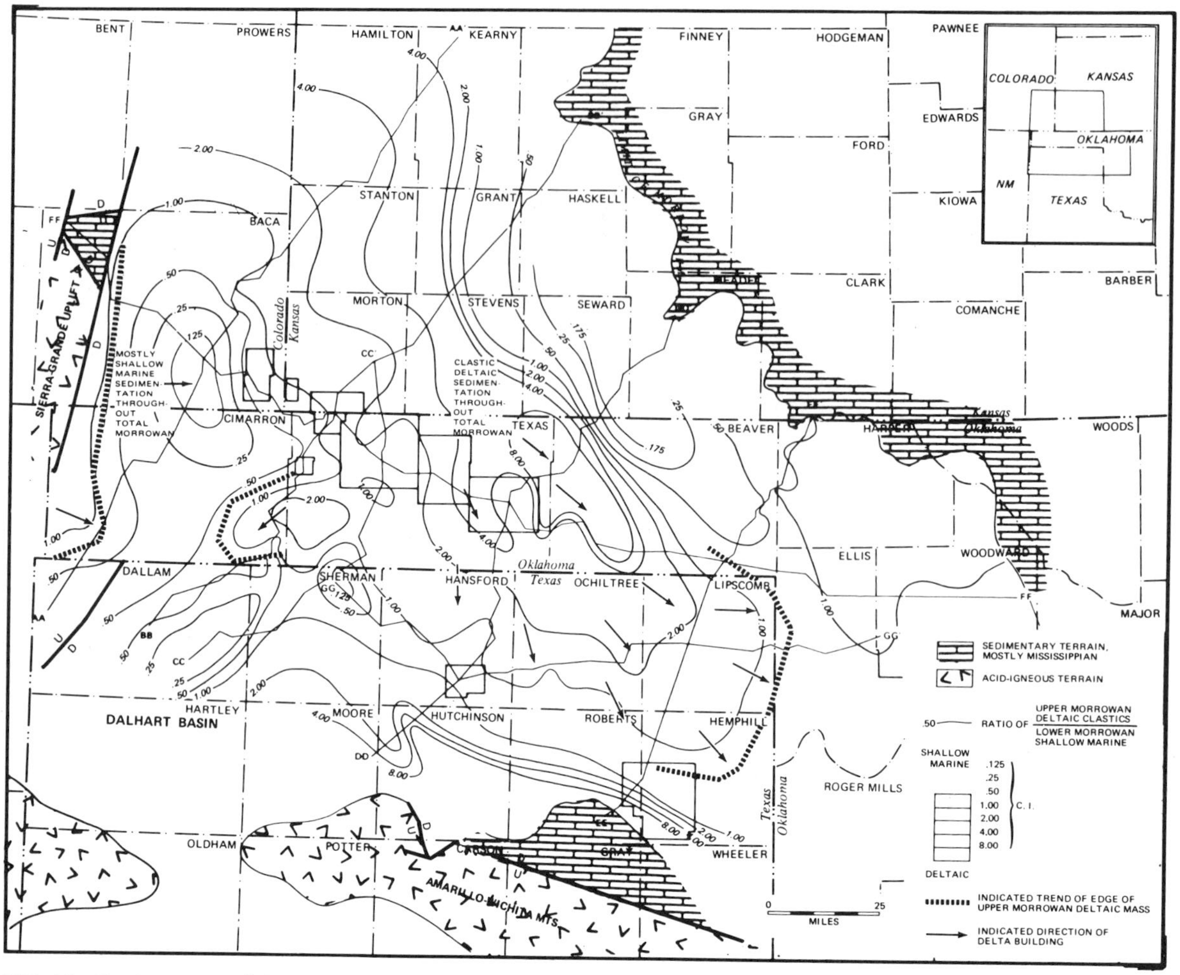

FIG. 42—Contour map of the ratio of upper Morrow deltaic clastics to lower Morrow nearshore marine deposits in the Anadarko basin.

factors must be considered in some detail: petrography and physical trend of the reservoir bodies.

To help determine the source of the upper Morrow reservoir deposits, petrographic thin sections from these deposits were examined. Stage B stream-mouth bars and stage C point bars were sampled in nine different areas throughout the basin. The main conclusions resulting from petrographic analysis were as follows:

(1) The source area contained much metamorphic material, characterized by metaquartzite.

(2) Metaquartzite grains were rare or absent in fine- to medium-grained sandstones but were common to abundant in coarse-grained sandstones, suggesting that the polycrystalline metaquartzite grains lost their identity through reduction in grain size.

(3) Metaquartzite fragments were the most common rock fragment and were the dominant type in all areas except the Mendota area, in some wells constituting more than 50% of the rock fragments.

(4) Quartz-muscovite schist fragments were noted in several samples.

(5) Euhedral inclusions of rutile, zircon, muscovite, apatite, tourmaline, sphene, and biotite were found in many quartz grains in all wells examined, suggesting a metamorphic source area.

(6) Sedimentary grains are the dominant rock fragment in stage C stream-mouth-

bar deposits of the Mendota area, including chert pebbles, shale fragments, quartz siltstones, and medium-grained quartz sandstones.

(7) Feldspar grains in the samples show a wide degree of weathering from almost fresh microclines in the northwest to badly altered clays elsewhere, the resulting kaolinite a product of an acid (probably fresh water) weathering environment.

The composition of the rock fragments in both the stream-mouth-bar and point-bar deposits indicates that the source terrain consisted of metasediments with a major contribution made by metaquartzite. Only material from the stage B bar-finger complex in the Mendota area departed from the general pattern. This particular deposit was derived from sedimentary rocks. The relative amounts of microcline, graphic granite, and granitoid fragments did not establish a clear-cut pattern, although they tended to increase slightly toward the west.

Figure 43 shows the position and graphically illustrates the petrography of some stage A and B stream-mouth-bar-finger deposits. An interpreted distributary-channel fill in the Spearman area is also concluded. Heavy dashed lines indicate the isopach trend of these deposits where control warrants the interpretation. Most of the deposits have a strong northwest-southeast trend and also have similar petrography, with the rock fragments containing much metamorphic material. The deposits in the Mendota area differ in trend and petrography. There the rock fragments contain a preponderance of sedimentary material, and the trend is northeast-southwest.

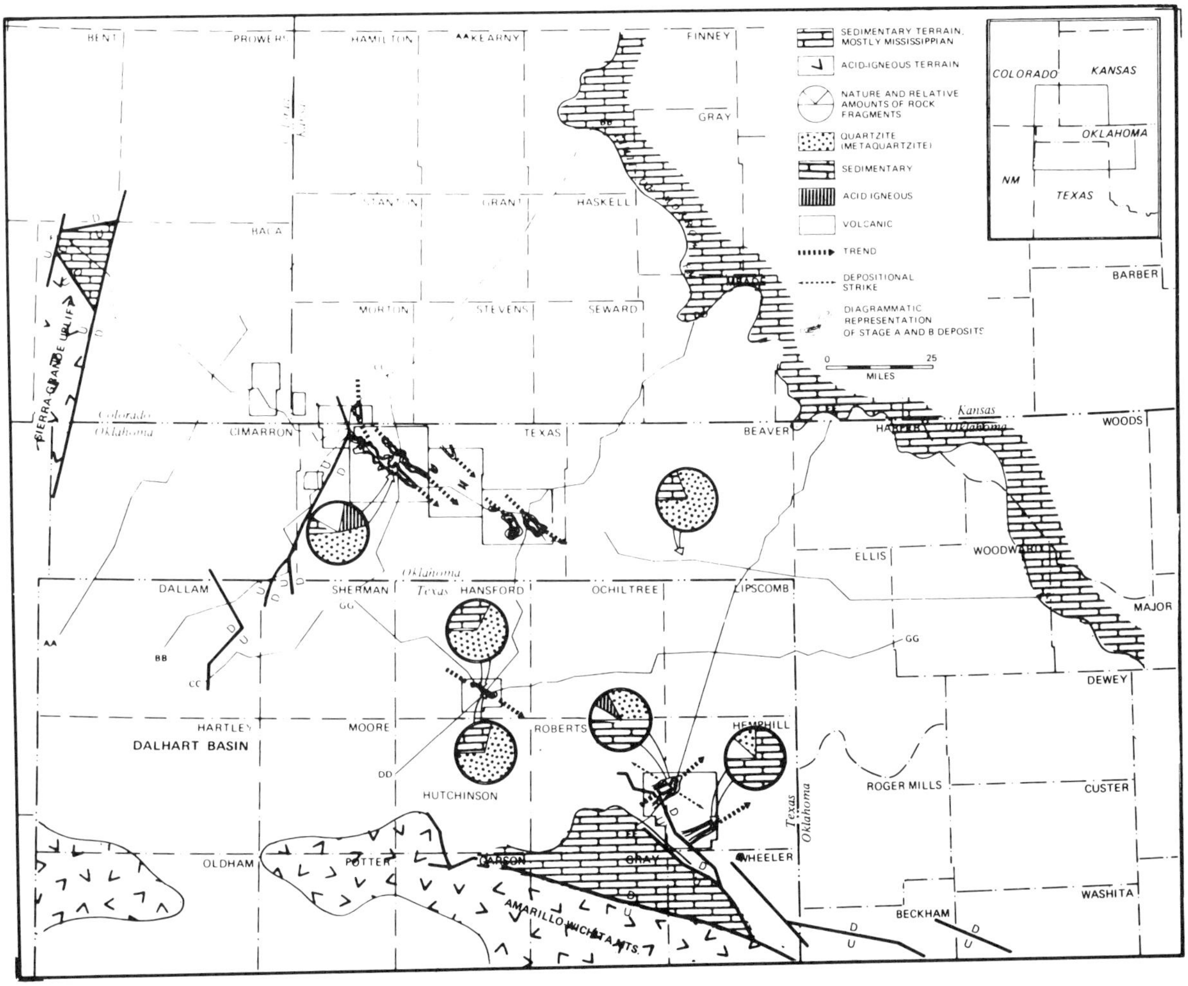

FIG. 43—Source (petrographic) and trend analysis of stage A and stage B upper Morrow stream-mouth deposits in the Anadarko basin.

Figure 44 shows the position and some petrography of the known stage C point-bar deposits. The rock fragments in most wells contain much metamorphic material, except in the Flank area, where igneous material was most abundant. This map likewise suggests a northwest-southeast trend of the meander belts and their porous reservoir sands.

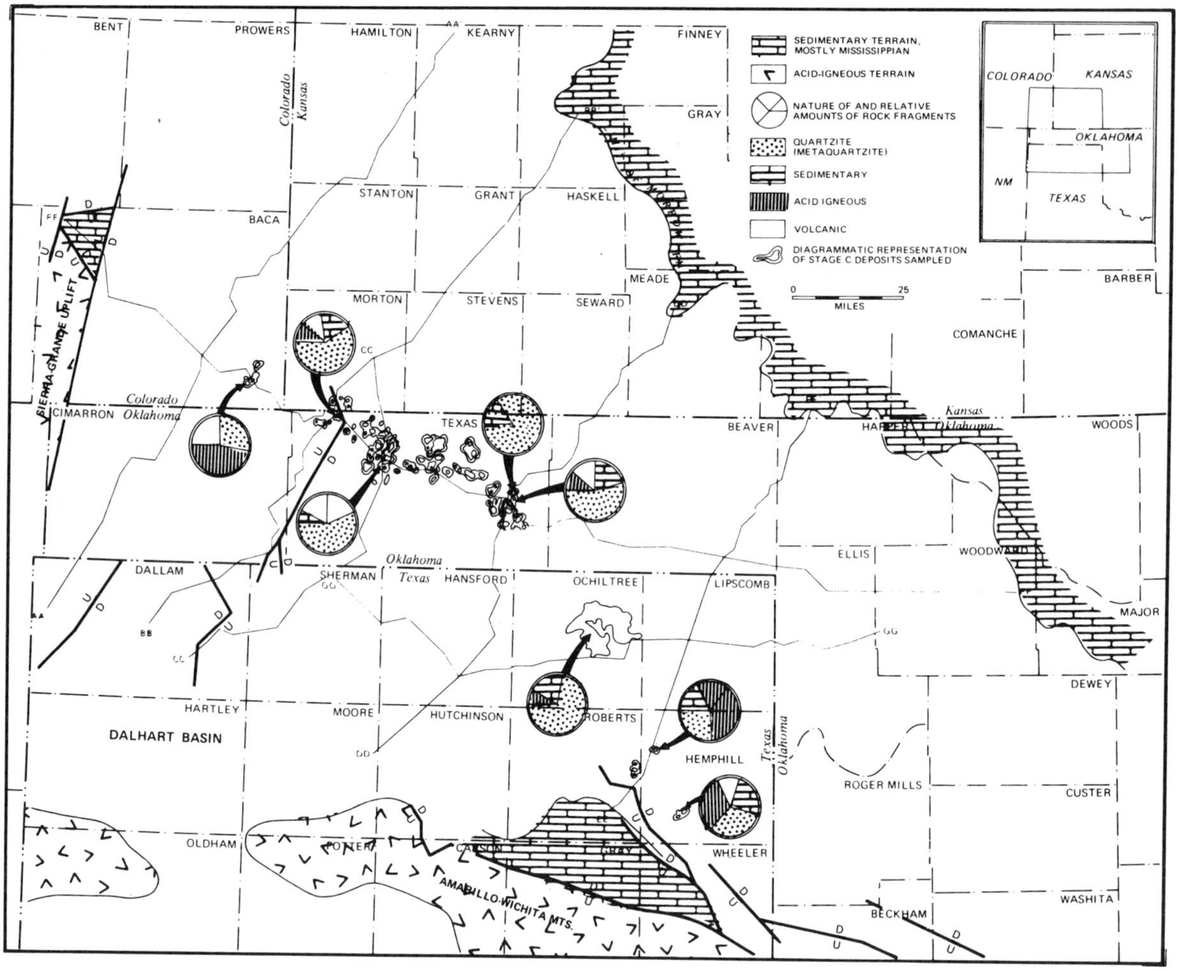

FIG. 44—Source (petrographic) and trend analysis of stage C upper Morrow point-bar deposits in the Anadarko basin.

Figure 45 is a compilation map showing the location of known upper Morrow coarse-grained deposits most of which are the result of stage C deposition. Although a more complex system is not evident, this map shows that the major trend is still northwest-southeast. An attempt can now be made to extend these meander belts into area where no environmental data are available and to use the concepts developed in areas of good control as a guide to further exploration and exploitation.

Paleogeographic Analysis. Since a metasediment terrain made significant contributions to the upper Morrow sediments, it is necessary to isolate areas where metasediments (metaquartzite) would be available during upper Morrow deposition. This can establish a source-depositional continuity which could be used to reconstruct the paleogeography. As stated before, if an area of deltaic sedimentation can be found, and its primary source area located, then there exists the potential of additional development of deltaic reservoirs between the two areas. This concept is based upon the supposition that the deltaic sediments of a particular basin are fed by one main stream system.

The concept of a clastic basin being filled by a multitude of small streams around its periphery is not compatible with present-day delta building. At all the Recent sites of deltaic sedimentation, the major amount of clastics is being introduced by one or, at most, two major stream systems. These stream systems usually drain a large area, particularly if they introduce large amounts of clastic material into the basin.

Denudation Rate and Sediment Volume. Some suppositions can be made concerning the approximate size of the drainage area which contributed sediment to the upper Morrow delta. These are based upon (1) the approximate quantity of deltaic material, (2) the probable size of the stream system (based on the size of the point-bar deposits, and (3) the known empirical relationships of stream characteristics.

Calculations based on regional cross-sections show that approximately 1,500 cu mi or 16 trillion tons of clastic material was deposited in the upper Morrow delta. Detailed analysis of the point-bar characteristics during stage C on the outer portion of the deltaic plain indicates that the stream system was slightly larger than the Brazos River near the Gulf of Mexico. Analogy with the Brazos and Colorado (Texas) Rivers and empirical relationship between thickness of point bars and stream width and depth show that the probable average annual discharge of the upper Morrow stage C stream was between 10,000 and 20,000 cu ft per sec and that the stream system added between 30,000,000 and 50,000,000 tons of sediment a year. The Colorado River of Texas has an average annual discharge of 4,374 cu ft per sec and has added material to its delta at the rate of 10,721,000 tons per year. The Brazos River, with an annual discharge of 8,099 cu ft per sec, is adding to its delta at the rate of 24,388,000 tons per year. The limiting rates of 30,000,000 and 50,000,000 tons per year gave 527,000 years as the longest and 316,000 years as the shortest time spans, respectively, in which the 1,500 cu mi of upper Morrow deltaic sediments could have accumulated.

Recent data from 12 river systems show that the tons per year contributed by each square mile of drainage area ranges from 105 to 808. The Colorado River (Texas) drainage area contributes 369 tons of sediment per square mile per year, whereas the Brazos River has 694 tons per year added by each square mile of its drainage area. The highest tons-per-square-mile-per-year figure (800) combined with the maximum time duration (527,000 years) gives a potential drainage area of 38,000 sq mi for the upper Morrow stream system. The lowest tons-per-square-mile-per-year factor (100) and the least number of years (316,000) give a probable drainage area of 480,000 sq mi for the same system. Average values indicate that the drainage area (source area) of the upper Morrow stream system was around 75,000 sq mi.

Empirical data on Recent meandering and braided streams show that meandering streams with discharges similar to that of the upper Morrow stream have slopes of 2.5 ft per mi or less and a probable total length of between 500 and 700 mi.

Identification of Source Area. Three main clues to the source area of the Morrowan deltaic deposits were as follows: (1) most of the deposits were probably sourced in an area having a significant amount of available metamorphic rocks (particularly metaquartzite), (2) the probable size of the drainage area was around 75,000 sq mi, and (3) the deposits themselves indicated a strong northwest-southeast trend. The only exception to these general conditions was in the Mendota area, where petrology and trend of bar fingers indicated a local source.

Figure 46 is a map of the late Morrowan paleogeography showing the four potential erosional source areas and the type of rocks available in them that could have contributed material to the upper Morrow delta. These are (A) the central Kansas uplift area, (B) the Amarillo-Wichita Mountains, (C) the Sierra Grande uplift, and (D) the positive area between the front range area and the Sioux uplift. The latter is called Frontrangia-Siouxia. The approximate area in square miles is also shown for areas (B), (C), and (D). All the potential source areas probably contributed some clastic material to the upper Morrow. This is particularly true in the early stages of development. Relatively short streams built "bar-finger fanlike" deposits around the periphery of the depositional basin. This condition is illustrated by the stage B stream-mouth-bar fingers in the Mendota area (Figs. 43, 45) which built out from the Amarillo-Wichita Mountains with a local source in the sedimentary terrain of the nearby mountains. This source is reflected in the petrography of the deposits. The stage C point bars in the same area

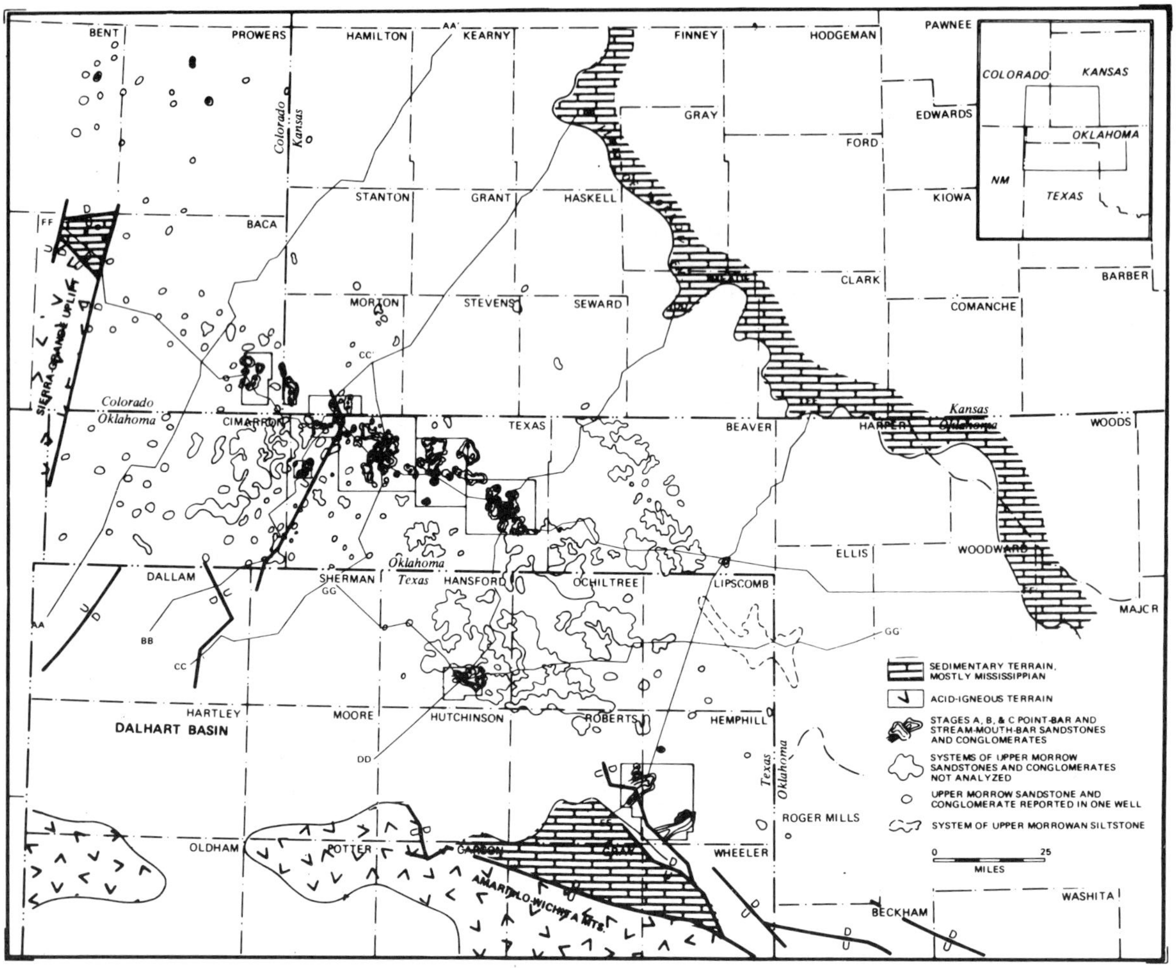

FIG. 45—Map showing location of upper Morrow deltaic reservoir deposits in the Anadarko basin.

have a more normal metamorphic-igneous source and may indicate the end of local "fan" development as positive areas become flatter, and the basin fills up. Except for relatively minor local sedimentation, however, the major upper Morrow deltaic sedimentation was probably the result of one major stream system. Its drainage area (source) was either in area (A), (B), (C), or (D) (Fig. 46).

Area (A) was large enough but is essentially a sedimentary terrain with minor metamorphic zones. Because of this, it is eliminated as the major source area. Area (B) has a variety of rock types available; however, metamorphic types are a minor part. Furthermore, it has an area of only 5,000 sq mi, too small to have sourced the upper Morrow deposits. Hence, this area can be eliminated. Area (C) is essentially an acid-igneous terrain with some sedimentary rocks and minor metamorphics. It is interesting to note that the known petrography of nearby deposits in southeast Colorado does not show any evidence of the sedimentary material. This area probably contributed material to the Dalhart basin and to local areas along its edge, but it is eliminated as the major source area because its drainage area of 18,500 sq mi is also too small to have sourced the major portion of the upper Morrow.

Area (D), on the other hand, consists of igneous rocks with a relatively large amount of metamorphic types present. Its drainage area is 80,000 sq mi, large enough to have easily provided the necessary material. The known trend of critical upper Morrow deltaic deposits is toward this area. Because of these reasons, area (D) is considered to be the source of upper Morrow deltaic deposits which built toward the southeast

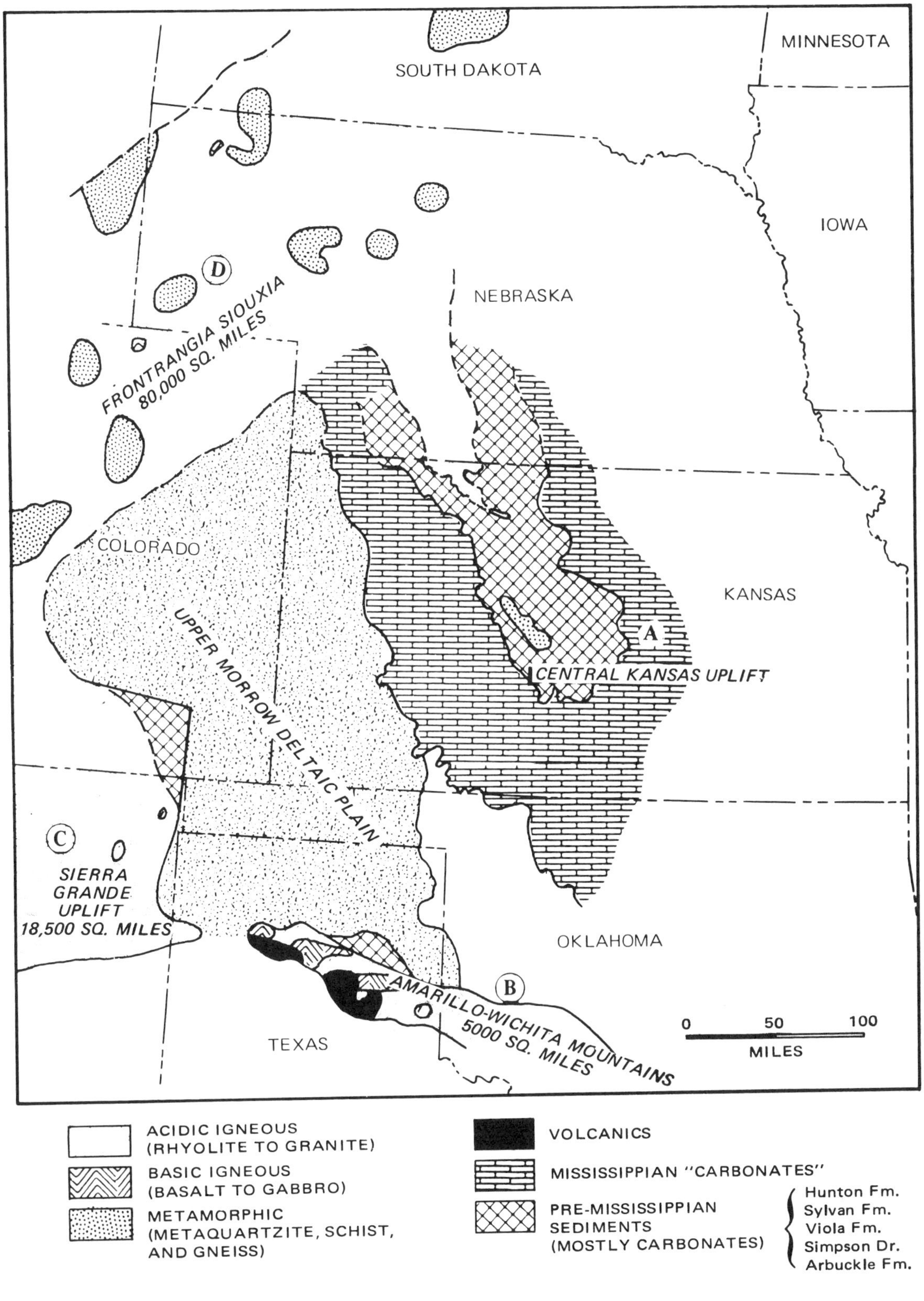

FIG. 46—Map showing type and location of source material available and approximate size of source areas during upper Morrowan stage C delta building.

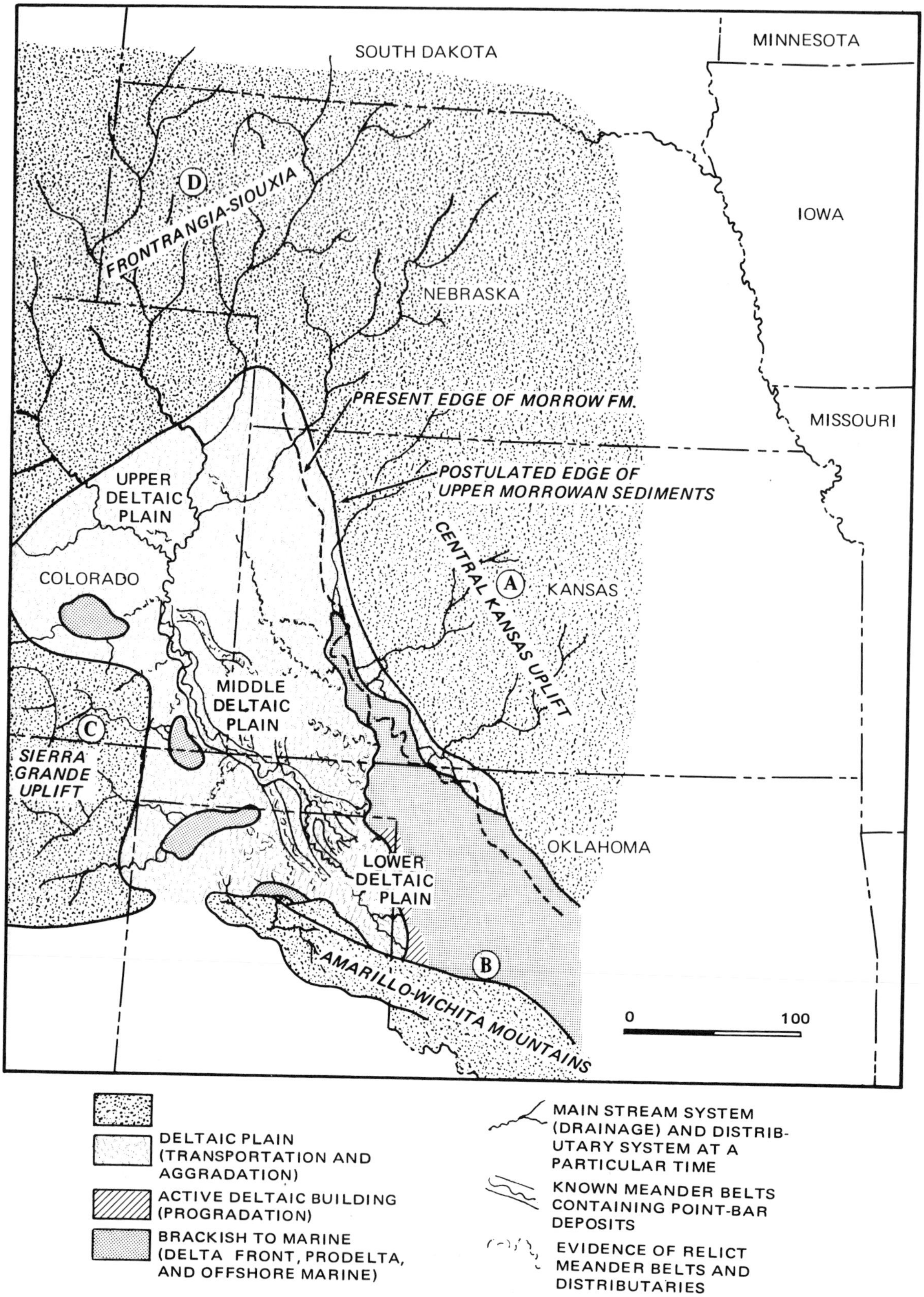

FIG. 47—Paleogeographic map during stage C, upper Morrowan time.

Upper Morrow Paleogeography. With this information, it is now possible to reconstruct the paleogeography of late Morrowan time. Figure 47 is a paleogeographic reconstruction of the upper Morrow stage C deltaic system. Known meander-belt areas are outlined, and hypothetical streams on the deltaic plain and in the source area are shown in order to illustrate the conditions affecting deposition. Minor stream systems developed in areas (A), (B), and (C). The main drainage area of Frontrangia-Siouxia (D) has a hypothetical network of streams which entered and crossed the upper deltaic plain, where they coalesced into a major stream. This stream picked up minor tributaries and slightly increased in size as it crossed the middle and lower plain. At the edge, it either entered the sea as a single large distributary or broke up into several smaller meandering distributaries. The change in elevation between the upper deltaic plain and the edge was probably not large, with most of the middle and lower deltaic plain not higher than 50 to 60 ft above sea level.

Subsidence was least in the upper deltaic plain and greatest in the lower deltaic plain. Active erosion occurred in the positive source areas. The upper deltaic plain acted more as a zone of transportation. As the streams meandered back and forth across this area, the sediments were continually reworked, and subsidence did not aid in burial. There, probably most earlier deltaic-front deposits were destroyed, and only the very last point-bar deposits remained. In the middle and lower deltaic plain, subsidence was great enough and accumulation of soft prodelta clay was sufficiently thick that some delta-front deposits sank and were preserved. In fact, there was vertical "stacking" of point-bar deposits in this area owing to subsidence. The streams on the deltaic plain often shifted course, which caused a change in the site of active delta building. Abandoned stream courses and deltaic lobes are shown in Figure 47 to illustrate this phenomenon.

The depth of water and fetch across the late Morrowan sea was not great, which resulted in marine processes of small magnitude. As the local sites of deposition shifted, the abandoned deltaic lobes were not significantly altered by these marine processes but tended merely to subside as part of the whole delta mass and were covered by deltaic-front fine-grained deposits (bay claystones, etc.).

Exploration Significance of Regional Geology — Reconstruction of the paleogeography of a source-stream-delta system should aid in exploration. Not only can the gross trend of reservoir bodies be established, but arguments can be advanced as to the existence of reservoirs in unexplored areas. This is particularly true of the upper Morrow, where the existence of deltaic sediments has been documented in the Oklahoma and Texas Panhandles. When an alluvial-deltaic system is recognized, extension of the system toward the source area should be an exploration objective. It can be argued that the major source of the deltaic clastics is the Frontrangia-Siouxia area of northeastern Colorado and western Nebraska. It should follow that there are deltaic-plain reservoirs (point-bar deposits) and possibly delta-front reservoirs (stream-mouth-bar deposits) in east-central and southeastern Colorado as well as in western Kansas. Although this area is relatively sparsely drilled to the Pennsylvanian, enough wells have penetrated upper Morrow sandstone bodies to indicate that this is true. Stage C point-bar deposits have been identified in cores from these areas.

The location of meander belts of the stage C stream courses would be of particular exploration interest. Although upper and middle deltaic-plain reservoirs would be smaller than those in the lower deltaic plain, they should be sufficiently well developed to be attractive exploration objectives, particularly if their meander belts could be determined. Because of less subsidence in these areas, there would be less vertical separation of reservoirs, which would lead to more continuity of water in reservoirs. There should be enough horizontal separation for many stratigraphic traps to occur.

Nothing has been said about the sediments lying above the stage C deposits. In most places, they represent delta-front and offshore-marine claystones which cover the deltaic deposits. This happened as the large supply of clastics in the source area stopped. During this period, the site of active delta building rapidly shifted northwest and toward the source area, and most of the stage C deltaic plain was covered by shallow-marine waters and younger delta lobes. This phenomenon would result in the last and youngest upper Morrow deltaic deposits occurring in southeastern and east-central Colorado. Reservoir beds of these deposits should also be expected there.

Basinward extension of known alluvial-deltaic deposits should also be considered an exploration objective. Deltaic-plain and delta-front deposits in the central Texas Panhandle and in the Oklahoma Panhandle may extend into western Oklahoma. Determination of the approximate basinward limit of deltaic-plain and delta-front environments would indicate prospective areas where associated point-bar and stream-mouth-bar deposits would occur.

When deltaic reservoirs are definitely identified, the exploration geologist should utilize all available knowledge to reconstruct the source-stream-delta system, because such a reconstruction may open up new avenues of exploration activity. This should be true whether there be 3 or 30,000 data points. Furthermore, the production geologist should endeavor to identify the facies types of any producing reservoirs so that his efforts at efficient production will be realized.

The American Association of Petroleum Geologists Bulletin
V. 57, No. 1 (January 1973), P. 162-189, 26 Figs., 2 Tables

Permeability-Porosity Patterns and Variations in Some Holocene Sand Bodies[1]

WAYNE A. PRYOR[2]
Cincinnati, Ohio 45221

Abstract Sandstone reservoirs are the results of long and commonly complex histories of geologic evolution. The combined processes of deposition, burial, compaction, diagenesis, and structural deformation yield final reservoir bodies of widely varied geometries, permeability-porosity characteristics, and structural configurations that are difficult to predict. In unraveling the evolution of sandstone reservoirs it is necessary to have detailed knowledge of their initial depositional characteristics and of the postdepositional modifications impressed upon them. This knowledge can provide a rational basis in predicting the characteristics of reservoir bodies away from areas of data control. Little information pertaining to the reservoir characteristics of freshly deposited sand bodies has been available. In an API sponsored study, permeability, porosity, and textural parameters were derived from 992 oriented and undisturbed sand samples of river bars, beaches, and dunes undergoing active sedimentation.

River point-bar samples have permeabilities ranging from 4 md to more than 500 darcys and average 93 darcys. Porosities in the river point-bars range from 17 to 52 percent and average 41 percent. Beach sand samples have a permeability range of 3.6 to 166 darcys and average 68 darcys. Porosities in beach sands range from 39 to 56 percent and average 49 percent. Permeability values in dune sands range from 5 to 104 darcys and average 54 darcys. Dune sand porosities range from 42 to 55 percent and average 49 percent. Permeability values in river-bar sands are extremely varied in comparison to those of beaches and dunes. In river bars, permeability decreases systematically downstream and bankward. Although of low variability, permeabilities on beaches are low on the beach faces, high on the beach crests, and variable on the beach-berm areas. Both river-bar and beach sands have well organized directional permeabilities, parallel with the length of the bodies in river bars and perpendicular to the length of the bodies in beaches. Dunes are characterized by low variability in permeability and porosity and show no significant patterns or trends.

There is greater variability within bedding and lamination packets than between them. In addition the boundary conditions between bedding and lamination packets are important factors in determining the effective reservoir characteristics of sand bodies, to the extent that a bedding unit of higher permeability completely surrounded by units of lower permeabilities will not demonstrate its ultimate through-flow capabilities, but will have an *effective permeability* influenced by and largely determined by the lower permeabilities of the bounding units. River-bar sand bodies have significantly different arrangement and variability between bedding units from those of beaches or dunes.

The ideal relations between permeability-porosity and textural parameters that have been set forth by various authors for artificially packed particles are demonstrated only slightly by these natural sands from different depositional environments. In all three depositional environments permeability increases with increase in grain size and porosity increases with increase in grain sorting. However, in river-bar sands permeability increases as grain sorting increases and porosity increases as grain size increases, just the opposite of the relations in beach-dune sands and in the artificially packed grain experiments. The underlying cause of these deviations is the different style of grain packing in the river-bar sands.

Introduction

Permeability and porosity are important characteristics of sand-reservoir bodies; their magnitudes, patterns, and variabilities significantly influence the migration, accumulation, and distribution of fluids and gases in the reservoirs, and just as significantly determine the abilities of reservoirs to release their fluids and gases to production stimulation.

The heterogeneous character of sand reservoirs is well known and much effort in exploration and production investigations is directed toward predicting the degree and patterns of the heterogeneities away from areas of data control. Predictions of reservoir character commonly are founded on studies of other similar reservoir bodies or on idealized models. However, sand reservoir bodies are the results of long and commonly complex histories of geologic evolution, and the combined processes of deposition, burial, compaction, diagenesis, and structural deformation yield final products of widely varied geometric, permeability, porosity, and structural configurations that are difficult to predict.

[1] Manuscript received, January 30, 1972; revised and accepted, June 7, 1972.

[2] Professor of Geology, University of Cincinnati.

This work was supported and encouraged by the American Petroleum Institute under API Project 91-B (Pryor, 1968). The Advisory Committee on Reservoir Heterogeneities was composed of P. H. Monagham, Esso Production Corp.; P. H. Braun, Atlantic Richfield Co.; C. H. Hewitt, Marathon Oil Co.; and F. J. Lucia, Shell Development Co. Their counsel and support were invaluable and are greatly appreciated. P. E. Potter, University of Cincinnati, reviewed this manuscript and gave many helpful comments. John Wiethe and Phil Ziegler, formerly students at the University of Cincinnati Geology Department, contributed many hours of assistance in the field and laboratory, as well as pertinent discussion of the results.

Many investigations and descriptions of buried, ancient sand reservoirs present permeability-porosity data and their relations to other important sediment properties; several notable examples are: Hutchinson *et al.* (1961), Horn (1964), Hewitt and Morgan (1965), Polasek and Hutchinson (1967), and Dodge *et al.* (1971). In addition, there are many descriptions and interpretations of permeability-porosity and their relations to textural and fabric properties of ideal particle masses and artificial sands (Graton and Fraser, 1935; Fraser, 1935; Krumbein and Monk, 1942; Wyllie and Spangler, 1952). However, these parameters and their relations to other sediment properties are not well known or documented for natural, recently deposited sands. The few notable studies of permeability and porosity relations in recent sediments (Fraser, 1935; Hamilton and Menard, 1956; Füchtbauer and Reineck, 1963) present significant, but widely scattered data from relatively few recent sediment samples and include no comprehensive, systematic, or detailed permeability and porosity investigations of sand bodies from specific depositional environments. Hence, it is obvious that knowledge of the initial depositional characteristics and the modifications impressed upon them by postdepositional processes is necessary to unravel the evolution of buried sand reservoirs and to enhance predictions of their heterogeneities.

In 1965, the API Advisory Committee on Reservoir Heterogeneity reaffirmed the need for this type of data and prompted the writer to undertake an investigation of the permeability-porosity heterogeneities of some recent sand bodies. The purpose of the investigation was to determine the patterns of permeability and porosity variations for recent sand bodies of several different origins, and to relate these heterogeneities to principal sedimentary parameters and to the formative conditions of the depositional environments. The purpose of this paper is to present the primary data derived from the investigation and to report in particular the patterns and variations of permeability-porosity in several recent sand bodies and their relations to textural variations.

Location

Alluvial, littoral, and eolian environments are considered to be important sites for the deposition of sand bodies that are representative of buried sand reservoirs. Hence, 3 river bars, 3 beaches, and 2 dune areas were selected for study. The river bars are on the Whitewater River, 15 mi west of Cincinnati, Ohio; on the Wabash River, 2 mi northeast of Grayville, Illinois; and in an abandoned course of the Mississippi River at the U.S. Corps of Engineers Lock Site near Jonesville, Louisiana. The beaches are on the east Texas Gulf Coast, 2 mi west of Sabine Pass, Texas; on the western end of Ship Island, 8 mi offshore from Gulfport, Mississippi; and on Santa Rosa Island, adjacent to the studied beach; and in St. Andrews State Park, near Panama City, Florida.

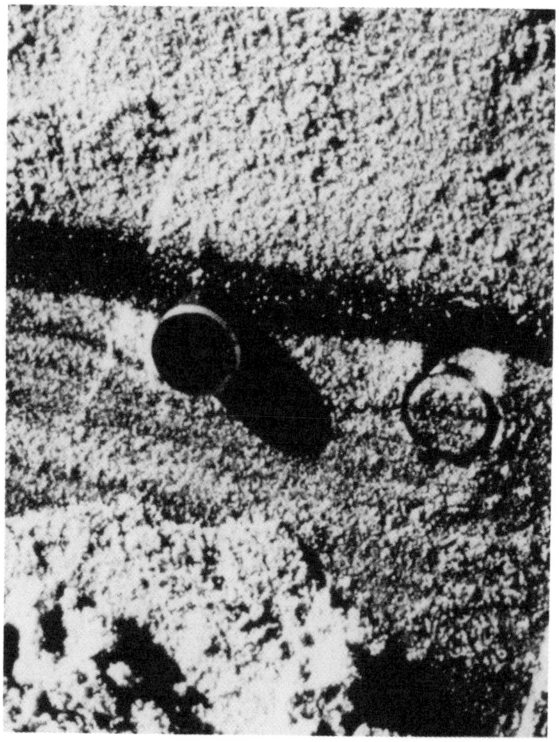

Fig. 1—Thin-wall sample tubes inserted into sand face of Ship Island beach trench.

Methods

Undisturbed and oriented samples were collected from these sand bodies for determination of permeability, porosity, and textural parameters. A nested samples pattern (Figs. 2, 3, 9, 13) was designed to acquire data at three levels of density within the deposits: (1) samples with at least a 10-ft spacing in a series of profile lines at least 100 ft apart and perpendicular to the length of the sand body; (2) a grid pattern of 25 samples on 5-ft centers; and (3) a closely spaced pattern within and among bedding units in trench faces, with sample spacings of less than 1 ft.

The samples were collected in thin-wall aluminum tubes (2.5 cm O.D. × 5.2 cm) by carefully pushing the tubes into the unconsolidated sediments (Fig. 1). In the profile line and grid patterns, samples were taken at a regular depth of 6 in. below the depositional surface to avoid the

FIG. 2—Aerial photograph of Wabash River bar, with locations of sample profiles, grid, and trench.

effects of eolian reworking. Attempts were made to collect only undisturbed samples. Because permeability in sands is related to the directional fabric of bedding and grain orientation (Mast and Potter, 1963), the samples were oriented parallel with the direction of bedding and lamination inclination to obtain maximum permeabilities for the given sample site. Bedding thickness and directional property orientation for each of the samples were recorded. An attempt was made to collect these sediment samples as soon as possible after they were deposited; in the case of river bars after flood periods, and in the case of beaches during the ebb phases of spring tides. The difficulties of collecting large numbers of undisturbed sand samples underwater were insurmountable and early attempts at this type of sampling were abandoned.

The permeabilities of the undisturbed samples were measured in a commercially available Constant Head-Falling Head Permeameter adapted to the dimensions of the thin-wall aluminum sample tubes. Water permeabilities, in Darcy units (darcys and millidarcys), were measured at relatively low head pressures (50 cm of water) to prevent disturbance of the unconsolidated sand samples by elutriation and repacking. The samples were uniformly treated by a standard procedure to minimize operational variations.

Porosity was measured by using a modification of Ludwick's Volumeter (Ludwick, 1956) and basing the calculations on the core volumes and grain volumes of the samples.

Textural parameters of the samples were determined by normal sieve analysis and Emory settling tube analyses and the principal parameters of mean diameter and standard deviation (Folk and Ward, 1957) were derived from the data.

Patterns of Permeability and Porosity

A total of 992 samples was collected from the three principal depositional environments: 480 from three river bars, 348 from three beaches, and 164 from three dune areas.

River point-bars—The point-bar studied on the Wabash River is shown (Fig. 2) at normal stage, several days after the peak stage of an August 1967 summer flood. The subaerial part of the bar is approximately 3,400 ft long, 500 ft wide, and at this stage level the river is an average of 500 ft wide and is 12 ft deep in the thalweg. The sediment distribution is characteristic for point-bars, with the material becoming finer and better sorted toward the downstream end and bankward. The mean grain size of the sampled sands on this bar is 0.500 mm (1.00 phi), the mean sizes range from about 1.0 mm to less than

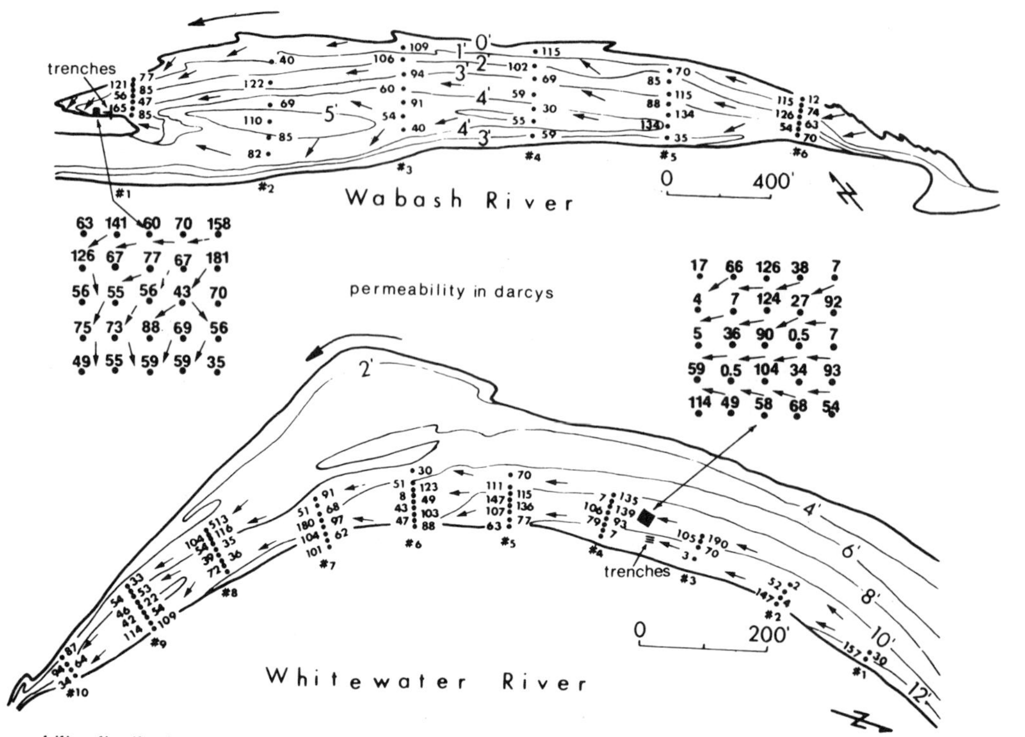

FIG. 3—Permeability distribution for Whitewater River bar and Wabash River bar. Arrows indicate directions of cross-stratification and maximum permeability directions.

0.30 mm. Size sorting (phi standard deviation) ranges from a well-sorted 0.32δ to a poorly sorted 1.60δ. The surface of the bar displays several well-developed, downstream curving sand-scale fronts with smaller ripple, megaripple, and sand-wave fields swarming on their stoss sides. These bed-form features increase in length and crest height toward the thalweg, as is normal on most point-bars. The point-bar on the Wabash River is represented by 167 samples: six profiles with 42 samples, one 20-ft square grid of 25 samples, and two intersecting-trench sections with 100 samples through a large sand-scale front.

The Whitewater River point-bar is approximately 3,000 ft long and 300 ft wide (Figs. 3, 4). The river is normally 300 ft wide and 4 ft deep in the thalweg. After summer floods only the lower 1,400 ft is composed of sand; the upper part is gravel and sandy gravel. The sand varies widely in grain size, from very fine sands to coarse gravelly granules, and the average grain size is about 0.420 mm (1.25 phi), however, mean sizes range from more than 1.0 mm to less than 0.25 mm. Size sorting is variable, ranging from a moderately sorted 0.55δto a poorly sorted 1.60δ. The sand is in the form of various types of ripples, sand waves, and one or more sand-scale fronts, which increase in size toward the thalweg. A total of 151 samples was collected on the Whitewater River bar: 10 profiles with 63 samples, one 20-ft square grid of 25 samples, and three trench sections through a sand wave with 63 samples.

On the Wabash River point-bar, permeabilities range from 4 md to 195.31 darcys, with a mean of 77.86 darcys (Fig. 3). Porosities range from 40 to 50 percent, with a mean of 44.95 percent. Overall values of variance for permeabilities and porosity are 1,544.49 and 2.76 respectively (Table 1). On the Whitewater River point-bar, the permeability values range from 160 md to 513.35 darcys, with a mean of 94.74 darcys. Porosities range from 17 to 52 percent, with a mean of 36.92 percent (Fig. 4). The extremely low porosities result from inclusion in the samples of some large pebbles. The permeabilities (Table 1) have an overall variance of 4,283.70 and porosity has an S^2 value of 58.98. Permeability values on both river bars (Fig. 3) decrease slightly in a downstream direction and increase toward the thalweg, and they are also more variable at the upstream positions. These trends are weak, but do exist. The porosity values (Fig. 4), however, show no significant trends. As most of the cross-stratification is inclined in a downstream direction, the trends of directional permeability are also in the downstream direction. The permeability-porosity data for the 5-ft grid samples (Figs. 3, 4) show relatively unorga-

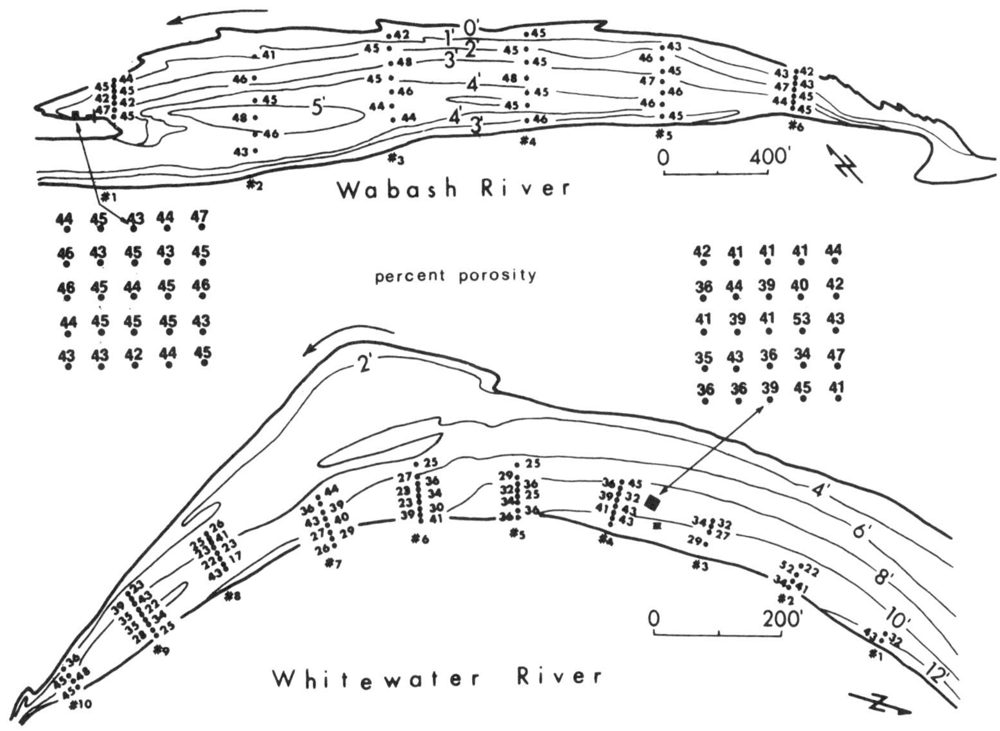

FIG. 4—Porosity distributions for Whitewater River bar and Wabash River bar.

nized patterns. The grid areas have widely varied distributions of permeability. The Wabash River bar grid sample has a mean permeability of 77.40 darcys with a variance of 1,257.41 (Table 1). The Whitewater River bar grid sample has a mean permeability of 50.55 darcys with a variance of 1,753.09. This high variability is the result of sampling different parts of ripples and sand waves, which within themselves have wide ranges of permeabilities and porosities. These variabilities and patterns of permeability-porosity values within bedforms are observed in the trench section samples (Figs. 5, 6) of the two major, different types of forms.

The trench sections *A* and *B* in Figure 5 have been cut through a "sand-scale front" on the downstream end of the Wabash River point-bar. This type of bedform feature is developed by sand avalanching over the crest of a sand ridge, in a conelike fashion, and coming to repose without eroding the toe area in advance of the feature. This results in planar cross-stratification with upward or convex laminations displayed in the transverse section (Fig. 5). This is a relatively important type of bedform for permeability character, in that the base of this depositional unit is not only nonerosional, but it is the depositional site of a nearly continuous carpet of silt and clay, which forms a significant permeability boundary. A change in permeability across this boundary can be seen in the bottommost vertical samples of trenches *A* and *B* (Fig. 5). The permeabilities are as low as 4 md, or more than 4 orders of magnitude lower than the permeabilities of the overlying planar cross-stratified unit. The planar-convex, cross-stratified unit is in turn erosionally overlain by a tangential-concave trough cross-stratified zone that results from the migration of sand, in the form of sandwaves, up the stoss side of the sand-scales. The permeability values in the planar cross-stratified unit are uniformly higher than those of the lower boundary or the overlying trough cross-stratified unit. The laminations in the planar unit usually are graded, and coarser and better sorted in the lower part. The upper unit usually is finer grained and less well sorted and also infiltrated by clay and silt materials, which effectively reduce permeabilities. The infiltration of the fines occurs when silt- and clay-laden water drains downward through the sediments, during the last stages of floodwater cover.

The trench sections *I-1,I-2,* and *I-3* (Fig. 6) have been cut longitudinally through a sand wave on the Whitewater River bar. The trenches were positioned in the midportion of this barchanlike bedform and are spaced 1 ft apart. Unlike the

Table 1. Permeability and Porosity, Means, Standard Deviations, and Variances

Sample Location	Permeability				Porosity		
	No.	Mean (Darcys)	Standard Deviation	s^2 Variance	Mean (%)	Standard Deviation	s^2 Variance
Wabash River Bar							
Total	167	77.86	39.30	1,544.49	44.95	1.66	2.76
Profile	42	79.21	30.36	921.73	44.85	1.49	2.22
Grid	25	77.40	35.46	1,257.41	44.41	1.17	1.37
Trench	100	78.88	42.94	1,843.84	45.22	1.81	3.27
Whitewater River Bar							
Total	151	94.74	65.45	4,283.70	36.92	7.68	58.98
Profile	63	81.95	69.98	4,897.20	33.76	7.78	60.53
Grid	25	50.55	41.87	1,753.09	40.78	4.02	16.16
Trench	63	124.95	54.01	2,917.08	38.57	7.41	54.91
Mississippi River Bar							
Total	30	6.39	5.76	33.17	44.98	2.37	5.62
Bedding Units	15	9.77	5.15	26.52	45.62	1.05	1.10
Boundary Units	15	3.02	4.13	17.05	44.35	3.05	9.30
Ship Island Beach							
Total	195	60.82	18.20	331.24	45.94	2.56	6.55
Profile	92	62.82	19.58	383.38	46.11	2.64	6.97
Grid	25	80.29	12.73	162.05	47.14	2.91	8.47
Trench	78	52.21	11.30	127.69	45.41	2.16	4.67
Santa Rosa Island Beach							
Total	245	75.34	19.73	389.27	49.61	2.87	8.23
Profile	90	76.62	22.96	527.16	48.24	2.31	5.62
Grid I	25	73.48	16.63	276.45	51.29	3.42	11.69
Grid II	25	72.42	15.23	231.95	48.26	2.11	4.45
Trench	105	75.61	18.27	333.79	50.75	2.57	6.61
Sabine Pass Beach							
Total	40	6.88	1.88	3.53	45.08	1.80	3.24
Grid	20	5.27	0.96	0.92	44.57	1.92	3.68
Trench	20	8.49	0.91	0.82	45.59	1.47	2.16
Santa Rosa Island Dune							
Trench	48	71.59	13.78	189.88	47.90	2.08	4.32
St. Andrew Park Dunes							
Total	116	36.17	10.55	111.20	50.82	2.85	8.12
Profile	61	35.89	11.16	124.54	49.07	2.50	6.25
Trench	55	36.58	9.89	97.81	52.83	1.59	2.53

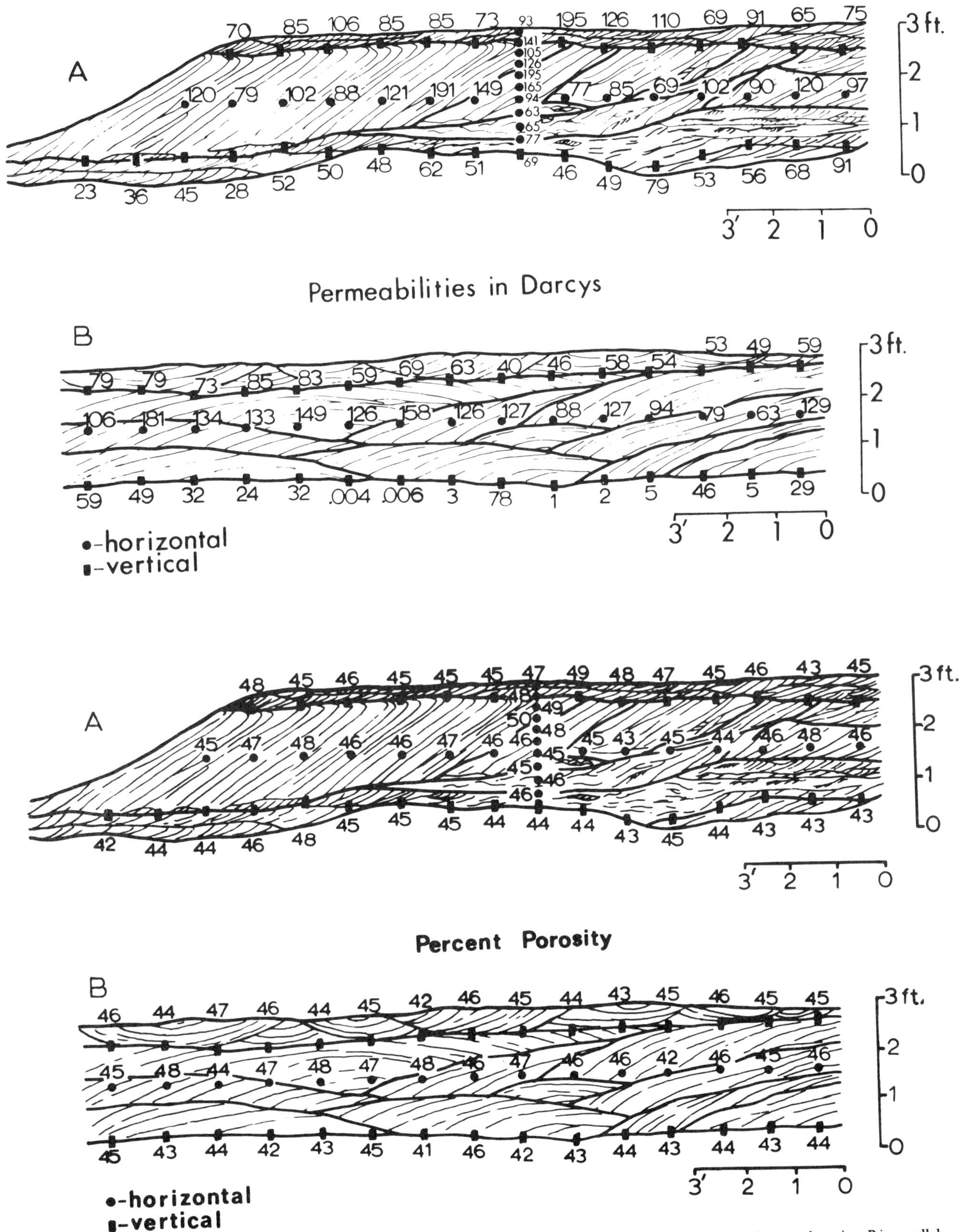

FIG. 5—Trench sections through sand-scale front on Wabash River bar. Section *A* is perpendicular to front and section *B* is parallel with front.

sand-scale front (Fig. 5), this feature has an erosional base and tangential-trough cross-stratification. The downstream end is at the right of the diagram and the permeability values decrease upward and in a downstream direction. In section *I-2,* the permeability values range from 190 darcys upstream to 200 md downstream, a change of nearly three orders of magnitude. The lower permeability values in the downstream position of this feature are caused by the infiltration of fine-grained materials. This downstream position is the location of the sand-wave scour pit, which becomes a small, muddy-water filled pond during the last stages of floodwater cover. When this clay-silt laden water evaporates and/or drains into the sand, the fine materials infiltrate or form a fine-grained surface layer, thus effectively reducing permeability.

In both planar and tangential cross-laminated units there is a downward increase in permeability within each of the inclined laminations, which is related to the downward increase in grain size in the laminations. There are wide variations in texture from one lamination to the next and hence wide permeability variations. However, the vertical and lateral variations within individual laminae are as great as those between them. As permeability is directional in cross-stratified sand (Mast and Potter, 1963), with maximum permeabilities parallel with the inclined laminations, consider here (Figs. 5, 6) the tortuosity patterns of lateral fluid transmission from one lamination to the other through these types of units in a subsurface reservoir.

The standard deviations and variances (S^2) presented in Table 1 show relatively high degrees of permeability variability, at all levels of sampling for both the Whitewater River and the Wabash River point-bars. The Whitewater samples, however, have S^2 values for permeability that are significantly higher than the Wabash samples. Porosity variability is also greater for the Whitewater River bar and is relatively low for the Wabash River bar. These differences are the result of variability in textural characteristics in the Whitewater River bar (Fig. 19).

Mississippi River bar samples were collected from an 1,100-year old point-bar sequence in an abandoned course of the river at Jonesville, Louisiana (Fig. 7). Thirty samples were collected in a vertical array from the upper 20 ft of a point-bar sequence that is at least 50 ft thick. Beginning at the bottom of this exposure is a tangential-trough cross-stratified unit 3 ft thick, followed by seven similar units, diminishing upward in thickness to less than 1 ft. The upper 6 ft of the sequence is represented by thin, tangential cross-stratified, ripple-stratified, and ripple-drift stratified units. The mean grain size of these sands is 0.125 mm (3.00 phi), but ranges from 0.16 mm to less than 0.88 mm. Size sorting (phi standard deviation) ranges from a well-sorted 0.31 δ to a moderately sorted 0.87δ (Fig. 7). The sands are friable, but show some degree of cementation and cohesiveness.

Permeability values range from 259 md to 18.79 darcys and average 5.96 darcys. S^2 values of 33.17 are given for total permeabilities in Table 1. Porosities range from 40 to 52 percent, average 44.98 percent, and have S^2 values of 5.62. Samples were taken parallel with the inclination of the cross-stratification in the central part of each bedding unit and vertically across the boundaries of each bedding unit (Fig. 7). This sampling pattern was designed to test the effects of boundary conditions on permeability. The permeabilities of the central parts of the sedimentary units average 9.77 darcys and range from 2.81 to 18.65 darcys, whereas samples through the boundaries have an average permeability of only 3.02 darcys and range from 259 md to 16.91 darcys. Most boundaries are erosional and the permeability reductions are caused by the differences in grain size between adjacent units and the break in continuity of grain arrangement between adjacent units. In the ripple and ripple-drift stratification zone, thin laminae of clay and silt commonly form the boundaries between units.

Comparison of absolute permeability values between the Mississippi River bar sequence and the recent point-bars of the Wabash and Whitewater Rivers is not practical, because of the diagenetic changes of compaction and cementation that the Mississippi River bar sediments already have undergone during the last 1,100 years. However, the general aspects of bedding-unit boundary conditions are comparable, as are the trends of permeability increase toward the thalweg positions.

Beaches—The beach on the western end of Ship Island (Fig. 8), is approximately 150 ft wide during low tide with a relief of about 3.5 ft. The beach is composed of a narrow, low-tide beach face, crest, and berm, and the dominant high-tide beach face, beach crest, broad landward-sloping beach berm, and back-berm trough. Low-lying, stabilized dunes, and a storm beach lie behind the beach. Symmetrical beach cusps are well developed and are spaced at approximately 35-ft intervals along both low- and high-tide faces. Sand laminations are well marked by concentrations of dark heavy minerals (Fig. 1). The bedding is inclined gulfward at low angles of less than 10° in the beach-face deposits and slope at

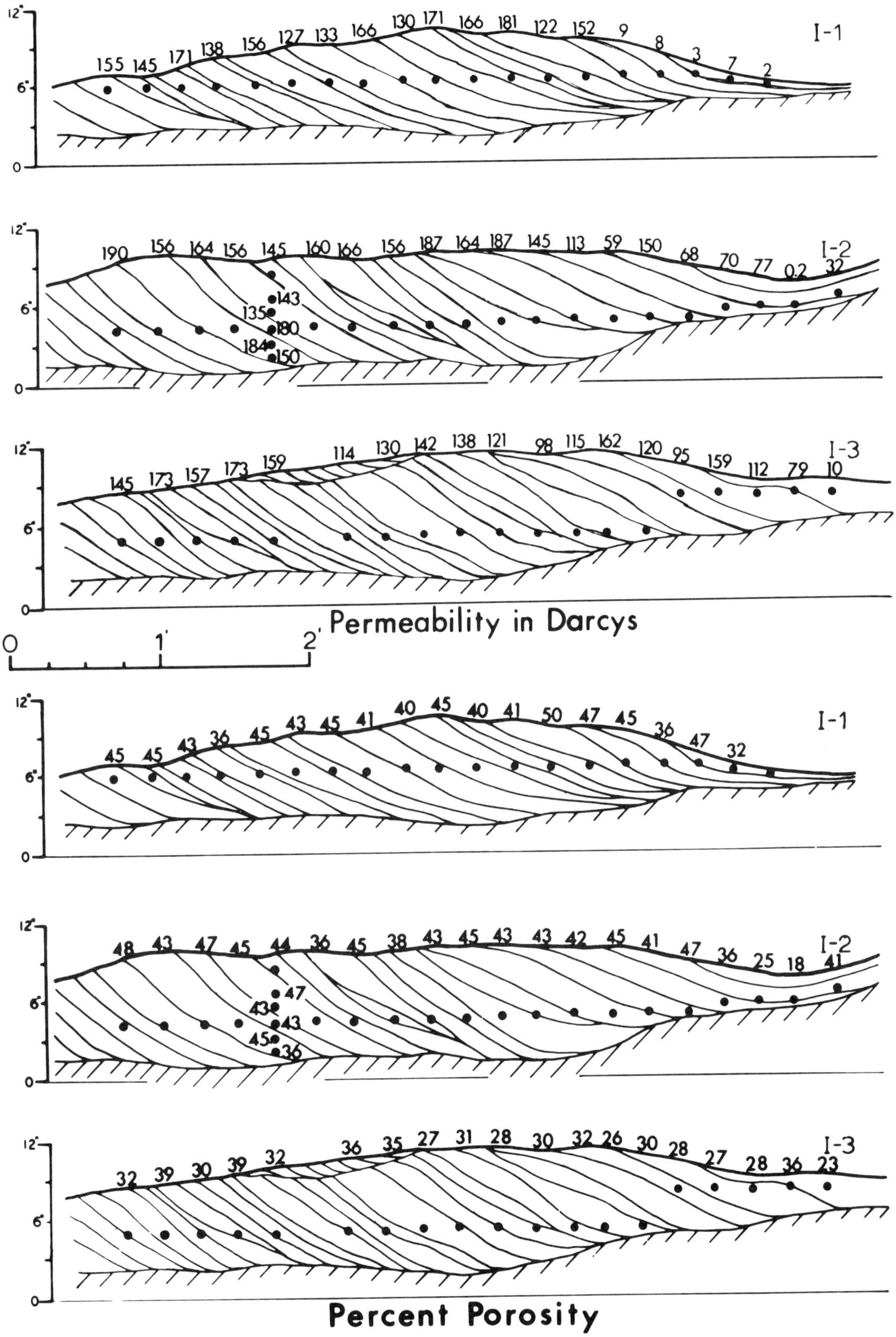

FIG. 6—Trench sections and sand wave on Whitewater River bar, with distribution of permeability and porosity data.

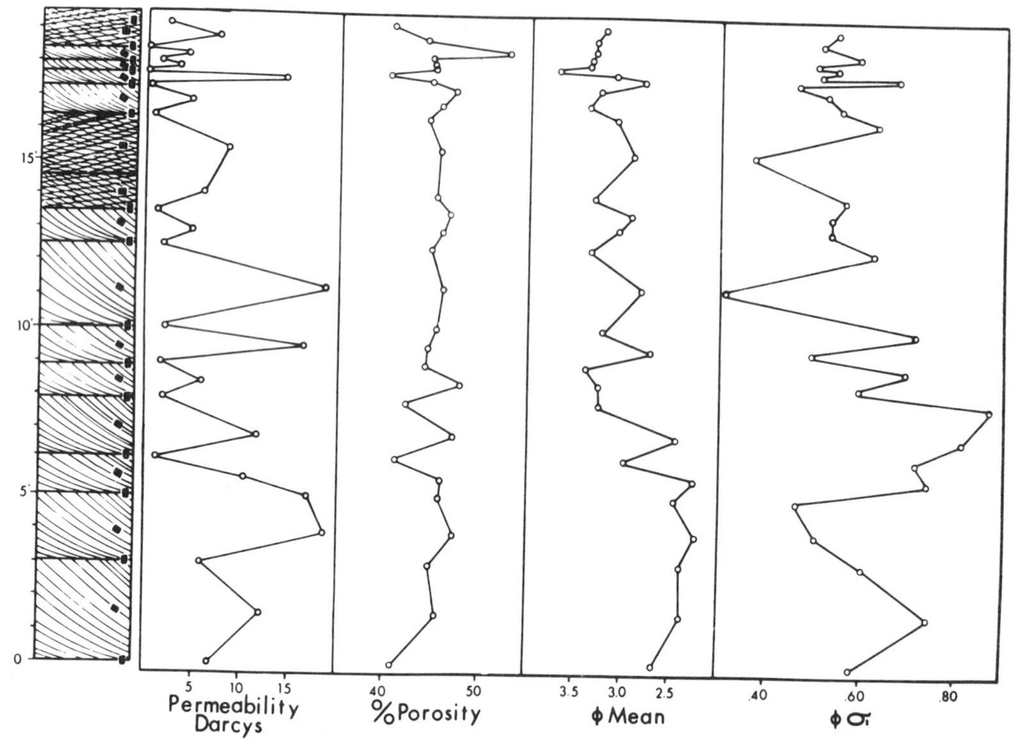

FIG. 7—Vertical profile of permeability-porosity, textural, and bedding data from abandoned Mississippi River bar deposits.

FIG. 8—Aerial photograph of Ship Island beach sample area.

even lower angles inland within the berm deposits. Small-scale, trough cross-stratification in the back-berm trough dip in directions parallel with the beach crest. A characteristic of these beach sands is the widespread extent of relatively thin laminations (see Figs. 11, 12, where individual laminations extend more than 25 ft along the base of a cusp unit). These laminations are formed in the swash zone of the cusp areas, and the less persistent laminations in the berm zone are formed by swash washover. On Ship Island beach, during late July 1967, the average grain size of sediments was 0.25 mm (2.00 phi), but ranged from 0.60 mm to less than 0.22 mm. The sands are well sorted and phi standard deviations only range from 0.20 to 0.58δ(Fig. 22). The beach on Ship Island is represented by 195 samples: 10 profiles with 92 samples, one 20-ft square grid of 25 samples, and three intersecting trench sections through the beach face and crest with 78 samples.

The sampled beach on Santa Rosa Island, Florida, is approximately 125 ft wide and has a relief of 6 ft. In late July and early August of 1967 there were three well-developed beaches—a low-tide beach, a high-tide beach, and a storm beach. The beach berms are narrow and slope shoreward at very low angles. Beach cusps are well developed on the low-tide and high-tide beach faces, with a spacing of approximately 15 ft. Dark heavy minerals are not abundant on this beach and the laminations are indistinct and difficult to observe. The steep beach faces give rise to relatively high angles of inclination (up to 35°) for the internal laminations of the sediments. The narrow cusps also result in less widely spread single laminations. On Santa Rosa Island the sand has an average grain size of 0.30 mm (1.75 phi), and ranges from 0.50 mm to less than 0.22 mm. The sands generally are well sorted ranging from phi standard deviations of 0.18 to 0.68δ. One significant characteristic of the sediments on the Santa Rosa Island beach is the "fluffy," loose nature of the sands on the beach crests and beach berms. The sands in these positions are so "fluffy" that inserting the thin-wall sample tubes, without compacting the sediment, was difficult. The fluffiness extended at least to the bottom of the 3-ft deep trench sections, although it was somewhat less than at the surface. The development of fluffy sands on the beach crests and beach berms was observed on both the Santa Rosa Island and Ship Island beaches. It was caused by the foamy, sand-laden surf water swashing over the beach crests and extending onto the berm zones, where the water drained out of the foamy suspension and a loosely packed sand lamination was deposited. This is distinctly different from the traction-load sand that is deposited on the beach faces. Directions of inclined laminae (Fig. 9) are seaward on the beach faces and landward on the berm areas. A total of 245 samples was collected from the Santa Rosa Island beach: 10 profiles with 90 samples, two 20-ft square grids of 25 samples each, and two intersecting trench sections across the beach face, crest, and berm with 105 samples.

Permeabilities on the Ship Island beach range from 30.58 to 166.57 darcys, with a mean of 60.82 darcys, and porosities range from 39 to 55 percent, with a mean of 45.94 percent. In Table 1 the S^2 values for permeability and porosity are 331.24 and 6.55, respectively. On the Santa Rosa Island beach, permeabilities range from 34.74 to 121.59 darcys, with a mean of 75.34 darcys, and porosities range from 44 to 56 percent, with a mean of 49.61 percent. The S^2 values for permeability and porosity are 389.27 and 8.23, respectively. Data for the profile and grid samples are presented in Figures 9 and 10, and for the trench sections in Figures 11 and 12. On both beaches, data from the profile samples generally show highest permeabilities on the beach crests, lowest on the beach faces, and most variable on the beach berms. Porosities also show this trend, with unusually high porosity values, related to the "fluffy" sands in the beach berm areas. The permeability-porosity and textural properties are relatively uniform in pattern along the length of the beaches.

The Ship Island beach-grid samples were taken across the crest of the high-tide beach and Figures 9 and 10 show rather uniform distributions of permeability and porosity. The Ship Island grid sample has a mean permeability of 80.29 darcys and a mean porosity of 47.14 percent (Table 1), and the S^2 values for permeability and porosity are 162.05 and 8.47, respectively. On Santa Rosa Island beach, grid I is located on the berm area and grid II is on the low-tide beach face. Grid area I has slightly higher permeability and porosity values than grid II and also higher degrees of variance in these properties (Table 1).

Three trenches were dug and sampled across a beach-face cusp area on Ship Island. Two of the trenches, *A* and *B* (Figs. 11, 12), are perpendicular to the shoreline and are intersected by trench *C*, which is parallel with the shore. In these trenches, laminations formed in the cusp and berm areas can be differentiated by directions of dip—seaward in the case of the cusp and shoreward in the berm. Permeabilities and porosities are rather uniform in these samples (variance = 127.69 and 4.67), ranging only from 36 to 87 darcys. This is especially notable in trench *C*, where a single, widespread lamination was sampled over a distance of 20 ft and had permeabilities ranging from only 33 to 53 darcys. Two intersecting trenches were dug and sampled across the high-tide beach face, crest, and berm on Santa Rosa Island (Figs. 11, 12). Although the permeability and porosity values are rather uniform (variance = 333.79 and 6.61), there is an important change in both parameters with depth. The near-surface samples have average permeabilities and porosities of 89 darcys and 52 percent, respectively, and the lowermost samples, at a depth of 3 ft, average 59 darcys permeability and 46 percent porosity. This permeability-porosity reduction is related directly to the loss of the fluffy or loosely packed texture with depth of burial.

The Sabine Pass beach was sampled during an extremely low storm-spring tide in late December 1966. A total of 40 samples was taken from the Sabine Pass beach: one 20-ft square grid of 20 samples, and one trench section through a beach cusp horn with 20 samples. A grid-pattern sample was taken on the nearly flat, rippled, tidal flat of the beach and a trench was dug and sampled through a cusp horn of the high-tide beach face. The average grain size on this beach is 0.105 mm (3.25 phi) with abundant, fine shell material marking the laminations. Permeability values on

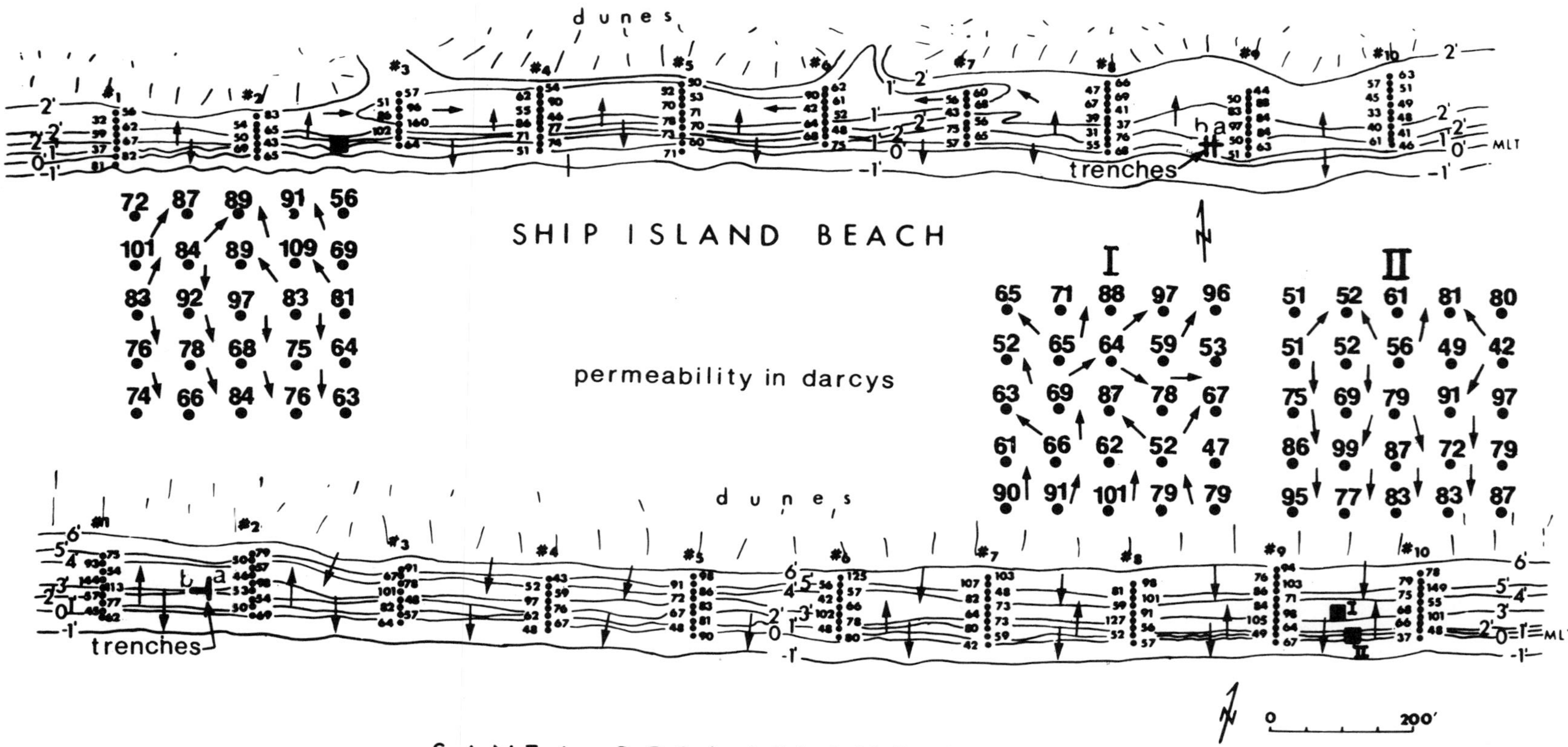

FIG. 9—Permeability distributions for Ship Island beach and Santa Rosa Island beach. Arrows indicate directions of cross-stratification and maximum permeability directions.

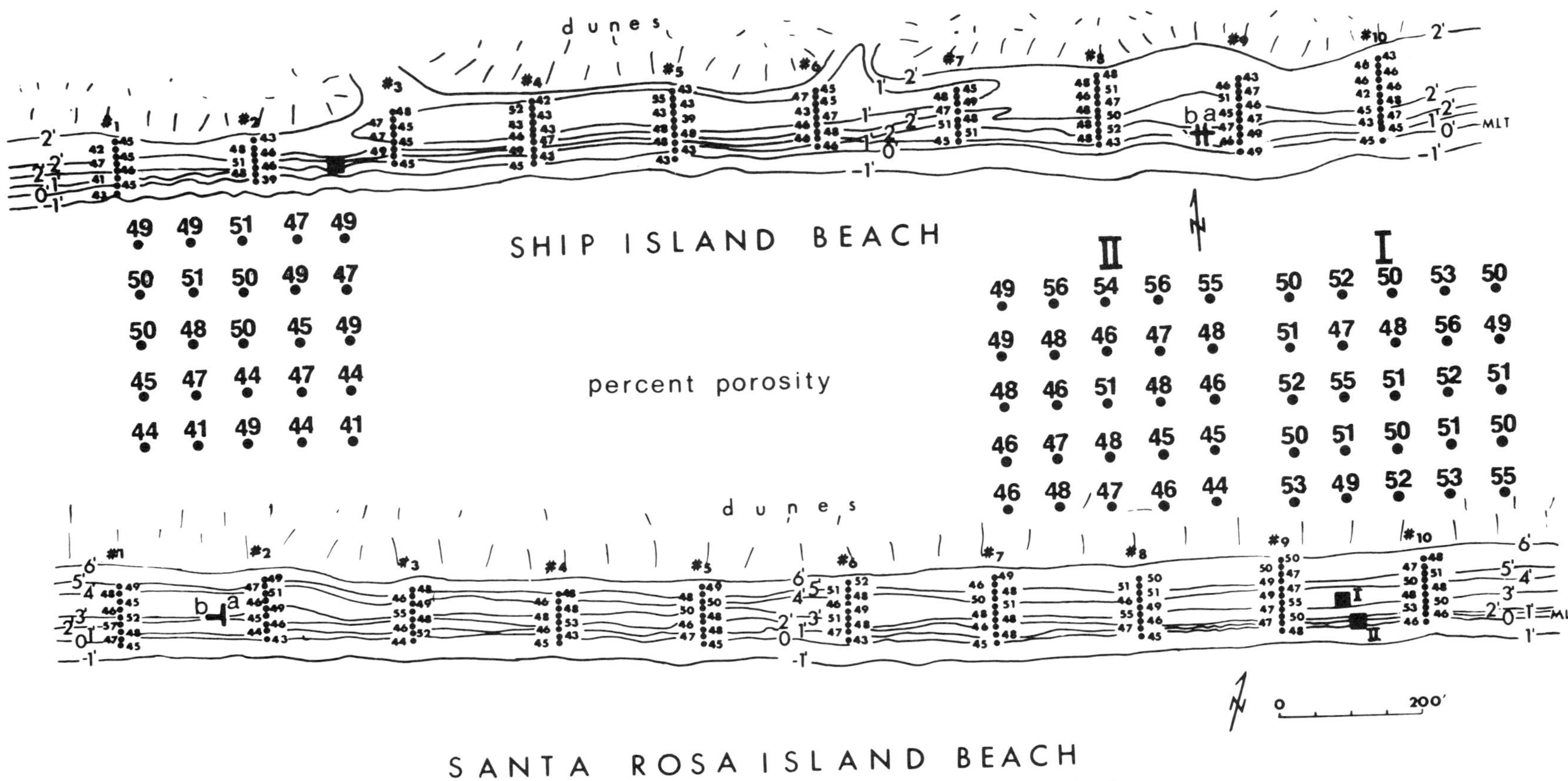

FIG. 10—Porosity distributions for Ship Island beach and Santa Rosa Island beach.

the Sabine Pass beach are comparatively low, ranging from 3.62 to 10.06 darcys, with a mean of 6.88 darcys. Porosities range from 40 to 50 percent, with a mean of 45.08 percent. In Table 1 the variability values of permeability and porosity are very low, 3.53 and 3.24, respectively. The grid samples on the tidal flat have lower permeability and porosity values than the trench samples. The mean values in the grid sample are 5.27 darcys and 44.51 percent, whereas in the trench the means are 8.49 darcys and 45.59 percent. The sample variability is very low, with an overall variance in permeability of 3.53 (Table 1). The trench-sample permeability values are slightly less variable than the grid samples.

Beaches are marked by a relatively low variability in permeability and porosity. The low degrees of variability that do exist are arranged perpendicular to the length of the beaches. Beach faces have lowest permeability and porosity values and are the least variable, beach crests have the highest permeabilities and porosities, and beach berms have the most variable permeabilities and porosities. Directional permeability is oriented seaward in the beach face zone and shoreward in the berm area just behind the beach crest. However, directional permeability is oriented parallel with the shoreline within the back-berm areas.

Dunes—On Santa Rosa Island, the selected dune was approximately 300 ft inland from the sampled beach. A vertical trench, 8 ft high and 5 ft wide, was excavated in the shoreward face of a low dune. A total of 48 samples was taken on 1-ft centers and 3-in. spacing along a vertical line (Figs. 13, 14). The average grain size of these dune sands is 0.340 mm (1.60 phi), and ranges only from 0.40 mm to less than 0.28 mm. Size sorting (phi standard deviation) also has a small range, from 0.30 to 0.48δ (well-sorted). Dips of the dune stratification range from horizontal to more than 25°. The permeability values of the Santa Rosa Island dune sample range from 46.41 to 104.82 darcys, with a mean of 71.59 darcys. Porosities range from 44 to 51 percent, with a mean of 47.90 percent. No significant distribution patterns are observed other than an extreme uniformity of values; variance is low for both permeabilities and porosities, 189.88 and 4.32, respectively.

In St. Andrews Park, dunes form a high cliff just behind the beach. A total of 116 samples was collected from these dunes. Sixty-one samples were taken in a series of 6 vertical profiles, spaced 400 ft apart, along the cliff face; the samples in each of the profiles are spaced a maximum of 2 ft apart. A trench 7 ft high and 7 ft wide was excavated on the dune face and 55 samples were taken on 1-ft centers, and on 3-in. spacing in one vertical line (Figs. 13, 14). The average grain size of these dune sands is 0.210 mm (2.25 phi), and has a narrow range of 0.27 to 0.14 mm. The sands are very well sorted with standard deviations ranging from 0.23 to 0.35δ. The permeability values for the St. Andrews dune samples range from 5.40 to 79.47 darcys, with a mean of 36.17 darcys. Porosities range from 42 to 55 percent, with a mean of 50.82 percent. Permeability and porosity values are quite uniform and show no significant patterns or trends. Permeability variances are relatively low (Table 1) with S^2 values of 111.30 for the total sample. The permeability variance for the trench sample is somewhat lower than that for the profiles.

Dunes are characterized by the low variability in permeability and porosity values.

Directional permeability—The relations among cross-stratification, grain-fabric orientation, and directional permeability have been demonstrated by Mast and Potter (1963). They found that permeability in bedded sandstones is greatest parallel with cross-stratification inclination and grain-fabric orientation. Hewitt and Morgan (1965) also determined that horizontal permeabilities in the Pennsylvanian Robinson sandstones in the Illinois basin were at a maximum parallel with the trough axis of the crossbedding, with minimum permeabilities of 85–95 percent of the maxima being observed at right angles to the crossbedding. Dodge *et al.* (1971, p. 1828), in a reservoir character study of some marine Cretaceous sandstones in Texas, concluded that permeabilities are higher parallel with the dip directions of crossbedding than normal to the dip directions.

Ten duplicate samples were collected perpendicular to bedding inclination from the sand bodies. The permeabilities of these samples were all within a range of 89–96 percent of the permeabilities of the adjacent parallel-to-crossbedding samples.

In the river point-bar samples, the directions of inclined cross-stratification crossbedding trough axis and ripple troughs are generally downstream. The directional variation is small on the upper and middle parts of the bars, but as seen by the directional arrows in Figure 3 they begin to swing toward the bank at the lower ends of the bars. Hence the directions of maximum permeability are generally parallel with the length of the river point-bar deposits.

The directional characteristics of beach deposits (Fig. 9) show that in the beach faces the laminae dip seaward, perpendicular to the length of

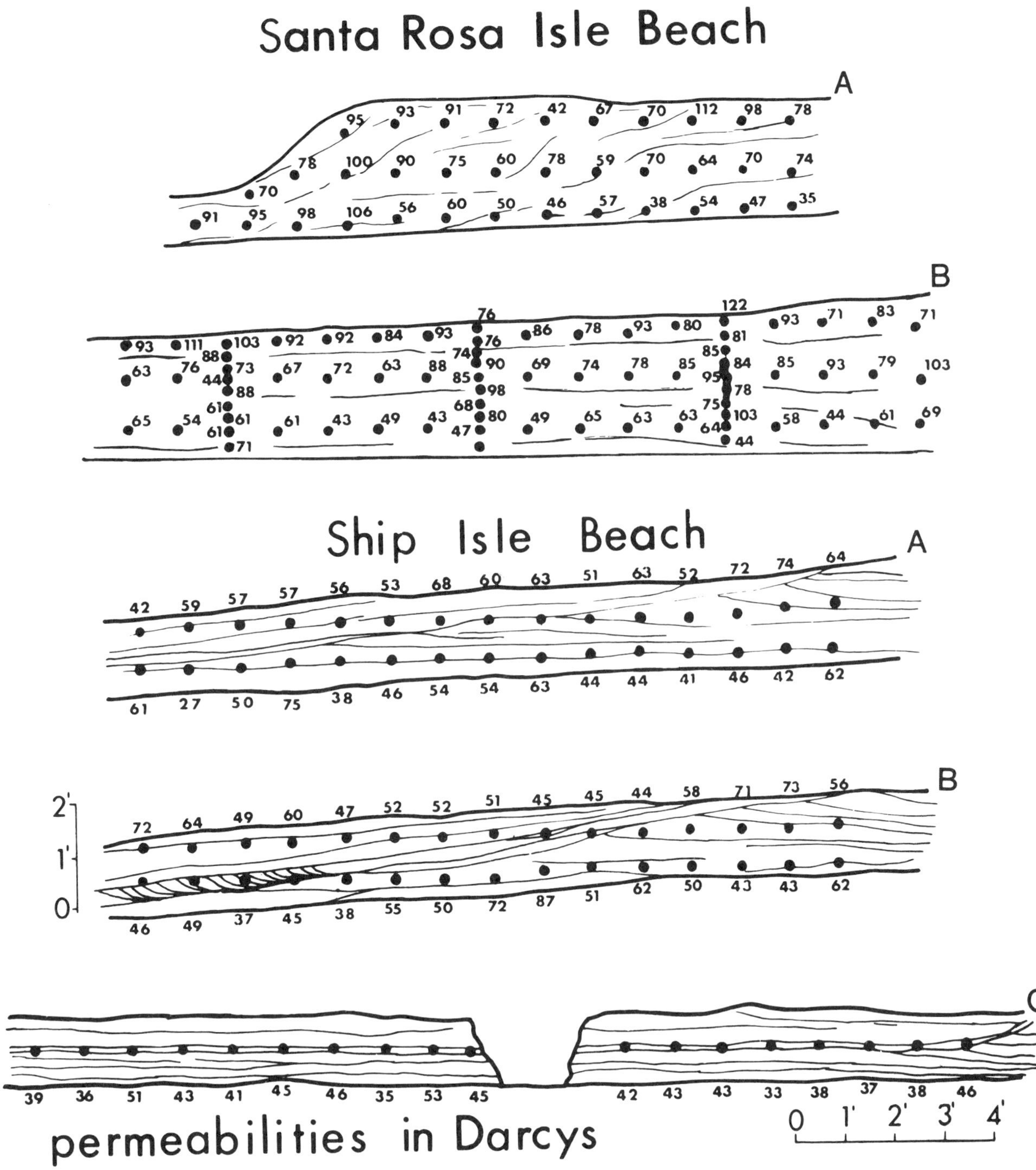

FIG. 11—Trench sections through beaches on Ship Island and Santa Rosa Island, giving distributions of permeability data. Locations of trenches are given in Figure 9.

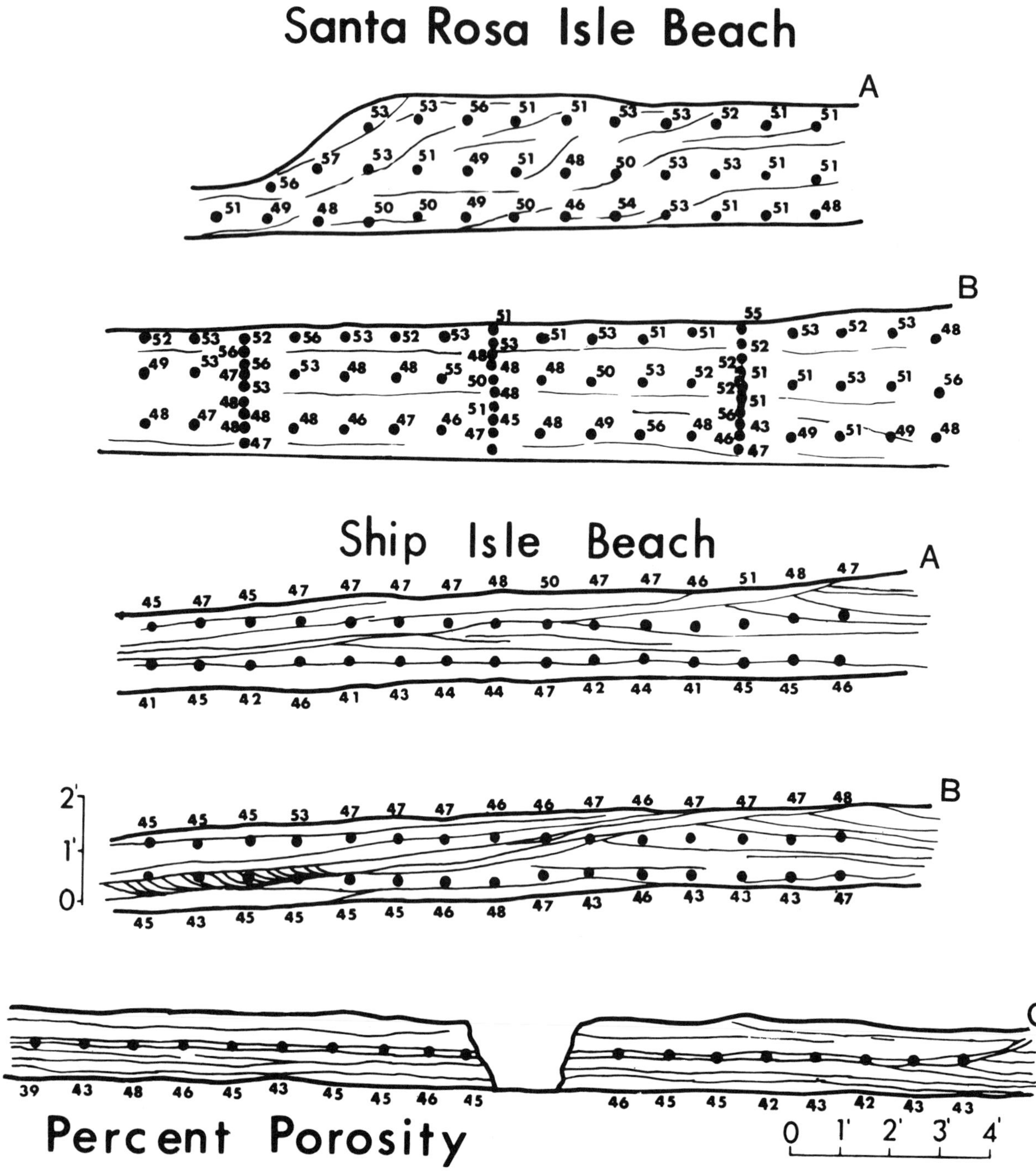

FIG. 12—Trench sections through beaches on Ship Island and Santa Rosa Island, giving distributions of porosity data.

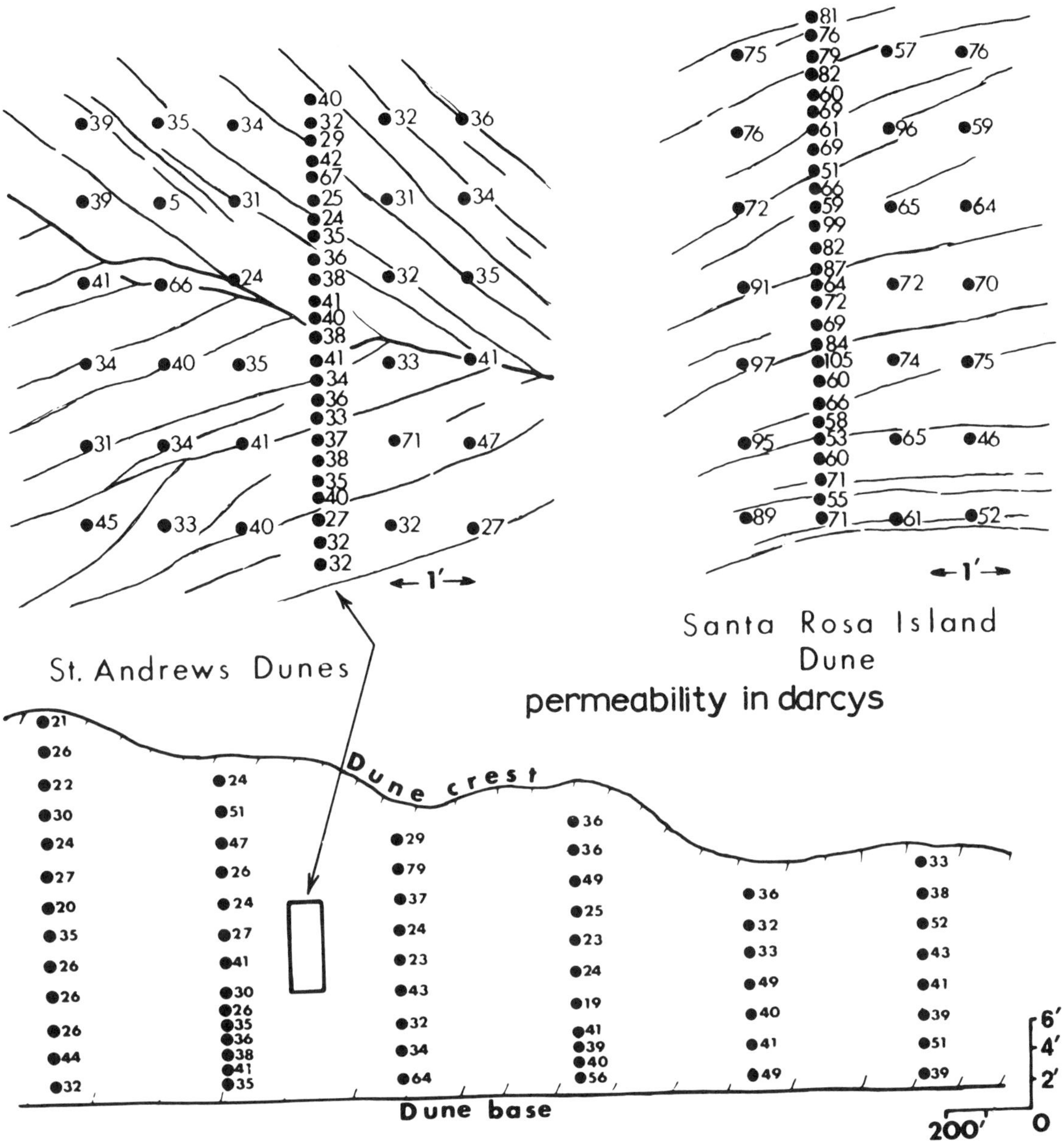

FIG. 13—Permeability distributions for dunes at Santa Rosa Island and St. Andrews Park.

the beach body. On the berm side of the beach crests the laminae dip shoreward; however, these are mixed with some crossbedding dips that parallel the length of the beach sand body. Hence beach sand bodies have maximum permeabilities that are oriented largely at right angles to the body and in diametrically opposing directions. *Directional permeabilities in beaches are more variable than in river point-bars.*

Crossbedding in the dune samples is extremely variable, locally shifting as much as 300°. However, the dominant direction is shoreward.

Character of bedding units—The fundamental building blocks of all sand bodies are the laminae and beds. These primary units have differing characteristics of texture, thickness, extent, and arrangement, which are dependent upon their depositional environments.

In beach sand bodies the beds and laminae are generally uniform in character, less than 1 in. thick, have a relatively wide extent, and are arranged into thick packets of parallel laminae. Truncation of the individual laminae and packets of adjacent laminae and packets is of relatively low frequency. Dune sand bodies are constructed of beds and laminae that also are relatively uniform in character and of wide extent, but are generally thicker than 1 in. The individual beds are arranged into very thick packets of steeply dipping crossbeds, in which truncation is abundant. Neither beach nor dune sand bodies have abundant clay or fine-grained interbeds that would inhibit permeability across bedding or laminae units.

In contrast to beach and dune sandbodies, river bar sand bodies are made up of laminae that are nonuniform in character; they are generally thin, of limited lateral extent, and have a markedly variable textural character. The individual beds are arranged into very thick packets of steeply dipping crossbeds, in which truncation is abundant. Neither beach nor dune sand bodies have abundant clay or fine-grained interbeds that would inhibit permeability across bedding or laminae units.

The most common laminae array includes the sinuous, tangential, trough cross-stratified packet and the broad, planar cross-stratified packet. The tangential, trough cross-stratified unit (Figs. 5, 6, 15) ranges from several inches to several feet in thickness and may extend longitudinally in a sinuous pattern for tens of feet. The planar, in places convex-upward, cross-stratified units are more widely spread and range from several inches to several tens of feet thick. Their base is nonerosional with planar foreset laminae commonly overlapping adjacent laminae (Fig. 15). Widespread major clay and silt units such as are present at the base of the cross-stratified unit in Figure 5 and shown diagrammatically in Figure 15, are common to river-bar sand bodies. These packets of cross-stratified sand are typified by a high variability in textural characteristic and permeability-porosity and by a high degree of truncation and intersection of adjacent packets. Silt and clay laminae are abundant along the truncation planes. Permeability increases downward in each dipping lamination (Fig. 15). Directional permeability is parallel with the length of the sinuous packets and in the direction of dip of the planar packets.

Means and variabilities—A comparison of the data and their variability for the different samples and at different sampling levels is given in Table 1. The most obvious differences are between the major sample groups. Most significant is the difference in permeability variability between the river-bar sands and the beach-dune samples. The river-bar permeabilities are one to three orders of magnitude in variance higher than the beach-dune permeability variances. This relation does not entirely hold for the porosities; the Whitewater River porosities are significantly more variable than the beach-dune porosities, but the Wabash River values are not. The relatively low porosity percentages and high variabilities for the Whitewater samples can be attributed to the poorly sorted and heterogeneous character of the gravelly sediments and especially to the inclusion of some relatively large particles in the samples.

The Sabine Pass beach permeabilities are not only the lowest of the studied samples, but their variability is also the lowest. Comparison of mean permeabilities between Ship Island and Santa Rosa Island beaches shows a slight absolute difference, but permeability variability is nearly identical. However, the Santa Rosa Island dune sample has a mean permeability only slightly lower than the nearby Santa Rosa Island beach sample, but the permeability variance is significantly lower. The permeability variability for both dune samples is similar and significantly lower than for the beach samples. Values of porosity variance for all samples, with the exception of the Whitewater River bar sample, are rather low and uniform.

Parameter Relations

Permeability versus porosity—Figure 16 shows the character of the permeability and porosity distributions for the various sample sites and for the total group of samples. Computations and experiments with artificially packed spheres by

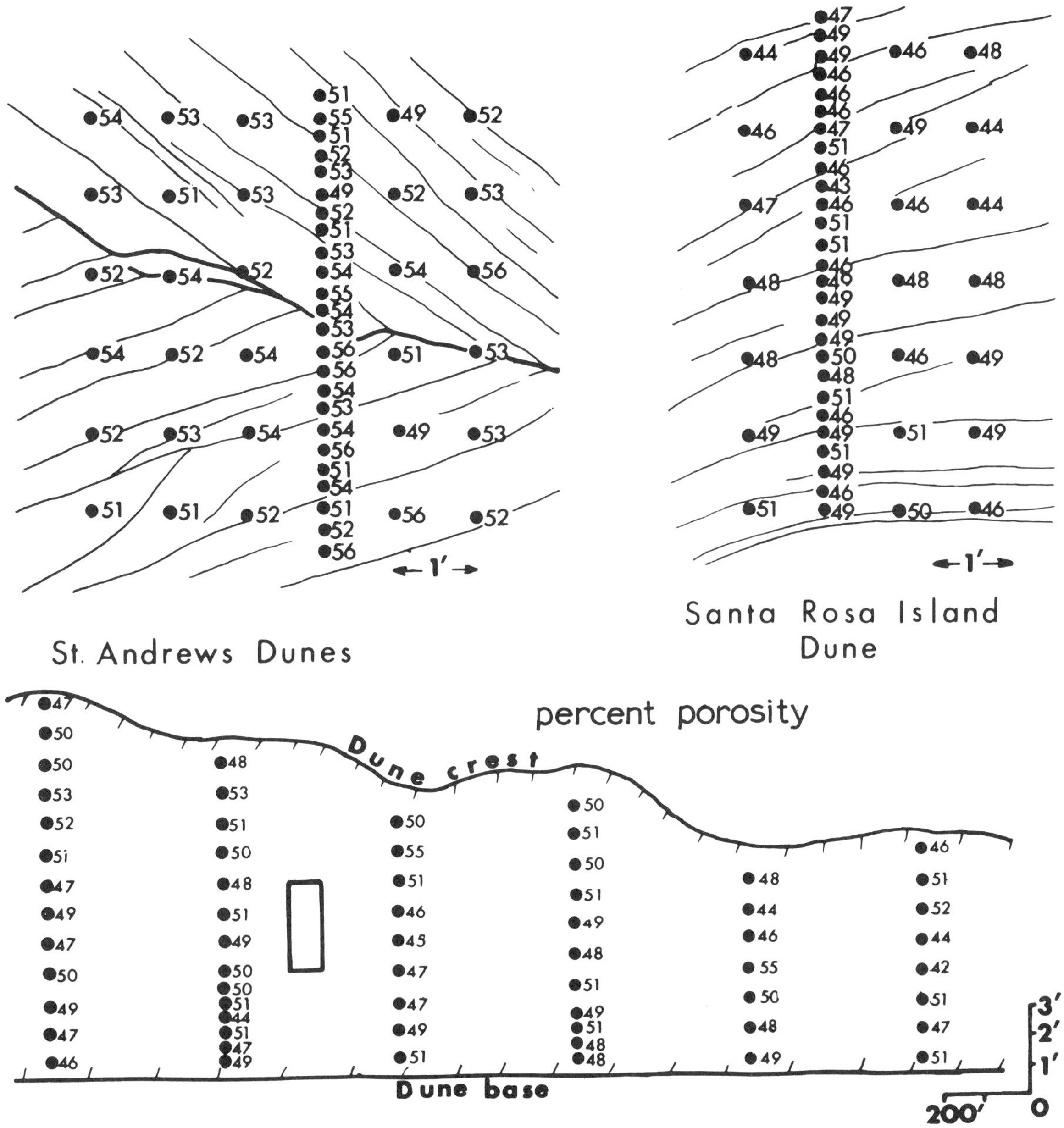

FIG. 14—Porosity distributions for dunes at Santa Rosa Island and St. Andrews Park.

Graton and Fraser (1935) show that a straight-line relation should exist between porosity and permeability, with permeability increasing as porosity increases. In studies of various sandstone reservoir rocks, Füchtbauer (1967) showed very clearly that permeability decreases as porosity decreases. However, in Figure 16 no clear relations between permeability and porosity for natural sands appear to exist. The trends shown by Füchtbauer (1967, Figs. 7-10) are related to much wider ranges in porosity than are present in these recent sands; the lower porosities of the older sediments resulting from burial compaction, cementation, and increased clay content, which also reduce permeabilities.

Permeability-porosity versus grain texture—The relations between permeability-porosity and grain-textural parameters for each of the sampled sand bodies are shown in Figures 17-24 with the relations for the samples combined in Figures 17 and 18. Mean grain sizes are given in phi units, where 0.0 phi equals 1.00 mm, 1.0 phi equals 0.5 mm, 2.0 phi equals 0.25 mm, *etc.* Sorting is presented in phi standard deviation units.

The relations between porosity and textural parameters, for the total group of samples show no significant trends. Fraser (1935) pointed out that theoretically grain size should have no effect on porosity, but that in natural grain assemblages as grain size decreases, porosity should increase, because of increased intergrain friction, adhesion, and bridging. This has been demonstrated clearly for sample groups with wide ranges of particle size by Hamilton and Menard (1956) and Füchtbauer and Reineck (1963), where recent sands with median diameters of approximately 0.7 mm have porosities of about 40 percent and clays with median diameters of less than 20 microns have porosities of nearly 80 percent. However, in the coarse- to fine-sand range these porosities are relatively uniform and show no tendency to increase or decrease with a change in median diameter (see von Engelhardt, 1960, Figs. 7-8). The data in Figure 17 also do not vary widely in grain size. This lack of a clear relation may be the result of the mixing of data from several different depositional environments. However, this is probably due to relatively small grain-size range in the samples used here. There is a very slight suggestion of a decrease in porosity with decrease of sorting in Figure 18. The reduction is relatively insignificant with the better sorted sands showing a slightly higher porosity and agreeing with the conditions that Fraser (1935) demonstrated experimentally with artificial mixtures of spheres.

Permeability on the other hand does show significant relations with grain size. As mean grain size increases, permeability clearly increases at a constant rate (Fig. 17), whereas the trends for permeability and sorting (Fig. 18) are poorly developed. The experimental work of Graton and Fraser (1935) and Fraser (1935) demonstrates the permeability-grain size relation in artificially packed spheres, where the throat-plane areas of the voids or the size of the intergranular channelways increase as the grain size increases, and hence the rate of flow increases. Krumbein and Monk (1942) determined that the coefficient of permeability in artificially packed sands varies as the square of the diameter and inversely as the log of the standard deviation. One therefore would expect to see some clear relation between permeability and decrease in sorting, caused by the filling of channelways with finer materials in poorly sorted sands (Fraser, 1935, Figs. 5, 6), or an increase in permeability with decreased sorting caused by the enlargement of channelways by the irregular, disruptive packing in poorly sorted sands (Fraser, 1935, Figs. 3-4). However, these possible trends are not demonstrated clearly in Figure 18. The data appear to be relatively scattered with only the weakly suggested increase in permeability with decrease in sorting, which Fraser demonstrated for artificial mixtures of spheres (Fraser, 1935, p. 971-973).

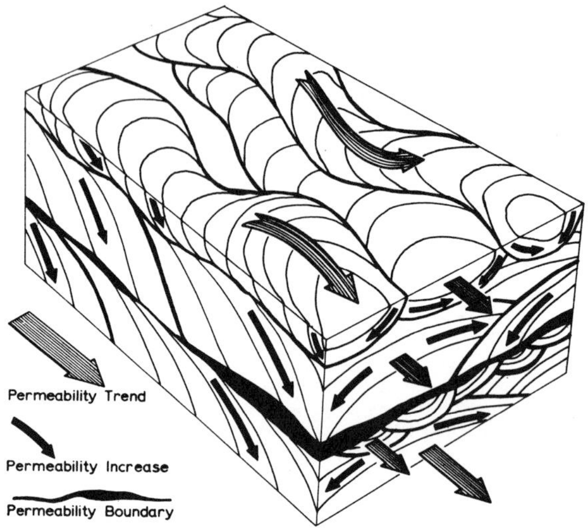

FIG. 15—Diagram of river-bar laminae packet with permeability characteristics.

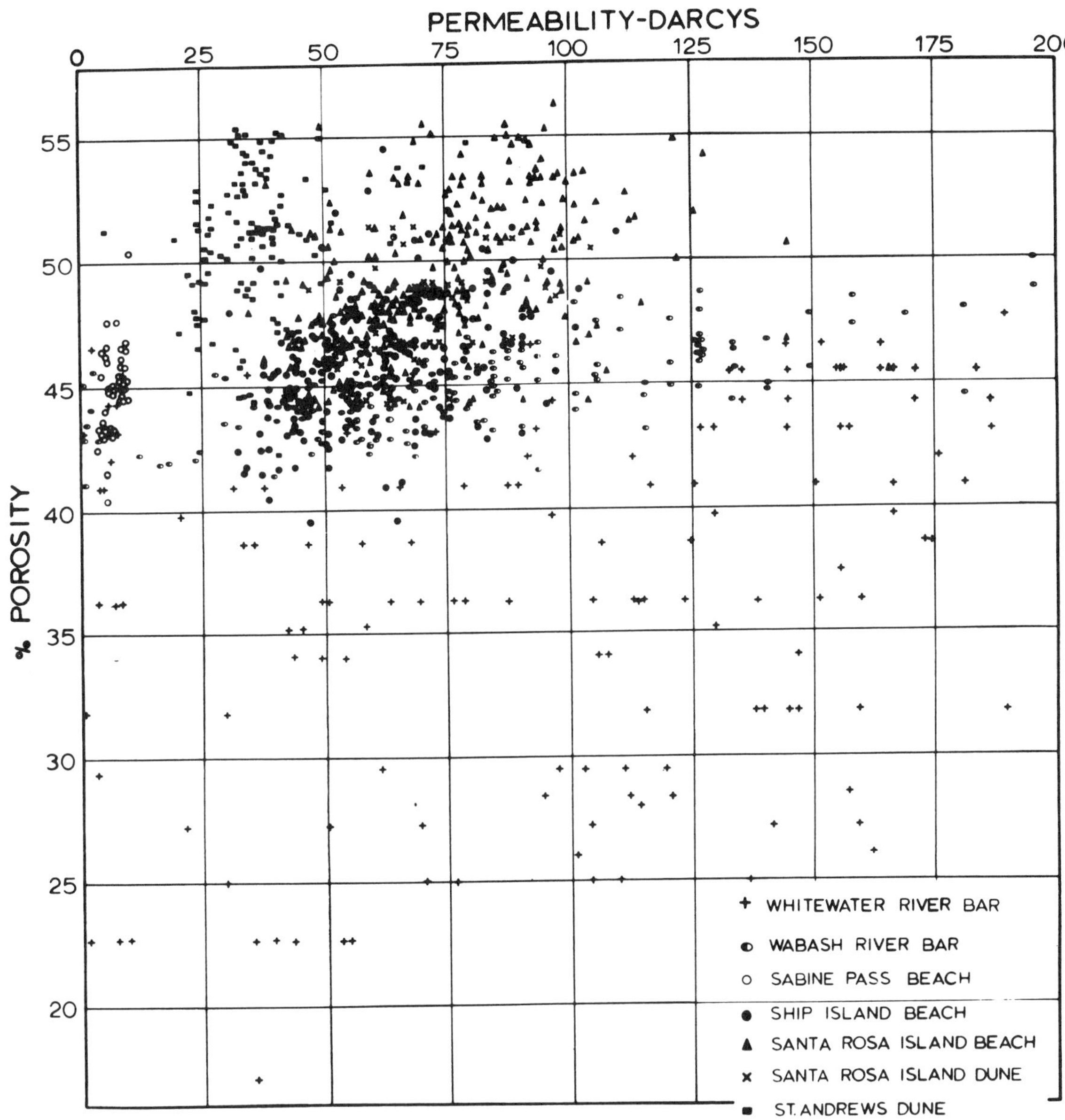

FIG. 16—Permeability versus porosity.

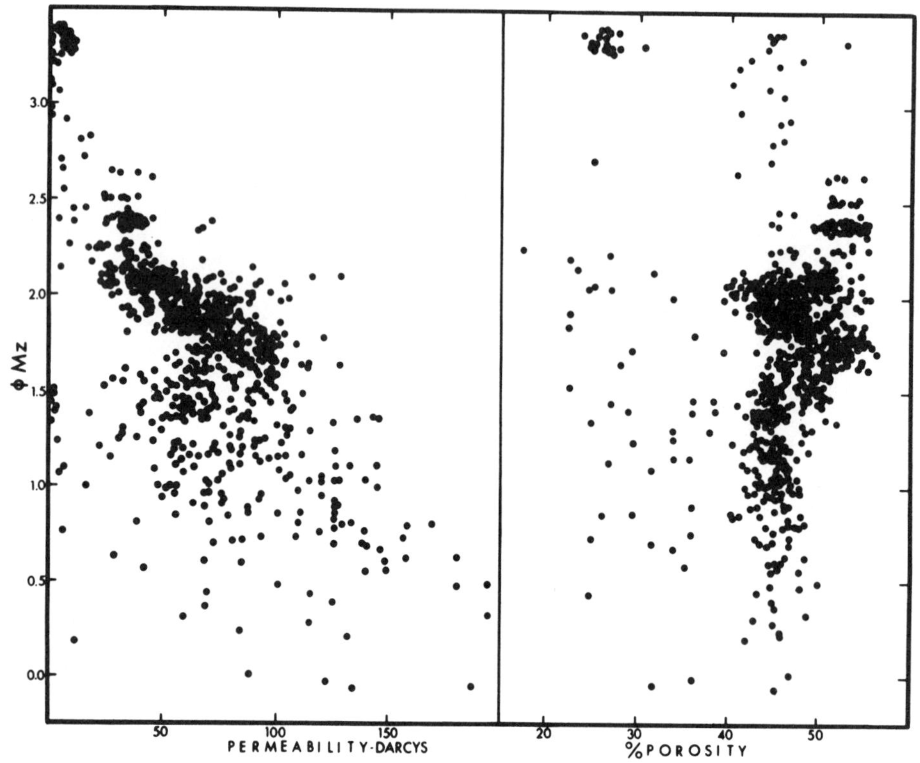

FIG. 17—Permeability-porosity versus mean grain size for all samples.

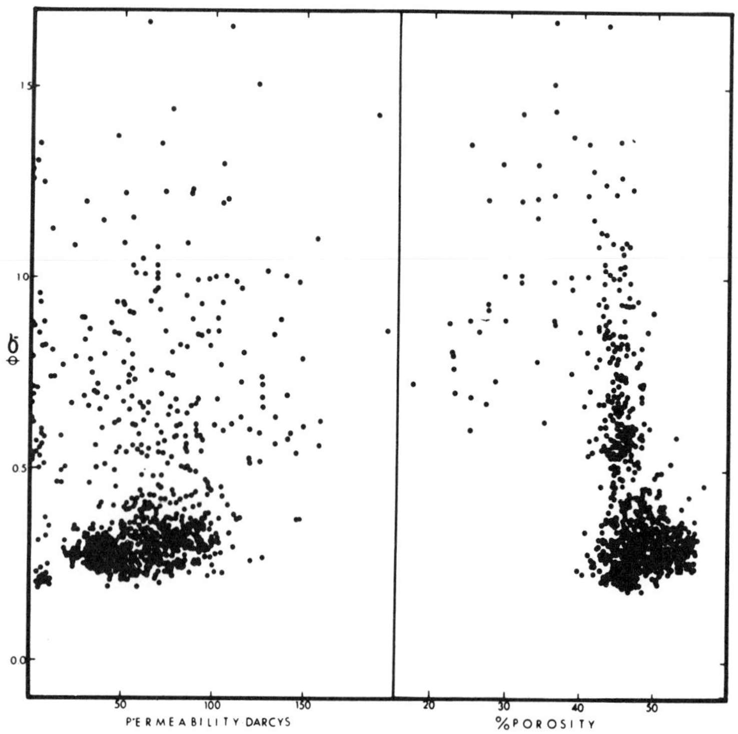

FIG. 18—Permeability-porosity versus phi inclusive graphic standard deviation for all samples.

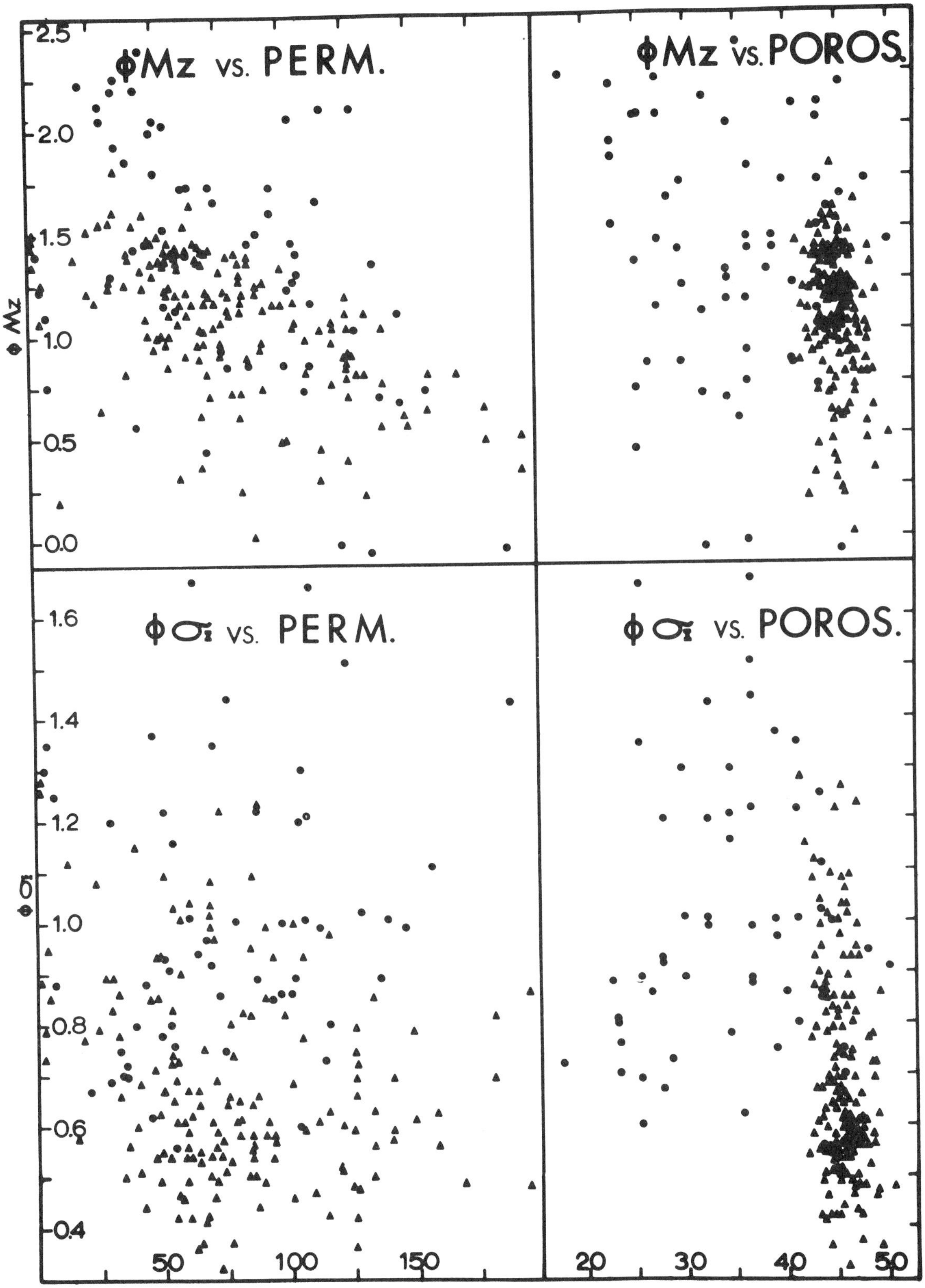

Fig. 19—Permeability and porosity versus textural parameters for Wabash River (triangles) and Whitewater River (circles) bars.

Figures 17 and 18 are the plottings of all the data from the three different environments of deposition and as such tend to obscure some relations that are environment related. When these data from each of the sample groups are examined individually (Figs. 19-24), significant trends and interenvironmental differences are exposed.

Figures 19 and 20 are the data plots for the river-bar samples which, in addition to showing trend differences between the river-bar data, show interesting and significant major differences between them and the beach-dune data (Figs. 21, 22, 24). Permeability increases with increase in median diameter and sorting. Porosity also appears to increase slightly with increase in median diameter and sorting. The permeability-mean diameter trends are clearly developed in all three sets of river-bar data, with the slope of the data distribution for the Wabash River samples being lower than the Whitewater or Mississippi samples. The permeability-sorting trends are developed weakly in the Wabash and Mississippi data, and the data are scattered and show no trends for the Whitewater sample. The porosity versus mean size-sorting trends in the Wabash data are well developed, but are poorly developed in the Whitewater and Mississippi data.

In contrast to the river-bar data, the beach-dune plots (Figs. 22-24) show that permeability increases as sorting decreases and that porosity increases as mean grain size decreases. An exception is the Sabine Pass beach sample (Fig. 21), where the permeability-sorting distribution is similar to that of the river-bar samples. The trends of permeability versus mean grain size and porosity versus sorting for beaches and dunes are similar to those of the river-bar data. Data distribution of each beach sample is slightly different from the others; slopes of permeability and porosity increases are steepest in the Sabine Pass beach sample and lowest in the Ship Island sample.

Permeability-porosity values and textural characteristics of each of the two dune samples are of uniform variability, so that individually no strong trends develop, but when both dune samples are plotted together enough variability exists to demonstrate significant trends. In Figure 24 permeability increases as mean size increases and sorting decreases, and porosity increases as mean size decreases and sorting increases.

River-bar, beach, and dune samples have the following in common: (1) permeability increases with increasing grain size, and (2) porosity increases with increase in sorting; both relations agree with the experiments of Fraser (1935) and the determinations of Krumbein and Monk (1942). The relations of porosity versus mean size and permeability versus sorting separate the data into two different groups: (1) the river bars, where porosity increases with increasing sorting and permeability increases with increasing sorting (the reverse of Fraser's finding), and (2) beaches and dunes, where porosity increases with decreasing mean grain size and permeability increases with decreasing sorting (both in accord with the findings of Fraser). When the data are plotted all together (Figs. 17, 18), the above relations are obscured. As shown in Table 2, the beach and dune sample relations are generally in accord with the relations between permeability-porosity and texture that Fraser (1935) and Graton and Fraser (1935) demonstrated for artificially packed mixtures of spheres. The river-bar sample relations, however, are the reverse of what is expected for permeability versus sorting and porosity versus grain size.

The ideal conditions set forth by Fraser (1935), Graton and Fraser (1935), and Krumbein and Monk (1942), for artificially packed quartz sands, are only weakly demonstrated for these natural sands from different depositional environments. Only the permeability-mean grain size relation is well developed.

In the experiments with artificial sediments, the investigators jarred the sediments in their containers until no further compaction was observable and maximum packing was attained. Although natural sands do not receive this treatment, the winnowing processes of the beach and dune environments result not only in better sorting of the sands, but also result in a closer packing geometry. The "dumping" processes of the fluvial environment yield a much looser, propped-open packing style than the beach-dune winnowing processes. Although no geometric data are available on the relative packing styles of sands from different depositional environments, their styles of packing may vary significantly and may be the cause of the differences shown in Table 2. Von Engelhardt and Pitter (1951) among others have shown that packing differences exert a strong control on variabilities in permeability and porosity.

To test the effect of depositional environment on packing, duplicate samples were selected from each of the major sample units. The samples were treated in the permeameter at excessively high head pressures (100 cm water), so that elutriation would cause repacking. Figure 25 shows the character of permeability and porosity change through time as repacking takes place. All samples show an initial rapid reduction in permeabil-

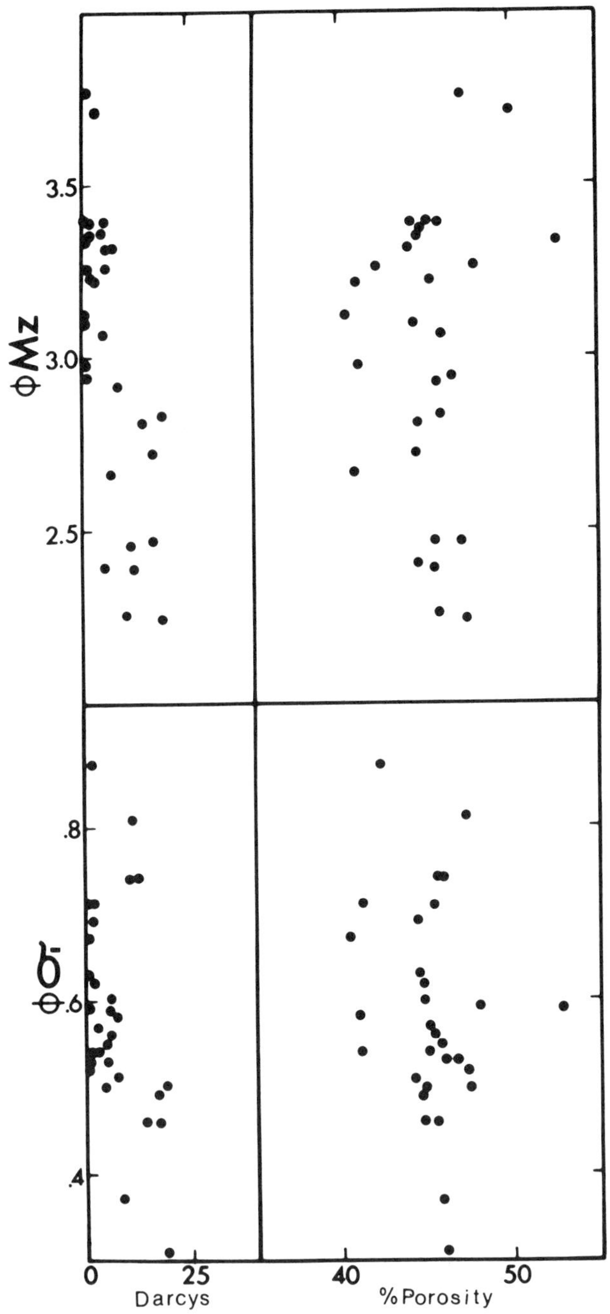

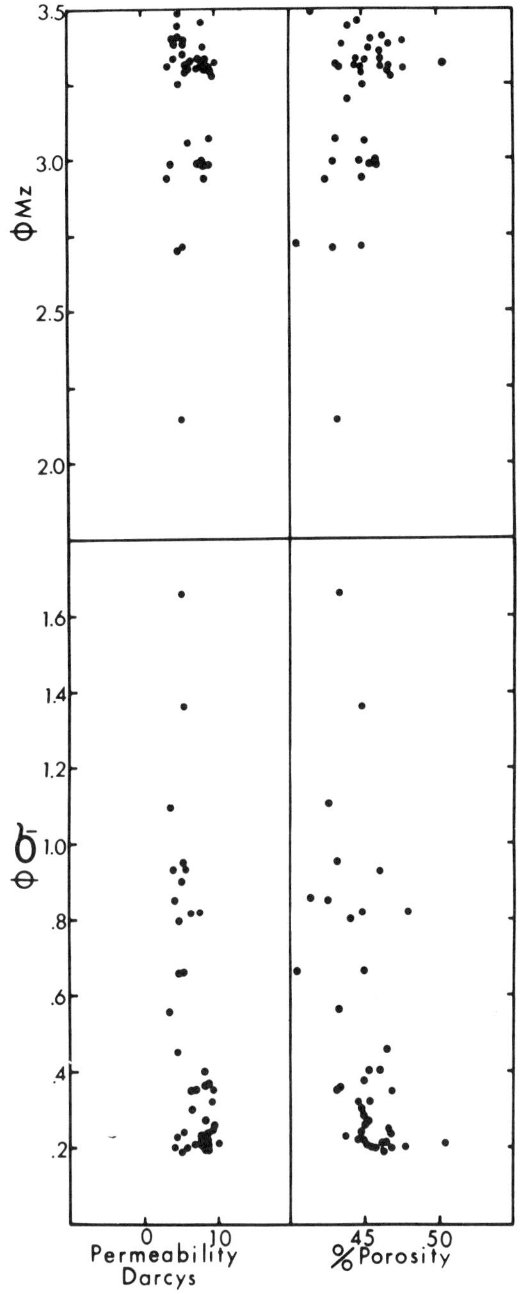

FIG. 20—(Left) Permeability and porosity versus textural parameters for Mississippi River bar.

FIG. 21—(Right) Permeability and porosity versus textural parameters for Sabine Pass beach.

ity during the first several hundred seconds of the test, followed by a leveling-off and stabilization. Porosities given for the undisturbed and the repacked samples show marked reductions. It is obvious from Figure 25 that the river-bar samples undergo the most extensive changes in packing and that beach and dune samples show the least. The "fluffy" beach crest sample from Santa Rosa Island also shows a rapid decrease in permeability and a significant reduction in porosity. The porosity decreases are accompanied, of course, by sample-volume reductions, further indicating repacking into a more dense particle arrangement. The 1,100-year old Mississippi River bar sand shows only a very small reduction of permeability and porosity through the test run, indicating that some postdepositional repacking and cementation already have taken place.

There appears to be a difference in the initial packing styles of river-bar sands and beach-dune sands; river-bar sands are more irregularly packed. This packing difference could explain the departure of permeability-porosity versus textural relations (Table 2) of river-bar sands from the ideal conditions of Fraser (1935) and Krumbein and Monk (1942).

Summary

In summary, recent sand bodies from river bars, beaches, and associated coastal dunes have

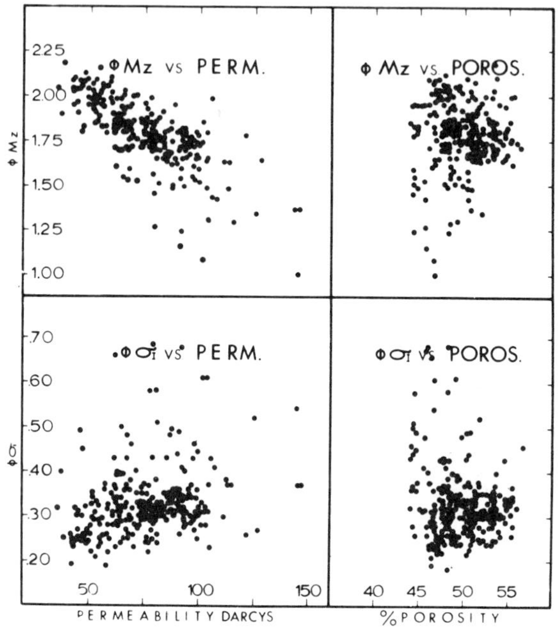

Fig. 23—Permeability and porosity versus textural parameters for Santa Rosa Island beach.

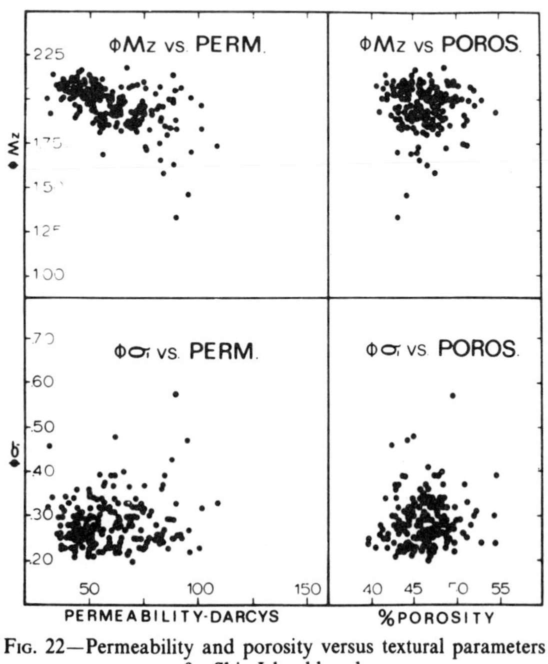

Fig. 22—Permeability and porosity versus textural parameters for Ship Island beach.

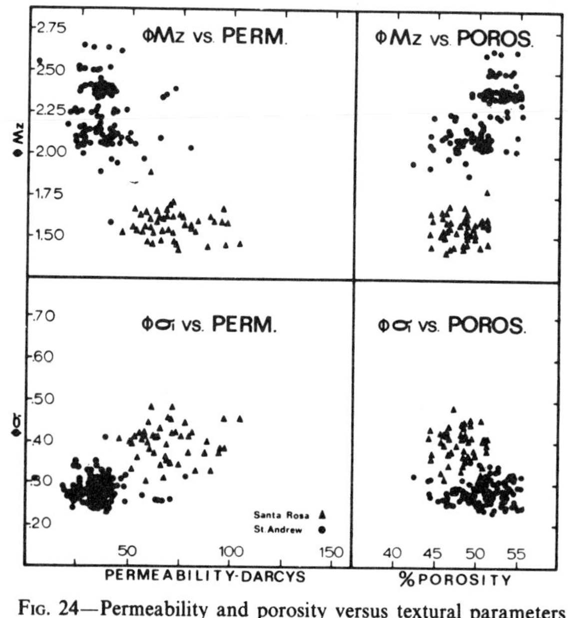

Fig. 24—Permeability and porosity versus textural parameters for Santa Rosa Island and St. Andrew Park dunes.

Table 2. Relations of Permeability and Porosity to Textural Parameters and Depositional Environments

	River Bar		Beach		Dune		Fraser (1935)	
	Perm.	Poros.	Perm.	Poros.	Perm.	Poros.	Perm.	Poros.
Grain Size increases	Increase	Increase	Increase	Decrease	Increase	Decrease	Increase	Decrease
Sorting increases	Increase	Increase	Decrease	Increase	Decrease	Increase	Decrease	Increase

relatively high permeabilities, ranging well above 1,000 md and averaging more than 50 darcys. Initial porosities of recent sand bodies are also relatively high, averaging well over 45 percent and ranging up to 56 percent.

Sand bodies from the various depositional systems exhibit well-organized and different patterns of permeability variations (Fig. 26). Permeabilities in river-bar sands are generally higher and more variable than in beaches and dunes and decrease systematically downstream and bankward. In beach sand bodies, permeability values have a small variability, but are relatively low on the beach faces, high on beach crests, and variable on the beach berms. Permeability-variation trends are parallel with the length of river-bar sand bodies and perpendicular to beach sand bodies. Dune sand bodies show no distinctive trend of permeability variation.

There are greater variations of grain characteristics and permeability-porosity within bedding and lamination subunits than between them; this is especially pronounced in river-bar sand bodies. Packets of laminae and beds in beach-dune sand bodies are more uniform in character than those in river-bar deposits. Clay laminae and other low-permeability units commonly are present between bedding and laminae packets in river-bar

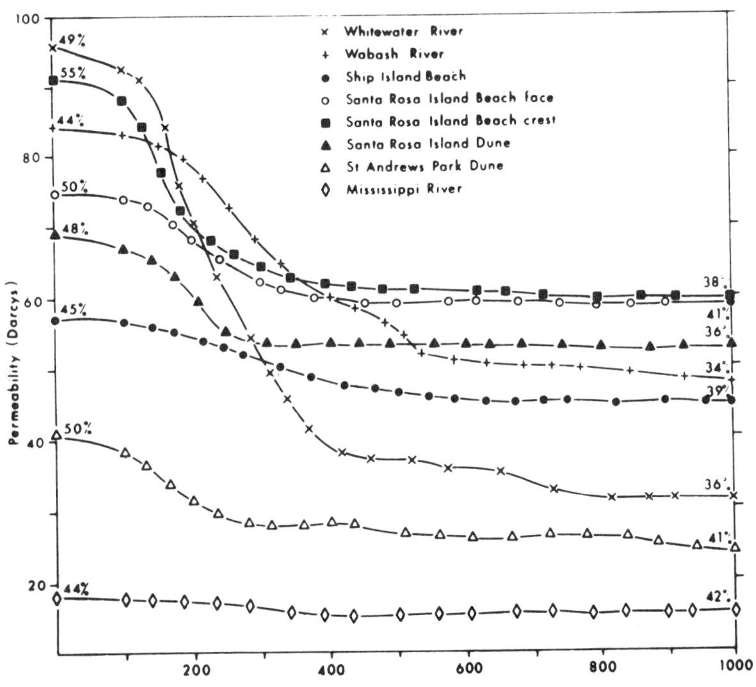

FIG. 25—Effects of repacking by elutriation on permeability and porosity of selected recent sands from different depositional environments.

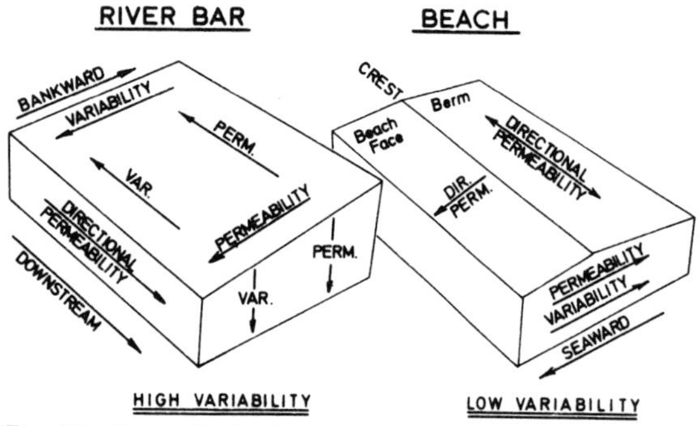

Fig. 26—Generalized relation of permeability trends in rivei bars and beaches.

sand bodies. In all deposits the boundary conditions between sediment packets are important factors in determining the effective reservoir characteristics of the sand bodies because sediment packets surrounded by other units of lower permeability will have effective permeabilities influenced by, and largely determined by, the lower permeabilities of the bounding units and hence will not demonstrate their ultimate through-flow capabilities.

The ideal relations between permeability-porosity and textural characteristics that have been shown by various authors for artificially packed particles are demonstrated only weakly by beach and dune sands and are not demonstrated by river-bar sands. The different styles of natural grain packing between river-bar sands and beach-dune sands are probably the causes of these deviations from the ideal models.

References Cited

Dodge, C. F., D. P. Holler, and R. L. Meyer, 1971, Reservoir heterogeneities of some Cretaceous sandstones: Am. Assoc. Petroleum Geologists Bull., v. 55, no. 10, p. 1814-1828.

Folk, R. L., and W. C. Ward, 1957, Brazos River bar—a study in the significance of grain-size parameters: Jour. Sed. Petrology, v. 27, no. 1, p. 3-26.

Fraser, H. J., 1935, Experimental study of the porosity and permeability of clastic sediments: Jour. Geology, v. 43, no. 8, pt. 1, p. 910-1010.

Füchtbauer, H., 1967, Influence of different types of diagenesis on sandstone porosity: 7th World Petroleum Cong. Proc., v. 2, p. 353-369.

——— and H. E. Reineck, 1963, Porosität und Verdichtung rezenter, mariner Sedimente: Sedimentology, v. 2, no. 4, p. 294-306.

Graton, L. C., and H. J. Fraser, 1935, Systematic packing of spheres, with particular relation to porosity and permeability: Jour. Geology, v. 43, no. 8, pt. 1, p. 785-909.

Hamilton, E. L., and H. W. Menard, Jr., 1956, Density and porosity of sea-floor surface sediments off San Diego, California: Am. Assoc. Petroleum Geologists Bull., v. 40, no. 4, p. 754-761.

Hewitt, C. H., and J. T. Morgan, 1965, The Fry *in situ* combustion test—reservoir characteristics: Jour. Petroleum Technology, v. 17, no. 3, p. 337-342.

Horn, D., 1964, Zur sedimentation des Dogger-beta-Hauptsandsteins im ostholsteinischen Juratrog: Meyniana, v. 14, p. 21-42.

Hutchinson, C. A., Jr., C. F. Dodge, and T. L. Polasek, 1961, Identification, classification, and prediction of reservoir nonuniformities affecting production operations: Jour. Petroleum Technology, v. 13, no. 8, p. 223-230.

Krumbein, W. C., and G. D. Monk, 1942, Permeability as a function of the size parameters of sedimentary particles: AIME Tech. Pub. 1492.

Ludwick, J. C., 1956, A volumeter for measuring porosity of incoherent sands: Jour. Sed. Petrology, v. 26, no. 3, p. 276-283.

Mast, R. F., and P. E. Potter, 1963, Sedimentary structures, sand-shape fabrics, and permeability, pt. 2: Jour. Geology, v. 71, no. 5, p. 548-565.

Polasek, T. L., and C. A. Hutchinson, Jr., 1967, Characterization of non-uniformities within a sandstone reservoir from a fluid mechanics standpoint: 7th World Petroleum Cong. Proc., México, D. F., v. 2, p. 397-407.

Pryor, W. A., 1968, Reservoir inhomogeneities of recent sandbodies: Am. Petroleum Inst. Research Project 91-B, Ann. and Final Rept., 53 p.

von Engelhardt, W., 1960, Der Porenraum der Sedimente: Berlin, Springer-Verlag, 207 p.

——— and H. Pitter, 1951, Über die Zusammenhänge zwischen Porosität, Permeabilität, und Korngrösse bei Sanden und Sandsteinen: Heidelberger Beitr. Mineralogie u. Petrographie, v. 2, no. 6, p. 477-491.

Wyllie, M. R. J., and M. B. Spangler, 1952, Application of electrical resistivity measurements to problems of fluid flow in porous media: Am. Assoc. Petroleum Geologists Bull., v. 36, no. 2, p. 359-403.

SANDSTONE DEPOSITIONAL ENVIRONMENTS

American Association of Petroleum Geologists Memoir 18,
Underground Waste Management and Environmental Implications, edited by T.D. Cook, copyright 1972, pp. 133-190.

Geometry of Sandstone Reservoir Bodies[1]

RUFUS J. LeBLANC[2]
Houston, Texas 77001

Abstract Natural underground reservoirs capable of containing water, petroleum, and gases include sandstones, limestones, dolomites, and fractured rocks of various types. Comprehensive research and exploration efforts by the petroleum industry have revealed much about the character and distribution of carbonate rocks (limestones and dolomites) and sandstones. Porosity and permeability of the deposits are criteria for determining their efficiency as reservoirs for fluids. Trends of certain sandstones are predictable. Furthermore, sandstone reservoirs have been less affected than carbonate reservoirs by postdepositional cementation and compaction. Fracture porosity has received less concentrated study; hence, we know less about this type of reservoir. The discussions in this paper are confined to sandstone reservoirs.

The principal sandstone-generating environments are (1) fluvial environments such as alluvial fans, braided streams, and meandering streams; (2) distributary-channel and delta-front environments of various types of deltas; (3) coastal barrier islands, tidal channels, and chenier plains; (4) desert and coastal eolian plains; and (5) deeper marine environments, where the sands are distributed by both normal and density currents.

The alluvial-fan environment is characterized by flash floods and mudflows or debris flows which deposit the coarsest and most irregular sand bodies. Braided streams have numerous shallow channels separated by broad sandbars; lateral channel migration results in the deposition of thin, lenticular sand bodies. Meandering streams migrate within belts 20 times the channel width and deposit two very common types of sand bodies. The processes of bank-caving and point-bar accretion result in lateral channel migration and the formation of sand bodies (point bars) within each meander loop. Natural cut-offs and channel diversions result in the abandonment of individual meanders and long channel segments, respectively. Rapidly abandoned channels are filled with some sand but predominantly with fine-grained sediments (clay plugs), whereas gradually abandoned channels are filled mainly with sands and silts.

The most common sandstone reservoirs are of deltaic origin. They are laterally equivalent to fluvial sands and prodelta and marine clays, and they consist of two types: delta-front or fringe sands and abandoned distributary-channel sands. Fringe sands are sheetlike, and their landward margins are abrupt (against organic clays of the deltaic plain). Seaward, these sands grade into the finer prodelta and marine sediments. Distributary-channel sandstone bodies are narrow, they have abrupt basal contacts, and they decrease in grain size upward. They cut into, or completely through, the fringe sands, and also connect with the upstream fluvial sands or braided or meandering streams.

Some of the more porous and permeable sandstone reservoirs are deposited in the coastal interdeltaic realm of sedimentation. They consist of well-sorted beach and shoreface sands associated with barrier islands and tidal channels which occur between barriers. Barrier sand bodies are long and narrow, are aligned parallel with the coastline, and are characterized by an upward increase in grain size. They are flanked on the landward side by lagoonal clays and on the opposite side by marine clays. Tidal-channel sand bodies have abrupt basal contacts and range in grain size from coarse at the base to fine at the top. Laterally, they merge with barrier sands and grade into the finer sediments of tidal deltas and mud flats.

The most porous and permeable sandstone reservoirs are products of wind activity in coastal and desert regions. Wind-laid (eolian) sands are typically very well sorted and highly crossbedded, and they occur as extensive sheets.

Marine sandstones are those associated with normal-marine processes of the continental shelf, slope, and deep and those due to density-current origin (turbidites). An important type of normal-marine sand is formed during marine transgressions. Although these sands are extremely thin, they are very distinctive and widespread, have sharp updip limits, and grade seaward into marine shales. Delta-fringe and barrier-shoreface sands are two other types of shallow-marine sands.

Turbidites have been interpreted to be associated with submarine canyons. These sands are transported from nearshore environments seaward through canyons and are deposited on submarine fans in deep marine basins. Other turbidites form as a result of slumping of deltaic facies at shelf edges. Turbidite sands are usually associated with thick marine shales.

[1] Manuscript received, March 17, 1972.

[2] Shell Oil Company. This paper is based on the writer's 30 years of experience in studies of modern and ancient clastic sediments—from 1941 to 1948, with the Mississippi River Commission, under the guidance of H. N. Fisk, and, since August 1948, with Shell Development Company and Shell Oil Company.

The writer is grateful to Shell Oil Company for permission to publish this paper, and he is deeply indebted to Alan Thomson for his critical review of the manuscript; he is also grateful to Nick W. Kusakis, John Bush, Dave C. Fogt, Gil C. Flanagan, and George F. Korenek for assistance in the preparation of illustrations and reference material; to Aphrodite Mamoulides and Bernice Melde for their library assistance; to Darleen Vanderford for typing the manuscript, and to Judy Breeding for her editorial assistance.

Numerous stimulating discussions of models of clastic sedimentation and the relationship of sedimentary sequences to depositional processes were held with Hugh A. Bernard and Robert H. Nanz, Jr., during the late 1940s and 1950s, when we were closely associated with Shell's early exploration research effort. The writer is particularly indebted to these two men for their numerous contributions, many of which are included in this paper.

The writer also wishes to thank W. B. Bull, University of Arizona, for his valuable suggestions concerning the alluvial-fan model of clastic sedimentation.

Introduction

Important natural resources such as water, oil, gas, and brines are found in underground reservoirs which are composed principally of the following types of rocks: (1) porous sands, sandstones, and gravels; (2) porous limestones and dolomites; and (3) fractured rocks of various types. According to the 1971 American Petroleum Institute report on reserves of crude oil and natural gas, sandstones are the reservoirs for about 75 percent of the recoverable oil and 65 percent of the recoverable gas in the United States. It is also estimated that approximately 90 percent of our underground water supply comes from sand and gravel (Walton, 1970).

Sandstone and carbonate (limestone and dolomite) reservoirs have been intensively studied during the past 2 decades; consequently, the general characteristics and subsurface distribution of these two important types of reservoirs are relatively well known in numerous sedimentary basins. The factors which control the origin and occurrence of fracture porosity have received less attention; thus, our knowledge and understanding of this type of reservoir are more limited.

The detection of subsurface porosity trends within sedimentary basins was recognized by the petroleum industry as one of its most significant problems, and for the past 2 decades it has addressed itself to a solution through extensive research. Largely as a result of this research, which is summarized below, our ability to determine trends of porous sedimentary rocks has progressed noticeably, especially during the past 10 years.

The amount of porosity and permeability present within sedimentary rocks and the geometry of porous rock bodies are controlled mainly by two important factors: (1) the environmental conditions under which the sediments were deposited and (2) the postdepositional changes within the rocks as a result of burial, compaction, and cementation. Postdepositional diagenetic processes have less effect on the porosity and permeability of sands and sandstones than they have on carbonate sediments; consequently, porosity trends are significantly more predictable for sandstones than for limestones and dolomites.

Organization of paper—The following two parts of this paper give a brief historical summary of the early research on clastic sediments and present a classification of environments of deposition and models of clastic sedimentation. A résumé of significant studies of modern clastic sediments—mainly by the petroleum industry, government agencies, and universities—follows. The main part of the paper concerns the sedimentary processes, sequences, and geometry of sand bodies which characterize each of the following models of clastic sedimentation: alluvial fan, braided stream, meandering stream, deltaic (birdfoot-lobate and cuspate-arcuate), coastal interdeltaic (barrier island and chenier plain), and marine (transgressive, submarine canyon, and fan).

Historical Summary of Early Research on Modern Clastic Sediments

Geologists are now capable of interpreting the depositional environments of ancient sedimentary facies and of predicting clastic porosity trends with a reasonable degree of accuracy (Peterson and Osmond, 1961; Potter, 1967; Rigby and Hamblin, 1972; Shelton, 1972). This capability stems from the extensive research conducted on Holocene sediments by several groups of geologists during the past 3 decades. Conditions which led to this research, and the most significant studies of clastic sedimentation which provided the models, criteria, and concepts necessary to make environmental interpretations, are summarized below.

During the late 1930s and early 1940s, petroleum geologists became aware that improved methods of stratigraphic interpretations were badly needed, and that knowledge and geologic tools necessary to explore for stratigraphic traps were inadequate. A detailed study made by the Research Committee of The American Association of Petroleum Geologists on the research needs of the industry ultimately led to the establishment of geologic research departments by major oil companies. By 1948, exploration research by the oil industry was in its early stages, and expansion proceeded rapidly thereafter.

Meanwhile, some very significant developments were occurring at Louisiana State University. H. V. Howe and R. J. Russell, together with their graduate students, had already published several Louisiana Geological Survey bulletins summarizing their pioneer work on the late Quaternary geology of southern Louisiana (Howe and Moresi, 1931, 1933; Howe *et al.*, 1935; Russell, 1936). Their early work on the Mississippi deltaic plain and the chenier plain of southwestern Louisiana is considered to be the beginning of the modern environmental approach to stratigraphy. Fisk became fascinated with the Howe and Russell approach, and he applied results of their research to his study of Tertiary sediments. The work of Fisk (1940) in central Louisiana, which included a study of the lower Red River Valley and part of the Mississippi Valley, attracted the attention of General Max Tyler, president of the Mississippi River Commission in Vicksburg. General Tyler engaged Fisk as a consultant and provided him with a staff of geologists to conduct a geologic investigation of the lower Mississippi River alluvial valley.

The Fisk (1944) report on the Mississippi Valley, which now has become a classic geologic document, established the relations between alluvial environments, processes, and character of sediments. The AAPG, recognizing the significance of this contribution, retained Fisk as Distinguished Lecturer, and the results and significance of his work became widely known. One of his most significant contributions came when, as the petroleum industry was getting geologic research under way, he was selected by a major oil company to direct its geologic research effort in Houston.

By 1950, a few major oil companies were deeply involved in studies of recent sediments. However, the small companies did not have staff and facilities to conduct this type of research, and American Petroleum Institute Project 51 was established for the purpose of conducting research on recent sediments of the Gulf Coast. Scripps Institution of Oceanography was in charge of the project, which continued for 8 years. Results of this research were available to all companies (Shepard *et al.*, 1960).

While the petroleum industry was conducting "in-house" research and supporting the API project, some significant research was being done by the U.S. Waterways Experiment Station in Vicksburg, Mississippi, and by the new Coastal Studies Institute at Louisiana State University under the direction of R. J. Russell. These two groups conducted detailed studies of recent sediments for many years, and results were made available to the petroleum industry.

By 1955, a fairly good understanding of processes of sedimentation and character of related sediments in several depositional environments had been acquired. Although the application of this wealth of knowledge to operational problems was very difficult, some useful applications nevertheless had been made by the middle 1950s, and it was generally agreed that the initial research effort was successful.

Since 1955, geologists all over the world have become involved in studying recent sediments and applying the results to research on older rocks. Geologists with the U.S. Geological Survey and several universities have conducted studies of alluvial fans, braided streams, and eolian deposits; and the oceanographic institutions, such as Scripps, Woods Hole, and Lamont, have investigated deep-marine sediments on a worldwide basis. Publication of papers on clastic sedimentation has been increasing rapidly. The first textbook on the geology of recent sediments cites more than 700 references, 75 percent of which have appeared since 1955 (Kukal, 1971). Many of these contributions, considered to be most significant to the current understanding of clastic sediments, are cited in this paper.

Models and Environments of Clastic Sedimentation

The realm of clastic sedimentation can be divided into several conceptual models, each of which is characterized by certain depositional environments, sedimentary processes, sequences, and patterns. What are considered to be some of the most common and basic models and environments[3] of clastic sedimentation, arranged in order from the periphery to the center of a depositional basin, are listed below and are shown on Figures 1–4.

- Continental
  - Alluvial (fluvial) models
    - Alluvial fan
    - Braided stream
    - Meandering stream (includes flood basins between meander belts)
  - Eolian (can occur at various positions within continental and transitional models)
- Transitional
  - Deltaic models
    - Birdfoot-lobate (fluvial dominated)
    - Cuspate-arcuate (wave and current dominated)
    - Estuarine (with strong tidal influence)
  - Coastal-interdeltaic models
    - Barrier-island model (includes barrier islands, lagoons behind barriers, tidal channels, and tidal deltas)
    - Chenier-plain model (includes mud flats and cheniers)
- Marine
  - (Note: Sediments deposited in shallow-marine environments, such as deltas and barrier islands, are

[3] The classification of depositional environments presented herein was initially developed by the writer and his colleague, Hugh A. Bernard, during the early 1950s (LeBlanc and Bernard, 1954) and was recently modified (Bernard and LeBlanc, 1965). For other classifications, refer to Laporte (1968), Selley (1970), Crosby (1972), and Kukal (1971).

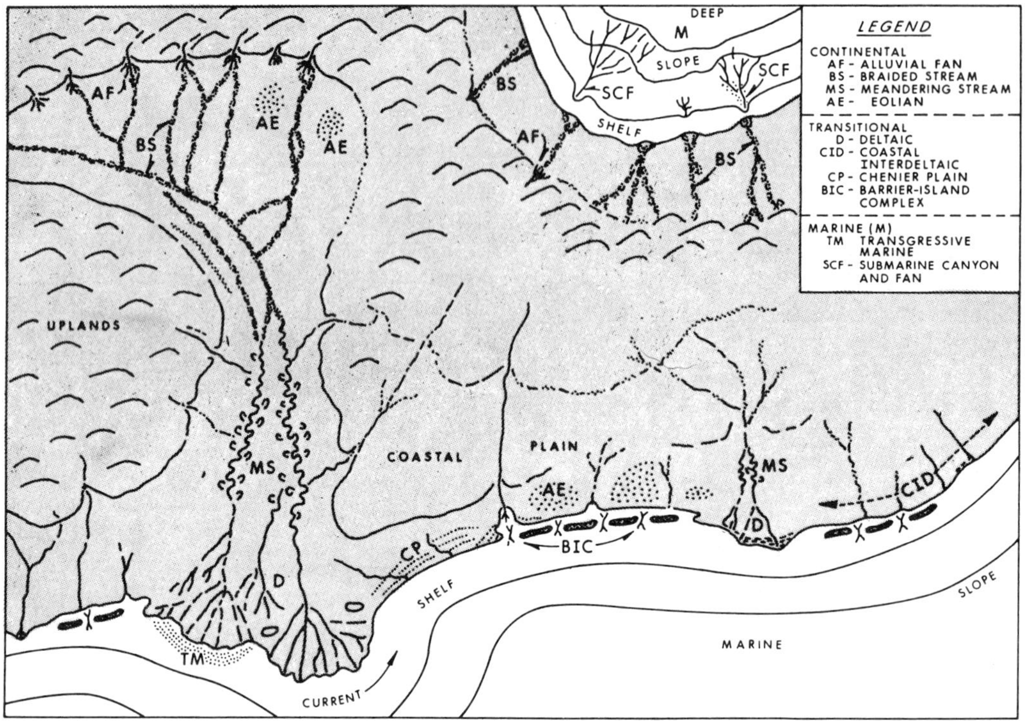

71-0363-4

FIG. 1—Some common models of clastic sedimentation. See Figures 2–4 for details.

included under the transitional group of environments.)
Transgressive-marine model
Submarine-canyon and submarine-fan model

RÉSUMÉ OF SIGNIFICANT STUDIES OF MODERN CLASTIC SEDIMENTATION

Alluvial Fans

Although much work has been done on alluvial fans, only a few papers discuss the relation of sedimentary sequences to depositional processes. Some of the more important contributions are by Rickmers (1913), Pack (1923), Blackwelder (1928), Eckis (1928), Blissenbach (1954), McKee (1957), Beaty (1963), Bull (1962, 1963, 1964, 1968, 1969, 1971), Hoppe and Ekman (1964), Winder (1965), Anstey (1965), Denny (1965, 1967), Legget *et al.* (1966), and Hooke (1967).

Braided Streams

Early papers on braided streams concerned channel patterns, origin of braiding, and physical characteristics of braided streams. Significant studies of this type were conducted by Lane (1957), Leopold and Wolman (1957), Chein (1961), Krigstrom (1962), Fahnestock (1963), and Brice (1964).

The relatively few papers on the relation of braided-stream deposits to depositional processes did not appear until the 1960s. Doeglas (1962) discussed braided-stream sequences of the Rhône River of France, and Ore (1963, 1965) presented some criteria for recognition of braided-stream deposits, based on the study of several braided streams in Wyoming, Colorado, and Nebraska. Fahnestock (1963) described braided streams associated with a glacial outwash plain in Washington. More recently, Williams and Rust (1969) discussed the sedimentology of a degrading braided river in the Yukon Territory, Canada. Coleman (1969) presented results of a study of the processes and sedimentary characteristics of one of the largest braided rivers, the Brahmaputra of Bangla Desh (formerly East Pakistan). N. Smith (1970) studied the Platte River from Denver, Colorado, to Omaha, Nebraska, and used the Platte model to interpret Silurian braided-stream deposits of the Appalachian region. Waechter (1970) has recently studied the braided Red River in the Texas Panhandle, and Kessler (1970, 1971) has investigated the Canadian River in Texas. Boothroyd (1970) studied braided streams associated with glacial outwash plains in Alaska.

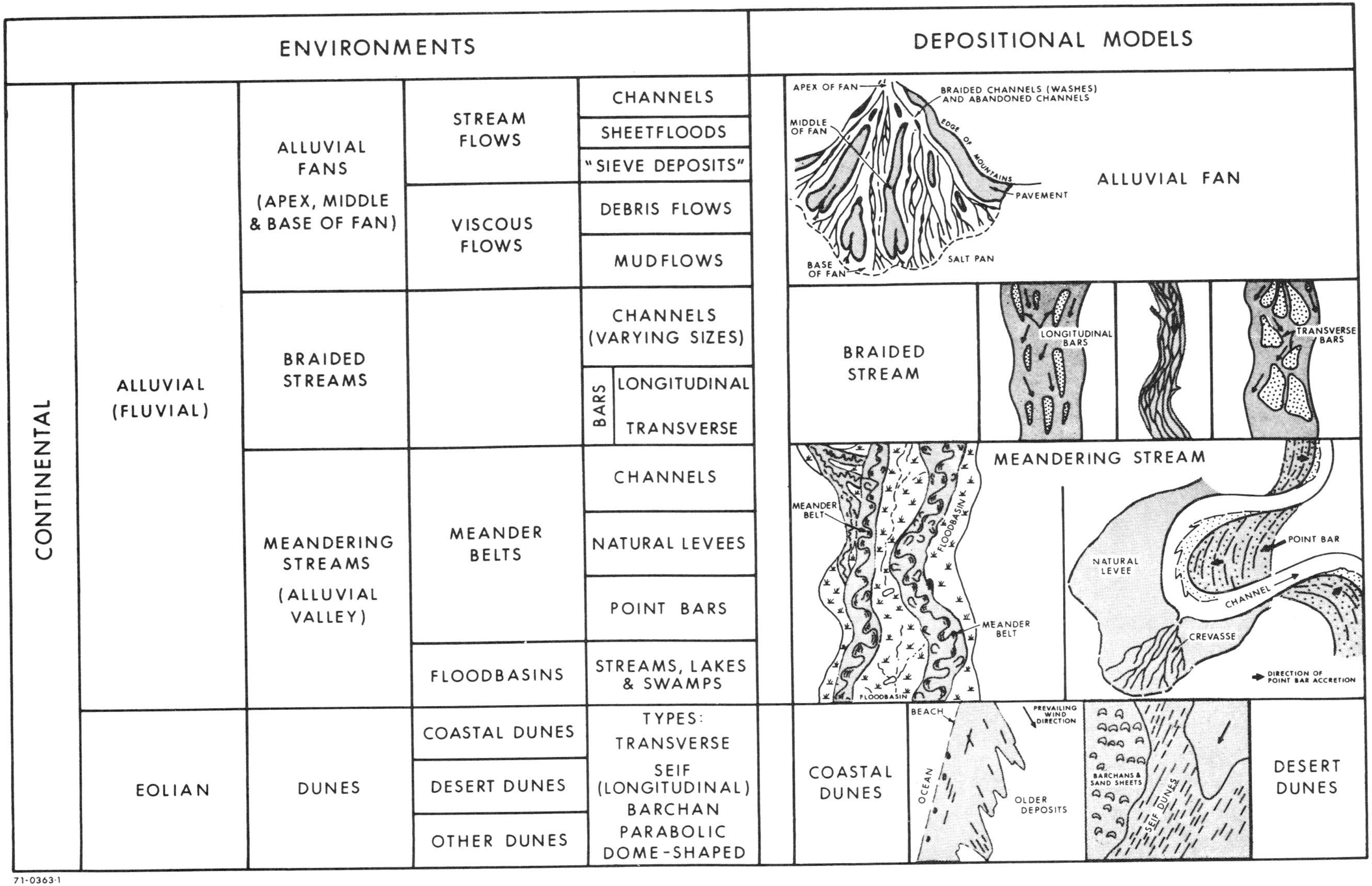

FIG. 2—Alluvial (fluvial) and eolian environments and models of clastic sedimentation.

ENVIRONMENTS						TYPES OF DELTAS (DEPOSITIONAL MODELS)
TRANSITIONAL	DELTAIC	UPPER DELTAIC PLAIN		MEANDER BELTS	CHANNELS	BIRDFOOT-LOBATE DELTA
					NATURAL LEVEES	
					POINT BARS	
				FLOODBASINS	STREAMS, LAKES & SWAMPS	
		LOWER DELTAIC PLAIN		DISTRIBUTARY CHANNELS	CHANNELS	CUSPATE-ARCUATE DELTA
					NATURAL LEVEES	
				INTER-DISTRIBUTARY AREAS	MARSH, LAKES, TIDAL CHANNELS & TIDAL FLATS	
		FRINGE	DELTA FRONT	INNER	RIVER-MOUTH BARS	ESTUARINE DELTA
					BEACHES & BEACH RIDGES	
					TIDAL FLATS	
				OUTER		
		DISTAL				

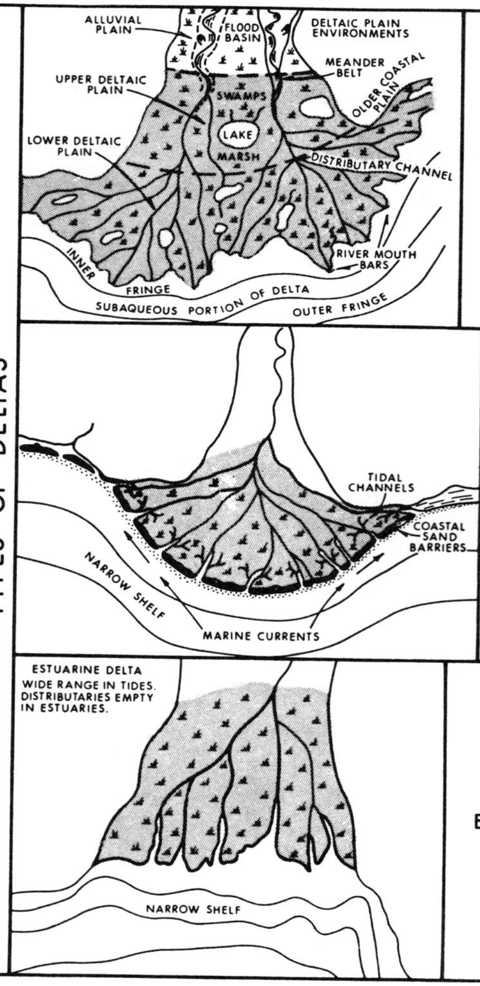

71-0363-2

FIG. 3—Deltaic environments and models of clastic sedimentation.

Meandering Streams

H. N. Fisk's studies of the Mississippi alluvial valley, conducted for the Mississippi River Commission during the period 1941–48, represent the first significant contribution on meandering stream environments and deposits. This pioneer effort provided geologists with knowledge of the fundamental processes of alluvial-valley sedimentation. Another study of a meandering stream, the Connecticut River, and its valley was made by Jahns (1947). Important work on alluvial sediments deposited by meandering streams was also done by Sundborg (1956) in Sweden, and by Frazier and Osanik (1961), Bernard and Major (1963), and Harms *et al.* (1963) on the Mississippi, Brazos, and Red River point bars, respectively. Thus, by 1963 the general characteristics of point-bar sequences, and the closely related abandoned-channel and flood-basin sequences, were sufficiently well established to permit geologists to recognize this type of sedimentary deposit in outcrops and in the subsurface.

Other important contributions were made by Allen (1965a) on the origin and characteristics of alluvial sediments, by Simons *et al.* (1965) on the flow regime in alluvial channels, by Bernard *et al.* (1970) on the relation of sedimentary structures to bed form in the Brazos valley deposits, and by McGowen and Garner (1970) on coarse-grained point-bar deposits.

Deltas

The early work by W. Johnson (1921, 1922) on the Fraser delta, Russell (1936) on the Mississippi delta, Sykes (1937) on the Colorado delta, and Fisk (1944) on the Mississippi delta provided a firm basis for subsequent studies of more than 25 modern deltas during the late 1950s and the 1960s.

Fisk continued his studies of the Mississippi delta for more than 20 years. His greatest contributions were concerned with the delta framework, the origin and character of delta-front sheet sands, and the development of bar-finger sands by seaward-migrating rivermouth bars.

Scruton's (1960) paper on delta building and the deltaic sequence represents results of API Project 51 on the Mississippi delta. Additional research on Mississippi delta sedimentation, sedimentary structures, and mudlumps was reported by Welder (1959), Morgan (1961), Morgan *et al.* (1968), Coleman *et al.* (1964), Coleman (1966b), Coleman and Gagliano (1964, 1965), and also by Kolb and Van Lopik (1966). Coleman and Gagliano (1964) also discussed and illustrated processes of cyclic sedimentation. The most recent papers on the Mississippi delta are by Frazier (1967), Frazier and Osanik (1969), and Gould (1970).

Studies of three small birdfoot deltas of Texas—the Trinity, Colorado, and Guadalupe—were made by McEwen (1969), Kanes (1970), and Donaldson (1966), respectively. In addition, Donaldson *et al.* (1970) presented a summary paper on the Guadalupe delta. These four contributions are valuable because each one presents photographs and logs of cores of complete deltaic sequences.

European geologists associated with the petroleum industry and universities also have made valuable contributions to our understanding of deltas. Kruit (1955) and Lagaaij and Kopstein (1964) discussed their research on the Rhône delta of southern France, Allen (1965c, 1970) summarized the geology of the Niger delta of western Africa, and van Andel (1967) presented a résumé of the work done on the Orinoco delta of eastern Venezuela. More recently, the Po delta of Italy was studied by B. Nelson (1970) and the Rhône delta of southern France by Oomkens (1970).

Other recent contributions on modern deltas are by Coleman *et al.* (1970) on a Malaysian delta, by R. Thompson (1968) on the Colorado delta in Mexico, and by Bernard *et al.* (1970) on the Brazos delta of Texas.

The deltaic model is probably the most complex of the clastic models. Although additional research is needed on this aspect of sedimentation, the studies listed have provided some valuable concepts and criteria for recognition of ancient deltaic facies.

Coastal-Interdeltaic Sediments

Valuable contributions to our knowledge of this important type of sedimentation have been made by several groups of geologists. In the Gulf Coast region, the extensive Padre Island–Laguna Madre complex was studied by Fisk (1959), and the chenier plain of southwestern Louisiana was studied by Gould and McFarlan (1959) and Byrne *et al.* (1959). The Galveston barrier-island complex of the upper Texas coast was investigated mainly by LeBlanc and Hodgson (1959), Bernard *et al.* (1959, 1962), and Bernard and LeBlanc (1965).

Among the impressive studies made by Europeans during the past 15 years are those by van Straaten (1954), who presented results of very

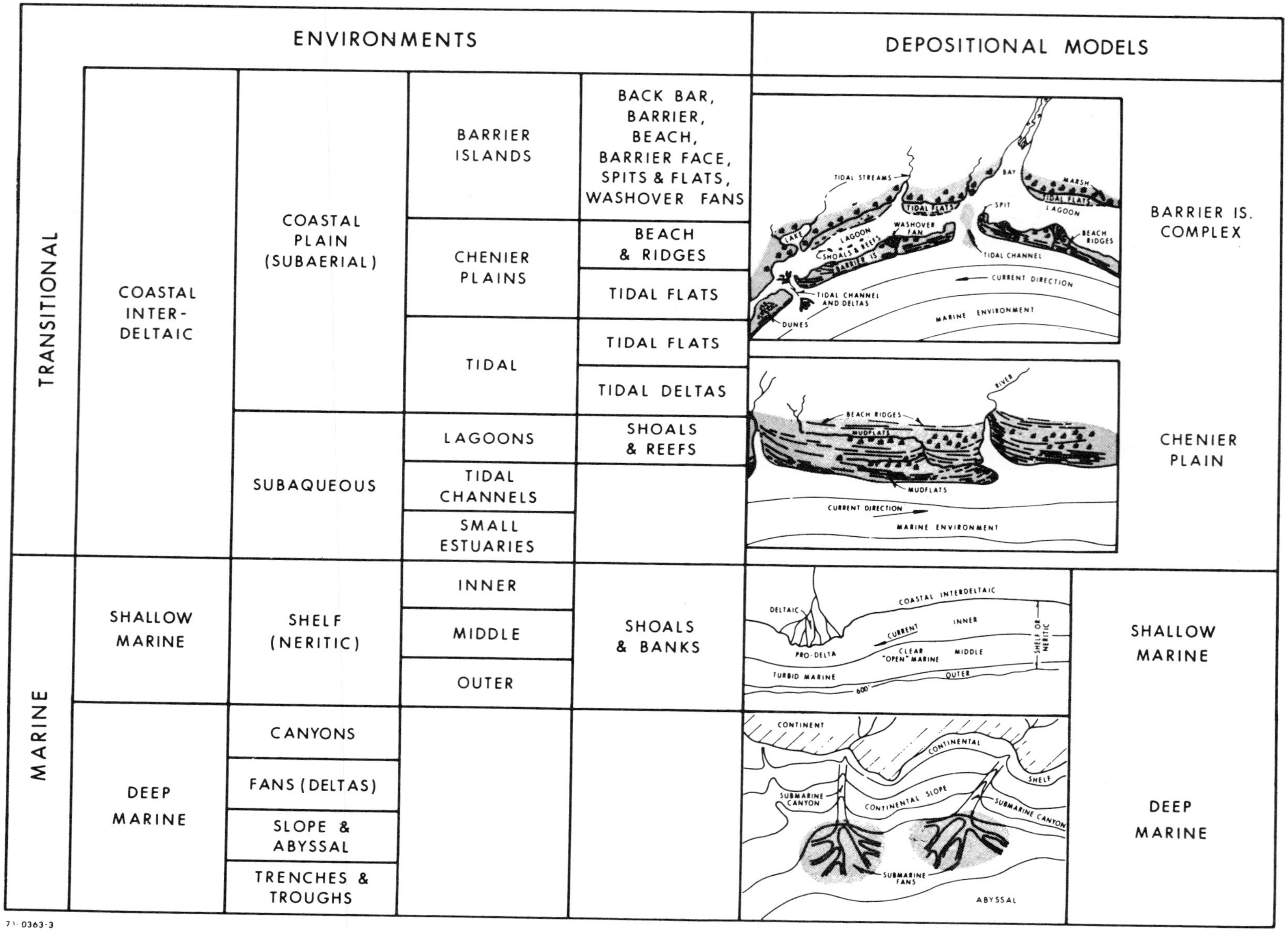

FIG. 4—Coastal-interdeltaic and marine environments and models of clastic sedimentation.

significant work on tidal flats, tidal channels, and tidal deltas of the northern Dutch coast, and by Horn (1965) and H. E. Reineck (1967), who reported on the barrier islands and tidal flats of northern Germany.

During the past several years, a group of geologists has conducted interesting research on the coastal-interdeltaic complexes which characterize much of the U.S. Atlantic Coast region. Hoyt and Henry (1965, 1967) published several papers on barriers and related features of Georgia. More recently, results of studies at the University of Massachusetts on recent coastal environments of New England were reported by Daboll (1969) and by the Coastal Research Group (1969).

In addition, Curray *et al.* (1969) described sediments associated with a strand-plain barrier in Mexico, and Potter (1967) summarized the characteristics of barrier-island sand bodies.

Eolian Sand Dunes

Prior to the middle 1950s, eolian depositional environments were studied principally by European geologists (Cooper, 1958). Since that time, the coastal sand dunes of the Pacific, Atlantic, and Gulf coasts of the United States, as well as the desert dunes of the United States and other countries, have been investigated by university professors and by geologists with the U.S. Geological Survey. Some of the most significant contributions, especially those concerned with dune stratification, are discussed in the section on the eolian model of clastic sedimentation.

Marine Sediments

Early work on modern marine sands, exclusive of those deposited adjacent to and related to interdeltaic and deltaic depositional environments, was conducted largely by scientists associated with Scripps, Woods Hole, and Lamont oceanographic departments. Several aspects of marine sediments were discussed by Trask *et al.* (1955), and the recent sands of the Pacific Ocean off California were studied by Revelle and Shepard (1939), Emery *et al.* (1952), and Emery (1960a). Stetson (1953) described the northwestern Gulf of Mexico sediments, and Ericson *et al.* (1952, 1955) and Heezen *et al.* (1959) investigated the Atlantic Ocean sediments. Later, Curray (1960), van Andel (1960), and van Andel and Curray (1960) reported results of the API project on the Gulf of Mexico. A few years later, results of the API project studies on the Gulf of California were reported by van Andel (1964) and van Andel and Shor (1964). Menard (1964) discussed sediments of the Pacific Ocean. For a more complete list of references to studies of recent marine sands, the reader is referred to Kuenen (1950), Guilcher (1958), Shepard *et al.* (1963), and Kukal (1971).

Much of the early research on modern marine environments was devoted to submarine canyons, fans, and basins considered by the investigators to be characterized mainly by turbidity-current sedimentation. Several scientists affiliated with Scripps and the University of Southern California published numerous papers on turbidites which occur in deep marine basins.

It is extremely difficult to observe the processes of turbidity-current sedimentation under natural conditions; consequently, the relations between sedimentary sequences and processes are still relatively poorly understood. Much of the research dealing with turbidity currents has been concerned with theory, laboratory models, and cores of deep-water sediments deposited by processes which have not been observed.

Alluvial-Fan Model of Clastic Sedimentation

Occurrence and General Characteristics

Alluvial fans occur throughout the world, adjacent to mountain ranges or high hills. Although they form under practically all types of climatic conditions, they are more common and best developed along mountains of bold relief in arid and semi-arid regions (Figs. 5, 6).

The alluvial-fan model has the following characteristics: (1) sediment transport occurs under some of the highest energy conditions within the entire realm of clastic sedimentation, (2) deposition of clastic sediment occurs directly adjacent to the areas of erosion which provide the sediments, and (3) deposits are of maximum possible range in size of clastic particles (from the largest boulders to clays) and are commonly very poorly sorted compared with other types of alluvial sediments (Fig. 5).

The size of individual alluvial fans is controlled by drainage-basin area, slope, climate, and character of rocks within the mountain range. Individual fans range in radius from several hundred feet to several tens of miles. Coalescing fans can occur in linear belts that are hundreds of miles long. Fan deposits usually attain their maximum thicknesses and grain size near the mountain base (apex of fan) and

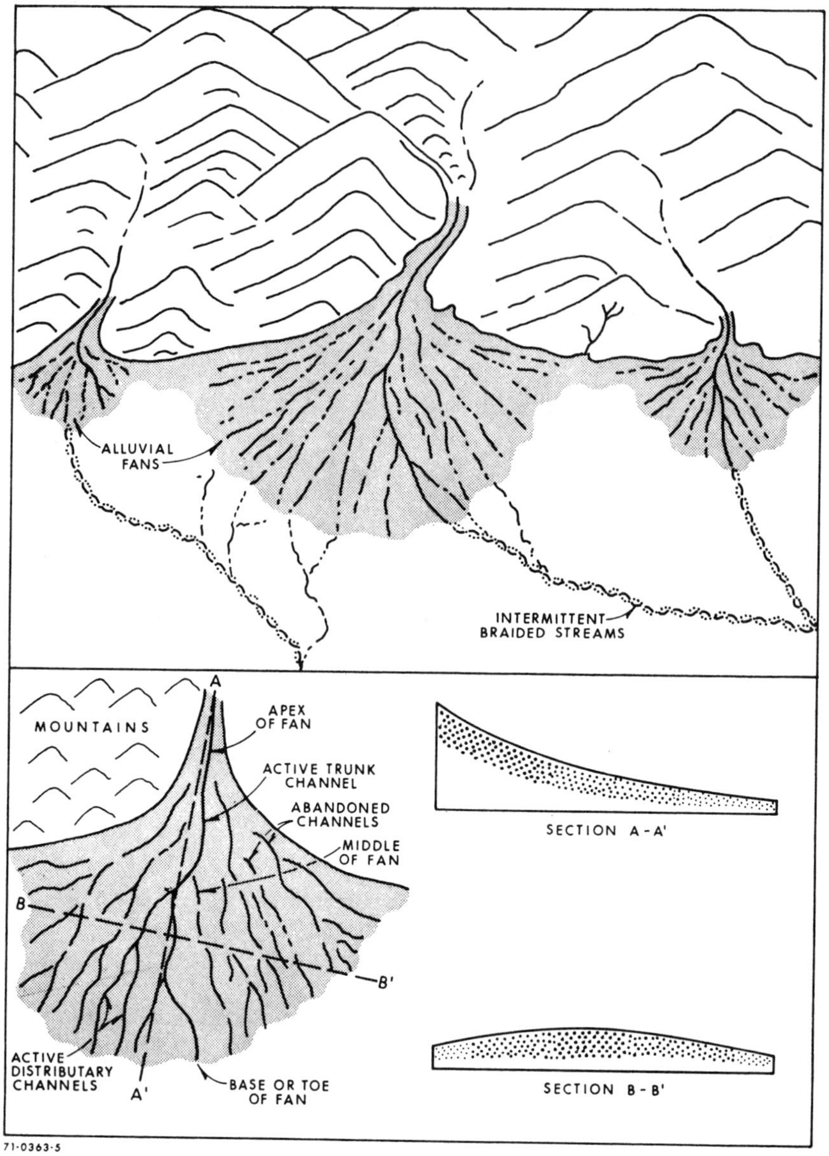

FIG. 5—Alluvial-fan model of clastic sedimentation.

gradually decrease in thickness away from the apex.

The alluvial-fan environments commonly grade downstream into braided-stream or playa-lake environments. In some areas, where mountains are adjacent to oceans or large inland lakes, alluvial fans are formed under both subaerial and submerged conditions. Such fans are now referred to as "Gilbert-type" deltas.

Alluvial-fan deposits form important reservoirs for groundwater in many areas, and adjacent groundwater basins are recharged through the fan deposits which fringe these basins.

Source, Transportation, and Deposition of Sediments

Tectonic activity and climate have a profound influence on the source, transportation, and deposition of alluvial-fan deposits. Uplift of mountain ranges results in very intensive erosion of rocks and development of a very high-gradient drainage system. The rate of weathering and production of clastic material is controlled mainly by rock characteristics and climate (temperature and rainfall).

Clastic materials are transported from source areas in mountains or high hills to alluvial fans by several types of flows: stream flows and sheetfloods and debris flows or mudflows. Sediment transport by streams is usually characteristic of large fans in regions of high to moderate rainfall. Mudflows or debris flows are more common on small fans in regions of low rainfall characterized by sudden and brief periods of heavy downpours.

Stream deposits—Streams which drain relatively small segments of steep mountain ranges have steep gradients; they may erode deep canyons and transport very large quantities of coarse debris. The typical overall stream gradient is concave upward, and the lowest gradient occurs at the toe of the fan (Fig. 5).

Hooke (1967) described a special type of stream-flow deposit, which he called "sieve deposits," on fans which are deficient in fine sediments. These gravel deposits are formed when water infiltrates completely into the fan. Bull (1969) described three types of water-laid sediments on alluvial fans: channel, sheetflood, and sieve deposits. Stream channels radiate outward from the fan apex and commonly are braided. The processes of channel migration, diversion, abandonment and filling, and development of new main channels and smaller distributary channels on the lower part of the fan surface are characteristic features. Most fan surfaces are characterized by one or a few active channels and numerous abandoned channels. Deposits on abandoned portions of gravelly and weathered fan surfaces are referred to as "pavement."

Alluvial-fan channel deposits have abrupt basal contacts and channel geometry; they are generally coarse. Bull (1972) described channel deposits as imbricated and massive or thick-bedded.

Heavy rainfall in mountainous source areas can result in floods on alluvial fans. The relatively shallow and wide fan channels are not capable of carrying the sudden influx of large volumes of water; consequently, the streams overtop their banks and flood part of the fan surface. The result is the deposition of thin layers of clastic material between channels. Bull (1969) reported sheetflood deposits to be finer grained than channel deposits, cross-bedded, and massive or thinly bedded.

Debris-flow deposits—Some workers refer to both fine-grained and coarse-grained types of plastic flowage in stream channels as mudflows, but others consider mudflows to be fine-grained debris flows. Examples of transportation and deposition of clastic sediments by mudflows were first described by Rickmers (1913) and Blackwelder (1928). The following conditions favor the development of mudflows: presence of unconsolidated material with enough clay to make it slippery when wet, steep gradients, short periods of abundant water, and sparse vegetation.

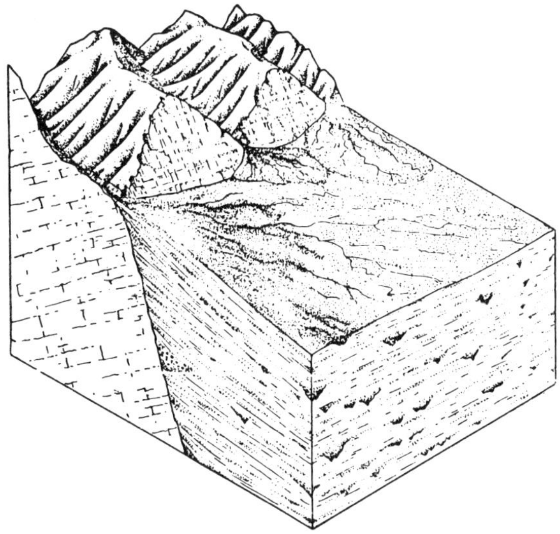

FIG. 6—Stratigraphic geometry of an alluvial fan. After Bull (1972).

Pack (1923) discussed debris-flow deposition on alluvial-fan surfaces. Debris flows occur as a result of very sudden, severe flooding of short duration. Beaty (1963) described eyewitness accounts of debris flows on the west flank of the White Mountains of California and Nevada. Debris flows follow channels, overtop the channel banks, and form lobate tongues of debris along channels. Debris-flow deposits are very poorly sorted, fine- to coarse-grained, and unstratified; they have abrupt margins. This type of deposit is probably most common on the upper parts of the fans between the apex and midfan areas.

Summary: Character and Geometry of Alluvial-Fan Deposits

Most of the alluvial-fan studies conducted thus far have been concerned primarily with the origin and general characteristics of fans and the distribution of sediments on the surfaces of fans. An exception is Bull's excellent summary paper (Bull, 1972), which contains significant data on the geometry of channel, sheetflood, debris-flow, mudflow, and sieve deposits. The abstract of Bull's paper is quoted below:

Alluvial fans commonly are thick, oxidized, orogenic deposits whose geometry is influenced by the rate and duration of uplift of the adjacent mountains and by climatic factors.

Fans consist of water-laid sediments, debris-flow deposits, or both. Water-laid sediments occur as channel, sheetflood, or sieve deposits. Entrenched stream channels commonly are backfilled with gravel that may be imbricated, massive, or thick bedded. Braided sheets of finer-grained sediments deposited downslope from the channel may be cross-bedded, massive, laminated, or thick bedded. Sieve deposits are overlapping lobes of permeable gravel.

Debris-flow deposits generally consist of cobbles and boulders in a poorly sorted matrix. Mudflows are fine-grained debris flows. Fluid debris flows have graded bedding and horizontal orientation of tabular particles. Viscous flows have uniform particle distribution and vertical preferred orientation that may be normal to the flow direction.

Logarithmic plots of the coarsest one percentile versus median particle size may make patterns distinctive of depositional environments. Sinuous patterns indicate shallow ephemeral stream environments. Rectilinear patterns indicate debris flow environments.

Fans consist of lenticular sheets of debris (length/width ratio generally 5 to 20) and abundant channel fills near the apex. Adjacent beds commonly vary greatly in particle size, sorting, and thickness. Beds extend for long distances along radial sections and channel deposits are rare. Cross-fan sections reveal beds of limited extent that are interrupted by cut-and-fill structures.

Three longitudinal shapes are common in cross section. A fan may be lenticular, or a wedge that is either thickest, or thinnest, near the mountains.

Ancient Alluvial-Fan Deposits

Some examples of ancient alluvial-fan deposits which have been reported from the United States, Canada, Norway, and the British Isles are summarized in Table 1, together with other types of alluvial deposits.

Braided-Stream Model of Clastic Sedimentation

Occurrence and General Characteristics

Braided streams occur throughout the world under a very wide range of physiographic and climatic conditions. They are common features on extensive alluvial plains which occupy a position in the clastic realm of sedimentation between the high-gradient alluvial-fan environment at the base of mountain ranges and the low-gradient meandering-stream model of sedimentation (downstream). In physiographic provinces characterized by mountainous areas adjacent to the sea, the braided-stream environment can extend directly to the coastline and thus constitute the predominant environment of alluvial deposition. In this type of situation, meandering streams do not exist (Fig. 7). The braided stream is also a common feature of glacial outwash plains associated with the fluvioglacial environment.

The braided-stream model is characterized by extremely variable rates of sedimentation in multiple-channel streams (Fig. 8), the patterns of which vary widely compared with meandering channels. Braided channels are usually wide and shallow; they contain numerous bars, are slightly sinuous or straight, and migrate at rapid rates. Stream gradients are high, are quite variable, and are less than those of alluvial fans but generally greater than those of meandering streams. Large fluctuations in discharge occurring over short periods of time are also common. The combination of steep gradients and high discharge rates results in the transporta-

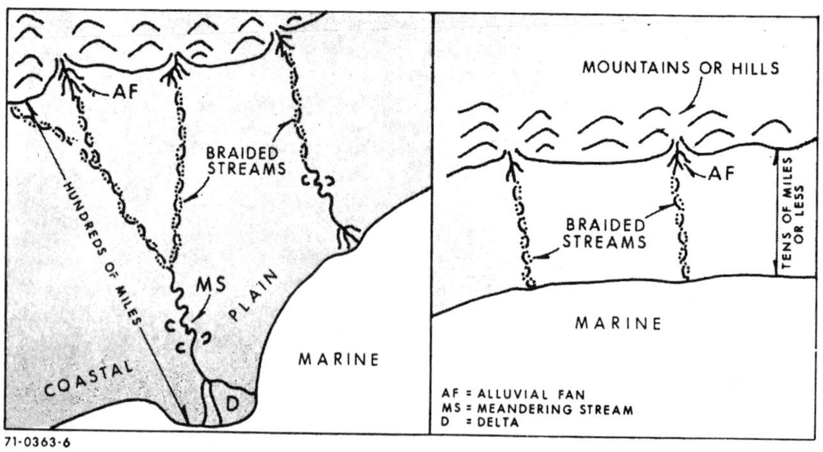

FIG. 7—Braided-stream model of clastic sedimentation.

tion and deposition of large amounts of coarse material, ranging from boulders to sand. Braided-stream deposits overall are finer than those of alluvial fans, coarser than those of meandering streams, and quite varied in stratification.

Source, Transportation, and Deposition of Sediments

Aggrading braided streams transport very large quantities of clastic material derived from a variety of sources, such as outwash plains, alluvial fans, mountainous areas, and broad plains. Unlike that of meandering streams, the bulk of the sedimentary load of most braided streams is transported as bed load. Rates of sediment transport and deposition are extremely variable, the maximum rate occurring during severe floods of short duration. High-gradient upstream segments of braided streams close to source areas are characterized by deposition of poorly sorted clastic sediments which

Table 1. Examples of Ancient Alluvial-Fan, Braided-Stream, and Meandering-Stream Deposits

Alluvial Fan	*Braided Stream*	*Meandering Stream*	*Composite*	*Author*
Arizona			Arizona	Melton, 1965
California				Crowell, 1954
California			California	Flemal, 1967
			California	Galehouse, 1967
Colorado				Boggs, 1966
			Colorado	Bolyard, 1959
		Colorado		Brady, 1969
			Colo. Plateau	Finch, 1959
			Colo. Plateau	Stokes, 1961
Colorado				Howard, 1966
Colorado				Hubert, 1960
Connecticut Valley				Klein, 1968
		Illinois		Hewitt *et al.*, 1965
		Illinois		Shelton, 1972
			Kansas	Lins, 1950
		Kansas		Shelton, 1971
	Llano Estacado			Bretz & Horberg, 1949
	Maryland	Maryland		Hansen, 1969
Massachusetts				Wessel, 1969
Massachusetts				Stanley, 1968
			Massachusetts	Mutch, 1968
		Michigan		Shideler, 1969
		Montana		Gwinn, 1964
	Mississippi	Mississippi		Berg & Cook, 1968
Montana	Montana			Gwinn & Mutch, 1965
			Montana	Shelton, 1967
Montana				Wilson, 1967, 1970
Montana				Beaty, 1961
			Nebraska	Exum & Harms, 1968
			Nebraska	Harms, 1966
	New York	New York		Buttner, 1968
	New Jersey, New York			Smith, 1970; Shelton, 1972
			North Dakota	Royse, 1970
			Oklahoma	Visher, 1965b
		Pennsylvania		Beutner *et al.*, 1967
	Pennsylvania			Smith, 1970
		Pennsylvania		Ryan, 1965
			Rhode Island	Mutch, 1968
S.W. USA				Bull, 1972
			Texas	Fisher & McGowen, 1969
Texas				McGowen & Groat, 1971
		Texas		McGowen & Garner, 1969; Shelton, 1971
		West Virginia		Beerbower, 1964, 1969
		Wyoming		Berg, 1968
Wyoming	Wyoming			Spearing, 1969
			Alberta	Byers, 1966
			Quebec	Dineley & Williams, 1968
Northeastern Canada				Klein, 1962
		Nova Scotia		Way, 1968
Northwest Territories	Northwest Territories	Northwest Territories		Miall, 1970
		England		Allen, 1964; Laming, 1966
Wales and Scotland				Bluck, 1965, 1967
		South Wales		Kelling, 1968
Norway				Nilsen, 1969
	Scotland			Williams, 1966, 1969
	Spain			Nagtegaal, 1966
	Spitsbergen	Spitsbergen		Moody-Stuart, 1966
		New South Wales		Conolly, 1965

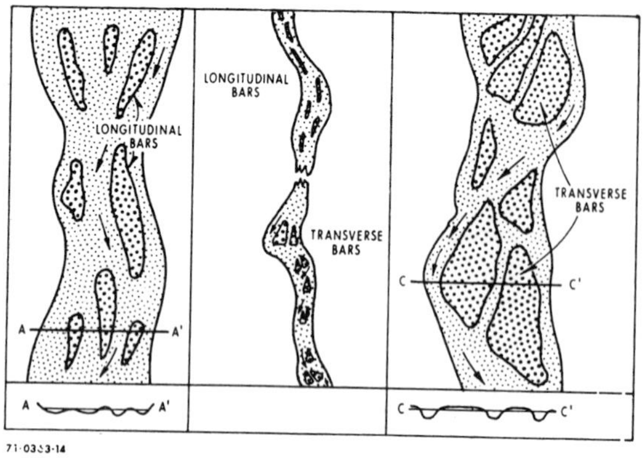

FIG. 8—Types of braided-stream channels and bars.

range in size from boulder to sand. Farther downstream, there is a gradual decrease in grain size and an increase in sorting.

The bed-load materials are transported under varying bed-form conditions, depending upon river stage. Coleman (1969) reported ripple and dune migration in the Brahmaputra River of Bangla Desh ranging from 100 ft to 2,000 ft (30–610 m) per day. Chein (1961) reported downstream movement of sandbars in the Yellow River of China to be as great as 180–360 ft (55–110 m) per day. (For comparison, the rate of bed-load movement in the meandering Mississippi is about 40 ft [12 m] per day.)

Process of channel division (braiding) by development of bars—The exact causes of channel division which results in the development of the braided pattern are not very well understood. Two methods in which channel division takes place have been described by Ore (1963) as follows:

> Leopold and Wolman (1957, p. 43–44), using results of both stream-table studies and observations of natural braided streams, discuss in some detail how channel division may take place. At any time, the stream is carrying coarser fractions along the channel center than at the margins, and due to some local hydraulic condition, part of the coarsest fraction is deposited. Finer material is, in part, trapped by coarser particles, initiating a central ridge in the channel. Progressive additions to the top and downstream end of the incipient bar build the surface toward water level. As progressively more water is forced into lateral channels beside the growing bar, the channels become unstable and widen. The bar may then emerge as an island due to downcutting in lateral channels, and eventually may become stabilized by vegetation. New bars may then form by the same process in lateral channels. These authors stress that braiding is not developed by the stream's inability to move the total quantity of sediment provided to it; as incapacity leads merely to aggradation without braiding. The condition requisite to braiding is that the stream cannot move certain sizes provided; that is, the stream is incompetent to transport the coarsest fraction furnished to a given reach. Observations for the present study substantiate the braiding process of Leopold and Wolman.
>
> Many features of streams, bars, and braided reaches result from changes in regimen (e.g., discharge, load, gradient), to a large extent representing seasonal fluctuations. Other features of bars result from normal evolution, and represent no change in regimen.
>
> The incipient longitudinal bar formed in a channel commonly has an asymmetric, downstream-pointing, crescentic shape. This coarse part is the "nucleus" of the bar, is coarser than successive additions to the downstream end, and largely retains its position and configuration as long as any part of the bar remains. During longitudinal bar evolution downstream of this incipient bar the water and its sediment load commonly sweep from one lateral channel diagonally across the downstream end of the bar, forming a wedge of sediment with an advancing front at its downstream edge. This wedge of sediment is higher at its downstream edge, both on the longitudinal bars described here, and where found as transverse bars to be considered later. The latter build up the channel floor, independent of longitudinal bar development, simply by moving downstream.
>
> After a certain evolutionary stage, bar height stops increasing because insufficient water for sediment transport is flowing over its surface, and deepening and widening of lateral channels slowly lower water level. From then on, the bar may be either stabilized by vegetation or dissected.
>
> Widening of a reach after bar deposition is in some cases associated with lateral dissection of the newly formed bar. Most erosion, however, apparently occurs on the outer channel margins. If water level remains essentially constant for long periods of time, lateral dissection may establish terraces along bar margins. A compound terrace effect may be established during falling water stages. The constant tendency of the stream to establish a cross-sectional profile of equilibrium is the basic cause of lateral cutting by the stream.
>
> Longitudinal bars which become awash during high-water stages may be dissected by small streams flowing transversely over their surfaces. In stream-table experiments, sediment added to a system eroding transverse channels on bar surfaces is first transported along lateral channels beside the bars. Eventually, these channels fill to an extent that sediment starts moving transversely over bar surfaces, and fills bar-top, transverse channels. The addition of sufficient sediment to fill lateral and bar-top channels often culminates in a transverse bar covering the whole bar surface evenly.
>
> Another process of braiding, in addition to that described by Leopold and Wolman, takes place in well sorted sediments, and involves dissection of transverse bars. This is in opposition to construction of longitudinal bars in poorly sorted sediment, the type of braiding discussed above. Both types may occur together geographically and temporally. During extended periods of high discharge, aggradation is by large tabular bodies of sediment with laterally sinuous fronts at the angle of repose migrating downstream. Stabilization of discharge or decrease in load after establishment of these transverse bars results in their dissection by anastomosing channels; bars in this case form as residual elements of the aggradational pattern.
>
> The transient nature of braided stream depositional surfaces is characteristic of the environment. The streams and depositional areas within the stream exhibit profound lateral-migration tendencies, especially during

periods of high discharge. Channel migration takes place on several scales. Individual channels erode laterally, removing previously deposited bars. They divide and coalesce, and several are usually flowing adjacent to one another concurrently within the main channel system. The whole channel system, composed of several flowing channels with bars between, also exhibits migrating tendencies.

Braided-stream deposits—Our knowledge of modern braided-stream deposits has increased substantially during the past several years as a result of studies of several rivers in Wyoming, Colorado, and Nebraska by Ore (1963, 1965); the Brahmaputra River of Bangla Desh (formerly East Pakistan) by Coleman (1969); the Platte River of Colorado and Nebraska by N. Smith (1970); the Red River of the Texas panhandle by Waechter (1970); the Canadian River of northwest Texas by Kessler (1970 1971); and the Copper River of Alaska by Boothroyd (1970). These studies revealed that braided-stream deposits are laid down principally in channels as longitudinal bars and transverse bars. Abandoned-channel deposits (channel fills) have been reported by Doeglas (1962) and Williams and Rust (1969).

According to Ore (1963, 1965), longitudinal-bar deposits occur mainly in upstream channel segments and transverse bars are more common in downstream segments; however, in some places these two types of bars occur together (Fig. 8). Longitudinal-bar deposits are lens-shaped and elongated in the downstream direction. Grain size decreases downstream from coarse to fine in an individual bar; deposits are poorly sorted and mainly horizontally stratified but laterally discontinuous. Transverse-bar deposits occur as long thin wedges and are highly dissected by channels. The downstream edges of transverse bars migrate to produce planar cross-stratification and some festoon crossbedding. Sediments of transverse bars are generally finer and better sorted than those of longitudinal bars.

N. Smith (1970) described some very significant relations between types of bars, stratification, and grain size in the Platte River. In the upstream segment in Colorado, the deposits consist mainly of longitudinal bars characterized by low-relief stratification, generally horizontally bedded but including some festoon crossbedding. The downstream channel segment in Nebraska is characterized by transverse-bar deposits consisting of better sorted, fine-grained sand with abundant tabular cross-stratification and some festoon crossbedding.

The Red River braided-stream sediments of West Texas consist of longitudinal-bar deposits with low-angle or horizontal stratification; they are deposited during waning flood stages (Waechter, 1970). Low-river-stage deposits consist mainly of migrating transverse-bar deposits (in channels) with tabular cross-stratification and some festoon crossbedding. The migration of very shallow channels results in stratification sets that are horizontal, tabular or lenticular, and laterally discontinuous.

Kessler (1970) reported longitudinal-bar deposits consisting mainly of fine sand in upstream reaches of the Canadian River in West Texas. Transverse-bar deposits are predominant in the downstream part of the area studied. Kessler (1971) also discussed individual flood sequences of deposits which contain parallel bedding and tabular and small-ripple cross-laminations. These sequences are covered by clay drapes and are laterally discontinuous.

Coleman (1969) presented the results of a significant study of one of the largest braided rivers of the world, the Brahmaputra in Bangla Desh. This river is 2–6 mi (3–9.5 km) wide and migrates laterally as much as 2,600 ft (790 m) per year; deposition of sediments in its channels during a single flood occurs in a definite sequence of change, ranging from ripples up to 5 ft (1.5 m) high that migrate downstream 400 ft (120 m) per day to sand waves 50 ft (15 m) high that migrate up to 2,000 ft (610 m) per day.

Williams and Rust (1969) presented results of a very detailed study of a 4-mi (6.5 km) segment of a degrading braided stream, the Donjek River of the Yukon Territory, Canada. They divided the bar and channel deposits, which range from coarse gravels to clays, into seven facies. Ninety-five percent of the bar deposits are of the longitudinal type and consist of gravel, sand, and some finer sediments. Abandoned-channel deposits consist of gradational sequences of gravels, sand, and clays that become finer upward.

Summary: Braided-Stream Deposits

Most of the sediments of modern braided streams studied during the past decade have been referred to by authors as transverse- or longitudinal-bar deposits. These sediments were deposited within braided channels during varying discharge conditions ranging from low water to flood stage. Thus, all longitudinal and transverse bars should be considered as a special type of bed form occurring within active braided channels.

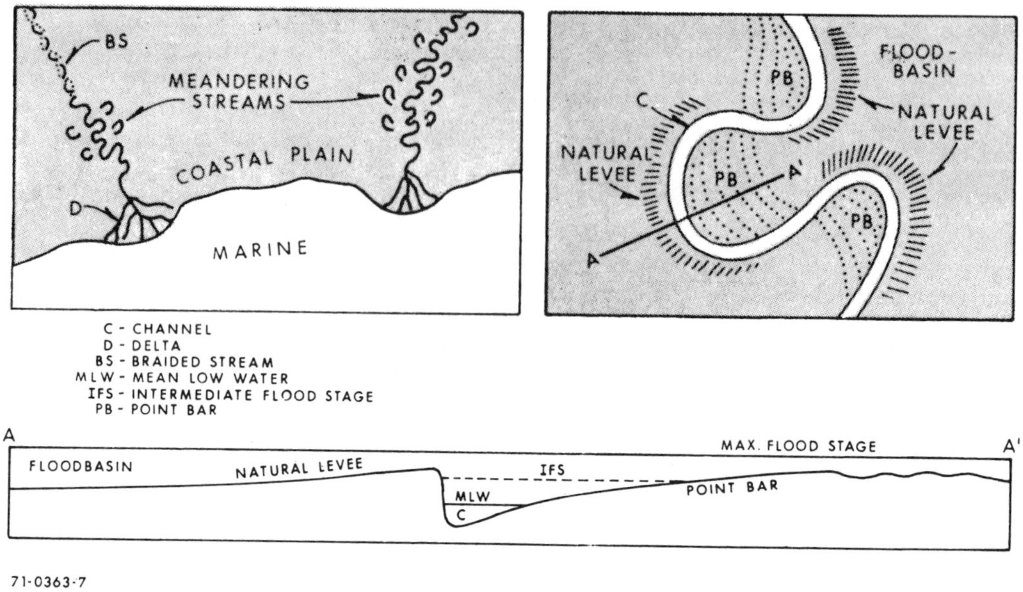

FIG. 9—Setting and general characteristics of meandering-stream model of clastic sedimentation.

Studies by Doeglas (1962) and Williams and Rust (1969) are significant because they describe abandoned-channel deposits. Doeglas discussed the methods of channel abandonment and described the channel-fill deposits as coarse grained, with channel or festoon laminations, in the upstream portions of abandoned channels, and as fine grained, silty, and rippled in the downstream portions of abandoned channels.

Ancient Braided-Stream Deposits

Some examples of ancient braided-stream deposits which have been reported from the United States, Spitsbergen, and Spain are summarized in Table 1.

Meandering-Stream Model of Clastic Sedimentation

Occurrence and General Characteristics

Meandering streams generally occur in coastal-plain areas updip from deltas and downdip from the braided streams. The axis of sedimentation is usually perpendicular to the shoreline (Fig. 9).

This model is characterized by a single-channel stream which is deeper than the multichannel braided stream. Meandering streams usually have a wide range in discharge (cu ft/sec) which varies from extended periods of low-water flow to flood stages of shorter duration. Flooding can occur one or more times per year and major flooding once every several years. The meandering channel is flanked by natural levees and point bars, and it migrates within a zone (meander belt) about 15 to 20 times the channel width. Channel segments are abandoned and filled with fines as new channels develop.

Source, Transportation, and Deposition of Sediments

Sediments are derived from whatever type of deposit occurs in the drainage area. Clays and fine silts are transported in suspension (suspended load), and coarser sediments such as sand, gravel, and pebbles are transported as bed load. Sediment transport and deposition during extended low-water stages are confined to the channel and can be nil or very slow. Maximum sediment transport occurs during rising flood stage when the bed of the channel is scoured.

The maximum rate of sediment deposition occurs during falling flood stages. Grain size depends on the type of sediment available to the channel; the coarsest sediments are deposited in the deepest part of the channel, and the finest sediments accumulate in floodbasins and in some parts of the abandoned channels.

Channel migration and deposition of point-bar sediments—The most important processes of sedimentation in the meandering-stream model are related to channel migration which occurs as a result of bank caving and point-bar accretion (Fig. 10). The process of bank caving occurs most rapidly during falling flood

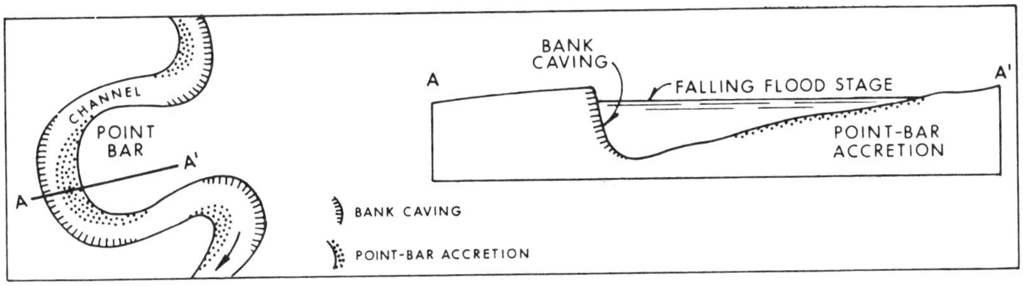

FIG. 10—Areas of bank caving and point-bar accretion along a meandering channel.

stage, when currents of maximum velocities are directed against the concave bank. Bank caving occurs at maximum rates in bends where the bed and bank materials are very sandy. Rates are much slower in areas where banks are characterized by clayey sediments (Fisk, 1947).

Deposition occurs on the convex bar (point bar) simultaneously with bank caving on the concave bank.

Bank caving and point-bar accretion result in channel migration and the development of the point-bar sequence of sediments (Fig. 11). The point bar is probably the most common and significant environment of sand deposition. The thickness of this sequence is governed by channel depths. Point-bar sequences along the Mississippi River attain thicknesses in excess of 150 ft (45 m). Medium-size rivers like the Brazos of Texas produce point-bar sequences that are 50 ft (15 m) thick (Bernard *et al.*, 1970).

Channel diversions and filling of abandoned channels—The process of channel diversion and channel abandonment is another characteristic feature of meandering streams. There are two basic types of diversion and abandonment: (1) the neck or chute cutoff of a single meander loop and (2) the abandonment of a long channel segment as a result of a major stream diversion (Fisk, 1947).

Meander loops which are abandoned as a result of neck or chute cutoffs become filled with sediment (Fig. 12A). The character of the channel fill depends on the orientation of the abandoned loop with respect to the direction of flow in the new channel. Meanders oriented with their cutoff ends pointing downstream (Fig. 12B) are filled predominantly with clays (clay plugs); those oriented with the cutoff ends pointing upstream are filled principally with sands and silts.

A major channel diversion is one which results in the abandonment of a long channel segment or meander belt, as shown in Figure 13. Channeling of flood water in a topographically low place along the bank of the active channel can rapidly erode unconsolidated sediments and create a new channel. This process can happen during a single flood or as a result of several floods. The newly established channel has a gradient advantage across the topographically lower floodbasin. A diversion can occur at any point along the channel.

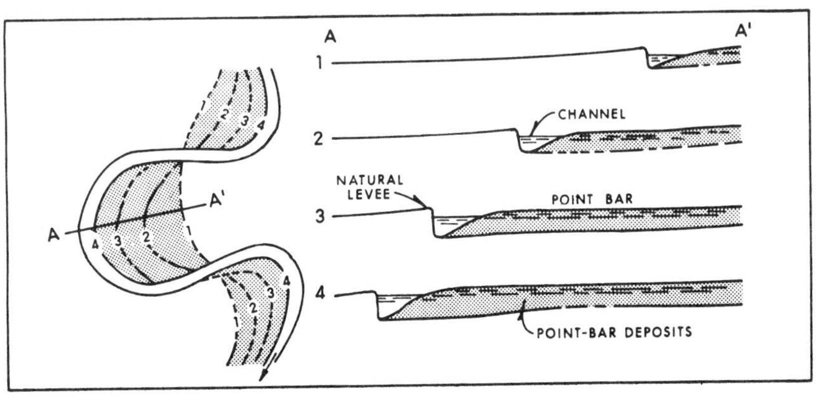

FIG. 11—Development of point-bar sequence of sediments.

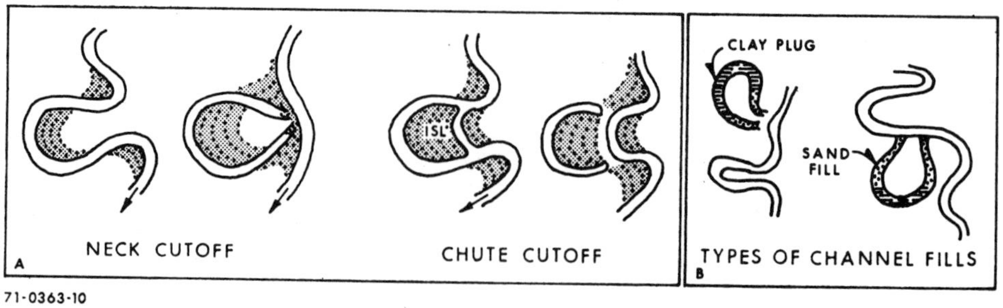

FIG. 12—Channel diversion, abandonment, and filling as a result of neck and chute cutoffs.

The character of the sediments which fill long channel segments is governed by the manner of channel diversion. Abrupt abandonment (during a single flood or a few floods) results in the very rapid filling of only the upstream end of the old channel, thus creating a long sinuous lake. These long, abandoned channels (lakes) fill very slowly with clays and silts transported by flood waters (Fig. 14, left).

Gradual channel abandonment (over a long period) results in very gradual channel deterioration. Diminishing flow transports and deposits progressively smaller amounts of finer sands and silts (Fig. 14, right).

Summary: Characteristics of Meander-Belt and Floodbasin Deposits

The meandering-stream model of sedimentation is characterized by four types of sediments: the point bar, abandoned channel, natural levee, and floodbasin. The nature of each of these four types of sediments and their interrelations are summarized in Figure 15.

Only two main types of sand bodies are associated with a meandering stream: the point-bar sands and the abandoned-channel fills. The former, which are much more abundant than the latter, occur in the lower portion of the point-bar sequence and constitute at least 75 percent of the sand deposited by a meandering stream. Coalescing point-bar sands can actually form a "blanketlike" sand body of very large regional extent. The continuity of sand is interrupted only by the "clay plugs" which occur in abandoned meander loops or in the last channel position of meander belts which have been abandoned abruptly.

Examples of ancient alluvial deposits of meandering-stream origin which have been reported in the literature are summarized in Table 1.

DELTAIC MODELS OF CLASTIC SEDIMENTATION

Occurrence and General Characteristics

Deltaic sedimentation occurs in the transitional zone between continental and marine (or inland seas and lakes) realms of deposition. Deltas are formed under subaerial and subaqueous conditions by a combination of fluvial and marine processes which prevail in an area where a fluvial system introduces land-derived sediments into a standing body of water.

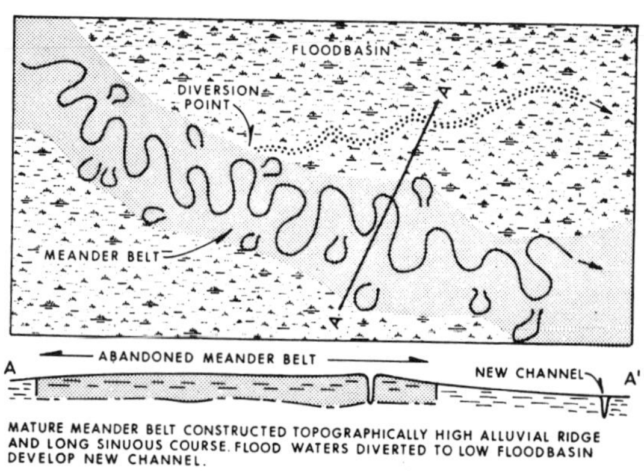

FIG. 13—Major channel diversion and abandonment of a meander belt.

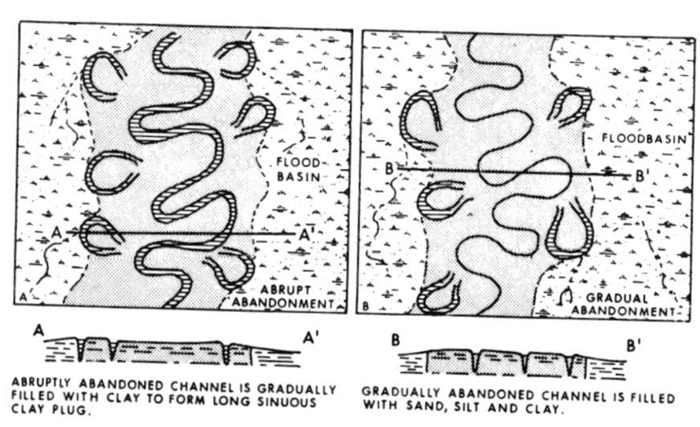

FIG. 14—Variations in character of abandoned channel fill typical of meander belts.

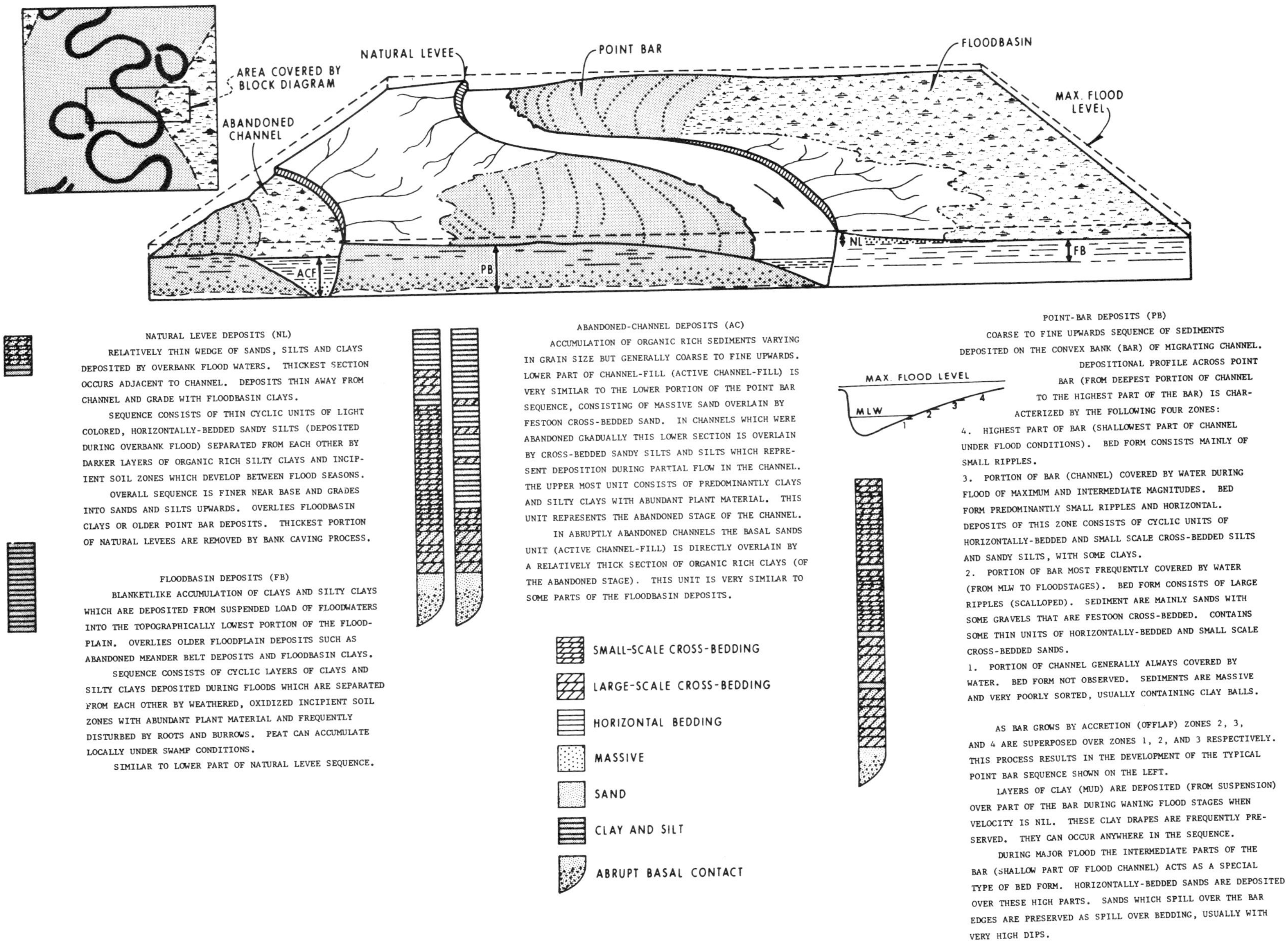

FIG. 15—Characteristics of meander-belt and floodplain deposits.

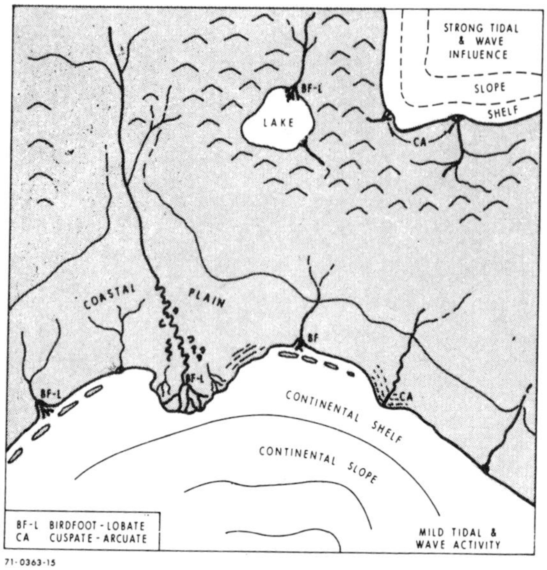

FIG. 16—Occurrence of deltaic models of clastic sedimentation.

Large deltas usually are associated with extensive coastal plains; however, all coastal plains do not include large deltas. The deltaic environment occurs downstream from the meandering-stream environment and is directly adjacent to, and updip (landward) from, the marine environment; it is flanked by the coastal-interdeltaic environment. Most large deltas occur on the margins of marine basins, but smaller deltas also form in inland lakes, seas, and coastal lagoons and estuaries (Fig. 16).

That portion of a delta which is constructed under subaerial conditions is called the "deltaic plain"; that portion which forms under water is called the "delta front," "delta platform," and "prodelta." The bulk of the deltaic mass is deposited under water.

Deltas are considered to be extremely important because they are the sites of deposition of sand much father downdip than the interdeltaic environment, as well as being the sites where clastic deposition occurred at maximum rates.

Source and Transportation of Sediments

Sediments deposited in large deltas are derived from extensive continental regions which are usually composed of rock types of varied compositions and geologic ages. Thus, the composition of deltaic sediments can be quite varied.

The sediment load of rivers consists of two parts: (1) the clays and fine silts transported in suspension and (2) the coarser silts and sands, and in some cases gravels, transported as bed load. The ratio of suspended load to bed load varies considerably, depending upon the rock types and climatic conditions of the sediment-source areas. The suspended load is generally much greater than the bed load.

The transportation of sediment to a delta is an intermittent process. Most rivers transport the bulk of their sediments during flood stages. During extended periods of low discharge, rivers contribute very little sediment to their deltas.

The extent to which deltaic sediments are dispersed into the marine environment is dependent upon the magnitude of the marine processes during the period that a river is in flood stage. Maximum sediment dispersal occurs when a river with a large suspended load reaches flood stage at the time the marine environment is most active (season of maximum currents and wave action). Minimum dispersal occurs when a river with a small suspended load (high bed load) reaches flood stage at a time when the marine environment is relatively calm.

Size of deltas—There is an extremely wide range in the size of deltas;[4] modern deltas range in area from less than 1 sq mi (2.6 sq km) to several hundreds of square miles. Some large deltaic-plain complexes are several thousand square miles in area. Delta size is dependent upon several factors, but the three most important are the sediment load of the river; the intensity of marine currents, waves, and tides; and the rate of subsidence. For a given rate of subsidence, the ideal condition for the construction of a large delta is the sudden large influx of sediments in a calm body of water with a small tidal range. An equally large sediment influx into a highly disturbed body of water with a high tidal range results in the formation of a smaller delta, because a large amount of sediment is dispersed beyond the limits of what can reasonably be recognized as a delta. Rapid subsidence enhances the possibility for a large fluvial system to construct a large delta.

Types of deltas—A study of modern deltas of the world reveals numerous types. Bernard

[4] Published figures on areal extent of deltas are based on size of the deltaic plain and do not include submerged portions of the delta, which in many cases are as large as or larger than the deltaic plains.

(1965) summarized some of the factors which control delta types as follows:

> Deltas and deltaic sediments are produced by the rapid deposition of stream-borne materials in relatively still-standing bodies of water. Notwithstanding the effects of subsidence and water level movements, most deltaic sediments are deposited off the delta shoreline in the proximity of the river's mouth. As these materials build upward to the level of the still-standing body of water, the remainder of deltaic sediments are deposited onshore, within the delta's flood plains, lakes, bays, and channels.
>
> Nearly 2,500 years ago, Herodotus, using the Nile as an example, stated that the land area reclaimed from the sea by deposition of river sediments is generally deltoid in shape. The buildup and progradation of deltaic sediments produces a distinct change in stream gradient from the fluvial or alluvial plain to the deltaic plain. Near the point of gradient change the major courses of rivers generally begin to transport much finer materials, to bifurcate into major distributaries, and to form subaerial deltaic plains. The boundaries of the subaerial plain of an individual delta are the lateral-most distributaries, including their related sediments, and the coast line. Successively smaller distributaries form sub-deltas of progressively smaller magnitudes.
>
> Deltas may be classified on the basis of the nature of their associated water bodies, such as lake, bay, inland sea, and marine deltas. Other classifications may be based on the depth of the water bodies into which they prograde, or on basin structure.
>
> Many delta types have been described previously. Most of these have been related to the vicissitudes of sedimentary processes by which they form. Names were derived largely from the shapes of the delta shorelines. The configuration of the delta shores and many other depositional forms expressed by different sedimentary facies appear to be directly proportional to the relative relationship of the amount or rate of river sediment influx with the nature and energy of the coastal processes. The more common and better understood types, listed in order of decreasing sediment influx and increasing energy of coastal processes (waves, currents, and tides), are: birdfoot, lobate, cuspate, arcuate, and estuarine. The subdeltas of the Colorado River in Texas illustrate this relationship. During the first part of this century, the river, transporting approximately the same yearly load, built a birdfoot-lobate type delta in Matagorda Bay, a low-energy water body, and began to form a cuspate delta in the Gulf of Mexico, a comparatively high-energy water body. Many deltas are compounded; their subdeltas may be representative of two or more types of deltas, such as birdfoot, lobate, and arcuate. Less-known deltas, such as the Irrawaddy, Ganges, and Mekong, are probably mature estuarine types. Others, located very near major scarps, are referred to the "Gilbert type," which is similar to an alluvial fan.

Additional studies of modern deltas are required before a more suitable classification of delta types can be established. J. M. Coleman (personal commun.) and his associates, together with the Coastal Studies Institute at Louisiana State University, are presently conducting a comprehensive investigation of more than 40 modern deltas. Results of their studies undoubtedly will be a significant contribution toward the solution to this problem.

Only three types of deltas will be considered in this report: the birdfoot-lobate, the cuspate-arcuate, and the estuarine.

Sedimentary Processes and Deposits of the Birdfoot-Type Delta

The processes of sedimentation within a delta are much more complex and variable than those which occur in the meandering-stream and coastal-interdeltaic environment of sedimentation. It is impossible to discuss these deltaic processes in detail in a short summary paper such as this; therefore, only a brief summary of the following significant processes is presented.

1. Dispersal of sediment in the submerged parts of the delta (from river mouths seaward);
2. Formation of rivermouth bars, processes of channel bifurcation, and development of distributary channels;
3. Seaward progradation of delta, deposition of the deltaic sequence of sediments, and abandonment and filling of distributary channels; and
4. Major river diversions, abandonment of deltas, and development of new deltas.

Dispersal and deposition of sediments—Riverborne sediments which are introduced in a standing body of water (a marine body or inland lakes and seas) are transported in suspension (clays and fine silts) and as bed load (coarse silts, sands, and coarser sediments). Most of the sands and coarse silts are deposited in the immediate delta-front environment as rivermouth bars and slightly beyond the bar-front zone. The degree of sand dispersal is, of course, controlled by the level of marine energy; however, in most birdfoot deltas, sands are not transported beyond 50-ft (15 m) water depths. Fisk (1955) referred to the sands deposited around the margins of the subaerial deltaic plain as "delta-front sands," and they are called "delta-fringe sands" herein.

The finer sediments (clays and fine silts), which are transported in suspension, are dispersed over a much broader area than the fringe sands and silts. The degree of dispersal is governed by current intensity and behavior. Accumulations of clays seaward of the delta-fringe sands are referred to as "prodelta" or "distal clays" (Fig. 17).

Channel bifurcation and development of distributary channels—Some of the most significant deltaic processes are those which result in the origin and development of distributary

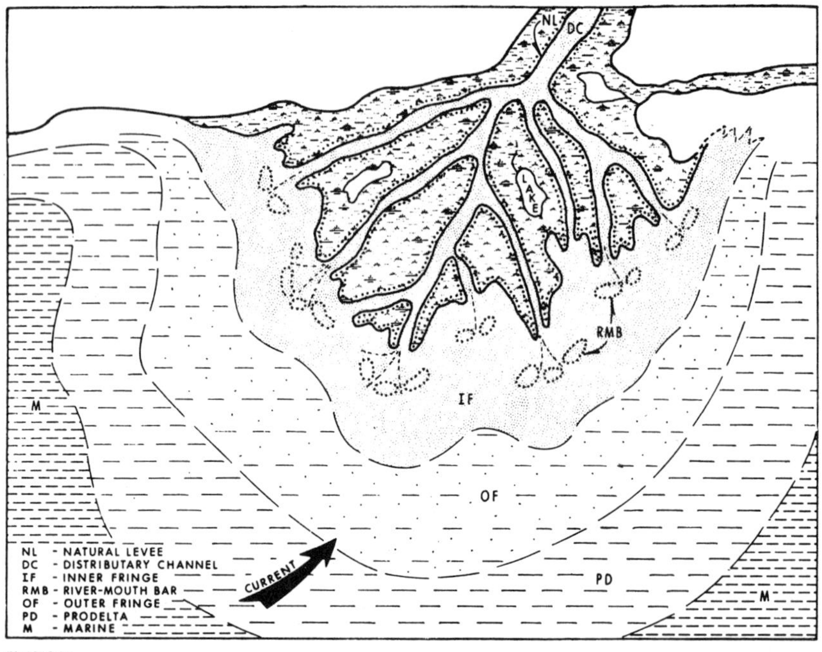

FIG. 17—Distribution of distributary-channel and fringe sands in a birdfoot-lobate delta.

channels. Welder (1959) conducted a detailed study of these processes in a part of the Mississippi delta, and Russell (1967a) summarized the origin of branching channels, as follows:

The creation of branching channels is determined by the fact that threads of maximum turbulence and turbulent interchange (Austausch; 1.2.3; 3.5) lie deep and well toward the sides of channels, particularly if they have flat beds (typical of clay and fine sediments in many delta regions). These threads are associated with maximum scour and from them, sediment is expelled toward areas of less turbulence and Austausch. Significant load is propelled toward mid-channel, where shoals are most likely to form.

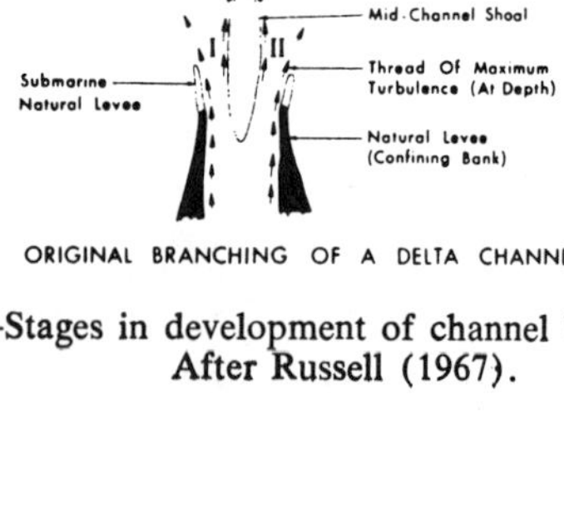

FIG. 18—Stages in development of channel bifurcation. After Russell (1967).

At its mouth, the current of a delta channel continues forward (as a result of momentum) and creates *jet flow* into the lake or sea it enters. After leaving the confinement imposed by fixed banks, however, the current flares marginally to some extent (widening the jet, reducing its velocity, and eventually dissipating its flow energy). Near the termination of confining banks the jet flow is concentrated and moves ahead into relatively quiet water. With flaring of jet flow comes an increase in spacing between threads of most intense turbulence and exchange. There is a tendency toward scour below each thread, but the exchange process sends most of the entrained material toward marginal quiet water on both sides (Fig. 8 [Fig. 18 of this paper]). Deposition creates a submarine natural levee on the outer side of each thread. Sediment is also attracted toward and deposited in the widening area of mid-channel water, where it builds a shoal. The channel divides around the shoal, creating two distributaries, each of which develops its own marginal threads of maximum turbulence, perpetuating conditions for other divisions below each new channel mouth. If not opposed by wave erosion and longshore currents, the subdivision continues in geometric progression (2, 4, 8, 16, etc.) as the delta deposit grows forward.

The marginal natural levees are submarine features at first and fish may swim across their crests. Later they grow upward, and for awhile become areas where logs and other flotsam accumulate and where birds walk with talons hardly submerged. Salt- or fresh-water-tolerant grasses invade the shallow water and newly created land, first along levee crests, later to widen as the levees grow larger. Salicornia and other plants become established pioneer trees such as willows, and eventually in the plant succession comes the whole complex characteristic of natural levees upstream. In tropical areas mangroves are likely to become the dominant trees.

A similar conversion exists in mid-channel, where the original shoal becomes land and either develops into a lenticular or irregular island or becomes the point of land at the head of two branching distributaries.

Progradation of delta and deposition of deltaic sequence—Fisk's discussion of the process of distributary-channel lengthening (progradation of delta seaward) is probably one of the most significant of his many contributions on deltaic sedimentation (Fisk, 1958). His description of this important aspect of delta development is presented below. (Stages in the development of a birdfoot-type delta are shown in Figure 19.)

Each of the pre-modern Mississippi River courses was initiated by an upstream diversion, similar to the one presently affecting the Mississippi as the Atchafalaya River enlarges (Fisk, 1952). Stream capture was a gradual process involving increasing flow through a diversional arm which offered a gradient advantage to the gulf. After capture was effected, each new course lengthened seaward by building a shallow-water delta and extending it gulfward. Successive stages in course lengthening are shown diagrammatically on Figure 2 [Fig. 19, this paper]. The onshore portion of the delta surface . . . is composed of distributaries which are flanked by low natural levees, and interdistributary troughs holding near-sea-level marshes and shallow water bodies. Channels of the principal distributaries extend for some distance across the gently sloping offshore surface of the delta to the inner margin of the steeper delta front where the distributary-mouth bars are situated. The offshore channels are bordered by submarine levees which rise slightly above the offshore extensions of the interdistributary troughs.

In the process of course lengthening, the river occupies a succession of distributaries, each of which is favorably aligned to receive increasing flow from upstream. . . . The favored distributary gradually widens and deepens to become the main stream . . . ; its natural levees increase in height and width and adjacent interdistributary troughs fill, permitting marshland development. Levees along the main channel are built largely during floodstage; along the distal ends of distributaries, however, levee construction is facilitated by crevasses . . . which breach the low levees and permit water and sediment to be discharged into adjacent troughs during intermediate river stages as well as during floodstage. Abnormally wide sections of the levee and of adjacent mudflats and marshes are created in this manner, and some of the crevasses continue to remain open and serve as minor distributaries while the levees increase in height. Crevasses also occur along the main stream during floodstages . . . and permit tongues of sediment to extend into the swamps and marshes for considerable distances beyond the normal toe position of the levee.

Distributaries with less favorable alignment are abandoned during the course-lengthening process, and their channels are filled with sandy sediment. Abandoned distributaries associated with the development of the present course below New Orleans vein the marshlands. . . . Above the birdfoot delta, the pattern is similar to that of the older courses . . . ; numerous long, branching distributaries diverge at a low angle.

Stream diversions, abandonment of deltas, and development of new deltas—Deltas prograde seaward but they do not migrate laterally, as a point bar does, for example. A delta *shifts position* laterally if a major stream diversion occurs upstream in the alluvial environment or in the upper deltaic-plain region (Fig. 20). Channel diversions were discussed in the section on the meandering-stream model.

Deltas, like meander belts, can be abandoned abruptly or gradually, depending upon the time required for channel diversion to occur. Once a delta is completely abandoned, all processes of deltaic sedimentation cease to exist in that particular delta. With a standing sea level, the sediments of the abandoned delta compact, and subsidence probably continues. The net result is the encroachment of the marine environment over the abandoned delta. This process has erroneously been referred to by some authors as "the destructive phase of deltaic sedimentation." The author maintains that the proper terminology for this process is "transgressive marine sedimentation." The two processes and their related sediments are significantly different, as the discussion of the transgressive marine model of sedimentation demonstrates (see the succeeding section on this model).

As the marine environment advances landward over an abandoned subaqueous delta front and the margins of the deltaic plain, the upper portion of the deltaic sequence of sediment is removed by wave action. The amount of sediment removed depends on the inland extent of the transgression and on the rate of subsidence. The front of the transgression is usually characterized by deposition of thin marine sand units. Seaward, sediments become finer and grade into clays. Thus, local marine transgressions which occur because of delta shifts result in the deposition of a very distinctive marine sedimentary sequence which is easily distinguished from the underlying deltaic sequence.

Concurrent with marine transgression over an abandoned delta, a new delta will develop on the flanks of the abandoned delta. Sedimentary processes in the new delta are similar to those described under the discussion of progradation of deltas.

Repeated occurrences of river diversions result in the deposition of several discrete deltaic masses which are separated by thin transgressive marine sequences (Fig. 20). Under ideal conditions, deltaic facies can attain thicknesses

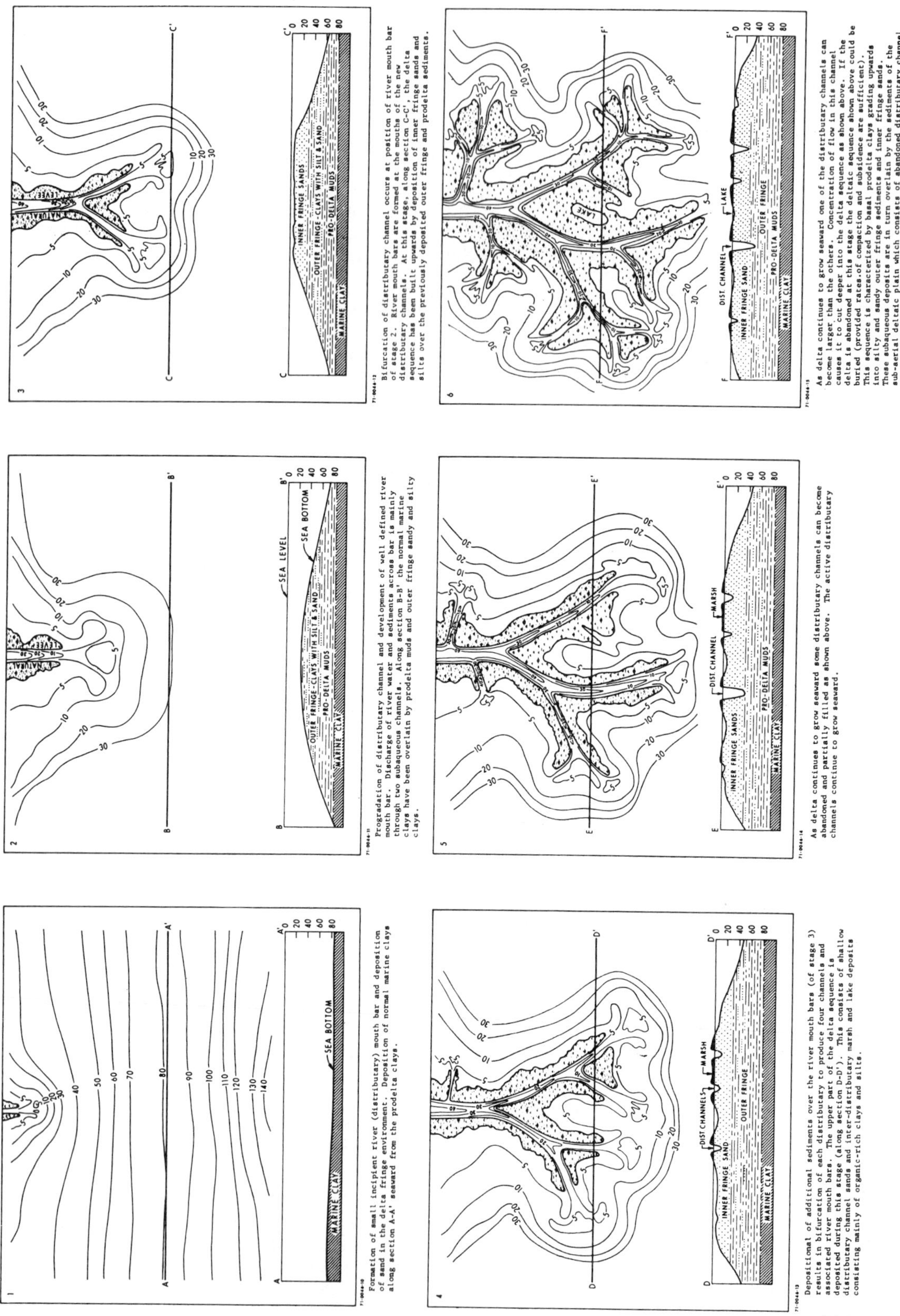

FIG. 19—Stages in development of birdfoot delta and deposition of deltaic sequence.

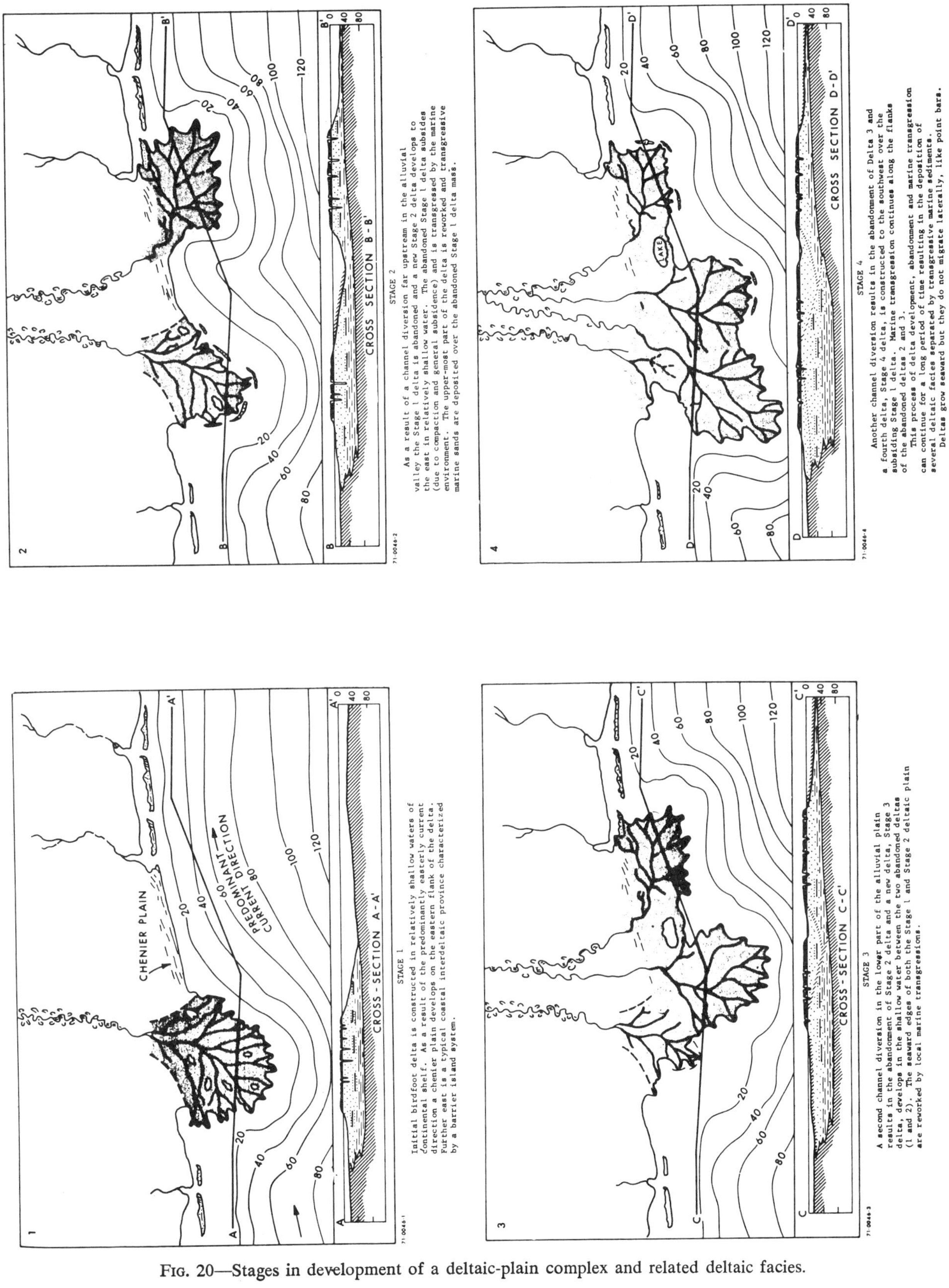

FIG. 20—Stages in development of a deltaic-plain complex and related deltaic facies.

of several hundreds of feet in a large sedimentary basin.

Sedimentary Processes and Deposits of the Cuspate-Arcuate Type of Delta

The shape of a delta is controlled by the influence of marine processes which are active against the delta front (Table 2). Russell (1967a) presented the following excellent summary of the modification of deltas by marine processes.

> The depositional processes characteristic of river mouths are opposed by marine processes that work toward removal of deposits. In a quiet sea or lake the geometric increase in number of distributaries is most closely approached. Below the most inland and earliest forking of the river, the delta builds out as a fan-shaped accumulation, with distributaries creating ribs with natural levees separating basins that widen and open toward the sea. The point deserving greatest emphasis is that the entire delta system originates underwater and only later becomes features visible as land.
>
> The ideal delta front is *arcuate* or has a *bird-foot* shape as viewed from the air or indicated on a chart. The latter pattern indicates a condition in which the deposition of load is dominant over the efforts of marine processes. It results from the forward growth of natural levees and the inability of longshore currents to carry away sediment about as rapidly as it is brought to the river mouth. The delta of the Mississippi is the largest and most typically cited example. Some talons of the foot extend out more than 20 miles and the basins between natural levees flank V-shaped marshes and bays up to about 1.5 fathoms deep. Many smaller bird-foot deltas occur in lakes and estuaries, where there is relatively little distance for fetch to generate high waves and where there are only feeble longshore currents.
>
> The arcuate-front deltas, such as those of the Nile and Niger, indicate sufficient wave action and removal of sediment by longshore currents to maintain relatively stable, smooth fronts. In some cases the momentum of jet flow is apparently sufficient to prevent much flaring, and a single pair of natural levees advances seaward to form a *cuspate* delta front, localized along a single channel. The Tiber, Italy, is the commonly cited example. The Sakayra River, on the Black Sea coast of Turkey has such a delta, but the reason is dominance of wave action. Ahead of it is a large area of shoal water with an extremely irregular system of channels and natural levees (changing so rapidly that a pilot keeps daily watch over them in order to guide boats back to the river mouth). Levees are prevented from growing up to sea level because wave erosion keeps them planed off to a depth of a few feet and because longshore currents entrain and transport sediment away effectively enough to prevent seaward growth of land area. This leaves but one channel mouth in a central position as a gently protruding single cusp.

The modern Brazos River delta of Texas (constructed since 1929) is a good example of a small modern arcuate delta which has been strongly influenced by marine processes. Bernard *et al.* (1970) discussed this small delta and its vertical sequence of sediments. Stages in the development of this type of delta are shown in Figure 21.

The modern Niger delta of western Africa is a classic example of a large arcuate-type delta that is highly influenced by marine processes and tidal currents. Allen (1965c, 1970) described the environments, processes, and sedimentary sequences of this interesting delta. On the basis of data presented by Allen, it is obvious that, although there are many similarities between the Niger delta and the birdfoot-type Mississippi delta, there are certainly some significant differences. For example, from the standpoint of sand bodies, the characteristics and geometry of the delta-fringe sands of the Niger are considerably different from those of the Mississippi. As indicated on Figure 22, a very large quantity of the sand that is contributed to rivermouth bars by the Niger is transported landward and deposited on the front of the deltaic plain as prominent beach ridges (this is a special form of delta-fringe sand according to the writer's deltaic classification). This process results in the development of a thick body of clean sands along the entire front of the deltaic plain.

Another important difference between the Niger and the Mississippi is the occurrence of a very extensive tidal-marsh and swamp environment on the Niger deltaic plain behind (landward of) the prominent beach ridges. This environment is characterized by a network of numerous small channels which connect with the main distributary channels. These channels, which are influenced by a wide tidal range, migrate rather freely and become abandoned to produce extensive point-bar deposits and many abandoned channel fills. In contrast, the Mississippi River distributary channels migrate very little and, hence, point-bar sands constitute only a small percentage of the deltaic deposits.

In summary, the Niger arcuate delta is characterized by prominent delta-fringe sands (which include the beach-ridge sands) occurring as a narrow belt along the entire front of the deltaic plain. Point-bar sand bodies are very common directly adjacent to and landward of the delta-fringe sands. The combination of high-level marine energy and strong tidal currents results in development of a relatively large quantity of distributary and point-bar (migrating channels) sands.

Table 2. Factors Which Influence Characteristics of Deltaic Deposits

Characteristics of Deltaic Deposits		Factors Which Influence Deltaic Sedimentation						
		Magnitude of Fluvial Processes (Sediment Influx)	Magnitude of Marine Processes (Wave Energy, Range in Tide & Currents)	Depth of Water	Subsidence Rate	Sea Level	Salinity of Water	Climate and Vegetation
Thickness of Deltaic Sequence		Large sediment influx required to produce thick sequence.	High wave and current energy tends to disperse sediments and reduce rate of sedimentation.	Water depths in basin controls max. thickness possible. Subsidence permits development of sequence thicker than depth.	Rapid subsidence produces thick sequence. Thin sequences occur in stable areas.	Rising sea level results in thicker sequences.		Abundant vegetation promotes development of thick organic rich layer in upper part of sequence.
Sand Content	Distributary Channel Fills	Percent of sand in channels depends on rapid or gradual diversion. Gradual abandonment produces more sand in fills.	Large tidal range can transport sand upstream near river mouths and increase sand content in fills.					
	Fringe	Amount of sand in fringe depends on sand load of river. Small for muddy rivers; large for rivers with high sand load.	Strong currents may transport sand along coast away from delta thus reducing sand content of fringe.					
Porosity and Permeability	Distributary Channel Fills		Large tidal range results in deposition of clays and salts in lower reaches of channels thus lowering porosity and permeability.					
	Fringe	High sand content. Rivers generally have best developed fringe sands.	Strong wave and current winnows fines and deposition of clean sand occurs. Dirty sands are deposited in areas of low marine energy.					
Geometry of Deltaic Mass			Strong marine currents can disperse sediment over extensive area to form broad subaqueous fringe and pro-delta mass.	Relatively thin blanket-like deltaic masses form in shallow water. Geometry influenced by bottom topography.	Thin blanket-like delta masses occur in stable areas. Local downwarps result in ladle-shape delta masses.	Rising sea level will produce on-lap fringe over marsh facies. Falling sea level causes strong off-lap of marsh over fringe.	Fine sediments in suspension flow over denser sea water and may be dispersed over broad areas by currents.	
Influence of Organisms on Sediments		Rapid influx of sediment and high turbidity greatly reduces organism population and degree of bioturbation.	Organisms have little influence on sediments in near shore zone with large wave energy.					
Number of Distributary Channel Fills			Strong wave and current action prevents normal development of river mouth bars necessary for bifurcation of channel.	Deltas constructed in shoal water have largest number of distributary channels.				
Thickness (Depth) of Dist. Channel Fills					Rapidly subsiding deltaic plain may result in thicker dist. channel fills.	Falling sea level may cause deepening of channels. Rising sea level causes shoaling.		
Rate of Sedimentation		Rates of sedimentation generally greatest in deltas of rivers with large sediment.	Strong currents can disperse sediments and reduce deposition rates.					Rapid growth of vegetation traps sediments and increases rate of deposition.
Thickness of Organic Rich Clays and Peats (Conc) in Upper Part of Sequence					Thick section of peat and organic clays can accumulate in a near sea level environment (with subsidence).	A rising sea level will result in thick section of organic material in upper portion of sequence.		Warm, wet deltas with dense vegetation are characterized by thick peats and organic rich clays.

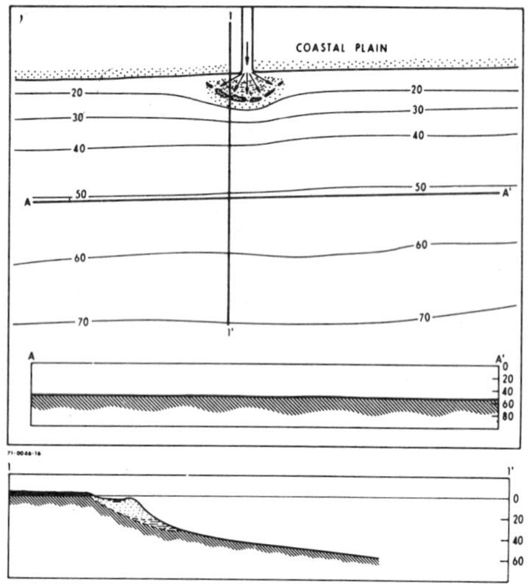

River diversion (at a point within alluvial valley or deltaic plain) results in the development of a new channel. During the initial stage of delta development a relatively small river mouth bar is formed in the marine environment. The distance of this bar from the original shoreline is approximately four times the width of the channel. This type of river mouth bar is generally breached by two or more subaqueous channels which radiate from the river mouth. The bar crest is characterized by sand and the bar-back area (which acts as a stilling basin) is characterized by clays and silts with minor amounts of sand. Seaward from the bar crest (the bar-front) the sediments grade from sands to silts and clays. Deposition along section 1-1 consists of normal marine muds.

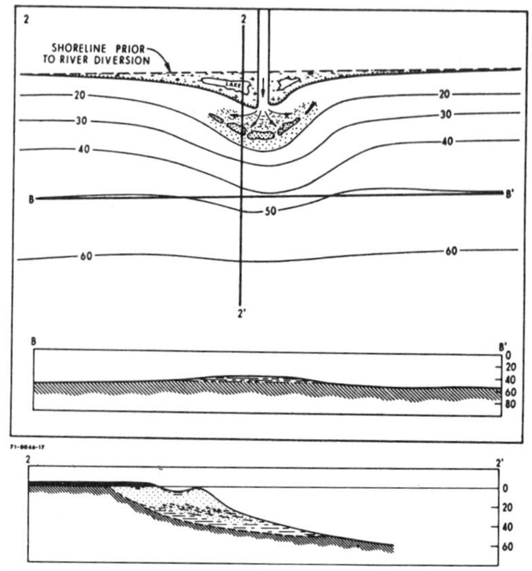

The initial river mouth bar (shown in stage 1) continues to grow (both vertically and seqward by accretion. If major flood occurs when marine environment is relatively quiet (non-storm period) the river mouth bar is built upwards to an elevation slightly above mean sea-level. Vegetation on newly formed subaerial bar traps fine-grained sediments and the bar becomes attached to the mainland. Thus the initial river mouth bar shown in Stage 1 becomes part of the deltaic plain. The river channel cuts across the old bar (of stage 1) and a new river mouth bar develops offshore. Deposition along section 2-2 consists of prodelta clays.

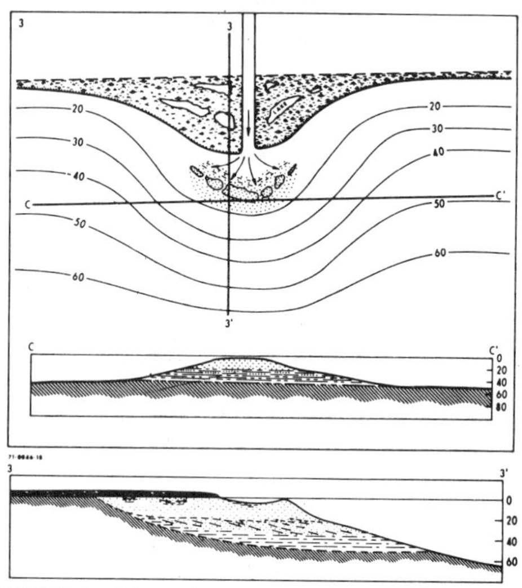

The stage 2 river mouth bar continues to grow seaward by accretion and vertically during flood stages. When bar crest reaches elevation slightly above mean sea-level it becomes attached to the stage 2 deltaic plain. The area of the stage 2 bar is separated from the stage 1 bar by shallow lakes. A new river mouth bar is constructed in the middle part of section 3-3, and delta fringe sands are deposited over the older outer fringe and prodelta sediments.

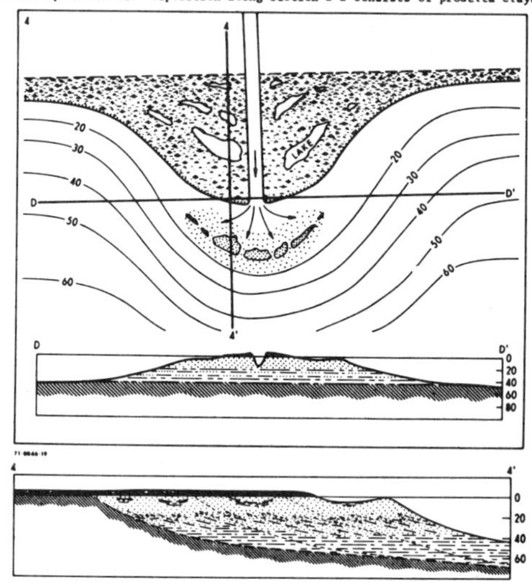

Delta continues to grow seaward and the uppermost portion of the delta sequence is deposited as the subaerial deltaic plain is constructed over the area along the center of section 4-4'. The rate at which this type of delta advances seaward is governed by the frequency and magnitudes of river floods and storms. If major floods occur frequently during periods characterized by relatively low marine enegry (non-storm) the delta will advance seaward rapidly. Severe current and wave action associated with major storms, especially during non-flood stages of the river, have a tendency to partially destroy the river mouth bars and thereby reduce the rate of seaward delta advance.

FIG. 21—Stages in development of a cuspate delta.

Sedimentary Processes and Deposits of Estuarine-Type Delta

Large deltas such as the Ganges, Amazon, and Colorado (in Gulf of California) are considered to be examples of estuarine-type deltas. Although our knowledge of these deltas is extremely limited, it is now reasonably well established that they are associated with extreme tidal conditions (up to 25 ft [8 m] at the mouth of the Colorado River). It is apparent that very strong tidal currents have a profound influence on the distribution of sediments. Sands are known to be transported for great distances in front of these deltas; however, the geometry of these sand bodies is unknown. Additional studies of this type delta are badly needed.

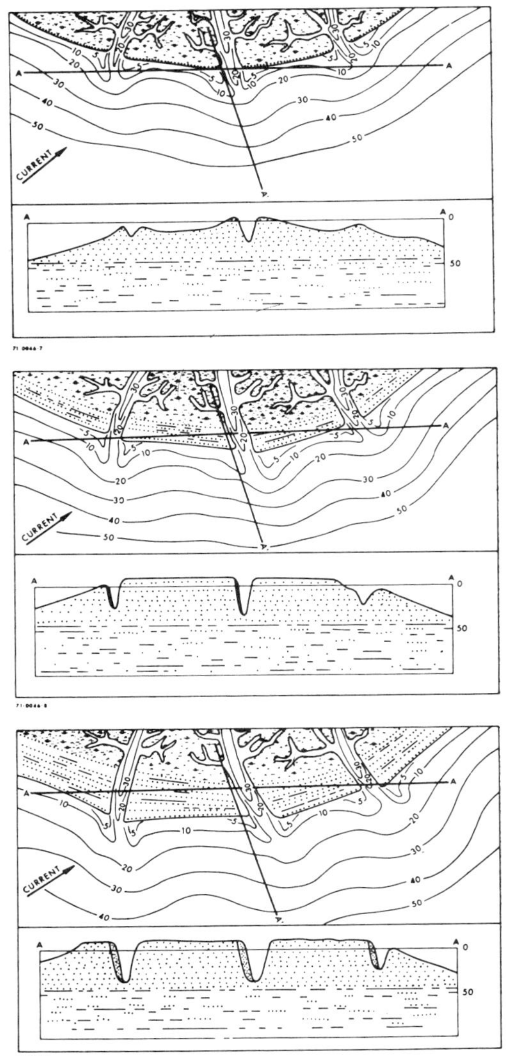

The three diagrams shown above illustrate stages in the development of a deltaic sequence which is characteristic of arcuate deltas. One of the most prominent features of this type of delta is the beach ridge complex on the delta front. This is a very special type of delta fringe sands. Note that as the delta advances seaward the subaqueous part of the delta sequence is locally replaced by distributary channels, which become deeper and wider as the delta advances seaward.

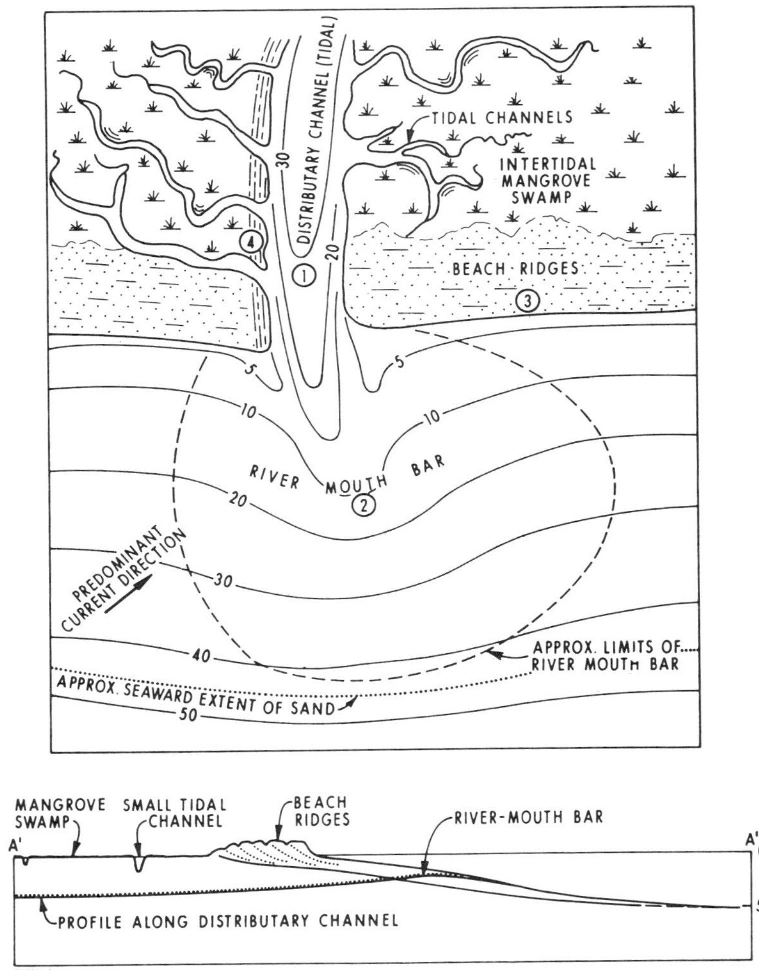

Arcuate deltas generally develop in areas where level of marine energy (currents, waves and tides) is very high as compared to the magnitude of the fluvial processes. Although river mouth bars develop as in other types of deltas, they are generally reworked and partially destroyed as rapidly as they are formed by strong marine currents and wave action, especially in situations where severe storm activity does not coincide with major flood periods. This situation precludes the full development of river mouth bars and resulting channel bifurcation.

Sands introduced to the marine environment by the distributary channel are initially deposited on or near the river mouth bars. These sands are subsequently transported to the beach environment by currents and waves. This leads to the development of prominent beach accretion ridges and the best developed delta fringe sands have the characteristics of beach deposits.

Strong littoral drift results in lateral migration (in same direction as marine current) of the lower ends of distributary channels. This lateral channel migration is accompanied by the deposition of sediments along the margins of the channel opposite the direction of migration (4).

If the tidal range is large the area behind the beaches of the delta front is generally characterized by a complex of tidal channels which empty into the main distributary channels. These channels are associated with marshes or swamps and they migrate to produce point bar sequences.

FIG. 22—Processes of deltaic sedimentation associated with an arcuate delta.

Table 3. Examples of Ancient Deltaic Deposits

Geographic Occurrence	*Author*
California	Todd and Monroe, 1968
Illinois	Lineback, 1968
	Swann *et al.*, 1965
Indiana	Hrabar and Potter, 1969
	Wier and Girdley, 1963
Iowa & Illinois	Laury, 1968
Kansas	Brown, 1967
	Hattin, 1965
Louisiana	Clark and Rouse, 1971
	Curtis, 1970
Michigan	Asseez, 1969
Miss., La. & Ala.	Galloway, 1968
Montana	Sims, 1967
Nebraska	Shelton, 1972
New Mexico	Schlee and Moench, 1961
New York	Friedman and Johnson, 1966
	Lumsden and Pelletier, 1969
New York & Ontario	Martini, 1971
New York	Wolff, 1967
North Dakota	Shelton, 1972
Ohio	Knight, 1969
	Lené and Owen, 1969
Oklahoma	Busch, 1953. 1971
	Shelton, 1972
	Visher *et al.*, 1971
Oregon	Dott, 1964, 1966
	Snavely *et al.*, 1964
Pa., W. Va., Ohio	Beerbower, 1961
	Ferm and Cavaroc, 1969
South Dakota	Pettyjohn, 1967
Texas	Brown, 1969
	Fisher and McGowen, 1969
	Gregory, 1966
	LeBlanc, 1971
	Nanz, 1954
	Shannon and Dahl, 1971
	Wermund and Jenkins, 1970
	Shelton, 1972
W. Va., Pa., Ohio	Donaldson, 1969
Wyoming	Barlow and Haun, 1966
	Dondanville, 1963
	Hale, 1961
	Paull, 1962
	Weimer, 1961b
Wyoming & Colorado	Weimer, 1965
Several states, U.S.A.	Fisher *et al.*, 1969
N. Appalachians	Ferm, 1970
Central Appalachians	Horowitz, 1966
Central Appalachians	Dennison, 1971
Upper Miss. embayment & Illinois basin	Pryor, 1960, 1961
Upper Miss. Valley	Swann, 1964
Okla., Iowa, Mo., Kans., Ill., Ind., Ky.	Manos, 1967
Okla. to Penn.	Wanless *et al.*, 1970
Central Gulf Coast	Mann and Thomas, 1968
Alberta, Canada	Carrigy, 1971
	Shawa, 1969
	Shepheard and Hills, 1970
	Thachuk, 1968
England	Allen, 1962
	Taylor, 1963
Ireland	Hubbard, 1967
Scotland	Greensmith, 1966

Summary: Deltaic Sand Bodies

There are three basic types of deltaic sand bodies: delta-fringe, abandoned distributary-channel, and point-bar sands. The relative abundance and general characteristics of these sand bodies in the three types of deltas considered herein are summarized below.

Birdfoot-type delta—The most common sands are those of the delta-fringe environment. These sands occur as relatively thin, widespread sheets, and they contain a substantial amount of clays and silts.

Abandoned distributary channels contain varied amounts of sand, probably composing less than 20 percent of the total delta sand content. These sand bodies are long and narrow, are only slightly sinuous, and are encased in the delta-fringe sands or prodelta clays, depending upon channel depths and the distance that the delta has prograded seaward.

Cuspate-arcuate type of delta—Delta-fringe sand complexes are wide (width of delta), though individual sand bodies are relatively narrow, and are generally much cleaner than delta-fringe sands of the birdfoot-type delta.

Distributary-channel sands and point-bar sands are much more common than in birdfoot-type deltas and can constitute up to 50 percent of the total sand content of the delta. These two types of sands are encased in delta-fringe and prodelta sediments.

Estuarine-type delta—Delta-fringe sands appear to be much more common than distributary and point-bar sands. They probably extend for great distances within the marine environment in front of the delta; however, their geometry remains unknown.

Ancient Deltaic Deposits

Deposits of deltaic origin have been reported from more than 40 states and from several foreign countries. Some examples are summarized in Table 3.

Coastal-Interdeltaic Model of Sedimentation

Setting and General Characteristics

This type of sedimentation occurs in long, narrow belts parallel with the coast where shoreline and nearshore processes of sedimentation predominate. The ideal interdeltaic deposit, as the name implies, occurs along the coast between deltas and comprises mud flats and cheniers (abandoned beach ridges) of the chenier-plain complex and the barrier-island–lagoon–tidal-channel complex (Fig. 23). It can also occur along the seaward edge of a coastal plain which is drained by numerous small streams and rivers but is devoid of any sizable deltas at the marine shoreline.

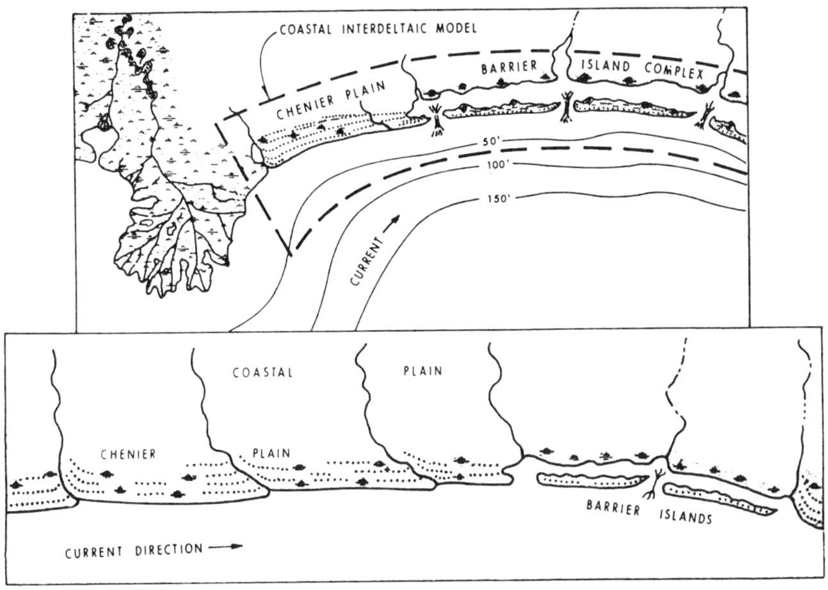

71-0363-17

FIG. 23—General setting and characteristics of coastal-interdeltaic model of clastic sedimentation.

Source and Transportation of Sediments

Most of the sediments deposited are derived from land, but minor amounts come from the marine environment. A portion of the sediment transported to the marine shoreline by rivers and smaller streams is dispersed laterally by marine currents for great distances along the coast. Clays and fine silts are carried in suspension, and sand is transported mainly as bed load or by wave action in the beach and nearshore zone. The suspended silt and clay load is dispersed at a rapid rate and is most significant in the development of the mud flats of the chenier plain. Lateral movement of the sand bed load occurs at a relatively slow rate and is most significant in the development of the cheniers and the barrier-island complex.

A minor amount of sediment can also be derived from adjacent continental-shelf areas if erosion occurs in the marine environment.

Sedimentary processes and deposits of chenier plain—Major floods result in the sudden large influx of sediments at river mouths. Much of the suspended load introduced to the coastal-marine environment is rapidly dispersed laterally along the coast by the predominant longshore drift. A considerable portion of this suspended load is deposited along the shoreline (on the delta flank) as extensive mud flats. This period of regressive sedimentation (progradation or offlap) occurs in a relatively short period when rivers are at flood stages (Fig. 24).

During long periods when rivers are not flooding, the supply of sediment to the coast is reduced considerably or is nil. Coastal-marine currents and wave action rework the seaward edge of the newly formed mud flat, and a transgressive situation develops. A slight increase in

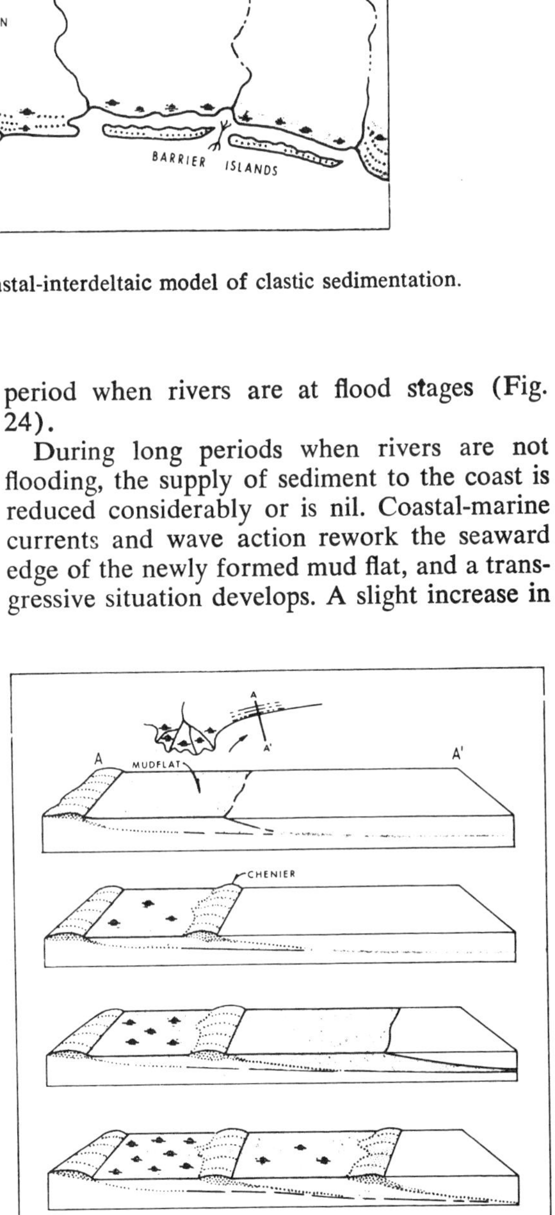

71-0363-18

FIG. 24—Stages in development of a chenier plain.

sand supply can result in a regressive situation, and the initial transgressive beach accumulation will grow seaward by regressive beach accretion to form a long, narrow, well-defined chenier on the seaward edge of the extensive mud flat.

Another period of river flooding develops another mud flat on the seaward edge of the chenier. During the subsequent nonflood season, the coastal-transgressive processes produce another beach ridge. Thus, over a long period, a chenier plain consisting of mud flat and beach ridges is constructed.

The width of a mud flat is varied and is dependent on the magnitude and duration of a river flood. The size of the chenier (height and width) is determined by two factors: duration of the nonflood season (absence of muds) and magnitude of coastal-marine processes, including storm tides and waves.

Small streams which drain to the coastline across a chenier plain contribute little sediment to the chenier-plain environment. The mouths of these streams are generally deflected in the direction of the littoral drift.

Sedimentary processes and deposits of barrier-island complex—The typical barrier-island complex comprises three different but related depositional environments: the barrier island, the lagoon behind the barrier, and the tidal channel–tidal deltas between the barriers.

The seaward face of a barrier island is primarily an environment of sand deposition. Coastal-marine energy (currents and wave action) is usually much greater than in the chenier–mud-flat regions. Sediments are transported along the coast in the direction of the predominant littoral drift. Coarser sands are deposited mainly on the beach and upper shoreface, and finer sands are deposited in the lower shoreface areas. Silts and clays are deposited in the lower shoreface zones on the adjacent shelf bottom—at depths greater than 40–50 ft (12–15 m). Storm tides and waves usually construct beach ridges several feet above sea level, depending on the intensity of storms, and also transport sandy sediments across the barrier from the beach zone to the lagoon.

Under ideal conditions, a barrier grows seaward by a beach-shoreface accretion process to produce a typical barrier-island sequence of sediments which grades upward from fine to coarse (Figs. 25, 27). The various organisms which live in the beach, shoreface, and adjacent offshore areas usually have a significant influence on the character of sedimentary structures.

Dry beach sand can be transported inland by the wind and redeposited as dune sand on the barrier, in the lagoon, or on the mainland across the lagoon.

Tidal channel–tidal delta—Tidal action moves a large quantity of water in and out of lagoons and estuaries through the tidal channels which exist between barrier islands. These channels are relatively short and narrow and vary considerably in depth. Maximum channel depths occur where the tidal flow is confined between the ends of barriers. The channel cross section is asymmetric: one side of the channel merges with the tidal flats and spit; the opposite side of the channel has abrupt margins against the barrier (Fig. 26).

As marine waters enter the lagoon or estuary system during rising tides, the inflow attains its maximum velocity in the deepest part of the confined channel. The tidal flow is dispersed as it enters the lagoon, and current velocities are greatly reduced. The result is the deposition of sediment in the form of a tidal delta which consists of a shallow distributary channel separated by sand or silt shoals. Similar tidal deltas are also formed on the marine side of the system by similar processes associated with the falling or outgoing tide.

The depth of tidal channels and the extent of tidal deltas are dependent on the magnitude of the tidal currents. The deepest channels and the largest deltas are associated with large lagoons and estuaries affected by extreme tidal ranges.

Tidal channels migrate laterally in the direction of littoral drift by eroding the barrier head adjacent to the deep side of the channel and by spit and tidal-flat accretion on the opposite side.

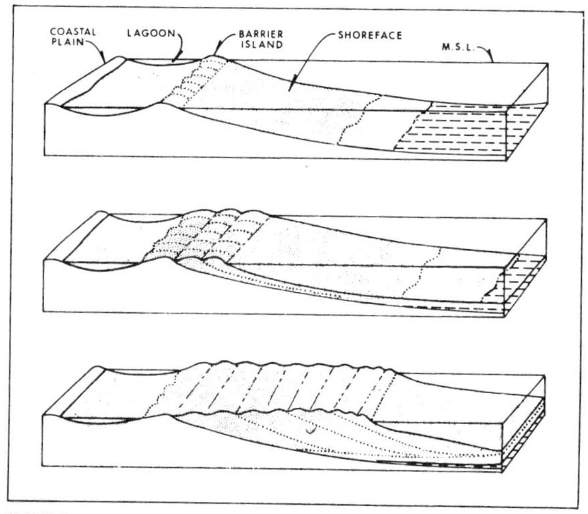

Fig. 25—Stages in development of a barrier island.

Lateral migration of the tidal system results in the deposition of the tidal-channel and tidal-delta sequences of sediments.

Summary: Characteristics of Coastal-Interdeltaic Deposits

The coastal-interdeltaic model of sedimentation is characterized by six distinct but related types of deposits: mud flat, chenier, barrier island, lagoon, tidal channel, and tidal delta. Characteristics of these deposits are summarized in Figure 27.

Three main types of sand bodies are associated with this model: barrier island, chenier, and tidal channel–tidal delta. The barrier-island sand body, which is the largest and most significant of the three, is long (usually tens of miles)and narrow (2–6 mi or 3–10 km), is oriented parallel with the coastline, and attains maximum thicknesses of 50–60 ft (15–18 m). The chenier sand bodies are very similar to those of the barriers; however, they are generally only about a third as thick. Tidal-channel sand bodies are oriented perpendicular to the barrier sands, and their thickness can vary considerably (less than, equal to, or greater than that of the barrier sands), depending on the depth of tidal channels.

Ancient Coastal-Interdeltaic Deposits

Examples of ancient coastal-interdeltaic deposits reported from 13 states are summarized in Table 4.

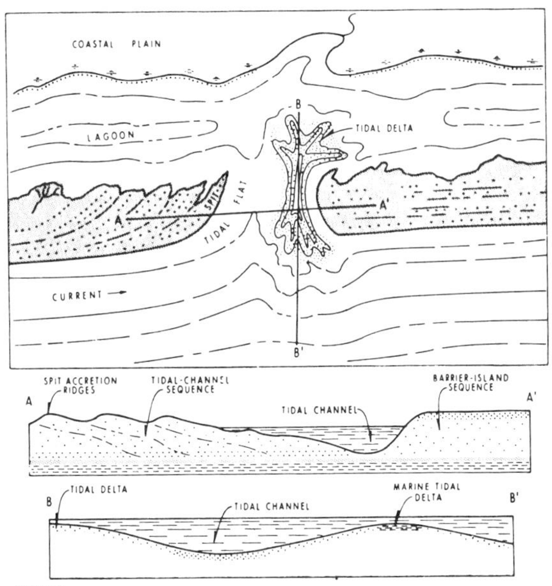

FIG. 26—Relation of tidal channels and tidal deltas to barrier islands.

Table 4. Examples of Ancient Coastal-Interdeltaic Deposits

Geographic Occurrence	*Author*
Colorado	Griffith, 1966
Florida	Gremillion *et al.*, 1964
Georgia	Hails and Hoyt, 1969
	MacNeil, 1950
Illinois	Rusnak, 1957
Louisiana	Sloane, 1958
Louisiana & Arkansas	Thomas and Mann, 1966
	Berg and Davies, 1968
Montana	Cannon, 1966
	Davies *et al.*, 1971
	Shelton, 1965
New Mexico	Sabins, 1963
New York	McCave, 1969
Oklahoma & Kansas	Bass *et al.*, 1937
	Boyd and Dyer, 1966
	Dodge, 1965
Texas	Fisher and McGowen, 1969
	Fisher *et al.*, 1970
	Shelton, 1972
	Harms *et al.*, 1965
	Jacka, 1965
Wyoming	Miller, 1962
	Paull, 1962
	Scruton, 1961
	Weimer, 1961a

Eolian Model of Sand Deposition

Occurrence and General Characteristics

A very common process of sedimentation is transportation and deposition of sand by the wind. Two basic conditions are necessary for the formation of windblown sand deposits: a large supply of dry sand and a sufficient wind velocity. These conditions are commonly present along coastlines characterized by sandy beaches and also in semiarid regions and deserts, where weathering and fluvial sedimentation produce a large quantitiy of sand (Fig. 28).

Under certain conditions, sands on the downstream parts of alluvial fans and along braided streams are transported and redeposited by the wind (Glennie, 1970). Sands originally deposited on point bars of meandering streams and along distributary channels of some deltas are also picked up by the wind and redeposited locally as dune sand. Similarly, sands deposited along beaches of the coastal-interdeltaic environments are redeposited by onshore winds as sand dunes on barrier islands or on the mainland. Thus, the eolian process of sand deposition is likely to occur within all models of clastic sedimentation discussed in the preceding sections.

Eolian Transport and Sedimentation

The complex processes of sand transport and deposition by the wind were studied and de-

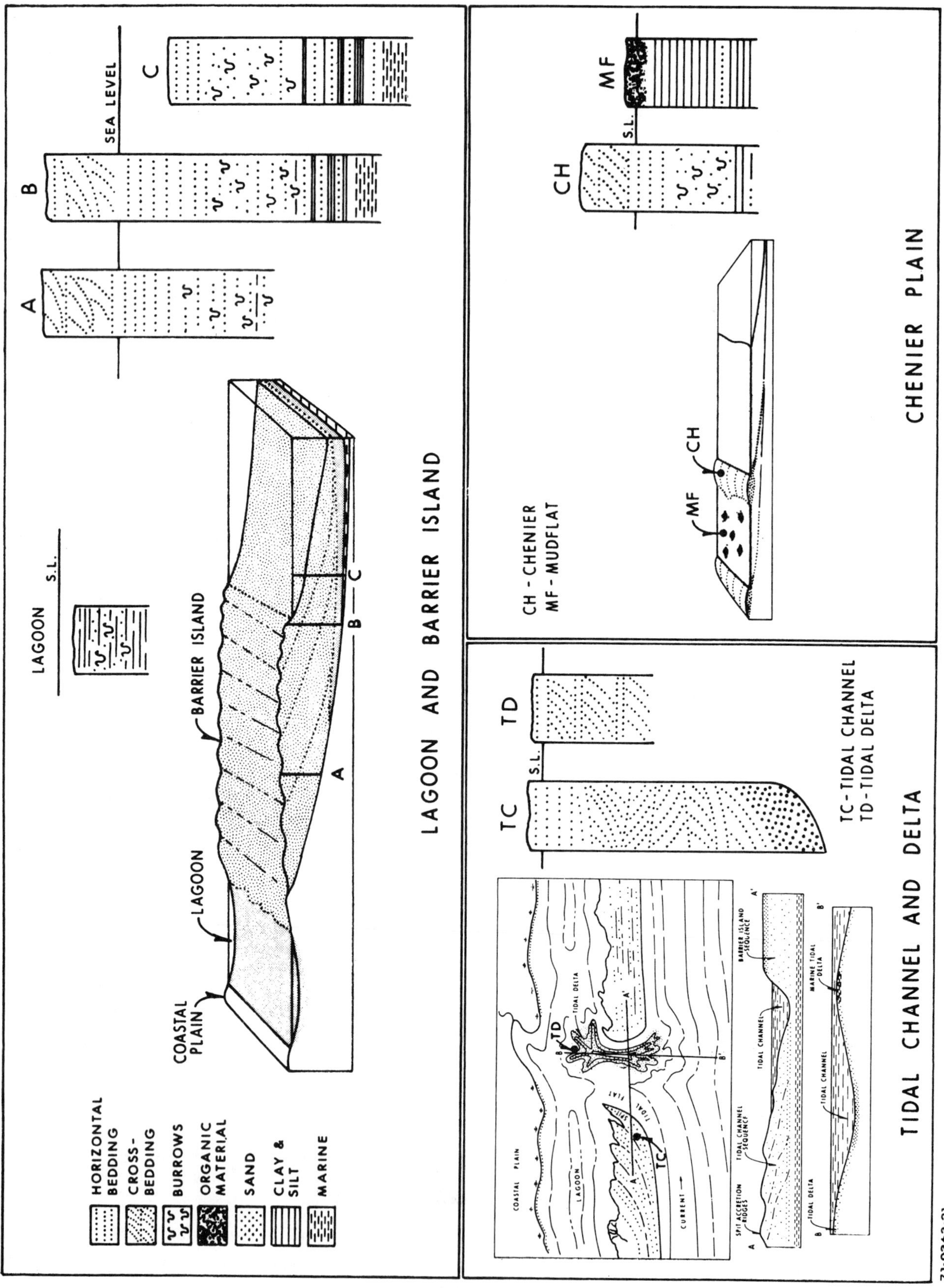

FIG. 27—Sedimentary sequences typical of chenier-plain and barrier-island complexes of coastal-interdeltaic model of clastic sedimentation.

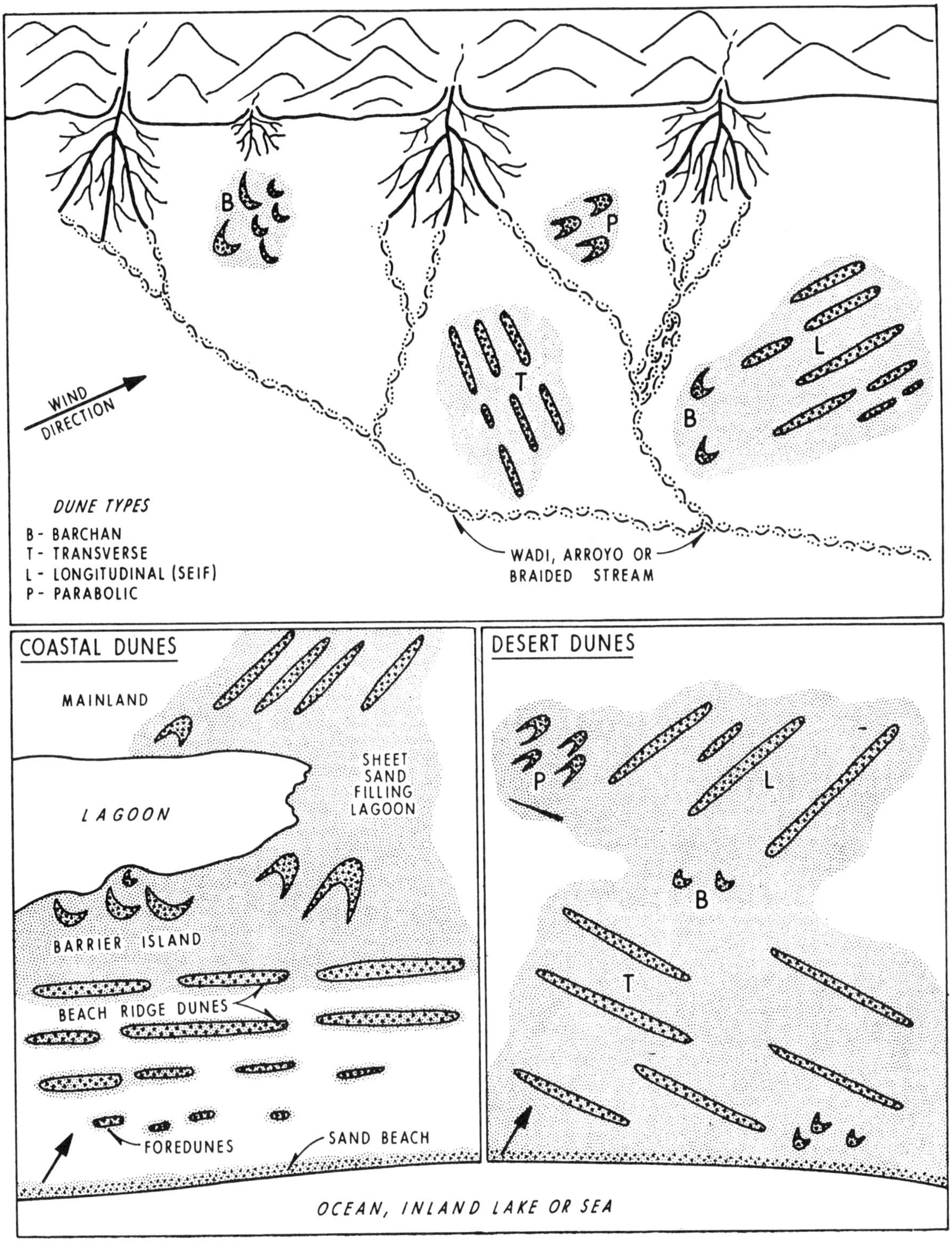

71-0363-22

Fig. 28—Occurrences of eolian sands in coastal and desert regions.

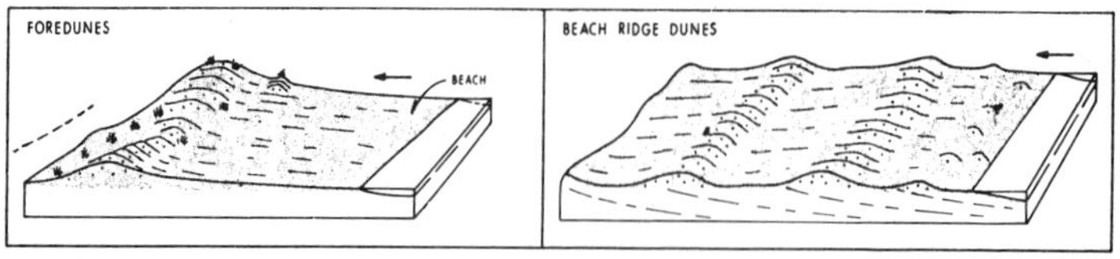

Low mounds and small elongate ridges a few feet high occurring adjacent and parallel to sand beach and shoreline, usually partly stabilized by vegetation.

Dunes against vegetation coalesce to form long, slightly sinuous ridge or series of ridges parallel to coastline. Closely associated with beach accretion ridges formed by wave action. Characteristic of barrier islands and shorelines on flanks of deltas.

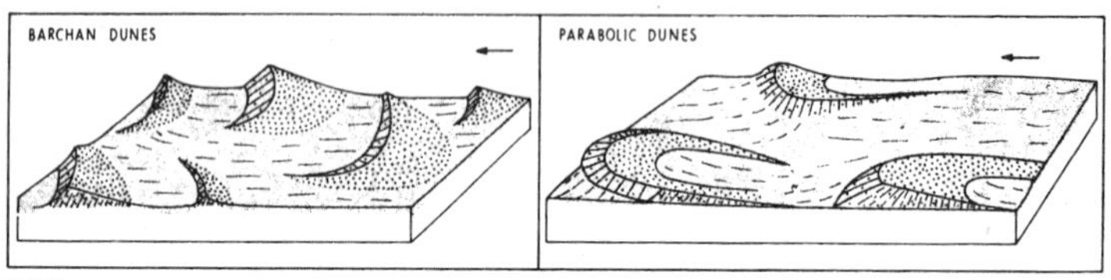

Concentric with steep slope on concave (leeward) side facing away from beach. Horns extend downwind. Can occur as scattered isolated dunes or several barchans can join to form sinuous ridge which resembles transverse dunes.

U-shaped with open end toward beach (windward) and steep side away from beach. Results from sand blowouts. Middle part moves forward (downwind) with respect to sides. Long arms usually anchored by vegetation.

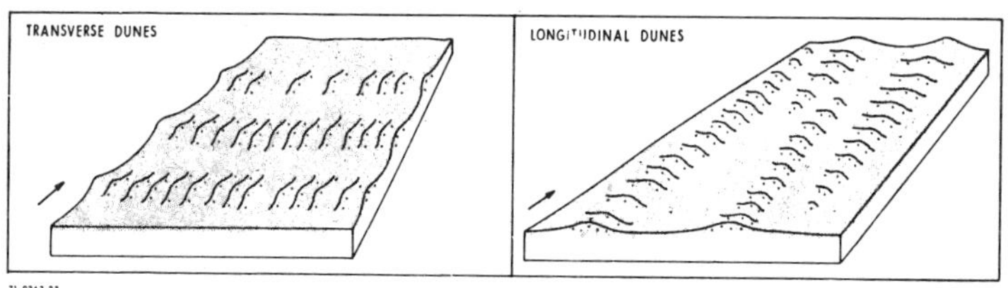

Dunes or ridges occur parallel or slightly oblique to coastline and elongated in direction perpendicular to effective wind direction. Generally symmetrical in cross section. Leeward side steep and windward side has very gentle slope.

Elongated parallel to wind direction and usually oblique or perpendicular to coastline. Cross section symmetrical. Separated from each other by flat areas. Seif dunes are special type of longitudinal dunes.

FIG. 29—Some common types of coastal dunes which also occur in deserts.

scribed by Bagnold (1941). Recently, Glennie (1970) summarized this type of sedimentation as observed under desert conditions.

The most common method of sand deposition is in the form of sand dunes. Many types of dunes have been recognized and described by numerous authors (Fig. 29). H. Smith (1954) presented the following classification and description of coastal dunes which can occur either under active or stabilized conditions.

1. Foredune ridges, or elongate mounds of sand up to a few tens of feet in height, adjacent and parallel with beaches.
2. U-shaped dunes, arcuate to hairpin-shaped sand ridges with the open end toward the beach.
3. Barchans, or crescentic dunes, with a steep lee slope on the concave side, which faces away from the beach.
4. Transverse dune ridges, trending parallel with or oblique to the shore, and elongated in a direction essentially perpendicular to the dominant winds. These dunes are asymmetric in cross profile, having a gentle slope on the windward side and a steep slope on the leeward side.
5. Longitudinal dunes, elongated parallel with wind direction and extending perpendicular or oblique to the shoreline; cross profile is typically symmetric.
6. Blowouts, comprising a wide variety of pits, troughs, channels, and chute-shaped forms cutting into or across other types of dunes or sand hills. The larger ones are marked by conspicuous heaps of sand on the landward side, assuming the form of a fan, mound, or ridge, commonly with a slope as steep as 32° facing away from the shore.
7. Attached dunes, comprising accumulations of sand trapped by various types of topographic obstacles.

McKee (1966) described an additional type, the dome-shaped dune, from White Sands National Monument, and Glennie (1970) described the seif dune of Oman, which is a special type of longitudinal dune. Many other dune types have been described; however, the above types appear to be the most common.

Studies of modern eolian sand bodies—Cooper (1958) reviewed the early studies of sand dunes, mainly by Europeans, and summarized the status of dune reseach in North America. Additional sand-dune studies in the United States since 1959 were made in Alaska by Black (1961), on the Texas coast by McBride and Hayes (1962), on the Georgia coast by Land (1964), in the Imperial Valley of California by Norris (1966), in coastal California by Cooper (1967), and in the San Luis Valley of Colorado by R. Johnson (1967). Additional studies outside the United States were made in southern Peru by Finkel (1959), in Baja California by Inman *et al.* (1966), in Libya by McKee and Tibbitts (1964), in Russia by Zenkovich (1967), and in Australia by Folk (1971).

During the past several years, some very important studies on eolian sands, which included detailed observations of internal dune structure and stratification in deep trenches cut through dunes, were made along the Texas coast by McBride and Hayes (1962), in White Sands National Monument by McKee (1966), along the Dutch coast by Jelgersma *et al.* (1970), and in the deserts of the Middle East by Glennie (1970). These authors presented photographs and sketches of various types of sedimentary structures exposed in trench walls and described their relations to dune types, wind regime, and grain-size distribution. These studies have provided some badly needed criteria for recognition of ancient eolian sands. The following summary of the geometry and general characteristics of modern eolian sand bodies was prepared largely from the references cited above.

Summary: Coastal Eolian Sand Bodies

Coastal eolian sand bodies, consisting of several types of dunes, are very long and quite narrow; they range in thickness from a few feet to a few hundreds of feet, and are aligned parallel with or oblique to the coastline. Because these sands are derived from beach deposits and form in vegetated areas, they commonly contain fragments of both shells and plants. They are characterized by high-angle crossbedding and are usually well sorted. The adjacent and laterally equivalent beach deposits are generally horizontally bedded and have some low-angle crossbedding.

Summary: Eolian Sand Bodies of Desert Regions

Desert eolian sand bodies differ from coastal eolian sands mainly in their distribution. The internal sedimentary structures and their relations to dune types are similar (Bigarella *et al.*, 1969). Seif dunes are products of two wind directions and appear to occur more commonly in desert areas. These dunes are characterized by high-angle crossbedding in two directions.

Ancient Eolian Deposits

Ancient eolian deposits have been reported from the Colorado Plateau by Baars (1961) and Stokes (1961, 1964, 1968), from the southwestern United States by McKee (1934), from England by Laming (1966), and from Brazil and Uruguay by Bigarella and Salamuni (1961). Criteria for recognition of eolian deposits have been summarized by Bigarella (1972).

Marine Clastic Sedimentation

Transportation and deposition of sand in the marine environment occur under a wide range of geologic and hydrologic conditions, ranging from those of the coastal shallow-marine environments to the deeper water environments of the outer continental shelves, the slopes, and the abyssal plains (Fig. 30).

As indicated in the Introduction, sands deposited under regressive (progradational) conditions within the coastal shallow-marine environments are considered herein as products of either the coastal-interdeltaic or the deltaic model of sedimentation. Other important shallow-marine sand bodies are produced as a result of marine transgressions.

During the past several years, studies made principally by the major oceanographic institutions on the modern deep-marine environments and research by petroleum geologists, university professors, and graduate students on ancient clastic sediments of various geologic ages have revealed that sand bodies of deep-marine origin are rather common throughout the world. Although most geologists now accept the fact that some sands are of deep-marine origin, our understanding of the various geologic processes which produce these sand bodies is relatively poor. The writer's personal experience with this type of clastic sedimentation is limited; however, on the basis of familiarity with the literature on marine sediments, it appears that most deep-marine sands are depos-

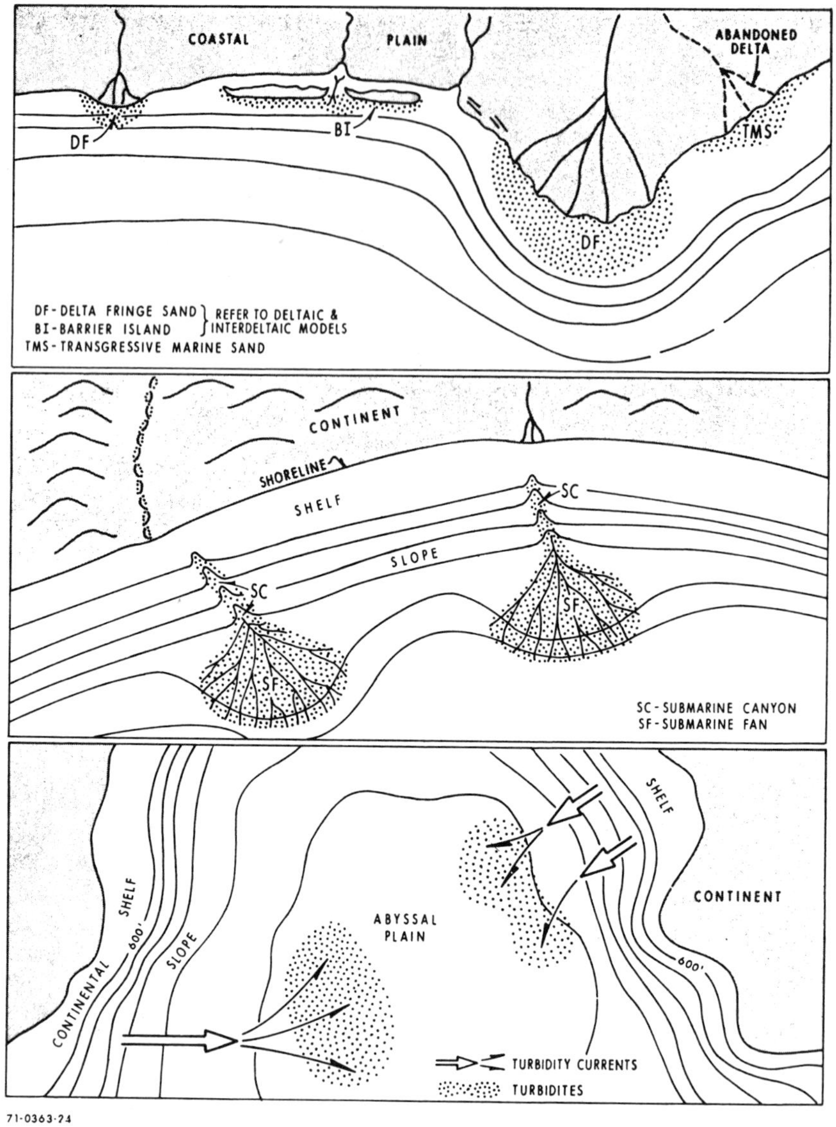

FIG. 30—Deposition of sand in marine environments.

ited under three principal types of environmental conditions: (1) on the outer shelf, the slope, and the continental rise, as a result of slumping, sliding, and tectonic activity such as earthquakes; (2) in abyssal plains, by density (turbidity) and bottom currents; and (3) in submarine canyons, fan valleys, and fans, by both bottom and density currents.

Only two of these several types of marine sands are discussed in this paper: (1) the shallow-marine sands deposited as a result of transgressive-marine sedimentation associated with the shifting of deltas and (2) the deep-marine sands deposited in submarine canyons, fan valleys, and fans.

Transgressive-Marine Model of Clastic Sedimentation

Setting and general characteristics—Deposition of clastic sediments during periods of marine transgressions (onlap) is a common process of sedimentation in most basins. There are two basic types of marine transgression: that which is associated with the shifting of deltas as a result of major river diversions during a period of standing sea level, and that which occurs as a result of a relative rise in sea level (due to subsidence of a coastal plain or eustatic rise in sea level). The inland and lateral extents of marine transgressions resulting from delta shifts are limited in size, depending on the dimensions of the abandoned deltaic plains. Marine transgressions resulting from relative changes in sea level extend over much broader regions and are commonly referred to as regional transgressions. Their dimensions are governed mainly by the topography of the coastal plain being transgressed and by the amount of relative rise in sea level. Thus, transgressive-marine deposition can occur locally over abandoned deltas or regionally over eolian, alluvial, interdeltaic, and deltaic deposits of a large part of a coastal plain.

Modern marine transgressions resulting from major changes in drainage and delta shifts have been described by several authors: Russell, 1936; Russell and Russell, 1939; Kruit, 1955; van Straaten, 1959; Scruton, 1960; Curray. 1964; Coleman and Gagliano, 1964; Rainwater, 1964; Coleman, 1966b; Scott and Fisher, 1969; L. Brown, 1969; and Oomkens, 1970.

Sources, transportation, and deposition of sediments—After a delta is abandoned because of upstream channel diversion, a very significant change occurs in conditions of sedimentation. The abandoned deltaic plain and subaqueous delta front no longer receive sediment and gradually subside owing to the compaction of the deltaic deposits. The seaward edge of the abandoned delta is attacked by marine wave and current action and recedes landward at relatively slow rates. As marine processes erode the upper part of the deltaic sequence, the sandy sediments within the sequence are winnowed and deposited along the advancing shoreline as barrier islands, beaches, and shallow-marine sands; finer sediments are deposited farther offshore. Thus, the transgressive-marine depositional profile is characterized by sands and shell material nearshore and by progressively finer sediments offshore. Over a period of time, as the transgression proceeds inland, the thin veneer of shallow-marine sands which are deposited over the underlying delta sediments is in turn overlain by marine silts and clays. Stages in the development of such a trangressive-marine sand body are illustrated in Figure 31.

Character of sediments—This type of sedimentation, although largely restricted in extent to abandoned deltas and adjacent and laterally

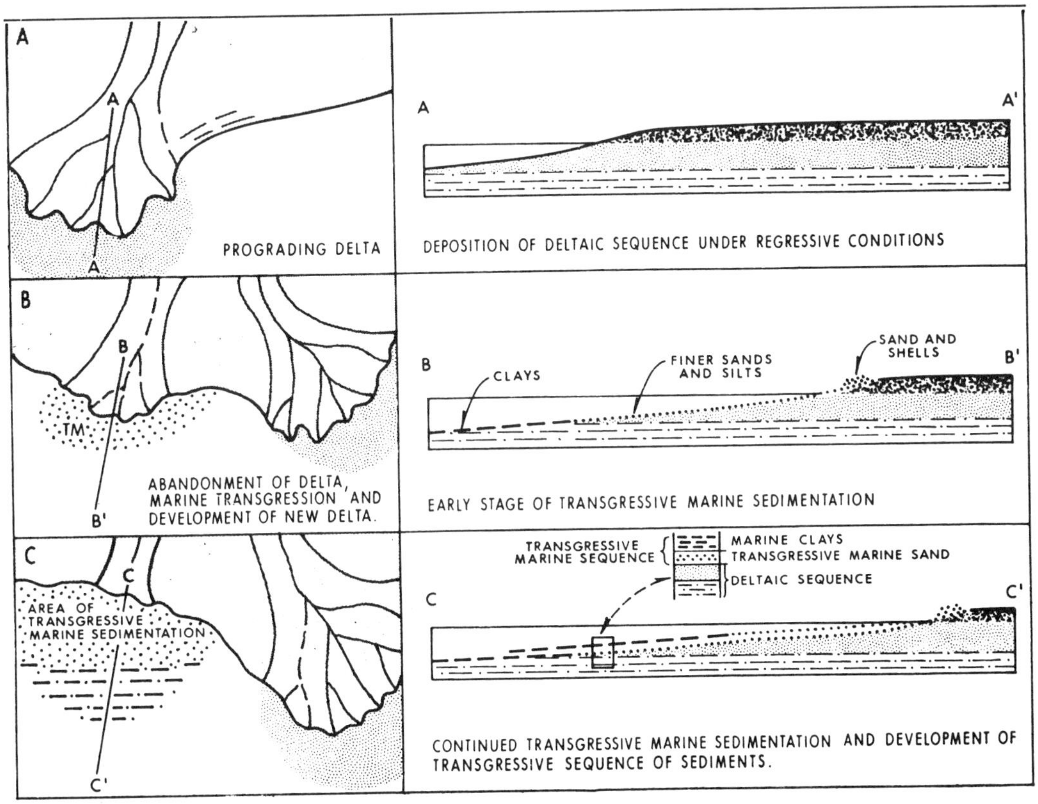

FIG. 31—Transgressive-marine sedimentation resulting from delta shifts.

equivalent interdeltaic and offshore-marine environments, is significant because it produces very diagnostic blanketlike layers of marine sediments (thin shallow-marine sand overlain by clays) which separate the individual deltaic units (Fig. 31). These layers usually provide the only good correlations within thick deltaic facies, and the marine shales act as impervious seals between deltaic sand bodies. The transgressive-marine sands containing calcareous shell material usually become cemented and thus do not form very efficient reservoirs.

Submarine Canyon-Fan Model of Clastic Sedimentation

Occurrence and general characteristics—The occurrence of modern and Pleistocene sands in deep-marine environments of the world is well documented as a result of numerous deep-sea investigations by oceanographic institutions during the past 20 years. Although there is much controversy regarding the origin of these sands, it is certain that such sands do exist. An analysis of the literature reveals that some of the most common deep-sea sands are those associated with submarine canyons and fans. (For a discussion of types of submarine canyons, troughs, and valleys, the reader is referred to Shepard and Dill, 1966.)

Submarine canyons and fans are common features associated with continental shelves, slopes, and rises. The canyons and fans off the Pacific Coast of the United States and Canada have received the most attention. Significant papers on these features off the coasts of Washington, Oregon, California, and Baja California, and off the Gulf and Atlantic coasts, are listed in the selected references. Also included are references to papers on canyons and fans in the Mediterranean, the Atlantic Ocean off Africa, and the Indian Ocean off Pakistan.

Characteristics and origin of submarine canyons have been discussed by numerous authors (for summary, see Shepard and Dill, 1966). Although it is still uncertain how some deep-sea canyons and valleys originated, it is now reasonably well established that a large number of canyons and fans are related to rivers, and that they were formed during stages of low sea level of the Quaternary Period (Figs. 32, 33). For

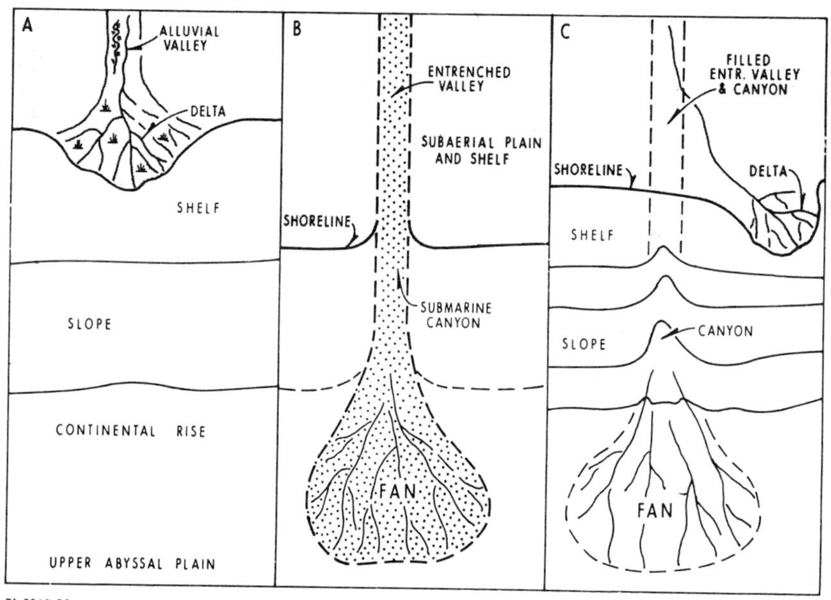

FIG. 32—Stages in development of submarine canyon and fan.

Stage A: Standing sea-level situation. Development of alluvial valley and delta and deposition of marine clays on shelf and slope. Base of aggrading river is well below sea level.

Stage B: Falling and low-sea-level situation. Development of entrenched-valley system on coastal plain and of submarine canyon offshore. Bases of entrenched valley (near coast) and of canyon are well below sea level. Rates of sedimentation are very high. Material removed by canyon-cutting and sediments flowing through canyon while it formed are deposited as extensive submarine fan.

Stage C: Rising and standing sea-level situation. Alluviation of entrenched-valley system and partial filling of canyon. Rates of sedimentation are greatly reduced after sea level reaches a stand. Slight modification of fan by normal-marine processes occurs.

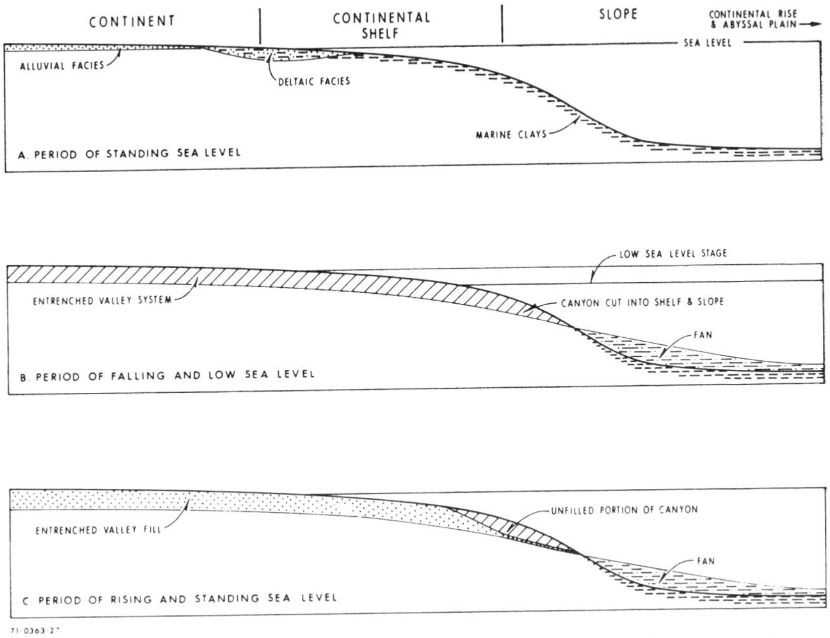

FIG. 33—Relation of submarine-fan deposits to submarine-canyon and entrenched-valley system.

example, the Mississippi Canyon off southern Louisiana is a continuation of the late Pleistocene Mississippi entrenched valley system (Fisk, 1944; Osterhoudt, 1946; Fish and McFarlan, 1955; and Bergantino, 1971). Also, the Astoria canyon and fan off Oregon are related to the Columbia River (Duncan and Kulm, 1970); the Newport Canyon is related to the Santa Ana River of California (Felix and Gorsline, 1971); the Congo Canyon connects with the Congo River (Heezen *et al.*, 1964); the Monterey and Soquel canyons and fans occur off the Great Valley of California (Martin and Emery, 1967); the Bengal deep-sea fan and the "Swatch-of-No-Ground" canyon occur off the Ganges River delta (Curray and Moore, 1971); and the Inguri canyon is related to a river flowing in the Caspian Sea (Trimonis and Shimkus, 1970). The *National Geographic* magazine maps of the Indian and Atlantic Ocean floors (Heezen and Tharp, 1967, 1969) show large fans off the Indus and Amazon Rivers and also off the Laurentian Trough, and the Hudson Canyon is associated with the Hudson River. Seismic reflection surveys between canyon heads (on shelves) and the coastline most probably will reveal more examples of canyons related to entrenched river valleys on land.

There is an extremely wide variation in the size of submarine canyon-fan systems. Some of the small ones off California studied by Gorsline and Emery (1959) include short canyons 5–10 mi (8–16 km) long and fan areas of about 50 sq mi (130 sq km). The largest canyon-fan systems studied thus far are those of the Congo, Ganges, and Rhône Rivers. The Bengal fan is 2,600 km long and 1,100 km wide; the Congo fan is more than 520 km long and 185 km wide; and one of the largest fans off the Pacific coast of the United States, the Delgade fan, is 300 km long and 330 km wide (Normark, 1970). Menard (1960) discussed the dimensions of several other fans.

Some very significant studies of deep-sea sands associated with canyons and fans—based on core, seismic reflection, and bathymetric data, and bottom observations and photography by divers—have been made during the past 3 years (Winterer *et al.*, 1968; Carlson and Nelson, 1969; Shepard *et al.*, 1969; Curray and Moore, 1971; Normark, 1970; Nelson *et al.*, 1970; Piper, 1970; Duncan and Kulm, 1970; and Felix and Gorsline, 1971).

Physiographic features.—Detailed bathymetric surveys over several canyons and fans of various sizes have revealed that these submarine features are characterized by physiographic features very similar to those of subaerial alluvial fans. The canyons are V-shaped and have steep walls and gradients. The surfaces of the fans are characterized by lower gradient distributary channels with natural levees and by topographically low interchannel areas. Some fans are crossed by relatively large fan valleys

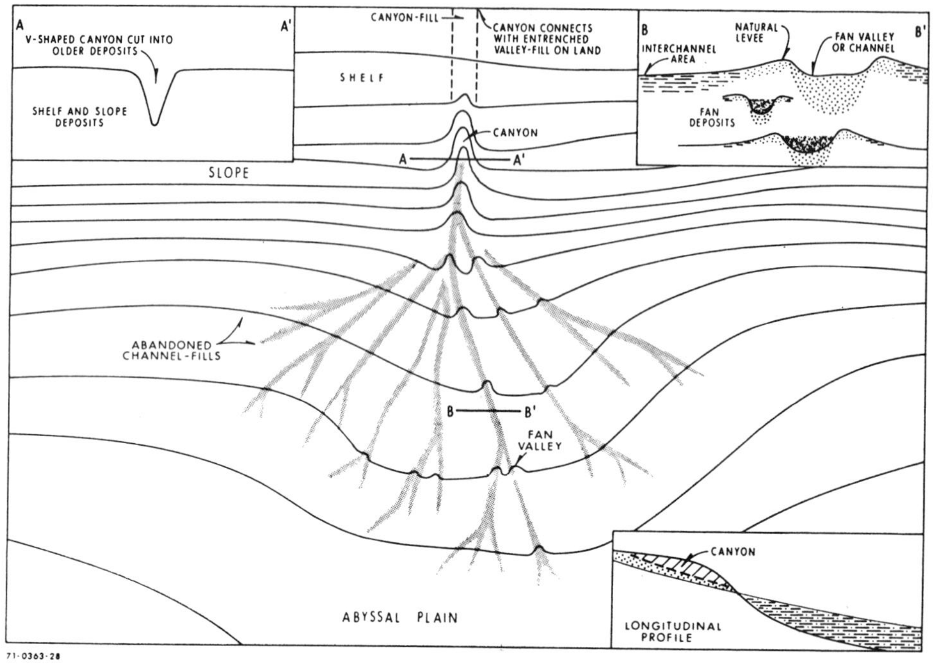

FIG. 34—Submarine-canyon and fan model of clastic sedimentation.

which also have natural levees. The principal physiographic features of a typical canyon and fan are illustrated in a generalized fashion in Figure 34.

The overall shape of a submarine fan can either be symmetrical or asymmetrical, depending on strong current directions and on the presence of high topographic features on the abyssal plains. Sizes of the distributary channels and natural levees are widely varied; the larger channels usually have the highest and broadest natural levees. The lower parts of fans merge with the abyssal plains. Channels on fan surfaces were probably formed by depositional processes. Erosional channels that have been reported probably represent an entrenchment stage, as is the case with subaerial alluvial fans.

Many canyons were cut across continental shelves and slopes, and the fans were constructed at the base of the slopes or on the continental rises. Some canyons presently do not extend landward across the continental shelves (*e.g.*, the Mississippi Canyon) because they have been filled with sediments. Seismic surveys reveal that this type of canyon was once connected with inland entrenched valley systems.

Longitudinal profiles of canyons and fans are concave upward. The steepest gradients occur in the upper (landward) portions of canyons, and the lowest gradients occur on the outer or lower portions of fans.

Depositional processes and character of sediments—It is absolutely certain that large quantities of sediment, including a significant amount of sand, have been transported through submarine canyons and deposited as submarine fans in deep-sea environments. The manner in which these sediments were transported, especially the sands, is much less certain. Nearly 2 decades ago, some very strong statements were made by oceanographers regarding the turbidity-current origin of both the canyons and the fan deposits. Although no one had actually seen or measured a turbidity current in a canyon or over a submarine fan, the turbidity-current concept was very popular with most oceanographers during the early 1950s. During the past 20 years, numerous additional observations have been made, but no one has yet seen a live turbidity current in a natural marine environment. On the basis of direct observations of the ocean bottom and sedimentary structures in cores, many oceanographers now believe that some submarine-fan sand deposits were transported mainly by normal bottom currents, especially during low stages of sea level of the Pleistocene. A typical example is the origin of the sand associated with the Mississippi cone in

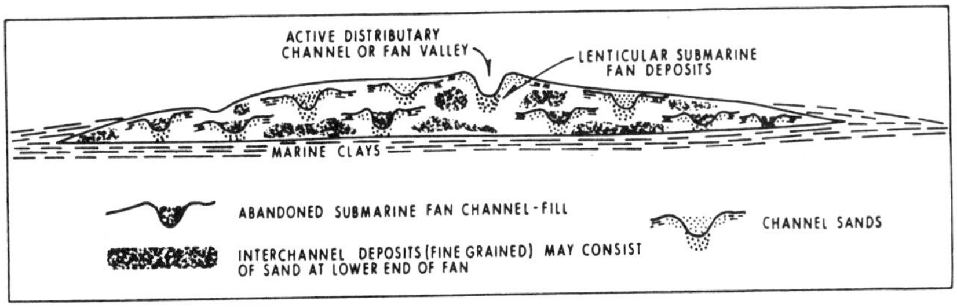

FIG. 35—Generalized distribution of submarine-fan deposits.

the Gulf of Mexico off southern Louisiana. Greenman and LeBlanc (1956) did not consider these sands to be of turbidity-current origin, but Ewing *et al.* (1958) were certain that the sands were transported and deposited by turbidity currents. Twelve years later, Huang and Goodell (1970) concluded, on the basis of detailed studies of sedimentary structures observed in numerous cores, that the sands are not of turbidity-current origin, but that the mechanisms of transport are bottom currents, differential pelagic settling, and mass movement by sliding and slumping. Walker and Massingill (1970) reported that part of the Mississippi cone sediments were recently involved in large-scale slumps. They presented evidence that one slump moved from near the mouth of the Mississippi Canyon southeastward for at least 160 n. mi. Thus, the origin of these deep-sea sands and many others remains a problem.

Regardless of the mechanisms of sediment transport through submarine canyons and of deposition of fans, the general nature and distribution of fan deposits have been determined for several fans. The coarsest and most poorly sorted sediments occur in canyons. Sands are common in distributary channels and fan valleys and on the lower parts of the open fan. Sandy sediments also occur on natural levees, but the interchannel areas are characterized by fine-grained sediments (Fig. 35). Core data from several fans indicate that sand bodies are usually thin and very lenticular, and are interbedded with fine-grained sediments.

For details concerning the sedimentary structures which characterize submarine-canyon and fan deposits, the reader is referred to Carlson and Nelson (1969); Shepard *et al.* (1969); Stanley (1969); Huang and Goodell (1970); and Haner (1971).

Horn *et al.* (1971) described the characteristics of sediments related to submarine canyons, fans, and adjacent abyssal plains of the northeast Pacific Ocean off Alaska, Canada, Washington, Oregon, and northern California. They interpreted sediments with a wide range in layer thickness, with graded and nongraded layers, and with sand in the basal parts of graded units to be proximal turbidites related to main routes followed by turbidity currents (probably channels). The finer grained sediments, mainly graded silts and clays, were interpreted as distal turbidites deposited beyond the main avenues of turbidite flows.

It is the opinion of the writer that many of the submarine canyons and related fans which now are found off rivers are the products of entrenchment (canyons) and deposition (fans) during stages of low sea level of the Pleistocene. Oceanographers who have studied several of these fan deposits have concluded that they are of Miocene to Pleistocene age. The geologic-age determinations were made on the basis of present sediment load of the related rivers and known thickness of fan deposits. This writer suggests that rates of sedimentation were probably several times greater during Pleistocene low-sea-level stages than at the present time (period of higher and standing sea level) and, consequently, that the fan deposits are probably chiefly of Pleistocene age.

Ancient examples of submarine canyon and fan deposits—Some examples of ancient deposits of submarine canyon and fan origin have been described from the Gulf coast by Osterhoudt (1946), Bornhauser (1948, 1960), Hoyt (1959), Paine (1966), and Sabate (1968); from California by Sullwold (1960), Martin (1963), Bartow (1966), Dickas and Payne (1967), Normark and Piper (1969), Piper and Normark (1971), Davis (1971), Fischer (1971), and Shelton (1972); from Canada by Hubert *et al.* (1970); from Europe by Walker (1966), Stanley (1967, 1969), and Kelling and Woollands (1969); and from Australia by Conolly (1968).

SELECTED REFERENCES

Al-Habeeb, K. H., 1970, Sedimentology of the flood-plain sediments of the middle Euphrates River, Iraq: Baghdad Univ., MS thesis.

Allen, J. R. L., 1962, Petrology, origin, and deposition of the highest lower Old Red Sandstone of Shropshire, England: Jour. Sed. Petrology, v. 32, p. 657–697.

——— 1964, Studies in fluviatile sedimentation: six cyclothems from the lower Old Red Sandstones, Anglo-Welsh basin: Sedimentology, v. 3, p. 163–198.

——— 1965a, A review of the origin and characteristics of Recent alluvial sediments: Sedimentology, v. 5, p. 89–191.

——— 1965b, Coastal geomorphology of eastern Nigeria: beach-ridge barrier-islands and vegetated tidal flats: Geologie en Mijnbouw, v. 44, p. 1–21.

——— 1965c, Late Quaternary Niger delta and adjacent areas: sedimentary environments and lithofacies: Am. Assoc. Petroleum Geologists Bull., v. 49, no. 5, p. 547–600.

——— 1970, Sediments of the modern Niger delta, *in* Deltaic sedimentation—modern and ancient: Soc. Econ. Paleontologists and Mineralogists Spec. Pub. 15, p. 138–151.

American Petroleum Institute, 1971: Petroleum facts and figures.

Anderson, G. S., and K. M. Hussey, 1962, Alluvial fan development at Franklin Bluff, Alaska: Iowa Acad. Sci. Proc., v. 69, p. 310–322.

Andrews, P. B., 1970, Facies and genesis of a hurricane washover fan, St. Joseph Island, central Texas: Texas Univ. Bur. Econ. Geology Rept. Inv. 67, 147 p.

Anstey, R. L., 1965, Physical characteristics of alluvial fans: U.S. Army Quartermaster Research and Eng. Command, Earth Sci. Div. Tech. Rept. ES-20, 109 p.

Arnborg, L. E., 1948, The delta of the Angerman River: Geog. Annaler, v. 30, p. 673–690.

Asseez, L. O., 1969, Paleogeography of Lower Mississippian rocks of Michigan basin: Am. Assoc. Petroleum Geologists Bull., v. 53, no. 1, p. 127–135.

Attia, M. I., 1954, Deposits in the Nile Valley and the delta: Cairo, Egypt Geol. Survey, 356 p.

Axelson, V., 1967, The Laitaure delta, a study of deltaic morphology and processes: Geog. Annaler, v. 49, p. 1–127.

Baars, D. L., 1961, Permian blanket sandstones of Colorado Plateau, *in* J. A. Peterson and J. C. Osmond, eds., Geometry of sandstone bodies: Am. Assoc. Petroleum Geologists, p. 179–219.

Bagchi, K., 1944, The Ganges delta: Calcutta, Calcutta Univ. Press, 157 p.

Bagnold, R. A., 1941, The physics of blown sand and desert dunes: New York, W. Morrow and Co., 265 p.

Bakker, W. T. J. N. P., and T. Edelman, 1965, The coast line of river-deltas: 9th Conf. on Coastal Engineering Proc., p. 199–218.

Barlow, J. A., Jr., and J. D. Haun, 1966, Regional stratigraphy of Frontier Formation and relation to Salt Creek field, Wyoming: Am. Assoc. Petroleum Geologists Bull., v. 50, p. 2185–2196.

Bartow, J. A., 1966, Deep submarine channel in upper Miocene, Orange County, California: Jour. Sed. Petrology, v. 36, no. 3, p. 700–705.

Bass, N. W., *et al.*, 1937, Origin and distribution of Bartlesville and Burbank shoestring oil sands in parts of Oklahoma and Kansas: Am. Assoc. Petroleum Geologists Bull., v. 21, p. 30–66.

Bates, C. C., 1953, Rational theory of delta formation: Am. Assoc. Petroleum Geologists Bull., v. 37, p. 2119–2162.

Beall, A. O., Jr., 1968, Sedimentary processes operative along the western Louisiana shoreline: Jour. Sed. Petrology, v. 38, p. 869–877.

Beaty, C. B., 1961, Boulder deposit in Flint Creek Valley, western Montana: Geol. Soc. America Bull., v. 72, p. 1015–1020.

——— 1963, Origin of alluvial fans, White Mountains, California and Nevada: Assoc. Am. Geographers Annals, v. 53, p. 516–535.

Beerbower, J. R., 1961, Origin of cyclothems of the Dunkard Group (Upper Pennsylvanian-Lower Permian) in Pennsylvania, West Virginia, and Ohio: Geol. Soc. America Bull., v. 72, p. 1029–1050.

——— 1964, Cyclothems and cyclic depositional mechanisms in alluvial plain sedimentation: Kansas Geol. Survey Bull. 169, v. 1, p. 31–42.

——— 1969, Interpretation of cyclic Permo-Carboniferous deposition in alluvial plain sediments in West Virginia: Geol. Soc. America Bull., v. 80, p. 1843–1848.

Berg, R. R., 1968, Point bar origin of Fall River sandstone reservoirs, northeastern Wyoming: Am. Assoc. Petroleum Geologists Bull., v. 52, no. 11, p. 2116–2122.

——— and B. C. Cook, 1968, Petrography and origin of lower Tuscaloosa sandstones, Mallalieu field, Lincoln County, Mississippi: Gulf Coast Assoc. Geol. Socs. Trans., v. 18, p. 242–255.

——— and D. K. Davies, 1968, Origin of Lower Cretaceous Muddy Sandstone at Bell Creek field, Montana: Am. Assoc. Petroleum Geologists Bull., v. 52, no. 10, p. 1888–1898.

Bergantino, R. N., 1971, Submarine regional geomorphology of the Gulf of Mexico: Geol. Soc. America Bull., v. 82, p. 741–752.

Bernard, H. A., 1965, A resume of river delta types: Am. Assoc. Petroleum Geologists Bull., v. 49, p. 334–335.

——— and R. J. LeBlanc, 1965, Resume of Quaternary geology of the northwestern Gulf of Mexico province, *in* Quaternary of the United States, Review volume for 7th Cong., Internat. Assoc. for Quaternary Research: Princeton, New Jersey, Princeton Univ. Press, p. 137–185.

——— and C. F. Major, 1963, Recent meander belt deposits of the Brazos River: an alluvial "sand" model: Am. Assoc. Petroleum Geologists Bull., v. 47, no. 2, p. 350.

——— R. J. LeBlanc, and C. F. Major, 1962, Recent and Pleistocene geology of southeast Texas, *in* Geology of the Gulf Coast and central Texas and guidebook of excursions, Geol. Soc. America 1962 Ann. Mtg.: Houston Geol. Soc., p. 175–224.

——— C. F. Major, and B. S. Parrott, 1959, The Galveston barrier island and environs—a model for predicting reservoir occurrence and trend: Gulf Coast Assoc. Geol. Socs. Trans., v. 9, p. 221–224.

——— *et al.*, 1970, Recent sediments of southeast Texas; a field guide to the Brazos alluvial and deltaic plains and the Galveston barrier island complex: Texas Univ. Bur. Econ. Geology Guidebook No. 11.

Beutner, E. C., *et al.*, 1967, Bedding geometry in a Pennsylvanian channel sandstone: Geol. Soc. America Bull., v. 78, no. 7, p. 911–916.

Bigarella, J. J., 1972, Eolian deposits—their characteristics, recognition, and importance, *in* J. K. Rigby and W. K. Hamblin, eds., Recognition of ancient

sedimentary environments: Soc. Econ. Paleontologists and Mineralogists Spec. Pub. 16, 340 p.
——— and R. Salamuni, 1961, Early Mesozoic wind patterns as suggested by dune bedding in the Botucatu Sandstone of Brazil and Uruguay: Geol. Soc. America Bull., v. 72, p. 1089–1106.
——— *et al.*, 1969, Coastal dune structure from Parana, Brazil: Marine Geology, v. 7, p. 5–55.
Bird, E. C. F., 1961, The coastal barriers of East Gippsland, Australia: Jour. Geology, v. 127, p. 460–468.
——— 1962, The river deltas of the Gippsland lakes: Royal Soc. Victoria Proc., v. 75, pt. 1, p. 65–74.
Black, R. F., 1961, Eolian deposits of Alaska: Arctic, v. 4, no. 2, p. 89–111.
Blackwelder, E., 1928, Mudflow as a geologic agent in semi-arid mountains: Geol. Soc. America Bull., v. 39, p. 465–483.
——— 1931, Desert plains: Jour. Geology, v. 39, p. 133–140.
Blissenbach, E., 1952, Relation of surface angle distribution to particle size distribution on alluvial fans: Jour. Sed. Petrology, v. 22, p. 25–28.
——— 1954, Geology of alluvial fans in semi-arid regions: Geol. Soc. America Bull., v. 65, p. 175–189.
Bluck, B. J., 1964, Sedimentation of an alluvial fan in southern Nevada: Jour. Sed. Petrology, v. 34, no. 2, p. 395–400.
——— 1965, The sedimentary history of some Triassic conglomerates in The Vale of Glamorgan, South Wales: Sedimentology, v. 4, p. 225–246.
——— 1967, Deposition of some upper Old Red Sandstone conglomerates in the Clyde area: a study in the significance of bedding: Scottish Jour. Geology, v. 3, no. 8, pt. 2, p. 139–167.
Boggs, S., Jr., 1966, Petrology of Minturn Formation, east-central Eagle County, Colorado: Am. Assoc. Petroleum Geologists Bull., v. 50, no. 7, p. 1399–1422.
Bolyard, D. W., 1959, Pennsylvanian and Permian stratigraphy in Sangre De Cristo Mountains between La Veta Pass and Westcliffe, Colorado: Am. Assoc. Petroleum Geologists Bull., v. 43, p. 1896–1939.
Bonham-Carter, G. F., and A. J. Sutherland, 1967, Diffusion and settling of sediments at river mouths: a computer simulation model: Gulf Coast Assoc. Geol. Socs. Trans., v. 17, p. 326–338.
Boothroyd, J. C., 1970, Recent braided stream sedimentation, south-central Alaska (abs.): Am. Assoc. Petroleum Geologists Bull., v. 54, no. 5, p. 836.
Bornhauser, M., 1948, Possible ancient submarine canyon in southwestern Louisiana: Am. Assoc. Petroleum Geologists Bull., v. 32, p. 2287–2290.
——— 1960, Depositional and structural history of northwest Hartburg field, Newton County, Texas: Am. Assoc. Petroleum Geologists Bull., v. 44, p. 458–470.
Bouma, A. H., 1965, Sedimentary characteristics of samples collected from some submarine canyons: Marine Geology, v. 3, no. 4, p. 291–320.
Boyd, D. R., and B. F. Dyer, 1966, Frio barrier bar system of south Texas: Am. Assoc. Petroleum Geologists Bull., v. 50, p. 170–178.
Brady, L. L., 1969, Stratigraphy and petrology of the Morrison Formation (Jurassic) of the Canon City, Colorado area: Jour. Sed. Petrology, v. 39, no. 2, p. 632–648.
Bretz, J. H., and L. Horberg, 1949, The Ogallala Formation west of the Llano Estacado: Jour. Geology, v. 57, no. 5, p. 477–490.
Brice, J. C., 1964, Channel pattern and terraces of the Loup Rivers in Nebraska: U.S. Geol. Survey Prof. Paper 422-D, 41 p.
Brickman, R., and L. J. Pons, 1968, A pedo-geomorphological classification and map of the Holocene sediments in the coastal plain of the three Guianas: Wageningen, Netherlands Soil Survey Institute, 40 p.
Brown, L. F., Jr., 1969, North Texas (eastern shelf) Pennsylvanian delta systems, *in* Delta systems in exploration for oil and gas: Texas Univ. Bur. Econ. Geology, p. 40–53.
Brown, S. L., 1967, Stratigraphy and depositional environment of the Elgin Sandstone (Pennsylvanian) in south-central Kansas: Kansas Geol. Survey Bull. 187, pt. 3, 9 p.
Buffington, E. C., 1952, Submarine "natural levees": Jour. Geology, v. 60, p. 473–479.
Bull, W. B., 1960, Types of deposition on alluvial fans in western Fresno County, California (abs.): Geol. Soc. America Bull., v. 71, p. 2052.
——— 1962, Relation of textural (cm) patterns to depositional environment of alluvial fan deposits: Jour. Sed. Petrology, v. 32, p. 211–216.
——— 1963, Alluvial fan deposits in western Fresno County, California: Jour. Geology, v. 71, p. 243–251.
——— 1964, Geomorphology of segmented alluvial fans in western Fresno County, California: U.S. Geol. Survey Prof. Paper 352-E, p. 59–129.
——— 1968, Alluvial fans: Jour. Geol. Education, v. 16, p. 101–106.
——— 1969, Recognition of alluvial-fan environments in stratigraphic record (abs.): Am. Assoc. Petroleum Geologists Bull., v. 53, p. 710.
——— 1972, Recognition of alluvial-fan deposits in the stratigraphic record, *in* J. K. Rigby and W. K. Hamblin, eds., Recognition of ancient sedimentary environments: Soc. Econ. Paleontologists and Mineralogists Spec. Pub. 16, 340 p.
Busch, D. A., 1953, The significance of deltas in subsurface exploration: Tulsa Geol. Soc. Digest, v. 21, p. 71–80.
——— 1971, Genetic units in delta prospecting: Am. Assoc. Petroleum Geologists Bull., v. 55, p. 1137–1154.
Buttner, P. J. R., 1968, Proximal continental rhythmic sequences in the Genesee Group (lower Upper Devonian): Geol. Soc. America Spec. Paper 106, p. 109–126.
Byers, P. R., 1966, Mineralogy and origin of the Eastend and Whitemud Formations of south-central and southwestern Saskatchewan and southeastern Alberta: Kingston, Ontario, Queen's University, Master's thesis.
Byrne, J. V., *et al.*, 1959, The chenier plain and its stratigraphy, southwestern Louisiana: Gulf Coast Assoc. Geol. Socs. Trans., v. 9, p. 237–260.
Cadwell, D. H., and J. H. Moss, 1969, Floodplain sediments and their movement during a high discharge, near Manheim, Pennsylvania: Pennsylvania Acad. Sci. Proc., v. 43, p. 185–186.
Cannon, J. L., 1966, Outcrop examination and interpretation of paleocurrent patterns of the Blackleaf Formation near Great Falls, Montana: Billings Geol. Soc. 17th Field Conf. Guidebook, p. 71–111.
Carlson, P. R., and C. H. Nelson, 1969, Sediments and sedimentary structures of the Astoria submarine canyon-fan system, northeast Pacific: Jour. Sed. Petrology, v. 39, p. 1269–1282.
Carrigy, M. A., 1971, Deltaic sedimentation in Atha-

basca tar sands: Am. Assoc. Petroleum Geologists Bull., v. 55, p. 1155–1169.
Casey, S. R., and R. B. Cantrell, 1941, Davis sand lens, Hardin field, Liberty County, Texas, *in* A. I. Levorsen, ed., Stratigraphic type oil fields: Am. Assoc. Petroleum Geologists, p. 564–599.
Cavaroc, V. V., 1969, Delineation of some outer deltaic plain subenvironments based on sedimentary properties in vertical section (abs.): Am. Assoc. Petroleum Geologists Bull., v. 53, no. 3, p. 710–711.
Challinor, J., 1946, Two contrasted types of alluvial deposits: Geol. Mag., v. 83, p. 162–164.
Chamberlin, T. K., 1964, Mass transport of sediment in the heads of Scripps submarine canyon, California, *in* R. L. Miller, ed., Papers in marine geology: New York, The Macmillan Co., p. 42–64.
Chawner, W. D., 1935, Alluvial fan flooding: the Montrose, California flood of 1934: Geog. Rev., v. 25, p. 255–263.
Chein, C., 1961, The braided river of the lower Yellow River: Sci. Sinica, v. 10, no. 6, p. 734–754.
Clark, R. H., and J. T. Rouse, 1971, A closed system for generation and entrapment of hydrocarbons in Cenozoic deltas, Louisiana Gulf Coast: Am. Assoc. Petroleum Geologists Bull., v. 55, p. 1170–1178.
Coastal Research Group, 1969, Field trip guidebook, coastal environments of northeastern Massachusetts and New Hampshire: Massachusetts Univ. Geol. Dept. Contr. 1-CRG, 462 p.
Coleman, J. M., 1966a, Ecological changes in the massive fresh-water clay sequence: Gulf Coast Assoc. Geol. Socs. Trans., v. 16, p. 159–174.
——— 1966b, Recent coastal sedimentation: central Louisiana coast: Louisiana State Univ. Coastal Studies Inst. Tech. Rept. 29, p. 1–73.
——— 1969, Brahmaputra River: channel processes and sedimentation: Sed. Geology, v. 3, no. 2–3, p. 129–239.
——— and S. M. Gagliano, 1964, Cyclic sedimentation in the Mississippi River deltaic plain: Louisiana State Univ. Coastal Studies Inst. Tech. Rept. 16, pt. G; also in Gulf Coast Assoc. Geol. Socs. Trans., v. 14.
——— and ——— 1965, Sedimentary structures: Mississippi River deltaic plain, *in* Primary sedimentary structures and their hydrodynamic interpretation: Soc. Econ. Paleontologists and Mineralogists Spec. Pub. 12, p. 133–148.
——— *et al.*, 1964, Minor sedimentary structures in a prograding distributary: Marine Geology, v. 1, p. 240–258.
——— *et al.*, 1970, Sedimentation in a Malaysian high tide tropical delta, *in* Deltaic sedimentation—modern and ancient: Soc. Econ. Paleontologists and Mineralogists Spec. Pub. 15, p. 185–197.
Collinson, J. D., 1968, Deltaic sedimentation units in the Upper Carboniferous of northern England: Sedimentology, v. 10, no. 4, p. 233–254.
——— 1968, The sedimentology of the Grindslow shales and the Kinderscant grit: a deltaic complex in the Namurian of northern England: Jour. Sed. Petrology, v. 39, no. 1, p. 194–221.
Conolly, J. R., 1965, Petrology and origin of the Hervey Group, Upper Devonian, central New South Wales: Geology Soc. Australia Jour., v. 12, pt. 1, p. 123–166.
——— 1968, Submarine canyons of the continental margin, East Bass Strait (Australia): Marine Geology, v. 6, p. 449–461.
——— and W. J. Cleary, 1971, Braided deep sea deltas and the origin of turbidite sands (abs.): Am. Geophys. Union Trans., v. 52, p. 244.
Cooper, W. S., 1958, Coastal sand dunes of Oregon and Washington: Geol. Soc. America Mem. 72, p. 1–169.
——— 1967, Coastal dunes of California: Geol. Soc. America Mem. 104, 131 p.
Corbeille, R. L., 1962, New Orleans barrier island, New Orleans area: Gulf Coast Assoc. Geol. Socs. Trans., v. 12, p. 223–229.
Credner, G. R., 1878, The deltas: Petermanns Geog. Mitt., v. 56, 74 p. (in German).
Cressey, C. B., 1928, The Indiana sand dunes and shorelines of the Lake Michigan basin: Geol. Soc. Chicago Bull. 8, p. 37.
Crickmay, C. H., 1955, Delta formation: Am. Assoc. Petroleum Geologists Bull., v. 39, p. 107–114.
——— 1960, Lateral activity in a river of northwestern Canada: Jour. Geology, v. 68, p. 377–391.
Croft, A., 1962, Some sedimentation phenomena along the Wasatch Mountain front: Jour. Geophys. Research, v. 67, p. 1511–1524.
Crosby, E. J., 1972, Classification of sedimentary environments, *in* J. K. Rigby and W. K. Hamblin, eds., Recognition of sedimentary environments: Soc. Econ. Paleontologists and Mineralogists Spec. Pub. 16, 340 p.
Crowell, J. C., 1952, Submarine canyons bordering central and southern California: Jour. Geology, v. 60, p. 58–83.
——— 1954, Geology of the Ridge Basin area: California Div. Mines Bull. 170, Map Sheet No. 7.
Culbertson, J. K., and D. R. Dawdy, 1964, A study of fluvial characteristics and hydraulic variables, middle Rio Grande, New Mexico: U.S. Geol. Survey Water-Supply Paper 1498-F, 74 p.
Curray, J. R., 1960, Sediments and history of Holocene transgression, continental shelf, northwest Gulf of Mexico, *in* F. P. Shepard *et al.*, eds., Recent sediments, northwest Gulf of Mexico: Am. Assoc. Petroleum Geologists, p. 221–266.
——— 1964, Transgressions and regressions, *in* Papers in marine geology: New York, The Macmillan Co., p. 175–203.
——— 1969a, Estuaries, lagoons, tidal flats and deltas, *in* New concepts of continental margin sedimentation: Am. Geol. Inst. Short Course Notes, Nov. 7–9, p. III–1 to III–30.
——— 1969b, Shore zone sand bodies: barriers, cheniers and beach ridges, *in* New concepts of continental margin sedimentation: Am. Geol. Inst. Short Course Notes, Nov. 7–9, p. II–1 to II–18.
——— and D. G. Moore, 1963, Facies delineation by acoustic-reflection: northern Gulf of Mexico: Sedimentology, v. 2, p. 130–148.
——— and ——— 1964, Holocene regressive littoral sand, Costa de Nayarit, Mexico, *in* Deltaic and shallow marine deposits: New York, Elsevier Publishing Co., p. 176–182.
——— and ——— 1971, Growth of the Bengal deep-sea fan and denudation in the Himalayas: Geol. Soc. America Bull., v. 82, p. 563–572.
——— *et al.*, 1969, Holocene history of a strand plain, lagoonal coast, Nayarit, Mexico, *in* Coastal lagoons, a symposium: Internat. Symposium on Coastal Lagoons Mem., Mexico, D. F., p. 63–100.
——— *et al.*, 1970, Late Quaternary sea-level studies in Micronesia: Carmarse expedition: Geol. Soc. America Bull., v. 81, p. 1865–1880.
Curtis, D. M., 1970, Miocene deltaic sedimentation,

Louisiana Gulf Coast, *in* Deltaic sedimentation—modern and ancient: Soc. Econ. Paleontologists and Mineralogists Spec. Pub. 15, p. 293–308.
Daboll, J. M., 1969, Holocene sediments of the Parker estuary, Massachusetts: Massachusetts Univ. Dept. Geology Contr. No. 3-CRG, 138 p.
Damuth, J. E., and R. W. Fairbridge, 1970, Equatorial Atlantic deep-sea arkosic sands and ice-age aridity in tropical South America: Geol. Soc. America Bull., v. 81, p. 189–206.
Davies, D. K., 1966, Sedimentary structures and subfacies of a Mississippi River point bar: Jour. Geology, v. 74, no. 2, p. 234–239.
——— *et al.*, 1971, Recognition of barrier environments: Am. Assoc. Petroleum Geologists Bull., v. 55, p. 550–565.
Davis, J. R., 1971, Sedimentation of Pliocene sandstones in Santa Barbara Channel, California (abs.): Am. Assoc. Petroleum Geologists Bull., v. 55, p. 335.
Davis, W. M., 1938, Sheetfloods and streamfloods: Geol. Soc. America Bull., v. 49, p. 1337–1416.
Dennison, J. M., 1971, Petroleum related to Middle and Upper Devonian deltaic facies in central Appalachia: Am. Assoc. Petroleum Geologists Bull., v. 55, p. 1179–1193.
Denny, C. S., 1965, Alluvial fans in the Death Valley region, California, Nevada: U.S. Geol. Survey Prof. Paper 466, 62 p.
——— 1967, Fans and pediments: Am. Jour. Sci., v. 265, p. 81–105.
Dickas, A. B., and J. L. Payne, 1967, Upper Paleocene buried channel in Sacramento Valley, California: Am. Assoc. Petroleum Geologists Bull., v. 51, no. 6, p. 873–882.
Dickinson, K. A., *et al.*, 1972, Criteria for recognizing ancient barrier coastlines, *in* J. K. Rigby and W. K. Hamblin, eds., Recognition of sedimentary environments: Soc. Econ. Paleontologists and Mineralogists Spec. Pub. 16, 340 p.
Dietz, R. J., and H. J. Knebel, 1971, Thou Sans Fond submarine canyon, Ivory Coast, Africa: Deep-Sea Research, v. 18, p. 441–447.
Dill, R. F., *et al.*, 1954, Deep-sea channels and deltas of the Monterey submarine canyon: Geol. Soc. America Bull., v. 65, p. 191–194.
Dillon, W. P., 1970, Submergence effects on a Rhode Island barrier and lagoon and inferences on migration of barriers: Jour. Geology, v. 78, no. 1, p. 94–106.
Dineley, D. L., and B. P. J. Williams, 1968, Sedimentation and paleoecology of the Devonian Escuminac Formation and related strata, Escuminac Bay, Quebec: Geol. Soc. America Spec. Paper 106, p. 241–264.
Dobby, E. H. G., 1936, The Ebro delta: Geog. Jour., v. 87, p. 455–469.
Dodge, C. F., 1965, Genesis of an Upper Cretaceous offshore bar near Arlington, Texas: Jour. Sed. Petrology, v. 35, no. 1, p. 22–35.
Doeglas, D. J., 1962, The structure of sedimentary deposits of braided rivers: Sedimentology, v. 1, p. 167–190.
Donaldson, A. C., 1966, Deltaic sands and sandstones, *in* Symposium on recently developed geologic principles and sedimentation of the Permo-Pennsylvanian of the Rocky Mountains: Wyoming Geol. Assoc. 20th Ann. Field Conf. Guidebook, p. 31–62.
——— 1969, Ancient deltaiç sedimentation (Pennsylvanian) and its control on the distribution, thickness and quality of coals, *in* A. C. Donaldson, ed., Some Appalachian coals and carbonates: models of ancient shallow-water deposition: West Virginia Geol. and Econ. Survey, p. 93–123.
——— *et al.*, 1970, Holocene Guadalupe delta of Texas Gulf Coast, *in* Deltaic sedimentation—modern and ancient: Soc. Econ. Paleontologists and Mineralogists Spec. Pub. 15, p. 107–137.
Dondanville, R. R., 1963, The Fall River Formation, northwestern Black Hills, lithology and geologic history: Wyoming Geol. Assoc.–Billings Geol. Soc. Joint Field Conf. Guidebook, p. 87–99.
Dott, R. H., Jr., 1964, Ancient deltaic sedimentation in eugeosynclinal belts, *in* Deltaic and shallow marine deposits: Amsterdam, Elsevier Pub. Co., p. 105–113.
——— 1966, Eocene deltaic sedimentation at Coos Bay, Oregon: Jour. Geology, v. 74, p. 373–420.
Duncan, J. R., and L. D. Kulm, 1970, Mineralogy, provenance and dispersal history of late Quaternary deep-sea sands in Cascadia basin and Blanco fracture zone of Oregon: Jour. Sed. Petrology, v. 40, p. 874–887.
Eckis, R., 1928, Alluvial fans of the Cucamonga district, southern California: Jour. Geology, v. 36, p. 224–247.
Emery, K. O., 1960a, Basin plains and aprons off southern California: Jour. Geology, v. 68, p. 464–479.
——— 1960b, The sea off southern California: New York, John Wiley and Sons, Inc.
——— *et al.*, 1952, Submarine geology off San Diego, California: Jour. Geology, v. 60, no. 6, p. 511–548.
Ericson, D. B., M. Ewing, and B. C. Heezen, 1951, Deep-sea sands and submarine canyons: Geol. Soc. America Bull., v. 62, p. 961–965.
——— ——— and ——— 1952, Turbidity currents and sediments in the North Atlantic: Am. Assoc. Petroleum Geologists Bull., v. 36, p. 489–511.
——— *et al.*, 1955, Sediment deposition in deep Atlantic, p. 205–220, *in* Arie Poldervaart, ed., Crust of the earth: Geol. Soc. America Spec. Paper 62, 762 p.
——— *et al.*, 1961, Atlantic deep-sea cores: Geol. Soc. America Bull., v. 72, p. 193–285.
Evans, G., 1965, Intertidal flat sediments and their environments of deposition in the Wash: Geol. Soc. London Quart. Jour., v. 121, p. 209–245.
Ewing, M., *et al.*, 1958, Sediments and topography of the Gulf of Mexico, *in* L. G. Weeks, ed., Habitat of oil: Am. Assoc. Petroleum Geologists, p. 995–1053.
Exum, F. A., and J. C. Harms, 1968, Comparison of marine-bar with valley-fill stratigraphic traps, western Nebraska: Am. Assoc. Petroleum Geologists Bull., v. 52, no. 10, p. 1851–1868.
Eymann, J. L., 1953, A study of sand dunes in the Colorado and Mojave Deserts: Los Angeles, Univ. Southern California, MS thesis, 91 p.
Fahnestock, R. K., 1963, Morphology and hydrology of a glacial stream—White River, Mount Rainier, Washington: U.S. Geol. Survey Prof. Paper 422-A, 70 p.
Felix, D. W., and D. S. Gorsline, 1971, Newport submarine canyon, California: an example of the effects of shifting loci of sand supply upon canyon position: Marine Geology, v. 10, p. 177–198.
Ferguson, J., 1863, Delta of the Ganges: Geol. Soc. London Quart. Jour., v. 19, p. 321–354.
Ferm, J. C., 1970, Allegheny deltaic deposits, *in* Deltaic sedimentation—modern and ancient: Soc. Econ. Paleontologists and Mineralogists Spec. Pub. 15, p. 246–255.
——— and V. V. Cavaroc, Jr., 1969, A field guide to Allegheny deltaic deposits in the upper Ohio Valley:

Ohio Geol. Soc. and Pittsburgh Geol. Soc.
Field, M. E., and O. H. Pilkey, 1971, Deposition of deep-sea sands: comparison of two areas of the Carolina continental rise: Jour. Sed. Petrology, v. 41, p. 526–536.
Finch, W. I., 1959, Geology of uranium deposits in Triassic rocks of the Colorado Plateau region: U.S. Geol. Survey Bull. 1074-D, 164 p.
Finkel, H. J., 1959, The barchans of southern Peru: Jour. Geology, v. 67, p. 614–647.
Fischer, P. J., 1971, An ancient (upper Paleocene) submarine canyon and fan: the Meganos Channel, Sacramento Valley, California (abs.): 67th Ann. Geol. Soc. America Cordilleran Section Mtg. Prog., v. 3, no. 2, p. 120.
Fisher, W. L., 1969, Gulf Coast Basin Tertiary delta systems, *in* Delta systems in exploration for oil and gas: Texas Univ. Bur. Econ. Geology, p. 30–39.
——— and L. F. Brown, Jr., 1969, Delta systems and oil and gas occurrence, *in* Delta systems in exploration for oil and gas: Texas Univ. Bur. Econ. Geology, p. 54–66.
——— and J. H. McGowen, 1969, Depositional systems in Wilcox Group (Eocene) of Texas and their relationship to occurrence of oil and gas: Am. Assoc. Petroleum Geologists Bull., v. 53, no. 1, p. 30–54.
——— *et al.*, 1969, Delta systems in the exploration for oil and gas: Texas Univ. Bur. Econ. Geology, 78 p.
——— *et al.*, 1970, Depositional systems in the Jackson Group of Texas—their relationship to oil, gas and uranium: Texas Univ. Bur. Econ. Geology Geol. Circ. 70–4.
Fisk, H. N., 1940, Geology of Avoyelles and Rapides Parishes, Louisiana: Louisiana Geol. Survey Geol. Bull. No. 18.
——— 1944, Geological investigation of the alluvial valley of the lower Mississippi River: Mississippi River Commission, Vicksburg, Mississippi.
——— 1947, Fine-grained alluvial deposits and their effects on Mississippi River activity: Mississippi River Commission, Vicksburg, Mississippi.
——— 1952, Geological investigation of the Atchafalaya basin and problems of Mississippi River diversion: Mississippi River Commission, Vicksburg, Mississippi, p. 1–145.
——— 1955, Sand facies of Recent Mississippi delta deposits: 4th World Petroleum Cong. Proc., Sec. I/C, p. 1–21.
——— 1958, Recent Mississippi River sedimentation and peat accumulation, *in* Ernest Van Aelst, ed., 4th Congres pour l'avancement des etudes de stratigraphie et de geologie du Carbonifere, Heerlen, 1958: Compte Rendu, v. 1, p. 187–199.
——— 1959, Padre Island and the Laguna Madre flats, coastal south Texas, *in* 2d Coastal Geog. Conf., April 6-9: Natl. Acad. Sci.–Natl. Research Council Pub., p. 103–151.
——— 1961, Bar-finger sands of the Mississippi delta, *in* J. A. Peterson and J. C. Osmond, eds., Geometry of sandstone bodies: Am. Assoc. Petroleum Geologists, p. 29–52.
——— and E. McFarlan, Jr., 1955, Late Quaternary deltaic deposits of the Mississippi River, *in* A. Poldervaart, ed., The crust of the earth: Geol. Soc. America Spec. Paper 62, p. 279–302.
——— *et al.*, 1954, Sedimentary framework of the modern Mississippi delta: Jour. Sed. Petrology, v. 24, no. 2, p. 76–99.
Flemal, R. C., 1967, Sedimentology of the Sespe Formation, southwestern California: Princeton, New Jersey, Princeton Univ., PhD thesis, 258 p.
Folk, R. L., 1971, Longitudinal dunes of the northwestern edge of the Simpson Desert, Northern Territory, Australia, Part 1. Geomorphology and grain size relationships: Sedimentology, v. 16, p. 5–54.
——— and W. C. Ward, 1957, Brazos River bar—a study of the significance of grain size parameters: Jour. Sed. Petrology, v. 27, p. 3–26.
Frazier, D. E., 1967, Recent deltaic deposits of the Mississippi River: their development and chronology: Gulf Coast Assoc. Geol. Socs. Trans., v. 17, p. 287–315.
——— and A. Osanik, 1961, Point-bar deposits, Old River locksite, Louisiana: Gulf Coast Assoc. Geol. Socs. Trans., v. 11, p. 121–137.
——— and ——— 1969, Recent peat deposits, Louisiana coastal plain: Geol. Soc. America Spec. Paper 114, p. 63–85.
Freeman, J. C., 1949, Strand-line accumulation of petroleum, Jim Hogg County, Texas: Am. Assoc. Petroleum Geologists Bull., v. 33, no. 7, p. 1260–1270.
Friedkin, J. F., 1945, A laboratory study of the meandering of alluvial rivers: U.S. Waterways Experiment Sta., Vicksburg, Mississippi, May.
Friedman, G. M., and K. G. Johnson, 1966, The Devonian Catskill deltaic complex of New York, type example of a "teutonic delta complex," *in* Deltas in their geologic framework: Houston Geol. Soc.
Galehouse, J. J., 1967, Provenance and paleocurrents of the Paso Robles Formation, California: Geol. Soc. America Bull., v. 78, p. 951–978.
Galloway, W. E., 1968, Depositional systems of lower Wilcox Group, north-central Gulf Coast basin: Gulf Coast Assoc. Geol. Socs. Trans., v. 18, p. 275–289.
Geijskes, D. C., 1952, On the structure and origin of the sandy ridges in the coastal zone of Surinam: Koninkl. Nederlandsch Aardrijksk. Genoot. Tijdschr., v. 69, no. 2, p. 225–237.
Gellert, J. F., 1968, The Yangtze River mouth and delta: Warsaw, Przegląd Geog., v. 40, p. 413–418 (in German).
Giddings, J. L., 1947, Mackenzie River delta chronology: Tree Ring Bull., v. 13, p. 26–29.
Gilbert, G. K., 1914, The transportation of debris by running water: U.S. Geol. Survey Prof. Paper 86.
Giles, R. T., and O. H. Pilkey, 1965, Atlantic beach and dune sediments of the southern United States: Jour. Sed. Petrology, v. 35, no. 4, p. 900–910.
Gilluly, J., 1927, Physiography of the Ajo region, Arizona: Geol. Soc. America Bull., v. 48, p. 323–348.
Glenn, J. L., and A. R. Dahl, 1959, Characteristics and distribution of some Missouri River deposits: Iowa Acad. Sci. Proc., v. 66, p. 302–311.
Glennie, K. W., 1970, Desert sedimentary environments—Developments in sedimentology, no. 14: Amsterdam, Elsevier Pub. Co., 222 p.
Gorsline, D. S., 1967, Sedimentologic studies of the Colorado delta: Southern California Univ. Geol. Rept. 67–1, 121 p.
——— and K. O. Emery, 1959, Turbidity-current deposits in San Pedro and Santa Monica basins off southern California: Geol. Soc. America Bull., v. 70, p. 279–290.
——— *et al.*, 1967, Studies of submarine canyons and fans off southern California: Southern Calif. Univ. Geol. Rept. 67–3, 27 p.
Gould, H. R., 1970, The Mississippi delta complex, *in* Deltaic sedimentation—modern and ancient: Soc.

Econ. Paleontologists and Mineralogists Spec. Pub. 15, p. 3–30.
——— and E. McFarlan, Jr., 1959, Geologic history of the chenier plain, southwestern Louisiana: Gulf Coast Assoc. Geol. Socs. Trans., v. 9, p. 261–270.
Gourou, P., 1967, The Zambezi River delta: Acta Geog. Lovaniensia, v. 5, p. 31–36 (in French).
Greenman, N. N., and R. J. LeBlanc, 1956, Recent marine sediments and environments of northwest Gulf of Mexico: Am. Assoc. Petroleum Geologists Bull., v. 40, p. 813–847.
Greensmith, J. T., 1966, Carboniferous deltaic sedimentation in eastern Scotland, a review and reappraisal, *in* Deltas in their geologic framework: Houston Geol. Soc.
——— and E. V. Tucker, 1969, The origin of the Holocene shell deposits in the chenier plain facies of Essex (Great Britain): Marine Geology, v. 7, p. 403–425.
Gregory, J. L., 1966, A lower Oligocene delta in the subsurface of southeast Texas: Gulf Coast Assoc. Geol. Socs. Trans., v. 16, p. 227–241.
Gremillion, L. R., *et al.*, 1964, Barrier and lagoonal sets on high terraces in the Florida panhandle: Southeastern Geology, v. 6, p. 31–36.
Griffith, E. G., 1966, Geology of Saber bar, Logan and Weld Counties, Colorado: Am. Assoc. Petroleum Geologists Bull., v. 50, p. 2112–2118.
Griggs, G. B., and L. D. Kulm, 1970, Sedimentation in Cascadia deep-sea channel: Geol. Soc. America Bull., v. 81, p. 1361–1384.
Guilcher, A., 1958, Coastal and submarine morphology: London, Methuen & Co., Ltd.: New York, John Wiley and Sons, Inc., 274 p.
——— 1959, Coastal sand ridges and marshes and their continental environment near Grand Popo and Ouidah, Dahomey: 2d Coastal Geog. Conf., U.S. Office Naval Research–Natl. Acad. Sci., Washington, D.C., p. 189–212.
Guy, H. P., *et al.*, 1966, Summary of alluvial channel data from flume experiments, 1956–1961: U.S. Geol. Survey Prof. Paper 462-I, p. 1–96.
Gwinn, V. E., 1964, Deduction of flow regime from bedding character in conglomerates and sandstones: Jour. Sed. Petrology, v. 34, no. 3, p. 656–658.
——— and T. A. Mutch, 1965, Intertongued Upper Cretaceous volcanic and nonvolcanic rocks, central-western Montana: Geol. Soc. America Bull., v. 76, p. 1125–1144.
Hack, J. T., 1941, Dunes of the western Navajo country: Geog. Rev., v. 31, p. 240–263.
Hails, J. R., 1964, The coastal depositional features of southeastern Queensland: Australia Geog., v. 9, p. 208–217.
——— and J. H. Hoyt, 1969, The significance and limitations of statistical parameters for distinguishing ancient and modern sedimentary environments of the lower Georgia coastal plain: Jour. Sed. Petrology, v. 39, no. 2, p. 559–580.
Hale, L. A., 1961, Late Cretaceous (Montanan) stratigraphy, eastern Washakie basin, Carbon County, Wyoming: Wyoming Geol. Assoc. 16th Field Conf. Guidebook, p. 129–137.
Hamilton, E. L., 1967, Marine geology of the abyssal plains in the Gulf of Alaska: Jour. Geophys. Research, v. 72, p. 4189–4213.
Haner, B. E., 1971, Morphology and sediments of Redondo submarine fan, southern California: Geol. Soc. America Bull., v. 82, p. 2413–2432.
Hansen, H. J., 1969, Depositional environments of subsurface Potomac Group in southern Maryland: Am. Assoc. Petroleum Geologists Bull., v. 53, no. 9, p. 1923–1937.
Hantzschel, W., 1939, Tidal flat deposits (Wattenschlick), *in* P. D. Trask, ed., Recent marine sediments: Am. Assoc. Petroleum Geologists, p. 195–206.
Happ, S. C., *et al.*, 1940, Some aspects of accelerated stream and valley sedimentation: U.S. Dept. Agriculture Tech. Bull., v. 695, 134 p.
Harms, J. C., 1966, Stratigraphic traps in a valley fill, western Nebraska: Am. Assoc. Petroleum Geologists Bull., v. 50, p. 2119–2149.
——— and R. K. Fahnestock, 1965, Stratification bed forms and flow phenomena (with an example from the Rio Grande), *in* Primary sedimentary structures and their hydrodynamic interpretation: Soc. Econ. Paleontologists and Mineralogists Spec. Pub. 12, p. 84–115.
——— *et al.*, 1963, Stratification in modern sands of the Red River, Louisiana: Jour. Geology, v. 71, p. 566–580.
——— *et al.*, 1965, Depositional environment of the Fox Hills sandstones near Rock Springs uplift, Wyoming: Wyoming Geol. Assoc. 19th Field Conf. Guidebook, p. 113–130.
Hattin, D. E., 1965, Stratigraphy of the Graneros shale (Upper Cretaceous) in central Kansas: Kansas Geol. Survey Bull. 178.
Heezen, B. C., and Marie Tharp, 1967, Indian Ocean floor: Natl. Geog. Mag., October.
——— and ——— 1969, Atlantic Ocean floor: Natl. Geog. Mag., June.
——— *et al.*, 1959, The floors of the oceans: Geol. Soc. America Spec. Paper 65, 122 p.
——— *et al.*, 1964, Congo submarine canyon: Am. Assoc. Petroleum Geologists Bull., v. 48, p. 1126–1149.
Hewitt, C. H., *et al.*, 1965, The Frye in situ combustion test—reservoir characteristics: Soc. Petroleum Engineers AIME Petroleum Trans., March, p. 337–342.
Holeman, J. N., 1970, The sediment yield of major rivers of the world: Water Resources Research, v. 4, no. 4, p. 737–747.
Hooke, R. L., 1967, Processes of arid-regional alluvial fans: Jour. Geology, v. 75, no. 4, p. 438–460.
Hoover, R. A., 1968, Physiography and surface sediment facies of a recent tidal delta, Harbor Island, central Texas coast: Austin, Texas, Texas Univ., PhD thesis, 223 p.
Hoppe, G., and S. R. Ekman, 1964, A note on the alluvial fans of Ladtjouagge, Swedish Lappland: Geog. Annaler, v. 46, p. 338–342.
Horn, D., 1965, Zur Geologischen entwicklung der sudlichen schlëimundung in Holozan: Meyniana, v. 15, p. 41–58.
Horn, D. R., *et al.*, 1971, Turbidites of the northeast Pacific: Sedimentology, v. 16, p. 55–69.
Horowitz, D. H., 1966, Evidence for deltaic origin of an Upper Ordovician sequence in the central Appalachians, *in* Deltas in their geologic framework: Houston Geol. Soc., p. 159–169.
Houbolt, J. J. H. C., and J. B. M. Jonker, 1968, Recent sediments in the eastern part of the Lake of Geneva (Lac Leman): Geologie en Mijnbouw, v. 47, p. 131–148.
Howard, J. D., 1966, Patterns of sediment dispersal in the Fountain Formation of Colorado: Mtn. Geologist, v. 3, p. 147–153.
Howe, H. V., and C. K. Moresi, 1931, Geology of

Iberia Parish: Louisiana Geol. Survey Geol. Bull. No. 1.
——— and ——— 1933, Geology of Lafayette and St. Martin Parishes: Louisiana Geol. Survey Geol. Bull. No. 3, 238 p.
——— *et al.,* 1935, Reports on the geology of Cameron and Vermilion Parishes: Louisiana Geol. Survey Geol. Bull. No. 6, 242 p.
Hoyt, J. H., 1964, High angle beach stratification, Sapelo Island, Georgia: Jour. Sed. Petrology, v. 32, p. 309–311.
——— 1967, Barrier island formation: Geol. Soc. America Bull., v. 78, p. 1125–1136.
——— 1969, Chenier versus barrier, genetic and stratigraphic distinction: Am. Assoc. Petroleum Geologists Bull., v. 53, p. 299–306.
——— and V. J. Henry, Jr., 1965, Significance of inlet sedimentation in the recognition of ancient barrier islands, *in* Sedimentation of Late Cretaceous and Tertiary outcrops, Rock Springs uplift: Wyoming Geol. Assoc. 19th Field Conf. Guidebook, p. 190–194.
——— and ——— 1967, Influence of island migration on barrier island sedimentation: Geol. Soc. America Bull., v. 78, p. 77–86.
——— and ——— 1971, Origin of capes and shores along the southeastern coast of the United States: Geol. Soc. America Bull., v. 82, p. 59–66.
——— and R. J. Weimer, 1963, Comparison of modern and ancient beaches, central Georgia coast: Am. Assoc. Petroleum Geologists Bull., v. 47, p. 529–531.
——— *et al.,* 1964, Late Pleistocene and Recent sedimentation on Georgia coast, *in* Deltaic and shallow marine deposits: New York, Elsevier Publishing Co., p. 170–176.
Hoyt, W. V., 1959, Erosional channel in the middle Wilcox near Yoakum, Lavaca County, Texas: Gulf Coast Assoc. Geol. Socs. Trans., v. 9, p. 41–50.
Hrabar, S. V., and P. E. Potter, 1969, Lower west Badeu (Mississippian) sandstone body of Owen and Green Counties, Indiana: Am. Assoc. Petroleum Geologists Bull., v. 53, no. 10, p. 2150–2160.
Huang, T., and H. G. Goodell, 1970, Sediments and sedimentary processes of eastern Mississippi cone, Gulf of Mexico: Am. Assoc. Petroleum Geologists Bull., v. 54, p. 2070–2100.
Hubbard, J. A. E. B., 1967, Facies patterns in the Carrowmoran Sandstone (Visean) of western County Sligo, Ireland: Geol. Assoc. Proc., v. 77, pt. 2, p. 233–254.
Hubert, C., *et al.,* 1970, Deep-sea sediments in the lower Paleozoic Quebec supergroup: Geol. Assoc. Canada Spec. Paper 7, p. 103–125.
Hubert, J. F., 1960, Petrology of the Fountain and Lyons Formations, Front Range, Colorado: Colorado School Mines Quart., v. 55, p. 1–242.
——— 1964, Textural evidence for deposition of many western Atlantic deep-sea sands by ocean-bottom currents rather than turbidity currents: Jour. Geology, v. 72, p. 757–785.
Inman, D. L., and R. A. Bagnold, 1963, Littoral processes, *in* The sea, v. 2, chap. 2: New York, John Wiley and Sons, p. 529–553.
——— *et al.,* 1966, Coastal sand dunes of Guerrero Negro, Baja California, Mexico: Geol. Soc. America Bull., v. 77, no. 8, p. 787–802.
Ives, R. L., 1936, Desert floods in Sonoyta Valley: Am. Jour. Sci., v. 32, p. 349–360.
Jacka, A. D., 1965, Depositional dynamics of the Almond Formation, Rock Springs uplift, Wyoming: Wyoming Geol. Assoc. 19th Field Conf. Guidebook, p. 81–100.
Jahns, R. H., 1947, Geological features of the Connecticut Valley, Massachusetts as related to recent floods: U.S. Geol. Survey Water-Supply Paper 996, 157 p.
Jelgersma, S., J. de Jong, and W. H. Zagwijn, 1970, The coastal dunes of the western Netherlands; geology, vegetational history and archeology: Geol. Stichting Med., new ser., no. 21, p. 93–154.
Johnson, A., 1965, A model for debris flow: University Park, Pennsylvania, Pennsylvania State Univ., PhD thesis, 233 p.
Johnson, R. B., 1967, The great sand dunes of southern Colorado: U.S. Geol. Survey Prof. Paper 575-C, p. C177–C183.
Johnson, W. A., 1921, Sedimentation of the Fraser River delta: Canada Geol. Survey Mem. 125, 46 p.
——— 1922, The character of the stratification of the sediments of the Recent delta of the Fraser River, British Columbia, Canada: Jour. Geology, v. 30, p. 115–129.
Kanes, W., 1969, Tidal inlets, tidal deltas, and barrier islands versus alluvial deposits: a discussion of sedimentary criteria, *in* A. C. Donaldson, ed., Some Appalachian coals and carbonates: models of ancient shallow water deposition: West Virginia Geol. and Econ. Survey, p. 259–263.
——— 1970, Facies and development of the Colorado River delta in Texas, *in* Deltaic sedimentation—modern and ancient: Soc. Econ. Paleontologists and Mineralogists Spec. Pub. 15, p. 78–106.
Kelling, G., 1968, Patterns of sedimentation in Rhondda Beds of South Wales: Am. Assoc. Petroleum Geologists Bull., v. 52, no. 12, p. 2369–2386.
——— and D. J. Stanley, 1970, Morphology and structure of Wilmington and Baltimore submarine canyons, eastern United States: Jour. Geology, v. 78, p. 637–660.
——— and M. A. Woollands, 1969, The stratigraphy and sedimentation of the Llandoverian rocks of the Rhoyader district, *in* A. Woods, ed., The Precambrian and lower Paleozoic rocks of Wales: Univ. Wales Press, p. 255–282.
Kessler, L. G., II, 1970, Bar development and aggradational sequences in the braided portions of the Canadian River, Hutchinson, Roberts and Hemphill Counties, Texas (abs.): 23d Ann. Rocky Mountain Sec. Geol. Soc. America Mtg. Prog., v. 2, no. 5, p. 338.
——— 1971, Characteristics of the braided stream depositional environment with examples from the South Canadian River, Texas: Earth Sci. Bull., v. 4, no. 1, p. 25–35.
King, C. A. M., 1959, Beaches and coasts: London, Edward Arnold Publications, Ltd., 403 p.
Klein, G. deV., 1962, Triassic sedimentation, Maritime Provinces, Canada: Geol. Soc. America Bull., v. 73, p. 1127–1146.
——— 1967, Comparison of Recent and ancient tidal flat and estuarine sediments, *in* Estuaries: Am. Assoc. Adv. Sci. Pub. 83, p. 207–218.
——— 1968, Sedimentology of Triassic rocks in the lower Connecticut Valley, *in* Guidebook for field trips in Connecticut, New England Internat. Collegiate Geol. Conf.: Connecticut Geol. and Nat. History Survey Guidebook 2, p. (c-1) 1–19.
——— 1970, Depositional and dispersal dynamics of intertidal sand bars: Jour. Sed. Petrology, v. 40, p. 1095–1127.

——— 1971, A sedimentary model for determining paleotidal range: Geol. Soc. America Bull., v. 82, p. 2585–2592.
——— and J. E. Sanders, 1964, Comparison of sediments from Bay of Fundy and Dutch Wadden Sea tidal flats: Jour. Sed. Petrology, v. 34, no. 1, p. 18–24.
Knight, W. V., 1969, Historical and economic geology of Silurian Clinton Sandstone of northwestern Ohio: Am. Assoc. Petroleum Geologists Bull., v. 53, no. 7, p. 1421–1452.
Kolb, C. R., 1962, Distribution of soils bordering the Mississippi River from Donaldsonville to Head of Passes: U.S. Army Corps Engineers, Waterways Expt. Sta. Tech. Rept. 3-601, p. 1–61.
——— 1963, Sediments forming the bed and banks of the lower Mississippi River and their effect on river migration: Sedimentology, v. 2, p. 227–234.
——— and R. I. Kaufman, 1967, Prodelta clays of southeast Louisiana: Internat. Marine Geotech. Research Conf. Proc., Monticello, Illinois, p. 3–21.
——— and J. R. Van Lopik, 1966, Depositional environments of the Mississippi River deltaic plain, southeastern Louisiana, *in* Deltas in their geological framework: Houston Geol. Soc., p. 17–61.
——— *et al.*, 1968, Geological investigation of the Yazoo basin, lower Mississippi Valley: U.S. Army Corps Engineers, Waterways Expt. Sta. Tech. Rept. 3–480.
Komar, P. D., 1969, The channelized flow of turbidity currents with application to Monterey deep-sea channel: Jour. Geophys. Research, v. 74, p. 4544–4558.
——— 1970, The competence of turbidity current flow: Geol. Soc. America Bull., v. 81, p. 1555–1562.
Krause, D. C., *et al.*, 1970, Turbidity currents and cable breaks in the western New Britain trench: Geol. Soc. America Bull., v. 81, p. 2153–2160.
Krigstrom, A., 1962, Geomorphological studies of Sandur plains and their braided river in Iceland: Geog. Annals, v. 44, p. 328–346.
Kruit, C., 1955, Sediments of the Rhône delta; I, Grain size and microfauna: Koninkl. Nederlandse Geol. Mijnb. Genoot. Verh., Geol. Ser., v. 15, p. 357–514.
Kuenen, Ph. H., 1950, Marine geology: New York, John Wiley and Sons, Inc., 568 p.
Kukal, Z., 1964, Diagnostic features of Paleozoic deltaic sediments of central Bohemia: Sedimentology, v. 3, p. 109–113.
——— 1971, Geology of Recent sediments: New York, Academic Press, 490 p.
Lagaaij, R., and F. P. H. W. Kopstein, 1964, Typical features on a fluvio-marine offlap sequence, *in* Developments in sedimentology, v. 1, Deltaic and shallow marine deposits: Amsterdam, Elsevier Publishing Co., p. 216–226.
Laming, D. J. C., 1966, Imbrication, paleocurrents, and other sedimentary features in the lower New Red Sandstone, Devonshire, England: Jour. Sed. Petrology, v. 36, no. 4, p. 940–959.
Land, L. S., 1964, Eolian cross bedding in the beach dune environment, Sapelo Island, Georgia: Jour. Sed. Petrology, v. 34, no. 2, p. 389–394.
Lane, E. W., 1957, A study of the shape of channels formed by natural streams flowing in erodible material: U.S. Army Corps Engineers, Missouri River Div., Omaha, Nebraska, Sediment Ser. 9, 106 p.
——— 1963, Cross-stratification in San Bernard River, Texas, point bar deposits: Jour. Sed. Petrology, v. 33, p. 350–354.
——— and E. W. Eden, 1940, Sand waves in the lower Mississippi River: Western Soc. Engineers Jour., v. 45, no. 6, p. 281–291.
Laporte, L. F., 1968, Ancient environments: Englewood Cliffs, New Jersey, Prentice-Hall, Inc., 116 p.
Laury, R. L., 1968, Sedimentology of the Pleasantview Sandstone, southern Iowa and western Illinois: Jour. Sed. Petrology, v. 38, no. 2, p. 568–599.
Lawson, A. C., 1913, The petrographic description of alluvial fan deposits: California Univ. Pubs. Geol. Sci., v. 7, p. 325–334.
LeBlanc, R. J., and H. A. Bernard, 1954, Resume of late Recent geological history of the Gulf Coast: Geologie en Mijnbouw, new series, v. 16c, p. 185–194.
——— and W. D. Hodgson, 1959, Origin and development of the Texas shoreline: Louisiana State Univ. Coastal Studies Inst., 2d Coastal Geog. Conf., p. 57–101; EPR Pub. 219, Shell E&P Research, Houston.
LeBlanc, R. J., Jr., 1971, Environments of deposition of the Yegua Formation (Eocene), Brazos County, Texas: Corpus Christi Geol. Soc. Bull., v. 11, no. 6, 70 p.
Legget, R. F., R. J. E. Brown, and G. H. Johnston, 1966, Alluvial fan formation near Aklavik, Northwest Territories, Canada: Geol. Soc. America Bull., v. 77, p. 15–30.
Lené, G. W., and D. E. Owen, 1969, Grain orientation in a Berea Sandstone channel at south Amherst, Ohio: Jour. Sed. Petrology, v. 39, no. 2, p. 737–743.
Leopold, L. B., and M. G. Wolman, 1957, River channel patterns: braided, meandering and straight: U.S. Geol. Survey Prof. Paper 282-B, 85 p.
——— and ——— 1960, River meanders: Geol. Soc. America Bull., v. 71, p. 769–794.
——— *et al.*, 1966, Channel and hillslope processes in a semi-arid area, New Mexico: U.S. Geol. Survey Prof. Paper 352-G.
Lineback, J. A., 1968, Turbidites and other sandstone bodies in the Borden Siltstone (Mississippian) in Illinois: Illinois Geol. Survey Circ. 425, p. 1–29.
Lins, T. W., 1950, Origin and environment of the Tonganoxie Sandstone in northeastern Kansas: Kansas Geol. Survey Bull., v. 86, p. 107–140.
Liteanu, E., and A. Pricajan, 1963, The geological structure of the Danube delta: Bucharest, Hidrobiologia, v. 4, p. 57–82 (in Rumanian).
Logvinenko, N. V., and I. N. Remizov, 1964, Sedimentology of beaches on the north coast of the Sea of Azov, *in* Deltaic and shallow marine deposits, developments in sedimentology, v. 1: Amsterdam, Elsevier Publishing Co., p. 244–252.
Lucke, J. B., 1934, A study of Barnegat Inlet, New Jersey and related shoreline phenomena: Shore and Beach, v. 2, p. 45–93.
Lum, D., and H. T. Stearns, 1970, Pleistocene stratigraphy and eustatic history based on cores at Waimanalo, Oahu, Hawaii: Geol. Soc. America Bull., v. 81, p. 1–16.
Lumsden, D. N., and B. R. Pelletier, 1969, Petrology of the Grimsby Sandstone (Lower Silurian) of Ontario and New York: Jour. Sed. Petrology, v. 39, no. 2, p. 521–530.
Lustig, L. K., 1963, Competence of transport in alluvial fans: U.S. Geol. Survey Prof. Paper 475-C, p. 126–129.
——— 1965, Clastic sedimentation in the Deep Springs Valley, California: U.S. Geol. Survey Prof. Paper 352-F.
MacIntyre, I. G., and J. D. Milliman, 1970, Physio-

graphic features on the outer shelf and upper slope, Atlantic continental margin, southeastern United States: Geol. Soc. America Bull., v. 81, p. 2577–2589.
MacNeil, F. S., 1950, Pleistocene shore lines in Florida and Georgia: U.S. Geol. Survey Prof. Paper 221-F.
Mammerickx, J., 1970, Morphology of the Aleutian abyssal plain: Geol. Soc. America Bull., v. 81, p. 3457–3464.
Mann, C. J., and W. A. Thomas, 1968, Ancient Mississippi River: Gulf Coast Assoc. Geol. Socs. Trans., v. 18, p. 187–204.
Manos, C., 1967, Depositional environment of Sparland cyclothem (Pennsylvanian), Illinois and Forest City basins: Am. Assoc. Petroleum Geologists Bull., v. 51, no. 9, p. 1843–1861.
Martin, B. D., 1963, Rosedale channel—evidence for late Miocene submarine erosion in Great Valley of California: Am. Assoc. Petroleum Geologists Bull., v. 47, p. 441–456.
——— and K. O. Emery, 1967, Geology of the Monterey Canyon, California: Am. Assoc. Petroleum Geologists Bull., v. 51, p. 2281–2304.
Martini, I. P., 1971, Regional analysis of sedimentology of Medina Formation (Silurian), Ontario and New York: Am. Assoc. Petroleum Geologists Bull., v. 55, p. 1249–1261.
Mathews, W. H., and F. P. Shepard, 1962, Sedimentation of Fraser River delta, British Columbia: Am. Assoc. Petroleum Geologists Bull., v. 48, no. 8, p. 1416–1444.
Matthews, J. L., 1966, Sedimentation of the coastal dunes at Oceano, California: Univ. California at Los Angeles, PhD thesis, 138 p.
McBride, E. F., and M. O. Hayes, 1962, Dune cross bedding on Mustang Island, Texas: Am. Assoc. Petroleum Geologists Bull., v. 46, p. 546–552.
McCave, I. N., 1969, Correlation using a sedimentological model of part of the Hamilton group (Middle Devonian), New York State: Am. Jour. Sci., v. 267, no. 5, p. 567–591.
McCloy, J. L., 1969, Morphologic characteristics of the Blow River delta, Yukon Territory, Canada: Baton Rouge, Louisiana, Louisiana State Univ., PhD thesis, 161 p.
McEwen, M. C., 1969, Sedimentary facies of the modern Trinity delta, *in* Galveston Bay geology: Houston Geol. Soc., p. 53–77.
McGee, W. J., 1897, Sheetflood erosion: Geol. Soc. America Bull., v. 8, p. 87–112.
McGowen, J. H., and L. E. Garner, 1969, Comparison of Recent and ancient coarse-grained point bars (abs.): Am. Assoc. Petroleum Geologists Bull., v. 53, no. 3, p. 731–732.
——— and ——— 1970, Physiographic features and stratification types of coarse-grained point bars: modern and ancient examples: Sedimentology, v. 14, p. 77–111.
——— and C. G. Groat, 1971, Van Horn Sandstone, West Texas—an alluvial fan model for mineral exploration: Texas Univ. Bur. Econ. Geology, 57 p.
McKee, E. D., 1934, The Coconino Sandstone—its history and origin: Carnegie Inst. Washington Pub. 440, Contr. Paleontology, p. 77–115.
——— 1939, Some types of bedding in the Colorado River delta: Jour. Geology, v. 47, p. 64–81.
——— 1957, Primary structures of some Recent sediments: Am. Assoc. Petroleum Geologists Bull., v. 41, p. 1704–1747.
——— 1966, Structures of dunes at White Sands National Monument, New Mexico: Sedimentology, v. 7, no. 1, 69 p.
——— and G. C. Tibbitts, 1964, Primary structures of a seif dune and associated deposits of Libya: Jour. Sed. Petrology, v. 34, no. 1, p. 5–17.
McKenzie, P., 1958, The development of sand beach ridges: Australian Jour. Sci., v. 20, p. 213–214.
Melton, M. A., 1965, The geomorphic and paleoclimatic significance of alluvial deposits in southern Arizona: Jour. Geology, v. 73, p. 1–38.
Menard, H. W., 1955, Deep-sea channels, topography, and sedimentation: Am. Assoc. Petroleum Geologists Bull., v. 39, p. 236–255.
——— 1960, Possible pre-Pleistocene deep-sea fans off central California: Geol. Soc. America Bull., v. 71, p. 1271–1278.
——— 1964, Marine geology of the Pacific: New York, McGraw-Hill Book Co., Inc., 271 p.
——— *et al.*, 1965, The Rhône deep-sea fan, *in* Submarine geology and geophysics: London, Butterworth, p. 271–286.
Miall, A. D., 1970, Continental marine transition in the Devonian of Prince of Wales Island, Northwest Territories: Canadian Jour. Earth Sci., v. 7, p. 125–144.
Miller, D. N., 1962, Patterns of barrier bar sedimentation and its similarity to Lower Cretaceous Fall River stratigraphy, *in* Symposium on Early Cretaceous rocks: Wyoming Geol. Assoc. 17th Field Conf. Guidebook, p. 232–247.
Milling, M. E., and E. W. Behrens, 1966, Sedimentary structures of beach and dune deposits: Mustang Island, Texas: Inst. Marine Sci. Pub., v. 11, p. 135–148.
Moody-Stuart, M., 1966, High and low sinuosity stream deposits, with examples from the Devonian of Spitsbergen: Jour. Sed. Petrology, v. 36, no. 4, p. 1102–1117.
Moore, D., 1966, Deltaic sedimentation: Earth Sci. Rev., v. 1, p. 87–104, Amsterdam, Elsevier Publishing Co.
Moore, D. G., 1970, Reflection profiling studies of the California borderland: structure and Quaternary turbidite basins: Geol. Soc. America Spec. Paper 107.
Moore, G. T., and D. O. Asquith, 1971, Delta: term and concept: Geol. Soc. America Bull., v. 82, p. 2563–2568.
Morgan, J. P., 1961, Mudlumps at the mouths of the Mississippi River, *in* Genesis and paleontology of the Mississippi River mudlumps: Louisiana Geol. Survey Geol. Bull. No. 35.
——— 1970, Depositional processes and products in the deltaic environment, *in* Deltaic sedimentation—modern and ancient: Soc. Econ. Paleontologists and Mineralogists Spec. Pub. 15, p. 31–47.
——— *et al.*, 1953, Occurrence and development of mud flats along the western Louisiana coast: Louisiana State Univ. Coastal Studies Inst. Tech. Rept. No. 2.
——— *et al.*, 1968, Mudlumps: diapiric structures in Mississippi delta sediments, *in* J. Braunstein and G. D. O'Brien, eds., Diapirism and diapirs: Am. Assoc. Petroleum Geologists Mem. 8, p. 145–161.
Muehlberger, F. B., 1955, Pismo beach–Point Sal dune field, California: Lawrence, Kansas, Univ. Kansas, Master's thesis, 106 p.
Muller, G., 1966, The new Rhine delta in Lake Constance, *in* M. L. Shirley, ed., Deltas in their geologic framework: Houston Geol. Soc., p. 107–124.
Mutch, T. A., 1968, Pennsylvanian non-marine sediments of the Narragansett Basin, Massachusetts and

Rhode Island: Geol. Soc. America Spec. Paper 106, p. 177–209.
Nagtegaal, P. J. C., 1966, Scour-and-fill structures from a fluvial piedmont environment: Geologie en Mijnbouw, v. 45, no. 10, p. 342–354.
Nanz, R. H., 1954, Genesis of Oligocene sandstone reservoir, Seeligson field, Jim Wells and Kleberg Counties, Texas: Am. Assoc. Petroleum Geologists Bull., v. 38, p. 96–117.
Nelson, B. W., 1970, Hydrography, sediment dispersal, and recent historical development of the Po River delta, Italy, *in* Deltaic sedimentation—modern and ancient: Soc. Econ. Paleontologists and Mineralogists Spec. Pub. 15, p. 152–184.
Nelson, C. H., 1968, Marine geology of Astoria deep-sea fan: Corvallis, Oregon, Oregon State Univ., PhD thesis, 287 p.
——— *et al.*, 1970, Development of the Astoria canyon-fan physiography and comparison with similar systems: Marine Geology, v. 8, p. 259–291.
Nichols, M. M., 1964, Characteristics of sedimentary environments in Moriches Bay, *in* Papers in marine geology: New York, The Macmillan Co., p. 363–383.
Nilsen, T. H., 1969, Old Red sedimentation in the Buelandet-Vaerlandet Devonian district, western Norway: Sed. Geology, v. 3, no. 1.
Normark, W. R., 1970, Growth patterns of deep sea fans: Am. Assoc. Petroleum Geologists Bull., v. 54, p. 2170–2195.
——— and D. J. W. Piper, 1969, Deep-sea fan-valleys, past and present: Geol. Soc. America Bull., v. 80, p. 1859–1866.
Norris, R. M., 1966, Barchan dunes of Imperial Valley, California: Jour. Geology, v. 74, p. 292–306.
Nossin, J. J., 1965, The geomorphic history of the northern Padang delta: Jour. Tropical Geog., v. 20, p. 54–64.
Nunnally, N. R., 1967, Definition and identification of channel and overbank deposits and their respective roles in flood plain formation: Prof. Geographer, v. 19, no. 1, p. 1–4.
Odem, W. I., 1953, Subaerial growth of the delta of the diverted Brazos River, Texas: Compass, v. 30, p. 172–178.
Oomkens, E., 1967, Depositional sequences and sand distribution in a deltaic complex: Geologie en Mijnbouw, v. 46E, p. 265–278.
——— 1970, Depositional sequences and sand distribution in the post-glacial Rhone delta complex, *in* Deltaic sedimentation—modern and ancient: Soc. Econ. Paleontologists and Mineralogists Spec. Pub. 15, p. 198–212.
Ore, H. T., 1963, Some criteria for recognition of braided stream deposits: Contr. Geology, v. 3, no. 1, p. 1–14.
——— 1965, Characteristic deposits of rapidly aggrading streams, *in* Sedimentation of Late Cretaceous and Tertiary outcrops, Rock Springs uplift: Wyoming Geological Assoc. 19th Field Conf. Guidebook, p. 195–201.
Osterhoudt, W. J., 1946, The seismograph discovery of an ancient Mississippi River channel (abs.): Geophysics, v. 11, p. 417.
Otvos, E. G., Jr., 1970, Development and migration of barrier islands, northern Gulf of Mexico: Geol. Soc. America Bull., v. 81, p. 241–246.
Pack, F. J., 1923, Torrential potential of desert waters: Pan-Am. Geologist, v. 40, p. 349–356.
Paine, W. R., 1966, Stratigraphy and sedimentation of Hackberry shale and associated beds of southwestern Louisiana: Gulf Coast Assoc. Geol. Socs. Trans., v. 16.
Paull, R. A., 1962, Depositional history of the Muddy Sandstone, Bighorn basin, Wyoming: Wyoming Geol. Assoc. 17th Field Conf. Guidebook, p. 102–117.
Peterson, J. A., and J. C. Osmond, eds., 1961, Geometry of sandstone bodies: Am. Assoc. Petroleum Geologists, 240 p.
Petrescu, G., 1948, Le delta maritime du Danube; son evolution physico-geographique et les problems qui s'y posent: Rumania, Univ. Iaşi Ann. Sci., 2d sec. (Sc. Nat.), v. 31, p. 254-303.
Pettyjohn, W. A., 1967, New members of Upper Cretaceous Fox Hills Formation in South Dakota, representing delta deposits: Am. Assoc. Petroleum Geologists Bull., v. 51, no. 7, p. 1361–1367.
Pezzetta, J. R., 1968, The St. Clair River delta: Univ. Michigan, PhD thesis, 168 p.
Phleger, F. B, 1965, Sedimentology of Guerrero Negro Lagoon, Baja California, Mexico, *in* 17th Submarine geology and geophysics—Colston Research Society Symposium Proc.; Bristol, England, Paper 171, p. 205–237: London, Butterworths.
——— 1969, Some general features of coastal lagoons, *in* Coastal lagoons, a symposium: Internat. Symposium on Coastal Lagoons Mem., Mexico, D F., p. 2–25.
——— and G. C. Ewing, 1962, Sedimentology and oceanography of coastal lagoons in Baja California, Mexico: Geol. Soc. America Bull., v. 73, p. 145–182.
Pierce, J. W., 1970, Tidal inlets and washover fans: Jour. Geology, v. 78, no. 2, p. 230–234.
——— and D. J. Colquhoun, 1970, Holocene evolution of a portion of the North Carolina coast: Geol. Soc. America Bull., v. 81, no. 12, p. 3697–3714.
Piper, D. J. W., 1970, Transport and deposition of Holocene sediment of La Jolla deep sea fan, California: Marine Geology, v. 8, p. 211–227.
——— and W. R. Normark, 1971, Re-examination of a Miocene deep-sea fan and fan valley, southern California: Geol. Soc. America Bull., v. 82, p. 1823–1830.
Potter, P. E., 1967, Sand bodies and sedimentary environments, a review: Am. Assoc. Petroleum Geologists Bull., v. 51, p. 337–365.
Price, W. A., 1947, Equilibrium of form and forces in tidal basins of coast of Texas and Louisiana: Am. Assoc. Petroleum Geologists Bull., v. 31, no. 9, p. 1619–1663.
——— 1954, Environment and formation of the chenier plain: Texas A&M Research Found. Proj. 63, Dept. Oceanography, Texas A&M College.
——— 1963, Patterns of flow and channeling in tidal inlets: Jour. Sed. Petrology, v. 33, no. 2, p. 279–290.
Priddy, R. R., *et al.*, 1955, Sediments of Mississippi Sound and inshore waters: Mississippi Geol. Survey Bull. 82.
Pryor, W. A., 1960, Cretaceous sedimentation in Upper Mississippi embayment: Am. Assoc. Petroleum Geologists Bull., v. 44, p. 1473–1504.
——— 1961, Sand trends and paleoslope in Illinois basin and Mississippi embayment, *in* J. A. Peterson and J. C. Osmond, Geometry of sandstone bodies: Am. Assoc. Petroleum Geologists, p. 119–133.
——— 1967, Biogenic directional features on several point bars: Sed. Geology, v. 1, p. 235–245.
Psuty, N. P., 1965, The geomorphology of beach ridges

in Tabasco, Mexico: Louisiana State Univ. Coastal Studies Inst. Tech. Rept. 30, p. 1–51.
Pugh, J. C., 1953, The Porto Novo-Badagri sand ridge complex: Nigeria, Univ. College of Ibadan Dept. Geography, Research Notes, no. 3, p. 1–14.
Rahn, P. H., 1967, Sheetflood, streamfloods, and the formation of pediments: Assoc. Am. Geographers Annals, v. 57, p. 593–604.
Rainwater, E. H., 1964, Transgressions and regressions in the Gulf Coast Tertiary: Gulf Coast Assoc. Geol. Socs. Trans., v. 14, p. 217–230.
Rapp, A., 1959, Avalanche boulder tongues in Lappland: Geog. Annaler, v. 41, p. 34–48.
Reineck, H. E., 1963, Sedimentgefuge im bereich der sudlichen Nordsee: Abhandl. Senckenberg: Naturf Gesell. 505, p. 1–136.
——— 1967, Layered sediments in tidal flats, beaches, and shelf bottoms of the North Sea, *in* Estuaries: Am. Assoc. Adv. Sci. Pub. 83, p. 191–206.
——— and I. B. Singh, 1967, Primary sedimentary structures in the Recent sediments of the Jade, North Sea: Marine Geology, v. 5, p. 227–235.
Reineck, H. R., 1972, Tidal flats, *in* J. K. Rigby and W. K. Hamblin, eds., Recognition of sedimentary environments: Soc. Econ. Paleontologists and Mineralogists Spec. Pub. 16, 340 p.
Revelle, R., and F. P. Shepard, 1939, Sediments off the California coast, *in* F. P. Shepard *et al.*, eds., Recent marine sediments: Am. Assoc. Petroleum Geologists, p. 245–282.
Rex, R. W., 1964, Arctic beaches, Barrow, Alaska, *in* Papers in marine geology: New York, The Macmillan Co., p. 384–400.
Rickmers, W. R., 1913, The Duab of Turkestan: New York, Cambridge Univ. Press, 563 p.
Rigby, J. K., and W. K. Hamblin, eds., 1972, Recognition of sedimentary environments: Soc. Econ. Paleontologists and Mineralogists Spec. Pub. 16, 340 p.
Rona, P. A., 1970, Submarine canyon origin on upper continental slope off Cape Hatteras: Jour. Geology, v. 78, p. 141–152.
Royse, C. F., Jr., 1970, A sedimentologic analysis of the Tongue River–Sentinel Butte interval (Paleocene) of the Williston basin, western North Dakota: Sed. Geology, v. 4, p. 19–80.
Ruhe, R. U., 1964, Landscape morphology and alluvial deposits in southern New Mexico: Assoc. Am. Geographers Annals, v. 54, p. 147–159.
Rusnak, G. A., 1957, A fabric and petrographic study of the Pleasantview Sandstone: Jour. Sed. Petrology, v. 27, no. 1, p. 41–55.
——— 1960, Sediments of Laguna Madre, Texas, *in* Recent sediments, northwest Gulf of Mexico: Am. Assoc. Petroleum Geologists, p. 153–196.
Russell, R. J., 1936, Physiography of lower Mississippi River delta, *in* Lower Mississippi River delta: Louisiana Geol. Survey Geol. Bull. 8, p. 1–199.
——— 1942, Geomorphology of the Rhône delta: Assoc. Am. Geographers Annals, v. 32, p. 149–254.
——— 1967a, River and delta morphology: Louisiana State Univ. Coastal Studies Inst. Tech. Rept. 52, 49 p.
——— 1967b, Aspects of coastal morphology: Geog. Annaler, v. 49, ser. A, p. 299–309.
——— and H. V. Howe, 1935, Cheniers of southwestern Louisiana: Geog. Rev., v. 25, no. 3, p. 449–461.
——— and R. D. Russell, 1939, Mississippi River delta sedimentation, *in* P. D. Trask, ed., Recent marine sediments: Am. Assoc. Petroleum Geologists, p. 153–177.
Ryan, D. J., 1965, Cross-bedding formed by lateral accretion in the Catskill Formation near Jim Thorpe, Pennsylvania: Pennsylvania Acad. Sci. Proc., v. 38, no. 2, p. 154–156.
Ryder, J. M., 1971, The stratigraphy and morphology of para-glacial alluvial fans in south-central British Columbia: Canadian Jour. Earth Sci., v. 8, p. 279–298.
Sabate, R. W., 1968, Pleistocene oil and gas in coastal Louisiana: Gulf Coast Assoc. Geol. Socs. Trans., v. 18, p. 373–386.
Sabins, F. F., Jr., 1963, Anatomy of stratigraphic trap, Bisti field, New Mexico: Am. Assoc. Petroleum Geologists Bull., v. 47, p. 193–228.
Saucier, R. T., 1964, Geological investigation of the St. Francis basin: U.S. Army Corps Engineers, Waterways Expt. Sta. Tech. Rept. 3–659.
——— 1967, Geological investigation of the Boeuf-Tensas basin, lower Mississippi Valley: U.S. Army Corps Engineers, Waterways Expt. Sta. Tech. Rept. 3–757.
Schlee, J. S., and R. H. Moench, 1961, Properties and genesis of "Jackpile" Sandstone, Laguna, New Mexico, *in* J. A. Peterson and J. C. Osmond, eds., Geometry of sandstone bodies: Am. Assoc. Petroleum Geologists, p. 134–150.
Scholl, D. W., *et al.*, 1970, The structure and origin of the large submarine canyons of the Bering Sea: Marine Geology, v. 8, p. 187–210.
Schumm, S. A., 1963, Sinuosity of alluvial rivers on the great plains: Geol. Soc. America Bull., v. 74, p. 1089–1100.
——— 1971, Fluvial paleochannels, *in* J. K. Rigby and W. K. Hamblin, eds., Recognition of sedimentary environments: Soc. Econ. Paleontologists and Mineralogists Spec. Pub. 16, 340 p.
Schwartz, M. L., 1971, The multiple causality of barrier islands: Jour. Geology, v. 79, no. 1, p. 91–94.
Scott, A. J., and W. L. Fisher, 1969, Delta systems and deltaic deposition, *in* Delta systems in exploration for oil and gas: Texas Univ. Bur. Econ. Geology, p. 10–39.
Scruton, P. C., 1960, Delta building and the deltaic sequence, *in* F. P. Shepard *et al.*, eds., Recent sediments, northwest Gulf of Mexico: Am. Assoc. Petroleum Geologists, p. 82–102.
——— 1961, Rocky Mountain Cretaceous stratigraphy and regressive sandstones: Wyoming Geological Assoc. 16th Field Conf. Guidebook, p. 241–249.
Selley, R. C., 1970, Ancient sedimentary environments: Ithaca, New York, Cornell Univ. Press, 237 p.
Shannon, J. P., Jr., and A. R. Dahl, 1971, Deltaic stratigraphic traps in West Tuscola field, Taylor County, Texas: Am. Assoc. Petroleum Geologists Bull., v. 55, p. 1194–1205.
Sharp, R. P., 1942, Mudflow levees: Jour. Geomorphology, v. 5, p. 222–227.
——— and L. H. Nobles, 1953, Mudflow in 1941 at Wrightwood, southern California: Geol. Soc. America Bull., v. 64, p. 547–560.
Shawa, M. S., 1969, Sedimentary history of the Gilwood Sandstone (Devonian), Utikuma Lake area, Alberta, Canada: Bull. Canadian Petroleum Geology, v. 17, p. 392–409.
Shelton, J. W., 1965, Trend and genesis of lowermost sandstone unit of Eagle Sandstone at Billings, Montana: Am. Assoc. Petroleum Geologists Bull., v. 49, p. 1385–1397.
——— 1967, Stratigraphic models and general criteria for recognition of alluvial, barrier-bar, and turbidity-

current sand deposits: Am. Assoc. Petroleum Geologists Bull., v. 51, no. 12, p. 2441–2461; also *in* Billings Geol. Soc. 17th Field Conf. Guidebook, p. 1–17.
——— 1972, Models of sand and sandstone deposits: Oklahoma Geol. Survey Bull. (in press).
Shepard, F. P., 1960, Gulf Coast barriers, *in* F. P. Shepard *et al.*, eds., Recent sediments, northwest Gulf of Mexico: Am. Assoc. Petroleum Geologists, p. 197–220.
——— F. B Phleger, and Tj. H. van Andel, 1960, Recent sediments, northwest Gulf of Mexico: Am. Assoc. Petroleum Geologists, 394 p.
——— 1965, Importance of submarine valleys in funneling sediments to the deep sea, *in* Progress in oceanography, v. 3: New York, The Macmillan Co., p. 321–332.
——— and E. C. Buffington, 1968, La Jolla submarine fan-valley: Marine Geology, v. 6, p. 107–143.
——— and R. F. Dill, 1966, Submarine canyons and other sea valleys: Chicago, Rand McNally, 381 p.
——— and D. G. Moore, 1955, Sediment zones bordering the barrier islands of central Texas coast, *in* Finding ancient shorelines: Soc. Econ. Paleontologists and Mineralogists Spec. Pub. 3, p. 78–96.
——— and ——— 1960, Bays of central Texas coast, *in* F. P. Shepard *et al.*, eds., Recent sediments, northwest Gulf of Mexico: Am. Assoc. Petroleum Geologists, p. 117–152.
——— *et al.*, eds., 1960, Recent sediments, northwest Gulf of Mexico: Am. Assoc. Petroleum Geologists, 394 p.
——— *et al.*, 1963, Submarine geology: New York, Harper and Row, 557 p.
——— *et al.*, 1969, Physiography and sedimentary process of La Jolla submarine fan and fan-valley, California: Am. Assoc. Petroleum Geologists Bull., v. 53, p. 390–420.
Shepheard, W. W., and L. V. Hills, 1970, Depositional environments, Bearpaw-Horseshoe Canyon (Upper Cretaceous) transition zone, Drumheller "Badlands," Alberta: Bull. Canadian Petroleum Geology, v. 18, p. 166–215.
Shideler, G. L., 1969, Dispersal patterns of Pennsylvanian sandstones in the Michigan basin: Jour. Sed. Petrology, v. 39, no. 3, p. 1229–1237.
Shirley, M. L., ed., 1966, Deltas: Houston Geol. Soc., 251 p.
Shuiyak, B. A., and V. L. Boldyrev, 1966, The processes of beach ridge formation: Oceanology (Engl. ed.), v. 6, no. 1, p. 88–94.
Siebold, E., 1963, Geological investigation of nearshore sand transport—examples of methods and problems from Baltic and North Seas, *in* Progress in oceanography, v. 1: New York, The Macmillan Co., p. 3–70.
Simons, D. B., and E. V. Richardson, 1961, Forms of bed roughness in alluvial channels: Am. Soc. Civil Engineers Proc., Jour. Hydraulics Div., v. 87, no. HY 3, May, p. 87–105.
——— 1962, The effect of bed roughness on depth-discharge relations in alluvial channels: U.S. Geol. Survey Water-Supply Paper 1498-E.
——— 1966, Resistance to flow in alluvial channels: U.S. Geol. Survey Prof. Paper 422-J, 69 p.
——— *et al.*, 1965, Sedimentary structures generated by flow in alluvial channels, *in* Primary sedimentary structures and their hydrodynamic interpretation: Soc. Econ. Paleontologists and Mineralogists Spec. Pub. 12, p. 34–52.
Sims, J. D., 1967, Geology and sedimentology of the Livingston Group, northern Crazy Mountains, Montana: Evanston, Illinois, Northwestern Univ., PhD thesis, 109 p.
Sioli, H., 1966, General features of the delta of the Amazon, *in* Scientific problems of the humid tropic zone deltas and their implications: UNESCO, Dacca Symposium Proc., p. 381–390.
Sloane, B. J., Jr., 1958, The subsurface Jurassic Bodcaw sand in Louisiana: Louisiana Geol. Survey Geol. Bull. 33.
Smith, A. E., Jr., 1971, Deltas of the world—modern and ancient, bibliography: Houston Geol. Soc. Delta Study Group, 42 p.
Smith, A. J., 1966, Modern deltas: comparison maps, *in* Deltas in their geologic framework: Houston Geol. Soc., p. 233–251.
Smith, H. T. U., 1954, Coastal dunes: Coastal Geog. Conf., February, U.S. Office Naval Research, p. 51–56.
Smith, N. D., 1970, The braided stream depositional environment: comparison of the Platte River with some Silurian clastic rocks, north central Appalachians: Geol. Soc. America Bull., v. 81, p. 2993–3014.
Snavely, P. D., Jr., *et al.*, 1964, Rhythmic-bedded eugeosynclinal deposits of the Tyee Formation, Oregon Coast Range, *in* Symposium on cyclic sedimentation: Kansas Geol. Survey Bull. 169, p. 461–480.
Spearing, D. R., 1969, Stratigraphy, sedimentation, and tectonic history of the Paleocene-Eocene Hoback Formation of western Wyoming: Ann Arbor, Michigan, Univ. Michigan, PhD thesis, 179 p.
Stanley, D. J., 1967, Comparing patterns of sedimentation in some modern and ancient submarine canyons: Earth and Planetary Sci. Letters, v. 3, p. 371–380.
——— 1968, Graded bedding-sole marking-graywacke assemblage and related sedimentary structures in some Carboniferous flood deposits, eastern Massachusetts: Geol. Soc. America Spec. Paper 106, p. 211–239.
——— 1969, Submarine channel deposits and their fossil analogs ("fluxoturbidites"), *in* New concepts of continental margin sedimentation: Am. Geol. Inst. Short Course Notes, Nov. 7–9, p. DJS-9-1–DJS-9-17.
——— 1970, Flyschoid sedimentation on the outer Atlantic margin off northeast North America, *in* Flysch sedimentology in North America: Geol. Assoc. Canada Spec. Paper No. 7, p. 179–210.
——— 1972, Submarine channel deposits, fluxoturbidites and other indicators of slope and base-of-slope environments in modern and ancient marine basins, *in* J. K. Rigby and W. K. Hamblin, eds., Recognition of sedimentary environments: Soc. Econ. Paleontologists and Mineralogists Spec. Pub. 16, 340 p.
——— and N. Silverberg, 1969, Recent slumping on the continental slope off Sable Island bank, southeast Canada: Earth and Planetary Sci. Letters, v. 6, p. 123–133.
——— *et al.*, 1971, Lower continental rise east of the middle Atlantic states: predominant sediment dispersal perpendicular to isobaths: Geol. Soc. America Bull., v. 82, p. 1831–1840.
Stetson, H. C., 1953, The sediments of the northwestern Gulf of Mexico—Part 1. The continental terrace of the western Gulf of Mexico: its surface sediments, origin and development, *in* Papers in physical oceanography and meteorology: Massachusetts Inst. Tech. Abs. Theses, v. 12, no. 4, p. 5–45.

Stokes, W. L., 1961, Fluvial and eolian sandstone bodies in Colorado Plateau, *in* J. A. Peterson and J. C. Osmond, eds., Geometry of sandstone bodies: Am. Assoc. Petroleum Geologists, p. 151–178.

——— 1964, Eolian varving in the Colorado Plateau: Jour. Sed. Petrology, v. 34, no. 2, p. 429–433.

——— 1968, Multiple parallel-truncation bedding planes—a feature of wind-deposited sandstone formations: Jour. Sed. Petrology, v. 38, no. 2, p. 510–515.

Sullwold, H. H., Jr., 1960, Tarzana fan, deep submarine fan of late Miocene age, Los Angeles County, California: Am. Assoc. Petroleum Geologists Bull., v. 44, p. 433–457.

Sundborg, A., 1956, The river Klarälven. A study of fluvial processes: Geog. Annaler, v. 38, p. 127–316.

Swann, D. H., 1964, Late Mississippian rhythmic sediments of Mississippi Valley: Am. Assoc. Petroleum Geologists Bull., v. 48, no. 5, p. 637–658.

——— *et al.*, 1965, The Borden Siltstone (Mississippian) delta in southwestern Illinois: Illinois Geol. Survey Circ. 386, 20 p.

Swift, D. J. P., 1969, Inner shelf sedimentation: processes and products, *in* New concepts of continental margin sedimentation: Am. Geol. Inst. Short Course Notes, Nov. 7–9, p. DS-4-1–DS-4-46.

——— and R. M. McMullen, 1968, Preliminary studies of intertidal sand bodies in the Minas Basin, Bay of Fundy, Nova Scotia: Canadian Jour. Earth Sci., v. 5, no. 2, p. 175–183.

Sykes, G., 1937, The Colorado delta: Carnegie Inst. Washington Pub. 460, 193 p.

Taylor, J. H., 1963, Sedimentary features of an ancient deltaic complex: the Wealden rocks of southeastern England: Sedimentology, v. 2, p. 2–28.

Thachuk, N. M., 1968, Geological study of the Middle Devonian Gilwood arkoses in the Nipisi area, Alberta: Petroleum Soc. of Canadian Inst. Mining and Metallurgy, 19th Ann. Tech. Mtg., Preprint No. 6828, 17 p.

Thom, B. G., 1967, Mangrove ecology and deltaic geomorphology: Tabasco, Mexico: Jour. Ecology, v. 55, p. 301–343.

Thomas, W. A., and C. J. Mann, 1966, Late Jurassic depositional environments, Louisiana and Arkansas: Am. Assoc. Petroleum Geologists Bull., v. 50, p. 178–182.

Thompson, R. W., 1968, Tidal flat sedimentation on the Colorado River delta, northwestern Gulf of California: Geol. Soc. America Mem. 107, 133 p.

Thompson, W. O., 1937, Original structures of beaches, bars, and dunes: Geol. Soc. America Bull., v. 48, p. 723–752.

Todd, T. W., 1968, Dynamic diversion: influence of longshore current-tidal flow interaction on chenier and barrier island plains: Jour. Sed. Petrology, v. 38, p. 734–746.

——— and W. A. Monroe, 1968, Petrology of Domengine Formation (Eocene) at Potrero Hills and Rio Vista, California: Jour. Sed. Petrology, v. 38, p. 1024–1039.

Trask, P. D., *et al.*, 1955, Recent marine sediments—a symposium: Soc. Econ. Paleontologists and Mineralogists Spec. Pub. 4, 736 p.

Treadwell, R. C., 1955, Sedimentology and ecology of southeast coastal Louisiana: Baton Rouge, Louisiana, Louisiana State Univ., unpub. thesis, p. 1–176.

Trimonis, E. S., and K. M. Shimkus, 1970, Sedimentation at the head of a submarine canyon: Acad. Sci. USSR, Oceanology (Engl. ed.), v. 10, p. 74–85.

Trowbridge, A. C., 1911, The terrestrial deposits of Owens Valley, California: Jour. Geology, v. 19, p. 709–747.

——— 1930, Building of Mississippi delta: Am. Assoc. Petroleum Geologists Bull., v. 14, p. 867–901.

van Andel, Tj. H., 1960, Sources and dispersion of Holocene sediments, northern Gulf of Mexico, *in* F. P. Shepard *et al.,* Recent sediments, northwest Gulf of Mexico: Am. Assoc. Petroleum Geologists, p. 34–55.

——— 1964, Recent marine sediments of Gulf of California, *in* Tj. H. van Andel and G. C. Shor, Jr., eds., Marine geology of the Gulf of California: Am. Assoc. Petroleum Geologists Mem. 3, p. 216–310.

——— 1967, The Orinoco delta: Jour. Sed. Petrology, v. 37, no. 2, p. 297–310.

——— and J. R. Curray, 1960, Regional aspects of modern sedimentation in northern Gulf of Mexico and similar basins, and paleogeographic significance, *in* F. P. Shepard *et al.*, eds., Recent sediments of northwest Gulf of Mexico: Am. Assoc. Petroleum Geologists, p. 345–364.

——— and G. C. Shor, Jr., eds., 1964, Marine geology of the Gulf of California: Am. Assoc. Petroleum Geologists Mem. 3, 408 p.

Van Lopik, J. R., 1955, Recent geology and geomorphic history of central coastal Louisiana: Louisiana State Univ. Coastal Studies Inst. Tech. Rept. 7, p. 1–89.

Vann, J. H., 1959, The geomorphology of the Guiana Coast: Louisiana State Univ. Coastal Studies Inst., 2d Coastal Geog. Conf. Proc., p. 153–187.

van Straaten, L. M. J. U., 1951, Texture and genesis of Dutch Wadden Sea sediments: 3d Internat. Cong. Sedimentology Proc., Netherlands, p. 225–244.

——— 1954a, Composition and structure of Recent marine sediments in the Netherlands: Leidsche Geol. Meded., pt. 19, p. 1–110.

——— 1954b, Sedimentology of Recent tidal flat deposits and the Psammites du Condroz (Devonian): Geologie en Mijnbouw, v. 16, p. 25–47.

——— 1959, Littoral and submarine morphology of the Rhône delta: Louisiana State Univ. Coastal Studies Inst., 2d Coastal Geog. Conf. Proc., p. 233–264.

——— 1961, Sedimentation in tidal flat areas: Alberta Soc. Petroleum Geologists Jour., v. 9, p. 203–226.

——— 1965, Coastal barrier deposits in south- and north-Holland . . . : Netherlands, Geol. Sticht. Meded., new series, no. 17, p. 41–75.

Venkatarathnam, K., 1970, Formation of the barrier spit and other sand ridges near Chilka Lake on the east coast of India: Marine Geology, v. 9, no. 2, p. 101–116.

Visher, G. S., 1965a, Use of vertical profile in environmental reconstruction: Am. Assoc. Petroleum Geologists Bull., v. 49, no. 1, p. 41–61.

——— 1965b, Fluvial processes as interpreted from ancient and Recent fluvial deposits, *in* Primary sedimentary structures and their hydrodynamic interpretation: Soc. Econ. Paleontologists and Mineralogists Spec. Pub. 12, p. 116–131.

——— 1968, Depositional framework of the Bluejacket-Bartlesville sandstone, *in* Guidebook to geology of the Bluejacket-Bartlesville sandstone, Oklahoma: Oklahoma City Geol. Soc., p. 32–51.

——— 1971, Physical characteristics of fluvial deposits, *in* J. K. Rigby and W. K. Hamblin, eds., Recognition of sedimentary environments: Soc. Econ. Paleontologists and Mineralogists Spec. Pub. 16, 340 p.

——— *et al.*, 1971, Pennsylvanian delta patterns and petroleum occurrences in eastern Oklahoma: Am. Assoc. Petroleum Geologists Bull., v. 55, p. 1206–1230.

Volker, A., 1966, The deltaic area of the Irrawaddy River in Burma, *in* Scientific problems of the humid tropic zone delta and their implications: UNESCO, Dacca Symposium Proc., p. 373–379.

Waechter, N. D., 1970, Braided stream deposits of the Red River, Texas panhandle (abs.): Geol. Soc. America Abs. with Programs, v. 2, no. 7, p. 713.

Walker, J. R., and J. V. Massingill, 1970, Slump features on the Mississippi fan, northeastern Gulf of Mexico: Geol. Soc. America Bull., v. 81, p. 3101–3108.

Walker, K. R., 1962, Lithofacies maps of Lower Mississippian clastics of eastern and east-central United States: Am. Assoc. Petroleum Geologists Bull., v. 46, p. 105–111.

Walker, R. G., 1966, Deep channels in turbidite-bearing formations: Am. Assoc. Petroleum Geologists Bull., v. 50, p. 1899–1917.

Walton, W. C., 1970, Groundwater resource evaluation: New York, McGraw-Hill Co., 664 p.

Wanless, H. R., *et al.*, 1970, Late Paleozoic deltas in the central and eastern United States, *in* Deltaic sedimentation—modern and ancient: Soc. Econ. Paleontologists and Mineralogists Spec. Pub. 15, p. 215–245.

Warme, J. E., 1966, Geology of Mugu Lagoon: Univ. California at Los Angeles, PhD dissert.

——— 1971, Paleoecological aspects of a modern coastal lagoon: California Univ. Pubs. Geol. Soc., v. 87, 134 p.

Way, J. H., Jr., 1968, Bed thickness analysis of some Carboniferous fluvial sedimentary rocks near Joggins, Nova Scotia: Jour. Sed. Petrology, v. 38, no. 2, p. 424–433.

Weidie, A. E., 1968, Bar and barrier-island sands: Gulf Coast Assoc. Geol. Socs. Trans., v. 18, p. 405–415.

Weimer, R. J., 1961a, Spatial dimensions of Upper Cretaceous sandstones, Rocky Mountain area, *in* J. A. Peterson and J. C. Osmond, eds., Geometry of sandstone bodies: Am. Assoc. Petroleum Geologists.

——— 1961b, Upper Cretaceous delta on tectonic foreland, northern Colorado and southern Wyoming (abs.): Am. Assoc. Petroleum Geologists Bull., v. 45, p. 417.

——— 1965, Late Cretaceous deltas, Rocky Mountain region (abs.): Am. Assoc. Petroleum Geologists Bull., v. 49, p. 363.

Welder, F. A., 1959, Processes of deltaic sedimentation in the lower Mississippi River: Louisiana State Univ. Coastal Studies Inst. Tech. Rept. No. 12, p. 1–90.

Wermund, E. G., and W. A. Jenkins, Jr., 1970, Recognition of deltas by fitting trend surfaces to Upper Pennsylvanian sandstones in north-central Texas, *in* Deltaic sedimentation—modern and ancient: Soc. Econ. Paleontologists and Mineralogists Spec. Pub. 15, p. 156–269.

Wessel, J. M., 1969, Sedimentary history of Upper Triassic alluvial fan complexes in north-central Massachusetts: Massachusetts Univ. Dept. Geology Contr. No. 2, 157 p.

Wier, C. E., and W. A. Girdley, 1963, Distribution of the Inglefield and Dicksburg Hills Sandstone members in Posey and Vanderburgh Counties, Indiana: Indiana Acad. Sci. Proc., v. 72, p. 212–217.

Wilde, P., 1965, Recent sediments of the Monterey deep-sea fan: Cambridge, Massachusetts, Harvard Univ., PhD dissert., 153 p.

Williams, G. E., 1966, Paleogeography of the Torridonian Applecross Group: Nature, v. 209, no. 5030, p. 1303–1306.

——— 1969, Characteristics and origin of a Pre-Cambrian pediment: Jour. Geology, v. 77, p. 183–207.

Williams, P. F., and B. R. Rust, 1969, The sedimentology of a braided river: Jour. Sed. Petrology, v. 39, no. 2, p. 649–679.

Wilson, M. D., 1967, The stratigraphy of the Beaverhead Group in the Lima area, southwestern Montana: Evanston, Illinois, Northwestern Univ., PhD dissert., 183 p.

——— 1970, Upper Cretaceous-Paleocene synorogenic conglomerates of southwestern Montana: Am. Assoc. Petroleum Geologists Bull., v. 54, p. 1843–1867.

Winder, C. G., 1965, Alluvial cone construction by alpine mudflow in a humid temperate region: Canadian Jour. Earth Sci., v. 2, p. 270–277.

Winterer, E. L., *et al.*, 1968, Geological history of the pioneer fracture zone with the Delgada deep-sea fan northeast Pacific: Deep-Sea Research, v. 15, no. 5, p. 509–520.

Wolff, M. P., 1967, Deltaic sedimentation of the Middle Devonian Marcellus Formation in southeastern New York: Ithaca, New York, Cornell Univ., PhD dissert., 231 p.

Wolman, M. G., and L. M. Brush, 1961, Factors controlling the size and shape of stream channels in coarse noncohesive sands: U.S. Geol. Survey Prof. Paper 282-G, p. 183–210.

——— and L. B. Leopold, 1957, River flood plains: some observations on their formation: U.S. Geol. Survey Prof. Paper 282-C, p. 87–107.

Wurster, P., 1964, Delta sedimentation in the German Keuper Basin, *in* Deltaic and shallow marine deposition: Amsterdam, Elsevier Publishing Co., p. 436–446.

Zenkovich, V. P., 1964, Formation and burial of accumulative forms in littoral and nearshore marine environments: Marine Geology, v. 1, p. 175–180.

——— 1967, Processes of coastal development: New York, Interscience Publishers—Division of John Wiley and Sons, Inc.

Distribution and Continuity of Sandstone Reservoirs—Part 1

R.J. LeBlanc, Sr., Shell Development

Introduction

Effective maximum exploitation of a hydrocarbon accumulation requires full understanding of the reservoir beds and the pore space they contain. Fundamental research on clastic sediments by the petroleum industry, universities, and government agencies has produced a wealth of published information on the distribution and continuity of sandstones. Criteria for recognizing the various types of sandstones utilizing conventional subsurface data, such as cores, cuttings, and well logs, are reasonably well established. Conceptual models for predicting or determining sand body distribution, continuity, and internal characteristics also are available. Any method of oil- or gas-field development that does not take advantage of these conceptual models and criteria must be considered negligent.

The distribution, continuity, and internal characteristics of sandstones are controlled primarily by original environments of deposition. Anyone who is seriously involved in determining the distribution and continuity of sandstones, whether on a local or regional basis, will profit substantially by studying the literature and developing a thorough understanding on these environments.

Therefore, this summary paper on the distribution and continuity of sandstone reservoirs consists of two parts. Part 1 consists of brief discussions of the three most significant sandstone depositional environments: (1) *continental* (alluvial fan, braided stream, meandering stream, and aeolian), (2) *transitional* (deltaic and coastal barrier islands, tidal channels, and aeolian and transgressive marine), and (3) *deep marine* (turbidities related to submarine canyon-fan systems). (See Fig. 1.)

These key sedimentary environments are reviewed through examination of modern examples and ancient analogs. Criteria for identifying sandstones from conventional data are presented with the conceptual models necessary for predicting their distribution and internal reservoir characteristics.

Part 2 is a bibliographic source document. It includes a genetic classification of environments of sand deposition and lists, by depositional environments and geographic occurrence, the significant literature on ancient sandstones. Some 250 selected references are included.

The scientist and engineer who together integrate appropriately the criteria, principles, and models of sandstone deposition presented here with other subsurface engineering data should benefit from the synergism bound to occur and, thus, should be better able to solve the practical reservoir problems.

Distribution and Continuity of Sandstone Reservoirs

Our present knowledge and understanding of the factors that control the distribution and continuity of sandstones are based on the research on Recent sediments and on the study of ancient clastic sediments.

The following discussion of the distribution and continuity of sandstone reservoirs describes those sand bodies believed to predominate in the spectrum of sandstone reservoirs.

Alluvial Fan and Braided Stream Sandstones

Sands of alluvial fan and braided stream origin are deposited under relatively high energy conditions, mainly along the front of mountains of high relief (Fig. 2). Deposition of sand in this type of environment takes place under two different conditions. Some fans are constructed almost entirely by braided stream sedimentation. Sediments derived from the mountain areas are funneled to the fan through canyons and then are dispersed by braided streams. Sands and coarser clastics generally are confined to the channels. However, during periods of extreme flooding on overbank sheet flow, another type of sedimentation results in deposition of sand over the entire fan surface. When this type of sedimentation prevails, especially in the middle to lower fan areas, sandstones with great lateral continuity are deposited. It is quite probable that most of the continental sheet sands, which are nearly devoid of clay bodies, were deposited in this class of environment.

Under certain conditions, sedimentation on alluvial fans occurs as a result of mud or debris flows. These flows usually follow the main channels on the fan to produce long, narrow bodies of extremely poorly sorted and unstratified sand or conglomeratic sand.

Some alluvial fan deposits, such as the Price River (Cretaceous age) in Utah, consist of interbedded sandstones, conglomerates, and mudstones (Fig. 3). It is quite probable that

This paper presents brief discussions of the three most important depositional environments of sandstone reservoirs. Criteria for identifying sandstones from conventional data are presented, along with conceptual models for predicting their distribution and internal characteristics.

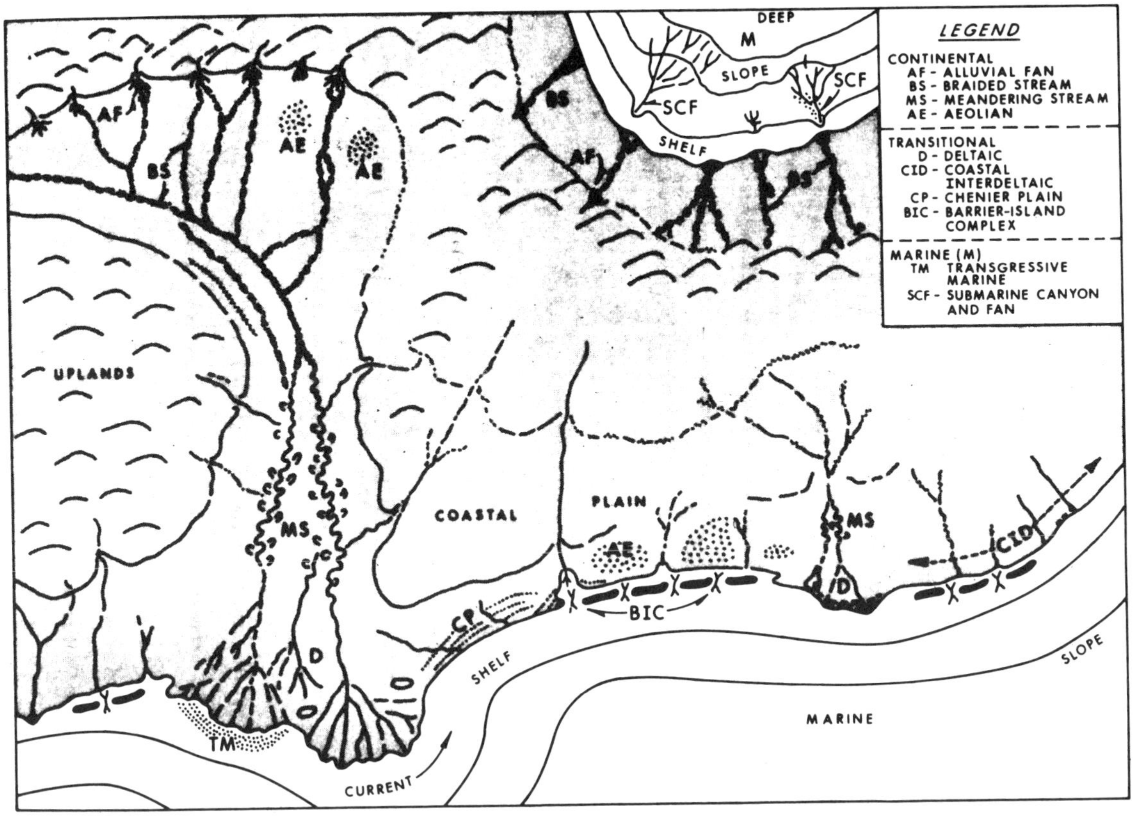

Fig. 1—Environments of clastic sedimentation. From Le Blanc.[126]

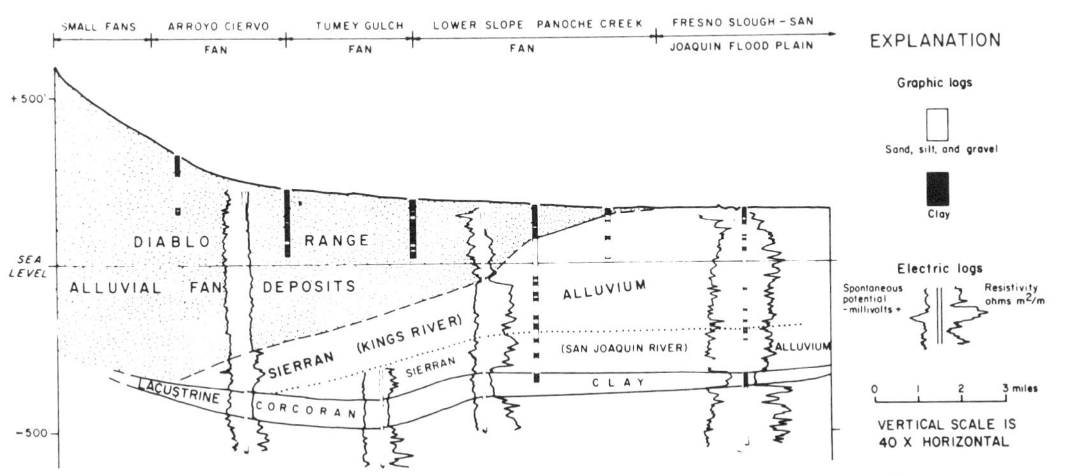

Fig. 2—Thick section of alluvial fan deposits adjacent to mountain front. From Bull.[243]

the deposition of these sandstones and conglomerates was confined to the fan channels. During periods of overbank flash flooding, fine-grained sediments were deposited in the interchannel areas. Thus, as channels were abandoned and new channels developed, the interchannel clay became interbedded with channel sand.

High- energy alluvial fan and braided stream deposits have not been studied as extensively as meandering stream sediments. However, on the basis of presently available information, it appears that there is great variability in the distribution and continuity of this type of sandstone. Some braided stream deposits of the lower fan can be distributed over very large areas and have extremely good lateral and vertical continuity. Sandstones related to the fans that receive a substantial amount of fine-grained sediments can be confined to fan channels and, thus, have poor lateral and vertical continuity.

Campbell[38]* has described an excellent example of fluvial sheet sandstone from the Morrison formation of northwestern New Mexico. The abstract of this important paper is quoted below and the distribution of the Morrison sandstones is shown in Figs. 4 and 5.

"The outcrops of the Westwater Canyon Member in the Morrison formation (Jurassic), northwestern New Mexico,

*See bibliography in "Distribution and Continuity of Sandstone Reservoirs — Part 2," for all references cited in Part 1.

Fig. 3—Interbedded channel conglomerates and sandstones and interchannel mudstones of alluvial fan origin, Price River (Cretaceous) of Utah.

show representative attributes of a fluvial sheet sandstone. This sandstone body consists of coalescing fluvial channel systems that in turn are composed of still smaller coalescing individual channels. The sheet is more than 63 miles wide transverse to the paleo-current direction and averages 200 feet in thickness; the downcurrent dimension is not preserved but may have been as much as 100 miles. Channel systems internal to the sheet average 7 miles wide and 50 feet deep. The individual channel deposits that form the systems average 600 feet wide by 12 feet thick.

"Any one vertical column through the sheet sandstone is usually more than 90 percent sandstone. Further, each vertical column consists of repetitive incomplete (top missing) fluvial sequences that show upward-decreasing grain sizes and scales of sedimentary structures.

"Fluvial sheet sandstone is probably deposited by aggrading and coalescing braided stream systems."

A large portion of the Castlegate sandstone (Cretaceous age), which crops out along the Book Cliffs of eastern Utah, is considered to be an excellent example of fluvial sheet sandstone deposited by braided streams downdip from alluvial fans. The Castlegate ranges from 200 to 500 ft thick and is very uniform in character for many miles (Fig. 6).

Fluvial sheet sandstones provide an ideal reservoir for hydrocarbons. Such a sandstone body is large enough to contain giant hydrocarbon reserves as well as an aquifer.

Sandstones of Meandering Stream Origin

The most common type of continental sandstones are those that are of meandering stream origin. This type of environment has been studied extensively and criteria for recognizing meandering stream deposits are extremely well developed.

There are two processes of sedimentation in the meandering stream model that are pertinent to the development of sand bodies: channel migration and channel abandonment. When channel migration occurs in the caving-bank area, there is a corresponding development of a point bar on the opposite side of the channel. Meandering streams migrate in zones (meander belts) that are 15 to 20 times the width of the channel. Fully developed Mississippi River meander belts are 15 to 20 miles wide and up to a few hundred miles long (Fig. 7).

Channel abandonments are of two types. The river abandons a single loop (oxbow lake) by cutoff or it abandons an

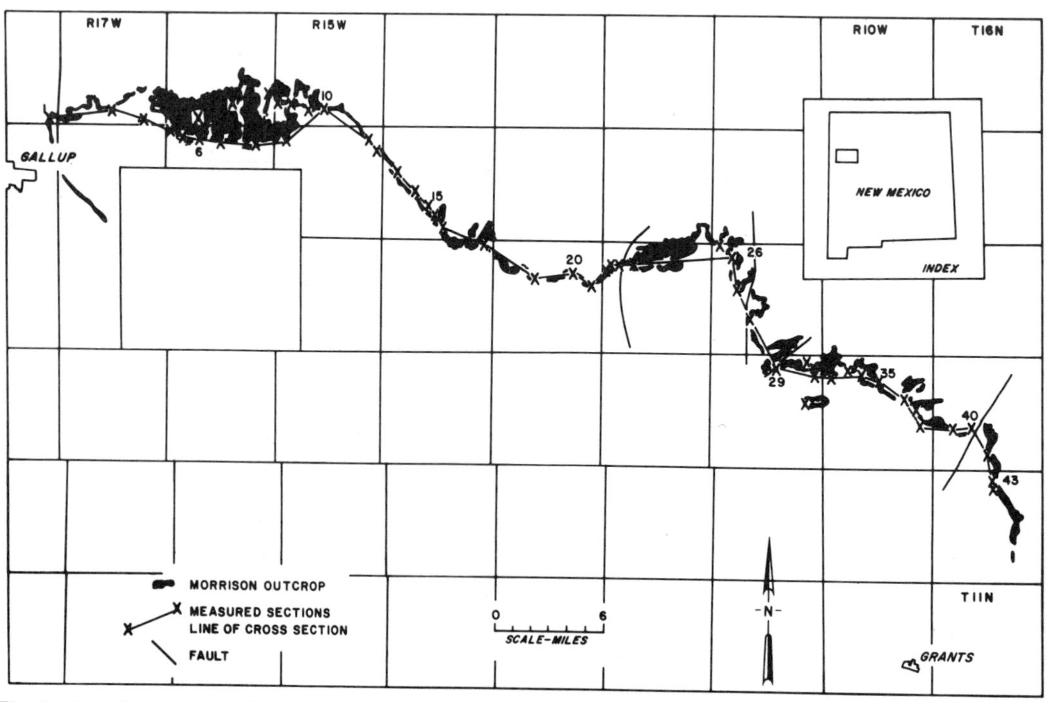

Fig. 4—Location map showing distribution of Morrison sandstones between Gallup and Grants, N.M. Refer to Fig. 5. From Campbell.[38]

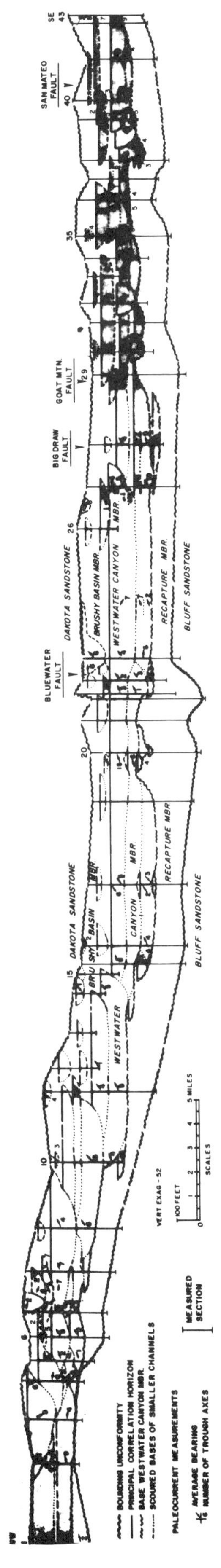

Fig. 5—Fluvial sheet sandstone of braided stream origin. Refer to Fig. 4 for location of this section. From Campbell.[38]

entire meander belt. Thus, the flood plain of a typical meandering stream is characterized by an active meander belt, several abandonment meander belts, and numerous oxbow lakes.

Most oxbow lakes are filled with fine-grained sediments that are called clay plugs. Fisk[77] described hundreds of clay plugs in the Mississippi Valley.

The abandonment of a meander belt can occur gradually or very abruptly. If abandonment occurs relatively fast, the last channel position within the belt will become filled with clays and silts, thus producing a long, sinuous clay plug along the entire abandoned meander belt. The last channel position of a gradually abandoned meander belt is usually filled with silty sands and sands.

Once a meander belt is abandoned and removed from the environment of sand transport, it is usually covered by flood basin clays, thus providing a seal over the sand body—this is discontinuity.

Some areas of the flood plain are never occupied by the meander belts. These environments (flood basins or backswamps) are topographically low and receive only fine-grained sediments. An example of the distribution of several meander belts and associated flood basin deposits is shown in Fig. 7.

The meandering stream produces extensive sheets of sand, in the form of coalescing bars, over a relatively short period of geological time. These sand sheets are not as continuous as those of braided stream origin described by Campbell.[38] The clay plugs in oxbow lakes and the last channels of rapidly abandoned meander belts comprise numerous discontinuities within the sheets of sand.

A point bar is characterized by a very distinctive vertical sequence (Fig. 8) consisting of four layers: a poorly sorted basal section of sand and silts, a zone of trough cross-bedded sand and gravel, a zone of horizontal and small-scale cross- bedded sand and silty sands, and an upper layer of finer-grained, small-scale, cross-bedded silts and sands.[242] The point bar sequence is as thick as the meandering channel that produced it.

All the sediments within the point bar sequence were deposited within the channel during flood stages of varying

Fig. 6—Braided stream sandstones in the Castlegate (Cretaceous) near Castlegate, Utah.

magnitudes. At the end of each flood cycle, when the stream velocity is nearly zero, a layer of clay is deposited on the bar in the form of a clay drape. This drape can cover the entire bar that is under water or only local depressions. Clay drapes of varying thicknesses, from 1/8 in. to more than 1 or 2 ft, are common within the point bar sequence. Most of the drapes occur in the upper half of the sequence, and they provide vertical discontinuities within the sequence. A point bar with several clay drapes usually is reflected on the SP log shape as a serrated bell.

Several well documented examples of hydrocarbon accumulations within point bar sandstone reservoirs have been published. Examples are shown in Figs. 9 and 10.

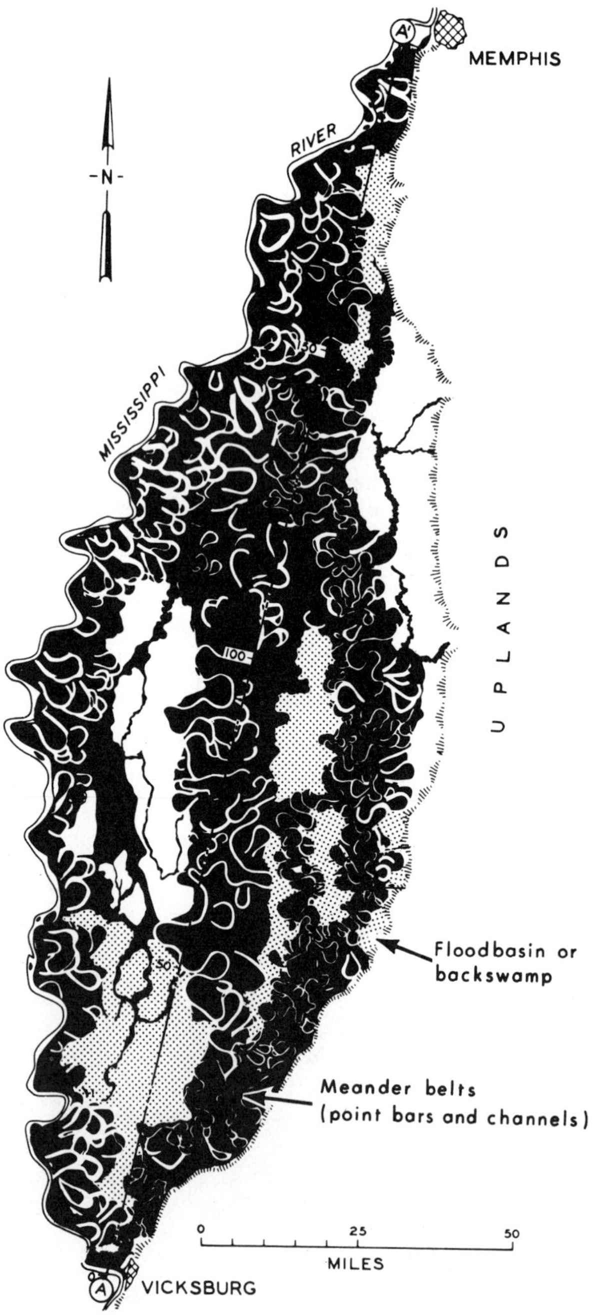

Fig. 7—Distribution of recent meander belt and flood basin deposits in the Mississippi Valley between Vicksburg, Miss., and Memphis, Tenn. Modified after Kolb *et al.*[121]

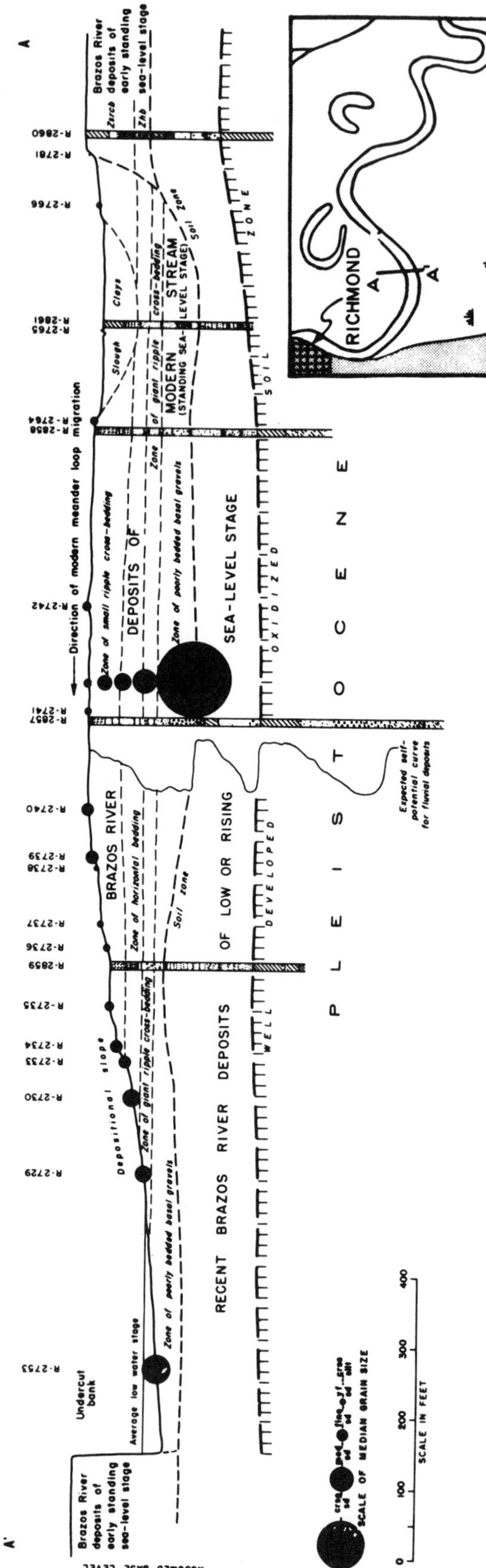

Fig. 8—Sequence of Recent point bar deposits of the Brazos River, southeast Texas. From Bernard *et al.*[242]

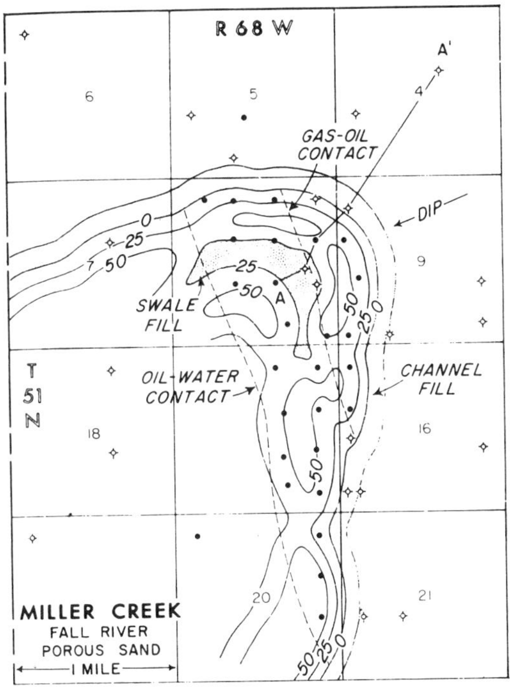

Fig. 9—Sandstone reservoir of point bar origin, Fall River sandstone of Wyoming. From Berg.[12] Refer to Fig. 10 for Cross-section A-A'.

Deltaic Sandstones

The most common and important hydrocarbon sandstone reservoirs are those of deltaic origin. Deltaic sands are important because they are very numerous and especially because of the position that they occupy within the sedimentary basin. With the exception of deep-sea submarine fan sandstones, the deltaic sands occur farther downdip than any other type of sand.

Deltaic sands are deposited in two types of environments: within distributary channels and within the delta front areas (Fig. 11). In this paper, the sands that are deposited in the subaqueous environment in front of the deltaic plain will be referred to as delta fringe sands or simply fringe sands.

Distributary Channel Sands. The distributary channels of the deltaic plain funnel sand to the delta front region. Most of this sand is transported downstream as bed load and, consequently, it is usually highly cross-bedded. Processes of deltaic sedimentation are such that a distributary channel, once it develops, usually does not exist for a long period of time. Modern deltaic plains are characterized by literally dozens of abandoned distributary channels but only a few active channels. For example, there are only two active distributaries on the Nile Delta.

A live distributary channel is full of water, but once the channel is abandoned it becomes filled with sediments. The character of the channel fill is governed by the manner of channel abandonment. Rapidly abandoned channels are filled mainly with fine-grained sediments; those that are abandoned gradually have more sands added in the fill. Generally, most distributary channels are not completely filled with sand; only the lower portions of the channel contain sand.

Some deltas are characterized by distributary channels that migrate, with the result that point bars are developed within the delta. Whenever this happens, the quality of sands in the delta is usually enhanced. Point bar sands are common in the Niger Delta, but very uncommon in the Mississippi Delta.

Sandstones of distributary channel origin usually make up only a small portion of the total sand package within a delta. The most common sandstones are those that are deposited as sheets in front of the deltaic plain, the fringe sands.

Delta Fringe Sands. Most of the sand that is introduced to the sea by distributary channels is deposited at the end of the channel in the form of river mouth bars. These are shifting sand shoals that present a hazard to navigation. Sandstones of this origin usually are highly cross-bedded.

Much of the sand that is deposited in the river mouth bar environment becomes dispersed by marine energy to form a fringe of sand in front of the deltaic plain (Fig. 11). These fringe sands normally extend seaward to water depths of about 50 ft.

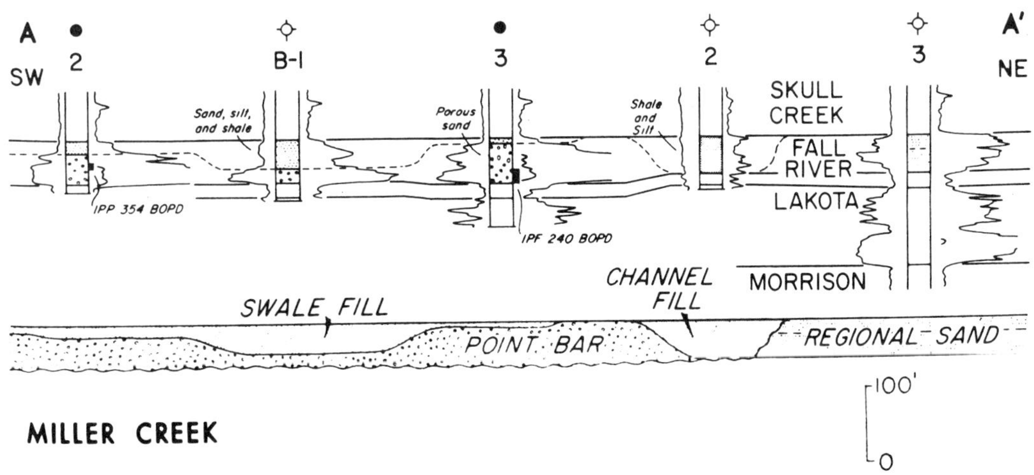

Fig. 10—Point bar sandstone reservoir and channel fill shale within the Fall River sandstone of Wyoming. From Berg.[12] Refer to Fig. 9 for location of cross-section.

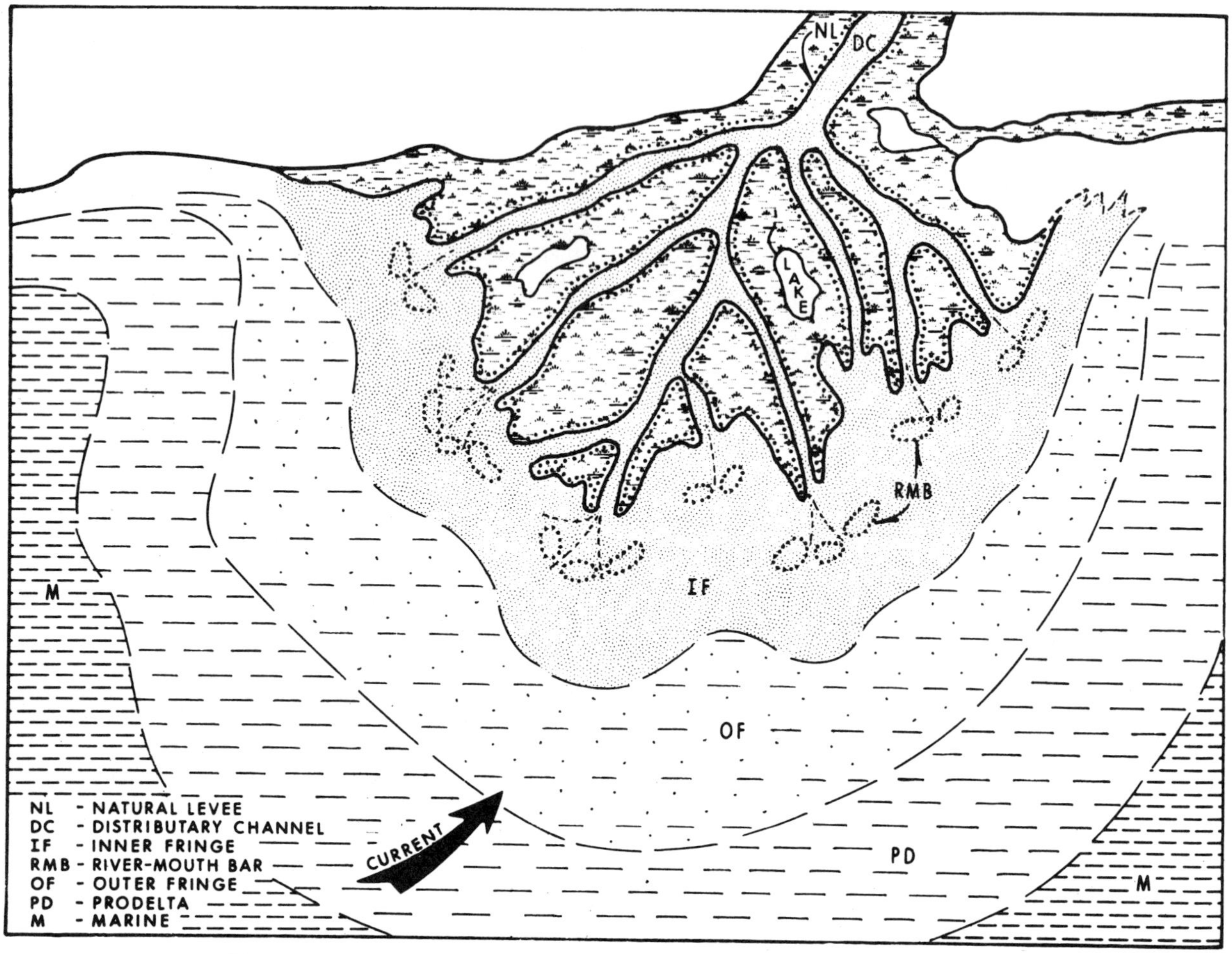

Fig. 11—Depositional environments typical of modern deltas. From Le Blanc.[126]

The variability of delta fringe sands is controlled by two main factors: (1) the amount and type of sand introduced to this environment by the distributary channels, and (2) the amount of energy in the marine environment (wave, tidal, current). If a delta with low sand content is constructed in a relatively quiet body of water, the delta fringe sands generally will be poor-quality reservoirs. On the other hand, if a delta with a high sand content is constructed in a body of water with relatively high levels of energy (such as waves and currents), the net result will be the deposition of clean, well sorted sand with excellent reservoir characteristics.

During the past 15 years, much attention has been given to different types of deltas. Deltas have been classified according to the shape of their deltaic plains, such as birdfoot, cuspate, and arcuate deltas. The most significant literature on deltas that has appeared in recent years is concerned with the range in character of deltas (Fig. 12). Coleman and Wright[42,43] are the outstanding authorities on this subject.

Relationship of Distributary Channels to Fringe. There is another aspect of deltaic process that must be considered: the relationship between distributary channel sandstones and delta fringe sandstones (Fig. 11). The most significant paper on this aspect of deltaic sedimentation was written by Fisk.[79]

Distributary channels are very shallow, usually less than 10 ft, at their distal ends where they meet the sea. Their depths increase upstream from the river mouths, and they do not end at the edge of the deltaic plain but continue in the subaqueous environment across the river mouth bars. Thus, the bases of the distributary channels do not have sharp contacts in this area. Rather, the sands of the basal part of the channels are transitional with the fringe sands.

As a delta progrades seaward, the distributary channel will cut into the fringe sands. If this movement continues for several miles, the distributary channels can cut through the entire delta sequence and into underlying marine shales. Thus, there is a very wide range in the thickness of distributary channel sandstones, depending on the size of the delta and the amount of delta progradation. Channel depths ranging from less than 10 ft to as much as 150 ft can be expected.

It was emphasized earlier that the thickness of a point bar sequence is controlled by the depth of the channel that produced it. The thickness of the delta sequence also is partly controlled by distributary channel depths; however, the most important factor is the depth of the water within which the delta was constructed.

The Delta Sequence. Two columnar sections are required to define the delta sequence. A vertical section at a point between distributary channels usually will have three distinct layers: basal prodelta muds, fringe sands, and the organic-rich upper layer of the deltaic plain. A vertical sec-

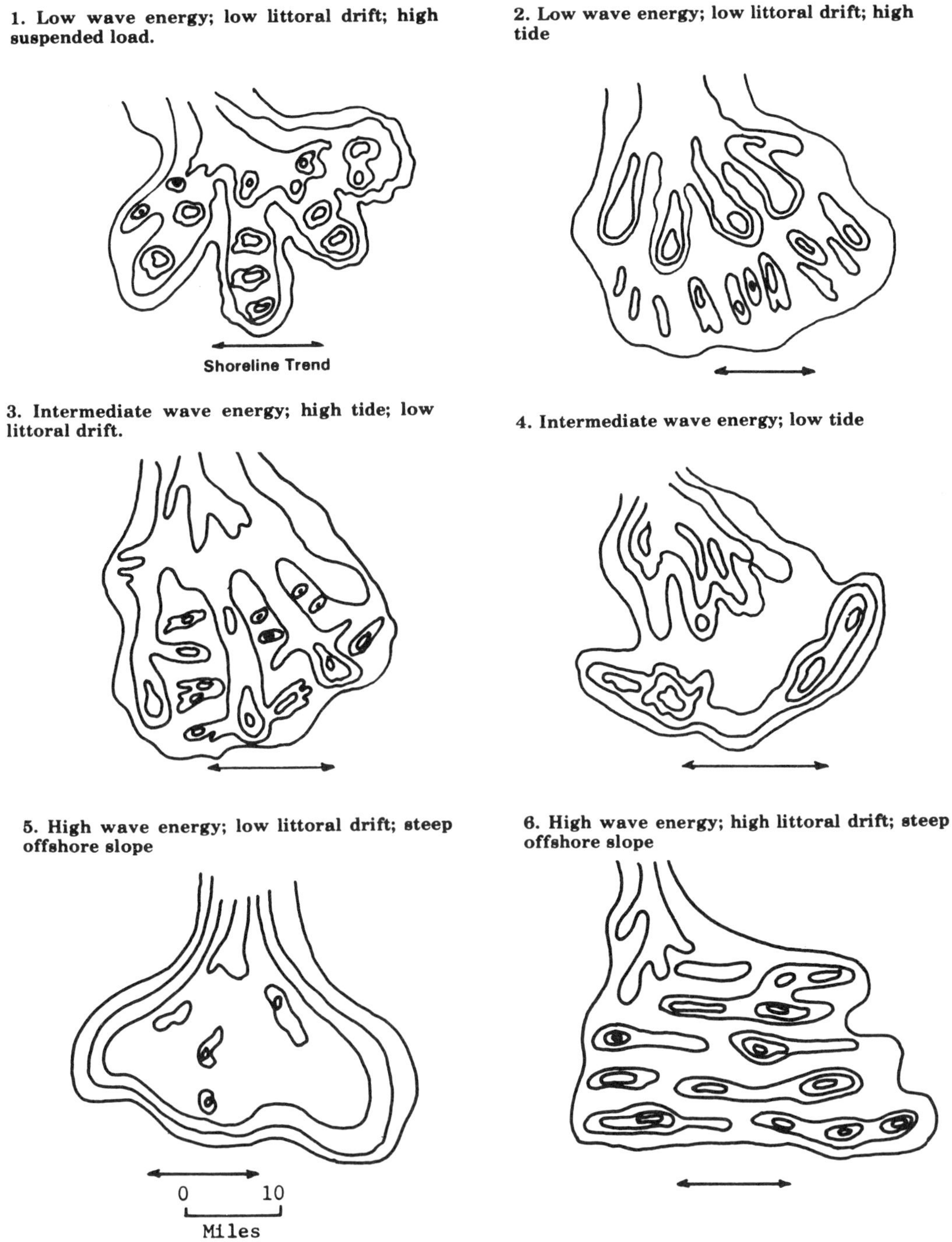

Fig. 12—Variations in net sand distribution within deltas. From Coleman and Wright.[42]

tion at a point within a distributary channel will show the channel fill resting on a portion of the delta sequence. The base of the distributary channel sandstone can rest on inner fringe sands, outer fringe sands, prodelta clays, or even on marine sediments below the delta.

SP Log Shapes. Deltaic sands are reflected by two basic SP log shapes. The most common shape will be a variety of funnels that represent the fringe and prodelta sediments (Fig. 13). The channel sandstones will be characterized by bell-shaped logs. If there is a considerable amount of clay layers in the channel fill, the bells will be serrated. Likewise, if there is a substantial amount of clay layers within the fringe sequence, the funnels will be serrated. Additional information on interpretation of log responses in deltaic sediments has been summarized by Gilreath and Stephens.[86]

Once the geologist recognizes that he is within a deltaic situation, it is a relatively simple procedure in many basins to interpret deltaic sandstones with log shapes, provided that he has a good understanding of the deltaic model of sedimentation.

Some examples of deltaic sandstone reservoirs are shown in Fig. 14.

Continuity of Deltaic Sands. Fringe sands have the widest distribution and the best lateral and vertical continuity.

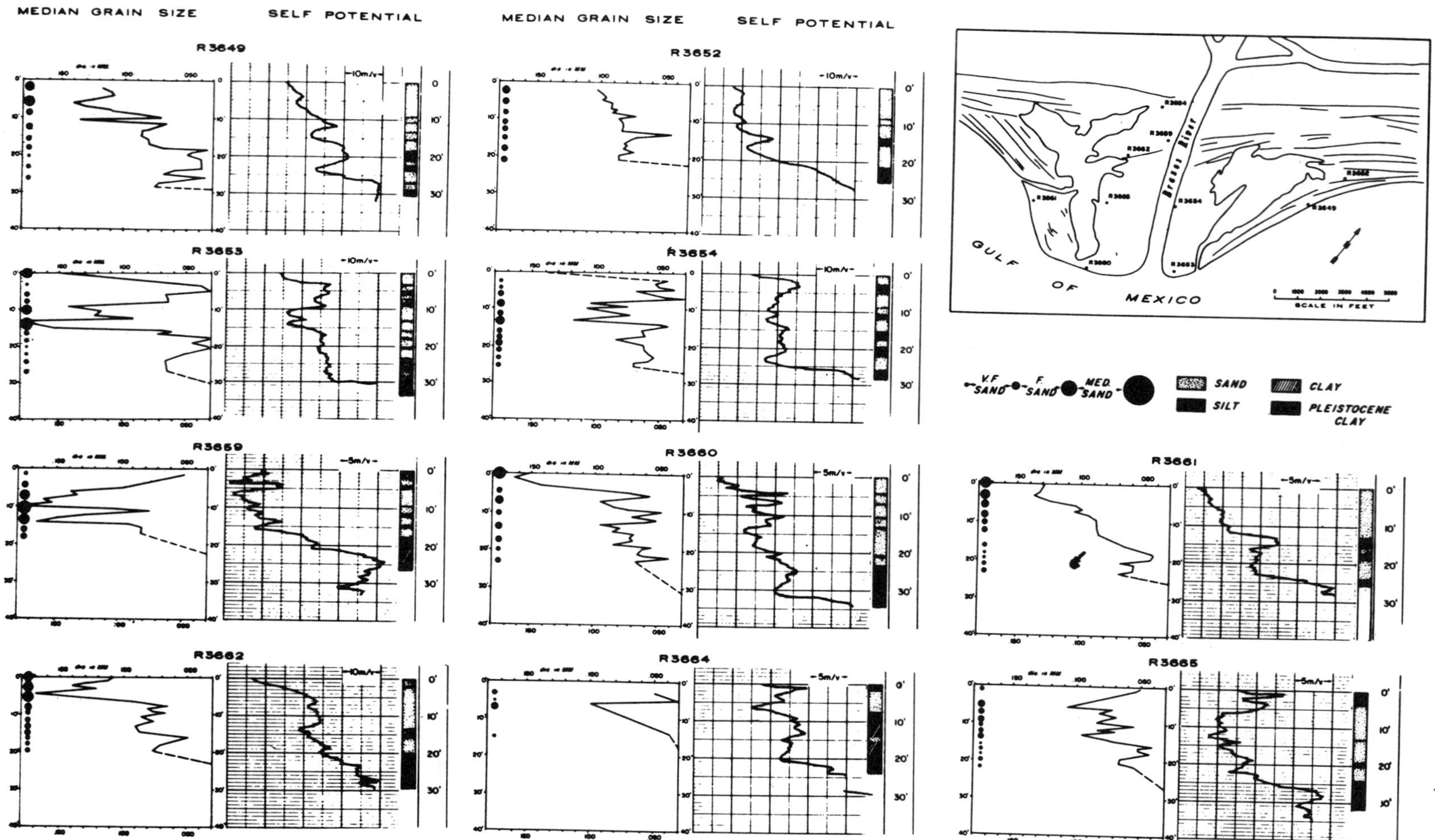

Fig. 13—SP log shapes of the modern Brazos River delta of southeast Texas. From Bernard *et al.*[242]

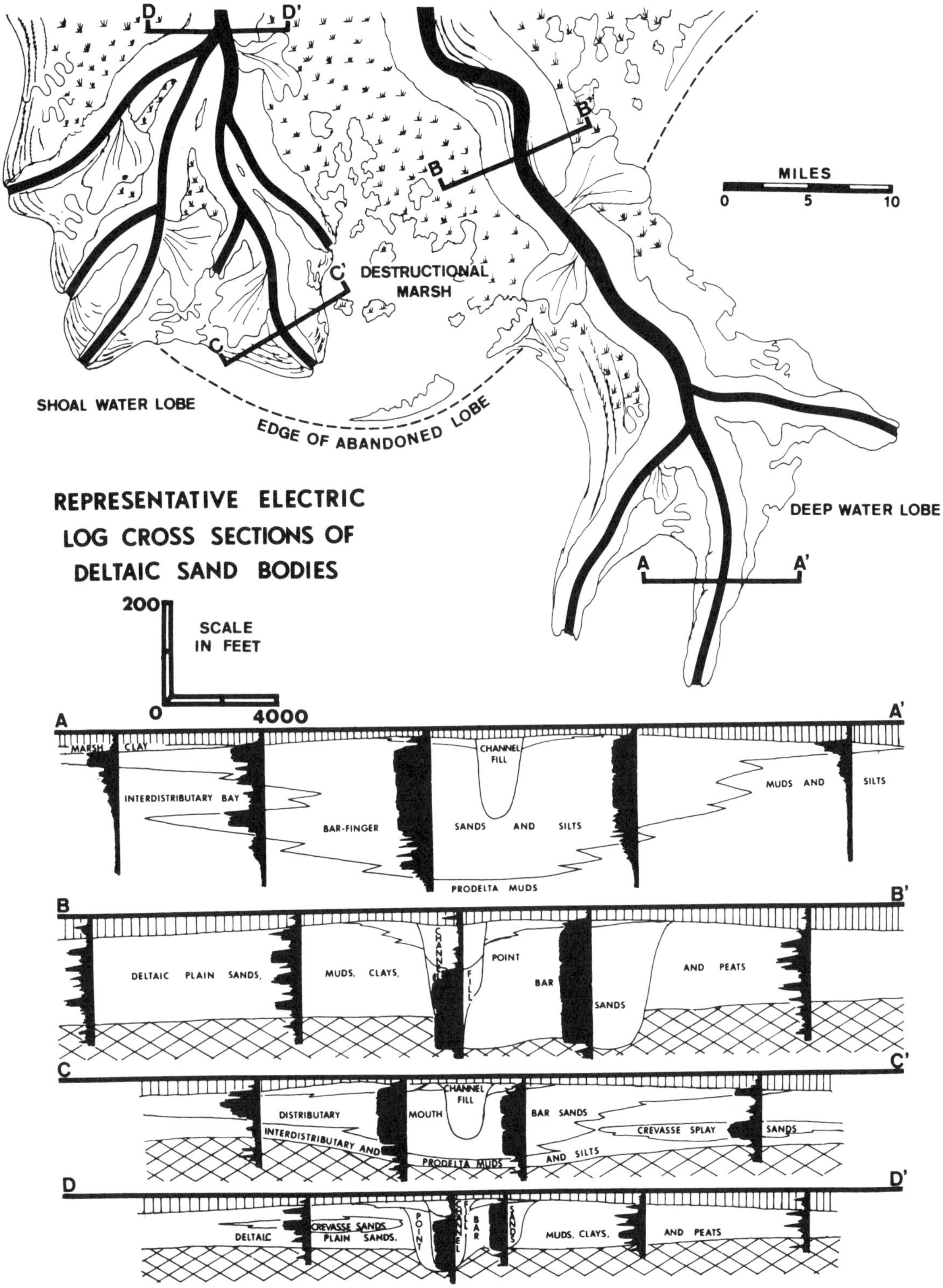

Fig. 14—Electric log cross-sections of deltaic sand bodies. From Galloway.[85]

Fig. 15—Pennsylvanian distributary channel (DC) sandstones overlying outer delta fringe (F) sandstones, eastern Oklahoma. Note continuity between these two types of sandstones.

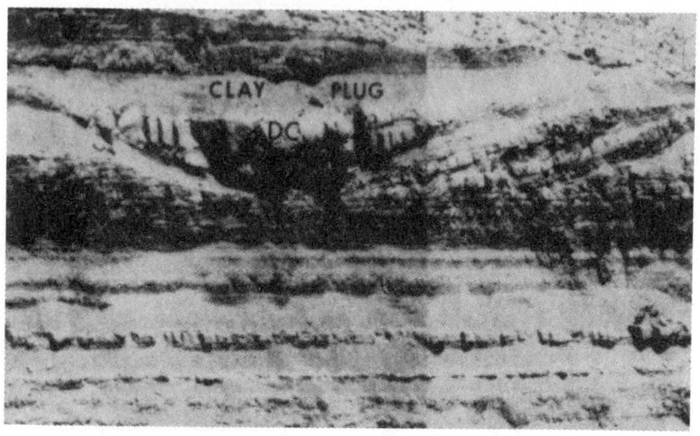

Fig. 16—Deltaic sediments within the Colton (Eocene) formation of Utah. Point bar (PB) and distributary channel (DC) sandstones overlie delta fringe (F) sandstones. Clay layer provides discontinuity between the channel and fringe sandstones.

These sands are in sharp contact with the laterally equivalent organic- rich deltaic plains clays (updip), but they grade gradually into silts and clays in a downdip direction. Distributary channel sandstones have much more limited distribution and the vertical continuity of these sands is quite variable, depending on the manner of channel abandonment.

Some distributary channel sandstones are in direct contact with fringe sands (Fig. 15), while others are separated from fringe sands by a layer of fine- grained sediments that acts as a permeability barrier between the two sand bodies (Fig. 16). Houser and Neasham[103] described variations in continuity and permeability of distributary channel and delta fringe sands of the Mississippi Delta.

Coastal Interdeltaic Barrier Island and Tidal Channel Sandstones

Barrier Island Sands. Criteria for recognition of sands related to barrier island complexes, and conceptual models for predicting the occurrences of this type of sand, were reasonably well established by 1960. Ancient barriers that exist in the subsurface are recognized very easily. They are difficult to find because they comprise only a small percentage of sands that occur in a basin. The recognition of barrier sands in the subsurface can be compared with the task of finding a needle in a large haystack. Although one knows exactly what the needle looks like, it is nevertheless hard to find because it is too small relative to the size of the haystack.

Barrier island sands provide excellent petroleum reservoirs. They occur as long, narrow, linear belts parallel to shorelines. Perhaps one of the most important features of this class of sand body is that these sands occur in situations that are ideal for trapping large quantities of hydrocarbons stratigraphically. Barrier sands usually have sharp contacts with updip lagoonal clays. Downdip, they pinch out into the silts and clays of the lower barrier shoreface (Fig. 17).

The thickness of a single barrier island sand body is controlled by the depth of inner continental shelf waters within which the barrier is formed. A good average thickness for barrier island sand is 50 ft. The barrier island sequence (Fig. 18) consists of four layers. The first layer is a lower interval of interbedded sands, silts, and clays, overlain by a thick zone of highly burrowed sand. The third layer consists of horizontally laminated sands of the beach and upper shore face. The uppermost layer consists of weathered oxidized sections of sediments that occur above sea level. In addition,

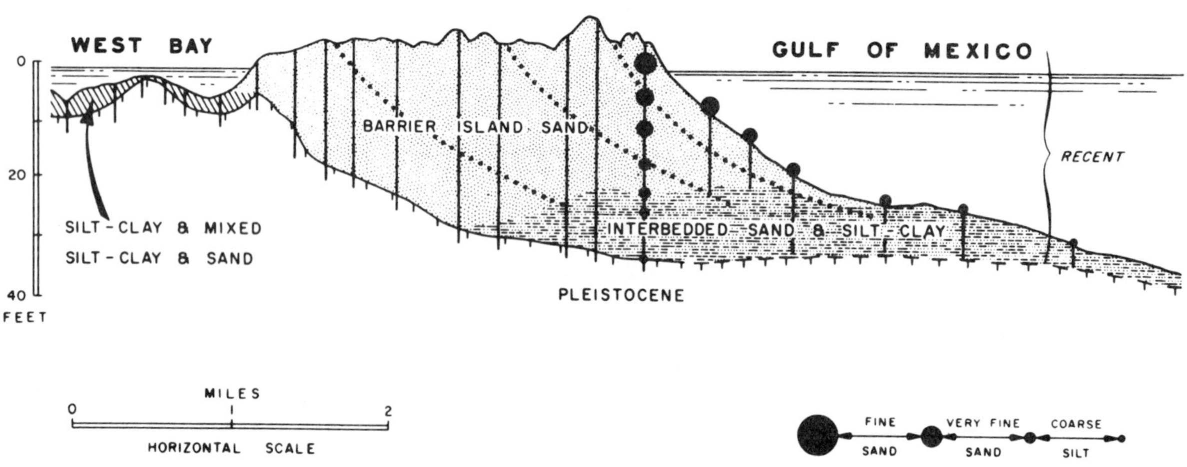

Fig. 17—Section across the Galveston barrier island showing relationship of barrier sands to lagoonal and marine fine-grained sediments. From Bernard *et al.*[242]

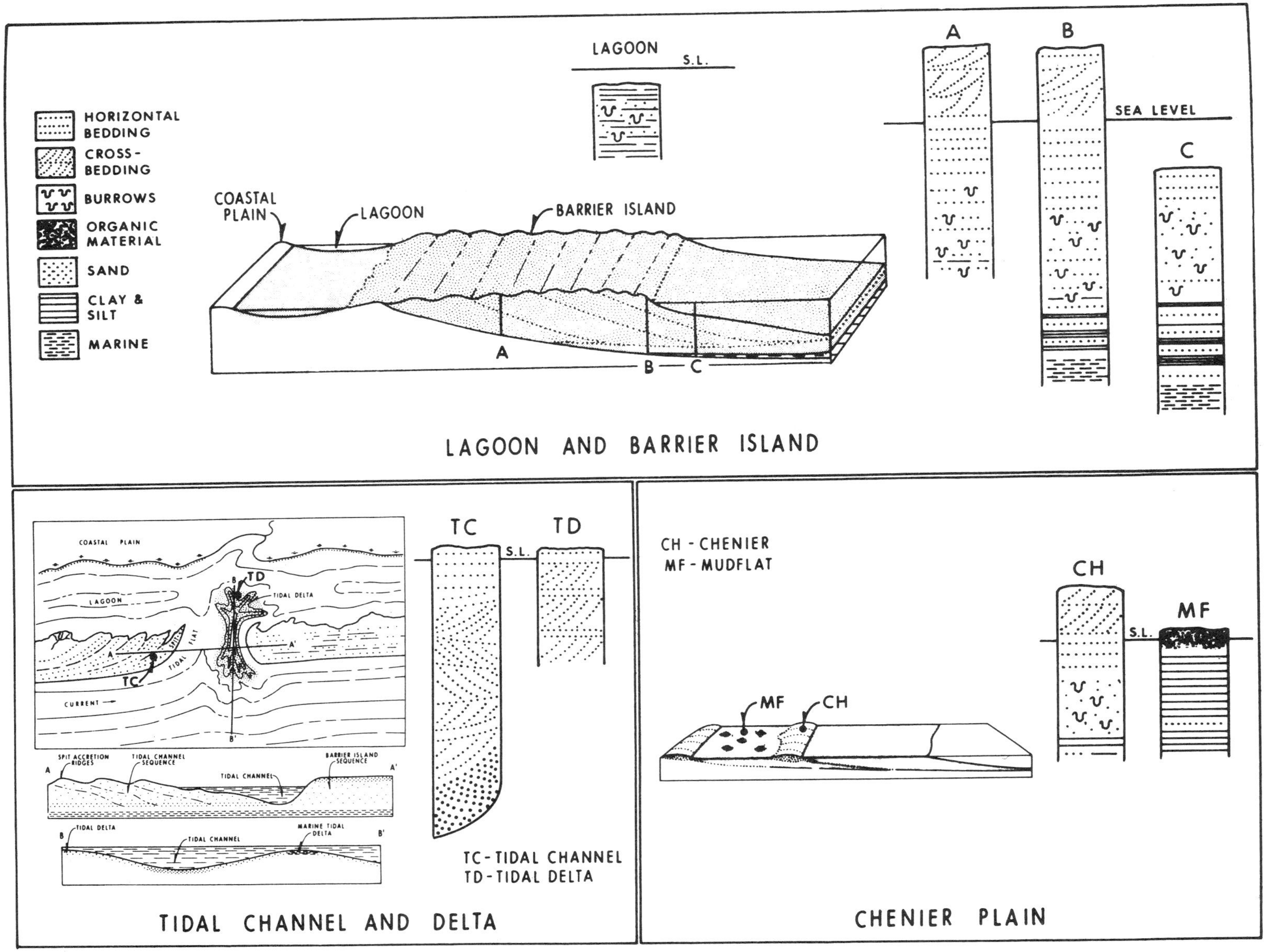

71-0363-21

Fig. 18—Barrier island, tidal channel, and chenier sands characteristic of the Texas-Louisiana coast. From Le Blanc.[126]

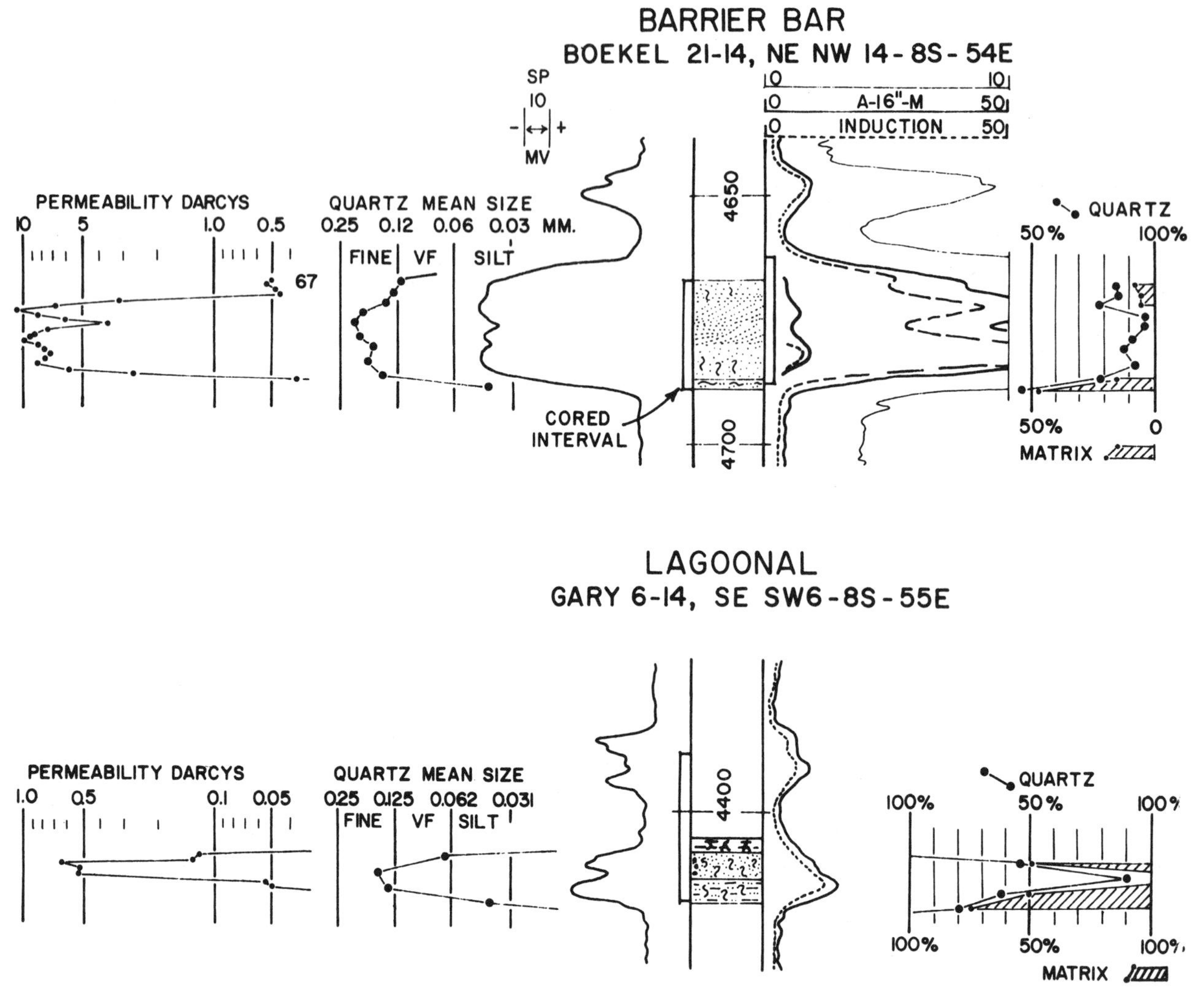

Fig. 19—Barrier island sandstone reservoir and lagoonal trap facies at Bell Creek field, Montana. From Davies and Berg.[55]

some barriers have a cover of cross-bedded windblown sand. A very significant characteristic of barrier sands is that they are nearly devoid of clays. Thus, they have excellent continuity along strike and dip directions.

The barrier island sand body is reflected on the SP log by a variety of log shapes. A log through the central part of the barrier will have a funnel shape (Fig. 19). A log through a barrier along its updip edge will be blocky because the lower zone of interbedded sediments is absent. Also, a log through the downdip portion of the barrier will be expressed by a thin funnel because the upper shore face and the layer of weathered sand are absent.

Barrier island sand bodies occur in the subsurface as individual units, or they may coalesce vertically to form a thick, linear sand body that is relatively narrow and parallel to the coast.

Tidal Channel Sandstones. The continuity of a barrier island along a coast line is interrupted by tidal channels, also called passes or inlets (Fig. 18). These tidal channels range in depth from a few feet to more than 100 ft. Their depths are controlled by the size of the estuaries that lie updip from the barriers. The larger the estuary, the deeper the channel.

Tidal channels usually migrate laterally along the coast. The direction of migration is governed by the direction of littoral drift. If two barriers are separated by a deep tidal channel that migrates, the net result will be the truncation of a portion or all of the barrier sequence by channel migration and the deposition of a tidal channel sequence of sediments on the opposite side (Fig. 18). Thus, the continuity of barrier sand, parallel to the coast line, will be interrupted by a tidal channel sand body that trends perpendicular to the barrier. These tidal channel sequences range from coarse to fine upward, have bimodal dip directions, and will appear on the SP log as bell-shaped. Thus, the continuity of barrier funnels will be interrupted by a series of tidal channel bells, but there will be no discontinuity of sand.

Relationship of Barrier Sands to Deltaic and Point Bar Sands

It is important to remember the positions that barrier island tidal channels occupy in a basin relative to the deltaic and point bar sands. Generally, barrier sand bodies occur on both flanks of a delta. In the case of large muddy deltas, like the Mississippi Delta, the barrier sands are separated from the delta by the chenier plain sands. Deltas occur downdip from the meander belt sandstones. It also should be remembered that the width of the deltaic plain is generally three to

Fig. 20—Cross-bedded aeolian sandstones of the Navajo (Jurassic), eastern Utah.

five times the width of alluvial valleys.

Another important relationship to remember is that barrier island sands do not occur as far downdip as the deltaic sands. Deltas usually protrude into the sea; hence, the sands are deposited much farther downdip than the barrier sands. Finally, barrier sands generally are seldom buried as deeply as the deltaic sands. One of the best examples of a well documented[14,55,56] barrier island reservoir is the Bell Creek field of southeastern Montana (Fig. 19). Fisher and McGowen[73] and Boyd and Dyer[28] described large Tertiary barrier island complexes of south Texas. For additional examples of coastal interdeltaic sandstones, refer to Table 4 in Part 2 of this paper.

Aeolian Sandstones

Most of the windblown sands are deposited in two different environmental situations. Sand dunes generally are common along most coastal regions characterized by barrier sands.[18,46,47] Significant dune fields also occur in inland or coastal desert environments.[87]

Aeolian sands usually are highly cross-bedded (Fig. 20) and form excellent reservoirs with wide distribution and excellent continuity. Glennie[88] described reservoirs of this type from the North Sea (Figs. 21 and 22).

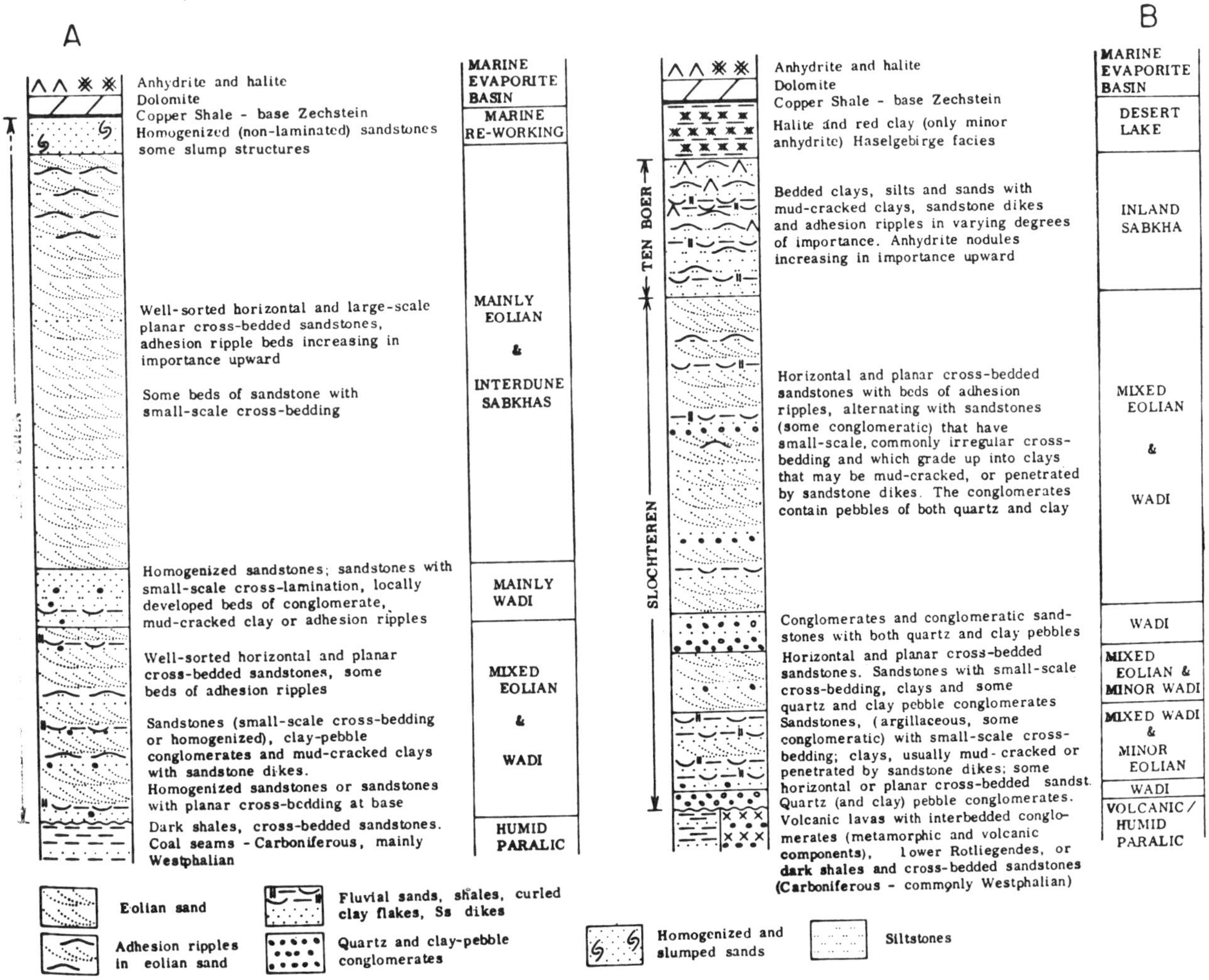

Fig. 21—Permian aeolian sandstone reservoirs of the North Sea. From Glennie[88]

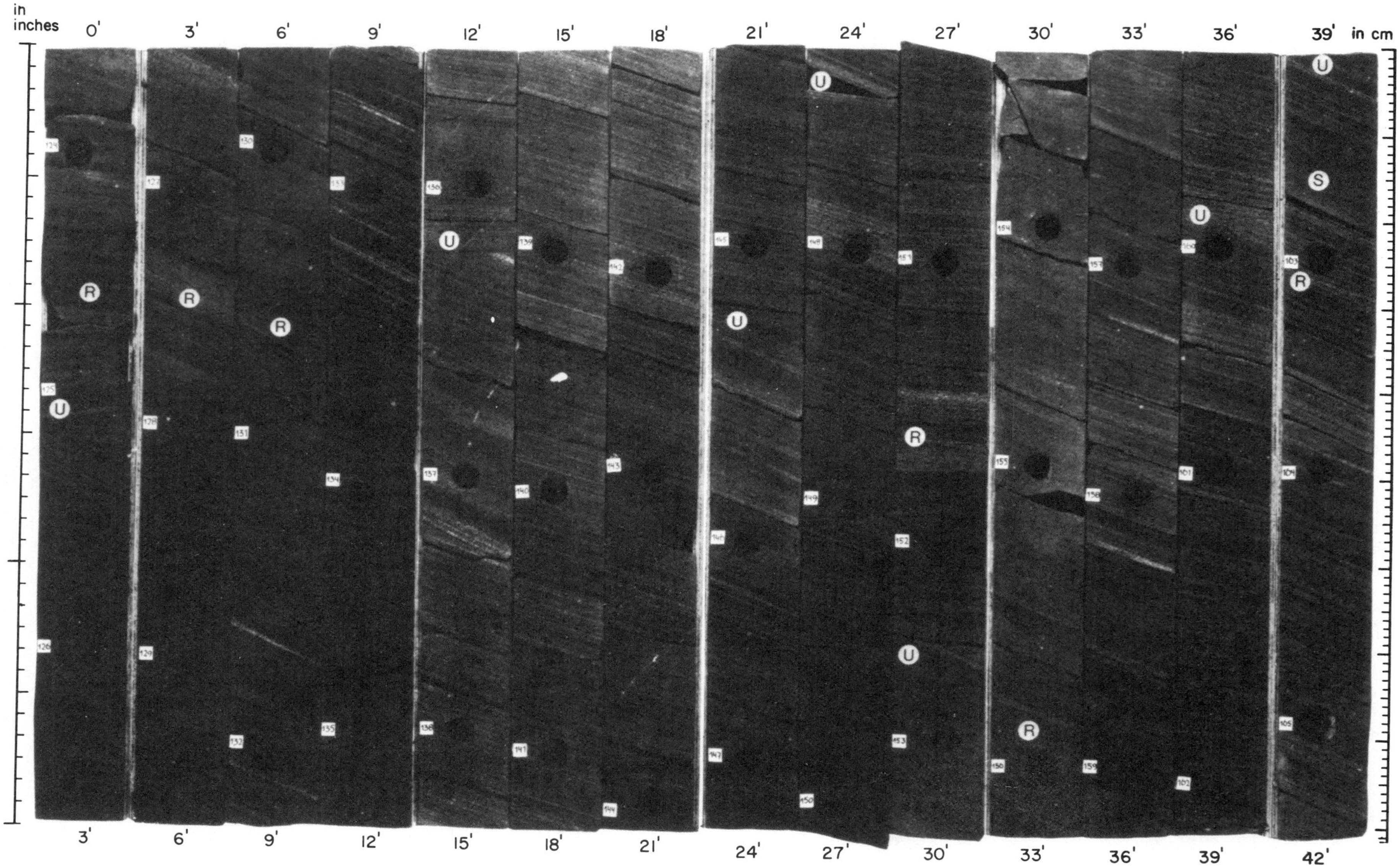

Fig. 22—Photograph of cored interval of Permian aeolian sandstone reservoir of the North Sea. From Glennie.[88]

Transgressive Marine Sandstones

Most of the deltaic and barrier sands are usually deposited under regressive conditions as the deltas and barriers grow seaward. The sands that are deposited as a result of the opposite situation, when the sea advances upon the land, are referred to here as transgressive marine sands.

Transgressions can occur as a result of a rising sea level, subsidence of the coastal plain, or simply because of delta abandonment during a period of standing sea level. Only one type of transgressive marine sand is discussed in the paper—that related to delta abandonment.

As the transgression progresses inland, a small portion of the upper layer of the delta sequence is removed by wave erosion and a relatively thin blanket of sand is deposited over the deltaic sediments. Farther seaward along the transgressive profile, silts and clays are deposited. Transgressive marine sands are usually very thin, seldom more than 10 ft, and have wide distribution and good continuity. As these sands become buried, they are cemented rapidly because of the large amount of shell material that they contain. These sands usually appear on the resistivity log as sharp spikes.

Transgressive marine sands are not important as reservoirs; however, because of their diagnostic characteristics and wide distribution, they are excellent marker beds for correlation purposes. It would be impossible to zone and correlate deltaic facies if these transgressive beds did not occur in the section.

Deep Marine Sandstones

Criteria for recognition of turbidities have been well established for more than 15 years. Hundreds of papers have been published on the general characteristics of turbidities of various ages throughout the world. The mechanism of turbidity current flow and, especially, an understanding of the mechanism required to trigger currents, remained vague for many years. An outcrop containing literally hundreds of discrete beds of sandstones interbedded with shale usually was explained as a result of gravity slumping of slope sediments or of earthquakes that caused slumping of slope sediments to form turbidity flows. Each sand bed was the product of a single slump or earthquake.

Recently, there was a very remarkable breakthrough in our understanding of turbidities. Mutti and Ricci-Lucchi[156] published a paper on turbidities of Italy. These two geologists recognized several distinct facies and three basic sequences, which they related to the submarine canyon fan model of clastic sedimentation. Thus, we now have a conceptual model for turbidite sedimentation, and for the first time in more than 30 years we can handle these deep-water sediments on the surface or in the subsurface in the same manner as the deltaic and other shallow- water sediments.

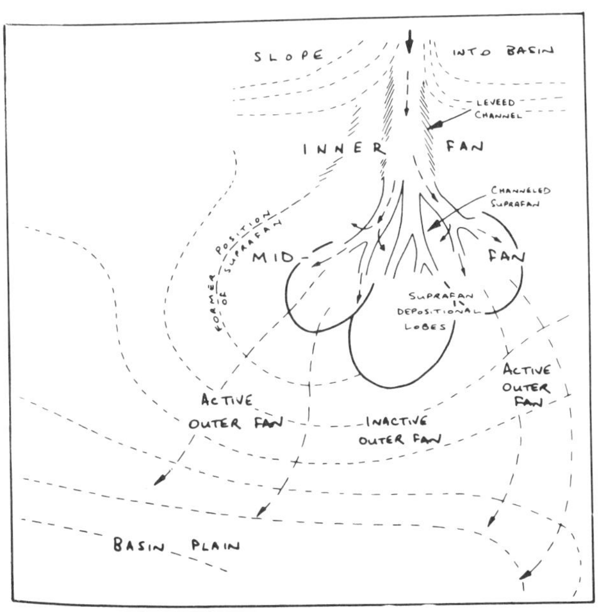

Fig. 23—Submarine fan model of clastic sedimentation. Modified from Walker and Mutti.[225]

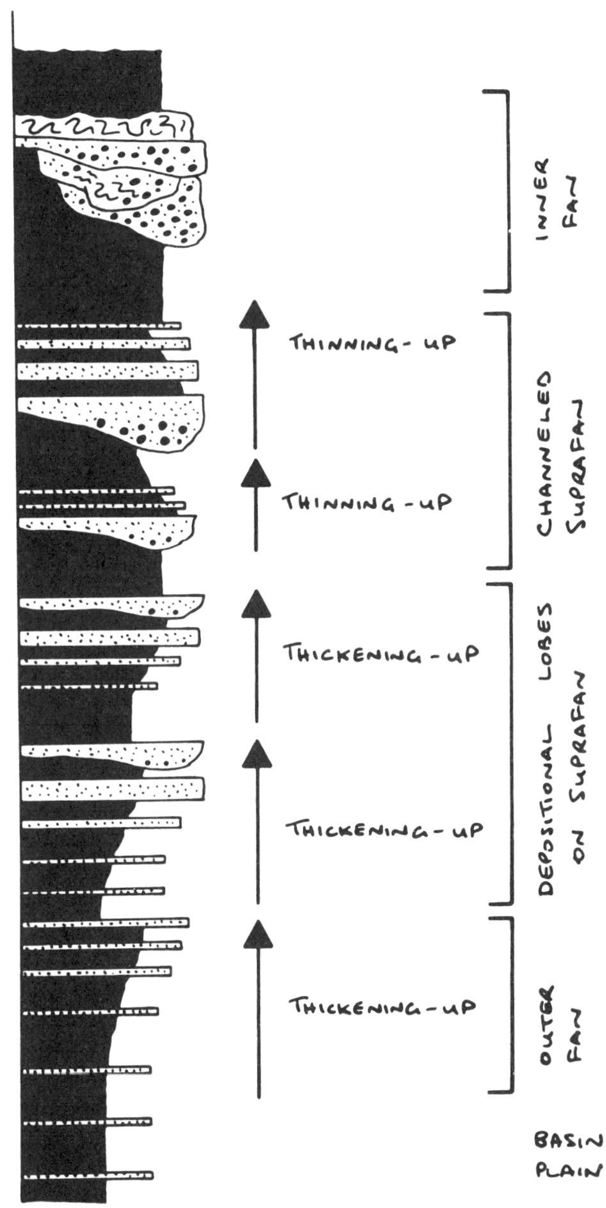

Fig. 24—Sequences of submarine fan channel and lobe sandstones. Modified from Walker and Mutti.[225]

Fig. 25—Three submarine fan channel sequences within the Jackfork (Paleozoic) of Arkansas. Note thinning of beds upward (to left) in each sequence.

Mutti and Ricci-Lucchi[156] and Walker and Mutti[225] recognized fan channel sequences, in which the beds thin upward ("thinning upward" sequence). They also described another sequence in which the beds thicken upward ("thickening upward" sequence). They related the fan channel sequences to the upper and middle fans and the thickening-upward sequences to the fan lobes (Figs. 23 and 24).

Fan channel sandstones have limited distribution, similar to delta distributary channels, but they are much longer simply because deep-sea fans are much larger than deltas. A photograph of three fan channel sequences in the Jackfork (Pennsylvanian) formation of Arkansas is shown in Fig. 25.

Fan lobe sandstones are probably distributed over much larger regions than any other type of sandstone. The lateral continuity of sand beds must be extensive. In many cases, fan lobe sandstones are interbedded with numerous shale beds. It is possible, therefore, that the vertical continuity of reservoirs will be rather poor. Most of our observations of fan lobe sandstones have been on outcrops whose lateral extent precludes the possibility of determining the extent of the lobes. This will have to await further study in the subsurface.

Fan lob sequences that occur in the Jackfork of eastern Oklahoma are shown in Fig. 26.

There is now general agreement among geologists that many thick sections of turbidities are of submarine canyon fan origin.

Fig. 26—Thick sequence of submarine fan lobe sandstones within the Jackfork (Paleozoic) of eastern Oklahoma. Note thickening of beds upward (to left).

Acknowledgments

I am very pleased for the opportunity to prepare and present this paper at the SPE-AIME 51st Annual Fall Technical Conference and Exhibition. I am grateful to Shell Development Co. for granting permission to share my experience and knowledge, especially during this period of energy crisis.

The genetic classification presented in Part 2 was initially developed by me and my former colleague, the late Hugh Allen Bernard, during the early 1950's.[128] Our original classification was modified through the years as we improved our knowledge and understanding of Recent sediments. The classification presented here is essentially the same we presented in 1965.[15]

I also am indebted to Doris M. Curtis of Shell Development Co., C.N. Tinker of Shell Oil Co., and Robert M. Sneider of Sneider and Meckel Associates for their critical review of this paper.

Distribution and Continuity of Sandstone Reservoirs—Part 2

R.J. LeBlanc, Sr.

This is a bibliographic source document supplementing Part 1 of this paper. It includes a genetic classification of environments of sand deposition and lists, by depositional environment and geographic occurrence, the significant literature on ancient sandstones.

Editor's Note: Part 1 of this paper (SPE 6137A) contained brief discussions of the three most important depositional environments of sandstone reservoirs: continental, transitional, and marine. Criteria for identifying sandstones using conventional data, along with the conceptual models used for predicting reservoir distribution and internal characteristics, were included. Part 2, presented here, contains a genetic classification and an extensive bibliography on sandstone reservoirs.

Resume of Studies of Ancient Sandstones

This second part is intended as a source document to supplement Part 1 of this paper. Our present knowledge and understanding of the factors that control the distribution and continuity of sandstones is based on the research on Recent sediments briefly reviewed in Part 1 and also on the study of ancient clastic sediments. While several groups of sedimentologists associated with the petroleum industry, government agencies, and universities were busy studying modern sands, another group of geologists was involved with the interpretation of depositional environments of ancient sandstones. Criteria and concepts derived from the study of Recent sediments were applied rapidly to the study of older rocks.

Presented here is a genetic classification (Table 1, Figs. 1 through 4) initially developed by me and the late Hugh Allen Bernard during the early 1950's.[128] Our original classification was modified through the years as we improved our knowledge and understanding of Recent sediments. The classification presented here is essentially the same one that we presented in 1965.

Some of the most significant studies of ancient clastic sediments of continental, transitional, and marine origin are summarized by depositional environments and geographic occurrence (Tables 2 through 7).

A selected bibliography is presented that covers our knowledge of the distribution and continuity of sandstone reservoirs.

Bibliography

1. Allen, J.R.L.: "Petrology, Origin, and Deposition of the Highest Lower Old Red Sandstone of Shropshire, England," *J. Sed. Petrology* (1962) 657-697.
2. Allen, J.R.L.: "Studies in Fluviatile Sedimentation: Six Cyclothems From the Lower Old Red Sandstones, Anglo-Welsh Basin," *Sedimentology* (1964) 163-198.
3. Asseez, L.O.: "Paleogeography of Lower Mississippian Rocks of Michigan Basin," *Bull.*, AAPG (1969) 127-135.
4. Baars, D.L.: "Permian Blanket Sandstones of Colorado Plateau," *Geometry of Sandstone Bodies,* AAPG (1961) 179-219.
5. Barlow, J.A., Jr., and Haun, J.D.: "Regional Stratigraphy of Frontier Formation and Relation to Salt Creek Field, Wyoming," *Bull.*, AAPG (1966) 2185-2196.
6. Bartow, J.A.: "Deep Submarine Channel in Upper Miocene, Orange County, California," *J. Sed. Petrology* (1966) 700-705.
7. Bass, N.W., Leatherock, C., Dillard,W.R., and Kennedy, L.E.: "Origin and Distribution of Bartlesville and Burbank Shoestring Oil Sands in Parts of Oklahoma and Kansas," *Bull.*, AAPG (1937) 30-66.
8. Beaty, C.B.: "Boulder Deposit in Flint Creek Valley, Western Montana," *Bull.*, GSA (1961) 1015-1020.
9. Beerbower, J.R.: "Origin of Cyclothems of the Dunkard Group (upper Pennsylvanian-Lower Permian) in Pennsylvania, West Virginia, and Ohio," *Bull.*, GSA (1961) 1029-1050.
10. Beerbower, J.R.: "Cyclothems and Cyclic Depositional Mechanisms in Alluvial Plain Sedimentation," *Bull.*, Kansas Geol. Survey (1964) 31-42.
11. Beerbower, J.R.: "Interpretation of Cyclic Permo-Carboniferous Deposition in Alluvial Plain Sediments in West Virginia," *Bull.*, GSA (1969) 1843-1848.
12. Berg, R.R.: "Point Bar Origin of Fall River Sandstone Reservoirs, Northeastern Wyoming," *Bull.*, AAPG (1968) 2116-2122.
13. Berg, R.R. and Cook, B.D.: "Petrography and Origin of Lower Tuscaloosa Sandstones, Mallalieu Field, Lincoln County, Mississippi, "*Trans.*, Gulf Coast Assn. Geol. Soc. (1968) 242-255.
14. Berg, R.R. and Davies, D.K.: "Origin of Lower Cretaceous Muddy Sandstone at Belt Creek Field, Montana," *Bull.*, AAPG (1968) 1888-1898.
15. Bernard, H.A. and Le Blanc, R.J.: "Resume of Quaternary Geology of the Northwestern Gulf of Mexico Province," *Quaternary of the United States*, Princeton U. Press, Princeton, N.J. (1965) 137-185.
16. Bernard, H.A., Major, C.F., and Parrott, B.S.: "The Galveston Barrier Island and Environs—A Model for Predicting Reservoir Occurrence and Trend," *Trans.*, Gulf Coast Assn. Geol. Socs. (1959) 221-224.
17. Beutner, E.C., Flueckinger, L.A., and Gard, T.M.: "Bedding Geometry in a Pennsylvanian Channel Sandstone," *Bull.*, GSA (1967) 911-916.

This is a bibliographic source document supplementing Part 1 of this paper. It includes a genetic classification of environments of sand deposition and lists, by depositional environment and geographic occurrence, the significant literature on ancient sandstones.

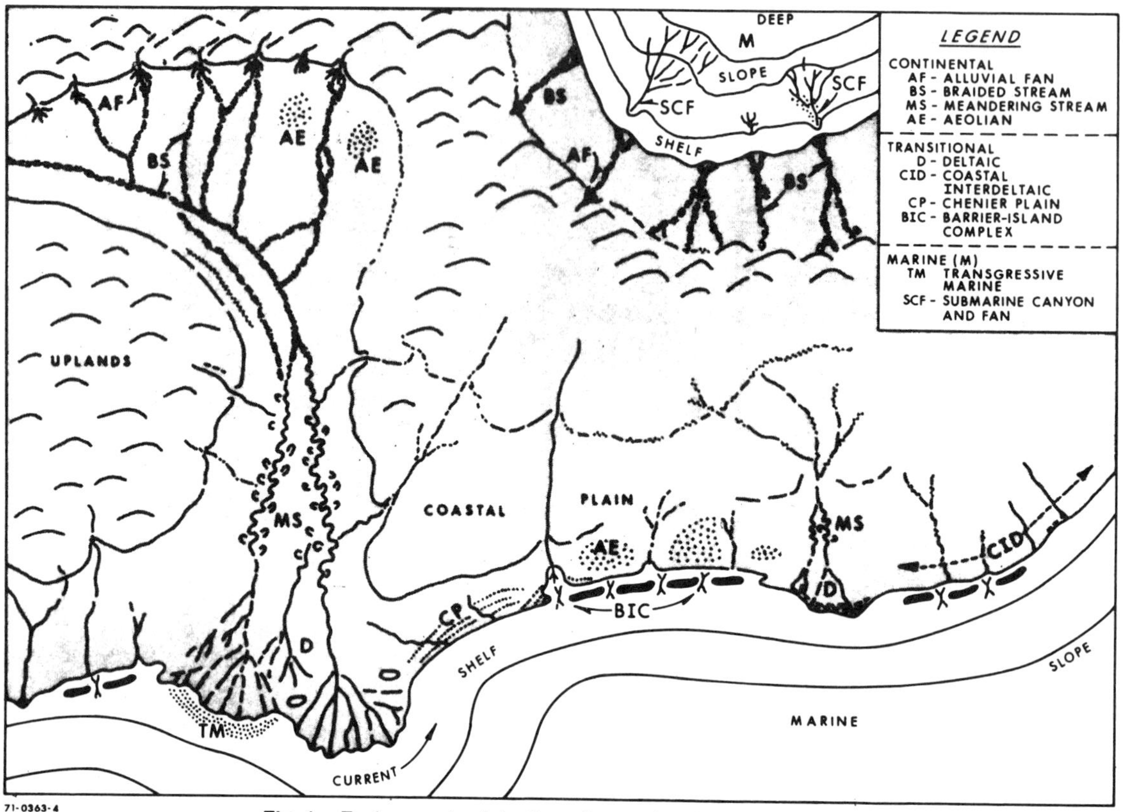

Fig. 1—Environments of clastic sedimentation. From Le Blanc.[126]

18. Bigarella, J.J.: "Eolian Deposits—Their Characteristics, Recognition, and Importance," *Recognition of Ancient Sedimentary Environments*, Spec. Pub. 16, Soc. Econ. Paleontologists and Mineralogists (1972).
19. Bigarella, J.J. and Salamuni, R.: "Early Mesozoic Wind Patterns as Suggested by Dune Bedding in the Botucatu Sandstone of Brazil and Uruguay," *Bull.*, GSA (1961) 1089-1106.
20. Blackwelder, E.: "Mudflow as a Geologic Agent in Semi-Arid Mountains," *Bull.*, GSA (1928) 465-483.
21. Bloomer, R.R.: "Depositional Environments of a Reservoir Sandstone in West- Central Texas," *Bull.*, AAPG (1976).
22. Bluck, B.J.: "The Sedimentary History of Some Triassic Conglomerates in The Vale of Glamorgan, South Wales," *Sedimentology* (1965) 225-246.
23. Bluck, B.J.: "Deposition of Some Upper Old Red Sandstone Conglomerates in the Clyde Area: A Study in the Significance of Bedding," *Scottish J. Geol.* (1967) Part 2, 139-167.
24. Boggs, S., Jr.: "Petrology of Minturn Formation, East-Central Eagle County, Colorado," *Bull.*, AAPG (1966) 1399-1422.
25. Bolyard, D.W.: "Pennsylvanian and Permian Stratigraphy in Sangre de Cristo Mountains Between La Veta Pass and Westcliffe, Colorado," *Bull.*, AAPG (1959) 1896-1939.
26. Bornhauser, M.: "Possible Ancient Submarine Canyon in Southwestern Louisiana," *Bull.*, AAPG (1948) 2287-2290.
27. Bornhauser, M.: "Depositional and Structural History of Northwest Hartburg Field, Newton County, Texas," *Bull.*, AAPG (1960) 458-470.
28. Boyd, D.R. and Dyer, B.F.: "Frio Barrier Bar System of South Texas," *Bull.*, AAPG (1966) 170-178.
29. Brady, L.L. "Stratigraphy and Petrology of the Morrison Formation (Jurassic) of the Canon City, Colorado, Area," *J. Sed. Petrology* (1969) 632-648.
30. Bretz, J.H. and Horberg, L.: "The Ogallala Formation West of the Llano Estacado," *J. Geol.* (1949) 477-490.
31. Brown, L.F., Jr.: "North Texas (Eastern Shelf) Pennsylvanian Delta Systems," *Delta Systems in Exploration for Oil and Gas*, U. of Texas, Bureau of Econ. Geol., Austin (1969) 40-53.
32. Brown, S.L.: "Stratigraphy and Depositional Environment of the Elgin Sandstone (Pennsylvanian) in South-Central Kansas," *Bull.*, Kansas Geol. Survey (1967).
33. Burke, K.: "Longshore Drift, Submarine Canyons and Submarine Fans in Development of Niger Delta," *Bull.*, AAPG (1972) 1975- 1983.
34. Busch, D.A.: "The Significance of Deltas in Subsurface Exploration," *Digest*, Tulsa Geol. Soc. (1953) 71-80.
35. Busch, D.A.: "Genetic Units in Delta Prospecting," *Bull.*, AAPG (1971) 1137-1154.
36. Buttner, P.J.R.: "Proximal Continental Rhythmic Sequences in the Genesee Group (Lower Upper Devonian)," Spec. paper 106, GSA (1968) 109-126.
37. Byers, P.R.: "Mineralogy and Origin of the Eastend and Whitemud Formations of South-Central and Southwestern Saskatchewan and Southeastern Alberta," MS thesis, Queen's U., Kingston, Ont. (1966).
38. Campbell, C.V.: "Reservoir Geometry of a Fluvial Sheet Sandstone," *Bull.*, AAPG (1976) 1009-1020.
39. Cannon, J.L.: "Outcrop Examination and Interpretation of Paleocurrent Patterns of the Blackleaf Formation Near Great Falls, Montana," Billings Geol. Soc. 17th Field Conf. Guidebook, Billings, Mont. (1966) 71-111.
40. Carrigy, M.A.: "Deltaic Sedimentation in Athabasca Tar Sands," *Bull.*, AAPG (1971) 1155-1169.
41. Clark, R.H. and Rouse, J.T.: "A Closed System for Generation and Entrapment of Hydrocarbons in Cenozoic Deltas, Louisiana Gulf Coast," *Bull.*, AAPG (1971) 1170-1178.
42. Coleman, J.M. and Wright, L.D.: "Variability in Modern River Deltas," *Trans.*, Gulf Coast Assn. Geol. Soc. (1973) 33-36.
43. Coleman, J.M. and Wright, L.D.: "Modern River Deltas: Variability of Processes and Sand Bodies," *Deltas, Models for Exploration*, Houston Geol.Soc. (1975) 99-149.
44. Conolly, J.R.: "Petrology and Origin of the Hervey Group, Upper Devonian, Central New South Wales," *Jour.*, Geol. Soc. Australia (1965) 123-166.
45. Conolly, J.R.: "Submarine Canyons of the Continental Mar-

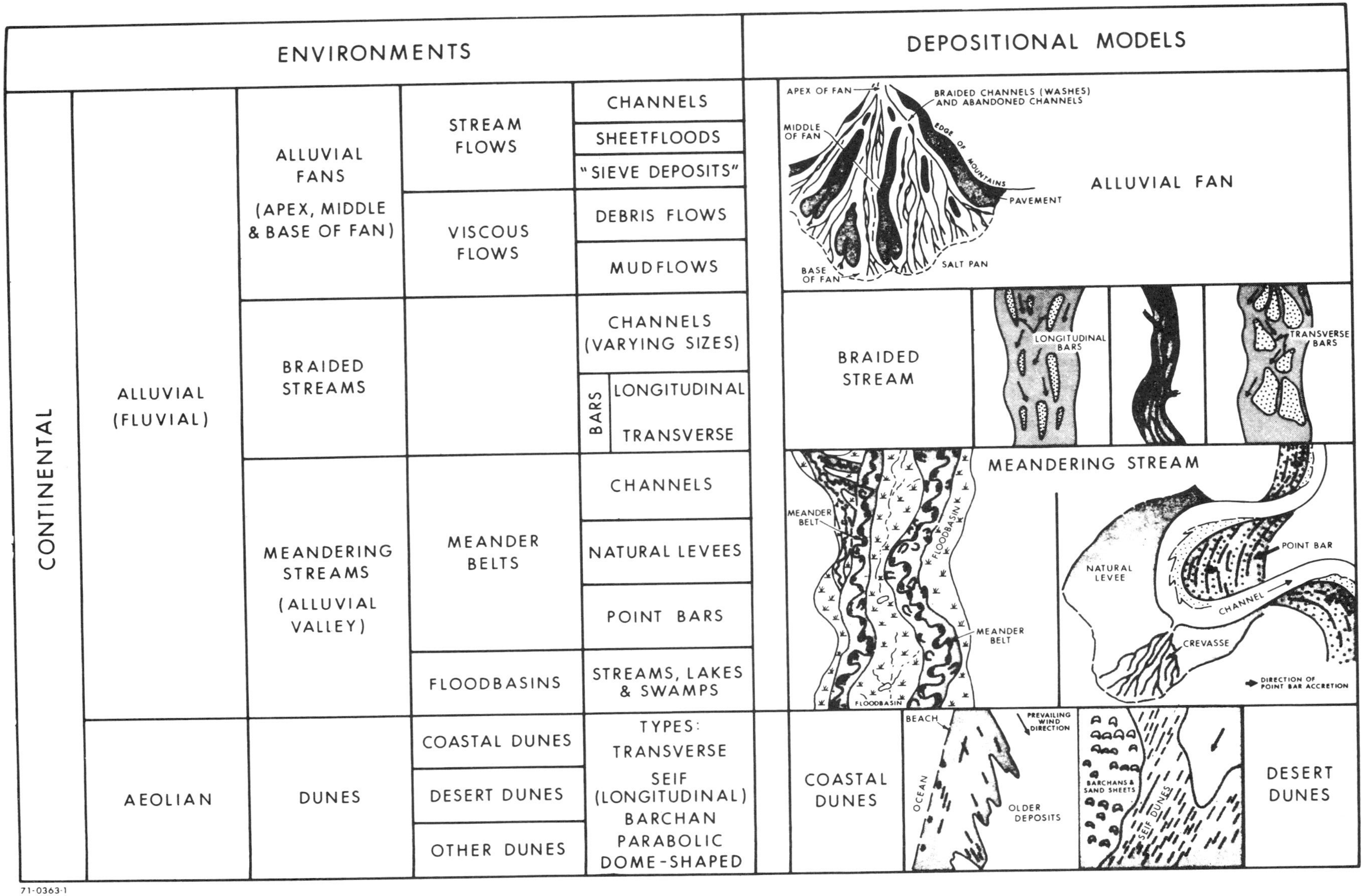

Fig. 2—Alluvial (fluvial) and aeolian environments of clastic sedimentation. From Le Blanc.[126]

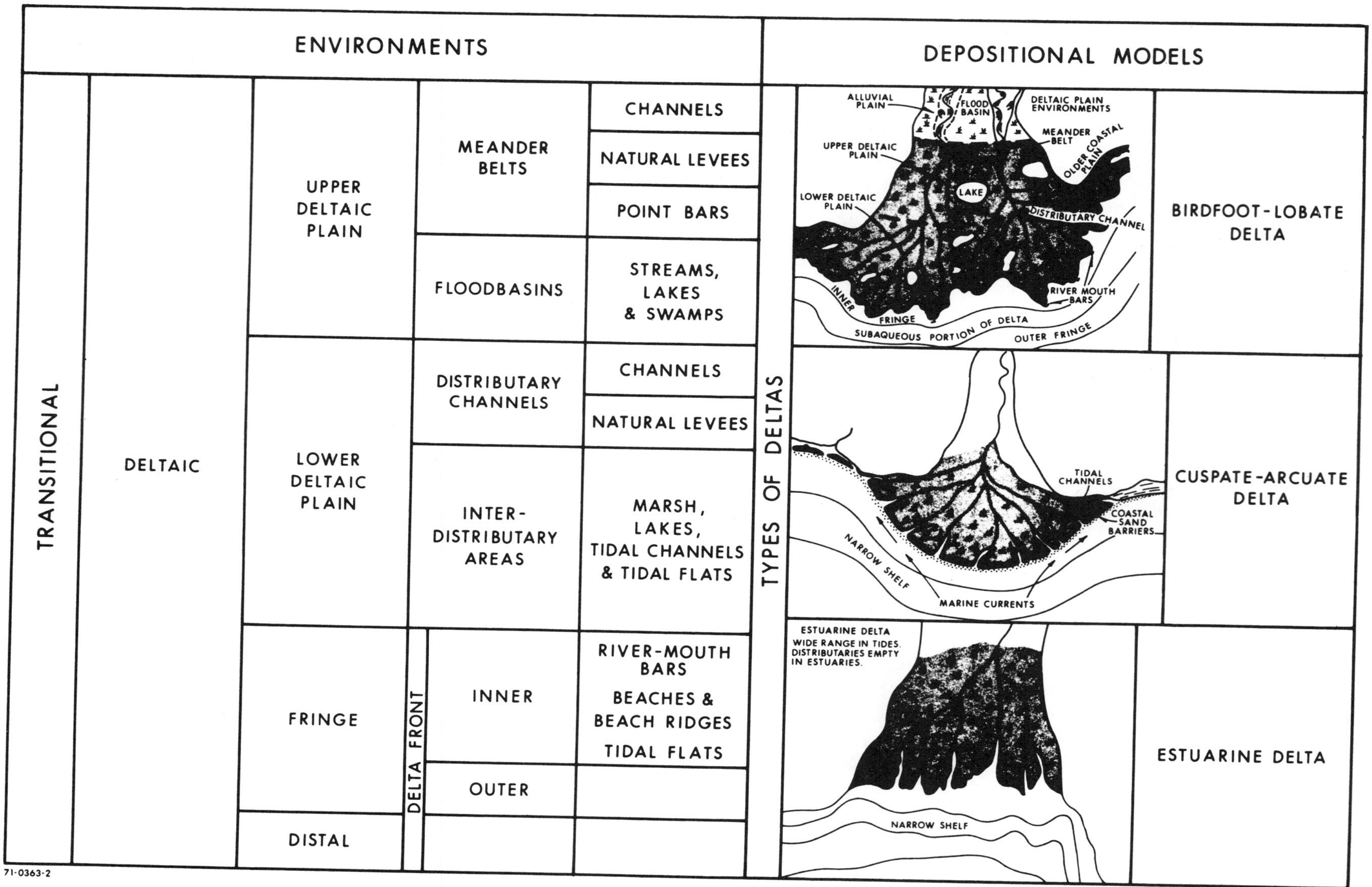

Fig. 3—Deltaic environments of clastic sedimentation. From Le Blanc.[126]

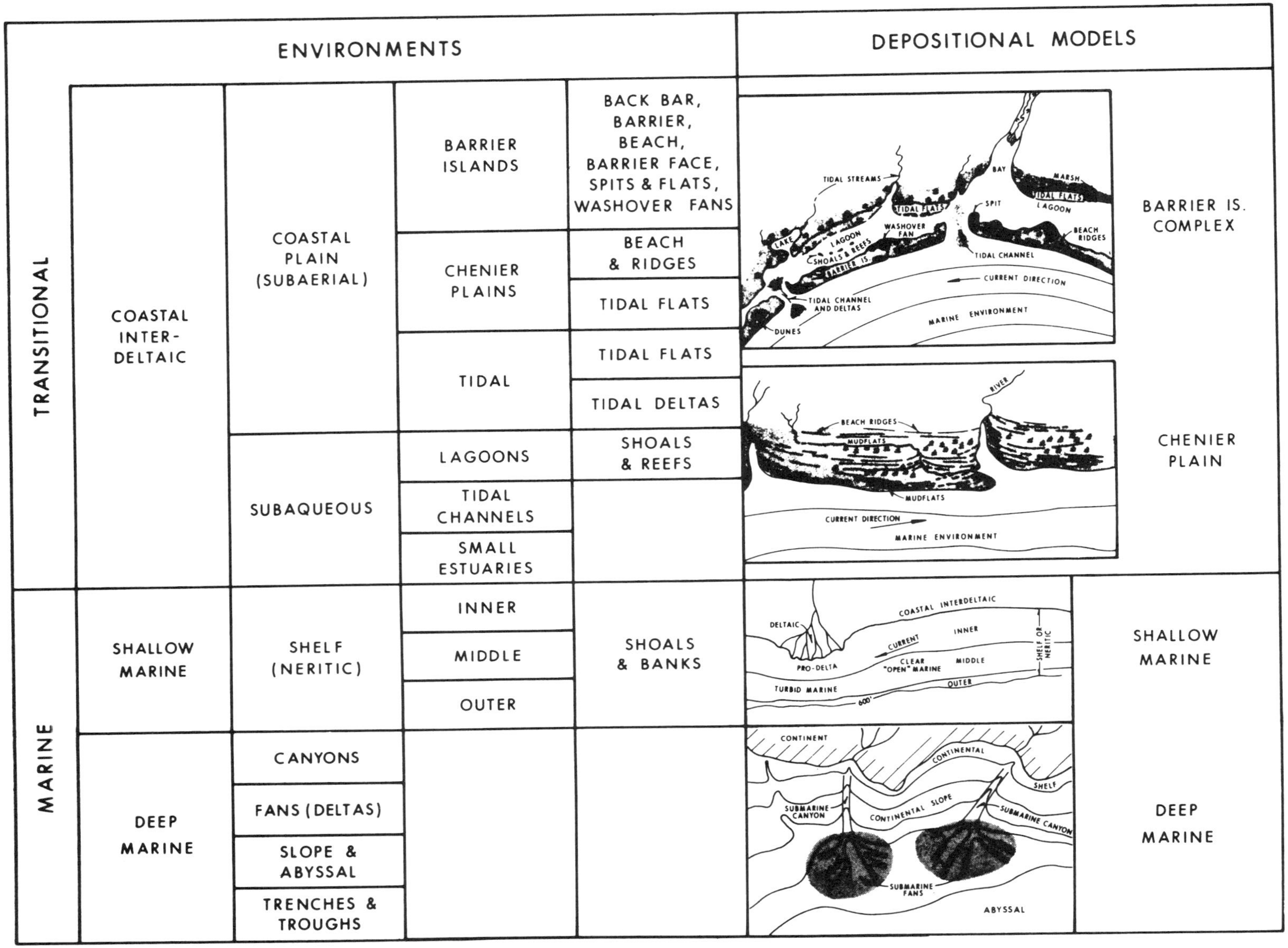

Fig. 4—Coastal interdeltaic and marine environments of clastic sedimentation. From Le Blanc.[126]

TABLE 2—EXAMPLES OF ANCIENT ALLUVIAL-FAN, BRAIDED-STREAM, AND MEANDERING-STREAM DEPOSITS

Alluvial Fan	Braided Stream	Meandering Stream	Composite	Author
Arizona			Arizona	Melton[149]
California				Crowell[50]
California			California	Flemal[82]
			California	Galehouse[84]
Colorado				Boggs[24]
			Colorado	Bolyard[25]
		Colorado		Brady[29]
			Colorado Plateau	Finch[71]
			Colorado Plateau	Stokes[204]
Colorado				Howard[104]
Colorado				Hubert[112]
Connecticut Valley				Klein[119]
		Illinois		Hewitt and Morgan[101]
		Illinois		Shelton[190]
			Kansas	Lins[133]
		Kansas		Shelton[190]
	Llano Estacado			Bretz and Horberg[30]
	Maryland	Maryland		Hansen[97]
Massachusetts				Wessel[234]
Massachusetts				Stanley[200]
			Massachusetts	Mutch[155]
		Michigan		Shideler[193]
		Montana		Gwinn[93]
	Mississippi	Mississippi		Berg and Cook[13]
Montana	Montana			Gwinn and Mutch[94]
			Montana	Shelton[189]
Montana				Wilson[239, 240]
Montana				Beaty[8]
			Nebraska	Exum and Harms[69]
			Nebraska	Harms[98]
	New York	New York		Buttner[36]
	New Jersey, New York			Smith,[196]
				Shelton,[190]
				Campbell[38]
			North Dakota	Royse[173]
			Oklahoma	Visher[220, 221]
		Pennsylvania		Beutner *et al.*[17]
	Pennsylvania			Smith[196]
		Pennsylvania		Ryan[176]
			Rhode Island	Mutch[155]
Southwest U.S.				Bull[243]
			Texas	Fisher and McGowen[73]
Texas				McGowen and Groat[146]
		Texas		McGowen and Garner,[144] Shelton[190]
		West Virginia		Beerbower[10, 11]
		Wyoming		Berg[12]
Wyoming	Wyoming			Spearing[198]
			Alberta	Byers[37]
			Quebec	Dinley and Williams[60]
Northeastern Canada				Klein[118]
		Nova Scotia		Way[228]
Northwest Territories	Northwest Territories	Northwest Territories		Miall[150]
		England		Allen[2]
				Laming[124]
Wales and Scotland				Bluck[22, 23]
		South Wales		Kelling[116]
Norway				Nilsen[161]
	Scotland			Williams[237, 238]
	Spain			Nagtegaal[157]
	Spitsbergen	Spitsbergen		Moody-Stuart[152]
		New South Wales		Conolly[44]

TABLE 3—EXAMPLES OF ANCIENT DELTAIC DEPOSITS

Geographic Occurrence	Author	Geographic Occurrence	Author
California	Todd and Monroe[215]	Texas	Gregory[90]
Colorado	Weimer[247, 248]		
Illinois	Lineback[132]	Texas	Le Blanc[130]
Illinois	Swann *et al.*[209]	Texas	Nanz[158]
Indiana	Hrabar and Potter[109]	Texas	Shannon and Dahl[186]
Indiana	Wier and Girdley[236]	Texas	Wermund and Jenkins[233]
Iowa and Illinois	Laury[125]	Texas	Shelton[190]
Kansas	Brown[32]	West Virginia, Pennsylvania, Ohio	Donaldson[62]
Kansas	Hattin[100]	Wyoming	Barlow and Haun[5]
Louisiana	Clark and Rouse[41]	Wyoming	Dondanville[63]
Louisiana	Curtis[53]	Wyoming	Hale[96]
Michigan	Asseez[3]	Wyoming	Paull[166]
Mississippi, Louisiana, Alabama	Galloway[85]	Wyoming	Weimer[229, 230]
Montana	Sims[194]	Wyoming, Colorado	Weimer[231]
Nebraska	Shelton[190]	Several states	Fisher *et al.*[74]
New Mexico	Schlee and Moench[181]	North Appalachians	Ferm[244]
New York	Friedman and Johnson[83]	Central Appalachians	Horowitz[102]
		Central Appalachians	Dennison[58]
New York	Lumsden and Pelletier[135]	Upper Mississippi embayment and Illinois Basin	Pryor[169, 170]
New York and Ontario	Martini[141]	Upper Mississippi Valley	Swann[208]
New York	Wolff[241]	Oklahoma, Iowa, Missouri, Kansas, Illinois, Indiana, Kentucky	Manos[139]
North Dakota	Shelton[190]		
Ohio	Knight[120]	Oklahoma to Pennsylvania	Wanless *et al.*[227]
Ohio	Lené and Owen[131]	Central Gulf Coast	Mann and Thomas[138]
Oklahoma	Busch[34, 35]	Alberta, Canada	Carrigy[40]
Oklahoma	Shelton[190]	Alberta, Canada	Shawa[187]
Oklahoma	Visher *et al.*[223]	Alberta, Canada	Shepheard and Hills[192]
Oregon	Dott[64, 65]	Alberta, Canada	Thachuk[212]
Oregon	Snavely *et al.*[197]	England	Allen[1]
Pennsylvania, West Virginia, Ohio	Beerbower[9]	England	Taylor[211]
Pennsylvania, West Virginia, Ohio	Ferm and Cavaroc[70]	Ireland	Hubbard[110]
South Dakota	Pettyjohn[167]	Scotland	Greensmith[89]
Texas	Brown[31]		
Texas	Fisher and McGowen[73]		

gin, East Bass Strat (Australia)," *Marine Geol.* (1968) 449-461.

46. Cooper, W.S.: "Coastal Sand Dunes of Oregon and Washington," Mem. 72, GSA (1958) 1-169.
47. Cooper, W.S.: "Coastal Dunes of California," Mem. 104, GSA (1967).
48. Cotter, E.: "Deltaic Deposits in the Upper Cretaceous Ferron Sandstone, Utah," *Deltas, Models for Exploration*, Houston Geol. Soc. (1975) 472-484.
49. Crimes, T.P.: "Sand Fans: Turbidites, Slumps and the Origin of the Bay of Biscay: A facies Analysis of the Guipuzcoan Flysch," *Paleogeogr., Paleoclim., Paleoecology* (1976) 1-15.
50. Crowell, J.C.: "Geology of the Ridge Basin Area," *Bull.* 170, California Div. Mines, Map Sheet No. 7 (1954).
51. Cruz, P., Verdugo-V., R., and Barcenas-P., R.: "Origin and Distribution of Tertiary Conglomerates, Veracruz Basin, Mexico," *Bull.*, AAPG (1977) 207-226.
52. Curray, J.R.: "Transgressions and Regressions," *Papers in*

Marine Geology, The Macmillan Co., New York (1964) 175-203.
53. Curtis, D.M.: "Miocene Deltaic Sedimentation, Louisiana Gulf Coast," *Deltaic Sedimentation—Modern and Ancient*, Spec. Pub. 15, Soc. Econ. Paleontologists and Mineralogists (1970) 293-308.
54. Curtis, D.M. and Echols, D.J.: "Eustasy in the Miocene? Interpretation of Stratigraphic Evidence From Gulf Coast (USA)," presented at Ninth Cong. Int'l. de Sedimentologie, Theme I, Indicateurs Sedimentologiques, Nice, France (1975) 37-41.
55. Davies, D.K. and Berg, R.R.: "Sedimentary Characteristics of Muddy Barrier-Bar Reservoir and Lagoonal Trap at Bell Creek Field," presented at 20th Annual Conf. of Montana Geol. Soc., Eastern Montana Symposium (1969) 97-105.
56. Davies, D.K., Ethridge, F.G., and Berg, R.R.: "Recognition of Barrier Environments," *Bull.*, AAPG (1971) 550-565.
57. Davis, J.R.: "Sedimentation of Pliocene Sandstones in Santa Barbara Channel, California," (abs.) *Bull.*, AAPG (1971) 335.
58. Dennison, J.M.: "Petroleum Related to Middle and Upper Devonian Deltaic Facies in Central Appalachia," *Bull.*, AAPG (1971) 1179-1193.
59. Dickas, A.B. and Payne, J.L.: "Upper Paleocene Buried Channel in Sacramento Valley, California," *Bull.*, AAPG (1967) 873-882.
60. Dineley, D.L. and Williams, B.P.J.: "Sedimentation and Paleoecology of the Devonian Escuminac Formation and Related Strata, Escuminac Bay, Quebec," Spec. Paper 106, GSA (1968) 241-264.
61. Dodge, C.F.: "Genesis of an Upper Cretaceous Offshore Bar Near Arlington, Texas," *J. Sed. Petrology* (1965) 22-35.
62. Donaldson, A.C.: "Ancient Deltaic Sedimentation (Pennsylvanian) and Its Control on the Distribution, Thickness and Quality of Coals," *Some Appalachian Coals and Carbonates: Models of Ancient Shallow-Water Deposition*, West Virginia Geol. and Econ. Survey (1969) 93-123.
63. Dondanville, R.R.: "The Fall River Formation, Northwestern Black Hills, Lithology and Geologic History," Wyoming Geol. Assn.-Billings Geol. Soc. Joint Field Conf. Guidebook (1963) 87-99.
64. Dott, R.H., Jr.: "Ancient Deltaic Sedimentation in Eugeosynclinal Belts," *Deltaic and Shallow Marine Deposits*, Elsevier Publishing Co., Amsterdam (1964) 105-113.
65. Dott, R.H., Jr.: "Eocene Deltaic Sedimentation at Coos Bay, Oregon," *J. Geol.* (1966) 373-420.
66. Eckis, R.: "Alluvial Fans of the Cucamonga District, Southern California," *J. Geol.* (1928) 224-247.
67. Emery, K.O., Butcher, W.S., Gould, H.R., and Shepard, F.P.: "Submarine Geology Off San Diego, California" *J. Geol.* (1952) 511-548.

TABLE 4—EXAMPLES OF ANCIENT COASTAL INTERDELTAIC DEPOSITS

Geographic Occurrence	Author
Colorado	Griffith[92]
Florida	Gremillion *et al.*[91]
Georgia	Hails and Hoyt[95]
	MacNeil[136]
Illinois	Rusnak[174]
Louisiana	Sloane[195]
Louisiana and Arkansas	Thomas and Mann[213]
	Berg and Davies[14]
Montana	Cannon[39]
	Davies *et al.*[56]
	Shelton[188]
New Mexico	Sabins[178]
New York	McCave[143]
Oklahoma and Kansas	Bass *et al.*[7]
	Boyd and Dyer[28]
	Dodge[61]
Texas	Fisher and McGowen[73]
	Fisher *et al.*[75]
	Shelton[190]
	Harms *et al.*[99]
	Jacka[114]
Wyoming	Miller[151]
	Paull[166]
	Scruton[184]
	Weimer[229, 230]
	Land[246]

TABLE 5—EXAMPLES OF ANCIENT AEOLIAN SANDSTONES

Geographic Occurrences	Author
Colorado Plateau, U.S.A.	Baars[4] Stokes[204-206]
Southwest U.S.A.	McKee[147]
England	Laming[124]
Brazil and Uruguay	Bigarella and Salamuni[19]
U.S.A., Great Britain, and South America	Bigarella[18]
North Sea	Glennie[88]
Wyoming	Steidtmann[202]

TABLE 6—EXAMPLES OF ANCIENT TRANSGRESSIVE MARINE SANDSTONES

Geographic Occurrence	Author
North Texas	Brown[31]
Gulf Coast	Curray[52] Rainwater[245] Scott and Fisher[182]

TABLE 7—EXAMPLES OF ANCIENT SUBMARINE CANYON AND FAN DEPOSITS

Geographic Occurrence	Author
Arkansas and Oklahoma	Thomson and Le Blanc[214]
California	Sullwold[207] Martin[140] Bartow[6] Dickas and Payne[59] Normark and Piper[162] Piper and Normark[168] Davis[57] Fischer[72] Shelton[190] Nelson and Nilsen[159] Van der Kamp *et al.*[217]
Gulf Coast	Bornhauser[26, 27] Hoyt[108] Paine[164] Sabate[177] Curtis and Echols[54]
Canada	Hubert *et al.*[111]
Europe	Walker[224] Stanley[199, 201] Kelling and Woollands[117] Kruit *et al.*[122]
Italy	Mutti and Ricci-Lucchi[156] Ricci-Lucchi[172]
Australia	Conolly[45]
Africa	Burke[33]
Worldwide summary	Whitaker[235]

68. Erlxeben, A.W.: "Deltaic and Related Carbonate Systems in the Pennsylvanian Canyon Group of North-Central Texas," *Deltas, Models for Exploration*, Houston Geol. Soc. (1975) 400-424.
69. Exum, F.A. and Harms, J.C.: "Comparison of Marine-Bar With Valley-Fill Stratigraphic Traps, Western Nebraska," *Bull.*, AAPG (1968) 1851-1868.
70. Ferm, J.C. and Cavaroc, V.V., Jr.: "A Field Guide to Allegheny Deltaic Deposits in the Upper Ohio Valley," Ohio Geol. Soc. and Pittsburgh Geol. Soc. (1969).
71. Finch, W.I.: "Geology of Uranium Deposits in Triassic Rocks of the Colorado Plateau Region," *Bull.* 1074-D, USGS (1959).
72. Fischer, P.J.: "An Ancient (Upper Paleocene) Submarine Canyon and Fan: The Meganos Channel, Sacramento Valley, California," (abs.) *Proc.*, 67th Annual GSA Cordilleran Section Meeting (1971) 120.
73. Fisher, W.L. and McGowen, J.H.: "Depositional Systems in Wilcox Group (Eocene) of Texas and Their Relationship to Occurrence of Oil and Gas," *Bull.*, AAPG (1969) 30- 54.
74. Fisher, W.L., Brown, L.F., Jr., Scott, A.J., and McGowen, J.H.: "Delta Systems in the Exploration for Oil and Gas," U. of Texas, Bureau of Econ. Geol., Austin (1969).
75. Fisher, W.L., Proctor, C.V., Jr., Galloway, W.E., and Nagle, J.S.: "Depositional Systems in the Jackson Group of Texas—Their Relationship to Oil, Gas and Uranium," Circ. 70-4, U. of Texas, Bureau of Econ. Geol., Austin (1970).
76. Fisk, H.N.: "Geological Investigation of the Alluvial Valley of the Lower Mississippi River," Mississippi River Commission, Vicksburg (1944).
77. Fisk, H.N.: "Fine-Grained Alluvial Deposits and Their Effects on Mississippi River Activity," Mississippi River Commission, Vicksburg (1947).
78. Fisk, H. N.: "San Facies of Recent Mississippi Delta Deposits," *Proc.*, Fourth World Pet. Cong., Rome (1955) Sec. I/C, 1-21.
79. Fisk, H.N.: "Recent Mississippi River Sedimentation and Peat Accumulation," *Fourth Congres pour l'Avancement des Etudes de Stratigraphie et de Geologie du Carbonifere*, Heerlen, Compte Rendu (1958) 187-199.
80. Fisk, H.N.: "Padre Island and the Laguna Madre Flats, Coastal South Texas," *Proc.*, Second Coastal Geog. Conf., NAS-NRC (1959) 103-151.
81. Fisk, H.N. et al.: "Sedimentary Framework of the Modern Mississippi Delta," *J. Sed. Petrology* (1954) 76-99.
82. Flemal, R.C.: "Sedimentology of the Sespe Formation, Southwestern California," PhD dissertation, Princeton U., Princeton, N.J. (1967).
83. Friedman, G.M. and Johnson, K.G.: "The Devonian Catskill Deltaic Complex of New York, Type Example of a 'Teutonic Delta Complex'," *Deltas in Their Geologic Framework*, Houston Geol. Soc. (1966).
84. Galehouse, J.J.: "Provenance and Paleo-currents of the Paso Robles Formation, California," *Bull.*, GSA (1967) 951-978.
85. Galloway, W.E.: "Depositional Systems of Lower Wilcox Group, North-Central Gulf Coast Basin," *Trans.*, Gulf Coast Assn. Geol. Soc. (1968) 275-289.
86. Gilreath, J.A. and Stephens, R.W.: "Interpretation of Log Responses in Deltaic Sediments," *Finding and Exploring Ancient Deltas in the Subsurface*, AAPG (1975) c-1 to c-31.
87. Glennie, K.W.: *Desert Sedimentary Environments- Developments in Sedimentology*, Elsevier Publishing Co., Amsterdam (1970).
88. Glennie, K.W.: "Permian Rotliegendes of Northwest Europe Interpreted in Light of Modern Desert Sedimentation Studies," *Bull.*, AAPG (1972) 1048-1071.
89. Greensmith, J.T.: "Carboniferous Deltaic Sedimentation in Eastern Scotland, a Review and Reappraisal," *Deltas in Their Geologic Framework*, Houston Geol. Soc. (1966).
90. Gregory, J.L.: "A Lower Oligocene Delta in the Subsurface of Southeast Texas," *Trans.*, Gulf Coast Assn. Geol. Socs. (1966) 227-241.
91. Gremillion, L.R., Tanner, W.F., and Huddlestun, P.: "Barrier and Lagoonal Sets on High Terraces in the Florida Panhandle," *Southeastern Geol.* (1964) 31- 36.
92. Griffith, E.G.: "Geology of Saber Bar, Logan and Weld Counties, Colorado," *Bull.*, AAPG (1966) 2112-2118.
93. Gwinn, V.E.: "Deductions of Flow Regime From Bedding Character in Conglomerates and Sandstones," *J. Sed. Petrology* (1964) 656-658.
94. Gwinn, V.E. and Mutch, T.A.: "Intertongued Upper Cretaceous Volcanic and Nonvolcanic Rocks, Central-Western Montana," *Bull.*, GSA (1965) 1125-1144.
95. Hails, J.R. and Hoyt, J.H.: "The Significance and Limitations of Statistical Parameters of Distinguishing Ancient and Modern Sedimentary Environments of the Lower Georgia Coastal Plain," *J. Sed. Petrology* (1969) 559-580.
96. Hale, L.A.: "Late Cretaceous (Montanan) Stratigraphy, Eastern Washakie Basin, Carbon County, Wyoming," Wyoming Geol. Assn. 16th Field Conf. Guidebook (1961) 129-137.
97. Hansen, H.J.: "Depositional Environments of Subsurface Potomac Group in Southern Maryland," *Bull.*, AAPG (1969) 1923-1937.
98. Harms, J.C.: "Stratigraphic Traps in a Valley Fill, Western Nebraska," *Bull.*, AAPG (1966) 2119-2149.
99. Harms, J.C., MacKenzie, D.B., and McCubbin, D.G.: "Depositional Environment of the Fox Hills Sandstones Near Rock Springs Uplift, Wyoming," Wyoming Geol. Assn. 19th Field Conf. Guidebook (19965) 113- 130.
100. Hattin, D.E.: "Stratigraphy of the Graneros Shale (Upper Cretaceous) in Central Kansas," *Bull.* 178, Kansas Geol. Survey (1965).
101. Hewitt, C.H. and Morgan J.T.: "The Fry In Situ Combustion Test-Reservoir Characteristics," *J. Pet. Tech.* (March 1965) 337-342; *Trans.*, AIME, **234**.
102. Horowitz, D.H.: "Evidence for Deltaic Origin of an Upper Ordovician Sequence in the Central Appalachians," *Deltas in Their Geologic Framework*, Houston Geol. Soc. (1966) 159-169.
103. Houser, J.F. and Neasham, J.W.: "Bed Continuity and Permeability Variations of Recent Deltaic Sediments," *Bull.*, AAPG (1976) 681.
104. Howard, J.D.: "Patterns of Sediment Dispersal in the Fountain Formation of Colorado," *Mtn. Geologist* (1966) 147-153.
105. Howe, H.V. and Moresi, C.K.: "Geology of Iberia Parish," *Bull. 1*, Louisiana Geol. Survey (1931).
106. Howe, H.V. and Moresi, C.K.: "Geology of Lafayette and St. Martin Parishes," *Bull. 3*, Louisiana Geol. Survey (1933).
107. Howe, H.V., Russell, R.J., McGuirt, J.H., Craft, B.C., and Stephenson, M.B.: "Reports on the Geology of Cameron and Vermilion Parishes," *Bull. 6*, Louisiana Geol. Survey (1935).
108. Hoyt, W.V.: "Erosional Channel in the Middle Wilcox Near Yoakum, Lavaca County, Texas," *Trans.* Gulf Coast Assn. Geol. Socs. (1959) 41-50.
109. Hrabar, S.V. and Potter, P.E.: "Lower West Badeu (Mississippian) Sandstone Body of Owen and Green Counties, Indiana," *Bull.* AAPG (1969) 2150-2160.
110. Hubbard, J.A.E.B.: "Facies Patterns in the Carrowmoran Sandstone (Visean) of Western County Sligo, Ireland," *Proc.*, Geol. Assn. (1967) 233-254.
111. Hubert, C., Lajoie, J., and Leonard, M.A.: "Deep-Sea Sediments in the Lower Paleozoic Quebec Supergroup," Spec. Paper 7, Geol. Assn. Canada (1970) 103-125.
112. Hubert, J.F.: "Petroloyg of the Fountain and Lyons Formations, Front Range, Colorado," *Quarterly*, Colorado School of Mines (1960) 1-242.
113. Humphreys, M. and Friedman, G.M.: "Late Devonian Catskill Deltaic Complex in North- Central Pennsylvania," *Deltas, Models for Exploration*, Houston Geol. Soc. (1975) 370-379.
114. Jacka, A.D.: "Depositional Dynamics of the Almond Formation, Rock Springs Uplift, Wyoming," Wyoming Geol. Assn. 19th Field Conf. Guidcbook (1965) 81-100.
115. Johnson, W.A.: "The Character of the Stratification of the Sediments of the Recent Delta of the Fraser River, British Columbia, Canada," *J. Geol.* (1922) 115- 129.
116. Kelling, G.: "Patterns of Sedimentation in Rhondda Beds of South Wales," *Bull.*, AAPG (1968) 2369-2386.
117. Kelling, G. and Woollands, M.A.: "The Stratigraphy and Sedimentation of the Llandoverian Rocks of the Rhoyader Dis-

trict," *The Precambrian and Lower Paleozoic Rocks of Wales*, U. of Wales Press (1969) 255-282.
118. Klein, G. deV.: "Triassic Sedimentation, Maritime Provinces, Canda," *Bull.*, GSA (1961) 1127-1146.
119. Klein, G. deV.: "Sedimentology of Triassic Rocks in the Lower Connecticut Valley," *Guidebook for Field Trips in Connecticut*, New England Int'l. Collegiate Geol. Conf., Connecticut Geol. and Nat. History Survey Guidebook 2. (1968) 1-19.
120. Knight, W.V.: "Historical and Economic Geology of Silurian Clinton Sandstone of Northwestern Ohio," *Bull.*, AAPG (1969) 1421-1452.
121. Kolb, C.R., Steinriede, W.B., Jr., Krinitzsky, E.L., Saucier, R.T., Mabrey, P.R., Smith, FL.L., and Fleetwood, A.R.: "Geological Investigation of the Yazoo Basin, Lower Mississippi Valley," Waterways Experimental Station Report 3-480, U.S. Army Corps Engineers (1968).
122. Kruit, C., Brouwer, J., and Ealey, P.: "A Deep-Water Sand Fan in the Eocene Bay of Biscay," *Nature Phy. Sci.* (1972) 59-61.
123. Kukal, Z.: *Geology of Recent Sediments*, Academic Press, New York (1971).
124. Laming, D.J.C.: "Imbrication, Paleo-Currents, and Other Sedimentary Features in the Lower New Red Sandstone, Devonshire, England," *J. Sed. Petrology* (1966) 940-959.
125. Laury, R.L.: "Sedimentology of the Pleasantview Sandstone, Southern Iowa and Western Illinois," *J. Sed. Petrology* (1968) 568-599.
126. LeBlanc, R.J.: "Geometry of Sandstone Reservoirs Bodies," *Underground Waste Management and Environmental Implications*, Mem. 18, AAPG (1972) 133-190.
127. Le Blanc, R.J.: "Significant Studies of Modern and Ancient Deltaic Sediments," *Deltas, Models for Exploration*, Houston Geol. Soc. (1975) 13-85.
128. Le Blanc, R.J. and Bernard, H.A.: "Resume of Late Recent Geological History of the Gulf Coast," *Geologie en Mijnbouw* (1954) 185-194.
129. LeBlanc, R.J. and Hodgson, W.D.: "Origin and Development of the Texas Shoreline," *Proc.*, Louisiana State U. Coastal Studies Inst., Second Coastal Geog. Conf. (1959) 57-101.
130. Le Blanc, R.J., Jr.: "Environments of Deposition of the Yegua Formation (Eocene), Brazos County, Texas," *Bull.*, Corpus Christi Geol. Soc. (1971).
131. Lené, G.W., and Owen, D.E.: "Grain Orientation in a Berea Sandstone Channel at South Amherst, Ohio," *J. Sed. Petrology*, (1969) 737-743.
132. Lineback, J.A.: "Turbidites and Other Sandstone Bodies in the Borden Siltstone (Mississippian) in Illinois," Circ. 425, Illinois Geol. Survey (1968) 1-29.
133. Lins, T.W.: "Origin and Environment of the Tonganoxie Sandstone in Northeastern Kansas," *Bull.*, Kansas Geol. Survey (1950) 107-140.
134. Lucke, J.B.: "A Study of Barnegat Inlet, New Jersey, and Related Shoreline Phenomena," *Shore and Beach* (1934) 45-93.
135. Lumsden, D.N. and Pelletier, B.R.: "Petrology of the Grimsby Sandstone (Lower Silurian) of Ontario and New York," *J. Sed. Petrology* (1969) 521-530.
136. MacNeil, F.S.: "Pleistocene Shore Lines in Florida and Georgia," Prof. Paper 221-F, USGS (1950).
137. Maldonado, A. and Stanley, D.J.: "The Nile Cone: Submarine Fan Development by Cyclic Sedimentation," *Marine Geol.* (1976) 27-40.
138. Mann, C.J. and Thomas, W.A.: "Ancient Mississippi River," *Trans.*, Gulf Coast Assn. Geol. Soc. (1968) 187-204.
139. Manos, C.: "Depositional Environment of Sparland Cyclothem (Pennsylvanian), Illinois and Forest City Basins," *Bull.*, AAPG (1967) 1843-1861.
140. Martin, B. D.: "Rosedale Channel—Evidence for late Miocene Submarine Erosion in Great Valley of California," *Bull.*, AAPG (1963) 441-456.
141. Martini, I.P.: "Regional Analysis of Sedimentology of Medina Formation (Silurian), Ontario and New York," *Bull.*, AAPG (1971) 1249-1261.
142. McBride, E.F., Weidie, A.E., and Wollenben, J.A.: "Deltaic and Associated Deposits of Difunta Group (Late Cretaceous to Paleocene), Parras and La Popa Basins, Northeastern Mexico," *Deltas, Models for Exploration*, Houston Geol. Soc. (1975) 485-522.
143. McCave, I.N.: "Correlation Using a Sedimentological Model of Part of the Hamilton Group (Middle Devonian), New York State," *Am. J. Sci.* (1969) 567-591.
144. McGowen, J.H. and Garner, L.E.: "Comparison of Recent and Ancient Coarse-Grained Point Pars," (abs.) *Bull.*, AAPG (1969) 731-732.
145. McGowen, J.H. and Garner, L.E.: "Physiographic Features and Stratification Types of Coarse-Grained Point Bars: Modern and Ancient Examples," Circ. 75-9, U. of Texas, Bureau of Econ. Geol., Austin (1975).
146. McGowen, J.H. and Groat, C.G.: "Van Horn Sandstone, West Texas—An alluvial Fan Model for Mineral Exploration," U. of Texas, Bureau of Econ. Geol., Austin (1971).
147. McKee, E.D.: "The Coconino Sandstone—Its History and Origin," Pub. 440, Carnegie Inst., Contr. Paleontology, Washington (1934) 77-115.
148. Meckel, L.D.: "Holocene Sand Bodies in the Colorado Delta Area, Northern Gulf of California," *Deltas, Models for Exploration*, Houston Geol. Soc. (1975) 239-265.
149. Melton, M.A.: "The Geomorphic and Paleoclimatic Significance of Alluvial Deposits in Souther Arizona," *J. Geol.* (1965) 1-38.
150. Miall, A.D.: "Continental Marine Transitin in the Devonian of Prince of Wales Island, Northwest Territories," *Cdn. J. Earth Sci.* (1970) 125-144.
151. Miller, D.N.: "Patterns of Barrier Bar Sedimentation and Its Similarity to Lower Cretaceous Fall River Stratigraphy," *Symposium on Early Cretaceous Rocks*, Wyoming Geol. Assn. 17th Field Conf. Guidebook (1962) 232-247.
152. Moody-Stuart, M.: "High and Low Sinuosity Steam Deposits, With Examples From the Devonian of Spitsbergen," *J. Sed. Petrology* (1966) 1102-1117.
153. Moore, G.T. and Fullam, T.J.: "Submarine Channel Systems and Their Potential for Petroleum Localization," *Deltas, Models for Exploration*, Houston Geol. Soc. (1975) 165-189.
154. Morgan, J.P.: "Mudlumps at the Mouth of the Mississippi River," *Genesis and Paleontology of the Mississippi River Mudlumps, Bull. 35*, Louisiana Geol. Survey (1961).
155. Mutch, T.A.: "Pennsylvanian Non-Marine Sediments of the Narragansett Basin, Massachusetts and Rhode Island," Spec. Paper 106, GSA (1968) 177-209.
156. Mutti, E. and Ricci-Lucchi, F.: "Le Torbiditi dell' Appennino Settentrionale: Introduzione all'Analisi di Facies," *Memorie della Societa Geologica Italiana* (1972) 161-199.
157. Nagtegaal., P.J.C.: "Scour-and-Fill Structures From a Fluvial Piedmont Environment," *Geologie en Mijnbouw* (1966) 342-354.
158. Nanz, R.H.: "Genesis of Oligocene Sandstone Reservoir, Seeligson Field, Jim Wells and Kleberg Counties, Texas," *Bull.*, AAPG (1954) 96-117.
159. Nelson, C.H. and Nilsen, T.H.: "Depositional Trends of Modern and Ancient Deep-Sea Fans," *Modern and Ancient Geosynclinal Sedimentation*, Spec. Pub. 19, SEPM (1974) 69-91.
160. Nelson, H.: "Late Pleistocene and Holocene Depositoinal Trends, Processes, and History of Astoria Deep-Sea Fan, Northeast Pacific," *Marine Geol* (1976) 129-173.
161. Nilsen, T.H.: "Old Red Sedimentation in the Buelandet-Vaerlandet Devonian District, Western Norway," *Sed. Geol.* (1969).
162. Normark, W.R. and Piper, D.J.W.: "Deep-Sea Fan Valleys, Past and Present," *Bull.*, GSA (1969) 1859-1866.
163. Pack, F.J.: "Torrential Potential of Desert Waters," *Pan-Am. Geologist* (1923) 349-356.
164. Paine, W.R.: "Stratigraphy and Sedimentation of Hackberry Shale and Associated Beds of Southwestern Louisiana," *Trans.*, Gulf Coast Assn. Geol. Socs. (1966).
165. Parsons, M.G.: "The Geology of the Laurentian Fan and the Scotia Rise," Mem. 4, Cdn. Soc. Pet. Geol. (1975) 155-167.
166. Paull, R.A.: "Depositional History of the Muddy Sandstone, Bighorn Basin, Wyoming," Wyoming Geol. Assn. 17th Field Conf. Guidebook (1962) 102-117.
167. Pettyjohn, W.A.: "New Members of Upper Cretaceous Fox

Hills Formation in South Dakota, Representing Delta Deposits," *Bull.*, AAPG (1967) 1361-1367.
168. Piper, D.J.W. and Normark, W.R.: "Re-Examination of a Miocene Deep-Sea Fan and Fan Valley, Southern California," *Bull.*, GSA (1971) 1823-1830.
169. Pryor, W.A.: "Cretaceous Sedimentation in Upper Mississippi Embayment," *Bull.*, AAPG (1960) 1473-1504.
170. Pryor, W.A.: "Sand Trends and Paleoslope in Illinois Basin and Mississippi Embayment," *Geometry of Sandstone Bodies*, AAPG (1961) 119-133.
171. Revelle, R. and Shepard, F.P.: "Sediments Off the California Coast," *Recent Marine Sediments*, AAPG (1939 245-282.
172. Ricci-Lucchi, F.: "Depositional Cycles in Two Turbidit Formations of Northern Apennines (Italy)," *J. Sed. Petrology* (1975) 3-43.
173. Royse, C.F., Jr.: "A Sedimentologic Analysis of the Tongue River-Sentinel Butte Interval (Paleocene) of the Williston Basin, Western North Dakota," *Sed. Geol.* (1970) 19-80.
174. Rusnak, G.A.: "A Fabric and Petrographic Study of the Pleasantview Sandstone," *J. Sed. Petrology* (1957) 41-55.
175. Russell, R.J.: "Physiography of Lower Mississippi River Delta," *Lower Mississippi River Delta, Bull. 8*, Louisiana Geol. Survey (1936) 1-199.
176. Ryan, D.J.: "Cross-Bedding Formed by Lateral Accretation in the Catskill Formation Near Jim Thorpe, Pennsylvania," *Proc.*, Pennsylvania Acad. Sci. (1965) 154- 156.
177. Sabate, R.W.: "Pleistocene Oil and Gas in Coastal Louisiana," *Trans.*, Gulf Coast Assn. Geol. Soc. (1968)) 373-386.
178. Sabins, F.F., Jr.: "Anatomy of Stratigraphic Trap, Bisti Field, New Mexico," *Bull.*, AAPG (1963) 193- 228.
179. Saucier, R.T.: "Geological Investigation of the St. Francis Basin," Tech. Report 3-659, Waterways Experimental Station, U.S. Army Corps of Engineers (1964).
180. Saucier, R.T.: "Geological Investigation of the Boeuf-Tensas Basin, Lower Mississippi Valley," Tech. Report 3-757, Waterways Experimental Station, U.S. Corps of Engineers (1967).
181. Schlee, J.S. and Moench, R.H.: "Properties and Genesis of 'Jackpile' Sandstone, Laguna, New Mexico," *Geometry of Sandstone Bodies*, AAPG (1961) 134-150.
182. Scott, A.J. and Fisher, W.L.: "Delta Systems and Deltaic Deposition," *Delta Systems in Exploration for Oil and Gas*, U. of Texas, Bureau of Econ. Geol., Austin (1969) 10-39.
183. Scruton, P.C.: "Delta Building and the Deltaic Sequence," *Recent Sediments, Northwest Gulf of Mexico*, AAPG (1960) 82-102.
184. Scruton, P.C.: "Rocky Mountain Cretaceous Stratigraphy and Regressive Sandstones," Wyoming Geol. Assn. 16th Field Conf. Guidebook (1961) 241-249.
185. Selley, R.C.: "Subsurface Environmental Analysis of North Sea Sediments," *Bull.*, AAPG (1976) 184-195.
186. Shannon, J.P., Jr., and Dahl, A.R.: "Deltaic Stratigraphic Traps in West Tuscola Field, Taylor County, Texas," *Bull.*, AAPG (1971) 1194-1205.
187. Shawa, M.S.: "Sedimentary History of the Gilwood Sandstone (Devonian), Utikuma Lake Area, Alberta, Canada," *Bull.*, Cdn. Pet. Geol. (1969) 392-409.
188. Shelton, J.W.: "Trend and Genesis of Lower-Most Sandstone Unit of Eagle Sandstone at Billings, Montana," *Bull.*, AAPG (1965) 1385-1397.
189. Shelton, J.W.: "Stratigraphic Models and General Criteria for Recognition of Alluvial, Barrier-Bar, and Turbidity-Current Sand Deposits," *Bull.*, AAPG (1967) 2441-2461; also in Billings Geol. Soc. 17th Field Conf. Guidebook, 1-17.
190. Shelton, J.W.: "Models of Sand and Sandstone Deposits," *Bull. 118*, Oklahoma Geol. Survey (1973).
191. Shepard, F.P., Phleger, F.B., and van Andel, T.H., eds.: *Recent Sediments, Northwest Gulf of Mexico*, AAPG (1960) 394 pp.
192. Shepheard, W.W. and Hills, L.V.: "Depositional Environments, Bearpaw-Horseshoe Canyon (Upper Cretaceous) Transition Zone, Drumheller 'Badlands,' Alberta," *Bull.*, Cdn. Pet. Geol. (1970) 166-215.
193. Shideler, G.L.: "Dispersal Patterns of Pennsylvanian Sandstones in the Michigan Basin," *J. Sed. Petrology* (1969) 1229-1237.
194. Sims. J.D.: "Geology and Sedimentology of the Livingston Group, Northern Crazy Mountains, Montana," PhD dissertation, Northwestern U., Evanston, Ill. (1967).
195. Sloan, B.J., Jr.: "The Subsurface Jurassic Bodcaw Sand in Louisiana," *Bull. 33*, Louisiana Geological Survey (1958).
196. Smith, N.D.: "The Braided Stream Depositional Environment: Comparison of the Platte River With Some Silurian Clastic Rocks, North Central Appalachians," *Bull.*, GSA (1970) 2993-3014.
197. Snavely, P.D., Jr., Wagner, H.C. and MacLeod, N.S.: "Rhythmic-Bedded Eugeosynclinal Deposits of the Tyee Formation, Oregon Coast Range," *Symposium on Cyclic Sedimentation, Bull. 169*, Kansas Geol. Survey (1964) 461-480.
198. Spearing, D.R.: "Stratigraphy, Sedimentation and Tectonic History ofthe Paleocene-Eocene Hoback Formation of Western Wyoming," PhD dissertation, U. of Michigan, Ann Arbor (1969).
199. Stanley, D.J.: "Comparing Patterns of Sedimentation in Some Modern and Ancient Submarine Canyons," *Earth and Planetary Sci. Letters* (1967) 371-380.
200. Stanley, D.J.: "Graded Bedding-Sole Marking-Graywacke Assemblage and Related Sedimentary Structures in Some Carboniferous Flood Deposits, Eastern Massachusetts," Spec. Paper 106, GSA (1968) 211-239.
201. Stanley, D.J.: "Submarine Channel Deposits and Their Fossil Analogs ('Fluxoturbidites')," *New Concepts of Continental Margin Sedimentation*, Am. Geol. Inst. Short Course Notes, Nov. 7-9, 1969, DJS-9-1 to DJS-9-17.
202. Steidtmann, J.R.: "Evidence for Eolian Origin of Cross-Stratification in Sandstone of Casper Formation, Southernmost Laramie Basin, Wyo.," *Bull.*, GSA (1974) 1835-1842.
203. Stetson, H.C.: "The Sediments of the North-Western Gulf of Mexico—Part 1. The Continental Terrace of the Western Gulf of Mexico: Its Surface Sediments, Origin and Development," *Papers in Physical Oceanography and Meteorology*, Abstract these, Massachusetts Institute of Technology, Cambridge (1953) 5-45.
204. Stokes, W.L.: "Fluvial and Eolian Sandstone Bodies in Colorado Plateau," *Geometry of Sandstone Bodies*, AAPG (1961) 151-178.
205. Stokes, W.L.: "Eolian Varving in the Colorado Plateau," *J. Sed. Petrology* (1964) 429-433.
206. Stokes, W.L.: "Multiple Parallel-Truncation Bedding Planes—A Feature of Wind-Deposited Sandstone Formation," *J. Sed. Petrology* (1968) 510-515.
207. Sullwold, H.H., Jr.: "Tarzana Fan, Deep Submarine Fan of Late Miocene Age, Los Angeles County, California," *Bull.*, AAPG (1960) 433-457.
208. Swann, D.H.: "Late Mississippian Rhythmic Sediments of Mississippi Valley," *Bull.*, AAPG (1964) 637-658.
209. Swann, D.H., Lineback, J.A., and Frund, E.: "The Borden Siltstone (Mississippian) Delta in Southwestern Illinois," Circ. 386, Illinois Geol. Survey (1965).
210. Sykes, G.: "The Colorado Delta," Pub. 460, Carnegie Inst., Washington (1937).
211. Taylor, J.H.: "Sedimentary Features of an Ancient Deltaic Complex: The Wealden Rocks of Southeastern England," *Sedimentology* (1963) 2-28.
212. Thachuk, N.M.: "Geological Study of the Middle Devonian Gilwood Arkoses int he Nipisi Area, Alberta," Preprint No. 6823 presented at Petroleum Soc. of CIM, 19th Annual Technical Meeting, 1968.
213. Thomas, W.A. and Mann, C.J.: "Late Jurassic Depositoinal Environments, Louisiana and Arkansas," *Bull.*, AAPG (1966) 178-182.
214. Thomson, A. and Le Blanc, R.J.: "Carboniferous Deep-Sea Fan Facies of Arkansas and Oklahoma," paper presented at GSA Annual Meeting (1975).
215. Todd, T.W. and Monroe, W.A.: "Petrology of Domengine Formation (Eocene) at Potrero Hills and Rio Vista, California," *J. Sed. Petrology* (1968) 1024- 1039.
216. Trask, P.D. *et al.*: "Recent Marine Sediments—A Symposium," Spec. Pub. 4, Soc. Econ. Paleontologists and Mineralogists (1955).

217. Van de Kamp, P.C., Harper, J.D., Conniff, J.J., and Morris, D.A.: "Facies Relations in the Eocene-Oligocene in the Santa Ynez Mountains, California," *J. Geol. Soc.*, London (1974) 545-565.
218. van Straaten, L.M.J.U.: "Composition and Structure of Recent Marine Sediments in The Netherlands," *Leidsche Geol. Meded.* (1954) 1-110.
219. van Straaten, L.M.J.U.: "Sedimentology of Recent Tidal Flat Deposits and the Psammites du Condroz (Devonian)," *Geologie en Mijnbouw* (1954) 25- 47.
220. Visher, G.S.: "Use of Vertical Profile in Environmental Reconstruction," *Bull.*, AAPG (1965) 41- 61.
221. Visher, G.S.: "Fluvial Processes as Interpreted From Ancient and Recent Fluvial Deposits," *Primary Sedimentary Structures and Their Hydrodynamic Interpretation*, Spec. Pub. 12, Soc. Econ. Paleontologists and Mineralogists (1965) 116- 131.
222. Visher, G.S. *et al.*: "Pennsylvanian Delta Patterns and Petroleum Occurrences in Eastern Oklahoma," *Bull.*, AAPG (1971) 1206-1230.
223. Visher, G.S. *et al.*: "The Coffeeville Format (Pennsylvanian) of Northern Oklahoma, a Model for Epeiric Sea Delta," *Deltas, Models for Exploration*, Houston Geol. Soc. (1975) 382-397.
224. Walker, R.G.: "Deep Channels in Turbidite-Bearing Formations," *Bull.*, AAPG (1966) 1899-1917.
225. Walker, R.G. and Mutti, E.: "Turbidite Facies and Facies Associations," *Turbidites and Deep Water Sedimentation*, SEPM Short Course, Anaheim, Calif. (1973) 1-38.
226. Walls, R.A.: "Late Devonian-Early Mississippian Subaqueous Deltaic Facies in a Portion of the Southeastern Appalachian Basin," *Deltas, Models for Exploration*, Houston Geol. Soc. (1975) 359-367.
227. Wanless, H.R., Baroffio, J.R., Gamble, J.C., Horne, J.C., Orlopp, D.R., Rocha-Campos, A., Souter, J.E., Trescott, P.C., Vail, R.S., and Wright, C.R.: "Late Paleozoic Deltas in the Central and Eastern United States," *Deltaic Sedimentation—Modern and Ancient*, Spec. Pub. 15, Soc. Econ. Paleontologists and Mineralogists (1970) 215-245.
228. Way, J.H., Jr.: "Bed Thickness Analysis of Some Carboniferous Fluvial Sedimentary Rocks Near Joggins, NOva Scotia," *J. Sed. Petrology* (1968) 424-433.
229. Weimer, R.J.: "Spatial Dimensions of Upper Cretaceous Sandstones, Rocky Mountain Area," *Geometry of Sandstone Bodies,* AAPG (1961).
230. Weimer, R.J.: "Upper Cretaceous Delta on Tectonic Foreland, Northern Colorado and Southern Wyoming," (abs.) *Bull.*, AAPG (1961) 417.
231. Weimer, R.J.: "Late Cretaceous Deltas, Rocky Mountain Region," (abs.) *Bull.*, AAPG (1965) 363.
232. Welder, F.A.: "Processes of Deltaic Sedimentation in the Lower Mississippi River," Tech. Report 12, Louisiana State U. Coastal Studies Inst., Baton Rouge (1959) 1-90.
233. Wermund, E.G. and Jenkins, W.A., Jr.: "Recognition of Deltas by Fitting Trend Surfaces to Upper Pennsylvanian Sandstones in North-Central Texas," *Deltaic Sedimentation—Modern and Ancient*, Spec. Pub. 15, Soc. Econ. Paleontologists and Mineralogists (1970) 156-269.
234. Wessel, J.M.: "Sedimentary History of Upper Triassic Alluvial Fan Complexes in North-Central Massachusetts," U. of Massachusetts, Dept. of Geol., Contr. No. 2 (1969).
235. Whitaker, J.H.M.: "Ancient Submarine Canyons and Fan Valleys," Spec. Pub. 19, SEPM (1974) 106-125.
236. Wier, C.E. and Girdley, W.A.: "Distribution of the Inglefield and Dicksburg Hills Sandstone Members in Posey and Vanderburgh Counties, Indiana," *Proc.*, Indiana Acad. Sci. (1963) 212- 217.
237. Williams, G.E.: "Paleogeography of the Torridonian Applecross Group," *Nature* (1966) 1303-1306.
238. Williams, G.E.: "Characteristics and Origin of a Pre-Cambrian Pediment," *J. Geol.* (1969) 183-207.
239. Wilson, M.D.: "The Stratigraphy of the Beaverhead Group in the Lima Area, Southwestern Montana," PhD dissertation, Northwestern U., Evanston, Ill. (1967).
240. Wilson, M.D.: "Upper Cretaceous- Paleocene Synorogenic Conglomerates of Southwestern Montana," *Bull.*, AAPG (1970) 1843-1867.
241. Wolff, M.P.: "Deltaic Sedimentation of the Middle Devonian Marcellus Formation in Southeastern New York," PhD dissertation, Cornell U., Ithaca, N.Y. (1967).
242. Bernard, H.A., Major, C.F., Jr., Parrott, B.S., and Le Blanc, R.J., Sr.: "Recent Sediments of Southeast Texas," U. of Texas, Bureau of Econ. Geol., Austin (July 1970).
243. Bull, W.B.: "Recognition of Alluvial-Fan Deposits in the Stratigraphic Record," *Recognition of Ancient Sedimentary Environments*, Spec. Pub. 16, Soc. Econ. Paleontologists and Mineralogists (1972).
244. Ferm, J.C.: "Allegheny Deltaic Deposits," *Deltaic Sedimentation— Modern and Ancient*, Spec. Pub. 15, Soc. Econ. Paleontologists and Mineralogists (1970).
245. Rainwater, E.H.: "Transgressions and Regressions in the Gulf Coast Tertiary," *Trans.*, Gulf Coast Assn. Geol. Socs. (1964) 217-230.
246. Land, C.B., Jr.: "Stratigraphy of Fox Hills Sandstone and Associated Formations, Rock Spring Uplift and Wamsutter Arch Area, Sweetwater County, Wyoming: A Shoreline Estuary Sandstone Model for the Late Cretaceous," *Quarterly*, Colorado School of Miens (1972) 69.
247. Weimer, R.J.: "A Guide to Uppermost Cretaceous Stratigraphy, Central Front Range, Colorado: Deltaic Sedimentation, Growth Faulting and Early Laramide Crustal Movement," *Mtn. Geologist* (1973) 53-97.
248. Weimer, R.J. and Land, C.B., Jr.: "First Guide to Dakota Group (Cretaceous) Stratigraphy, Golden-Morrison Area, Colorado," *Mtn. Geologist* (1972) 241-267.

Paper (SPE 6137B) was accepted for publication April 26, 1977. Paper was first presented at the SPE-AIME 51st Annual Fall Technical Conference and Exhibition, held in New Orleans, Oct. 3-6, 1976.

The American Association of Petroleum Geologists Bulletin
Vol. 53, No. 1 (January, 1969), P. 30-54, 12 Figs.

Depositional Systems in Wilcox Group (Eocene) of Texas and Their Relation to Occurrence of Oil and Gas[1]

W. L. FISHER[2] and J. H. McGOWEN[2]
Austin, Texas 78712

Abstract Regional investigation of the lower part of the Wilcox Group in Texas in outcrop and subsurface indicates seven principal depositional systems: (1) Mt. Pleasant fluvial system developed updip and in outcrop north of the Colorado River; (2) Rockdale delta system, present primarily in the subsurface, chiefly between the Guadalupe and Sabine Rivers; (3) Pendleton lagoon-bay system in outcrop and subsurface largely on the southern flank of the Sabine uplift; (4) San Marcos strandplain-bay system in outcrop and subsurface mainly on the San Marcos arch; (5) Cotulla barrier-bar system in the subsurface of South Texas; (6) Indio bay-lagoon system developed updip and in outcrop in South Texas; and (7) South Texas shelf system, an extensive system entirely within the subsurface. The Rockdale delta system consisting of large lobate wedges of mudstone, sandstone, and carbonaceous deposits, is the thickest and most extensive of the lower Wilcox depositional systems. It grades updip to the thinner terrigenous facies of the Mt. Pleasant fluvial system. Deposits of the Rockdale delta system were the source of sediments redistributed by marine processes and deposited in laterally adjacent marine systems. Delineation of depositional systems and, more specifically, delineation of component facies of the several systems permit establishment of several regional oil and gas trends which show the relation of producing fields and distribution of potentially productive trends.

Introduction

The Wilcox Group (early Eocene) is a thick sequence of predominantly terrigenous clastic sedimentary rocks which is volumetrically a significant part of the large terrigenous fill of the western Gulf Coast province. It is an economically important group of rocks, providing oil, gas, and freshwater reservoirs, as well as deposits of lignite, ceramic clay, and industrial sand.

The study area is approximately 40,000 sq mi in the Texas Gulf coastal plain, extending from the Sabine River on the east to the Rio Grande on the south in a 100- to 300-mi-wide belt including outcrop and subsurface. Subsurface control consisted of approximately 2,500 electric logs, supplemented by lithologic control based on outcrop descriptions, and cuttings and cores of strategically located wells. Fifty stratigraphic dip sections and 15 strike sections were prepared at spacings of 8 to 10 mi to determine three-dimensional geometry and relations of principal lithologic units. Emphasis in outcrop study was on the mapping of principal rock units, interpretation of primary sedimentary structures, and determination of fossil composition.

This paper is part of a continuing, larger regional study of the Wilcox Group in Texas which has been under way at the Bureau of Economic Geology for the past three years. Herein, consideration is restricted to the lower part of the Wilcox Group, including the sequence underlain by marine mudstone and claystone of the Midway Group and normally overlain by marine or lagoonal mudstone of the middle part of the Wilcox Group. The Carrizo Formation and the upper part of the Wilcox Group are not considered.

Terminology and Depositional Systems

The stratigraphic terminology used by the writers is informal. Reference is made to defined, formal stratigraphic units (chiefly in outcrop) where applicable. The terminology used derives from the concept of depositional systems or complexes, which has been employed recently, for example, in the study of modern

[1] Manuscript received and accepted, February 2, 1968. Parts of this paper were presented by W. L. Fisher to the Gulf Coast Association of Geological Societies, San Antonio, Texas, 1967. Present paper is republished with minor modifications from paper by Fisher and McGowen in Gulf Coast Assoc. Geol. Socs. Trans., v. 17, p. 105–125, 1967. This paper was nominated by the Gulf Coast Association of Geological Societies for republication in the AAPG *Bulletin;* publication authorized by the Director, Bureau of Economic Geology, The University of Texas at Austin, and by the Gulf Coast Association of Geological Societies.

[2] Bureau of Economic Geology, The University of Texas at Austin.

The writers thank M. J. Casey, L. F. Brown, Jr., and P. T. Flawn, Bureau of Economic Geology, and Alan J. Scott, Department of Geology, The University of Texas at Austin, for reading the original manuscript and offering many helpful suggestions. Gerald H. Baum, Texas Water Development Board, made available the extensive well-log library of that agency; numerous Gulf Coast operators provided records of specific wells. Peter U. Rodda and J. Stewart Nagle, Bureau of Economic Geology, The University of Texas at Austin, have been helpful in many ways during course of the project. Illustrations were prepared under direction of J. W. Macon, Bureau of Economic Geology, The University of Texas at Austin.

sediments and depositional environments of the South Texas coast (Hayes and Scott, 1964) and in reference to the Frio barrier bar system of South Texas (Boyd and Dyer, 1964). In studies of modern sediments emphasis generally is placed not only on description of specific environments and two-dimensional distribution of sediments and environments, but also on analysis and integration of facies; fabrication of facies tracts and three-dimensional depositional models of known depositional complexes or systems are used. These approaches are significant in the interpretation of ancient sediments, especially in regional analysis. The writers place more emphasis on lithologic and fossil composition, sedimentary structures, and distribution, geometry, and relations of component facies than on regional correlation and persistence of locally established formal stratigraphic units.

The principal unit here employed is the depositional system, recognized by specific criteria and designated by a genetic term, *e.g.*, delta system, fluvial system, barrier-bar system, *etc.* Components of a depositional system are referred to simply as facies, *e.g.*, the delta-front facies or distributary-channel facies of a delta system. Smaller scale components are designated by such terms as units, beds, or deposits. Basically, the concept of depositional systems in ancient sediments, fabricated principally from integration of facies, involves recognition of large-scale natural genetic elements, comparable to modern depositional systems readily apparent from physiographic characteristics.

Principal Depositional Systems

Seven principal depositional systems are recognized in the lower part of the Wilcox Group of Texas (Fig. 1). The dominant depositional element is the large Rockdale delta system, which comprises large lobate wedges of mudstone, sandstone, and carbonaceous deposits, chiefly lignite (Fig. 1). Maximum lower Wilcox sedimentation, with sequences as much as 5,000 ft thick, took place during deposition of this system. It is present principally in the subsurface and grades up the depositional slope and up the structural dip to thinner terrigenous clastic facies of the Mt. Pleasant fluvial system, present in outcrop north of the Colorado River. A large and probably main part of the original fluvial system was deposited updip of the present outcrop, and only a part of the system is preserved in the East Texas basin.

Southwestward transport of sand from the delta system, principally by longshore currents during periods of delta destruction, and the presence of local embayments marginal to the delta system resulted in distinct depositional systems laterally associated with the delta system. These systems include, on the east, the Pendleton lagoon-bay system developed in embayments marginal to the delta system; on the southwest, the San Marcos strandplain-bay system developed proximal to the delta system; and the Cotulla barrier bar and complementary Indio lagoon-bay systems developed more distally from the main area of deltation. Contemporaneous sedimentation in systems marginal and lateral to the delta system was relatively slow, and the sequences generally are less than one-fourth the thickness of the delta system. Gulfward from the Cotulla barrier bar system and partly lateral to the Rockdale delta system is the South Texas shelf system, marked chiefly by mudstone and relatively minor sandstone.

Mt. Pleasant Fluvial System

The Mt. Pleasant fluvial system includes most of the lower part of the Wilcox Group in outcrop north of the Colorado River and in the northern half of the Sabine uplift (Figs. 1, 2). It extends down structural dip to a line running approximately from central Bastrop County through central Lee, northern Burleson, southern Robertson, central Leon, northern Houston, southern Cherokee, and central Nacogdoches Counties. South and downdip from this line the system grades into the Rockdale delta system (Fig. 3); laterally it grades into embayed marine facies developed marginally to the fluvial and delta systems. Three fairly distinct facies are preserved, recognition of which is based on (1) sandstone-body geometry, (2) sandstone-isolith pattern, (3) internal sedimentary structures, (4) vertical sequence in structures and textures, (5) grain size, and (6) relative amounts of channel and overbank deposits (Figs. 1–3). These facies reflect different fluvial environments.

Tributary-Channel Facies

In northernmost Wilcox outcrops, from central Van Zandt County northeast to Texarkana, and in outcrop in the northern part of the Sabine uplift, sandstone deposits are present in definite, separated areas (*e.g.*, Mt. Pleasant, Reilly Springs, Simms deposits); claystone and mudstone beds crop out in intervening areas. Isoliths of sandstone units outline elongate bodies with an overall dendritic tributary pat-

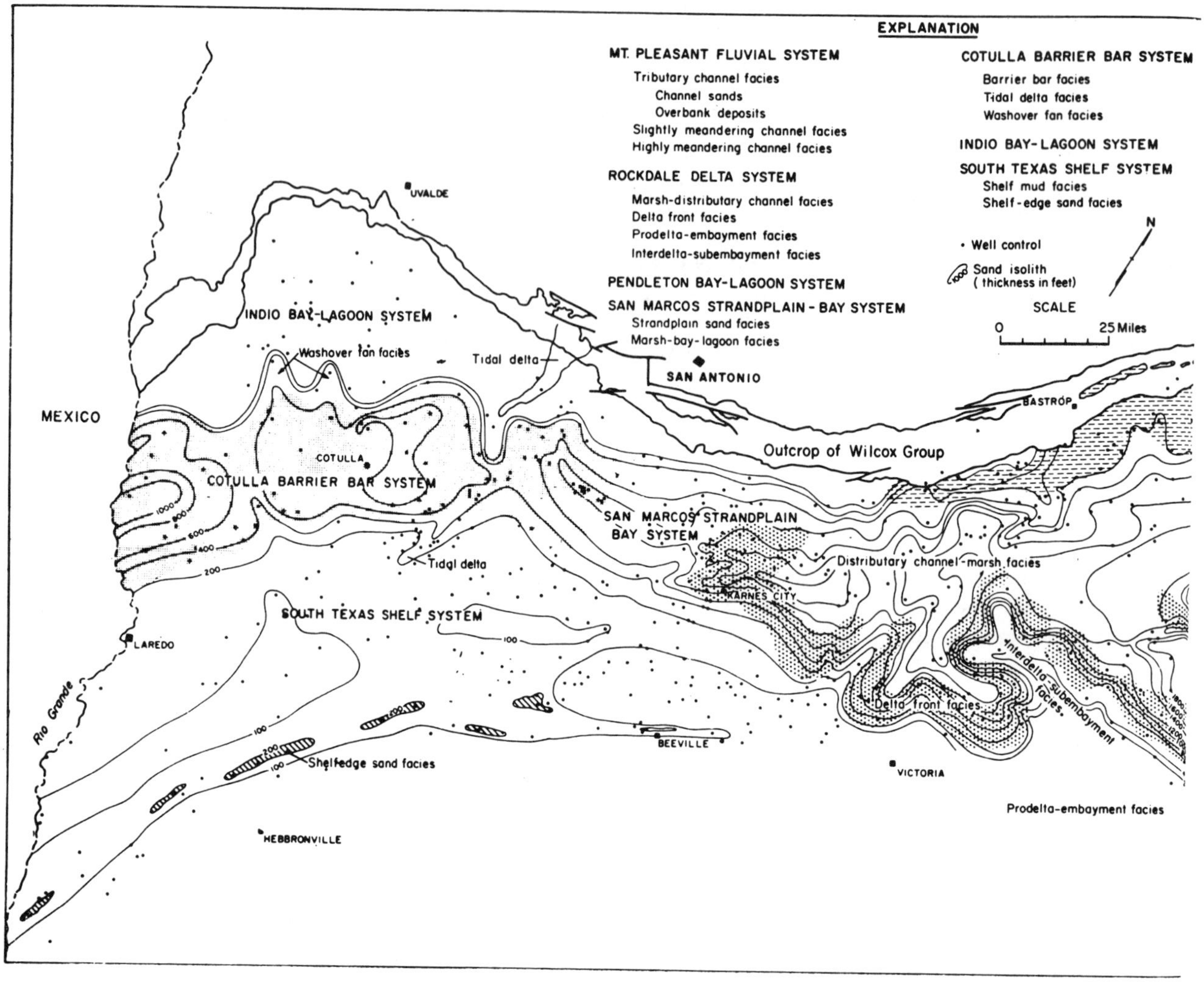

FIG. 1.—Principal depositional systems,

tern (Fig. 2). The fluvial facies defined chiefly by this pattern extends from the outcrop southward in the subsurface to a line from southwestern Smith County to west-central Rusk County. The thickness of the tributary-channel facies ranges from about 300 to 800 ft and commonly increases downdip and toward the axis of the East Texas basin; the facies consists of about 40 percent sandstone or channel deposits and about 60 percent mudstone or overbank deposits. The overbank (or nonchannel) mudstone beds are mostly laminated, light to moderate gray, and contain minor amounts of carbonaceous matter. Individual sandstone units range from 5 to about 80 ft in thickness and are channel deposits with flat tops and convex-downward bases. Bases of channels are characterized by channel lag of granules to fine pebbles, and a few angular clay pebbles. Sandstone in lower parts of the channel deposits contains trough or festoon cross-beds and in upper parts contains wedge sets of cross-beds. Upper parts of channel sandstone are horizontally bedded. Sandstone in this facies is mostly fine grained to locally medium grained; mean sand grain size averages about 0.14 mm, and range in mean is about 0.11 to 0.25 mm; channel sandstone typically becomes finer upward. Channels are cut into underlying and laterally adjacent overbank mudstone layers. Sinuous, elongate, narrow lignite deposits commonly are found laterally adjacent to channel sandstone deposits; they apparently accumulated as peat bogs in abandoned stream courses. Claystone beds within and adjacent to sandstone channel deposits are chiefly kaolinitic; feldspar is mostly weathered. Claystone of overbank deposits, by contrast, is largely nonkaolinitic—mostly illitic, chloritic, and montmorillonitic; feldspar of overbank mudstone deposits gener-

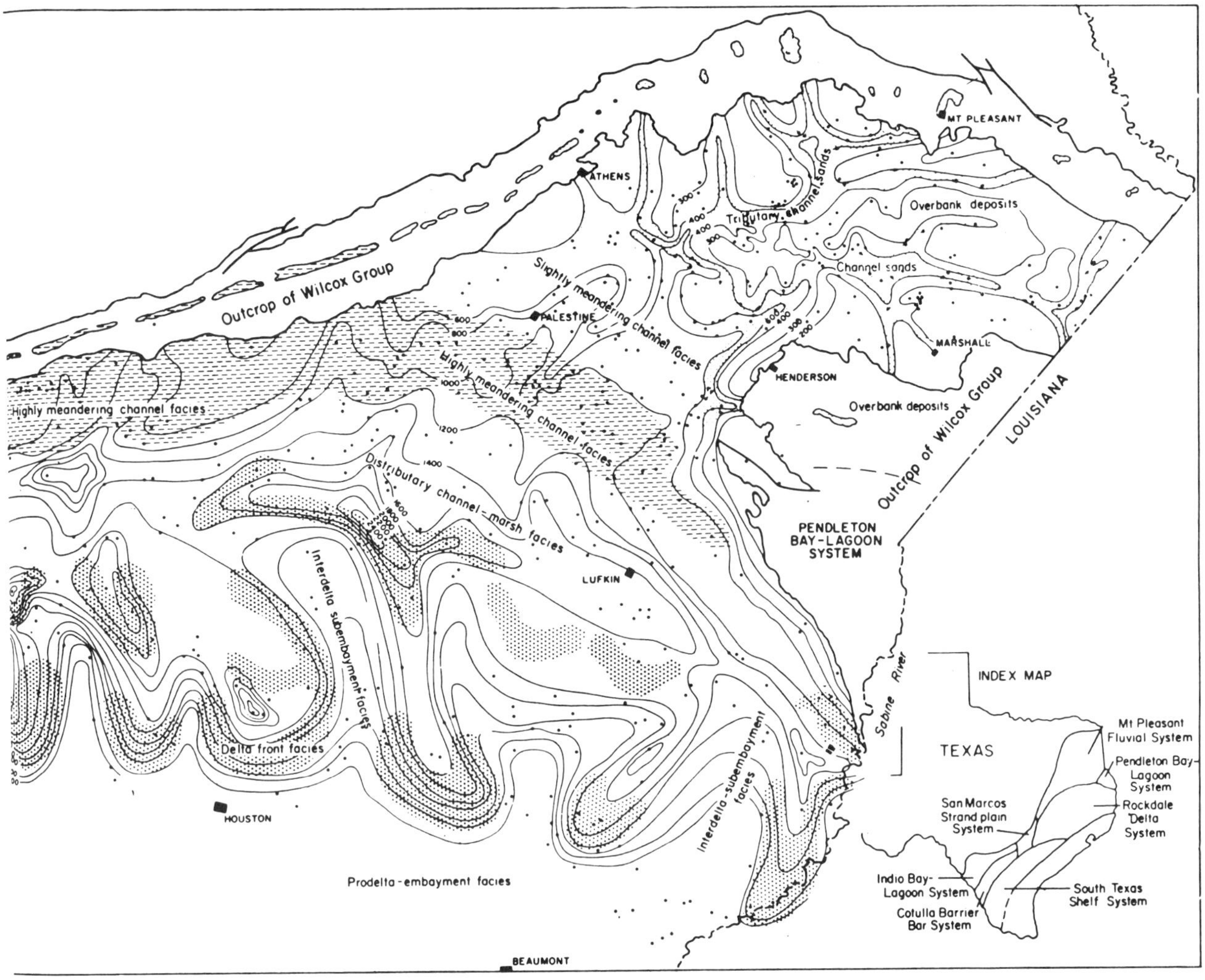

lower part of Wilcox Group (Eocene), Texas.

ally is unweathered. Ceramic clay plants and numerous abandoned lignite mines of East Texas are adjacent to the local Wilcox sandstone deposits because of the close association of kaolinitic clay and lignite to the channel deposits.

Slightly Meandering Channel Facies

Down the depositional slope in the fluvial system a transitional facies is characterized by uniform to slightly dendroid sandstone isolith patterns (Figs. 1, 2), which are less dendritic than those in the tributary-channel facies. This change in pattern probably represents a slightly greater meandering of the original stream system. Channel sandstone deposits are more numerous within this facies than in the tributary-channel facies; they make up about 60 percent of the total facies, and overbank deposits compose 40 percent. The facies ranges in thickness from about 500 to 800 ft; it crops out from central Freestone County, through western Henderson County, into southern Van Zandt County and extends downdip in the subsurface to a line from central Freestone County through central Anderson and central Cherokee Counties, and into northwestern Nacogdoches County. In addition to containing a higher percentage of channel deposits, individual sandstone bodies in this facies are slightly more extensive, and mean grain size is somewhat larger than that of the tributary-channel facies. Primary sedimentary structures of sandstone and relation of channels to clay-mineral distribution and lignite deposits are similar for the two facies.

Highly Meandering Channel Facies

A distinct facies is developed in the Mt. Pleasant fluvial system just downslope deposi-

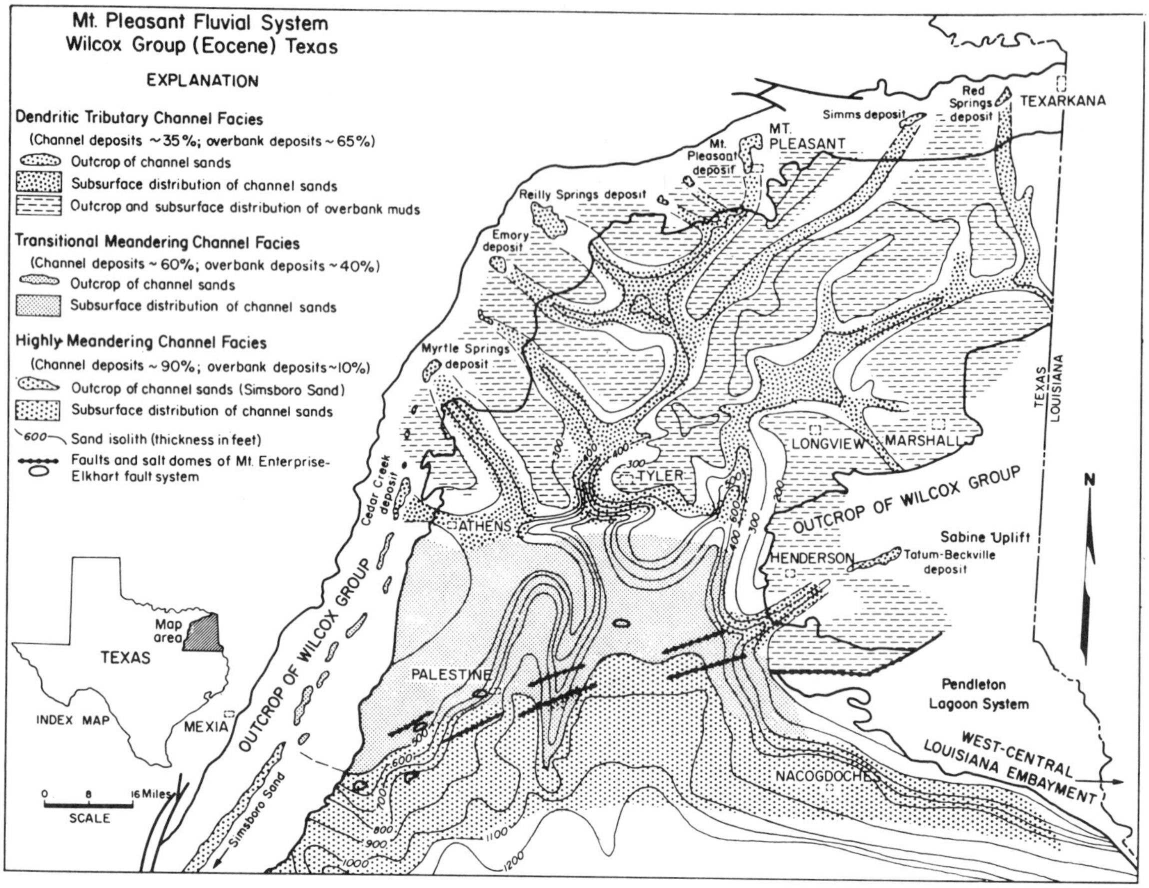

FIG. 2.—Outcrop and subsurface distribution, Mt. Pleasant fluvial system, Wilcox Group, Texas. For well control, see Figure 1.

tionally and downdip structurally from the Mt. Enterprise-Elkhart fault zone and associated salt domes on the southwest (Figs. 1, 2). Within this facies sandstone predominates and overbank mudstone composes less than 10 percent. Sandstone isolith lines generally conform and sandstone in the facies (*e.g.,* Simsboro sandstone outcrop) forms areally persistent units. For such units as these, Pettijohn *et al.* (1965) suggested the term multilateral. The bodies are made up of multiple channels which trend at various angles to the strike of the sandstone isolith lines. Aggregate thickness of sandstone in the facies ranges from 600 to 1,100 ft. Individual sandstone units, comprising numerous superposed and multilateral channel deposits, are as much as 30 mi wide. Areas along the strike between the more extensive sandstone units are made up of greater amounts of overbank mud deposits. Sand grains of the facies have a mean average size about 0.22 mm and are much coarser than sand grains in upslope fluvial facies. Sandstone channel deposits that compose the multilateral sandstone bodies extend down and laterally into underlying and adjacent channel sandstone. Channels are broadly convex downward and are characterized by a basal, locally conglomeratic zone with festoon or trough cross-bedding which represents channel-lag deposition. This zone is overlain by a thin unit with tabular cross-beds, in turn overlain by a zone of wedge-set cross-beds (epsilon bedding of Allen, 1963). The tabular and wedge cross-bedded sandstones probably represent lateral deposition as the toe and lower parts of point bars. The horizontally bedded and rippled zones, characteristic of the upper parts of many point-bar deposits, are rarely preserved in this facies, as the channel-lag and lower point-bar deposits are succeeded disconformably by channel deposits of a similar sequence. Within the preserved channel sequence there is only slight, if any, upward fining. The relatively small amount of clay in this facies is very kaolinitic. Lignite deposits, typically associated

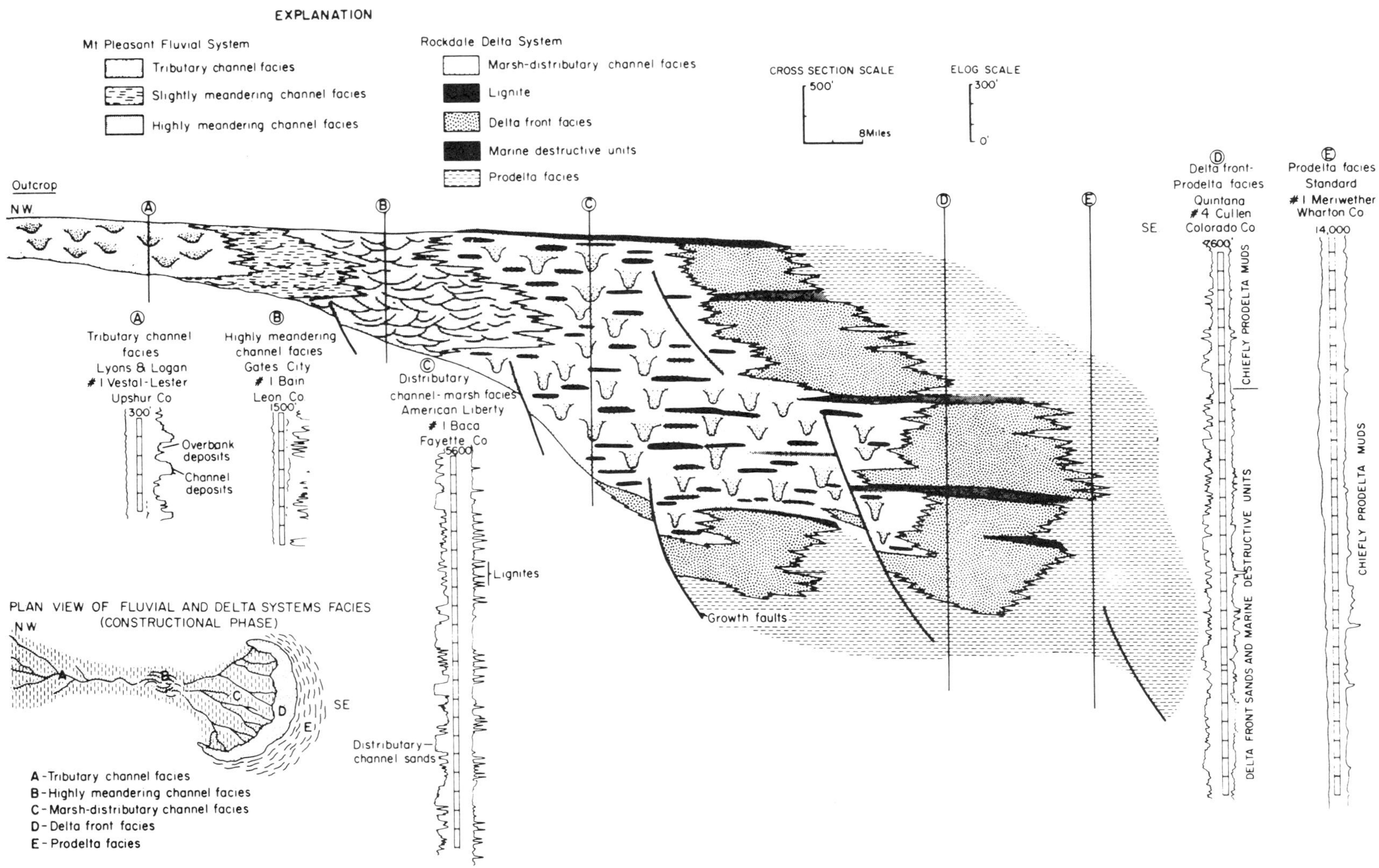

FIG. 3.—Diagrammatic stratigraphic dip section of Mt. Pleasant fluvial system and Rockdale delta system shows relation and character of principal component facies.

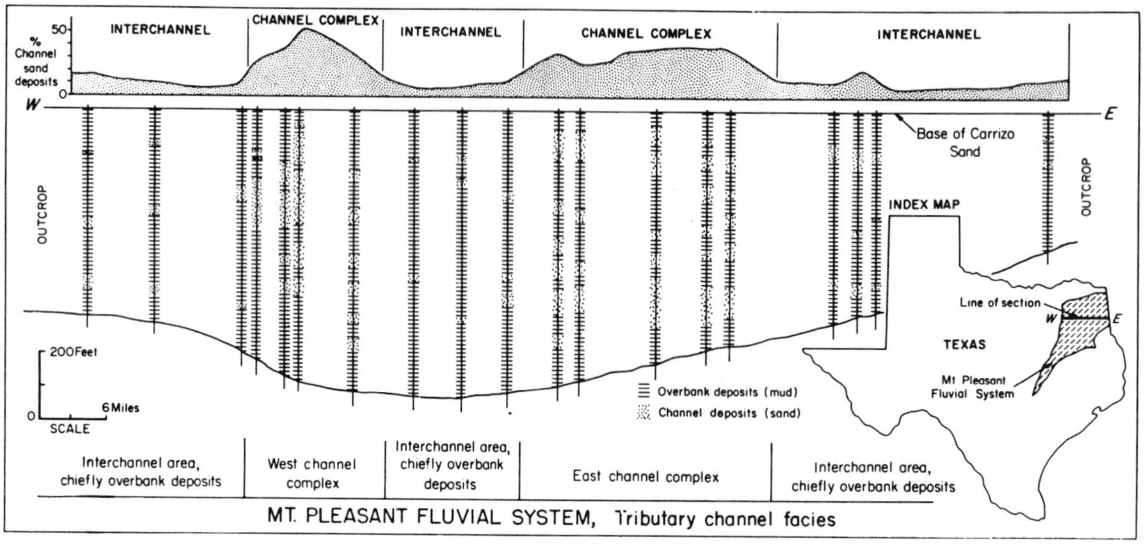

FIG. 4.—Diagrammatic stratigraphic strike section of Mt. Pleasant fluvial system, Wilcox Group, Texas.

with channel deposits in the other fluvial facies, are not preserved in the highly meandering channel facies. Lateral continuity and permeability of the sandstone make this facies a significant aquifer.

The highly meandering channel facies is the farthest downslope facies of the preserved part of the Mt. Pleasant fluvial system, grading downslope depositionally or southeastward to the distributary channel-marsh-swamp facies of the Rockdale delta system. It thus heads the delta system, extending in outcrop from central Bastrop County northward to southern Freestone County, and in subsurface from central Bastrop County northeastward to western Nacogdoches County (Fig. 1). Undoubtedly, it originally extended a considerable distance northwest of the present outcrop, probably coincident with the general courses of the main Wilcox stream systems. In central Bastrop and western Nacogdoches Counties the fluvial facies changes abruptly to the marginally embayed marine mudstone of the San Marcos strandplain-bay and Pendleton lagoon-bay systems, respectively. Within narrow belts marking these transitions, fluvial sandstones of the facies, which were modified by marine processes, have very well-sorted and fine grains and were locally redeposited as beach, barrier-bar, and strandplain deposits.

Multistory and Multilateral Sandstone Bodies

A characteristic feature of sandstone bodies in the fluvial facies containing relatively high proportions of overbank muds (tributary-channel and slightly meandering channel facies) is vertical persistence or stacking (Fig. 4). Successive sandstone bodies are superposed roughly, and complementary interchannel areas are void of significant channel sand deposits. This feature, termed multistorying by Feofilova (1954), has been noted in several facies containing channel sand deposits (*e.g.*, Mueller and Wanless, 1957; Friedman, 1960; Potter, 1963; Brown *et al.*, 1967). There is variation in degree of stacking, but vertical persistence is sufficient to show a fluvial drainage pattern of composite sequences similar to patterns of individual sandstone bodies. Where point-bar sequences are developed as parts of these facies, the sequences are characteristically complete. In contrast, sandstone bodies in fluvial facies containing low proportions of overbank muds (highly meandering facies) are markedly persistent laterally. Lateral and vertical repetition of channel sandstone units results in multilateral sandstone bodies as much as 30 mi wide. The channel units in these multilateral bodies commonly show an incomplete point-bar sequence; the coarse-grained channel-lag and lower point-bar deposits are bounded erosively vertically and laterally by similar channel sequences. Controlling factors in the formation of these distinct fluvial sandstone units apparently are differences in stream regime, differences in amounts of mud and sand in the system affording different degrees of compaction, and regional structure and tectonics.

Rockdale Delta System

The Rockdale delta system (as in the nomenclatural designation of Echols and Malkin,

1948) is the dominant depositional system of the lower Wilcox Group in Texas, comprising areally about 60 percent and volumetrically about 80 percent of the known lower Wilcox (Fig. 1). It was the ultimate depositional site of most sediments transported through the fluvial system and the principal source of sediments redistributed and deposited in associated barrier-bar, strandplain, lagoon-bay, and shelf systems.

The Rockdale delta system is downdip structurally from the fluvial system. The southern limit is marked by a line extending from north-central Karnes County north to central Bastrop County, and the eastern limit is marked by a line from central Nacogdoches County through northern San Augustine and Sabine Counties to the Sabine River. The downdip limit is beyond present well control (Fig. 1). Maximum thicknesses of 3,500–5,000 ft coincide with the central axis of the system which is chiefly along drainage of the modern Brazos River and roughly coincident with structural axis of the Houston embayment. Laterally and along the present structural strike the delta system generally, but not uniformly, thins to about 1,500 ft in the southwestern part of the system and to about 1,000 ft in the eastern part.

Criteria for recognition of this part of the lower Wilcox Group as a delta system include (1) composition and boundary relations of component facies, (2) external geometry of skeletal sand facies, (3) position in relation to laterally associated depositional systems, and (4) contrast in rates of deposition, with thick, rapidly deposited terrigenous clastic sequences (constructional units) alternating with thin, slowly deposited, relatively persistent marine and marsh deposits (destructional units).

Principal Facies

Four regional facies define the Rockdale delta system. Downslope depositionally and downdip structurally from the highly meandering facies of the Mt. Pleasant fluvial system, the lower Wilcox consists of an extensive facies of sandstone units alternating with mudstone, siltstone, and numerous beds of lignite (Fig. 3). Farther downdip this facies changes rather abruptly to thick sandstone, which in turn grades down dip and along structural strike to thick, uniform mudstone sequences. The alternating sandstone and mudstone-lignite sequences are interpreted to be distributary-channel and interdistributary deposits accumulated on a delta plain; the sandstone facies represents delta-front deposits laid down chiefly as distributary-mouth bars; the downdip thick mudstone sequence represents prodelta deposition. Mudstone sequences along strike of the delta-front sandstone facies are interpreted to be interdelta facies.

Distributary channel-marsh-swamp facies.—The facies consisting of alternating sandstone and mudstone-siltstone-lignite units is an areally extensive facies of the Rockdale delta system (Fig. 3). It contrasts markedly with the immediately upslope fluvial facies, chiefly in sandstone-body geometry and relative proportion of channel and nonchannel deposits. The fluvial facies contains cross-bedding structures and textural features indicative of channel-lag and lateral point-bar deposition in highly meandering streams. The result is relatively persistent, uniform, multilateral sandstone units with very few nonchannel deposits preserved. The sandstone changes downdip structurally and downslope depositionally to isolated, elongate, narrow sandstone bodies interspersed in nonchannel mudstone and associated lignite. In these respects it resembles the tributary-channel and slightly meandering facies of the Mt. Pleasant fluvial system. These facies differ significantly, however, in that the channels in the delta system are thicker, are symmetrical in cross section, and show a distributary rather than a tributary pattern in plan. Symmetry in cross section, thickness, and compactional features of underlying mudstone suggest accumulation of the sandstone as bedload deposits in relatively straight, stabilized streams, in which accumulation by downward accretion was facilitated by compactional subsidence of the channel sandstone in a framework of interchannel mudstone. Such features characterize certain modern delta-plain distributary channels and contrast with the lateral deposition in point bars of meandering, very sinuous streams. Mudstone and siltstone, composing about 60 percent of this facies, accumulated as subaerial levee, crevasse-splay, lacustrine, and flood-basin interdistributary deposits.

Two kinds of lignite are present within the distributary channel-marsh facies, both distinct from the elongate lignite deposits of the fluvial system. One lignite, which is tabular in shape but only local in extent, probably represents peat accumulation in small interdistributary depressions similar in origin to interdistributary peat in the Mississippi delta system, as discussed by Fisk (1960) and Frazier (1967). A second type is also tabular in shape but much more extensive; this type developed as regional, landward facies of marine delta destruction. Other distinctions between fluvial and delta-plain lignites are made on the basis of grade and composition. Fluvial lignites are more variable in composition than delta lignites and contain high percentages of woody materials consistent with a swamp origin. Delta lignites are slightly higher in grade and have higher specific gravity, and the relatively low percentages of woody material are more characteristic of marsh growth. Regional variations in Wilcox lignites (Fisher, 1963) reflect differences in depositional origin.

Delta-front facies.—Downdip the distributary channel-marsh-swamp facies changes to a thick sandstone facies in which lignite is absent and mudstone content is low (Fig. 3). The sandstones mostly have fine grains, generally much better sorted than those of fluvial and distributary channel sandstones. Commonly, the sandstone contains thin, multidirectional ripple cross-laminae and other current features. Plant and detrital lignite fragments are present locally. Individual sandstone units may be 200 ft or more thick; in plan, external geometry is lobate or lunate. The units are interpreted to be distributary-mouth bars representing delta-front deposition. The high percentage of sandstone in this facies resulted from deposition of bedload of delta-plain distributary channels debouching into relatively quiet marine water; the muds were carried in suspension farther seaward and accumulated as prodelta deposits. Relative cleanness, good sorting, and multidirectional current features indicate winnowing and reworking of sands in this environment by both channel and open-water currents under shoal conditions in the delta-front environment. Redistribution of these sands and lateral coalescence of the constructive delta-front deposits formed a fairly continuous delta-front sandstone facies. The relative amount of redistribution and lateral coalescence within the delta-front facies in the different Rockdale deltas resulted in delta lobes ranging from elongate to rounded.

Prodelta facies.—The farthest downdip sections of lower Wilcox penetrated by wells generally consist of relatively uniform, thick sequences of dark mudstone (Fig. 3). They contain moderately large amounts of very finely comminuted and disseminated organic matter, laminations marked commonly by slight textural and color variations, and a marine fauna exceedingly low in number and species diversity. This mudstone facies interfingers with, and grades updip to, the delta-front sandstone facies and represents deposition from suspension seaward from the delta front and river mouth as prodelta facies. Thickness and downdip extent of the facies are difficult to determine, primarily because few wells have penetrated significant thickness of the lower Wilcox in the downdip areas. However, the few deep wells indicate that the facies is probably the thickest in the delta system; approximately 4,000–5,000 ft has been recorded in some deep wells.

Interdelta, subembayment facies.—Sequences in areas between principal delta lobes are similar in some respects to the prodelta facies (Fig. 1). They consist principally of mudstone, much of which is laminated, but are thinner than the prodelta facies and in addition contain several thin, burrowed mudstone units or thin units of fossiliferous siltstone and very fine-grained sandstone with small amounts of glauconite. These thin units accumulated more slowly than the laminated mudstone, in areas not directly affected by deltation. The interdelta facies grades updip structurally to marsh facies of the delta plain and laterally along strike to delta-front sandstone facies.

Delta-Destruction Units

Delineation of a thick terrigenous clastic sequence such as the Rockdale delta system into component and individual deltas is a time-consuming task. Basically, it involves recognition and correlation of thin marine units formed in the process of marine destruction of a constructive delta mass. During periods of great sediment influx, the tract of delta facies (distributary channel-marsh-swamp, delta front, prodelta) is established and, during progradation and basinal and compactional subsidence, a thick terrigenous sequence accumulates. With cessation of sediment influx due either to diminishment of the source area or, more commonly, to upstream avulsion resulting in abandonment of the previous drainage system and establishment of a new network, compaction of the water-saturated sediments of the constructed delta mass exceeds the rate of deposition. The delta lobe, or at least marginal parts of it, is subjected to marine encroachment involving reworking of the constructive delta surface and underlying delta sediments. Rate of sedimentation during destruction is very low and as a consequence units are relatively thin. In the Rockdale delta system, delta-destruction units consist of very well-sorted, fine-grained, slightly glauconitic, commonly bioturbated sandstone, shell detritus, and mudstone either burrowed or containing a marine fauna. These units were deposited chiefly on the lower parts of the abandoned delta plain. Landward destruction of the plain involved extensive marsh growth, coincident with compactional subsidence, which resulted in areally extensive lignite deposits. This lignite contrasts with the thinner, local lignite bodies that formed as interdistributary peat. The delta-destruction units apparently represent deposition in environments comparable with delta-destruction environments in the modern Mississippi system. The marine sandstone originated under conditions similar to those of the Chandeleur Islands, and the burrowed or fossiliferous mudstone originated as open-bay deposits similar to those of Breton Sound.

Recognition of delta-destruction units and their facies relations in delta systems should be followed by studies of their areal extent by careful correlation of closely spaced wells in a large area. By this means the delta system can be subdivided and smaller delta units mapped. Determination of sand content of such subdivisions permits isolith mapping of skeletal elements of the delta (distributary-channel and delta-front facies) and, with mapping of component facies, allows delineation of specific delta lobes. Most delta-destructive units are restricted areally to a specific, underlying, constructive delta lobe. Because delta construction is a relatively rapid process and delta destruction a slow process, destruction of one delta lobe may persist during the construction, subsidence, and subsequent initiation of destruction of an adjacent lobe; thus destructional units of two or more lobes may be partly contemporaneous. Recognition of widespread destructional units permits larger scale, though only approximate, subdivision of delta systems.

Individual Deltas in Rockdale Delta System

The writers have mapped and delineated 16 principal deltas within the lower Wilcox Rockdale delta system (Fig. 5). These have the component facies and facies pattern heretofore described. Striking, and significant in another context, are the vertical stacking of individual deltas and the persistence of interdelta subembayments within the delta system. Further, the Rockdale deltas show a marked coincidence with modern coastal-plain streams (Fig. 5).

Maps of several delta-destruction units have made possible the delineation of three main phases of deltation in the Rockdale system, in addition to the outlining of principal individual deltas. The initial phase of deltation resulted in three relatively small deltas situated roughly along drainage of the modern Colorado, Brazos, and Trinity Rivers (Fig. 5). They range in area from 200 to 350 sq mi, and thicknesses are on the order of 300 to 500 ft. A second or intermediate period of delta building resulted in the largest, thickest, and most prograded deltas of the system. The larger deltas of this phase, as much as 2,000 sq mi in area, were constructed along the axes of the modern Guadalupe, Colorado, and Brazos Rivers; the smaller deltas, 500–700 sq mi in area, developed along the axes of the modern Sabine and Neches Rivers. The Brazos and Colorado deltas (Fig. 5) are the thickest, with maximum thickness of 1,500 ft, exclusive of the prodelta facies. A terminal phase involved delta construction in all areas of intermediate-phase deltation, as well as construction of a large delta situated primarily in Angelina, Houston, and Trinity Counties (Angelina delta). These terminal deltas are comparable in size and thickness with the underlying, intermediate-phase deltas, but do not extend as far downdip structurally.

Although individual deltas within the Rockdale system are varied in size and sediment volume, a more significant variation is in the shape of the delta lobes (Fig. 6). One type, shown by the intermediate-phase delta constructed in the area of the Trinity River, the four intermediate and terminal deltas constructed in the area of the Neches and Sabine Rivers, and, to some extent, the three initial-phase deltas, has the elongate, narrow shape of the modern birdfoot delta and certain initial deltas of older complexes delineated by Frazier (1967) in the Mississippi delta system. A second type, shown by the other deltas of the Rockdale system, has a generally lobate or rounded shape similar to that of premodern, abandoned deltas of the Mississippi system. If a delta line is drawn in the manner in which bayline maps commonly are made (Lowman, 1949), the elongate, narrow birdfoot deltas of the Trinity, Neches, and Sabine Rivers extend considerably beyond the line; thus, progradation into deeper waters, as in the modern deep-water birdfoot delta of the Mississippi system (Fig. 6), is suggested. These deltas are associated with a sequence relatively high in mudstones, a feature that strengthens the analogy with the sand-poor, deep-water birdfoot delta of the Mississippi. The initial-phase deltas similarly represent progradation over un-

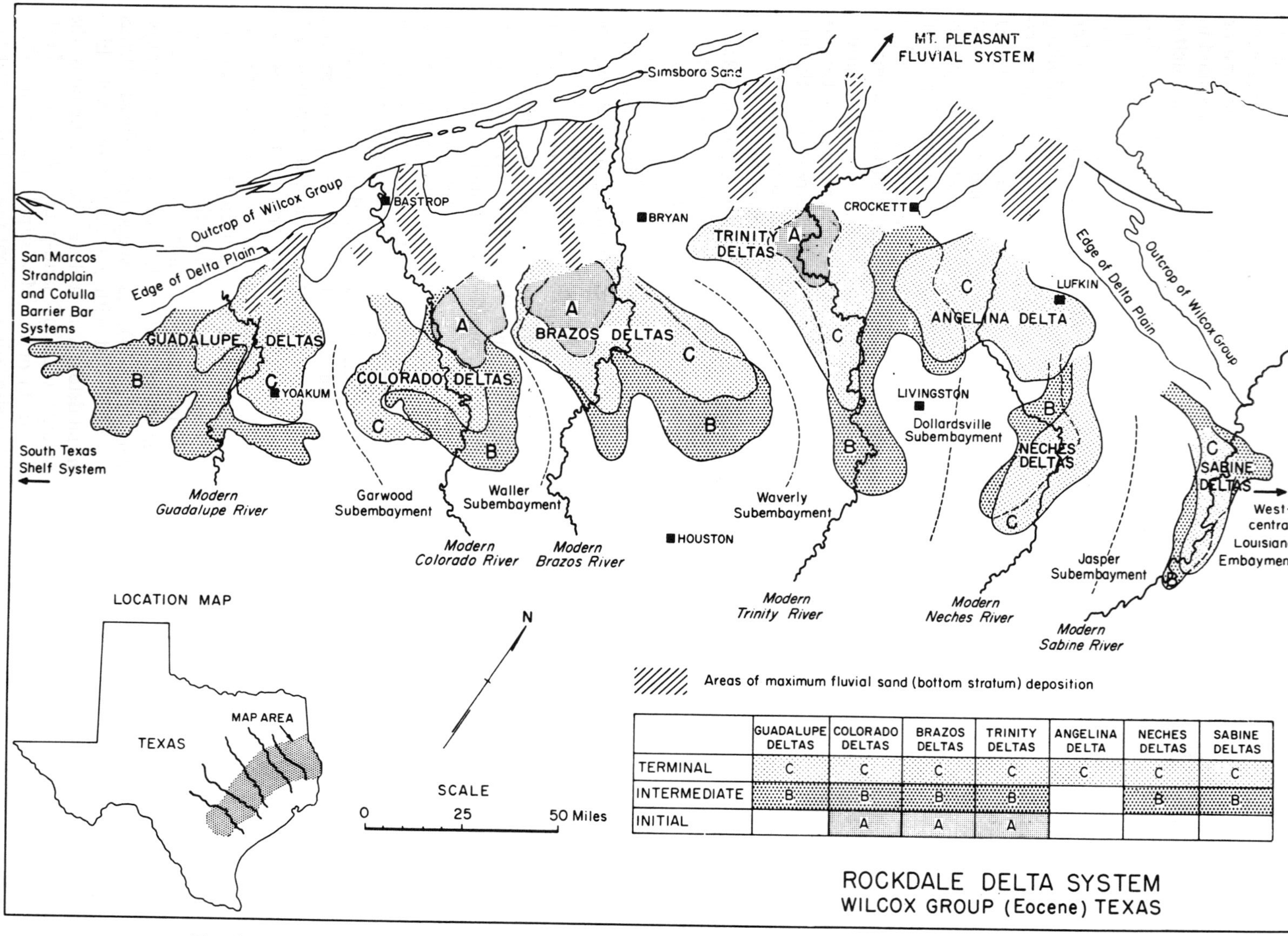

FIG. 5.—Distribution of principal deltas, Rockdale delta system, Wilcox Group, Texas; lobal outline drawn at distal margin of delta front facies. For well control, See Figure 1.

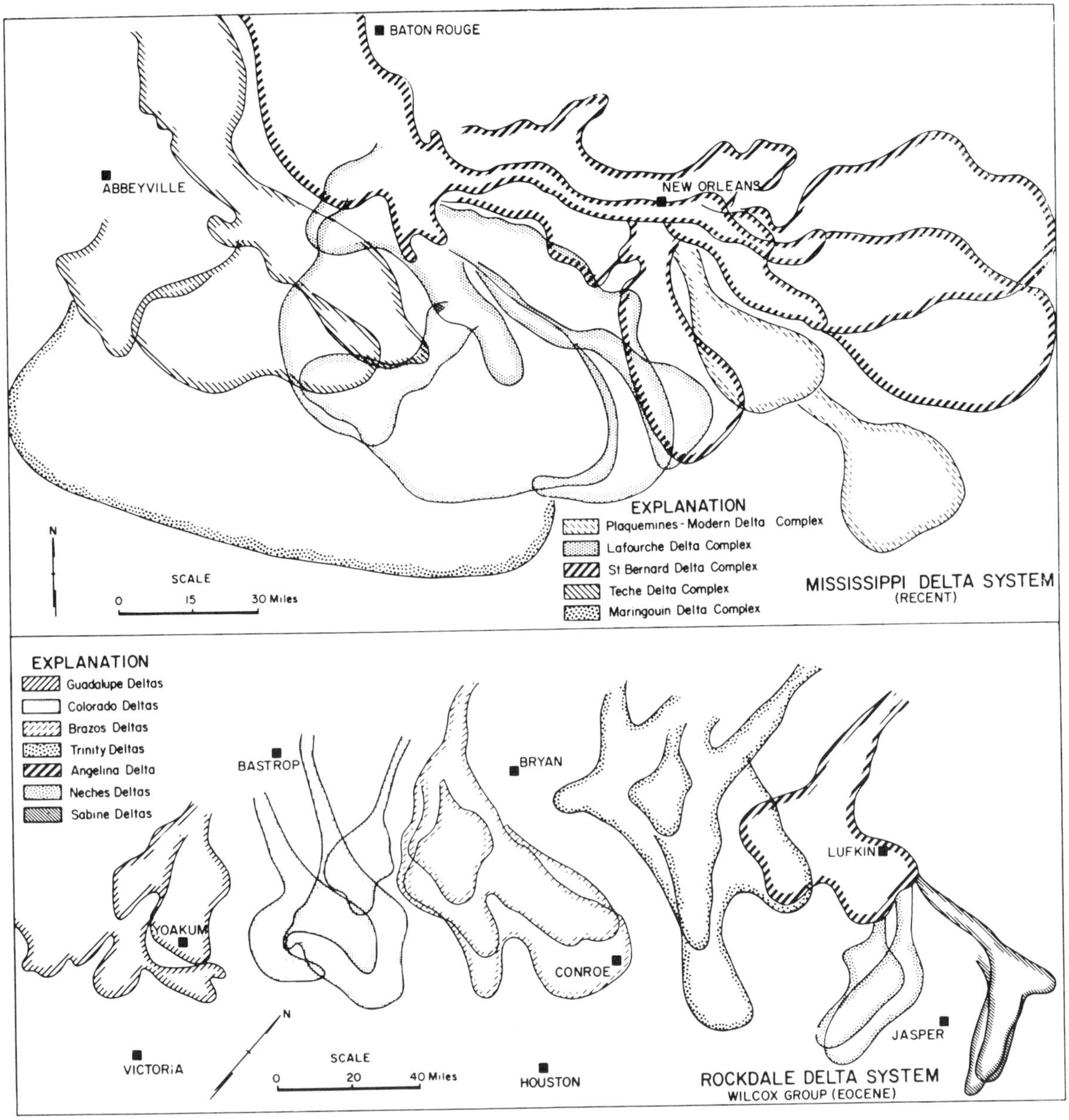

FIG. 6.—Comparison of size, distribution, and arrangement of principal delta lobes of the Holocene Mississippi delta system, southeastern Louisiana (modified from Frazier, 1967), and Rockdale delta system, Wilcox Group (Eocene, Texas).

derlying thick mudstone. By contrast, the lobate or rounded deltas of the Rockdale system consist of relatively large proportions of sandstone, a characteristic that suggests construction as shoal-water deltas analogous to the sand-rich, lobate, shoal-water deltas of the modern Rhone (Kruit, 1955) and premodern Mississippi (Fisk, 1955, 1961; Frazier, 1967).

Vertical stacking or approximate superposition of individual deltas within the system, persistence of interdelta subembayments, and coincidence of Rockdale deltas and modern drainage have been mentioned. The controlling factor or factors, like those of the multistoried channel sandstone bodies in upslope fluvial facies of the Mt. Pleasant system, are not fully understood. Such features apparently are common in fluvial systems containing relatively high percentages of mud and may be common in delta systems, although few ancient delta systems have been mapped and studied. Maps of delta systems in the Pennsylvanian of the Eastern Interior basin (Wanless *et al.*, 1963) and the modern and premodern deltas of the

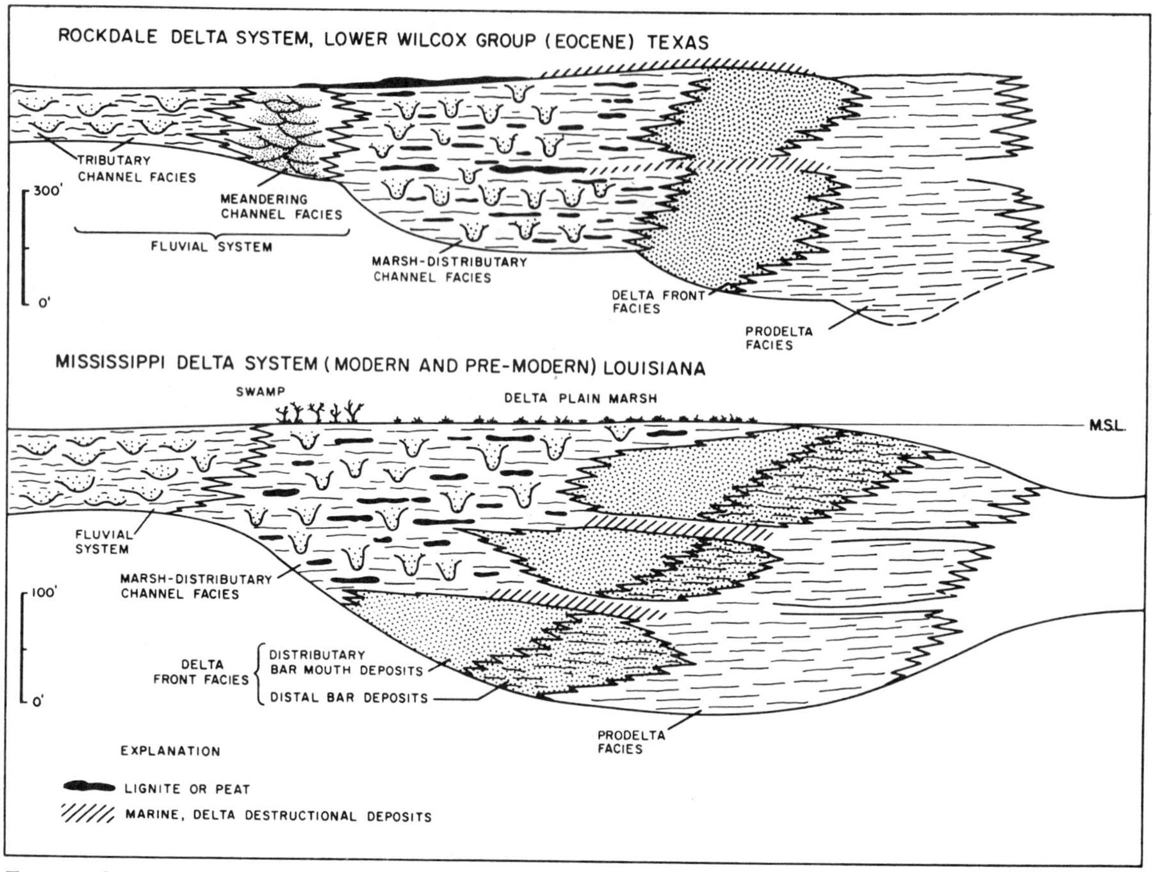

FIG. 7.—Comparison of component facies of Mississippi delta system (modified from partly hypothetical cross section of Coleman and Gagliano, 1964) and Rockdale delta system, Wilcox Group, Texas.

Mississippi system (Kolb and van Lopik, 1966; Frazier, 1967; Fig. 6) indicate some vertical stacking, though perhaps not as pronounced as that in the Rockdale delta system. Significant in this regard is not only the approximate coincidence of Rockdale deltas and modern coastal-plain drainage, but also the coincidence of other Eocene delta systems (*e.g.*, Yegua and Fayette) and Pleistocene delta systems along these axes. Certainly, compactional subsidence is a possible factor, although the more important control probably is tectonic. Structural control is indicated by the presence of growth-fault zones coincident with the updip limits of the thicker facies of the Rockdale delta system (Fig. 3).

Comparison with Mississippi Delta System

The Eocene Rockdale and the Holocene Mississippi delta systems, both built into the Gulf Coast basin, are similar in shape, occurrence, position of shoal-water lobate deltas and deep-water birdfoot deltas, and vertical stacking of individual deltas. The two systems are comparable in area (Fig. 6). A most significant similarity of the two systems is the distribution of component facies (Fig. 7), in which the high-sandstone delta-front facies is bounded up-delta or landward by a relatively high-mudstone and organic facies (distributary-channel and marsh-swamp facies) and off-delta or seaward by a high-mudstone prodelta facies. These detrital lenses are bounded vertically by marine and landward-equivalent destructional units. The significance of bounding facies in delta sequences has been emphasized by Coleman and Gagliano (1964); delineation of component facies, together with determination of external geometry of the skeletal elements (distributary-channel and delta-front sandstone bodies) and delineation of delta-destruction units, are judged to be the best regional criteria for recognition of ancient deltas or delta systems.

Pendleton Lagoon-Bay System

Marginal to the Rockdale delta system and along the southern part of the Sabine uplift (chiefly Shelby and northern parts of San Augustine and Sabine Counties), the Wilcox Group is predominantly a mudstone facies. This facies includes parts or all of several formally defined stratigraphic units (Pendleton, Marthaville, Hall Summit, Lime Hill, Converse, Cow Bayou, Dolet Hills, and Naborton Formations). It is present in outcrop and in subsurface. The facies, designated informally as the Pendleton lagoon-bay system (Fig. 1), is approximately 1,000 ft thick in Texas. It is comparable in thickness with fluvial facies along the strike, but thicker than updip fluvial facies and thinner than downdip delta facies.

The Pendleton system consists chiefly of laminated to locally bioturbated mudstone, and local lenses and beds of very fine- to fine-grained, sparingly glauconitic, massive to cross-bedded sandstone and fine-grained, broken, massive to cross-bedded, lignitic sandstone. Certain of the mudstones are gypsiferous; clay-ironstone concretions are common. Lignite is locally common, but relatively impure and in thin, discontinuous beds. The system contains a few fairly discrete units with marine fossils. Faunal composition is chiefly molluscan, with a few foraminifers and ostracods; fossils are present mostly in the sparingly glauconitic units. Species diversity is moderate to low, at least for fossiliferous units, but not so low as to suggest very abnormal salinity. Mollusks comprise largely a bank and mud fauna, characteized by several thick-shelled forms. The fauna of the Pendleton system has been described by Barry and LeBlanc (1942).

The system is underlain by marine clay of the Midway Group (Porters Creek Formation) and overlain by sandstone of the Carrizo Formation. A large part of the facies is contemporaneous with delta facies (Sabine deltas of the Rockdale system) on the west and south, although the upper part is contemporaneous with other Wilcox depositional systems not considered here. The Pendleton system grades west and north into the various fluvial facies of the Mt. Pleasant system; it extends east into Louisiana for an undetermined distance.

The lithologic and biologic composition of the Pendleton system and its situation marginal to the Rockdale delta system suggest that deposition took place in low-energy lagoons, lakes, bays, and mudflats flanking the area of maximum deposition in the delta system, much like the marginal environments of the Mississippi delta system. Deposition was chiefly marine and circulation ranged from restricted to open; the system opened seaward primarily on the southeast, and there was greater brackish-water and freshwater influence on the west and north, where the system grades into fluvial facies of the Mt. Pleasant system.

San Marcos Strandplain-Bay System

In outcrop south of the Colorado River, the highly meandering channel sandstone facies of the Mt. Pleasant system changes abruptly to a facies primarily of mudstone which locally contains marine fossils (Fig. 1). This facies persists in outcrop through southern Bastrop, Caldwell, Guadalupe, and northern Wilson Counties and extends a short distance downdip into the subsurface. It includes the Hilbig facies of Claypool (1933). The faunal and lithologic composition of the facies is very similar to that of the Pendleton system. A similar origin is suggested, with deposition in embayments and on mudflats along the opposite margins of downslope fluvial facies of the Mt. Pleasant system and delta facies of the Rockdale system. In outcrop, mud deposits of this facies are continuous below the downslope facies of the Mt. Pleasant system (highly meandering channel facies or Simsboro sandstone) about as far north as midway between the Brazos and Trinity Rivers, where the mudstone grades into fluvial and delta facies of earliest Rockdale deltation (initial phase, Fig. 5). The northern extent of this mudstone facies includes at least parts of units formally designated the Hooper and Seguin Formations. Southwestward, displacement of the facies is contemporaneous with large-scale deltation in the lower Wilcox Rockdale system (intermediate and terminal phases, Fig. 5).

Downdip and marginal to the shoal-water deltas of the southwestern part of the Rockdale system (Guadalupe deltas), mudstone grades into a facies consisting of interbedded sandstone and mudstone. The sandstone units generally have fine, well-sorted grains, are calcareous and sparingly glauconitic, and commonly contain *Ophiomorpha* burrows, a few detrital shell fragments, and small armored mud balls. They are present as elongate bodies parallel with the depositional strike and as tabular or sheet sandstone characterized by low-angle cross-bedding. Individual sandstone and mudstone units range up to 50 ft in thickness. Mudstone units are similar to those of the updip mudstone facies, but are commonly pyritic and slightly gypsiferous. Carbonaceous materials are present locally as thin lignitic stringers. This interbedded sandstone and mudstone facies is approximately 1,200 ft thick, and is present chiefly in southwestern Gonzales, Wilson, southwestern Karnes, and eastern Atascosa Counties. Composition of the facies and its position marginal to shoal-water deltas of the Rockdale system suggest an origin at least mechanically analogous to that of the Holocene chenier system of southwestern Louisiana. The mud accumulated in low-energy environments ranging from mudflat to open bay, contemporaneous with periods of active deltation. During marine destruction of the adjacent sand-rich, shoal-water deltas, sands were redistributed laterally by prevailingly southwesterly longshore currents; elongate and sheet sands were deposited in a strandplain marginal to the delta system. Transgression and regression of relatively thin marine units in this facies thus were related to periods of active deltation and delta destruction. Subsurface sandstone units such as the Poth and Bartosch are typical of this facies. In reference to their origin and area of principal occurrence, the updip mudstone and downdip sandstone-mudstone facies along the southwestern margin of the Rockdale delta system are designated the San Marcos strandplain-bay system.

Cotulla Barrier-Bar and Indio Lagoon-Bay Systems

Southwestward, sheet and elongate sandstone bodies of the San Marcos system grade into facies of elongate sandstone bodies arranged parallel with the depositional strike. The sandstone bodies generally have fine, very well sorted grains and are locally glauconitic; internal structures are not known because of lack of cores. Thickness of individual sandstone units ranges as much as 100 ft; aggregate sandstone thickness is 400–1,000 ft for the facies. Sandstone isolith lines show an extensive elongate sandstone unit oriented parallel with the depositional strike. This sandstone belt trends northeast-southwest in western Atascosa, southern Frio, northwestern LaSalle, southeastern Dimmit, and northwestern Webb Counties (Fig. 8). Updip from the sandstone belt, in southern Bexar, northern Atascosa, northern Frio, Zavala, western Dimmit, and northwestern Webb Counties and extending to the outcrop, is a predominantly mudstone facies, including most of the Indio Formation (Fig. 8). The principal lithologic component of this facies is mud, although locally sand is predominant. Mudstone is mostly gray to buff, calcareous to locally gypsiferous, and pyritic, and some is glauconitic or dolomitic. Structurally, it ranges from distinctly laminated to extensively burrowed. Marine fossils are chiefly thick-shelled mollusks; a few units contain large amounts of a very few species of foraminifers, mostly textulariids and globigerinids. Concentric mounds with internal structures, probably of algal origin, commonly are associated with the mudstone. Lignite beds of this facies are thin, discontinuous, and commonly impure, in contrast with the lignite of the fluvial and delta systems. Sandstone in this facies is of varied texture, but most has fine, moderately well-sorted grains; it is developed locally in relation to distinct depositional features, including tidal deltas, washover fans, and small-scale delta and fluvial units.

The lower Wilcox sandstone and mudstone facies of South Texas are judged to have originated as extensive barrier-bar and complementary lagoon-bay systems and are designated here, respectively, the Cotulla barrier-bar system and the Indio lagoon-bay system. The sand was transported laterally from shoal-water deltas of the Rockdale system on the northeast by prevailing southwesterly longshore currents. In the San Marcos strandplain system, situated intermediate to the delta and barrier-bar systems, the adjacent delta system served not only as a source of sand but also, through periods of active and inactive deltation, as a control for the distribution of the sheet sand. Accumulation of sand in the barrier-bar system was too distant from the delta system for the latter to have had direct control on sandstone-body geometry. The principal distinction in sand accumulation between these two systems is that sheet or tabular units formed in the San Marcos and elongate bodies formed in the Cotulla. Similar patterns of sand dispersal are present in modern and premodern marginal-marine systems southwest of the Mississippi delta system. Accumulation of sand in relatively narrow elongate belts in the barrier-bar system, as opposed to sheet-sand accumulation in the strandplain, facilitated development of extensive, landward mud facies—the Indio lagoon-bay system. Development of a distinctive lagoon-bay system is thus a complement of the barrier-bar system.

Other features substantiate interpretation of the South Texas, lower Wilcox facies as barrier-bar and lagoon-bay systems. First, sandstone-isolith patterns depicting the lagoonal side of the barrier-bar system are generally more irregular than those of the seaward side; further, the transition from sandstone to mud-

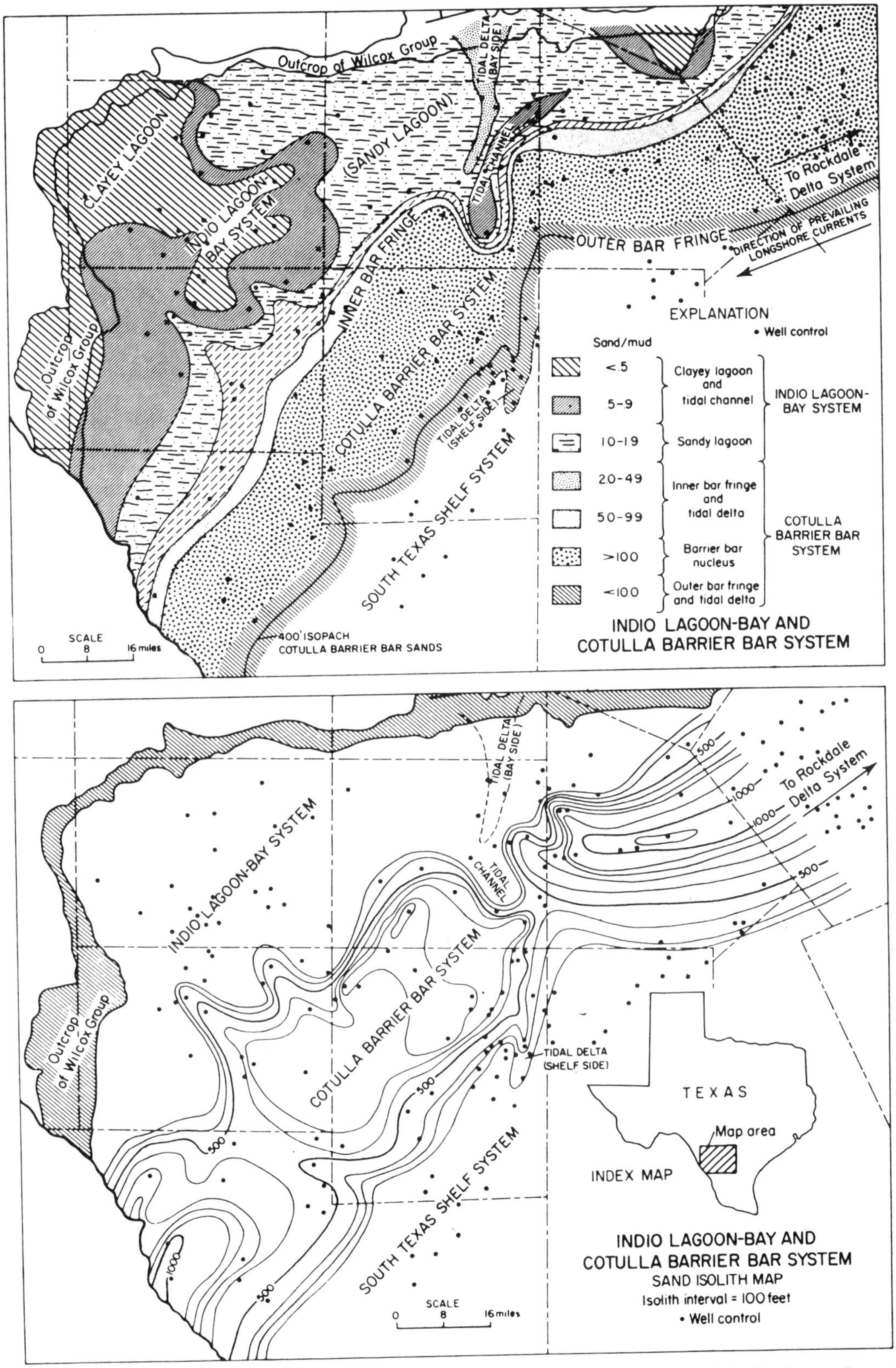

FIG. 8.—Sand-mud ratio map and sand isolith map of Cotulla barrier-bar system and Indio lagoon-bay system, Wilcox Group, Texas.

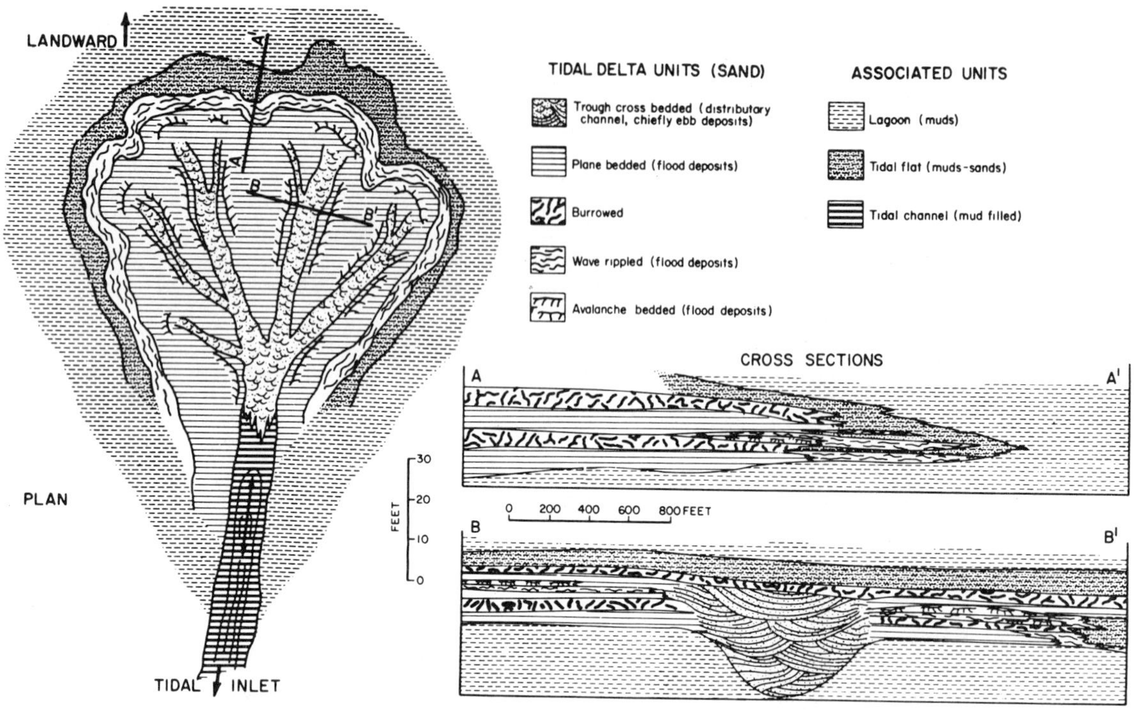

FIG. 9.—Principal rock units, lagoon-side tidal delta, Indio lagoon-bay system, Wilcox Group, southeastern Medina (outcrop) and northeastern Frio (subsurface) Counties, Texas.

stone is more abrupt on the lagoonal than on the seaward side (Fig. 8). Such features are typical of modern barrier-bar systems. Second, in southeastern Frio and northeastern LaSalle Counties, sandstone isolith lines of the Cotulla barrier-bar system are indented sharply on the lagoon side; a less pronounced indentation is present on the seaward side. Position and configuration of these indentations suggest the former presence of a tidal channel, a feature common to modern barrier bars. Flood and ebb tidal surges move through a channel or inlet in the bar, accreting sand in the form of tidal deltas both within the lagoon and onto the foreshore or shelf side of the bar. Presence of an anomalous thickness of sandstone in the Indio lagoon-bay system updip from the terminus of the inferred tidal channel (in northeastern Frio and southeastern Medina Counties and extending into outcrop), as well as sedimentary structures and units within the sandstone bodies, indicate accumulation as a lagoonal tidal delta (Fig. 8). Principal units of the lagoonal tidal deltas developed in outcrop of southern Medina County include flood and ebb deposits (Fig. 9). Deposition occurred both within tidal distributary channels and on interchannel flats. Channel deposits are trough or festoon cross-bedded sandstone, with axes of cross-bed sets trending generally south; deposition was principally during ebb flow. Toward the barrier bar, and in the subsurface, the tidal channel becomes relatively straight and nondistributive; this part of the channel is filled with mudstone and apparently was filled after the channel became inactive. Deposits of the interchannel flats are chiefly plane-bedded sandstone grading marginally through megarippled sandstone to tidal-flat and lagoonal mudstone; locally, thin, avalanche-bedded sandstone is associated with the plane-bedded deposits. Bedding, structures, and directional features of the interchannel deposits indicate that deposition took place mainly during flood surge. Plane-bedded sandstone shows graded bedding with a vertical increase in the amount of clay. The upper units are extensively burrowed, apparently as the result of destruction while parts of the tidal delta became emergent. Lobate extension of sandstone on the foreshore side of the bar and just south of the inferred tidal channel (east-central LaSalle County) probably represents a delta constructed by ebb-tide transport of sand through the tidal channel. On the assumption that these features originated as a tidal channel and complementary tidal deltas, the position of the deltas in respect to the channel, as well as the shape and configuration of the tidal channel, substantiate the inference of a prevailing southwestward transport of sediment by longshore currents. Lobate lagoonward extensions of sandstone in northeastern and north-central Dimmit County (Fig. 8), without associated isolith indentations indicative of tidal channels, possibly represent deposition as washover fans, with sand carried into the lagoon by storm surges. This feature is common in modern barrier bars (Andrews, 1966).

South Texas Shelf System

Downdip and seaward of the Cotulla barrier-bar system, a predominantly mudstone system characterizes the lower Wilcox of South Texas. This system extends across about 3,000 sq mi in southwestern Karnes, northern Goliad, northern Bee, north Live Oak, McMullen, northwestern Duval, Webb, Zapata, and western Starr Counties (Fig. 1). It is relatively uniform in thickness through much of its extent, ranging from about 1,000 to 1,500 ft, but thickens northeastward toward the Rockdale delta system and southward along the Rio Grande. Principal lithologic component is mudstone; aggregate thickness of sandstone in the system is generally less than 200 ft (Fig. 1). Individual sandstone units are varied in number, but commonly 8–10 are present. Thickness ranges generally from 10 to a maximum of 100 ft; average thickness of sandstone units is about 30 ft. Mudstones are of varied composition, but generally are light or medium gray to olive or buff, and laminated to extensively burrowed; marine fossils and glauconite pellets are common. Mudstones are commonly calcareous, though some are pyritic or carbonaceous. The sandstone is very fine to fine grained, well sorted to muddy, slightly glauconitic, and commonly contains detrital shell fragments. Faunal diversity is moderate; composition is chiefly molluscan, but thick-shelled forms are not predominant as in other lower Wilcox marine facies. Foraminifers are numerous and a few echinoderms are present.

The comparative uniformity of the system, the predominantly mudstone composition, marine fauna, and position relative to other depositional systems suggest that deposition took place on an extensive, shallow-water continental shelf; hence the system is designated the South Texas shelf system. Part of the Falls City Formation, as proposed and described by Hargis (1962), is included in this system. Vertical variations in abundance and diversity of marine fossils, amount of glauconite, extent of burrowing, and presence of sandstone units probably are related to variations in rate of deposition coincident with filling and foundering of the shelf; these cycles presumably relate to phases of delta construction and destruction in the Rockdale delta system along strike on the northeast.

Origin of the relatively thin sandstone units in the South Texas shelf system is not clear. Their external geometry is not well known, but present control suggests that there are both laterally persistent sheet sandstones, more numerous updip toward the Cotulla barrier-bar system, and elongate, NE-SW-trending sandstone bodies. The source of sand in the South Texas shelf system, like that of updip sandstone systems, probably was the Rockdale delta system on the northeast, but mode of transport and dispersal is unknown. Relatively sharp indentations in sandstone isolith lines of the southwestern lobes of the Guadalupe deltas of the Rockdale delta system (Fig. 1) suggest that submarine channeling of the deltas may have occurred during periods of delta destruction, with possible dispersal of sand across the shelf by southward-flowing turbidity currents. (Submarine channeling in the Wilcox as reported by Hoyt in 1959 is associated with an upper Wilcox system.) A more probable explanation of the origin of these sandstone units, is accumulation as barrier bars during periods of initial marine destruction of gulfward-projecting delta lobes and subsequent to shelf filling coincident with principal delta construction. The origin thus is similar to that of the Cotulla barrier-bar system, but there is only minor accumulation downdip and principal accumulation updip in the Cotulla system. Sheet sandstone of the shelf system may be related to transgression associated with periods of continued delta destruction; shelf-edge marine reworking of foundered barrier-bar sands during this period was likely. Extensive mud deposition in this system presumably was contemporaneous with periods of active deltation in the Rockdale delta system, although mudstones of the two systems are compositionally distinct. The amount of mud deposition on the shelf was small in comparison with the amount of contemporaneous prodelta mud deposition. Differences in composition and amount of mud in the two systems probably are related to marked differences in rates of accumulation.

Sand-Dispersal Patterns

The dispersal pattern of coarser fraction sediments (generally sand) in a terrigenous clastic system provides significant clues to the origin of specific facies and also provides a basis for the integration of various facies and systems.

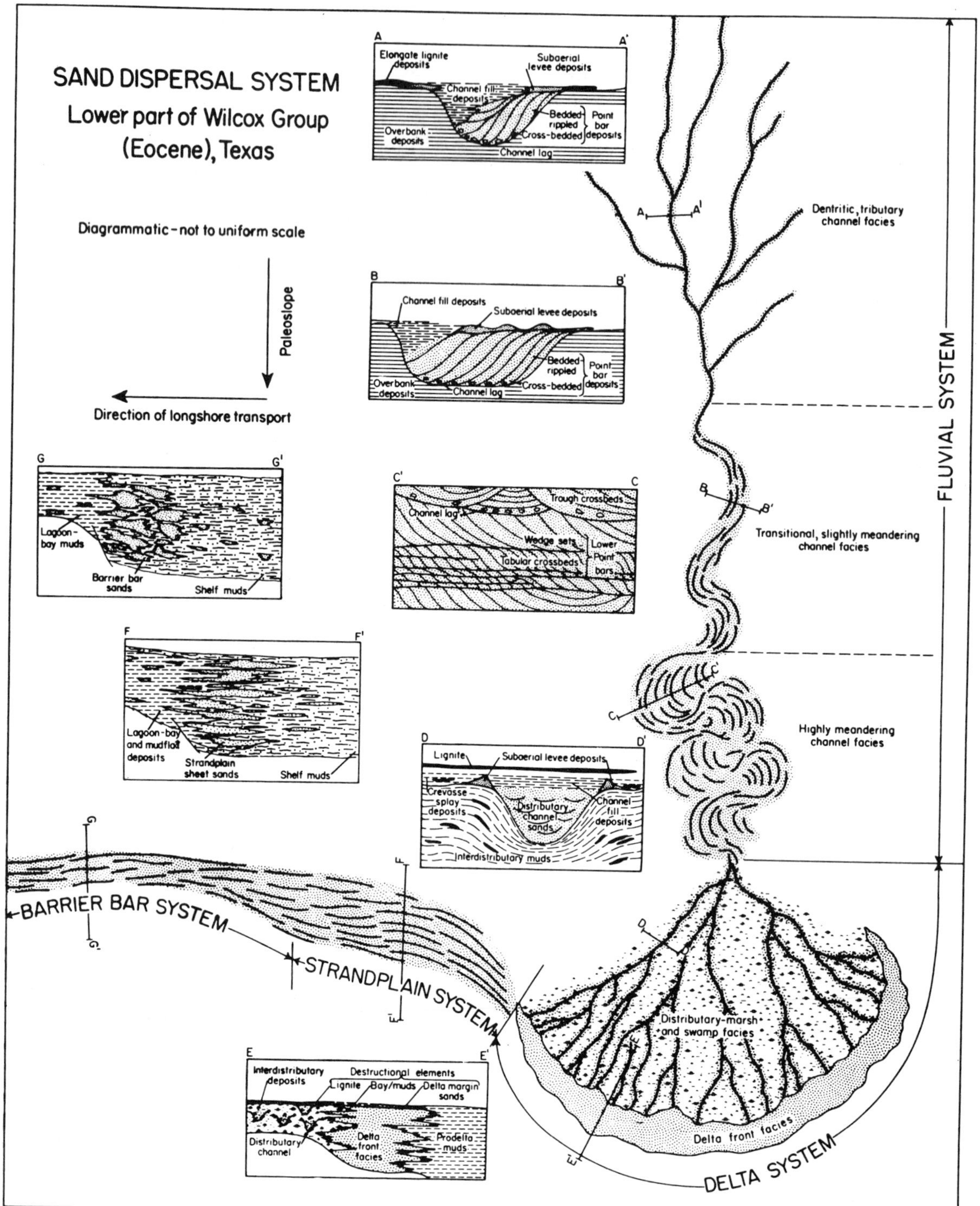

FIG. 10.—Idealized sand-dispersal system in various depositional systems, Wilcox Group, Texas.

Transport, dispersal, and deposition of the coarser fraction sediments normally involve higher energy elements, and thus the geometry of sandstone bodies effectively outlines skeletal units of the system. Within the various depositional systems of the lower Wilcox, several modes of sand transport and deposition occurred; each resulted in sandstone units of distinct geometry, structure, and texture (Fig. 10). The pattern involves dispersal through several fluvial facies, in which only a relatively small part of the sand was permanently stored or preserved, and through separate delta facies in which most deposition involved permanent storage, especially in the deep-water, elongate deltas. Marine redistribution of a relatively small part of the sand deposited in the delta-front facies, chiefly in shoal-water deltas, resulted in accumulation of sand in marginal strandplain, barrier-bar, and shelf systems (Fig. 10). Sandstone bodies of the fluvial system and of distributary channels of the delta system have regional trends normal to the depositional strike; sandstone bodies of the delta-front facies and those in the shelf system roughly parallel the depositional strike; those of the barrier-bar and strandplain systems closely parallel the depositional strike.

The highest upslope element preserved in the lower Wilcox sand-dispersal pattern, apparently only a minor part of the original fluvial system, consists of elongate channel sandstone with dendritic tributary distribution; ratio of nonchannel or overbank mudstone to channel sandstone is high. Downslope depositionally, the ratio of nonchannel mudstone to sandstone decreases, and individual sandstone bodies increase in width as a result of deposition by increasingly meandering streams. A zone of highly meandering stream deposition, as judged from the dominance of a point-bar sequence, resulted in a fluvial facies composed chiefly of extensive, multilateral sandstone bodies with very little associated nonchannel mudstone. This fluvial facies is present downslope and as the preserved part of the main fluvial system. Upslope this facies is bounded by a fault system (present Mt. Enterprise-Elkhart system) or threshold (Tyler basin threshold) which influenced deposition in other Cenozoic and Mesozoic units. It suggests the possible presence of a fall line, or at least a local gradient change that resulted in increased sand deposition. Once such a zone was established, weak bank conditions of streams traversing it would contribute to a high degree of meandering or lateral shifting. A similar high-sandstone fluvial facies in sand-dispersal patterns of Upper Pennsylvanian units in the Eastern Interior basin is reported by Potter (1963).

The next downslope element in the sand-dispersal pattern consists of elongate sandstone bodies in a distributive network; these are associated with relatively large amounts of nonchannel mudstone (Fig. 10). Channel sand accumulation was chiefly by downward accretion through compaction, as indicated by the nonerosive bases of the distributary channels and compactional features of underlying mudstone.

Principal deposition of sand in the delta system occurred in the delta-front facies, built up of distributary-mouth bars. Where delta-front sands form elongate, narrow sand facies (bar-finger sands of Fisk, 1961) in relatively high-mud frameworks and under deeper water conditions (birdfoot deltas), sand storage is generally permanent. This is apparently true of the elongate delta fronts characteristic in the eastern part of the Rockdale system where they are not associated with marginal barrier-bar and strandplain sands; interdelta subembayment facies contain very few sandstone bodies. Where the delta front is constructed as a lobate to rounded facies of sand-rich, shoal-water deltas (typical of deltas in the southwestern part of the Rockdale system), some of the sand may be redistributed laterally by marine destruction of the deltas and transported by longshore currents. In such situations two main types of sandstone bodies accumulated, though both have a common source. Adjacent to the delta system, accumulation generally resulted in relatively thin, areally persistent sheet sands which made up a strandplain system. Down the longshore current and more distal from the delta system, sand accumulated in more narrowly restricted, elongate barrier bars with the complementary development of a distinct landward lagoon-bay system.

A final pattern of sand dispersal, not shown on Figure 10 but possibly a part of lower Wilcox depositional systems, is transport and deposition by turbidity currents. Sharp indentations in sandstone isolith lines of certain of the shoal-water deltas in the Rockdale system (Guadalupe and Colorado deltas) may reflect mud-filled, submarine canyons that were cut and subsequently filled during specific stages of delta construction and destruction. Such a feature has been documented in a higher Wilcox system (Hoyt, 1959), and submarine channeling is well known from several modern delta systems. Comparable features in lower Wilcox systems are probable. Under such conditions delta sands eroded during channeling would be carried down depositional slope by turbidity currents and deposited as fans at the first major flattening of the slope. Mode of dispersal would be similar to that outlined by Walker (1966, Fig. 18) for near-source submarine fans associated with the Pennine delta in the upper Carboniferous of northern England. Verification of such fans in the lower Wilcox awaits the drilling of more deep downdip wells.

Depositional Systems and Basin Tectonics

The writers have used the concept of depositional systems not only as an approach to interpreting certain rock units and facies of the Wilcox Group, but also as a basis for establishing criteria for recognizing large-scale, regionally developed depositional units. Many sedimentary basins, and notably the Gulf Coast basin, contain very thick and large masses of terrigenous clastic rocks. Application of diagnostic criteria for recognition of specific depositional environments and the utilization of detailed depositional models are facilitated first by outlining major depositional systems involved in the basin fill.

The nature of a sedimentary basin, especially its structural or tectonic attitude, is critical to the concept of depositional systems because of its effect on (1) volume of sediment or magnitude of sequences involved, and (2) variation in distribution of component facies. First, in a relatively unstable basin where subsidence is generally a response to sediment load (as is judged to be the case of the Gulf Coast basin during Wilcox deposition), sites of deposition of particular facies tend to be maintained. Thus a barrier-bar system, for example, can be constructed of several individual barrier bars, the main features of which are similar to those of the larger system. Essential difference between the two is size. In magnitude, the Cotulla barrier bar system here recognized in the lower Wilcox of South Texas is closely comparable to other barrier-bar systems recognized elsewhere in the Gulf Coast basin—Frio barrier-bar system (Neogene) of South Texas (Boyd and Dyer, 1964) and Terryville barrier bar (Jurassic) of southern Arkansas and northern Louisiana (Thomas and Mann, 1966).

Second, the tectonic nature of the basin markedly influences the distribution of component facies within a depositional system, although composition and character of component facies of a given depositional system may be very similar in different structural basins. For example, delta systems developed in relatively stable basins, such as the Pennsylvanian of North Texas (Eastern Interior basin) and the western interior United States, commonly show fairly extensive progradation. Sections within such systems contain a vertical succession of component facies with prodelta mudstone overlain by delta-front sandstone, in turn succeeded by delta-plain and fluvial facies. By contrast, delta systems formed in rapidly subsiding, unstable basins, such as the Gulf Coast basin, commonly do not contain a pronounced vertical succession of facies, but rather have pronounced lateral facies that generally persist vertically. Within any particular part of the delta system, therefore, a vertical sequence generally shows only a single component facies of the system.

Oil and Gas Occurrence

A well-established generalization in Gulf Coast petroleum geology is that occurrence of oil and gas is controlled regionally by depositional facies, and local entrapment is by structures coincident with depositional facies that contain optimum reservoirs. The environmental control of oil and gas occurrence in terrigenous clastic rocks has been emphasized particularly by Rainwater (1963). Distribution of depositional facies in the context of the regional system of which they are a part explains control of known accumulations and provides a basis for outlining regional, potentially productive trends. In the lower Wilcox of the Texas Gulf Coast, five or six main oil- and gas-producing trends (Fig. 11) are defined in relation to depositional systems established herein: (1) de'ta trend (chiefly delta-front constructional and marine destructional sandstone), (2) strandplain trend, (3) barrier-bar trend (including tidal deltas), (4) shelf trend, (5) shelf-edge trend, and (6) a possible submarine-fan trend not yet defined.

Oil and gas accumulation in the lower Wilcox from north-central Karnes County northeastward to the Sabine River is controlled closely by distribution of specific deltas of the Rockdale delta system. Principal reservoirs are delta-front sandstone and closely associated marine-reworked sandstone from delta destruction; known fields are largely coincident with a relatively narrow belt determined by the more distal fringes of delta-front sandstone of specific delta lobes (Fig. 11). Interfingering of delta-front sandstone with adjacent marine prodelta mudstone provides the multipay reservoirs typical of many of the large delta-front fields (*e.g.*, Falls City, Sheridan, Columbus, Lake Creek, New Ulm, and Quicksand Creek). Marine clay deposited over delta-front sand during stages of marine destruction of delta lobes locally provides an effective seal for sandstone reservoirs. Numerous large-scale structures along the structural trend of lower Wilcox pro-

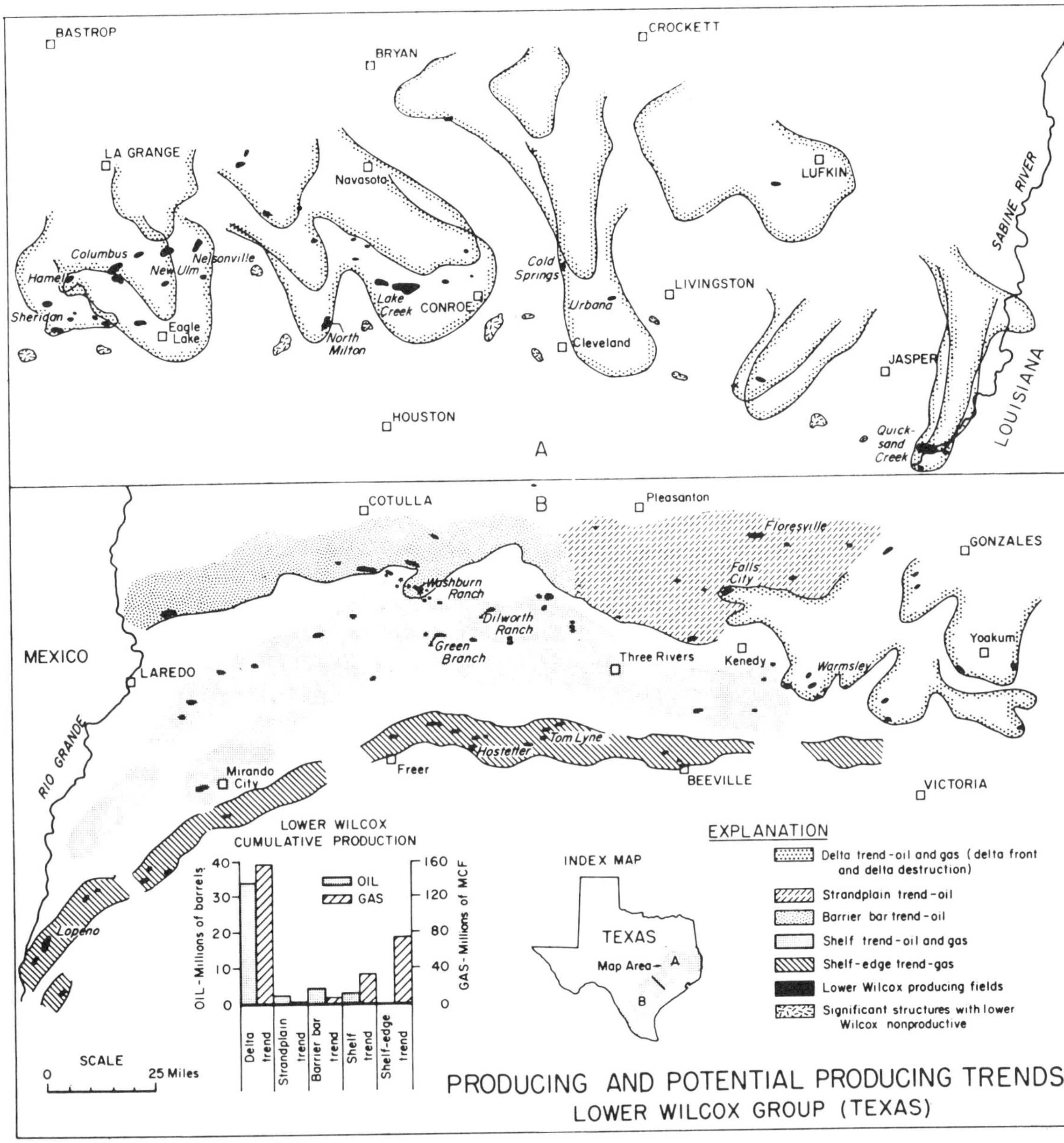

FIG. 11.—Productive and potentially productive oil and gas trends, lower part of Wilcox Group, Texas. For well control, see Figure 1.

duction within the delta system are notably nonproductive because of their location in sandstone-poor interdelta subembayment facies (Fig. 11). Growth faults and diapiric structures are associated with zones of thick deposition, such as the delta front; structures associated with these features are locally important traps.

Marginal to the delta trend is the standplain trend, controlling oil accumulation chiefly in the lower Wilcox of northwestern Gonzales, Wilson, northwestern Karnes, and eastern Atascosa Counties. Areally extensive sheet sandstone interbedded with dominantly marine mudstone forms a wide productive trend, which includes the farthest updip lower Wilcox production.

Sandstone units on the seaward side of the Cotulla barrier-bar system constitute the barrier-bar trend which extends in a narrow zone from northern McMullen County southwestward across LaSalle County through western Webb County. Most of the established production from this trend is in east-central LaSalle

County, associated mainly with a shelf-side tidal delta of the Cotulla system. Provided that entrapping structures are present, additional discoveries in the barrier-bar trend are likely; significantly, the trend is at shallow depth. Production has been chiefly oil; however, some significant gas wells have been completed recently in this trend in Webb County.

Several lower Wilcox fields have been developed in the sandstone reservoirs of the South Texas shelf system. With the exceptions of Green Branch and Dilworth Ranch, however, fields are small. Individual sandstone units within this extensive system are sporadic in distribution and not easily definable; many sandstone units are muddy with low permeability. Production has been primarily gas with small amounts of oil.

A final, main lower Wilcox trend is referred to here as the shelf-edge trend (Fig. 11). Sandstone bodies of this trend are part of the South Texas shelf system, though their origin is not fully understood. They form a fairly well-defined, narrow zone extending from the southwestern margin of the Rockdale delta system through central Bee, central Live Oak, northwestern Duval, southeastern Webb, eastern Zapata, and western Starr Counties. Productive sandstone units presumably originated as barrier bars similar to elongate sandstone bodies updip on the remaining part of the shelf system and in the Cotulla system. Their characteristics and downdip occurrence suggest that shelf-edge reworking and cleaning may have been a result of the higher energies associated with such a zone.

If the presence of submarine-fan deposits is indicated by deep drilling, an additional lower Wilcox trend may be established. Most likely area of such sandstone development would be downdip of the major shoal-water deltas of the Rockdale system, chiefly in Goliad, Victoria, Jackson, and Wharton Counties. Identification of such a trend will require extensive deep drilling.

Comparison of Lower Wilcox and Northwest Gulf of Mexico Depositional Systems

Comparison of the lower Wilcox with the Holocene of northwest Gulf of Mexico shows many similarities both in the kinds and distribution of depositional systems and in the component facies of these systems (Fig. 12). This is expected, because both are developed in the same structural basin. Prominent in the northwest Gulf is the extensive Mississippi delta system made up of five distinct delta complexes (Fig. 6) and at least 16 subdeltas (Frazier, 1967), with total area of about 20,000 sq mi. The lower Wilcox Rockdale delta system consists of seven delta complexes and at least 16 subdeltas (Fig. 5) with a total area of about 24,000 sq mi. Similarities in component facies and shapes of specific deltas in the two systems have been noted.

The Mississippi delta system is fed by the large, interior-draining fluvial system of the Mississippi River. Much of the analogous fluvial system of the lower Wilcox (Mt. Pleasant fluvial system) extended inland from present outcrop and is not preserved. On the basis of comparable size of the Rockdale and Mississippi delta systems, a large, interior-draining fluvial system is presumed to have been present during lower Wilcox deposition. Maximum thickness of the preserved part of the Mt. Pleasant fluvial system is between the modern Colorado and Trinity Rivers, and presumably the larger streams entered the Gulf in that area. The principally preserved part of the Mt. Pleasant system in northeast Texas was probably a minor or secondary part of the system, with streams comparable to modern streams east of the Mississippi River such as the Pearl and Tangipahoa Rivers.

Marginal to the area of maximum development of the gulfward-extending Mississippi River and delta systems are relatively low-energy basins or depressions in which freshwater to marine deposition occurs (Lake Pontchartrain, Lake Borgne, and part of Mississippi Sound on the east, Vermillion Bay, West Cote Blanche Bay, White Lake, and Grand Lake on the west). Lower Wilcox analogues include the predominantly marine mudstone systems of the Pendleton lagoon-bay system developed along the east flank of the Rockdale and Mt. Pleasant systems and the updip part of the San Marcos system on the southwest or opposite side.

A net southwestward longshore transport of sediments from both the modern Mississippi and the Wilcox Rockdale delta systems formed comparable depositional systems along strike of these systems. The southwestern Louisiana chenier system, adjacent to the Mississippi delta system, is at least mechanically analogous to the San Marcos strandplain system, which developed just down the longshore drift of the Rockdale delta system. More distally along the strike of the Holocene Mississippi delta system is a barrier-bar system, of which the modern Bolivar-Galveston barrier bar is a part. This system is considered analogous to the bar-

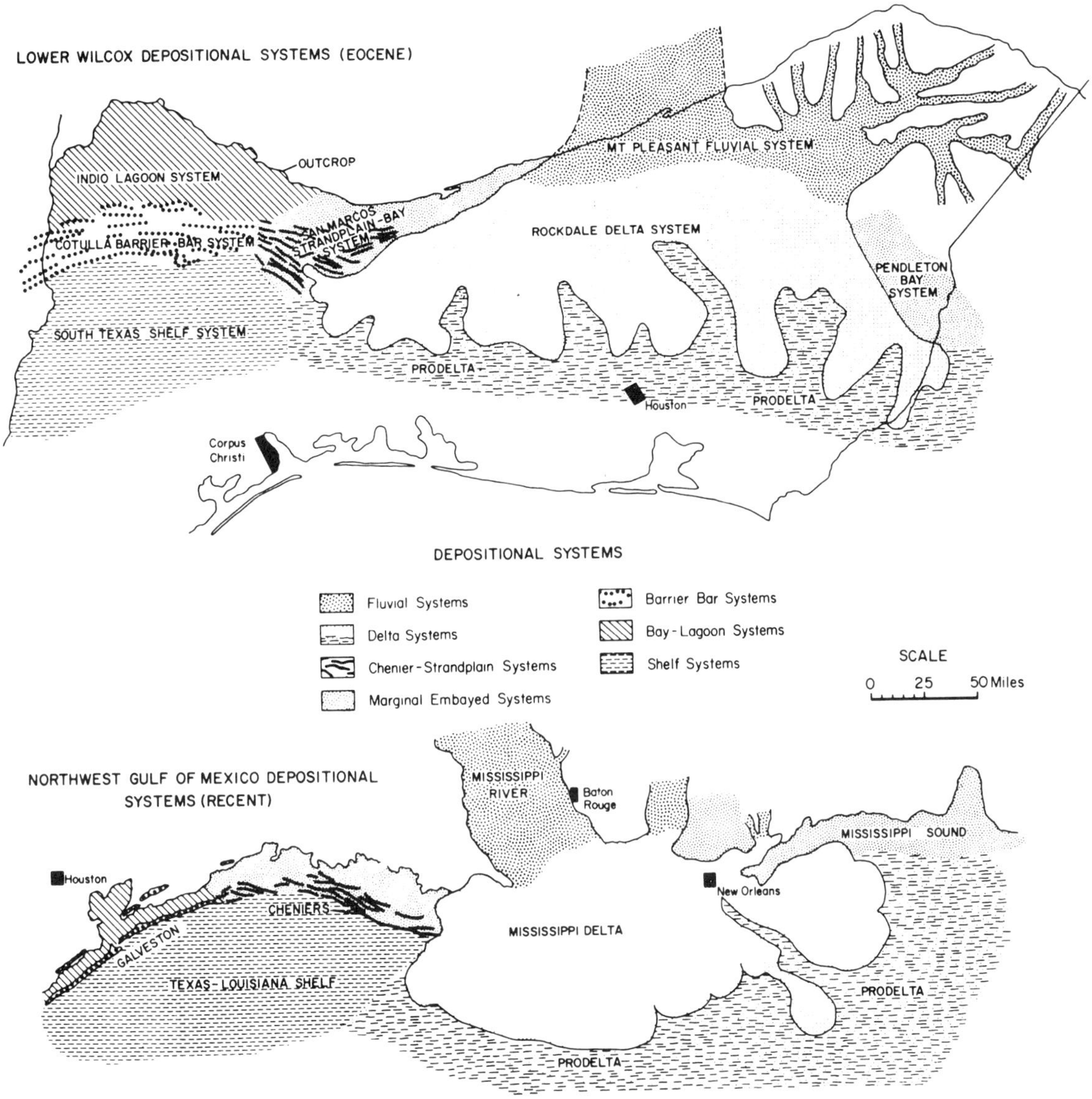

FIG. 12.—Comparison of lower Wilcox (Eocene) and northwestern Gulf of Mexico (Holocene) depositional systems.

riers making up the Cotulla barrier-bar system of the South Texas lower Wilcox. Landward complement of the modern barrier bars is a bay-lagoon system, represented at present by Galveston Bay. Deposits of this system are the modern analogue of the Indio lagoon-bay system. The extensive Louisiana-Texas continental shelf, situated gulfward of the barrier bar and chenier system and partly marginal to the Mississippi delta system, is the modern counterpart of the lower Wilcox South Texas shelf system.

References Cited

Allen, J. R. L., 1963, The classification of cross-stratified units, with notes on their origin: Sedimentology, v. 2, p. 93–114.

Andrews, P. B., 1966, Facies and genesis of a hurricane washover fan, St. Joseph Island, central Texas coast: Unpub. Ph. D. dissert., Texas Univ. at Austin, 238 p.

Barry, J. O., and R. J. LeBlanc, 1942, Lower Eocene faunal units of Louisiana: Louisiana Geol. Survey Bull. 23, 208 p.

Boyd, D. R., and B. F. Dyer, 1964, Frio barrier bar system of South Texas: Gulf Coast Assoc. Geol. Socs. Trans., v. 14, p. 309–322.

Brown, L. F., *et al.*, 1967, Role of compaction in development of geometry of superposed elongate sandstone bodies (abs.): Am. Assoc. Petroleum Geologists Bull., v. 51, p. 455–456.

Claypool, C. B., 1933, The Wilcox of central Texas: Abs. of Ph. D. thesis, Illinois Univ., 12 p.

Coleman, J. M., and S. M. Gagliano, 1964, Cyclic sedimentation in the Mississippi River deltaic plain: Gulf Coast Assoc. Geol. Socs. Trans., v. 14, p. 67–80.

Echols, D. J., and D. S. Malkin, 1948, Wilcox (Eocene) stratigraphy, a key to production: Am. Assoc. Petroleum Geologists Bull., v. 32, p. 11–33.

Feofilova, A. P., 1954, The place of alluvium in sedimentation cycles of various orders and time of its formation: Akad. Nauk SSSR Geol. Inst. Trudy, v. 151, Coal ser., no. 5, p. 241–272.

Fisher, W. L., 1963, Lignites of the Texas Gulf coastal plain: Texas Univ. Bur. Econ. Geology Rept. Inv. 50, 164 p.

——— and J. H. McGowen, 1967, Depositional systems in the Wilcox Group of Texas and their relationship to occurrence of oil and gas: Gulf Coast Assoc. Geol. Socs. Trans., v. 17, p. 105–125.

Fisk, H. N., 1955, Sand facies of recent Mississippi delta deposits: 4th World Petroleum Cong. Proc., Sec. 1–C, p. 377–398.

——— 1960, Recent Mississippi River sedimentation and peat accumulation: 4th Internat. Cong. Carboniferous Stratigraphy and Geology, Heerlen 1958, Compte Rendu, p. 187–199.

——— 1961, Bar-finger sands of the Mississippi delta, *in* Geometry of sandstone bodies: Am. Assoc. Petroleum Geologists, p. 29–52.

Frazier, D. E., 1967, Recent deltaic deposits of the Mississippi River: their development and chronology: Gulf Coast Assoc. Geol. Socs. Trans., v. 17, p. 287–311.

Friedman, S. A., 1960, Channel-fill sandstones in the middle Pennsylvanian rocks of Indiana: Indiana Geol. Survey Rept. Prog. 23, 59 p.

Hargis, R. N., 1962, Stratigraphy of the Carrizo-Wilcox of a portion of South Texas and its relationship to production: Gulf Coast Assoc. Geol. Socs. Trans., v. 12, p. 9–25.

Hayes, M. O., and A. J. Scott, 1964, Environmental complexes, South Texas coast: Gulf Coast Assoc. Geol. Socs. Trans., v. 14, p. 237–240.

Hoyt, W. V., 1959, Erosional channel in the middle Wilcox near Yoakum, Lavaca County, Texas: Gulf Coast Assoc. Geol. Socs. Trans., v. 9, p. 41–50.

Kolb, C. R., and J. R. van Lopik, 1966, Depositional environments of the Mississippi River deltaic plain—southeastern Louisiana, *in* Deltas in their geologic framework: Houston Geol. Soc., p. 17–61.

Kruit, C., 1955, Sediments of the Rhone delta; I, Grain size and microfauna: Koninkl. Nederlandsch Geol. Mijnbouwk Gen. Verh., Geol. Ser. v. 15, p. 397–499.

Lowman, S. W., 1949, Sedimentary facies in the Gulf Coast: Am. Assoc. Petroleum Geologists Bull., v. 33, p. 1939–1997.

Mueller, J. C., and H. R. Wanless, 1957, Differential compaction of Pennsylvanian sediments in relation to sand-shale ratios, Jefferson County, Illinois: Jour. Sed. Petrology, v. 27, p. 80–88.

Pettijohn, F. J., *et al.*, 1965, Geology of sand and sandstone: Indiana Geological Survey and Dept. Geology, Indiana Univ. Conf., 205 p.

Potter, P. E., 1963, Late Paleozoic sandstones of the Illinois basin: Illinois Geol. Survey Rept. Inv. 217, 92 p.

Rainwater, E. H., 1963, The environmental control of oil and gas occurrence in terrigenous clastic rocks: Gulf Coast Assoc. Geol. Socs. Trans., v. 13, p. 79–94.

Thomas, W. A., and C. J. Mann, 1966, Late Jurassic depositional environments, Louisiana and Arkansas: Am. Assoc. Petroleum Geologists Bull., v. 50, p. 170–182.

Walker, R. G., 1966, Shale grit and Grindslow shales: transition from turbidite to shallow water sediments in the upper Carboniferous of northern England: Jour. Sed. Petrology, v. 36, p. 90–114.

Wanless, H. R., *et al.*, 1963, Mapping sedimentary environments of Pennsylvanian cycles: Geol. Soc. America Bull., v. 74, p. 437–486.

The American Association of Petroleum Geologists Bulletin
V. 66, No. 6 (June 1982), P. 649-688, 25 Figs., 3 Tables

Frio Formation of Texas Gulf Coastal Plain: Depositional Systems, Structural Framework, and Hydrocarbon Distribution[1]

WILLIAM E. GALLOWAY,[4] DAVID K. HOBDAY,[3] and KINJI MAGARA[4]

ABSTRACT

The Frio Formation (Oligocene-Miocene) is one of the major Tertiary progradational wedges of the Texas Gulf coastal plain and has yielded nearly 6 billion bbl of oil and 60 tcf of gas. The Frio, and its updip equivalent, the Catahoula Formation, consists of deposits of two large fluvial and associated deltaic systems, centered in the Houston and Rio Grande embayments. Structural history in the Houston embayment is dominated by syndepositional deformation of underlying Jurassic salt; mobilization of thick, undercompacted prodelta and slope muds characterized the tectonic evolution of the deltaic sequence in the Rio Grande embayment. These two major deltaic depocenters are separated by a vertically stacked, strike-parallel coastal barrier and strandplain system. Underlying, interbedded, and transgressive shelf, prodelta, and continental slope mudstone sequences provide principal source and sealing facies. Sparse organic geochemical data, regional thermal gradients, and distribution patterns of hydrocarbons show that large volumes of oil and gas have likely been generated within and effectively expulsed upward and landward from thermally mature, normally to moderately undercompacted sequences of bounding mudstone facies.

All Frio depositional systems contain major, geologically defined, hydrocarbon-producing plays. Oil and gas field productivity data show a log normal frequency distribution for all but the largest fields. Per volume productivity, production styles, and types of hydrocarbons within each of the ten recognized plays reflect available source rock types, differing thermal and compaction histories, and variable reservoir and trap configurations that characterize each depositional system. Systematic distribution patterns of both physical and chemical properties of produced hydrocarbons can be related to source facies, regional thermal regime, and post generation modification by (1) continued maturation, thermal cracking, and deasphalting, (2) migration and attendant chromatographic separation, and (3) bacterial alteration.

INTRODUCTION

The Frio Formation of Oligocene-Miocene age is one of the principal progradational wedges of the Texas Gulf Coast Tertiary basin. Several hundred feet of vertical aggradation of the ancient coastal plain accompanied progradation of a continental platform wedge more than 15,000 ft (4,500 m) thick and up to 50 mi (80 km) wide. This wedge, which consists of interfingering marine and nonmarine sands and shales, has produced more than 16 billion BOE[5] of hydrocarbons, making it one of the most important petroleum reservoirs of the Gulf coastal basin.

More than 50 years of intensive and increasingly deeper exploration and development make the Frio Formation a mature exploration target that may well have yielded the most of its contained and producible hydrocarbon reserves. However, even a small remaining fraction of undiscovered petroleum would constitute a major target for continued exploration and an energy resource of regional significance.

Objective

The objective of this report is twofold: (1) Compile the geologic data base necessary for a thorough assessment of factors influencing the distribution and characteristics of oil and gas within the Frio Formation. (2) Integrate, as appropriate, the geologic framework description and the known distribution of petroleum accumulations as a basis for defini-

[1] Manuscript received, August 3, 1981; accepted, January 26, 1982.
This article is extracted from a report on a project entitled "Resource evaluation of the Frio Formation (Frio MSU—Texas Gulf Coast province)" and jointly funded by the Texas Bureau of Economic Geology and the Division of Oil and Gas Resources, U.S. Geological Survey.
Publication of this paper is authorized by the director of the Bureau of Economic Geology, The University of Texas at Austin.

[2]Bureau of Economic Geology, The University of Texas at Austin, Austin, Texas 78712.
[3] Formerly at the Bureau of Economic Geology; current address, The University of Sydney, Sydney, New South Wales 2006, Australia.
[4] Formerly at the Bureau of Economic Geology; current address, petroleum consultant, Austin, Texas 78731.

The writers acknowledge the help and support provided by numerous individuals and organizations. Geochemical data were supplied by J. L. Clayton, M. Smith, and L. C. Price of the U.S. Geological Survey, J. A. Momper and R. L. Ames of Amoco Production Co., and L. C. Bonham and R. W. Jones of Chevron Oil Field Research Co. Statistical data for oil and gas production compiled by Petroleum Information Corp. were made available through the help of R. F. Mast and T. Dyman of the U.S. Geological Survey. Engineering data on oil and gas fields were supplied by T. Garland, J. Hicks, M. Carrales, Jr., and H. Rahman of the Energy Information Administration. M. M. Dodge and J. S. Posey contributed freely to assembling the data base and in locating or providing additional information. Drafting, typing, and manuscript preparation were performed by personnel of the Bureau of Economic Geology. W. L. Fisher, R. A. Morton, and L. F. Brown, Jr., reviewed the manuscript. Research assistants Mark Helper, Diana Morton, Nathan Smith, and Victor Gavenda contributed to data gathering and synthesis.

[5] The volume of natural gas here and throughout the report is converted to liquid equivalents on an approximate Btu basis of 1 bbl equals 6 mcf.

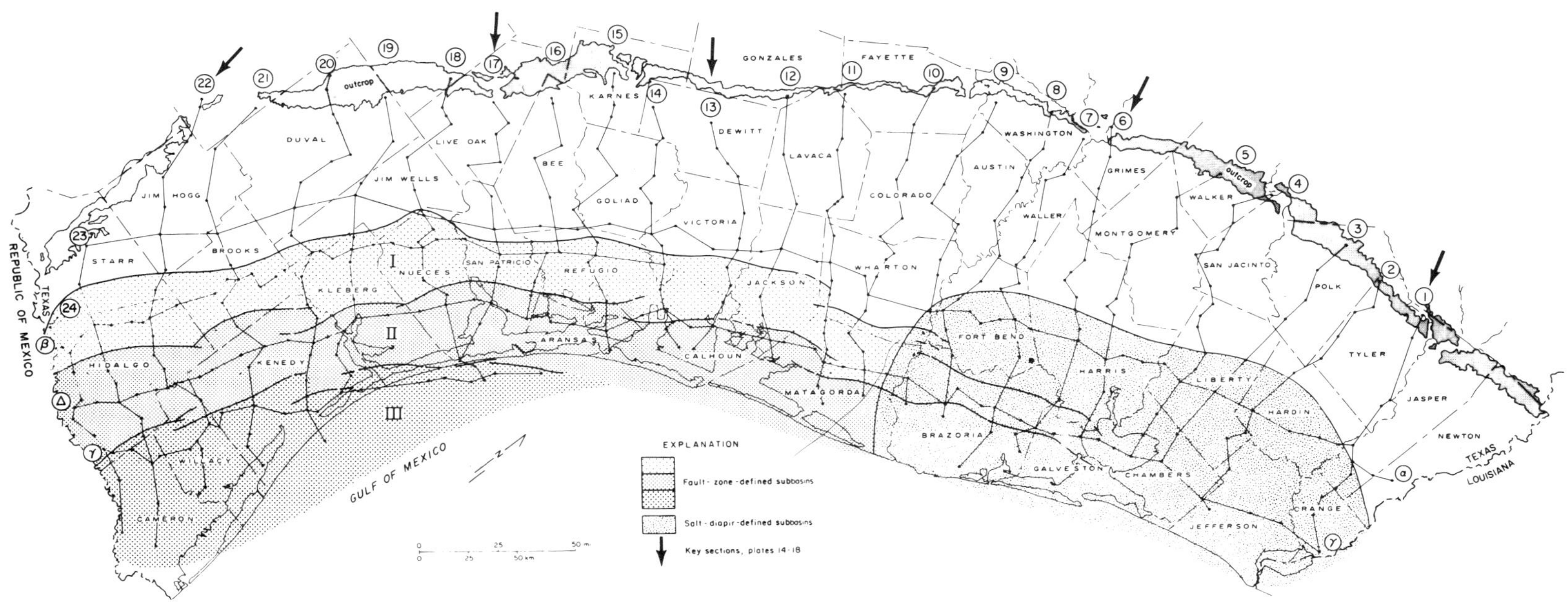

FIG. 1—Index map showing location of study area and grid of dip and strike cross sections used for establishing correlation framework, log facies mapping, and temperature and pressure analyses. Frio Formation was deposited in several regional subbasins defined by major growth fault belts (I-III) and extent of underlying bedded salt.

tion and comparative resource evaluation of principal producing Frio plays. The hydrocarbon play, which is the working geographic subdivision applied in this analysis, is a geologically homogeneous area characterized by its production style or suite of styles (White, 1980).

Methodology

The approach used involved four component elements: (1) regional delineation of the Frio Formation and its updip equivalent, the Catahoula Formation; (2) compilation of a systematic inventory of known hydrocarbon distribution; (3) synthesis and interpretation of the geologic framework; and (4) delineation and quantitative description of hydrocarbon-producing plays.

Regional delineation of the Frio Formation was based on a grid of 3 strike- and 24 dip-oriented cross sections (Fig. 1) modified from regional sections of Dodge and Posey (1982). The conventional subsurface nomenclature was integrated with the regional correlation network.

The *hydrocarbon inventory* required manual collation of data from public sources, including the International Oil Scouts Association (1978) *Texas Yearbook 1978*, the Railroad Commission of Texas (1978) annual report, published field papers, and tabulated calculations of field reserves compiled by the Office of Applied Analysis, Energy Information Administration, Department of Energy. The resultant inventory provides systematic information on all Frio fields that have produced more than 1 million BOE of hydrocarbons, including year of discovery, number of producing wells, depth, thickness, and hydrocarbon gravity of producing zones, cumulative production of oil, condensate, casing-head gas and gas, calculated reserves of gas and oil, and additional significant producing units.

The *geologic framework studies* included regional mapping of thickness, structural features and trends, and sand distribution, as well as delineation of principal depositional systems and component facies associations, discrimination of controls on reservoir quality, determination of the volume, quality, and distribution of potential source facies, and evaluation of thermal and fluid flow history and concomitant hydrocarbon expulsion and migration efficiency.

Play delineation integrated the geologic and production studies to outline geographically and describe key aspects of the major hydrocarbon plays that characterize the Frio MSU (Major Stratigraphic Unit).

The careful delineation of geologically defined plays and consequent data synthesis and projection within relatively homogeneous populations should provide not only the best basis for improved resource estimation, but also produce a comparative index of exploratory potential for different portions of a major producing unit in a large, mature, hydrocarbon province. Significantly, different plays within the Frio may have greater geologic diversity than exhibited by comparable plays in totally different major stratigraphic units (for example, some Frio plays may be more analogous to certain Wilcox plays than to other plays within the Frio).

DESCRIPTIVE GEOLOGIC FRAMEWORK

The Frio Formation was deposited along the northern margin of the Gulf Coast Tertiary basin, an extracratonic basin characterized by rapid subsidence in areas of sediment loading. The Tertiary infill accumulated to great thicknesses where the continental platform began to build toward the Gulf beyond the underlying Mesozoic shelf margin and onto transitional to oceanic crust (McGookey, 1975). Major progradational wedges, such as the Frio Formation, consist of an updip and relatively shallow section of interbedded continental and marginal marine sandstone and mudstone underlain by several thousand feet of undercompacted slope and basinal mudstone. Resultant instability due to rapid loading of plastic, water-saturated sediment at the crest of a gently inclined subaqueous slope resulted in large-scale, syndepositional down-to-the-coast faulting and intrastratal deformation.

Three major structural provinces can be defined on the basis of syndepositional structural style and nature of the diapiric material caught up in the deformational process (Fig. 2). The broadly defined Houston embayment of east Texas is characterized by salt diapirism and associated faulting and large salt withdrawal subbasins (Bebout et al, 1978). Across the San Marcos platform and southward toward the Rio Grande embayment, underlying salt is thin or absent, and long, linear belts of growth faults and associated shale ridges, massifs, and diapirs dominate the structural style (Bruce, 1973; Bishop, 1978). In the Rio Grande embayment, large but more discontinuous belts of growth faults and deep-seated shale ridges and massifs occur. In addition to the strike-parallel fault trends, subsidiary, oblique trends of deep, low-relief anticlinal structures are particularly prominent in this embayment (Fig. 2).

The effect of structural dislocation decreases upsection within the major wedges. Consequently, updip portions of the Frio are little affected by faulting that is dominantly manifested in the underlying Eocene section, and dips are gentle and uniformly basinward. As the Frio section thickens gulfward, major fault and diapiric displacement extends up into and through the unit, and structure becomes increasingly complex and segmented. The Vicksburg flexure, a uniquely continuous, narrow fault zone characterized by extreme vertical displacement of the underlying section, forms the updip limit of significant structural deformation of the Frio (Loucks, 1978).

Stratigraphic Correlation and Relationships

Lateral and vertical stratigraphic relationships of the Frio as defined in this report are illustrated in Figure 3. In both the Houston and Rio Grande embayments, the outcrop and shallow subsurface Catahoula Formation is the updip equivalent of the Frio Formation. The Frio is underlain by the Oligocene Vicksburg Formation, which is thickest and best developed within the Rio Grande embayment. Also, a thin mud-rich unit, called the Frio Clay, is mapped at outcrop in south Texas. Contrary to older published sections, this thin unit is now believed to correlate in part to both the Vicksburg and the lowest Catahoula-Frio of the deep subsurface (Baker, 1978; Galloway and Kaiser, 1980). The confusion of names is a relic of repeated redefinition of type sections during the early 1900s. Downdip, the Frio is regionally overlain by an onlapping transgressive marine shale wedge, the Anahuac Shale.

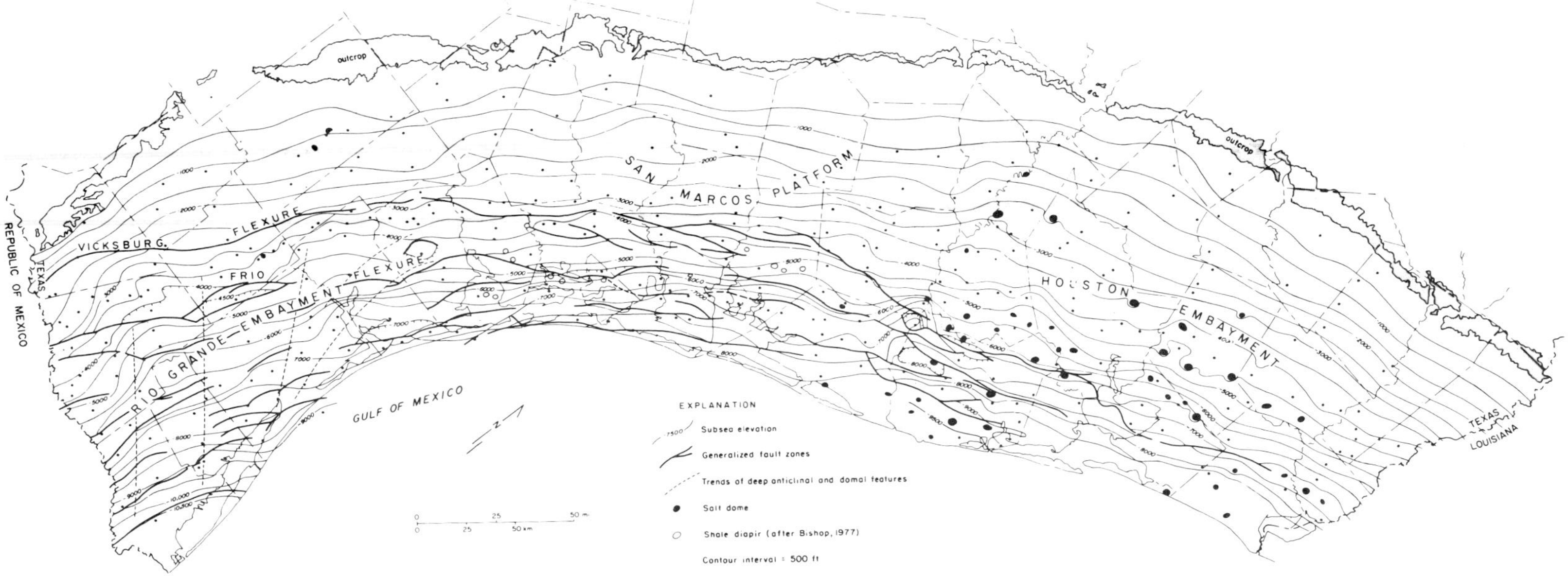

FIG. 2—Frio structure map. Contours are drawn on top of Frio Formation. Major fault zones and diapirs affecting all or part of Frio interval are indicated schematically. Generalized fault trends and location of diapiric structures derived from published and unpublished sources.

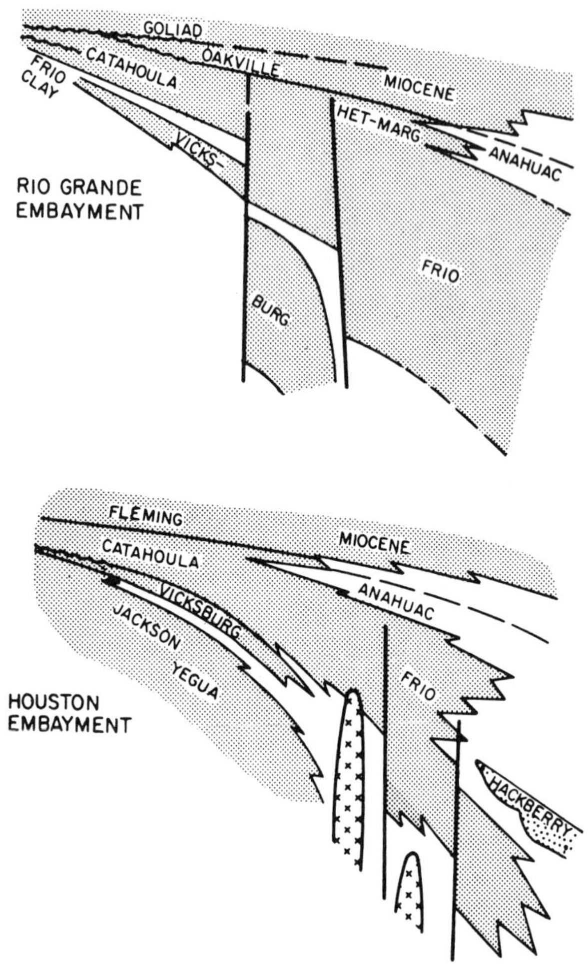

FIG. 3—Schematic Frio stratigraphic correlation sections for Houston and Rio Grande embayments, Texas Gulf coastal plain, showing commonly applied nomenclature and principal bounding units.

In the Rio Grande embayment, the updip extent of the marine shelf incursion was limited by the continued active influx of coarse sediment, commonly called the *Heterostegina* and *Marginulina* sands. Renewed progradation of the Miocene Oakville (south Texas) and equivalent lower Fleming (east Texas) terminated the Anahuac transgression. The westernmost edge of the Hackberry facies of Louisiana extends into the Houston embayment and lies within middle Frio equivalent marine facies (Fig. 3).

Within the global framework of sea level cycles defined by Vail et al (1977), progradation of the Frio wedge was initiated at or slightly before a major fall. Deposition then continued through a period of generally slowly rising sea level punctuated by sharp, temporary rises during early Miocene time. Following the Anahuac transgressive event, renewed progradation and seaward advance of the shoreline predated the middle Miocene highstand by several million years.

The grid of stratigraphic cross sections was correlated with outcrop, regional depositional units such as the Anahuac, and paleontologic zones in key wells to provide an internally consistent framework for further subdivision and mapping of the Frio. Analysis of the intermediate depth strike section $\beta\beta'$ suggested a regionally correlative lithostratigraphic subdivision of the MSU into upper, middle, and lower operational units of subequal thickness (Fig. 4). Projection of these broadly defined operational units downdip established their relation to paleontologic markers.

The cross sections show that the gradual coastward thickening of the updip Catahoula-Frio is accentuated across the Vicksburg flexure, and it increases markedly at the next downdip belt of faults. The documented thickness exceeds 10,000 ft (3,000 m) in the Houston embayment and 12,000 ft (3,600 m) in the Rio Grande embayment.

Correlation of the three operational units further defines the nature of the basinward expansion of the Frio. In central and southern portions of the coastal plain, three fault-defined depositional subbasins are recognized (Fig. 1). The updip basin (I) is characterized primarily by thickening of the lower Frio operational unit across the underlying Vicksburg flexure. Subbasin II lies downdip of the well-integrated series of major growth faults that constitute the Frio flexure. The lower Frio expands tremendously, and it rapidly loses sand content farther downdip. The middle Frio also displays significant expansion. The third subbasin (III) is well developed only within the Rio Grande embayment and lies downdip of a somewhat diffuse, irregular zone of growth faults. Within this subbasin, the middle Frio operational unit expands greatly, and the lower Frio is rarely penetrated by drilling. The subbasins do not extend as recognizable entities into the salt dome province of the Houston embayment. Regional strike sections were intentionally designed to lie within each of the major fault-defined subbasins (Fig. 1).

Primary Facies Mapping

Several hundred wells were projected into the cross-section framework, correlated, and used in the preparation of a suite of primary facies maps that form the basis of further interpretive mapping, geologic reconstruction, and volumetric calculations. Primary facies maps, including net sand isolith maps, sand percentage maps, and electric log facies maps, were constructed for each of the three intervals. The complete suite of maps is reproduced in Galloway et al (1982). Only the three isolith maps and the middle Frio sand percent and log facies maps are reproduced here as Figures 5 to 9. The preparation of the isolith and percentage maps was based on summation of total sand, interpreted from electrical logs, within each of the operational units. Log facies maps were generated by establishing a suite of type log segments representative of patterns typically encountered in the Frio (Fig. 9) and by visually classifying all logs used on the cross sections using the type segments for guidance. Log facies can be grouped as aggradational (sharp-based, blocky, or upward-decreasing deflections), progradational (thin to thick transitional lower deflections on sand bodies), or mixed suites (Fig. 9). The most common log facies for each operational unit was then transposed to the map view to provide an overview of geographic distributions. The lateral and vertical distributions and the relationships of progradational and aggradational sequences provide important criteria for the in-

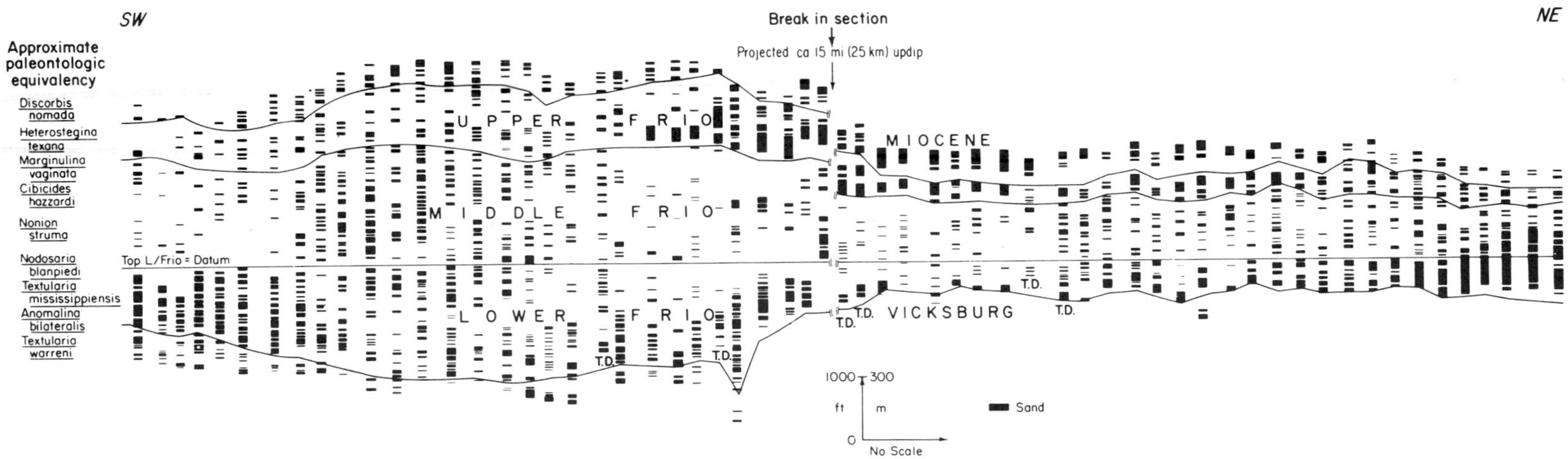

FIG. 4—Regional strike section $\beta\beta'$ displays lithostratigraphic basis for subdivision of Frio section into three operational units for facies mapping. Downdip correlations with the marine section defines relationships of unit boundaries to conventional micropaleontologic zones.

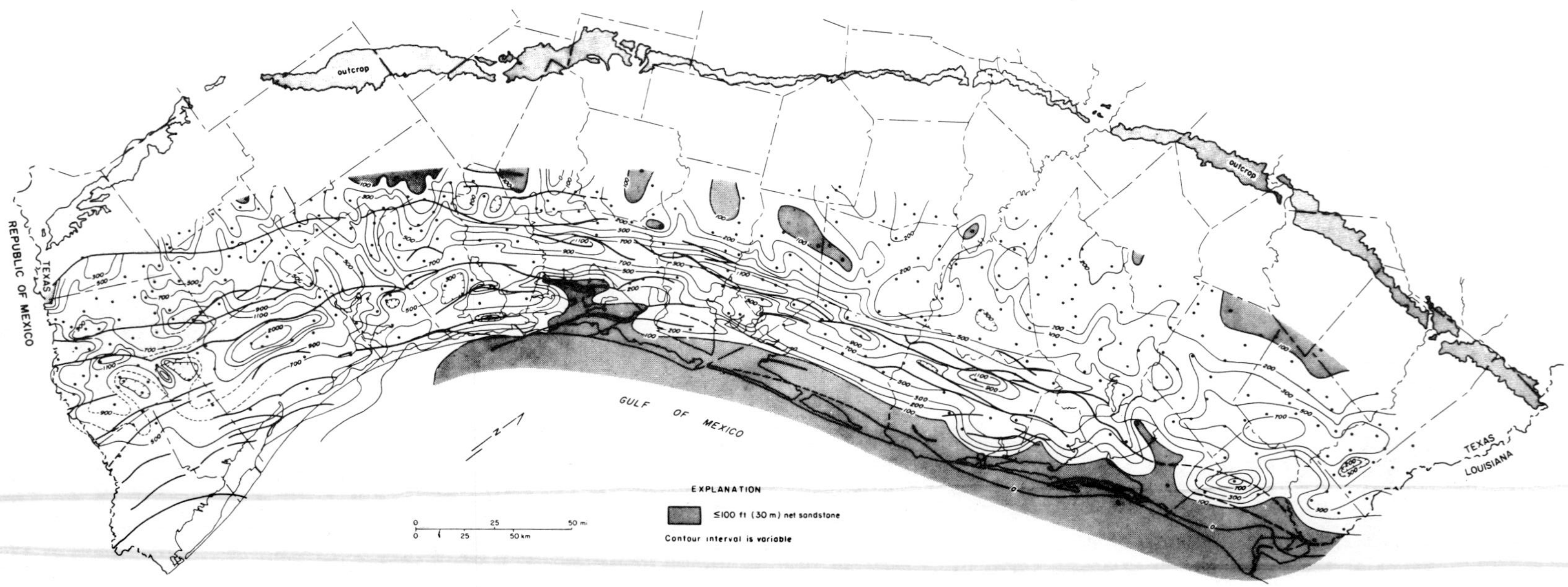

FIG. 5—Net sandstone isolith map of lower Frio operational unit.

terpretation of depositional environment and history (Frazier, 1974).

Comparison of the primary facies maps for each operational unit reveals several important features.

1. All maps show a general areal parallelism of trends.

2. Total sand, as expected, is particularly sensitive to major changes in gross thickness associated with expansion into fault-defined or salt-withdrawal subbasins.

3. Sand percentage best reflects primary facies and reservoir trends, particularly where abrupt thickness changes resulted from syndepositional structural movement.

4. Upper Frio trends are displaced inland relative to the lower and middle Frio intervals which show a generally coastward progression, particularly in south Texas.

5. Three geometrically different contour patterns characterize each map. In both the Rio Grande and Houston embayments, dip-oriented trends expand downdip into irregular, lobate patterns. Between, extending across the San Marcos arch, facies boundaries and contours are strongly strike oriented, particularly near the coast.

DEPOSITIONAL SYSTEMS

Division of the Frio MSU into a number of distinct but spatially related depositional systems (Fig. 10) was accomplished primarily on the basis of sand distribution and trends as delineated by regional facies maps of the upper, middle, and lower Frio operational units (Figs. 5 to 8) and by the areal and vertical distribution of log facies as defined by maps and key cross sections (Figs. 9, 11, 12, and 13). Fundamental procedures in the recognition and delineation of these genetic units followed the approach of Fisher and McGowen (1967) but used a greater density of well control.

Major progradational complexes, designated the Houston and Norias delta systems, are centered on the Houston and Rio Grande embayments, respectively (Fig. 10). The Houston delta system is characterized by numerous small, laterally dispersed lobes, in contrast to the more clearly defined, vertically stacked lobes of the considerably larger Norias delta system. These delta systems in east and south Texas were supplied by fluvial systems of a different character. The east Texas Chita-Corrigan fluvial system consisted of several rivers carrying a mixed load of sand, silt, and clay, whereas the Gueydan system to the south was a single, considerably coarser grained river (Galloway, 1977).

Separating these two major structurally controlled fluvial-deltaic facies tracts was an area in central Texas spanning the San Marcos arch that was traversed by relatively minor coastal plain streams. Deposition of the Moulton/Choke Canyon/Flatonia stream-plain system in this area was dominated by silt and mud accumulation, with widely spaced, minor belts of coarser grained stream deposits that are best developed in its western half. Shore-zone sediments of central Texas were derived in part along strike from adjacent deltaic depocenters, and together with reworked sediments of the small interdeltaic streams, they built up the substantial system of Frio coastal barriers and beach ridges first documented by Boyd and Dyer (1964). This trend is termed the Greta/Carancahua barrier-strand-plain system in the present study. The similar, but smaller Buna strand-plain/barrier system developed on the eastern flank of the Houston delta system and extends toward the Louisiana border. Updip from this zone of accreting beach ridges was an extension of the Chita-Corrigan fluvial system. Fluvial sedimentation was progressively less active eastward where muddy coastal lake and stream-plain deposits are abundant (Fig. 10).

Basinward Frio elements include a shelf platform initially inherited from Jackson and Vicksburg deposition, and a broad unstable prograding slope, which in east Texas was deeply incised and embayed. The extensive, deep-water Hackberry embayment of Louisiana continued into Orange and Jefferson Counties, and was possibly represented in other smaller slope canyons and embayments along the Texas coast (Stuckey, 1964). Although the 2,000 ft (600 m) thickness of Hackberry in Jefferson County represents a reduction from the maximum in Louisiana, both the "Lower Hackberry Sand" and "Upper Hackberry Shale" are recognized. Together these correspond approximately to the middle Frio operational unit of the present study.

Smaller, strike-aligned intraslope basins elsewhere along the Texas coast were a product of growth faulting. These intraslope basins locally trapped large quantities of sediment transported by slumping and density flow.

Gueydan Fluvial System

The Gueydan fluvial system records the entry of a major extrabasinal river into the Gulf basin along the axis of the Rio Grande embayment (Galloway, 1977). The system is characterized by coarse-grained, bed-load channel fills, and sections commonly contain from 15 to more than 40% sand.

Location and boundaries.—The mapped extent of the main drainage axes of the system averages 75 mi (120 km) in width and trends 50 to 70 mi (80 to 110 km) downdip, where it merges with the Norias delta system. In the middle and upper Frio sequence, fluvial facies extend across the Vicksburg flexure. The system is bounded along strike by sand-poor stream-plain facies.

Facies description.—The principal component facies of the Gueydan fluvial system include bed- and mixed-load channel-fill sand and conglomeratic sand and laterally associated crevasse-splay sand and flood-plain mud and siltstone (Galloway, 1977). Channel-fill sequences are 10 to 30 ft (3 to 9 m) thick, and commonly stack, producing sand bodies 50 to 100 ft (15 to 30 m) thick (Fig. 11). Channel sand units also amalgamate laterally, forming sand-rich belts up to several miles wide. Log patterns are typically blocky, with sharp bases and abrupt to gradational tops (aggradational log facies a, Fig. 9).

Overbank splay and flood-plain sequences consist of massive mudstone with thin, discontinuous sand beds and lenses. Aggradational log facies b-type sequences typify such deposits.

Structural style.—Fluvial sequences dip gently basinward. Low relief, broad domes and anticlines, as well as some minor fault displacement occur, particularly where the system overlies deep-seated Eocene or Oligocene growth fault trends. Structural modification is greatest where the Gueydan system crosses the Vicksburg flexure.

Reservoir and trap configurations.—Hydrocarbons are produced where fluvial and splay sands encased in the more mud-rich flood-plain sequences cross anticlinal structures. Production from such large, low-relief features is typically

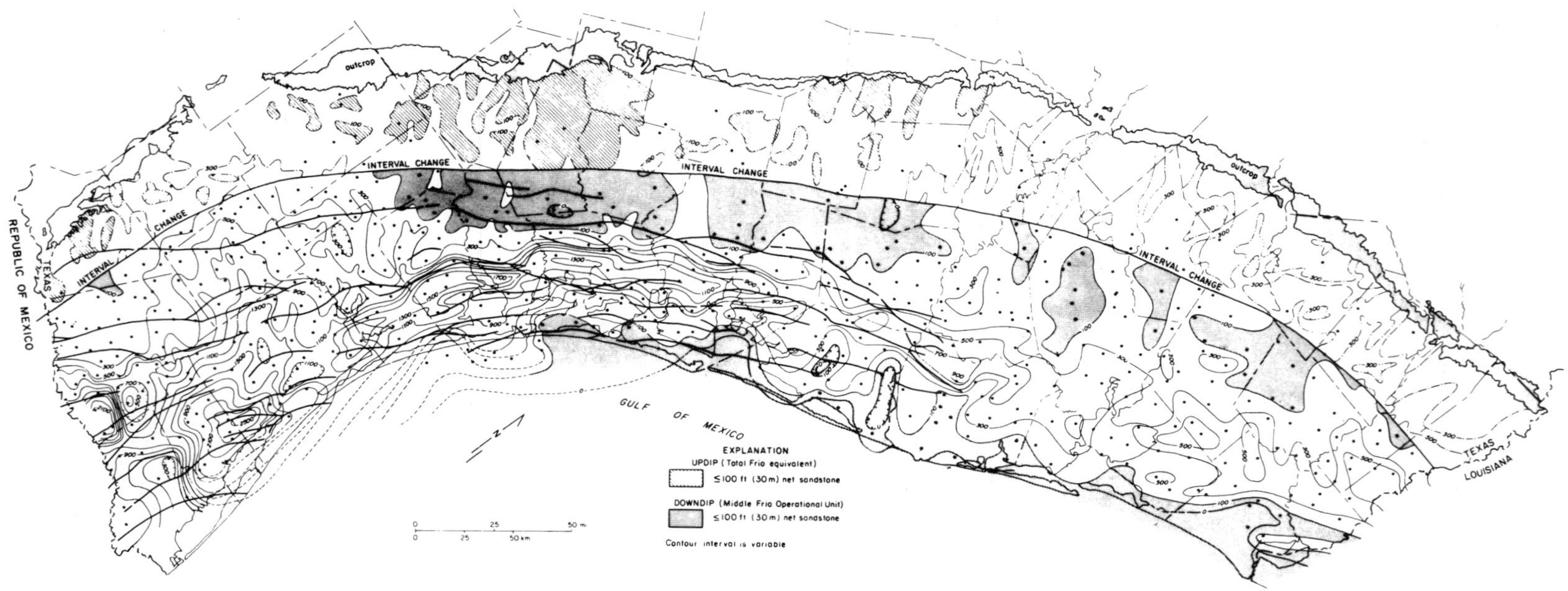

FIG. 6—Net sandstone isolith map of middle Frio operational unit. Updip, broken contours show sand content of undivided, Frio-equivalent Catahoula Formation.

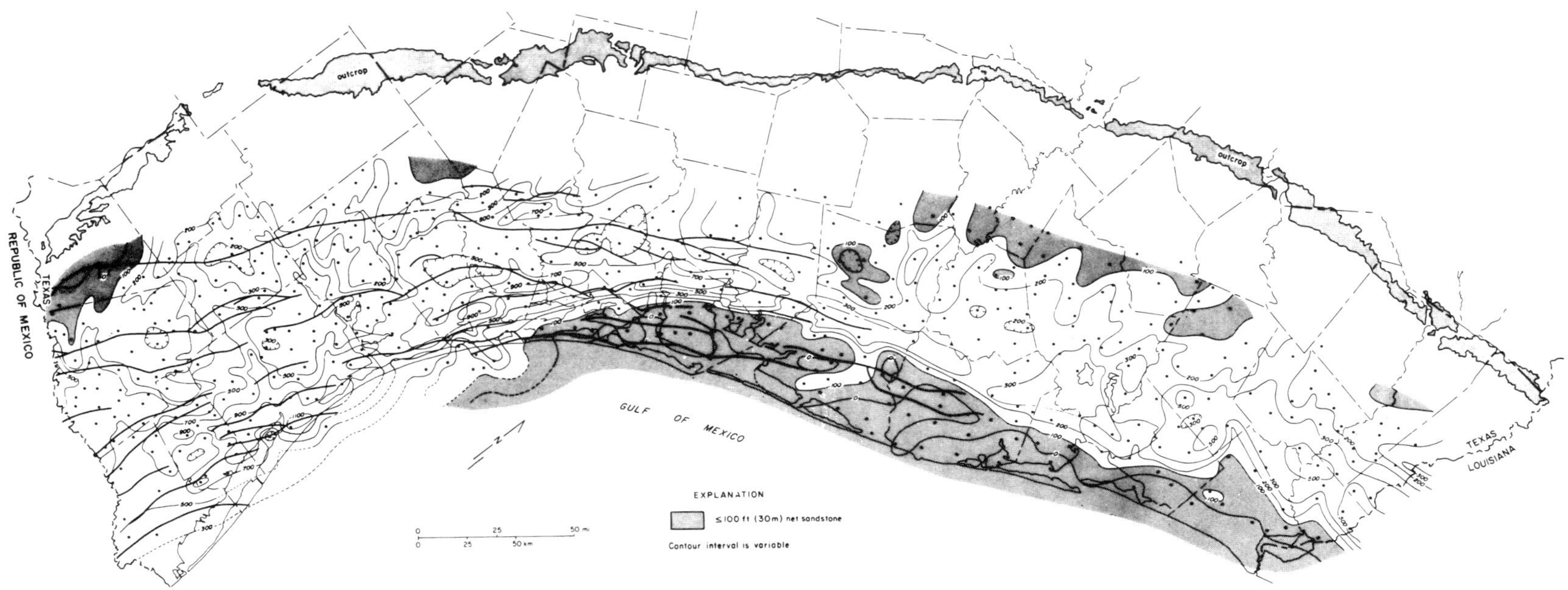

FIG. 7—Net sandstone isolith map of upper Frio operational unit.

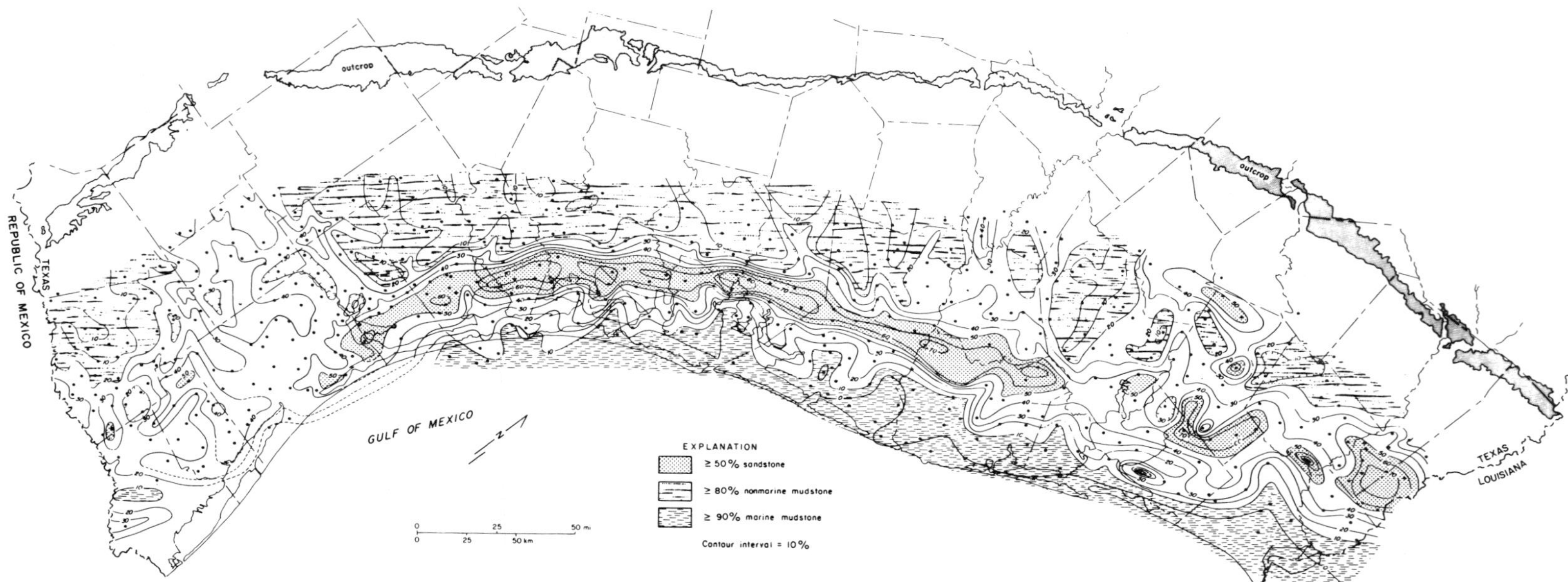

FIG. 8—Sandstone percent map of middle Frio operational unit.

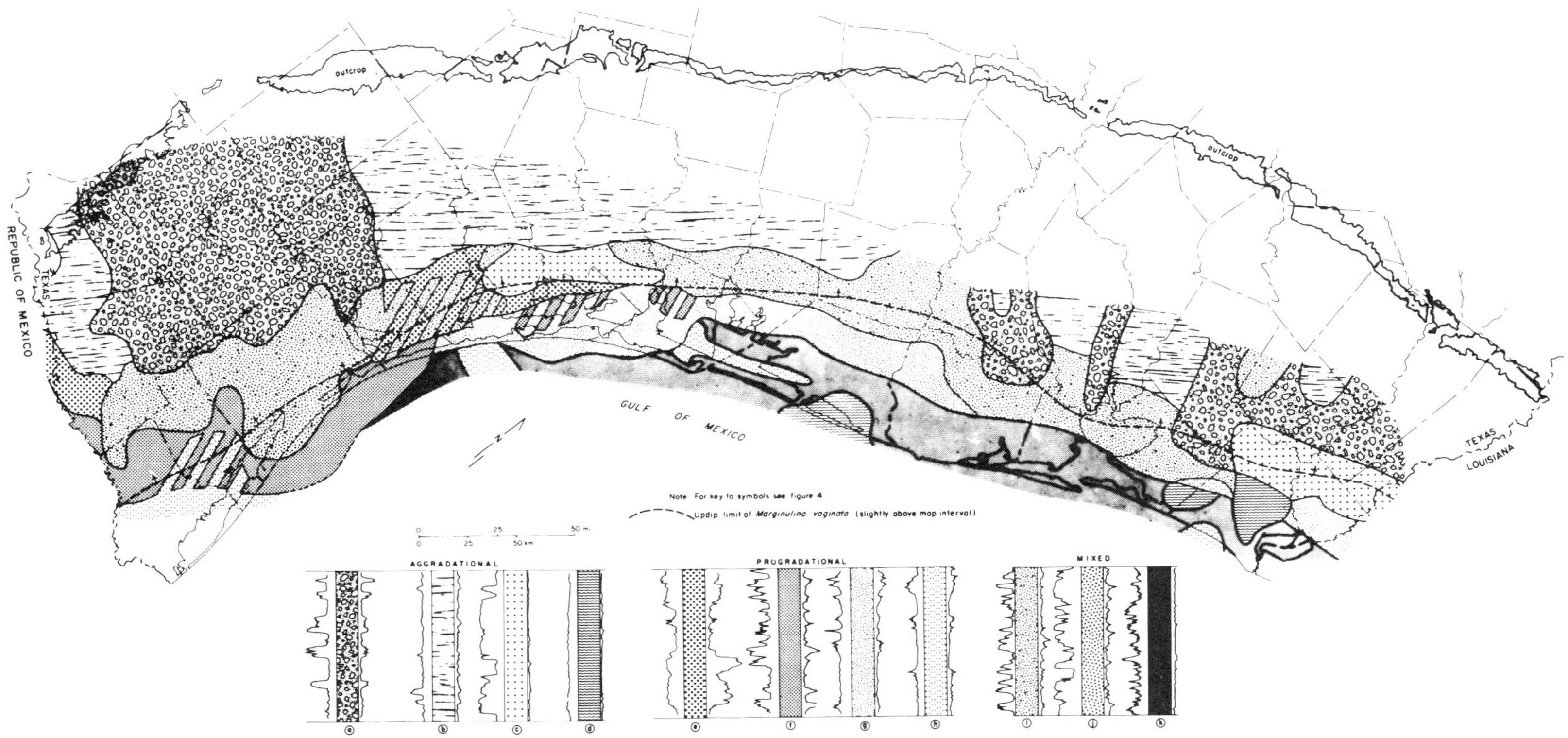

FIG. 9—Electric log (SP and short-normal resistivity) facies map of middle Frio operational unit. Log facies types are defined on basis of dominant vertical sequence (aggradational, progradational, or mixed), thickness of lithologic sequences, and abundance of sandstone. Aggradational facies include (a) multiple blocky sand bodies within sand-rich sequence, (b) thin, fining-upward to blocky sands within sand-poor sequence, (c) thick, blocky sand bodies in sand-rich sequence, and (d) massive mudstone. Progradational sequences are characterized by common coarsening-upward bases to sand bodies, and include (e) thick sand bodies within sand-rich sequence, (f) multiple thin to medium, erratically coarsening-upward sand units, (g) multiple thin sand units, and (h) thick, well-developed coarsening-upward units within mud-rich section. Mixed log facies types include (i) composite sequences of coarsening- and fining-upward, thin to medium sand bodies, (j) blocky, thick sand bodies within sand-dominated sections, and (k) thinly interbedded sand and mud forming thick packages that display both transitional and abrupt unit boundaries.

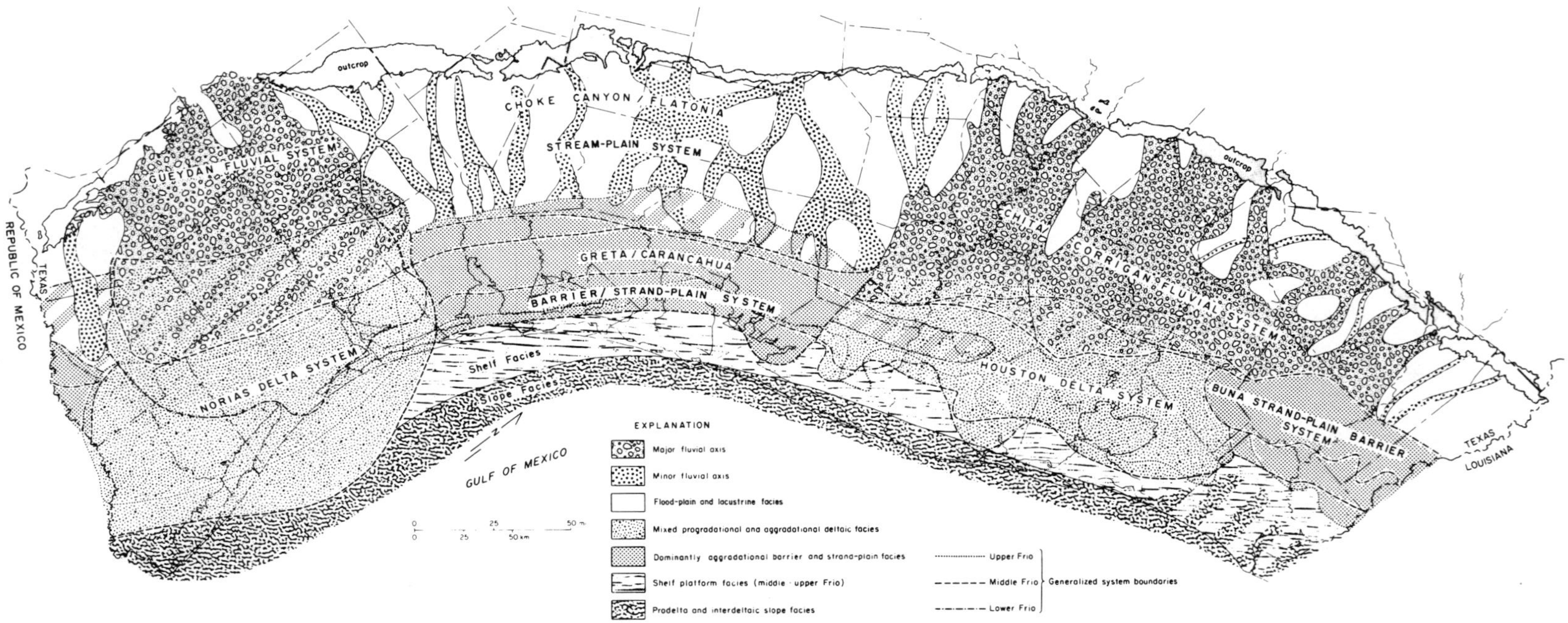

FIG. 10—Depositional systems of Frio Formation.

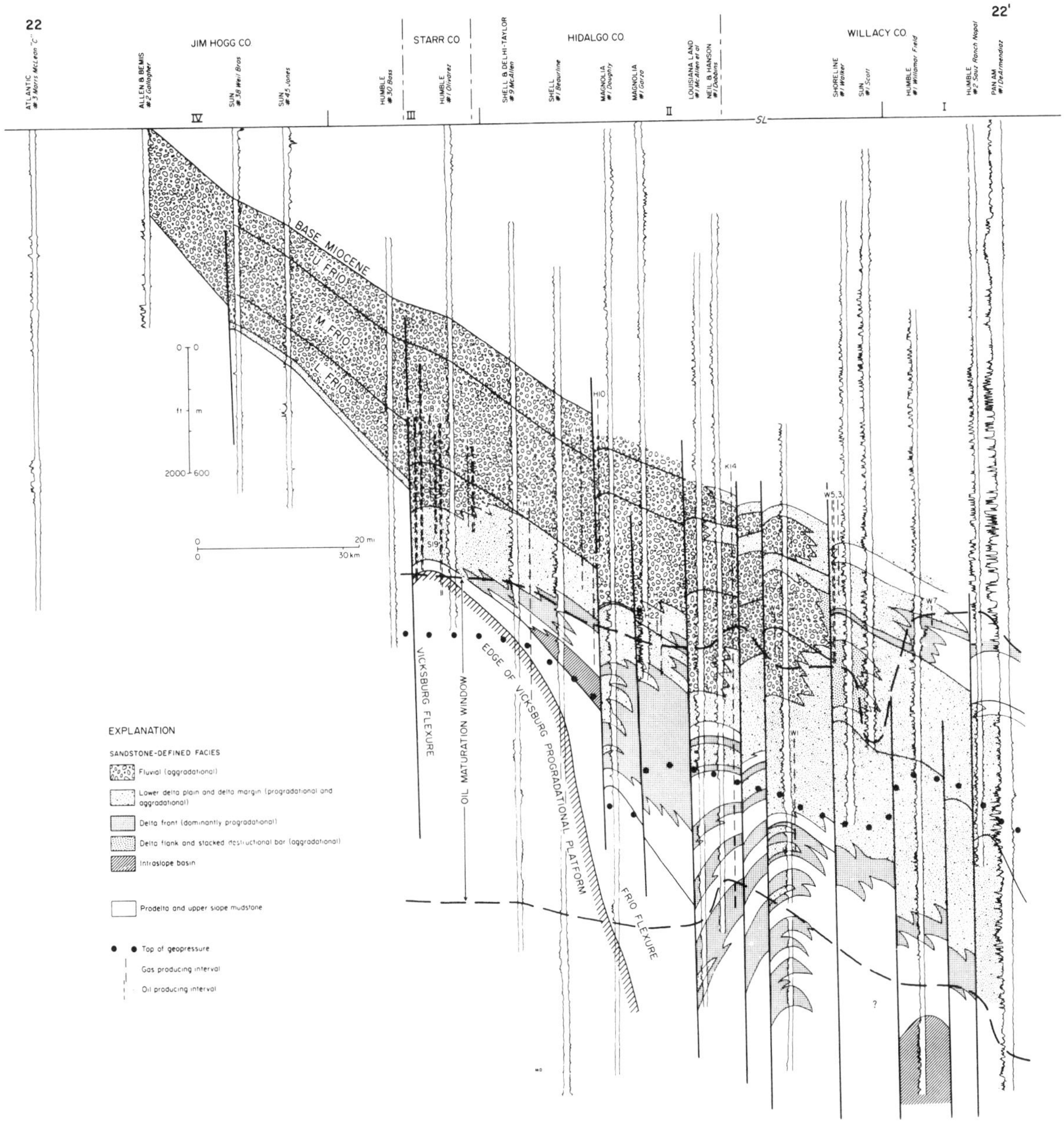

FIG. 11—Stratigraphic dip section 22-22′ through Gueydan fluvial system and Norias delta system. In addition to correlation markers and interpreted depositional facies assemblages, section shows (1) approximate vertical distribution of oil and gas reservoirs in fields that lie on or adjacent to section, (2) approximate top and base of hydrocarbon liquid window, and (3) top of geopressured zone. For location of section, see Figure 1.

complicated by faulting and by stratigraphic changes. Some stratigraphic entrapment also occurs.

Norias Delta System

The Norias delta system constitutes the main Frio depocenter in the south Texas coastal plain. The sand content ranges from 25 to 40 % of a Frio section that is over 12,000 ft (3,700 m) thick. Deposition of the system resulted in a progradation of the continental margin more than 60 mi (95 km) basinward, primarily during deposition of the lower and middle Frio operational units. This major offlapping episode was terminated by the Anahuac transgression, but the rate of sediment supply to the Norias system was sufficient to severely limit updip incursion of transgressive marine shelf facies (Fig. 11). The upper Frio *Heterostegina-Marginulina* deltaic complexes continued to prograde locally across the Frio platform in the face of regional onlap (Agagu, 1975).

Location and boundaries.—Unlike the Houston delta system, the Norias delta is the product of shifting drainage axes of a single large fluvial system. Consequently, lateral boundaries of the system remained fairly fixed through time (Fig. 10). The maximum width of the system exceeds 125 mi (200 km), and seismically derived velocity mapping (Domenico, 1980) indicates that sand-rich deltaic facies extend up to 10 or more mi (16 or more km) offshore from Kenedy County. Downdip, and at great depth, deltaic facies, including genetically associated prodelta and upper slope deposits, grade into deep-water slope deposits of the offlapping continental platform. Northward, along strike, the system grades into the Greta/Carancahua barrier/lagoon system. Rapid and extensive progradation of the Norias delta at the south end of the barrier-strand-plain coastal bight is probably responsible for the development of a relatively sediment-starved delta margin embayment, which was barred by strike-reworked coastal barrier sandstones, thus producing a barrier-lagoon couple. To the northeast, slow progradation of the Houston delta system produced a marginal strand-plain lacking well-developed lagoonal or embayment facies separating laterally reworked strandline sands from subaerial coastal plain deposits. An analogous barrier-strand-plain complex appears to flank the Norias delta system to the south, but its characteristics are poorly defined by present mapping, which terminates at the Rio Grande.

Facies description.—Regional sand distribution patterns, as well as local sand-body configurations (Weise et al, 1982), indicate deposition of the Norias delta system as a mosaic of wave-modified lobate to wave-dominated cuspate delta lobes (terminology of Galloway, 1975). Component facies assemblages include delta plain, delta front, prodelta and upper slope, and delta flank sequences (Fig. 11).

Delta-plain and closely associated reworked delta and splay margin sandstones include deposits of distributary channel, crevasse splay, and destructional bar and beach ridge environments. Distributary and larger splay channel fills form blocky to fining-upward, coastward bifurcating sand bodies 20 to 60 ft (6 to 20 m) thick (Nanz, 1954). Several such units commonly stack, forming thick belts 1 to 3 mi (1.5 to 5 km) wide, which also contain interbedded splay and destructional sands. Larger splays form laterally discontinuous, coarsening-upward sequences interbedded with delta plain and interdistributary bay mudstone. Mixed progradational and aggradational patterns represented by log facies i typify the facies assemblage. Bedded organic debris is notably absent in the Norias delta plain facies, probably a reflection of the dry, subarid middle to late Tertiary paleoclimate that characterized south Texas (Galloway, 1977). Sandstones and shales do, however, contain abundant detrital carbonaceous debris.

Delta-front facies include dominantly progradational sequences of channel-mouth bar, laterally equivalent wave-reworked delta margin, and subaqueous delta-front splay deposits, all of which are interbedded with and cap prodelta mud and siltstone. In updip, or shallower positions within the main body of the Norias system, delta-front sand bodies are commonly less than 50 ft (15 m) thick and stack to produce repetitive cycles (log facies f, Fig. 9). In deeper, or more distal settings where the deltas prograded directly onto continental slope and into rapidly subsiding growth-fault subbasins, the overall vertical sequence increases to as much as several hundred feet, and cyclic progradation of successive delta lobes across active growth faults produced composite sequences several thousands of feet thick (wells 24 and 25, Fig. 11). Large-scale slumping, faulting, and soft-sediment deformation, as observed at the modern Mississippi delta front (Coleman and Prior, 1980), dominated Norias delta fronts. The thickness of distal delta-front splays (initiated by sporadic slumping, debris flow, or flood discharge) increases greatly in proportion to the wave-reworked channel-mouth bar or delta-margin sandstone cap, producing thick coarsening-upward sequences characterized by highly erratic or serrated log patterns (log facies h, Fig. 9). Across some of the largest shelf-edge growth faults, where genetically equivalent sequences may thicken by an order of magnitude, remobilization of delta-margin sediment by slumping and density current flow may have decoupled significant quantities of sand from the delta front, moving it down the prodeltaic upper slope into isolated, strike-elongate intraslope subbasins (Fig. 11). Downslope damming by intraslope basins, and consequent limited basinward extent of frontal splay and slump sands, is an inherent product of growth-fault mechanics (Dailly, 1976; Crans et al, 1980). Logs in such highly expanded delta-front and intraslope basin settings include both blocky and highly serrated patterns, as shown in Figure 9, log facies k.

Delta-flank facies include thick, up to 100 ft (30 m) or more, massive sand bodies characterized by transitional coarsening-upward bases and located primarily along the flanks of the Norias delta system. Such sand bodies are dominantly strike parallel to lobate and continuous over several miles. Sand-rich progradational log facies patterns typify the assemblage, which is interpreted to consist of wave-reworked, stacked coastal barrier and thick delta-front sandstones deposited along the flanks of maximum progradation and fluvial influence during delta building. The facies grades laterally into massive stacked barrier sandstones of the interdeltaic coastline.

Structural style.—Sediments of the Norias delta system are segmented by multiple belts of coastward-dipping growth faults. Faulting is spatially related to the development of deep shale ridges and dip-reversals or "roll over" on the downthrown side. The greatest fault activity accompanied deposition of delta-front and prodelta facies; thus the age of fault development tends to decrease basinward, reflecting the offlapping pattern of sedimentation (Fig. 11).

Reservoir and trap configuration.—Petroleum production from distributary channel, crevasse-splay, delta-front, and delta-flank sandstones occurs throughout the delta system. Interbedded shelf or thick prodelta shales are required to provide the necessary seal. Traps include shale-cored rollover anticlines, faulted anticlines, and less commonly, closure against faults on their upthrown side.

Chita/Corrigan Fluvial System

The Chita/Corrigan mixed-load fluvial system constitutes a broad belt approximately 150 mi (225 km) wide and extending more than 50 mi (80 km) into the subsurface across the upper Texas coastal plain. Maps and sections suggest the contemporaneous existence of at least three main drainage

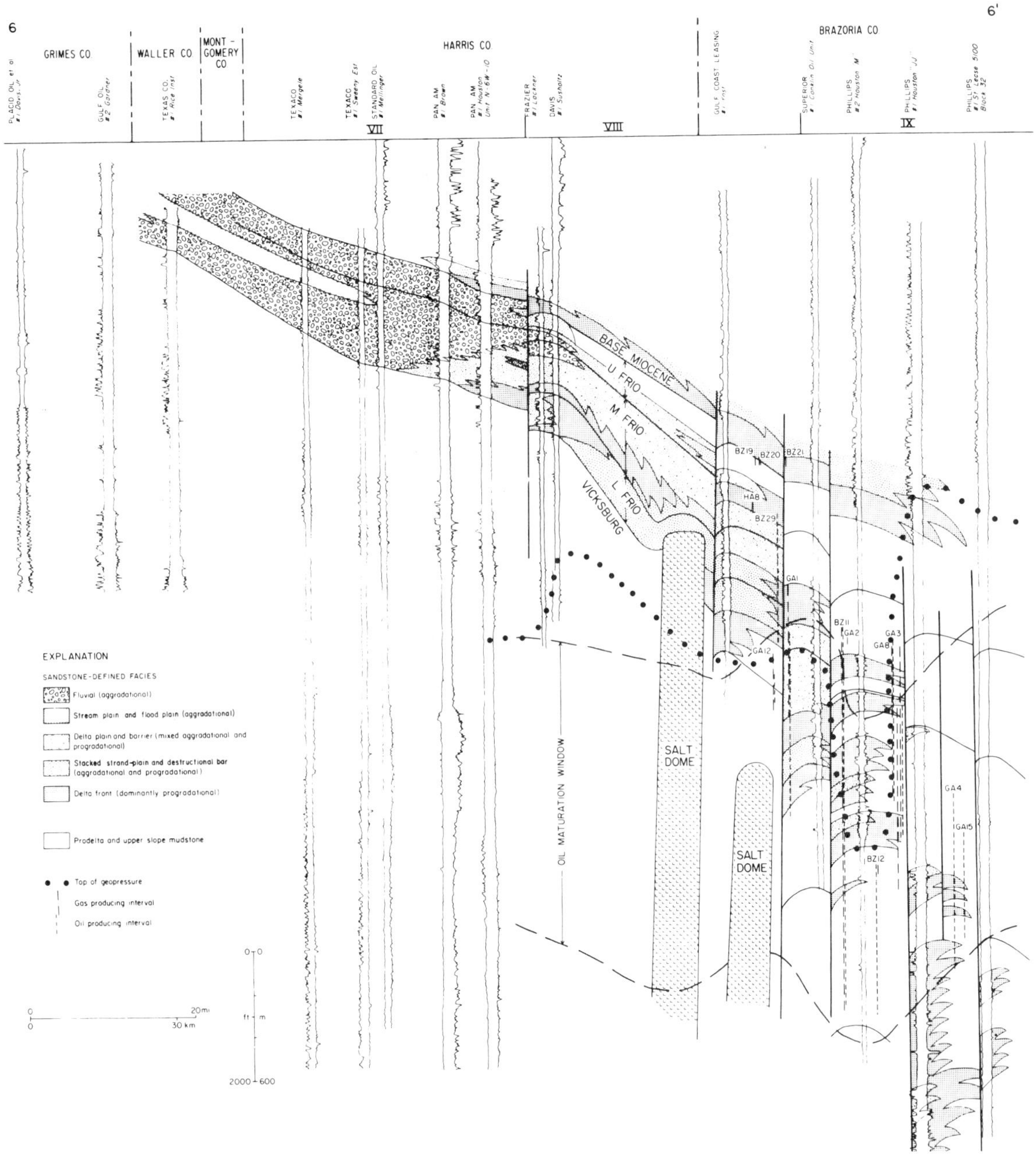

FIG. 12—Stratigraphic dip section 6-6′ through Chita/Corrigan fluvial system and Houston delta system. For location of section, see Figure 1.

axes (Galloway, 1977). Component facies include meanderbelt and crevasse-splay sand bodies encased within floodplain and interchannel lake mudstone, and produce type a and b aggradational log facies sequences. The regional coastward dip of the system is little affected by structure except in the immediate vicinity of a few shallow piercement salt domes.

Houston Delta System

Several minor, laterally coalescent and vertically repetitive deltaic cycles make up the Houston delta system (Fig. 10), the main locus of terrigenous accumulation in the Frio Formation of east Texas. The system was fed by streams probably comparable in size to the modern Trinity, Brazos, and Colorado rivers, but unlike these modern analogs, the deltas periodically extended beyond the main strandline into the open Gulf, and thus they were subject to the brunt of deep-water wave attack. Marine processes redistributed deltaic sand along strike, supplying strand-plain systems downdrift. Furthermore, there was constant switching in the position of the delta lobes, accompanied by destructional marine reworking and inundation of the abandoned site. Consequently, the rate of coastal progradation was slow, and, in contrast to the comparatively abrupt northern margin of the Norias delta system, there is a very broad transition zone with the Greta/Carancahua barrier/strand-plain system (Fig. 10). A further consequence of the limited fluvial influx was the extensive submergence of the Houston delta system by the Anahuac transgression (Fig. 12). The thickness of the total Frio interval corresponding to the Houston delta system ranges between 1,800 and 7,500 ft (550 and 2,300 m), with the sand content decreasing distally from a local maximum of 60 to 10% (Fig. 8).

Location and boundaries. —The Houston delta system extends 130 mi (200 km) along strike and has a width of 30 to 45 mi (50 to 70 km). It underlies parts of nine countries, but its most typical and consistent development is centered in southern Harris County. Whereas the boundaries with the adjacent barrier and strand-plain systems fluctuated substantially, the updip and downdip boundaries were relatively constant, shifting within narrow limits in response to both local and regional patterns of transgression and regression. The resulting superposition of successive progradational sequences within a coast-parallel tract contrasted with the broadly prograding Norias deltaic headland.

Facies description.—Vertical log patterns and the areal distribution of sandstone bodies reflect deposition by frequently shifting delta systems of limited extent. Updip delta systems probably corresponded to the wave-dominated arcuate category of Galloway (1975), whereas a lobate configuration was characteristic of units deposited either during episodes of maximum progradation or in areas of accelerated subsidence associated with salt withdrawal (Bebout et al, 1978). Facies of the Houston delta system include delta plain, back barrier, strand plain and destructional bar, delta front, and prodelta-upper slope (Fig. 12).

Strand-plain and destructional bar facies comprise blocky, strike-oriented sandstones of log facies c (Fig. 9) between 50 and 1,000 ft (15 and 300 m) thick, which lie above, or updip of delta-front facies. They resemble sandstones of the broad interdeltaic tracts, but they are an integral part of the Houston delta system. This facies is notably absent in the most regressive deltaic units. The origin of the strand-plain and destructional bar facies was related to pronounced marine modification of deltaic deposits, probably in part contemporaneous with the progradation of arcuate deltas. The considerable thickness of several of these blocky sandstones is too great to be accounted for by progradation alone. It is more likely that transgressive reworking resulted in aggradation of the destructive bar sands and the development of stacked strand-plain sands along strike.

Delta plain and back-barrier facies are closely associated, and they are distinguished from one another only on the basis of their position relative to equivalent seaward facies, and to a lesser extent on the basis of the great strike continuity of certain back-barrier units. Environments represented include distributary channel and splay channel (blocky or upward-fining sandstones with little lateral continuity along strike), splay subdeltas (thin, upward-coarsening units with moderate strike continuity, which are commonly stacked in multiples totaling several hundred feet in thickness), and destructive barrier sandstones, including washover fans and flood tidal deltas (thin sandstones that are normally gradationally based and strike elongate, and which grade landward into logoonal shales). Together, these units, characterized by log facies i (Fig. 9), constitute a predominantly aggradational succession which gradationally overlies a progradational delta sequence. In contrast to older Tertiary delta systems, such as that of the lower Wilcox (Fisher and McGowen, 1967), bedded lignite is absent.

Delta-front and prodelta-slope facies are represented by upward-coarsening sequences or partial sequences. Two variants are recognized on the basis of sequence thickness, position relative to major growth faults, and sand body geometry. These originated on the shallow shelf and deeper unstable shelf edge, respectively.

Updip progradational cycles of the stable shelf are thin, 50 to 150 ft (15 to 45 m), and vertically repetitive. Log facies g (Fig. 9) represents purely progradational units, whereas log facies f represents a combination of delta-front progradation and delta-plain, strand-plain, and destructional bar aggradation. Some of the sandstone units are broadly lobate, but most are strike oriented as a result of effective longshore redistribution of sand by marine processes acting on arcuate delta fronts.

Thick, growth-faulted sequences of the unstable shelf edge attain a maximum of 1,000 ft (300 m), and on the downthrown side of major growth faults, they are superimposed to a thickness of 3,000 ft (900 m) or more (Fig. 12, log 18). The greater part of these sequences is made up of shale; this shows an upward increase in the proportion of erratic, thin sandstone units, probably representing flood-generated or gravity-resedimented units of the upper prodelta and distal mouth bar. Capping sandstone units of these shelf-edge deltas are relatively thin, averaging between 40 and 100 ft (15 and 30 m), and tend to be lobate or elongate perpendicular to the paleoshoreline (Bebout et al, 1978).

Structural style.—A very complex structural framework includes deep salt anticlines, shallow salt diapirs and associated radial fault patterns, salt-withdrawal basins, and growth faults. Once initiated, growth faulting continued to displace successively higher strata, even though additional loading may have been insignificant; this was due to the inherent instability of the buoyant salt at depth. Continued fault displacement and ready initiation of new faults produced a closely spaced pattern of strike-parallel, broadly arcuate fractures. This is distinct from the more widely spaced, clearly defined regional fault systems of south Texas. Isolated, circulat to ovoid areas of anomalously thick sand accumulation may represent salt-withdrawal basins. Alternatively, some may be related to growth faulting.

Reservoir and trap configurations.—Delta-front and delta-plain sandstones coalesce along strike to form semi-persistent reservoirs. Reservoir continuity in the dip direction is severely limited by facies changes, which for the most part were related to growth faulting. Locally abrupt changes in sand body geometry and thickness are associated with salt diapirs.

The majority of producing traps are related to faulted anticlines above deep salt ridges, or to growth-fault traps, where they are sealed against either the upthrown or downthrown block. Stratigraphic factors contribute to trapping in areas of abrupt facies change, but in all cases, they are subordinate to structure in determining the configuration of the major traps.

Buna Strand-Plain/Barrier System

Location and boundaries.—The Buna strand-plain/barrier system is restricted to portions of Liberty, Jefferson, Hardin, Jasper, Newton, and Orange counties, and continues eastward into Louisiana. Shore-zone progradation continued at a limited rate throughout Frio deposition, even within the upper Frio interval when elsewhere there was a tendency toward shoreline retreat. More importantly, however, was the process of aggradation, producing thick sequences of stacked shore-zone sands.

Facies description.–Strike-oriented sands with a relatively small proportion of finer sediments are the distinguishing feature of the Buna strand-plain/barrier system. Blocky log patterns are common, as are irregularly upward-coarsening sandy successions up to 500 ft (150 m) thick, some of which include subordinate upward-fining packages.

The dominant sandy facies are stacked, aggradational shoreface and beach deposits. Abundant sand carried along strike from the Houston delta system, and possibly from other comparable delta systems in Louisiana, produced a lateral gradation from shallow-water arcuate deltas through strand/plain to barrier/lagoon shorelines. The latter is distinguished by moderate updip development of shore-parallel back-barrier shales with sporadic sands introduced along the seaward margin. The strand-plain coast immediately east of the Houston delta system received some sediment directly from major, but relatively short-lived, fluvial influx.

Subordinate finer grained facies, which vertically separate thick packages of strand-plain and barrier sands, accumulated during episodes of regression in marsh, slough, and logoonal environments, and during episodes of minor transgression in lower shoreface and shelf environments. The blocky shore-zone sands are multistoried and imply prolonged in-situ aggradation (Fisher, 1969; Winker, 1980). Additional variations between upward-coarsening and upward-fining packages are attributable to minor shifts of the strandline position occasioned by variations in rates of sediment input, local subsidence rate, or possibly eustatic processes.

Fine-grained lagoonal facies show complex interfingering along the landward margin of the barrier trend. Sandy beds within the predominantly shale succession probably represent flood-tidal delta and washover deposits.

Offshore, shelf-slope equivalents of the Buna strand-plain deposits include the Texas portion of the Hackberry slope canyon-embayment system. The Hackberry wedge includes erratic upward-coarsening log patterns and elongate basinward-diverging sandstones which led Halbouty and Barber (1961) to interpret a progradational deltaic origin for the lower Hackberry sandstones of Port Acres field, Jefferson County. Subsequent studies by Paine (1971) of equivalent strata in Louisiana established a deep-water origin (Brown and Fisher, 1977), and this has recently been extended by Berg and Powers (1980) into east Texas. In Louisiana, elongate slope channel sandstones, with the typical internal sequence recording waning flow conditions, merge basinward into blanketlike fan deposits, which are thickest over topographic and structural depressions in the sub-Hackberry unconformity. These sandstones grade up into shales containing a deep-water fauna. In east Texas, turbidite channel-fill units are characterized by stacked massive sandstones. Interbedded sandstones and shale along the channel margins merge into interchannel overbank shales with thin turbidite sandstones (Berg and Powers, 1980). In some places, the coeval deeper water fan deposits are not present in the drilled succession but are anticipated downdip. Other canyons may have filled with an onlapping succession of turbidite channel sandstones overlain by more distal fan deposits.

Structural style.—The dominant structures are a combination of growth faults, low-amplitude anticlines underlain by salt, and shallow salt diapirs. There is not the major disruption of sedimentation patterns characteristic of the salt withdrawal basins adjoining on the west.

Reservoir and trap configurations.—The strand-plain and barrier reservoir sandstone are thick, with great lateral continuity along strike, and high values of porosity and permeability. They lack seals, however. Traps are, therefore, largely confined to major growth faults and salt domes, with some deep anticlinal traps.

Choke Canyon/Moulton/Flatonia Stream-Plain System

The middle portion of the Texas Frio coastal plain consists of a broad sand-poor apron containing a multiplicity of thin, narrow, dip-oriented mixed-load to bed-load fluvial channel-fill and meander-belt sand bodies encased in siltstone and mudstone sequences deposited in flood basin, broad open shallow lake, and tributary settings (Galloway, 1977). The section dips gently basinward with minor, local faulting and low-relief anticlinal structures associated with deep-seated Eocene growth fault zones. Numerous shallow fields produce from such low-relief structures and from stratigraphic traps. The system is characterized by aggradational log facies pattern b.

Greta/Carancahua Barrier/Strand-Plain System

Location and boundaries.—The Greta/Carancahua barrier/strand-plain system corresponds in large degree with the barrier-bar system first documented by Boyd and Dyer (1964). It comprises a linear sandstone belt separating marine from brackish water shales, and it extends from Nueces County in the west, where it is gradational with the Norias delta system, to the Houston delta system in the east. The barrier system thus coincides regionally with the updip sand-poor stream-plain and coastal lake system. The locus of sand accumulation shifted 10 to 20 mi (15 to 30 km) basinward during lower to middle Frio deposition before experiencing a transgressive reversal in the upper Frio. As in the example of the Buna strand-plain/barrier system, vertical upbuilding of sands was the main depositional pattern.

Facies description.—Massive, blocky, stacked barrier and strand-plain sands are present through intervals of as much as 4,000 ft (2,500 m), with thinner, irregular shaly interbeds most abundant near the base and top. In cross sction (Fig. 13), the sands pinch out abruptly seaward into shaly shelf facies containing neritic foraminiferal assemblages (Martin, 1969).

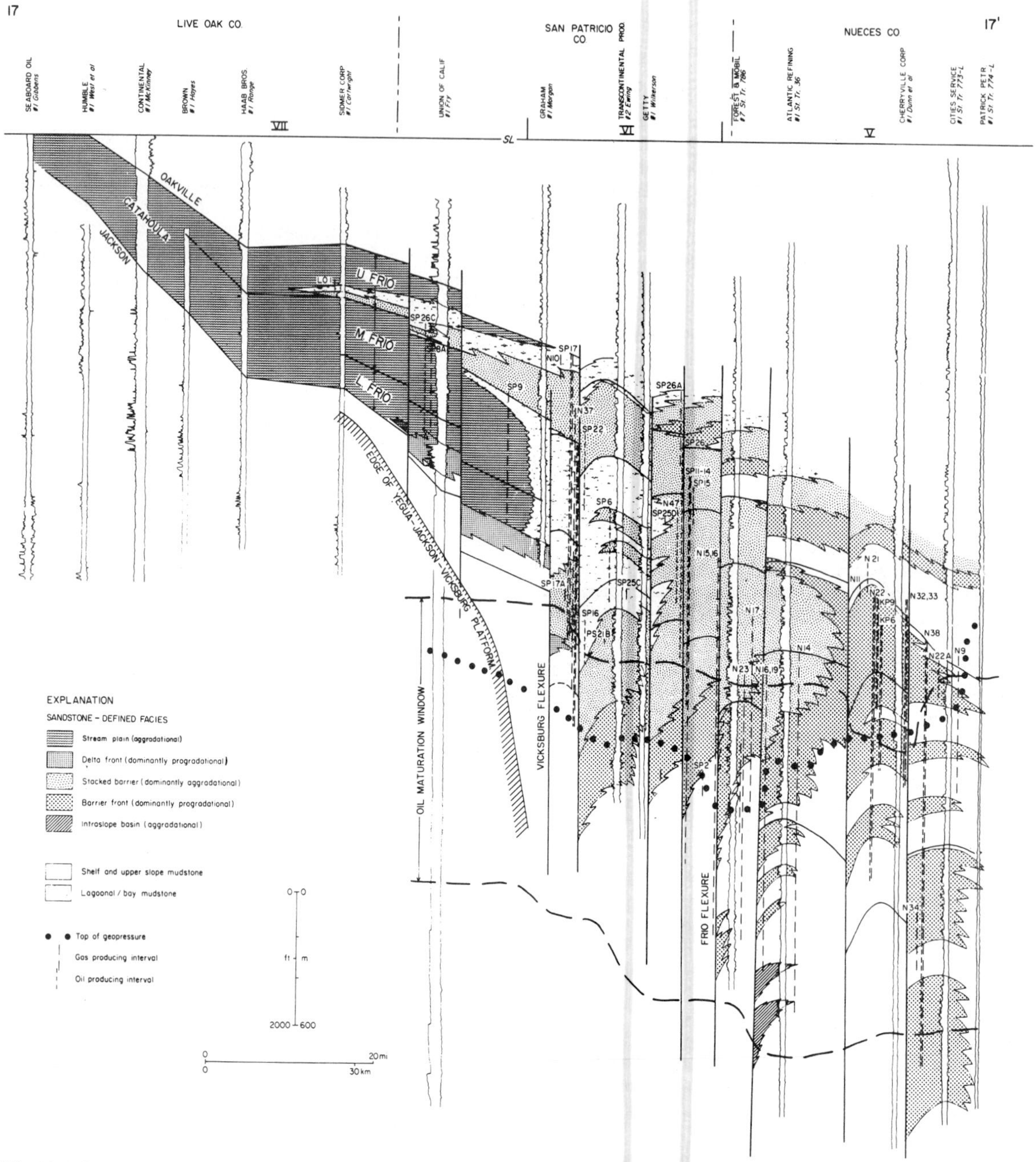

FIG. 13.—Stratigraphic dip section 17-17′ through Choke Canyon-Flatonia stream-plain and Greta-Carancahua barrier strand-plain system. Stacked barrier and updip lagoonal facies are particularly well developed in this area of Frio coastline. For location of section, see Figure 1.

Landward, the gradation into back-barrier lagoon, bay, or lower coastal plain fluvial-deltaic mudstone is spread over a broader area, particularly in the upper parts most affected by marine transgression.

Shoreline conditions remained remarkably uniform during lower and middle Frio deposition. To the south, a wave-dominated barrier coast, probably resembling the Padre-Mustang-Matagorda Island complex of modern Texas barriers, underwent significant aggradation. Northward, many small streams fed sediment directly into the Gulf, where sands were combined with strike-reworked sediment to form a complex assemblage of strand-plain and delta destructional facies. A large proportion of these sands was preserved during ensuing transgressions because of continued subsidence. Analogous multistoried sands of the Ingleside strandline trend (Pleistocene of Texas) are as much as 5 mi (8 km) wide and 80 ft (25 m) thick (Winker, 1979). Frio deposition was characterized by a stable shoreline position over a very much longer time period. The modern Texas coast has entered an aggradational phase (Morton, 1977), but little vertical growth of stable barriers has occurred in the limited time span of a few thousand years. The lenticular and tongue-shaped sands along the landward margin of the barrier trend probably are stacked washover and flood-tidal delta deposits. As in the other coastal facies of the Frio Formation, extensive sheet sands are present near the top. In the southern part of the system, aprons of sand and silt deposited in lower shoreface and shelf settings by wave action and storm-generated currents (Berg and Powell, 1976) produce barrier front facies sequences that may exceed 1,000 ft (300 m) in thickness in rapidly subsiding growth-fault bounded subbasins (Fig. 13).

Boyd and Dyer (1964) were of the opinion that the barrier and strand-plain sands were derived from the delta system (Norias) centered on the Rio Grande embayment. It is equally possible that, like the modern Padre Island coast (Morton, 1977), longshore drift systems converged from the southwest and northeast, supplying sand from the Houston delta system as well. Littoral convergence would have ensured adequate supply of sand over the considerable distance represented and would account for the relatively uniform development of shoreline sands between the two delta systems. Sand mineralogy (Bebout et al, 1978) supports such convergence and indicates little material from the Norias delta system was transported northward beyond San Patricio County.

Structural style.–The dominant structural control was growth faulting. This augmented the tendency of littoral processes to develop a marked strike facies orientation. The greatest vertical stacking of sand occurs between the Vicksburg and Frio flexures. The axis of the system, however, crosses to the downdropped side of the Frio flexure as it approaches the Norias deltaic headland during deposition of middle and upper Frio (Figs. 6 and 7). Parallelism of structural and depositional trends suggests a strong interaction between fault growth and localization of specific depositional environments.

Reservoir and trap configurations.—The paucity of hydrocarbon seals has severely limited oil and gas accumulation in the thick, multistoried barrier-strand-plain sandstone core. Stratigraphic factors are important both landward and seaward of the thickest sandstone development, where a number of stratigraphic and combination traps are present. Structure plays a key role, and trapping is most commonly related to simple and faulted anticlines, with rare fault sealing.

Frio Depositional History

The Frio progradational wedge developed in response to pronounced subsidence of the inherently unstable platform edge, which accommodated significant volumes of terrigenous sediments. Areally distinct patterns of growth faulting, diapirism, and associated sedimentation resulted from differences in the strata underlying the Frio Formation. Salt diapirism, salt-withdrawal basins, and complex fault arrangements characterize the Houston embayment. The San Marcos arch and Houston embayment areas are underlain primarily by shale and show a more orderly pattern of sequential growth faulting, which defines three elongate subbasins initiated successively gulfward. Thickening of individual units across growth faults is most marked in the outermost subbasin of the Rio Grande embayment.

The updip, time-equivalent Catahoula Formation is up to 1,000 ft (300 m) thick, and accumulated on a progradational platform inherited from Yegua, Jackson, and Vicksburg deposition. The large integrated Gueydan fluvial system that originated beyond the state boundary flowed down the axis of the Rio Grande embayment. In contrast, two or three separate river networks constituted the Chita/Corrigan fluvial system in the Houston embayment; net sand isopachs show, however, that the positions of the fluvial axes changed substantially with time (Galloway, 1977).

A semiarid climate prevailed in the south. Coupled with the widespread destruction of vegetation by ash emanating from the volcanic terrane of northwest Mexico, and attendant flash floods, the dry climate favored low sinuosity bed-load channel systems flanked by broad, ash-laden crevasse splays (Galloway, 1977). Paralleling the modern climatic zonation, relatively humid conditions existed in east Texas. Sinuous, mixed-load channels draining a mixed sedimentary hinterland were separated by large lacustrine basin and flood-plain areas (Galloway, 1977). Small, localized streams flowing across the San Marcos arch drained basin margin exposures of carbonate-rich Cretaceous strata.

Downdip gradation from fluvial to deltaic, barrier, and strand-plain deposits occurs below a present-day depth of approximately 2,000 ft (600 m). Patterns of coastal progradation varied according to local structural controls. In east Texas, continued regional subsidence was accentuated by salt withdrawal and platform-edge foundering. Infilling of the Rio Grande embayment proceeded across successive fault-bounded subbasins, each of which underwent slope aggradation followed by progradation of deltaic complexes over the platform constructed thereby.

The main processes contributing to deep-water aggradation were the accumulation of pelagic clays and terrigenous muds and gravity resedimentation of shallow-water deposits. Mass movement of delta front deposits is thought to have been responsible for irregularly interbedded slope sandstones off both the Norias and Houston delta systems. Delta-front mudflows (Roberts, 1980) and slumps (Coleman and Pryor, 1980) were undoubtedly important mechanisms for downslope transport, particularly in the most regressive deltaic units containing a relatively large component of fine-grained sediment. According to Roberts, the high water content of up to 80%, together with the gas liberated by the abundant organic matter in these fine-grained sediments, produces excess pore-water pressures which contribute to slope failure. Strike-parallel intraslope basins related to growth faulting and mud or salt ridges may have trapped large quantities of delta-front sediment transported by slumping, mass flow, and turbidity currents.

Mass flow processes comparable to those operating on the modern Mississippi delta slope may have been responsible for initiating erosion of the large submarine canyons in the outer neritic to deep-water Hackberry embayment, a prominent feature during middle Frio deposition. Bathyal depths of accumulation are established by the foraminiferal assemblage in the Hackberry shale. The major portion of the embayment was located in Louisiana, but it extended through much of Orange and Jefferson counties, Texas, where canyons in the outer shelf and slope of east Texas were at least 600 ft (180 m) deep and 3.5 mi (5.6 km) wide (Berg and Powers, 1980). Erosionally carved canyons were filled by onlapping turbidity channel and overbank deposits, suggesting that major submarine-fan sand deposits would be anticipated at greater depth (Paine, 1971; Brown and Fisher, 1977). Headward retreat of gully systems by slumping (Roberts, 1980) updip of the major canyons would have tapped large volumes of shelf, and possibly even shoreface, sand. This would account for the stacking of individual 10 ft (3 m) massive turbidite channel fill sands to total thicknesses of up to 200 ft (60 m) along the major dip-oriented axes. Canyon-filling shales are distinguished from adjacent units of comparable lithology on the basis of their microfaunal content. Other smaller deep-water embayments probably existed along the Texas coast during Frio deposition. Stuckey (1964) identified a broad mudstone-filled reentrant extending through coastal Brazoria and Matagorda counties, corresponding in position to the outer shelf and slope off the Houston delta system.

The Norias delta system constituted the most actively prograding coastal sector. Deposition of the lower Frio operational unit was characterized by shoal-water arcuate or wave-modified lobate deltas that produced a massive, sandy facies sequence extending basinward to the approximate position of the Frio fault zone (Fig. 5). Beyond this, highly expanded shelf-edge delta sequences were deposited. These deltas probably assumed a more fluvially dominated lobate geometry in plan view because of rapid subsidence and consequent reduced marine redistribution of sand (Galloway, 1975). This was followed by middle Frio progradation (Fig. 6) of the deltaic platform by as much as 40 mi (65 km). By upper Frio (Fig. 7), the shelf edge had built gulfward to eastern Cameron County and locally beyond the modern shoreline.

Regional marine transgression and onlap caused the Norias delta shoreline to retreat by only 25 mi (40 km), and delta progradation persisted locally during this regionally transgressive interlude, spreading mud across the foundered platform. Marine redistribution of sediment was accentuated and arcuate delta geometries likely prevailed (Curtis, 1970; Galloway, 1975), interspersed with destructional bar and strand plain, and grading into a barrier/lagoon environment in the delta-margin embayment to the north.

The delta-margin embayment of south Texas merged across the San Marcos arch, with a broad strand plain flanking the Houston delta system. This Greta/Carancahua barrier/strand-plain system constituted a prominent zone of strike-parallel aggradation brought about by a combination of structural and depositional factors. Growth faults maintained a consistent coast-parallel trend, defining narrow, linear zones. Convergent littoral drift provided large volumes of sand from the deltaic headlands on either side. Shoreline conditions along the Greta/Carancahua trend remained remarkably uniform during lower and middle Frio deposition, resulting in the development of an exceptionally thick but relatively narrow, vertically homogeneous sandy succession. Minor progradation occurred during lower and middle Frio deposition, with landward retreat of the progressively less well-defined shore-zone sand belt during upper Frio deposition.

The Houston delta system of east Texas, like the Norias delta, was characterized by elongate to lobate geometries during the most regressive phases and more arcuate forms during episodes of general transgression and shoreline retreat. Because of the greater structural complexity of the salt-floored Houston embayment, the systematic patterns of delta-front progradation and retreat observed in deposits of the Norias delta system are not developed. Furthermore, because of the more dispersed, multiple drainage network which supplied sediment at several points along the coast, and the constant switching of delta depocenters, there was a tendency toward imbricate stacking of facies along a regional strike-parallel belt. Vertically repetitive upward-coarsening sequences and partial sequences are characteristic. Maximum thickness of individual prodelta-delta front sequences approaches 1,000 ft (300 m) in shelf-edge progradational deposits of the lower Frio; thinner platform-delta sequences 50 to 300 ft (30 to 185 m) thick typify the middle and upper Frio operational units. Complex aggradational delta-plain deposits cap many of the individual or stacked progradational sequences. Delta-plain facies include blocky distributary sandstones and thin upward-coarsening bay-fill and crevasse-splay subdelta packages. Much thicker (250 to 1,000 ft; 75 to 300 m) blocky sandstones, which occupy the same stratigraphic position as the delta-plain deposits, record the development of multistoried wave-reworked sands of recurrent delta destructional phases.

During deposition of the lower Frio, the shelf edge was approximately coincident with the eastern extension of the Vicksburg fault zone, although this is not well defined in east Texas. Localized shelf-edge deltas built seaward over muddy, aggrading slope environments, which were subject to the sporadic influx of resedimented delta front sands. The greatest volumes of distributary mouth-bar sand were preserved coincident with the seaward terminations of major fluvial axes, for example, along the trend through Wharton and central Matagorda County.

By middle Frio deposition, the shelf edge had been extended seaward, and generally shallower water conditions prevailed. The main locus of coarse-grained deposition had shifted to the east, where sandy, marine-modified deltas were supplied by large fluvial channel systems flowing through Fort Bend, Harris, and Liberty counties. The middle Frio expands westward, and it contains a greater proportion of mud. Small, lobate shoal-water deltas built rapidly to the shelf edge, but they were largely reworked after abandonment into coast-parallel bars.

Depositional style of the upper Frio Houston delta system was strongly influenced by Anahuac marine transgression. Successive transgressions extended farther landward but were separated by episodes of delta progradation. The delta lobes became smaller as they retreated landward, and wave-reworking along the encroaching shoreline produced thick, time-transgressive blanket sands (such as the Greta sand) near the top of the Frio Formation.

The Buna strand-plain/barrier system marginal to the Houston delta system on the east resembles the Greta-Carancahua system in most respects but shows evidence of frequent minor coastal regressions and retreats superimposed

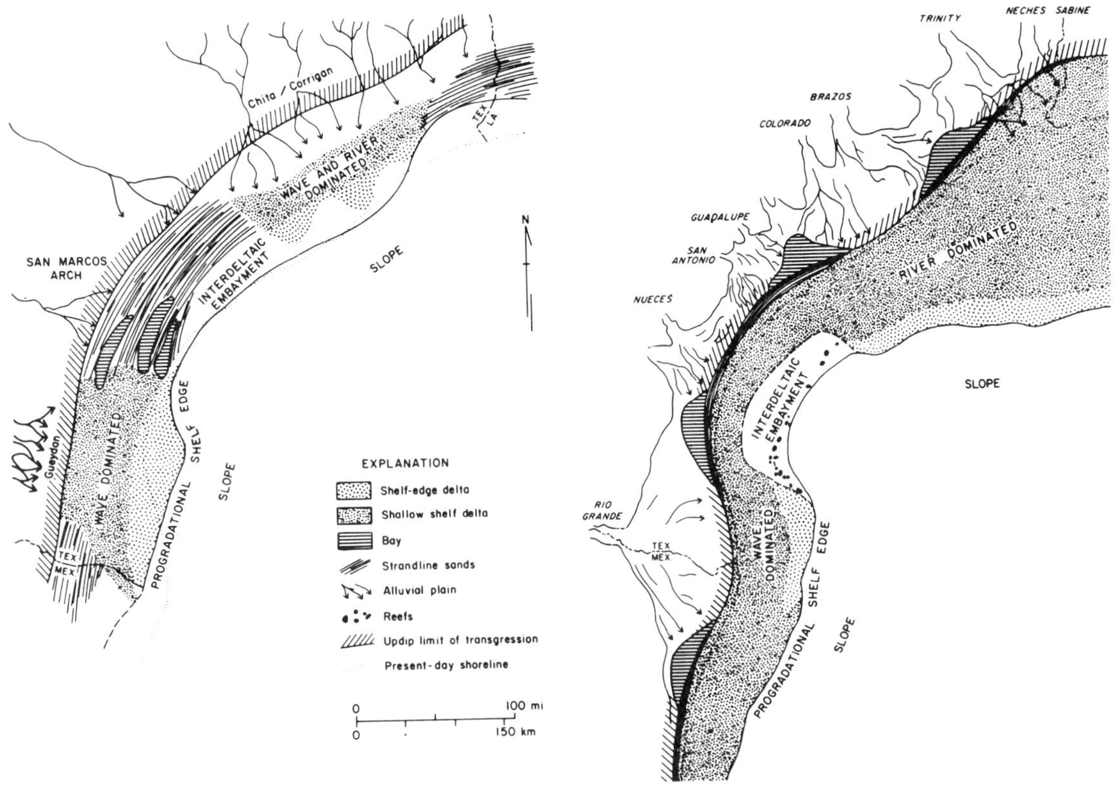

FIG. 14—Comparative Frio Formation and late Pleistocene (Winker, 1979, 1980) paleogeography of Texas coastal plain and shelf. Pleistocene deltaic deposition occurred primarily during progradational phase; fluvial, strand-plain, and back-barrier deposition characterized aggradational and transgressive intervals. During Frio deposition, aggradation also occurred concurrently with progradation as loci of sediment input shifted across peripheries of broad, advancing deltaic headlands.

on the generally aggradational arrangement of blocky sands. These oscillations probably stemmed from variations in alongshore sediment supply resulting from changing positions of the delta lobes, and from a fluctuating local supply of sediment from coastal plain streams.

This easternmost segment of the Frio shoreline was subject to lower rates of subsidence than elsewhere along the Texas coast, favoring prolonged reworking by waves. Wave-dominated arcuate deltas, in which marine reworking was contemporaneous with progradation, developed locally during lower Frio deposition but were volumetrically less important than strike-fed strand-plain systems. Water depths decreased following these initial episodes of delta progradation, and the remainder of Frio deposition was characterized by slower accumulation of sand in a stable shore-zone setting. Areas of back-barrier mud accumulation were subject to sporadic influx of washover and flood-tidal delta sands.

Log patterns in the Miocene succession above the major Anahuac transgressive shale wedge indicate a continuation of the Buna strand-plain/barrier style of sedimentation but with considerable expansion in the thickness of the various facies as the Miocene depocenter shifted eastward into Louisiana.

Comparison of Frio and Late Pleistocene Paleogeography

The geographic distribution of environments constructed for the Frio Formation (Fig. 14a) bears a striking resemblance to the late Pleistocene (Winker, 1979, 1980) and modern coastal plain and shelf of Texas (McGowen et al, 1977; Morton and McGowen, 1980). The tectonic framework, climatic zonation, drainage network distribution, and coastal and nearshore hydrologic regime were closely comparable to the present. A primary difference is that early to middle Tertiary sea levels were not subject to the short-term glacio-eustatic oscillations characteristic of the Quaternary. These glacio-eustatic sea level changes were faster than regional subsidence rates (Winker, 1979) except near the shelf edge. Conversely, local basinal factors were more important than the slower tectono-eustatic changes during Frio deposition. The same depositional phases—aggradational, progradational, and transgressive—are nevertheless represented in the Frio and Quaternary successions, marking a similar response to shifting shorelines and depocenters, albeit on a different time scale. This time factor has resulted in major differences in the thickness of units occupying the same paleogeographic zone.

The modern Texas coast, having recently recovered from a rapid postglacial rise in sea level, has reached a stage of general stability, although minor transgression still prevails over much of its length (Morton, 1977). The aggradational phase has commenced, but has not yet proceeded significantly. A more advanced stage of development is recorded in the late Pleistocene Beaumont Formation (Winker, 1979), which was deposited under conditions practically identical to those of the present day. The greater thickness, broader lateral extent, and the more fully developed range of depositional phases in the Beaumont Formation make it an ideal analog for comparison with the Frio Formation (Fig. 14).

Winker's synthesis of the Beaumont Formation shows that a characteristic sedimentary architecture is associated with each of the three depositional phases. He further divides the progradational phase into shallow-shelf and shelf-edge components.

Shallow-shelf deposits of the progradational phase developed in response to the falling sea level accompanying the onset of the Wisconsin glaciation. Progradation was rapid, generating thin, upward-coarsening sequences that were best developed in the wave-dominated deltas of the Rio Grande embayment. Deltas in east Texas were multilobate and river-dominated. Rivers became entrenched during the progradational phase, resulting in a limited volume of fluvial accumulation.

Frio drainage patterns and distribution of delta types were strikingly similar to the late Pleistocene (Fig. 14), but slower rates of relative subsidence permitted fluvial deposits to spread laterally and coalesce within the Rio Grande and Houston embayments. Rates of growth faulting were slow, and delta sequences are characteristically thin but areally extensive.

Shelf-edge deposits of the progradational phase of the Beaumont Formation were characterized by rapid vertical accumulation in a narrow zone along the shelf margin, a pattern closely duplicating that of Frio deposition. The same three factors contributed to rapid subsidence in both cases: growth faulting, salt and mud withdrawal, and isostatic loading.

The aggradational phase of the Beaumont Formation followed Sangamon transgression. Major sand bodies developed during this phase are barrier, strand-plain, and fluvial deposits. Strand plains developed along the flanks of deltaic headlands, and barriers formed in marginal embayments. The same patterns prevailed during Frio deposition, except that the aggradational phase was not so clearly defined because significant aggradation also occurred contemporaneously with downdip progradation and transgression. Thick, multistoried Frio barrier/strand-plain successions record reworking during local recurrent transgressive events produced by shifting of deltaic depocenters and rapid subsidence.

The transgressive phase of the late Pleistocene was characterized by slow sediment influx and sediment reworking. Numerous thin laterally extensive shale units within the Frio Formation correspond to this phase and contain a neritic fauna and pinch out updip. The considerably thicker Anahuac Shale wedge marks the incursion of a comparatively sediment-starved transgressive shelf, which terminated the Frio depositional cycle.

SANDSTONE COMPOSITION AND POSTDEPOSITIONAL DIAGENETIC MODIFICATION

Sandstone mineralogy differs markedly between the Norias and Houston delta systems (Loucks et al, 1977, 1979a), reflecting contrasting source terranes drained by the Gueydan and Chita/Corrigan rivers. Quartz, feldspar (including both plagioclase and orthoclase), and lithic grains (dominantly volcanic and carbonate rock fragments) are subequal constituents of the mineralogically immature Norias sandstones. In contrast, sandstones of the Houston delta system are quartz-rich with subordinate feldspar and minor rock fragments, which consist of mixed volcanic and metamorphic types. Interdeltaic coastal sandstones of the Greta/Carancahua system reflect strike-reworking of deltaic sands and input of locally derived quartzofeldspathic sands by small rivers of the stream plain. The finer grain size and greater mechanical reworking of strike-transported strandline sands enhanced quartz and feldspar content at the expense of rock fragments.

Reservoir quality varies systematically among the major depositional systems and closely reflects compositional trends. Geographic variability of sandstone consolidation gradients (Fig. 15) is dramatic. At an equivalent depth of 10,000 ft (3,000 m), sandstones of the mineralogically immature Norias delta system retain 8% less average total porosity than do sandstones of the Houston system, although average grain size and sorting are comparable. Permeability data show similar trends (Loucks et al, 1979a, p. 75). In the Norias system, permeabilities rarely exceed 1 md below 15,000 ft (4,500 m), whereas measured whole-core permeabilities of as much as one darcy occur at comparable depths in the Houston system.

The loss of permeability and porosity with depth results primarily from physical compaction and chemical diagenesis. Principal cements, which fill pore space and degrade reservoir quality, include calcite, silica (in the form of quartz overgrowths), and, in deeply buried sandstones, iron-rich carbonate minerals and kaolinite pore fill. As emphasized by Loucks et al (1979b), formation of secondary leached porosity becomes increasingly important in the preservation of high-quality reservoirs at depth. Within the oil-productive window, more than 50% of the preserved porosity is of secondary origin. A major grain and cement leaching event occurs at a depth of burial that coincides closely with the oil generation window, thus providing optimum conditions for petroleum migration into the enhanced reservoirs (Loucks et al, 1979b). The poorer reservoir quality of the Norias delta system sandstones is a product of extreme carbonate cementation at intermediate burial depths and ubiquitous postleaching precipitation of kaolinite and iron-rich carbonate minerals at greater depths (Loucks et al, 1979a).

HYDROCARBON OCCURRENCE AND PRODUCTION

The Frio has produced nearly 6 billion bbl of oil and 60 tcf of gas, for a total of more than 16 billion BOE of hydrocarbons. Included in this figure is a substantial contribution from very old fields discovered and developed during the early part of the century, as well as significant new discoveries of the last decade. In addition, sands of the Frio contain an estimated 264 tcf (44 billion BOE) of dissolved methane (Gregory et al, 1980). Conservatively, then, the Frio contained about 60 billion BOE of hydrocarbons.

The hydrocarbon inventory developed for this study con-

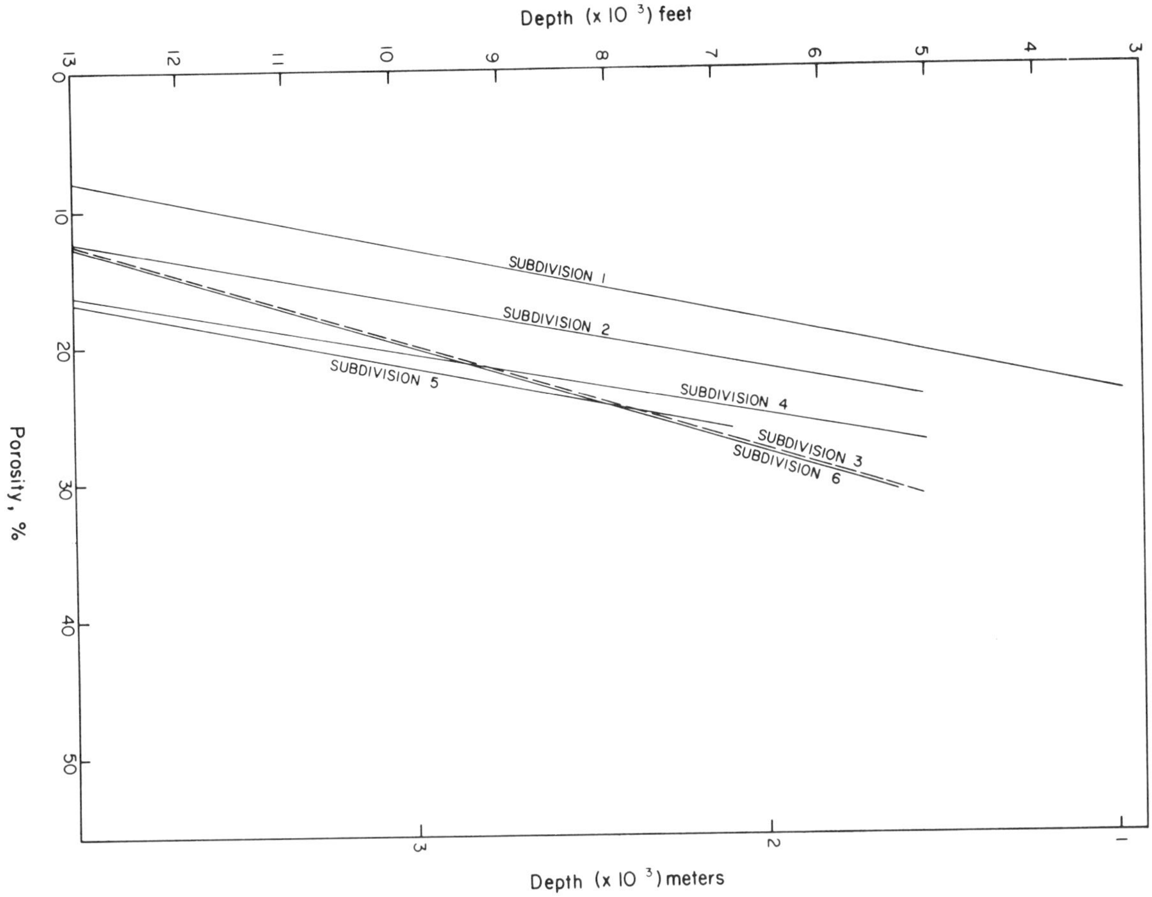

FIG. 15—Comparison of porosity versus depth trends in downdip portions of Frio Formation. Correspondence of data subdivisions and Frio depositional systems is as follows: subdivision 1 = Norias delta system; subdivision 2 = Norias to Greta/Carancahua transition; subdivision 3 = Greta/Carancahua barrier/strand-plain system; subdivisions 4 and 5 = Houston delta system; subdivision 6 = Houston to Buna transition (from Gregory et al, 1980).

sidered only fields containing over 1 million bbl of oil or gas equivalent. We estimate that the resultant data base includes more than 95% of the known Frio production, an estimation consistent with the observation that the majority of producible hydrocarbons typically occurs in large fields. Errors introduced by deletion of smaller fields are less than the uncertainty introduced by the incomplete and contradictory record keeping characteristic of the early production history of many of the largest and most prolific fields.

Play Recognition and Description

Hydrocarbon production in the Frio MSU can be grouped into ten geologically defined plays (Fig. 16), seven of which contain the majority of the fields. Principal defining paramenters of plays include the depositional system of facies assemblage of the reservoirs, dominant structural style and the type of hydrocarbon produced. Key attributes of each play are summarized in Table 1. Although genetic stratigraphy, structural style, and production are typically interrelated, some plays are defined by the dominance of one variable. For example, plays V and VI are separated, primarily because they occur on the marine and landward sides, respectively, of the axis of the Greta/Carancahua barrier/strand-plain system. Play III is dominated by a single major structural feature, the Vicksburg flexure. Plays VIII and IX are distinguished by the relative importance of salt domes in their structural style. Play II is characterized by the fact that it produces almost exclusively gas.

Quantitative geologic attributes of each play are listed in Table 2. In addition to total area, total rock volume and volume of sandstone and mudstone were determined by planimetry of appropriate isopach and isolith maps. Cross sections were used to estimate the percentage of the section lying below the 200°F (94°C) isotherm and within the hydrocarbon maturation window. The approximate volume of thermally mature mudstone was then calculated using this estimate. Structural features, including miles of faulting, number of salt piercement diapirs, and number of closures, were determined from commercial and published structural maps. These data provide at best a qualitative comparison, however, as mapping horizons and their positions within the

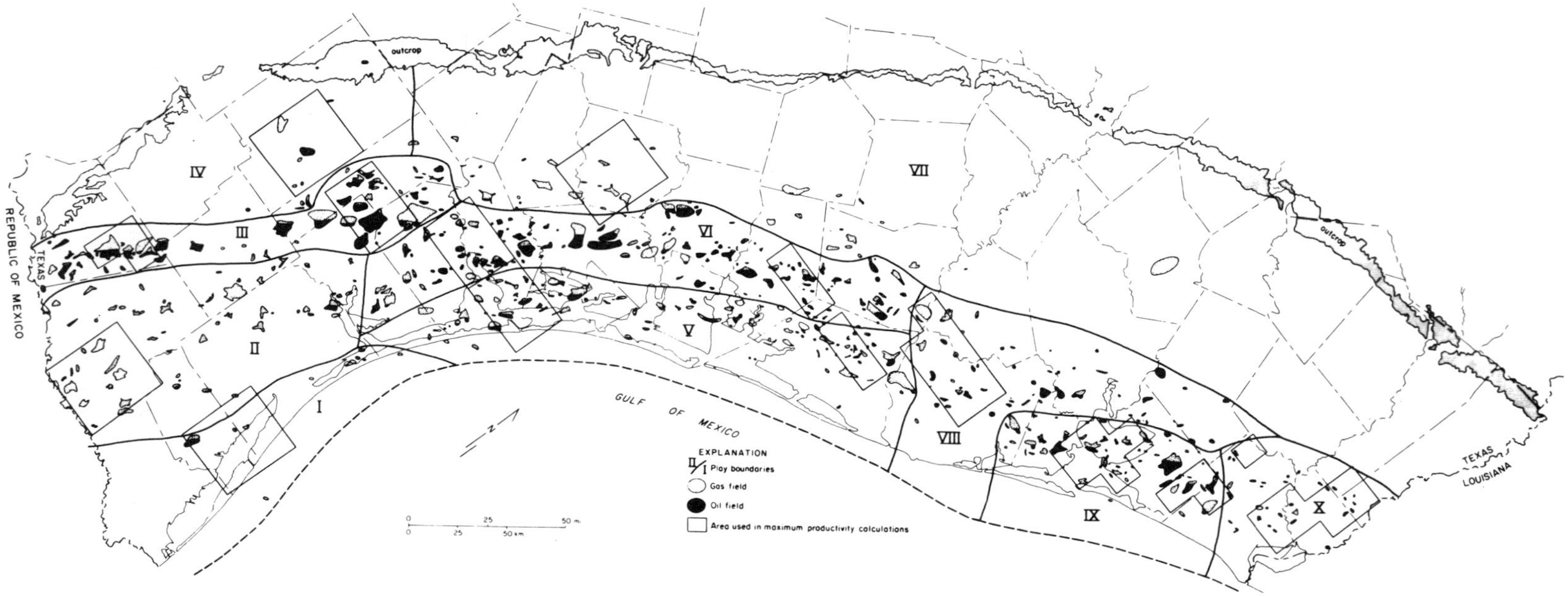

FIG. 16—Major producing petroleum fields and defined plays (I-X) of Frio Formation.

Table 1

Play	Productive System(s)	Structural Style	Exploration Maturity	Frontiers	Limiting Factors
I	Norias Delta System	Growth fault & shale ridge	Immature	Depth & OCS	(1) Reservoir quality (2) Migration efficiency
II	Norias Delta System	Growth fault & shale ridge	Mature	Depth	(1) Reservoir quality (2) Migration efficiency
III	Norias and Gueydan Systems	Vicksburg Flexure	Supermature	None	(1) Well density
IV	Gueydan Fluvial System	Inherited low amplitude fault & anticline	Mature	None	(1) Well density (2) Thin/shallow section (3) Lack of indigenous, mature source
V	Greta-Carancahua Barrier/ Strand-Plain System	Growth fault & shale ridge	Mature	Depth & downdip (limited)	(1) Reservoir distribution/ volume (2) Well density
VI	Greta-Carancahua Barrier/ Strand-Plain System	Growth fault & shale ridge	Supermature	None	(1) Well density
VII	Moulton Stream-Plain & Chita/ Corrigan Fluvial Systems	Inherited low amplitude fault & anticline	Mature	None	(1) Well density (2) Thin, shallow section (3) Lack of indigenous, mature source
VIII	Houston Delta System	Salt diapirism & associated faulting	Mature	Depth	(1) Reservoir distribution/ volume?
IX	Houston Delta System	Growth fault & minor salt diapirism	Mature	Depth	(1) Reservoir distribution/ volume
X	Buna Strand-Plain System	Growth fault & salt diapirism	Mature	Depth?	(1) Reservoir distribution/ volume

Frio differ from play to play. Average sandstone porosity was estimated from regional Frio consolidation profiles in Loucks et al (1979a). Average drilling density, expressed as the number of well penetrations of middle Frio per square mile, was estimated from county well depth audits prepared by Petroleum Information, Inc. Interpreted source rock quality, likely source facies, and the average of available total organic carbon analyses provided by company and public sources complete the geologic description.

A hydrocarbon inventory of each play tabulated cumulative production of oil, associated gas, and nonassociated gas (Fig. 17 and Table 3). Cumulative production statistics were used as the primary data base for most further calculations, because they are historical figures subject to the least amount of interpretation or extrapolation. Total produced hydrocarbons for each play was calculated by using the conversion factor based on energy equivalency of 6 mcf of gas per barrel of oil. Other conversions might be used, but energy equivalency establishes a yardstick for comparison of the produced and potential *total energy content* of each play volume. The total can be readily disaggregated following various calculations by using known or assigned proportions of gas to oil.

The combination of volumetric data and hydrocarbon production produces a series of play yield factors (Table 3 and Fig. 18) indexed to volumes of total rock and of reservoir rock. These figures are minimums in the sense that they are based generally on the 80 to 90% of the producible hydrocarbon reserve that has actually been extracted. However, visual examination of data shows that similar calculations based on interpreted total producible or total in-place hydrocarbons would have little effect on the relative ranking. Both whole-rock and reservoir yield factors display an order-of-magnitude range. For comparison, White et al (1975) calculated a reservoir yield factor of approximately 3.1×10^6 BOE/mi^3 for the gas-prone Oligocene-Miocene section of Louisiana. We agree with their conclusion that production per volume of reservoir is the best index of play hydrocarbon productivity.

Table 2

Play	Area (mi²)	ROCK VOLUME: Rock volume (mi³)	Sandstone (mi³)	Mudstone (mi³)	Percent that is thermally mature†	Thermally mature mudstone (mi³)	STRUCTURE: Miles of faulting	Number of salt diapirs	Number of mapped closures	Average sand porosity (%)	Average well density (wells/mi²)‡	Source rock quality (avg. TOC)	Principal potential source rock facies
I	3,120	6,360 (depth ≤ 20,000')	900+	5,460	95±	5,000	570	0	28	18	0.04	0.3	Prodelta and slope mudstone
II	4,420	6,800	1,900	4,900	80	3,900	1,010	0	64	18	0.11	0.3	Prodelta and slope mudstone
III	2,534	1,760	370	1,390	6	80	270	1	24	26	1.53	0.3	Delta plain mudstone
IV	4,310	1,625	200	1,425	0	0	minimal	2	not mapped	27	1.02	0.4	Flood-plain mudstone
V	5,300	8,090	640	7,450	95±	7,000	970	2	41	25	1.13	0.2	Shelf and upper slope mudstone
VI	4,088	3,195	1,120	2,075	31	650	690	1	35	30	1.85	0.3	Shelf and lagoonal mudstone
VII	21,868	6,378	1,286	5,092	0	0	minimal	11	not mapped	30	1.12	0.4	Flood-plain mudstone
VIII	3,740	2,731	793	1,938	60	1,170	125	58	58	31	1.20	0.3	Prodelta, shelf, and upper slope mudstone
IX	2,630	2,980	189	2,791	87	2,435	111	35	35	29	2.2	0.3	Prodelta, shelf, and upper slope mudstone
X	1,750	1,400	252	1,148	44	500	16	12	12	33	0.9	0.3	Shelf and upper slope mudstone
TOTALS	53,760	41,319	7,650	33,669		20,735	3,762+	122	297+				

† Present temperature ≥ 200° F (90°C)
‡ Average penetration of upper half of total Frio interval

Play Stratigraphy, Structural Relationships, and Hydrocarbon Production

Each of the ten plays is characterized by a distinct association of structural features, sedimentary facies, and hydrocarbon production, but the relative importance of specific criteria varies from one play to another. The plays vary considerably in size and thickness of the Frio Formation. Updip plays extend to the Catahoula outcrop; downdip plays extend gulfward to the limit of well coverage, which is largely depth imposed. This corresponds approximately to the distal extremities of known reservoir sands, although these may persist at greater depth.

Play I is a poorly explored, gas-prone division corresponding to the upper slope, prodelta, and distal delta front of the Norias delta system (Table 1). The proportion of reservoir sandstone is low, and the bulk of the shale source rock is thermally mature (Table 2). These mudrocks, of prodelta and slope origin, therefore constitute a potential hydrocarbon source, but source rock quality is poor, with an average of only 0.3% TOC. Reservoir quality is also poor because of fine grain size and advanced diagenetic destruction of porosity and permeability. Migration efficiency was low because of limited interfingering of sandstone and mudstone, and low permeability values.

Dominant structures are growth faults and shale ridges, with the associated anticlines acting as hydrocarbon traps. Reservoir facies include marine-modified delta front sandstones within the upper Frio. The producing interval lies directly below the Anahuac shale, which provides a seal. The play has been sparsely drilled, with an average density of only 0.04 wells/mi² (Table 2).

Play II corresponds to the main body of the Norias delta system and its area of interfingering with the Gueydan fluvial system (Table 1). It is characterized by production of nonassociated gas, but little oil. Play II suffers from the same diagenetically imposed reservoir limitations as play I, and source rock quality and migration efficiency are likewise poor. Potential mudstone source rock makes up three-quarters of the total volume, and 80% is thermally mature. Underlying deltaic and slope mudstones of the Vicksburg Formation constitute an additional potential source. The major structural features are growth faults and shale ridges. Shale-cored rollover anticlines and faulted anticlines are the most important traps. Some hydrocarbons are trapped by closure against faults on the upthrown side. There are subordinate stratigraphic traps. Reservoir facies vary with stratigraphic position. Thick, upward-coarsening, wave-reworked delta front sandstone reservoirs are most important in the lower Frio, giving way to fluvial sandstones in the middle and upper Frio operational units. Drilling density is low, averaging 0.11 wells/mi² (Table 2).

Play III is a gas-prone play located over the Vicksburg flexure, and has reached the supermature stage of exploration. It includes part of the Gueydan fluvial system, extending into the farthest updip positions of the Norias delta system, and into coastal lake and stream-plain systems along strike. Eighty percent of the total rock volume is mudstone, of which only 6% is thermally mature. Average TOC content is only 0.3%. Subjacent Vicksburg deltaic and interdeltaic coastal mudstones and Jackson Group marine shales constitute most likely sources for the hydrocarbons. Reservoir porosity averages 26% and is thus greatly improved over the more deeply buried downdip plays. Migration efficiency is likewise enhanced.

Dip-oriented fluvial channel-fill sand bodies are the domi-

Table 3

Play	Cumulative Production					Percent Already Produced		Remaining Proved Reserve				
Play	Oil (10^6bbl)	Associated gas (10^3mmcf)	Nonassociated gas (10^3mmcf)	Total gas (10^3mmcf) and liquid equivalent in (10^6BOE)	Total hydrocarbons (10^6boe)	Oil	Gas	Average oil recovery (%)†	Oil (10^6bbl)	Associated gas (10^3mmcf)	Nonassociated gas (10^3mmcf)	Average $\frac{\text{BOE gas}}{\text{bbl oil}}$
I	29	281	185	466 / 80	107	~60	~60	34	3.6	9.6	303	2.7
II	50	1	6,530	6,531 / 1,090	1,140	—	81	34	0	0	1,600	22
III	917	7,690	7,110	14,800 / 2,470	3,390	96	91	34	28	322	1,500	2.7
IV	52	2	77	79 / 15	65	100	95	34	1	0.04	6	0.2
V	519	4,200	7,700	11,900 / 1,980	2,500	98	83	44	11	89	2,400	3.8
VI	2,500	4,750	5,850	10,600 / 1,765	4,230	85	90	44	440	836	1,200	0.7
VII	41	45	886	931 / 155	191	82	83	46	8	8.2	199	4.8
VIII	1,620	1,110	2,440	3,550 / 590	2,210	77	85	48	490	336	630	0.4
IX	857	2,120	4,150	6,270 / 1,045	1,900	89	88	48	110	272	870	1.2
X	236	465	1,530	1,995 / 330	567	96	83	48	9	17.7	420	1.4
TOTAL	6,821	20,664	36,458	57,122 / 9,520	16,300				1,100	1,890	9,130	1.4

* Calculated energy equivalent in bbls of oil (6 mcf = 1 BOE)

† Figures apply to entire production within the RR Commission District in which the play lies

‡ Exclusive of a giant field lying within the play

nant reservoir facies, along with some reworked deltaic sands in the lower Frio. Anticlines, faulted anticlines, and fault closure along the Vicksburg flexure are the most important traps. Stratigraphic features contribute to trapping in approximately one half of the fields; this results from the abrupt facies changes along the Vicksburg flexure. The play is densely drilled, averaging 1.53 wells/mi^2. The nonproductive gap in the middle of the strike-elongate play corresponds to the sand-rich axis of the Gueydan fluvial system, and it probably reflects the near absence of continuous shales necessary for trap sealing.

Play IV is characterized by minor oil-dominant production in the updip Gueydan fluvial system (Table 2). This thin, shallow interval contains a low proportion of sandstone. The mudstones average 0.4% TOC, but have undergone inadequate thermal maturation for oil generation. The play is therefore entirely lacking in indigenous hydrocarbon liquids. Inherited low-amplitude faults and anticlines are the major structures. Reservoir facies are mainly fluvial sandstones, and faulted anticlines and stratigraphic traps are most important. Drilling density is moderately high.

Play V comprises a gas-prone trend lying mainly within the Greta/Carancahua barrier/strand-plain system and adjacent shelf. It has reached a mature stage of exploration, with 1.13 wells/mi^2 (Table 2). Shelf and upper slope mudstones average only between 0.2 and 0.4% TOC, but over 95% is thermally mature. Underlying Vicksburg and Jackson marine shales constitute additional potential sources. Sandstones downdip of the barrier/strand-plain axis constitute the most important reservoirs. These have an average porosity of 25% and are interbedded with mudrocks to provide maximum efficiency of hydrocarbon migration. Traps are simple and faulted shale-cored anticlines, and growth faults with closure on the upthrown block. The produced hydrocarbons show a well-defined thermal zonation (Fig. 13), with gas dominating at greater depth.

Play VI, the most important in terms of oil and total hydrocarbon production, is another intensively explored

Table 3 (continued)

Yield Factors (f)														
Production/Total Volume - 10^6BOE/mi^3						Production/Reservoir Volume - 10^6BOE/mi^3								
Entire Play			Best Area			Entire Play			Best Area					
Oil	Gas	Total	Oil	Gas	Total	Oil	Gas	Total	Oil	Gas	Total	Production/mile of faulting (10^6BOE/mi)	Production/structural closure (10^6BOE/trap)	Production/volume of thermally mature mudstone (10^6BOE/mi)
0.005	0.012	0.017	0.016	0.042	0.06	<0.033	<0.086	<0.12	<0.08	< 0.22	< 0.3	0.2	4	0.02
0.007	0.16	0.17	0.009	0.198	0.21	0.026	0.57	0.60	0.039	0.86	0.9	1.1	18	0.3
0.52	1.40	1.93	0.92	2.47	3.40	2.48	6.68	9.16	5.06	13.64	18.7	12.5	140	42
0.032	0.008	0.04	0.088	0.023	0.11	0.26	0.067	0.33	0.95	0.24	1.2	NA	NA	NA
0.064	0.25	0.31	0.076	0.30	0.37	0.81	3.09	3.90 (2.2)‡	0.41	1.58	2.0	2.5	60	0.3
0.78	0.55	1.32	0.56	0.40	0.95	2.23	1.55	3.78 (3.1)‡	1.71	1.19	2.9	6.1	121	6.5
0.006	0.025	0.03	0.017	0.067	0.08	0.032	0.12	0.15	0.064	0.24	0.3	NA	NA	NA
0.59	0.22	0.81	0.56	0.21	0.77	2.04	0.74	2.78	2.64	0.96	3.6	17.7	41	1.9
0.29	0.35	0.64	0.60	0.73	1.33	4.53	5.50	10.03	4.13	5.01	9.2	17.1	58	0.8
0.17	0.24	0.41	0.33	0.46	0.79	0.94	1.31	2.25	1.46	2.04	3.5	35.4	52	1.1
0.17	0.23	0.40				0.89	1.24	2.13						0.8

trend, with 1.85 wells/mi^2. It is situated updip of the main barrier/strand-plain axis, and , like play V, it shows favorable alternation of source and reservoir rocks. However, the source mudstones of shelf and lagoonal origin have a low organic carbon content (Table 2) and only one-third of the total mudstone volume is thermally mature. Major hydrocarbon sources likely include underlying Eocene marine shelf and slope mudrocks.

Reservoir sandstones are highly porous, averaging 30% porosity. The main producing facies are mixed progradational-aggradational units and stacked blocky sandstone units of barrier and strand-plain origin, and thick successions of irregularly alternating sandstone and shale that probably represent back-barrier deposits. Many of the thickest, unbroken sandstones are unproductive, because they lack adequate seals. Growth faults and shale diapirs are well developed. Simple and faulted anticlines are the dominant traps, with some stratigraphic and combination traps.

Play VII constitutes a large but relatively unimportant hydrocarbon-producing area extending across the Moulton/Choke Canyon/Flatonia stream-plain and Chita/Corrigan fluvial systems, and incorporating the updip limits of the Greta/Carancahua and Buna barrier and strand-plain systems. Flood-plain mudstone contains higher than average organic carbon, but none is thermally mature. Low-amplitude faults and anticlines inherited from the underlying Wilcox Group are the most important traps. The area has been drilled to a density of 1.12 wells/mi^2 primarily as a result of exploration for deeper targets.

Play VIII extends from the downdip edge of the Chita-Corrigan fluvial system through the Houston delta system. It is distinguished by the abundance of salt diapirs (Table 2), many of which are large and profoundly influenced structure, sedimentation, and hydrocarbon production patterns. It is an oil-prone area, with a large volume of total produced hydrocarbons (Table 3). There is a substantial proportion of sandstone with high porosity. Prodelta, shelf, and upper slope mudstone source rocks are 60% thermally mature, but they have a low organic carbon content. Potential source rocks include interbedded Frio prodelta, upper slope and shelf mudstones, and underlying Jackson shelf and Yegua prodelta mudstones. Dominant structures are salt diapirs with their associated fault patterns, large salt-withdrawal basins, and growth faults.

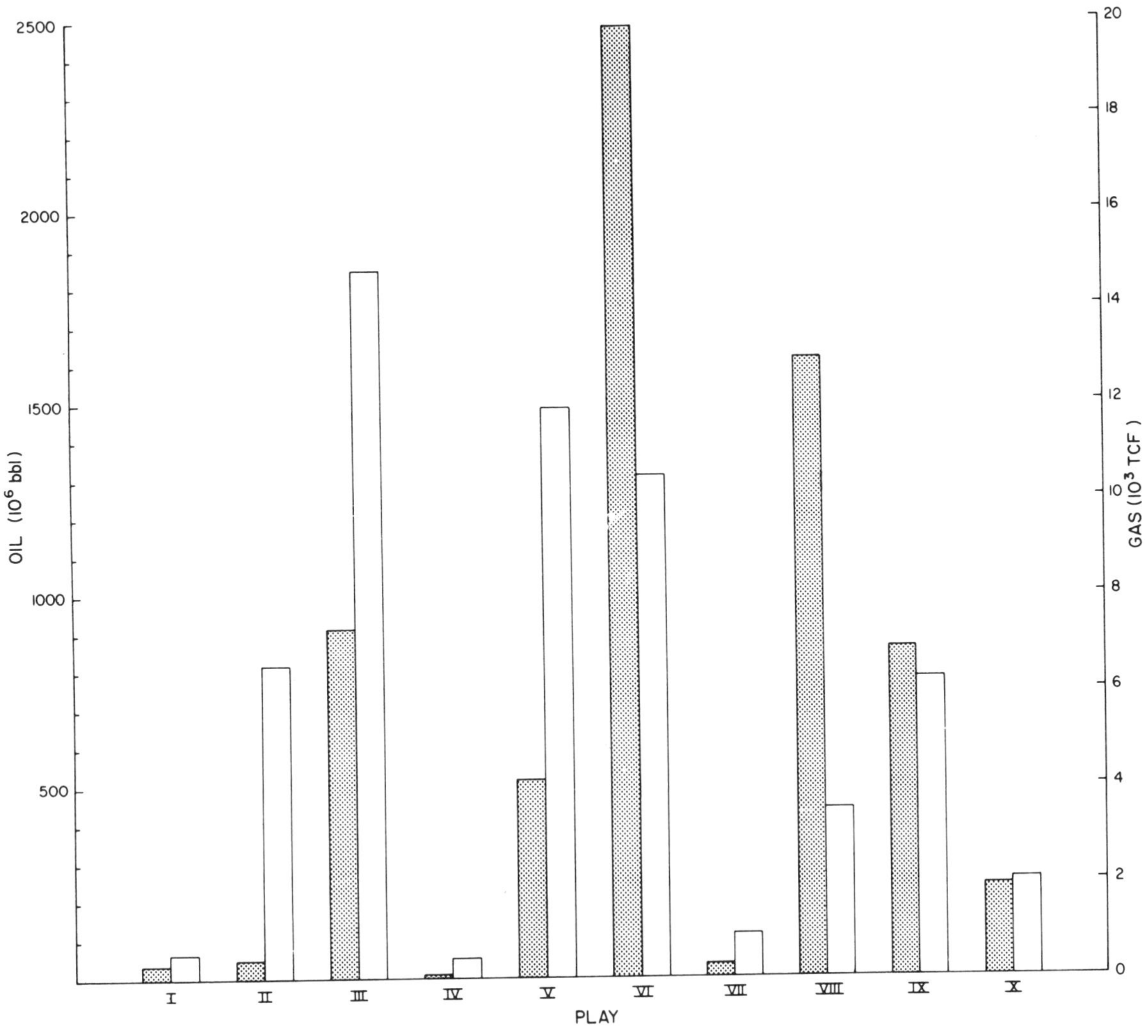

FIG. 17—Cumulative production of oil and gas for each of ten Frio plays.

Hydrocarbons are present in faulted anticlines adjacent to salt diapirs and in fault traps sealed against the upthrown block. Reservoir facies range from basinward, mudstone-dominated successions with upward-coarsening or randomly interbedded sandstones in the lower and middle Frio, to channel-fill and reworked delta front deposits in the middle and upper Frio.

Play IX differs from play VIII in having fewer large salt diapirs and salt-withdrawal basins. It corresponds to part of the Houston delta system together with equivalent shelf and slope facies. The proportion of reservoir sandstone is exceptionally low, amounting to only 6%, but this sandstone retains good porosity at depth. Mudstone source rocks are about 85% thermally mature, but again they are characterized by low levels of organic carbon typical of prodelta, shelf, and upper slope mudstones. Underlying Eocene shelf and deltaic mudstones are likely contributors as well.

Reservoir facies include mudstone-rich progradational sequences and slope channel-fill in the lower Frio, with fluvial channel-fill and reworked delta front sandstones in the middle and upper Frio. Traps include faulted anticlines above deep salt ridges, simple anticlines, and growth faults. Play IX is the most intensively drilled of any, with an average of 2.2 wells/mi^2.

Play X is a gas-prone area of limited production. Sandstones, which constitute 18% of the total volume, have exceptionally high porosity (Table 2). Frio shelf and upper slope mudstone source rocks are 44% thermally mature. Underlying Eocene deltaic and shelf deposits are probable contributors as well. Repetitive interfingering of source and reservoir rocks within the Frio provides an ideal avenue for hydrocarbon migration and collection. Growth faults and salt diapirs both influenced patterns of sedimentation and hydrocarbon trapping.

Reservoir facies are extremely diverse and include slope turbidite sandstones in the middle Frio (Hackberry facies), the seaward margins of the Buna strand-plain/barrier systems, and updip channel complexes. Hydrocarbon traps range from salt domes to simple rollover anticline and faulted anticlines. Stratigraphic traps are present in the middle Frio, where

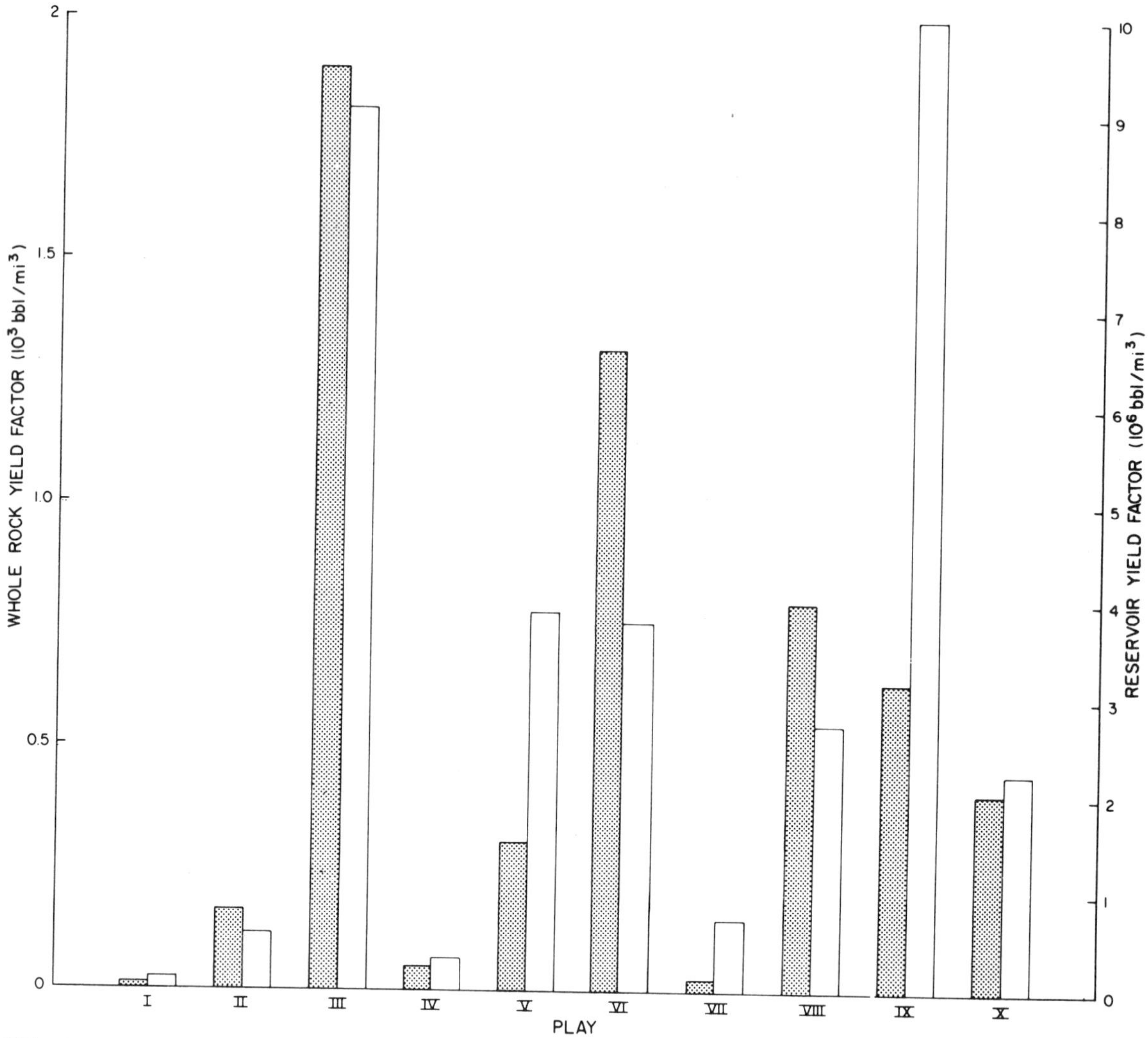

FIG. 18—Total hydrocarbon yield factor calculated on basis of volume of total rock and total sandstone (reservoir rock) contained within each of the ten Frio plays. Yield factor referenced to reservoir volume provides best index of play productivity.

deep-water sandstones pinch out updip against a scoured basal unconformity. The area has been moderately explored, with an average of 0.9 wells/mi^2.

FRIO SOURCE ROCKS AND MATURATION HISTORY

In addition to establishing reservoir distribution and geometry, depositional environments control the areal and stratigraphic extent, richness, and organic matter type of potential source facies. Within the Frio, prodelta, slope, open shelf, delta platform, and lagoonal mudstones all constitute potential source facies that are volumetrically important and are interbedded with productive reservoir sandstones. A survey of the organic richness and thermal evolution of associated and subjacent mudrocks is required to qualify the potential for internal source of Frio plays.

Gulf Coast Tertiary source rocks are known to be generally lean, and the Frio is typical in that regard. In all, TOC data collected from various industry and government sources for 140 Frio samples averaged slightly below 0.3 wt. % organic carbon. Few samples exceeded the 0.4% lower limit for significant Gulf Coast petroleum source rocks suggested by Dow (1978). No obvious patterns in regional source richness are apparent (Galloway et al, 1982). Data did not, however, permit detailed correlation with depositional facies.

Information on Frio organic matter type is particularly scarce. Samples from the DOE-General Crude Pleasant Bayou 1 and 2 wells, which penetrated the basinward margin of the Houston delta system in Brazoria County, show that herbaceous and woody organic matter dominates, indicating a gas-prone source (Brown, 1979). Relatively small quantities of oil-prone amorphous organic matter were also found. It is likely that similar woody and herbaceous organic material characterizes fluvial and delta-influenced Frio mudstone, whereas amorphous marine kerogen occurs more abundantly in marine-dominated facies. Preliminary results for sample

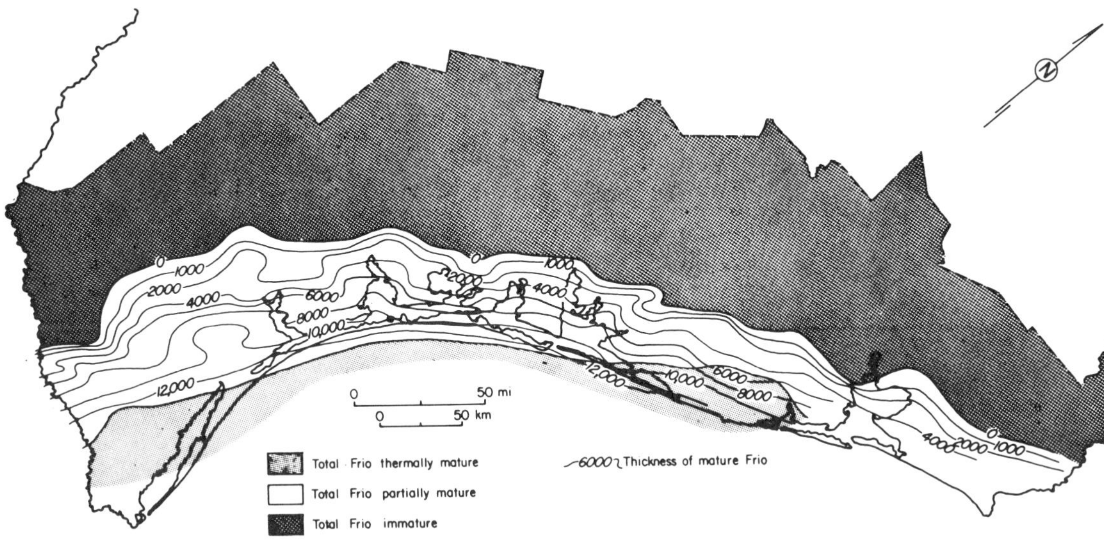

FIG. 19—Thickness of Frio strata that lie below top of the oil generation window, as approximated by the 200°F (94°C) isotherm. Significant volumes of mature source rock lie only within downdip portions of deltaic and barrier/strand-plain systems.

sets from two additional wells support this assumption.

The depth of the petroleum generation zone, the oil maturation window, is a product of burial time and temperature. Dow (1978) presented composite profiles of vitrinite reflectance (R_o) for the Cretaceous through Pliocene-Pleistocene sequences of the Louisiana Gulf Coast. According to his plot of Oligocene sediments, the temperature that corresponds to the onset of oil generation (about 0.5% R_o) is approximately 210°F (99%C). Some oil generation could start slightly earlier. The upper limit for oil generation (R_o =1.0%) occurs at about 300°F (148°C); at higher temperature, oil would have begun conversion to wet gas, and would be largely converted to gas at 1.35% R_o, or at a temperature of 330°F (166°C). Generally then, for Frio sediments, the oil window should lie approximately between the 100 and 300°F isotherms, assuming thermal gradients have not fluctuated greatly since Oligocene time. Gas and distillate would be generated at hotter temperatures. For the DOE-General Crude well, vitrinite reflectance data indicate a general correspondence between R_o values and the present isotherms (Galloway et al, 1982), placing the oil window from about 9,000 through 16,500 ft (3,000 through 5,000 m).

Using the 200 to 300° isotherms (calculated from corrected bottom-hole temperature measurements) as a guide, the oil maturation zone was plotted on each of the regional dip sections (Figs. 11 to 13, for example), and a regional isopach of the thickness of thermally mature Frio section was constructed (Fig. 19). Much of the Frio section is mature within plays I, II, V, VIII, IX, and X. Highly prolific plays III and VI, by contrast, contain little thermally mature Frio section.

In summary, although Frio mudstones contain low percentages of organic carbon and are dominated by gas-prone woody and herbaceous organic matter types, the volumes of potential source rock lying within or below the oil maturation window are immense. In the more deeply buried downdip plays, the possibility of internal source must be considered viable. In shallow plays, a deeper, extrinsic hydrocarbon source is required.

Patterns of Hydrocarbon Distribution

Examination of geographic and stratigraphic distributional patterns of Frio hydrocarbons within the context of the geologically homogeneous plays reveals several trends and relationships pertinent to understanding both petroleum origin and exploration potential. Generalized patterns suggested by data compilations incorporating all Gulf Coast Tertiary reservoirs are more clearly defined and can be more closely correlated with other variables such as depositional facies, depth of burial, or thermal history.

Field size distribution.—Both oil and gas field size, as defined by cumulative production statistics, suggest a log normal frequency distribution over all but the largest size ranges (Fig. 20). More than half of the total number of Frio fields have individually produced less than 8 million BOE. These same fields, however, account for less than 10% of the cumulative Frio production. A plot of the proportional production of gas and oil for each size class (Fig. 20) shows a bimodal pattern. A broad peak produced by numerous fields in the 32 to 128 BOE size range is separated by a slight sag from a second peak produced by a few 256 million-plus BOE fields. Deviations from well-developed frequency and production trend lines at this giant and near-giant end of the field size spectrum support the presumption that such fields are statistically unpredictable, geologically unique occurrences.

Disaggregation of the production versus field size data for the seven highly productive Frio plays (Fig. 21) shows that the bimodal pattern is particularly pronounced in plays VIII through X of the upper coastal plain, is apparent for the major hydrocarbon type in plays V and VI of the middle coastal plain, and is absent in south Texas plays II and III. The exact size class of the trough varies from play to play. This com-

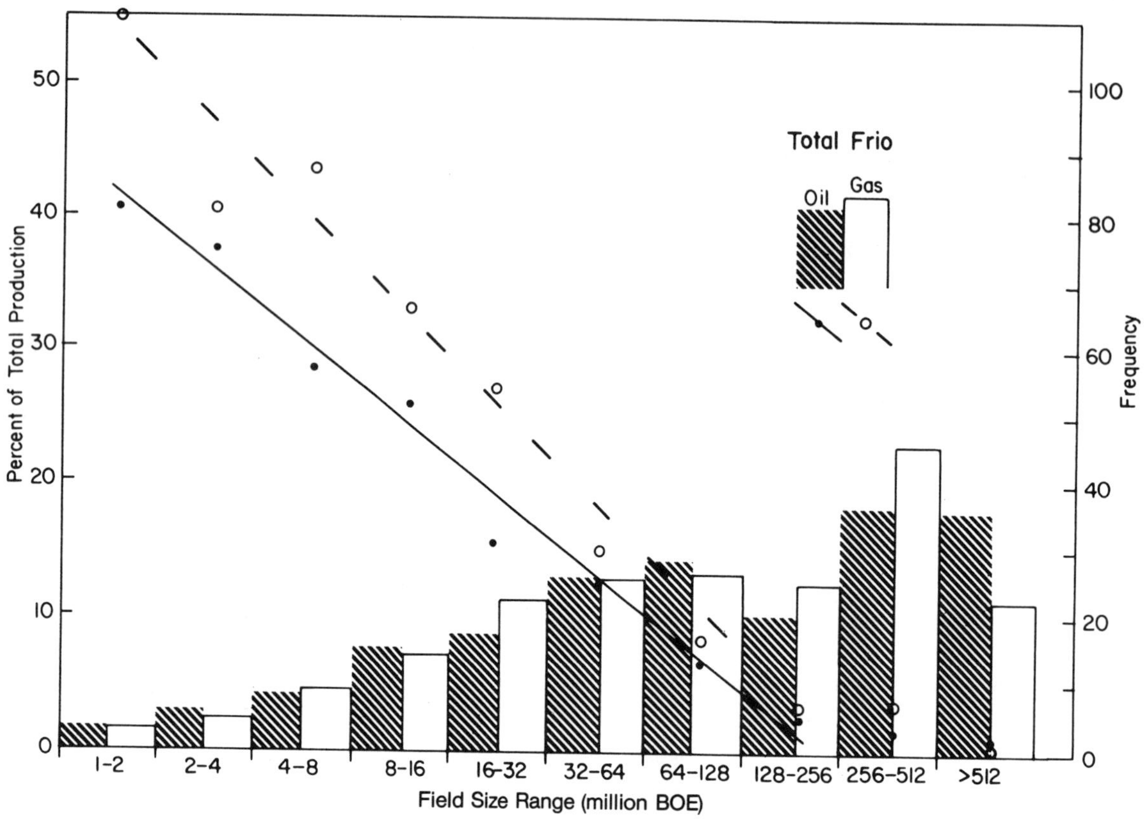

FIG. 20—Field size distribution for Frio Formation. Greatest percentage of total produced hydrocarbons occurred in the 64 to 128 and 256 to 512 million BOE size classes. As commonly assumed, number of fields within increasingly large size categories decreases linearly on semilog scale, indicating log normal distribution for all but the few largest fields.

parison suggests that both the complex and long-lived faulting associated with salt mobilization is particularly efficient at sporadically integrating large volumes of source rock into a single anomalously large drainage system. Regional growth faulting and shale diapirism may rarely produce such an anomalous integrated drainage network within sand-rich Frio facies of the Greta-Carancahua barrier-strand-plain system, but it produces a more continuous spectrum of smaller drainage compartments in moderate to sand-poor sequences of the Norias delta system.

Burial depth distribution.—The present depth of Frio reservoirs closely approximates maximum burial depth. Thus the present depth distribution is primarily determined by the depth to the zones of oil and gas generation and by the extent of vertical migration. Plots of data from several thousand reservoirs confirm generally accepted generalizations (Tissot and Welte, 1978, p. 401-403) and reveal some second order regional trends (Fig. 22).

1. Oil production significantly overlaps both wet and dry gas production. Only in the salt-floored Houston embayment, however, do wet and dry gas-producing intervals significantly overlap. In that area, dry gas occurs at unusually great depths.

2. The average depth of oil reservoirs (based on visual estimation of original data plots) ranges from about 5,500 ft (1,700 m) in the Rio Grande embayment to 7,300 ft (2,300 m) in the Houston embayment (Fig. 22). Concomitantly, the deepest significant occurrence of oil increases from about 9,000 ft (2,800 m) to almost 12,000 ft (3,700 m). Plays V and VI of the middle coastal plain display intermediate values.

3. In all areas, the average depth of condensate production averages 1,500 to 2,000 ft (450 to 600 m) below average oil depth (Fig. 22).

Petroleum density distribution patterns.–Because density measurements, in the form of API gravity, exist for several thousand Frio oils and distillates, this parameter is useful for examining spatial variation of liquid hydrocarbon composition. For each play, plots of API gravity versus depth for liquids produced from oil or gas reservoirs (as classified by the Railroad Commission of Texas) were plotted against depth. Representative examples of two such plots are shown in Figure 23. As expected, play V, which is downdip, exhibits greater depths of production and more distillate-productive reservoirs. Play VI, which is updip, possesses more oil reservoirs at generally shallower depths. Data points form two distinctive groupings on each plot, one for distillates and one for oils. These groupings can be enclosed within an envelope that excludes only a few highly divergent points. Superposition of the two sets of plots shows that the envelopes overlap; each play, in effect, revealing about two-

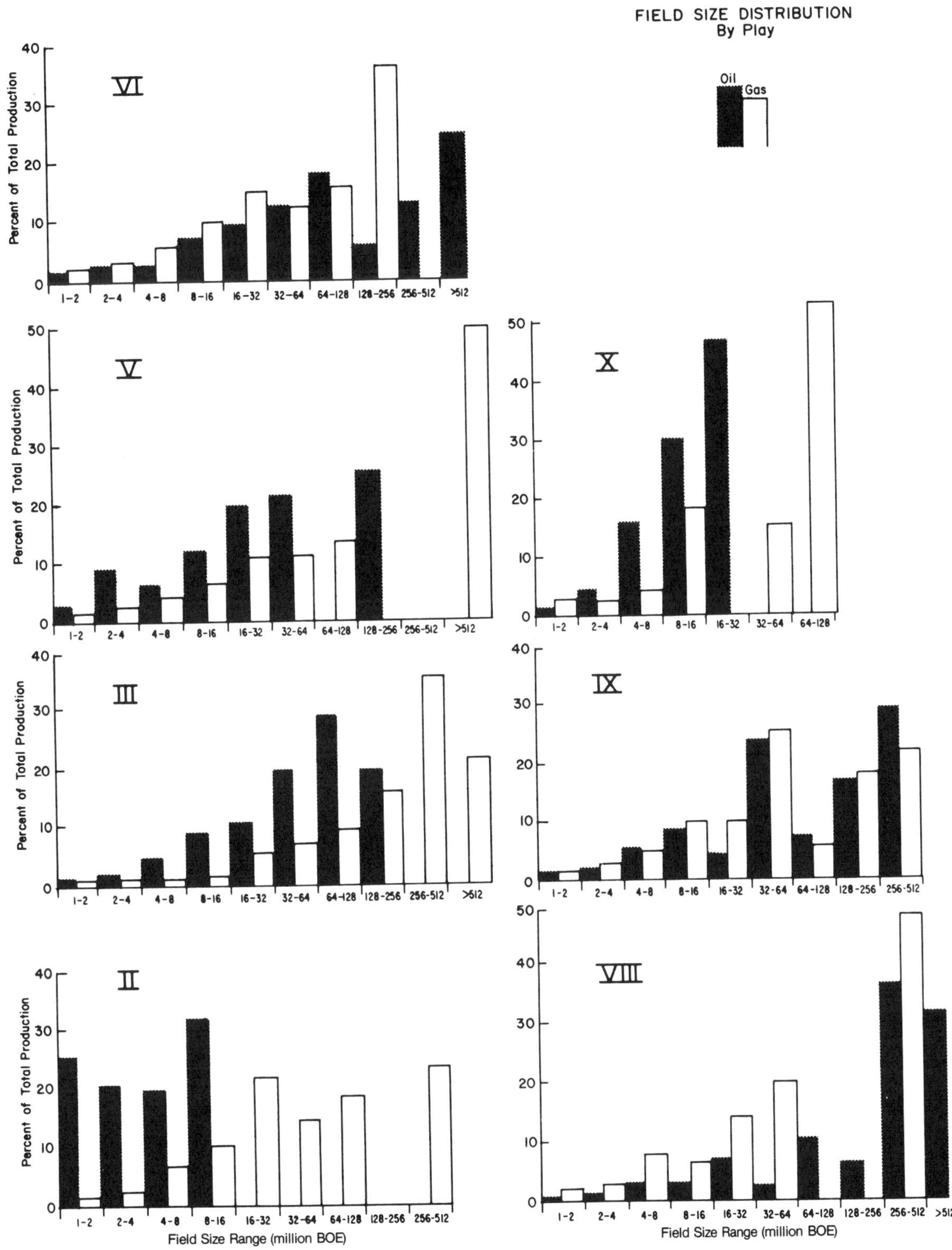

FIG. 21—Field size distributions within seven significant Frio plays.

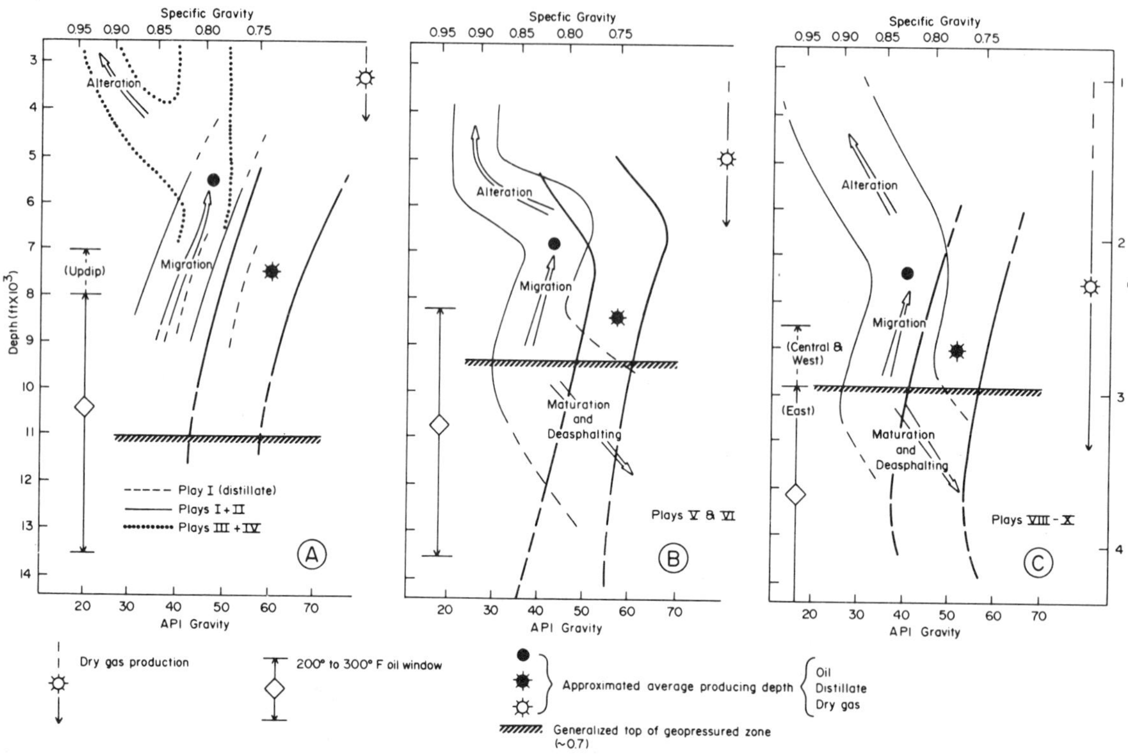

FIG. 22—Generalized envelopes showing systematic variation in density of oils and condensates with depth. Envelopes include over 90% of several tens to hundreds of data points and reveal consistent vertical trends in petroleum geochemistry. Data are grouped into lower (A), middle (B), and upper (C) coastal plain plots to maximize depth range on each plot. Well symbols indicate average depth of oil, condensate, and dry gas reservoirs in each area. Average vertical extent of hydrocarbon liquid window and top of geopressure are also indicated.

thirds of a complex vertical trend that characterizes hydrocarbons of the Greta/Carancahua system reservoirs (Fig. 22B).

Cross plots for the remaining major plays can similarly be combined to produce generalized graphs for produced hydrocarbons of the Norias and Gueydan systems (Rio Grande embayment) and of the Houston and Buna systems (Houston embayment; Fig. 22A and C). Liquids of plays I through IV exhibit the greatest contrasts. Play I distillates are unusually dense, play II produces dominantly light to very light distillate, and plays III and IV produce light to heavy oil and dry gas (Fig. 22A). The heterogeneity of depth-density trends indicates different hydrocarbon sources likely dominated in different parts of the south Texas Frio, a conclusion supported by other geologic and geochemical attributes of the different plays. In contrast, combination of plays VIII, IX, and X reveals considerable data overlap and similar depth-related trends (Fig. 22C). The envelopes for oil and gas reservoirs resemble their counterparts in plays V and VI, indicating uniformity of hydrocarbon source types. Indeed, a similar suite of Frio and older mudstone facies are possible hydrocarbon sources for all of these plays. Vertical trends found in all three areas indicate regional modification of the migrating or trapped hydrocarbon liquids by depth-dependent processes.

Additional compositional data for about 60 Frio oils is contained in analyses tabulated by Coleman et al (1978). Although Frio oils are low in sulfur, sulfur content does increase in direct proportion to increasing density as would be expected (Fig. 24). In addition, a density versus sulfur cross-plot highlights two sets of anomalous data. All very low sulfur oils (<0.05 wt. %) come from play III or closely adjacent portions of plays II, V or VI. Although data points are few, these oils may define a second distinct linear correlation function on the cross plot (Fig. 24). At the other extreme, two oils from play I are very high in sulfur and plot far off of the main trend (Fig. 24). Oils from the remaining plays show no systematic variations within the generalized linear envelope. Thus, variations in sulfur content support the interpretation of multiple sources for oils in the Norias delta and Gueydan fluvial systems and of a more uniform source or combination of sources for hydrocarbons in the other Frio depositional systems. Of equal importance, trends in measured API gravity reveal parallel trends in sulfur content and can be interpreted accordingly.

Examination of the three generalized density-depth envelopes for oil reservoirs (Fig. 22) shows that each contains at least two of three recurrent trend segments. Dense (and therefore relatively sulfurous) oils become lighter at depth until a relatively high API gravity between 35 and 50° is reached. Below this level, density gradually increases downward until oils display an average API gravity of 30 to

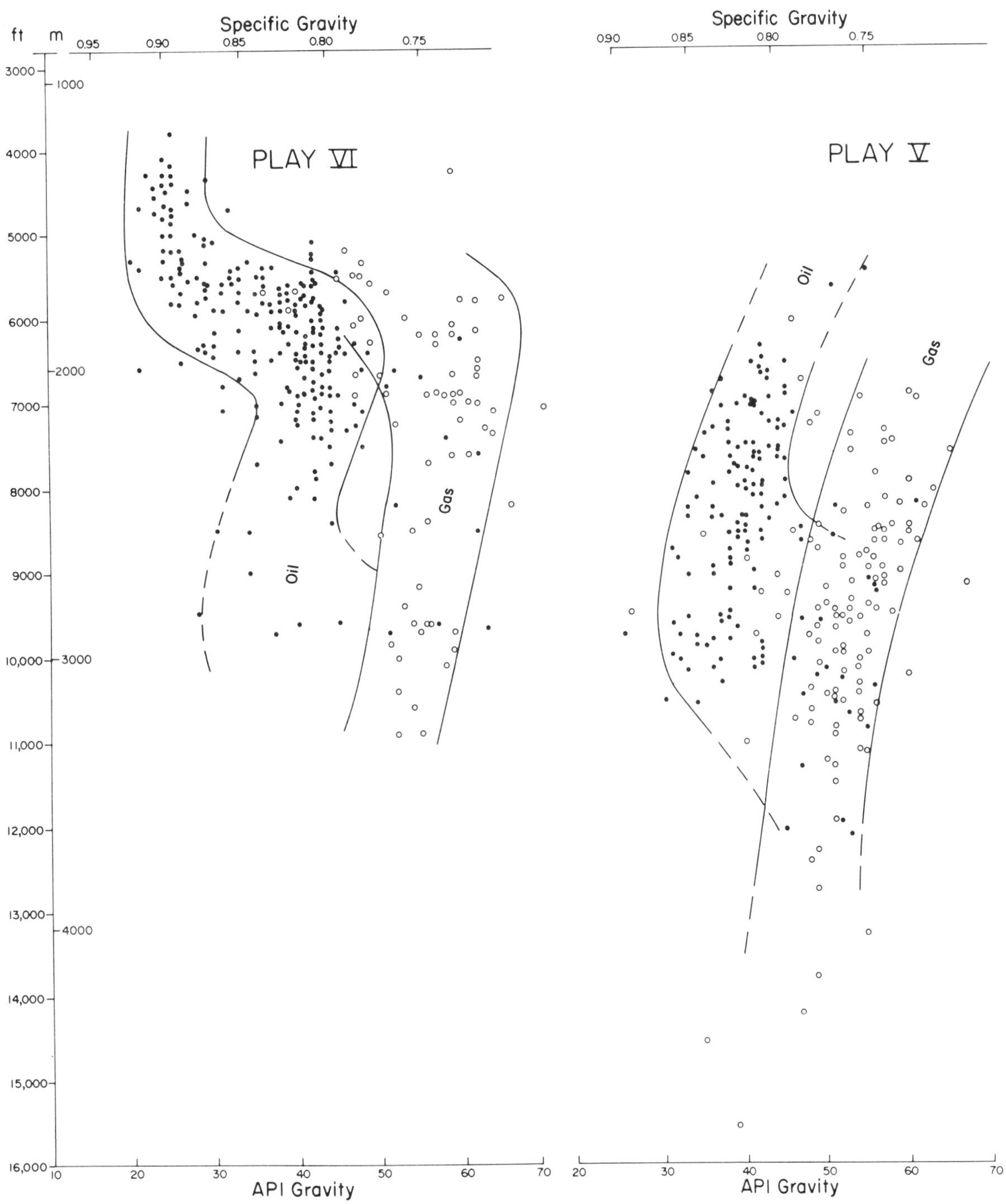

FIG. 23—Representative plots of hydrocarbon liquids density versus depth for plays V and VI. Data are typical of plots from other major plays that were used to construct Figure 22.

40° at the approximate top of the oil generation zone. Below the top of the oil window, density is highly variable but generally decreases, and few oils are found below the inferred depth of peak liquid generation (Fig. 22). Liquids from gas reservoirs show mainly the trend of increasing density with depth and merge with remnants of the oil trend in the middle to lower part of the liquid window. Only in the Greta/Carancahua system do distillates extend to sufficiently shallow depths to show an inflection back toward greater density.

Vertical distribution of the three zones and their relationship to the oil window suggests three processes have acted to

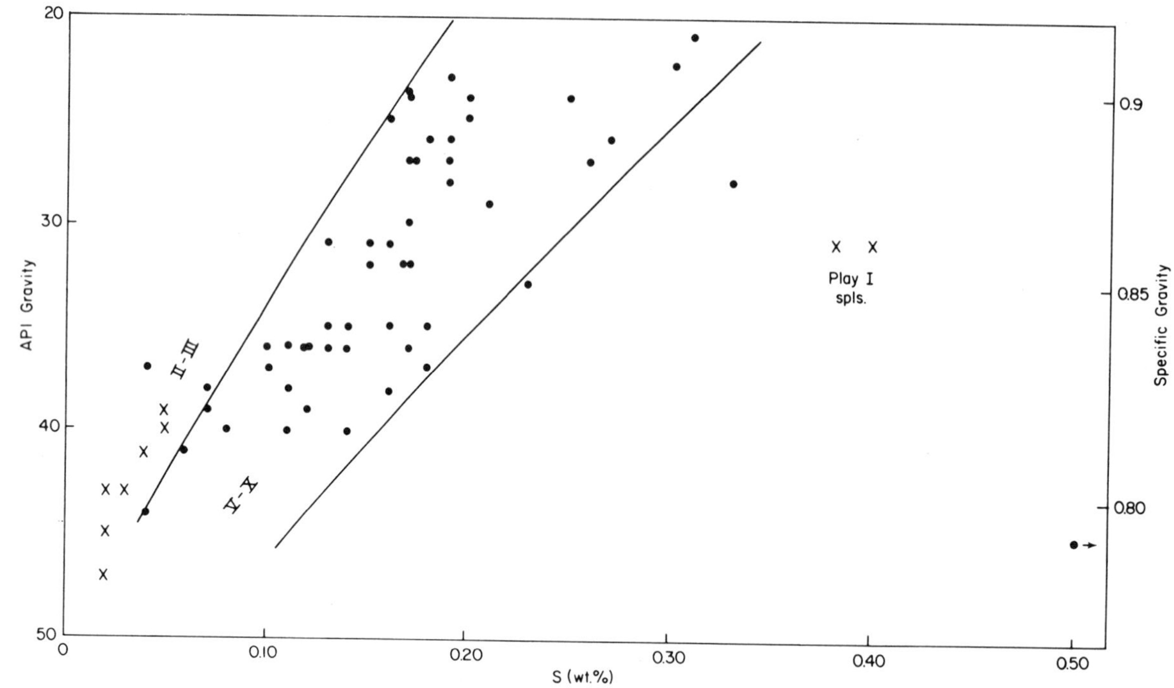

FIG. 24—Cross plot of weight percent sulfur versus density for Frio oils shows strong positive correlation. In addition, extremely low sulfur crudes from plays II and III define separate fields, suggesting geochemically different source. Two oils from play I are anomalously high in sulfur (data from Coleman et al, 1978).

modify original crude oils that were generated as source mudrocks matured: (1) continued maturation, thermal cracking, and deasphalting, (2) migration and attendant chromatographic separation, and (3) bacterial alteration.

Maturation and deasphalting.—Oils initially generated and trapped within the liquid window were carried to greater depths and hotter temperatures as subsidence continued. Continued thermal maturation and increasing concentration of light fractions result in precipitation of the heavy ends with consequent removal of sulfur and decrease in density. Gas generation becomes proportionally dominant, and gas with light distillate is a likely end product. Depending on stratigraphic and geographic location, deep Frio reservoirs that exhibit the maturation trend have been embedded within the oil generation zone for 10 and 20 million years (assuming thermal gradients have remained reasonably constant), allowing ample time for thermal cracking and deasphalting to proceed.

Migration and chromatographic separation.—If peak generation is assumed to occur near the thermal midpoint of the generation window, then the center of mass of reservoired oils lies approximately 4,500 ft (1,400 m) above the level of maximum generation, thus indicating considerable upward migration. For the interval extending from 2,000 to 3,000 ft (600 to 900 m) above the top of the oil window, the common trend for all oils is one of decreasing density and sulfur content. This upward density decrease, which averages 5 to 10 API units, is most likely a record of the migration process. The same trend is even more apparent for distillates, which occur over a much broader interval between zones of active thermal cracking and shallow alteration. Chromatographic separation of hydrocarbons is a well-known phenomena and would favor greater migration of lighter fractions (Nagy, 1960). Such a process was suggested for Mahakam delta oils by Combaz and De Matharel (1978).

Bacterial alteration.—Although each envelope is somewhat different, all three plots in Figure 22 indicate alteration of some or all shallow oils to a denser, more sulfur-rich, 20 to 30 °API gravity crude. The increased density is accompanied by an increase in the Correlation Index for the C_4 to C_7 fraction, which can be interpreted as indirect evidence for an increase in naphthene and aromatic fractions at the expense of normal paraffins (Coleman et al, 1978). Together these data and interpretations imply anaerobic bacterial alteration of the oils (Bailey et al, 1973). Bacterial alteration, in turn, requires some flushing of the reservoirs by meteoric waters. The increase in depth and pervasive impact of the alteration process northeast along the coastal plain indirectly supports this interpretation. The known depth of freshwater penetration within Oligocene and Miocene aquifers similarly increases from southwest to northeast, and such a pattern has likely persisted through late Tertiary time as the modern climatic gradient was well established during deposition of the Frio (Galloway, 1977). Carothers and Kharaka (1980) propose meteoric intrusion and petroleum alterations within the Gulf Coast Miocene section based on depletion of dissolved organic acids at reservoir temperatures below 160°F (70°C). Evamy et al (1978) describe bacterial degradation of Tertiary Niger delta oils within reservoirs at temperatures of up to 180°F (80°C). Temperature and extent of meteoric flushing would limit the depth range of bacterial alteration. Maximum depths of altered Frio oils are in fact near the 150°

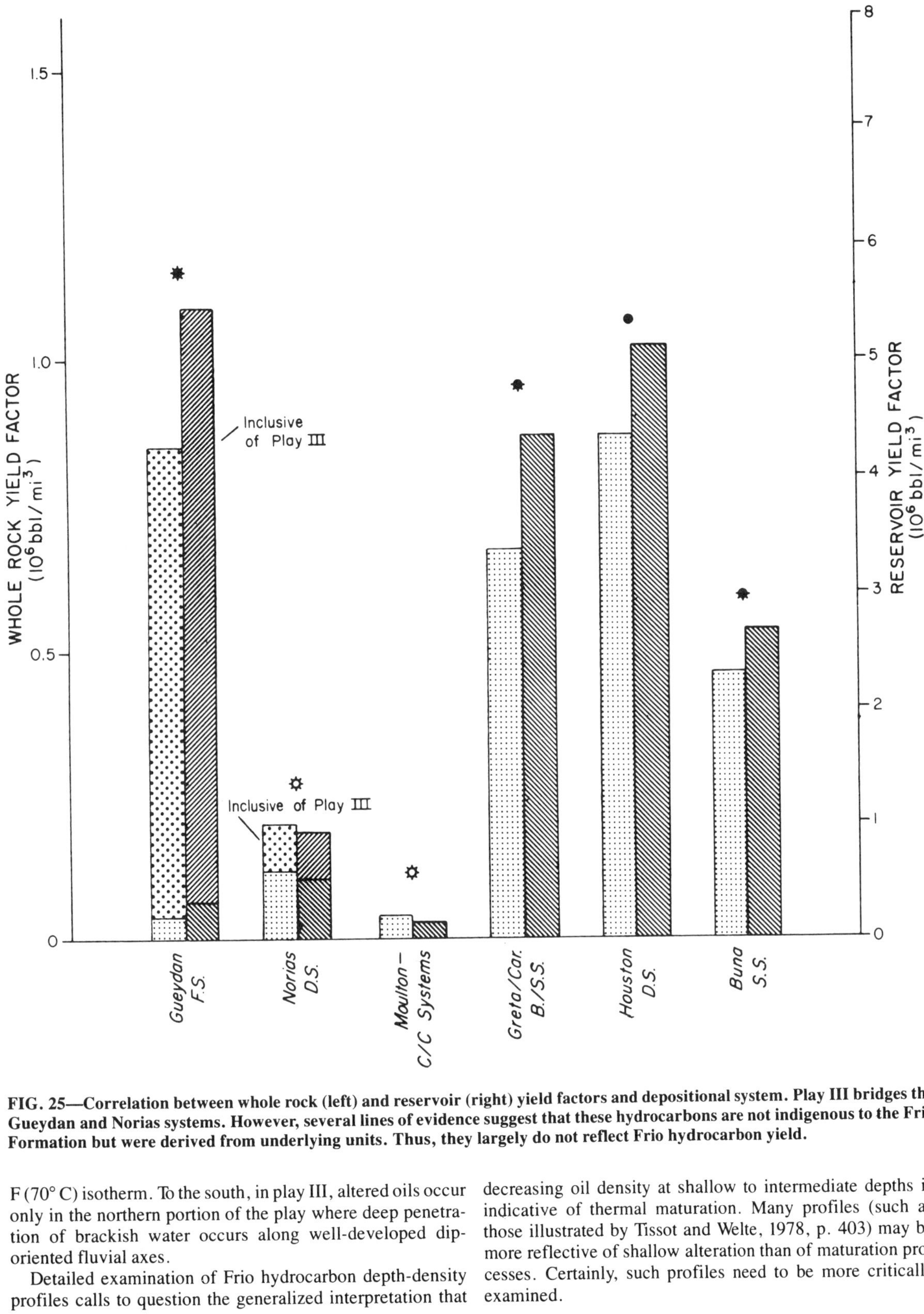

FIG. 25—Correlation between whole rock (left) and reservoir (right) yield factors and depositional system. Play III bridges the Gueydan and Norias systems. However, several lines of evidence suggest that these hydrocarbons are not indigenous to the Frio Formation but were derived from underlying units. Thus, they largely do not reflect Frio hydrocarbon yield.

F (70° C) isotherm. To the south, in play III, altered oils occur only in the northern portion of the play where deep penetration of brackish water occurs along well-developed dip-oriented fluvial axes.

Detailed examination of Frio hydrocarbon depth-density profiles calls to question the generalized interpretation that decreasing oil density at shallow to intermediate depths is indicative of thermal maturation. Many profiles (such as those illustrated by Tissot and Welte, 1978, p. 403) may be more reflective of shallow alteration than of maturation processes. Certainly, such profiles need to be more critically examined.

Effects of temperature and pressure regimes.—Over 90% of the oil reservoirs lie above the top of geopressure. However, the plot for plays I through IV (Fig. 22A) indicates the correlation with temperature is more consistent. Where the top of the oil window is several thousand feet higher than the top of geopressure, the oil basement rises accordingly. The liquid window does lie largely within the geopressure zone in most of the oil-rich plays (Fig. 22B and C). Consequently, effective expulsion of oils generated within this zone, which is otherwise characterized by poor fluid drainage, is required.

As with oil, almost all condensates occur substantially above their presumed depth of origin. Dry gas shows the least evidence of a correlation with temperature or pressure, particularly in plays VIII through X.

Correlations with depositional facies.—Recognition of facies-controlled variations in petroleum properties is complicated by the effects of migration, bacterial alteration, and further maturation. Some possible associations are evident, however. The extremely sulfur-poor, gas-rich hydrocarbons of plays II and III likely reflect nonmarine sources within the Frio Norias delta system or in analogous facies of the underlying Vicksburg. Humic, type III kerogen would dominate prodelta and delta platform mudstones of the delta system, and thick aggradational deltaic sequences now lie within the liquid window (see cross section in Figure 11). In contrast, marine organic matter would be more abundant in interdeltaic shelf and slope muds, which are commonly interbedded with or underlie Frio reservoirs of the remaining principal plays. Marine organic matter contains oil-prone type II kerogen and produces relatively more sulfur-rich hydrocarbons. Occurrence of play I oils directly below the transgressive marine Anahuac shale, which may act both as a source and seal, perhaps accounts for their atypical sulfur content.

CONCLUSIONS

The Frio Formation of the Texas Gulf coastal plain is a major progradational clastic wedge capped by a regional transgressive marine tongue. It consists of the deposits of seven principal depositional systems, including the Gueydan fluvial and Norias delta systems of the Rio Grande embayment, the Chita/Corrigan fluvial, Houston delta, and Buna strand-plain systems of the Houston embayment, and the interdeltaic Moulton/Choke Canyon/Flatonia stream-plain and Greta/Carancahua barrier/strand-plain systems.

Structural elements, which include regional growth fault zones, salt diapirs and interdiapiric salt-withdrawal basins, and shale diapirs and ridges, are the product of depositional loading of undercompacted slope and basinal muds or of bedded salt. Structural features and their growth histories reflect depositional patterns and the nature of the deeply buried plastic, deformable material. Thus each of the three principal progradational systems is characterized by a distinctive variation on the general structural theme.

Sand composition, pore fluid evolution, and burial history vary significantly among the various depositional systems. Consequently, deep reservoir quality, which is controlled primarily by burial diagenetic modification of porous sandstones, is related to the depositional framework. Porosity and permeability destruction with depth severely limit production potential of mineralogically immature sandstones of the Norias delta system.

Integration of structural, stratigraphic, and production data defines ten producing plays within the Frio Formation. Each play is characterized by geologic similarities and a common production style. Seven of the plays contained nearly all of the more than 16 billion bbl of oil and liquid-equivalent gas produced from the Frio.

Hydrocarbon productivity, whether indexed to total rock volume, total reservoir volume, or total volume of thermally mature mudstone, varies widely from play to play and among the various depositional systems. Richest plays have produced approximately 10 million BOE of hydrocarbons/mi^3 of sandstone. The Gueydan fluvial system is the most prolific hydrocarbon producing system on a per-reservoir volume basis (Fig. 25). However, most of the production centers around the Vicksburg flexure, and geologic and geochemical considerations suggest that the hydrocarbons originated far below the Frio section. The two major delta systems reveal both the best and worst per-volume hydrocarbon yields of more deeply buried Frio systems capable of generating indigenous petroleum (Fig. 25). Thus, even though there are distinct correlations of productivity with major depositional elements, other factors, including structural style and relation of the oil window to pressure and potential source facies, play commanding roles in determining hydrocarbon productivity of the systems.

Vertical patterns of hydrocarbon distribution are determined by the stratigraphic position and organic composition of thermally mature source mudstone facies, the depth and thickness of the zone of petroleum liquids generation and its relationship to the geopressure zone, and to systematic geochemical modifications produced by migration, continuing thermal maturation, or biodegradation. Vertical distribution of oil reservoirs relative to the oil generation window suggests upward migration averaged 3,000 to 4,000 ft (900 to 1,200 m). Extensive vertical migration has minimized direct correlation of depositional environment and hydrocarbon productivity. Indigenous Frio hydrocarbons were derived from source rocks exhibiting organic carbon contents that average significantly below the cutoff of 0.4% commonly applied in geochemical qualification of source potential.

REFERENCES CITED

Agagu, O. K. 1975, Depositional characteristic of the Frio Formation, subsurface south Texas: Master's thesis, Univ. Texas at Austin.

Bailey, N. J. L., et al, 1973, Alteration of crude oil by waters and bacteria—Evidence from geochemical and isotope studies: AAPG Bull., v. 57, p. 1276-1290.

Baker, E. T., Jr., 1978, Stratigraphic and hydrogeologic framework of part of the coastal plain of Texas: U.S. Geol. Survey Open File Rept. 77-712, 32 p.

Bebout, D. G., R. G. Loucks, and A. R. Gregory, 1978, Frio sandstone reservoirs in the deep subsurface along the Texas Gulf Coast: Texas Univ. Bur. Econ. Geology Rep. Inv. 91, 92 p.

Berg, R. R., and R. R. Powell, 1976, Density-flow origin for Frio reservoir sandstones, Nine Mile Point field, Aransas County, Texas: Gulf Coast Assoc. Geol. Socs. Trans., v. 26, p. 310-319.

——— and B. K. Powers, 1980, Morphology of turbidite-channel reservoirs, lower Hackberry (Oligocene), southeast Texas: Gulf Coast Assoc. Geol. Socs. Trans., v. 30, p. 41-48.

Bishop, R. S., 1978, Mechanism for emplacement of piercement diapirs: AAPG Bull. v. 62, p. 1561-1583.
Boyd, D. R., and B. F. Dyer, 1964, Frio barrier bar system of south Texas: Gulf Coast Assoc. Geol. Socs. Trans. v. 14, p. 309-322.
Brown, L. F., Jr., and W. L. Fisher, 1977, Seismic-stratigraphic interpretation of depositional systems: Examples from Brazilian rift and pull-apart basins, *in* C. E. Payton, ed., Seismic stratigraphy—Applications to hydrocarbon exploration: AAPG Memoir 26, p. 213-248.
Brown, S. W., 1979, Hydrocarbon source facies analysis, Department of Energy and General Crude Oil Company, Pleasant Bayou No. 1 and 2 wells, Brazoria County, Texas, *in* Proceedings M. H. Dorfman and W. L. Fisher, eds., Fourth United States Gulf Coast Geopressured-Geothermal Energy Conference: Research and Development, v. 1, p. 132-152.
Bruce, D. H., 1973, Pressured shale and sediment deformation: mechanism for development of regional contemporaneous faults: AAPG Bull., v. 57, p. 878-886.
Carothers, W. W., and Y. K. Kharaka, 1980, Stable carbon isotopes of HCO_3 in oil field waters—Implications for the origin of CO_2: Geochim. et Cosmochim. Acta, v. 44, p. 323-332.
Coleman, H. J. et al, 1978, Analyses of 800 crude oils from United States oilfields: Technical Information Center, U.S. Dept. Energy Rept. BETC/RI-78/14, 447 p.
Coleman, J. M., and D. B. Prior, 1980, Deltaic sand bodies: AAPG, Continuing Education Note Ser., no. 15, 171 p.
Combaz, A., and M. De Matharel, 1978, Organic sedimentation and genesis of petroleum in Mahakam delta, Borneo: AAPG Bull., v. 62, p. 1684-1695.
Crans, W., G. Mandl, and J. Haremboure, 1980, On the theory of growth faulting: a geomechanical delta model based on gravity sliding: Jour. Petroleum Geology, v. 2, no. 3, p. 265-307.
Curtis, D. M., 1970, Miocene deltaic sedimentation, Louisiana Gulf Coast, *in* Deltaic Sedimentation, Modern and Ancient: SEPM Spec. Pub. 15, p. 293-308.
Dailly, G. C., 1976, A possible mechanism relating progradation, growth faulting, clay diapirism and overthrusting in a regressive sequence of sediments: Bull. Canadian Petroleum Geology, v. 24, no. 1, p. 92-116.
Dodge, M. M., and J. S. Posey, 1982, Structural cross sections, Tertiary formations, Texas Gulf Coast: Texas Univ. Bur. Econ. Geology, in press.
Domenico, S. N., 1980, Velocity and lithology: Notes, AAPG/SEG Stratigraphic Interpretation of Seismic Data School, September 23-27, AAPG, Tulsa, Oklahoma, 62 p.
Dow, W. G., 1978, Petroleum source beds on continental slopes and rises: AAPG Bull., v. 62, p. 1584-1606.
Evamy, D. D., et al, 1978, Hydrocarbon habitat of Tertiary Niger delta: AAPG Bull., v. 62, p. 1-39.
Fisher, W. L., 1969, Facies characterization of Gulf Coast basin delta systems, with some Holocene analogs: Gulf Coast Assoc. Geol. Socs. Trans., v. 19, p. 239-261.
——— and J. H. McGowen, 1967, Depositional systems in the Wilcox Group of Texas and their relationship to occurrence of oil and gas: Gulf Coast Assoc. Geol. Socs. Trans., v. 17, p. 105-125.
Frazier, D. E., 1974, Depositional-episodes: their relationship to the Quaternary stratigraphic framework in the northwestern portion of the Gulf basin: Texas Univ. Bur. Econ. Geology Geol. Circ. 74-1, 28 p.
Galloway, W. E., 1975, Process framework for describing the morphologic and stratigraphic evolution of deltaic depositional systems, *in* M. L. Broussard, ed., Deltas: Houston Geol. Soc. p. 87-98.
——— 1977, Catahoula Formation of the Texas coastal plain: depositional systems, composition, structural development, ground-water flow history, and uranium distribution: Texas Univ. Bur. Econ. Geology Rep. Inv. 87, 59 p.
——— and W. R. Kaiser, 1980, Catahoula Formation of the Texas coastal plain: origin, geochemical evolution, and characteristics of uranium deposits: Texas Univ. Bur. Econ. Geology Rept. Inv. 100, 81 p.
———D. K. Hobday, and K. Magara, 1982, Frio Fromation of the Texas Gulf coastal plain—depositional systems, structural framework, and hydrocarbon origin, migration, distribution and exploration potential: Texas Univ. Bur. of Econ. Geology Rept. Inv., in press.
Gregory, A. R., et al, 1980, Volume and accessibility of entrained (solution) methane in deep geopressured reservoirs—Tertiary formations of the Texas Gulf Coast: Texas Univ. Bur. Econ. Geology Contract Rept. 390 p.
Halbouty, M. T., and T. D. Barber, 1961, Port Acres and Port Arthur fields, Jefferson County, Texas: Gulf Coast Assoc. Geol. Socs. Trans., v. 11, p. 225-234.
International Oil Scouts Association, 1978, Texas oil and gas development, *in* Texas Yearbook 1978: Austin, v. 6, 360 p.
Loucks, R. G., 1978, Sandstone distribution and potential for geopressured geothermal energy production in the Vicksburg Formation along the Texas Gulf Coast: Gulf Coast Assoc. Geol. Socs. Trans., v. 28, p. 239-271.
——— D. G. Bebout, and W. E. Galloway, 1977, Relationship of porosity formation and preservation to sandstone consolidation history—Gulf Coast lower Tertiary Frio Formation: Gulf Coast Assoc. Geol. Socs. Trans., v. 27, p. 109-120.
———M. M. Dodge, and W. E. Galloway, 1979a, Sandstone consolidation analysis to delineate areas of high-quality reservoirs suitable for production of geopressured geothermal energy along the Texas Gulf Coast: Texas Univ. Bur. Econ. Geology, unpublished report prepared for U.S. Dept. Energy, 98 p.
——— ——— ——— 1979b, Importance of secondary leached porosity in lower Tertiary sandstone reservoirs along the Texas Gulf Coast: Gulf Coast Assoc. Geol. Socs. Trans., v. 29, p. 1-8.
Martin, G. B., 1969, The subsurface Frio of South Texas: Stratigraphy and depositional environments as related to the occurrence of hydrocarbons: Gulf Coast Assoc. Geol. Socs. Trans., v. 19, p. 489-501.
McGookey, D. P., 1975, Gulf Coast Cenozoic sediments and structure: An excellent example of extra-continental sedimentation: Gulf Coast Assoc. Geol. Socs. Trans., v. 25, p. 104-120.
McGowen, J. H., L. E. Garner, and B. H. Wilkinson, 1977, The Gulf shoreline of Texas. Processes, characteristics, and factors in use: Texas Univ. Bur. Econ. Geology Geol. Circ. 77-3, 27 p.
Morton, R. A., 1977, Historical shoreline changes and their causes, Texas Gulf Coast: Gulf Coast Assoc. Geol. Socs. Trans., v. 27, p. 352-364.
———and J. H. McGowen, 1980, Modern depositional environments of the Texas coast: Texas Univ. Bur. Econ. Geology Guidebook 20, 167 p.
Nagy, B., 1960, Review of the chromatographic "plate" theory with reference to fluid flow in rocks and sediments: Geochim. et Cosmochim. Acta, v. 19, p. 289-296.
Nanz, R. H., Jr., 1954, Genesis of Oligocene sandstone reservoir, Seeligson field, Jim Wells and Kleberg Counties, Texas: AAPG Bull., v. 38, p. 96-117.
Paine, W. R., 1971, Petrology and sedimentation of the Hackberry sequence of southwest Louisiana: Gulf Coast Assoc. Geol. Socs. Trans., v. 21, p. 37-55.
Railroad Commission of Texas, 1978, Annual report of the Oil and Gas Division: Austin, 613 p.
Roberts, H. H., 1980, Sediment characteristics of Mississippi River delta-front mudflow deposits: Gulf Coast Assoc. Geol. Socs. Trans., v. 30, p. 485-496.
Stuckey, C. W., 1964, The stratigraphic relationship of the Hackberry, Abbeville, and Harang faunal assemblages: Gulf Coast Assoc. Geol. Socs. Trans., v. 14, p. 209-216.
Tissot, B. P., and D. H. Welte, 1978, Petroleum formation and occurrence: New York, Springer-Verlag, 538 p.
Vail, P. R., R. M. Mitchum, Jr., and S. Thompson, III, 1977, Seismic stratigraphy and global changes of sea level, part 4: global cycles of relative changes of sea level, *in* C. E. Payton, ed., Seismic stratigraphy—Applications to hydrocarbon exploration: AAPG Memoir 26, p. 83-97.
Weise, B. R., et al, 1982, Geologic studies of geopressured and hydropressured zones in Texas—Test well site selection: Contract Rept. prepared for Gas Research Institute, contract 5011-321-0125, Texas Univ. Bur. Econ. Geology, in press.
White, D. A., 1980, Assessing oil and gas plays in facies-cycle wedges: AAPG Bull., v. 64, p. 1158-1178.
——— et al, 1975, Assessing regional oil and gas potential: AAPG Studies Geology 1, p. 143-159.
Winker, C. D., 1979, Late Pleistocene fluvial-deltaic deposition, Texas coastal plain and shelf: Master's thesis, Univ. Texas at Austin, 187 p.
——— 1980, Depositional phases in late Pleistocene cyclic sedimentation, Texas coastal plain and shelf, with some Eocene analogs, *in* B. F. Perkins and D. K. Hobday, eds., Middle Eocene coastal plain and nearshore deposits of East Texas: a fieldguide to the Queen City formation and related papers: Gulf Coast Sec. SEPM, p. 46-66.

THE AMERICAN ASSOCIATION OF PETROLEUM GEOLOGISTS BULLETIN

JULY 1976 VOLUME 60, NUMBER 7

Reservoir Geometry of a Fluvial Sheet Sandstone[1]

C. V. CAMPBELL[2]
Houston, Texas 77001

Abstract Fluvial sheet sandstone provides an ideal reservoir for hydrocarbons. Such a sandstone body is large enough to contain giant hydrocarbon reserves as well as an aquifer that could flush the hydrocarbons.

The outcrops of the Westwater Canyon Member in the Morrison Formation (Jurassic), northwestern New Mexico, show representative attributes of a fluvial sheet sandstone. This sandstone body consists of coalescing fluvial-channel systems that in turn are composed of still smaller coalescing individual channels. The sheet is more than 63 mi (100 km) wide transverse to the paleocurrent direction and averages 200 ft (61 km) in thickness; the downcurrent dimension is not preserved but may have been as much as 100 mi (160 km). Channel systems internal to the sheet average 7 mi (11 km) wide and 50 ft (15 m) deep. The individual channel deposits that form the systems average 600 ft (183 m) wide by 12 ft (4 m) thick.

Any one vertical column through the sheet sandstone is usually more than 90 percent sandstone. Further, each vertical column consists of repetitive incomplete (top-missing) fluvial sequences that show upward-decreasing grain sizes and scales of sedimentary structures.

Fluvial sheet sandstone probably is deposited by aggrading and coalescing braided-stream systems.

INTRODUCTION

A common type of petroleum reservoir is a sheet-sandstone body that is a complex of multilateral fluvial channels. Oil-productive examples include parts of the Pennsylvanian Bartlesville (Bluejacket) Sandstone in Oklahoma (Saitta B. and Visher, 1968; Visher et al, 1971), the Cretaceous Woodbine in East Texas (Oliver, 1971, p. 8), and the Eocene Wilcox Group of the Gulf Coast of Central Texas (Fisher and McGowen, 1967, p. 110). All of these examples represent delta-plain deposits within a fluvial-deltaic environmental framework.

However, fluvial-sheet sandstone also is a component of several other environmental frameworks. The Upper Cretaceous Castlegate Sandstone in Utah is an example from a beach framework where the sheet of fluvial sandstone is separated from beaches by coastal swamps and marshes (Hale, 1959, p. 60, pl. III; Van de Graaff, 1972, p. 562). An example of a sheet of fluvial sandstone that interfingers with tidal-flat and estuarine deposits is the Misoa-Mirador Formation in the Maracaibo basin, Venezuela (Brondjik, 1967a, b; Walton, 1967; and writer's observations). Examples of sheet sandstones from wholly continental depositional frameworks include the Triassic Shinarump, Moss Back, and Gartra Members of the Chinle Formation in the southern Rocky Mountains (McCormick and Picard, 1970; Stewart et al, 1972). Other examples include large areas of the "Bundsandstein" (Lower Triassic) in Germany (Wurster, 1968; Backhaus, 1971) and the "Nubian" of North Africa (McKee, 1962).

In this paper usage of the term "multilateral sand body" follows Potter's original definition (1967, p. 338), but usage of the term "sheet sandstone" differs from the sheet sandstone of most geologists, who follow Potter (1963, p. 15; 1967, p. 338). Potter uses "multilateral sand body" for those sand bodies that consist of a complex of laterally coalescing depositional cycles. The cycles are the channel systems subsequently described in this paper. These coalescing channel systems form a hydrologically continuous sheet of

[1]Manuscript received, December 3, 1975; accepted, January 30, 1976.

[2]Exxon Production Research Co.

Special thanks are due W. C. Duke for his able field assistance and P. E. Melancon and J. W. Boettcher for their advice and stimulating discussions.

fluvial sandstone that covers 1,000 or more sq mi (2,600 sq km; arbitrary minimum). Widths and lengths are measured in tens of miles. Shapes in plan may be equidimensional or rectangular. Thicknesses range from 60 to 1,000 ft (18 to 305 m) but usually fall within the range of 100 to 300 ft (30 to 90 m). Except along their edges or in uncommon holes in the sheet, the stratigraphic interval of sheet sandstone may consist of more than 90 percent sandstone. The minimum-size sheet sandstone as here described shows all gradations to noncoalescing channel systems; the maximum area of a sheet sandstone may exceed 25,000 sq mi (65,000 sq km).

Such a sheet of fluvial sandstone—the Westwater Canyon Member of the Morrison Formation (Jurassic) in northwestern New Mexico (Fig. 1)—is the subject of this paper. Even though the Westwater Canyon Member does not produce petroleum and is from a wholly continental association of environments, its internal multilateral channel structure seems typical of sheets of fluvial sandstone. Further, the cliff-face exposures of the Westwater Canyon make an excellent field laboratory for studying the external geometries of such sandstone bodies. Thus, geometric and vertical-sequence attributes of the Westwater Canyon may be of use in evaluating petroleum reservoirs that are sheet-like bodies of fluvial sandstone.

Along the cliff faces of the Westwater Canyon Member, stratigraphic and facies relations were mapped by tracing the tops of separate sandstone bodies between measured sections (Fig. 2). When a top terminated laterally by nondeposition or intersection with a younger incised sandstone body, a stratigraphically higher or lower top was traced. Continuation of this procedure along the outcrop places the different sandstone bodies in their relative time-stratigraphic relations, for the depositional tops of all bodies are parallel (Fig. 3) according to many field measurements. The tops of the sandstone bodies were traced by a combination of walking, observation through binoculars, study of aerial photographs, and aerial reconnaissance at the level of the cliff faces.

More than 80 sections were measured with emphasis on locations where lateral changes were abrupt. Lithologies were described in the field for each measured section, and the layering characteristics were described according to the principles noted by Campbell (1967). In addition, paleocurrent measurements were recorded where suitable structures for measurement were present.

After first placing the Westwater Canyon Member in its geologic setting, this paper describes the distribution, external geometry, internal components, vertical sequence of textures and structures, and paleocurrent measurements. The

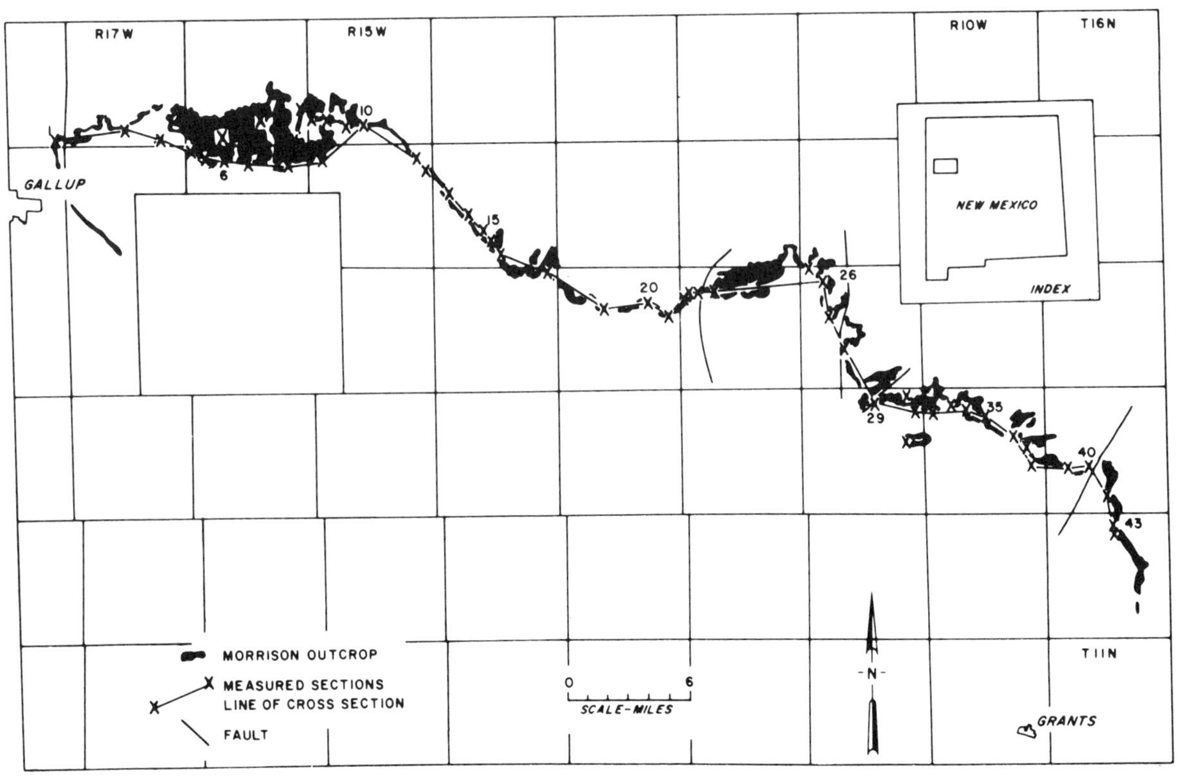

FIG. 1—Location map, Gallup to Grants, northwestern New Mexico.

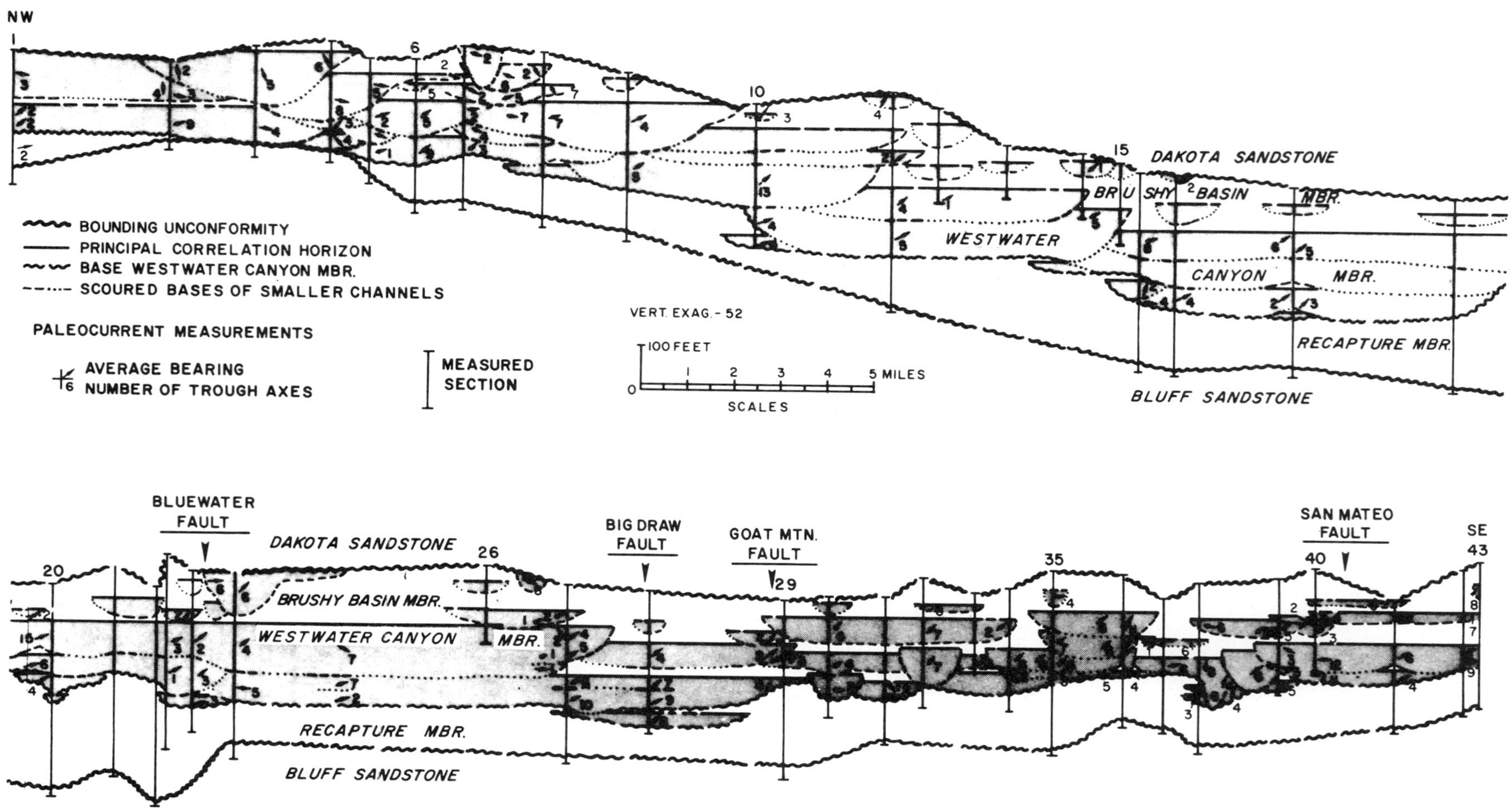

FIG. 2—Stratigraphic cross section showing relations of Westwater Canyon Member to enclosing rocks. This section is approximately transverse to paleocurrent directions and shows cross-sectional dimensions of channel systems that compose continuous sheet of Westwater Canyon sandstone.

FIG. 3—Photograph showing parallel tops of Westwater Canyon channel systems. Parallel tops are marked on photograph in middle of Morrison Formation, which is overlain by cliff-forming Dakota Formation and underlain by moundy topography of Bluff Formation.

paper concludes with a discussion of the possible origins of sheets of fluvial sandstones.

GEOLOGIC SETTING

The Morrison Formation is the uppermost unit of the sequence of Upper Jurassic, stable-shelf, nonmarine rocks in northwestern New Mexico (Fig. 4). This formation extends from the study area (Fig. 1) northward through the Rocky Mountains to the border between the United States and Canada, westward into central Utah, and eastward into central Kansas. The limits of the Morrison are mostly truncated edges (Peterson, 1972). Marine Upper Jurassic rocks appear mostly disconnected with the Morrison but are present on the Gulf Coast, in British Columbia, Canada, and in California, Oregon, and Washington (Imlay, 1957).

Upper Jurassic rocks in the study area are all continental deposits. Jurassic deposition began with the red-bed Entrada Sandstone (Fig. 4), which is largely eolian with probable interdune lake and fluvial deposits. Limestone and gypsum-anhydrite of the Todilto Formation probably accumulated in a shallow lake that formed on the Entrada Sandstone; the Todilto is unconformable on the Entrada. The even, parallel-bedded, red-bed Summerville siltstones overlie the Todilto, and these siltstones may represent the final filling of the Todilto lake or some other as yet unidentified type of continental deposit. Even, parallel-bedded Bluff Sandstone overlies and appears gradational with the Summerville. This lower Bluff Sandstone was deposited in an as yet unidentified type of depositional environment. However, the upper Bluff Sandstone is of eolian origin and is overlain unconformably by the Morrison Formation. The basal Morrison, the Recapture Member, represents fluvial and flood-plain deposits. The middle or Westwater Canyon Member is a sheet-like complex of multilateral fluvial-channel deposits; and the sheet-like complex grades laterally northward and vertically into probable lake deposits, the Brushy Basin Member at the top of the Morrison. Above the unconformity at the top of the Morrison, Dakota beds onlap and overlap Morrison rocks. The Dakota rocks are fluvial in the western part of the study area but become increasingly marine eastward.

DESCRIPTION OF CROSS SECTION

The cross section in Figure 2 extends for 63 mi (100 km) along the outcrop of the Morrison Formation (Fig. 1), and the datum for the section is the mapped succession of parallel tops of overlapping channel-filling sandstone bodies. In viewing the cross section, one must remember that the vertical exaggeration is 52×; thus, the sandstone body has a truly sheet-like cross section.

Figure 2 shows the distribution of the Westwater Canyon Member based on the cliff-face exposures in the study area (Fig. 1), and from these exposures this member has been mapped

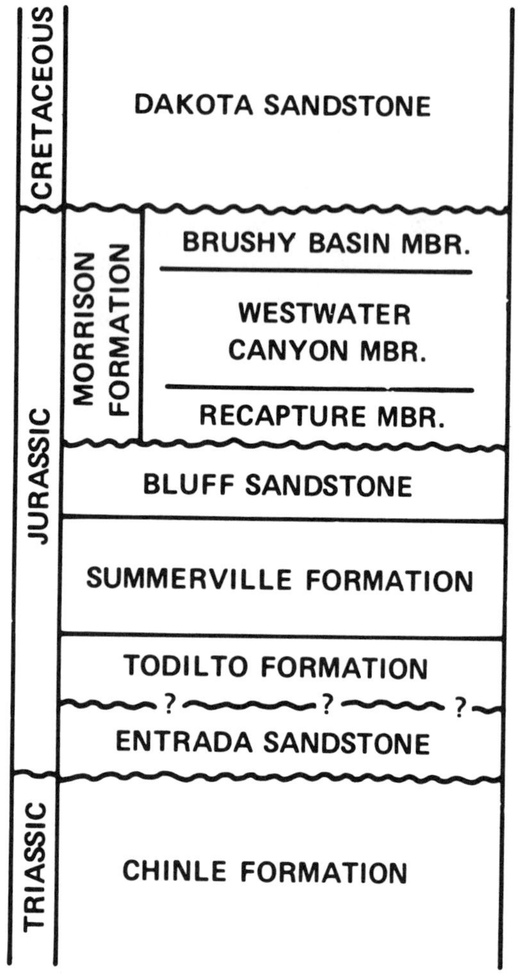

FIG. 4—Stratigraphic column, Jurassic rocks, northwestern New Mexico.

northwestward for 150 mi (240 km) and northeastward for 100 mi (160 km; Craig et al, 1955, Fig. 28). In the Gallup area the Westwater Canyon Member is truncated and overlapped by the Cretaceous Dakota Formation (Fig. 1), and this truncated edge extends southeastward to a point south of Grants (Granger, 1968, p. B65); consequently the original southwestward limit of the Westwater Canyon Member is unknown, but it originally covered an area greater than 20,000 sq mi (52,000 sq km).

The Westwater Canyon Member is approximately 200 ft (61 m) thick within the study area (Fig. 2). Northward and northwestward this member exceeds 300 ft (91 m) in thickness (Craig et al, 1955, Fig. 28; Saucier, 1967, Fig. 2). Southeastward the thick single-sandstone unit of the Westwater Canyon divides into narrower and thinner channel-filling sandstone bodies (Fig. 2), and this change suggests that it is losing its hydrologic continuity. Northwestward along the cross section, tops of successive channel-filling sandstones step up, indicating that the westernmost sandstone bodies are the youngest. Each successively younger body interfingers eastward with the Brushy Basin Member, but the sandstone bodies show essentially no gradation upward into the shale of the Brushy Basin Member. This shale, interpreted as a lake deposit, is greenish-gray or gray claystone that commonly contains floating silt and sand grains and lenses and nodules of limestone. The shale composes even, parallel beds with laminations parallel with the bedding surfaces.

As the channel-filling sandstones of the Westwater Canyon Member step up in time, they also overlap the Recapture Member onto the Bluff Sandstone. The base of the Westwater Canyon Member is always a sharply defined scoured surface that commonly must be traced as much as a quarter of a mile (0.4 km) to demonstrate truncation of underlying beds. The underlying and lateral beds, the Recapture Member interpreted as flood-plain deposits, consist of alternating grayish-red and greenish-gray siltstone; and these lithologies are present in even and wavy parallel beds that have been homogenized by the combined activity of roots and burrowing animals. The Recapture also contains channel-filling lenses of sandstone that have essentially the same texture as the Bluff Sandstone, which is fine-grained, well sorted, and subrounded to rounded.

Internally, the sheet of Westwater Canyon sandstone consists of two scales of channel-filling deposits. The largest, here termed channel systems, are the channel-filling bodies having the parallel tops shown in Figure 2. These channel systems range in width from 1 to 21 mi (1.6 to 34 km; average 7 mi or 11 km) and in thickness from 20 to 200 ft (6 to 61 m; average 50 ft or 15 m). Because of their great width compared with depth, most of the channel systems composing the Westwater Canyon appear tabular in cross section with abrupt thinning only at the channel edges. Separate channel systems commonly are marked by contrasting overall colors in outcrops, such as grayish-red versus salmon-pink versus buff.

The second scale of channel-filling deposit is the individual channel deposit that is the building block of the channel systems (Fig. 5). Each younger individual channel deposit fills scours into older ones, and consequently only parts of the individual channel deposits are preserved. Original dimensions of the individual channel deposits are width, 100 to 1,200 ft (30 to 366 ft; average 600 ft or 183 m), and thickness, 3 to 20 ft (1 to 6 m; average 12 ft or 4 m).

FIG. 5—Photograph showing individual channels composing channel systems, marked 1, 2, and 3, and curved scoured bases of readily identifiable individual channels, also marked.

Figure 2 also shows isolated channel systems in the Brushy Basin Member. These down-current offshoots of Westwater Canyon channel systems are narrower and commonly deeper than systems in the Westwater Canyon Member. These Brushy Basin systems range in width from 200 ft to 3 mi (61 m to 5 km) and in thickness from 3 to 100 ft (1 to 30 m).

VERTICAL SEQUENCE OF SEDIMENTARY STRUCTURES AND TEXTURES

The vertical sequence of sedimentary structures in the Westwater Canyon Member shows repeated top-missing fluvial sequences with occasional complete sequences, especially at the top of channel systems (Fig. 6). The complete fluvial sequence may begin with (1) a thin basal siltstone or mudstone that commonly contains conglomerate clasts eroded from the channel walls (Fig. 7), (2) a thin lens of pebble conglomerate, or (3) coarser grains over finer grains across an extensive scour surface. Above the basal unit, cross-laminated trough beds are characteristic. Many of the troughs are broad (20 to 50 ft or 6 to 15 m wide) and shallow (1 to 3 ft or 0.3 to 1 m deep), and many troughs show asymmetric lateral filling rather than symmetric filling. The thicknesses of the cross-laminated trough beds decrease upward. Next, although not always present, are one or two beds with parallel bedding surfaces that enclose laminae parallel with the bedding surfaces or low-angle (less than 10° dip) cross laminae. Above the parallel beds (or small-scale cross-laminated trough beds if the parallel beds are absent) current-ripple-laminated beds top the sequence. However, the complete fluvial sequence is uncommon because of removal of its top during scouring by successively younger channels; and only incomplete or top-missing fluvial sequences commonly are repeated in vertical succession.

Textural differences, although studied only in the field, show distinct grading on three scales. First is the upward decrease in grain size in each cross-laminated trough bed. Then, as the trough beds decrease in size, their overall grain size decreases; and this overall gradation continues into the current-ripple-laminated unit at the top of complete fluvial sequences. Finally, overall gradation within channel systems is expressed by generally finer grain sizes in successively younger individual channel deposits. Exceptions to these patterns of grading are present but are not particularly common.

Grain sizes in the Westwater Canyon Member range from pebbles to siltstone, but sandstone predominates. Basal pebble conglomerates are

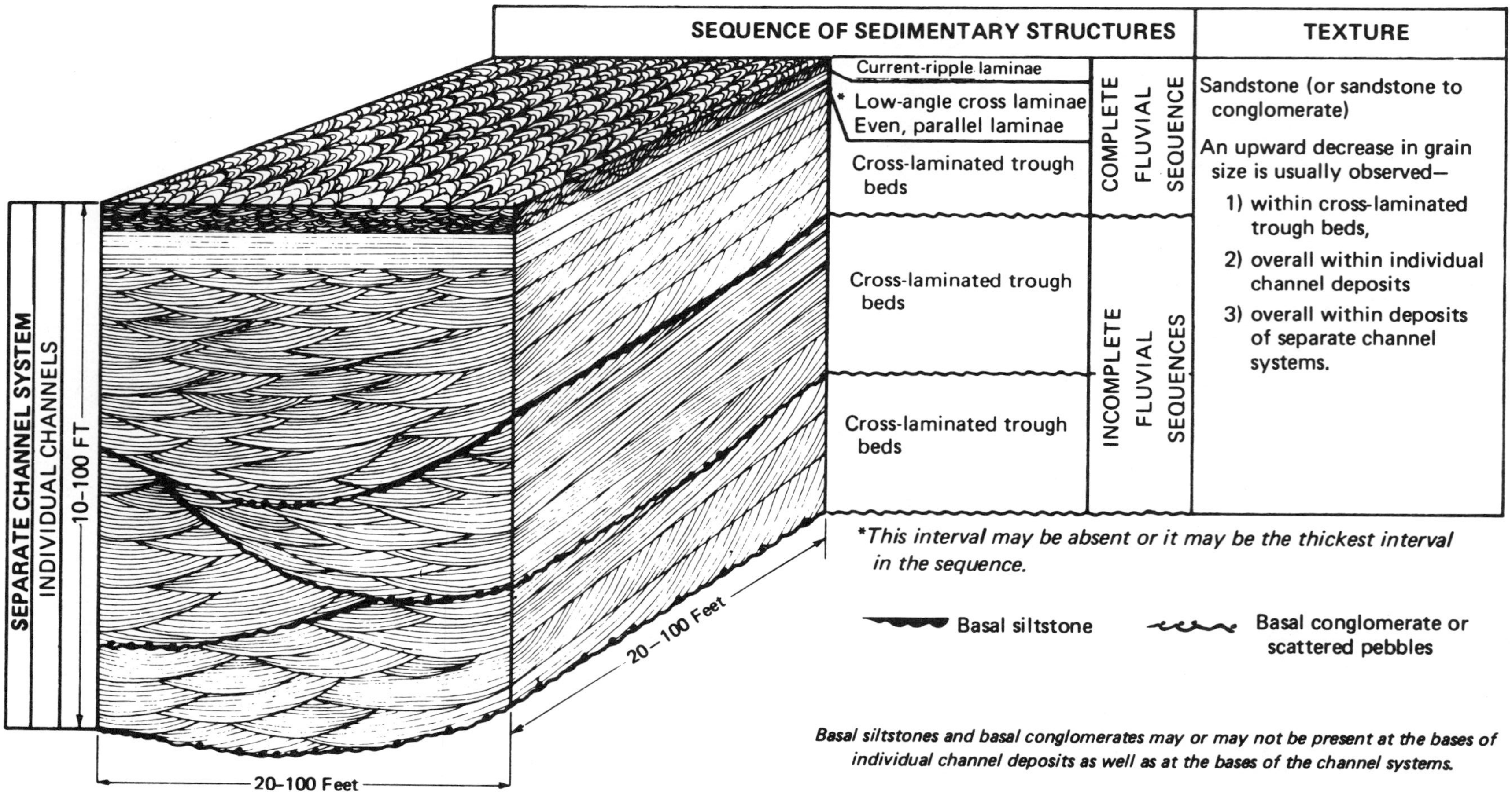

FIG. 6—Sequence of sedimentary structures and textures.

FIG. 7—Photograph showing basal siltstones and flood-plain remnants. Basal siltstones are above marked bases of individual channels. Flood-plain remnant marked *A* consists of thin, even parallel beds composed of siltstone.

most abundant in the western part of the study area; basal siltstone with pebbles and occasional cobbles seems more common in the east. Elsewhere grain-size variations depend on the position of the sample within vertical sequences. For instance, a channel system having coarse-grained sandstone at the base usually will be fine grained or very fine grained at the top. Within channel systems, preserved parts of lower individual channel deposits may grade upward from coarse grained to medium grained, and higher individual channels from medium grained to fine grained. The uppermost individual channel deposits usually are fine grained or very fine grained at their tops. More than 90 percent of any one vertical column through most locations on a sheet sandstone is sandstone.

On the basis of data from four boreholes, Westwater Canyon sandstones average about 33 percent fine, 58 percent medium, and 9 percent coarse (Santos, 1970, p. 13). In addition, four samples of Westwater Canyon sandstone from the study area have the following grain compositions: quartz, 46 to 63 percent; feldspar, 24 to 42 percent; and volcanics, 8 to 15 percent (Cadigan, 1967, p. 75). The feldspar is divided about equally between potassic feldspar and plagioclase. The sandstone varies from friable to hard; the hard samples usually are cemented by calcite, or sometimes clay matrix; and both friable and hard varieties may contain iron oxide cement.

Channel systems uncommonly enclose tabular remnants of flood-plain deposits, which usually terminate laterally with near vertical scour surfaces (Fig. 7). These remnants show the same bedding, internal structures, and lithologies as Recapture flood-plain deposits. However, brown to black sandstone thought to be the B zone of a soil profile marks the top of the channel systems and lies directly below leached siltstone and mudstone at the base of the remnants.

PALEOCURRENTS

Paleocurrent directions were measured where axes of troughs were present, and the average bearing for the direction of plunge of troughs for localities within channel systems is shown on Figure 2. An attempt was made to measure only one trough from each separate individual channel at the sample locality. The bearing of plunge of troughs rather than dip of cross laminae was measured, because the direction of plunge of troughs provides a superior (in terms of less dispersion) and more economical (in terms of less required measurements) index of paleocurrents (Dott, 1973, p. 783).

The average bearings of paleocurrents (Fig. 2) indicate a northeasterly trend, but occasional averages are directed either northwesterly or southeasterly. Within each channel system the bearings are remarkably similar; this correspondence suggests that nearly all of the individual channels composing a channel system flowed in the same direction with little meandering. Sinuous rather than meandering stream patterns also are indicated by the few individual channel courses that could be traced by the edges of channel systems; one edge was traced for 2 mi (3 km).

Origin of Sheets of Fluvial Sandstone

Conceivably, sheets of fluvial sandstone could form by the coalescing of either point bars or braided-stream deposits. Of these two possibilities, the latter seems better to fit the features of the Westwater Canyon Member.

A point-bar origin is ruled out for several reasons. First, the pattern in plan of the individual channels that compose the channel systems seems sinuous rather than meandering; both paleocurrent measurements and channel courses traced in limited exposures suggest sinuous or straight individual channels. Second, according to Leopold and Langbein (1966, p. 60), streams meander only where their banks consist of fine-grained cohesive material; almost all Westwater Canyon individual channels cut into noncohesive sandstone. However, the much larger channel systems cut at least in part into mudstone, but the separate channel systems are usually many times wider than 20 times the individual channel width cited as the width of meander belts (LeBlanc, 1972, p. 148). Finally, the general absence of siltstone and mudstone within the Westwater Canyon suggests sand-choked high-gradient channels rather than low-gradient meandering channels that fill partly with sand and partly with clay plugs.

Comparison of the Westwater Canyon Member with deposits in modern braided streams is not wholly satisfactory. Most of the described modern examples are conglomeratic rather than sandy and are associated either with glacial outwash (Krigström, 1962; Fahnestock, 1963, 1970; Carryer, 1966; Williams and Rust, 1969; McDonald and Banerjee, 1970; Boothroyd, 1972; Bluck, 1974; Gustavson, 1974; Smith, 1974) or with alluvial fans (Denny, 1965; Hooke, 1967; Beaty, 1970; Bull, 1972). Modern sandy braided-stream deposits include (1) the alluvial fan of the Kosi River (Gole and Chitale, 1966) and (2) the braided reaches of the present Yellow River (Chien, 1961) and the Brahmaputra River (Coleman, 1969). All of these examples are deposits by aggrading streams. In addition, sandy braided-stream deposits that accumulated in degrading river valleys and thus are confined laterally have been described by many (Doeglas, 1962; Ore, 1963, 1965; Sarkar and Basumallick, 1968; Collinson, 1970; Kessler and Cooper, 1970; Smith, 1970, 1971; Waechter, 1970; Kessler, 1971; Shelton and Noble, 1974). No one of these examples describes all of the critical attributes of the Westwater Canyon, and consequently, the aggrading braided-stream deposits of the Westwater Canyon are identified by comparison with a combination of attributes from different braided-stream habitats: (1) area, (2) thickness, (3) internal multilateral channel structure, (4) vertical sequence of textures and structures, and (5) enclosing facies.

1. The width (70 mi or 112 km) of the alluvial fan of the Kosi River approximates the width (63 mi or 100 km) of the Westwater Canyon Member in the study area. Further the Kosi fan is about 70 mi (112 km) long and covers more than 4,000 sq mi (10,400 sq km). This fan apparently has formed since 1731 (Gole and Chitale, 1966).

2. The few recorded thicknesses for modern sandy braided-stream deposits generally fall between the average 200 ft (61 m) for the Westwater complex and 50 ft (15 m) for internal channel systems. Kessler (1971, p. 33) noted that most modern braided-channel deposits would be preserved as lateral sand bodies 100 ft (30 m) or less in thickness. Ore (1964, p. 13) also suggested that braided-stream deposits are sheet-like and generally less than 100 ft (30 m) thick. Coleman (1969, p. 163) reported that the average depth of sand across the 8-mi (13 km) wide channel of the Brahmaputra is about 60 ft (18 m) with 120 ft (37 m) at node points.

3. The channel shifting in braided streams described by many authors suggests the internal multilateral channel structure in the Westwater Canyon Member. Small channel fills, which range from 3 to 20 ft (1 to 6 m) deep and average 12 ft (4 m), within the Westwater Canyon channel systems have counterparts in the Cimarron River (average 15 ft or 5 km deep; Shelton and Noble, 1974) and in the Durance and Ardeche Rivers (3 to 30 ft or 1 to 6 m; Doeglas, 1962). Also Westwater Canyon channel systems, which range from 1 to 20 mi (1.6 to 32 km) wide and average 7 mi (11.2 km), have modern braided analogs in the

Kosi River (4 to 10 mi or 6 to 16 km wide; Gole and Chitale, 1966, p. 115), in the Brahmaputra River (2 to 10 mi or 3 to 16 km wide; Coleman, 1969, p. 162), and in the Yellow River (17 mi or 27 km between dikes; Chien, 1961, p. 734).

4. The vertical sequence of textures and sedimentary structures in the Westwater Canyon Member is like the fining-upward cycle described by Allen (1964, 1965), Visher (1965), and Potter (1967). These authors illustrated but did not emphasize the vertical multiple-channel character of the cycles. Also Visher (1965, Fig. 13) suggested that the fining-upward sequence identifies point bars, and essentially the same cycle or vertical sequence was described by Harms et al (1963, Fig. 5) and by Bernard and Major (1963) and Bernard et al (1970, illustrations) from present point bars. Many other workers seem to identify the fining-upward cycle with point bars rather than with braided-stream deposits (for example, Sarkar and Basumallick, 1968; Van de Kamp, 1973), but Costello and Walker (1972) reported fining-upward cycles in Pleistocene braided-outwash gravels and sands, Gustavson (1974) described fining-upward lenses from present glacial outwash, and Cant (1973) interpreted Devonian braided-stream deposits that contain fining-upward cycles.

Further, illustrations of modern braided-stream deposits suggest that fining-upward cycles thick enough to compose a component of the Westwater Canyon multiple-fluvial sequence are in present braided-stream deposits. For example, Doeglas (1962, Fig. 30) illustrated cross-laminated trough beds underlying even, parallel beds having laminations parallel with bedding surfaces; and current-ripple laminae overlie the even, parallel beds. Moreover, large-scale cross-laminated trough beds overlain by current-ripple laminae were shown by Coleman (1969, Fig. 39A) and Hartshorn and Ashley (1972, Fig. 5). Also, even, parallel beds composed of laminae paralleling bedding surfaces are overlain by current-ripple laminae in other photographs (Coleman, 1969, Fig. 39D; Smith, 1970, Fig. 10).

Nearly all students of modern braided streams cite longitudinal or transverse bars (or both) as characteristic, and identifying these bars in ancient deposits seems critical to a braided-stream interpretation. In the Westwater Canyon Member these bars seem to be represented by partly preserved cross-laminated trough beds 20 to 50 ft (6 to 15 m) or more wide. These trough beds commonly look like cross-laminated planar beds in sections parallel with the current direction and like even, parallel beds composed of laminae parallel with bedding surfaces in sections transverse to the current direction. Blodgett (1974) noted a similar preservation of linguoid bars in Oligocene braided-stream deposits. In addition, we should not expect preservation of bar morphologies in ancient braided streams owing to the constant shifting of channels and the coincident formation and destruction of bars (see Krigström, 1962; Ore, 1965, p. 197; Smith, 1974).

Consequently, the "fining-upward cycle" seems to be present in both braided-stream deposits and point bars and is not a distinguishing feature of either of these types of deposits (also see Moody-Stuart, 1966, p. 1105). More conclusive distinction between these types of deposits depends finally on their differing external geometries and internal stream patterns (Kessler and Cooper, 1970, p. 269) together with the presence of clay plugs in point-bar deposits.

5. The facies enclosing the Westwater Canyon Member probably are flood-plain deposits below and laterally and lake deposits downcurrent; upcurrent deposits have been eroded. Thus the Westwater Canyon Member is interpreted as an apron of coalescing braided-stream deposits (possibly distal alluvial-fan deposits) marginal to a freshwater lake.

Conclusions

Fluvial sheet sandstone, such as the Westwater Canyon Member of the Morrison Formation, consists of coalescing multilateral fluvial-channel deposits. They are large enough to reservoir giant hydrocarbon reserves together with an aquifer large enough to flush the hydrocarbons. The most favorable environmental setting for oil-productive sheet sandstones is marginal marine. However, such sheets must be in favorable structural settings, where porosity and permeability have been preserved, and in association with source rocks.

References Cited

Allen, J. L., 1964, Studies in fluviatile sedimentation: six cyclothems from the lower Old Red Sandstone, Anglo-Welsh basin: Sedimentology, v. 3, p. 163-198.

——— 1965, Fining-upwards cycles in alluvial successions: Geol. Jour., v. 4, p. 229-246.

Backhaus, E., 1971, Red beds of the "Bundsandstein" (Lower Triassic) in the Odenwald and Pfalz-Vosges areas, *in* Sedimentology of parts of central Europe: 8th Internat. Sedimentological Cong., Heidelberg, Germany, Guidebook Excursions, p. 105-122.

Beaty, C. R., 1970, Age and estimated rate of accumulation of an alluvial fan, White Mountains, California, U.S.A.: Am. Jour. Sci., v. 268, p. 50-77.

Bernard, H. A., and C. F. Major, Jr., 1963, Recent meander belt deposits of the Brazos River: an alluvial "sand" model (abs.): AAPG Bull., v. 47, p. 350.

——— ——— B. S. Parrott, and R. J. LeBlanc, Sr.,

1970, Recent sediments of southeast Texas. A field guide to the Brazos alluvial and deltaic plains and the Galveston barrier island complex: Texas Univ. Bur. Econ. Geology Guidebook 11, [83] p.
Blodgett, R. H., 1974, A braided stream model for the White River Group in Wyoming and Nebraska (abs.): Geol. Soc. America Abs. with Programs, v. 6, p. 247.
Bluck, B. J., 1974, Structure and directional properties of some valley sandur deposits in southern Iceland: Sedimentology, v. 21, p. 533-554.
Boothroyd, J. C., 1972, Coarse-grained sedimentation on a braided outwash fan, northeast Gulf of Alaska: Univ. South Carolina ONR Tech. Rept. No. 6-CRD, 127 p.
Brondjik, J. F., 1967a, The Misoa and Trujillo Formations: Asoc. Venezolana Geología, Minería y Petróleo Bol. Inf., v. 10, p. 3-19.
——— 1967b, "Eocene" formations in the southwestern part of the Maracaibo basin: Asoc. Venezolana Geología, Minería y Petróleo Bol. Inf., v. 10, p. 35-50.
Bull, W. B., 1972, Recognition of alluvial-fan deposits in the stratigraphic record, *in* Recognition of ancient sedimentary environments: Soc. Econ. Paleontologists and Mineralogists Spec. Pub. 16, p. 63-83.
Cadigan, R. A., 1967, Petrology of the Morrison Formation in the Colorado Plateau region: U.S. Geol. Survey Prof. Paper 556, 113 p.
Campbell, C. V., 1967, Lamina, laminaset, bed and bedset: Sedimentology, v. 8, p. 7-26.
Cant, D. J., 1973, Devonian braided stream deposits in Battery Point Formation, Gaspé Est, Quebec: Maritime Sediments, v. 9, p. 13-20.
Carryer, S. J., 1966, A note on the formation of alluvial fans: New Zealand Jour. Geology and Geophysics, v. 9, p. 91-94.
Chien, N., 1961, The braided stream of the lower Yellow River: Sci. Sinica, v. 10, p. 734-754.
Coleman, J. M., 1969, Brahmaputra River: channel processes and sedimentation: Sed. Geology, v. 3, p. 129-239.
Collinson, J. D., 1970, Bedforms of the Tana River, Norway: Geog. Annaler, v. 52A, p. 31-56.
Costello, W. R., and R. G. Walker, 1972, Pleistocene sedimentology, Credit River, southern Ontario: a new component of the braided river model: Jour. Sed. Petrology, v. 42, p. 38-400.
Craig, L. C., et al, 1955, Stratigraphy of the Morrison and related formations, Colorado Plateau region, a preliminary report: U.S. Geol. Survey Bull. 1009-E, p. 125-168.
Denny, C. S., 1965, Alluvial fans in the Death Valley region, California and Nevada: U.S. Geol. Survey Prof. Paper 466, 62 p.
Doeglas, D. J., 1962, The structure of sedimentary deposits of braided rivers: Sedimentology, v. 1, p. 167-190.
Dott, R. H., Jr., 1973, Paleocurrent analysis of trough cross stratification: Jour. Sed. Petrology, v. 43, p. 779-783.
Fahnestock, R. K., 1963, Morphology and hydrology of a glacial stream—White River, Mt. Rainier, Washington: U.S. Geol. Survey Prof. Paper 422-A, p. A1-A70.
——— 1970, Channel form, pattern, and activity of gravel-bearing streams (abs.): Geol. Soc. America Abs. with Programs, v. 2, p. 548-549.
Fisher, W. L., and J. H. McGowen, 1967, Depositional systems in Wilcox Group (Eocene of Texas) and their relation to occurrences of oil and gas: Gulf Coast Assoc. Geol. Socs. Trans., v. 17, p. 105-125 (*also in* AAPG Bull., v. 53, p. 30-54, 1969).
Gole, C. V., and S. V. Chitale, 1966, Inland delta building activity of Kosi River: Am. Soc. Civil Engineers Proc., Jour. Hydraulics Div., HY 2, no. 4722, p. 111-126.
Granger, H. C., 1968, Localization and control of uranium deposits in the southern San Juan basin mineral belt, New Mexico, a hypothesis: U.S. Geol. Survey Prof. Paper 600-B, p. B60-B70.
Gustavson, T. C., 1974, Sedimentation on gravel outwash fans, Malaspina Glacier foreland, Alaska: Jour. Sed. Petrology, v. 44, p. 374-389.
Hale, L. A., 1959, Intertonguing Upper Cretaceous sediments of northeastern Utah–northwestern Colorado, *in* Symposium on Cretaceous rocks of Colorado and adjacent areas: Rocky Mtn. Assoc. Geologists Guidebook, 11th Field Conf., p. 55-66.
Harms, J. C., D. B. Mackenzie, and D. G. McCubbin, 1963, Stratification of modern sands of the Red River, Louisiana: Jour. Geology, v. 71, p. 566-580.
Hartshorn, J. H., and G. M. Ashley, 1972, Glacial environment and processes in southeastern Alaska: Univ. Massachusetts Coastal Research Center, Tech. Rept. 4-CRC, 69 p.
Hooke, R. LeB., 1967, Processes on arid-region alluvial fans: Jour. Geology, v. 75, p. 438-460.
Imlay, R. W., 1957, Paleoecology of Jurassic seas in the western interior of the United States, *in* Treatise on marine ecology and paleontology: Geol. Soc. America Mem. 67, v. 2, p. 469-504.
Kessler, L. G., II, 1971, Characteristics of the braided stream depositional environment with examples from the South Canadian River, Texas: Earth Sci. Bull., v. 4, p. 25-35.
——— and F. G. Cooper, 1970, Channel sequences and braided stream development in the South Canadian River, Hutchinson, Roberts, and Hemphill Counties, Texas: Gulf Coast Assoc. Geol. Socs. Trans., v. 20, p. 263-273.
Krigström, A., 1962, Geomorphological studies of sandur plains and their braided rivers in Iceland: Geog. Annaler, v. 44, p. 328-346.
LeBlanc, R. J., 1972, Geometry of sandstone reservoir bodies, *in* Underground waste management and environmental implications: AAPG Mem. 18, p. 133-190.
Leopold, L. B., and W. B. Langbein, 1966, River meanders: Sci. American, v. 214, p. 60-70.
McCormick, C. D., and M. D. Picard, 1970, Petrology of Gartra Formation (Triassic), Uinta Mountains area, Utah and Colorado: Jour. Sed. Petrology, v. 39, p. 1484-1508.
McDonald, B. C., and I. Banerjee, 1970, Sediments and bed forms on a braided outwash plain: Canadian Jour. Earth Sci., v. 8, p. 1282-1301.
McKee, E. D., 1962, Origin of the Nubian and similar sandstones: Geol. Rundschau, v. 52, p. 551-587.

Moody-Stuart, M., 1966, High- and low-sinuosity stream deposits, with examples from the Devonian of Spitsbergen: Jour. Sed. Petrology, v. 36, p. 1102-1117.
Oliver, W. B., 1971, Depositional systems in the Woodbine Formation (Upper-Cretaceous), northeast Texas: Univ. Texas Bur. Econ. Geology Rept. Inv. 73, 28 p.
Ore, H. T., 1963, Some criteria for recognition of braided stream deposits: Wyoming Univ. Contr. Geology, v. 3, p. 1-14.
——— 1965, Characteristic deposits of rapidly aggrading streams, *in* Sedimentation of Late Cretaceous and Tertiary outcrops, Rock Springs uplift: Wyoming Geol. Soc. Guidebook, 19th Field Conf., p. 195-201.
Peterson, J. A., 1972, Jurassic system, *in* Geologic atlas of the Rocky Mountain region, United States of America: Rocky Mtn. Assoc. Geol., p. 177-189.
Potter, P. E., 1963, Late Paleozoic sandstones of the Illinois basin: Illinois State Geol. Survey Rept. Inv. 217, 92 p.
——— 1967, Sand bodies and sedimentary environments: a review: AAPG Bull., v. 51, p. 337-365.
Rust, B. R., 1972, Structure and process in a braided stream: Sedimentology, v. 18, p. 221-245.
Saitta B., S., and G. S. Visher, 1968, Subsurface study of the southern portion of the Bluejacket delta, *in* A guidebook to the geology of the Bluejacket-Bartlesville Sandstone, Oklahoma: Oklahoma City Geol. Soc., p. 52-68.
Santos, E. S., 1970, Stratigraphy of the Morrison Formation and the structure of the Ambrosia Lake district, New Mexico: U.S. Geol. Survey Bull. 1272-E, 30 p.
Sarkar, S. K., and S. Basumallick, 1968, Morphology, structure, and evolution of a channel island in the Barakar River, Barakar, West Bengal: Jour. Sed. Petrology, v. 38, p. 747-754.
Saucier, A. E., 1967, The Morrison Formation in the Gallup region, *in* Guidebook Defiance-Zuni-Mt. Taylor region, Arizona and New Mexico: New Mexico Geol. Soc., 18th Field Conf., p. 138-144.
Shelton, J. W., and R. L. Noble, 1974, Depositional features of braided-meandering stream: AAPG Bull., v. 58, p. 742-749.
Smith, N. D., 1970, The braided stream depositional environment: comparison of Platte River with some Silurian clastic rocks, north-central Appalachians: Geol. Soc. America Bull., v. 81, p. 2993-3014.
——— 1971, Transverse bars and braiding in the lower Platte River, Nebraska: Geol. Soc. America Bull., v. 82, p. 3407-3420.
——— 1974, Sedimentology and bar formation in the upper Kicking Horse River, a braided outwash stream: Jour. Geol., v. 82, p. 205-223.
Stewart, J. H., F. G. Poole, and R. F. Wilson, 1972, Stratigraphy and origin of the Chinle Formation and related Upper Triassic strata in the Colorado Plateau region: U.S. Geol. Survey Prof. Paper 690, 336 p.
Van de Graaff, F. R., 1972, Fluvial-deltaic facies of the Castlegate Sandstone (Cretaceous), east-central Utah: Jour. Sed. Petrology, v. 42, p. 558-571.
Van de Kamp, P. C., 1973, Holocene continental sedimentation in the Salton basin, California: a reconnaissance: Geol. Soc. America Bull., v. 84, p. 827-848.
Visher, G. S., 1965, Fluvial processes as interpreted from ancient and recent fluvial deposits, *in* Primary sedimentary structures and their hydrodynamic interpretation: Soc. Econ. Paleontologists and Mineralogists Spec. Pub. 12, p. 116-132.
——— S. Saitta B., and R. S. Phares, 1971, Pennsylvanian delta patterns and petroleum occurrences in eastern Oklahoma: AAPG Bull., v. 55, p. 1206-1230.
Waechter, N. B., 1970, Braided stream deposits of the Red River, Texas Panhandle (abs.): Geol. Soc. America Abs. with Programs, v. 2, p. 713.
Walton, W. M., 1967, The informal units of the subsurface Eocene: Asoc. Venezolana Geología, Minería y Petróleo Bol. Inf., v. 10, p. 21-30.
Williams, P. F., and B. R. Rust, 1969, The sedimentology of a braided river: Jour. Sed. Petrology, v. 39, p. 649-679.
Wurster, P., 1968, Paläogeographie der deutschen Trias und die palägeographische Orientieruing der Lettenkohle in Südwestdeutschland: Eclogae Geol. Helvetiae, v. 61, p. 157-166.

Reprinted by permission of Elsevier Scientific Publishing Company. Published in *Sedimentary Geology*, V. 21, No. 2 pp. 129-147.

Sedimentary Geology, 21 (1978)

STUDIES IN FLUVIATILE SEDIMENTATION: AN EXPLORATORY QUANTITATIVE MODEL FOR THE ARCHITECTURE OF AVULSION-CONTROLLED ALLUVIAL SUITES

J.R.L. ALLEN

Sedimentology Research Laboratory, Department of Geology, The University, Reading (Great Britain)

(Received December 29, 1977)

ABSTRACT

Allen, J.R.L., 1978. Studies in fluviatile sedimentation: an exploratory quantitative model for the architecture of avulsion-controlled alluvial sites. Sediment. Geol., 21: 129—147.

A river crossing a coastal plain is assigned a zone of influence within which it may abruptly move and construct new channel sand-bodies following avulsion. Assuming uniform and steady subsidence at a fixed site, the alluvium formed beneath the zone is constructed over a series of equal time-steps, at the start of each of which an avulsion occurs, and during each of which an increment of sediment is accumulated beneath the plain. Each increment comprises an overbank sequence of a thickness proportional to subsidence rate and avulsion period, laterally equivalent to a generally weakly multi-storey sand body of a thickness proportional to channel depth, avulsion period, and subsidence rate. The choice of sand-body width is determined by stream size and mode of behaviour. The new position of the river after an avulsion is chosen using random-number tables combined with rules for the avoidance of older, relief-creating sand bodies. The model otherwise ignores sequence compaction, and so is valid only when operated for comparatively short periods.

In accordance with a theoretical model of sand-body connectedness, in which uniform bodies are regularly packed, experimental sand bodies are virtually unconnected in those alluvial suites containing 50% or more of overbank fines. The degree of connectedness grows very rapidly as the proportion of sand bodies increases above 50%. The use of rules for avoidance does not prevent the experimental suites from being somewhat disordered internally, as shown by the wide variation in the composition of one-dimensional profiles through individual sequences. The results obtained from the model suggest that coarsening-upward alluvial suites of coastal origin may owe their character as much to progressively decreasing subsidence as to any independent decline in channel sinuosity.

INTRODUCTION

An understanding of the architecture of alluvial suites underpins the successful exploitation of many hydrocarbon resources (Campbell, 1976), particularly when secondary or tertiary recovery is contemplated, and is vital to the efficient detection and exploitation of metalliferous ores accumulated in river palaeochannels (Craig et al., 1955; Mullens and Freeman, 1957). But lack of evidence on the size, shape, and relationships of alluvial sand bodies often means that recourse must be had to models in order to obtain the necessary insight into the composition and structure of these suites. Quantitative models for this purpose have hitherto been lacking.

In an earlier paper (Allen, 1974a) semi-quantitative models were presented for the architecture of alluvial suites of a type well known in the Old Red Sandstone (Siluro-Devonian) of southern Britain, which accumulated on an extensive coastal plain. The models were discussed in this paper with particular reference to the occurrence and significance of pedogenic and other superficial carbonate accumulations, but they clearly had implications for many other problems of the architecture of alluvial sequences. In building these models, use was made of available quantitative data on rivers and their processes, while such partly random events as the construction of new channel belts after river avulsion were simulated by assigning channel position with the aid of dice or random-number tables. The ideas explicit or implied in these models have since been extended, partly in connection with postgraduate teaching, and in the present paper we give an exploratory quantitative model for the structure of alluvial suites formed in a coastal plain environment as the result of river avulsion. The model is of the simplest kind, with only avulsion being treated as a random process. Steady subsidence is taken as the factor permitting an alluvial suite to build up, while compaction is ignored except for the effects of buried sand bodies on surface relief. Hence the model would seem valid only for periods that are small on a geological time-scale (order 10^5—10^6 years). Leeder (1977) has independently presented a quantitative model for alluvial suites, but in the context of valley filling and with overbank sedimentation as the sole control on build-up. As in the present model, and some of Allen's (1974a) earlier ones, avulsion is assumed to be the process causing channels to alter position bodily.

SPECIFICATION OF MODEL

Coastal plain of alluviation

We imagine a linear coastal plain of alluviation (Fig. 1A) of uniform width and slope lying between uplands and the sea. As in the Cenozoic-Quaternary Gulf of Mexico Coastal Plain (Bernard and LeBlanc, 1965), differential subsidence is taking place with the consequence that the plain and an imaginary hinge line parallel with the coastline are slowly moving seaward. Along any spatially fixed section AB parallel with the coastline and seaward of the hinge, a uniform and constant rate of subsidence, R, is considered to prevail in the short term. The model will show how a sequence of alluvium is built up as the plain subsides beneath AB, the products of its operation being a vertical two-dimensional section of the strata, normal to the transport direction. We finally assume that the entire region is uniform in geological compo-

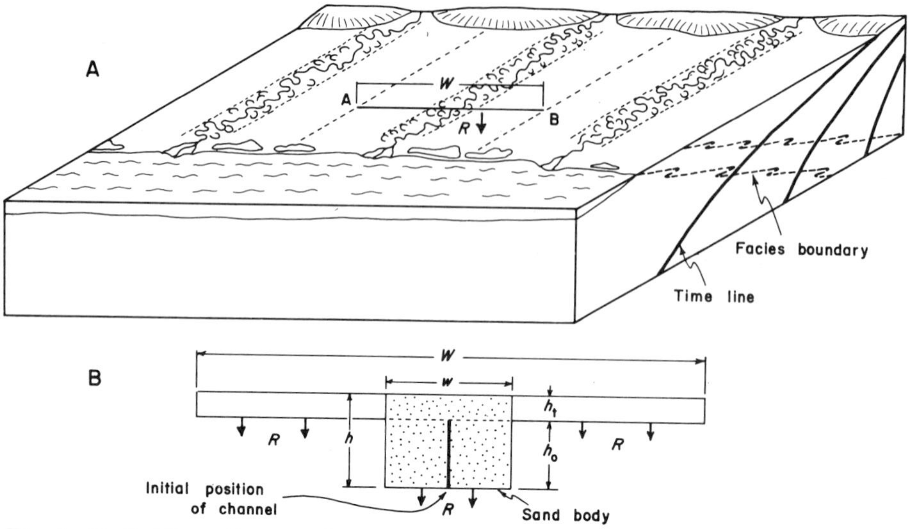

Fig. 1. Definition diagram for rivers building an alluvial suite on a coastal plain during subsidence-controlled offlap. A. General character of plain and zones of river influence. B. Details of sediment increment constructed between one avulsion and the next (fine-grained overbank sediments shown blank).

sition and structure, and also in physiography and climate. Rivers of a uniform character may then be regarded as crossing the plain at a constant longitudinal spacing. Each river is considered to have a zone of influence of width W on the plain, within which it alone is responsible for the channel sand-bodies accumulated.

Rivers

No restrictions beyond uniformity and constancy are placed on the rivers with respect to their aqueous and sediment discharges, calibre of load, channel pattern, and rate of local channel movement through bank caving and scour. But as drainage basin area and stream length both increase with discharge (Hack, 1957), a value should be assigned to W consistent with these other quantities. Larger streams should have wider zones of influence as well as, on the average, producing larger sand bodies.

Each river is assumed periodically to change position within its zone of influence as the result of avulsion, the sudden abandonment by the whole river of an existing channel for a new site under the stimulus of an extreme hydrological event. The albeit meagre evidence (Fisk, 1944, 1947; Russell, 1954; Anderson, 1961) suggests that the period of avulsion, P, under conditions of stable base-level is in the order of 10^3 years. To retain simplicity we will assume that P is constant and uniform, and that only channel position after an avulsion is subject to a random process, qualified as will be explained below. The frequency of avulsion within each zone of influence, measured as the number of events per unit width and time, is inversely related to both P and W. This frequency will be denoted by F. The alluvial sequence accumulated beneath AB in the model is therefore built up in a series of equal time-marching steps, each step being of duration P in time and resulting, beneath those parts of the floodplain removed from the sand body actively being formed, in the accumulation of a layer of overbank fines RP units thick.

Sand bodies

In the intervals between avulsion, the laterally migrating channel of each river sweeps out a zone of constant width w within which a sand body is accumulated (Fig. 1B). The width of this zone may if desired be set equal to and identical with the instantaneous channel width, but is more likely to be selected so as to express the existence of a meander-belt complex, where "meander" here implies the lateral and downstream wanderings of low-sinuosity braided as well as the conventionally meandering channels.

The sand body thickness presents more of a problem. We suggest that the thickness, h, of the body which has been accumulated by the end of one period of avulsion comprises two parts, h_0 and h_t (Fig. 1B). The first part, h_0, is equivalent to the instantaneous channel depth, assumed to be constant throughout the history of the river, regardless of its possible secular changes of sinuosity at the site of the section AB. This may be artificial, as there is some evidence that maximum channel depth is a complex function of sinuosity (Allen, 1977). To accept a channel depth dependent on sinuosity, however, would unnecessarily complicate a model intended to be exploratory, although unquestionably being a desirable feature of a more realistic treatment. It will of course also be appreciated that with avulsion as the cause of channel shifting, the channel at the start of each new time-step is incised to a depth of h_0 below the surface of the floodplain as it existed at the end of the immediately preceding step (Fig. 1B). At the end of each time-step, the increment of sediment deposited on the alluvial plain is h units thick at the site of the sand body and RP units thick elsewhere. The second component, h_t, of the final sand-body thickness represents sediment accumulated beneath the level of the floor of the active channel as it rises, relative to the base of the alluvial sequence, between one avulsion and the next (Fig. 1B). Hence, under our previous assumptions, $h_t = RP$. This component represents a possible multi-storey element within the sand body, since the time-scale of local bank migration for most rivers can generally be taken as small compared with the avulsion period, whence the river may be supposed rapidly to comb across the width w (Wolman and Leopold, 1957). In a vertical section through AB in Fig. 1, each sand body is therefore regarded as having a rectangular cross-section of width w and uniform thickness $h = h_0 + h_t$. Whereas h_0 depends only on the river regime, h_t is a function partly of the tectonic environment.

Where will the river move after an avulsion? In the absence of faulting, folding, climatic change, and base-level change acting in some combination to cause terracing, the relief and slope of an alluvial plain are extremely small except close to an active channel, where an alluvial ridge may be formed as the result of levee-building and crevassing. Considering this factor alone, each river after avulsion should avoid the immediate neighbourhood of its former channel, on account of the alluvial topography existing there, but should have little preference for any one site over another within the remainder of its zone of influence. But differential compaction of the overbank fines around buried sand bodies in the growing alluvial suite will also help to shape the topography of the alluvial plain above (Brown, 1975), with the result that the river on avulsion will also tend to avoid sites of inherited relief. The effect diminishes, however, the more deeply buried is the sand body, and may bear a complex relationship to the sand/mud ratio in the existing alluvium. Taking a sand body at a fixed depth, we should expect its effect to be least when the sand/mud ratio is high (the buried mud bodies are isolated and contained, and the sand compacts but little) and also when the ratio is low (the mud is abundant and continuous, and lateral migration from above and below sand bodies is both rapid and easy). The body should afford the strongest surface relief when mud and sand are about equally abundant, since the sand is virtually incompressible but the mud is present in substantial quantity and is partly constrained.

These considerations permit a new site for the river after avulsion to be chosen from amongst designated sites using a random number table, but with the proviso that two avoidance factors are included in this process. One must account for the relief of the alluvial ridge abandoned at the start of the time-step, and the other the relief inherited from older sand bodies. Assuming that w is ranged symmetrically about the initial position of the channel immediately after an avulsion (Fig. 1B), the first effect is accounted for by the arbitrary but not unreasonable rule that the new sand body may touch laterally but not overlap (erode) its immediate predecessor. The effect of older bodies is accommodated by the rule that a given sand body may not overlap (erode) any earlier body whose top lies less than nRP units below the surface of the alluvial plain at the start of the time-step, where $n = 1, 2, 3$, etc. is left for the modeller to choose.

It is important to understand that compaction is not otherwise taken into account in the preceding model. This is justified by the restriction of the operation of the model to short periods, when generally an alluvial suite composed of 10—100 m of sediment (uncompacted) will be generated under realistic subsidence rates.

QUANTITATIVE DESCRIPTION OF MODEL SEQUENCES

General

It seems best to treat alluvial sequences generated by the model as far as possible quantitatively, so that useful comparisons can be drawn with real alluvial suites. Amongst the more important properties of the two-dimensional sections of alluvium generated beneath AB in Fig. 1A that can be quantified are: (1) the number-density of the sand bodies, (2) their area-density, (3) the connectedness of the bodies, and (4) the variability of one-dimensional profiles through the sequence, that is, the variability of such as bore-hole sections or conventional lithostratigraphic logs.

Sand-body density

Under the preceding assumptions and rules, the *number-density*, N, of sand bodies accumulated in the model during a time-period t is:

$$N = \frac{F(W - w)t}{RWt + Wh_0} \tag{1a}$$

where F is the frequency of avulsion, equal to $1/P(W - w)$. The numerator is the number of bodies produced during the period, and the denominator the cross-sectional area of the alluvial suite generated. The term $(W - w)$ appears in the numerator in recognition of the finite zone of influence of each river. Avulsion must occur with a zone of width $(W - w)$ if the sand bodies due to one river are not to stray partly into the zones of influence of immediately adjacent streams. However, when w becomes very small compared with W, we may write:

$$N = \frac{Ft}{Rt + h_0} \tag{1b}$$

approximately. Using the above avoidance factors, the number of sand bodies present in an experimental alluvial suite at the end of the period t is always equal to the number of bodies introduced during that period, even though some of the bodies, because of reworking by later ones, may be reduced in cross-sectional area.

In the same way, the *total area-density*, A, of sand bodies accumulated beneath AB in a period t is:

$$A = \frac{F(W - w)thw}{RWt + Wh_0} \tag{2a}$$

where the numerator is the product of the number of bodies produced with the cross-sectional area of the individual body, and the denominator is the total cross-sectional area of the alluvium generated. When w is very small compared with W, we again have:

$$A = \frac{Fthw}{Rt + h_0} \tag{2b}$$

approximately. The total area-density ranges up from zero and may exceed unity, because it measures the sum of the areas of every sand body introduced into the alluvial suite compared with the final cross-sectional area of the suite, no account being taken of the reworking of earlier bodies by later ones. To account for reworking, we define the *residual area-density*:

$$a = \frac{\text{Total area of sand bodies preserved in cross-section of alluvial suite}}{\begin{array}{c}\text{Cross-sectional area of alluvial suite accumulated in same time}\\ \text{(sand bodies plus floodplain fines)}\end{array}} \qquad (3)$$

which should be capable of estimation from the experimental alluvial suites. Clearly a cannot exceed unity, whereas A may range to very high values. Provided that A is in the order of unity or smaller, we should expect the two measures of area-density to be roughly equal, since the use of avoidance factors means that the bodies tend to a regular two-dimensional spacing and to a uniform degree of erosion of one by another.

The denominators in eqs. 1a, b and 2a, b include a term in h_0, where h_0 is equivalent to the depth of the river channel. This is done so that the model can operate satisfactorily at very small subsidence rates. Even if the rate were zero, we would expect the river to create after sufficient avulsions a sheet of sediment of width W and depth h_0. It is important only to remember that, when the subsidence rate is non-zero, the increment Wh_0 need not be wholly formed of alluvium deposited by the river that built the overlying part of the suite.

Sand-body connectedness

The connectedness of sand bodies is a complex attribute dependent on their packing in space, the frequency and extent to which they touch, and the degree to which they are in mutual communication.

As a complex property, connectedness in the model can be expressed quantitatively in a number of different ways. We can define it in terms of the two-dimensional *coordination*, *C*, of the bodies, that is, the average number of bodies with which an individual body is in immediate contact in the plane of the section through AB. Connectedness can also be described by identifying clusters of sand bodies that are mutually connected in the plane of the section but which, as clusters in this section, are each fully embedded in overbank fines and so lie isolated from other clusters. Such a two-dimensional cluster may be called a *domain*, and an aspect of the connectedness of the bodies may be described in terms of the two-dimensional *domain size*, *D*, the average number of bodies in the domains. Clearly, the domain size is a measure of the scale of an alluvial hydrocarbon reservoir. The spatial extent to which contiguous sand bodies are in contact could significantly affect the ease with which pore fluids pass from one body to another in a domain. This aspect of connectedness is described by the two-dimensional *average fractional contact*, *K*, the average fraction of the sand-body perimeter in contact with immediately surrounding bodies. Each of the coordination, domain size, and average fractional contact can be measured from experimentally constructed alluvial suites. Each should increase with growing sand content in the alluvium, a property which can be estimated from one-dimensional sequences. Because of the equivalence of linear, areal and volumetric proportions in heterogeneous materials (Chayes, 1956), it should be possible for given experimental parameters to obtain empirical relationships between C, D, K and sand content.

By extending the analogy between sand-body and sediment-particle packings implicit in the treatment of coordination, it seems possible to obtain some theoretical guidance as to the form of the relationship between the residual area-density and the average fractional contact of an alluvial suite. Fig. 2 shows three regular modes of packing of sand bodies of uniform rectangular cross-section of width w and thickness h. Regular as opposed to random modes are chosen partly because of the uniformity of the experimental bodies, and partly because the use of avoidance factors means that clustering of the bodies is minimized. In each mode an increase of packing density is obtained by a simple translation, in mode 1 parallel with the long edges of the bodies, in mode 2 parallel with their diagonals, and in mode 3 parallel with the short edges. If sand bodies were packed in contact in infinitely extensive arrays according to these modes, the geometry of each array could be described in terms of relationships within the "unit cells"

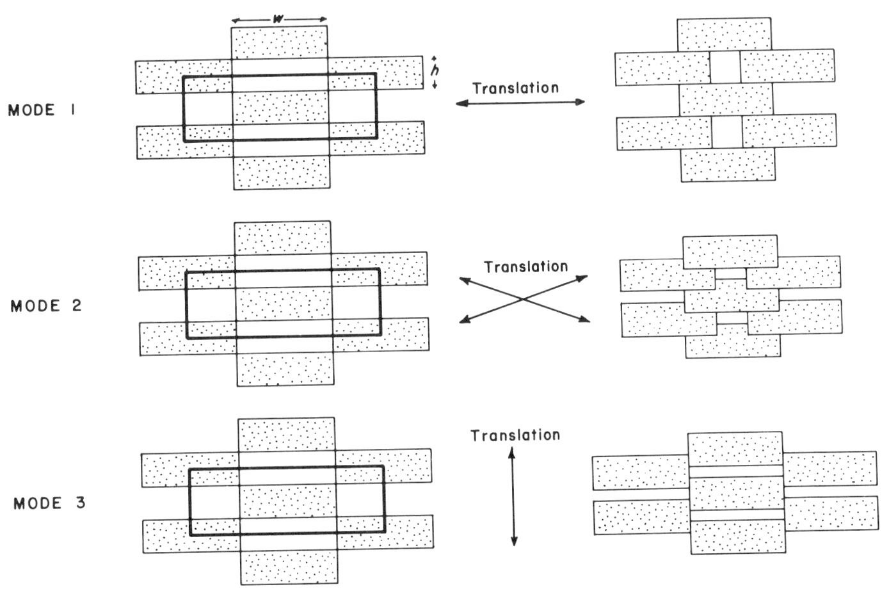

Fig. 2. Model showing modes of regular two-dimensional packing of uniform sand bodies of width w and thickness h. The heavy lines in the left-hand column of packings define the "unit cell" by means of which the character of the packing can be calculated. The cell is in each case formed by joining the centres of the four bodies clustered around a central body.

shown by the heavy lines in Fig. 2. In each cell let a fraction $2k_1$ of each long edge of each sand body be in contact with neighbours, and let each short edge be in contact with other bodies to the extent of the fraction $2k_2$. For the general case, the residual area-density is related to k_1 and k_2 by:

$$a = \frac{(1 - 2k_1k_2)}{(2 - 2k_1 - 2k_2 + 2k_1k_2)} \tag{4}$$

and we have the definition:

$$K = \frac{2(k_1w + k_2h)}{w + h} \tag{5}$$

relating the average fractional contact to k_1, k_2, w and h.

This model based on regular packings has interesting properties. When k_1 and k_2 are simultaneously equal to zero, the residual area-density in each packing mode is equal to one-half. Therefore whatever their precise arrangement, we should not expect many sand bodies to be in contact in an alluvial suite consisting of less than 50% of such bodies. As k_1 and k_2 each approach their limit of one-half, the residual area-density for the three modes approaches its upper limit of unity. An increase in sand/mud ratio should, therefore, mean an increase in the average fractional contact and, since a degree of disorder must exist in the packing of real bodies, in the coordination and domain size. It is thought possible that sand bodies in real alluvial suites may pack similarly to mode 2 in Fig. 2, for which $k_1 = k_2 = \frac{1}{2}K$ and K approaches an upper limit of unity as a approaches unity. In the other modes, as well as in the general case, K is less than unity when a has reached its upper limit. We should note that these conclusions stem from a consideration of the properties of the unit cell alone, which means that they are strictly true only for infinitely large alluvial suites. Real suites are finite and some may be of a quite small size, for example, those filling palaeovalleys. The relations between a, K, k_1 and k_2 for them should differ slightly from those given by eqs. 4 and 5, because the peripheral sand bodies are less closely packed than in the unit cell.

EXPERIMENTAL RESULTS

General

Six experiments covering a wide range in subsidence rate were made using the model (Table I, Figs. 3—8). The ratio of sand-body width to channel depth was held constant at 500. This was felt to be a reasonable choice in the light of Schumm's (1968) data on moderately sized rivers, allowing the channel belts of broad, shallow low-sinuosity streams, as well as the matured

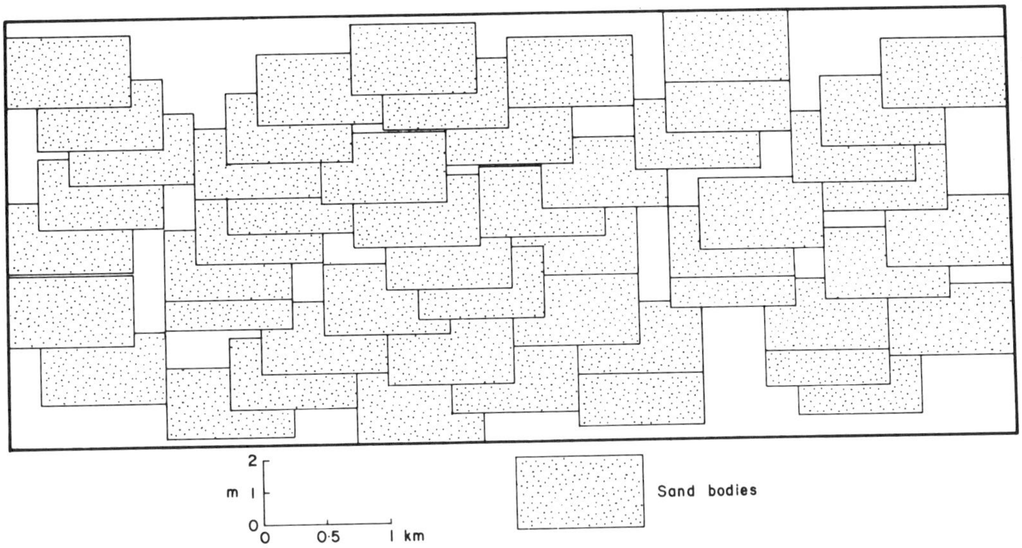

Fig. 3. Alluvial suite obtained during experiment 1 (Table I). Note that if sand-body complexes are defined in vertical profile as lying between overbank fines, then most complexes consist of a number of internal sequences (parts of sand bodies) each defined below by an erosional contact. Since the sand bodies are individually weakly multi-storey, a hierarchy of multi-storey elements may be said to be present in each complex. See Tables II—IV.

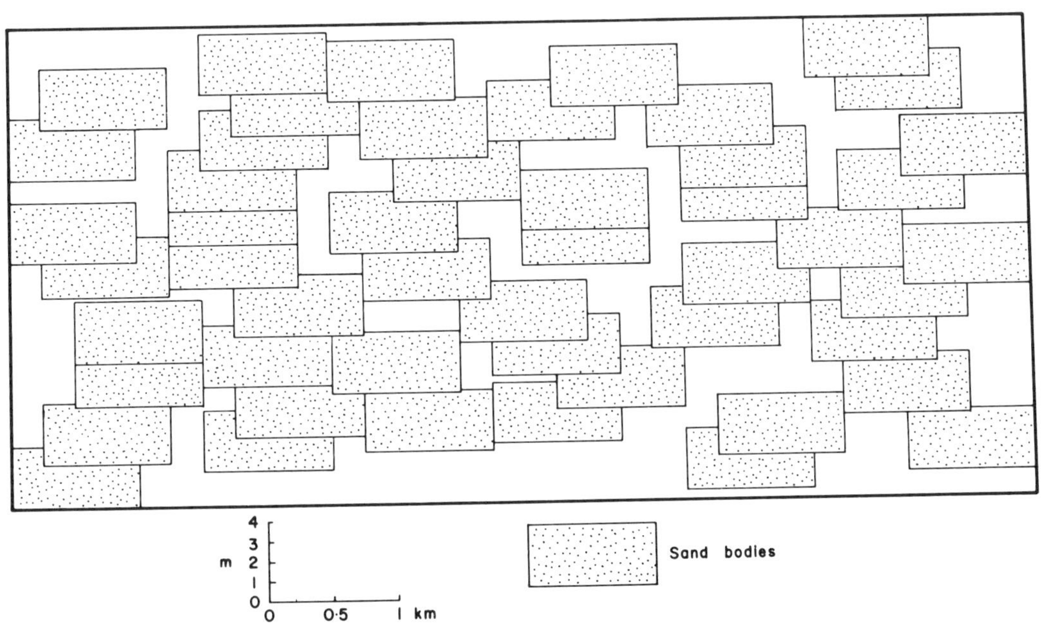

Fig. 4. Sand bodies obtained during experiment 2 (Table I). Overbank fines are now more in evidence than in experiment 1 (Fig. 3) but many sand-body complexes would again include a hierarchy of multi-storey elements. See Tables II—IV.

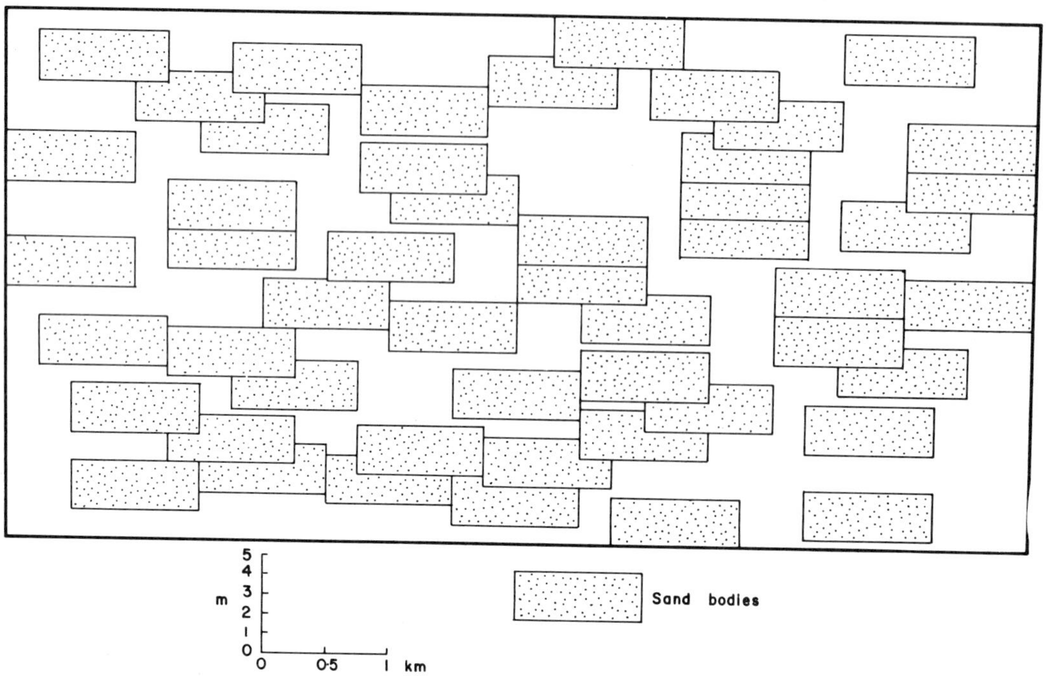

Fig. 5. Alluvial suite obtained during experiment 3 (Table I). Sand bodies here have about the same abundance as overbank fines. Complexes of sand bodies are no longer predominant in vertical profiles. See Tables II—IV.

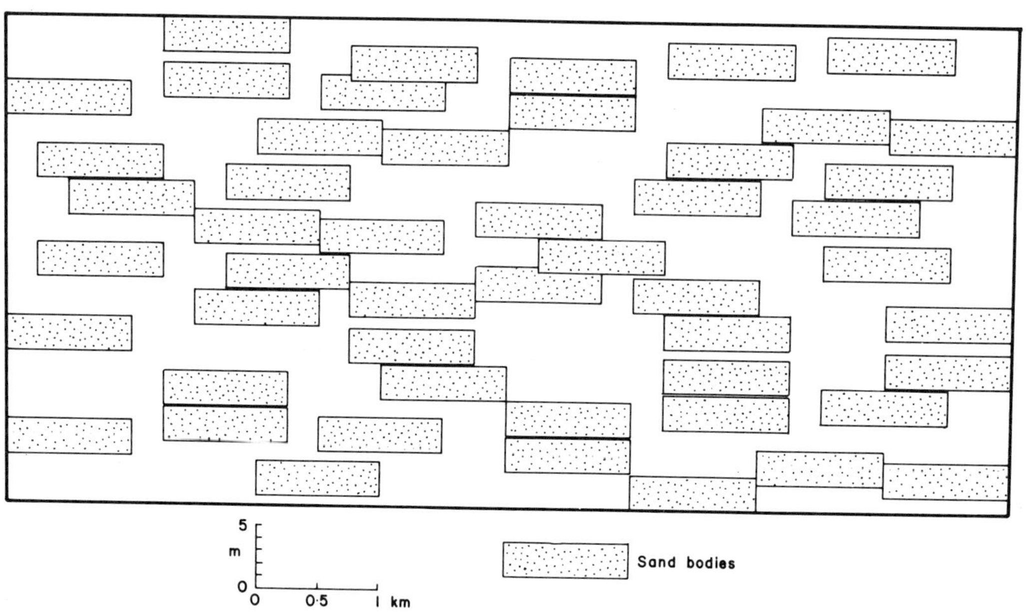

Fig. 6. Alluvial suite obtained during experiment 4 (Table I). Overbank fines now dominate over sand bodies and few bodies are in contact. There is only one sand-body complex. See Tables II—IV.

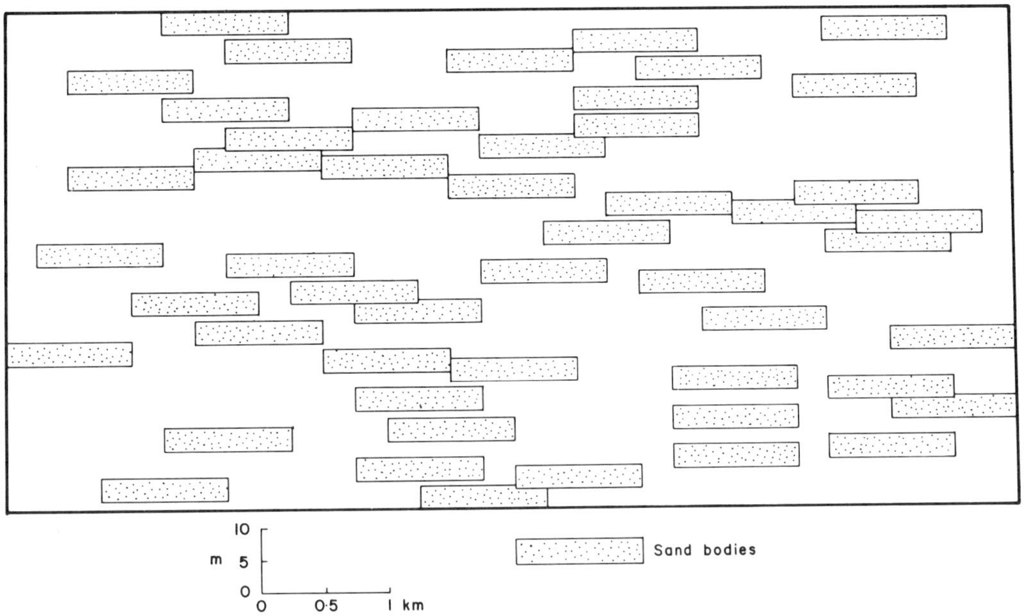

Fig. 7. Alluvial suite obtained during experiment 5 (Table I). Again overbank fines are dominant. There are few sand-body complexes and few sand bodies in contact. See Tables II—IV.

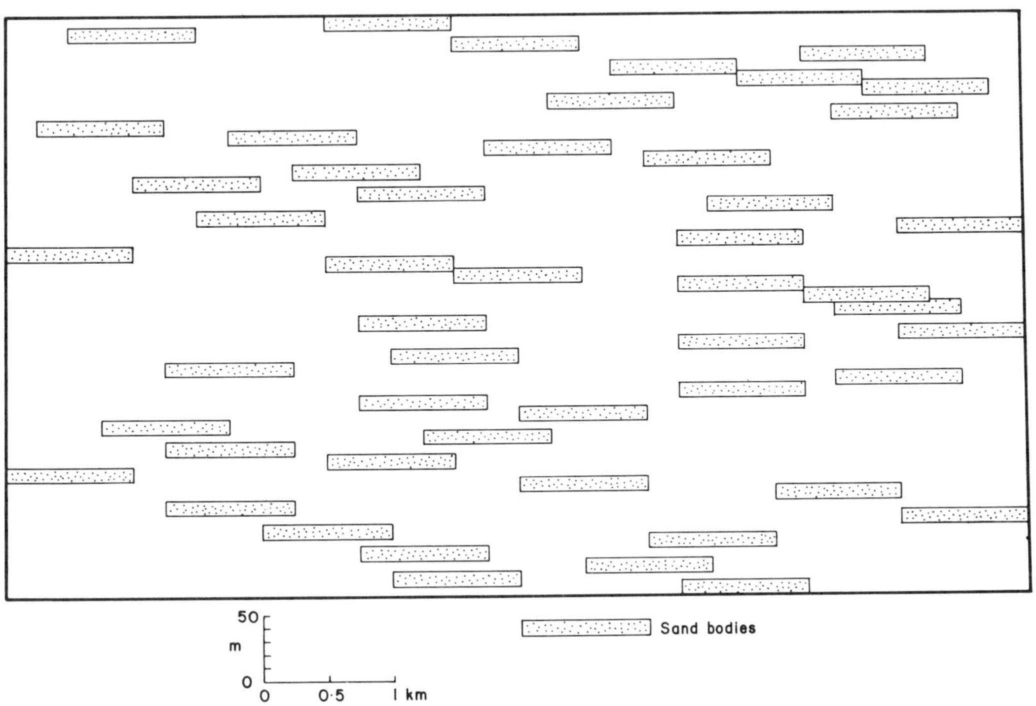

Fig. 8. Alluvial suite generated during experiment 6 (Table I). Overbank fines are predominant. Contacts between sand bodies and sand-body complexes are few. Note that some vertical profiles include long sections of overbank fines laterally equivalent to many sand bodies. Individual sand bodies are strongly multi-storey under the high subsidence rate prevailing in the experiment, in contrast to the bodies in experiments 2—5 which are all weakly multi-storey. See Tables II—IV.

TABLE I

General experimental conditions and results

Experiment	R (m yr^{-1})	h (m)	N (m^{-2})	A	a (measured)	K (measured)
1	$1.146 \cdot 10^{-4}$	2.229	$4.64 \cdot 10^{-4}$	1.035	0.756	0.615
2	$1.592 \cdot 10^{-4}$	2.318	$3.49 \cdot 10^{-4}$	0.809	0.660	0.587
3	$2.457 \cdot 10^{-4}$	2.491	$2.35 \cdot 10^{-4}$	0.586	0.539	0.327
4	$3.519 \cdot 10^{-4}$	2.704	$1.68 \cdot 10^{-4}$	0.453	0.448	0.0251
5	$7.653 \cdot 10^{-4}$	3.531	$7.95 \cdot 10^{-4}$	0.281	0.274	0.0700
6	$4.430 \cdot 10^{-3}$	10.859	$1.40 \cdot 10^{-5}$	0.153	0.152	0.0015

In all experiments P = 2000 years, t = 100,000 years, W = 8000 m, and w = 1000 m.

meander belts of the relatively deeper high-sinuosity rivers to be accommodated. The rules for vertical sand-body avoidance are: $n = 3$ in experiments 1 and 2; $n = 2$ in experiments 3 and 4; $n = 1$ in experiments 5 and 6.

Density of sand bodies

Sand-body density as measured by the number-density (N), total area-density (A), and residual area-density (a) increases steadily with decreasing subsidence rate (Table I, Figs. 3—8). As was expected at the lower rates, the residual area-density falls below the total area-density due to the partial reworking of earlier by later sand bodies. A boundary effect which becomes increasingly important with decreasing subsidence rate is apparent in Figs. 3—8, the sand bodies being least dense around the borders of the cross-sections. Along the vertical edges, at the margins of the zone of influence of the river, the effect extends inward about one-half of a sand-body width, and is related to the rule that the sand bodies do not overlap this zone of influence. The effect along the horizontal edges has an extent of one sand-body thickness or less, and is related to the starting and stopping of the build-up of alluvium, that is, to the finiteness of the sequence in the flow of time. The values given for the residual area-density therefore underestimate the sand-body density in what might be called the core of each suite.

Coordination and domain size

Tables II and III give data respectively on the coordination (C) of the experimental sand bodies and on the number and average size (D) of the sand-body domains. Provided that the residual area-density is less than approximately 0.5 (experiments 4—6), the bodies are virtually unconnected within the fine-grained overbank sediments and contribute to many domains, none of which is large. The degree of coordination and the domain size increase very rapidly, however, as the residual area-density is increased above

TABLE II

Frequency of number of contacts per sand body

Experiment	Number of contacts per body							C	Total number of bodies
	0	1	2	3	4	5	6		
1	0	5	10	15	15	4	1	3.12	50
2	0	11	19	13	4	3	0	2.38	50
3	5	16	17	12	0	0	0	1.72	50
4	31	14	5	0	0	0	0	0.48	50
5	28	18	2	2	0	0	0	0.56	50
6	42	6	2	0	0	0	0	0.20	50

0.5. These results are consistent with the regular packing model summarized in Fig. 2. According to this model, there are no contacts between sand bodies until the residual area-density is increased to 0.5, above which the coordination is constant at 4. Table II shows that the coordination at the lowest subsidence rate is 3.12, when approximately 75% of the area of the cross-section is occupied by sand bodies (Fig. 3). At this high degree of coordination, there is but one domain, and each sand body is in communication with every other.

Average fractional contact

The measured values of the average fractional contact listed in Table I are also broadly consistent with the regular packing model (Fig. 2). According to eq. 4, $K = 0$ for $a \leqslant 0.5$, the table listing very small values for the average fractional contact in the last three runs, because of the partly random arrangement of the experimental bodies. For $a > 0.5$, K increases in a manner dependent on the packing mode (Fig. 9). The graph shows that, in this higher range, both the mode 1 and the mode 2 packings generally under-

TABLE III

Characteristics of domains

Experiment	Number of domains	D	Number of bodies in largest domain
1	1	50	50
2	3	16.7	46
3	11	4.55	12
4	37	1.35	4
5	35	1.43	6
6	45	1.11	3

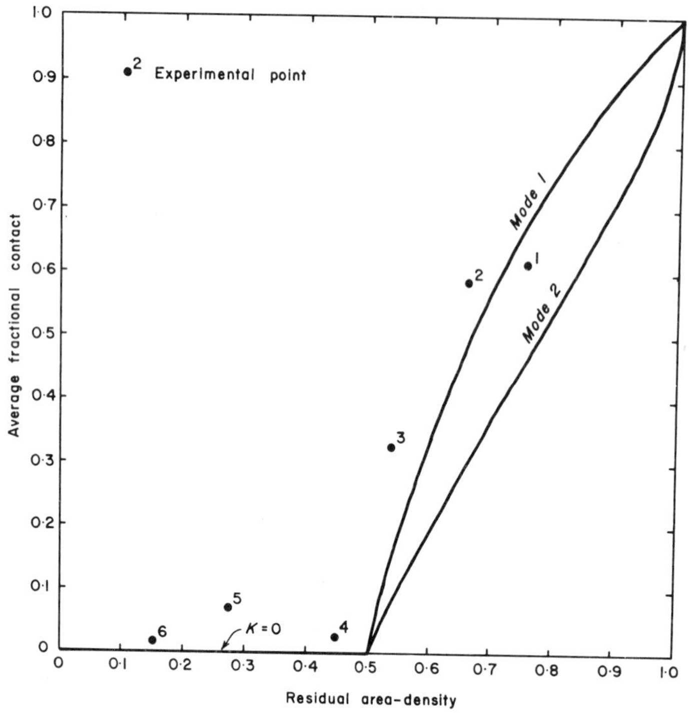

Fig. 9. Measured values of the average fractional contact (K) and residual area-density (a) in terms of the theoretical model of regular sand-body packing (see eqs. 4, 5).

estimate K, perhaps as a consequence of the boundary effects noted, though the observed and theoretical forms of variation are in broad agreement. There is no evidence that the suites depart significantly from being isotropic.

Variability of alluvial suites

Although the use of avoidance factors brings a measure of order to the experimental alluvial suites, a glance at Figs. 3—8 will show that each suite remains very variable in lithological composition, one vertical profile being compared with another. The variability is quantified in Table IV, showing the frequency with which sand-body intersections occur in a total of 28 vertical profiles through each alluvial suite. The number of intersections ranges from as small as three per profile, with four intersections as the usual lower limit, to as large as ten per profile. These results suggest that the sand content of an alluvial suite is unlikely to be reliably estimated by a single vertical profile through the suite, such as a borehole or conventional log, unless the section has cut so many sand bodies that the effects of clustering have been smoothed out. As in the earlier models (Allen, 1974a), most suites include thick sections of overbank sediment individually equivalent to numerous sand bodies.

TABLE IV

Frequency of vertical profiles intersecting stated number of sand bodies

Experiment	Number of sand-body intersections in vertical profile										Total number of intersections
	1	2	3	4	5	6	7	8	9	10	
1	0	0	0	3	5	8	6	4	1	1	178
2	0	0	0	3	4	5	10	3	1	2	185
3	0	0	0	0	4	9	10	2	3	0	187
4	0	0	2	2	3	5	6	8	1	1	184
5	0	0	0	3	6	4	11	1	3	0	182
6	0	0	0	4	7	8	1	1	5	2	179

COMPARISON WITH THE OLD RED SANDSTONE

Viewed in reverse order and as parts of an ascending stratigraphical succession, Figs. 3—8 suggest how a thick alluvial suite might appear that had been accumulated beneath an alluvial plain involved in progressive offlap of the type that is affecting the Gulf of Mexico Coastal Plain (Fig. 1A). The sequence would coarsen upwards as the hinge-line advanced ever closer toward AB, fixed as regards its horizontal position, and the rate of subsidence at this line of section was progressively reduced. In addition, once the hinge-line had advanced through AB, an unconformity would be developed and the upper parts of the alluvium would be eroded.

The Devonian-early Carboniferous "Catskill" facies of the Appalachian Mountains of North America (Allen and Friend, 1968) and the Lower Old Red Sandstone of Wales and the Welsh Borderland (Allen, 1962a, 1963, 1965, 1974a, b; Allen and Dineley, 1976) are examples of such thick coarsening-upward alluvial suites, each of which is terminated by an unconformity. In the lower part of the Catskill facies, channel sandstones and overbank mudstones are about equally common amongst the repetitive fining-upwards sequences. Mudstones are rare in the higher parts, however, which consist largely of thick sandstones. These include numerous horizons of intraformational conglomerate and show channelling, both suggestive of the erosional contacts between successive sand bodies as depicted in Figs. 3 and 4. Similar features typify the Siluro-Devonian Lower Old Red Sandstone in Wales and the Welsh Borderland. In such formations as the Sandstone and Marl Group, the Porth y Mor Beds, and the Ledbury Formation, alluvial sandstone bodies account for less of the section than mudstones attributable to overbank environments. In stratigraphically higher formations such as the Brownstones and the Woodbank Group, the dominant rocks are thick sandstones with many indications that they are divisible between numerous once more complete channel sandstone bodies. At these levels there is much evidence for the reworking of overbank mudstones and calcretes, ranging from thick intraformational conglomerates to widely disseminated sand-size grains (Allen, 1962a, b).

There has been a tendency to interpret coarsening-upward alluvial suites as recording the gradual replacement of high-sinuosity by low-sinuosity streams, either because of an increase of channel slope and calibre of discharged sediment harmonious with the basinward migration of environmental belts (e.g. Allen and Friend, 1968), or in response to tectonic uplift in the source areas of the rivers or to climatic change (e.g. Leeder, 1973). The progressive upward reduction in the proportion of overbank fines is seen as consistent with the "bed-load" character of low-sinuosity streams, their presumed inability to create thick overbank sequences, and with their extreme channel instability. The model sketched above suggests that the upward-coarsening of an alluvial suite can also be explained by the secular decrease in the subsidence rate at a place fixed in space during the *steady* development of a region of tectonic instability. A reduction in the subsidence rate at a fixed site during offlap merely means a relative increase in the extent of reworking by the rivers, which are not themselves required to change substantially in character in the process, though there could be some sympathetic response to the alteration in bank sand/mud ratio. There is a tendency for channel sinuosity to increase with increasing distance downstream from the head of a river, but whether a significant change of sinuosity can occur within the more limited scope of the depositional zone an alluvial plain seems unlikely, to judge from maps of modern examples. So far as the Lower Old Red Sandstone is concerned, there is evidence for low-sinuosity streams at most levels within the grossly upward-coarsening succession. The deposits of high-sinuosity streams certainly are present in the lower, muddier parts, but the proportion of sandstone bodies showing indisputable evidence of lateral deposition is invariably low. The sand/mud ratio of an alluvial suite may be an unreliable guide to channel type.

Although the case is irrelevant to the Lower Old Red Sandstone, Figs. 3—8 may alternatively be viewed as they stand, as suggesting how an alluvial suite might appear that had accumulated during a period of increasing subsidence rate, perhaps leading up to actual marine transgression. Grossly, this suite should fine upward. If subsidence had followed a period of stability or mild uplift, the suite should surmount an unconformity, the vertical pattern of stratigraphical relationship and sand-body content being the inverse of that prompted by offlap. The lowest member, immediately above the unconformity surface, could well be a laterally extensive sheet sandstone, formed as rivers repeatedly reworked their earlier deposits under conditions of negligible or very weak subsidence.

CONCLUSIONS

(1) A simple exploratory model of the possible architecture of an alluvial suite accumulated beneath a coastal plain has been constructed, on the basis

of a steady subsidence rate and partly random channel-shifting through avulsion. Compaction in the sequence is ignored, other than to recognize that a river after avulsion will avoid areas of positive relief caused by older shallow-lying sand bodies.

(2) For the short operating periods (10^5—10^6 years) for which the model is thought valid, the character of an alluvial suite depends on the size of the zone of influence of the chosen river, the scale and mode of behaviour of the river, the avulsion period, and the subsidence rate. If all variables except subsidence rate are held constant, the character of the suite becomes a function only of this rate.

(3) The character of an alluvial suite may be described in terms of the number-density and area-density of sand bodies, where the latter is equivalent in a statistically isotropic suite to the average sand/mud ratio afforded by one-dimensional vertical profiles (boreholes, conventional logs).

(4) The connectedness of the sand bodies may be described by the coordination (average sand-body contacts per body), domain size (average number of bodies in bounded cluster of bodies), and the average fractional contact (fraction of sand-body perimeter in contact with adjacent bodies).

(5) For sand-body contents of less than 50%, the degree of connectedness is very small, in accordance with the predictions of a theoretical model of connectedness based on regular packings of uniform bodies. Above a content of 50%, the connectedness increases very rapidly. For example, at a content of 75.6%, the experimental coordination is 3.12 contacts per body, the bodies (total of 50) form but one domain, and an average of 61.5% of the body perimeter touches neighbouring bodies.

(6) The experimental results suggest that coarsening upward alluvial suites such as the Lower Old Red Sandstone could have been formed during a period of gradually decreasing subsidence (at a fixed site) accompanying offlap, rather than in response to externally controlled changes of stream pattern and regime from high to low sinuosity. The model also suggests that alluvial suites which fine upward above unconformities may be due to an increasing subsidence rate after stability or mild uplift.

REFERENCES

Allen, J.R.L., 1962a. Petrology, origin, and deposition of the highest Lower Old Red Sandstone of Shropshire, England. J. Sediment. Petrol., 32: 657—697.

Allen, J.R.L., 1962b. Intraformational conglomerates and scoured surfaces in the Lower Old Red Sandstone of the Anglo-Welsh Cuvette. Geol. J., 3: 1—20.

Allen, J.R.L., 1963. Depositional features of Dittonian rocks: Pembrokeshire compared with the Welsh Borderland. Geol. Mag., 100: 385—400.

Allen, J.R.L., 1965. Sedimentation and palaeogeography of the Old Red Sandstone of Anglesey, North Wales. Proc. Yorks. Geol. Soc., 35: 139—185.

Allen, J.R.L., 1974a. Studies in fluviatile sedimentation: implications of pedogenic carbonate units, Lower Old Red Sandstone, Anglo-Welsh outcrop. Geol. J., 9: 181—208.

Allen, J.R.L., 1974b. Sedimentology of the Old Red Sandstone (Siluro-Devonian) in the Clee Hills area, Shropshire, England. Sediment. Geol., 12: 73—167.

Allen, J.R.L., 1977. Changeable rivers: some aspects of their mechanics and sedimentation. In: K.J. Gregory (Editor), River Channel Changes. Wiley, London, pp. 15—45.
Allen, J.R.L. and Dineley, D.L., 1976. The succession of the Lower Old Red Sandstone (Siluro-Devonian) along the Ross—Tewkesbury Spur Motorway (M.50), Hereford and Worcester. Geol. J., 11: 1—14.
Allen, J.R.L. and Friend, P.F., 1968. Deposition of the Catskill facies, Appalachian region: with notes on some other Old Red Sandstone basins. Geol. Soc. Am. Spec. Pap., 106: 21—74.
Anderson, B., 1961. The Rufiji Basin, Tanganyika. VII. Soils of the Main Irrigable Areas. FAO Report to Government of Tanganyika, 125 pp.
Bernard, H.A. and LeBlanc, R.J., 1965. Résumé of the Quaternary geology of the Northwestern Gulf of Mexico Province. In: H.E. Wright and D.G. Frey (Editors), The Quaternary of the United States. Princeton University Press, Princeton, N.J., pp. 137—185.
Brown, L.F., 1975. Role of sediment compaction in determining geometry and distribution of fluvial and deltaic sandstones. In: G.V. Chilingarian and K.H. Wolf (Editors), Compaction of Coarse-grained Sediments, 1. Elsevier, Amsterdam, pp. 247—292.
Campbell, C.V., 1976. Reservoir geometry of a fluvial sheet sandstone. Bull. Am. Assoc. Pet. Geol., 60: 1009—1020.
Chayes, F.J., 1956. Petrographic Modal Analysis. Wiley, New York, N.Y., 115 pp.
Craig, L.C., Holmes, C.N., Cadigan, R.A., Freeman, R.A., Mullens, T.E. and Weir, G.W., 1955. Stratigraphy of the Morrison and related formations, Colorado Plateau region, a preliminary report. U.S. Geol. Surv. Bull., 1009-E: 125—168.
Fisk, H.N., 1944. Geological Investigation of the Alluvial Valley of the Lower Mississippi River. Mississippi River Commission, Vicksburg, Miss., 78 pp.
Fisk, H.N., 1947. Fine Grained Alluvial Deposits and their Effects on Mississippi River Activity. Mississippi River Commission, Vicksburg, Miss., 82 pp.
Hack, J.T., 1957. Studies of longitudinal stream profiles in Virginia and Maryland. U.S. Geol. Surv. Prof. Pap., 294-B: 97 pp.
Leeder, M.R., 1973. Sedimentology and palaeogeography of the Upper Old Red Sandstone in the Scottish Border Basin. Scott. J. Geol., 9: 117—144.
Leeder, M.R., 1977. Artificially generated sections through subsiding alluvial plains, with special reference to controls upon channel sandstone extent and 'interconnectedness'. Program and Abstracts, First International Symposium on Fluvial Sedimentology, Calgary, October 1977, p. 18.
Mullens, T.E. and Freeman, V.L., 1957. Lithofacies of the Salt Wash Member of the Morrison Formation, Colorado Plateau. Geol. Soc. Am. Bull., 68: 505—526.
Russell, R.J., 1954. Alluvial morphology of Anatolian rivers. Ann. Assoc. Am. Geogr., 44: 363—391.
Schumm, S.A., 1968. River adjustment to altered regimen — Murrumbidgee River and paleochannels, Australia. U.S. Geol. Surv. Prof. Pap., 598: 65 pp.
Wolman, M.G. and Leopold, L.B., 1957. River floodplains: some observations on their formation. U.S. Geol. Surv. Prof. Pap., 282-C: 87—107.

Reprinted from the *Geological Society of America Bulletin*, V. 71, No. 6 (June 1960), pp. 769-793, by permission of the Geological Society of America.

RIVER MEANDERS

BY LUNA B. LEOPOLD AND M. GORDON WOLMAN

ABSTRACT

Most river curves have nearly the same value of the ratio of curvature radius to channel width, in the range of 2 to 3. Meanders formed by meltwater on the surface of glaciers, and by the main current of the Gulf Stream, have a relation of meander length to channel width similar to rivers. Because such meanders carry no sediment, the shapes of curves in rivers are evidently determined primarily by the dynamics of flow rather than by relation to debris load.

Velocity distributions along river curves provide a generalized picture of flow characteristics. Evidence on flow resistance in curved channels suggests that a basic aspect of meander mechanics may be related to the distribution of energy loss provided by a particular configuration or curvature. No general theory of meanders is as yet satisfactory, however; in fact, present evidence suggests that no single theory will explain the formation and characteristics of all meanders and that few of the physical principles involved have yet been clearly identified.

RÉSUMÉ

Dans la plupart des courbes de rivières la valeur du rapport entre le rayon de courbure et la largeur du lit, de l'ordre de grandeur de 2 à 3, est presque toujours la même. Les méandres formés par l'eau de fonte à la surface des glaciers et par le principal courant du Gulf Stream, montrent une relation entre la longueur du méandre et la largeur du chenal semblable á celle des rivières. Puisque de tels méandres ne transportent pas de sédiments, on peut en conclure que la forme des courbes de rivières est determinée principalement par la dynamique de l'écoulement d'eau plutôt que liée à la charge sédimentaire.

La distribution des vélocités le long des courbes de rivières donne une image générale des caracteristiques d'écoulement. Ce qu'on sait de la résistance à l'écoulement dans les chenaux courbes suggère que la résistance minimum est un facteur de base dans la mécanique des méandres. Toutefois, aucune théorie générale des méandres n'est encore complètement satisfaisante. En fait, les données actuelles suggèrent qu'aucune théorie ne suffira pas à expliquer tous les méandres, et que quelques-uns seulement des principes de physique en jeu ont déjà été reconnus.

ZUSAMMENFASSUNG

Die meisten Flußwindungen haben etwa dasselbe Wertverhältnis von Krümmungsradius zu Bettbreite, ungefähr zwei zu drei. Mäanderwindungen, die durch Schmelzwasser auf der Oberfläche der Gletscher entstehen, und solche, die durch die Hauptströmung des Golfstromes bedingt sind, haben ein Verhältnis von Mäanderlange zu Bettbreite, welches dem der Flüsse ähnlich ist. Da solche Mäander kein Sediment mit sich führen, sind die Formen der Flußwindungen offensichtlich in erster Linie durch die Dynamik der Strömung bedingt, weniger durch die Gegebenheiten des mitgeführten Materials.

Die Verteilung der Strömungsgeschwindigkeit entlang den Flußwindungen zeigt ein verallgemeinertes Bild der Strömungsmerkmale. Was über den Strömungswiderstand in gewundenen Flußbetten beobachtet wurde, läßt die Vermutung aufkommen, daß der geringste Widerstand eine wichtige Ursache für Mäanderbildungen ist. Tatsachlich lassen die gegenwärtig beobachteten Tatsachen vermuten, daß vorläufig noch keine einzige Theorie alle Mäanderbildungen erklaren kann, und daß nur wenige der physikalischen Gesetzmäßigkeiten, die dabei eine Rolle spielen, bisher gefunden worden sind.

МЕАНДРЫ РЕК

Люна Б. Лиополд и М. Гордон Уолмэн

Резюме

Большинство речных излучин обладает почти одинаковой величиной отношения радиуса кривизны к ширине канала, колеблющейся в порядке от 2 до 3. Меандры, образованные талой водой на поверхности ледников и излучины главного течения Гольфштрема, обладают отношением длины излучины к ширине канала, подобным тому, которое наблюдается в реках. Ввиду того, что такие меандры не содержат осадков, формы излучин в реках, очевидно, определяются в первую очередь динамикой течения, нежели зависимостью от наличия кластического материала.

Распределение скорости течения вдоль речных излучин дает обобщенную характеристику течения. Доказательством сопротивления течения в криволинейных каналах наводит на мысль о том, что основным фактором в механике образования меандр является минимальное сопротивление. Еще не выработана общая удовлетворительная теория о происхождении меандр. В действительности, современные данные дают возможность сказать, что отдельная теория не даст объяснения происхождения всех типов меандр. Еще не изучены многие физические факторы, влияющие на происхождение меандр.

CONTENTS

TEXT

ILLUSTRATIONS

TABLE

General Statement

The most characteristic features of all stream channels, regardless of size, are the absence of long straight reaches and the presence of frequent sinuous reversals of curvature. These bends, whether or not they display regular reversals and sufficient symmetry to warrant the name meander, tend to be scaled versions of a given set of proportions. The proportions are determined by three dimensions: the repeating distance or length of the curve, the width of the channel, and the radius of curvature. In brief, the size of the bend appears to be proportional to the size of the river; large rivers have large bends, and small rivers have small bends.

An adequate theory of meanders should explain why all river bends seem to be of proportional sizes, and by what mechanical or physical principle a bend becomes adjusted in size to the stream discharge. Such a theory should also explain why some rivers meander and others do not. It should explain why meandering rivers occur on ice with high velocity of flow and no sediment load, and why tortuous channels are so prevalent in tidal estuaries. A meander theory should also explain how chance perturbations of the flow may persist and lead to the development of a sinuous pattern.

The present paper does not provide any theory. An attempt is made to analyze and integrate what appear to be the most prevalent essential characteristics of meandering channels in nature. At least some of these prevalent characteristics are probably basic to the mechanics of meandering channels which are as yet poorly understood. These include the external features of the channel, as well as the distribution of velocity and the pattern of flowing water. Such a summary may prove useful in the eventual definition of the necessary and sufficient conditions for meandering of channels in diverse natural environments.

There is a large published record on the occurrence and physiographic setting of meandering channels, as well as on the form and theory of river curves. It is not our purpose here to cite this record exhaustively, but rather to sift out those facts which appear to us most pertinent, emphasizing at the same time that opinions may differ as to what is pertinent.

Our studies of meandering streams and of meander mechanics lead us to write a somewhat different kind of a paper than might usually be expected in a "review" series. Many new observations are available, some previously unpublished, some recently published or in process of publication. We hope it will be of some service, especially to the reader who has an interest but not a specialty in the subject, to have these new observations summarized and related to earlier observations. We believe that the status of the problem will be clearer as a result, and these observations are, in our opinion, of some importance in indicating directions toward which future work might profitably be directed.

We then attempt to show that the problem of river meanders can be broken down into two parts. The first concerns the mechanical and hydraulic processes that govern the form, size, and probably the occurrence of meanders. The second includes the physiographic history of a particular channel and the formation, maintenance, and possible dissolution of a meandering pattern. Although these two parts are almost inseparable they do represent quite different ways of viewing the meander problem. We hope it will be clear from this review that many important physiographic questions cannot be answered with satisfaction because there is insufficient knowledge at present about meander mechanics.

The authors are much indebted to J. Hoover Mackin and W. W. Rubey for many constructive suggestions and for a critical reading of an early draft of the manuscript.

Geometry and Patterns of Channel Bends

Because most channels are not straight but sinuous, the forms which they display in plan view are, in detail, infinitely varied. In our experience it is unusual for a reach of natural stream to be straight for a distance exceeding 10 channel widths (Leopold and Wolman, 1957, p. 53). Yet certain common features are discernible. These features, common to many individual bends, probably have some significance in the mechanics or hydraulics of meanders, but this significance is only poorly understood. The geometric characteristics of a channel seen in plan view are defined by Figure 1.

We use the term "arc distance" to mean the length measured along the channel center line from one point of inflection to the next. "Sinuosity" we use to mean the ratio of arc distance to half the meander length.

The geometry of meander curves has been the object of extensive statistical study, examples of which are papers by Jefferson (1902), Inglis (1937; 1949, Pt. 1, p. 143), Bates (1939), and Leopold and Wolman (1957, p. 58). An additional independent sample in the appendix includes measurements of rivers in which the presence of a well-defined flood plain indicates that the channel form and pattern have developed reasonably free of bedrock control. These data for rivers of moderate to large size were derived from published topographic maps and for small streams from plane-table maps made by the authors. The appendix also includes some data from published laboratory studies of meandering streams in erodible materials.

Studies of the geometry of the patterns of meandering rivers give generally similar results.

The most consistent correlations are between meander length, channel width, and radius of curvature, as shown by Figure 2, plotted from the data in the appendix. Because of the variability in the dimensions of channels in nature, considerable scatter of plotted data is expected. Despite the scatter the relations among the factors appear to hold through a very large range of stream size, from laboratory streams a foot wide to the Mississippi River nearly a mile wide. Various authors have expressed the relations for rivers in alluvial flood plains by regression equations, a few of which are presented in Table 1. Table 1 also includes relations derived from the data in the appendix.

The exponents in the regression equations are all so close to unity that the relations between meander length (wave length), amplitude, radius of curvature, and channel width may be considered linear. Among different groups of data fairly consistent values are obtained for the coefficient relating meander length to channel width; meander length ranges from 7 to 10 times the channel width. This ratio holds in general not only for meanders on a broad flood plain but also characterizes sinusoidal or meandering valleys incised into consolidated materials (Dury, 1958, p. 109).

A marked inconsistency is exhibited, however, in the ratio of amplitude to channel width. Amplitude correlates only poorly with meander length. We tentatively interpret these findings to mean the amplitude of

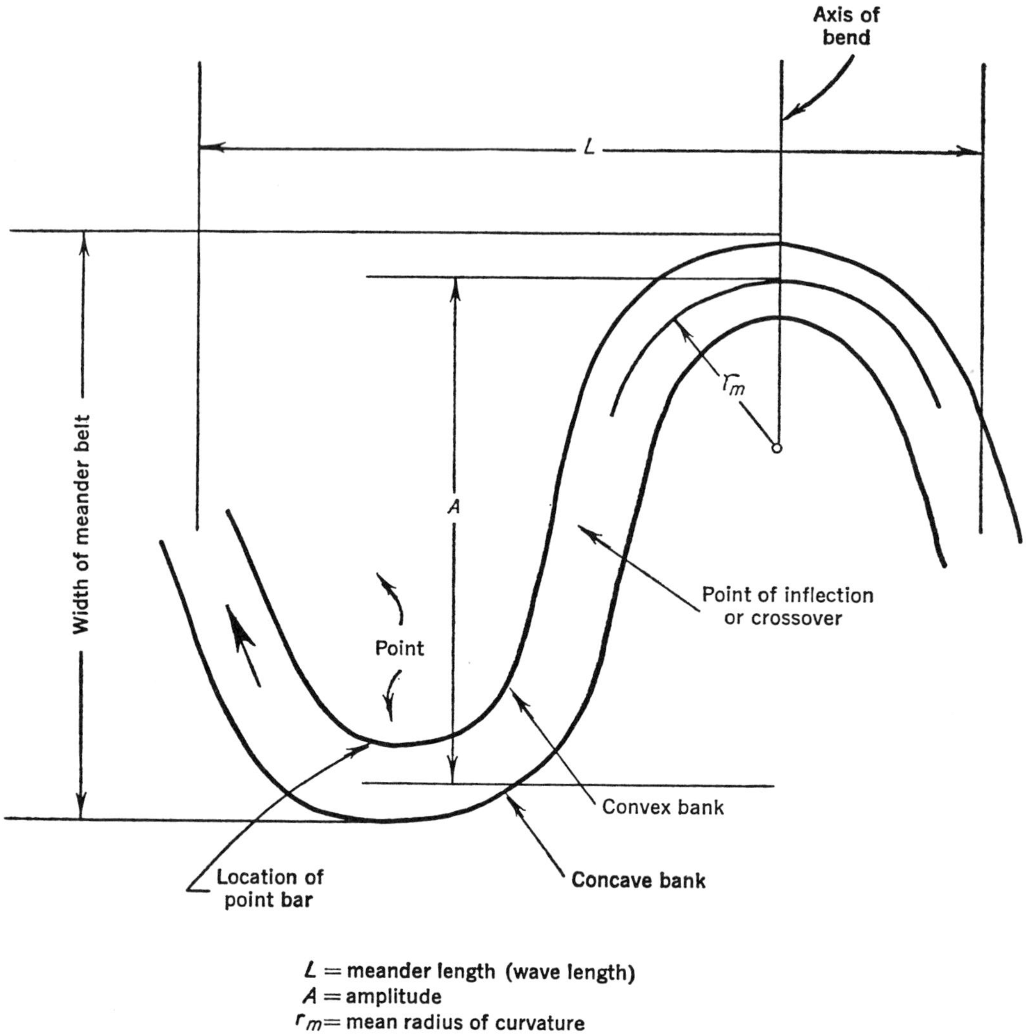

FIGURE 1.—SKETCH TO DEFINE TERMS USED IN DESCRIBING GEOMETRIC CHARACTERISTICS OF A MEANDERING CHANNEL

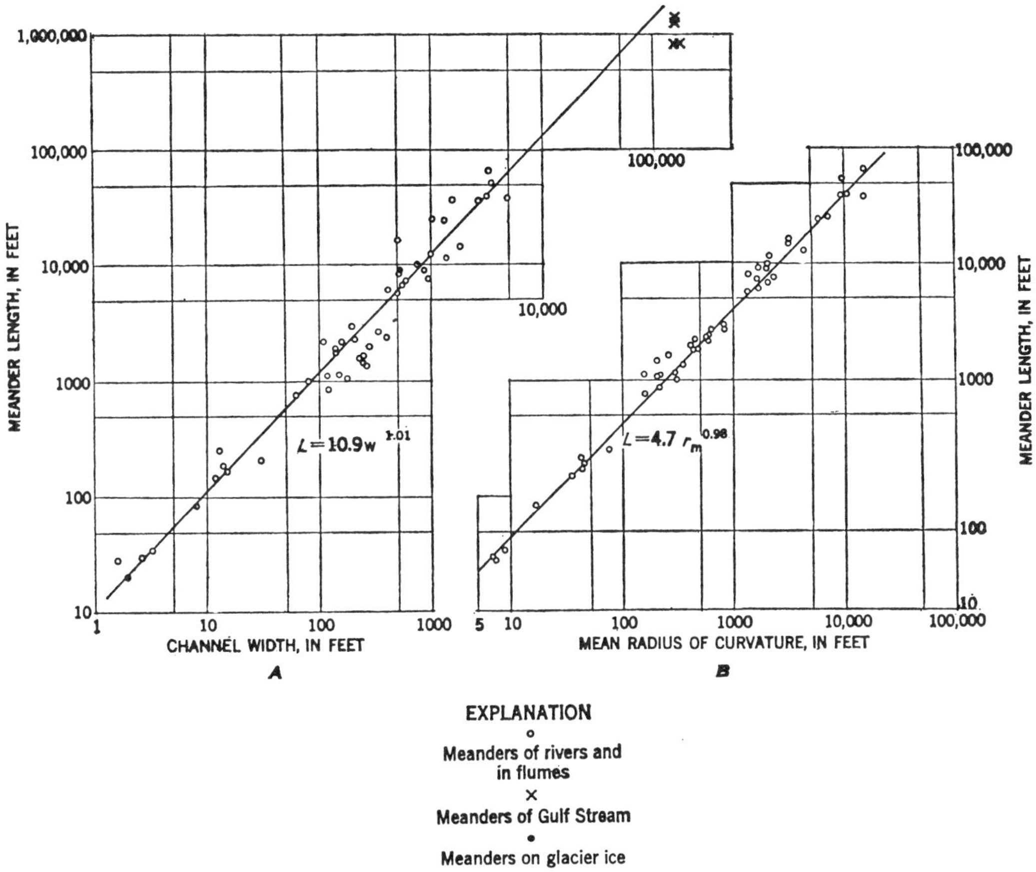

FIGURE 2.—RELATIONS BETWEEN MEANDER LENGTH AND CHANNEL WIDTH (LEFT DIAGRAM) AND MEAN RADIUS OF CURVATURE (RIGHT DIAGRAM)

TABLE 1.—EMPIRICAL RELATIONS BETWEEN SIZE PARAMETERS FOR MEANDERS IN ALLUVIAL VALLEYS

	Meander length to channel width	Amplitude to channel width	Meander length to radius of curvature
Inglis (1949, pt. 1, p. 144, Jefferson data)	$L = 6.6w^{0.99}$	$A = 18.6w^{0.99}$	..
Inglis (1949, pt. 1, p. 149, Bates data)	..	$A = 10.9w^{1.04}$	..
Leopold and Wolman (this paper)	$L = 10.9w^{1.01}$	$A = 2.7w^{1.1}$	$L = 4.7r_m^{0.98}$

meander loops is determined more by erosion characteristics of stream banks and by other local factors than by simple hydraulic principles.

The U. S. Waterways Experiment Station studies (Friedkin, 1945, p. 15) showed that in uniform material amplitude did not progressively increase nor did meander loops cut off as the meanders migrated downstream. Friedkin interpreted this to mean that "natural cut-offs result from local differences in the erodibility of bank materials (p. 16)."

These considerations appear to support the concept that a limit to the amplitude of meanders is provided by the formation of cut-offs. The rate of downstream migration relative to the rate of growth of the amplitude of bends probably depends on the distribution of shear against the stream banks, which in turn is a function of channel shape. The latter is determined at least partly by the characteristics of the flood-plain alluvium and the moving debris.

We believe relations between channel width,

meander length, and radius of curvature, in contrast to amplitude, are to a great extent independent of bed and bank materials and are related in some unknown manner to a more general mechanical principle.

Examining the correlations presented above one can write

$$L = 10.9w^{1.01} = 4.7r_m^{0.98}.$$

Assuming the exponents to be unity and solving for the ratio of radius of curvature to width,

$$r_m/w = \frac{10.9}{4.7} = 2.3.$$

Computing the value of r_m/w river by river from the data in the appendix rather than equating the least-square regression lines, it is found that in the sample of 50 rivers the median value is 2.7, mean 3.1, and two-thirds of the cases lie between 1.5 and 4.3. About one quarter of the values lie between 2.0 and 3.0. The tendency for similarity in the ratio of curvature to width makes all rivers look quite similar on planimetric maps, as suggested by Figure 3. In fact when one inspects a planimetric map of a river without first glancing at the map scale it is not immediately obvious whether the river is large or small owing to this tendency for similar ratio of radius to width regardless of river size.

As will be mentioned later, the value of the ratio of radius of curvature to channel width has an important bearing on resistance to flow, and this appears to offer some clue as to the hydraulic basis for the observed geometric similarity among channels of different sizes.

Curves in Sediment-Free Channels

Meandering channels are carved in glacier ice by streams of meltwater flowing on the glacier surface. To obtain measurements of such channels we went to one of the Dinwoody glaciers on the east side of the crest of the Wind River Range in Wyoming (Lat. 43°12′, Long. 109°38″). The meandering channels on this small valley glacier can be seen in Plate 1. Although concentrated primarily upon the steeper slope downstream, meandering channels began on the relatively flat portion of the profile, that is, near the point of inflection from concave to convex portions of the surface profile. The channels extended downstream to the convex-upward part of the glacier snout.

Some of the meandering channels were incised 6 to 12 feet into the ice, and those which appear on Figure 1 of Plate 1 are of this kind. Satisfactory measurements of these incised meanders could not be obtained, but the channel shown in Figure 2 of Plate 1 incised 1 to 2 feet, appeared to be comparable to the more deeply incised streams.

Comparison of the meandering stream with incipient channels being formed by water flowing over the ice surface led us to conclude that an irregular sheet of water gradually concentrated in a shallow flat channel at some place where local factors favored melting. The meandering form developed as the flowing water gradually melted a channel downward. Channels in the ice only a foot wide and 2 inches deep had no well-developed meander form, but another channel of similar width but 1 foot deep showed a well-developed sinusoidal pattern.

Although occasional rocks and pebbles occurred in the channel none were in motion at time of observation. In some bends a few pebbles had collected on the inside of the curve in the position of the point bar in ordinary streams. The surface of the glacier in the locality was strewn with cobbles and rocks in a random manner owing to the fact that the channel measured was near the edge of a poorly developed medial moraine.

In several places terrace remnants were cut into the ice and abandoned as downcutting progressed. No pebbles were seen on these terrace remnants. The meanders appeared to be cut by melting alone with no effective help from abrasion by transported pebbles.

Over-all dimensions (in feet) and flow parameters of the channel pictured in Figure 2 of Plate 1 were as follows:

Channel width	2.3
Channel incision below ice surface	1.0
Width of surface of flowing water	2.0
Mean depth of flowing water, d_m	0.25
Average velocity from measurements of flow, v	5.3 ft/sec.
Average meander length	20
Average channel slope, s	0.023
Discharge	2.4 cfs
Froude number $v/\sqrt{gd_m}$ (where g is the acceleration of gravity)	1.9

The Froude number, 1.9, was well within the range of supercritical or shooting flow. In shooting flow it is usual to observe surface wave trains and the diagonal crisscross of

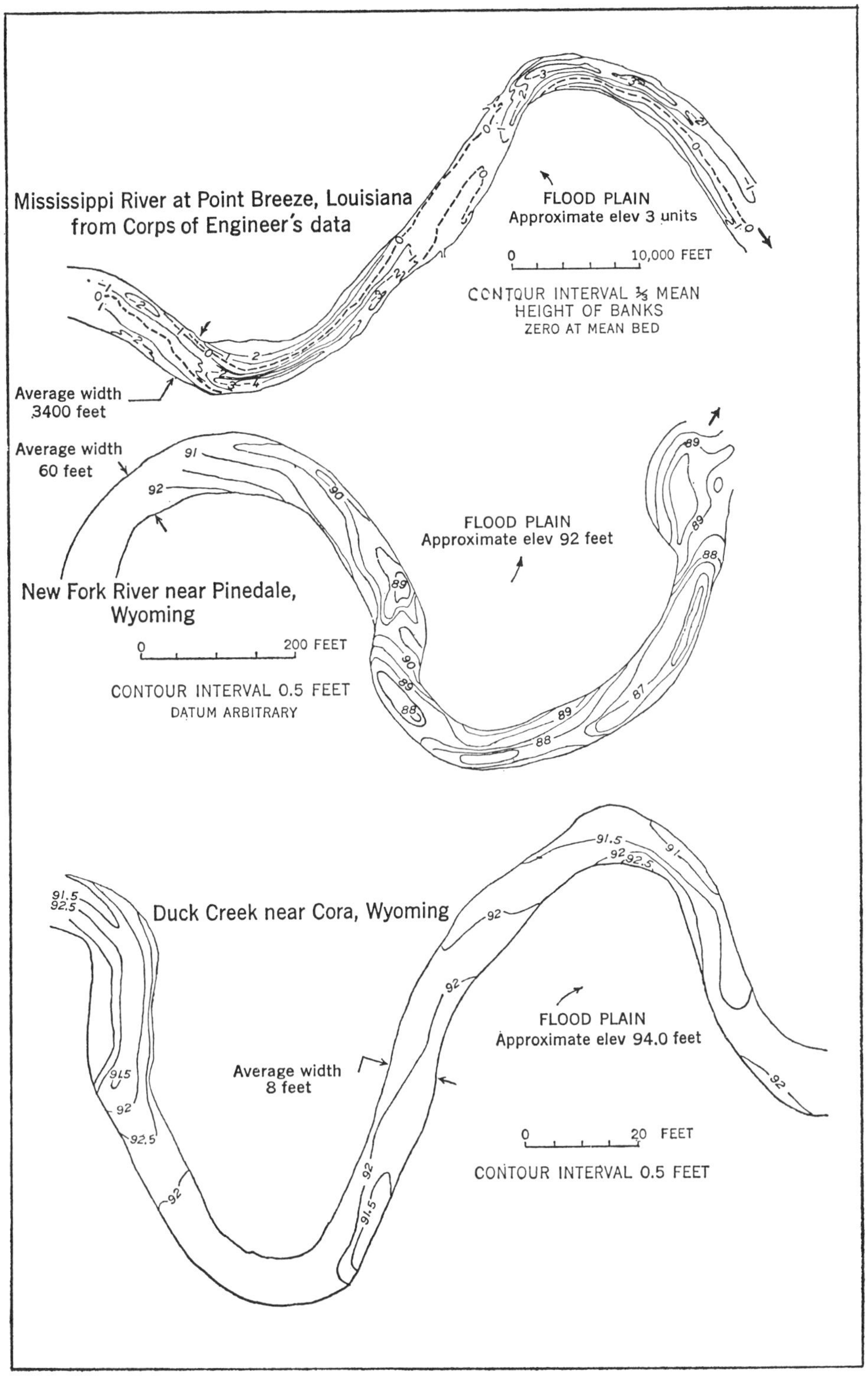

FIGURE 3.—PLANIMETRIC MAP AND BED TOPOGRAPHY OF A MEANDER OF THE MISSISSIPPI RIVER AT POINT BREEZE, LOUISIANA, THE NEW FORK NEAR PINEDALE, WYOMING, AND DUCK CREEK NEAR CORA, WYOMING

Scales are chosen so that meander length is equal on the printed page.

surface waves reflected off the channel sides. If these were present they were masked by the violent turbulence, although they were probably the cause of a marked hump in water-surface elevation extending along the channel center line in some places. The water was conspicuously superelevated on the outside of the curves.

The measurements of meander length and width of this stream cut in ice are plotted on Figure 2. The meander length of channels developed in ice bears the same relation to channel width as in ordinary meandering streams. Measurements made by D. G. Anderson of meanders of small streams on the ice of Chamberlain Glacier, Alaska, confirm this conclusion (Oral communication to Leopold, 1958).

Data on meanderlike phenomena from another source deserve mention as possibly having some relation to the present problem. In the Gulf stream of the North Atlantic oceanographers have found bands of relatively high speed which have a meandering pattern in plan view. Stommel (1954, p. 887) described these as "horizontal eddy and meander formations," and Von Arx (1952, p. 213) used the term "meandering current." These currents appear both in cross sections of the velocity field and in planimetric maps of the temperature field. Maps of the temperature field at a depth of 200 meters have been published by Stommel (1954, Fig. 2) and by Fuglister (1955, chart 3a). From some of these published maps rough measurements of meander length may be obtained and, less satisfactorily, estimates of current width. Better estimates of current width are obtainable from velocity cross sections published by Von Arx (1952, Fig. 2) and Worthington (1954, Fig. 9).

Four measurements of the width and meander length were possible from the published data, and these are plotted on Figure 2. Although they fall slightly below the line drawn through the points representing river data, the graph suggests that meandering currents in the Gulf Stream bear certain analogies to river meanders. Although frictional flow in rivers should not be compared to geostrophic flow of an ocean current, the observed similarity in meander dimensions appears worthy of attention.

The importance of meanders in ice and in ocean currents lies in the fact that both flow systems exhibit meanderlike phenomena in the absence of sediment debris. The establishment of meander length–width relations similar to those of sediment-laden rivers suggests that sediment alters or affects but does not cause the meander pattern. (For a contrasting view *see* Matthes, 1941.)

Channel Cross Sections and Longitudinal Profiles

From a study of bends of the Garonne River, Fargue (1908, p. 25) stated as a general rule of river behavior that the shallowest sections occur downstream from the crossover, and the deepest sections downstream from the axis of the bend or point of maximum curvature. The topographic maps of sample meanders in Figure 3 show in a general way the features described by Fargue and many workers after him. These same figures, however, also indicate that the location of deeps and shallows may be highly variable. For example, data from 25 bends on Buffalo Creek near Buffalo, New York, indicate that the deep points are usually found downstream from the point of maximum curvature but do occur as well upstream from this point (Parsons, 1959). On a reach of the Popo Agie River near Hudson, Wyoming, containing four successive bends of amazing symmetry deeps occurred upstream from the point of maximum curvature and not downstream.

In the reach of the Mississippi River shown in Figure 3 the deepest portion occurs at the axis of curvature in one bend but somewhat downstream from this axis in the other bend. In the New Fork the deepest portion in one bend occurs practically at the point of inflection. The reach on Duck Creek more nearly conforms to Fargue's generalization.

There seems little doubt that the depth of water at a bend of a river is closely correlated

---

Plate 1.—MEANDERS ON DINWOODY GLACIER, WYOMING

Figure 1.—In middle foreground are meandering channels of meltwater streams carved in the glacier ice. View southwesterly from Sentinel Peak; photograph by Mark F. Meier

Figure 2.—Channel carries meltwater from glacier. The water flowed at high velocity and was violently agitated.

FIGURE 1

FIGURE 2

MEANDERS ON DINWOODY GLACIER, WYOMING

with the bend curvature. Using 103 measurements of mean depth and corresponding radii of curvature on the River Elbe, Leliavsky (1955, p. 118) shows that the two factors bear a simple linear relation and that depth increases inversely as a function of radius of bend.

The point of inflection in a river curve is closely associated with a shallow portion of the reach, or a depositional bar on the bed. The occurrence of a bar or riffle in the bed of a straight river reach should correspond morphologically therefore to the point of inflection of a meandering channel. It is on this basis that twice the distance between successive riffles in a straight reach was compared with the wave length of meanders (Leopold and Wolman, 1957, p. 55).

Not only do the riffles in straight channels tend to be in a position analogous to comparable shallows in a meandering reach, but there also is some tendency for the shape of the bars to be the same. In some straight reaches successive gravel or sand bars which constitute the riffles assume the form of lobate wedges sloping alternately toward one bank and then the other. In some meanders the bar near the point of inflection slopes sharply across the channel toward the deep near the concave bank downstream. This is apparent on the topographic map of the New Fork (Fig. 3) and is accentuated at low flow when water flows diagonally across the channel down the steepest local bed slope. At high flow there is rapid divergence as the cross-sectional area expands.

The longitudinal profile of bars, whether associated with the point of inflection of a meander, or a riffle in a straight channel, is usually moundlike rather than dunelike and is asymmetric in cross section. Perhaps ideally the bar has a dunelike profile with a short, steep downstream face. The topography of the bend in the Mississippi River (Fig. 3) suggests this, but many are less regular.

The bar is asymmetric in cross section in a different sense on upstream and downstream faces. In a meander, as one moves downstream toward the point of inflection, the surface of the bar slopes laterally toward the concave bank, but immediately having passed the crest of the bar the asymmetry changes, and the surface of the bar slopes toward the opposite bank.

For this reason the channel cross section in a meandering reach is symmetrical only for a very short distance. We estimate that on the average the distance through which the cross section is essentially symmetrical is only about one-tenth wave length. This short distance can be seen on topographic maps of Figure 3.

The importance of the configuration of the bars and their relation to the deeps lies in the relation of bed topography to bed shear. River pilots and others know that in large rivers the crossings (points of inflection) tend to scour at low flow and fill at high flow; the bends (pools) tend to fill at low flow and scour at high (Straub, 1942, p. 617; Lane and Borland, 1954, p. 1075).

Local bed shear must be greater at points of scour than where filling occurs, but direct measurements of bed shear are few, and little is known in detail about the relation of bed topography to stress on the bed.

Lane and Borland (1954, p. 1079) explain the observed sequence of scour and fill as a consequence of the different shape of cross section in pool and riffle. They say that the crossing or riffle has a larger cross-sectional area than the pool during high stages and a smaller area during low stages. Laursen and Toch (1954, p. 1085) state that width often tends to be greater at a crossing than at a bend. The expansion in width they believe would cause deposition on the wide crossing while scour occurs in the narrower bend.

We have made measurements of slope and of depth over pool and riffle separately at various stages of flow in an attempt to compute approximately the variation of mean stress. At low flow the bend is a still pool over which the water has a relatively flat slope. In contrast, in the riffle the water tends to plunge over the steep downstream face of the bar. As stage increases, the slope over the pool or bend increases, while that over the bar decreases until near bankfull stage when the water surface attains a continuous uniform slope, and all obvious surface effects of the shallow bar have been erased.

With increasing stage, depth increases in both bend and crossing. The product of depth and slope, then, increases with stage more rapidly in pool than at the riffle or crossing.

In the one stream measured by us, Seneca Creek near Dawsonville, Maryland, the computed mean shear is greater in the riffle at low flow and greater in the pool at high flow (Leopold, in press). This observation is in qualitative agreement, then, with the concept of scouring in the crossing at low flow and in the bend at high flow. The analysis is incomplete, however, inasmuch as variations in the distribution of velocity in successive cross sections have not been included in the computation of shear, and additional observations

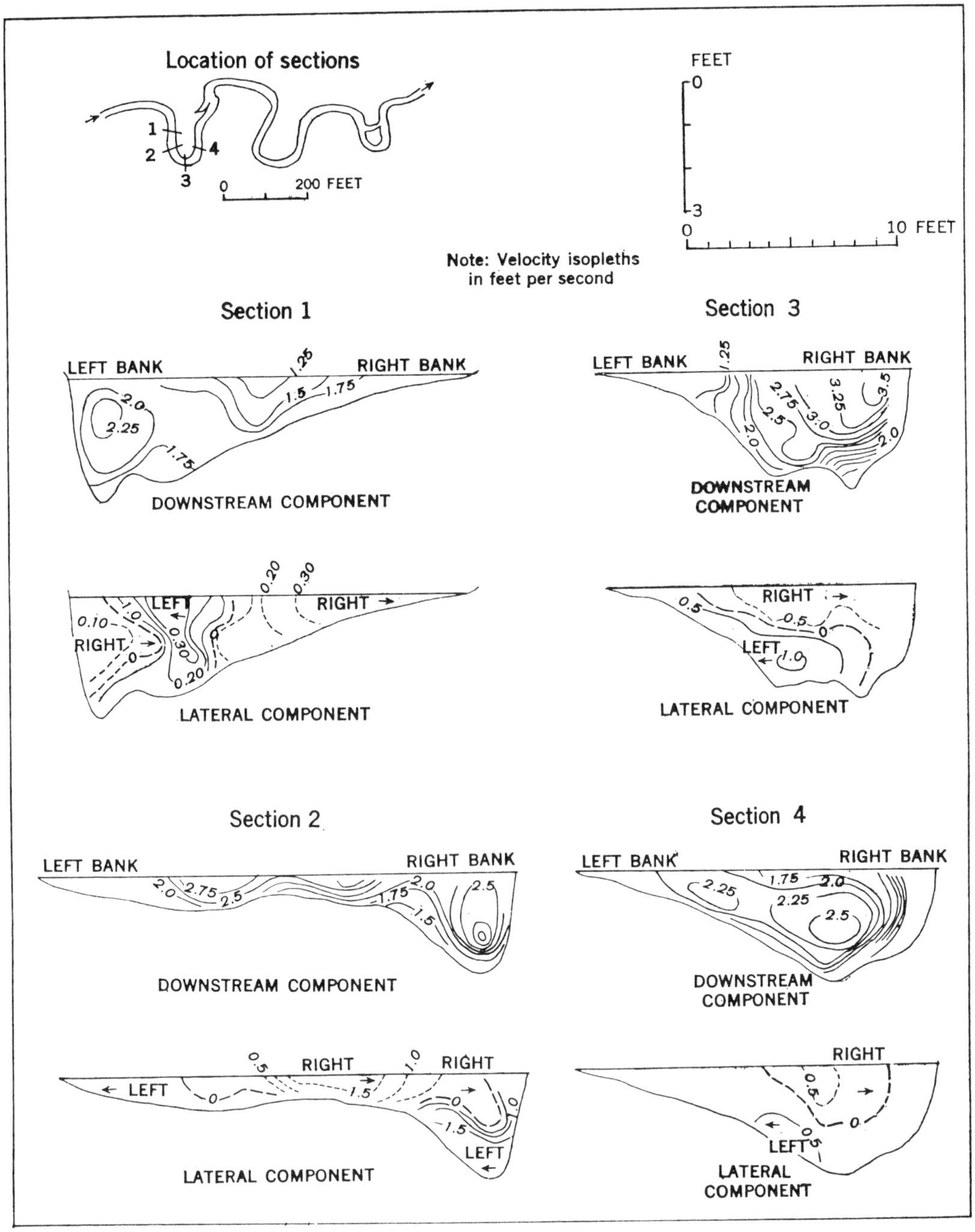

FIGURE 4.—LATERAL AND DOWNSTREAM COMPONENTS OF VELOCITY AT VARIOUS CROSS SECTIONS IN A BEND, BALDWIN CREEK, NEAR LANDER, WYOMING

elsewhere are necessary before any generalization can be made. Mean depth and slope, even over a short reach, probably do not measure the true shear exerted on any local area of bed. Direct measurement of shear stress in various parts of a bend at various stages is needed.

DYNAMIC AND FLOW CHARACTERISTICS

Pattern of Flow in a Meander Bend

Relatively few detailed measurements of the distribution of flow in a natural meander have been published (Blue *et al.*, 1934; Eakin, 1935; Leliavsky, 1955, p. 96–100; Van Til and Tops,

1953, p. 19–23). The velocity distributions which follow are intended both as additions to the slim store of data and as illustrations of the general characteristics of flow in bends in natural channels. Figure 4 shows downstream and transverse components of velocity at the axis of the bend on Baldwin Creek near Lander, Wyoming. Point velocities were measured by Price current meter and horizontal angles by a vane attached to the wading rod. Measured angles were used in conjunction with the measured velocity to compute the lateral components of velocity. Measurements were made at a high but less than bankfull stage.

At least in narrow channels a cross-channel velocity component is directed toward the convex bank or point bar near the bed, and toward the concave bank near the surface. Continuity requires, then, that surface water plunge downward near the concave bank and that some bed water emerge at the surface near the convex bank.

This circulatory motion in the cross-sectional plane of a channel, as clearly explained by Thomsen (1879), results from the larger centrifugal force on fast-moving surface parcels than on slower-moving ones near the bed. The motion gives to an individual water parcel a path resembling a helix.

Despite the tendency for circulatory motion in the stream cross section, surface stream lines (Fig. 5) show that no single parcel of water crosses completely from the convex to the concave bank. The maximum cross-channel motion does not exceed perhaps two-thirds of the channel width in any given meander bend. This accounts for the observation by Friedkin (1945, p. 5) in laboratory models that material eroded from one bank tends to deposit on a point bar downstream on the same side of the stream.

Matthes (1941, p. 634) stated that in wide, shallow rivers helical circulation does not exist. It is our opinion that in rivers of large width/depth ratio the helical motion exists, but not as a single rotating cell involving the whole channel (Nemenyi, 1946). The existence of several cells of circulation in the Tamm Bend data from the Mississippi can be seen in the distribution of the lateral components of velocity published by Eakin (1935). As in the atmosphere, circulation cells of large magnitude tend to break down into smaller cells. The "polar-front" cell and the "trade-wind" cell of Rossby (1941, p. 610) are directly driven by dynamic forces which we consider analogous to the direct helical motion in river bends due to the combined effect of frictional and centrifugal forces. Rossby's "frictionally driven" circulation cell in the middle latitudes furnishes, in our opinion, an analogy to those circulation cells seen in the Eakin data of the Mississippi which rotate in a direction opposite to the helical motion expected in a single cell occupying the entire cross section.

It is well documented that the water surface is superelevated near the concave bank of a channel bend (Blue *et al.*, 1934; Eakin, 1935; Leliavsky, 1955, p. 123; and others). The amount of the superelevation is proportional to

$$\frac{v^2 w}{g r_m}$$

where v is the mean velocity, w is channel width, r_m the mean radius of curvature, and g the acceleration of gravity (Leliavsky, 1955, p. 122).

This superelevation is a consequence of the curved path of water flowing around a bend and is a correlative of the helical motion.

Generalized Picture of Flow in a Meander

Measurements in meandering streams and in curved flumes (for example Mockmore, 1944, p. 569) allow the construction of a generalized picture of the flow pattern in a meander (Fig. 5). The isometric view of the two principal components of velocity at various positions in the bend show the main features. The scale is such that superelevation of the water surface in the bend does not show on the diagrams but is implied by the velocity distribution.

At the crossover or point of inflection of the bend (Fig. 5, section 1), the cross-sectional shape is not completely symmetrical but is slightly deeper near the bank which was concave in the bend immediately upstream. Downstream from the crossover the section (2) becomes approximately symmetrical, but, because this section is in the bend, some cross-current component exists in accord with the curving stream lines.

The velocity in a meander crossover is not symmetrically distributed. As would be expected, proceeding downstream from the axis of the bend the thread of maximum velocity is much closer to the concave bank than to the center of the channel. The high velocity, moreover, continues to hug this side through the point of inflection of the curve.

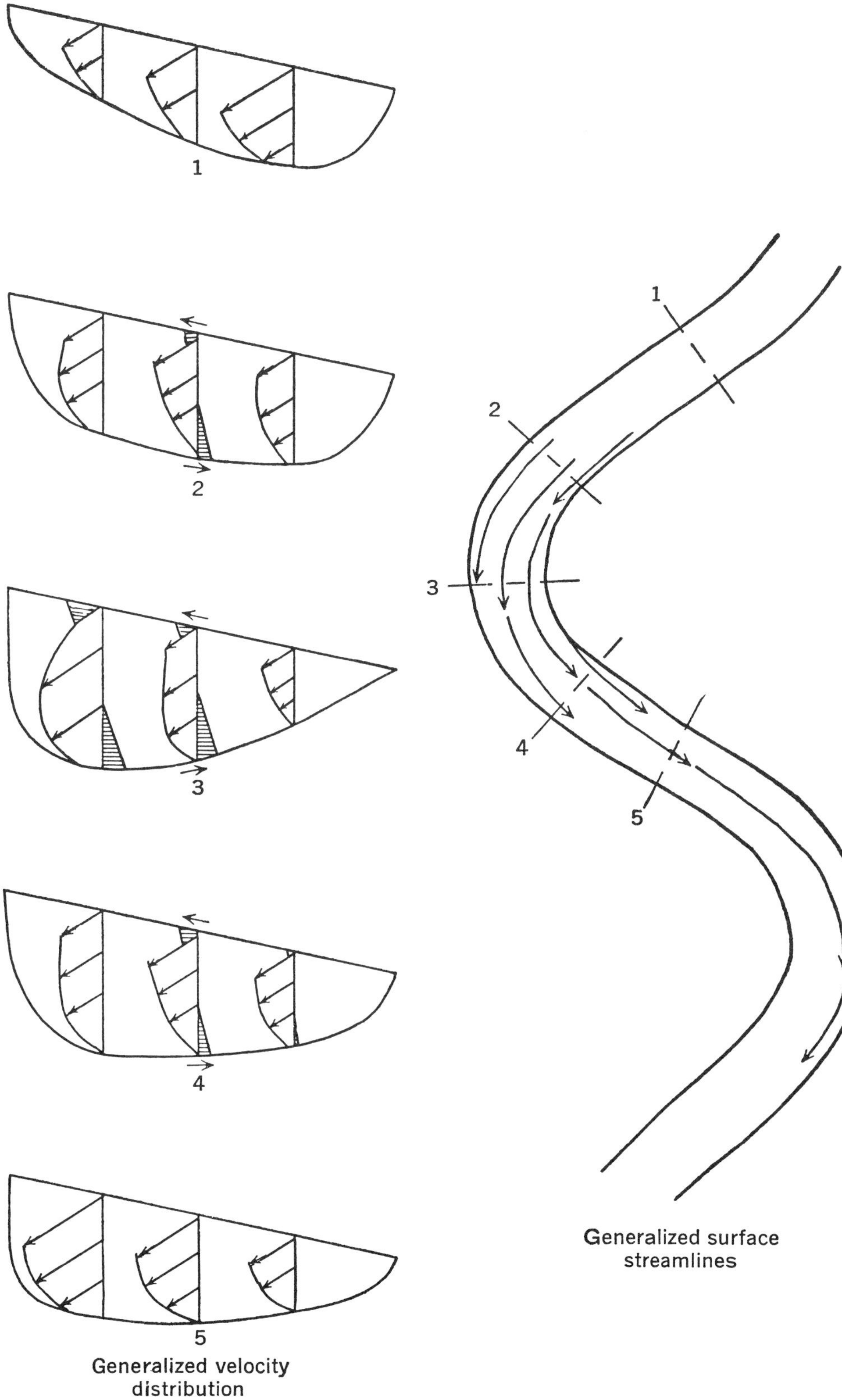

FIGURE 5.—ISOMETRIC VIEW OF GENERALIZED DIAGRAM OF FLOW DISTRIBUTION IN A MEANDER

Showing downstream (open parabolas with arrows) and lateral (closely lined areas) components of velocity as vectors, and surface stream lines. All sections viewed from a changing position to the left of and above the individual section.

The asymmetry in the velocity field persists downstream from the point of reversal of curvature, at least as far as the most nearly rectangular and symmetrical channel cross section.

Downstream from the symmetrical cross section the maximum downstream velocity approaches the concave bank. The highest velocity at any point occurs near the concave bank just downstream from the axis of bend. Individual filaments of water accelerate and decelerate along a stream line. The maximum velocity is not always associated with the same parcels of water.

Measured distributions of velocity show also that the maximum-point velocity in the bend is depressed below the water surface. These observations agree with the measurements made by Mockmore (1944, p. 596) in a curved flume.

An effect of increasing discharge in a meander reach is to move downstream the position on the concave bank where the thread of maximum velocity impinges. The stream lines of highest velocity do not touch the concave bank but approach it and then swerve into a course more or less parallel to that bank.

Channel Shape and Movement in Relation to the Flow Pattern

The slight lack of congruence of stream-line curvature with bank curvature leads to the tendency for the locus of point-bar deposition to occur downstream from the axis of the bend. A building point bar tends to concentrate caving of the bank downstream from the axis of bend and thus accounts for the progressive down-valley movement of meander bends. This same tendency exists in all river curves, even in those lacking sufficient symmetry to be called meanders, and this accounts in part for normal movement of river channels and the consequent construction of flood plains.

Material slumping into the bed by bank caving is caught in the transverse component and carried toward the middle of the stream. If the location from which it is derived is far enough downstream in the bend, insufficient length of reach remains for the material to get all the way across the channel. Instead it moves through the crossover reach without having crossed the channel. Once into the reversed curve the particle moving near the bed is drawn toward the same bank from which it started rather than to the opposite bank. This does not rule out the likelihood that much fine material which is not carried near the bed of the stream will be deposited at a downstream location on the side opposite to the one from which it was derived.

The plunging or diving motion of surface water near the concave bank is one factor tending to depress the point of maximum velocity below the surface near the axis of the bend. Bank friction also has a similar effect. Because the highest velocity gradient on both bed and bank occurs where the thread of highest velocity is nearest the boundary, this maximum velocity gradient is equivalent to maximum shear stress. Its location accounts, then, for deepening of the bend reach near the concave bank. This scour, together with building of the point bar, explains the triangular shape of the cross section.

Increasing the depth of water in the channel reduces the vertical velocity gradient. In a channel that is very deep relative to its width, the effect of bed friction becomes relatively small, and the velocity is nearly constant with depth except close to the bed. Thus the helical circulation becomes negligible. Flow in a curved channel of great depth and small width approaches potential flow—that is, the downstream velocity varies across the channel inversely as the radius of curvature of the stream lines. This was demonstrated by Wattendorf (1935, p. 574) in a deep, narrow, curved flume. In the absence of the circulation in cross section the thread of high velocity hugs the convex bank and does not cross the channel. Thus the mechanism for building the point bar is probably absent in channels of extremely small width–depth ratio.

Effect of Flow Pattern on Deposition and Erosion

The vigorous cross currents near the bed can transport considerable quantities of bed material from the concave to the convex side of the channel. On a meander bend of Baldwin Creek (Fig. 4) depth-integrated samples of suspended load were taken at each of five verticals spaced across the stream and repeated in each of five cross sections along a meander wave. A nearly uniform concentration of sediment existed across the section in a crossover. In the sections located in the curved reach the concentration increased markedly near the convex bank.[1] A similar distribution of sediment was observed by Eakin (1935,

---

[1] Dr. John P. Miller assisted the senior author in obtaining these measurements.

p. 471) in his samples taken at Tamm Bend on the Mississippi River. This increased concentration near the convex bank is attributed to the fact that water in this location has, to some extent at least, come laterally in a path close to the bed and is comprised, therefore, of water elements with relatively high concentrations of sediment.

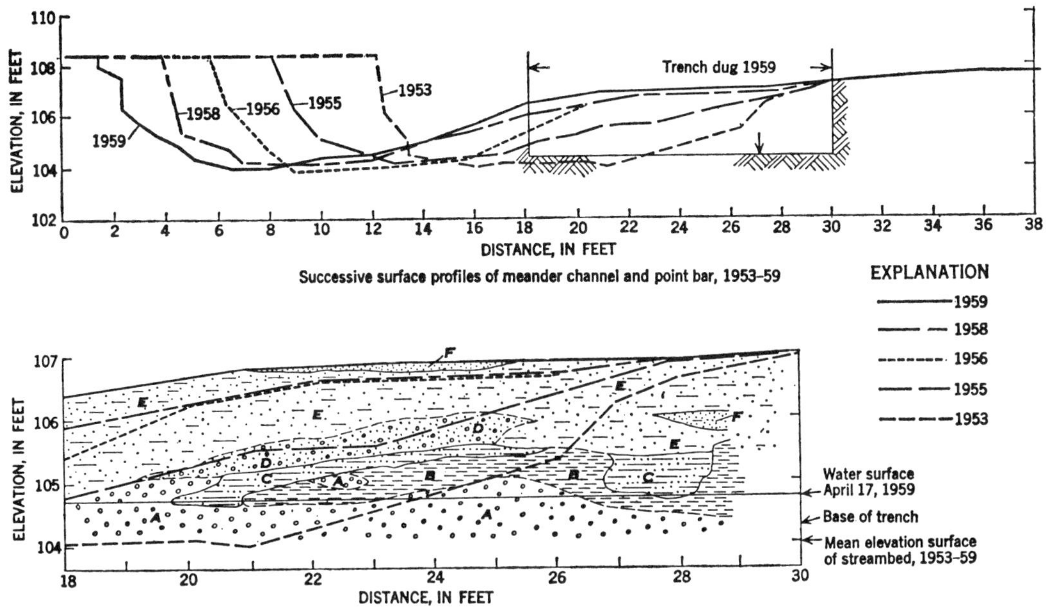

Large scale cross-section of stratigraphy in relation to profiles

FIGURE 6.—STRATIGRAPHY OF POINT BAR IN RELATION TO MEASURED CROSS-SECTIONAL PROFILES DURING SIX YEARS OF OBSERVATION, WATTS BRANCH, 1 MILE NORTHWEST OF ROCKVILLE, MARYLAND

Key to stratigraphic units is as follows:
A Gravel, mostly 3–8 mm with considerable 8–20 mm and a few 64 mm
B Olive-gray clayey silt, with organic matter, small mica flakes
C Orange-brown, mottled sandy silt with some clay, lenses of leaflike organic mterial, mica
D Coarse sand, brown-stained, with pebbles up to 8 mm, some fine roots *in situ,* lenses of silt
E Brown sandy silt gradually changing upward in places into fine sand
F Fine sand with some silt

Although there is some evidence that rifling in a spirally welded pipe tends to increase the ability of a flow to transport sediment over that of a cast-iron pipe (M. L. Albertson, oral communication to authors), evidence from preliminary experiments (L. M. Brush, oral communication to Wolman) indicates that transport in a meandering channel is much reduced over that in a straight channel of similar dimensions and boundary roughness. This effect of curvature on transport is in accord with the added energy losses accompanying increased curvature.

The cross-channel component of bed flow toward the convex bank is probably the principal mechanism for the building of a point bar. The path of particles which are deposited on a building point bar has, of course, only a small component across the stream compared with a large down-stream component. The lobate form of a point bar may be compared to a sector of a very flat cone. A bed particle, taking a path along a chord of this cone, may climb the inclined plane or ramp presented by the cone. Thus, even some of the coarser debris of the channel bed will be deposited on the point bar at an elevation considerably higher than the average level of the stream bed.

Variation in discharge would lead one to expect that a thin layer of coarse material may be deposited on the point bar at one time and at another time a layer of fine material may be deposited upon the coarse. On the average, fine material will be carried higher up the bar than will coarse material, so the point bar would show on the average a rough gradation of size from coarse near the base to fine near the flood-plain level. In many instances, then, a growing point bar would not merely lap over the coarse material on the stream bed.

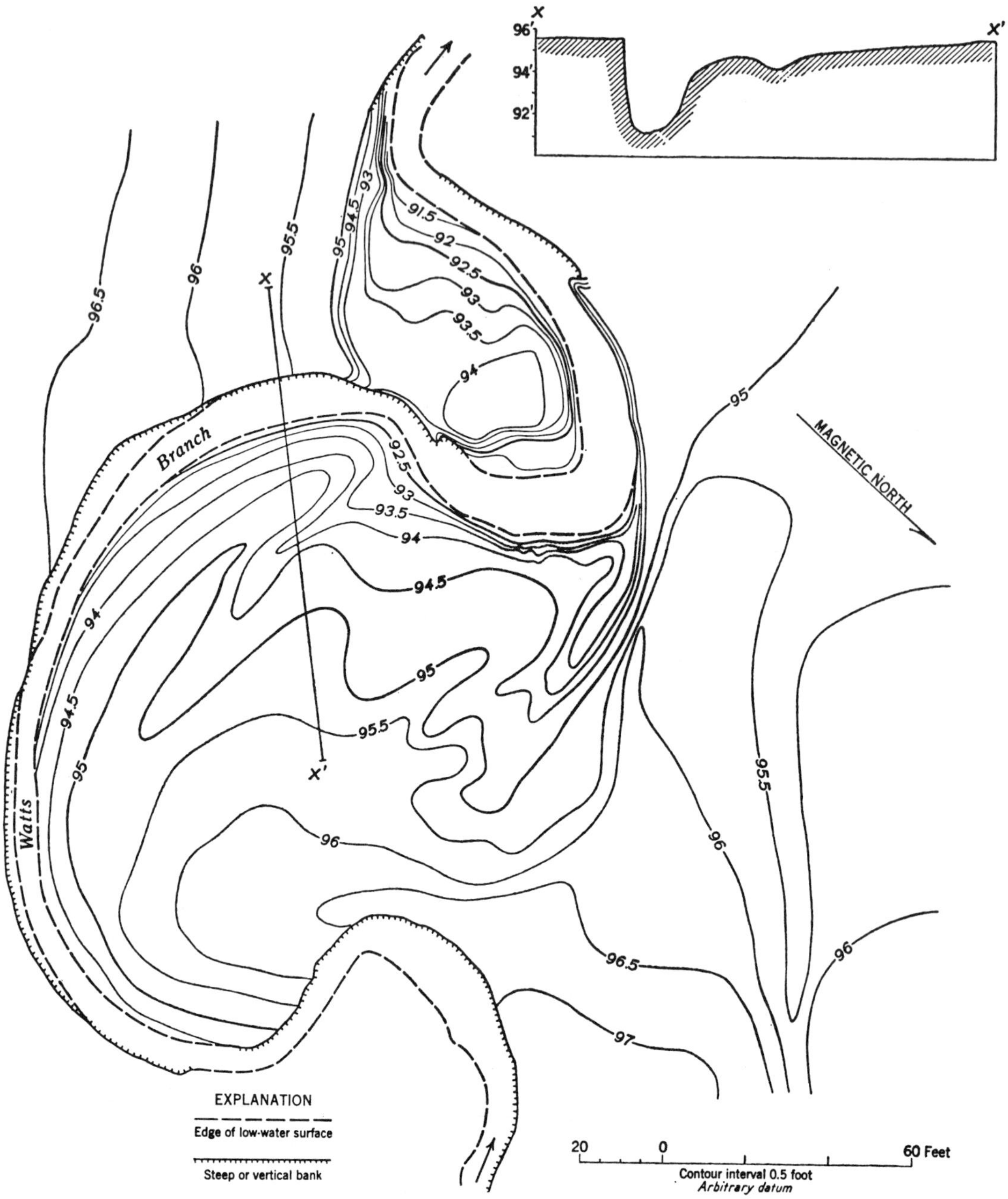

FIGURE 7.—MAP AND CROSS SECTION OF A TYPICAL POINT BAR OF THE FLOOD PLAIN OF WATTS BRANCH, 1 MILE NORTHWEST OF ROCKVILLE, MARYLAND
After Wolman and Leopold (1957)

An unusual opportunity to verify this reasoning was provided by a series of cross sections made in a meandering reach of Watts Branch near Rockville, Maryland. During 6 years of observation the stream has moved laterally a distance equal to one channel width. At the end of this period of observation a trench was dug in the point bar normal to the channel. The stratigraphy revealed in this trench superimposed on known surface positions is sketched in Figure 6. This trench was located on the right side of the stream on section x-x′ of Figure 7.

Approximate contact surfaces between materials of different textures are more or less parallel to past surface profiles. Coarse material characteristic of the bed is observed in a position well above the channel bed. In addition, there is no doubt that deposition occurred up to the level of the water surface at bankfull stage. In the same period deposition over the flat flood plain by overbank flow has been too small to measure. These observations appear to confirm the authors' (Wolman and Leopoldt 1957) hypothesis that point-bar building i, the primary process of flood-plain developmens in flood plains of this type.

Bank erosion is greatly influenced by wetting of the bank materials. Arroyos cut in fine-grained alluvium experience most bank cutting after, not during, flow. The wetting causes later slumping (Leopold and Miller, 1956, p. 5). Bank erosion is also enhanced by return seepage of water which infiltrated the banks during high flow. Upon lowering the stage the balancing pressure of the water in the channel is released, and the banks slump or collapse (Inglis, 1949, pt. 1, p. 152). A study of bank cutting in Watts Branch near Rockville, Maryland, showed that a combination of bank wetting and ice-crystal formation promoted the greatest bank erosion (Wolman, 1959, p. 214). Although the largest discharges occurred in summer, the winter provided more thorough soil wetting which, in combination with freeze and thaw, led to maximum bank erosion.

As bank erosion occurs in the bend of a meander, over a period of time it is usual for an approximately equal amount of deposition to occur on the opposite bank. This general equality of deposition and erosion is the reason width and cross-sectional area remain about the same as the channel moves laterally across the flood plain.

Meander Mechanics and Physiographic Problems

General Discussion

The vagaries of nature provide endless opportunities for perturbations in the flow—local bank erosion, chance emplacement of a boulder, fallen trees, or blocks of other vegetation—any one of which would alter the path of a straight channel. Thus one need hardly inquire why a stream channel is not straight. On the other hand, a random succession of chance perturbations might be expected merely to result in random bends of different patterns. Although this situation describes many channels, the existence of beautifully symmetrical meander bends and the remarkable similarity of bends in rivers of different sizes and physiographic settings must be explained. There is not yet available any theory or dynamical principle which explains qualitatively the characteristic patterns common to meandering channels. In the absence of such a general principle, however, attempts have been made to explain at least qualitatively how symmetrical successive bends begin and grow, and how size of bends is related to the size of the river.

Initiation and Development of Meanders

A large body of experience and literature on river regulation has been built up by European engineers on rivers in Europe, India, and Africa. This experience has covered a far greater span of time and of field conditions than has American practice. Regarding meanders, Leliavsky recently summarized the salient concepts developed (1955, esp. p. 111–141). Apparently the consensus of these workers is that the effect of helical flow is the dominant factor. Leliavsky expresses it (p. 128) as follows:

> "For some reason or other, a small abrasion . . . in the . . . bank of a straight channel is supposed to have taken place. The water particles moving alongside the eroded portion of the bank follow a curved trajectory and develop, consequently, a centrifugal force. This force, in turn, gives rise to a local helicoidal current, which intensifies the original abrasion and works its way deeper and farther into the shore, . . . until the whole channel becomes finally involved in the process and the eroded material accumulates on the opposite bank. This, then, is the birth of a meander."

Prus-Chacinski (1954) also argues that helical flow is the basic mechanism leading to meandering. He showed, further, that, by introducing an artificial secondary flow at the entry to the first bend, it is possible to produce various kinds of secondary circulations in the next successive bend which, in turn, affect the circulation in the next bend, and so on. He demonstrates that the downstream effects of a given circulation pattern are quite persistent, often through several successive bends. The "cause" of meandering Prus-Chacinski ascribes to any disturbance which produces an initial secondary circulation.

It seems clear that helical flow may play an important role in the process of deposition on a point bar. A building point bar helps concentrate shear against the concave bank and thus promotes bank caving and channel movement. Even denying that helical flow exists in wide rivers, Matthes' concept of meander development is closely allied with that just mentioned.

Matthes (1941, p. 633) indicates that bank cutting and orderly transfer of sediment to its place of deposition on point bars is a principal requirement for meandering. He observed that material tends to be deposited on point bars on the same side of the channel as that of the concave bank from which it was eroded. Thus, by bank erosion of a concave bank and concurrent deposition on a point bar across the channel, the channel will move laterally and downstream. Thus, the meander wave may migrate downstream generally maintaining its configuration if the materials are uniform. Friedkin's concept of the process (1945, p. 4) is essentially similar to that of Matthes (1941).

Only a few observers have studied this process of meander formation and development in laboratory channels. Quraishy (1944) described the development of a sinuous channel from an initially straight one in a sand-bed flume. Of particular interest was the formation of a series of small dunelike ripples of sand which he called "skew shoals." These developed on the initially flat bed after the sand grains had been in motion for a short time. These skew shoals were regularly spaced and alternated in position along the channel sides. After their full development they obstructed the path of water and forced it to assume a sinuous course.

This kind of mechanism is highly suggestive because it produced alternating bends, spaced uniformly along an initially straight channel at distances depending on the spacing of depositional barlike features on the bed. Although the Quraishy experiments do not indicate how the shear force of flowing water interacts quantitatively with bed debris to govern size and spacing of the dunelike features, it implies that the geometry of meander waves might be a function of bed forms.

The coincidence of the spacing of pools and riffles in straight reaches and meander lengths in rivers of comparable size might well be the result of a principle of bar or dune formation, as yet unknown, allied in some manner to skew shoal formation. Wolman and Brush (in press) found, however, that in noncohesive sands similar skew shoals formed only at supercritical flow (Froude number, based on mean depth and velocity, greater than 1.0). Although their wide shallow channels did not meander, Friedkin (1945, p. 4) produced a meandering channel from an initially straight one and concluded that bank erosion alone was sufficient to initiate meandering as long as channel widening and shallowing did not proceed too rapidly. In the study by Wolman and Brush (in press) helical flow was observed in the channel bends produced from the skew shoals.

Werner (1951) among others[2] has attempted to develop a general equation to describe meander formation and geometry. He (1951, p. 899) has suggested that the initiation of meanders is caused by "some local impulse or disturbance" which creates a transverse oscillation in the straight stream channel. Expressing velocity in terms of roughness, slope, and depth, as given by the Manning equation (p. 900), he derives an equation in which meander length is a function of initial width, channel roughness, slope, depth, and a coefficient proportional to sediment load.

An assumption is made that sediment prolongs the period of oscillation, inasmuch as any mechanism relying principally on inertial oscillations of the water within the channel width (seiche effects) gives wave periods much too short to account for meanders. The equation agrees qualitatively with respect to the interrelationships of some parameters. It indicates, for example, that meandering will be inhibited at high slopes, a conclusion which appears to be borne out in natural rivres (Leopold and Wolman, 1957, Fig. 46). On the other hand, the equation does not appear to describe quantitatively the observed relation of meander length to channel width. Further, if sediment load is not a necessary condition of meandering, the equation is invalid, for, when the load is zero, meander length is zero, and hence meandering should not occur.

Hjülstrom (1942, p. 252; 1949, p. 84) presented an expression for meander length based on seiche theory. He computed wave length as a product of seiche period and wave celerity, period being that of a seiche having a fetch equal to the width of the meander belt and water depth equal to that of the river. Wave celerity he considered to be the mean down-valley velocity of the river. The wave length so determined was also made a function of a coefficient of turbulent friction.

The resulting expression provided Hjülstrom a basis for reasoning about the effects of changing discharge on wave length through the interacting effects of water depth, velocity,

[2] The Coriolis force or the lateral deflection due to the rotation of the earth has frequently been cited as the cause of meandering, but calculations by Ludin (1926) and others indicate that the virtual radius of curvature attributable to the Coriolis force in a stream with a velocity of 3 feet per second would be on the order of 8 miles at a latitude of 60°.

and turbulence, and his argument showed keen insight and understanding of field conditions (1949, p. 86–88). But the formula has the disadvantage of making meander length dependent on wave amplitude; in our opinion, measurement data do not demonstrate this.

Although a simple comprehensive expression is still wanting, it appears that the forces determined by the velocity distribution, including the helical circulation, are all that is necessary to account for (1) the shape of the cross section in a meander, (2) the depositional and erosional pattern, and (3) the progressive down-valley migration of the meander. These observations are not new, but it is important to emphasize the following idea. Although point-bar formation and associated erosion of the opposite bank are necessary if a straight channel is to develop curves, the concept of helical flow, as Leliavsky (1955, p. 128) recognized, does not seem to explain how the secondary circulation determines the characteristic dimensions or proportions of meandering channels. The existence of meanders on glacier ice also implies that erosion and deposition may be a collateral, not the governing principle of meander development and movement. Because the hydraulic or mechanical significance of the pattern of curvature is closely tied to the fundamental physiographic questions, we consider this aspect of meander mechanics in the following section.

Problem of Channel Equilibrium

The preceding discussion was concerned principally with the initiation and development of a meander, elements that can be discussed in terms of a short reach of river or a single meander wave. There are a host of broader problems of channel adjustment to external controls which might be thought of as physiographic problems for want of a more specific term.

To begin, one might ask how meandering of a channel relates to the fundamental process of stream adjustment and stream equilibrium. It is generally believed that channel equilibrium is constantly approached, although rarely attained, by a process of continual adjustment. To use the words of Rubey (1952, p. 129),

> ". . . . with changing conditions, the stream is constantly cutting or filling and modifying its slope, velocity and cross section so as eventually to accomplish the imposed work with the least expenditure of energy."

If, indeed, a principle of least work is involved—for it is not yet proven—how does the development of a meandering pattern help accomplish this objective? Is the reduction of gradient achieved by increasing length relative to a straight channel necessary because of an excess of energy? Or, as was once believed, is meandering the aimless wandering of a channel too sluggish to accomplish any work of erosion? Does a river reach a stage at which vertical erosion is negligible and thence expend its excess energy on lateral erosion by meandering?

Under what changes of conditions would a river change its pattern from meandering to nonmeandering or vice versa? What would be the effect of an increase or decrease in discharge or in sediment load from the drainage basin?

Opinions on some of these questions have been published, but data or measurements are meager. Existing data may answer a particular question but do not explain why the observed result was obtained. In the following paragraphs some of these questions are considered along with related observations from the literature. Results of recent work, where applicable, are cited, and we suggest what appears to us some of the directions in which future work is needed.

Many of these queries relate to a single, fundamental question—what has the meandering pattern to do with energy expenditure? Inglis (1949, p. 158) states that,

> "Meandering is Nature's way of damping out excess energy during a wide range of varying flow conditions, the pattern depending on the grade of material, the relation between discharge and charge (load) and the rate of change of discharge and charge."

Schoklitsch (1937, p. 149) earlier stated what appears to be the same idea, that meander formation

> "might be due to the fact that the slope in such stretches is too great and is not in equilibrium with the size of the bed-sediment grains."

Water-surface slope of a river is a measure of energy expenditure. Whereas the equilibrium slope is closely related to the size of particles on the bed, there are other factors involved. Hack (1957, p. 61) confirms quantitatively the generally held belief that channel slope is controlled to a great extent by bed grain size and discharge, but to what extent channel shape enters is still unclear. As Rubey explains

(1952, p. 131), channel shape may adjust jointly with slope. At constant discharge if the principal effect of bed particle size is on channel roughness or frictional drag, then other forms of drag need to be considered. Total channel resistance is materially influenced by form drag of various kinds, pools and riffles, bars and dunes, and channel curvature.

There is reason from the hydraulic standpoint to believe that meandering may in part be a function of frictional drag and thence energy loss. More energy loss occurs per unit of length in a curved than in a straight channel of the same depth and cross section, owing to eddying, secondary circulation, or increased rate of shear. These eddy losses result from deflection of the water to a new direction as it moves around a curve or bulge. It is known from hydraulic experiments in pipes (King, 1954, Fig. 87) that energy loss first decreases and then increases with a decrease in the ratio radius of curvature/pipe diameter. Although any curved channel offers greater resistance to flow than a straight one, the minimum increase in resistance is about 40 per cent, and this applies within a narrow range of the ratio of radius of curvature/diameter when that ratio has a value of 2 to 3.

It was pointed out earlier that in meandering channels the comparable ratio, radius of curvature/channel width, is relatively conservative. Values of this ratio also tend to fall in the range of 2 to 3.

Bagnold (in press) suggests an explanation of this minimum increase in resistance at this value. He postulates that, as radius of curvature is reduced and becomes about 2 to 3 times the channel width, an eddy or zone of reverse flow forms just downstream from the bulge or convex bank. With the appearance of the reverse eddy the local width is constricted, and there is a local increase in effective radius of curvature and a net decrease in energy loss.

To the extent that further work confirms the indication that the modal value of this ratio is in the range 2–3, meanders tend to be characterized by a geometric pattern which happens to offer the smallest energy loss of any configuration of curved channel. The significance of this observation is unknown, but it suggests that some principle related to energy conservation does operate in the meander mechanism.

Assuming that a bend actually does tend to develop a configuration such that the energy loss due to the bend is a minimum, it does not follow that the total energy expenditure has been minimized. When a given discharge falls through some specified vertical distance, a certain amount of energy is transferred from potential form to some other form. If the water does not accelerate (velocity remains about constant downstream), then this potential energy is expended as work or heat. The same amount of energy is spent whether the water moves in a straight channel on a steep slope or in a longer curved channel at a smaller gradient. The energy expenditure per foot of channel length is smaller in the longer curved channel than in the shorter straight one. The question, then, is how much energy is utilized per foot of channel length and in what form it is used.

This energy may be spent in moving particles of debris, or it is otherwise dissipated into heat. The energy may be spent in removing particles from the bank and transporting them (bank erosion) or in transporting bed or suspended particles. If this energy is concentrated in such a manner that more of it goes into moving particles from one place than from another, then local scour will occur there and deposition elsewhere. For a channel to be in equilibrium scour must balance fill within the reach in question, and, further, the energy must be so expended that the net amount of debris coming into the reach must equal the net amount carried out of the reach.

It is generally agreed that meandering channels are often stable or in quasi-equilibrium. They may be so even though, over a period of time, a meander wave moves gradually downstream. The slope, discharge, and channel shape tend to become adjusted so that the above requirements are fulfilled. Adjustments in channel shape occur through erosion or deposition which in turn affect velocity, depth, and width. Specific hydraulic requirements relating depth, slope, velocity, and total resistance, including resistance offered by bed configuration (form resistance), bed and bank grains (skin resistance), and channel curvature (a particular kind of form resistance) must also be maintained.

Meandering is one way in which erosion and deposition may change the distribution, location, and amount of energy expenditure per unit of channel boundary. By lengthening the channel between two points at different elevations, the energy expenditure per foot of length is reduced. By bank erosion, point-bar building, and by scour and fill, the channel cross section is adjusted, and the energy expenditure is redistributed. Presumably abrupt discontinuities in the rate of energy expenditure in a reach of channel are less compatible with conditions of equilibrium than is a more or less continuous or uniform rate of energy loss. It may well be that a meandering channel is most stable when the energy loss due to curvature is at a minimum. Such a conclusion is perhaps implied in the modal distribution of values of the ratio of radius of curvature to width in natural streams.

If we view curvature as simply one method of altering the distribution of energy expenditure in a given length of channel, it is clear that the pattern of meandering will respond to changes in discharge and load. It is well known that an increase in discharge in a meandering channel will increase the channel width and will increase the size of the meander bends. Decrease in discharge will gradually reverse the process.

In the natural rivers, geologic and stratigraphic evidence clearly demonstrates that during late Pleistocene time an increasing discharge markedly decreased the gradient of Öster-Dal River and enlarged the width of channel as well as the size of meander beds (Wenner and Lannerbro, 1952, p. 108). During this gradual degradation of the valley floor the meandering pattern persisted. Thus the meander pattern where it exists in nature appears to be a persistent attribute of the river.

Change in load will cause aggradation or degradation and thus change in channel slope and size of bend (Friedkin, 1945, p. 7–9); an increase in slope will produce an increase in meander length and amplitude. Schoklitsch (Shulits, 1935, p. 644–646) and Bagnold (1960) have postulated that at high discharges sediment transport is a function of the rate of work done per foot of channel length, or power intensity. In a meandering channel in equilibrium increasing tightness of bend (curvature) through its effect on the rate of energy expenditure decreases the rate of transport. For equilibrium, then, a balance must be maintained between curvature and transport quite apart from any change in intensity of energy loss brought about by a change in length.

Direction for Future Work

No wholly adequate explanation of meandering is yet available. Probably no single simple mechanism will suffice to explain all aspects of meandering. Although there is general agreement on the manner in which bank erosion and point-bar formation are related to the orderly transfer of sediment which is basic to meandering, no physical or mechanical principle has been identified which explains qualitatively the size and geometry of meander curves.

There is need to investigate oscillatory forces which might explain more adequately the manner in which an initial bulge or depression in a stream bank leads to a symmetrical reversal of curvature.

Although the available velocity distributions in channel bends do permit general descriptions of the flow, many characteristics, including orientation and position of the helical patterns, are as yet poorly defined. More detailed measurements in natural channels are required to define the loci of energy losses and their relation to flow resistance and localized erosion of bed and banks.

The way in which a natural channel distributes the energy loss as between boundary friction, form resistance, curvature, and transport is little understood. Without such understanding it is virtually impossible to explain or predict the behavior of a meandering channel. Although some studies of energy losses in curved channels have been made (Allen, 1939; Leopold *et al.*, in press), it appears to us that laboratory as well as field studies on the distribution of energy expenditure in straight and curved alluvial channels are needed. It would be particularly desirable to map the distribution and magnitude of boundary shear in bends of different patterns but with similar cross section and depth. These observations must in turn be related to the mechanics of sediment transport in meandering channels of different patterns.

The principal unsolved problem with regard to the pattern of flow and its relation to erosion and deposition is in the area of the mechanics of sediment transport. Present theories are inadequate to explain the transport of heterogeneous sizes under the variety of conditions found in nature. The stress needed to induce and maintain motion is probably different for scattered rocks on a sand bed, uniform grading from sand to cobbles, graded cobbles and boulders without sand, or a few isolated boulders on a cobble bed. At the same time, objective criteria are needed to describe the ability of varying bank materials to withstand erosive stresses. A quantitative explanation of the meandering process will require balancing the erosive stress produced by tangential shear of the flowing water and the comparable resisting stress provided by the bank material.

It is particularly important that critical field observations be tied to theoretical and laboratory studies. A few fundamental concepts can unify a vast amount of empirical observation if the observations include those critical measurements which make possible an adequate test of theory.

BIBLIOGRAPHY

Allen, Jack, 1939, The flow in a tortuous stretch of a river and in a scale model of the same: Jour. Inst. Civil Engineers, v. 11, p. 115–132

Bagnold, R. A., 1960, Sediment discharge and stream power—a preliminary announcement: U. S. Geol. Survey Circ. 421, 15 p.

—— in press, Some aspects of the shape of river meanders: U. S. Geol. Survey Prof. Paper 282

Bates, R. E., 1939, Geomorphic history of the Kickapoo region, Wisconsin: Geol. Soc. America Bull., v. 50, p. 819–879

Blue, F. L., Jr., Herbert, J. K., and Lancefield, R. L., 1934, Flow around a river bend investigated: Civil Eng., v. 4, p. 258–260

Dury, G. H., 1958, Tests of a general theory of misfit streams: Inst. of British Geographers Trans., pub. 25, p. 105–118

Eakin, H. M., 1935, Diversity of current-direction and load-distribution on stream-bends: Am. Geophys. Union Trans., pt. II, p. 467–472

Fargue, L., 1908, La forme du lit des riviéres á fond mobile: Gauthier-Villars, Paris, 187 p.

Friedkin, J. F., 1945, A laboratory study of the meandering of alluvial rivers: U. S. Waterways Expt. Sta., 40 p.

Fuglister, F. C., 1955, Alternative analyses of current surveys: Deep-Sea Research, v. 2, p. 213–229

Hack, J. T., 1957, Studies of longitudinal stream profiles in Virginia and Maryland: U. S. Geol. Survey Prof. Paper 294–B, p. 45–97

Hjulstrom, F., 1942, Studien über das mäanderproblem: Geog. Annaler, B. XXIV, p. 233–269

—— 1949, Climatic changes and river patterns, *in* Glaciers and climate: Geog.Annaler, h. 1–2, p. 83–89

Inglis, C. C., 1937, The relationships between meander belts, distance between meanders on axis of stream, width, and discharge of rivers in flood plains and incised rivers: Ann. Rept. (Tech.), Central Board of Irrigation (India) 1938–39, p. 49

—— 1949, The behavior and control of rivers and canals: Central Waterpower Irrigation and Navigation Res. Sta., Poona, Res. Pub. 13, v. 1 and 2, 486 p.

Jefferson, M., 1902, Limiting width of meander belts: Nat. Geog. Mag., v. 13, p. 373–384

King, H. W., 1954, Handbook of hydraulics, 4th ed.: New York, McGraw-Hill Book Co., Inc., 400 p.

Lane, E. W., and Borland, W. M., 1954, River-bed scour during floods: Am. Soc. Civil Engineers Trans., v. 119, p. 1069–1079

Laursen, E. M., and Toch, A., 1954, Discussion of Lane and Borland, River-bed scour during floods: Am. Soc. Civil Engineers Trans., v. 119, p. 1084–1087

Leliavsky, Serge, 1955, An introduction to fluvial hydraulics: London, Constable and Co., 257 p.

Leopold, L. B., in press, The gravel bar: some observations on its role in channel morphology: Internat. Geol. Cong., Copenhagen

Leopold, L. B., and Miller, J. P., 1956, Ephemeral streams, hydraulic factors and their relation to the drainage net: U. S. Geol. Survey Prof. Paper 282A, p. 1–37

Leopold, L. B., and Wolman, M. G., 1957, River channel patterns—braided, meandering, and straight: U. S. Geol. Survey Prof. Paper 282B, p. 39–85

Leopold, L. B., Bagnold, R. A., Wolman, M. G., and Brush, L. M. Jr., in press, Flow resistance in sinuous and irregular channels: U. S. Geol. Survey Prof. Paper 282

Ludin, A., 1926, Influence of the rotation of the earth on rivers: Die Wasserkraft, v. 21, p. 216, partial translation by Sam Shulits, 1958

Matthes, G. H., 1941, Basic aspects of stream meanders: Am. Geophys. Union Trans., v. 22, p. 632–636

Mockmore, C. A., 1944, Flow around bends in stable channels: Am. Soc. Civil Engineers Trans., v. 109, p. 593–628

Nemenyi, P. F., 1946, Discussion of Vanoni, V., Transportation of suspended sediment by water: Am. Soc. Civil Engineers Trans., v. 111, p. 116–125

Parsons, D. A., 1959, Observations of flood flow effects on channel boundaries: U. S. Dept. Agriculture, Agric. Research Service, mimeographed

Prus-Chacinski, T. M., 1954, Patterns of motion in open-channel bends: Assoc. Internat. d'Hydrologie, pub. 38, v. 3, p. 311–318

Quraishy, M. S., 1944, The origin of curves in rivers: Current Sci., v. 13, p. 36–39

Rossby, C. G., 1941, The scientific basis of meteorology: U. S. Dept. Agriculture Yearbook, p. 599–656

Rubey, W. W., 1952, Geology and mineral resources of the Hardin and Brussels quadrangles (in Illinois): U. S. Geol. Survey Prof. Paper 218, 175 p.

Schoklitsch, Armin, 1937, Hydraulic structures, v. 1: Am. Soc. Mech. Engineers, 504 p.

Shulits, Sam, 1935, The Schoklitsch bed-load formula: The Engineer, v. 139, p. 644–646, 687

Stommel, H., 1954, Circulation in the North Atlantic Ocean: Nature, v. 173, p. 886–893

Straub, L. G., 1942, Mechanics of rivers, p. 614–636 *in* Meinzer, O. E., Hydrology: New York, Dover Publications, Inc., 712 p.

Thomsen, J., 1879, On the origin of windings of rivers in alluvial plains: Royal Soc. London Proc., v. 25, p. 5–6

Van Til, K., and Tops, J. W., 1953, Results of measurements in some bends of the Netherlands Rhine branches: 18th Internat. Navigation Cong., Sec. I, Ques. 1d, p. 15–25

Von Arx, W. S., 1952, Notes on the surface velocity profile and horizontal shear across the width of the Gulf Stream: Tellus, v. 4, p. 211–214

Wattendorf, F. L., 1935, A study of the effect of curvature on fully developed turbulent flow: Royal Soc. London Proc., ser. A, v. 148, p. 565–598

Wenner, C. G., and Lannerbro, R., 1952, The meander field at Mora against the background of the geology of the Siljan basin and the history of the Mora district: Ymer., h. 2, p. 81–109

Werner, P. W., 1951, On the origin of river meanders: Am. Geophys. Union Trans., v. 32, p. 898–902

Wolman, M. G., 1959, Factors influencing erosion of a cohesive river bank: Am. Jour. Sci., v. 257, p. 204–216

Wolman, W. G., and Brush, L. M., Jr., in press, Laboratory study of equilibrium channels in non-cohesive sands: U. S. Geol. Survey Prof. Paper 282

Wolman, W. G., and Leopold, L. B., 1957, River flood-plains—some observations on their formation: U. S. Geol. Survey Prof. Paper 282C, p. 87–109

Worthington, L. V., 1954, Three detailed cross sections of the Gulf Stream: Tellus, v. 6, p. 116–123

U. S. Geological Survey, Washington, D. C.; The Johns Hopkins University, Baltimore, Md.

Manuscript Received by the Secretary of the Society, October 29, 1959

Publication Authorized by the Director, U. S. Geological Survey

Reprinted by permission of the Society of Economic Paleontologists and Mineralogists. Published in SEPM Special Publications 2 (November 1951), pp. 76-107.

SEDIMENTARY HISTORY OF THE VENTURA BASIN, CALIFORNIA, AND THE ACTION OF TURBIDITY CURRENTS

M. L. NATLAND AND PH. H. KUENEN
Richfield Oil Corporation and Geologisch Instituut, Groningen, Netherlands

ABSTRACT

Foraminiferal evidence shows that the water depth of the Ventura Basin was several thousands of feet during most of its upper Tertiary sedimentary history. The coarse sandstones between the shales were deposited by turbidity currents of high density, and the conglomerates were emplaced by highly mobile submarine slides.

INTRODUCTION

One of the present authors (Natland) has, for many years, been occupied with studying the geology of the Ventura Basin in southern California. Among others, he compared the benthonic Foraminifera of thousands of systematically collected Pliocene and Pleistocene outcrop samples to those from Recent bottom samples taken off the west coast of North America, mainly off southern California. This enabled him to demonstrate that the depth of the ancient sea floor was about 5,000 feet during the lower Pliocene, and has gradually decreased until terrestrial deposition prevailed toward the end of the Lower Pleistocene. During this time, nearly 20,000 feet of sediment had been deposited on the gradually subsiding substratum. The evidence has been summarized elsewhere and will be published in detail in the near future.

However, because of the extensive interstratification of coarse sands and conglomerates with the foraminiferal shales, Natland long questioned the validity of his own results until the accumulated evidence became overwhelming. There are geologists who still continue to place more value on these coarse deposits, as pointers to shallow water, than on the faunal evidence for great depths collected by Natland (1933).

Dr. Ian Campbell, of the California Institute of Technology, Pasadena, California, kindly pointed out in 1948 that here was apparently conflicting evidence that might be reconciled on the basis of the turbidity current mechanism as advocated in a general way by Kuenen at the International Geological Congress in London that year.

The present authors decided to view some of the field evidence together and, accordingly, made two short field trips to the Ventura Basin during the early summer of 1950.

Description of Localities Examined

The six localities especially studied for this report are shown on figure 1. A brief description of them follows:

Locality 1.—At the top of the marine Pleistocene in Hall Canyon. It is composed of well-sorted, clean, cross-bedded, beach sand with abundant megafossils of a beach-dwelling type.

Locality 2.—Upper Pliocene in Hall Canyon. Composed of alternating sands, shales, and coarse conglomerates, with a few thin beds of carbonized wood. Abundant Foraminifera (*Uvigerina peregrina, Pseudoparrella pacifica, Bolivina spissa*) indicate a depositional environment greater than 1,000 feet and shallower than 2,500 feet.

Locality 3.—Alternating thin beds of upper Pliocene sands and shale, exposed on the west bank of the Ventura River, about 600 feet south of the mouth of Diablo Canyon. The beds studied occur immediately above and below the six-inch, white tuff bed composed of ash and

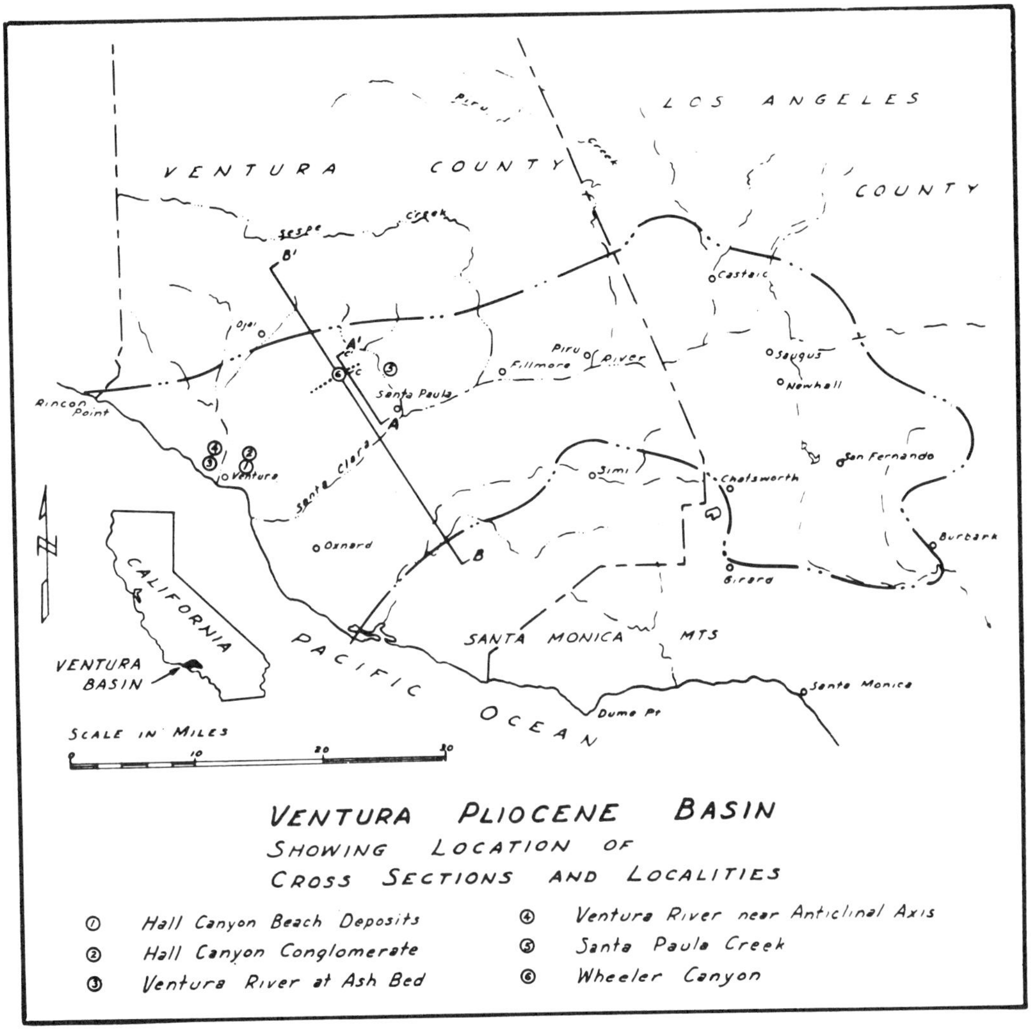

FIG. 1.—Ventura Pliocene Basin.

glass shards. These beds are approximately 200 feet stratigraphically lower than those at Locality 2. Abundant Foraminifera indicate that these beds were deposited in water deeper than 1,000 feet and shallower than 2,500 feet.

Locality 4.—Thick upper Pliocene sandstones exposed 1,000 feet south of the axis of the Ventura Anticline. Depth of water at the time of deposition, as indicated by Foraminifera, was from 1,500 to 2,500 feet.

Locality 5.—Hard sandstone reefs, in laminated middle Pliocene shale, exposed in bottom of Santa Paula Creek at its juncture with Mud Creek. Indicated environmental depth of accumulation, 2,500 to 4,000 feet.

Locality 6.—Composed of 300 feet of coarse, lenticular, lower Pliocene conglomerate beds, with interbedded siltstone and ripple-marked sandstone. These beds crop out at the head of the west fork of Wheeler Canyon. Foraminifera indicate that the conglomerates accumulated in water deeper than 4,000 feet.

At the above localities, evidence was found indicating that the coarse sands had been deposited by turbidity cur-

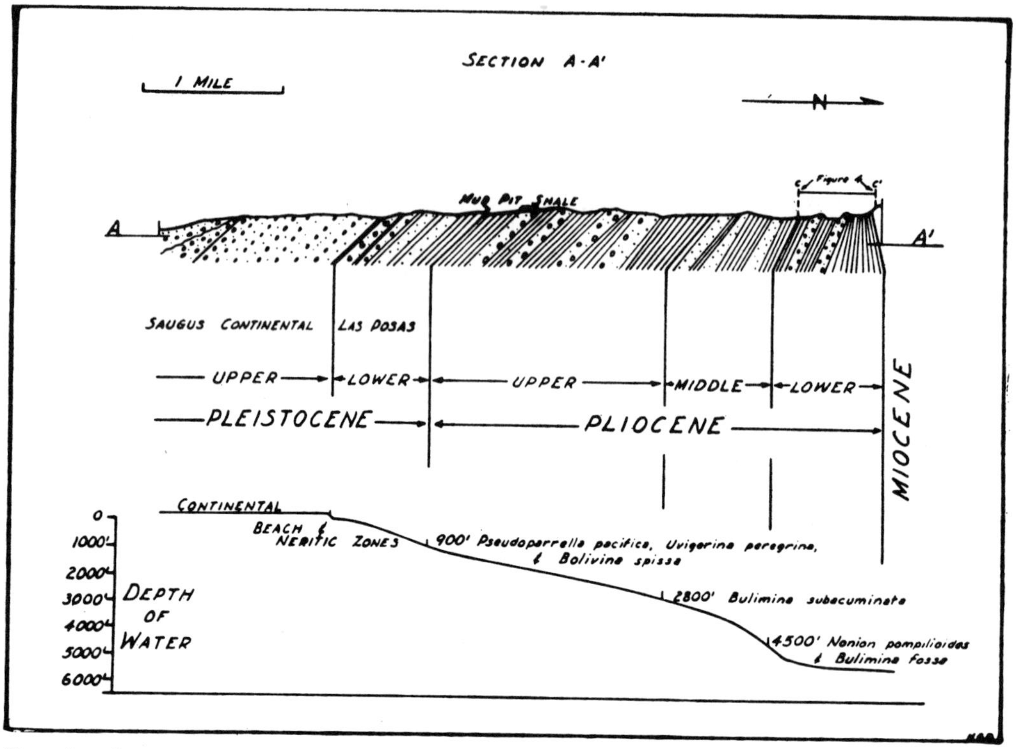

FIG. 2.—Cross section of Ventura Pliocene-Pleistocene, the location of which is shown on figure 1. Probable depth of water during deposition of beds is indicated below.

rents, whereas the conglomerates appear to have resulted from submarine sliding or highly-mobile flows. In a following section, the paleogeography formerly deduced by Natland will be briefly outlined as a setting for the action of the turbidity currents and slides. The evidence for the action of these will be treated in some detail.

ACKNOWLEDGMENTS

The writers wish to thank the staffs of the sedimentary laboratories of the Hancock Foundation at the University of Southern California, Los Angeles, California, and the Scripps Institution of Oceanography, at La Jolla, California, for kindly providing the mechanical analyses of the samples of graded sandstones. The identification of the Pliocene plant remains, by Dr. D. I. Axelrod of the University of California at Los Angeles, California, is highly appreciated. Thanks are due Helen Duggan and Harold Paris, of Long Beach, California, for the drafting of the plates and for editorial help.

BIOSTRATIGRAPHY AND STRUCTURE OF THE VENTURA BASIN

General Statement

The biostratigraphy and structure of the Ventura Basin is extremely varied and complicated and, therefore, cannot be fully treated here. Consequently, only those features having a definite bearing on the Pliocene sedimentation problems at hand will be discussed. The location and extent of the Ventura Basin is shown on figure 1. This basin began to form as a part of a large downwarp which affected the western half of southern California at the close of Oligocene time. The sag continued in a regional way to the close of Miocene time, when, as a result of the uplifting of the Santa Ynez, Topa Topa, San Gabriel, and the Santa Monica Mountains, the large basin was

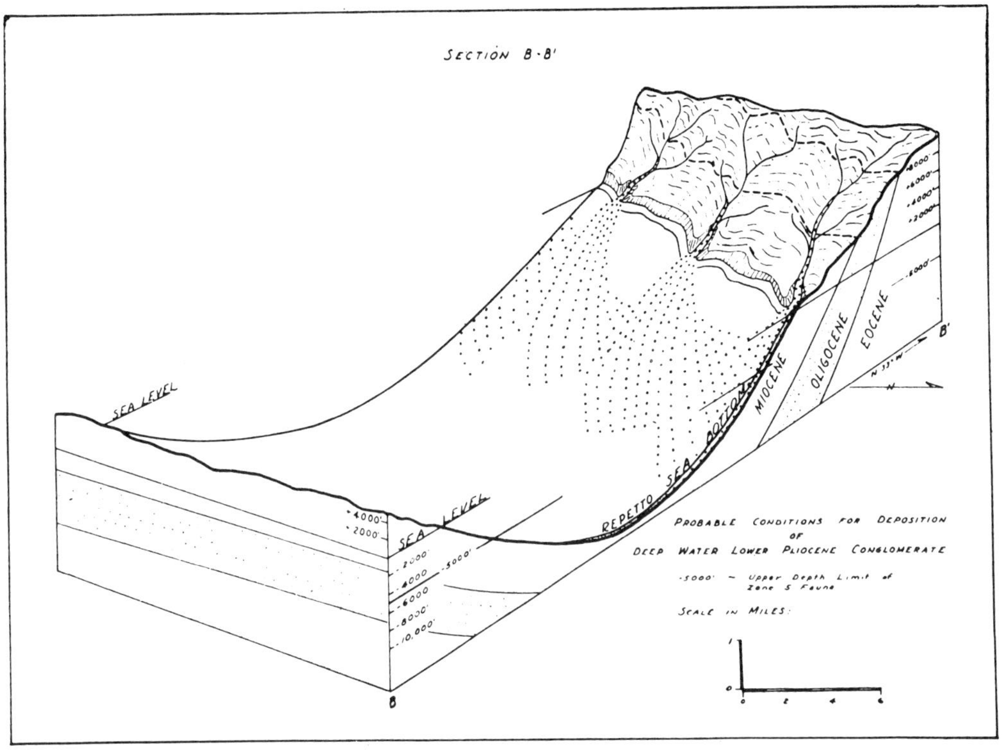

FIG. 3.—Block diagram illustrating the environmental conditions under which the lower Pliocene sediments accumulated.

separated into several smaller ones such as the Santa Maria, Los Angeles, and Ventura.

During lower Pliocene time, as indicated by faunal evidence (Natland, 1933), the Ventura Basin was covered by approximately 5,000 feet of marine water in the central portion. From lower Pliocene time to Recent, the basin continued to subside, but the rate of subsidence was gradually exceeded by the rate of sedimentation and, consequently, it shallowed and filled to its present configuration.

Eocene to Miocene

The section from Eocene to Miocene is approximately 26,000 feet thick (Reed and Hollister, 1936) and crops out along most of the northern side of the basin.

The Eocene is composed of about 10,000 feet of marine and continental indurated sandstone (Matillija, Coldwater) and shales (Cozydell) which hold up the higher elevations of the Santa Ynez and Topa Topa Mountains. Next about 6,000 feet of variegated Oligocene (Sespe) sandstones and shales of continental origin was deposited.

During Miocene time the area was again submerged and, by the close of this epoch, a widespread blanketing series of massive mudstones, sandstones, and siliceous shales—totaling 10,000 feet—had been deposited. As previously stated, the widespread Miocene basin was divided by uplifts into several smaller Pliocene basins, one of which was the Ventura Basin.

Ventura Basin Pliocene

In the downwarp which continued between the uplifting Santa Ynez, Topa Topa, and Santa Monica Mountain ranges, approximately 13,000 feet of beds were deposited during the Pliocene. Following is a description of these sediments:

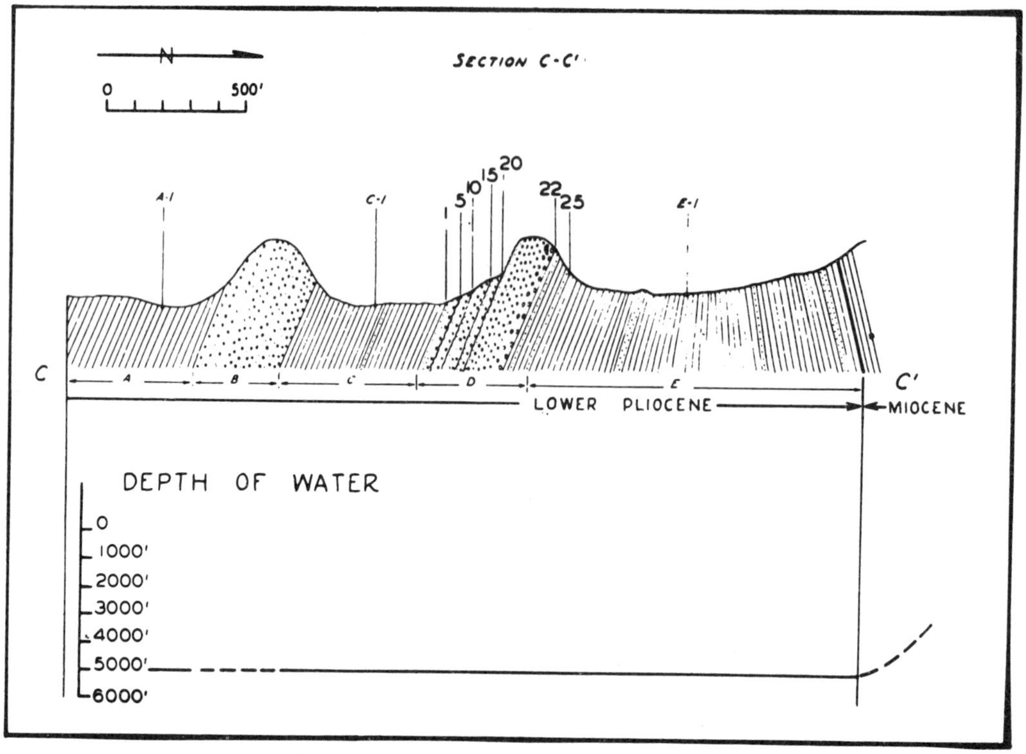

FIG. 4.—Cross section showing position of outcrop samples taken across the lowest of the lower Pliocene conglomerates. Depth of water during depositions, is indicated below.

Lower Pliocene (Repetto)

This member is 2,900 feet thick in Wheeler Canyon and is composed of siltstone, shale, sandstone, and conglomerate. The beds made up of finer sediments are traceable over most of the western part of the Ventura Basin. However, the sand and conglomerates are lenticular, especially the conglomerates, which are seldom traceable in outcrop for more than a mile. They thin rapidly toward the south in the subsurface. These conglomerate beds, in general, are confined to the northern slope of the basin (fig. 3), and their points of maximum thickness are probably related to the location of Pliocene river mouths. The lowest Pliocene conglomerate bed has been singled out for special study.

This conglomerate, at Locality 6 (figs. 1–5), is deserving of special discussion because it was deposited under conditions which, by present standards, are considered to be unusual. This bed is about 325 feet thick. It is very lenticular, grading easterly to sand and silt within a quarter of a mile. Toward the west it is more continuous but varies greatly in thickness and in percentage of coarse constituents. The lower half is composed of very coarse material, with boulders exceeding 2 feet in diameter (fig. 5). The upper half contains interbedded conglomerate and sand, with a few beds of ripple-marked sandstone and fine silt. These beds were sampled, and the Foraminifera contained therein were classified and plotted graphically on figure 6. The ranges of the species on this chart indicate no abrupt change of fauna, as one might expect had the basin shallowed to depths considered usual for the deposition of conglomerate. The same fauna extends from the shale below—through the conglomerate—to the shale above, without any appreciable change. However, there is considerable change in the

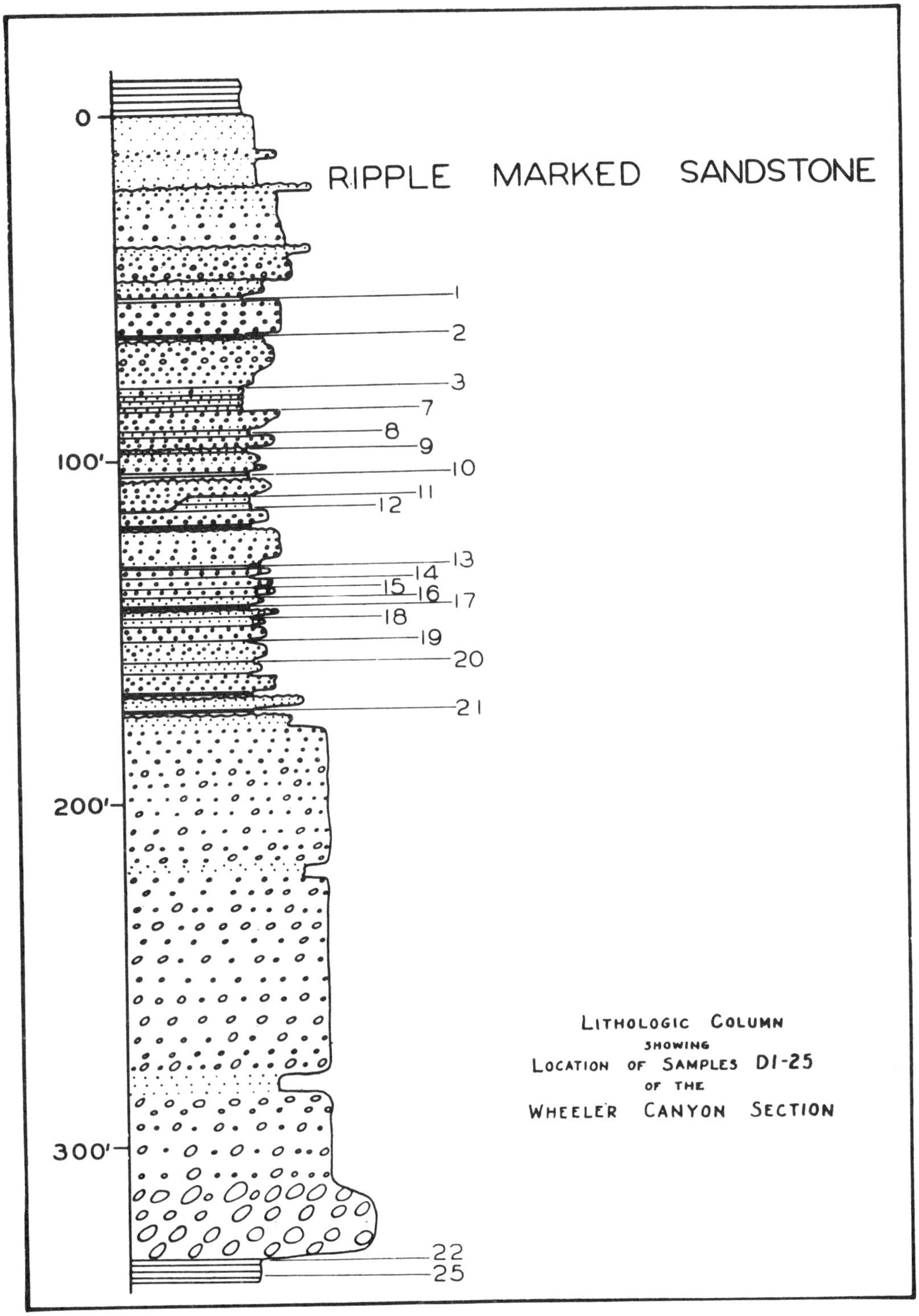

FIG. 5.—Columnar representation of the lithologic character of the lowest Pliocene conglomerate, showing location of samples described in figure 6.

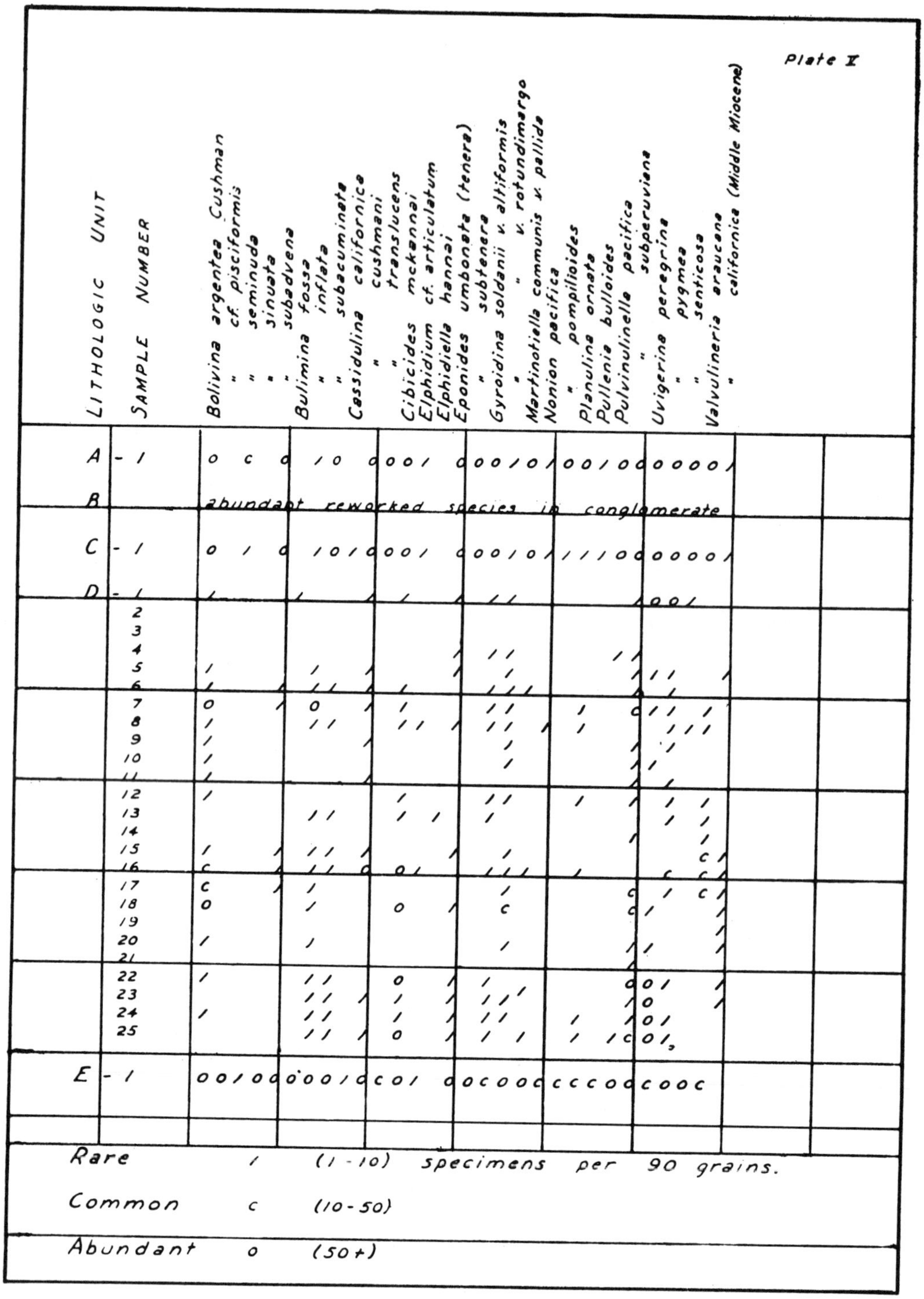

FIG. 6.—Stratigraphic ranges of foraminiferal species found in samples collected from the lower Pliocene, as shown on figures 4 and 5.

relative abundance and regularity with which the species occur through the conglomerate. This is only natural when one considers the relative rates of accumulation of shale and conglomerate. The *Elphidiums* listed on the chart are usually found in depths of water ranging from shore to 300 feet. Their presence here with deep-water species can be explained most satisfactorily by assuming that they were transported by turbidity flows from shallow to deep water.

The block diagram (fig. 3) illustrates the probable condition under which the lower Pliocene conglomerates were deposited. Note the exposed and eroding Miocene to Eocene beds along the uplifted northern border. Streams running down this elevated region carried sediments to the sea, where—if the concentration was high enough and the gradient of the bottom was sufficiently steep—they immediately continued their flow down the submarine slope. However, if the density of the suspended material was too low and the slope too gentle, only the silt and clay fractions were able to proceed immediately down the submarine slope, leaving the sand and coarse material to accumulate on the beach. Longshore currents then moved this coarse material to points where the currents were impeded by some promontory where it was deposited. When these deposits became sufficiently thick, or the angle of repose was exceeded by additional tilting, the sands then slid seaward in a massive turbidity flow.

A. Heim, R. Moser, and A. Bürkli-Ziegler (1888) described a landslide which may have been similar in action to those which produced the Ventura Basin lower Pliocene conglomerates. This landslide occurred in the year 1887 and carried part of the city of Zug, Switzerland, into Zug Lake. Fortunately, a very accurate contour map of the lake bottom had been made before the slide occurred. A similar survey of the lake bottom, made after the slide, revealed the configuration of the material transported far out into the lake. A. O. Woodford, of Pomona College, California, has published a review of the article describing the landslide at Zug (Woodford, 1951).

In an earlier publication (Natland, 1933) the ecology of the Ventura Pliocene has been discussed in some detail. However, as a result of much additional work on the bathymetric ranges of Recent Pacific Foraminifera, some revision in the depth limits of the zones now appears necessary. The bathymetric ranges of species in 150 samples collected off Panama—from shore to a depth of 4,140 meters—have revealed that the species *Bulimina fossa* (*rostrata*) and *Nonion pompilioides* (*umbilicatula*), characteristic of Recent Zone V and the lower Pliocene, begin to appear between 4,000 and 5,000 feet. In order to check these new data, a section was collected off La Jolla, California, in the spring of 1948. The course of this section was carefully chosen to avoid steep slopes where slumping might occur, and closed basins with shallow sills where abnormal thermal gradients exist. The shallowest ranges of the Zone V species in this section also began between 4,000 and 5,000 feet, which suggests that the lower Pliocene, characterized by these same deep-water species, was deposited in depths ranging downward from 4,000 to 5,000 feet, rather than from 6,500 feet downward as originally suggested. This discrepancy has undoubtedly resulted from the unfortunate use of samples, in the 1933 report, which were collected from closed basins where normal oceanic circulation does not exist. A colleague, Robert W. Crouch, is now studying the relationships between abnormal thermal gradients in these basins and the ranges of foraminiferal species. At present, his work points to temperature as being a dominant factor controlling the bathymetric ranges of the species, and that a temperature necessary to produce an assemblage similar to the lower Pliocene fauna is about 2.5° C. As Mr. Crouch will soon publish the results of his investigation, no further details will be given at this time.

If temperature is a dominant control,

one might argue that during lower Pliocene time the ocean was much colder, with 2.5° C. at the surface permitting species which now live below 4,000 feet to be found in shallow water and, therefore, the coarse sands and conglomerates in the lower Pliocene may have accumulated in an environment usually held appropriate for this type of deposit.

However, there is much evidence which strongly indicates that there has been little change in the thermal gradients of the offshore waters since lower Pliocene time. First, if the maximum ocean temperature was 2.5° C. in the shallow water, glacial conditions would have quite certainly existed on shore, and there are no recognizable glacial features to indicate that glacial conditions existed during the lower Pliocene.

Second, a collection of leaves was obtained from the lower Pliocene conglomerate, Sample Locality D 8, figure 5. These leaves, which occurred in a yellow-brown, rather soft sand, were identified by D. I. Axelrod, of the University of California at Los Angeles. He recognized species of avocado, cottonwood, live oak, a legume, and a sumac. In summarizing the climatic implications of this flora, he states: "From the evidence at hand it appears that members of oak woodland, thorn forest and chaparral vegetation were in the area. As gauged from similar living vegetation, the climate was semi-arid and probably somewhat like that now found in the foothills of central Sonora where rainfall ranges from 15 to 20 inches yearly, is of biseasonal distribution, and where frosts are rare or absent." This evidence clearly points to an onshore climate during Pliocene time little different from that of Ventura today, except that it was probably a little warmer and certainly not glacial. It is well known that ocean surface temperatures, to a large extent, control the onshore climate, and as the lower Pliocene climate is indicated by the flora to be about the same as it is today, it logically follows that the surface temperature of the sea was about the same. It is, therefore, reasonable to conclude that the marine thermal gradients were similar; consequently, we would expect a 2.5° C. temperature, necessary to produce the lower Pliocene fauna, to occur at depths between 4,000 and 5,000 feet.

Third, in the lower Pliocene, Lore R. David[1] has found scales of fish, the like of which today are known only from abyssal depths.

Fourth, W. P. Woodring (1938) found that the lower Pliocene (Repetto) molluscan fauna points to environmental depths of from 2,000 to 4,000 feet. This is not as deep as indicated by the Foraminifera. At present, no explanation is at hand to bring the two lines of evidence into closer conformity. However, the Foraminifera are vastly more abundant in the lower Pliocene than mollusks and, therefore, they perhaps may be more reliably used as depth indicators. In summary, it appears from several lines of evidence that the lower Pliocene sediments—including the conglomerates—were deposited in water deeper than 4,000 to 5,000 feet.

Middle Pliocene

The conformable group of beds usually referred to this division is lithologically indistinguishable from the lower Pliocene. In the Wheeler Canyon area the formation is approximately 3,200 feet thick and is composed of shales, mudstone, sands, and conglomerate. The formation is recognizable over most of the western half of the basin. Locality 5, situated on Santa Paula Creek, as shown on figure 1, belongs in this unit. The top of the unit is taken at the highest occurrence of *Bulimina subacuminata* and the base, at the first occurrence of *Nonion pompilioides*.

Bulimina subacuminata, in so far as known, is the most significant ecologic marker in the middle Pliocene. In the Recent section it generally appears at a depth between 2,000 and 3,000 feet. It

[1] Oral communication, 1948.

continues abundantly through most of the lower Pliocene. The graded bedding at Locality 5, therefore, is presumed to have accumulated in waters ranging from 2,000 to 4,000 feet deep.

Upper Pliocene

In Wheeler Canyon this formation is 6,800 feet thick. The upper part of it is usually called the "Mud Pit Shale," which is predominantly a massive, gray to rust-colored mudstone, becoming more fissile toward the base where it grades imperceptibly into the middle Pliocene. The top is marked by the highest in-place occurrence of *Pseudoparrella pacifica*, *Uvigerina peregrina*, *Bolivina spissa*, and *Bolivina interjuncta*. All of these species range down into the lower Pliocene, and the first three are abundant in the upper Miocene. Offshore, at present, these species seldom range shallower than 900 feet. The fauna present at Localities 1 to 4, which fall within this unit, indicate depths of deposition between 1,500 and 2,000 feet.

Lower Pleistocene

Lithologically, the 2,400 feet of beds included within the unit are more sandy and are generally of buff color instead of the gray color of the Pliocene. These beds are composed principally of sandy silts, sands with occasional megafossil beds, and conglomerates with well-rounded Eocene cobbles and boulders. The basal portion is characterized by *Cassidulina limbata* and *Cassidulina tortuosa*, and the upper part by *Elphidiella hannai* and *Elphidium spinatum*. The *Cassidulinas* in the lower part indicate that these beds were deposited in water ranging from 125 to 900 feet, while the *Elphidiums* in the upper part indicate a depositional environment of from shore to 125 feet. At the top of this formation is approximately 100 feet of clean, well-sorted, cross-bedded sand, abundant with mollusks which today live in the Eulittoral Zone.

Upper Pleistocene

This formation, of continental origin, is 3,600 feet thick in Wheeler Canyon. It appears to rest conformably on the lower Pleistocene and is composed of buff-to-yellow-colored silts, sands, and conglomerates. These sediments are abundant with a mixture of foraminiferal species redeposited from older formations. A few remains of indigenous *Equus* are also present.

In summary, on the basis of the present bathymetric ranges of the same species which mark the divisions of the Ventura Pliocene and Pleistocene, it is reasonable to conclude that this basin was deeper than 4,000 to 5,000 feet during the lower Pliocene and gradually shallowed, to be capped by a thick series of continental beds. During this time, the central part of the basin subsided an accumulative total of approximately 14,000 feet.

SIGNIFICANCE OF REDEPOSITED FORAMINIFERA AND SEDIMENTS

The distribution of Foraminifera and distinctive lithologic types redeposited from older formations into younger ones clearly points to a progressive uplifting of the northern margin of the basin and a sagging of the central part. The lower Miocene conglomerate beds are composed principally of well-rounded Eocene sandstone boulders and siliceous Miocene shale fragments. Such restricted Miocene species as *Valvulineria californica* and *Siphogenerina collomi* are found redeposited in this member. Even the basal lower Pliocene was unroofed by upper Pliocene time and was contributing such species as *Nonion pompilioides* and *Bulimina fossa* to the upper Pliocene. This influx of lower Pliocene species appears abruptly above the highest thick marine conglomerate in the upper Pliocene, suggesting that perhaps the deposition of the conglomerate was caused by the same orogenic movements which uplifted the lower Pliocene to an erosional position. Middle and upper Pliocene species, in turn, are found abundantly redeposited in the continental beds. The block diagram (fig. 3) illustrates the condition of marginal uplift and central subsidence. This process of marginal uplift and central subsidence is all important to the method of sedimentation under discussion, because it produced a sea bottom which was constantly, or periodically, being tilted. This tilting, which increased the gradient, may well have created the conditions necessary to send the numerous turbidity flows of fine sediments and sand and landslides of conglomerate on their way down the slopes of the basin.

EMPLACEMENT OF THE SANDS

The faunal evidence as to great depths of the Ventura Basin has been doubted, because of the intimate interstratification of fine and coarse sandstones with the lutites containing deep-water Foraminifera throughout the entire section.

If the coarse deposits actually did indicate shallow depths, two alternative pictures could be imagined: (1) rapid and great variations in depth took place, or (2) we are dealing throughout with shallow-water deposits.

In the present case, one could hardly go so far as to maintain that each of the thousands of coarse layers represents a renewed, vast regression. That would not only mean vertical displacements of hundreds to thousands of feet, but, each time, the floor of the basin must have reverted to the same depth, except for the gradual progressive shoaling throughout its history. It is no less unprofitable to speculate on sea level having provided the unstable element.

The alternative—that of continued shallow depths—is hardly more satisfactory. In the first place, ripple marks, scour channels, shallow-water mollusks, large-scale cross bedding and other markers for shallow-water conditions, are almost entirely absent in the sandstones. Where they do occur they are not of a nature tending to confirm shallow-water conditions. On the contrary, a more satisfactory explanation of all these features can be offered on the basis of turbidity currents, as will be pointed out below.

In the second place, the fine siltstones and shales—interbedded with the sandstones—are remarkably regular in thickness and show fine lamination, thus also indicating undisturbed bottom conditions.

The upward transition to the sandstones is abrupt in the extreme. However, there is definitely no erosion of the fine laminae by current or wave action. Absolute tranquility must have prevailed on the basin floor, except for occasional small-scale sliding and, especially, the sudden and repeated introduction of the sand deposits. In other words, no disturbance of the bottom waters occurred, other than that associated with the advent of a new coarse stratum.

In the third place, the faunal evidence makes no sense if the distribution of benthonic Foraminifera were not controlled by the very same factors as now regulate the distribution of Recent species off southern California. The same foraminiferal associations are found in the fossil and the Recent environments, and there is no arbitrary mixing. In the sandstones, the scarce tests correspond either to present shallow environments with only occasional deep-water species, or they were derived by reworking from older Tertiary deposits which had evidently been raised above sea level and were being denuded at that time. Hence, the materials of the coarse deposits were carried from shallow environments to their final resting place among the deep-water lutites. They were deposited so quickly that no bottom fauna became embedded in them.

The interbedded shales, on the other hand, contain tests from one of the Recent populations associated with a certain temperature and depth, and lack representatives typical of both shallower or deeper Recent environments. Moreover, passing down the section, the indicated depth is found to change gradually without sudden jumps, whereas along the strike there is hardly any variation in depth in the area under this study.

Had shallow depths prevailed and the Foraminifera not yet acquired the "temperature-depth" habit they now follow, there should be found an arbitrary intermixing of types instead of the observed, careful, mutual avoidance of species from different Recent biozones.

It is worth remembering that the problem of how the sand and gravel were transported and deposited could not be solved by assuming more conservative estimates of the depth. Wave and current action capable of moving pebbles are restricted to a few dozen meters depth and would be as much out of place on a basin floor at 1,000 feet depth as at 4,000 feet. Hence, no advantage would be gained by reducing the estimates, and nothing warrants such an attempt. On the contrary, the difficulty of accounting for the various features of the sediments, to be described below, would only be augmented.

On previous pages, several independent lines of evidence have rather conclusively shown that, during the lower Pliocene to Pleistocene history of the Ventura Basin, the floor was 4,000 to 5,000 feet deep to begin with and gradually shoaled to a few hundred feet, the marine deposits finally giving way to beach and terrestrial sediments. This great depth having been established beyond reasonable doubt, one is faced with the problem of how to explain the emplacement of the sandstones in this deep environment.

The bottom was not agitated by currents, as will be established in greater detail presently. Neither did wave action reach such depth. Even if the bottom slope were as high as 1 in 10, the coastal distance must have been 10 to 15 miles. This is much too far out for floods, surface currents or wind to have carried grains of 1 mm diameter. Yet, many sandstones contain grains of much larger size—some, even pebbles of 1 or 2 inches diameter. Obviously, normal processes of transportation are entirely inadequate. The writers are convinced that the mechanism of turbidity currents not only offers a satisfactory explanation of the transport and deposition of the coarse material, but accounts for a number of other observed features. Admittedly, the picture cannot be completed in all detail, and the mechanism—especially the starting off—of these currents is still a matter of pure speculation.

The observed features and their bearing on the mechanism of sedimentation will be treated under headings "A–H." It will be found that the picture of the action of turbidity currents presented by Kuenen and Migliorini (1950), and by Kuenen in this symposium, is confirmed and amplified in several respects, but that the general conditions differed significantly in the Ventura Basin and the Tertiary geosyncline of the Apennines, in that the general scale appears to have been smaller in the present case.

A. Indications of the absence of wave action and of normal bottom currents.

1. With the exception of one or two doubtful cases, no ripple marking is found anywhere in connection with the sandstones or in the intervening shales.

2. No scour channels were observed, although—here and there—small irregularities in the bedding occur. These are probably due to slump and drag phenomena.

3. Cross bedding is extremely rare in the sandstones and, in so far as observed, never clearly developed. In the silty parts of intervening shales, small-scale wavy cross bedding—such as is normally associated with current ripples—was observed a few times. These occurrences are so haphazard and rare that they cannot be attributed to normal bottom stirring. It is suggested that they were formed by dilute turbidity currents such as may be expected to form now and then, especially along the margins of more dense turbidity flows.

4. Many of the shales and siltstones

FIG. 7.—Aerial photograph taken February 22, 1941, off Del Mar, California, showing muddy water from Escondido, San Diequito, and Sorrento Creeks, spreading fan-shape over the surface of the Pacific Ocean. Maximum seaward spread of fan, approximately 1 mile.

show perfectly flat, paper-thin laminae, each of these traceable for many feet. These must have formed in quiet water.

B. Indications of moderate size of each unit.

1. The sandstone beds are of moderate thickness. Where earlier work has indicated members several hundred feet thick, these were found to consist of a number of separate and distinct units, none of which appears to exceed 4 or 5 feet in thickness.

2. Viewed at a distance, the bedding is of remarkable uniformity in dip and thickness. At each locality the sandstones tend to be of roughly-equal thickness, but where in some cases the average is 1 foot or even more, in others it is no more than a few inches. Closer examination shows, however, that both along the strike and dip some sandstones of a few inches in thickness tend to wedge out in a few dozen feet, and others of a foot and more in thickness may show distinct variations in thickness or even wedge out within the limits of a single exposure (50 feet or so).

3. Larger units comprising a number of thick sandstone beds have been traced by drilling over distances of 15 miles, but even these dwindle when followed toward the present coast line, that is, away from their probable source.

The moderate size of each individual bed militates against any attempt to explain the transport of material by surface currents. Aerial photographs taken during flash floods (fig. 7) show how the fresh water spreads out in a semi-circular fan on the surface of the sea, as a fresh water surface layer on the ocean.

If a deposit were dropped on the sea floor, reaching many miles from the coastline in deep water, it would necessarily cover a great lateral extent as well. On the other hand, turbidity currents of limited extent might occur, especially if there were sufficient bottom slope to guide them. Hence, this mechanism can account for beds of limited extent oc-

FIG. 8.—'Pull-apart' structure (for sketch see fig. 9), Locality 5, Santa Paula Creek.

curring far from the contemporary shoreline.

C. Indications of bottom slope.

1. All exposures visited showed some distortion of bedding, from slight undulations to marked contortions. Usually only one or two beds have partaken in each example of slipping. In other cases, several laminae were found to be involved at a time. The thickness of the contorted units varied from less than an inch to 2 feet. At Locality 5, for instance, highly contorted beds are locally piled together, forming a typical slump structure.

2. 'Pull-apart' structures. Repeatedly, structures were encountered showing that certain beds, both of sand and clay, had been stretched and torn apart into

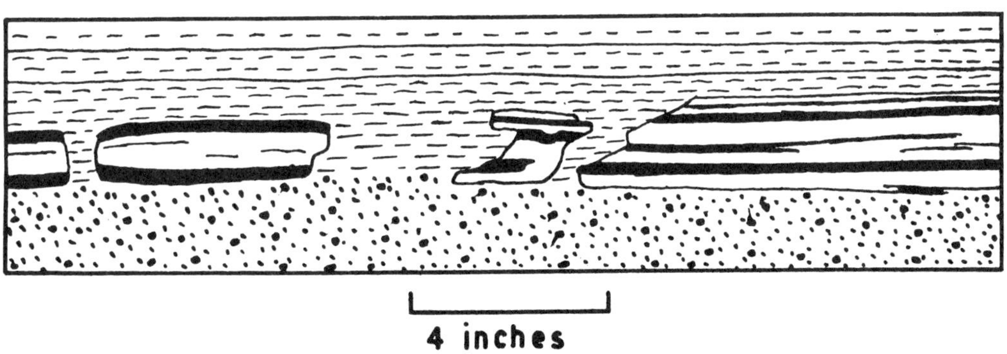

FIG. 9.—'Pull-apart' developed in laminated bed of clay lying on gravel bed and covered by silts. The structure of the latter in the gaps could not be ascertained. Hence, it is not known whether case (a) or (b) of figure 10 is represented. Locality 5, Santa Paula Creek.

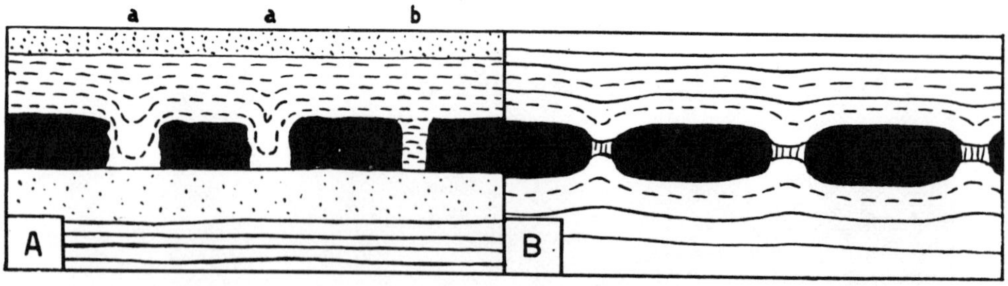

FIG. 10.—Schematic representation of 'pull-apart' (A), and boudinage (B). On the latter, the brittle layer thins toward the gaps where secondary matter tends to accumulate. The beds on both sides bend into the gaps. In the 'pull-apart' gaps, generally, only the upper beds bend down as at (a) if the covering bed antedates the structure, or the gap is filled in before the covering bed is deposited as at (b).

slabs. The intervening cracks had generally been filled in from the top and, in some cases, possibly from below. Although there is a certain resemblance to "boudinage," there is in the present case less regularity, not much thinning, and less flow from below.

3. Clay pebbles. All transitions were encountered between relatively rare, solitary 'pull-aparts,' via multiple 'pull-aparts' with the separate slabs still aligned or tilted, to rather frequent, angular clay fragments or beds of rounded clay pebbles. Sandy 'pull-aparts' are easily disintegrated and do not appear to have formed pebbles. All observed soft rock pebbles consisted of shale or silty shale. Laminated and bent shale fragments testify to the unconsolidated state of the materials at the time of deposition. The particles range in size from minute flakes to chunks several inches across or even 2 feet long. They are found isolated or in groups, in the lower parts or at higher levels of the sand beds; in other cases they occur packed close together, as distinctive conglomeratic beds.

The most obvious explanation of these structures is to attribute them to contemporaneous sliding on a gently sloping sea floor, and the present writers are convinced that this is the true cause of many of them. On the other hand, the

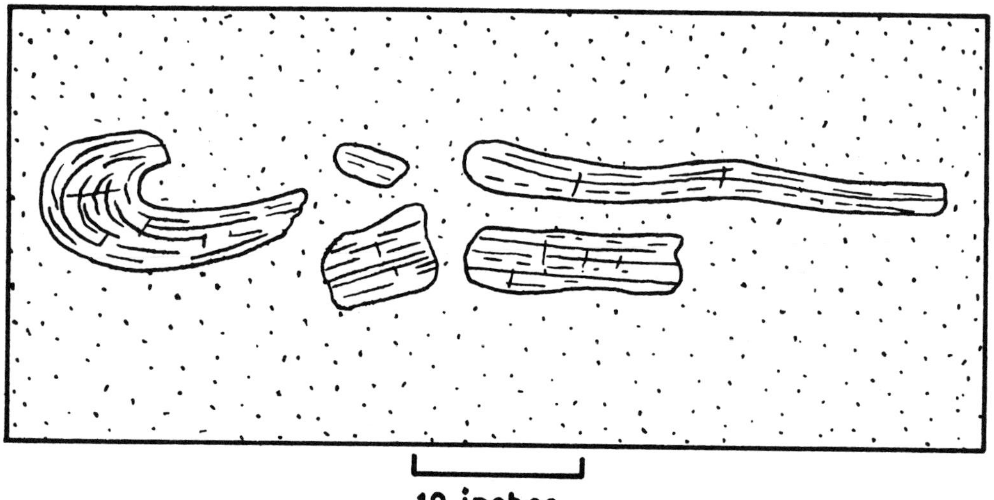

FIG. 11.—Bent chunks of clay in a sandy bed, evidently formed from 'pull-apart.' Locality 5, Santa Paula Creek.

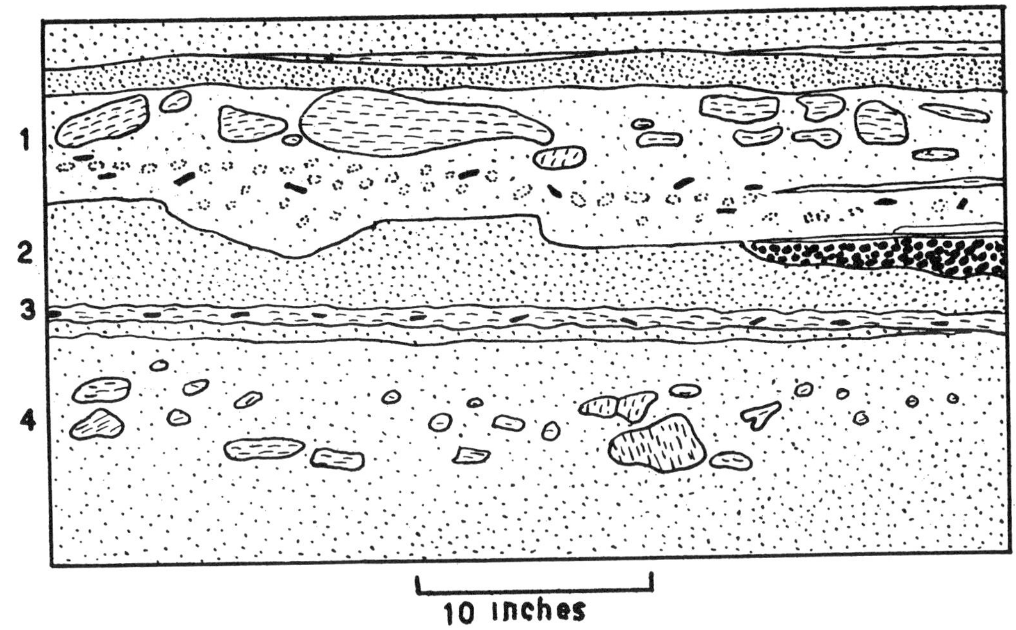

FIG. 12.—Irregular type of bedding. Bed (1) contains large clay pebbles in upper part and, in the center, many diatomaceous shale pebbles and carbonized wood. Bed (2) is eroded in places, and to the right it is covered by a fine conglomerate. Bed (3) is shaly, with carbonized wood. Bed (4) is sandy, with many irregular clay pebbles. Locality 2, Hall Canyon.

beds deposited by experimental turbidity currents were occasionally found to present marked folds where the bottom of the ditch was horizontal and the surface of the deposit sloped only one degree. This observation tends to indicate that these distortions of the artificial beds are not due to sliding. Possibly, the drag exerted by a turbidity current on its bed consisting of unconsolidated, highly mobile sediment may suffice to distort these beds in a manner strongly recalling true slumping under the influence of gravity.

Both the experimental distortions and those encountered in the field are characterized by the absence of clear orientation or alignment, thus leaving one in doubt as to the direction of movement.

In the graded series of the Macigno sandstones, Migliorini observed—besides plenty of typical slump structures—a different formation which he calls crinkled bedding (stratificazione arricciata; Migliorini, 1950). As these structures die out upwards and show overhanging bedding planes in the most intensenly distorted parts, and as no gaps are formed, he concludes that neither ripple marking nor post-depositional distortion can explain them. He offers the suggestion that the deformations are caused by the extrusion of water from the lower part of the bed while the upper part is being deposited from the turbidity current. This explanation is supported by the finding of vertical cylindrical structures in the lower part of the beds below the bulges of the crinkled zone. The rising water is thought to have passed up through the sand along these cylinders until it met a less permeable stratum, causing the latter to bulge upwards while depositing fine sediment below.

The upward dying-out of the crinkling, as described by Migliorini, is also visible in the laminated upper part of a graded Ventura sandstone (fig. 13).

The main objection to the mechanism he suggests is that the water would form a laccolith while lifting the impervious laminae. Even if we admit that a lac-

FIG. 13.—"Crinkled" bedding at upper limit of graded sandstone. Covered by dark bed of silty sand. Locality 4, Ventura River.

colith, and not a sill, was thus produced with stretching, but without rupture of the containing cover, the only addition to the mass of the bulge after draining is the fine sediment carried along by the water. This could form only a thin lens and it could not fail to stand out as a separate mass with a sharp and flat lower surface. The present authors are, therefore, inclined to assume that extrusion of water by settling cannot be the sole cause of the crinkling, although it may have been a subsidiary phenomenon.

Yet another mechanism may be invoked, the local sinking of an overburden into a hydroplastic substratum. This process appears to have been active in forming the "pockets" described below and will be treated in connection with these, but it may also have contributed to the formation of crinkled bedding and slump structures.

The distorted bedding is obviously still an unsolved problem, the explanation of which must be left to future research. Probably several causes combine to form the observed features, and it may also well be that we are actually dealing with more than one distinct, though outwardly similar, phenomenon, each the result of a different cause.

As to the clay pebbles, these can also be explained either by sliding or by the drag of the turbidity currents, or by a combination of the two.

4. There are, however, other indications of bottom slope. The relatively small extent of each separate graded bed or slump structure is not easily reconciled with the picture of a distant origin of the current, these then having reached the point in question by traveling over a horizontal sea floor. The impression gained is more that of local origin and action of the turbidity currents—this, in turn, implying a bottom slope sufficient for starting the movements.

5. The paleontological evidence for bottom slope. Thomas L. Bailey (1935) described the lateral change in the upper

Pliocene and lower Pleistocene faunas in outcrops exposed in Peppertree Canyon, near the mouth of Wheeler Canyon and about 6 miles northeast of Ventura, California. He was able to trace a distinct lithologic member in a north-to-south direction around the eastward plunge of the Ventura anticline, so that no question exists as to time stratigraphic correlation. However, on the south limb of the anticline Foraminifera typical of the uppermost part of Biozone IV (Natland, 1933) were found, indicating a depositional environment greater than 900 feet. In the same horizon, on the north side of the anticline, Foraminifera of Biozone III were found, indicating a depositional environment of approximately 200 feet. A slope of about 5 degrees southerly, away from the source area, is necessary to produce the 700-foot elevation difference in the 11,000 feet between the two localities.

D. Grading.

In all localities visited, the great majority of the sandstones showed grading from coarse at the bottom to fine at the top. No instances were observed, among the hundreds of beds examined, in which the grading was complete, and in the opposite sense, although departures from the ideal case do occur. In some beds no grading is seen, not so much by complete mixing of all sizes as by excellent sorting throughout, perhaps caused by good sorting of the sand in the source material. Here and there a sandy bed shows, toward its top, distinct parallel lamination or even poorly developed cross-bedding with slight angles of slope. Although the bed as a whole is graded, the supply appears to have shown pulsations and to have taken some time before local irregularities of the sea floor became filled in and smoothed off.

By far the most frequent case is that of medium- to coarse-grained sand resting immediately on top of the preceding shale and grading upward into very fine sand or silt. This, in turn, is followed more or less abruptly by a thin stratum of shale, in some places with a narrow zone of transition.

Coloring in the shale brings out fine laminae in many places. This lamination is sometimes seen to be caused by slight variations in grain size, perhaps caused by lateral spreading of dilute clouds from a turbidity current passing at a distance from the site in question, and possibly in some cases to transport in surface waters after flash floods, followed by settling to the bottom.

Mechanical analyses of three well-graded sandstones from the Ventura Basin were made to compare with those of beds deposited by experimental turbidity currents. The former range in thickness from 2 to 4 feet, whereas the latter were less than an inch thick. Yet the essential similarity is striking.

In beds I and II, with coarse lower parts, there is a marked tendency for the Trask sorting coefficient to decrease upwards and for the Trask coefficient of skewness to increase. Bed III is perfectly but slightly graded and shows no marked variation in sorting or skewness.

Of the artificial beds, "C" was deposited close to the origin of the turbidity current, and this may explain why the sorting and skewness are different at various levels. "A" and "B" were taken a few meters farther from the origin, where one may expect less influence from arbitrary pulsations in the current. They are in all respects similar to bed III.

These data are insufficient to establish general laws concerning sorting and skewness in graded beds. Only by examination of a large number of analyses from a variety of beds—preferably, also, from several sedimentary basins and Recent deposits—could it be ascertained whether there is a general tendency for the sorting to improve from the bottom upwards, the lower parts to be skewed to coarse, and the upper to be symmetrical or skewed to fine.

Less commonly, the bed starts with fine- or even medium-sized gravel, with cobbles up to 3 inches in largest diameter, and grades upward quite regularly to the sand and silt fractions. Sometimes the

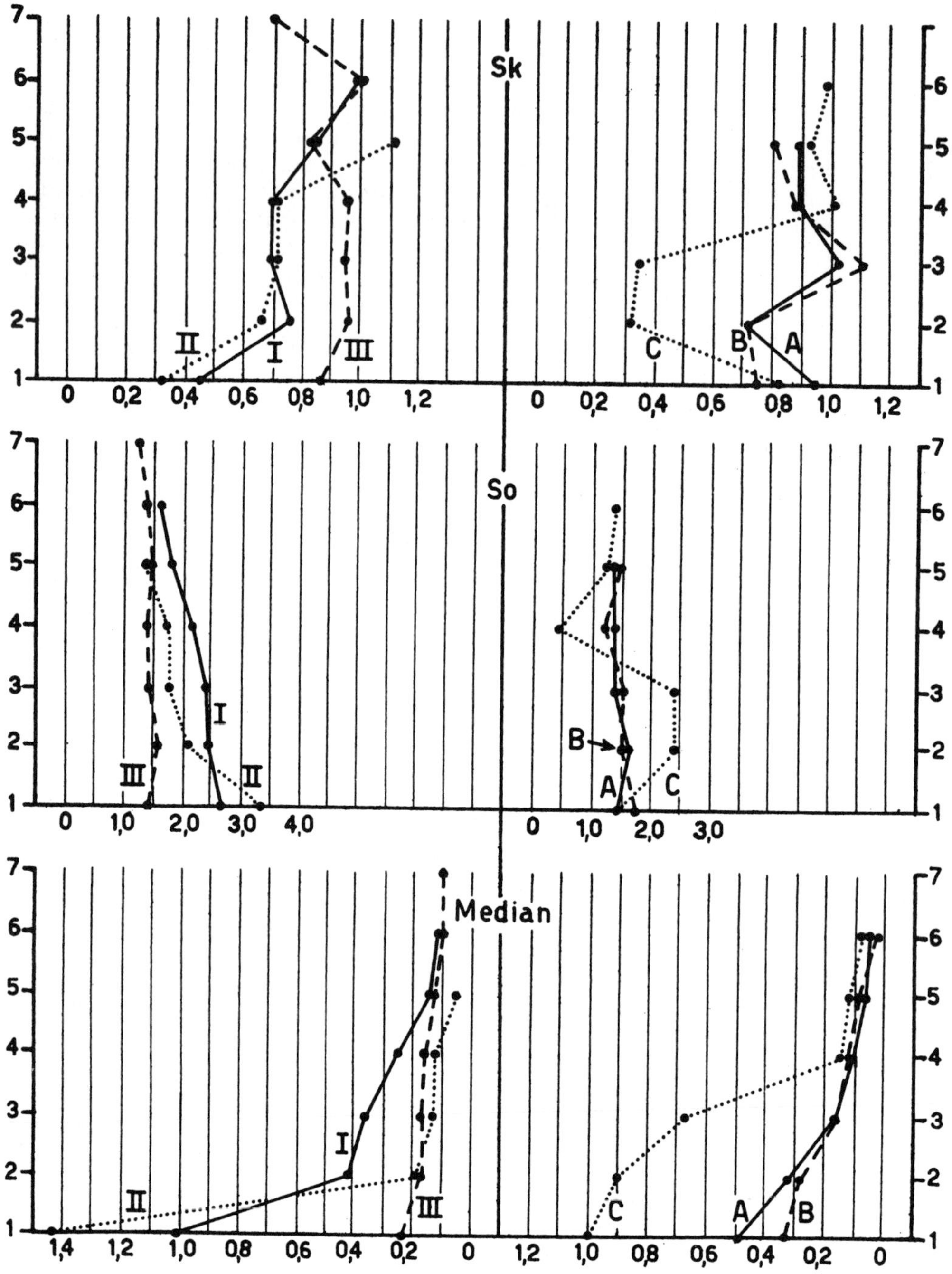

FIG. 14.—Mechanical analyses of three natural graded beds: (I) Ventura River, Locality 4, thickness 4 ft; (II) Ventura River, Locality 4, thickness 4 ft; (III) Hall Canyon, Locality 2, thickness 2 ft; all sampled roughly at equally spaced intervals from bottom (1) to top (5, 6 or 7); and of three beds (A, B, C) deposited by experimental turbidity currents (for histograms of latter, see Kuenen and Migliorini, 1950, p. 107).

FIG. 15.—Lower part of a very coarse bed, 3 feet thick. Outside the photograph it grades upward into fine sand without pebbles. Locality 5, Santa Paula Creek.

pebbles appear a few inches above the coarse sand forming the lower limit of the bed, or the pebbles are concentrated in the middle portions of the bed. Even then, the coarse grains come in more abruptly than they die out upward, showing that we are dealing with a deposit of a turbidity current advancing in two successive impulses, the second carrying coarser material than the first.

Broadly speaking, the coarseness of the grain size and the thickness of the bed increase together, but there are beds of 2 inches in thickness with fine gravel, of two-thirds foot with medium gravel and, conversely, sand beds 3 feet thick with only very coarse sand.

The pebbles are derived from a great variety of rocks; schists, sediments, and igneous rocks. Special mention should be made of diatomaceous shale pebbles. Occasionally they are found together with other types of pebbles, but in many beds they are the only constituent of large size. Now and again one finds a single large, subangular chunk (up to 8 inches across). The remarkable feature of these fragments is that they do not follow the grading of the normal rock material, but may lie embedded in the medium-grained sand in the middle of a bed.

The specific gravity of a diatomaceous shale pebble, determined in saturated condition, was found to be 1.65. From this it follows that the density of the flow at the time the pebbles were deposited must have been less than 1.65 because, otherwise, they would have floated.

Theoretically, it should be possible to calculate the density of the flow by comparing the size of the largest sand grains (about 1 mm diameter) deposited at the same time as pebbles of 4 inches in diameter and having a density of 1.65. According to Stokes' law, the pebbles and sand grains should settle at the same velocity in a liquid of about 1.64 density.

However, this does not take account of the viscosity and turbulence, nor of the fact that the few large pebbles extend

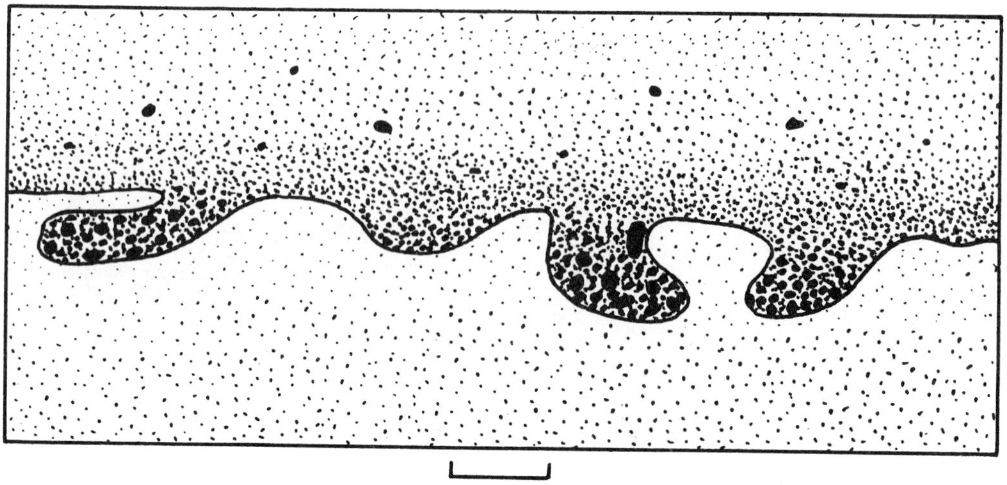

FIG. 16.—Typical pockets, filled with gravel grading upward imperceptibly into the coarse, sandy base of a graded sandstone. This bed lies on the fine sandy surface of another graded sandstone. Locality 5, Santa Paula Creek.

above the bottom when they are dropped, whereas the sand grains form a smooth sheet. Hence, the turbidity current must have shown a considerably lower density than 1.64 at the time the pebbles were finally embedded.

On the other hand, the simultaneous deposition of small sand grains and large pebbles definitely points to a medium of much higher density than pure water. Moreover, a few inches of sand were nearly always deposited before the pebbles came to rest. The experimental investigations showed that the density of a flow is highest at its head. Hence, the frontal portions of the turbidity currents must have been of higher density than the parts depositing the diatomaceous pebbles. The head of the flows is, therefore, estimated to have shown densities of the order of $1\frac{2}{3}$ and the middle portions of $1\frac{1}{3}$. It is considered highly improbable that the currents had maximum densities below $1\frac{1}{3}$. Admittedly, these estimates are only first approximations.

The clay pebbles, some of which are warped, must have been in a plastic condition and, therefore, contained a large percentage of water. Hence their density must also have been less than 2.0. The case represented in figure 12, where the clay pebbles are larger and lie at a higher level in the bed than the diatomaceous pebbles, indicates that the density was appreciably lower than that of the latter material.

E. Pockets, depressions, and protuberances.

A remarkable and highly significant feature repeatedly observed is the development of pockets by which some beds extend downward into the preceding stratum. As far as our observations go, it was invariably coarse sands and gravels that showed this structure, and generally the pockets were formed in the fine sandy surface of graded beds, seldom in shale beds. The pockets range in size from 1 to several inches deep and across. They do not appear to be elongated to any marked degree. In shape, all transitions were observed between slight dimples to deep pouches with a narrow upward opening. The internal surface may form a smooth, symmetrical indenture or a complicated, irregular cavity. Occasionally, a single pocket is observed, but most usually they occur in groups, spaced at irregular intervals along the base of a bed.

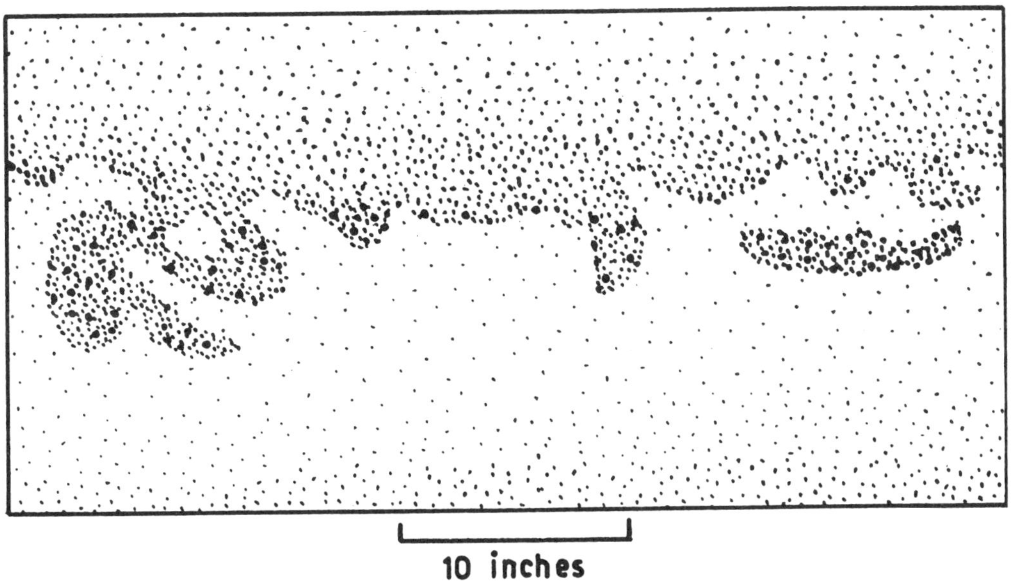

FIG. 17.—Highly irregularly pocketed lower surface of a coarse-graded sandstone, lying on the fine sandy surface of another graded sandstone. Locality 4, Ventura River.

In all cases noted by the present authors, the contents of the pockets were similar to that of the lower margin of the bed they belong to, but often considerably coarser and then grading upward, inside the pocket, to merge imperceptibly with the lower surface of the next bed.

The complicated shape of the pockets, with protuberances from the walls into the pocket, proves that they cannot represent original hollows in the lower bed which became subsequently filled. Only a well-indurated sandstone could sustain such shapes without collapsing, yet the condition must have been that of a wet and unconsolidated sand. It appears, therefore, that the pockets were developed by the coarse material they contain, settling on a sand which was in a highly mobile condition and then sinking into the sand by differential loading.

One of the experimental beds shows development of the same pocket structure. In a former paper these were referred to as irregular undulations (Kuenen, 1950, p. 109). By careful scraping away of the lower bed, the true shape has later been exposed in the dried out deposit and was found to greatly resemble the natural pockets (fig. 18).

The process can be imagined as follows. An advancing turbidity current, containing a small number of grains con-

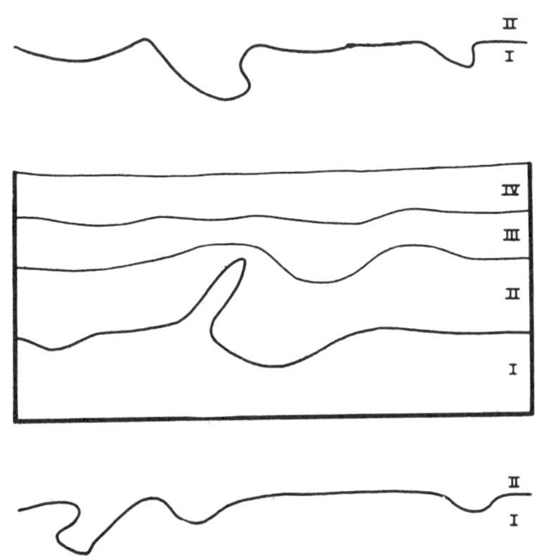

FIG. 18.—Section through deposits of experimental turbidity currents at right angles to direction of flow, showing pocket structure (exp. B, Kuenen and Migliorini, 1950). The shape of the bedding plane below II, at distances of 1 inch both ways from the section, is also shown.

siderably larger than the bulk of its load, must, at a certain point, start to deposit some of these larger grains in minor indentures. As soon as pebbles start to collect somewhere and to sink into the floor, further coarse material should tend to be captured. In this manner, the exceptionally large grains become entrapped in separate pockets. This process will continue while the head of the flow, where these pebbles are contained, is passing. A moment later, the chief body of the flow begins to deposit its load of coarse sand in much greater mass and in an even bed over the entire surface. Differential loading thus ceases, as well as the development of the pockets.

This process explains why: (1) the pockets contain exactly the same types of rock material as the overlying bed, differing only in size; (2) why the upward grading ends in grains of the same size as the lower limit of the covering bed, instead of finer or coarser material; and (3) it likewise accounts for the shapes of the pockets which depend on the relative amounts of pebbles, the mobility of the indented layer, etc.

Upward bulges surrounding the pockets were not observed. This may be due to erosion while the bulges were forming, to inaccuracy of observation, or to both.

The pockets here described appear to be of the same nature as Shrock's "flow casts" (Shrock, 1948, p. 156), but the latter are described as always forming at the lower surface of sand deposited on either clay or peat, whereas the present pockets are formed in fine sand; the filling, moreover, is graded gravel. Emery (1950) recently reported large pocket-structures developed where beach sand has sunk into an underlying marsh deposit of sandy silt.

Attention may also be called to an interesting paper by Macar and Antun (1950), who describe pseudo-nodules (slump balls) from the Devonian of the Ardennes. They show how these were formed from gradually developing basins, a few inches to several feet across, which sank into the substratum while sedimentation proceeded, like micro-geosynclines. These may develop until showing reversed limbs, but minute slides into the basins demonstrate how these presented surface depressions while being formed. Hence the basins cannot be due to differential loading but must have been formed under the influence of slump folding. The close association with typical slump structures and the various stages of development from undulating beds—via separate basins to deep, almost closed pockets—testify to the combined action of gravity sliding and vertical subsidence. The even more complicated slump balls described by Kuenen (1948), from the Carboniferous of Wales, evidently developed through the pockets being taken up and carried along in slump sheets.

The present authors are of the opinion that pockets, flow casts, crinkled bedding, slump balls, and slump structures, are formed by the contemporaneous action of gravity. The pockets and flow casts are formed by subsidence under unequal loading, but with crinkled bedding a certain amount of horizontal sliding may have been added. Pseudo-nodules (slump balls) result from combined vertical and horizontal movements, and in slump structures the sliding predominates or acts alone. There is probably no sharp division between these products of gravity action, and the relative importance of the two types of movement will vary from case to case.

It is typical of all these structures to show no polarity indicating the sense of movement. Folds dying out upwards, even complicated ones, are frequently encountered among typical slump structures (see, for instance, Kuenen, 1948, Pl. XXV, fig. 1). Apparently, this curious feature can develop without concomitant sedimentation. Lack of competence in the highly mobile upper portions of the sliding mass may be invoked as explanation. In some cases, compaction and extrusion of water, as advocated by

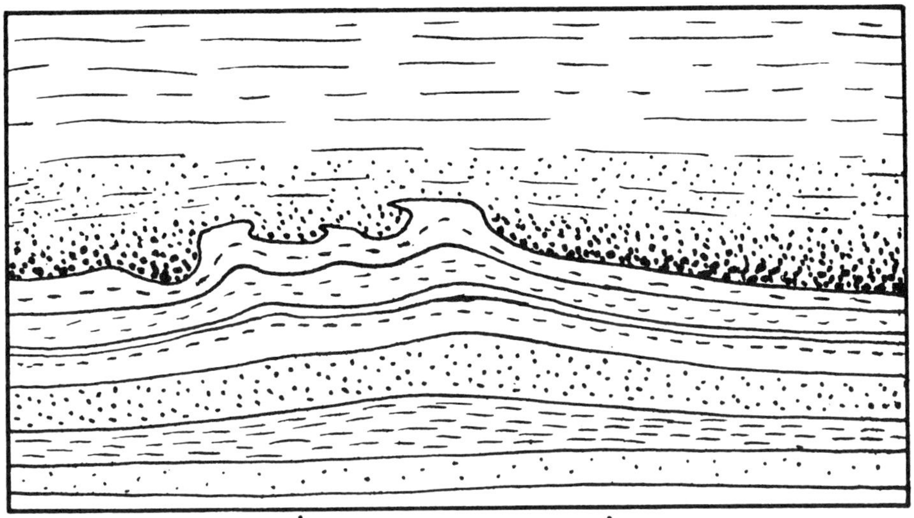

FIG. 19.—Local bulge of silty clays, covered by a graded sandstone bed with gravel in the depressions. Locality 3, Ventura River.

Migliorini, may form an additional factor.

Before leaving the subject of contemporaneous contortions, it may be pointed out that many of the periglacial structures ("Brodelboden," cryoturbate structures) are due to the same causes. Solifluction is merely a type of slumping. The complicated contortions, injections, and pocket structures result from local subsidence and/or sliding, sometimes assisted by frost expansion. The intensive saturation of the surface strata overlying permanently frozen ground results in high mobility, just as in the subaqueous processes where lack of consolidation offers suitable material for the action of gravity.

Clear cases of the opposite development, the localized updoming of the floor between larger areas of load, were also observed in the shape of irregular upward protuberances of lower beds which, at least in part, may be due to upwelling combined with drag exerted by the upper bed as it was being deposited under the action of the turbidity current (fig. 19). But these features can also be attributed to erosion and sliding, with concomitant deposition of the next bed.

The various structures described under this heading are significant in proving the unconsolidated nature and marked mobility of the beds at the time the next one was deposited. The lutites appear to have been, on the whole, slightly less mobile, although not compacted to any marked degree.

F. Denting of bedding planes.

One single instance of surface denting by the pebbly bottom of the next bed was encountered a short distance below the 6-inch ash bed at Locality 3 (fig. 20). The upper bed is 8 inches thick, excellently graded from pebbles to silt. The well-rounded pebbles are one-half to 1 inch in diameter and lie, on their flat sides, on well-laminated silts. The pebbles have dented the deposit of laminated silts to a depth of 1 inch, each successive lamina downward showing less deformation. The upper surface shows pits about one-half inch deep, and in cross section the silt is seen to have pressed up into the interstices between the pebbles. Differential compaction of the silt, subsequent to the covering, can account for only a fraction

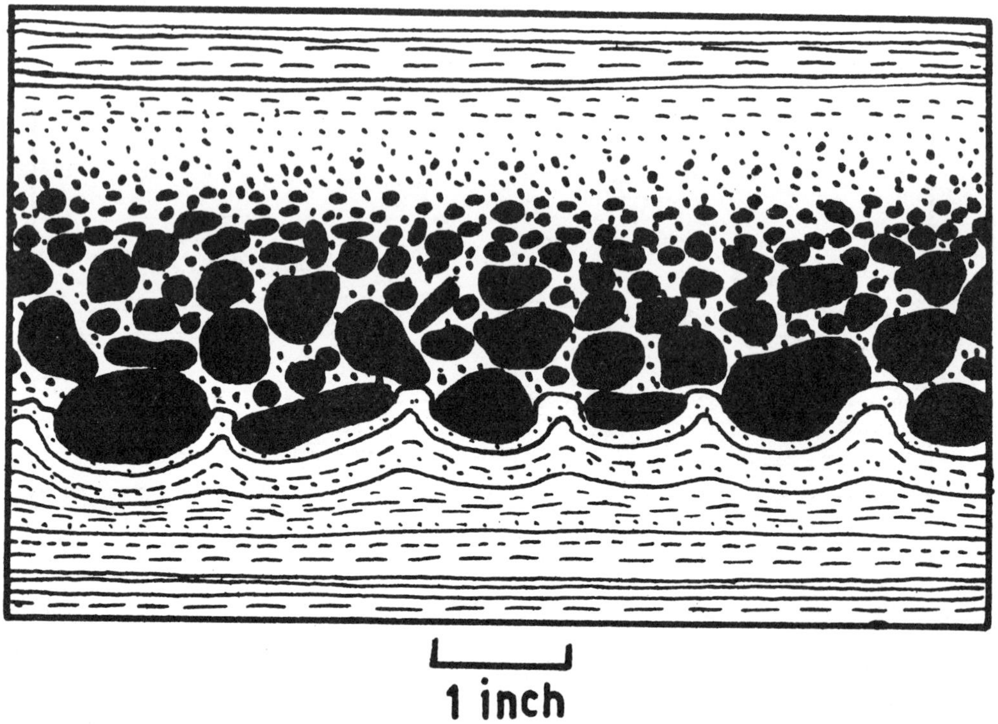

FIG. 20.—Schematic drawing of highly graded bed with pebbles at lower surface (actually the bed is somewhat thicker). The underlying sandstones have bulged up between the pebbles. Locality 3, Ventura River.

FIG. 21.—Very coarse graded bed. The underlying silty clay is dented by the pebbles (for sketch, see fig. 20). Locality 3, Ventura River.

of the deformation. The thickness of the deformed layer is 50 per cent less under the pebbles than between them. A silt capable of so high a degree of compaction could not fail to be deformed plastically under the mere weight of the pebbles. Hence the denting we now observe must be attributed mainly to plastic deformation, to which subsequent differential compaction added only a small amount.

Here is a striking illustration of the remarkable properties of turbidity currents. They carried medium-sized pebbles across a bed consisting of laminated silt, depositing them in a stratum of uniform thickness which graded upward into finer gravel and sand. Yet, in spite of the high competency of the current, testified by the pebbles of an inch in diameter, the sandy silt was not eroded, although the latter was so mobile as to be thoroughly pitted by the weight under water of a bed 8 inches thick.

G. Carbonaceous matter.

Both at Localities 2 and 4 (as in many other exposures not examined during the recent visit), appreciable quantities of carbonized plant remains are incorporated in a few of the beds exposed. Sticks up to several inches long and an inch thick have been found. It appears that in each case a brush fire must have scorched the land adjacent to the basin. The next flash flood would then sweep the country practically clean of carbonized debris and carry it out to sea with a large volume of sediment. The result would be either a direct deposit in the neighborhood of the coast, or the development of a turbidity current from the densely laden river waters, and a graded bed out in deep water. The very marked concentration of the carbonized remains in only a few of the many beds shows that, once produced by fire, practically all of this material was soon carried out to sea in one flood. There is no heralding by a few pieces in preceding beds, no arrival of stragglers in subsequent laminae. As do the fossil leaves mentioned previously, this points to a climate equivalent to that now present in southern California where, in dry seasons, the moderate brush growth is occasionally ignited by lightning and the resulting charred debris is swept away by the accompanying cloudburst. Although the carbonization of the wood and plant material specifically referred to was probably produced by fire, it must be admitted that some may have been carbonized by a process of slow oxidation after entombment. However, cylindrical sticks are found which lack the flattened shapes resulting from compaction during slow carbonization.

H. Questionable ripple marking and cross bedding (not associated with the course conglomerates).

Irregularities, of the size of ripples, on bedding planes are widespread in the Ventura Basin. Only in one or two cases were these found to show a rhythmic character and general shape such that the origin might be attributed to ripple marking. The best instance of the latter was discovered a couple of feet below the dented bed described above, at Locality 3 (fig. 22). The interspace is some 10 to 15 inches, four or five troughs of an inch deep being exposed. The crests are smoothly rounded, the troughs somewhat sharper. The subsequently deposited lutite accumulated faster in the troughs and the structure was buried in less than an inch thickness. Whether this feature represents ripple marking, "pull-aparts," markings formed by the rolling of a large boulder, or activity of organisms, could not be ascertained in the available time.

Small scale cross bedding, such as is normally associated with small current ripples, is not infrequent in the fine laminated sands, especially at Locality 5, but the authors were not able to discover sufficient rhythmic character to be sure whether true ripples had been developed.

In several instances the cross bedding was developed in a stratum a couple of inches thick, covering a bed with slump structures or other irregularities. It ap-

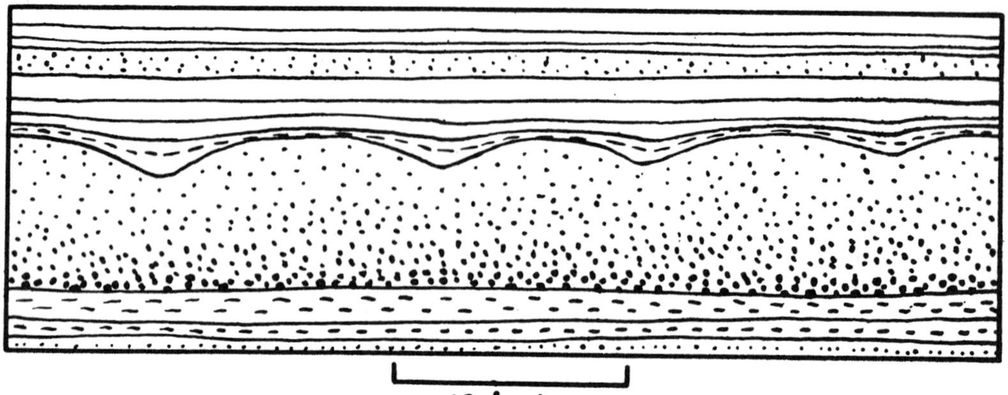

FIG. 22.—Rhythmic denting of graded sandstone, possibly ripple marks filled in by laminated clay Locality 3, Ventura River.

pears as if the depositing currents formed the cross bedding when flowing over an irregular surface until a smooth bedding plane had been developed.

Somewhat larger scale cross bedding in sand layers is poorly developed at a few spots. However, nothing comparable to current bedding, as normally developed in river deposits or other shallow-water sands, was observed.

Summary

The features associated with the sands show that they were deposited at considerable depth on a floor where no appreciable currents swept the bottom. The grain size and coastal distance were too great, the extent of separate beds too small, for surface currents to have carried the sediment and dropped it at the point of deposition. Contorted bedding, dented bedding planes, "pull-aparts," clay pebbles, and warped clay slabs testify to the unconsolidated nature of the beds on which the sands were deposited. The depositing currents formed neither ripple marks, cross bedding, nor scour channels. All these features are readily explained by the mechanism of turbidity currents. The excellent grading of the great majority of the sand beds, often starting with fine- or even medium-sized gravel, strongly supports this explanation.

The turbidity currents were of limited volume and extent, not infrequently of a pulsating nature. The density was high, probably in excess of 1.50 at the head. The basin floor probably sloped slightly away from the source; that is, toward the south.

EMPLACEMENT OF THE COARSE CONGLOMERATES

Besides the many graded sandstones containing pebbles at their base, coarse conglomerates with large boulders occur in a few deep-water horizons, besides in the beach and terrestrial beds. The deep-water conglomerates were examined at Locality 2 in Hall Canyon and Locality 6, Wheeler Canyon.

At Locality 2 the fragments range in size from pebbles to large boulders, 2 to 3 feet in diameter. They are very well-rounded, occasionally show pholad borings, and consist of a large variety of crystalline and sedimentary rocks. The pholad-bored rocks undoubtedly rolled from shallow, near-shore areas to depths of approximately 1,500 feet. About a dozen beds are well exposed over a stratigraphic thickness of about 300 feet between graded sandstones and shales. The conglomerates are seen to range in thickness from several inches to 3 feet. Most beds are of roughly constant thickness as far as exposed, but at least one, which is 2 feet thick, wedges out

FIG. 23.—Photograph showing attitude and character of conglomerates at Locality 6. Note ripple marks on the right side.

upward over a distance of some 20 feet. The conglomerates are poorly consolidated and, generally speaking, the larger the boulders, the thicker the bed. Otherwise, there is no sorting as to size and no grading. There is, however, a marked tendency for each bed to show a limited assortment of rock types. This sorting as to composition, but much less as to size, is hard to explain. Possibly several streams crossing different areas of source rock supplied the materials more or less separately. This would imply the absence of a common lower reach where mixing would occur. Also, there cannot have been prolonged accumulation on a beach or foreshore.

The beds immediately below the series containing conglomerates show the most pronounced irregularity observed in the Ventura Basin. It appears likely that the bottom slope was somewhat above the average, and turbidity currents were of small extent. The proximity of the coast in the relatively shallow water (1,000 feet) may have brought about general conditions of the nature of foreset delta beds with relatively moderate slope.

At Locality 6 the conglomerates occur in two horizons, one 500 feet above the other (figs. 2, 23), the maximum thickness for both attaining 300 feet. Along the strike, both sequences, as previously stated, are quite lenticular. Although faults may be present to complicate the structure, there can be no doubt as to this primary disappearance along the strike. The resemblance to the conglomerates already described is fairly close, but the boulders do not attain quite such large sizes, no sorting as to composition was observed, and the beds are remarkably regular in thickness—many continuing down dip for at least several hundreds of feet, as far as the exposure can show. Here and there, however, part of an underlying sandy bed is missing, so that the conglomerate suddenly steps down to greater thickness (fig. 24). The upper surface is not influenced and it is obvious that the wide gap in the lower bed, or beds, was de-

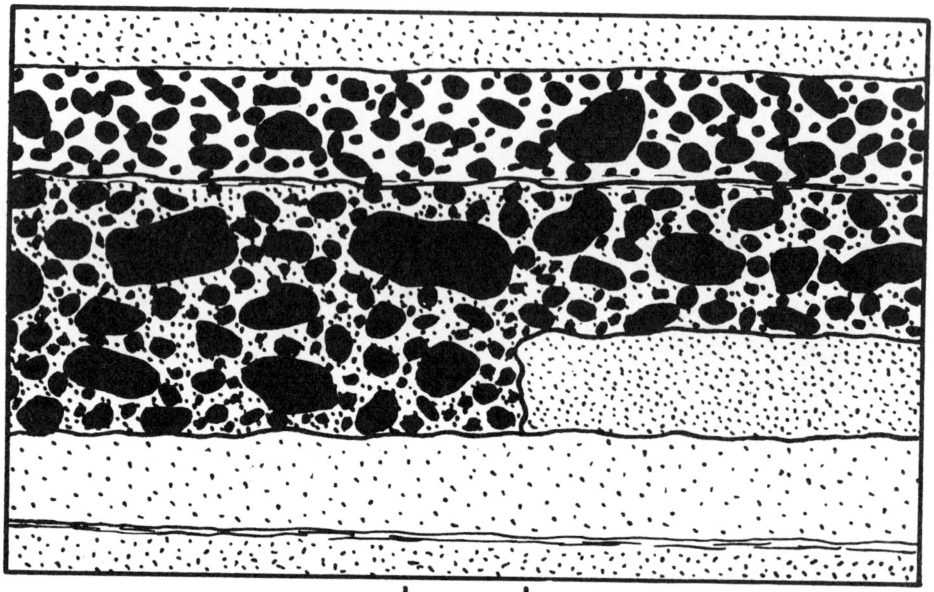

FIG. 24.—Schematic drawing of two conglomerate beds, the lower one stepping down across the underlying sandstone. Locality 6, Wheeler Canyon.

veloped during, and by the emplacement of, the covering conglomerate. If formed earlier, the edges could not have remained so sharp; if developed later, the surface of the conglomerate would also have been distorted. There is no evidence that the emplacement of the sand was accompanied by erosion of the conglomerates.

Although the largest cobbles or boulders may show a slight tendency to occur either in the middle or at the lower side of some beds, there is no sorting or grading. The larger fragments are probably in contact with each other, the interstices being occupied by fairly clean sand. The sandstones and shales, with which they are interbedded, show the normal degree of grading.

The enumerated features do not point toward the activity of turbidity currents, nor does it appear probable that the cobbles and boulders rolled down a submarine slope individually. As the maximum angle of repose is different for various grain sizes, the shales, sandstones, and conglomerates would wedge out either up- or down-slope. The rolling or sliding cobbles would have disturbed and eroded the sands and lutites over which they passed. No imbricate structures, which are typical of conglomeratic, deltaic, foreset beds, were observed. We are thus led to infer that submarine landslides or sandy mudflows were responsible for the emplacement.

The localized occurrence points to the accumulation taking place off a river mouth, not off a cobble beach of any great lateral extent. The slope on which the series of beds was laid down must have been equivalent to that of the surrounding sea floor. The normally graded sand beds and shales, and the absence of slump structures, certainly does not fit the picture of an abnormally steep slope. The structural slope, being at present equal to that of the beds in the continuation along the strike, points to the same conclusion. We are, therefore, not dealing with the foreset beds of a delta. The normal bottom slopes of the Ventura Basin during sedimentation are not known accurately, but the depth range

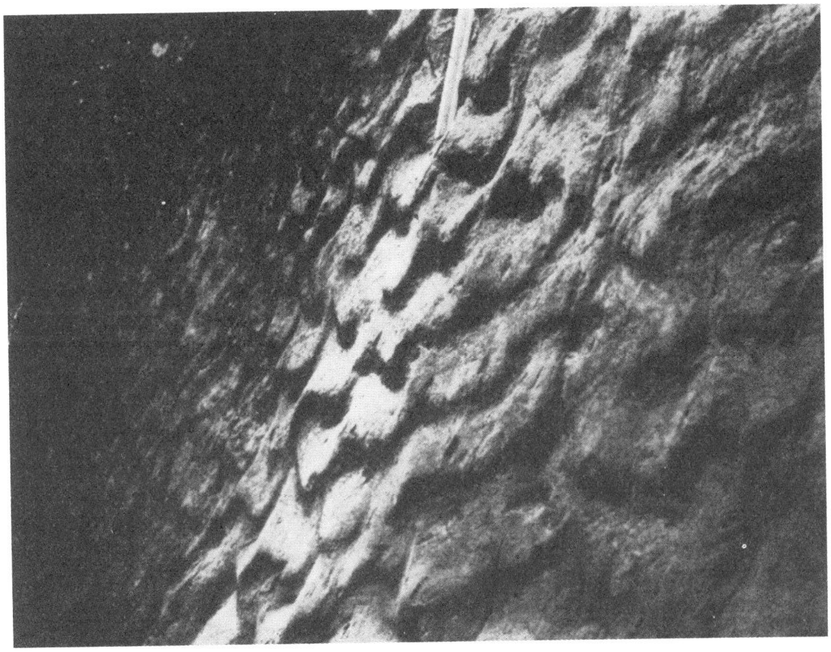

FIG. 25.—Photograph showing close view of current ripple marks at Locality 6. Note cuspate, asymmetrical shape and linear arrangement.

and horizontal extent together point to approximately a few degrees as the maximum that can be assumed. Even if the slope higher up were considerably steeper, the locality at which the conglomerates accumulated must have been situated several miles offshore, at least. The limited spread along the strike confirms the conclusion that we are not dealing with a deltaic formation.

The available field evidence is not sufficient to determine whether the conglomerates accumulated in a valley, or as a wide fan in front of such a feature, or that the mud flows proceeded down a flat slope. Possibly, detailed studies might reveal significant features to solve this problem of a fossil submarine valley. There is at least one line of evidence that could be followed which might prove significant; namely, the occurrence of ripple marks in between the conglomerates.

Ripple Marks in the Conglomerate Series

In a number of horizons between the deep-water conglomerates of Wheeler Canyon, large fields of asymmetrical, cuspate ripple marks are preserved. In all examples observed, the steeply dipping bedding plane is entirely covered with an extremely regular pattern of parallel ridges (fig. 25). The largest area exposed is roughly 1,000 square feet. Obviously, these fields must originally have been quite extensive, possibly covering the entire extent of the bedding plane. Moreover, the pattern in all cases shows roughly the same orientation, with the ridges nearly parallel to the structural strike and with the steep side facing what is thought to have been the original down-slope. The continuity and steepness of the ridges varies somewhat, but the ripple length is generally about the same, some 6 inches from crest to crest. The pattern is developed in thinly bedded, fine-grained deposits which occur between the conglomerates and is usually capped by a half-inch or so of fine silt or clay.

The asymmetrical shape of the ripples indicates that they were produced by currents. These ripple marks are restricted to the area of the conglomerates and are absent on most bedding planes even there, although conditions for preservation appear to have been favorable in the Ventura Basin, as fine laminations, slump structure, etc., have been perfectly preserved. This strongly indicates that the currents forming the ripples were not a normal feature of the circulation in the deep, ancient basin, or even the result of occasional basin-wide occurrences such as tsunamis, but that they were restricted to the same environment as the conglomerates and occurred only occasionally. This leads us to suggest that the ripple-forming currents were dilute turbidity currents, of sufficient velocity to move a light bed load by traction, but not able to lift an appreciable amount of sediment from the sea floor. Outside the area of the conglomerates, such currents did not develop. The only reason the authors can suggest is that the conglomerates were deposited in or at the lower end of a submarine valley. The walls of this valley, by hemming in a dilute turbidity current, would prevent lateral spreading. In this manner, a large hydraulic radius would be maintained, resulting in a higher velocity than on a wide slope where a spreading and thinning dilute turbidity current would creep slowly down slope, unable to ripple its floor.

CONCLUSIONS

1. The faunas indicate that the lower Pliocene was laid down in water 4,000 to 5,000 feet deep, and that—from that time to the present—a gradual shallowing to sea level, of the waters over the basin, took place.

2. The conglomerates and graded sands of the lower Pliocene were deposited at depths greater than 4,000 feet. The sands and conglomerates of younger formations were deposited in progressively shallower water toward the Recent.

3. The distribution of redeposited foraminifera in the section points to a continually uplifting northern marginal area and a downwarping central area. The tilting accompanying this process created slopes favorable for the generation of turbidity currents and landslides.

4. The emplacement of the graded sands in the deep water appears to have been accomplished by the action of turbidity flow, rather than by other currents of non-turbid origin.

5. The conglomerates of the Pliocene were deposited by submarine landslide action.

REFERENCES

Bailey, T. L., *1935*. Lateral change of fauna of the lower Pleistocene: *Geol. Soc. America Bull.*, vol. 46, pp. 489–502.

Emery, K. O., *1950*. Contorted Pleistocene strata at Newport Beach, California: *Jour. Sedimentary Petrology*, vol. 20, pp. 111–115.

Heim, A., Moser, R., and Bürkli-Ziegler, A., *1888*. Die Catastrophe Von Zug, 5 July 1887: Zurich, pp. 1–60.

Kuenen, Ph. H., *1937*. Experiments in connection with Daly's hypothesis on the formation of submarine canyons: *Leidsche Geol. Med.*, vol. 8, pp. 327–335.

———, *1948*. Slumping in the Carboniferous rocks of Pembrokeshire: *Geol. Soc. London Quart. Jour.*, vol. 104, pp. 365–385.

———, *1950*. Turbidity currents of high density: *Rept. 18th Int. Geol. Congress*, part 8, London, pp. 44–52.

———, and Migliorini, C. I., *1950*. Turbidity currents as a cause of graded bedding: *Jour. Geology*, vol. 58, pp. 91–127.

Macar, P., and Antun, P., *1950*. Pseudo-nodules et glissement sous-aquatique dans l'Emsien inférieur de l'Oesling: *Ann. Soc. Géol. de Belgique*, vol. 73 B, pp. 121–150.

Migliorini, C. I., *1950*. Dati a conferma della risedimentazione delle arenarie del macigno: *Atti. Soc. Toscana, Sci. Nat. Mem.*, vol. 57, ser. A, pp. 3–15.

Natland, M. L., *1933*. Depth and temperature distribution of some Recent and fossil Foraminifera in the southern California region: *Bull. Scripps Inst. Oceanography*, La Jolla, California.

Pettijohn, F. J., *1950*. Turbidity currents and graywackes—a discussion: *Jour. Geology*, vol. 58, pp. 169–171.
Reed, R. D., and Hollister, J. S., *1936*. Structural evolution of southern California: Am. Assoc. Petroleum Geologists publication.
Shrock, R. R., *1948*. *Sequence in layered rocks:* McGraw-Hill Book Co., Inc., New York, 507 pp.
Woodford, A. O., *1951*. Stream gradients and Monterey sea valley: *Geol. Soc. America Bull.*, vol. 62, pp. 811–813.
Woodring, W. P., *1938*. Lower Pliocene mollusks and echinoids from the Los Angeles Basin, California: *U. S. Geol. Survey Prof. Paper 190.*

Deltaic Environment Reservoir Types and Their Characteristics

R. M. Sneider, SPE-AIME, Sneider and Meckel Assocs., Inc.
C. N. Tinker, SPE-AIME, Shell Oil Co.
L. D. Meckel, Sneider and Meckel Assocs., Inc.

Introduction

The need to increase ultimate recovery requires that the engineer and earth scientist work together to depict accurately the character and geometry of a reservoir. This synergistic relationship results in a better understanding of the quality, distribution, and continuity of the reservoir and its contained fluids.

As most known accumulations of oil and gas trapped in terriginous (land-derived) sediments are contained in sandstones deposited in deltas, there is an increasing need to understand these reservoir types better. Knowledge of (1) the type of reservoir potentially available, (2) the distribution and quality of pore space in terms of porosity, permeability, and capillary pressure properties, and (3) the location of barriers to flow, both internal and external, should enhance greatly the ultimate recovery from this important source of oil and gas.

The need for information about these reservoirs prompted this summary paper, which is based solely on and distilled from information available in published sources.

In view of the spectrum of environments in which deltas, both ancient and modern, are known to exist, it would be foolhardy in this summary to attempt a review of the myriad types available. Therefore, only the processes responsible for creating the end members of a continuum of deltaic sedimentation and the characteristics of reservoirs created by these processes are presented. With this information and enough data, the inquiring geological/engineering team should be able to build on this foundation and depict accurately any specific reservoir in the deltaic environment.

0149-2136/78/0011-6701$00.25

Principles of Deltaic Sedimentation

Deltas are abundant in the geologic record and are known to exist in ancient, Recent, and modern settings. For example, 32 large deltas are forming at this time (Fig. 1). Countless others are in various stages of growth throughout the world.

In 1912, Joseph Barrell described a delta as "a deposit partly subaerial (above sea level) built by a river into or against a body of permanent water." Three prerequisites for deltaic sedimentation are apparent from this apt description (Fig. 2): (1) drainage basin, a source of sediment, (2) river, to move the material, and (3) receiving basin, to store and rework it. As a delta forms, the outer and lower parts are constructed below water, and the upper and inner surfaces become land "reclaimed" from the sea. Therefore, even though deltas are visualized commonly as subaerial, their subaqueous portions are truly significant.*

*Deltas formed entirely in a subaerial environment are called "alluvial fans." When similar principles apply, these are not discussed here because of significant differences in many aspects of reservoir continuity.

Processes responsible for creating end members of a continuum of deltaic sedimentation and the characteristics of reservoirs resulting from these processes are presented. Deltas tend to be either high-energy sand deltas or low-energy mud deltas. In either case, reservoirs generally are confined to the bar-fringe sands and/or to distributary channel sands. Frequently, reservoir quality is discontinuous.

A delta is created by an iterative process of progradation (built outward) in which sediment literally builds on itself until a particular pathway is no longer available and another delta forms in a different location. Thus, any given delta is a composite reflecting conditions at hand from the time of inception until ultimate abandonment of a particular deposition center.

The process, given the right conditions (large source, good distribution system, low dispersal potential), can be remarkably creative and very fast. For example, in just the past century, the modern "bird-foot" portion of the Mississippi River has grown spectacularly from a system of isolated distributary channels to an area of sediment covering tens of miles (Fig. 3). The bird foot is the coalescent result of several lobes of sediment, each initiated at a different time (Fig. 4). Similarly, the modern Mississippi Delta complex was formed only during the past 5,000 years and now encompasses hundreds of square miles (Fig. 5).

As conditions of source, distribution, and dispersal vary, so do the deltas they produce. However, based on the dispersal energy of the receiving basin relative to the volume of sediment being introduced, deltas can be placed in two general classes or end members and characterized as follow.

1. High-energy or sand deltas have few active distributary channels; distributary channels are meandering; and shorelines are continuous sand. Modern examples are the Nile, Rhone, and Brazos-Colorado rivers (Fig.6).

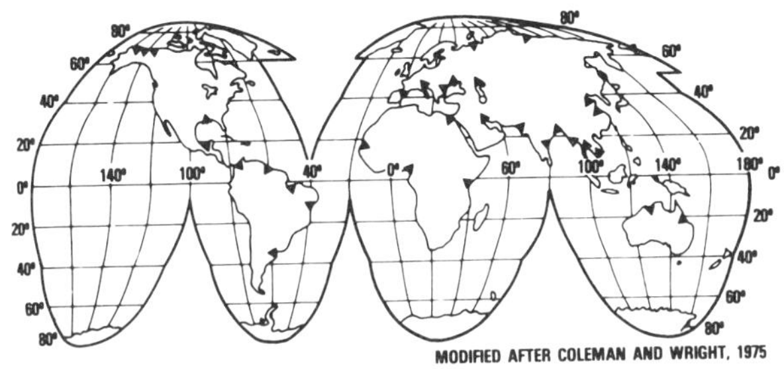

Fig. 1—Major modern deltas.

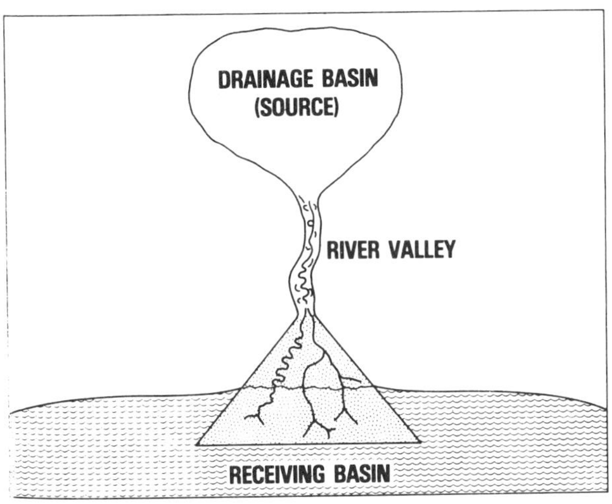

Fig. 2—Delta prerequisites.

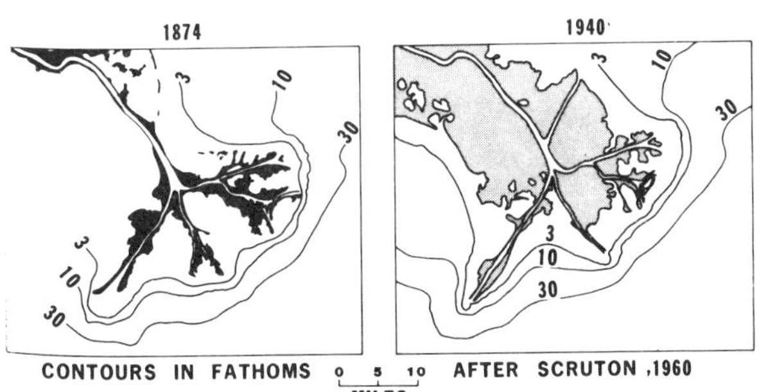

Fig. 3—Modern Mississippi delta growth.

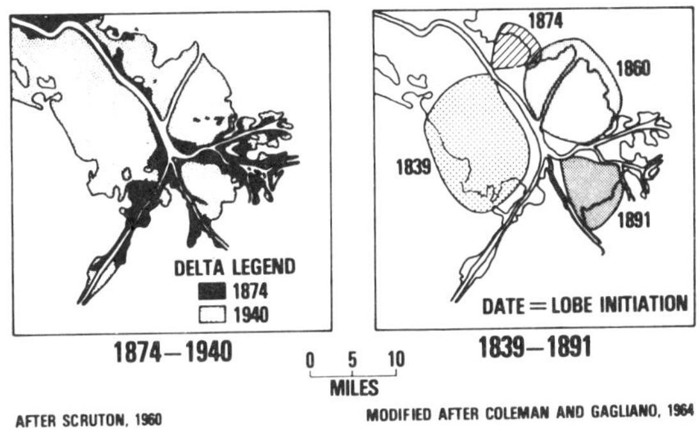

Fig. 4—Modern Mississippi delta growth from 1839-1940.

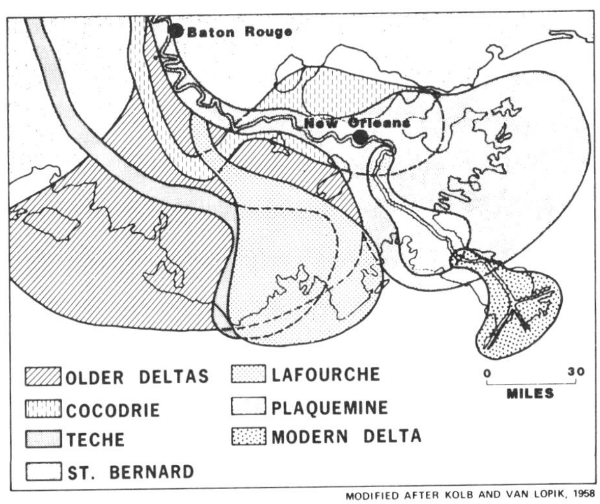

Fig. 5—Mississippi delta growth during the past 5,000 years.

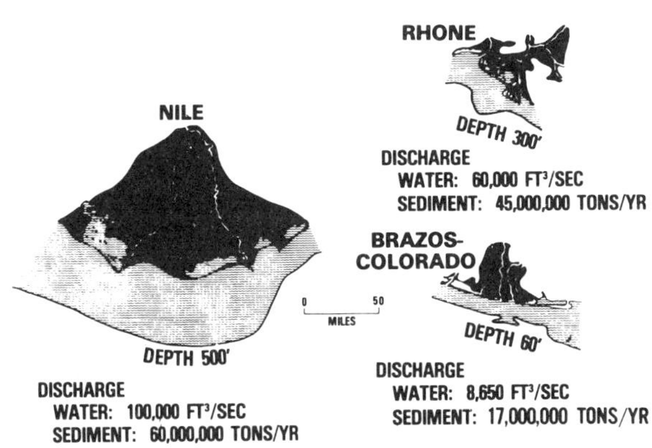

Fig. 6—Sand deltas.

2. Low-energy or mud deltas have many bifurcating (branching) distributary channels; distributary channels are straight to sinuous; and shorelines are discontinuous sands and muds. In addition to the modern Mississippi River, modern examples are the Orinoco, Lena, and Volga rivers (Fig.7).

These characteristics directly result from the volume and type of sediment available relative to the dispersal forces acting on it. Each characteristic bears directly on the quality, distribution, and continuity of reservoirs that may be created in the system.* For example, the nature and direction of forces molding the modern Niger delta (a high-energy sand delta) are illustrated in Fig. 8.

When such forces dominate the sediment load being introduced, fines (silt and clay) are winnowed, leaving a coarser-grained sedimentary environment on which the delta continues to build. This affects the pattern of the distributary system (a few long meandering channels active at any one time), the shape of the delta front and, thus, the continuity of sand deposition. In sand deltas, shorelines are frequently arcuate, and barrier bars are well preserved and continuous. The channels contain many point bars and, thus, throughout the complex there are many sand-on-sand contacts.

When these forces are not enough to dominate the sediment charge, the fines create a muddy or clay environment on which the delta builds. As a result, distributary channels are straight and bifurcate frequently (clay is more difficult to incise than sand), and the delta front is lobe-like, causing frequent interruption of the shoreline or fringe (bar) sands, which are relatively thin and discontinuous. The

*The impact of compaction, diagenesis, structure, and faulting, each of which bear on trap creation, are not covered here.

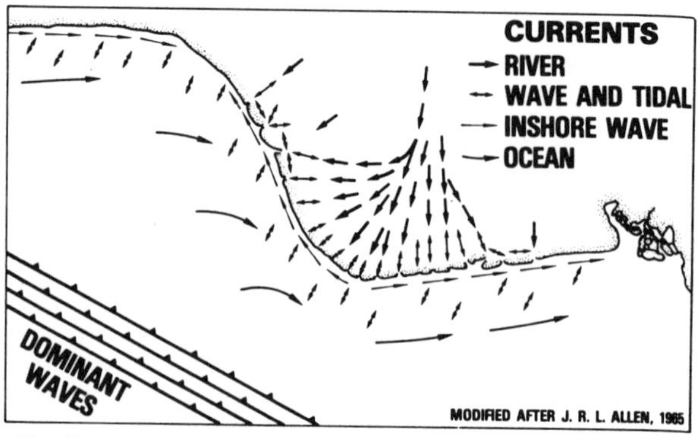

Fig. 8—Nature and direction of forces molding the Niger delta.

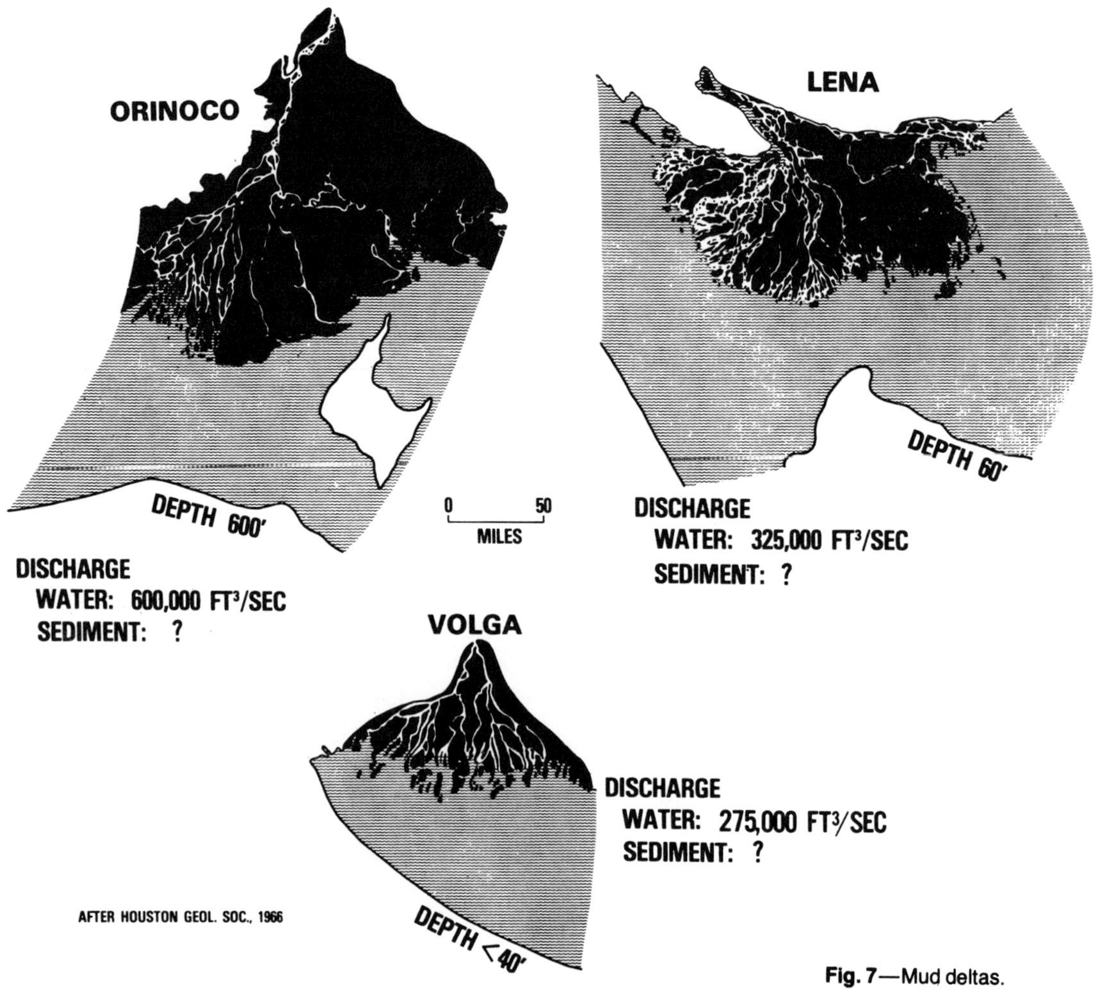

Fig. 7—Mud deltas.

channels usually are filled with sand (especially at the bottom) and clay-sized material. Thus, sand-on-sand contacts are not as frequent as in a sand delta.

Any delta system (whether high or low energy) develops two parts as it grows. One part is a regressive sequence of sediments (produced as the shoreline advances seaward) building upward from the bottom. The base of this sequence usually contains fine-grained, pro-deltaic silts and clays. The upper portions (near shore) can be highly variable, depending on the available dispersal energy, and range from thick, coastal barrier-bar sands to thin fringe sands to mud tidal flats and marshes (Fig. 9 dip section).

The other part of the delta system is its distributary channels. These may be meandering if cutting into a sand environment or relatively straight when cutting into clay. The resulting channel fill may be entirely sand or nearly all shale. Some mix of sand at the base and shale or silt at the top is most common (Fig. 9 strike sections).

Thus, the deltaic environment has two potential types of reservoirs—channel and bar sands. The rest of the delta sequence (silts, clays, and organic material) seals the reservoirs and frequently provides a prolific source of hydrocarbons to charge them.

Characteristic Reservoir Types

No two deltas are exactly alike in their distribution and continuity of reservoir and nonreservoir rock. However, many generalizations can be made about distribution and continuity from a synthesis of the numerous studies of deltas in Recent environments, outcrops, and the subsurface.

Before discussing reservoir/nonreservoir distribution and continuity, a review of the geologic factors controlling reservoir types (pore space) is worthwhile. The following geologic processes acting on a sediment during and after its deposition determine reservoir-rock quality and pore space.

Depositional		Diagenetic Processes
Texture	Grain size	Compaction
	Sorting	Cementation
	Grain shape	
	Grain packing	Solution
	Grain roundness	
Mineral composition		Fracturing

The most important parameter controlling reservoir quality in most unfractured reservoirs is the *size and sorting of the grains.* Studies of unconsolidated to slightly consolidated sands show that grain size and sorting correlate directly with porosity, pore size, and permeability. Even many moderately to well consolidated reservoir rocks in which compaction and/or cementation have reduced significantly original pore space, the porosity, pore size, and permeability correlate with grain size and sorting. Except for reservoir rocks that are tightly compacted or highly cemented, texture is the important control of reservoir quality.

In deltaic deposits, the interplay among the type and amount of sediment brought into a basin and the forces acting on and dispersing the sediment control the texture and, thus, the initial pore space.

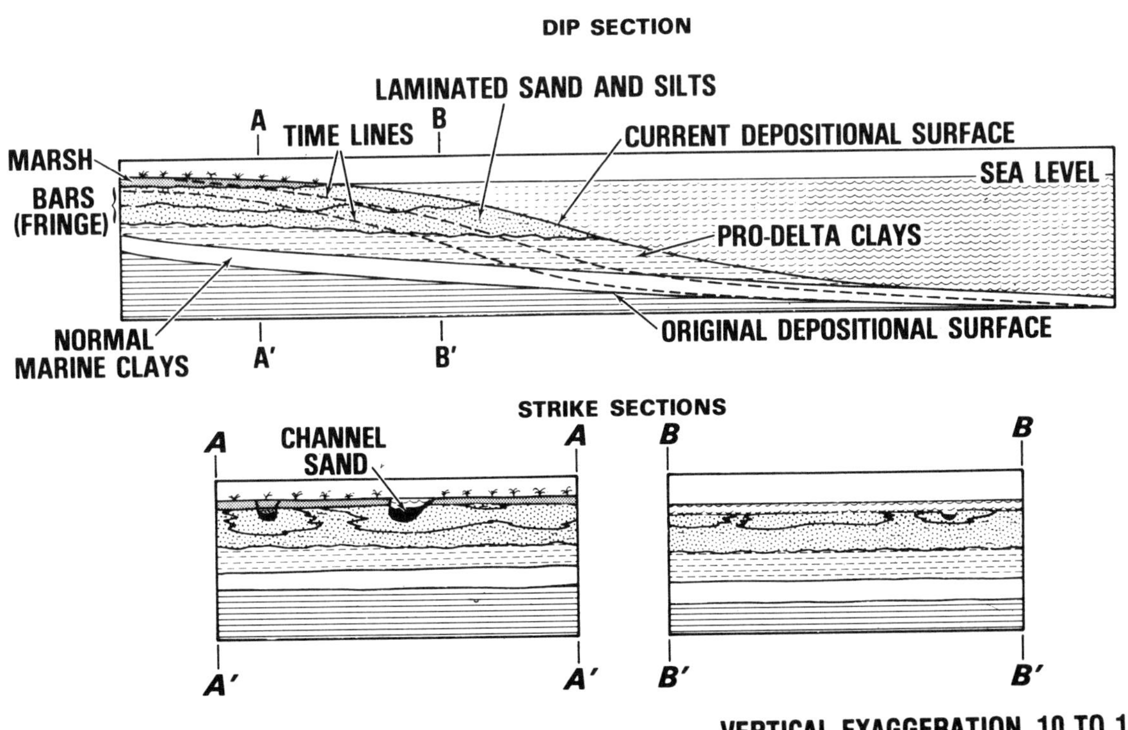

Fig. 9—Delta growth for dip and strike sections.

The Deltaic Sequence

Most reservoirs in deltas are in the channel and bar (fringe) deposits. Fig. 10 illustrates typical sequences and relationships for the reservoir and nonreservoir deposits formed by a delta building into a marine basin. The over-all, complete vertical sequence of deposits from top to bottom is (1) marsh (nonreservoir clay, silt, coal with a few reservoir sands of limited extent); (2)inner bar (reservoir sand with some clay/silt intercalations); (3) outer bar (reservoir sand with many clay/silt intercalations); (4) prodelta (nonreservoir clay/silt); and (5) marine (nonreservoir clay).

Channels of various dimensions cut down through the marsh-bar prodelta-marine sequence. The geometry of the channel deposits is a function primarily of (1) size of the delta, (2) position of the channel in the delta, (3) type of material being cut into, and (4) to some degree, the forces at the mouth of the channels distributing the sediments. For example, the channels of the modern Mississippi River generally are less than 600 ft wide and 35 ft deep where sediment is being dispersed into the Gulf of Mexico (e.g., Southwest Pass or Southeast Pass), but upstream on the delta plain, the Mississippi River is 2,000 to 3,000 ft wide and has cut down 150 to 200 ft (e.g., at New Orleans). Channels can be filled with up to 90% sand or clay/silt.

Conceptually, the best reservoir-quality sand in the bar deposits is at or near the top, and quality deteriorates gradually downward and laterally away from the main sand buildup. In contrast, the best reservoir-quality sand in the channels is in the lower part, and sand quality decreases vertically upward.

An understanding of the expected distribution and continuity of reservoir-quality sand, clay/silt interbeds, and pore space in subsurface deltaic deposits can be gained best by comparing the previously described delta end members—sand deltas vs mud deltas.

High-Energy Sand Deltas

Sand deltas, like the modern Nile and Niger deltas, are characterized by nearly continuous sand bars cut by a few meandering distributary channels. This delta type shows excellent continuity of sand in the bars and channels, shown by the subsurface example from Nigeria (Figs. 11 and 12).

Bar Sand Continuity. These sands are continuous reservoirs. Continuous silt/clay interbeds are common only in the lower part of the bar deposit and decrease in thickness and areal extent vertically upward. In a bar deposit, grain size increases and sorting improves vertically upward, so that the coarsest and best sorted sand occurs at the top of the deposit parallel to the shoreline or trend of the bar. Over-all grain size, sorting, and sand thickness decrease in both landward and seaward directions. Figs. 11 and 12 show that the reservoir from the Nigerian example actually is composed of two separate bar deposits and a channel-fill deposit.

Channel Sand Continuity. Channels in a sand delta are typically the result of meandering streams, which create point-bar deposits. The coarsest grain sizc and poorer sorting occurs in the lower part of the channel. The coarser-grained sediments in the lower portion are overlain by progressively finer-grained sediments toward the top. The sands in the lower portion of the channel are continuous. Some local discontinuities are formed by thin clay/silt interbeds that are usually less than 1 to 2 in. thick, covering less than 10 ft in areal extent. The amount and thickness of silt/clay interbeds increase toward the top of many channel deposits in sand deltas. Some channel deposits are filled completely with sand having excellent continuity, and essentially lack clay/silt beds with any appreciable areal extent (Fig. 12).

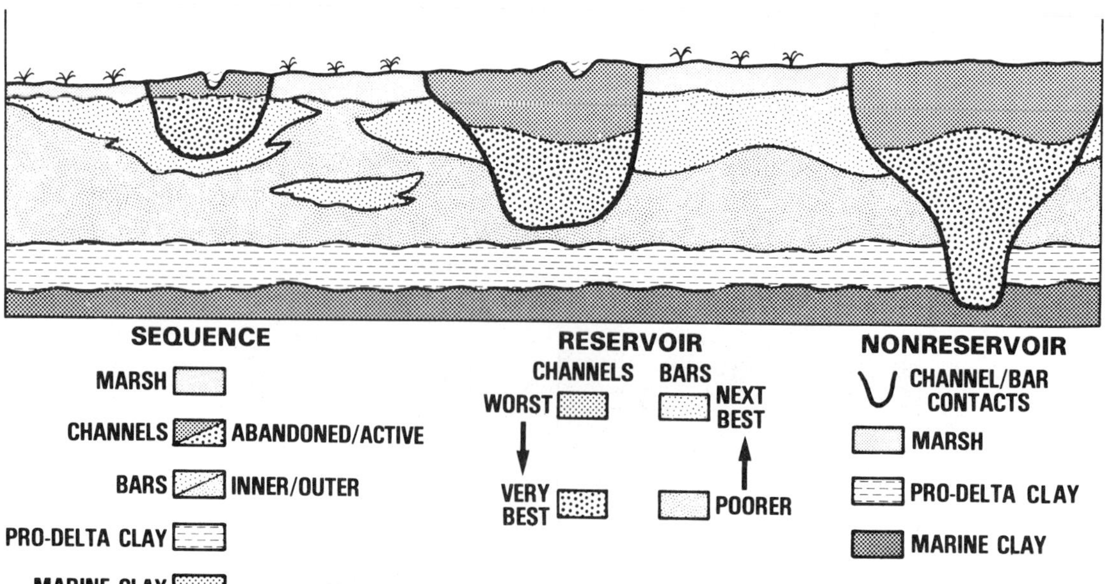

Fig. 10—Deltaic environment reservoir character and continuity.

In many channels, the amount of interstitial clay/silt and the number of interbeds increase near the channel boundary. Observations of channel boundaries in outcrops show that very thin (less than 6 to 8) in.) clay/silt beds line the boundary surface in such a way that the lining acts as a significant barrier to fluid flow.

There may exist a clay/silt barrier ("clay plug") formed originally by deposition in the last water-filled portion of the channel after the channel was abandoned. This plug can be situated anywhere in the channel, but is always on top of reservoir-quality sand. When present, this plug of nonreservoir material acts as a barrier to lateral fluid flow.

Pore-Space Distribution and Continuity. Initial pore space is controlled principally by grain size and sorting. For unconsolidated to moderately consolidated reservoirs, the general relationship between texture and pore space (porosity, pore size, and permeability) is illustrated in Fig. 13.

In the bar deposits, permeability increases vertically upward throughout the bar and is highest at or near the top. Fig. 14d is a typical bar permeability profile. Permeability profiles should be similar parallel to the trend of the bar. Permeability is highest at the top, near the shore, and decreases progressively seaward. Porosity should be well connected throughout the bar, except in the lower part.

Channel permeability profiles are illustrated in Figs. 14a, 14b, and 14c. In the lower portion of the sand, porosity and permeability will be continuous. Permeability decreases vertically upward, and the silt/clay interbeds increase in number, thickness, continuity, and areal extent.

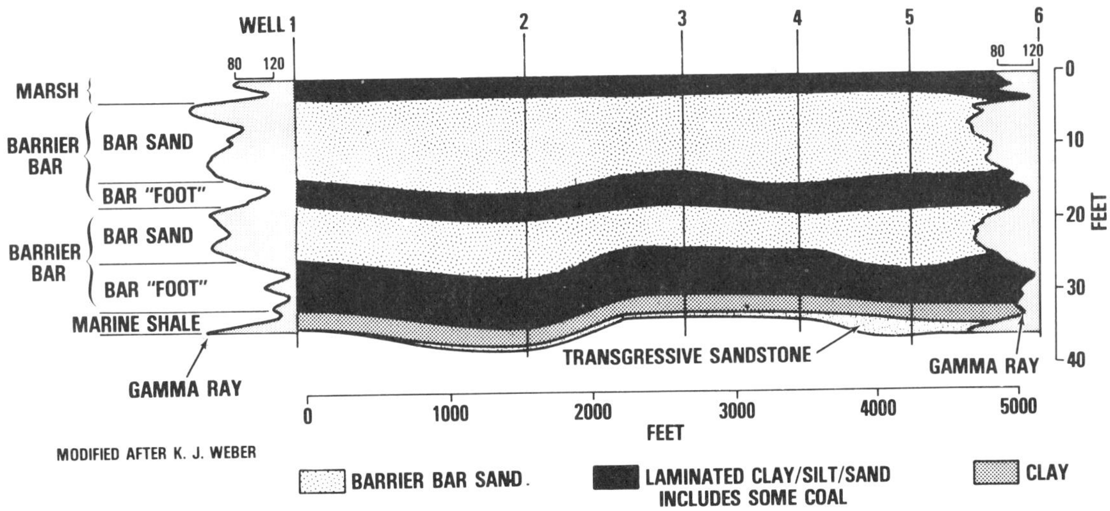

Fig. 11—Continuity in sand-delta system, Afiesere Field, Niger delta.

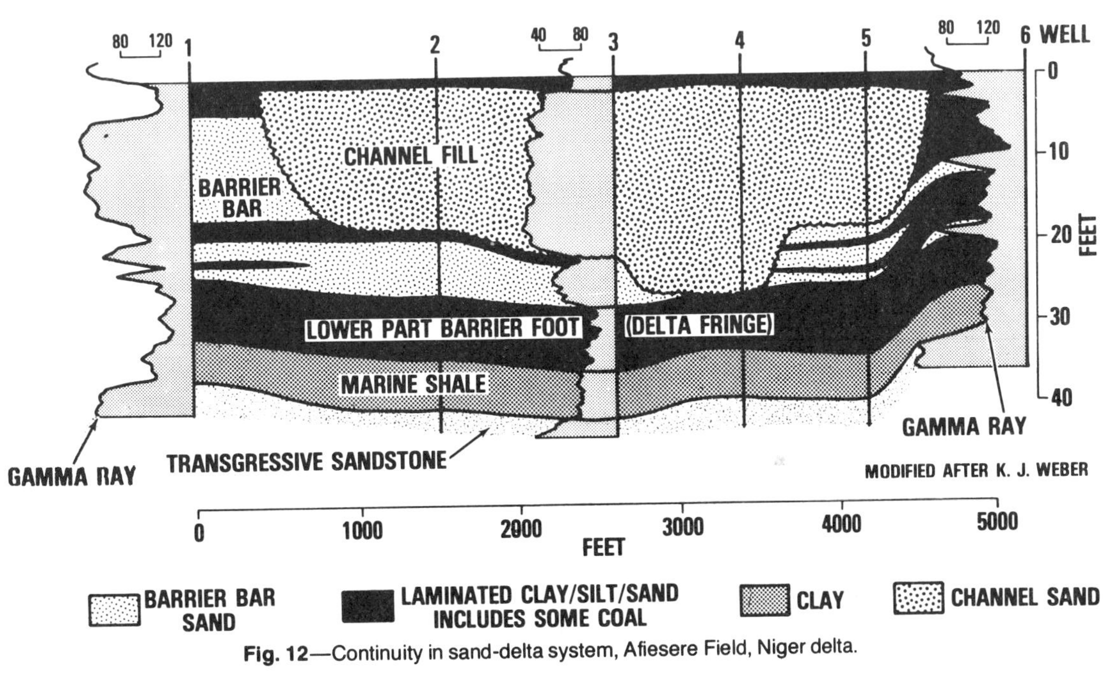

Fig. 12—Continuity in sand-delta system, Afiesere Field, Niger delta.

TYPE	TEXTURE		PORE SPACE		"CAPILLARITY"		CONTINUITY
	GRAIN SIZE	SORTING	POROSITY	PORE SIZE	PERMEABILITY	Sw	
CHANNELS Top	Finest	Best	Highest	Very Fine	Lowest	Highest	Deteriorates Upward
	↑	↑	↑	↑	↑	↑	↑
Bottom	Coarsest	Poorest	Lowest	Large	Highest	Lowest	Best
BARS Top	Coarsest	Best	Highest	Large	Highest	Lowest	Best
	↓	↓	↓	↓	↓	↓	↓
Bottom	Finest	Poorest	Lowest	Very Fine	Lowest	Highest	Deteriorates Downward

Fig. 13—Deltaic reservoir conceptual relationships.

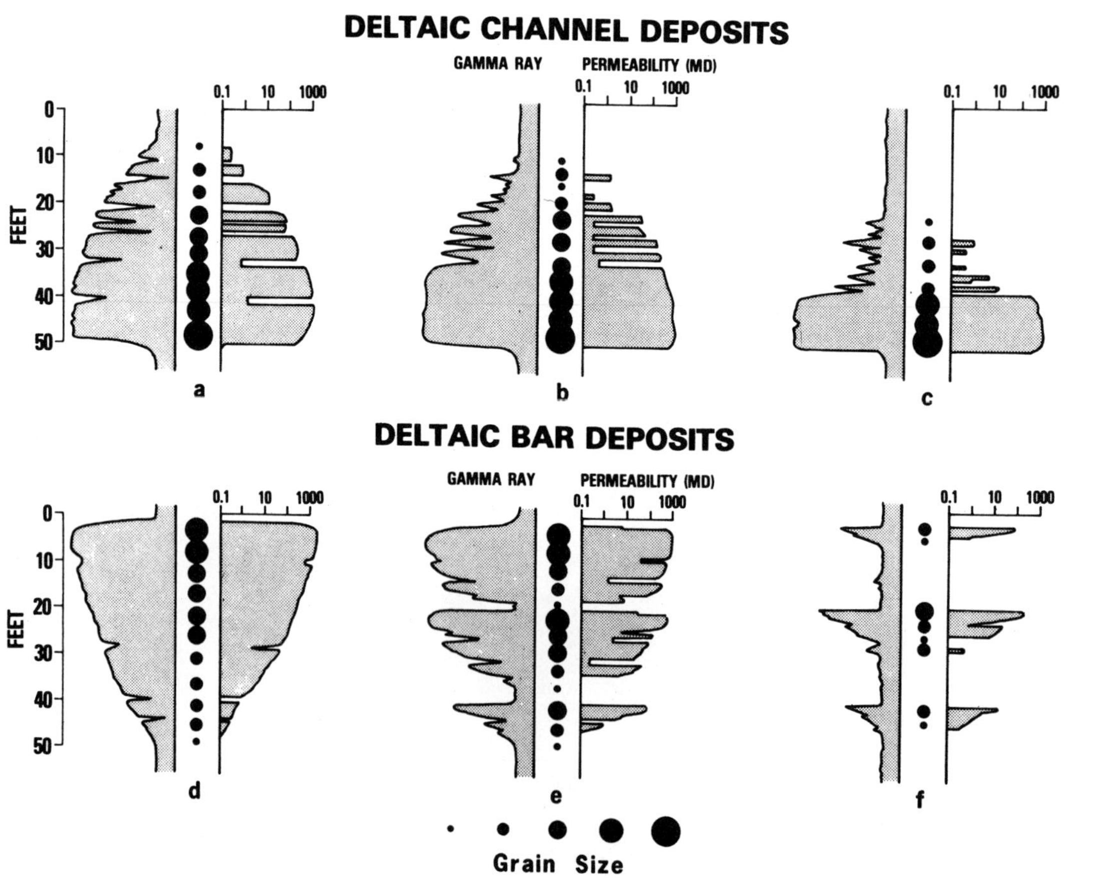

Fig. 14—Idealized permeability profiles and log responses.

Low-Energy, Mud Deltas

Mud deltas have shorelines with discontinuous sand (bars) and mud deposits that result from deposition from numerous separate, bifurcating distributary channels. The major differences between mud and sand deltas are (l) in a mud delta, the clay, silt, and sand are carried into a basin through numerous channels during floods, and (2) the clay/silt is not winnowed from the sand before the influx of sediment from the next flood. Thus, the sands are more discontinuous, and the clay/silt interbeds in the system are more numerous and continuous.

Bar-Sand Continuity. Figs. 15 and 16 illustrate in cross-section form the channel-bar relationship of two mud deltas in south Louisiana. The less continu-

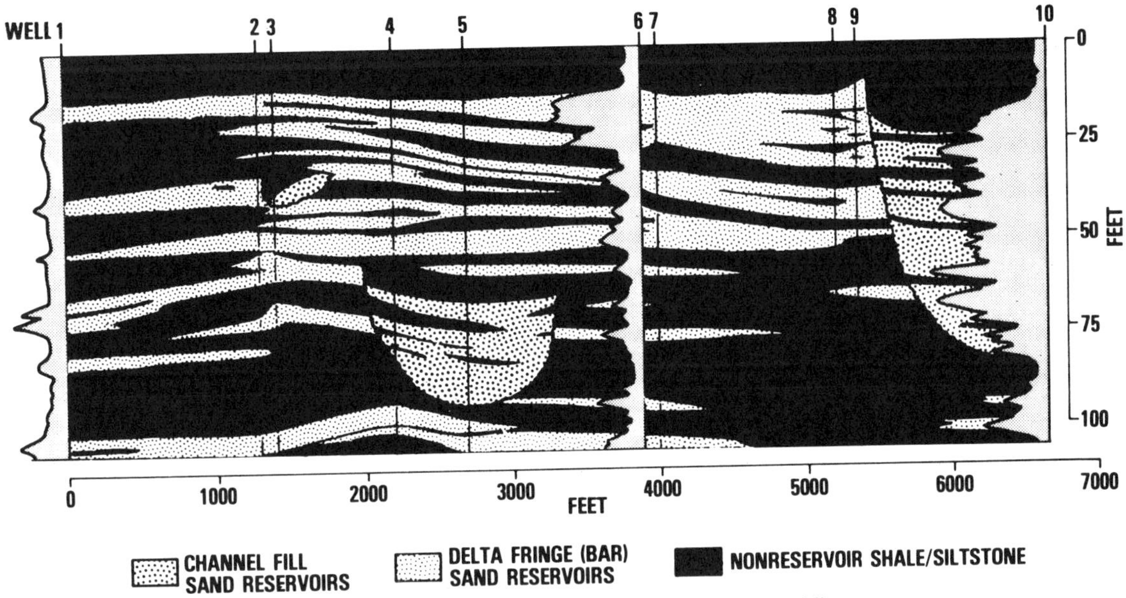

Fig. 15—Continuity in mud-delta system in south Louisiana Miocene.

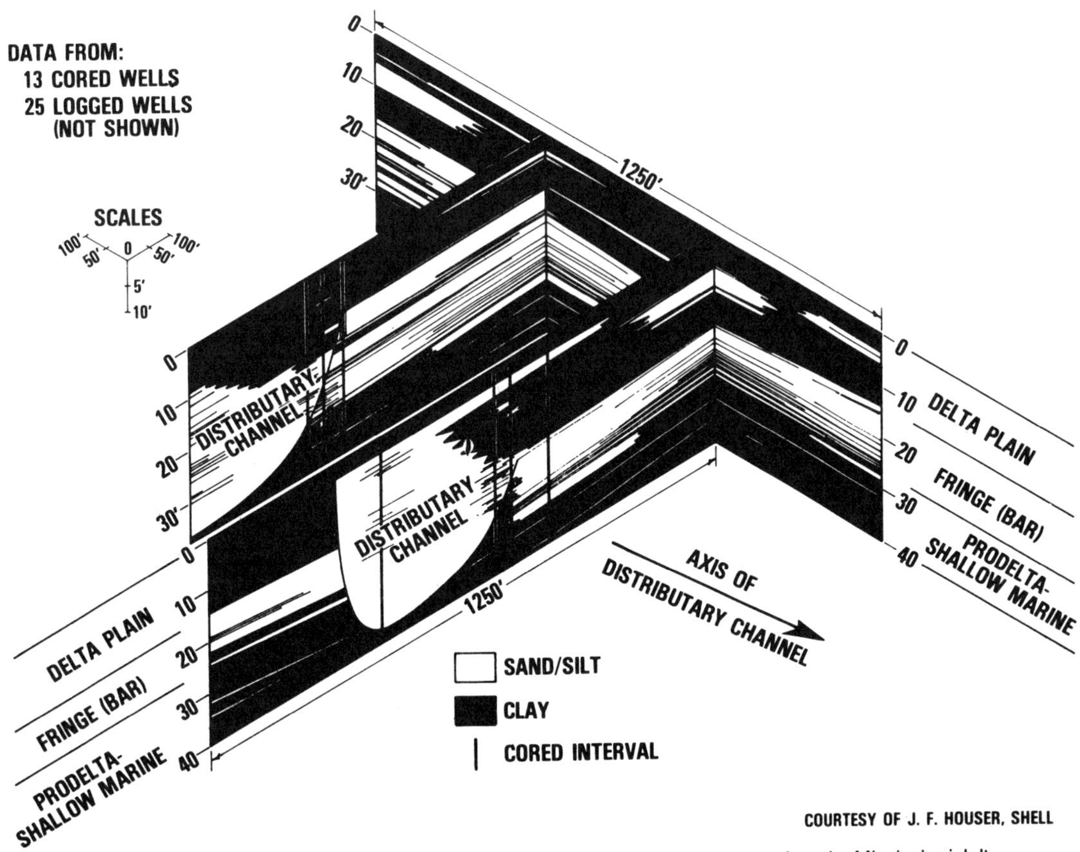

Fig. 16—Geometry and continuity of reservoir and nonreservoir units for Recent Lafourche Mississippi delta.

ous nature of the sands, because of the more continuous clay/silt interbeds, is evident, especially in the bar deposits (Fig. 5). Many clay/silt interbeds (usually less than 1 in. thick) are numerous and continuous over hundreds or thousands of feet, especially in the lower half of the bar (Fig. 16).

In the fringe or bar deposits, the grain size increases, sorting improves vertically upward, and silt/clay interbeds decrease in number, thickness, and areal extent. In individual bars, reservoir quality decreases laterally away from the sand deposition pathways. The best reservoir-quality rock (coarsest and best sorted) is at or near the top of the bar; quality decreases vertically downward and in a seaward direction. Because of the numerous, shifting distributary channels in a mud delta, individual pulses of sand (bars) are separated commonly by widespread clay/silt beds ranging from less than 1 ft to more than 12 ft thick.

Channel Sand Continuity. Sand always occurs in the lower part of distributary channels. Some sands locally have very thin clay/silt beds, generally less than 1 ft thick and from less than a few feet to a few tens of feet in maximum areal extent. These clay/silt interbeds are not barriers to vertical flow. The amount and thickness of sand and clay/silt above the continuous sand depend on how fast and where the channel is abandoned. With rapid abandonment or cutting off many miles away, good basal-channel sand grades rapidly upward into silt-clay/organic-rich impermeable sediment. The slower the abandonment or the closer to the point where the channel cuts off, the more sand filled the channel. Over all, as the channel fills, the number, thickness, and areal extent of the silt/clay interbeds increase vertically upward.

Along the channel boundary, thin clay/silt beds line the contact so that this lining acts as a fluid-flow barrier.

Pore-Space Distribution and Continuity. Figs. 14e and 14f are typical bar permeability profiles. In the bar deposits, permeability increases vertically upward and is highest at or near the top. Permeability decreases laterally away from the center of the bar buildup (the mouth of the distributary). The number, thickness, continuity, and areal extent of the clay/silt interbeds increase laterally away and downward from the bar buildup. Thus, pore space is best and porosity continuous only in the upward part of the bar.

Figs. 14b and 14c are typical permeability profiles for mud-delta distributary channels. The basal-channel sands are continuous throughout the distributary. The amount and quality of sand present above this continuous channel-sand layer depends on the location and rate of channel abandonment. With slow abandonment, or near the channel cutoff point, the upper part of the channel exhibits a gradual upward decrease in permeability and an increase in clay/silt interbeds. With rapid abandonment, or away from the channel cutoff point, the upper sequence comprises low-permeability thin sands with abundant clay/silt interbeds.

Summary

Most accumulations of oil and gas trapped in terrigenous sediments are contained in sandstones deposited in deltas. Deltas abound in the geologic record and are known to be created by an iterative process of progradation into a body of water. Based on the dispersal energy in the receiving basin relative to the volume of sediment introduced, two general classes of deltas exist—high-energy or sand deltas and low-energy or mud deltas. Each end member has characteristics bearing directly on the quality, distribution, and continuity of reservoirs in the delta. Each delta, whether high- or low-energy, is composed of (l) a regressive sequence of sediments in which the potential reservoirs are bar (fringe sands), and (2) distributary channels filled (at least partially) with sand. Thus, two prolific types of reservoir exist in the deltaic environment-channel sands and bar (fringe) sands.

The most important factor controlling reservoir quality is the initial size and sorting of grains. Thus, the best reservoir-quality sand in bar deposits is at or near the top; the sand then deteriorates downward and laterally away from the main sand buildup. The best reservoir-quality sand in channels is in the bottom and it then deteriorates upward. As might be expected, high-energy sand deltas have the best continuity of both bars and channel sands, while the low-energy mud deltas display much more discontinuity of sands because of the more continuous clay/silt interbeds.

Therefore, to exploit effectively these prolific sources of hydrocarbons, early determination of the original depositional setting of the delta as it bears on the type of reservoir potentially available, the distribution and quality of pore space, and the location of barriers to flow is essential. Effective application of this knowledge will enhance greatly the ultimate recovery from this important source of oil and gas.

Acknowledgments

We thank J. F. Houser, Shell Development Co., and Shell Oil Co. for providing Fig. 16. We also acknowledge J. G. Richardson of Exxon Production Research Co., who articulated the need for this study, and R. C. Merrill and M. L. Mardick of Shell Oil Co. for their encouragement and support. In addition, we give particular recognition and thanks to those who introduced us to the subject, willingly shared their knowledge, and encouraged us to continue building on that knowledge—the late Hugh Allen Bernard, and our friend, colleague, and mentor, Rufus J. LeBlanc, Sr.

Bibliography

1. Allen, J. R. L.: "Late Quaternary Niger Delta, and Adjacent Areas: Sedimentary Environments and Lithofacies," *Bull.*, AAPG (1965) **49**, 547-600.

2. Barrell, Joseph: "Criteria for the Recognition of Ancient Delta Deposits," *Bull.*, GSA (1912) **23**, 377-446.

3. Bernard, H.A. : "A Resume of River Delta Types," *Bull.*, AAPG (1965) **49**, 334-335.

4. Bernard, H. A. and LeBlanc, R. J., Sr.: "Resume of Quaternary Geology of the Northwestern Gulf of Mexico Province," *Quaternary of the U. S.*, Princeton U. Press, Princeton, NJ (1965) 137-185.

5. Bernard, H. A., Major, C. F., Jr., Parrott, B. S., and LeBlanc, R. J., Sr.: "Recent Sediments of Southeast Texas: A Field Guide to the Brazos Alluvial and Deltaic Plains and the Galveston Barrier Island Complex," *Guidebook No. 11*, Bureau of Economic Geology, U. of Texas, Austin (1976).

6. Busch, D. A. : "The Significance of Deltas in Subsurface Exploration," *Tulsa Geol. Soc. Digest* (1953) **21**, 71-80.

7. Busch, D. A.: "Genetic Units in Delta Prospecting," *Bull.*, AAPG (1971) **55**, 1137-1154.

8. Coleman, J. M.: *Deltas: Processes of Deposition and Models for Exploration,* AAPG, Tulsa (1976).

9. Coleman, J. M. and Gagliano, S. M.: "Cyclic Sedimentation in the Mississippi River Deltaic Plan," *Trans.*, Gulf Coast Assn. Geol. Soc. (1964) **14.**

10. Coleman, J. M. and Wright, L. D.: "Variability of Modern River Deltas," *Trans.*, Gulf Coast Assn. Geol. Soc. (1973) **23**, 33-36.

11. Coleman, J. M. and Wright, L.D.: "Modern River Deltas: Variability of Processes and Sand Bodies," *Deltas,* Houston Geol. Soc. (1975) 99-149.

12. Donaldson, A. C.: "Deltaic Sands and Sandstones," *Proc.*, Symposium on Recently Developed Geologic Principles and Sedimentation of the Permo-Pennsylvanian of the Rocky Mountains, Wyoming Geol. Assn. 20th Annual Conference (1966) 31-62h.

13. Fisher, W. L., McGowen, J. H., Brown, L. F., Jr., and Groat, C. G.: "Environmental Geologic Atlas of the Texas Coastal Zone—Galveston-Houston Area," Bureau of Economic Geology, U. of Texas, Austin (1972).

14. Fisk, H. N.: "Sand Facies of Recent Mississippi Delta Deposits," *Proc.,* Fourth World Pet. Cong., Rome (1955) Sec. 1/c, 377-398.

15. Fisk, H. N.: "Recent Mississippi River Sedimentation and Peat Accumulation," *Proc.,* Fourth Congress l'Avancement des etudes de Stratigraphie et de Geologie du Carbonifere, Heerlen, Compte Render (1958) **1**, 187- 199.

16. Fisk, H. N., McFarlan, Edward, Jr., Kolb, C. R., and Wilbert, L. J.: "Sedimentary Framework of the Modern Mississippi Delta," *J. Sed. Petrology* (1954) **24**, 76-99.

17. Fisk, H. N.: "Bar-Fringe Sands of the Mississippi Delta," *Geometry of Sandstone Bodies,* AAPG, Tulsa (1961) 29-52.

18. Kolb, C. R. and Van Lopik, J. R.: "Depositional Environments of the Mississippi River Deltaic Plain— Southeastern Louisiana," *Deltas in their Geological Framework,* Houston Geol. Soc. (1966) 17-61.

19. LeBlanc, R. J., Sr.: "Significant Studies of Modern and Ancient Deltaic Sediments," *Deltas,* Houston Geol. Soc. (1975) 13-85.

20. McGowen, J. G., Brown, L. F., Jr., Evans, T. J., Fisher, W. L. and Groat, C. G.: "Environmental Geologic Atlas of the Texas Coastal Zone—Bay City-Freeport Area," Bureau of Economic Geology, U. of Texas, Austin (1976).

21. Meckel, L. D.: "Holocene Sand Bodies in the Colorado Delta Area, Northern Gulf of California," *Deltas,* Houston Geol. Soc. (1975) 239-265.

22. Meckel, L. D. and Sneider, R. M.: "Application of Subsurface Data to Delta Exploration," *Finding and Exploring Ancient Deltas in the Subsurface,* AAPG, Tulsa (1975) B1-B11.

23. Rainwater, E. H.: "The Geological Importance of Deltas," *Deltas in Their Geologic Framework,* Houston Geol. Soc. (1966) 1-15.

24. Rainwater, E. H.: "Petroleum in Deltaic Sediments," *Deltas,* Houston Geol. Soc. (1975) 3-11.

25. Scruton, P.C.: "Delta Building and Deltaic Sequence," *Recent Sediments, Northwest Gulf of Mexico,* AAPG, Tulsa (1960) 82-102.

26. Shannon, J. P., Jr., and Dahl, A. R.: "Deltaic Stratigraphic Traps in West Tuscola Field, Talor County, Texas," *Bull.*, AAPG, (1971) **55**, 1197-1205.

27. Sneider, R. M.: "Some Oil Occurrences in the Tar Springs (Mississippi) Delta, Illinois," *Bull.*, AAPG (1965) 350.

28. Swann, D. H.: "Late Mississippian Rythmic Sediments of Mississippi Valley," *Bull.*, AAPG (1964) **48**, 637-658.

29. Van Andel, T. H.: "The Orinoco Delta," *J. Sed. Petrology* (1967) **37**, 297-310.

30. Weber, K. J.: "Sedimentological Aspects of Oil Fields in the Niger Delta," *Geol. en Mijnbouw* (1971) **50**, 559-576.

31. Wright, L. D. and Coleman, J. M.: "Variations in Morphology of Major River Deltas as Functions of Ocean Wave and River Discharge Regimes," *Bull.*, AAPG (1973) **57**, 370-398.

Original manuscript received in Society of Petroleum Engineers office Sept. 20, 1977. Paper accepted for publication Feb. 23, 1978. Revised manuscript received Sept. 14, 1978. Paper (SPE-6701) first presented at the SPE-AIME 52nd Annual Fall Technical Conference and Exhibition, held in Denver, Oct. 9-12, 1977.

SANDSTONES

Reproduced by permission of the Geological Society from *Sedimentology: Recent Developments and Applied Aspects*, P.J. Brenchley, editor, pp. 189-220.

Clastic diagenesis

S. D. Burley, J. D. Kantorowicz & B. Waugh

SUMMARY: Diagenesis is the sum of those processes by which originally sedimentary clastic assemblages attempt to reach equilibrium with their environments. The subject has rapidly evolved, over the last 20 years, after several decades of routine petrographical analysis, to become a discipline of sophisticated analytical geochemistry. This progression may be traced through the evolution of depositional facies models, the growth of theoretical geochemistry, and the development and application of quantitative analytical techniques. Consequently, the study of diagenesis today involves the integration of data gathered from a range of interrelated disciplines.

The publication in recent years of such integrated studies, particularly from the Texas Gulf Coast of the U.S.A., has led to a number of wide ranging models being established. These models allow an interpretation of the sequence of authigenic minerals in terms of their relationship to the depositional environment or surface chemistry (eogenesis), burial or subsurface conditions (mesogenesis) and weathering or reexposure to surface conditions (telogenesis).

More specifically, once the pre-depositional controls on diagenesis have been established, it is possible to relate the inferred eogenetic mineral assemblages to exact geochemical sedimentary environments. Sedimentary mineral assemblages chemically are characterized by relative instability and so tend to interact with interstitial pore waters. Thus, in non-marine environments, mineral authigenesis may reflect arid-oxidizing, or wet-reducing pore-water conditions, and in the marine environment, either oxidizing or reducing pore waters.

During mesogenesis different processes become important. Elevated temperatures add energy to the reacting system, lowering reaction barriers and increasing reaction rates. Furthermore, widespread pore fluid migration at depth, transporting large quantities of solute, is likely to impart major, regional diagenetic changes to sediments. That there is a remarkable degree of consistency in deep burial settings suggests that depth-related processes conform to a predictable pattern.

Additional geochemical instability is introduced into the diagenetic system during uplift and exposure to the telogenetic realm. Minerals formed during burial, at elevated temperatures and pressures, and from concentrated formation waters, may become unstable in oxidizing, meteoric waters.

Diagenetic research requires the complete dissection of sedimentary rocks and, subsequently, the quantitative chemical and mineralogical analysis of their individual components. Such an approach to diagenetic studies, when related to an assessment of their paragenesis, may eventually lead to a predictive, integrated model for the evolution of sedimentary clastic rocks.

The purpose of this review is to highlight only the more significant research of the previous 20 years which we feel has advanced diagenesis from a prolonged era of Sorbian petrographical description into a period of multidisciplinary analytical geochemistry. These advances may be discussed in several distinct categories, namely the evolution of depositional facies models, the development and application of sophisticated analytical techniques, and the generation of theoretical geochemical models. Also important in providing an impetus to the subject has been the realization by some oil companies that an understanding of diagenesis may ultimately enhance recovery from hydrocarbon reservoirs. Additionally, we describe and discuss the results of recent studies which illustrate the approach which we feel may further advance our understanding of clastic diagenesis.

In spite of the number of books and papers written on the subject in the last 20 years, there is no universally accepted definition of diagenesis. Authors vary in restricting diagenesis to processes following deposition, or to processes prior to lithification. However, a case can also be made for extending diagenesis to include weathering, and the processes involved in altering rocks exposed at the earth's surface. We favour the latter approach, and would, therefore, describe the modifications made to rocks prior to reburial beneath unconformities as well as their complete destruction during continued weathering as diagenetic.

In addition to being variously defined in an absolute sense, diagenesis is also variously defined in a relative sense—i.e. with respect to metamorphism. Heroux *et al.* (1979) provided an up-to-date appraisal of the indices which may be used to separate metamorphism from diagenesis. Wisely however, they like many other authors, and indeed ourselves, do not attempt a rigorous distinction between the two. Burial diagenesis and metamorphism reflect a continuum of geological processes within which there is no single criterion for distinguishing between the two that can be used for all sediments. There is, therefore, no consensus as to what diagenesis is, when it starts, or when it stops. We do not propose a rigid definition of the subject because considerable advances can been made by the application of techniques and theories from loosely interrelated disciplines. Consequently, it is preferable perhaps to examine all aspects of all sediments, rather than to prejudice one's understanding by preselecting those areas on which to concentrate study.

In prefacing a review of clastic diagenesis it is, however, important to point out how vaguely defined the subject is. This is because clastic diagenesis does not lend itself to a straightforward 'State of the Art' review. Numerous subdivisions and classifications have been proposed but none adopted universally. Conflicting terminology abounds, different methods of routine investigation are pursued by those groups with a different approach, and finally, whilst the identification of authigenic minerals and diagenetic modification is now universally accepted, it is probably fair to say that we have made little progress towards understanding exactly how these processes take place (cf. Bjørlykke 1979; Blatt 1979; Boles & Franks 1979b; Land & Dutton 1979; Wood & Hewett 1982; Haszeldine *et al.* 1984).

The terminology of Schmidt & McDonald (1979a), subdividing diagenesis into eogenesis, mesogenesis, and telogenesis, is used in this review. Although this terminology is a genetic scheme and is, therefore, strictly an interpretative classification, it is preferred to the schemes which merely divide diagenetic processes into early and late stages. These often make little reference to the role of surface chemistry in diagenesis and therefore are not readily applicable to studies designed to elucidate the relationship between diagenesis and the depositional pore-water chemistry. Also, other classifications often exclude weathering by making reference to 'after deposition' or 'during burial', and we would not exclude weathering so readily.

The plates accompanying this review are intended to exhibit the range of textures and diagenetic products described in the text. Plates have also been chosen to demonstrate the application of specific techniques to the resolution of diagenetic problems. This selection is not intended to be exhaustive, it is hoped that the references herein may satisfy readers who wish further to investigate particular aspects.

Towards integrated modelling

Clastic diagenesis is low-temperature geochemistry; it is everything that contributes to making up a clastic rock, from the weathering of its source until its metamorphism. We are aware of many excellent studies on isolated aspects of a clastic sedimentary rock, but too few integrated or holistic ones. In this review we intend highlighting and promoting an integrated approach to the study of clastic diagenesis. Consequently, before briefly describing historically how the subject has developed, it is necessary to outline the philosophy and the style of investigation which is currently advancing our understanding of diagenesis.

The integrated approach involves initially establishing *all* the conditions which influenced the composition of the original clastic sedimentary assemblage, as well as its depositional pore waters. Secondly, by 'taking the rock apart', determining whether the constituents are detrital or authigenic and in what order the latter were formed. Thirdly, evaluating the entire burial history of both the sediment itself as well as any other sediments within the basin that are likely to influence pore-water composition in the subsurface. Finally, using the information above, combined with existing geochemical models, to explain how all the constituents were formed and, if possible, when. With the resulting information it may be possible to establish whether specific diagenetic modifications are related to any particular aspect of the history of the sediment, such as a distinct source area (see Stalder 1975), the depositional environment, burial temperatures and pressures, or hydrocarbon migration.

Historically, clastic rocks were originally investigated by mineralogists and petrologists. Only more recently, however, have sufficient data become available in related fields to allow some of the diagenetic processes involved to be understood more fully. These data result from research into topics as diverse as recent oceanic sediments, soils and aquifer properties, incorporating investigations of biological, chemical and hydrological processes. Results from many of these studies are included in the following sections, either specifically or within the more general reviews referenced. Unfortunately, much of this research is outside the scope of this paper and as a consequence can be mentioned only in passing.

The DSDP projects have produced a wealth of data concerning the mineralogy, biology and chemistry of surficial deep-sea sediments (see Warme *et al.* 1981, and references therein). Similarly, Berner and his co-workers in the

FOAM Group have studied shallow marine sediments intensively in recent years (see Krom & Berner 1981). The general results of these and other studies were concisely reviewed by Curtis (1978, 1980). Recently the work of hydrologists has become significant in view of our poor understanding of how material is transferred, and how large volumes of cement are produced in the subsurface. Their results suggest simple models of fluid flow through sediments may not be applicable (Wood & Hewett 1982; see also Domenico 1977 for a review). In addition the considerable volume of data collected by soil scientists in recent decades is now being utilized by sedimentologists searching for suitable analogues for their observations of the ancient sedimentary record. Dixon & Weed (1977) and Fitzpatrick (1980) exemplify pedology *per se*, whilst the investigations by Nahon, Tardy and their co-workers illustrate its possible application in a geological context (see Tardy 1971; Gac & Tardy 1980; Nahon *et al.* 1980). Wilson (1983) illustrated how study of a range of sediments may significantly contribute to analysis of the sedimentary record.

The early days

Petrography, the description of rocks, began in the 1800s with the work of Sorby (see Folk 1965). Early work though, was often mainly descriptive and interpretation of diagenetic processes did not really begin until the 1960s. Thus, by the time quantitative geochemical explanations began to appear in the literature, features as diverse as quartz overgrowths and berthierine ooliths had been recorded for up to a century (Phillips 1881).

Although our understanding of diagenesis is mostly the result of very recent advances, some notable contributions to the subject were made on the basis of detailed petrography alone. Some of the descriptions and interpretations have been superseded with time although they are significant even today in having influenced several decades of interpretation. For example, Lerbekmo (1957) described authigenic montmorillonite from andesitic Tertiary sandstones in California. Its formation was attributed to dissolution of the andesitic detritus, irrespective of the depositional environment. By contrast Carrigy & Mellon (1964) and Shelton (1964) described authigenic clays in a variety of occurrences. They concluded that there was a relationship between the clay mineralogy and depositional environment.

Other workers, however, biased future thinking towards the subsequently controversial topic of 'pressure solution'. Both Waldschmidt (1941) and Taylor (1950) attributed the loss of porosity with depth in sandstones to solution and stylolitization. The textures they described were subsequently observed to be between overgrowths, thus demanding both a reconsideration of accepted depth/porosity gradients, and more significantly, a reevaluation of the problematical source of silica in diagenesis (see Blatt 1979).

The advent of sedimentary geochemistry

Diagenesis can be resolved into a number of distinct thermodynamic processes. The classic work of Garrels & Christ (1965) showed how mineral thermodynamics can be applied to sedimentary systems and its importance for understanding diagenetic processes. Although there are many shortcomings in the application of thermodynamics and kinetics to diagenetic studies, careful use of some of the concepts is an invaluable aid to defining and understanding some of the processes that take place during diagenesis (Berner 1971; Curtis 1978; Hutcheon 1981).

The basis for diagenetic reactions is conveniently summarized by the 'law' of mineral stability (Keller 1969), which simply states that 'minerals are in equilibrium with their surroundings only in the environment in which they form'. This fundamental principle seems to have been generally underrated by sedimentologists in the past although it has been appreciated in soil science for many years. Goldich (1938) observed that the persistence of common rock-forming minerals during weathering was generally the reverse of the sequence of minerals crystallizing as Bowen's reaction series during magmatic fractionation. Minerals stable at the highest temperatures and pressures were apparently least stable under surface conditions and rapidly underwent alteration. This relationship was considered to result from the relative cation–oxygen bond strengths in the various minerals (Keller 1954). However, Curtis (1976) demonstrated that the persistence of a particular mineral at surface temperature and pressure conditions is largely a function of total released energy liberated during breakdown into new mineral products. Providing there is a net free energy loss in a diagenetic reaction then it should proceed in the diagenetic environment. Furthermore, the greater the free energy loss, the greater the tendency to react, and hence the more unstable the mineral. Thus, not only is mineral stability a function of the diagenetic environment, but also of the rate at which the minerals react.

In diagenetic studies this general concept has widespread applications. On deposition an original detrital mineral assemblage comprises a mixture of minerals formed under widely differing temperature–pressure regimes, usually higher than those prevailing at the earth's surface. Therefore, in the sedimentary environment detrital minerals are often either metastable or unstable with respect to their pore waters. Typically, the detrital assemblage is unstable as a whole and so during diagenesis may react with the ambient environment and move towards equilibrium. Kinetic models, however, indicate that despite favourable free energies, reactions need not proceed spontaneously because of an energy barrier which has to be overcome in the course of the reaction, so that it is possible for phases to persist metastably. Geological reaction barriers arise from several causes which may be physical, biological or chemical in origin (Berner 1980). In essence, the dominance of metastability during early diagenesis is a direct consequence of the slowness of chemical reaction rates at low temperatures (Berner 1971; Curtis 1978;

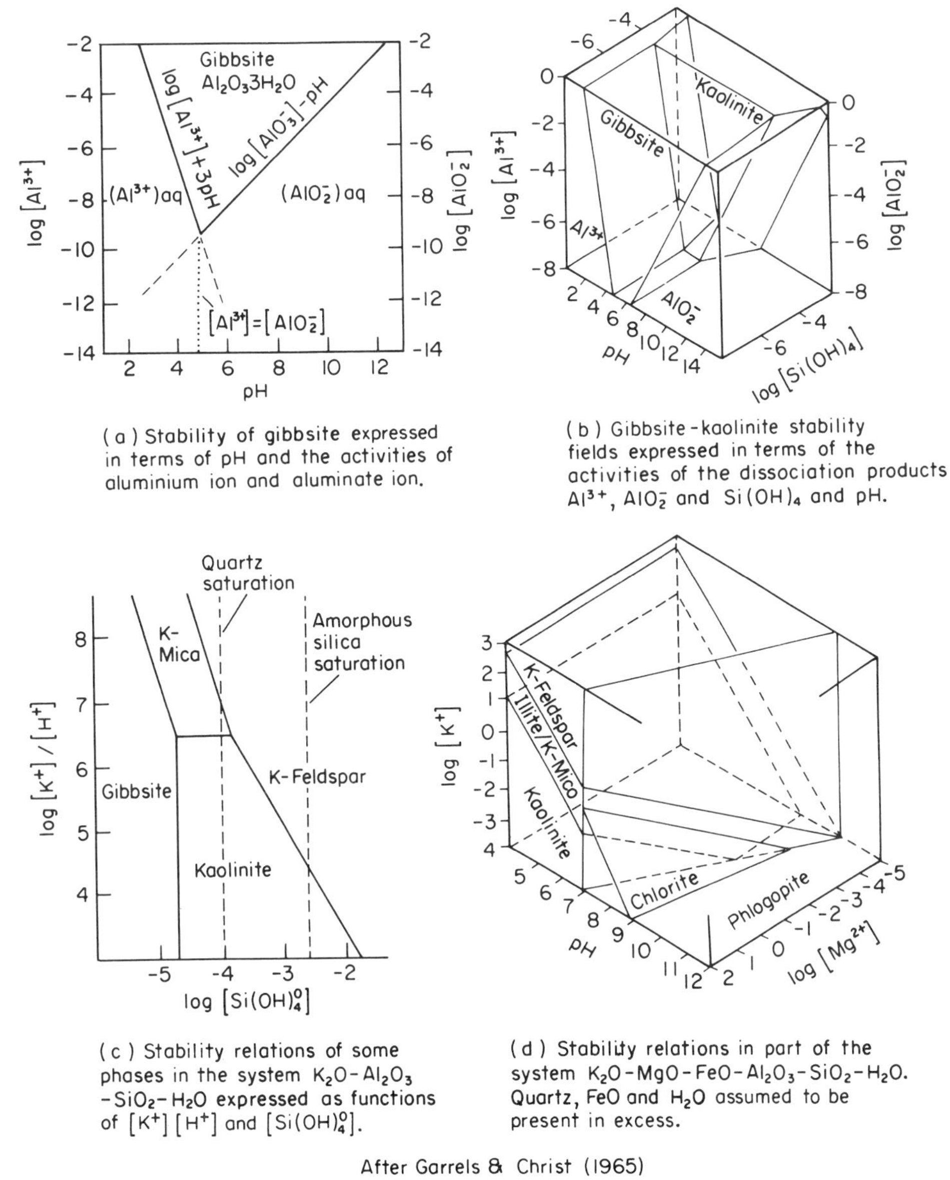

FIG. 1. The development of complex stability diagrams from a simple, two-dimensional, three-component system to multicomponent, three-dimensional diagrams (after Garrels & Christ 1965). (c) is a particularly useful plot because the ratio of $(K^+)/(H^+)$ in the hydrolysis reactions which define the stability fields is unity so that the complex stability relations can be expressed in terms of a two-dimensional plot.

Lippman 1982). Thus, despite complex interaction with pore waters, it is unlikely that true thermodynamic equilibrium is ever reached during early diagenesis at near-surface temperatures and pressures (Curtis 1978), although Hutcheon (1981) suggests that authigenic minerals may reach equilibrium with their pore waters during the later, deep burial stages of diagenesis.

Useful qualitative constraints can be placed on the chemical behaviour of species in solution by plotting the stability fields of minerals as a function of the activities of the relevant ions dissolved in solution. Such activity diagrams, largely developed by Garrels & Christ (1965) and expanded by Helgeson *et al.* (1969), serve to illustrate the relationships of the various minerals in equilibrium with the pore waters that bathe them, and provide a graphic summary of the mineral sequences that might be expected to precipitate if equilibrium were attained. The stability of a mineral in aqueous solution depends upon the temperature, the total pressure and the concentration of the chemical components in the co-existing aqueous solution. Not only are stability field boundaries temperature/pressure dependent, but the limiting hydrolysis reactions depend on free energy values, many of which are poorly known. Furthermore, most theoretical thermodynamic calculations also assume that equilibrium states are approached or attained in a closed system. Activity coefficients have been determined experimentally in very dilute aqueous solutions and essentially at low, near-surface pressures. The application of theoretical thermodynamic models to open geological systems at elevated temperatures and pressures in concentrated basinal oilfield brines with a dynamic flux of interstitial pore waters therefore requires extreme caution and an appreciation of the shortcomings. Despite the potential complexity, and the present lack of detailed understanding of high salinity, temperature and pressure thermodynamics, general stability relations and trends can be conveniently expressed using activity diagrams, providing they are used within their limitations. Figure 1 illustrates the development of multicomponent activity diagrams, whilst Figs 2 and 3, provide specific examples of the usefulness of such diagrams.

There are now many published studies which use thermodynamic stability models to explain diagenetic assemblages. Almon *et al.* (1976) used diagenetic cement assemblages to reconstruct the effective pore water composition which prevailed during cementation of Cretaceous sandstones from Montana. Similarly, Merino (1975) calculated the distribution of aqueous species in interstitial pore waters of the Eocene McAdams Sandstone of California and demonstrated, using activity diagrams, that the interstitial pore waters are in equilibrium with the diagenetic minerals. Also, Nesbitt (1980) showed that deep brines in the Carboniferous Illinois basin have equilibrated with the diagenetic mineralogy. Stoessell & Moore (1983) investigated the relationship between mineral diagenesis and reservoir fluid compositions in several Texas Gulf Coast reservoirs where the diagenetic sequence was already relatively well

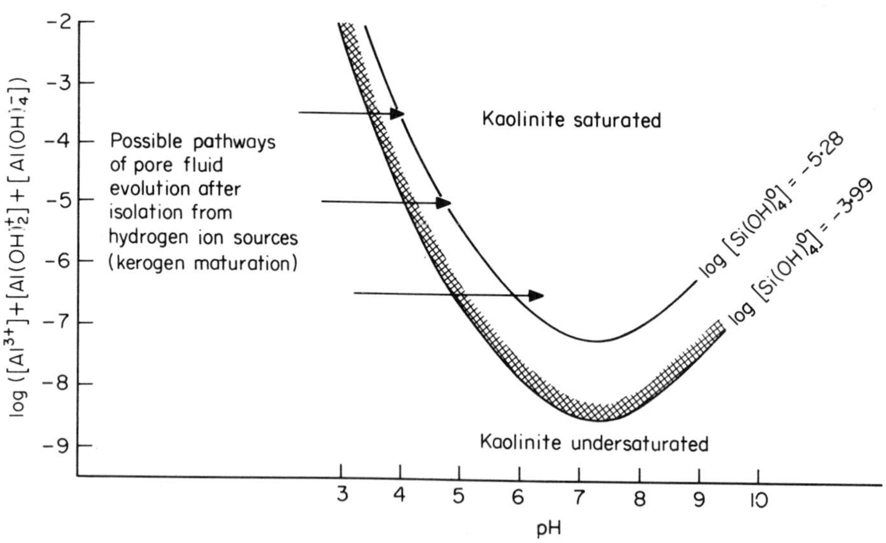

FIG. 2. The stability of kaolinite relative to the solubility of silicon and aluminium species as a function of pH (from Curtis 1983a). Pore fluids containing aluminium species must precipitate kaolinite if the pH of the solution rises.

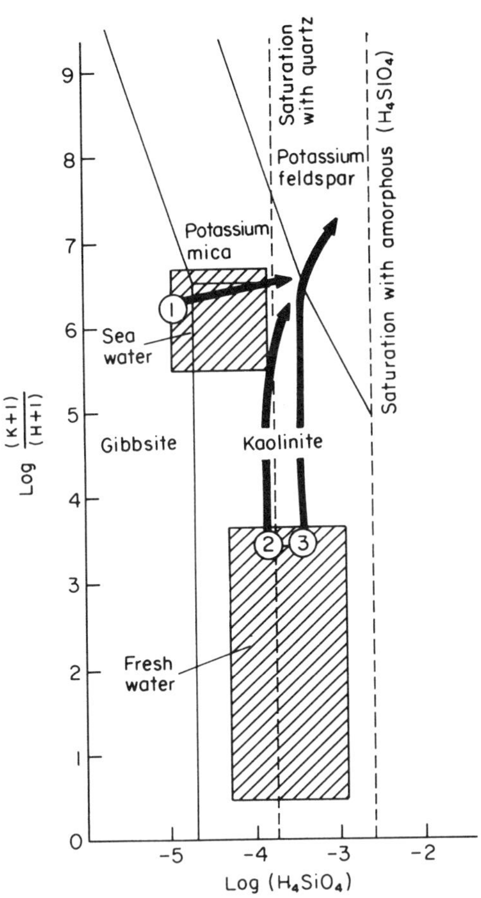

FIG. 3. Practical use of the stability diagram shown in Fig. 1c, showing the stability fields of kaolinite, illite and potassium feldspar together with approximate fields of marine pore waters and meteoric waters. Pathway 1 illustrates the evolution of marine pore waters through reaction with detrital minerals. Pathways 2 and 3 show equilibration of more acidic freshwater with detrital minerals. Pathway 3 involves more quartz precipitation.

known (see 'mesogenetic realm'). Using the thermodynamic data base of Helgeson together with several assumptions regarding stabilities of authigenic minerals they were able to propose a realistic model for the origin of present-day Gulf Coast reservoir fluids.

By contrast Curtis (1983a) used thermodynamic stability fields and kinetics to explain petrographical observations. The association of late-stage secondary porosity development and pore-filling kaolinite during burial diagenesis has been noted by several authors (Loucks *et al.* 1977; Lindquist 1977; Land & Dutton 1978; Schmidt & McDonald 1979a, b) and is readily explained in Fig. 2. Acidic pore waters generated at depth through thermal decarboxylation of organic matter are capable of dissolving and transporting large quantities of metal solutes, including aluminium. If separated from the source of acidity during a compactional drive, the pH of these acid solutions must rise as carbonate dissolution occurs. As the solubility of kaolinite decreases rapidly with increasing pH, dissolution of calcite should cause the precipitation of kaolinite.

The scanning electron microscope can be used as a qualitative guide to determining geochemical precipitation and dissolution processes. Precipitation of a solid substance from an aqueous solution can be considered as two basic processes, nucleation and crystal growth (Berner 1980). Both the rate of crystal nucleation and crystal growth are a function of the degree of supersaturation of the solution. At low levels of supersaturation, crystal nucleation is very slow and excess dissolved material is consumed by crystal growth on a limited number of nuclei. This results in a high degree of crystallinity of the authigenic phase, and will also tend to favour growth on existing mineral phases as syntaxial overgrowths. In contrast, with high solute concentrations and a high level of supersaturation, the rate of nucleation may be so fast that virtually all the excess solute is rapidly precipitated as nuclei (in order of 10–100 Å), effectively decreasing the saturation so that little solute remains for crystal growth. The results will tend to be very poorly crystallized, often fine-grained precipitates, with a high degree of disorder.

The actual processes of dissolution of solids by an aqueous solution involves the removal of ionic species by either diffusion transport or by surface reaction. Observation of partially dissolved crystals with the scanning electron microscope helps to elucidate the dissolution rate-controlling mechanism. Slow partial dissolution of mineral phases by surface reaction will result in the formation of crystallographically controlled features such as etch pits and contrasts with non-specific corrosion resulting from more rapid transport-controlled dissolution.

Energetically, the precipitation of overgrowths can also be considered as an attempt by detrital grains to become more stable and crystal-like (Hurst & Stout 1980). Rounded or angular grains are crystallographically anhedral and characterized by high surface area to mass ratios. During diagenesis the energetic response will tend to lower both the surface free energy and surface area by precipitation of stable authigenic overgrowths, so that the detrital grain approaches the low surface free-energy state of a euhedral crystal. Similarly, dissolution is influenced by the surface free-energy properties of grain and crystal surfaces (Hurst 1981). Low-index crystal faces possess low surface free energy and are least likely to suffer corrosion or dissolution, whilst high-index crystal faces, face edges and corners, abraded detrital surfaces and fractures all have high surface free energies and are therefore susceptible to dissolution and corrosion.

The above account of sedimentary geochemistry is out of necessity, brief and simplified. To summarize, diagenesis from a geochemical viewpoint may be considered as the sum of those processes by which originally sedimentary assemblages attempt to reach equilibrium with their environment, although it is unlikely that sediments ever reach equilibrium in stages with successively changing pore-water chemistry, temperature and pressure. Diagenetic processes are controlled by the activities of dissolved ions, Eh and pH, temperature, pressure and interaction with organic systems. Not only does the original sedimentary assemblage play an active role during diagenesis but so does the interstitial fluid phase. The reaction products of diagenetic processes, released into solution may fundamentally change the pore-water composition. A detrital assemblage moving towards equilibrium with its interstitial pore-water environment may therefore become unstable and undergo further dissolution and precipitation reactions as a result of either *in situ* pore-fluid evolution, or after the introduction of formation waters. Thus, it is important to remember that diagenetic processes take place in a dynamic, evolving aqueous system, so that the availability of water within the sediment, and the ability to move water through the sediment, are as important in diagenesis as the actual pore-fluid composition.

Depositional and pre-depositional influences

Many of the variables which affect diagenesis exert an influence on the sedimentary assemblage before and during its accumulation. The complexity of isolating a single factor is discussed by Hayes (1979). Not only can the composition, texture and geometry of the original sedimentary deposit be expected to exert a controlling influence but also the gross physical and chemical environment. The depositional environment is likely, therefore, to be an important factor. As a review of this topic is to be found elsewhere in this volume we do not discuss it here in any detail. Needless to say, in order to suggest likely depositional pore-water chemistry, as well as any likely local variations in ground-water conditions, it is necessary to establish, in considerable detail, the probable depositional environment. Consequently, and especially in the field of core studies rather than outcrop-based projects little progress was made until depositional-facies models were detailed during the last 20 years (see Reading 1978; Walker 1979; and Anderton, this volume).

The depositional environment is in turn a function of other variables. Climate may play a significant role in affecting the degree of surficial weathering in a particular environment, which will determine the stability of minerals and hence the chemistry of the pore waters. Detrital mineralogy can also be affected by weathering in the source area, transport distance and provenance. These in turn may be affected by the gross tectonic setting (Siever 1979).

The influence of such variables unfortunately does not end with the eogenetic pore-water system. A specific depositional environment will determine the sediment texture, sand to mud ratio, and overall geometry. Hence, porosity-permeability characteristics, and interconnectivity of sand bodies in the subsurface may all be related to the original depositional system. Furthermore, the style and rate of burial will also be related to the sedimentation rate and especially the tectonic setting. Clearly both the tectonics and sedimentology of the depositional basin affect diagenesis during and after burial, and an appreciation of the original setting is important to any diagenetic investigation.

Techniques

Petrography (see Figs 4–7)

Thin-section studies, employing optical microscopy, still form the basis for understanding textural relationships between detrital and authigenic minerals. Furthermore, quantities derived from modal analyses provide a basis for sandstone classification and modelling of the relationship between detrital and authigenic components and porosity (see Pettijohn *et al.* 1973). Mineral identification may be refined by use of suitable staining techniques, particularly for carbonate cements (Dickson 1965) where optical properties of individual minerals are very similar (Fig. 6c), and for feldspars (Bailey & Stevens 1960) where distinguishing potassium feldspar from plagioclase is often difficult. Recently petrography has been supplemented by more sophisticated analytical techniques such as XRD and cathodoluminesence.

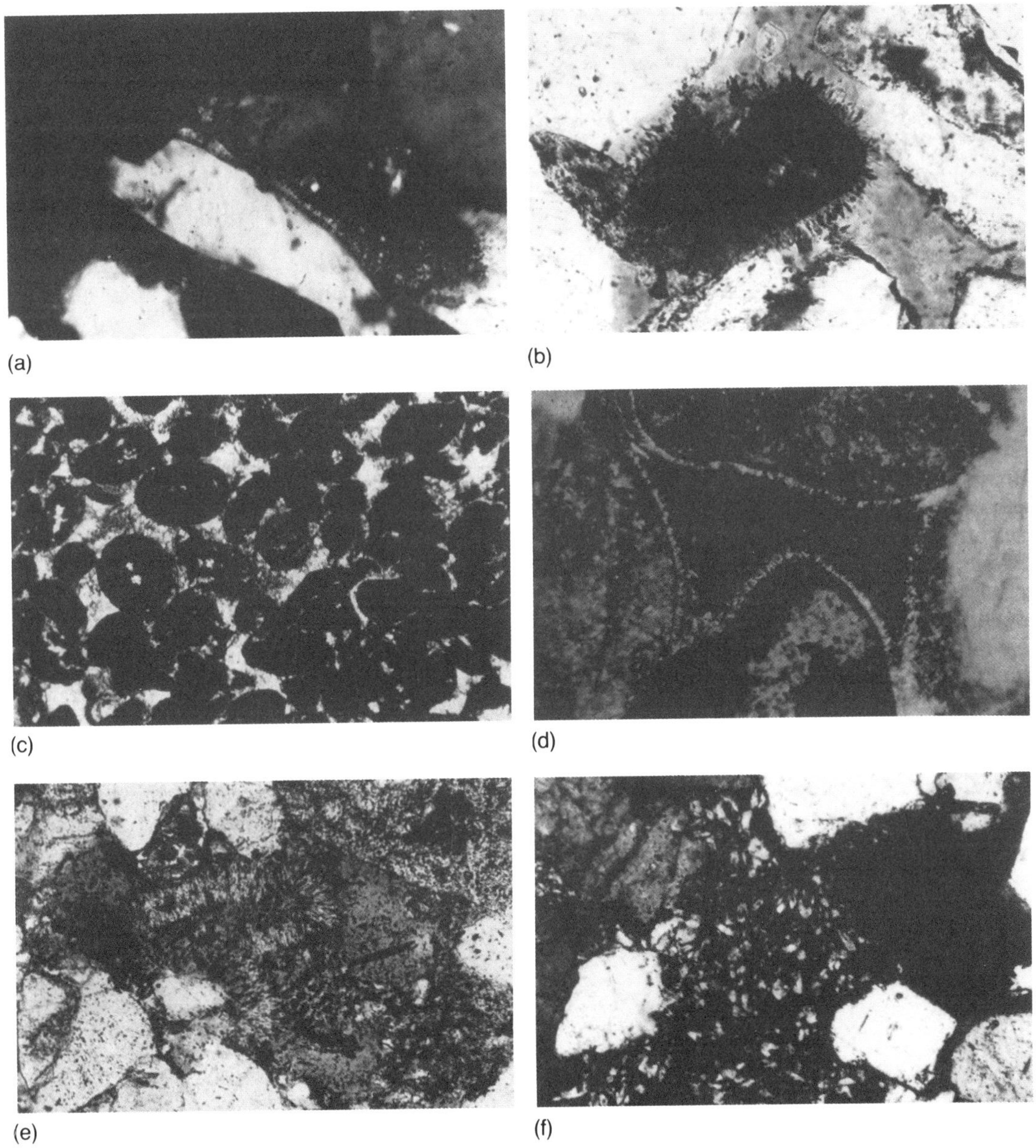

FIG. 4. Colour photomicrographs of authigenic clays. (a) Grain-coating chlorite vermiculite. Clay inhibits overgrowths and allows preservation of intergranular porosity. Vermiculite in this case is inferred to have formed during recent weathering of chlorite. (×70.) (b) Rim of radial authigenic smectite around an altered detrital volcanic grain. Adjacent quartz grains in contrast are coated with mechanically infiltrated haematite-stained clay and quartz overgrowths strongly suggesting the ions necessary for the formation of the smectite were provided by the host grain. (×180.) (c) Berthierine ooliths, partially oxidized in places, and cemented with siderite. (×18.) (d) Radial illite. Thick coating of radial illite growing on a variety of host grains but inhibiting subsequent overgrowths. (×180.) (e) Large stacked crystal aggregate of authigenic vermiform kaolinite within an oversized pore probably a result of grain dissolution. The growth of such large kaolinite aggregates may be related to a flux of meteoric water and may be correspondingly rapid (see Fig. 8e). (×70.) (f) Hugely oversized pore filled with blocky dickite aggregates. Note the corroded nature of quartz overgrowths suggesting that dickite postdates the removal of a carbonate cement (see Fig. 8f). (×70.)

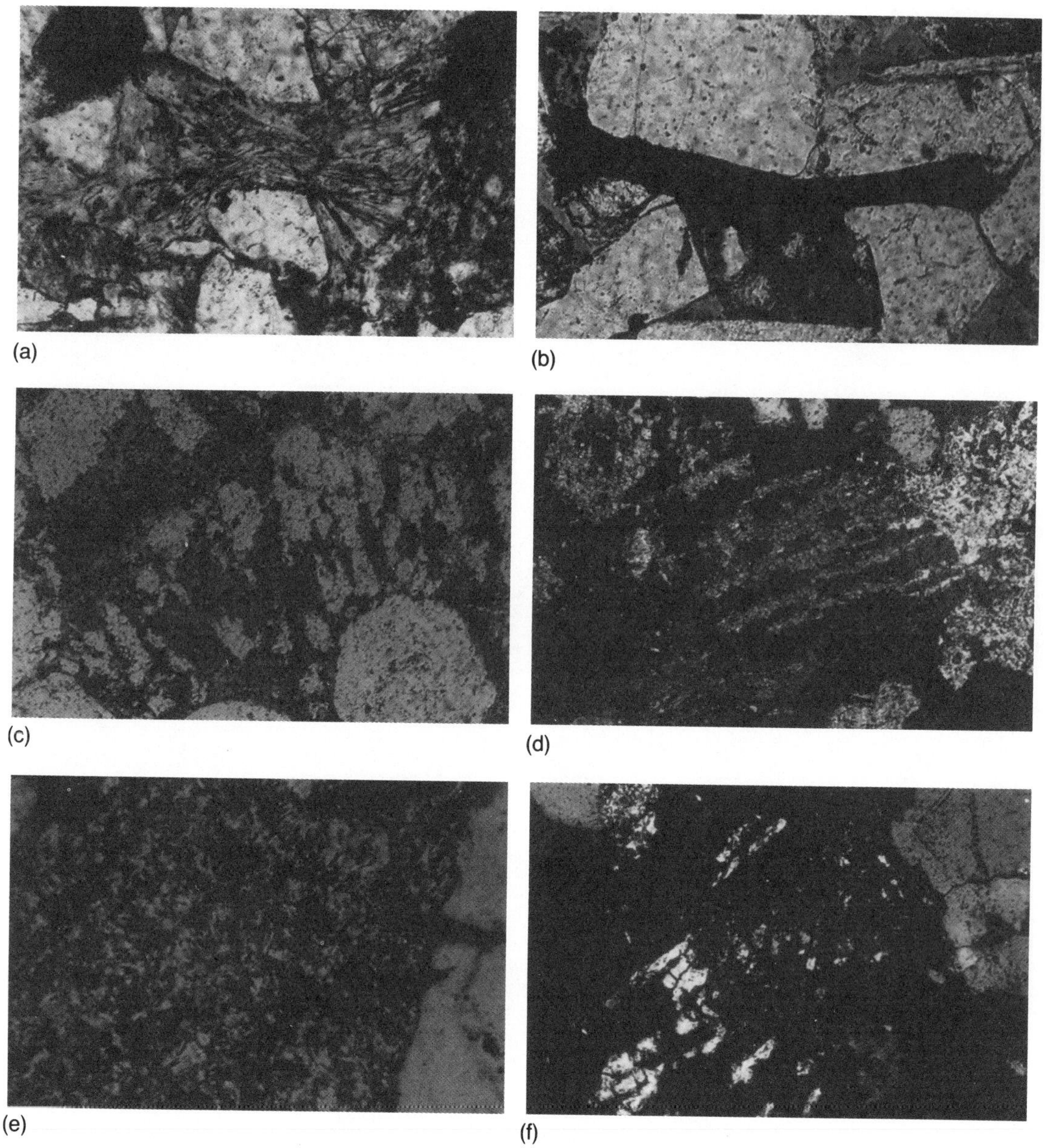

FIG. 5. Colour photomicrographs of replacement textures. (a) Mechanical and chemical alteration of a detrital muscovite grain to kaolinite. The intense basal cleavage is accompanied by extensive potassium loss clearly demonstrable by microprobe analysis. (× 200.) (b) Similar potassium loss in biotite under oxidizing conditions also results in the oxidation of octahedrally co-ordinated ferrous iron in the biotite lattice. The iron precipitates as haematite on the altering biotite which presumably acts as a nucleus. (×180.) (c) Feldspar appears to be unstable in a variety of diagenetic environments and the alteration products depend on several factors including the type of feldspar and the ambient geochemical environment. Here a potassium feldspar grain has been intensely replaced by a pervasive authigenic kaolinite cement. (×70.) (d) Solutions which precipitate carbonate cements are commonly highly corrosive to framework silicates, and feldspar is particularly susceptible to such alteration. This iron-rich ankerite cement has extensively corroded detrital quartz grains (left top, right top) detrital feldspar (centre) and quartz overgrowths (right top). (×70.) (e) Earlier formed clay minerals can become unstable subsequently. Kaolinite is being replaced by strongly birefringent illite. (×70.) (f) Authigenic lath-shaped crystals of barytes associated with and replacing an altered detrital plagioclase feldspar grain containing barium. The association suggests that whilst sulphate for the authigenic barytes was probably derived from the enclosing interstial pore waters at least some barium was provided by the altering grain which acted as a nucleus for barytes growth. (×70.)

FIG. 6. Colour photomicrographs of carbonate and sulphate cements. (a) Siderite spheruliths, exhibiting cruciform extinction being comprised of elongate crystals of siderite radiating from an amorphous nucleus. Siderite spheruliths occur commonly in suboxic soil profiles and grow displacing and replacing surrounding clay particles. (×35.) (b) Nodular, non-ferroan calcite cement associated with calcrete development in coarse-grained alluvial sandstones. Note the displacive growth textures (sand grains concentrated outside the nodule) and the intensive grain replacement and corrosion. (×18.) (c) Complex sequential carbonate cementation in a sandstone, highlighted by staining techniques. Initial dolomite rhombs (unstained carbonate left centre, cloudy core) are enclosed by non-ferroan calcite cement (pink stain) and later ferroan calcite (blue stain). (×70.) (d) Well-developed ankerite (stained blue, identified by microprobe analysis) partially cementing late stage mesogenetic secondary porosity in hydrocarbon reservoir. (×70.) (e) Dolomite cement associated with and replacing a highly altered and subsequently oxidized ferromagnesian grain suggesting possible source of magnesium ions from altered grain. Dolomite cement was probably originally zoned (with calcite ?) and more soluble zones have been leached. (×70.) (f) Pore filling, brightly birefringent poikilotopic anhydrite cement associated with either environmentally controlled early diagenesis in arid sediments or as a late-stage cement precipitated from brines derived from dissolution of evaporite minerals. (×180.)

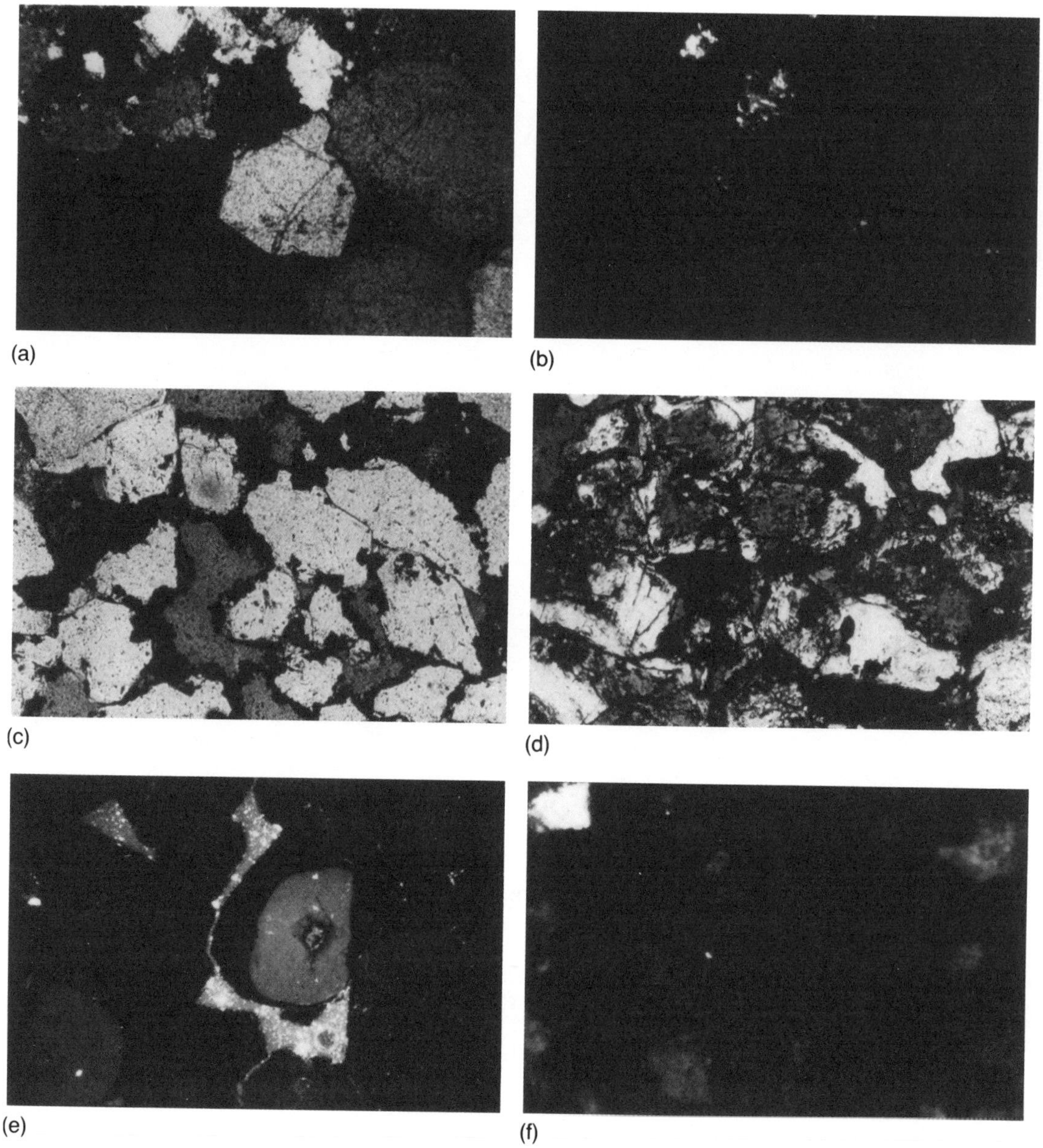

FIG. 7. Colour photomicrographs of authigenic cements and porosity. (a) and (b) Crossed polars and cathodoluminescence pair illustrating complex authigenic mineral precipitation sequence not normally visible using only optical microscopy. A thin black radial chlorite surrounds the rounded detrital grain. Two phases of quartz overgrowth cementation are separated by a phase of kaolinite precipitation (blue luminescence). (×115.) (c) Extensive intergranular and enlarged intergranular secondary porosity created by the dissolution of a replacive siderite cement. The siderite is now partially oxidised to goethite. Compare the quartz grain and overgrowth surfaces where replacement occurs, with pore walls enclosing subsequently precipitated dickite (see (f) and Fig. 4f). (×70.) (d) Intense dissolution of detrital feldspar grains producing extensive mesogenetic secondary porosity. (×70.) (e) Polyphase quartz overgrowth cementation in a quartz-rich sandstone. Note also dissolution at grain contacts. (×70.) (f) Extensive pore-filling dickite (bright blue) within occasionally enlarged intergranular porosity. Note the variety of quartz grain colours suggesting a complex source area, or sedimentary recycling. Note also the irregular nature of overgrowth surfaces (black). (×70.)

X-ray diffraction (XRD)

The introduction of this technique enabled the structural identification of minerals previously only identified on the basis of their thin-section properties—see Brindley (1951) for a contemporary review and Brindley & Brown (1980) for an up-to-date consensus. X-ray diffraction (XRD) is now widely applied to both the analysis of mudrocks and sandstones. Analysis of the powdered whole rock provides much information on the main sedimentary components of fine-grained clastics, although it is limited in its usefulness in sandstones without supportive techniques (see Almon & Davies 1978). Of much more value in diagenetic studies is XRD analysis of the separated fine fraction from both mudrocks and sandstones for determining detailed clay mineralogy. Consider the difficulty in distinguishing the range of clays exhibited in the figures on the basis of their morphology alone. The application of techniques such as XRD to diagenetic problems is well exemplified by the studies of depth-related mudrock mineralogy from the Texas Gulf Coast (Burst 1959, 1969; Perry & Hower 1970; Hower *et al.* 1976).

Clay-mineral studies documenting similar depth-related transformations are now recorded from other major sedimentary basins—e.g. Canadian Arctic (Foscolos & Powell 1979a, b); and North Sea (Pearson *et al.* 1982). X-ray diffraction has also been utilized to identify authigenic clays following their widespread recognition in sandstones (see Lerbekmo 1957; Shelton 1964; Carrigy & Mellon 1964). More recently, clay fraction XRD has supported studies distinguishing detrital and authigenic clays (see Wilson & Pittman 1977), elucidating complex clay mineral reactions (Boles & Franks 1979a) and highlighting specific clay mineral details (Almon *et al.* 1976; Guven *et al.* 1980, McHardy *et al.* 1982).

Scanning electron microscopy (SEM)

The SEM has revolutionized studies of sandstone diagenesis and is widely used to enhance two-dimensional (2-D) thin-section observations. The extreme depth of focus and wide range of magnification (with a resolution some 500 times greater than that of the optical microscope) make the SEM ideal for elucidating the three-dimensional (3-D) morphology of detrital and authigenic minerals and the nature of cement-porosity relationships (see Figs 8–10).

Early applications of the SEM in diagenetic studies were largely of pore throat geometry (Weinbrandt & Fatt 1969) and quartz overgrowth morphology (Waugh 1970; Pittman 1972). With the realization that authigenic clays were widespread in sandstones, the SEM has been used extensively to characterize clays detected by XRD (Borst & Keller 1969; Wilson & Pittman 1977). Extreme caution, however, needs to be taken concerning the identification of clay minerals. A wide range of morphologies and textures may be exhibited by clay minerals of the same structural group (Keller 1976; Wilson & Pittman 1977; Tompkins 1981) and cannot therefore be identified on the basis of SEM alone. This situation has been partially resolved with the development of solid-state energy dispersive X-ray analysers (EDS) which can be attached to SEM and provide qualitative chemical identification (e.g. Guven *et al.* 1980). Back-scatter SEM analysis and X-ray image mapping have enabled further refinements of the distribution of elements in sandstones to be made. Critical-point drying should dramatically affect understanding of the role of clays in reservoir behaviour (McHardy *et al.* 1982).

Electron-microprobe analysis

Problems inherent in the SEM/EDS technique such as the takeoff angle or 3-D geometry, can be resolved largely by examining either polished thin sections or polished blocks with a perfectly flat surface, thereby overcoming absorption of X-rays, etc. This has been fully utilized in the electron microprobe (EM) which allows quantitative chemical analyses to be made of individual minerals using a fully focussed 10–0.5 μm electron beam. The EM is particularly well suited to the analysis of mineral transformation (Velde 1977; Walker *et al.* 1978; Boles 1981). or of zonation in cementing minerals (Keighin & Fouch 1981).

The EM has also been of particular significance in identifying specific carbonate cements such as ankerite (Boles 1978) which would have been difficult to characterize structurally (Irwin 1981). Microprobe analyses provide the only way of characterizing many phases individually (Velde 1977; Merino & Ransom 1982) although problems of mechanical mixtures may arise because of the resolution of the optical system (Curtis *et al.* 1984). However, this may be resolved by use of more sophisticated high-resolution electron microscopes (Huggett 1984). The electron microscope is of most value in determining the chemistry of authigenic minerals such as feldspars which can be readily optically resolved (Waugh 1978; Ali & Turner 1982).

Stable isotope analysis

It is now well known that stable isotope analysis can provide a wealth of information on

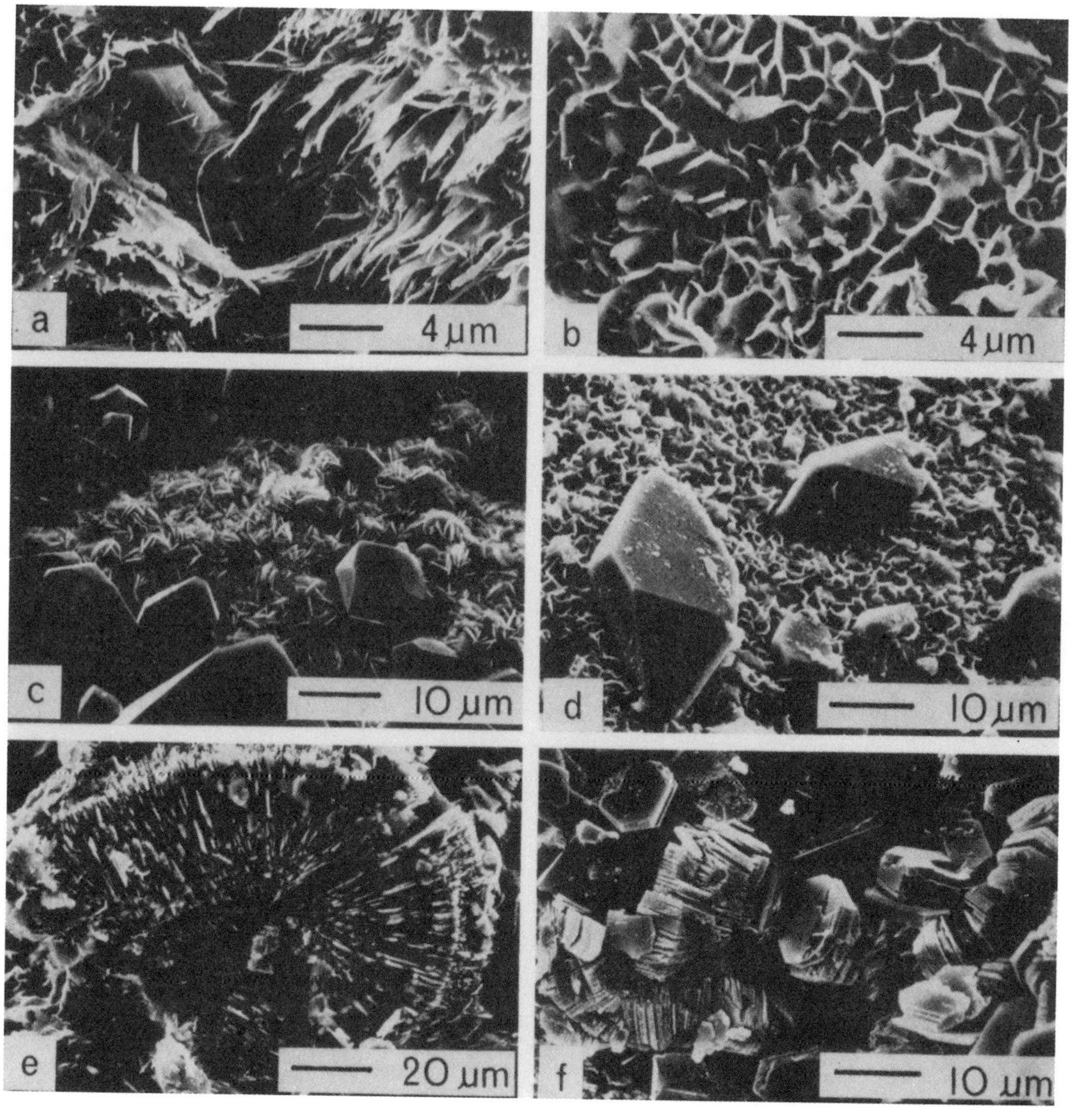

FIG. 8. Scanning electron micrographs of authigenic clays (all clays identified from 2μm clay fraction XRD analysis). (a) Elongate, lath-shaped, radially orientated illitic clay with a small expandable component (10%) and low crystallinity index. (b) Well-developed boxwork texture of ordered illite-smectite mixed layer clay, typical of eogenetic diagenesis in hot, dry red beds. (c) Iron-rich chlorite crystals (14 Å, chamosite *s.s.*) coating detrital grains and partially inhibiting development of quartz overgrowths. (d) Mixed-layer chlorite-smectite clay forming a pervasive grain coating and also inhibiting quartz overgrowth development. Note how the clay mineral morphology resembles the illite-smectite mixed-layer clay. (e) Large crystal aggregate of vermiform kaolinite. The vermiform aggregate is composed of many small stacked crystals of kaolinite. (f) Euhedral crystals of dickite with a high crystallinity partially enclosed by late quartz overgrowth development. Note the thickness of individual crystals compared with their diameter.

cementing minerals in sedimentary rocks (Hudson 1977). The majority of studies however have investigated carbonate rather than clastic sediments. Nonetheless, analysis of carbon, oxygen, and sulphur isotopes has refined our understanding of clastic diagenesis significantly in certain specific fields.

Principally, carbon- and oxygen-isotope analysis is employed to investigate the conditions

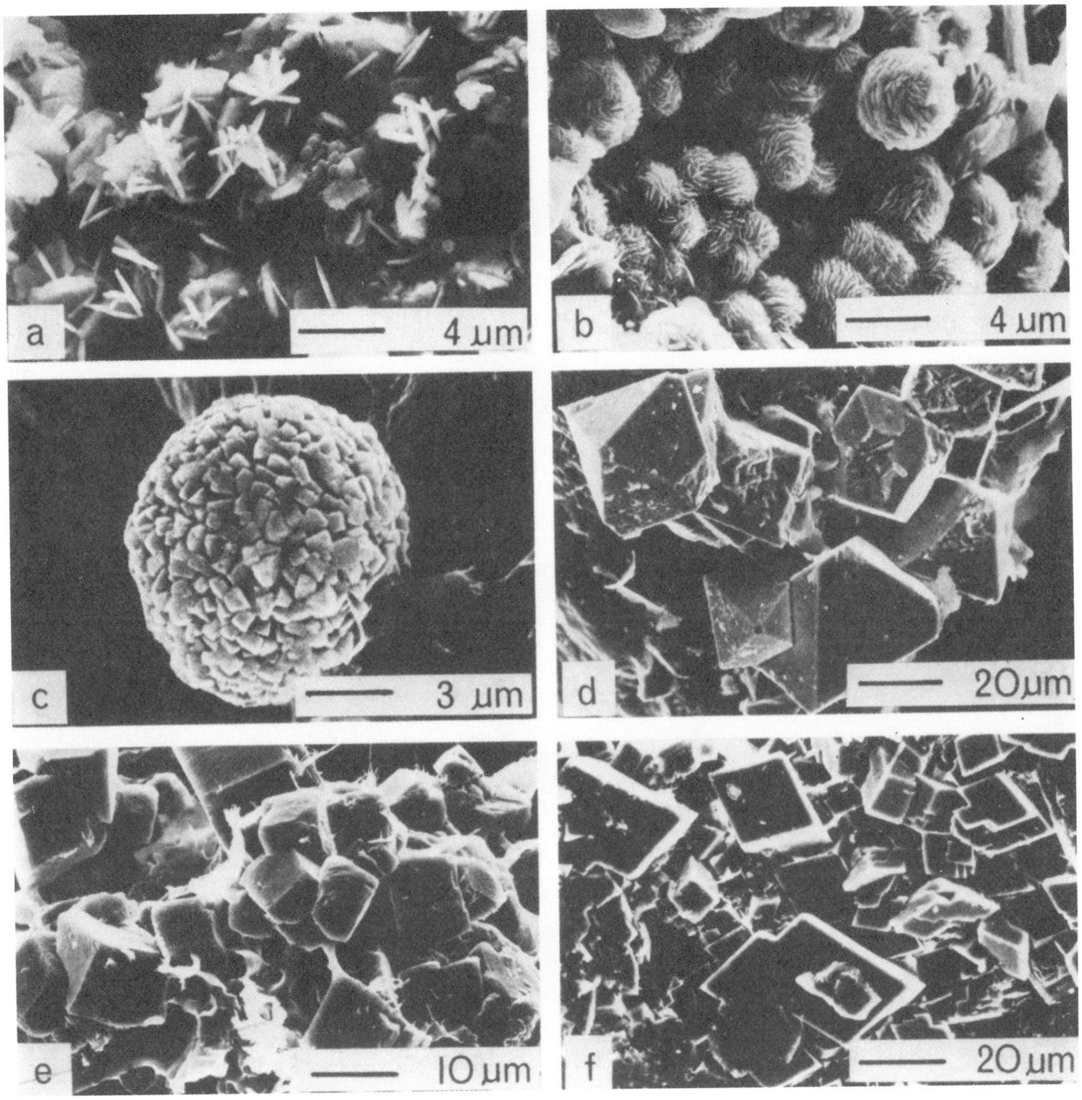

FIG. 9. Scanning electron micrographs of authigenic cements. (a) Specularite crystals of authigenic haematite mixed with authigenic clay. Such authigenic haematite is characteristically produced during eogenesis of hot, dry red beds. (b) Rosettes of highly crystalline haematite, which extensively cements a Triassic Sandstone. What controls the difference in haematite growth forms between (a) and (b)? (c) Framboid of authigenic pyrite crystals developed in a Carboniferous mudstone. (d) Euhedral octahedra of mesogenetic pyrite from a gas reservoir. Is the difference in growth form a reliable indicator of geochemical environment? (e) Non-ferroan dolomite crystals associated with illite-smectite authigenic clay interpreted to be an early pedogenic development. (f) Highly crystalline, isotopically light, calcian ankerite with well developed crystal faces. Compared with Fig. 5a is the difference in crystal form and development a function of chemistry or rate of growth?

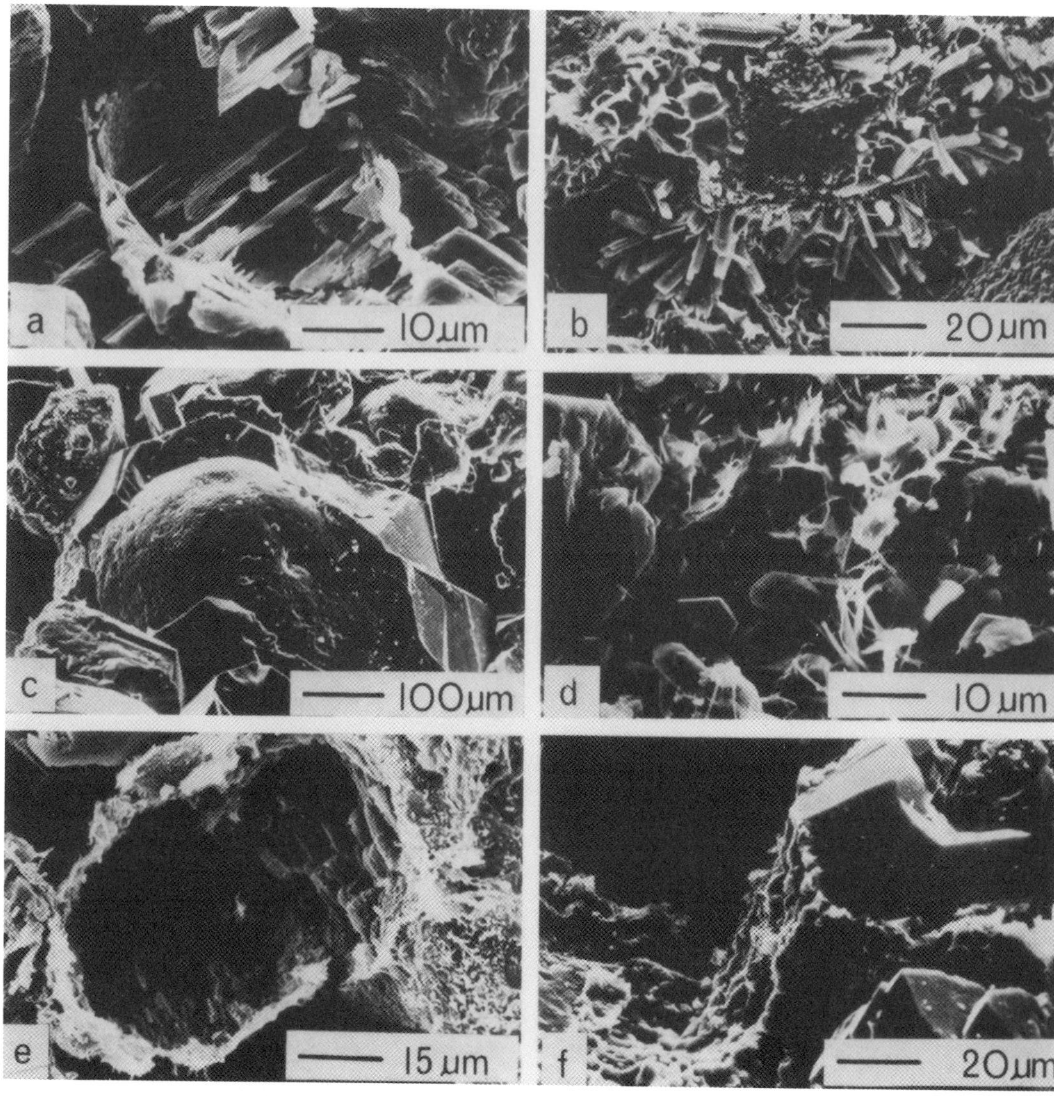

FIG. 10. Scanning electron micrographs of authigenic cements and porosity. (a) Intense dissolution of potassium feldspar leaving only thin clay rim which originally coated the detrital feldspar grain. Authigenic overgrowths and 'ingrowths' occur either side of the relic clay rim. Is the dissolution–precipitation process a response to high instability of the original high temperature feldspar with resultant precipitation of a low-temperature polymorph? (b) An unusual elongate, prismatic development of feldspar overgrowths around a detrital grain. What controls the morphology of feldspar overgrowths? (c) Typical development of euhedral quartz overgrowths (low surface free energy) enclosing rounded detrital grain (high surface free energy). (d) A cluster of small prismatic crystalls of authigenic quartz lining a dissolution void. Is crystal growth the result of precipitation from high levels of silica saturation? (e) Secondary porosity through total grain dissolution in a sandstone with a quartz grain framework preventing compaction. (f) Extensive intergranular secondary porosity through removal of a widespread carbonate phase (siderite in this example) which has previously corroded quartz overgrowths. Non-specific corrosion by transport-controlled dissolution?

under which minerals formed, and is an especially useful technique in distinguishing between non-marine and marine pore waters (e.g. Tan & Hudson 1974). However, changing pore-fluid compositions during burial also may be revealed, in both mudrocks (Irwin *et al.* 1977) and sandstones (Boles 1978; Land & Dutton 1978), the result basically of 'taking the rock apart' as originally advocated by Hudson. Irwin & Hurst (1983) have recently illustrated the significance of this technique in elucidating the nature of evolving pore-fluid compositions during diagenesis. Moreover, theoretical models allow the approximate temperature of mineral authigenesis to be deduced from stable isotope data (e.g. Milliken *et al.* 1981). Also, the study of modern calcretes has helped confirm their identification in the ancient sedimentary record (e.g. Salomons *et al.* 1978).

In addition, sulphur-isotope analysis (Coleman 1977), although applied primarily to ore deposits (e.g. Boast *et al.* 1981) may be used to assess the origin of sulphates both as cements in sandstones (e.g. Taylor 1983) and in both early and late diagenetic pyrite (e.g. Coleman & Raiswell 1981; Hudson 1982).

Cathodoluminescence (CL)

The luminoscope has evolved as a powerful petrographic tool in carbonate studies since its development 20 years ago (Smith & Stenstrom 1963; Long & Agrell 1965). Sippel (1968), Sibley & Blatt (1976) and others have applied the luminoscope in studies of silica authigenesis. However its more widespread use by clastic petrologists has been limited. Cathodoluminescence (CL) may be applied usefully in diagenetic studies to elucidate possible zonation of diagenetic minerals, and to study textures produced during development of secondary porosity. Little work has been published in these fields. This is because the majority of coarse clastic rocks are composed of quartz, which compared to non-ferroan calcites does not luminesce particularly well. Consequently, early workers were only able to record photographically their results. Sippel (1968) for example, employed 3000 ASA film with up to 20 min exposures. This problem has been overcome by recent research which has shown that quartz is more easily excited at lower temperatures and therefore emits more intense luminescence under cold stage conditions. Alternatively, Zinkernagel (1978) has developed a luminoscope with a hot cathode gun which emits more radiation than the 'cold' cathodes employed previously, excites quartz intensely and allows observation of its luminescence. A more sophisticated hot cathode luminoscope developed at the University of Bern (Ramseyer, 1983) also produces intense luminescence in clastic rocks. Repeated zonation in quartz overgrowths and clay inclusions are easily recognized by their characteristic luminescence colours (Fig. 7a, b, e). Complex replacement textures may be interpreted (Fig. 7f) and quantification of grain-to-grain dissolution is possible.

Miscellaneous techniques

In recent years, various additional techniques have been applied to specific diagenetic problems. Studies of fluid inclusions in quartz overgrowths and carbonate cements have been used to determine the timing and conditions of cementation or mineralization (see Roedder 1979). Pittman & Duchatko (1970) describe the use of epoxy-resin casts in studying pore geometry. Age dating has been applied in both sandstone and shale studies, to pinpoint the timing of mineral authigenesis (e.g. Perry 1974; Aronson & Burtner 1982). A range of laboratory techniques derived from soil-science investigations has aided the analysis of fine-grained sediments and, moreover, of amorphous constituents in fine- and coarse-grained clastic rocks (see Jackson 1956; Dixon & Weed 1977). Finally, Ixer *et al.* (1979) illustrate the use of reflected light microscopy in an investigation of opaque cementing minerals.

Diagenetic models

A review of published studies indicates that three broad schemes of diagenetic modification occur in the eogenetic environment, whilst a variety occur during both mesogenesis and telogenesis. Although no single factor *controls* diagenesis, the three eogenetic schemes appear to be largely a function of the depositional pore-water chemistry.

Eogenetic regime

Firstly, the marine environment, with dominantly alkaline pore waters (White 1965) constitutes one end member of the scheme. Non-marine environments, however, include a wide variety of original depositional pore-water systems, in which two extremes may be recognized. Warm and wet, often sub-tropical to temperate settings, where intense weathering is typically developed in the source area, have either initially acidic or anoxic pore waters with high concentrations of dissolved species (Tardy 1971; Gibbs 1977). In contrast, semi-arid to arid environments are characterized by low precipitation and little source-area weathering. Surface depositional pore waters in such desertic settings are typically dilute and slightly alkaline (Cooke & Warren 1973). As Hurst & Irwin (1982) concluded, the environment of deposition will, therefore, tend to create a broad division of diagenetic modifications in sandstones. The stability of detrital minerals is thus a function of pore-water chemistry. Despite this pertinent observation, few recent diagenetic studies detail, or even take into account, the depositional environment.

Of equal importance in determining the stable eogenetic assemblage is the potential contribution of dissolved species from the associated mudrocks. Mudrock diagenesis is, however, predominantly controlled by the amount of organic matter present during burial (Curtis 1977). In this respect, the depositional environment also creates a broad division of diagenetic mineralogy. In most marine and non-marine environments with warm, wet conditions organic matter becomes involved in diagenetic reactions and contributes reaction products to the pore waters. However, both in semi-arid environments or in a depositional framework with a slow rate of burial, organic matter is rapidly oxidized near the sediment–water interface and effectively removed from the reacting system. This not only has important implications in the eogenetic regime but can also impart a long-lasting influence on diagenesis throughout the mesogenetic regime when depth-related mudrock reactions may release significantly concentrated solutions into the enclosed sandstones.

Marine environments

Seawater is typically slightly alkaline with a salinity of around 35 parts per thousand (White 1965). However, there is little potential for chemical reaction between the common detrital mineral phases, and normal marine pore waters (Garrels & Christ 1965). Consequently, most detrital minerals should be metastable following their deposition in marine environments. Hence, the diagenetic processes which operate in both fine- and coarse-grained clastic sediments result either from changes in pore-water chemistry during initial burial, or the reaction of amorphous material and the less stable detritus (Curtis 1980).

Diagenetic processes in marine mudrocks commence immediately following deposition. These processes are well documented in studies of various recent sediments (e.g. Curtis 1978; Froelich *et al.* 1979; Berner 1981; Krom & Berner 1981). Close to the sediment–water interface the bacterial oxidation of organic matter releases bicarbonate ions into solution. These readily react with and dissolve the often amorphous or fine-grained clastic detritus, particularly weathering products such as clay minerals coated with iron sesquioxides (Greenland 1975). It is likely that this amorphous material is rapidly removed from the system by these reactions close to, or at, the sediment–water interface (Curtis 1980; cf. van Elsberg, 1978). Vertical permeability in fine-grained sediments is generally low. Consequently, whilst some reaction products may diffuse into the sea, the majority will either be fixed within the mudrock itself or during shallow burial compaction, expelled into any enclosed sandbodies (Meade 1966). Following aerobic bacterial processes sulphate-reducing bacteria liberate further bicarbonate ions into the pore waters and continue to reduce pH by the production of hydrogen sulphide. Eventually, fermentation processes dominate in the sediment column, releasing bicarbonate ions and methane. By-products of bacterial metabolic processes are thus released into the depositional pore waters and as a result authigenic minerals form rapidly at or near the sediment surface. Pyrite may precipitate by reaction between residual detrital iron oxides and hydrogen sulphide (Fig. 9c). Iron- and manganese-bearing calcite and dolomite may locally precipitate as concretions (Irwin *et al.* 1977) whilst the bicarbonate may significantly alter pore-water composition to effect the dissolution of amorphous material. Thus, the interaction of organic and inorganic detrital components with marine pore waters, modified by the reaction products of bacterial metabolism, probably causes rapid diagenetic modifications of the initial sediment (Coleman & Raiswell 1981).

Marine sandstones are characterized by an early authigenic suite of minerals which includes illite, interstratified illite-smectite, chlorite, potassium-feldspar overgrowths, quartz, pyrite and various carbonate cements (Fig. 4d; Almon *et al.* 1976; Hawkins 1978; Storey 1979; Tillman & Almon 1979; Hurst & Irwin 1982; Kantorowicz 1984). Furthermore, most workers record clay authigenesis, associated with quartz and feldspar overgrowths, followed by carbonate precipitation, as the typical order of cementation, suggesting that eogenetic marine diagenesis conforms to a predictable pattern. It is likely, however, that the paragenesis of these various minerals reflects not only the composition and chemistry of the seawater but also the original sedimentary mineralogy and the early diagenetic reactions in adjacent mudrocks.

Quartz arenites and subarkoses, with a generally stable detrital assemblage, contain authigenic illite or illite-smectite formed following grain dissolution or directly from seawater, followed by quartz overgrowths (Elverhøi & Bjørlykke 1978). Precipitation is probably the result of an increase in the ionic concentration of the seawater as silica- and alumina-bearing solutions migrate from adjacent shales to combine with potassium from the original interstitial seawater (Fig. 3; Harder 1974; Kastner & Siever 1979). With a much less stable original mineralogy, as in volcaniclastic sediments, the potential for reaction with the pore waters is increased and more ions can be added to the reacting system during grain dissolution. Thus, dissolution of volcaniclastic detritus, possibly stimulated by the reaction products of mudrock diagenesis, leads to increased activity of the iron and magnesium species in solution. Consequently chlorite or chlorite-smectite clays tend to precipitate in preference to more illitic clays (Fig. 4a; Almon *et al.* 1976; Dutton 1977; Galloway 1979; Wescott 1983). Illitic clays observed today are likely to have formed as disordered, more smectitic, end-members which have evolved in time. Clay mineral authigenesis is typically followed by quartz and feldspar overgrowth precipitation (Fig. 11). It is likely that this reflects the geochemical evolution of the system, with more easily formed minerals forming rapidly, whilst those requiring more ordered crystal growth forming subsequently. The diagenesis of aluminosilicate minerals in marine sediments is arrested ubiquitously by carbonate cementation (e.g. Stanton 1977). Calcite, or less commonly dolomite or siderite precipitation may involve either carbonate and bicarbonate ions originally in solution, or those released from the bacterial degradation of organic matter in the surrounding mudrocks (Coleman & Raiswell 1981).

In addition to a variety of authigenic minerals, marine clastic sedimentary assemblages are widely reported to contain a range of green sheet-silicates (Porrenga 1967; Knox 1970; James & van Houten 1979; Bhattacharyya & Kakimoto 1982). In the past, and indeed in some contemporary publications these are loosely referred to as either 'glauconite' or 'chamosite'. However, more recent research indicates that both the chemistry and the mineralogy of these sheet silicates is extremely diverse (see Odin & Matter 1981; Brindley 1982). The glauconite or glaucony minerals comprise a complex range of iron-rich, potassium-poor smectites. The complexity of recent glaucony compared with the relative purity of the mica end-member reported from ancient deposits confirms the general suggestion that clay mineral authigenesis is a relatively rapid process which, given a wide range of ions available, may initially form a chemically complex and poorly crystalline mineral. During diagenesis, this smectitic material may become more potassium-rich, and evolve or mature into an end-member which is chemically and mineralogically a potassium mica.

There appears to be a genetic relationship between glaucony and 'chamosite' (or berthierine *s.s.*, see below), the former occurring further from river mouths and deltas (Fig. 12). However, the origin of 'chamosite' is problematical, and consequently its use as a marine indicator is in doubt. Many 'chamosites' reported from the literature have now been shown to be 7 Å layer silicates, and, mineralogically therefore berthierines. In addition, chamosite is currently used to describe 2:1 14 Å layer silicates, the iron-rich chlorites (Bailey 1980). Consequently, it is often difficult to establish which mineral is being described in the literature. It is also suggested that berthierine *s.s.* evolves during diagenesis and alters to form chamosite *s.s.* (Iijima & Matsumoto, 1982). However, chamosite authigenesis may take place independently causing both minerals to occur together without any genetic relationship.

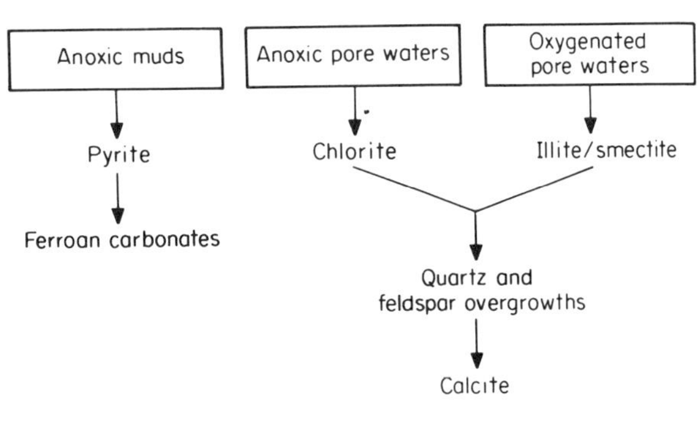

FIG. 11. Schematic diagenetic pathways in marine sediments.

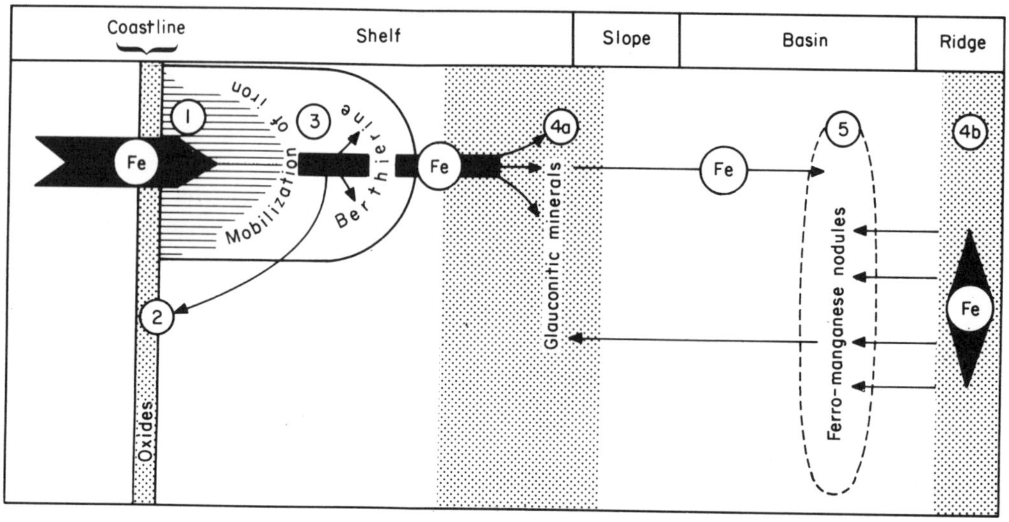

FIG. 12. Geochemical pathways of iron in the marine environment, (after Odin & Matter 1981). Iron accumulates in the coastal zone and inner shelf (1) where chemical precipitation of iron and oxidation of detrital iron (2) take place. Mobilization of iron results in authigenic growth of berthierine and (3) reorganization of detrital clay micelles. Further out on the shelf authigenic growth of glauconite (4) takes place in the deep basin. Where reducing conditions dominate, ferrous nodules (5) grow. Ooliths may be derived or accreted in place but berthierine is essentially reworked and altered lateritic or kaolinitic soil.

Furthermore, the origin of thc ooliths which berthierine characteristically forms (Fig. 4c) is currently controversial. Theories proposed include replacement of carbonate ooliths, *in situ* growth on the sea floor, and alteration of derived iron-coated clay ooliths (Knox 1970; Kimberley 1979, 1980a, b; Bradshaw *et al.* 1980; Bhattacharyya & Kakimoto 1982). The association of berthierine ooliths with deltas (Fisher & McGowen 1967; James & van Houten 1979; Odin & Matter 1981), the widely reported occurrence of ferruginous ooids and clay pellets on modern tropical alluvial plains (Nahon *et al.* 1980, and references therein), and the geochemical constraint of ferrous iron requiring reducing conditions (Curtis & Spears 1968), suggest that clay authigenesis is probably related to seawards size fractionation of kaolinite and smectite.

Vegetated (warm wet) non-marine environments

Petrological studies of a variety of alluvial sediments suggest that diagenesis was influenced by initially acidic or neutral, typically anoxić, pore waters. These studies are mostly of sediments deposited in well-vegetated and hence equatorial or tropical, even temperate climates (see Hemingway 1968; Hawkins 1978; Staub & Cohen 1978; Almon & Davies 1979; Edwards 1979; Storey 1979; Keighin & Fouch 1981; Huggett 1984). The diagenesis of these sediments is inferred to have been influenced significantly by the bacterial degradation of organic matter. Consequently, although the wide-ranging effects of grain dissolution and mineral precipitation may be seen throughout, the majority of the reactions involved probably resulted from changes in pore-fluid composition influenced strongly by bacterial activity (Curtis 1977, 1978). In addition, most authigenic minerals reflect the relative paucity of potassium, magnesium and sulphate ions in solution in freshwater (White 1965; Berner 1971).

In these warm, wet and vegetated conditions, organic matter is not completely oxidized at the sediment surface. Indeed, significant quantities survive to impart to the mudrocks present, a distinct black colour which remains throughout their burial. Subsequently, bacterially controlled processes operate, degrading this organic matter and significantly influencing ground-water chemistry immediately after deposition. These reactions initially involve aerobic bacteria, removing oxygen from the pore waters, causing the release of bicarbonate ions into solution and as a result lowering the pH of the solution. Such acidic conditions may also stimulate further organic and inorganic reactions which may subsequently affect the pore-water composition (Curtis 1977). Indeed, similar processes are likely to have operated throughout the sedimentary basin, affecting not only the composition of the original detrital clastic assemblage, but also that of the depositional groundwaters (Thomas 1974). Nonetheless, many of the studies listed above report early iron-chlorite authigenesis and hence, by contrast, suggest that pore waters were neutral and anoxic rather than acidic and oxidizing (Velde 1977; Harder 1978). It is equally possible that these requirements may also be satisfied by the continued diagenesis and interaction of organic matter with pore waters. This is because following the depletion of oxygen, anaerobic bacterial processes commence. Consequently, and in the absence of sulphate ions in solution (Berner 1971), ferric iron sesquioxides are reduced, releasing hydroxyl ions and causing a rise in pH (Curtis 1977).

Eogenetic modifications to quartz arenites and subarkoses under more acidic conditions are often similar and unspectacular (Fig. 13). Variations generally only occur where abundant rock fragments were deposited or where the detrital sediments were finer grained. Quartz overgrowths (Fig. 7c), pore-filling kaolinite (Figs 4e and 8e), altered (neomorphosed) muscovite (Fig. 5a) and dissolved feldspar grains are typically reported from quartz arenites (see Fig. 3). Both the quartz and the kaolinite may have precipitated directly from ground waters (Tardy 1971). In addition, silica and alumina released during the dissolution of feldspar grains may have contributed to these authigenic phases (Curtis & Spears 1971; Berner & Holdren 1979). This is especially true of crevasse-splay sandstones. These are often colonized by plants and as a result leached by acidic fluids, causing removal of feldspars and detrital micas whilst kaolinite precipitates in the underlying muds. The dissolution of feldspar grains and fine-grained clays, as well as alteration of muscovite may be explained by their relative instability in acidic conditions (Fanning & Keramides 1977; Huang 1977; Lin & Clemency 1981). Similarly, silica is relatively insoluble at low pH with the result that it is readily precipitated (Blatt 1966). In some sandstones diagenesis is effectively complete by this stage, with wholesale quartz overgrowth cementation. This is perhaps best described as a wet version of Wopfner's (1978) tropical silcrete model. In more arkosic sediments more authigenic kaolinite is often reported, suggesting that a relationship may exist between detrital mineralogy and clay authigenesis. Further confirmation of this model may be found in the quartz overgrowths (Breese 1960), and large vermiform aggregates of kaolinite plates (Fitzpatrick 1980) which have been observed in modern tropical soils.

In both quartz arenites and with the introduction of more iron-rich labile grains into the detrital sedimentary assemblage, further diagenetic variations may be observed. Chlorite rather than kaolinite is often reported as the first authigenic clay. This chlorite is invariably iron-rich and technically therefore chamosite *s.s.* (Fig. 8c). In addition, siderite rather than calcite cements are often found. Here it may be inferred that bacterial processes rapidly depleted oxygen with the result that anoxic conditions prevailed before significant diagenetic modifications to the sediments had occurred. Several studies confirm the importance of the nature of particular authigenic minerals. Chlorite occurs as a pore lining, inhibiting further quartz overgrowths and enhancing the preservation potential of intergranular porosity. Kaolinite on the other hand occurs as a pore filling, creating a rock with different reservoir properties. The significance of this distinction is that only very local variations in ground-water chemistry are required to produce either of the two situations (Kantorowicz 1984). Consequently, the reservoir character of such sediments is difficult to predict (see Fig. 13).

Siderite authigenesis often marks the last diagenetic modification that may be related to the degradation of organic matter, and consequently to depositional pore-water chemistry (Fig. 13). However, the role of iron in clastic diagenesis is somewhat problematical. Many authigenic sequences include iron-bearing

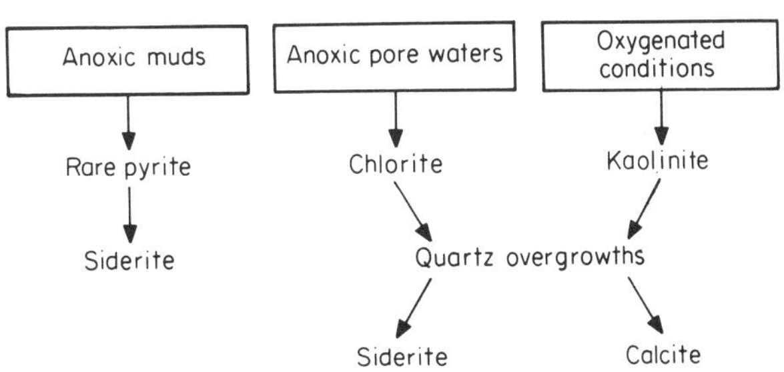

FIG. 13. Schematic diagenetic pathways in warm, wet non-marine sediments.

minerals but lack any obvious sources of iron. Nonetheless, Greenland (1975) and Picard & Felbeck (1976) suggest that iron may be transported into basins either coating clays, or complexed with organic polymers. In this context Gac & Tardy (1980) describe iron-coated clay micelles from Lake Chad, whilst Gibbs (1977) estimates that 6–7% of the material presently transported by the Amazon and Yukon Rivers is iron and that some 90% of this is transported as solid particles or as grain coatings. Gibbs (1977), Nahon *et al.* (1980) and Odin & Matter (1981) all related the mobility and derivation of iron to erosion of lateritic weathering products. This material is most likely to accumulate on flood plains as a result of overbank flooding. As a result sphaerosiderite-dominated soil horizons may develop, whilst at depth ferrous and sideritic concretions may form as detrital iron sesquioxides are mobilized (Fig. 6a; Coleman & Ho 1968; Retallack 1976; Curtis *et al.* 1972; Gould & Smith 1979).

Within the stabler sedimentary assemblages recorded in most of the above examples, diagenesis involves some dissolution of the aluminosilicate framework, as well as alteration to a variety of clay minerals, authigenesis of clays, and precipitation of both overgrowths and pore filling occasionally, replacive cements. These processes involve the interaction of the sediments themselves, particularly the organic matter present, and also the depositional pore waters. In a few extreme cases, the breakdown of abundant labile grains, such as volcaniclastics, overrides the subtler affects outlined above. Potassium, calcium, and magnesium ions may be released into solution during grain dissolution and result in the precipitation of complex diagenetic assemblages characterized by smectites and zeolites (Galloway 1974; Vessel & Davies 1981).

Although the studies reported here infer acidic or anoxic ground-water conditions they often involve essentially lateritic weathering assemblages. However, the diagenesis of similar sedimentary clastic assemblages described below is inferred to form red beds, with the eogenetic mineral assemblage dominated by haematite and illite/smectite authigenesis. The essential differences appear to reflect both the original climate and the original basin geometry and burial rate. Deposition and rapid burial often associated with tectonically active areas will create different conditions to those on stable cratons with slower sedimentation rates or in other tectonic settings with prolonged exposure to surface oxidation (Curtis 1977; Coleman *et al.* 1979; see Fig. 15). It is significant, therefore, to note that many of the rocks detailed above are from the Jurassic or Carboniferous of Western Europe (Figs 4a, c, 6a, 7c, 8c and 9c), or the Tertiary of the Gulf Coast. More significantly these studies concentrate on sediments inferred to have been deposited as basin centres and in areas with a consistently high water table.

Desertic (hot dry) non-marine environments

Red-coloured sediments have for many years been interpreted as being of continental origin. Furthermore, the cause of the red coloration has been well documented to be due to the presence of ferric oxide (see Fig. 9a; Grabau 1920; Berner 1969). The actual quantity of iron is small, often totalling less than 1% of the sediment (Walker & Honea 1969) and it is typically no greater than in non-red sediments. Clearly, high iron contents are not needed to produce red sediments and virtually all sediments would appear to have the potential to form red beds.

The origin of the haematite, however, has been the subject of considerable debate. Typically, red-bed sequences are late to post-orogenic fanglomerate deposits which grade laterally into fluvial sandstones and basin centre playa lake sediments. Mineralogically, they are often highly feldspathic, indicating a source area of crystalline granitic or metamorphic basement. Historically, two opposing theories have developed to account for this common association and the development of haematite in red beds. Earliest ideas suggested that red coloration was caused by iron liberated during the decomposition of hornblende and mica (Lyell 1852). Subsequent workers considered the haematite to form in lateritic soils in tropical, humid climates under deep weathering conditions (Krynine 1935, 1949). However, because ancient red beds are frequently associated with evaporites, Krynine (1950) suggested that red river alluvium, derived from lateritic soils, was transported from the tropical uplands and deposited as large, red alluvial fans in desertic basins. Despite widespread acceptance of this theory (see van Houten 1961) there was no proven modern analogue to such transported red soils. Indeed according to Walker (1967) the recent alluvium from Mexico quoted by Krynine was actually greyish-brown. Red coloration in the alluvium only developed on much older soils that had been exposed to prolonged oxidation.

To develop this concept of reddening increasing with age, Walker and his co-workers undertook detailed studies of Recent, Pleistocene and Pliocene deposits of the Sonoran Desert, and Baja California, and compared the results with studies of late Palaeozoic

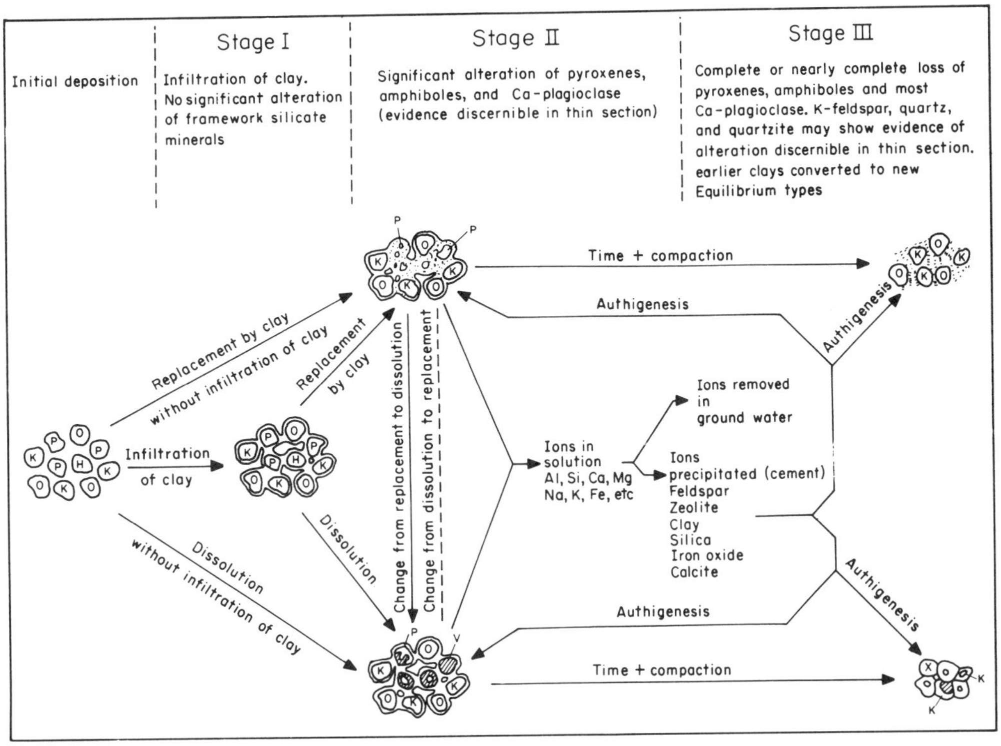

FIG. 14. Schematic diagram illustrating observed and inferred diagenetic alterations in first cycle, arkosic hot, dry (desertic) sediments (after Walker *et al.* 1978).

red-bed sediments of Colorado (see Walker 1967; Walker *et al.* 1967; Walker & Honea 1969; summarized in Walker *et al.* 1978; Fig. 14). Essentially, they were able to show that, regardless of age or location, the haematite pigment in red beds is a product of the *in situ* diagenetic alteration of iron-bearing grains.

In the modern alluvial fans of the Sonoran desert the sediments are distinctly non-red in colour, whilst the Pleistocene counterparts are brownish and the older Pliocene sediments are a characteristic brick-red colour. It appears that the reddening process takes place *in situ*, and is the result of several distinct post-depositional processes. Initially, the clay content of this coarse alluvium is negligible. However, it gradually increases with age to as much as 25% in the Pliocene sediments. Mechanical infiltration of clay contributes significantly to this increase (Fig. 4b). Clay also forms as a replacement of unstable detrital ferromagnesian grains, and is typically smectitic in composition (Fig. 8b). Some of this clay is unstable in the desert vadose environment and alters to liberate iron (Fig. 9a). During these various processes ions may be liberated into solution. In the oxidizing environment of the hot desert, iron rapidly becomes ferric and either enters the newly forming clay mineral lattice, or forms a fine coating on individual clay particles (Fig. 9a, b). This iron is subsequently converted to an oxide, amorphous hydrated goethite in the Pleistocene sediments, which in the Pliocene deposits has aged to crystalline haematite. Selective or total dissolution of unstable silicate grains (Figs 7d and 10a, c) such as volcaniclastics, also liberates many ions into solution which again may be precipitated as a result of the local pore-water chemistry or remain in solution and be removed from the reacting system (Fig. 14). In the associated dune fields a similar series of reddening processes develops.

Walker also recognized the widespread development of red soils in sediments of the Sonoran Desert, although the youngest ones are of late Pleistocene age. Fine-grained material in the playa lakes also becomes red with time. Here the iron-oxide clay coatings age to haematite whilst the clay may undergo post-depositional alteration, yielding additional iron which ultimately forms more haematite.

In addition to iron, various other chemical species are released into solution by the replacement and dissolution of detrital grains. Although in detail the exact chemical pathways involved in these hydrolysis reactions are not fully known they may be approximated by some of the 'weathering' reactions compiled by Curtis

(1976). The dissolution of framework silicates (feldspar) rock fragments and ferromagnesian grains liberates silica, alumina, alkali and alkali-earth elements into the interstitial pore waters as hydrated ions and complexes. Much of the released silica, alumina and labile elements is precipitated as pore-lining authigenic clay (Figs 8b, d). In the Sonoran Desert the most abundant clay is randomly interstratified illite-smectite with 80–95% expandable layers.

Sediments with high detrital feldspar contents release sufficient quantities of potassium into the pore waters to precipitate potassian intermediate sanidine as authigenic overgrowths, stable in the diagenetic environment (Fig. 10b; see also Waugh 1978; Kastner & Siever 1979; Ali & Turner 1982). Any excess silica forms authigenic quartz occurring as both tabular, and prismatic, discrete, euhedral crystals in addition to incipient overgrowths (Fig. 10c). More locally, where the detrital sediments are volcaniclastic in composition, the zeolite mineral clinoptilolite is a common cement. The ions necessary for all these authigenic cements can be generated by dissolution and replacement reactions taking place within the original sediment (Fig. 14). The remaining elements may be concentrated to form caliche deposits, or as evaporites within the playa lakes of enclosed basins (Figs 6b, f and 9e).

Walker's studies were important for several reasons, particularly in the red-bed context. The occurrence in a desert region with an arid climate, of red soils formed through diagenetic processes, illustrates conclusively that red beds need not only form in areas of lateritic weathering in moist tropical climates. However, red soils may form in such tropical settings. Both in the Orinoco Basin (Walker 1974) and in Columbia (van Houten 1972) although modern alluvium is brown, older sediments have reddened due to their prolonged exposure to favourable oxidizing conditions. Walker, therefore, concluded that there is no justification for using red beds as indicators of palaeoclimate, either in the source area or depositional basin. Chemical and mineralogical analyses reveal no obvious criteria that can be used to differentiate ancient sediments with a red coloration being formed diagenetically in moist climates from those found in hot desert climates. Thus, all sediments which contain sufficient free iron are capable of being oxidized to produce red sedimentary rocks, but they will only do so if the interstitial pore-water environment favours the formation and preservation of ferric iron oxides (see Fig. 15).

There are many examples of ancient analogues to the above red-bed model. Walker (1975) considered the late Palaeozoic red beds of the Western U.S. Interior and the Triassic Moenkopi Formation of Colorado as classic examples of diagenetically produced red beds. Nagtegaal (1969) described similar red beds from post-Hercynian Carboniferous to Lower Triassic sediments in Spain, whilst Pedersen & Anderson (1980) proposed the same diagenetic origin for the red Triassic sandstones of Jutland. Here in Britain the Old Red Sandstone (Turner 1980), Permian (Waugh 1978) and Triassic (Burley, 1984) all exhibit diagenetic features typical of desertic red beds, although the cautionary remarks of Turner & Archer (1977) and Wilson & Duthie (1981) should be recalled. The Rotliegendes of the southern North Sea is also considered as a classic example of desertic red beds and a wealth of literature has been written on it (Glennie *et al.* 1978; Seeman 1979; Rossel 1982).

In general, these studies concentrate on the petrology and paragenesis of authigenic minerals. They reveal a general trend which is comparable to that observed in studies of recent sediments. Namely, detrital grains are commonly coated with haematite and either tangential or radial developments of illite/smectite and illite clays. These initial developments are succeeded by quartz and feldspar overgrowths although these are often thinly spread and the sandstones may remain friable and loosely cemented. Calcite and dolomite cementation which occurs locally is sometimes related to calcrete formation, or evaporative conditions, respectively.

In contrast, red beds considered to form in a humid, tropical climate are described from deltaic sediments in the Difunta Group of Mexico (McBride 1974) and from the late Carboniferous Coal Measures of North Staffordshire (Besly & Turner 1983) and of South Wales (Archer 1965). These red beds are typically associated with palaeosols—see Turner (1980) for a summary. Frequently the red coloration is restricted to the mudstone units which are colour mottled red, purple, green and yellow. Calcrete nodules, spherulitic siderite, pyrite and roots are present at several horizons within the mudstones. Although Walker (1974) considered that the origin of red beds in both desert and moist tropical climates was broadly similar, the above examples suggest there are distinct differences. In these clay-rich, fine-grained tropical mudrocks the *in situ* reddening of detrital ferric hydroxides appears to be the dominant process. This is brought about by the seasonal lowering of water tables and the development of well-drained oxygenated

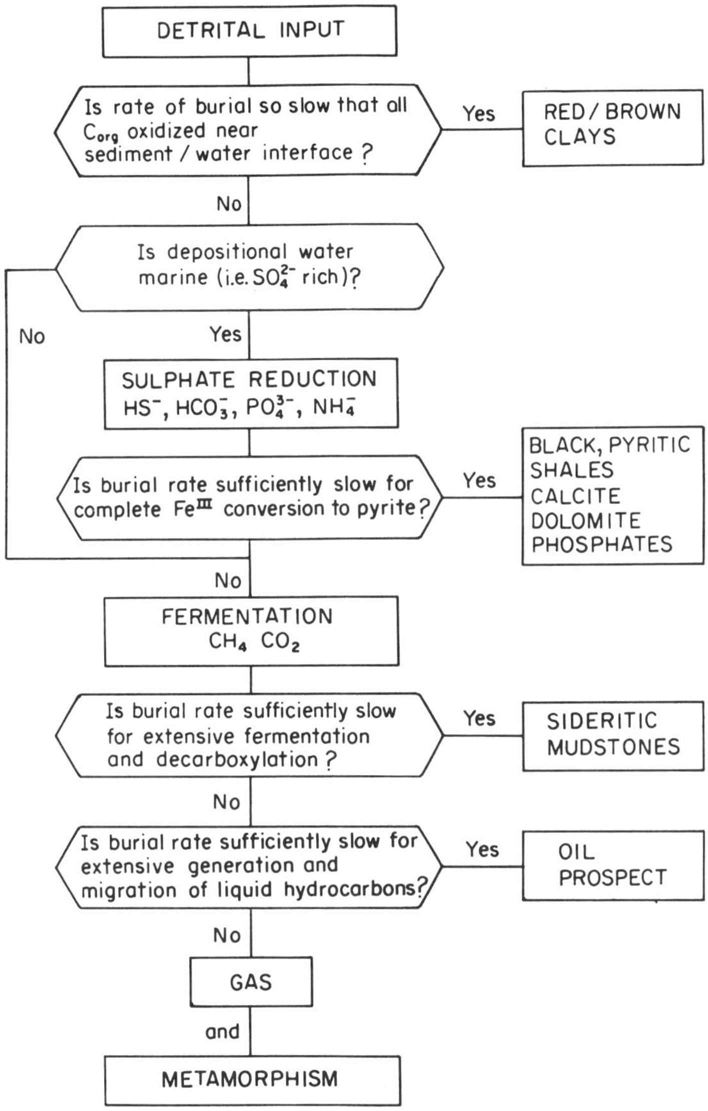

FIG. 15. Schematic diagram to illustrate the interrelations between sediment mineralogy, mud source rock potential and rate of burial (after Curtis 1977).

conditions, allowing the oxidation of organic matter and eventually development of haematite (McBride 1974; Besly & Turner 1983). A further contribution to the red coloration results from the rapid oxidation of finely disseminated pyrite producing ferric oxyhydroxides which redden with age (Berner 1969).

Mesogenetic regime

The sediment is gradually removed from the influence of depositional pore waters during burial. Although difficult to define precisely, the mesogenetic regime begins when the sediment is effectively sealed from the predominant influence of surface agents in the interstitial pore water, and persists to the onset of metamorphism or uplift and re-exposure.

Following the eogenetic modifications, regardless of original depositional environment, the sedimentary assemblage will comprise a mixture of those phases that were initially stable (or metastable) and new authigenic phases which have reached stability (or metastability) through reaction with the pore waters. To effect a further change in the diagenetic assemblage requires a change in the physicochemical conditions of temperature, pressure or pore-fluid chemistry. However, the solubility of many mineral phases increases with increasing temperature so that pore fluids will be capable of dissolving more ionic species. Also kinetically, increasing temperature adds energy to the reacting system. Reaction rates will therefore increase and reaction barriers can be more easily overcome so that metastable phases will react more readily.

This is well illustrated by the relationship between clay mineral transformations and burial temperatures which has been studied by many workers in several sedimentary basins. Probably the most extensively documented reaction is the conversion of smectites to illites through an intermediate mixed-layer illite-smectite with increasing depth and temperature, particularly well studied in the Tertiary of the Gulf Coast of the U.S.A. (e.g. Burst 1959, 1969; Perry & Hower 1970; Hower *et al.* 1976; Boles & Franks 1979a; summarized by Boles 1981 and Hower 1981). In detail the reaction of smectite to illite is complex and does not simply take place by cation exchange, but also involves aluminium substitution for silicon, reduction of octahedral iron and eventually the loss of both octahedral magnesium and iron (Hower *et al.* 1976).

Boles & Franks (1979a) suggested that the reaction does not retain the 2:1 structure of the original smectite. The aluminium needed for the reaction is concentrated in the smectite causing the collapse of the 2:1 structure forming new illitic layers and liberating silica. Any potassium present in the original smectite would contribute to the newly forming illite, the remainder coming from external sources.

However, the reaction may take place in a different manner. Hower *et al.* (1976) suggest that the original 2:1 smectite layer structure remains intact and the reaction essentially involves ionic substitution. Nadeau *et al.* (1984) reviewed the smectite-illite transformation with the aid of the TEM. They concluded that illite crystals are amalgamated 10 Å smectite layers. Solid solution recrystallization does not occur, hence this model may need to be re-assessed.

Boles (1981) considered that the smectite to illite transformation is not only significant in shale diagenesis but also has far-reaching implications for sandstone cementation (see Figs 16 and 17). Detailed petrographical studies of Gulf Coast sediments (Boles 1978; Boles & Franks 1979a; Land & Milliken 1981; Milliken *et al.* 1981) indicate that a variety of cements occur in the associated sandstones. These include quartz overgrowths, late ferroan carbonates—ferroan calcite, ferroan dolomite and ankerite (Figs 5d, 6d and 9f)—chlorite and extensive albitization of plagioclase. The breakdown of smectite to illite releases silica, calcium, sodium, iron and magnesium into the pore waters which may be capable of transporting dissolved ionic species from mudrocks into sandstones. Silica so

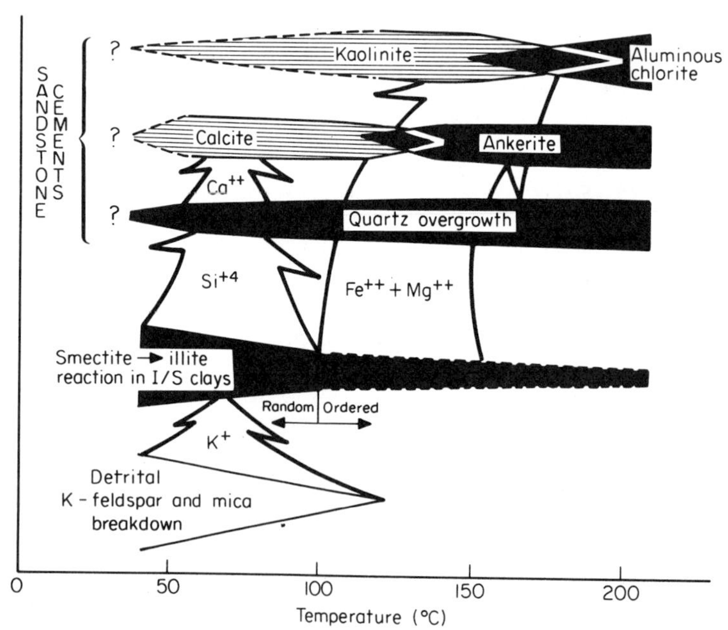

FIG. 16. The influence of the mixed layer illite-smectite clay depth reaction on sandstone cementation and mineral phases in the Wilcox Sandstones of the Texas Gulf Coast (after Boles & Franks 1979a). Vertical arrows depict ion transfer between illite-smectite reactions and mineral phases in sandstones.

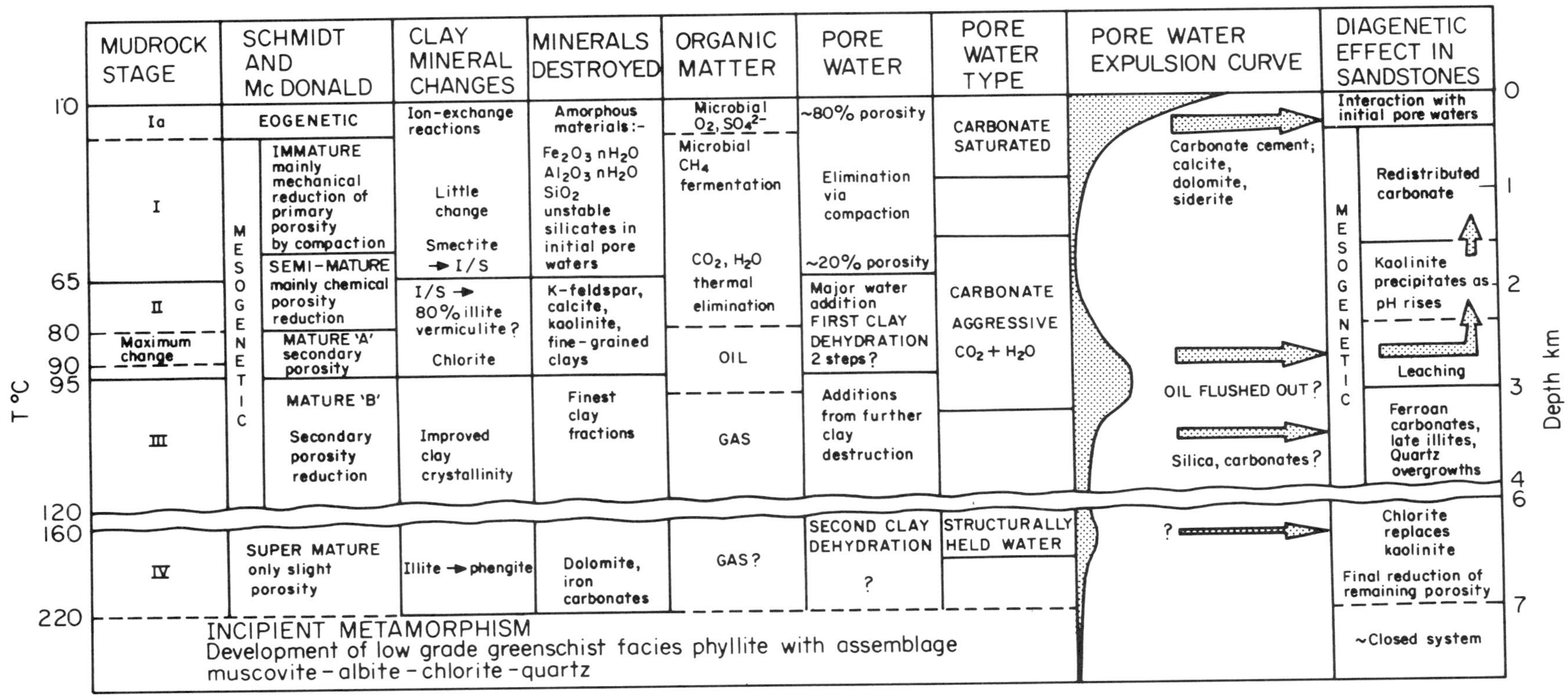

FIG. 17. Summary of depth-related diagenetic stages and important reactions in mudrocks which may contribute solutes to sandstones, via the associated trends in pore water evolution and expulsion (modified from Curtis 1983b, with additional data from Powers 1967; Perry & Hower 1970; Walker *et al.* 1978; Boles & Franks 1979a; Foscolos & Powell 1979a; Schmidt & McDonald 1979a).

released may precipitate as quartz overgrowths (Figs 7a, b, e and f), whilst iron and magnesium may contribute to the formation of chlorite from kaolinite and ferroan dolomite or ankerite from pre-existing calcites. Sodium liberated may contribute to the albitization of plagioclase.

This strongly implies that mass transfer of ionic species occurs between mudrocks and sandstones during mesogenetic diagenesis. There are other related lines of evidence to support the concept of mass transfer in the deep subsurface. Concomitant with the depth-related changes taking place in the mineralogy is the evolution and expulsion of the interstitial pore waters. Undisturbed muds, when initially deposited, have porosities in excess of 80%, whilst well sorted sands may have porosities of 50% (Burst 1969). During early burial most of the eogenetic interstitial pore waters are expelled by mechanical compaction. However, with elevated temperature, many diagenetic reactions occur which involve dehydration and will, therefore, liberate water. Hydrous or amorphous minerals rapidly dehydrate and, in smectitic clays undergoing transformation to illite, large quantities of interlayer water are expelled (Burst 1969). This smectitic clay dehydration is actually a complex process and probably takes place in at least two steps (Perry & Hower 1972). The initial water loss is associated with the collapse of smectite interlayers. Other complications arise where vermiculite is present as dehydration may be delayed until much later in diagenesis (Foscolos & Powell, 1979b). The destruction of clay minerals and organic matter also liberates structurally held water, even at considerable burial depths.

Throughout burial diagenesis therefore, in addition to evolution of the interstitial pore waters through diagenetic reactions, potentially large volumes of water may be added to the reacting system through the breakdown of the clay minerals and organic material. In addition, organic matter undergoes a series of depth-related changes following microbial degradation (Tissot *et al.* 1974) Bacterial formation may extend to depths of 1 km after which either increasing temperature or the exhaustion of suitable organic material prevents the activity of bacteria. With continued burial abiotic reactions dominate and thermal decarboxylation results in the liberation of carbon dioxide to depths of around 3 km and corresponding temperatures of 100°C. Further increase in temperature results in the generation of hydrocarbons.

All these depth-related processes may conveniently interact in the deep subsurface to produce a repeated pattern of sandstone diagenesis (Fig. 17). Throughout burial, pore waters expelled from mudrocks will tend to migrate upwards under a pressure gradient. Carbon dioxide liberated during thermal decarboxylation of organic matter produces acidic solutions. These migrating acidic pore waters can potentially carry high concentrations of aluminium and metal cations (Curtis 1983a). Water movement will tend to follow the more permeable channels in a sedimentary sequence and will therefore migrate into major sandbodies. Here, acidic pore waters will be capable of generating widespread secondary porosity by the removal of carbonate cements (see Figs 2 and 17 for explanation; and Figs 7d, f and 10e, f). These solutions may also precipitate extensive kandites (Figs 4f, 7f and 8f). Many authors now recognize deep mesogenetic secondary porosity in sandstone reservoirs (see Schmidt & McDonald 1979a, b, for a synopsis; Loucks *et al.* 1977; Lindquist 1977; Land & Dutton 1978; Milliken *et al.* 1981). An essential precursor to the development of this porosity is the presence of an early, framework-supporting cement. Given a stable silicate framework, the creation of deep mesogenetic porosity at depth clearly has important implications for petroleum reservoirs. Furthermore, smectite dehydration, supplying a water source at depth, and the release of chemical species that may precipitate higher in the section, may provide a mechanism for the migration of petroleum and formation of a 'diagenetic seal' causing hydrocarbon entrapment and overpressuring (Burst 1969; Bonham 1980).

Much of the above data has been accumulated from the Gulf Coast of the U.S.A. This is a clay-dominated sequence which has undergone continuous and active burial since the late Cretaceous and therefore has a simple burial history. Maximum burial depths, temperatures and pressure are known *in situ* (Boles 1981) and water migration is known to be up dip, largely through a compactional drive. However, the depositional environment is often ignored or given little consideration and little attention has been paid to the eogenetic processes that may have taken place. When studying depth-dependent mineralogical changes, it is important to bear in mind that the starting mineralogy in shales at different levels is likely to vary considerably (see Hower 1981).

Similar depth-related changes to those recorded from the Gulf Coast are known in other deep basins. In NE British Columbia and the NW Territories of Canada the smectite to illite transformation has been studied by Powell *et al.*

(1978). However, their results suggest that the burial depths at which clay dehydration occurs are attained prior to oil generation and that the water expelled by this process does not aid in oil migration. Foscolos & Powell (1979b) considered that the actual destruction of clay minerals (mostly expandables and kaolinite) releases crystal lattice water later during diagenesis which may assume an important role in petroleum migration. This second clay dehydration occurs after oil has become cracked to gas because of the presence of a calcium-vermiculite component in the mixed layer clays which is stable to higher temperatures (see Fig. 17).

Considerably less work has been published on deep-burial diagenesis in the U.K. Pearson *et al.* (1982) demonstrated that the smectite clay dehydration reaction is taking place in the early Tertiary sediments of the northern North Sea Viking Graben. The Kimmeridge Clay in the Jurassic, which is widely regarded as the main source for North Sea oil, has undergone smectite-illite ordering and Pearson *et al.* (1982) considered that water released during illitization would have been available at the time of hydrocarbon generation. Rossel (1982) envisaged that a similar depth-related clay mineral transformation has occurred in the Rotliegendes of the southern North Sea. Eogenetic clay mineral suites comprised a smectitic phase derived from the alteration of iron-bearing silicates. Under deep-burial conditions this smectite and kaolinite became unstable and converted to illite (e.g. Fig. 8a), although the source of potassium is envisaged to be overlying Zechstein halite. The occurrence and dominance of illite at depth in the North Sea has been noted by several authors (Fig. 5e; Sommer 1978; Hancock & Taylor 1978; McHardy *et al.* 1982) but there is little general agreement on its cause or relative timing.

Depth-related diagenetic assemblages are recorded from Triassic (Burley 1984) and Carboniferous sandstones (Hawkins 1978; Huggett 1984) in several basins in the U.K. Interpretation of the diagenetic history of these sediments is hindered by the complex burial history which includes tectonic inversion. However, late cements include isotopically light ankerite, quartz overgrowths and illite with associated secondary porosity (Figs 5e, f, 6d and 9d, f). That there is a remarkable degree of consistency in deep-burial diagenetic assemblages from different tectonic and environmental settings suggests that depth-related physico-chemical conditions conform to a predictable pattern. Widespread pore-fluid migration at depth from compaction of mudrocks, clay mineral reactions or large-scale migration of formation water is likely to impart changes in sediments on a regional scale, resulting in a uniformity of diagenetic texture and mineral assemblages.

Although the main theme of this account of mesogenesis has dealt with chemical reactions and mineralogical changes it would be inappropriate to conclude without reference to the physical processes which may be important during burial diagenesis. Following deposition, clastic sediments are compressed and compacted by the weight of the overlying sediments which results in a reorganization of grain fabric and a general reduction in porosity.

Fine- and coarse-grained clastics react in different ways. Mud-dominated sequences may be compacted to 10% of their original thickness, coincidentally reducing intergranular porosity from 80% to perhaps 5% and hence expelling considerable quantities of water into adjacent sands. This process may be important both during the most rapid compaction at shallow depths and later during burial when temperature-related reactions release absorbed water into pore spaces. During compaction the orientation of particles in muds changes from random to orientated parallel to basal spacings. Pore water is initially easily expelled in the early stages of compaction, as both porosity and horizontal permeability are relatively high. With continued compaction, when the mudrock is reduced to around 50% of its original volume, further pore-water escape is hindered by decreasing horizontal permeabilities, whilst adsorbed water becomes more tightly bonded. At depths of up to 1 km, water expulsion in mudrocks is largely driven by gravitational displacement, but clay mineral diagenesis and aquathermal pressuring are the main contributing agencies at depth.

In coarser-grained sequences pressure brings about several porosity-reducing adjustments. During initial mechanical compaction grains move closer together. Rotation and grain slippage, brittle grain deformation and fracturing, and plastic deformation of ductile grains may all be important. Nagtegaal (1978) demonstrated the effect of primary mineralogy on the stability of various sandstone frameworks. Lithic arenites, with high contents of relatively soft and volcanic fragments, are particularly susceptible to mechanical compaction and plastic deformation. In contrast, quartz arenites and arkosic arenites typically exhibit high framework stability to compaction effects (Fig. 10e, f).

Telogenetic regime

Rocks uplifted to the earth's surface and subsequently either altered and reburied, or weathered and eroded, may exhibit a wide range of diagenetic modifications. This is because minerals formed during burial at elevated temperatures and pressures, and from concentrated formation waters, are often unstable in oxidizing and occasionally acidic conditions (e.g. Berner *et al.* 1980). Such modifications may be illustrated with reference to records of both ancient sediments exposed beneath unconformities, and a range of rocks being altered or weathered at or near the earth's surface today.

Taylor (1978) described the alteration of organic matter, ferrous minerals and sulphides in Westphalian Carboniferous sediments in the Lake District. He related these various reactions to oxidation during exposure during the early Permian. These observations are entirely consistent with theoretical models derived to explain modern weathering reactions (Curtis 1976). Hence organic matter, preserved under the originally euxinic conditions during burial, was oxidized and the carbon completely removed during early Permian exposure. In addition, iron carbonates were oxidized to iron oxides, eventually producing haematite, whilst the oxidization of sulphides resulted in the formation of sulphates such as gypsum.

It is suggested by a number of authors that telogenetic modifications occur in Middle Jurassic clastic sediments in Brent Group reservoirs from the northern North Sea. Brent Group sandstones are variously described as flushed by freshwater both in the Middle Jurassic and subsequently during uplift and subaerial exposure beneath Upper Jurassic or Lower Cretaceous unconformities (e.g. Hancock & Taylor 1978; Sommer 1978; Bjørlykke *et al.* 1979; Giles & Marshall, in press). Exposure to acidic freshwater in these conditions is proposed to explain the presence of kaolinite or dickite, occurring both as a pore filling (Fig. 8f) and as a replacement of detrital feldspars (Fig. 5c). In addition, this may explain the improved porosity that often occurs up dip beneath unconformities in these reservoirs. Nevertheless, a number of problems exist with this general theory. Its two basic premises appear to be that kaolinite can only form from the dissolution of feldspars, and freshwater is the only potential source of acidity. However, a review of the occurrences of kaolinite and dickite reveals firstly that they may form in quartz arenites and, moreover, in limestones (Shelton 1964; Dickson & Coleman 1980; Burchette & Britten, this volume). Theoretically, kaolinite may also be produced as the result of mudrock diagenesis (Curtis 1983a), whilst freshwater is not necessarily acidic (see above). Even if such meteoric waters were available, Curtis (1983b) does not believe 'freshwater' has the chemical potential to create the secondary porosity observed. In addition, freshwater is not the only source of acidic pore waters available, as acidic solutions are believed to emanate from mudrocks during their burial, especially when they contain organic matter (e.g. Parker 1974; Tissot *et al.* 1974; Pearson *et al.* 1982; Rossel 1982). An additional point overlooked by many adherents of the 'freshwater flushing' school is that in actively forming sedimentary basins water movement is predominantly upwards driven by compactional processes (see mesogenesis section). With a change of tectonic regime to a situation in which the sedimentary pile remains static or is uplifted considerable quantities of water may be driven into and through the basin. The model of Schmidt & McDonald (1979a) was based largely on data from the former situation whilst many examples of freshwater penetration quote ancient sediments presently at or near the earth's surface. This topic is currently controversial and will probably be debated vigorously by the opposing schools of thought for some time to come.

The alteration or weathering of rocks at the earth's surface represents not only the end of one cycle but also the beginning of another in clastic diagenesis. This topic demonstrates particularly well the important contribution that has been made to our understanding of diagenetic processes by soil scientists. Pedogenic studies are particularly relevant because the alteration of detrital grains at the surface in soil profiles may take place by similar processes to those which affect grains subsequently. Hence, an understanding of the ways in which detrital minerals are modified or the mechanism by which pedogenic (authigenic) minerals form allows the conditions during sediment diagenesis to be inferred. Dixon & Weed (1977) provided comprehensive coverage of this subject, illustrating both the minerals and the processes involved as well as the variety of techniques employed. In addition, Curtis (1976) and Berner (1981 and references therein) both studied the thermodynamics of these processes. Their work and its application demonstrates the importance of an integrated approach to diagenesis.

Numerous studies of surface or near-surface alteration could be reviewed in this section. For example, Nishiyama *et al.* (1978) describe a Triassic chlorite-bearing red shale from Toyoma in Japan. At outcrop this shale is found to contain an irregularly interstratified chlorite–vermiculite rather than pure chlorite. They concluded that alteration of chlorite to vermiculite is most probably a weathering process (cf. Fig. 4a). Glasmann (1982) described the alteration of andesite from soils in the Cascades, illustrating both the more labile minerals within the rock, and their alteration products. Augite and plagioclase were both observed to alter to smectites, which grow as globules initially and act as nuclei for subsequent fibrous growth (cf. Guven *et al.* 1980). These studies are particularly useful in attempts to establish the original detrital mineralogy of greywackes and other digenetically mature sediments. Benson & Teague (1982) described the alteration of Miocene basalts from the Pasco Basin, Washington. They reported a paragenetic sequence comprising nontronite (an iron-rich smectite), clinoptilolite, and silica, with minor amounts of mordenite replacing the clinoptilolite. More significantly the alteration of basalts to these various minerals can be related to low-temperature processes, in ground-water conditions similar to those occurring today. Consequently, they proposed that zeolite formation is possible within 1000 m of the earth's surface and without elevated temperatures.

A detailed hydrogeochemical study by Walton (1981) demonstrated that dilute acidic ground waters (pH 5–6) in the Triassic Budleigh Salterton Pebble Beds aquifer of south Devon was capable of extensively leaching haematite and carbonate from the sediments. Furthermore, detrital feldspar in these sandstones has been replaced *in situ* by an intimate association of illite, kaolinite, and authigenic potassium feldspar. This telogenetic alteration is readily explained by the modern ground-water chemistry as the Pebble Bed waters are presently saturated with respect to both kaolinite and illite, containing considerable quantities of dissolved alumina. Thus, under surface conditions of temperature, pressure and modern pore-water chemistry, the Triassic mineral assemblage is unstable and has undergone extensive reaction with the ground waters in a shift towards equilibrium.

Volcaniclastic sediments

The role of volcaniclastic grains in clastic diagenesis has been mentioned briefly in the sections above. However, volcaniclastic sediments are especially significant because of the extreme results of their alteration when they dominate a clastic sedimentary assemblage. Volcaniclastic material is relatively unstable at the earth's surface. Consequently its constituents react readily with their pore waters and dissolve resulting in the precipitation of a variety of authigenic minerals. Vessel & Davies (1981), for example, report the generation of 20% authigenic clay within 3–5 years of deposition in volcaniclastic sediments from Guatemala. The speed of diagenesis of volcaniclastic sediments is illustrated by the results of studies from the geological record. The potential reactivity of sediments of this type was originally recognized by Cummins (1962) who proposed a diagenetic rather than detrital origin for the matrix of greywackes. This proposal was subsequently confirmed experimentally by Whetten & Hawkins (1970). Galloway (1974) reported how the reaction of predominantly ferromagnesian sediments controls their diagenesis with little reference to likely variations in depositional pore-water chemistry. Moreover, whilst initial reactions formed calcite and then chlorite, reactions at higher temperatures and pressures, and involving both the remaining detritus and initial diagenetic minerals, resulted in zeolite formation (cf. Benson & Teague 1982). Because of their textural transformations during diagenesis dominantly volcaniclastic sediments have poor reservoir potential (Nagtegaal 1978). This results from the wholesale dissolution and breakdown of detrital grains, which destroys the sediment's original framework, reducing the potential for preservation of any intergranular pore space. Volcaniclastic sediments suffer perhaps the most extreme diagenetic modifications, manifesting little or no textural or mineralogical similarities to the original assemblage.

Discussion and conclusions

The past 20 years have seen significant changes in the approaches taken to studying clastic rocks. Similarly, important advances have been made in our understanding of the diagenesis of these rocks. Diagenesis has evolved from a purely petrographical microscopic-based subject into a broad discipline encompassing all aspects of sedimentary geochemistry. This change may be traced from the 1950s through the establishment of predictive depositional-facies models, and the development of sophisticated quantitative analytical techniques, to more recent integration of data from numerous related disciplines such as organic geochemistry, pedology and hydrology. As a result it is now almost possible to establish the entire history of a rock from the source of its detrital components, via the depositional environment, with the associated (possibly climatically influenced) early cementation and through deep burial and changing pore-fluid composition and eventually into weathering following uplift. During diagenesis, distinct realms may be recognized in which pore-fluid composition is influenced by the depositional environment (eogenesis), or by formation waters (mesogenesis), or finally by the reintroduction of surficial agents during uplift and exposure (telogenesis). Within each of these realms differing styles of diagenesis may be recognized. Eogenetic processes are dominated by the relatively rapid dissolution and reaction of amorphous and unstable components resulting in the formation of characteristic minerals, such as haematite. Mesogenetic processes dominantly operate within the framework established previously with most significant changes influenced by solutions introduced from elsewhere in the basin. Telogenesis, by contrast, involves the alteration of minerals often formed under drastically different conditions and may eventually lead to the wholesale destruction of the rock.

A number of important points may be emphasized by contrasting the various eogenetic assemblages, and by comparing the general style of eogenetic modifications with those occurring subsequently. Although there are similarities in the relative order of diagenetic relations during eogenesis, the diversity of products of any initial reactions suggests that truly predictive models may be extremely hard to establish (see Almon & Davies 1979; Storey 1979; Besly & Turner 1983). Nonetheless, the inherent metastability of many of these assemblages results in rocks being 'fingerprinted' by the interstitial water conditions prevailing immediately following deposition. Furthermore, because of the relative instability of the original sedimentary assemblage the changes which may occur during eogenesis are often more drastic than any subsequent modifications. Consequently, with the obvious exception of volcaniclastic sandstones, mesogenetic modifications are generally restricted to the precipitation and dissolution of authigenic mineral phases within the solid framework established during eogenesis.

In each of the three eogenetic regimes outlined above a sequence of clay authigenesis, quartz and feldspar overgrowth cementation and carbonate cementation generally occurs. Within this general pattern there is however considerable diversity. Firstly, the clay mineralogies outlined above encompass illitic and smectitic clays in dry non-marine settings, kaolinite and chlorite in wet ones, and chlorite or smectitic clays in marine depositional environments. There is clearly some considerable overlap between these various eogenetic regimes. Consequently, distinct assemblages may form, and these may be explained in terms of the known depositional history. However recognition of distinct assemblages does not allow any ready interpretation of the depositional environment.

As well as the authigenesis of clay minerals, equally rapid eogenetic precipitation, dissolution and replacement reactions, which may be observed at the sediment surface today in modern analogues, are interpreted to take place. These lead to the formation of calcretes and haematite in dry settings, lateritic and sideritic soils in wet non-marine environments, whilst in the sea glauconite and berthierine ooliths form in addition to a range of incipient ferrous and calcareous concretions within muds rich in organic matter. These various processes generally have the effect of fixing many of the previously amorphous constituents of the sedimentary assemblage. They also remove from solution many of the cations liberated during the dissolution of igneous or metamorphic rock fragments. The subsequent formation of quartz overgrowths demonstrates that pore waters at this time must be excessively supersaturated with respect to silica, more than sufficient being present to allow its continued precipitation.

During mesogenesis a range of effects may occur. These vary according to the original detrital mineralogy, with the more pronounced effects being felt where relatively unstable grains survive eogenetic reactions, and form an integral part of the rock framework during its burial. With a more stable mineralogy—e.g. in quartz arenites—early cementation with quartz overgrowths often creates a rigid framework. The stability of this framework allows the repeated precipitation and dissolution of pore-filling cements during burial without the gross composition of the rock being affected. With the introduction of more lithic components or feldspar grains into the sandstone, reactions during burial may be more effective creating significant grain-dissolution porosity (Heald & Larese 1973). Alternatively in less well-cemented rocks, feldspar or lithic grains are compressed or compacted leading to a gradual reduction in intergranular porosity (McBride 1978). With greater percentages of these labile grains, wholesale destruction of the detrital grains often occurs during eogenesis. Nonetheless, any grains surviving into mesogenesis are usually altered during this phase, with the result that the original depositional framework of the rock may cease to exist whilst the mineralogical components may have changed from those present initially (e.g. Galloway 1974). The recognition of the potential relationship between detrital mineralogy and reservoir rock potential during burial also has important implications for hydrocarbon exploration.

The relatively high initial geochemical instability of detrital minerals in the eogenetic environment will tend to produce a shift toward sediment–pore-water equilibrium during early diagenesis. To produce significant changes in the resulting stable (or metastable) assemblage would require a change in the gross physical and chemical environment. It seems probable therefore that pressures and temperature, and pore waters introduced into the sediments from elsewhere in the basin, are likely to play increasingly important roles during mesogenesis. The remarkable degree of consistency recorded in deep-burial diagenetic assemblages (crystalline illites, chlorite, isotopically light ferroan carbonates, quartz cements) and common secondary porosity, from a variety of environmental and tectonic settings, does suggest that depth-related diagenetic processes conform to a pattern that may be predictable (see Fig. 17). Clearly, large volumes of pore fluids do migrate at depth in the subsurface, and the evidence for widespread solute transport is convincing.

To date, therefore, the study of diagenesis has enabled general models for the evolution of sediments deposited in a range of environmental and tectonic settings to be developed. As a result of these investigations a wide range of diagenetic mineral associations and textures can be interpreted, and some diagenetic processes are now understood. However, there are many aspects of diagenesis that remain poorly understood and provide scope for much research. There are many situations, for example, in which the diagenetic history of a clastic sediment is too complex to unravel with standard techniques. Continued application of more sophisticated techniques, such as stable isotope analysis, fluid inclusion studies and cathodoluminescence, should help resolve this. Many diagenetic processes are not fully understood, and here the use of geochemical modelling and high-resolution quantitative analytical techniques can be expected to increase our knowledge. Hydrogeology is a field which should also make significant contributions towards our understanding of subsurface fluid flow and sediment/water interactions over the next few years.

In conclusion, diagenesis has come of age as a discipline in the last 20 years, although it still remains a long way from being a truly predictive science. We hope that this admittedly biased review will provide a useful introduction to diagenesis for workers in diverse disciplines. We also hope that we may stimulate other workers to confirm or refute the models we have described, many of which cannot be regarded as the consensus of opinion at present.

ACKNOWLEDGEMENTS: The research which forms the basis of the review was carried out whilst all the authors were in receipt of NERC financial assistance at the University of Hull. The technical and academic staff of the Department of Geology are thanked for their high-quality support and stimulating encouragement. We are also grateful to the Geological Institute at the University of Berne and AGAT Consultants Inc., for continued assistance during the preparation of the manuscript. Special mention must be made of Julia Davidson for typing and editing innumerable versions of the manuscript.

A review paper by definition utilizes the research of other workers. We should like to thank the numerous people who have listened patiently to our ideas and whose helpful discussions have clarified our thinking. Mistakes of fact and opinion are our own.

Karl Ramseyer, of the University of Berne, took the luminescence pictures, Alan Mitchell, of Britoil, kindly provided the colour slides of the Rotliegendes. Photographic plates were prepared by Andy Werthemann at the University of Berne. Finally, we extend our gratitude to our reviewers and editors and to the expert staff of Blackwell Scientific Publications and Alden Press for their patience in piecing together this multifarious contribution.

References

ALI, A. D. & TURNER, P. 1982. Authigenic K-feldspar in the Bromsgrove Sandstone Formation (Triassic) of Central England, *J. sedim. Petrol.* **52**, 187–98.

ALMON, W. R. & DAVIES, D. K. 1978. Clay technology and well stimulation *Trans. Gulf-Cst Ass. geol. Socs*, **28**, 1–6.

—— & —— 1979. Regional diagenetic trends in the Lower Cretaceous Muddy Sandstone, Powder River Basin. *In*: SCHOLLE, P. A & SCHLUGER, P. R. (eds) *Aspects of Diagenesis. Spec. Publs Soc. econ. Miner. Paleont., Tulsa*, **26**, 379–400.

——, FULLERTON, L. B. & DAVIES, D. K. 1976. Pore space reduction in Cretaceous sandstones through chemical precipitation of clay minerals. *J. sediment. Petrol.* **46**, 89–96.

ARCHER, A. A. 1965. Red beds in the Upper Coal Measures of the Western part of the South Wales Coalfield. *Bull. geol. Surv. Gt Br.* **23**, 57–64.

ARONSON, J. L. & BURTNER, R. L. 1982. K-Ar dating of illitic clay in sandstone reservoirs and timing of petroleum migration. *Am. Ass. Petrol. Geol. Bull.* **66**, 1442 (abstract).

BAILEY, E. H. & STEVENS, R. E. 1960. Selective staining of K-feldspar and plagioclase on rock slabs and thin section. *Am. Miner.* **45**, 1020–5.

BAILEY, S. W. 1980. Summary of recommendations of AIPEA nomenclature committee. *Clay Miner.* **15**, 85–93.

BENSON, L. V. & TEAGUE, L. S. 1982. Diagenesis of basalts from the Pasco Basch, Washington. I. Distribution and composition of secondary mineral phases. *J. sedim. Petrol.* **52**, 595–613.

BERNER, R. A. 1969. Goethite stability and the origin of red beds *Geochim. cosmochim. Acta*, **33**, 267–73.

—— 1971. *Principles of Chemical Sedimentology*, McGraw-Hill, New York. 240 pp.

—— 1980. *Early Diagenesis: A Theoretical Approach*, Princeton University Press, Princeton, New Jersey. 241 pp.

—— 1981. Kinetics of weathering and diagenesis. *In*: LASAGA, A. C. & KIRKPATRICK, R. J. (eds) *Kinetics of Geochemical Processes. Reviews in Mineralogy*, **8**, 111–34. Mineralogical Society of America,

—— & HOLDREN, G. J. 1979. Mechanism of feldspar weathering. II. Observations of feldspars from soils. *Geochim. cosmochim Acta*, **43**, 1173–86.

——, SJOBERG, E. L. & VELBEL, M. A. & KROM, M. D. 1980. Dissolution of pyroxenes and amphiboles during weathering. *Science*, **207**, 1205–6.

BESLY, B. M. & TURNER, P. 1983. Origin of red beds in a moist tropical climate (Etruria Formation, Upper Carboniferous, U.K.). *In*: WILSON, R. C. L. (ed.) *Residual Deposits: Surface Related Weathering Processes and Materials. Geol. Soc. London Spec. Publ.* **11**, 131–47.

BHATTACHARYYA, D. P. & KAKIMOTO, P. K. 1982. Origin of ferriferous ooids: and SEM study of ironstone ooids and bauxite pisoids. *J. sedim. Petrol.* **52**, 845–57.

BJØRLYKKE, K. 1979. Cementation of sandstone—discussion. *J. sediment. Petrol.* **49**, 1358–60.

——, ELVERHOI, K. A. & MALM, A. O. 1979. Diagenesis in Mesozoic sandstones from Spitsbergen and the North Sea, a comparison. *Geol. Rdsch.* **68**, 1152–71.

BLATT, H. 1966. Diagenesis of sandstones: processes and problems. *Wyoming Geol. Ass. Ann. Conf.* **65**, a–o.

—— 1979. Diagenetic processes in sandstones. *In*: SCHOLLE, P. A. & SCHLUGER, P. R. (eds) *Aspects of Diagenesis. Spec. Publs Soc. econ. Miner. Paleont., Tulsa*, **26**, 141–57.

BOAST A. M., COLEMAN, M. L. & HALLS, C. 1981. Textural and stable isotopic evidence for the genesis of the Tynagh Base Metal deposit, Ireland, *Econ. Geol.* **76**, 27–55.

BOLES, J. R. 1978. Active ankerite cementation on the subsurface Eocene of south-west Texas. *Contrib. Mineral. Petrol.* **68**, 13–22.

—— 1981. Clay diagenesis and effects on sandstone cementation (case histories from the Gulf Coast Tertiary). *In*: LONGSTAFFE, F. J. (ed.) *Clays and the Resource Geologist. Min. Ass. Can., Short Course Hdbk*, **7**, 148–68.

—— 1982. Active albitisation of plagioclase, Gulf Coast Tertiary. *Am. J. Sci.* **282**, 165–80.

—— & FRANKS, S. G. 1979a. Clay mineral diagenesis in Wilcox of south-west Texas: implications of smectite diagenesis in sandstone cementation. *J. sedim. Petrol.* **49**, 55–70.

—— & —— 1979b. Cementation of sandstones—reply. *J. sedim. Petrol.* **49**, 1362.

BOHNAM, L. C. 1980. Migration of hydrocarbons in compacting basins. *Am. Ass. Petrol. Geol. Bull.* **64**, 549–67.

BORST, R. L. & KELLER, W. D. 1969. Scanning electron micrographs of API reference clay minerals and other selected samples. *Proc. Int. Clay Conf.* **1**, 871–901.

BRADSHAW, H. J., JAMES, S. J. & TURNER, P. 1980. Origin of oolitic ironstones—discussion. *J. sedimen. Petrol.* **58**, 295–9.

BREESE, D. E. 1960. Quartz overgrowths as evidence of silica deposition in soils. *Aust. J. Sci.* **23**, 18–20.

BRINDLEY, G. W. (ed.) 1951. *X-ray Identification and Crystal Structures of Clay Minerals*, Mineralogical Society, London. 345 pp.

—— 1982. Chemical composition of berthierines. *Clays Clay. Miner.* **30**, 153–5.

—— & BROWN, C. (eds) 1980. *Crystal structures of clay minerals and their x-ray identification. Mineralogical Society Monograph*, **5**, Mineralogical Society, London. 455 pp.

BURLEY, S. D. 1984. Distribution and origin of authigenic minerals in the Triassic Sherwood Sandstone Group, U.K. *Clay Miner.* **19**, 403–40.

BURST, J. F. 1959. Post diagenetic clay mineral environmental relationships in the Gulf Coast Eocene. *Clays Clay Miner.* **6**, 327–41.

—— 1969. Diagenesis of Gulf Coast clayey sediments and its possible relation to petroleum migration. *Am. Ass. Petrol. Geol. Bull.* **53**, 73–93.

CARRIGY, M. A. & MELLON, G. B. 1964. Authigenic clay mineral cements in Cretaceous and Tertiary sandstones of Alberta *J. sedim. Petrol* **34**, 461–72.

COLEMAN, J. M. & HO, C. 1968. Early diagenesis and compaction in clays. Abnormal Subsurface pressures. *Louisiana Coastal Studies Institute Technical Report 62 AD No. 687550*, pp. 23–50. Louisiana State University, Louisiana.

COLEMAN, M. L. 1977. Sulphur isotopes in petrology. *J. geol. Soc. London*, **133**, 595–608.

—— & RAISWELL, R. 1981. Carbon, oxygen and sulphur isotope variations in concretions from the Upper Lias of N.E. England. *Geochim. cosmochim. Acta*, **45**, 329–40.

——, CURTIS, C. D. & IRWIN, H. 1979. Burial rate a key to source and reservoir potential. *Wld Oil*, **5**, 83–92.

COOKE, R. U. & WARREN, A. 1973. *Geomorphology in Deserts*, Batsford Ltd London. 374 pp.

CUMMINS, W. A. 1962. The Greywacke Problem. *Lpool Manchr geol. J.* **3**, 57–72.

CURTIS, C. D. 1976. Stability of minerals in surface weathering reactions: a general thermochemical model. *Earth Surf. Proc.* **1**, 63–70.

—— 1977. Sedimentary geochemistry: environments and processes dominated by involvement in an

aqueous phase. *Phil. Trans. R. Soc.* **A286**, 353–71.
—— 1978. Possible links between sandstone diagenesis and depth related geochemical reactions occurring in enclosing mudstones. *J. geol. Soc. London*, **135**, 107–17.
—— 1980. Diagenetic alteration in black shales. *J. geol. Soc. London*, **137**, 189–94.
—— 1983a. A link between aluminium mobility and destruction of secondary porosity. *Am. Ass. Petrol. Geol. Bull.* **67**, 380–4.
—— 1983b. Geochemical studies on development and destruction of secondary porosity. *Geol. Soc. London Spec. Publ.* **12**, 113–25.
—— & SPEARS, D. A. 1968. Formation of sedimentary iron minerals. II. *Econ. Geol.* **63**, 262–70.
—— & SPEARS, D. A. 1971. Diagenetic development of kaolinite. *Clays Clay Miner.* **19**, 219–27.
——, PETROWSKI, C. & OERTEL, G. 1972. Stable carbon isotopes within carbonate concretions: a clue to place and time of formation. *Nature, Lond.* **235**, 98–100.
——, IRELAND, B. J., WHITEMAN, J. A., MULLANEY, R. & COBLEY, T. 1984. Stability of authigenic clay minerals: new evidence from analytical transmission electron microscopy. *Clay Miner.* **19**, 471–81.
DICKSON, J. A. D. 1965. A modified staining technique for carbonates in thin section *Nature, Lond.* **205**, 587.
—— & COLEMAN, M. L. 1980. Changes in carbon and oxygen isotope composition during limestone diagenesis. *Sedimentology*, **27**, 107–18.
DIXON, J. E. & WEED, S. B. (eds) 1977. *Minerals in Soil Environments*, Soil Science Society of America, Madison, Wisconsin. 948 pp.
DOMENICO, P. A. 1977. Transport phenomena in chemical rate processes in sediments. *Ann. Rev. Earth planet. Sci.* **5**, 287–317.
DUTTON, S. P. 1977. Diagenesis and secondary porosity distribution in deltaic sandstone, Strawn Series (Pennsylannian) north-central Texas. *Trans. Gulf-Cst Ass. Geol. Socs*, **27**, 272–7.
EDWARDS, M. B., 1979. Sandstone in Lower Cretaceous Helvetiafjellet Formation, Svalbard. Bearing on reservoir potential of Barents Shelf. *Am. Ass. Petrol. Geol. Bull.* **63**, 2193–203.
ELSBERG, J. N. VAN 1978. A new approach to sediment diagenesis. *Can. Pet. Geol. Bull.* **26**, 57–86.
ELVERHØI, A. & BJØRLYKKE, K. 1978. Sandstone diagenesis—Mesozoic rocks from southern Spitsbergen. *Norsk Polarinst. Arb.* **1977**, 145–57.
FANNING, D. S. & KERAMIDES, V. Z. 1977. Micas. *In*: DIXON, J. N. & WEED, S. B. (eds) *Minerals in Soil Environments*, pp. 195–258. Soil Science Society of America, Madison, Wisconsin.
FISHER, W. L. & MCGOWEN, J. H. 1967. Depositional Systems in the Wilcox Group of Texas and their relationship to occurrence of oil and gas *Trans. Gulf-Cst Ass. geol. Socs*, **17**, 105–25.
FITZPATRICK, E. A. 1980. *Soils: Their Formation, Classification and Distribution*, 2nd edn, Longman, London. 353 pp.
FOLK, R. L. 1965. Hentry Clifton Sorby (1826–1908). The founder of petrology. *J. Geol. Educ.* **8**, 43–7.
FOSCOLOS, A. E. & POWELL, T. G. 1979a. Catagenesis in shales and occurrence of authigenic clay in sandstones, North Sabine Well H-49, Canadian Arctic Islands. *Can. J. Earth Sci.* **16**, 1309–14.
—— & —— 1979b. Mineralogical and geochemical transformation of clays during catagenesis and their relation to oil generation. *Can. Soc. Pet. Geol. Mem.* **6**, 153–72.
FROELICH, P. N., KLINKHAMMER, G. P., BENDER, M. L., LUEDTKE, N. A., HEATH, G. R., CULLEN, D., DAUPHIN, P., HAMMOND, D., HARTMAN, B. & MAYNARD, V. 1979. Early oxidation of organic matter in pelagic sediments of the estern equatorial Atlantic: suboxic diagenesis. *Geochim. cosmochim. Acta*, **43**, 1075–90.
GAC, J. Y. & TARDY, Y. 1980. Geochemistry of a tropical landscape on granitic rocks: the Lake Chad Basin. *In*: CAMPBELL, A. (ed.) *Proceedings 3rd International Symposium on Water–Rock Interaction*, 8–10. Alberta Research Council.
GALLOWAY, W. E. 1974. Deposition and diagenetic alteration of sandstones in northeast Pacific arc-related basins: implications of greywacke genesis. *Bull. geol. Soc. Am.* **85**, 379–390.
—— 1979. Diagenetic control of reservoir quality in arc-derived sandstones; implications for petroleum exploration. *In*: SCHOLLE, P. A. & SCHLUGER, P. R. (eds) *Aspects of Diagenesis. Spec. Publs Soc. econ. Miner. Paleont., Tulsa*, **26**, 251–62.
GARRELS, R. M. & CHRIST, C. L. 1965. *Solutions, Minerals and Equilibria*, Harper and Row, New York, 450 pp.
GIBBS, R. J. 1977. Transport phases of transition metals in the Amazon and Yukon Rivers. *Bull. geol. Soc. Am.* **88**, 829–43.
GILES, M. R. & MARSHALL, J. D. in press. Processes controlling secondary porosity development. *Clay Miner.*
GLASMANN, J. R. 1982. Alteration of andersite in wet, unstable soils of Oregon's western Cascades *Clays Clay Miner.* **30**, 253–63.
GLENNIE, K. W., MUDD, G. C. & NAGTEGAAL, P. J. C. 1978. Depositional environment and diagenesis of Permian Rotliegendes sandstones in Leman Bank and Sole Pit areas of the U.K. Southern North Sea. *J. geol. Soc. London*, **135**, 25–34.
GOLDICH, S. S. 1938. A study in rock weathering. *J. Geol.* **46**, 17–58.
GOULD, K. W. & SMITH, J. W. 1979. The genesis and isotopic composition of carbonates associated with some Australian coals. *Chem. Geol.* **24**, 137–50.
GRABAU, A. W. 1920. *Textbook of Geology. I. General Geology*, Heath and Co., Boston. 864 pp.
GREENLAND, D. J. 1975. Charge characteristics of some kaolinite-iron hydroxide complexes. *Clay Miner.* **10**, 407–16.
GUVEN, N., HOWER, W. F. & DAVIES, D. K. 1980. Nature of authigenic illites in sandstone reservoirs. *J. sediment. Petrol.* **50**, 761–6.

HANCOCK, N. J. & TAYLOR, A. M. 1978. Clay mineral diagenesis and oil migration in the Middle Jurassic Brent Sand Formation. *J. geol. Soc. London*, **135**, 69–72.

HARDER, H. 1974. Illite mineral synthesis at surface temperatures. *Chem. Geol.* **14**, 241–53.

—— 1978. Synthesis of iron-layer silicate minerals under natural conditions. *Clays Clay Miner.* **26**, 65–72.

HASZELDINE, R. S., SAMSON, I. M. & CORNFORD, C. 1984. Quartz diagenesis and convective fluid flow: Beatrice Oilfield, U.K., North Sea. *Clay Miner.* **19**, 331–41.

HAWKINS, P. J. 1978. Relationship between diagenesis, porosity reduction and oil emplacement in late Carboniferous sandstone reservoirs: Bothamsall Oilfield, East Midlands. *J. geol. Soc. London*, **135**, 7–34.

HAYES, J. B. 1979. Sandstone diagenesis—the hole truth. *In*: SCHOLLE, P. A. & SCHLUGER, P. R. (eds) *Aspects of Diagenesis. Spec. Publs. Soc. econ. Miner. Paleont., Tulsa*, **26**, 127–9.

HEALD, M. T. & LARESE, R. E. 1973. The significance of the solution of feldspar in porosity development. *J. sediment. Petrol.* **43**, 458–60.

HELGESON, H. C., GARRELS, R. M. & MACKENZIE, F. T. 1969. Evaluation of irreversible reactions in geochemical processes involving minerals and aqueous solutions. II. Applications. *Geochim. cosmochim. Acta*, **33**, 455–81.

HEMINGWAY, J. E. 1968. Sedimentology of coal bearing strata. *In*: MURCHISON, D. & WESTOLL, T. S. (eds) *Coal and Coal Bearing Strata*, 43–69. Oliver and Boyd, Edinburgh.

HEROUK, Y., CHAGNON, A. & BERTRAND, R. 1979. Compilation and correlation of major thermal maturation indicators. *Am. Ass. Petrol. Geol. Bull.* **63**, 2128–44.

HOUTEN, F. B. VAN 1961. Climatic significance of red beds. *In*: NAIRN, A. E. M. (ed.) *Descriptive Palaeoclimatology*, 89–139. New York Interscience, New York.

HOUTEN, F. B. VAN 1972. Iron and clay in tropical savanna alluvium, North Columbia, a contribution to the origin of red beds. *Bull. geol. Soc. Am.* **83**, 2761–72.

HOWER, J. 1981. Shale diagenesis. *In*: LONGSTAFFE, F. J. (ed.) *Clays and the Resource Geologist. Min. Ass. Can., Short Course Hdbk*, **7**, 60–80.

——, ESLINGER, E. V., HOWER, M. E. & PERRY, E. A. 1976. Mechanism of burial and metamorphism of argillaceous sediment. I. Mineralogical and chemical evidence. *Bull. geol. Soc. Am.* **87**, 725–37.

HUANG, P. M. 1977. Feldspars, olivines, pyroxénes and amphiboles. *In*: DIXON, J. B. & WEED, S. B. (eds) *Minerals in Soil Environments*, 553–602. Soil Science Society of America, Madison, Wisconsin.

HUDSON, J. D. 1977. Stable isotopes and limestone lithification. *J. geol. Soc. London*, **133**, 637–60.

—— 1982. Pyrite in ammonite-bearing shales form the Jurassic of England and Germany. *Sedimentology*, **29**, 639–67.

HUGGETT, J. M. 1984. Controls of mineral authigenesis in Coal Measures Sandstones of the East Midland, U.K. *Clay Miner.* **19**, 343–57.

HURST, A. R. 1981. A scale of dissolution for quartz and its implications for diagenetic processes in sandstones. *Sedimentology*, **28**, 451–60.

—— & IRWIN, H. 1982. Geological modelling of clay diagenesis in sandstones. *Clay Miner.* **17**, 5–22.

—— & STOUT, J. H. 1980. Dissolution and growth of quartz grains from sandstone. *Int. Ass. Sedimentologists. 1st Europ. Mtg Bochum*, pp. 151–4 (abstract).

HUTCHEON, I. 1981. Application of thermodynamics to clay minerals and authigenic mineral equilibria. *In*: LONGSTAFFE, F. K. (ed.) *Clays and the Resource Geologist. Min. Ass. Can., Short Course Hdbk*, **7**, 169–92.

IIJIMA, A. & MATSUMOTO, R. 1982. Berthierine and chamosite in coal measures of Japan. *Clays Clay Miner.* **30**, 264–74.

IRWIN, H. 1981. On calcic dolomite-ankerite from the Kimmeridge clay. *Min. Mag.* **44**, 105–7.

—— & HURST, A. R., 1983. Applications of geochemistry to sandstone reservoir studies. *Geol. Soc. London Spec. Publ.* **12**, 127–46.

——, CURTIS, C. D. & COLEMAN, M. L. 1977. Isotopic evidence for source of diagenetic carbonates formed during burial of organic-rich sediments. *Nature, Lond.* **269**, 209–13.

IXER, R. A., TURNER, P. & WAUGH, B. 1979. Authigenic iron and titanium oxides in Triassic red beds (St Bees Sandstone), Cumbria, Northern England. *Geol. J.* **14**, 179–92.

JACKSON, M. L. 1956. Soil chemical analysis—advanced course. Department of Soil Science, University of Wisconsin, Madison, Wisconsin.

JAMES, H. E. & HOUTEN, F. B. VAN 1979. Miocene goethite and chamositic oolites, northeastern Columbia. *Sedimentology*, **26**, 125–34.

KANTOROWICZ, J. D. 1984. The origin, nature, and distribution of authigenic clay minerals from Middle Jurassic Ravenscar and Brent Group Sandstones. *Clay Miner.* **19**, 359–75.

KASTNER, M. & SIEVER, R. 1979. Low temperature feldspars in sedimentary rocks. *Am. J. Sci.* **279**, 435–79.

KEIGHIN, C. W. & FOUCH, T. D. 1981. Depositional environments and diagenesis of some non-marine Upper Cretaceous reservoir rocks. Uinta Basin, Utah. *In*: ETHRIDGE, F. G. & FLORES, R. M. (eds) *Recent and Ancient Non-marine Depositional Environments: Models for Exploration. Spec. Publs Soc. econ. Miner. Paleont., Tulsa*, **31**, 109–25.

KELLER, W. D. 1954. Bonding energies of some silicate minerals. *Am. Miner.* **30**, 783–93.

—— 1969. *Chemistry in Introduction Geology*, Luca Bros, Columbia, Missouri. 108 pp.

—— 1976. Scan electron micrographs of kaolins collected from diverse environments of origin. I. *Clays Clay Miner.* **24**, 107–33.

KIMBERLEY, M. M. 1979. Origin of oolitic iron formations. *J. sediment. Petrol.* **49**, 111–32.

—— 1980a. Origin of oolitic iron formations—reply.

J. sediment. Petrol. **50**, 299–302.
—— 1980b. Origin of oolitic iron formations—reply. *J. sedim. Petrol.* **50**, 1003–4.
KNOX R. W. O'B. 1970. Chamositic ooliths from the Winter Gill Ironstone (Jurassic) of Yorkshire, England *J. sedim. Petrol.* **40**, 1216–25.
KROM, M. D. & BERNER, R. A. 1981. The diagenesis of phosphorus in a nearshore marine environment. *Geochim. cosmochim. Acta*, **45**, 207–16.
KRYNINE, P. D. 1935. Arkose deposits in the humid tropics. *Am. J. Sci.* **39**, 353–63.
—— 1949. The origin of red beds. *NY Acad. Sci. Trans.* **II**, 60–68.
—— 1950. Petrology, stratigraphy and origin of the Triassic sedimentary rocks of Connecticut. *Conn. Geol. Nat. Hist. Survey Bull.* **73**, 239 pp.
LAND, L. S. & DUTTON, S. P. 1978. Cementation of a Pennsylvannian deltaic sandstone: isotopic data. *J. sedim. Petrol.* **48**, 1167–76.
—— & —— 1979. Cementation of sandstones—reply. *J. sedim. Petrol.* **49**, 1359–61.
—— & MILLIKEN, K. L. 1981. Feldspar diagenesis in the Frio Formation, Brazoria County, Texas, Gulf Coast. *Geology*, **9**, 314–18.
LERBEKMO, J. F. 1957. Authigenic montmorillonoid cement in andesitic sandstones *J. sedim. Petrol.* **27**, 298–305.
LIN, F. C. & CLEMENCY, C. V. 1981. The kinetics of dissolution of muscovite at 25°C and 1 atm. CO_2 partial pressure. *Geochim. cosmochim. Acta*, **45**, 571–6.
LINDQUIST, S. J. 1977. Secondary porosity development and subsequent reduction overpressed Frio Formation (Oligocene) south Texas. *Trans Gulf-Cst Ass. geol. Socs*, **27**, 99–107.
LIPPMAN, F. 1982. The thermodynamics status of clay minerals. *Developments in Sedimentology*, **35**, 475–486.
LONG, J. V. P. & AGRELL, S. O. 1965. The cathodoluminescence of minerals in thin section. *Min. Mag.* **34**, 318–26.
LOUCKS, R. G., BEBOUT, D. G. & GALLOWAY, W. E. 1977. Relationship of porosity formation and preservation to sandstone consolidation history. *Trans. Gulf-Cst Ass. geol. Socs*, **27**, 109–20.
LYELL, C. 1852. *Manual of Elementary Geology*. Murray, London.
MCBRIDE, E. F. 1974. Significance of colour in red, green, purple, olive-brown and grey beds of Difunta Group, northwestern Mexico, *J. sedim. Petrol.* **44**, 760–73.
—— 1978. Porosity loss in sandstone by ductile grain deformation during compaction. *Trans. Gulf-Cst Ass. geol. Socs*, **28**, 323–5.
MCHARDY, W. J., WILSON, M. J. & TAIT, J. M. 1982. Electron microscope and X-ray diffraction studies of filamentous illitic clay from sandstones of the Magnus Field. *Clay Miner.* **17f**, 23–9.
MEADE, R. H. 1966. Factors influencing the early stages of the compaction of clays and sands—review. *J. sedim. Petrol.* **36**, 1085–101.
MERINO, E. 1975. Diagenesis in Tertiary sandstones from Kettleman, North Dome, California. I. Diagenetic mineralogy *J. sedim. Petrol.* **45**, 320–36.
—— & RANSOM, B. 1982. Free energies of formation of illite solid solutions and their compositional dependence *Clays Clay Miner.* **30**, 29–39.
MILLIKEN, K. L., LAND, L. S. & LOUCKS, R. G. 1981. History of burial diagenesis determined from isotopic geochemistry. Frio Formation, Brazoria Country, Texas. *Am. Ass. Petrol. Geol. Bull.* **65**, 1397–413.
NADEAU, P. H., TAIT, J. M., MCHARDY, W. J. and WILSON, M. J. 1984. Interstratified XRD characteristics of physical mixtures of elementary clay particles. *Clay Miner.* **19**, 67–76.
NAGTEGAAL, P. J. C. 1969. Sedimentology, palaeoclimatology and diagenesis of post Hercynian continental deposits of the south central Pyrenees, Spain. *Leid. Geol. Meded*, **42**, 143–238.
—— 1978. Sandstone framework instability as a function of burial diagenesis. *J. geol. Soc. London*, **135**, 101–5.
NAHON, D., CARROZI, A. V. & PARRON, C. 1980. Lateritic weathering as a mechanism for the generation of ferruginous ooids. *J. sedim. Petrol.* **50**, 1287–98.
NESBITT, H. W. 1980. Characterization of mineral-formation water interactions in Carboniferous sandstones and shales of the Illinois sedimentary basin. *Am. J. Sci.* **280**, 607–30.
NISHIYAMA, T., OINUMA, K. & SATO, M. 1978. An interstratified chlorite-vermiculite in weathered red shale near Toyoma, Japan, (Abstract) *Proc. 6th Int. Clay Conf., Oxford*, p. 157.
ODIN, G. S. & MATTER, A. 1981. De glauconiarum origin. *Sedimentology*, **28**, 611–41.
PARKER, C. A. 1974. Geopressures and secondary porosity in the deep Jurassic of Mississippi. *Trans. Gulf-Cst Ass. geol. Socs*, **24**, 69–80.
PEARSON, M. J. WATKINS, D. & SMALL, J. S. 1982. Clay diagenesis and organic maturation in Northern North Sea sediments. *Dev. Sedimentol.* **35**, 665–75.
PEDERSEN, G. K. & ANDERSON, P. F. 1980. Depositional environments, diagenetic history and source areas of some Bunter Sandstones in Northern Jutland. *Danm. geol. Unders. Arborg,* **1979**, 69–93.
PERRY, E. A. 1974. Diagenesis and the K-Ar dating of shale and clay minerals. *Bull. geol. Soc. Am.* **85**, 827–30.
—— & HOWER, J. 1970. Burial diagenesis in Gulf Coast pelitic sediments. *Clays Clay Miner.* **18**, 165–178.
—— & —— 1972. Late stage dehydration in deeply buried pelitic sediments *Am. Assoc. Petrol. Geol. Bull.* **56**, 2013–21.
PETTIJOHN, F. J., POTTER, P. E. & SIEVER, R. 1973. *Sand and Sandstone*, Springer Verlag, Berlin. 618 pp.
PHILLIPS, 1881. On the constitution and history of grits and sandstones. *Q. Jl geol. Soc. London*, **37**, 6–25.

Picard, G. L. & Felbeck, G. T. jr 1976. The complexation of iron by marine humic acid *Geochim. cosmochim Acta*, **40**, 1347–50.

Pittman, E. D. 1972. Diagenesis of quartz in sandstones as revealed by SEM *J. sedim. Petrol.* **42**, 507–19.

—— & Duschatko, R. W. 1970. Use of pore casts and SEM to study pore geometry. *J. sedim. Petrol.* **40**, 1153–7.

Porrenga, D. K. 1967. Glauconite and chamosite as depth indicators in the marine environment. *Mar. Geol.* **5**, 495–501.

Powell, T. G., Foscolos, A. E., Gunther, P. R. & Snowdon, L. R. 1978. Diagenesis of organic matter and fine clay minerals: a comparative study. *Geochem. cosmochim. Acta*, **42**, 1121–97.

Powers, M. C., 1967. Fluid release mechanisms in compacting marine mudrocks and their importance in oil exploration *Am. Assoc. Petrol. Geol. Bull.* **51**, 1240–54.

Ramseyer, K. 1983. A new cathodoluminescence microscope and its application to sandstone diagenesis. *11th Int. Cong. Sed. (I.A.S.) Hamilton* p. 885 (abstract).

Reading, H. G. (ed.) 1978 *Sedimentary Environments and Facies*, Blackwell Scientific Publications, Oxford. 557 pp.

Retallack, G. J. 1976. Triassic palaeosols in the Upper Narrabeen Group of New South Wales. I. features of the palaeosols. *J. geol. Soc. Aust.* **23**, 383–99.

Roedder, E. 1979. Fluid inclusion evidence on the environment of sedimentary diagenesis, a review. *In*: Scholle, P. A. & Schluger, P. R. (eds) *Aspects of Diagenesis. Spec. Publs Soc. econ. Miner. Paleont., Tulsa*, **26**, 89–107.

Rossel, N. C. 1982. Clay mineral diagenesis in Rotliegend aeolian sandstones of the Southern North Sea. *Clay Miner.* **17**, 69–77.

Salomons, W., Gouldie, A. S. & Mook, W. G. 1978. Isotopic composition of calcrete deposits from Europe, Africa and India. *Earth Surf. Proc.* **3**, 43–57.

Schmidt, V. & McDonald, D. A. 1979a. The role of secondary porosity in the course of sandstone diagenesis. *In*: Scholle, P. A. & Schluger, P. R. (eds) *Aspects of Diagenesis. Spec. Publs Soc. econ. Miner. Paleont., Tulsa*, **26**, 175–207.

—— & —— 1979b. Texture and recognition of secondary porosity in sandstones. *In*: Scholle, P. A. & Schluger, P. R. (eds) *Aspects of Diagenesis. Spec. Publs Soc. econ. Miner. Paleont., Tulsa*, **26**, 209–25.

Seeman, U. 1979. Diagenetically formed interstitial clay minerals as a factor in Rotliegend Sandstone reservoir quality in the Dutch Sector of the North Sea. *J. Petrol. Geol.* **1**, 55–62.

Shelton, J. W. 1964. Authigenic kaolinite in sandstone. *J. sedim. Petrol.* **34**, 102–11.

Sibley, D. F. & Blatt, H. 1976. Intergranular pressure solution and cementation of the Tuscorora Orthoquartzite. *J. sediment. Petrol.* **46**, 881–96.

Siever, R. 1979. Plate-tectonic controls on diagenesis. *J. Geol.* **87**, 126–55.

Sippel, R. F. 1968. Sandstone petrology, evidence from luminescence petrography. *J. sedim. Petrol.* **38**, 530–54.

Smith, J. V. & Stenstrom, R. C. 1963. Electron excited luminescence as a petrologic tool. *J. Geol.* **73**, 627–35.

Sommer, F. 1978. Diagenesis of Jurassic sandstones in the Viking Graben. *J. geol. Soc. London*, **135**, 63–7.

Stalder, P. J. 1975. Cementation of Pliocene–Quaternary fluviatile clastic deposits in and along the Ocean Mountains. *Geologie Mijnb.* **54**, 148–56.

Stanton, G. D. 1977. Secondary porosity in sandstones of the Lower Wilcox (Eocene), Kames County, Texas. *Trans. Gulf-Cst Ass. geol. Soc.* **27**, 197–207.

Staub, J. R. & Cohen, A. D. 1978. Kaolinite-enrichment beneath coals: a modern analog, Snuggedy Swamp, South Carolina. *J. sediment. Petrol.* **48**, 203–10.

Stoessell, R. K. & Moore, C. H. 1983. Chemical constraints and origins of four groups of Gulf Coast reservoir fluids. *Am. Ass. Petrol. Geol. Bull.* **67**, 896–906.

Storey, S. R. 1979. Clay-carbonate diagenesis of deltaic sandstones—Basal Belly River Formation (Upper Cretaceous) Central Alberta, Canada (Abs) *Am. Ass. Petrol. Geol. Bull.* **63**, 534.

Tan, F. C. & Hudson, J. D. 1974. Isotopic studies of the palaeoeology and diagenesis of the Great Estuarine Series (Jurassic) of Scotland. *Scott. J. Geol.* **10**, 91–128.

Tardy, Y. 1971. Characterisation of the principal weathering types by the geochemistry of waters from European and African crystalline massifs. *Chem. Geol.* **7**, 253–71.

Taylor, B. J. 1978. Westphalian. *In*: Moseley, F. (ed.) *The geology of the Lake District. Yorkshire Geological Society Occasional Publication*, **3**, 180–9.

Taylor, J. M. 1950. Pore-space reduction in sandstone. *Am. Ass. Petrol. Geol. Bull.* **34**, 701–16.

Taylor S. R. 1983. A stable isotope study of the Mercia Mudstones (Keuper Marl) and associated sulphate horizons in the English Midlands. *Sedimentology*, **30**, 11–31.

Thomas, M. F. 1974. *Tropical Geomorphology. A Study of Weathering and Landform Development in Warm Climates*, MacMillan Press Ltd, London. 332 pp.

Tillman, R. W. & Almon, W. R. 1979. Diagenesis of Frontier Formation offshore bar sandstones, Spearhead Ranch Field, Wyoming. *In*: Scholle, P. A. & Schluger, P.R. (eds) *Aspects of Diagenesis. Spec. Publs Soc. econ. Miner. Paleont., Tulsa*, **26**, 337–78.

Tissot, B., Durand, B., Espitalic, J. & Combaz, A. 1974. Influence of nature and diagenesis of organic matter in formation of petroleum *Am. Ass. Petrol. Geol. Bull.* **58**, 499–506.

Tompkins, R. E. 1981. Scanning electron microscopy

of a regular chlorite/smectite (corrensite) from a hydrocarbon reservoir sandstone *Clays Clay Miner.* **29**, 233–5.

TURNER, P. 1980. *Continental red beds. Dev. Sedimentol.* **29**, 562 pp.

—— & ARCHER, R., 1977. The role of biotite in the diagenesis of red beds from the Devonian of northern Scotland. *Sed. Geol.* **19**, 241–51.

VELDE, B. 1977. *Clays and clay minerals in natural and synthetic systems. Dev. Sedimentol.* **21**, 218 pp.

VESSEL, R. K. & DAVIES, D. K. 1981. Non-marine sedimentation in an active fore arc basin. *In*: ETHRIDGE, F. G. & FLORES, R. M. (eds) *Recent and Ancient Non-marine Depositional Environments: Models for Exploration. Spec. Publs Soc. econ. Miner. Paleont.*, Tulsa, **31**, 31–45.

WALDSCHMIDT, W. A. 1941. Cementing materials in sandstones and their probable influence on migration and accumulation of oil and gas. *Am. Ass. Petrol. Geol. Bull.* **25**, 1839–79.

WALKER, R. G. (ed.) 1979. Facies models. *Geoscience Canada Reprint Series No. 1 Geol. Assoc. Canada*, 211 pp.

WALKER, T. R. 1967. Formation of red beds in modern and ancient deserts. *Bull. geol. Soc. Am.* **79**, 353–68.

—— 1974. Formation of red beds in moist tropical climates: a hypothesis. *Bull. Geol. Soc. Am.* **85**, 633–8.

—— 1975. Diagenetic origin of red beds. *In*: FALKE, M. (ed.) *The Continental Permian in Central, West and South Europe. NATO A.S.I. Series C*, **22**, 240–82.

——, RIBBE, P. H. & HONEA, R. M. 1967. Geochemistry of hornblend alteration in Pliocene red beds, Baja California, Mexico. *Bull. geol. Soc. Am.* **78**, 1055–60.

—— & HONEA, R. M. 1969. Iron content of modern deposits in the Sonoran desert: a contribution to the origin of red beds. *Bull. geol. Soc. Am.* **80**, 534–44.

——, WAUGH, B. & CRONE, A. J. 1978. Diagenesis in first cycle desert alluvium of Cenozoic age, southwestern United States and northwestern Mexico. *Bull. geol. Soc. Am.* **89**, 19–32.

WALTON, N. R. G. 1981. A detailed hydrogeochemical study of groundwaters from the Triassic sandstone aquifer of south west England. *Rep. Inst. Geol. Sci* No. 81/5.

WARME, J. E., DOUGLAS, R. G. & WINTERER, E. L. (eds) 1981. *The Deep Sea Drilling Project: A Decade of Progress. Spec. Publs Soc. econ. Miner. Paleont., Tulsa*, **32**, 564 pp.

WAUGH, B. 1970. Petrology, provenance, and silica diagenesis of the Penrith Sandstone (lower Permian) of northwest England. *J. sedim. Petrol.* **40**, 1226–1240.

—— 1978. Authigenic K-feldspar in British Permo-Triassic sandstones. *J. geol. Soc. London*, **135**, 51–6.

WEINBRANDT, R. M. & FATT, I. 1969. A scanning electron microscope study of the pore structure of sandstone. *J. Petrol. Tech.* **21**, 543–8.

WESCOTT, W. A. 1983. Diagenesis of Cotton Valley Sandstone (Upper Jurassic) East Texas: implications for tight gas formation pay recognition. *Am. Ass. Petrol. Geol. Bull.* **67**, 1002–13.

WHETTEN, J. T. & HAWKINS, J. W. 1970. Diagenetic origin of greywacke matrix minerals. *Sedimentology*, **15**, 347–61.

WHITE, D. E. 1965. Saline water of sedimentary rocks. *In*: YOUNG, A. & GALLEY, J. E. (eds) *Fluids in Subsurface Environments. Mem. Am. Ass. Petrol. Geol.* **4**, 342–66.

WILSON, M. D. & PITTMAN, E. D. 1977. Authigenic clays in sandstones: recognition and influence on reservoir properties and palaeoenvironment analysis. *J. sedim. Petrol*, **47**, 3–31.

WILSON, M. J. & DUTHIE, D. M. L. 1981. Some aspects of intrastratal alteration in the Old Red Sandstone. *Scott. J. Geol.* **23**, 57–64.

WILSON, R. C. L. (ed.) 1983. *Residual Deposits: Surface Related Weathering Processes and Materials. Geol. Soc. London Sp. Publ.* **11**, 258 pp.

WOOD, J. R. & HEWETT, T. A. (1982). Fluid convection and mass transfer in porous sandstones—a theoretical model. *Geochim. cosmochim. Acta*, **46**, 1707–14.

WOPFNER, H. 1978. Silcretes of Northern South Australia and adjacent regions. *In*: LANGFORD-SMITH, T. (ed.) *Silcrete in Australia*, 93–141. Department of Geography, University of New England, Australia.

ZINKERNAGEL, U. 1978. Cathodoluminescence of quartz and its applications to sandstone petrology. *Contrib. Sedimentol.* **8**, 69 pp.

S. D. BURLEY, Geologisches Institut, Universität Bern, Baltzerstrasse-1, 3012 Bern, Switzerland.

J. D. KANTOROWICZ, Koninklijke/Shell Exploratie en Produktie Laboratorium, Volmerlaan 6, Rijswijk ZH, The Netherlands.

B. WAUGH, AGAT Consultants Inc., 1215 18th Street, 260 Cable Building, Denver, Co 80202, USA.

Reprinted by permission of the Society of Economic Paleontologists and Mineralogists. Published in *Aspects of Diagenesis,* P.A. Scholle and P.R. Schulger, editors, SEPM Special Publication No. 26, pp. 159-173.

SEPM SPECIAL PUBLICATION NO. 26, P. 159–173, MARCH 1979

POROSITY, DIAGENESIS AND PRODUCTIVE CAPABILITY OF SANDSTONE RESERVOIRS

EDWARD D. PITTMAN
Amoco Production Co., Research Center, Tulsa, Oklahoma 74102

ABSTRACT

Four basic types of porosity occur in sandstones: intergranular, dissolution, micro and fracture. The first three types are related to rock texture and can be considered end members of a ternary classification diagram. Fracture porosity may be associated with any other porosity type.

All sandstones initially have intergranular porosity, which, if not destroyed, often is associated with good permeability, large pore apertures, and prolific hydrocarbon production. Dissolution porosity results from leaching of carbonate, feldspar, sulfate, or other soluble material. Sandstone reservoirs with dissolution porosity range from excellent to poor, depending on amount of porosity and interconnection of pores. Isolated dissolution pores result in low permeability. Sandstones with significant amounts of clay minerals have abundant microporosity, high surface area, small pore apertures, low permeability, high irreducible water saturation, and an increased sensitivity to fresh water. Fracture porosity, which contributes no more than a few percent voids to storage space, will enhance the deliverability of any reservoir. Open fractures, either natural or induced, are essential for economic deliverability rates from reservoirs with essentially only micropores or isolated dissolution pores.

Porosity type and/or pore geometry change with diagenesis: macropores become micropores, minerals dissolve to create voids, and pores are partly to completely occluded by precipitation of minerals. It is important to have an understanding of pore geometry, that is the size, shape, and distribution of pores in a reservoir. Pore geometry influences the type, amount, and rate of fluid produced.

Porosity type seldom is homogeneous in rocks. As a result, log interpretation problems may occur in sandstones containing significant micropores and interconnected macropores. Micropores may hold irreducible water while macropores may hold producible oil, gas, or water, depending on height above the oil- or gas-water contact. Log calculations may indicate high water saturation and a nonproductive interval, although the reservoir may be capable of water-free hydrocarbon production because the water is not producible.

INTRODUCTION

The success of massive hydraulic fracture treatments and the scramble for new hydrocarbon reserves has enabled the oil industry to complete wells in sandstone reservoirs that a few years ago were considered uneconomic. When dealing with these low-quality reservoirs, it is important to have an understanding of pore geometry, that is the size, shape, and distribution of pores. In other words, a three-dimensional representation of interconnected void space in the rock. Pore geometry influences the type, amount, and rate of fluid produced. Fluid factors, such as viscosity, also play a role but will not be covered in this paper.

Purpose of this paper is to discuss the basic types of porosity that occur in sandstones and to show examples that illustrate the importance of understanding pore geometry.

RESERVOIR CHARACTERISTICS OF POROSITY TYPES

Three types of porosity, excluding fractures, occur in sandstones: intergranular, dissolution and micro. These three types of porosity can be considered end members of a ternary diagram representing total matrix porosity (Fig. 1). Fracture porosity will be considered later as a secondary feature that enhances other porosity types. Any of the porosity end members may be hydrocarbon reservoirs under certain conditions. Generally, permeability and pore aperture size increase toward the intergranular porosity pole. Irreducible water, i.e., water held in the rock by the physical attraction of the rock for the liquid (capillary pressure), increases toward the microporosity pole.

Diagenesis may cause a change of porosity types. For example, precipitation of authigenic clay minerals in pores and pore throats increases surface area, creates microporosity, decreases effective intergranular macroporosity and lowers permeability. Of course, any porosity type may be partly to completely occluded by precipitation of mineral cements. Diagenesis may also create void space through dissolution of soluble components of the rock.

Intergranular porosity.—Intergranular pores occur among detrital sand grains. The best sandstone reservoirs from viewpoint of storage porosity and deliverability have predominantly intergranular porosity. Sandstone reservoirs with intergranular porosity range from a bottom cutoff for pay of about 5% to totally unconsolidated sand with porosities exceeding 40%.

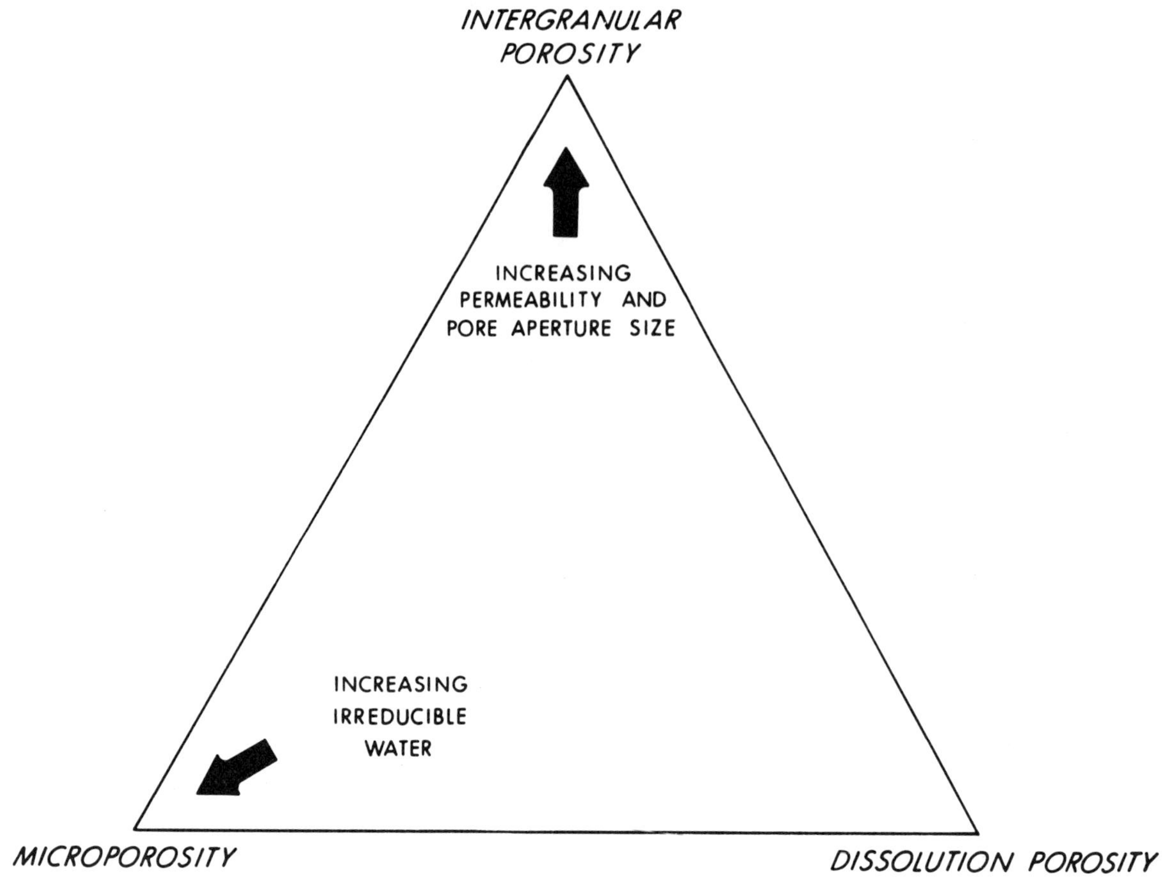

FIG. 1.—Ternary classification diagram of matrix related porosity. Fracture porosity represents a fourth porosity type.

All sands start with primary intergranular pores that are modified in size and shape, or even destroyed by diagenetic processes. Secondary intergranular porosity can form through dissolution of carbonate cement (see Schmidt and McDonald, this publication).

The Pennsylvanian Tensleep Sandstone of Wyoming is an example of a quartz arenite with intergranular porosity (photomicrograph, Fig. 2). This example has 15.3% porosity and 280 md permeability. A mercury injection curve reveals large, well-sorted pore apertures (approximately 20 μm breakthrough): desirable characteristics for a reservoir rock (Fig.2).

The Tensleep sample has quartz overgrowths and dolomite cement (Figs. 2, 3A). A scanning electron micrograph of a pore cast, which is an epoxy resin replica of the pore space (Pittman and Duschatko, 1970), for this reservoir shows that the intergranular voids are interconnected in a three-dimensional network (Fig. 3B).

Dissolution porosity.—Dissolution pores result from removal of carbonates, feldspars, sulfates, or other soluble materials (Rowsell and DeSwardt, 1975; Schmidt, *et al.,* 1976; Lindquist, 1976; Stanton and McBride, 1976). Soluble components may be: (1) detrital grains; (2) authigenic mineral cement; or (3) replacement minerals. Individual sandstones may contain dissolution pores of single or multiple origins.

Sandstone reservoirs with only dissolution porosity range from poor to excellent, depending on amount of porosity and interconnection of pores. Sandstones with only dissolution porosity may have essentially no measurable matrix permeability if the soluble grains are disseminated because upon leaching they form isolated pores with microporous interconnections. Good permeability exists if the soluble material is abundant enough to be in contact because of grain packing or interconnection of mineral cement. Excellent reservoirs develop where carbonate cement has been dissolved to form secondary intergranular porosity (see Schmidt and McDonald, this publication).

In my opinion, porosity created by the dissolution of feldspar (or dissolution of carbonate that has replaced feldspar) is common worldwide in

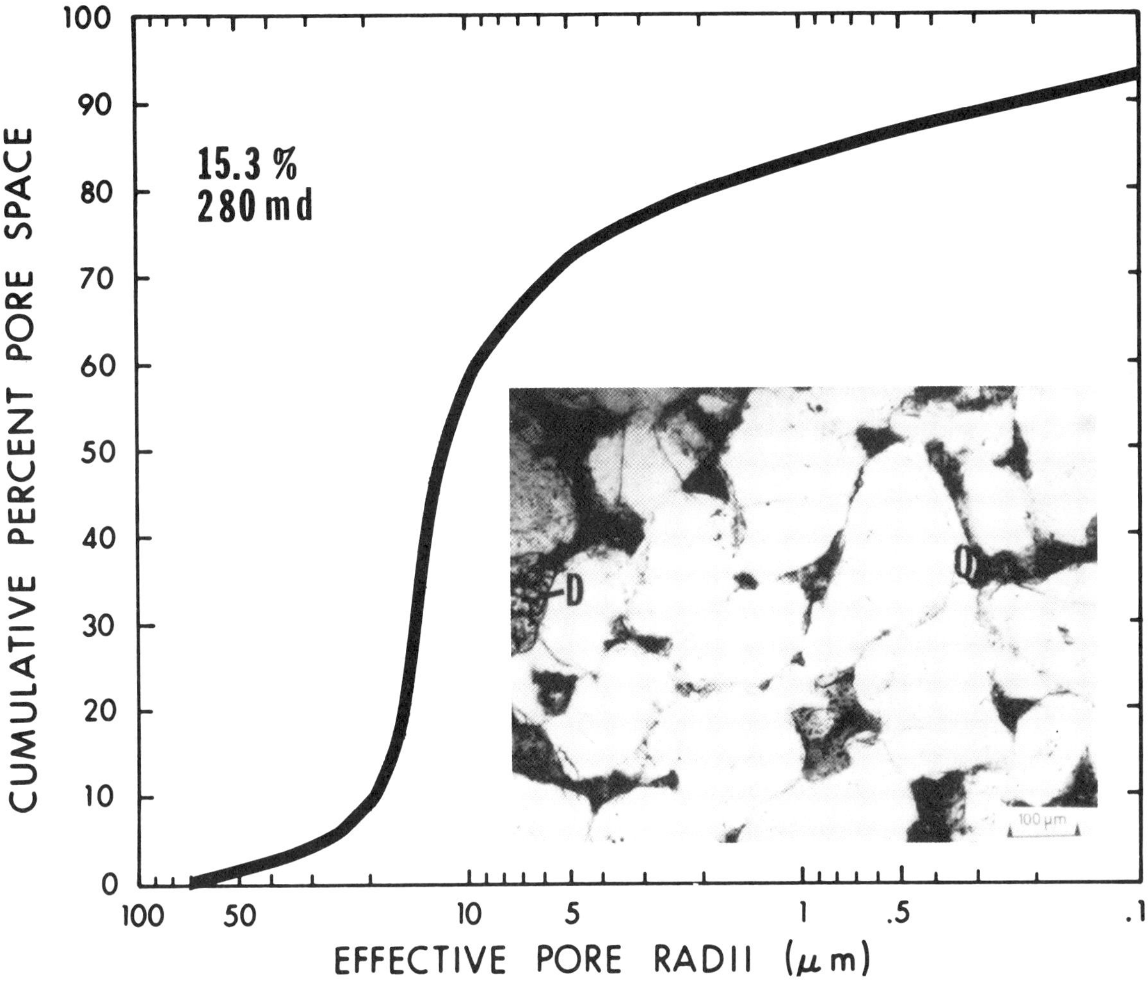

FIG. 2.—Mercury injection curve and photomicrograph of Tensleep Sandstone (Pennsylvanian), Wyoming. Dark gray areas among grains represent void space occupied by epoxy resin. Grains are cemented by quartz overgrowths (Q) and dolomite (D). Mercury injection curve reveals relatively coarse and well-sorted pore apertures indicative of a good reservoir.

sandstones of all ages. Dissolution may affect portions of feldspar grains (Fig. 4A–C), or essentially the entire grain (Fig. 4D–E). If feldspar is abundant in the sandstone, considerable porosity can originate in this manner.

Morgan and Gordon (1970) appear to have first discussed the importance of feldspar dissolution on reservoir rock performance. Mineralogy of dissolved feldspar and impact of dissolution pores on reservoirs were described by Shenhav (1971), Heald and Larese (1973), and Larese (1974). Feldspars commonly affected by dissolution are sanidine, orthoclase, and plagioclase. Leached feldspars are often turbid, due to incipient sericite or vacuoles. In contrast, microcline, which often coexists in sandstone with partly dissolved other feldspar, is usually clear and lacks any sign of dissolution. Presumably, this is explainable through the relative temperature of formation-stability series for the potash feldspar group (Deer *et al.,* 1962). Sanidine and orthoclase form at higher temperatures and possess more disordered structures than microcline. Thus, they are not as stable under diagenetic conditions. Plagioclase feldspar also forms at higher temperatures than microcline.

When the soluble material has been totally removed, identification is a matter of conjecture based on mold size and shape. For example, in Figure 4F, a pore filling material has been removed to create a pore in an otherwise tightly quartz-cemented sandstone. A carbonate mineral is a likely choice, perhaps as cement around a fossil fragment.

An example of a sandstone with isolated dissolution porosity is the Silurian Tuscarora from

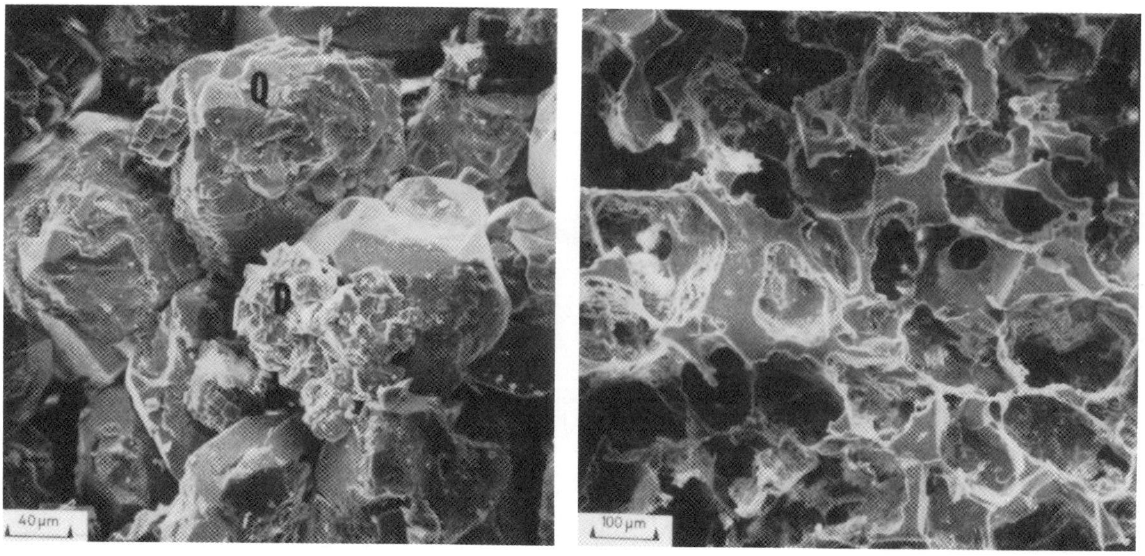

A B

FIG. 3.—Scanning electron micrographs of Tensleep Sandstone. *A*, Rock has obvious pores despite some quartz overgrowths (Q) and dolomite cement (D). *B*, Pore cast of specimen in "A" reveals that the pores are interconnected three dimensionally and that the apparent pore apertures are many times smaller than the large intergranular pores.

the Appalachian Province (photomicrograph, Fig. 5). This example has 10.0% porosity but only 0.05 md permeability. The mercury injection curve reveals small pore throats (breakthrough of mercury at about 0.5 μm) caused by abundant quartz overgrowths and possibly pressure solution (Fig. 5). In this example, there is no remnant material to provide a clue as to what was dissolved to produce the pores. Scanning electron microscope examination of the rock and particularly the pore cast confirms that the rock has small pore apertures (Figs. 6A, B). Pore interconnections are sparse and thin compared to the Tensleep example discussed earlier.

Microporosity.—Generalizing, microporosity —as distinguished from macroporosity—may be defined as pores with pore-aperture radii less than 0.5 μm. Pores with very small pore apertures usually are associated with small pores, but relatively large pores with very small pore apertures are not uncommon.

Micropores occur among clay minerals or at pore-throat restrictions in sandstone. Argillaceous sandstones commonly have significant microporosity, regardless of whether the clay is authigenic or allogenic in origin. Based on my observations, authigenic clay minerals and associated microporosity are common in sandstone reservoirs.

An argillaceous sandstone with only microporosity has essentially no measurable matrix permeability. A rock of this type with small pore apertures and high surface area will also have high irreducible water saturation.

In a water-wet system, it is possible for hydrocarbons to displace water in micropores. To do this, the height of the hydrocarbon column must be sufficient to provide adequate buoyancy pressure to overcome capillary pressure, which tends to hold water in the rock. I do not know of a totally microporous sandstone reservoir where this has happened. However, the Ekofisk chalk

FIG. 4.—Dissolution porosity in sandstones. *A–C*, Partial dissolution of detrital feldspar grains in sandstones. Medium to dark gray area in photomicrographs represents blue epoxy in pores. *D, E*, Nearly complete dissolution of feldspar grains. Only small scraps of feldspar remain in "D". In "E", the authigenic feldspar rim is conspicuous despite the near absence of a rounded detrital core. *F*, A small cavity or vug in a sandstone. The size and shape suggests that something larger than the quartz sand grains was removed. A calcareous fossil, perhaps surrounded by calcite cement, is a likely possibility.

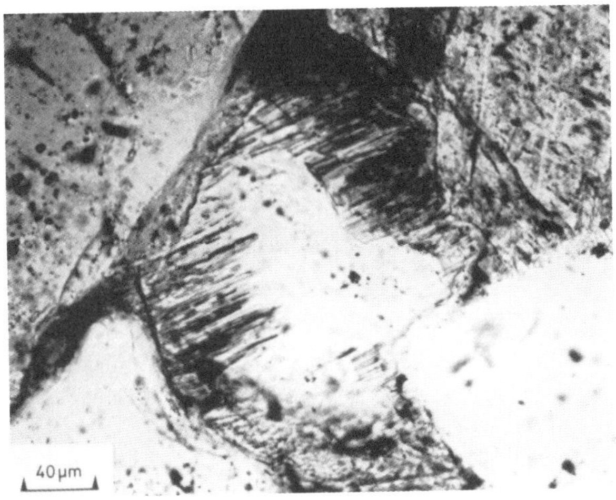

A

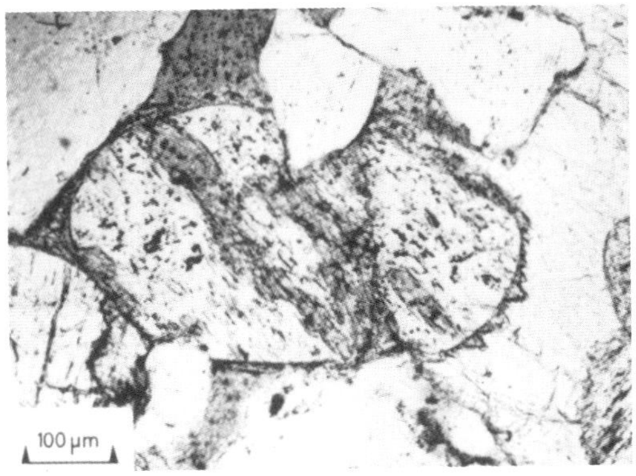

B

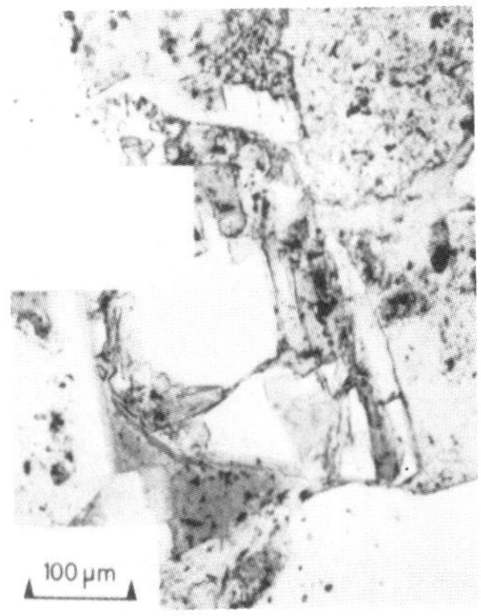

C

D

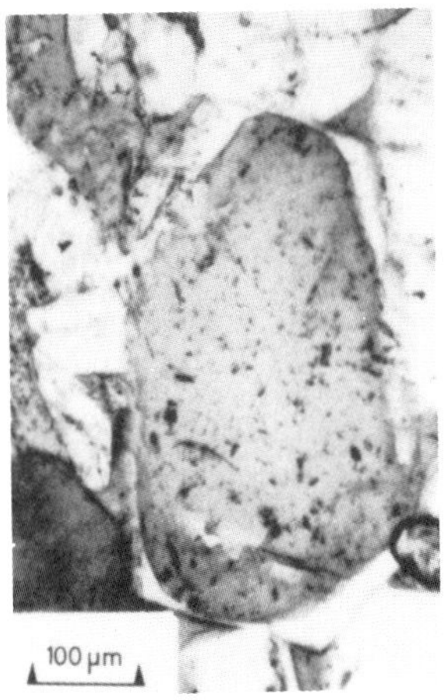

E

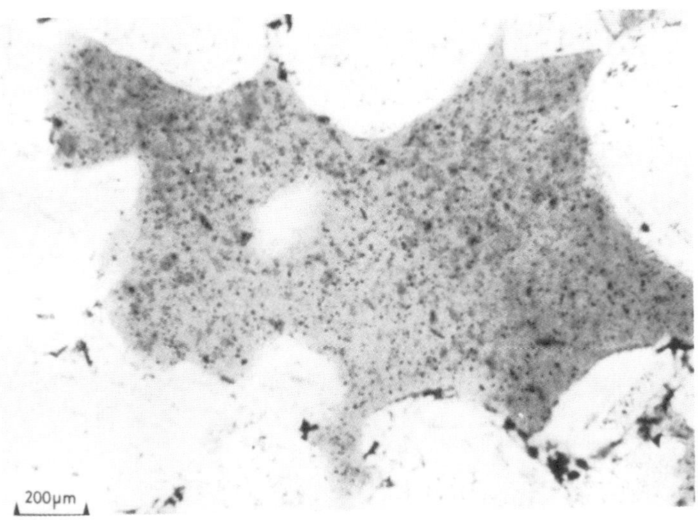

F

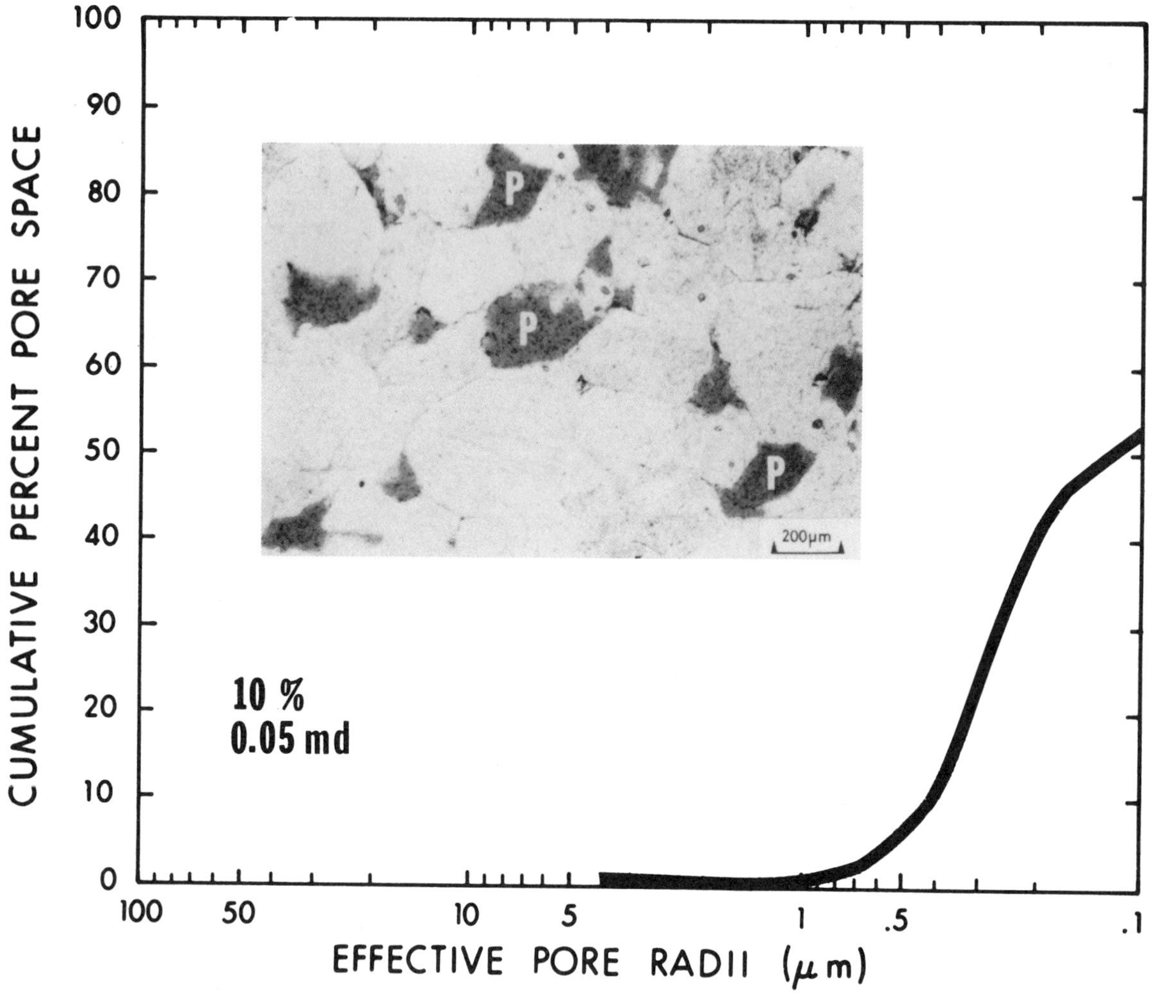

FIG. 5.—Mercury injection curve and photomicrograph of Tuscarora Sandstone (Silurian), Appalachian Province. Pores (P) are dark colored and appear to be isolated, but are actually interconnected by micropores. The mercury injection curve indicates the rock has small pore apertures because mercury did not break through until about 0.5 µm. A reservoir with these characteristics needs fractures, either natural or induced, to provide an economic deliverability rate.

reservoir in the North Sea, which has over 600 feet of oil column, is an example of this principle in carbonates.

The Mesaverde Sandstone, shown in the photomicrograph of Figure 7, has 11.1% porosity (all microporosity) and less than 0.05 md permeability. In this example, allogenic and authigenic clay minerals fill intergranular pore space. Micropores occur among the clay particles. Ninety-eight percent of the pore-aperture radii are less than 0.5 µm and 69% are less than 0.1 µm (Fig. 7). Breakthrough of mercury occurs at approximately 0.2 µm. Examination of scanning electron micrographs of this sandstone and its pore cast confirm the small pore apertures (Figs. 8A, B). This type of rock would be expected to have high water saturation due to irreducible water. In order for this particular sandstone to become a reservoir, irreducible water would have to be displaced by hydrocarbon and the rock fractured either naturally or artificially.

ROLE OF FRACTURES

Fracture porosity, which contributes no more than a few percent voids to storage space, will enhance the deliverability of any sandstone reservoir. Intergranular porosity, unless of poor quality, does not need associated fractures to be a suitable reservoir because the pores are interconnected. A sandstone reservoir with significant micropores or isolated dissolution pores typically needs fractures, either natural or induced, to

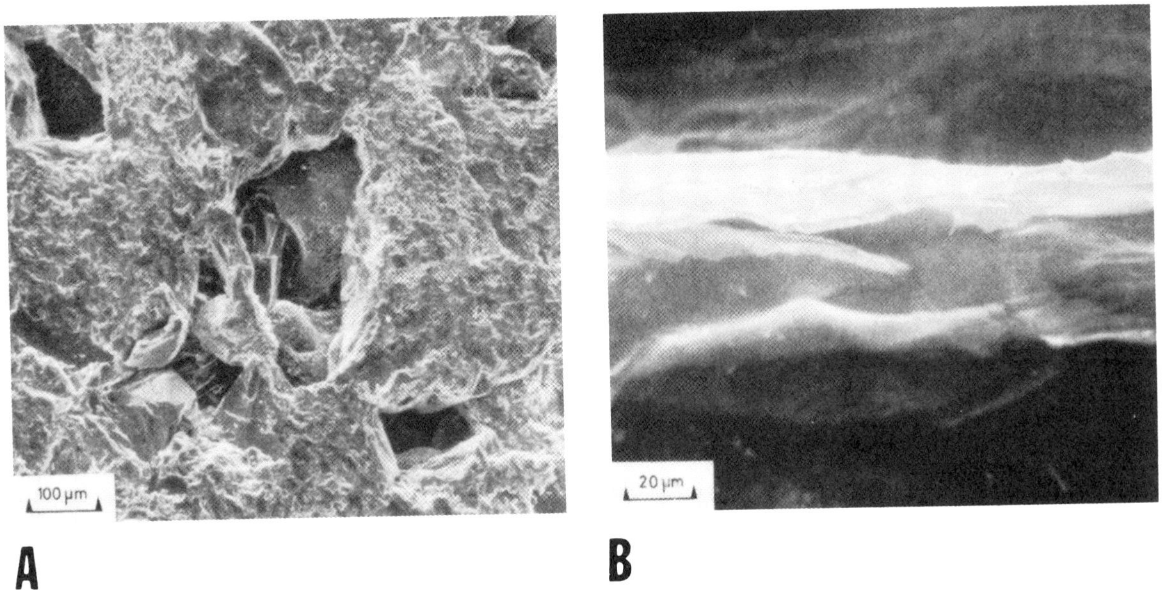

FIG. 6.—Scanning electron micrographs of Tuscarora Sandstone. *A*, Relatively large dissolution pore in the center has no obvious interconnections to other pores. *B*, Pore cast showing apparent pore aperture represented by curved plastic sheet, which is less that 1 μm thick in places.

provide permeability and economic deliverability rates.

Natural fractures may represent the entire storage porosity and permeability of a reservoir. Porosity due only to fractures is rarely greater than a few percent and usually is less than one percent. Reservoirs of this type characteristically have a high initial flow rate followed by rapid depletion.

Natural fractures may provide the necessary permeability to make productivity more attractive for a reservoir with low matrix permeability. A reservoir with a well-developed natural fracture system would be expected to yield a strong initial flow of hydrocarbons with a gradual decline as matrix porosity is depleted.

Sometimes natural fractures occur in reservoirs with good matrix porosity and permeability. In this situation, the fractures improve the productive capability and recovery efficiency of the reservoir (assuming coning of water is not a problem).

Induced fractures can provide the necessary permeability when matrix porosity is present but matrix permeability and effective natural fractures are absent. A hydraulically fractured reservoir of this type yields high initial hydrocarbon production with a rapid early decline followed by a slower later decline as the matrix yields fluid to the fracture.

ESTIMATION OF POROSITY TYPES

By combining study of thin sections prepared from rocks impregnated with colored epoxy and scanning electron microscopy, it is possible to make reliable estimates of the different porosity types. Intergranular pores can be readily identified in thin section by their position among detrital grains. Dissolution pores are recognizable in thin section by their location, size, and shape outline with respect to the rock framework and in some cases by the presence of a soluble precursor. Micropores are best recognized in scanning electron micrographs.

EXAMPLES OF SANDSTONE RESERVOIRS WITH MIXED POROSITY TYPES

Second Wall Creek Sandstone.—The Second Wall Creek Sandstone, Wyoming, has significant amounts of intergranular and dissolution porosity, but only minor microporosity. Data for a selected suite of samples from four wells plot along the intergranular-dissolution side of the ternary diagram (Fig. 9). Porosity ranges from 15.4 to 23.3% and permeability from 3.98 to 115 md. Higher permeability samples have a tendency to plot nearer to the intergranular pole, which illustrates the generalization expressed earlier. Intergranular pores interconnect dissolution pores to provide an effective macroporous network (photomicrograph, Fig. 10). Micropores occur among authi-

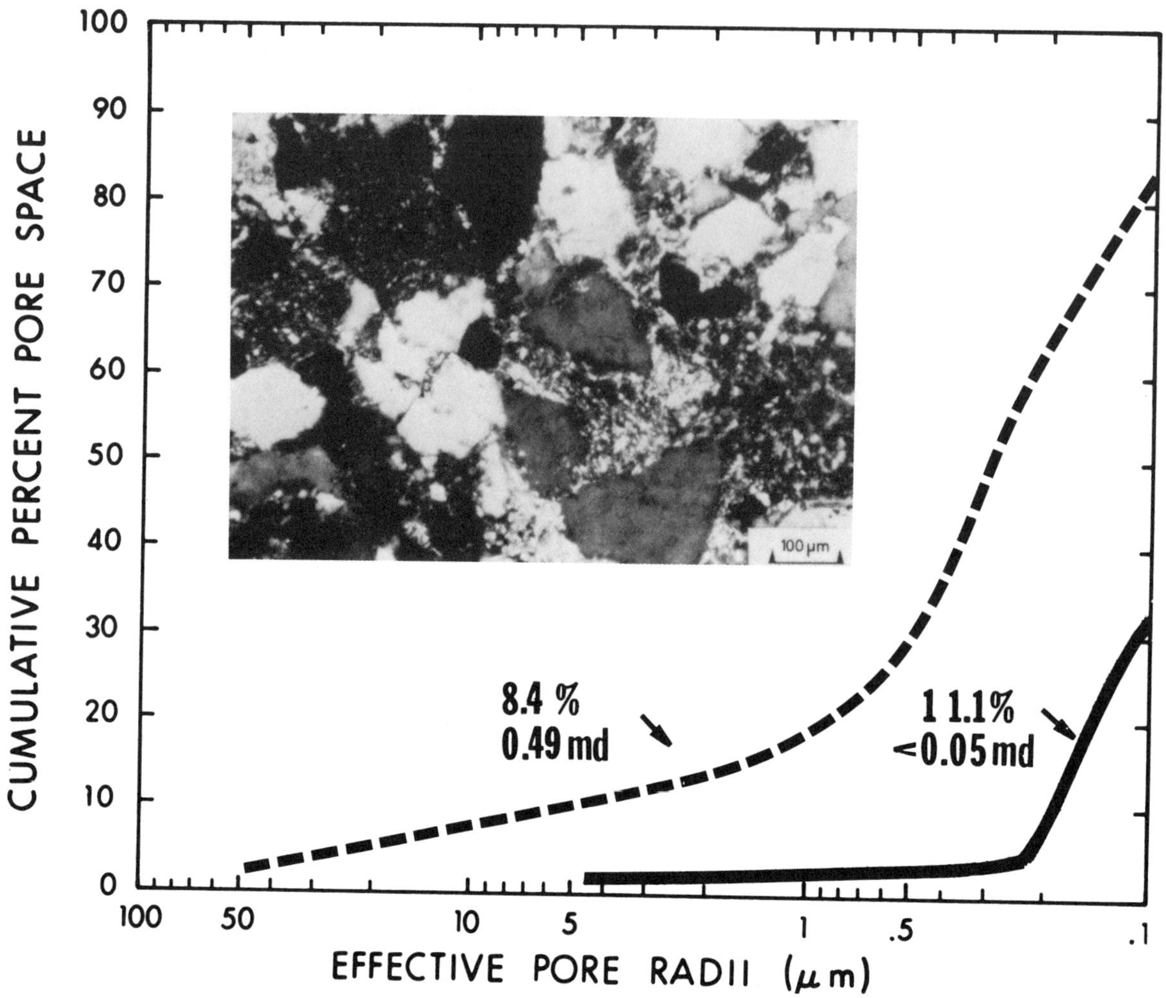

FIG. 7.—Mercury injection curves and photomicrograph of sandstone from Mesaverde (Cretaceous), Wyoming. Intergranular porosity is filled by clay, which has reduced pore apertures. The mercury injection curve (solid line) for the sample corresponding to the photomicrograph indicates that the break in the curve is at about 0.2 µm. This particular rock did not yield hydrocarbons. The dashed curve is for a Mesaverde sandstone that was productive. Note the coarser pore apertures and higher permeability although the porosity is even lower than for the other samples.

genic clays, but are not abundant enough to hold significant amounts of irreducible water. This is reflected in the mercury injection curve, which shows relatively large pore apertures (breakthrough at approximately 5 µm; Fig. 10).

Dissolution porosity in the Second Wall Creek Sandstone apparently originated from leaching of potash feldspar and plagioclase or from leaching of carbonate that replaced the feldspars. Detrital feldspar grains are partially to completely removed, but authigenic feldspar overgrowths are essentially unaltered (Figs. 10, 11). Based on scanning electron microprobe analyses, a compositional difference exists between authigenic rims and detrital cores of feldspars in the Second Wall Creek Sandstone (Fig. 12). In all cases, the rims are essentially K-rich potassium feldspar or Na-rich plagioclase compared to the cores. This compositional difference between rim and core may explain why the authigenic rim composed of more stable feldspar is preserved, while the detrital nuclei composed of less stable feldspar have been dissolved.

Permian Sandstone, South Africa.—Permian sandstones from South Africa have intergranular, dissolution, and microporosity (Fig. 13). These data plot slightly to the left and below data for the Second Wall Creek (Fig. 9). Samples of Permian Sandstone with significant amounts of microporosity and dissolution porosity have lower

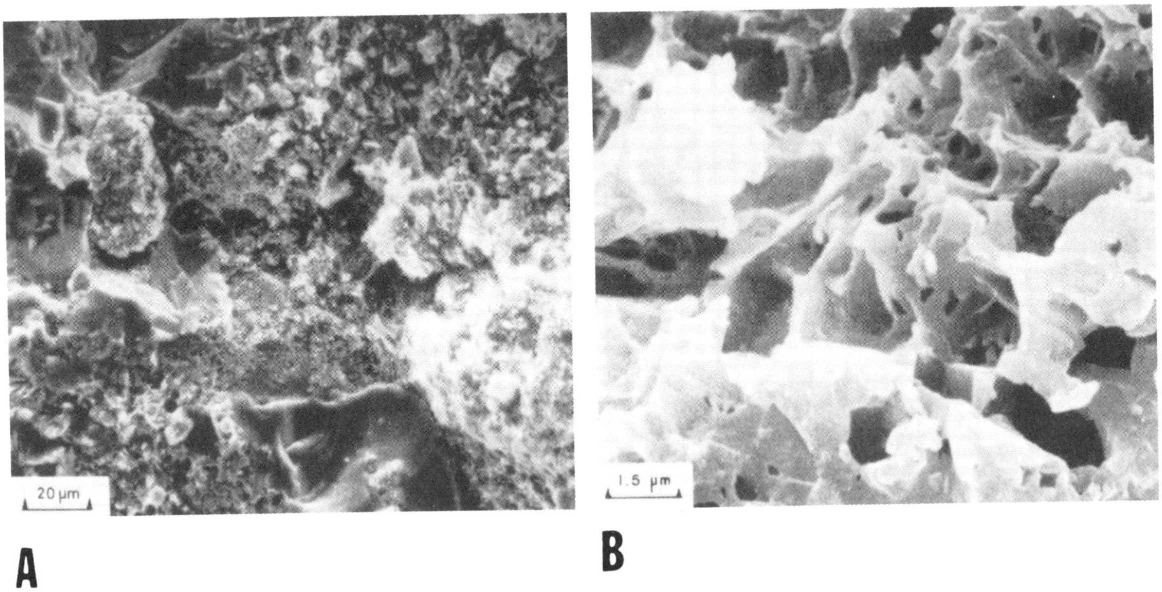

FIG. 8.—Scanning electron micrographs of Mesaverde Sandstone. *A*, Rock contains so much clay that individual sand grains are indistinct. *B*, Pore cast of this rock reveals an extremely fine pore network among the clay particles. Some pore apertures are only about 0.1 μm across.

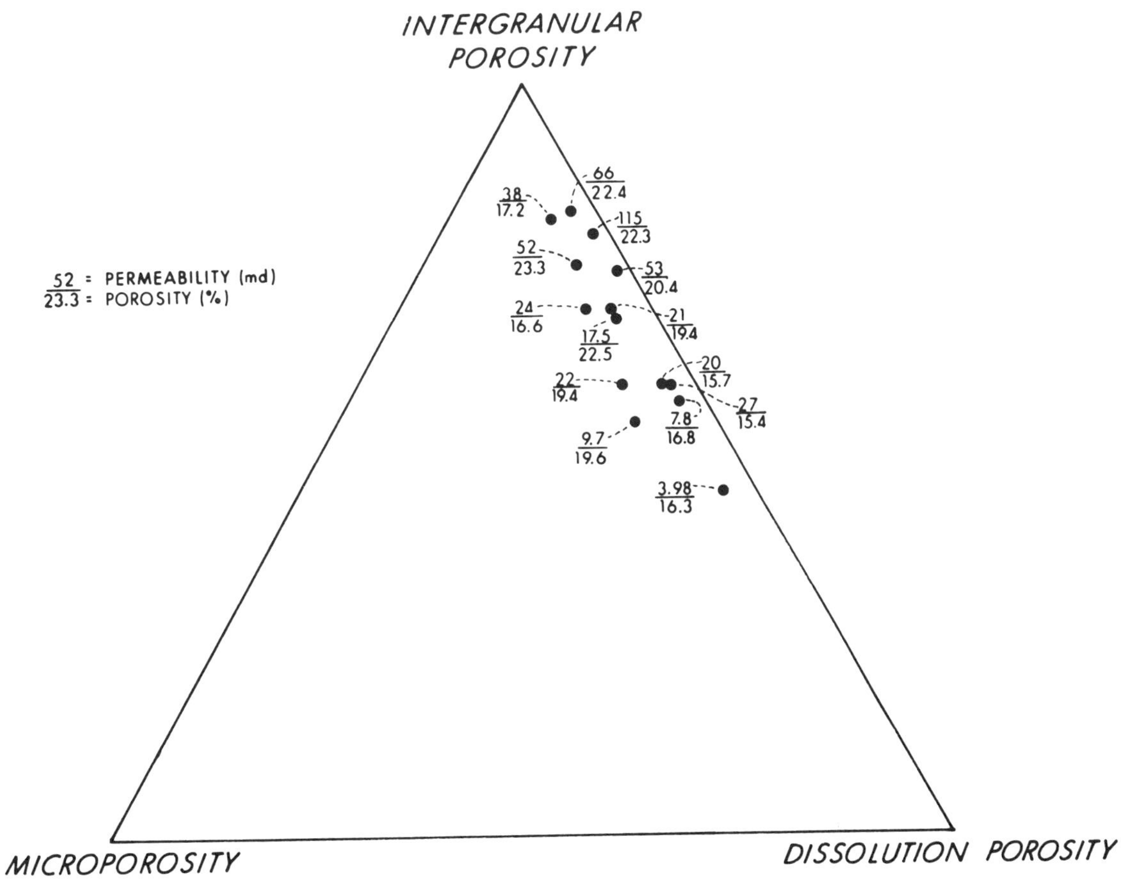

FIG. 9.—Triangular plot of porosity types from four wells in the Second Wall Creek Sandstone (Cretaceous), Wyoming. Note the tendency for permeability to increase towards the intergranular pole.

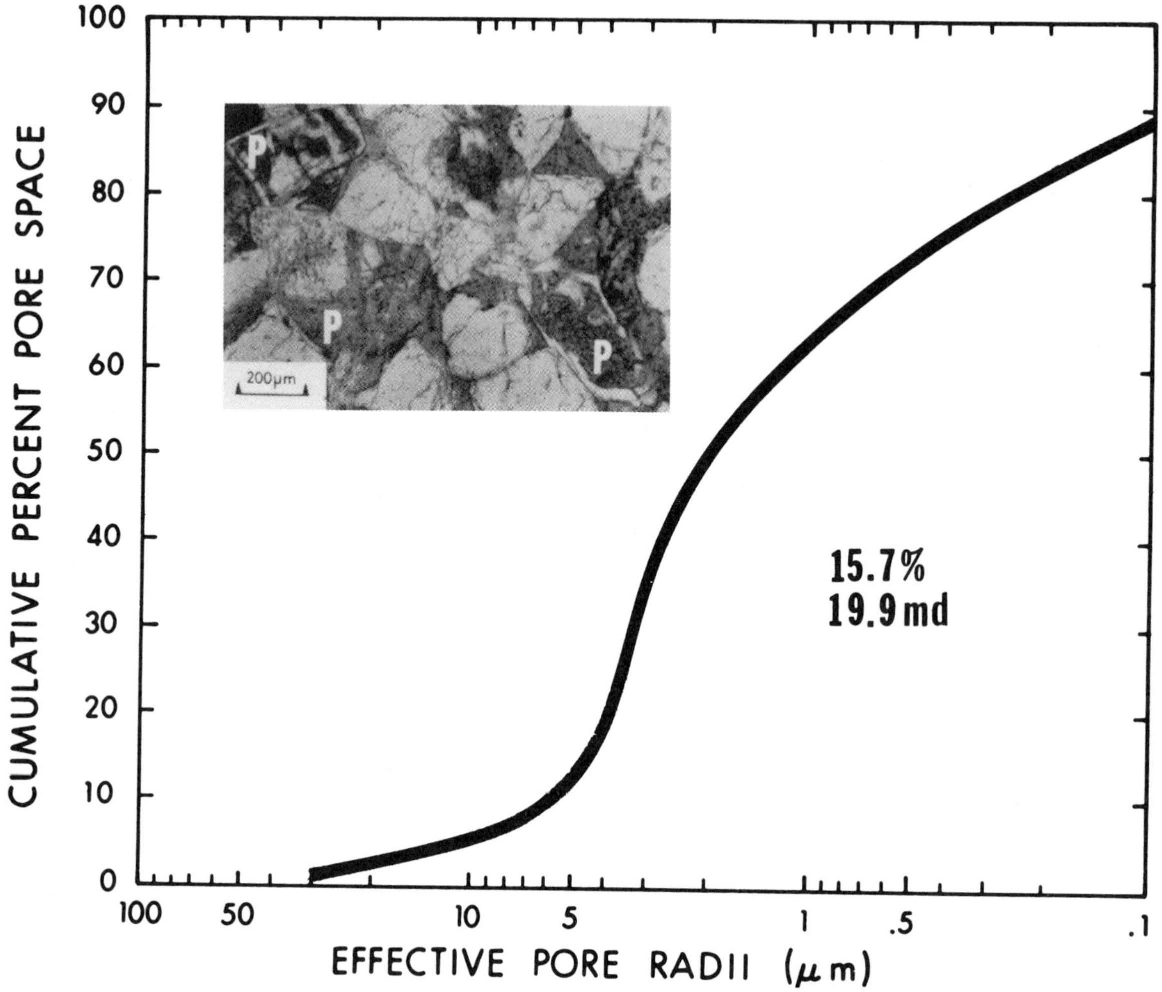

FIG. 10.—Mercury injection curve and photomicrograph of Second Wall Creek Sandstone, Wyoming. Dissolution pores (P) caused by dissolution of feldspar, and intergranular pores are visible in the micrograph. The pore system is well interconnected in contrast to the Tuscarora example (Fig. 5). The mercury injection curve indicates a relatively coarse pore system with breakthrough at about 5 μm.

permeabilities than samples containing predominantly intergranular porosity. In this case, dissolution porosity, which is more variable than microporosity, appears to have a pronounced influence on permeability. Note the significant increase in permeability moving in the direction of the intergranular pole of the diagram.

Muddy Formation.—Sandstones of the Cretaceous Muddy Formation, Denver-Julesburg

FIG. 11.—Scanning electron micrograph showing pore formed by dissolution of feldspar. Authigenic feldspar rim is well preserved. Kaolinite (K) occurs within the pore and as coating on detrital grain in foreground; Second Wall Creek Sandstone, Wyoming.

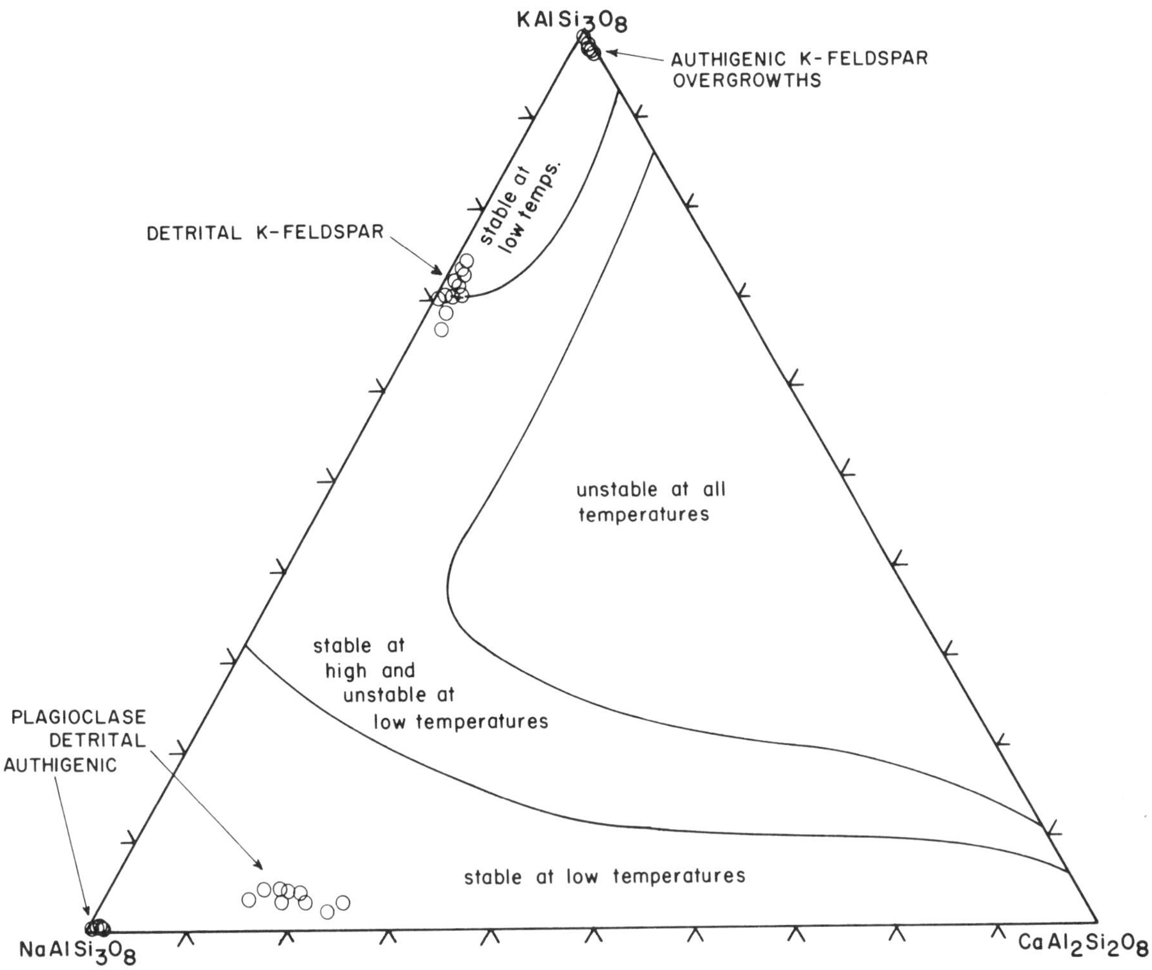

FIG. 12.—Composition of feldspars shown as mean values for each grain or component of grain (detrital nucleus and authigenic overgrowth), based on 127 analyses of 27 grains in 2 thin sections of Second Wall Creek Sandstone. Stability fields from Winchell and Winchell (1951, p. 264).

Basin, often have a mixture of porosity types. Figure 14 shows two groups of data: one near the microporosity pole and the other towards the intergranular pole. All but one sample contains the three porosity end members; however, dissolution porosity is less abundant than intergranular or microporosity. Samples plotting near the microporosity pole have low permeability (generally less than 1 md) caused by small pore apertures associated with abundant authigenic clay minerals. Porosity, however, is adequate for a reservoir (5.6 to 23.6%). For comparison, samples grouped in the upper part of the diagram have significantly better permeability (3.31 to 161 md) although the porosity is about the same (7 to 24.3%). This reflects the larger and better interconnected pores associated with the intergranular porosity.

The photomicrographs of Figure 15 illustrate the differences between the two groups of sandstones in Figure 14. The sample from the upper data group has essentially no clay, is cemented by quartz, and has obvious pores (Fig. 15A). The sample from the lower data group has abundant authigenic clay, which has reduced pore-aperture size and permeability (Fig. 15B). The obvious grain size difference between samples in 15A and B is coincidental. Very fine- and fine-grained sandstones occur in both groups of data.

Rocks plotting in the data group near the microporosity pole require a massive fracture treatment to make an economic well. Rocks in the upper data group do not. In most cases, logs alone do not provide definitive data to distinguish between these two distinctly different types of sandstone. A microlog would provide a measure of invasion and, therefore, matrix permeability for rocks in the upper group of data. However, these logs are not run often.

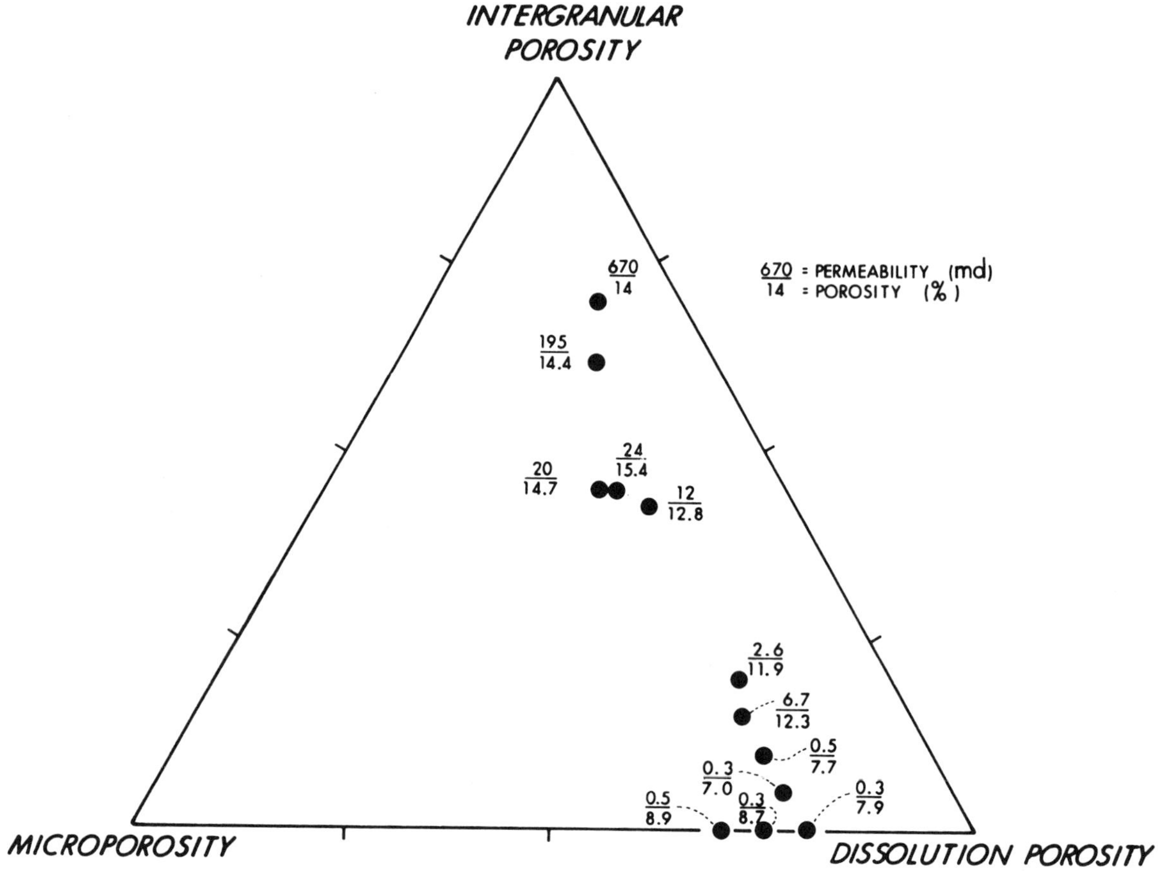

FIG. 13.—Triangular plot of porosity types for Permian sandstones, South Africa. There is a strong tendency for permeability to increase in the direction of the intergranular porosity pole.

Muddy Sand reservoirs often have small pore apertures and low permeability, but adequate porosity for storage of hydrocarbons (dashed curve, Fig. 16). This sample, which is from an oil reservoir, has 14% porosity and 0.34 md permeability. The sandstone has small pore apertures because of authigenic clay and required a massive hydraulic fracture treatment to be economically productive. About 50% of the pore apertures in this reservoir rock have a radii greater than 0.28 μm.

The solid curve in Figure 16 is an example of better reservoir rock from the Muddy Formation. Pore apertures are larger and the rock has 19.5% porosity and 119 md. permeability. Rock of this type does not require fracture treatment to make a well.

The Muddy Formation is a good example of how pore geometry may control fluid flow and water saturation. Intergranular macropores hold mobile fluids, such as oil, gas, and some water, while micropores hold irreducible water, which is relatively immobile. Log calculations may indicate water saturations are too high for a productive interval. However, the reservoir may be capable of essentially water-free production because the water is predominantly irreducible. Artificial fractures often are needed to provide commercial productive rates because of small pore apertures and low permeability. A sample situated near the

FIG. 15.—Scanning electron micrographs showing differences between two groups of samples plotted on Fig. 13. *A*, This example, from the upper data group, has little if any clay, quartz cement, and much larger pore apertures. Porosity is 19.5% and permeability 119 md. *B*, Example representative of the data group near the microporosity pole has abundant authigenic clay, which has reduced porosity, pore apertures, and permeability. This example has 14% porosity and 0.34 md. The difference in grain size between A and B is coincidental: both data groups contain fine- and very fine-grained sandstones.

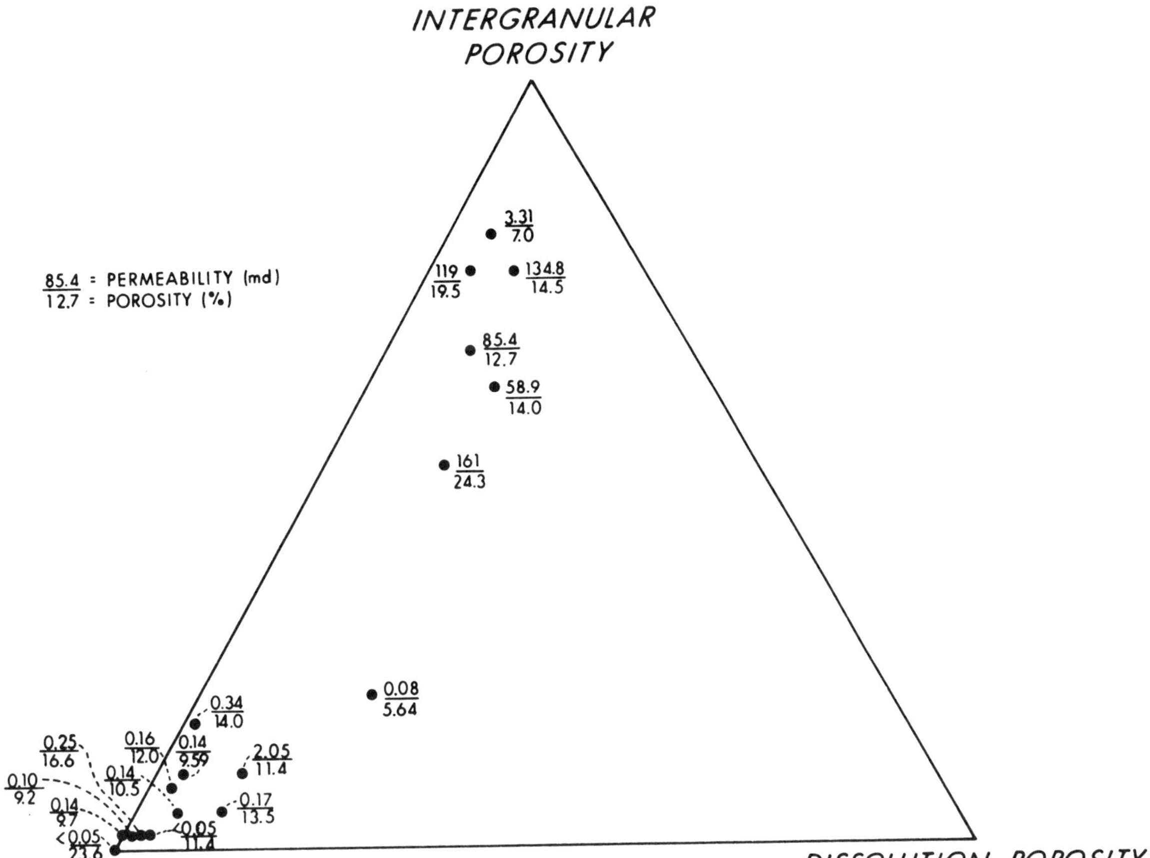

FIG. 14.—Triangular plot of porosity types for Muddy Sandstone (Cretaceous), Colorado and Wyoming. There are two groups of data: one near the microporosity pole and the other in the upper part of the triangle. Both groups have comparable porosities (up to 24%), but considerably different permeabilities. Data near the microporosity pole have permeabilities less than 1 md (except for one sample), whereas data in the upper group range from 3.31 to 161 md. Rocks near the microporosity pole require massive hydraulic fracture treatment the other rocks do not.

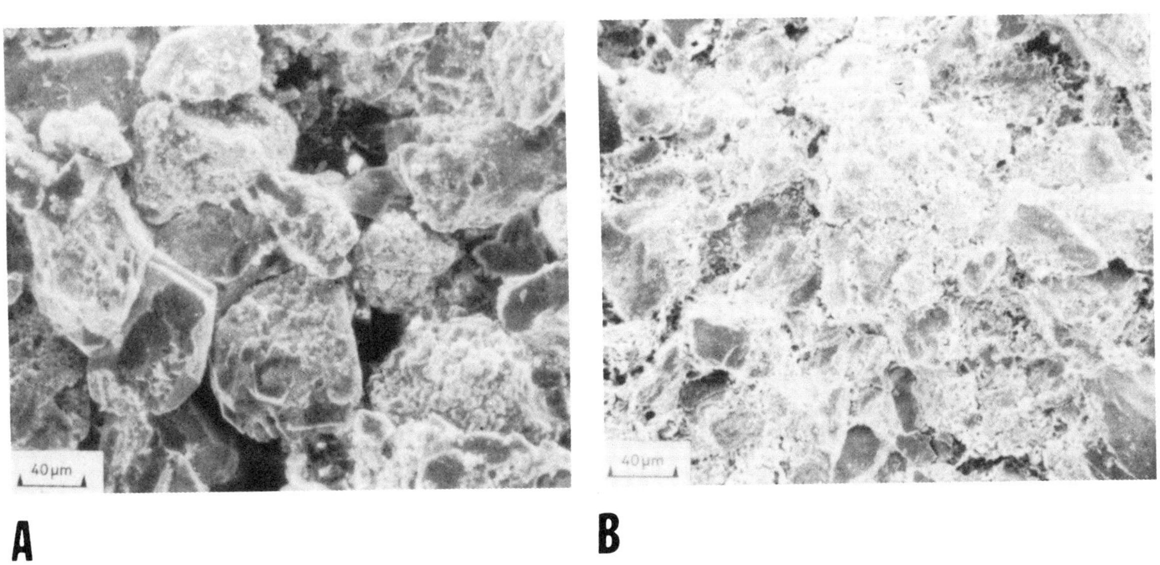

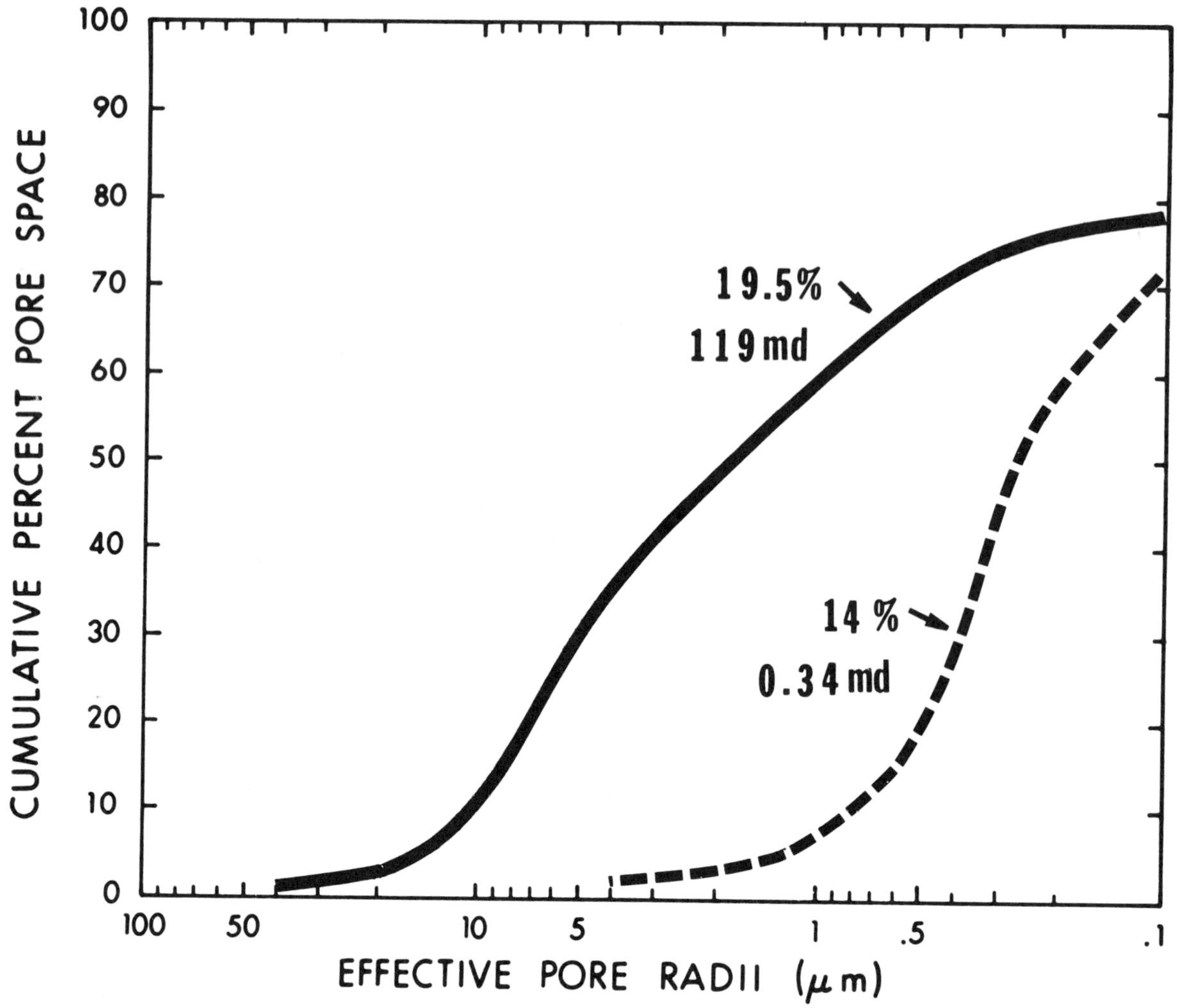

FIG. 16.—Mercury injection curves for two Muddy Sand reservoirs illustrated in Fig. 15. The solid curve is from the coarser, essentially clay-free sandstone of Fig. 15A. This sample has relatively large pore apertures, which is reflected in the good permeability (119 md). The dashed curve was derived from the finer sandstone (Fig. 15B), which has abundant authigenic clay that reduces pore apertures. The rock has 14% porosity, but only 0.34 md permeability. This sandstone required a fracture treatment to make an economic well.

microporosity-intergranular porosity base line nearer to the intergranular porosity pole (for example, 50% microporosity) could have relatively high irreducible water saturations, but with a better interconnected macropore system would not require artificial fractures for permeability.

CONCLUSIONS

1. Sandstone reservoirs often have pores of diverse origin and variable size.
2. Pore geometry influences the type, amount, and rate of production from reservoirs.
3. The best sandstone reservoirs, in terms of volume and flow rate, have intergranular porosity.
4. Potential log calculation problems for water saturation exist for rocks with abundant microporosity.
5. Fractures, either natural or induced, are required to provide economic deliverability rates for reservoirs with abundant microporosity and/or isolated dissolution porosity.
6. More cores will be needed in the future for petrographic study because available logs often cannot discriminate between important differences in pore geometry.
7. Study of porosity in rocks can help explain reservoir behavior and aid in anticipating potential reservoir problems.

ACKNOWLEDGEMENTS

I thank Amoco Production Company for permission to publish this paper, and also express gratitude to R. R. Thompson, H. D. Winland, D. J. Hartmann, J. B. Thomas, and R. A. Nelson for the helpful discussions and/or critical review of the manuscript.

REFERENCES

DEER, W. A., HOWIE, R. A., AND ZUSSMAN, J., 1962, Rock forming minerals, v. 4, framework silicates: New York, Wiley, 435 p.

HEALD, M. T., AND LARESE, R. E., 1973, The significance of solution of feldspar in porosity development: Jour. Sed. Petrology, v. 43, p. 458–460.

LARESE, R. E., 1974, Petrology and stratigraphy of the Berea Sandstone in the Cabin Creek and Gay-Fink Trends, West Virginia: PhD. Diss., Univ. West Virginia, 246 p.

LINDQUIST, SANDRA, 1976, Leached porosity in overpressured sandstones—Frio Formation (Oligocene), south Texas [abs.]: Am. Assoc. Petroleum Geologists Bull., v. 60, p. 1612–1613.

MORGAN, J. T., AND GORDON, D. T., 1970, Influence of pore geometry on water-oil relative permeability: Jour. Petroleum Technology, October 1970, p. 1199–1208.

PITTMAN, E. D., AND DUSCHATKO, R. W., 1970, Use of pore casts and scanning electron microscope to study pore geometry: Jour. Sed. Petrology, v. 40, p. 1153–1157.

ROWSELL, D. M., AND DE SWARDT, A. M. J., 1974, Secondary leaching porosity in Middle Ecca sandstones: South Africa Geol. Soc. Trans., v. 77, p. 131–140.

SCHMIDT, V., MACDONALD, D. A., AND PLATT, R. L., 1976, Pore geometry and reservoir aspects of secondary porosity in sandstones [abs.]: CIM-CSPG Joint Meeting on Enhanced Recovery, June 7–11, 1976, Program and Abs., p. 106–107.

SHENHAV, H., 1972, Lower Cretaceous sandstone reservoirs, Israel; petrography, porosity, permeability: Am. Assoc. Petroleum Geologists Bull., v. 55, p. 2194–2224.

STANTON, G. D., AND MCBRIDE, E. F., 1976, Factors influencing porosity and permeability of Lower Wilcox (Eocene) sandstone, Karnes County, Texas [abs.]: Am. Assoc. Petroleum Geologists and Soc. Econ. Paleontologists and Mineralogists Ann. Meeting Abs., v. 1, p. 119.

WINCHELL, A. N., AND WINCHELL, HORACE, 1951, Elements of optical mineralogy: New York, John Wiley and Sons, 551 p.

Reprinted by permission of the Society of Economic Paleontologists and Mineralogists. Published in *Journal of Sedimentary Petrology,* V. 47, No. 1 (March 1977), pp. 3-31.

AUTHIGENIC CLAYS IN SANDSTONES: RECOGNITION AND INFLUENCE ON RESERVOIR PROPERTIES AND PALEOENVIRONMENTAL ANALYSIS[1]

MICHAEL D. WILSON and EDWARD D. PITTMAN
Exxon Production Research Co., Houston, Texas
Amoco Production Co., Tulsa, Oklahoma

ABSTRACT: Thin-section/scanning electron microscope studies of thousands of sandstones representing many ages, compositions, and depositional environments indicate authigenic clays are far more common than previously recognized. Authigenic clays occur as pore linings, pore fillings, pseudomorphous replacements, and fracture fillings. Sandstones also contain allogenic clays. Allogenic clays originate as terrigenous material (dispersed matrix, sand-sized floccules, sand- to cobble-sized mud or shale clasts) or are introduced subsequent to deposition as a result of bioturbation or infiltration. An authigenic origin can be established on the basis of clay composition, structure, morphology and distribution, and sandstone textural properties. However, no individual criterion is an infallible indicator of authigenic origin. The most reliable criteria are a) delicacy of clay morphology, which precludes sedimentary transport, b) occurrence of the clay as pore linings absent only at grain contacts, and c) composition radically different than associated allogenic clays. Distinctions between authigenic and allogenic clays become difficult if the latter are recrystallized or if either is extensively deformed by burial or tectonism. Studies of authigenic clays utilizing a combination of thin-section, scanning electron microscope with nondispersive elemental analyzer, and x-ray diffraction show that each of the major clay groups exhibit a limited number of distinctive morphologies. Smectite occurs as highly wrinkled or honeycomb-like pore linings, the individual flakes of which are not resolvable. Illite forms pore-lining overlapping flakes whose edges tend to curl away from the grain surface and from which highly elongate lath-like projections may extend. Mixed-layer smectite/illite morphologies resemble those of both smectite and illite. Chlorite occurs primarily as pore-lining pseudohexagonal flakes with a cardhouse, honeycomb, or rosette arrangement. Kaolinite and dickite most commonly form pore-filling books of stacked pseudohexagonal flakes. Occasionally, they form thin pore-lining sheets of overlapping pseudohexagonal flakes. Authigenic clays are a major control on reservoir quality. Permeability and water saturation are particularly sensitive to the relative abundance of clays. Interpretations of depositional environment and provenance based on sandstone composition and texture may be inaccurate if the presence of authigenic clay is overlooked.

INTRODUCTION

Purpose

This paper draws attention to the importance of authigenic clay in sandstones. Clays in sandstones are allogenic or authigenic. Allogenic refers to clay minerals originating outside of the rock of which they now constitute a part. Authigenic clay minerals are formed or regenerated in place.

Many geologists are unaware of the widespread distribution of authigenic clay. We examined over three thousand sandstones of different age (Cambrian through Pliocene), composition (quartz, arkosic, and lithic arenites), depositional environment (fluvial to deepwater turbidite), and geographic distribution and found that authigenic clays are present in most samples.

More specifically, this paper 1) illustrates the morphologies of allogenic and authigenic clays; 2) lists criteria for distinguishing between allogenic and authigenic clays; and 3) discusses the effects of authigenic clay on sedimentologic interpretations and reservoir properties of sandstones.

Previous Work

The bulk of the geologic literature on sandstones gives little indication of their content of authigenic clay. The major American reference works, including those of recent vintage (Blatt *et al.*, 1982; Pettijohn *et al.*, 1972), devoted only cursory attention to this subject. Foreign reference texts are more informative. Füchtbauer and Müller (1970, p. 119-129) described sequences of diagenetic alterations which affect sandstones of various compositions under differing geochemical conditions. Clay neoformation (precipitation from solution as opposed to regeneration of preexisting clays) and replacement of detrital grains are typical in many of these sequences. Supporting references are made to articles by Russian, German, and American authors. Similarly, Millot (1970) claimed kaolinite is a common diagenetic product and that both sericite and chlorite are characteristically developed during diagenesis, particularly under conditions approaching those of metamorphism. Millot supported his statements by references to studies by French and

TABLE 1.—*Distribution of Authigenic Clay in Siliciclastic Sandstones*

Age	Number of Samples	Number of Formations	Number (Percent) Containing Authigenic Clay	Number (Percent) Without Diagenetic Clay		
				Pores Filled With Allogenic Clay	Pores Filled With Carbonate Or Evaporite Cement	Pores Partially Filled With Overgrowths
Tertiary	225	18	201 (89.4)	8 (3.6)	10 (4.5)	6 (2.5)
Mesozoic	329	21	310 (94.3)	2 (0.6)	5 (1.5)	12 (3.6)
Paleozoic	231	16	205 (88.8)	4 (1.7)	22 (9.5)	0 (0.0)
Total	785	55	716 (91.3)	14 (1.8)	37 (4.7)	18 (2.2)

Russian authors. Extensive referencing of the Russian literature by Sarkisyan (1972) suggested that authigenic clays are common in many Mesozoic and Paleozoic hydrocarbon-bearing sandstones of various sedimentary basins in Russia.

A few studies by American and Canadian geologists have suggested that authigenic clays are common in sandstones. Shelton (1964) compiled a list of occurrences of authigenic kaolinite and Hayes (1970) did the same for authigenic chlorite. Carrigy and Mellon (1964) and Carrigy (1971) concluded that authigenic clays (smectite, chlorite, and kaolinite) are abundant and widely distributed in sandstones of divergent compositions and depositional environments in the thick Cretaceous and Early Tertiary sequence of Alberta. Galloway (1974) found that diagenetic chlorite and smectite were very common in the compositionally immature Tertiary sandstones of the Bristol Bay, Queen Charlotte, Grays Harbor and Chehalis Basins of the northeast Pacific region. Walker (1967, 1973, and 1974) and Walker and Waugh (1973) demonstrated that the Fe-rich clay matrix in arkosic sediments was derived from intrastratal alteration of hornblende and biotite.

Study of 785 samples by the senior author (Wilson and Tillman, 1974) over a two-year period indicated that authigenic clay is present in approximately 90% of the samples (Table 1). In the remaining samples without authigenic clay, 20% contain abundant allogenic clays, 50% contain carbonate and evaporite cement and 20% retain appreciable porosity with only minor silica cement.

The size, distribution, and optical character of authigenic clays often cause them to go unnoticed during cursory petrographic examination. Optical properties of common authigenic clays and photomicrographic examples of each are presented in Carrigy and Mellon (1964, p. 462-465).

Recent papers have made use of the scanning electron microscope to interpret an authigenic origin for illite and kaolinite (Stalder, 1973); kaolinite, chlorite, and illite (Sarkisyan, 1971, 1972); chlorite (Pittman and Lumsden, 1968; Hayes, 1970); smectite, chlorite, and kaolinite (Flesch and Wilson, 1974); smectitie, illite, kaolinite, and chlorite (Pittman, 1974; Wilson, 1974) and glauconite (Borst and Keller, 1969; Odin, 1972; Hein *et al.*, 1974). Although Gaida and others (1973) did not specifically discuss the origin of clay in sandstone, it is obvious that they interpret the clay to be authigenic in some scanning electron micrographs.

Some geologists assume the texture of a sandstone directly reflects the environmental conditions at the time of deposition. Implicit in this approach is the concept that all clay minerals in sandstones are detrital in origin as exemplified by many of the older schemes for sandstone classification (Gilbert in Williams *et al.*, 1955; Pettijohn, 1957).

Cummins (1962) first expressed the view that the matrix in graywacke was diagenetic rather than primary in origin. One of Cummins' arguments against a primary matrix was the failure to find a modern sediment of the graywacke type. Hollister and Heezen (1964) examined 118 sand cores from the deep sea and found only two cores that contained what they considered to be a graywacke type sand with a primary matrix. Buller and McManus (1973) used textural analysis to conclude that diagenetic disintegration of about 30% of the contained sand-size mineral grains would convert a typical modern turbidite into a graywacke. Brenchley (1969) concluded that even well sorted, shallow-water sandstones can attain the typical graywacke texture by diagenetic decomposition of unstable components. Based on petrographic evidence, Lovell (1972), Reimer (1972), and Galloway (1974) interpreted the matrix in graywackes as diagenetic. Audley-Charles (1967 and 1968) argued for a primary matrix in the Mio-Pliocene sandstones of the Viqueque Formation, Timor, while Rahmani (1968) countered with an argument favoring a diagenetic origin for the matrix in these rocks.

Laboratory experiments support the viewpoint that clays can form diagenetically. Whetten and Hawkins (1970, 1972), and Whetten (1971) showed that clays are produced by diagenetic changes at temperatures of 250-300 °C in loose sand packs of Columbia River sediment (65-70% glassy volcanic rock fragments). Keen *et al.* (1973) produced smectite experimentally from alteration of tholeitic basalt in sea water at 150-250 °C and 1 Kb. Divis and McKenzie (1975) produced diagenetic clay experimentally from the primary detrital components of feldspathic sands at 200 °C.

Procedures

Mineralogical identification of clays reported in this study was based on comparison of scanning

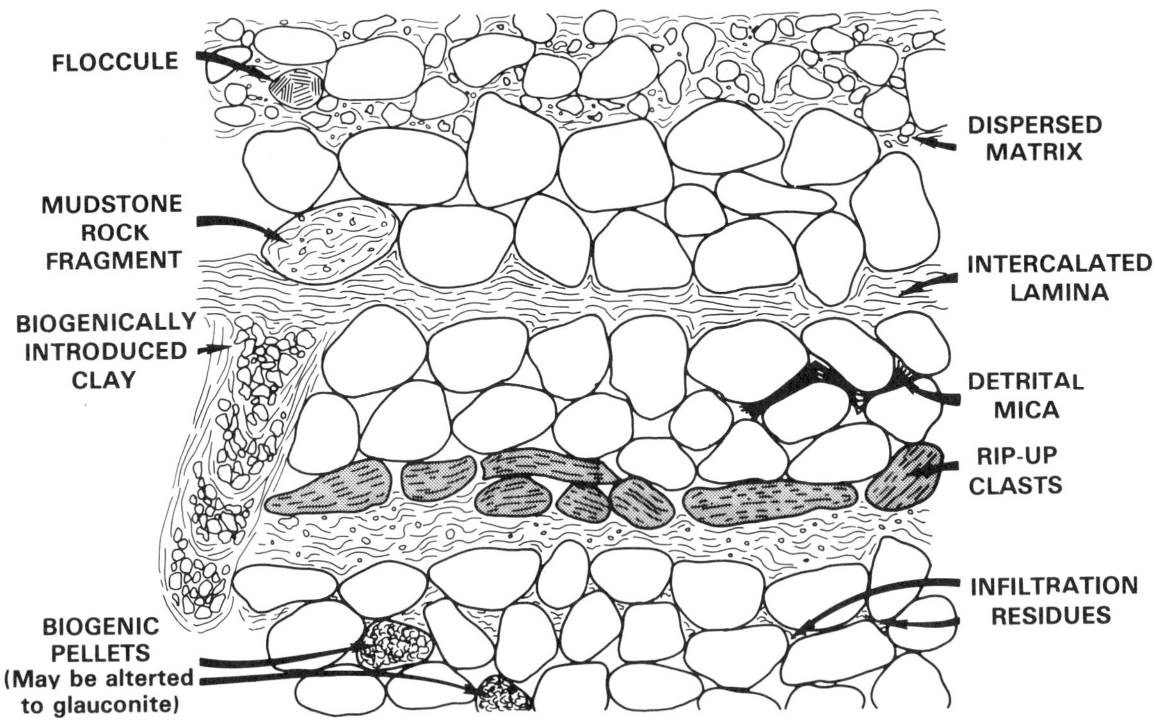

FIG. 1.—Modes of occurrence of allogenic clay in sandstones. The authors have not observed clay floccules in sandstones but grains interpreted as such have been described by Pryor and Van Wie (1971).

electron micrographs, X-ray diffraction analyses, thin section petrography, and nondispersive detector analyses. Nondispersive detector analyses were made on enough samples to provide confidence in identification of clays based on morphology seen in scanning electron micrographs.

All thin sections were prepared from rocks impregnated with epoxy. Carrigy and Mellon (1964, p. 462) stressed the care required to preserve authigenic clays during thin section preparation. It is not unlikely that many petrographers have used thin sections in which authigenic clays have been eliminated by the wet grinding of unimpregnated samples.

Samples for scanning electron microscopy were coated with a gold-palladium alloy. Carbon or gold-palladium was used as a coating for samples examined with the nondispersive detector. The relative abundance of the following elements were used to aid in identifying clay minerals: (1) kaolinite - aluminum; (2) illite - potassium; (3) chlorite - iron and/or magnesium; (4) smectite - magnesium, calcium, and sodium.

Clay composition is best determined by X-ray diffraction analysis, provided a relatively pure clay fraction can be separated from the sand fraction. To avoid contamination with fragments of nonclay grains produced by disaggregation, we recommend gentle crushing using mortar and pestle, followed by immersion in a high intensity ultrasonic bath for several hours. Decanting the suspended material followed by air drying generally provides sufficient quantity for one or more oriented slides. Many of the authigenic clays liberated by this technique may be greater than 2 μm in maximum dimension.

Care must be taken during sample preparation to avoid contamination by allogenic clays present as thin laminae, clay "rip-up" clasts, biogenic pellets, and burrow filling or linings. Similarly, coarse muscovite and biotite often are concentrated along thin laminae and represent a fraction that does not occur in the authigenic clay assemblage.

MODES OF OCCURRENCE OF CLAY IN SANDSTONES

Allogenic clays are formed prior to deposition and are mixed with the sand fraction during or immediately following deposition. Authigenic clays develop subsequent to burial and include both new and regenerated forms. Clays that are presumed to form at or near the depositional interface can be classed as authigenic or separated into a special group (hamrolytic) if desired. Clays generally are degraded during weathering, erosion, and transport and generated or regenerated during burial diagenesis.

Allogenic (also referred to as detrital) clay types are described below and diagrammed in Figure 1.

Allogenic

Syndepositional

Clays introduced during deposition occur as clay-sized particles, as well as silt- to boulder-sized clay aggregates (Table 2). The size, shape, and dis-

TABLE 2.—*Modes of Clay Occurrence in Sandstones*

Allogenic (detrital)	Authigenic
Syndepositional	Neoformed
Dispersed matrix	Pore linings
Intercalated laminae	Pore fillings
Mudstone and shale rock fragments (ancient)	Replacements
Mud rip-up clasts (contemporaneous)	Fracture and vug fillings
Floccules	Regenerated (can occur in any of the allogenic or neoformed modes)
Biogenic pellets (transported)	
Introduced shortly after deposition	
Burrow linings and fillings	
Infiltration residues	

tribution of the clay may be controlled by physical or biogenic processes.

Individual clay particles may be dispersed as matrix in a sandstone. Terrigenous clay and silt matrix is deposited simultaneously with the sand-size debris. Clays in sandstones also are concentrated in thin laminae (e.g., clay flasers) formed as a result of variations in competency of the depositional medium.

Detrital clay also forms sand-sized or larger aggregates of clay minerals. One type is clay or mud "rip-up" clasts, which are derived from erosion of layers deposited penecontemporaneously with the sandstones in which they are found. Clays also may flocculate to form sand-sized pellets, as suggested by Pryor (1971). Reworked fragments of older rocks such as shale or mudstone may be difficult to distinguish from "rip-up" clasts and biogenic pellets.

Biogenic clay or mud pellets are produced by ingestion and excretion of mud by organisms. These pellets may be retained in burrows or discharged and transported as detrital particles (Pryor, 1975). Biogenic pellets generally are ovoid in outline and occasionally exhibit a regular internal structure. Because pellets have a low grain density (clay and organic material) and bulk density (pellets have internal porosity), they should be coarser then syndepositional quartz, feldspar, and most rock fragments. Alteration of biogenic pellets to glauconite at the sediment-water interface is a common occurrence.

Soft aggregates of clay minerals tend to physically disintegrate during transport and are easily deformed during compaction to form what appears to be matrix. Even well lithified argillaceous grains can be squeezed into pores such that the original grain shape and texture become indistinct.

Introduced Shortly After Deposition

Reworking of sediments by organisms is an important mechanism for mixing mud and sand. Howard and Reineck (1972), in an X-radiography study of sediments off the Georgia coast, have shown that extensive homogenization of mud and sand is common. Other studies by Howard (1969) and Frey and Howard (1972) showed the importance of bioturbation in modifying texture. Through bioturbation, clays may line burrows or be irregularly dispersed.

Infiltration clays move into the interstices of sand by downward or lateral migration of water. Through experimental work, Crone (1974) has shown that such infiltrated clays occur as follows: (1) flocculated aggregates randomly distributed in voids; (2) coatings composed of platelets oriented parallel to grain surfaces; (3) ridges on grains consisting of platelets generally perpendicular to grain surfaces. Walker and Crone (1974) showed that sandstones and conglomerates of desert alluvium, which are essentially clay-free when deposited, become increasingly clay-rich with increasing age, and that the clay originates by mechanical infiltration from influent stream flow. Infiltration of clays similar to that described by Crone occurs in the zone of drilling mud invasion in well cores, and adjacent to the well bore, affording further evidence that infiltration is an expectable process.

Authigenic

All the major varieties of clay minerals are known to form authigenically in sandstone either as a direct precipitate from formation waters (neoformation) or through reactions between precursor materials and the contained waters (regeneration). The major types of neoformed clays and clays produced by their regeneration are discussed below and illustrated in Figure 2. Regenerated clays may take the form of any of their allogenic or neoformed precursors.

Pore Linings (also referred to as clay films, or grain coatings)

Pore linings are formed by clay coatings deposited on the surfaces of framework grains, except at points of grain-to-grain contact. The individual clay particles or aggregates commonly exhibit a preferred orientation normal to or parallel to the detrital grain surface. Pore linings grow outward from the grain surfaces and may merge with the linings or adjacent grains. Pore lining clays may be enveloped during subsequent cementation by such materials as quartz and feldspar overgrowths.

Pore Fillings

Pore-filling clays plug interstitial pores and the individual flakes, or aggregates of flakes, exhibit no apparent alignment relative to the detrital grain surfaces. Pore linings and pore fillings of clay are the major cementing agents in many friable sandstones.

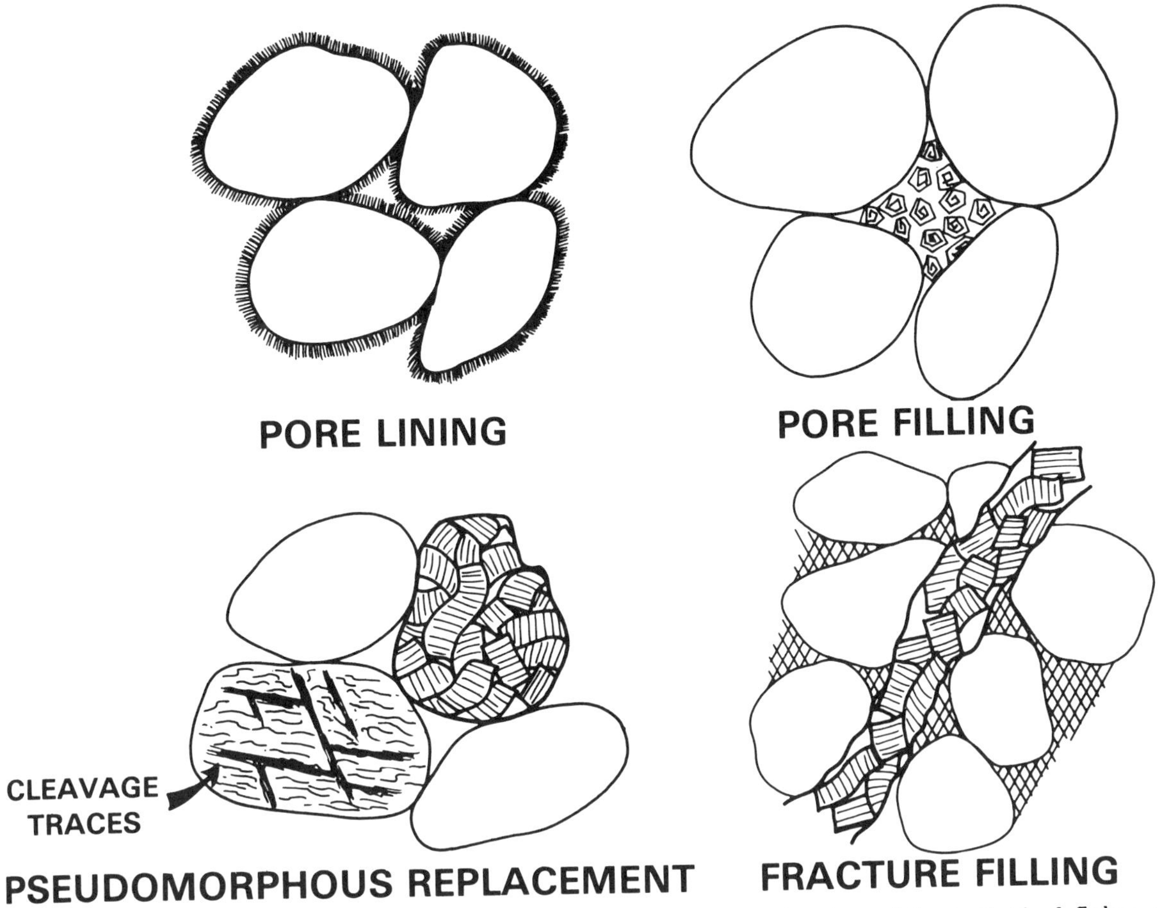

FIG. 2.—Modes of occurrence of authigenic clay in sandstones. The sizes of the individual flakes or aggregates have been exaggerated for purposes of illustration.

Replacements

Replacement clays are those that partially or completely replace detrital grains or infill voids left by dissolution of detrital grains. In some cases, textures of the replaced host grain are preserved. The grain types most commonly replaced are volcanic glass and basic and intermediate rock fragments, pyroxenes, amphiboles, calcic plagioclase, sanidine, and carbonate rock fragments or fossil debris.

Fracture and Vug Fillings

Rarely clays fill fractures or vugs which transect a series of detrital grains.

CRITERIA FOR DISTINGUISHING AUTHIGENIC CLAY

Over the last 30 years, numerous criteria have been proposed for differentiation of authigenic clay from clays of other origins (usually considered detrital). The criteria discussed permit differentiation between neoformed clays (and products of their regeneration) and allogenic clays. Allogenic clays which have experienced extensive regeneration are very difficult to distinguish from neoformed clays using these criteria. These criteria were based primarily on petrographic examination and X-ray diffraction analysis. However, morphology and many textural relationships are better observed with the scanning electron microscope. This instrument is used widely in combination with the standard methods of analysis.

The criteria listed below are based on a combination of experience and a thorough search of the literature on sandstone diagenesis. Some criteria are more reliable than others (Table 3), but none of the criteria should be considered infallible. Exceptions to most criteria can be found and will be so noted. Diligent search and application of several criteria will usually provide sufficient evidence to make a firm conclusion.

The criteria are placed into five categories—composition, morphology, structure, texture and distribution. The first three are intrinsic properties of the clay particles, whereas the last two refer to their relationships with the host rock.

Composition

(1) Authigenic clays can attain a high degree of purity. This may be reflected by their chemical composition (Keller, 1970, p. 792), their uniformity of color and texture, (Lerbekmo, 1961, p. 198) and

TABLE 3.—*Reliability of Criteria for Authigenic Origin of Clay*

Criteria	Reliability	Frequency of Occurrence
Composition		
1) Absence of impurities	O	C
2) Clay assemblage monomineralic	X	R
3) Significantly different from associated shales	+	U
4) Concentric color zoning	O	R
Morphology		
5) Crystal outlines	+	C
6) Delicate projections	+	C
7) Undeformed	O	C
8) Pseudomorphous replacement	+	R
Structure		
9) Low-temperature polytypes	+	?
10) High degree of crystallinity	O	U
Texture		
11) Gap in grain-size distribution (silt fraction lacking	X	C
12) Large particle size	O	C
Distribution		
13) Pore linings missing at grain contacts	+	C
14) Scattered pore fillings	X	C
15) Fracture fillings	+	R
16) Absent in early diagenetic concretions	+	C
17) Cover diagenetic components formed at earlier stage	+	C
18) Laminae with abrupt lateral terminations	+	R
19) Radial alignment of individual flakes	O	C
20) Medial sutures	O	R
21) Bridges between detrital grains near points of contact	+	C

X—unreliable; O—generally reliable; +—very reliable; R—rare; C—common; U—ubiquitous; ?—unknown

by their transparency (Dickinson, 1970, p. 702). Many authigenic clays exhibit these features. However, clouding by minor impurities such as finely crystalline diagenetic pyrite or anatase and staining by iron oxides or hydroxides may negate use of this feature. The impure nature of allogenic clays can be attributed to degradation during erosion and transport and mixing with nonclay impurities during deposition. In many sandstones the size of the authigenic clay flakes may be too small for detailed evaluation of this parameter.

(2) Some authigenic clay mineral suites are monomineralic. This can be attributed to generation of such clays under a limited range of subsurface physical and chemical conditions. Because they are derived from a diverse series of rock types and soils, allogenic clays are generally mixtures of two or more clays.

Although few authigenic clay mineral assemblages are truly monomineralic, many contain only traces of other clays. On the other hand, many sandstones contain two authigenic clays and sandstones with three authigenic clays also are known. Some allogenic clay suites may be essentially monomineralic, although most of these may have been produced by early or late stage diagenetic regeneration (e.g., fire-clays and deeply buried illitic shales).

(3) Authigenic clay suites in sandstones often differ radically from detrital clay suites in closely associated mudstone laminae or beds (Glass *et al.*, 1956; Glass, 1958; Weaver, 1960; Smoot, 1960; Füchtbauer and Goldschmidt, 1963, p. 100; Shelton, 1964, p. 103; Bayliss *et al.*, 1965, p. 424; Triplehorn, 1970; Wilson *et al.*, 1972). This relationship is best determined by comparison of X-ray diffraction patterns. However, in thin section relatively pure authigenic clay commonly abuts a detrital clay lamina of distinctly different optical character (color, birefringence). Authigenic and allogenic assemblages differ primarily because the former are generated within a narrow range of physicochemical conditions compared to the latter. The allogenic clays may have been generated in rocks or soils under conditions ranging from surface weathering through hydrothermal alteration to high grade metamorphism. This criterion is considered very reliable, provided the suites are significantly different. Because the differing clay suites usually are intimately interbedded, variations in provenance can be eliminated as a likely cause. Commonly, the authigenic suite may be dominated by a clay type that is a trace component or is lacking in the detrital assemblage (Fig. 3).

An authigenic origin of a clay assemblage is not ruled out if only slight differences occur. The clay mineral assemblage produced by diagenesis of framework grains, particularly in volcanic provinces, may closely match the clay assemblage produced by weathering of these same rock types in the source terrain. It also should be noted that the authigenic clay assemblages within an individual sandstone bed may differ if an oil-water contact is crossed (Kulbicki and Millot, 1960; Smoot, 1960; Füchtbauer and Goldschmidt, 1963; Prozorovich, 1967).

(4) Authigenic pore-lining clays may exhibit color zoning concentric with pore boundaries (Mellon, 1964, p. 793; Dickinson, 1970, p. 702). This feature is relatively uncommon and is restricted to sandstones with thick chlorite or smectite pore linings (Fig. 4). We can think of no reason to expect allogenic clays to exhibit such a characteristic, even if modified by diagenetic processes. However, if samples exhibiting this feature were collected at surface outcrops, the possibility of staining as a result of near-surface weathering should be considered.

Morphology

(5) Authigenic clays often exhibit crystalline habits (Fig. 5A). This is one of the first (Ross and Kerr, 1931, p. 73; Smithson and Brown, 1954, p. 180) and most widely used criterion for distinguish-

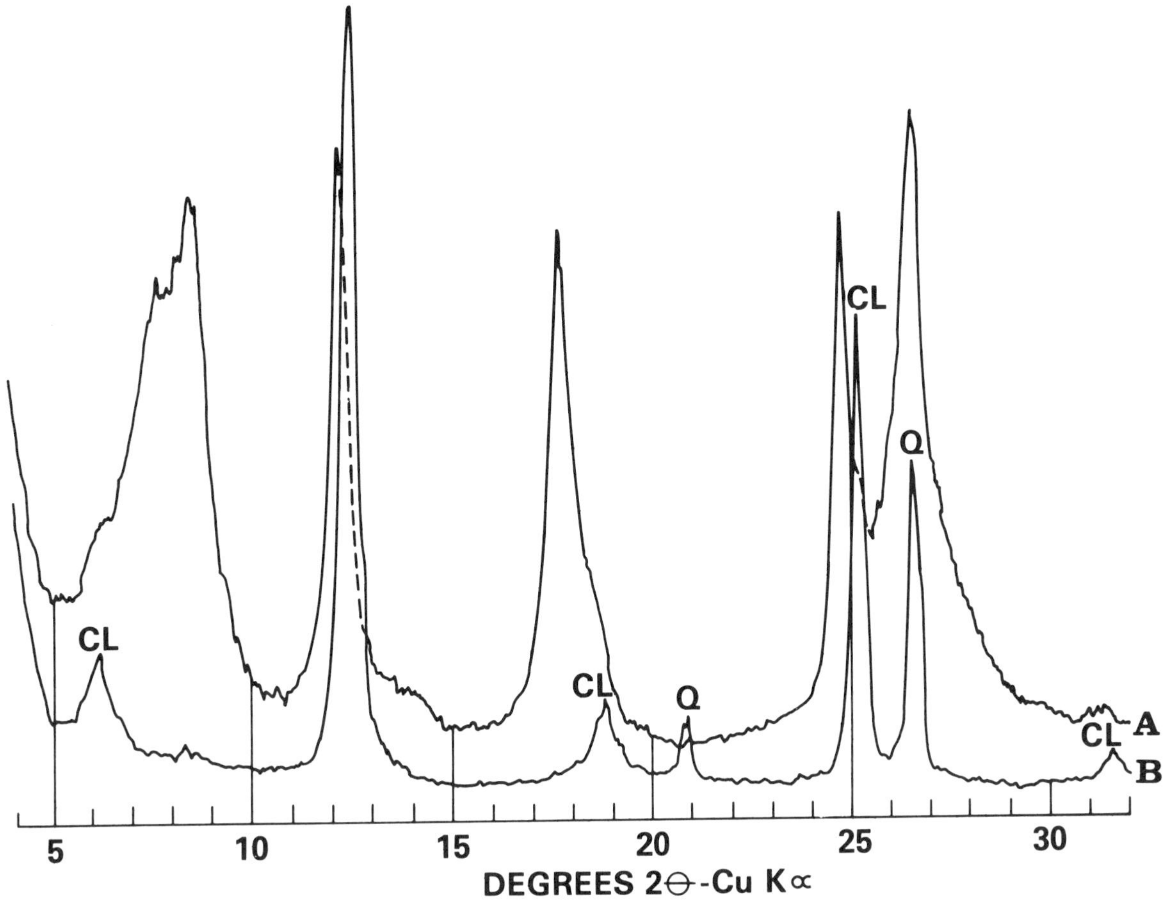

FIG. 3.—Comparison of <2 μm clay assemblages from a three inch shale bed (A) and from a four inch horizontally laminated sandstone bed (B) two feet below in the Upper Cretaceous Woodbine Formation of Texas. The detrital clay suite from the shale contains poorly crystallized mixed-layer smectite/illite and kaolinite and very minor amounts of chlorite. As evidenced by the extremely sharp diffraction peaks, the clay suite from the sandstone contains only well crystallized chlorite (CL). Minor quartz (Q) also is present.

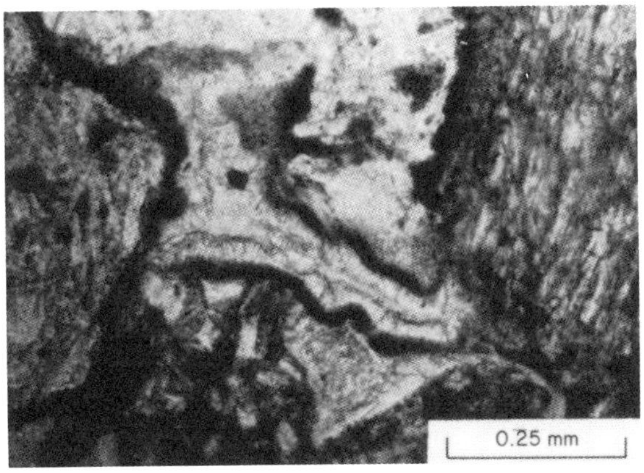

FIG. 4.—Pore-lining smectite exhibiting concentric color zonation parallel to the detrital grain surface. The outer rim is greenish and the central portion is white. Surrounding grains are volcanic rock fragments. Miocene Tyonek Formation, Alaska.

ing authigenic clays and is considered to be very reliable. The delicacy of the clay flakes precludes extended transport. This criterion is particularly applicable to chlorite, kaolinite and dickite. Some of these also occur in the form of delicate rosette or vermicular aggregates. Although the outlines of individual flakes of smectite are generally impossible to distinguish, the aggregate morphology may be very regular and easily distinguished from allogenic morphologies.

Authigenic mixed-layer smectite/illite and illite with obvious crystal outlines have not been observed. This observation conflicts with those of Rex (1966) who claimed to have found perfect mica crystals in many sandstone units. It is possible the needle-like crystals described by Rex occur primarily as the elongate projections (referred to in the next section) on larger irregular mica flakes. If recrystallization has not occurred, clays of undisputed detrital origin invariably exhibit irregular

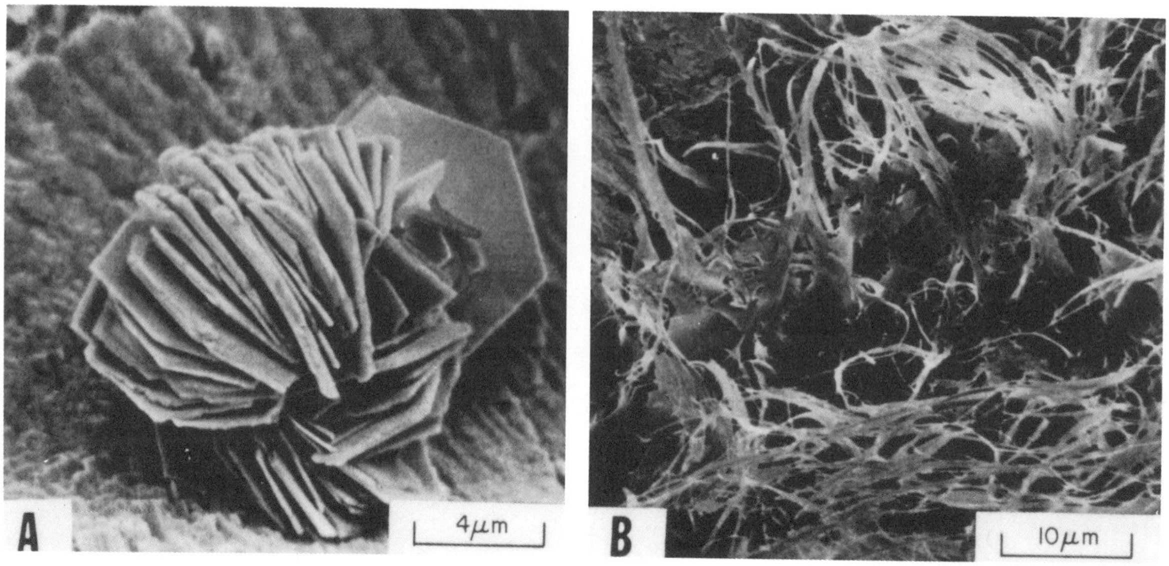

FIG. 5.—(A) Kaolinite with pseudohexagonal idiomorphic form. Crystalline growth habits are considered to be reliable indicators of authigenic origin. Eocene Frio Sand, Texas. (B) Authigenic illite with long lath-like projections. This illite is attached to the surface of a detrital grain and has a sheet-like appearance at the base that splits into delicate, tapering, lath-like structures. Permian Rotliegendes Sandstone, North Sea.

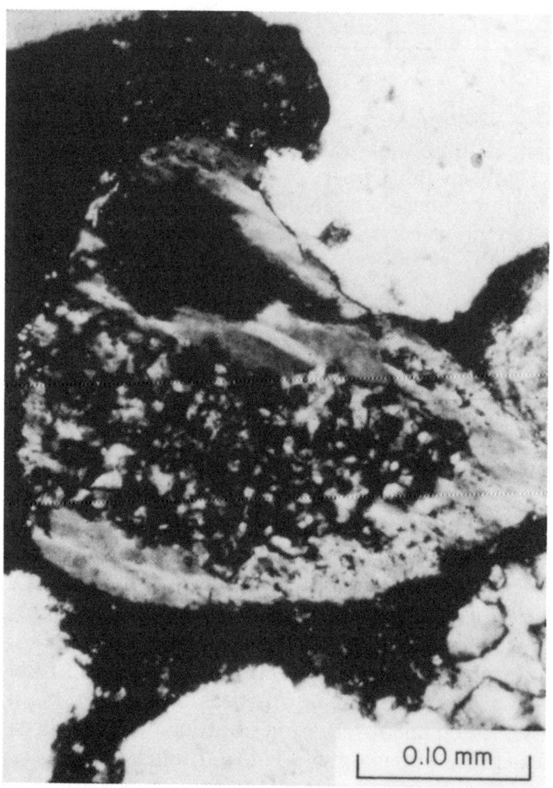

FIG. 6.—Microcline almost entirely altered to microcrystalline kaolinite. Preservation of a few isolated remnants with identical optical orientations provide evidence of the detrital grain mineralogy. Cretaceous Muddy Sandstone, Wyoming.

outlines. Distinction between well-crystallized neoformed clay and recrystallized matrix may be somewhat difficult and is best done utilizing other criteria presented here, particularly those involving composition.

(6) Some authigenic clays exhibit delicate spine or lath-like projections (Fig. 5B). Mixed-layer smectite/illite and illite exhibit projections from a fraction of a micron to 30 microns long. Such projections preclude transport of the clay as a detrital particle. However, if the projections are stubbier and more ragged in appearance, they become less distinguishable from irregularities along the edges of detrital grains, in which case differentiation from recrystallized matrix requires alternate criteria.

(7) Authigenic clays are not deformed by compaction (Gilbert, 1949, p. 6). Allogenic clays and micas often conform to the outlines of the adjacent coarser detrital grains as a result of deformation during compaction. This criterion is generally reliable but may be useless where sandstones have been subjected to directed or overburden stresses sufficient to disrupt the grain-cement framework. The latter should be expected in sandstones that have been deeply buried, such as those now being explored by an increasing number of deep wells.

(8) Authigenic clays can form pseudomorphs after some detrital grains (Heald, 1950, p. 629). Feldspar can be partially to completely replaced by or dissolved and later infilled by chlorite, kaolinite, or illite (Fig. 6). Chlorite and smectite often parti-

FIG. 7.—(A) Sandstone containing iron oxide stained smectite/illite pore linings and irregularly distributed pore-filling clots of kaolinite. Pore-filling kaolinite clots (arrows) appear as clogged pores. The distinct break in the grain size distribution between the sand and clay size material and high degree of sorting of the sand-sized grains indicates a diagenetic origin for the clay. (B) A closeup of a pore plugged by kaolinite. Jurassic Aztec Sandstone, Nevada.

ally or completely replace volcanic rock fragments, particularly volcanic glass.

The physically unstable character of these clay replacements preclude sedimentary transport prior to deposition. However, unless some features or structures of the original detrital grain are preserved, authigenic origin is difficult to establish. Faint traces of cleavage or twinning in feldspar may be preserved by the alignment of some of the authigenic clay crystals. Feldspar laths in volcanic rock fragments whose matrix has been replaced by chlorite may preserve relict volcanic rock textures (e.g., intergranular, trachytic).

Unstable mineral grains or rock fragments, replaced pseudomorphously by clays in the weathering zone, are physically weak. Such grains cannot survive more than slight transportation or reworking. When they occur, such grains represent first-cycle material deposited extremely close to the source area.

Structure

(9) Low-temperature clay polytypes are typically authigenic (e.g., Types Ia and Ib chlorite and 1M illite) (Brown and Bailey, 1962; Hayes, 1970; Triplehorn, 1967; Sarkisyan, 1972; and Wilson *et al.*, 1972). An authigenic origin for low-temperature polytype clays in sandstones is indicated if clays in intercalated detrital shale or mudstone laminae or beds consist predominantly of high-temperature polytypes. However, not all authigenic clays are low-temperature polytypes. Hayes (1970, p. 298) has observed authigenic IIb chlorite in some ironstones and sandstones and suggests its presence may be the result of high temperatures associated with deep burial or proximity to igneous intrusions.

(10) Authigenic clays exhibit a high degree of crystallinity (Shelton, 1964; Sarkisyan, 1972; Wilson *et al.*, 1972). The vast majority of authigenic clays that we have studied produce sharp X-ray peaks (as defined by measures of peak width, such as that used by Weaver, 1960, p. 1514). This feature is well illustrated in Figure 3.

However, mixed-layer clays produce broad diffuse peaks and are notable exceptions to this rule. As noted by J. Hayes (written communication) the relationship between peak sharpness and crystallinity is meaningless in a mixed-layer system.

Hayes (1970, p. 294) hypothesizes the existence of a sequence of development for authigenic chlorites, with crystallinity increasing as the "energy level of the diagenetic environment increases."

Texture

(11) Sandstones containing authigenic clay often exhibit a distinct break in their grain size distribution (Shelton, 1964, p. 103; Triplehorn, 1967, p. 883). Correcting for the obvious authigenic clay fraction, the sand mode generally is relatively well sorted (Figs. 7A and 7B). The silt component may be absent or present only in very small amounts. However, it is not unusual for allogenic and authigenic clay to occur together in sandstones.

Sandstones which contain more than a few percent allogenic clay commonly exhibit a unimodal distribution continuous from the sand through the clay fractions. However, some tuffaceous sandstones containing clay produced by the diagenetic alteration of fine vitric grains also may yield a unimodal grain size distribution.

(12) Authigenic clay particles in sandstones are larger than clay particles in adjacent detrital clay

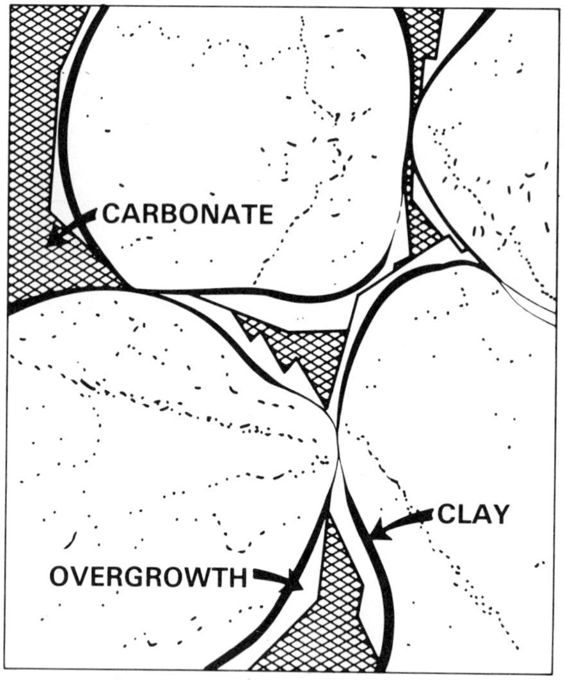

FIG. 8.—Sequential filling of pores by three diagenetic components. It is assumed the pores are filled inwards, the sequence being determined by the principle of superposition. Clay occurs as thin partial coatings immediately adjacent to the detrital grain surfaces and is the earliest of the three components to form. Idiomorphic quartz overgrowths, the next to form, attach to the quartz nuclei and cover the clay coatings. Carbonate cement fills the remainder of the pores.

laminae or mudstone beds (Lerbekmo, 1961, p. 198). Authigenic kaolinite typically occurs in booklets 2-10 μm across, although some attain long dimensions (parallel with the c-axis) of up to 2500 μm. Authigenic chlorite flakes are generally 2-10 μm across, and flakes up to 150 μm diameter have been observed. Though most allogenic clay particles are less than 5 μm across, many chlorite and muscovite flakes reach maximum dimensions of 100 to 1000 μm. If the latter are disregarded based on alternate criteria (see criteria 1, 2, 5, 7, 9) this relationship is reliable.

Distribution

Many of the criteria listed below are based on several concepts regarding the possible spatial arrangements of diagenetic materials with respect to detrital sand-sized grains and to one another. The validity of these assumptions has been tested using the criteria discussed previously. First, and of prime importance, is the assumption that pores are filled from the grain surface inwards. This is a reasonable assumption, inasmuch as the cementing agents require a supporting substrate; they cannot precipitate in the void centers. Cement in the centers of pores are younger than those adjacent to the grain surface (Fig. 8). In effect, this can be considered the "principle of superposition of cements." Thus, late stage cements cover or infill around earlier cements but neither replace them (exceptions to this are noted below) nor translate them by force of crystallization.

In some instances, clays have been observed both on the surface and within detrital grains, such as feldspar, volcanic rock fragments and carbonate clasts. It is assumed that such relatively unstable detrital grains often are coated, and partially dissolved and infilled during a single diagenetic stage.

(13) Some diagenetic clays line pores. Allogenic clay commonly is present at grain-to-grain contacts and may fill pores completely. In contrast, many authigenic clays form thin coatings on grain surfaces. Furthermore, such clays are absent at grain-to-grain contacts (Heald, 1950, p. 627), and there is no evidence that the clay was initially present and later pushed aside by compaction of the framework grains (Fig. 9A). This criterion is usually very reliable, although exceptions exist. Where clay growth is extensive, the grain coatings may extend outward, completely filling the pores. Much of the evidence for authigenic origin may be lost unless concentric banding or medial sutures (see criterion 4 and 20) are present. Detrital sand grains may rotate during compaction; this causes pore lining clays formed during an earlier stage of diagenesis to be shifted to points of grain-to-grain contact.

Clay coatings apparently coat all detrital grains regardless of their composition. This was demonstrated by Hayes (1970) who observed that chlorite grain coatings are present on nonsilicate grains such as fossil fragments.

In many sandstones studied by the authors, pore linings did not completely coat every grain. Secondary overgrowths develop only where such gaps are present (Zimmerle, 1963; Horn, 1965; Pittman, 1972; Heald and Larese, 1974).

(14) Authigenic clays occur as scattered pore fillings. Some authigenic clays —kaolinite and dickite in particular—generally form *irregularly* distributed clots from a fraction of a millimeter to several centimeters across (Fig. 9B). On the other hand, allogenic clays, which also fill pores, invariably are distributed *evenly* throughout their zone of occurrence (infiltration clays may be an exception) or are concentrated in thin zones paralleling the depositional lamination. One must take care to avoid confusing pore-filling authigenic clays and ductile detrital grains such as clay pellets or phyllite (Carrigy and Mellon, 1964). Dickinson (1970) introduced the term pseudomatrix for interstitial matrix formed by deformation of weak detrital grains. This pseudomatrix sometimes can be distinguished from authigenic clays in that the former consist of a variety of texturally or compositionally distinct matrix types. Additional criteria useful in recognizing pseudomatrix were listed by Dickinson (1970, p. 702).

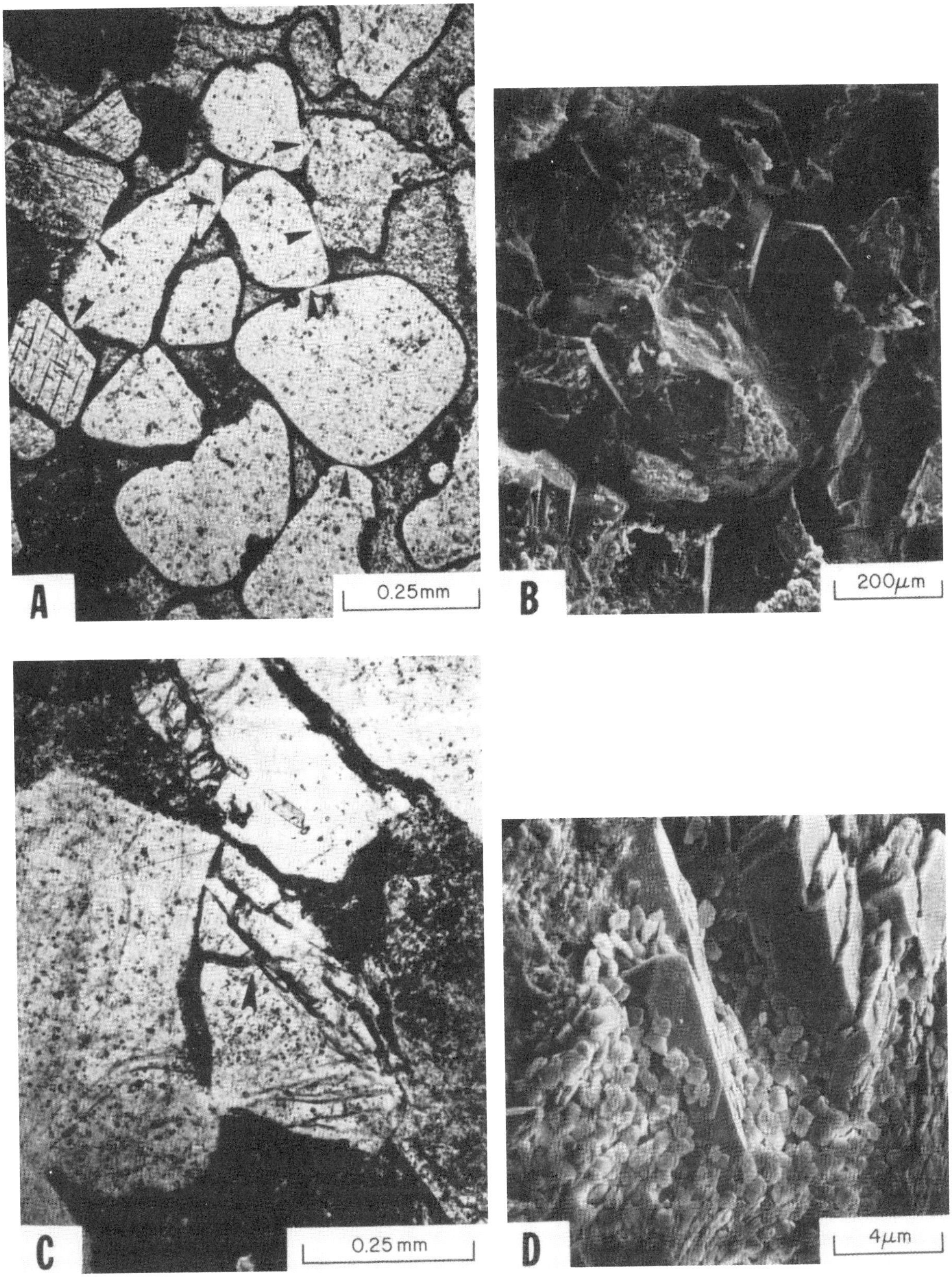

FIG. 9.—(A) Chlorite pore linings (dark rims) covering detrital grains but absent at grain contacts (arrows). Calcite fills the remaining pore space. Jurassic Morrison Formation, New Mexico. (B) Clusters of kaolinite booklets plugging scattered pores in a quartz arenite cemented at an earlier stage by well-developed quartz overgrowths. Cambrian Flathead Quartzite, Wyoming. (C) Chlorite present on detrital grain surfaces as well as along fractures within a grain (arrow). Unnamed Miocene unit, California. (D) Pseudohexagonal flakes and books of kaolinite partially covering feldspar overgrowths. The flakes and books tend to be aligned parallel to the overgrowth surfaces. Jurassic Morrison Formation, New Mexico.

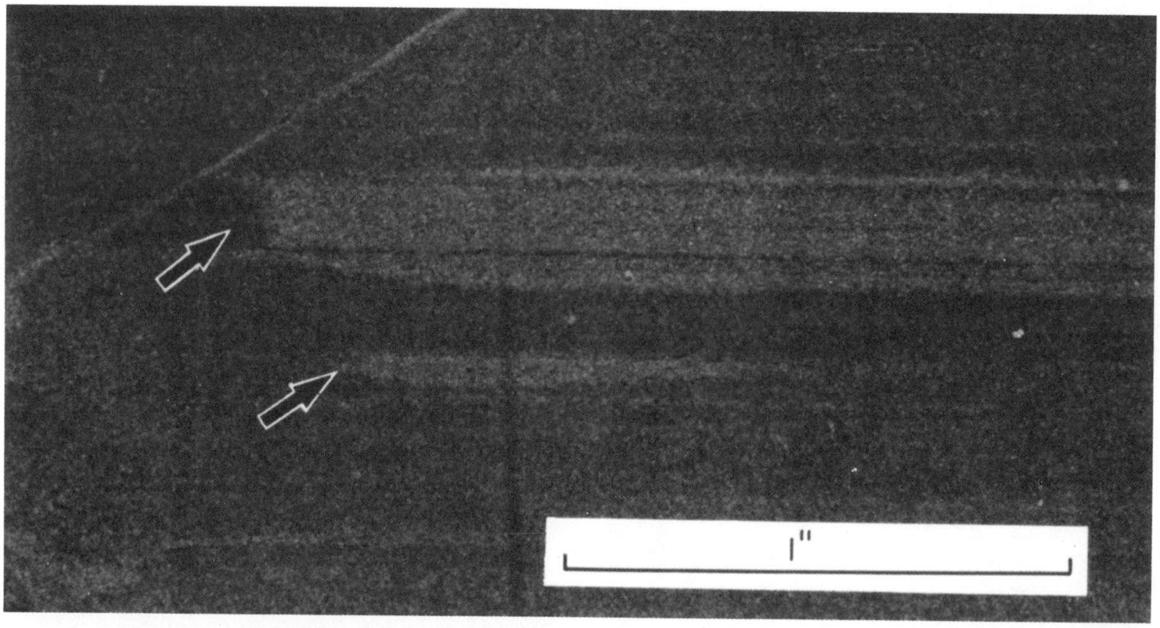

FIG. 10.—Abrupt lateral termination of zones containing pore-lining chlorite (arrows). Eocene Frio Sand, Texas. Photo courtesy of D. Horowitz.

(15) Clays present along fractures are authigenic (Pye, 1944, p. 74). Authigenic clays have been observed along fractures in individual grains as well as on the adjacent grain surfaces (Fig. 9C). This observation and the fact that the segments of the fractured grain commonly are optically discontinuous rule out prediagenetic replacement. A lack of fracturing in adjacent grains would support the interpretation that such limited fracturing is the result of early compaction. Kaolinite has been reported to occur along multigrain fracture surfaces by Shelton (1964).

(16) Authigenic clays are absent in those portions (i.e., concretions or laminae) of a sandstone thoroughly cemented during very early diagenesis, and present in the uncemented portions. Examples in which carbonate cements preclude the formation of authigenic clay are reported by Heald (1956), Brenchley (1969), and Galloway (1974). One might attribute the lack of clay in the cemented zones to a relatively late-stage replacement by the cementing material, but this interpretation is contradicted by the open grain packing of the cemented zones. This criterion is difficult to apply to surface samples inasmuch as it can be claimed that the tightly cemented zones are protected from the effects of surface weathering, which would introduce clay into a previously clay-free sandstone. However, if clay films of similar composition and structure are present in subsurface samples of the same unit, weathering is less likely to be the factor controlling clay development.

(17) Clays coating materials formed at an earlier stage of diagenesis are themselves diagenetic in origin (Heald, 1950, p. 628; Thompson, 1970, p. 605). Assuming that clays are not replaced or shoved aside by force of crystallization, clays covering other authigenic components must therefore be authigenic in origin. Examples of kaolinite covering quartz and feldspar overgrowths are common, and this feature is often apparent from both petrographic and SEM observation (Fig. 9D). It must be noted that invasion of a core by drilling mud commonly deposits clay and other particulate matter over diagenetic cements.

(18) Authigenic clays occur in layers with abrupt lateral terminations (Fig. 10) (Horowitz, oral communication). If the possibility that such clay zones are the result of small scale cross stratification or bioturbation is eliminated, an authigenic origin is the most reasonable explanation.

(19) Individual flakes of some authigenic clays are aligned radially with respect to detrital grain surfaces (c-axis aligned tangential to grain surface) (Dickinson, 1970, p. 702). Chlorite and smectite (and mixed-layer combinations of chlorite, smectite, and vermiculite?) are the only clays known to the authors that exhibit this arrangement (Fig. 11A). The radial arrangement indicates the clay particles grow from the grain surface outward into the pores. There is no obvious reason to expect recrystallized matrix to exhibit this characteristic. This arrangement could be destroyed by later compaction as observed by Carrigy (1971, p. 48).

(20) Some authigenic clays exhibit medial sutures (Dickinson, 1970, p. 702). Authigenic clays which line pores may grow outward from the detrital grain surfaces and intersect along sutures near the centers of the pores (Fig. 11B). The sutures indicate growth was from the edge of the pore inward and is not observed within pores filled by allogenic matrix.

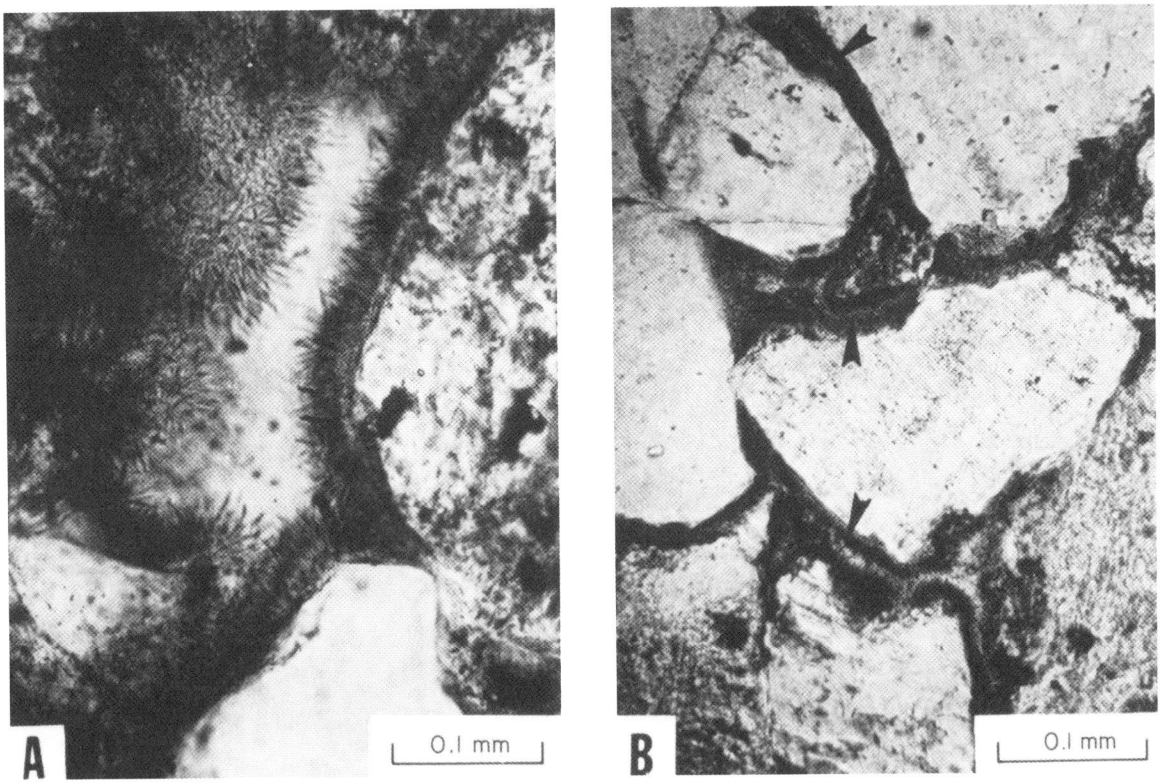

FIG. 11.—(A) Radial alignment of chlorite flakes on detrital grain surfaces suggesting development outward into the pores during diagenesis. Note the lack of any indication of deformation of clay adjacent to grain contacts. (B) Faint sutures (arrows) formed by the juncture of authigenic pore-lining clays growing inwards from opposing sides of pores. Miocene Bear Lake Formation, Alaska.

(21) Certain authigenic clays (smectite, mixed-layer smectite/illite, illite) may form bridges up to 50 μm long between detrital grain surfaces near points of grain contact (Fig. 12). There is no obvious way in which allogenic clays can be deposited in this arrangement.

CHARACTERISTICS OF THE COMMON FORMS OF CLAY

Allogenic Clay

Laminae of clay in sandstone are composed of individual clay particles that have a strong preferred orientation imposed by the platy nature of the particles. For example, when viewed across the stratification, the chlorite lamina in Figure 13A has an undulating appearance, swirling around and enclosing sand grains. The individual chlorite flakes, which have an irregular, rounded outline, can be seen better looking down on stratification (Fig. 13B).

Illite laminae are similar to chlorite in the packing of flakes. However, individual illite flakes have a ragged appearance with stubby projections dissimilar to the spine-like edges that develop in authigenic smectite/illite and illite.

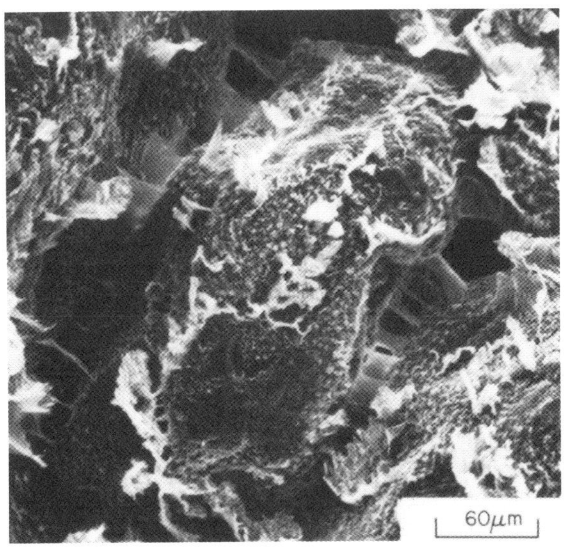

FIG. 12.—Smectite forming web-like pillars near contacts between detrital grains. Smooth areas on grain in center of photo are areas of contact with adjacent grains removed during sample preparation. Shredded ends of some large clay flakes are the result of breakage of the clay bridges by plucking out of detrital grains. Jurassic Morrison Formation, New Mexico.

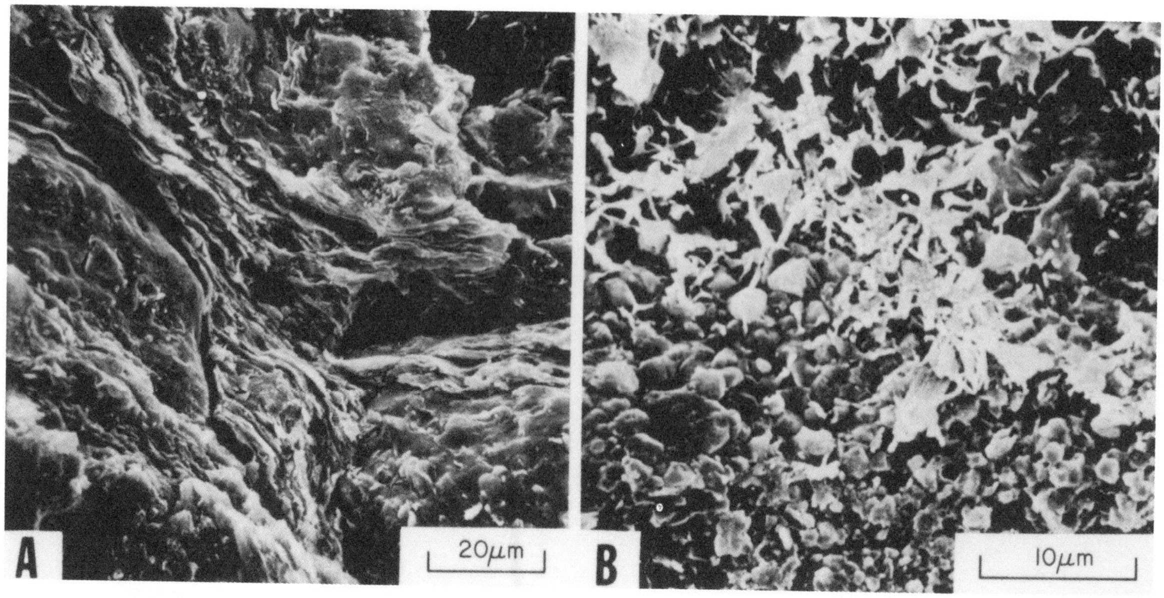

FIG. 13.—(A) Scanning electron micrograph showing detrital clay (chlorite) lamina in a sandstone. Platy clay particles impart a preferred fabric to the layer. Note how the lamina splits to flow around the sand grain in lower right of micrograph. Compare with Figure 13B which shows a different orientation. (B) Same sample as Figure 15A but looking down on the lamination so that the outlines of individual clay particles are visible. These chlorite detrital particles have an irregular but rounded outline. Where a void exists in the rock (upper center), authigenic illite has formed. Note the contrast between the crystal form and habits of the two clay minerals. Cretaceous Muddy Sand, Colorado.

House-of-cards structures, typical of clay floccules, do not appear to be preserved in ancient sandstones and shales, presumably because of reorientation of clay platelets with overburden pressure.

In thin section, clay laminae commonly are a very dark color, contain organic contaminates, and are homogeneously mixed with silt.

Authigenic Clay

Kaolinite and Dickite

Kaolinite and dickite are the most commonly recognized authigenic clays in sandstones (e.g., Fothergill, 1955, p. 305; Karpova, 1969, p. 7) because they commonly are coarse enough to be readily visible in thin section.

We have been unable to differentiate between kaolinite and dickite on the basis of scanning electron microscopy. They are readily distinguished by X-ray diffraction. Authigenic kaolinite-dickite always occurs as pseudohexagonal plates, most commonly as pore fillings, but also as pore linings. The characteristics of the major types of authigenic clays are summarized in Table 4.

Stacked Plates (books)—The most common habit of kaolinite-dickite is a face-to-face (c-axis) stacking of pseudohexagonal plates (Fig. 14A). Individual plates generally range from 3 to 20 μm in diameter. In most cases kaolinite fills only scattered pores in sandstones (Fig. 9B), but occasionally all pores are plugged. In all situations the clay fabric is identical: stacks of pseudohexagonal clay platelets randomly arranged with respect to each other. Grooves, or notches, in kaolinite flakes and books may be indications of twinning as suggested by Mansfield and Bailey (1972). Such indentations are less common in dickite which always has an extremely well-crystallized appearance. Sarkisyan (1972, p. 4) illustrated and classified similar appearing kaolinite as allogenic. However, it is difficult to explain the crystalline morphology by an allogenic origin; thus we prefer an authigenic origin.

Vermicular—A less common occurrence of kaolinite is a delicate vermicular growth, a sequence of stacked pseudohexagonal plates that may extend the length of a pore (Fig. 14B). This form of authigenic kaolinite is typically associated with book-type morphologies.

Overlapping Plates—Rarely, kaolinite occurs as overlapping pseudohexagonal plates coating sand-sized detrital grains (Fig. 9D). These plates are less than 1 μm thick and range from 2 to 10 μm in diameter. Often the detrital grain is completely covered by kaolinite

Chlorite

This mineral is the most variable in form among the authigenic clays. It has been reported as occurring in plates attached to detrital sand grains (Pittman and Lumsden, 1968; Hayes, 1970), but also

TABLE 4.—*Characteristics of Authigenic Clays*

	Morphology of Individual Flakes	Form of Aggregates	Relationship to Sand Size Detrital Grains	Thickness of Coating or Long Dimension of Aggregates (Microns)	Special Features
Kaolinite and Dickite	pseudohexagonal	stacked plates (book)	pore filling	2–2500 (generally 2–20)	flakes notched or embayed (twinned?)
	pseudohexagonal	vermicule	pore filling	10–2500 (generally 20–200)	flakes notched or embayed (twinned?)
	pseudohexagonal	sheet	pore filling	0.1–1	flakes notched or embayed (twinned?)
Chlorite	pseudohexagonal	plates (2-dimensional card-house)	pore lining	2–10	
	curled equidimensional with rounded edges	honeycomb	pore lining	2–10	
	equidimensional with angular or lobate edges	rosette or fan	pore lining and pore filling	4–150 (generally 4–20)	
	fan-shaped fibrous bundles	cabbagehead	pore lining and pore filling	8–40	
Illite	irregular with elongate spines	sheet	pore lining	0.1–10	bridging between sand grains
Smectite	not recognizable	wrinkled sheet or honeycomb	pore lining	2–12	bridging between sand grains
Mixed-layer smectite/illite	subequant with stubby spines	imbricate sheet to ragged honeycomb	pore lining	2–12	bridging between sand grains

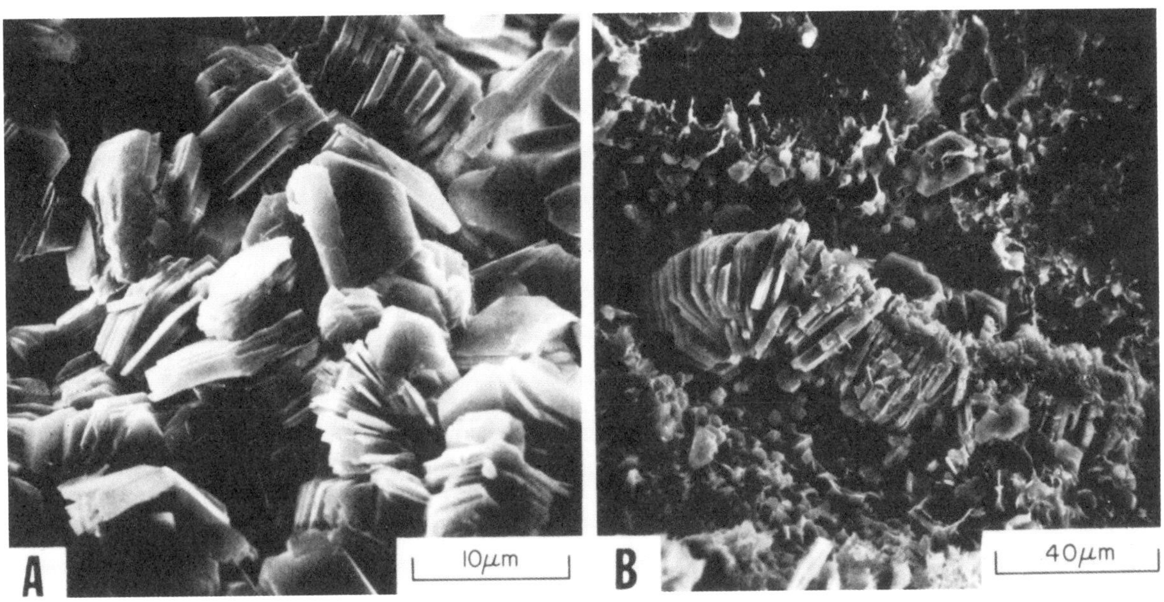

FIG. 14.—(A) Stacked plates of kaolinite in porous sandstone. Note the face-to-face arrangement and pseudohexagonal outlines of individual plates. Eocene Frio Sand, Texas. (B) Vermicular authigenic kaolinite in porous sandstone. The kaolinite also occurs as individual isolated plates. Sand grains are coated with illite which apparently formed prior to the kaolinite. Permian Rotliegendes Sandstone, North Sea.

occurs as rosettes, honeycombs, or cabbagehead-like growth.

In thin section, chlorite pore linings may be obvious if well developed (Fig. 11A). If the lining is relatively thick, it may appear either pale green or essentially colorless. Small crystals are often difficult to detect because of their low relief and birefringence.

Nondispersive detector analyses on the four growth habits of chlorite show the following relative concentrations of iron: plates and rosettes > honeycomb > cabbagehead.

Plates—Chlorite most commonly develops as individual idiomorphic crystals usually attached on edge to the surface of sand grains (Fig. 15A). The crystals appear attached to sand grains along their longest dimension, which ranges from approximately 2 to 10 μm. Often it is difficult to determine the size of the shortest dimension because of the dense packing and orientation of the plates.

These crystals appear to have a face-to-edge orientation with respect to each other (two dimensional cardhouse arrangement). This is similar to the structure of some clay floccules (Van Olphen, 1963); however, we believe these crystals have an authigenic origin.

Honeycomb—This growth also consists of plates, but they are arranged in a cellular pattern analogous to a honeycomb. The crystals are attached on edge to the detrital sand grains, and commonly curve and intersect to form the distinctive cellular pattern seen in Figures 15B and 15C.

Rosettes—Authigenic chlorite also forms rosette and fan-shaped clusters, which most commonly develop as pore linings but also occur as pore fillings (Fig. 15D). The crystals usually have lobate edges. This aids in distinguishing them from kaolinite, which they may resemble superficially. Individual rosettes or fans generally are 5-20 μm in diameter but may reach 150 μm. Individual crystals typically range from 4 to 18 μm in diameter.

Pores of a sandstone can be packed with rosette chlorite, which causes a drastic reduction in permeability, although the porosity may still be appreciable.

Cabbagehead—The rarest growth habit of authigenic chlorite is the cabbagehead. At low magnifications they appear to be small, equidimensional grains usually attached to sand-sized grains (Fig. 15E). At higher magnifications, concentric ribbon- or sheet-like layers of chlorite with complex orientations are visible (Fig. 15F). Individual structures range from approximately 8 to 40 μm. Most commonly, the cabbagehead structures form pore linings, but occasionally they occur as pore fillings.

Illite

In thin section, illite, because of a higher birefringence, is more easily detected than kaolinite-dickite or chlorite. Pore linings are readily visible using crossed polarizers (Fig. 16A). Illite generally occurs as irregular flakes with lath-like projections, the most delicate growth habit of all the authigenic clay minerals. This growth form varies in appearance depending on the development of lath-like projections. The authigenic growths, attached to the surface of sand grains, develop as sheets that curl away from the point of attachment. Edges of the sheets project lath-like appendages that commonly are relatively short, as shown in Figure 16B. This growth form can be confused with mixed-layer illite/smectite. Occasionally, the sheets of illite may develop relatively long, delicate-appearing, lath-like projections, which are commonly curled and may be up to 30 μm long and range from 0.5 to 2 μm in width (Fig. 5B). Commonly, these lath-like growths bridge the gap between adjacent detrital sand grains.

Smectite

Authigenic smectites have two basic growth habits: a crinkly coating on detrital sand grains (Fig. 17A) and a cellular structure (Fig. 17B) similar to honeycomb chlorite (Fig. 15C). Thickness of the clay layer ranges from 2 to 12 μm. Thick layers appear to be more deeply crenulated than thin layers. Presumably the clay coatings would be less crenulated, or crinkly, in a hydrated state. Individual smectite crystals cannot be resolved under the SEM. Smectite may bridge the gap between adjacent detrital sand grains, as shown in Figure 12.

Mixed Layer Illite/Smectite

Mixed-layer clays have characteristics similar to participating clay minerals. If illite-rich, the structure may be essentially indistinguishable from illite and takes the form of sheets with short lath-like digitate edges (Fig. 18). A smectite-rich mixed-

→

FIG. 15.—(A) Chlorite coating sand grain as seen with the SEM. Individual plates are attached on edge to the detrital sand grain. Pennsylvanian Spiro Sand, Oklahoma. (B) Honeycomb growth form of chlorite as coating on sand grains. Dark spots on grains represent points of contact between adjacent grains, some of which were removed during sample preparation. See Figure 15C for an enlargement of chlorite. (C) Honeycomb chlorite that has formed as a coating on a sand grain. Individual chlorite crystals curve and intersect to create the honeycomb appearance. Jurassic Norphlet Sand, Florida. (D) Chlorite in the form of rosettes occurring as a pore filling. Note the lobate edges which contrast to the pseudohexagonal outline of kaolinite that it superficially resembles. Miocene Discorbis Sand, Louisiana. (E) Low magnification view showing chlorite in the form of small cabbagehead structures attached to sand-sized detrital grains. See Figure 15F for a magnified view of cabbageheads. (F) Cabbagehead form of chlorite. Note the individual layers of clay particles that comprise the structure. Eocene Frio Sand, Texas.

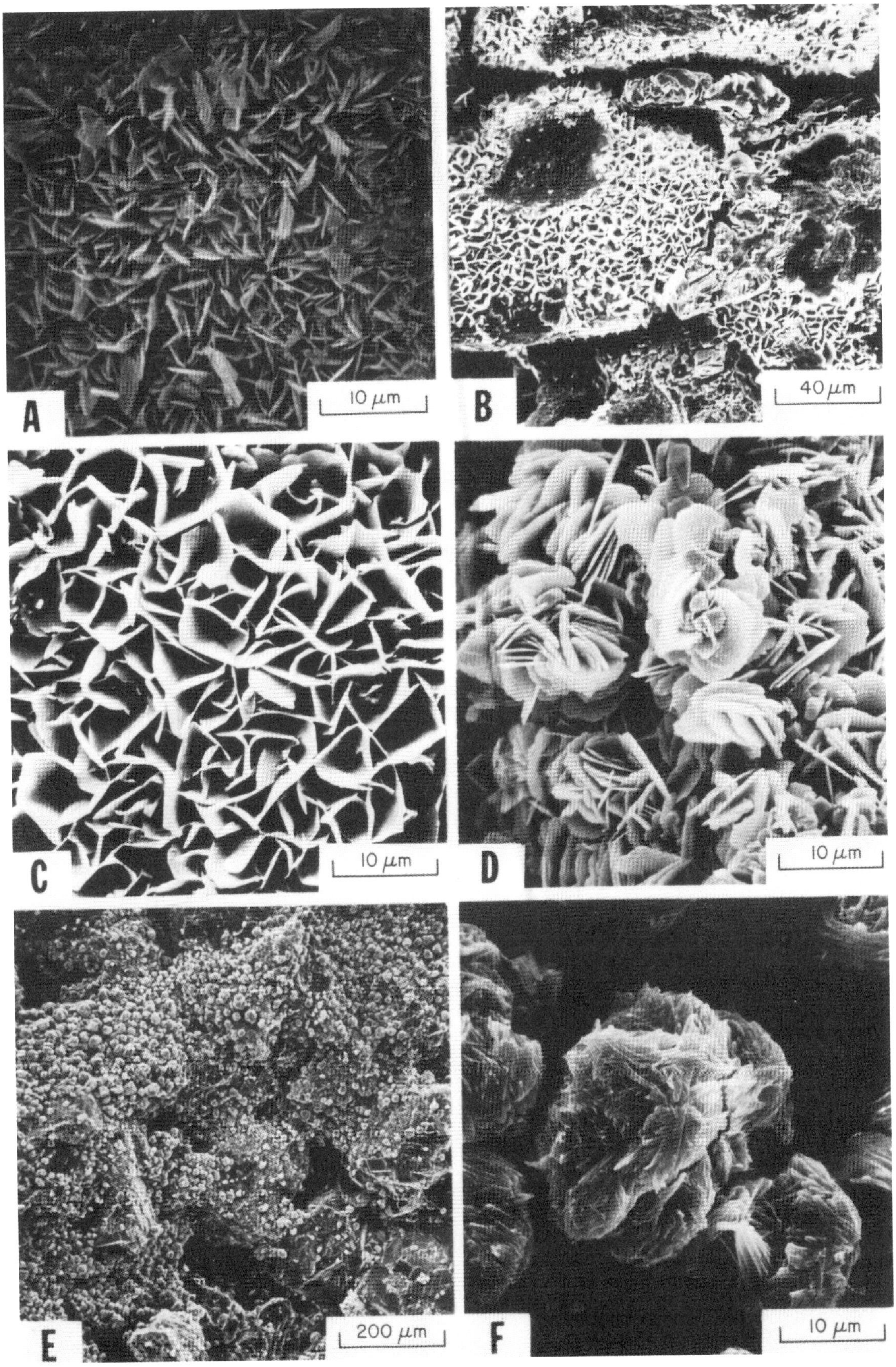
A
10 μm
B
40 μm
C
10 μm
D
10 μm
E
200 μm
F
10 μm

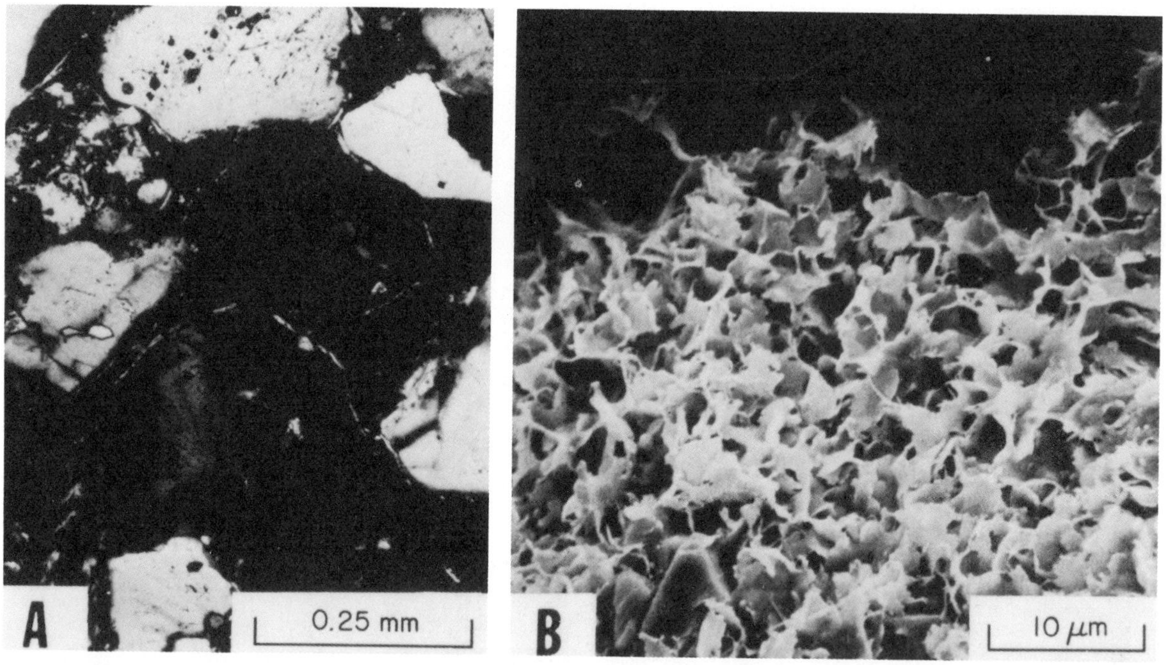

FIG. 16.—(A) Photomicrograph showing illite as a high birefringent coating on sand grains. Permian Rotliegendes Sand, North Sea. (B) Sand grains coated by authigenic illite with short lath-like projections. Quartz overgrowths without a coating of illite occur in dark area at top of micrograph. Cretaceous Muddy Sandstone, Colorado.

layer clay may resemble the crinkly form of smectite and differ only by the presence of spiny projections. Distinguishing illite and smectite from mixed-layer illite/smectite based just on scanning electron microscopy can be misleading.

Glauconite

Based on the study of a small number of samples, this mineral exhibits a wide range of morphologies generally resembling smectite, mixed-layer smectite/illite, or illite. Our present limited understanding of this morphological diversity prevents detailed discussion at the present time.

MINERALS OR MINERALOIDS WITH CLAY-LIKE MORPHOLOGY

Some minerals and mineraloids are known to exhibit morphologies in scanning electron micrographs that resemble one or more clay minerals. Listed below are those observed by the authors. Identification of clay types on the basis of morphology alone is dangerous; other minerals or mineraloids, as yet unrecognized, also may exhibit strong similarities to diagenetic clay minerals.

Cliachite

Seen in thin section, this amorphous aluminum hydroxide forms brownish concentric pisolitic or botryoidal coatings up to 100 μm in diameter on detrital grains. SEM examination shows spherulitic, or mound-like masses, of curled flaky particles (Flesch and Wilson, 1974). The overall appearance is somewhat like that of a marigold (Fig. 19A). If identification were based on morphology alone, cliachite possibly could be mistaken for chlorite or smectite. Spectral analysis of cliachite utilizing an X-ray energy spectrometer or a microprobe will indicate the absence of silica and thus prevent confusion with clay minerals. Nondispersive spectral analysis indicates that the cliachite in Figure 22A contains only aluminum and a minor amount of manganese (elements of atomic number below that of sodium, 12, are not resolved). The cliachite in this example is associated with both gibbsite and kaolinite, which are the only minerals represented by peaks on an X-ray diffractogram of $<$ 2 μm material.

Cristobalite

Generation of cristobalite by alteration of biogenic opaline tests has been recognized in a variety of cherts, chalks and siliceous clays (Wise and Kelts, 1972; Wise *et al.*, 1972; Wise and Weaver, 1973; Lancelot, 1973; Oehler, 1975). Cristobalite (Fig. 19B) forms spherulitic or hemispherical clusters 3 to 12 μm in diameter. The individual particles are relatively flat and have a regular edge to face arrangement. Cristobalite could be mistaken for

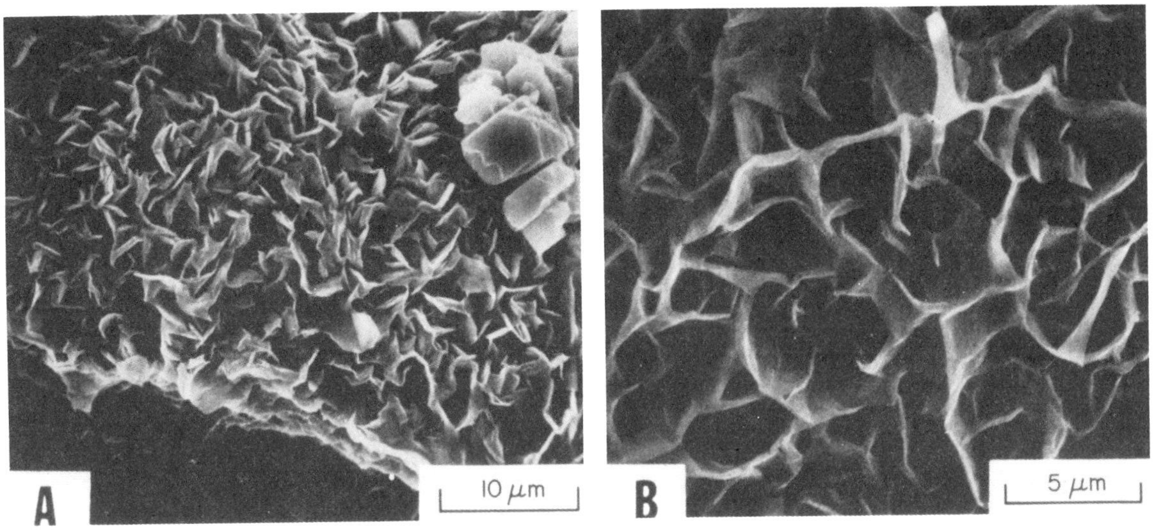

FIG. 17.—(A) Authigenic smectite occurring as a grain coating. The highly crenulated surface may be the result of the dehydrated state. Pseudohexagonal plates of kaolinite overlie the smectite near the upper right margin of the micrograph. Cretaceous Second Wall Creek Sandstone, Wyoming. (B) Authigenic smectite with a honeycomb-like appearance that superficially resembles the honeycomb chlorite shown in Figure 15C. However, the chlorite structure is composed of individual crystals whereas individual crystals of the smectite cannot be resolved. Jurassic Norphlet Formation, Alabama.

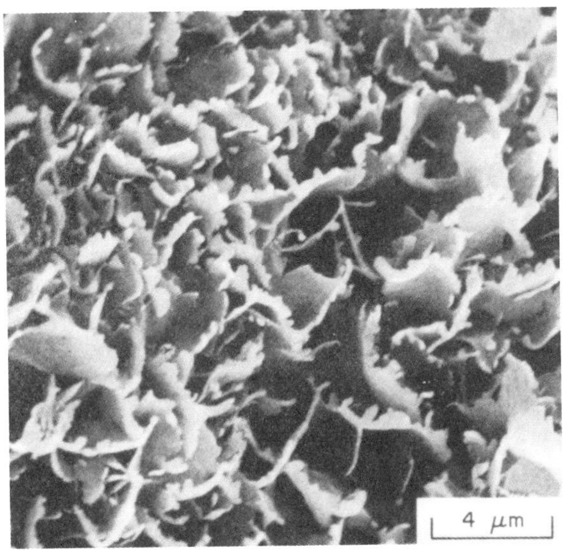

FIG. 18.—Authigenic mixed-layer smectite/illite. Mixed-layer clay minerals appear to take on the characteristics of the participating minerals. This example resembles the illite shown in Figure 16B. Distinguishing mixed-layer smectite/illite from illite or smectite may be unreliable if only the SEM is used. Cretaceous Mesaverde Group, Colorado.

rosette chlorite aggregates (Fig. 15D). These can be differentiated with ease if a spectral analysis can be made by nondispersive analyzer attachment to the SEM or by microprobe. Because of their lack of biogenic opaline tests, siliciclastic sandstones are unlikely to contain cristobalite.

Iron Oxide

Scattered small 15 to 30 μm grain coating reddish ovoid grains were observed in thin sections of Aztec sandstone (Jurassic) from the Muddy Mountains of southern Nevada. These particles are too small and of insufficient quantity for routine chemical or X-ray diffraction analysis. It is most likely this material is iron oxide. SEM examination indicates the particles in question are rounded packets of flakes, which appear to be slightly warped (Fig. 19C). This morphology resembles kaolinite which is also present in these samples. However, kaolinite books (and dickite) exhibit a distinct pseudohexagonal outline, and individual flakes are not distorted. Also spectral analysis indicates that the unknown grains contain only iron (deep penetration by X-ray energy spectrometer beam was used to insure that the material analyzed was not a thin surface coating on a particle of entirely different composition).

EFFECTS OF AUTHIGENIC CLAY ON SEDIMENTOLOGIC AND PETROGRAPHIC INTERPRETATIONS AND RESERVOIR PROPERTIES

The presence of even minor amounts of authigenic clay in sandstones may have a profound effect on many important properties of sandstones.

Effects Gas Exploration and Exploitation

Authigenic clays control petrophysical properties relating to reservoir quality, particularly per-

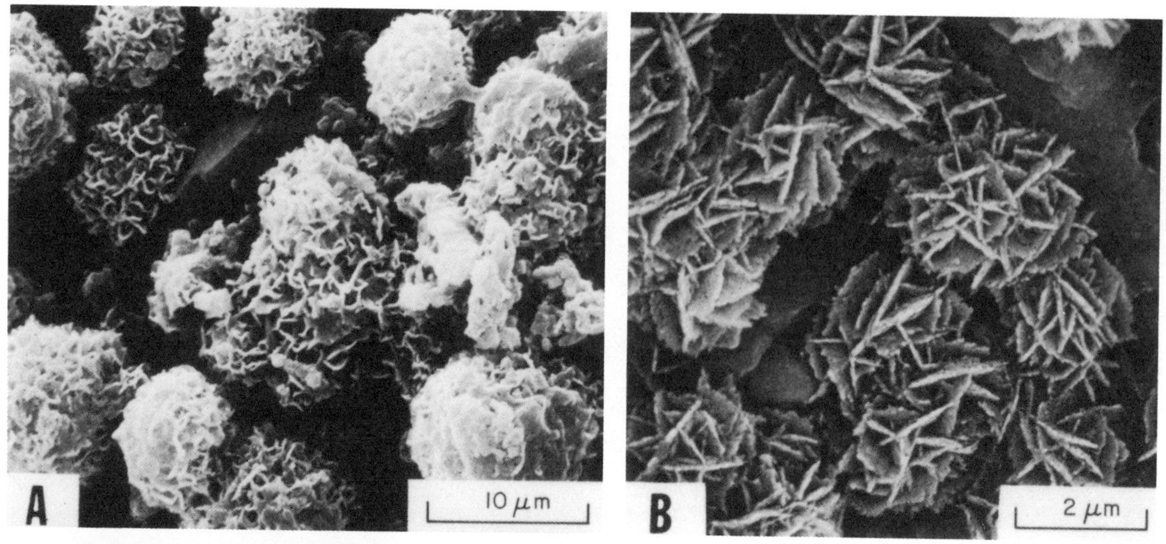

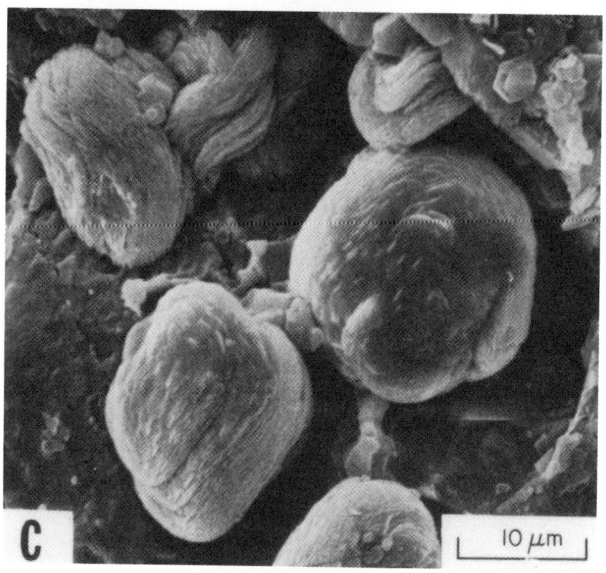

FIG. 19.—(A) Irregular spheroidal aggregates of curled flaky material identified as cliachite by optical and spectrometer analyses. The morphology of these cliachite grains can be easily mistaken for pore-lining smectite or chlorite. Jurassic Morrison Formation, New Mexico. (B) Rosettes of diagenetic cristobalite formed of flat individual plates with ragged edges. These spheroidal aggregates strongly resemble chlorite rosettes but flakes have serrate edges not observed on chlorite flakes. Cretaceous chalk, North Sea. (C) Rounded flaky aggregates typical of iron oxide(?) grains. Also present are pseudohexagonal books of kaolinite which often appear to partially cover the iron oxide(?) grains. The rounded outline and distortion of the flakes distinguishes the iron oxide(?) grains from kaolinite books. Jurassic Aztec Sandstone, Nevada.

meability and water saturation. A sporadic occurrence of authigenic clays in pores (Fig. 9B) certainly has less influence on permeability than clay that lines pores (Figs. 15A, 16B, and 5B) and restricts pore throats. Stalder (1973) showed that shifts in trend of the porosity-permeability relationship for the Rotliegendes Sandstone can be related to variations in authigenic clays.

Pore size and shape are modified by clay minerals, either occurring as infilling of pores, grain coatings reducing pore apertures, or complex combinations of these. For example, the Frio Sandstone (Eocene, Willamar West Field, Texas) has only authigenic kaolinte that sporadically plugs pores whereas the Morrow Formation (Pennsylvanian) Carlsbad Field, New Mexico contains both pore-lining chlorite and pore-filling kaolinte.

Micropores occurring among clay particles are capable of holding irreducible water, that is, water tightly held because of the physical attraction of the solid for the liquid. Some authigenic clays have more micropores and a greater surface area than others. Data from Gaida *et al.* (1973) show that low permeability and high surface area arc associated with illite having lath-like projections. A reservoir containing abundant authigenic clays with associated micropores (Fig. 5B) would be expected to have a long transitional oil-water or gas-water contact because of the absorbed water associated with high surface area.

Certain authigenic clays may create problems during drilling, stimulation, and production of hydrocarbons as a result of their ability to block pores by expansion or migration, or to trap drilling mud particulate matter.

Paleoenvironmental Analysis

The textural properties of sandstones (cumulative curve shape, moment measures of size, sorting, skewness, etc.) are often used as paleoenvironmental indicators. If authigenic clays are mistaken for allogenic clays, the resulting paleoenvironmental interpretations may be grossly incorrect.

Syndepositional clay matrix material has generally been considered to typify fluvial and turbidite environments and to be absent or minimal in environments where energy expenditure is sufficiently high (dune, beach, tidal channel) (Pettijohn *et al.*, 1972, Chap. 11; Blatt *et al.*, 1972, p. 312). Davies and Ethridge (1974) present evidence suggesting matrix content is one of the most potent petrographic criteria in the discrimination of various fluviodeltaic and marine subenvironments in Holocene sands and ancient sandstones.

Misidentification of authigenic clay is most likely in ancient turbidite sandstones, particularly graywackes, where pore sorting of the sand- and silt-sized debris suggests the clay fraction is primary, and subsequent modification by tectonism and burial or low grade regional metamorphism partially obliterates initial composition and texture.

Perhaps less commonly recognized is the fact that a high content of matrix clay may occur locally in beach sands (Friedman, 1967) and may be low or absent in many fluvial sands. A study by Moss (1972) indicated that interstitial silt and clay content is very low or lacking in Holocene river sands and flume sands deposited in the "coarse" ripple, dune, and transitional flow stages. Exceptions are thin, silty, and clayey sand laminae intercalated with clean sands as a result of fluctuations in sediment load or by concentration in the troughs of dune bed forms. Sands deposited in these three flow stages form the bulk of many ancient fluvial sandstones. Clay and silt content can be relatively high in sands deposited in the other flow stages (rheological, referred to as plane bed phase of Simons *et al.*, 1961, and fine ripple) investigated by Moss.

Our petrographic observations indicate most fluvial channel sandstones as well as most turbidite, beach, dune, and tidal sands contain little, if any, detrital clay (excluding bioturbation).

On the other hand, sandstones deposited in any depositional environment may contain significant amounts of authigenic clay, regardless of compositional maturity. For example, authigenic clay is locally a prominent component of such highly quartzose sandstones as the Pennsylvanian Morrow of southeastern New Mexico, and Lower Cretaceous Hosston Formation of southern Mississippi, which lack any trace of possible precursor detrital grains. These sandstones provide conclusive evidence that the presence of authigenic clays is not necessarily correlated to the relative abundance of unstable detrital grains. The dissolved reactants must have been derived predominantly from sources external to the sandstone body.

Though the abundance of allogenic clay may be a useful paleoenvironmental indicator, the distinction between detrital and authigenic clay may be difficult to make.

Interpretations of paleoenvironments based on sieve and settling tube grain size analyses may be seriously incorrect if diagenetic or disaggregation processes significantly alter the textural properties of a sandstone (Wilson and Tillman, 1974; Mousinho De Meis and Amador, 1974).

The effect of authigenic clay on measured sandstone textures is well illustrated by a sample from the Aztec Sandstone in the northern Muddy Mountains of southern Nevada. The depositional environment has been interpreted as an eolian dune based on the presence of a thick, massive sequence of large scale (up to 25 to 30′ thick) trough crossbeds, lack of trace or body fossils, and absence of intercalations of other lithologies. In thin section, authigenic clay with a high birefringence occurs as thin (2 to 5 μm), iron oxide stained smectite/illite pore linings, along with scattered patches of colorless, pore-filling kaolinite books with very low birefringence. The less than 2 μm fraction is 5.7% by weight of the whole rock. Minor silt-sized hematite(?) and partially dissolved feldspar grains are also present. Other than a very few small quartz overgrowths, no cementing agents other than clay were observed.

After disaggregation, the sample was sieved at $1/4\phi$ intervals. The cumulative curve and summary statistics for this sandstone are shown in Figure 20. Prior to correction for the presence of post-depositional silt and clay, the cumulative curve has a distinct break at 3.5ϕ and the moment measures indicated the sample was moderately sorted, highly skewed, and had a high peakedness. The size of the finest detrital material was determined by thin section study and all sieve fractions finer than this (less than 0.0625 mm or greater than 4ϕ) were subsequently deleted from the original size distribution. The cumulative curve resulting from this operation is devoid of any obvious breaks (Fig. 20). Sorting has improved significantly (from moderately to very well-sorted) and skewness and kurtosis are reduced to very low values. Clearly the presence of authigenic clay had a profound effect on grain size distribution and summary statistics of this sandstone. Similar results and effects can be expected in many other sandstones. Failure to recognize the importance of diagenetic effects on sandstone grain-size properties can lead to serious errors, e.g., in paleoenvironmental interpretation. The high silt and clay content of the Aztec Sandstone sample hardly suggests an eolian origin. Once the diagenetic materials are discounted, the textural properties of the sample are compatible with an eolian interpretation.

Sandstone Composition and Interpretation of Provenance

Diagenetic processes can cause significant changes in sandstone composition. The most common diagenetic alterations are complete or partial dissolution or replacement of chemically unstable

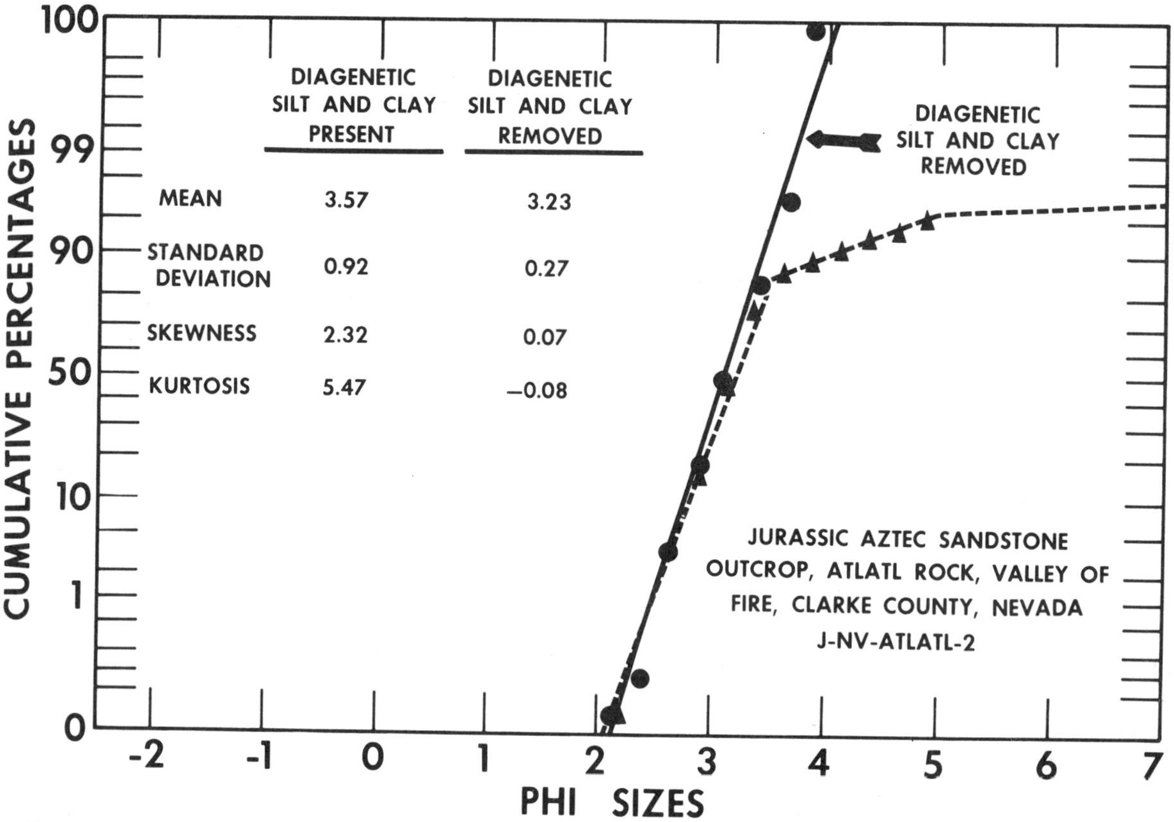

FIG. 20.—Comparison of grain size cumulative curves and moment measures of a sandstone before and after removal of authigenic clay and silt. The size of the smallest detrital particles present is estimated from petrographic examination.

detrital grains and introduction of pore-filling cements. Most cementing agents are easily recognized, and their presence can be ignored in estimating original detrital composition. However, when dissolution or replacement of unstable grains nears completion, it becomes very difficult to establish the nature of the altered detrital grains or even recognize that alterations have occurred.

The effect of diagenetic destruction of feldspar on sandstone composition and thus on interpreted source rock types and areas is illustrated with an example from the Norphlet Formation of southern Mississippi. Plagioclase grains, suggestive of feldspar outlines rather completely altered to illite while potassium feldspars are unaffected. The evidence that plagioclase was once present is (1) alignment of illite flakes along what appears to be cleavage traces; (2) sharp-angled corners to many illitic grains, suggestive of feldspar outlines rather than the irregular or rounded outlines of ductile clay pellets; (3) occurrence of quartz-potassium feldspar-illite aggregates looking very much like "granitic" rock fragments; and (4) presence of partially illitized plagioclase in presumably equivalent sandstones in an adjacent well. The illitic grains are likely to be misinterpreted as clasts of mica schist, shale, or penecontemporaneous clay. If included as lithic fragments, these sandstones plot as lithic arenites (Fig. 21). However, if the illitic grains are correctly identified as altered plagioclase, these sandstones plot as feldspathic arenites. Other examples of increased mineralogic maturity resulting from alteration of unstable detrital grains to clay include Moench and Schlee (1967), Walker (1973), Wilson and Tillman (1974), and Flesch and Wilson (1974).

Misidentification of the altered plagioclase might lead one to interpret the source rock types as being different from those supplying debris to adjacent areas. Alternatively, one might relate the absence of plagioclase to a spatial or temporal variation in the climate or the degree of weathering as controlled by paleogeographic conditions.

ACKNOWLEDGMENTS

The senior author thanks Cities Service and Exxon Oil Companies for permission to publish this article and is especially grateful to Dr. Frank Wantland (Cities Service Oil) for his encouragement in pursuing the study of diagenetic clays. The junior author thanks Amoco Production Company for release of information contained in Geologic Research Report F72-G-28. Mr. Allen White and Mr. William Fugate (Cities Service Oil) and Mr. Jack Nash (Amoco Production Co.) assisted with

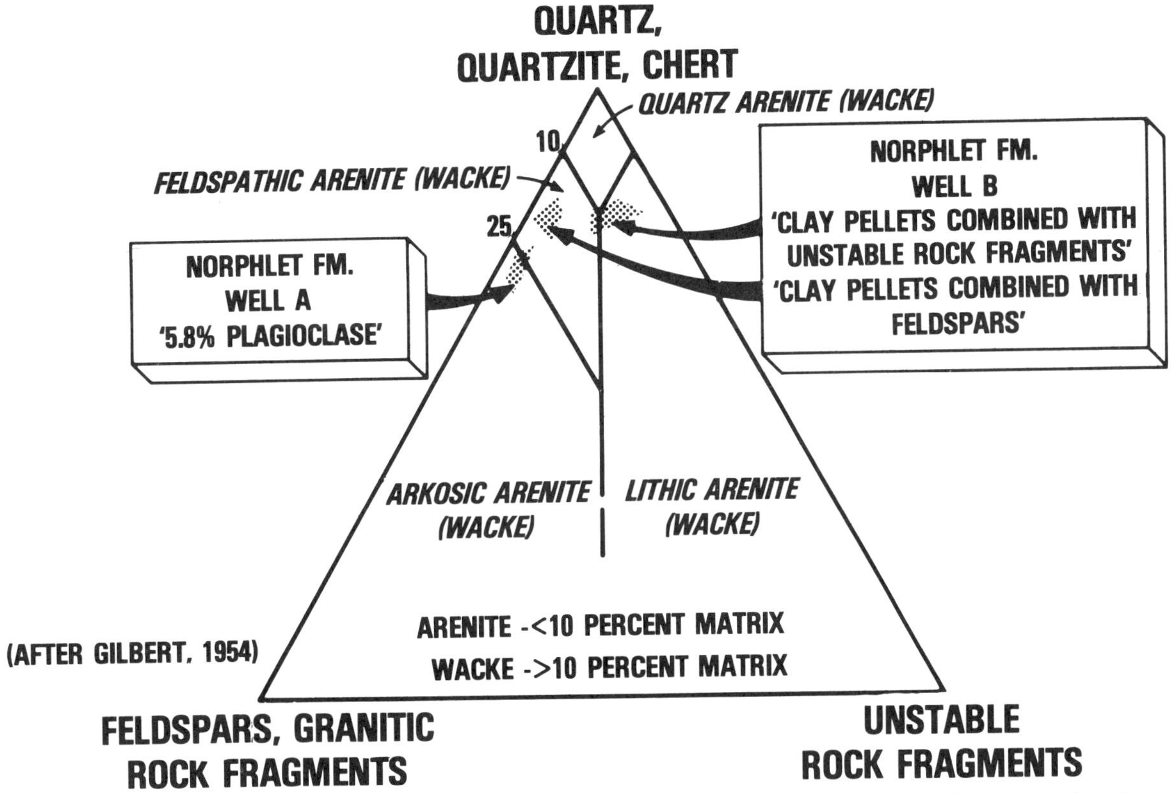

FIG. 21.—Shift in composition of a series of Jurassic Norphlet sandstones (Well A) resulting from misidentification of illitized plagioclase. Partially illitized plagioclase grains in an adjacent well (Well B) provide a guide to the proper identification of the completely illitized grains.

scanning electron microscope and X-ray diffraction analyses. Drs. Jack Hayes, John Whetten, Wayne Pryor, and Charles Mansfield read the manuscript and contributed many useful comments.

REFERENCES

Audley-Charles, M.G, 1967, Greywackes with a primary matrix from the Viqueque Formation (Upper Miocene-Pliocene), Timor: Jour. Sed. Petrology, v. 37, p. 5-11.

——— , 1968, Greywackes with a primary matrix from the Viqueque Formation (Upper Miocene-Pliocene), Timor: Reply to discussion by Riyadh A. Rahmani: Jour. Sed. Petrology, v. 38, p. 273.

Bayliss, P., F.C. Loughman, and J.C. Standard, 1965, Dicktite in the Hawkesbury Sandstone of the Sydney Basin, Australia: Am. Mineral., v. 50, p. 418-426.

Blatt, H., G. Middleton, and R. Murray, 1972, Origin of sedimentary rocks: Prentice-Hall, Englewood Cliffs, N.J., 634 p.

Borst, R.L., and W.D. Keller, 1969, Scanning electron micrographs of API reference clay minerals and other selected samples: Proc. Intl. Clay Conf., v. 1, p. 871-901.

Brenchley, P.J., 1969, Origin of matrix in Ordovician graywackes, Berwyn Hills, North Wales: Jour. Sed. Petrology, v. 39, p. 1297-1301.

Brown, B.E., and S.W. Bailey, 1962, Chlorite polytypism: I. Regular and semi-random one-layer structures: Amer. Min., v. 47, p. 819-850.

Buller, A.T., and J. McManus, 1973, Modes of turbidite deposition deduced from grain-size analyses: Geol. Mag., v. 109, p. 491-500.

Carrigy, M.A., 1971, Lithostratigraphy of the uppermost Cretaceous (Lance) and Paleocene strata of the Alberta Plains: Res. Council Alberta Bull. 27, 153 p.

——— , and G.B. Mellon, 1964, Authigenic clay mineral cements in Cretaceous and Tertiary sandstones of Alberta: Jour. Sed. Petrology, v. 34, p. 461-472.

Crone, A.J., 1974, Experimental studies of mechanically-infiltrated clay matrix in sand: Geol. Soc. America, Abs. with Programs, v. 6, no. 7, p. 701.

Cummins, W.A., 1962, The greywacke problem: Liverpool Manchester Geol. Jour., v. 3, p. 51-71.

Davies, D.K., and F.G. Ethridge, 1975, Sandstone composition and depositional environment: Amer. Assoc. Petroleum Geol. Bull., v. 59, p. 239-264.

Dickinson, W.R., 1970, Interpreting detrital modes of greywacke and arkose: Jour. Sed. Petrology, v. 40, p. 695-707.

Divis, A.F., and Judith McKenzie, 1975, Experimental authigenesis of phyllosilicates from feldspathic sands: Sedimentology, v. 22, p. 147-155.

Flesch, G., and M.D. Wilson, 1974, Petrography of Morrison Formation (Jurassic) sandstone of the Ojito Springs Quadrangle, Sandoval Co., New Mexico. Twenty-fifth New Mexico Geol. Soc. Field Conf. Guidebook, Central-Northern New Mexico, p. 197-210.

Folk, R.L., 1968, Petrology of sedimentary rocks: Hemphill's Austin, Texas, 170 p.

——— , P.B. Andrews, D.W. Lewis, 1970, Detrital sedimentary rock classification and nomenclature for use in New Zealand: New Zealand Jour. of Geology and Geophysics, v. 13, p. 937-968.

Fothergill, C.A., 1955, The cementation of oil reservoir sands and its origin: 4th World Petroleum Cong., Sec. 1/B, paper 1, p. 301-314.

Frey, R.W., and J.D. Howard, 1972, Georgia Coastal Region, Sapelo Island, U.S.A., Sedimentology and Biology. VI. Radiographic study of sedimentary structures made by beach and offshore animals in aquaia: Senckenbergiana Maritima, v. 4, p. 169-182.

Friedman, G.M., 1967, Dynamic processes and statistical

parameters compared for size frequency distributions of beach and river sands: Jour. Sed. Petrology, v. 37, p. 327-354.
Füchtbauer, H., and H. Goldschmidt, 1963, Beobachtungen zur Tonmineral Diagenese: Proc. Inter. Clay Conf. (Stockholm), p. 99-111.
———, and G. Müller, 1970, Sediment petrologie, (Teil II - Sediment sedimentgesteine): E. Schweizerbart'sche Verlagsbuchhendlung, Stuttgart, 709 p.
Gaida, K.H., W. Ruhl, and W. Zimmerle, 1973, Rasterelektronenmikroskopische Untersuchen des Porenraumes von Sandsteinen: Erdoel Erdgas Zeitschrift, v. 89, p. 336-343.
Galloway, W.E., 1974, Depositional and diagenetic alteration of sandstone in northeast Pacific arc-related basins: Implications for graywacke genesis: Geol. Soc. America Bull., v. 85, p. 379-390.
Gilbert, C.M., 1949, Cementation of some California Tertiary reservoir sands: Jour. Geol., v. 57, p. 1-17.
Glass, H.D., 1958, Clay mineralogy of Pennsylvanian sediments in southern Illinois: Proc. Fifth Natl. Conf. on Clays and Clay Minerals, p. 227-241.
———, P.E. Potter, and R. Sevier, 1956, Clay mineralogy of some basal Pennsylvanian sandstones, clay, and shales: Amer. Assoc. Petroleum Geol. Bull., v. 40, p. 750-754.
Hayes, J.B., 1970, Polytypism of chlorite in sedimentary rocks: Clays and Clay Minerals, v. 18, p. 285-306.
Heald, M.T., 1950, Authigenesis in West Virginia sandstones: Jour. Geol., v. 58, p. 624-633.
———, 1956, Cementation of Simpson and St. Peter Sandstones in parts of Oklahoma, Arkansas, and Missouri: Jour. Geol., v. 64, p. 16-30.
———, and R.E. Larese, 1974, Influence of coatings on quartz cementation: Jour. Sed. Petrology, v. 44, p. 1269-1274.
Hein, J.R., A.O. Allwardt, and G.B. Griggs, 1974, The occurrence of glauconite in Monterey Bay, California, diversity, origins, and sedimentary environmental significance: Jour. Sed. Petrology, v. 44, p. 562-571.
Hollister, C.D., and B.D. Heezen, 1964, Modern graywacke-type sands: Science, v. 146, p. 1523-1574.
Horn, V.D., 1965, Diagenese and porosität des Dogger-beta-Hauptsandsteins in den Olfeldern Plón-Ost und Preetz: Erdöl-Kohle-Erdgas-Petrochemie, v. 18, p. 249-255.
Howard, J.D., 1969, Radiographic examination of variations in Barrier Island Facies, Sapelo Island, Georgia: Trans. Gulf Coast Assoc. Geol. Soc., v. XIX, p. 217-232.
———, H. Reineck, 1972, Georgia Coastal Region, Sapelo Island, U.S.A., Sedimentology and Biology, IV. Physical and Biogenic Sedimentary Structures of the Nearshore Shelf: Senckenbergiana Maritima, v. 4, p. 81-123.
Karpova, G.V., 1969, Clay mineral post-sedimentary ranks in terrigenous rocks: Sedimentology, v. 13, p. 5-20.
Keene, J.B., D.A. Clague, and R.K. Nishimori, 1973, Experimental hydrothermal alteration of tholeiitic basalt in seawater; resultant mineralogy and textures: Geol. Soc. America, Cordilleran Section Annual Mtg., Abstracts with Programs, p. 65-66.
Keller, W.D., 1970, Environmental aspects of clay minerals: Jour. Sed. Petrology, v. 40, p. 788-813.
Kieke, E.M., and D.J. Hartmann, 1973, Scanning electron microscope application to formation evaluation: Gulf Coast Assoc. Geol. Soc. Trans., v. 23, p. 60-67.
Kulbicki, G., and G. Millot, 1960, L'evolution de la fraction argileuse des grès pétroliers cambro-ordoviciens du Sahara central: Bull. Serv. Carte Géol. Als. Lor., v. 13, p. 147-156.
Lancelot, Y., 1973, Chert and silica diagenesis in sediments from the central Pacific: Initial Reports of the Deep Sea Drilling Project, U.S. Gov. Printing Office, Washington, D.C., v. 17, p. 377-405.
Lerbekmo, J.F., 1961, Porosity reduction in Cretaceous sandstones of Alberta: Alberta Soc. Petroleum Geol. Jour., v. 9, p. 192-199.
Lovell, J.P.B., 1972, Diagenetic origin of graywacke matrix minerals: a discussion: Sedimentology, v. 19, p. 141-143.
Mansfield, C.F., and S.W. Bailey, 1972, Twin and pseudotwin intergrowths in kaolinite: Amer. Min., v. 57, p. 411-425.
Mellon, G.B., 1964, Discriminatory analysis of calcite- and silicate-cemented phases of the Mountain Park Sandstone: Jour. Geol., v. 72, p. 786-809.
Millot, G., 1970, Geology of clays (translated by Forrand, W.R. and Paquet, H.): Springer-Verlag, New York, 429 p.
Moench, R.H., and J.S. Schlee, 1967, Geology and uranium deposits of the Laguna District, New Mexico: U.S. Geol. Survey Prof. Paper 519, 113 p.
Moss, A. J., 1972, Bed-load sediments: Sedimentology, v. 18, p. 159-219.
Mousinho De Meis, M.R., and E.S. Amador, 1974, Note on weathered arkosic beds: Jour. Sed. Petrology, v. 44, p. 727-737.
Odin, G.S., 1972, Observations nouvelles sur la structure de la glauconie en accordeon ("vermicular pellets"); description du processus de genese de ces granules por neoformation: Sedimentology, v. 19, p. 285-294.
Oehler, J.H., 1975, Origin and distribution of silica lepispheres in porcelandite from the Monterey Formation of California: Jour. Sed. Petrology, v. 45, p. 252-257.
Pettijohn, F.J., 1957, Sedimentary rocks (2nd ed.), New York, Harper, 718 p.
———, P.E. Potter, and R. Sevier, 1972, Sand and sandstone: Springer-Verlag, New York, 618 p.
Pittman, E.D., 1972, Diagenesis of quartz in sandstones as revealed by scanning electron microscopy: Jour. Sed. Petrology, v. 42, p. 507-519.
———, 1974, Origin of clay in sandstone-evidence from scanning electron microscopy: Amer. Assoc. Petroleum Geol. and Soc. Econ. Paleon. and Min. Ann. Mtgs. Abstracts, v. 1, San Antonio, Texas, p. 71.
———, and D. N. Lumsden, 1968, Relationship between chlorite coatings on quartz grains and porosity, Spiro sand, Oklahoma: Jour. Sed. Petrology, v. 38, p. 668-670.
Prozorovich, G.E., 1967, On the distribution of clay minerals and small chemical elements in the clay fractions from the Neocomian rocks of the western Surgut oil fields (western Siberia): Rept. USSR Acad. Sci., v. 174, p. 948-951.
Pryor, W.A., 1975, Biogenic sedimentation and alteration of argillaceous sediments in shallow marine environments: Geol. Soc. America Bull., v. 86, p. 1244-1254.
———, and Wm. Van Wie, 1971, The "sawdust sand"—an Eocene sediment of floccule origin: Jour. Sed. Petrology, v. 41, p. 763-769.
Pye, W.D., 1944, Petrology of Bethel Sandstone of South-Central Illinois: Amer. Assoc. Petroleum Geol. Bull., v. 28, p. 63-122.
Rahmani, R.H., 1968, Graywackes with a primary matrix from the Viqueque Formation (Upper Miocene-Pliocene), Timor: Comment on a paper by M.G. Audley-Charles: Jour. Sed. Petrology, v. 38, p. 271-272.
Reimer, T.O., 1972, Diagenetic reactions in early Precambrian graywackes of the Barberton Mountain Land (South Africa): Sed. Geol., v. 7, p. 263-282.
Rex, R.W., 1966, Authigenic kaolinite and mica as evidence for phase equilibria at low temperatures: Clays and Clay Minerals, v. 13, p. 95-104.
Ross, C.S., and P.F. Kerr, 1931, The kaolin minerals: U.S. Geol. Survey. Prof. Pap., 165E, p. 151-170.
Sarkisyan, S.G., 1971, Application of the scanning electron microscope in the investigation of oil and gas reservoir rocks: Jour. Sed. Petrology, v. 41, p. 289-292.
———, 1972, Origin of authigenic clay minerals and their significance in petroleum geology: Sed. Geol., v. 7, p. 1-22.
Shelton, J.W., 1964, Authigenic kaolinite in sandstone: Jour. Sed. Petrology, v. 34, p. 102-111.
Simons, D.B., E.V. Richardson, and M.L. Albertson, 1961, Flume studies using medium sand (0.45 mm): U.S. Geol. Surv. Water Supply Pap., 1498A, 76 p.
Smithson, F., and G. Brown, 1954, The petrography of dickitic sandstones in North Wales and northern England: Geol. Mag., v. 91, p. 177-188.

Smoot, T.W., 1960, Clay mineralogy of pre-Pennsylvanian sandstones and shales of the Illinois Basin, Part III—Clay minerals of various facies of some Chester Formations: Ill. Geol. Surv. Circ. 293, p. 1-19.

Stalder, P.J., 1973, Influence of crystallographic habit and aggregate structure of authigenic clay minerals on sandstone permeability: Geol. en Mijnbouw, v. 52, p. 217-220.

Thompson, A.M., 1970, Geochemistry of color genesis in red-bed sequence, Juniata and Bald Eagle Formations, Pennsylvania: Jour. Sed. Petrology, v. 40, p. 599-615.

Todd, T.W., 1968, Paleoclimatology and the relative stability of feldspar minerals under atmospheric conditions: Jour. Sed. Petrology, v. 38, p. 832-844.

Triplehorn, D.M., 1967, Occurrence of pure well-crystallized 1M illite in Cambro-Ordovician sandstone from Rhourde El Baguel Field, Algeria: Jour. Sed. Petrology, v. 37, p. 879-884.

———, 1970, Clay mineral diagenesis in Atoka (Pennsylvanian) sandstones, Crawford County, Arkansas: Jour. Sed. Petrology, v. 40, p. 838-847.

Van Olphen, H., 1963, An introduction to clay colloid chemistry: Interscience, John Wiley and Sons, N.Y., p. 89-107.

Walker, T.R., 1967, Formation of red beds in modern and ancient deserts: Geol. Soc. America, Bull., v. 78, p. 353-368.

———, 1973, Intrastratal alterations in the Fountain Formation (Pennsylvanian age), Denver area, Colorado: Geol. Soc. America, Abs. with Programs (Southeastern Section), v. 5, no. 6, p. 521-522.

———, 1974, Formation of red beds in moist tropical climates: A hypothesis: Geol. Soc. America Bull., v. 85, p. 633-638.

———, and A.J. Crone, 1974, Mechanically-infiltrated clay matrix in desert alluvium: Geol. Soc. America, Abs. with Programs, v. 6, no. 7, p. 998.

———, and B. Waugh, 1973, Intrastratal alteration of silicate minerals in Later Tertiary fluvial arkose, Baja, California, Mexico: Geol. Soc. America Abs. with Programs, v. 4, p. 853.

Weaver, C.E., 1960, Possible uses of clay minerals in search for oil: Amer. Assoc. Petroleum Geol. Bull., v. 44, p. 1505-1518.

Whetten, J.T., 1971, Diagenetic origin of matrix minerals: laboratory and field evidence: Intl. Sedimentol. Congr., 8th, Heidelberg, Abst. p. 109.

———, and J.W. Hawkins, Jr., 1970, Diagenetic origin of graywacke matrix minerals: Sedimentology, v. 15, p. 347-361.

———, and ———, 1972, Diagenetic origin of graywacke matrix minerals: A reply: Sedimentology, v. 19, p. 144-146.

Williams, H., F.J. Turner, and C.M. Gilbert, 1955, Petrography: Freeman and Co., San Francisco, 406 p.

Wilson, M.D., 1974, Common forms of diagenetic clay in sandstones: Amer. Assoc. Petroleum Geol. and Soc. Econ. Paleon. and Min. Ann. Mtgs. Abstracts, v. 1, San Antonio, Texas, p. 100-101.

———, and R.W. Tillman, 1974, Diagenetic destruction of feldspar and genesis of clay: their influence on sandstone classification and grain size analysis: Geol. Soc. America Abstracts with Programs, v. 5, p. 130-131.

Wilson, M.J., D.C. Bain, W.J. McHardy, and M.L. Berrow, 1972, Clay-mineral studies on some Carboniferous sediments in Scotland: Sed. Geol., v. 8, p. 137-150.

Wise, S.W., and K.R. Kelts, 1972, Inferred diagenetic history of a weakly silicified deep sea chalk: Gulf Coast Assoc. Geol. Soc. Trans., v. 22, p. 177-203.

———, and F.G. Bennett, and F.M. Weaver, 1972, Chemically precipitated sedimentary cristobalite and the origin of chert: Eclogae Geol. Helv., v. 65, p. 157-163.

———, and F.M. Weaver, 1973, Origin of cristobalite-rich Tertiary sediments in the Atlantic and Gulf Coastal Plain: Gulf Coast Assoc. Geol. Soc. Trans., v. 23, p. 305-323.

Zimmerle, V.W., 1963, Zur Petrographie und Diagenese des Dogger-beta-Hauptsandsteins im Erdölfeld Plön-Ost: Erdöl u. Kohle, v. 16, p. 9-16.

Reprinted by permission of the Society of Economic Paleontologists and Mineralogists. Published in *Aspects of Diagenesis,* P.A. Scholle and P.R. Schulger, editors, SEPM Special Publication No. 26, pp. 175-207.

SEPM Special Publication No. 26, p. 175–207, March 1979

THE ROLE OF SECONDARY POROSITY IN THE COURSE OF SANDSTONE DIAGENESIS

VOLKMAR SCHMIDT AND DAVID A. MCDONALD[1]
Petro-Canada, Calgary, Alberta, and Mobil Oil Canada Ltd., Calgary, Alberta

ABSTRACT

Secondary porosity plays an important role in the diagenesis of some sandstones. The volume of secondary porosity equals or exceeds that of primary porosity in the sandstones of many sedimentary basins worldwide, and a significant percentage of the world's reserves of natural gas and crude oil are contained in secondary sandstone porosity. Prudhoe Bay Field and the Jurassic fields of the North Sea are examples of the many giant hydrocarbon accumulations in secondary sandstone porosity.

Chemical, physical, physicochemical, biochemical and biophysical processes result in secondary sandstone porosity through leaching and shrinkage of rock constituents, or through the opening of fractures and porous burrows and borings. Secondary sandstone porosity can originate anywhere in the sedimentary crust: (1) before effective burial in the environment of deposition (eogenetic)[2]; (2) at any depth of burial above the zone of metamorphism (mesogenetic)[2]; and (3) during exposure following a period of burial (telogenetic)[2]. Secondary porosity may occur in sandstones of any mineralogical or textural composition and of any Phanerozoic age. It is most common in sandstones that have undergone relatively long lasting, deep burial and have lost their primary porosity.

Most of the secondary porosity in ancient sandstones originated as a result of mesogenetic leaching of the carbonate minerals calcite, dolomite and siderite. This decarbonatization removes depositional carbonate constituents and diagenetic carbonate such as cements or replacements. Most of the mesogenetic decarbonatization results from the decarboxylation of organic matter in strata adjacent to the sandstone during the course of organic maturation. The process of decarboxylation leads to the generation of carbon dioxide which, in the presence of water, produces carbonic acid. This acid reacts with the carbonate minerals.

In most instances it is possible to differentiate microscopically between primary and secondary sandstone porosity. Thus it is possible to trace the loss of primary porosity during burial. In the presence of water and hydrostatic pressure, primary sandstone porosity cannot exist beyond specific limits of temperature-time exposure except for a small volume of irreducible lamellar porosity between grains. The limiting temperature-time exposure increases with increasing mineralogical stability of the sandstones and, subordinately, with increasing grain size.

The mesodiagenesis of sandstones can be divided into four stages: (1) immature—mechanical compaction; (2) semi-mature—chemical compaction of primary porosity; (3) mature—only secondary porosity present; and, (4) supermature—no effective primary or secondary porosity.

Decarbonatization may create considerable quantities of secondary porosity during semi-mature mesodiagenesis. However, the average addition of carbonate to the sandstone in this diagenetic stage usually exceeds the average carbonate removal.

Decarbonatization culminates during mature mesodiagenesis at which stage it greatly outweighs carbonatization. Much secondary sandstone porosity, therefore, originates after effective primary porosity has been lost. Fractures and irreducible lamellar porosity apparently provide sufficient access for decarbonatizing fluids to start the leaching process even in sandstones of low permeability.

Enormous volumes of carbonate move upwards in solution from diagenetically mature sandstones and are, at least in part, reprecipitated in immature and semi-mature sandstones. Within a subsiding prism of clastic sediments much of the carbonate content is being recycled upwards in this fashion and sandstones at shallower depths are being enriched in carbonate.

Primary migration of hydrocarbons commonly follows closely after the secondary porosity has been formed, because in the maturation of organic matter, the main phase of hydrocarbon generation follows after the culmination of decarboxylation. This close association of source and reservoir in time and space favours the accumulation of hydrocarbons in secondary porosity.

In the presence of pore water, secondary porosity is gradually reduced during deep burial although at a much slower rate compared to primary porosity.

The main geological and economic significance of secondary sandstone porosity is that it extends the depth range for effective sandstone porosity far below the generally accepted depth limit for effective primary porosity. Generation and primary migration of hydrocarbons occurs mainly below the range of effective primary porosity.

[1]Present address: Petro-Canada, Calgary, Alberta.

[2]Diagenesis terminology adopted from Choquette & Pray (1970).

The path of primary migration and the site of accumulation of hydrocarbons are, therefore, commonly controlled by the distribution of secondary porosity.

INTRODUCTION

The subject of secondary porosity in sandstones received little attention in the published literature prior to 1975. Most workers considered only a very small portion of sandstone porosity to be of secondary origin. Except for fractures, most accounts of secondary porosity described outcropping sandstone formations where invading meteoric water had removed soluble constituents (Sedimentation Seminar, 1969; Hraber & Potter, 1969). Krynine (1941) indicated that leaching of carbonate in the Oriskany Sandstone may possibly have occurred in the subsurface. Proshlyakov (1960) was probably the first author to convincingly report significant amounts of sandstone porosity that formed in the subsurface in the presence of saline formation-water. The first detailed discussion of the possible processes which create secondary porosity in the subsurface was published by Savkevic (1971). Rowsell and De Swardt (1973) observed secondary porosity of probable subsurface origin in the Middle Ecca sandstones of Pennsylvanian age in the northern Karoo basin of South Africa. They were the first authors outside Russia to apply the concepts developed by Russian geologists concerning mesogenetic porosity. Until 1976, these Russian and South African studies went largely unnoticed in North America. Unaware of the above studies, several groups of North American geologists have, during the last several years, recognized the common occurrence of secondary sandstone porosity that originated in the subsurface.

Recently, results of these investigations have appeared in the literature and several workers have described occurrences from the subsurface where the porosity originated after burial (Schmidt, 1976; Stanton & McBride, 1976; Alcock & Benteau, 1976; Pittmann, this volume; Hayes, this volume). Hayes and the authors of the present paper recognized the widespread occurrence of secondary sandstone porosity as early as 1970. Subsequently the authors refined the criteria for recognizing secondary porosity (Schmidt et al., 1977) and studied the phenomenon of secondary sandstone porosity in various geological settings in Canada, Alaska and Europe. A study of the Parsons Sandstone of the Mackenzie Delta (Schmidt, *ibid.*) provided several new concepts on the origin of secondary porosity in the subsurface. McBride and his students have applied these concepts and criteria very successfully when investigating secondary porosity in the subsurface of southern Texas (Stanton & McBride, 1976; Lindquist, 1976).

As the authors' studies progressed, it became apparent that much secondary porosity previously has been misidentified as primary porosity, and their views on the abundance of secondary porosity have been changing with increasing information (Fig. 1). At the present time it is certain that at least one-third of total sandstone porosity is of secondary origin. It appears possible that secondary porosity in sandstone is more abundant than primary sandstone porosity. Undoubtedly, secondary sandstone porosity is of major geological significance. Its origin, stability, and distribution form important aspects of sandstone diagenesis.

The study of secondary sandstone porosity is still very much in its infancy. However, the available information is sufficient to bring the emerging role of secondary porosity in sandstone diagenesis into perspective. This paper is intended to provide a preliminary account of the occurrence of secondary porosity in sandstones and to follow its evolution through the course of diagenesis. In this way a useful stepping stone will be established for other workers who may be interested in further investigating this very important subject.

GEOLOGICAL OCCURRENCE AND CAUSE

Textural Origin.—We recognize five principal classes of secondary sandstone porosity that can be differentiated on the basis of process of origin and fabric relationships (Fig. 2): (1) fractures, (2) shrinkage voids, (3) voids after sedimentary material, (4) voids after authigenic cement, and (5) voids after authigenic replacement minerals. The texture of secondary sandstone porosity and the criteria for its recognition are discussed in more detail in a separate paper of this publication (Schmidt and McDonald).

Secondary porosity can originate from fracturing, shrinkage, and dissolution only if the grain or crystal framework of the rock is sufficiently strong to prevent immediate collapse. There is a limit to the amount of secondary porosity that can be formed in a given sandstone under specific conditions of stress because the supportive strength of the rock framework will ultimately fail leading to reduction of porosity by collapse.

Usually textural origin and abundance of secondary porosity in sandstones with little or no primary porosity can be reliably determined using petrographic criteria. The textural relationship between secondary porosity and sandstone host provides evidence of the diagenetic changes that affected the rock prior to and following the

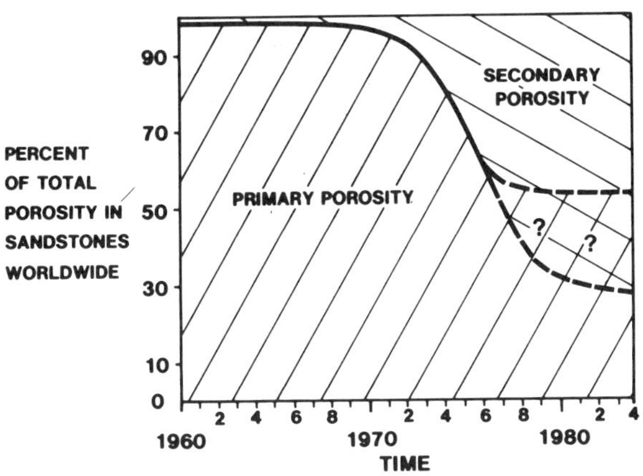

FIG. 1.—Change of interpretation of nature of sandstone porosity.

formation of the secondary porosity. For example, secondary pores which cut across welded grain fabrics indicate that welding occurred before leaching. On the other hand, secondary pores that are lined by bitumen originated prior to hydrocarbon migration. Thus diagenetic sequences can be established and fitted to the geological history of the sandstones.

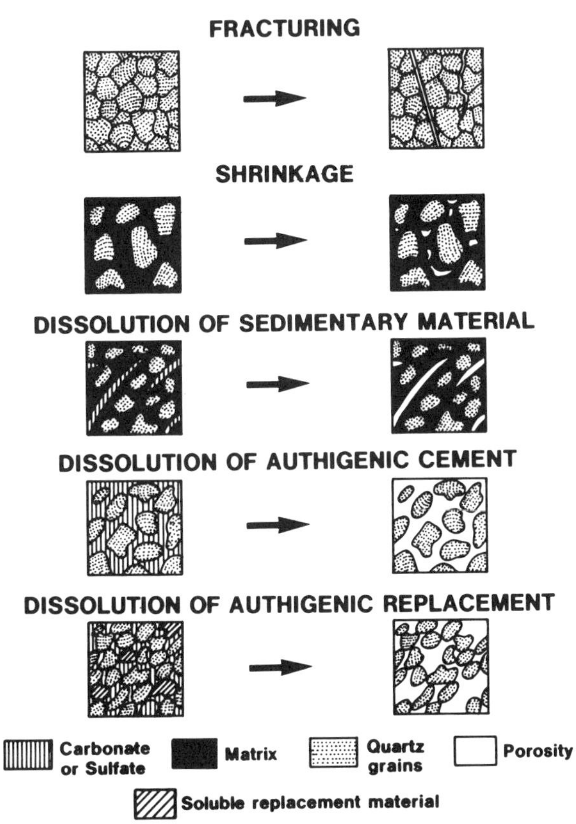

FIG. 2.—Textural origin of secondary sandstone porosity.

Regimes of diagenesis.—Secondary sandstone porosity is formed and preserved, changed and destroyed in almost any conceivable diagenetic environment. The system of diagenetic regimes that Choquette and Pray (1970) formulated for the study of porosity in carbonate rocks is equally appropriate for investigation of secondary porosity in sandstones and has been adopted in this paper. *Eodiagenesis* is defined as the regime at or near the surface of sedimentation where the chemistry of the interstitial water is mainly controlled by the surface environment prior to effective burial (effective burial is burial under strata that seal the sandstone from a predominant influence of surface agents on the chemistry of the interstitial water). *Mesodiagenesis* is the subsurface regime during effective burial. *Telodiagenesis* represents the regime at or near the surface after effective burial. The amounts of secondary porosity originating in each diagenetic regime differ greatly (Fig. 3).

A minor portion of secondary porosity evolves during eodiagenesis, mainly as a result of dissolution of sedimentary and eogenetic constituents, although fracturing and shrinkage may be subordinate contributors. In some instances eogenetic processes created considerable secondary porosity, for example, in calcareous quartz sandstones of Pleistocene and Holocene age from the Trucial Coast (unpublished petrographic study by Schmidt).

By far the largest amount of secondary porosity is produced during mesodiagenesis, and most of it can be attributed to leaching of sedimentary, eogenetic, and mesogenetic carbonate constituents (Hayes, this volume; Schmidt, 1976). Fractures form an important, although subordinate, portion of mesogenetic sandstone porosity. Mesogenetic shrinkage porosity plays a minor role and much of the shrinkage occurs during shallow burial.

Only a minor share of the total sandstone porosity originates during telodiagenesis, mainly as the result of leaching and, subordinately, due to fracturing. This we conclude from the comparison of outcrop samples and subsurface cores of numerous sandstone formations, and from the stratigraphic consideration that only a small percentage of sandstones have undergone a period of telodiagenesis.

Diagenetic processes.—The individual diagenetic processes that result in secondary porosity vary significantly in their quantitative importance (Fig. 4).

Fracturing of rocks, grains and intergranular constituents generally creates only insignificant to minor pore volume. However, rock fractures can, on occasion, form the major porosity type in some sandstone with low matrix porosity.

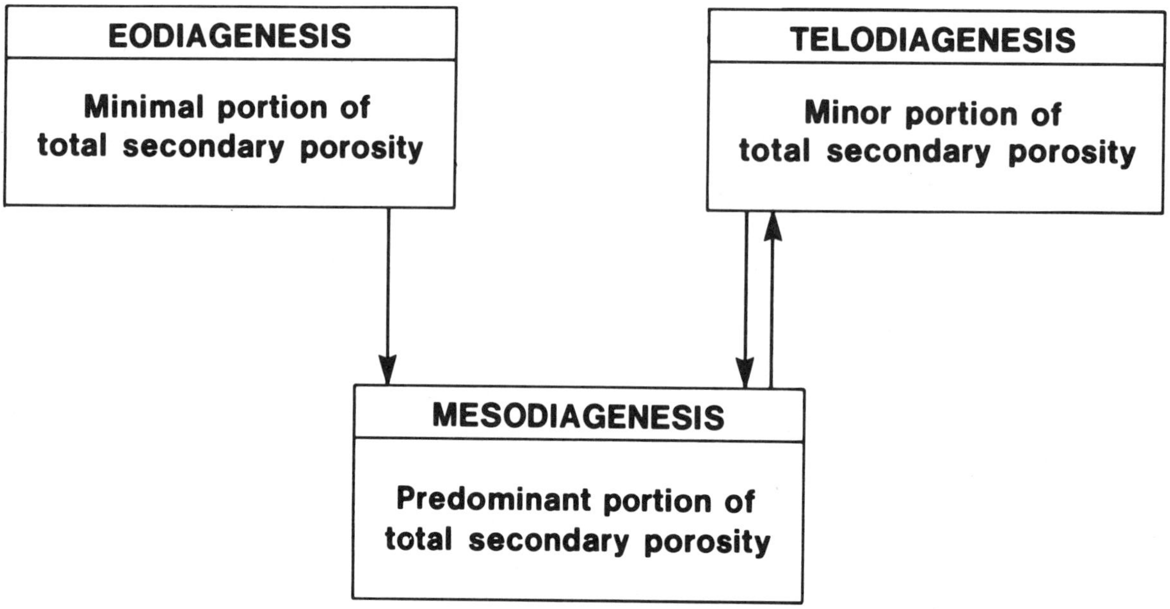

FIG. 3.—Diagenetic regimes of development of secondary sandstone porosity.

DIAGENETIC PROCESSES		PORTION OF SECONDARY POROSITY
	ROCK FRACTURING	* MINOR
	GRAIN FRACTURING	MINOR
	SHRINKAGE	* MINOR
DISSOLUTION OF	CALCITE	MAJOR
	DOLOMITE	MAJOR
	SIDERITE	MAJOR
	SULFATE	* MINOR
	OTHER EVAPORITES	* MINOR
	SILICATE	VERY MINOR
	OTHER NONSILICATE	VERY MINOR

* MAY BE OF MAJOR IMPORTANCE IN INDIVIDUAL STRATIGRAPHIC UNITS

FIG. 4.—Diagenetic processes creating secondary sandstone porosity.

Shrinkage, if present, usually supplies little porosity although exceptions do occur.

The vast majority of the total volume of secondary sandstone porosity results from the dissolution of carbonate minerals. Calcite, dolomite, and siderite each are the precursors of secondary porosity on a major scale (Figs. 5A–C, 6). The dissolved carbonate constituents mainly consist of cements, but also include substantial percentages of sedimentary and replacive carbonate.

The sulfate minerals such as anhydrite, gypsum, barite, and celestite commonly are subject to leaching (Fig. 5D). This may occur during any stage of diagenesis. The secondary porosity resulting from the removal of sulfates is minor in most sandstone formations. However, it can be of major importance in some instances, for example, in Cambrian sandstones of the Northwest Territories (McDonald, in progress).

The dissolution of nonsulfate evaporite minerals such as halite and sylvite appears to create only a small fraction of secondary sandstone porosity because of the relatively infrequent occurrence of these evaporite minerals in sandstones. Porosity of this origin, however, is easily misidentified and may be much more common than is presently realized. Hartman (1968) recognized a case of leaching of halite that created the reservoir for the Pelahatchie Field in the Gulf Coast region of the United States. Therefore, dissolution of the evaporite minerals can, in some instances, develop secondary porosity of major importance. It may occur either in the subsurface or near the surface in contact with meteoric waters.

Direct dissolution of silicate minerals such as quartz or feldspar generally creates little or no secondary porosity. Dissolution of silicate minerals as a rule occurs along grain contacts without creating porosity. However, Hayes (this volume) reports a case of significant leaching of silicate minerals in Tertiary sandstones of South Central Alaska. Heald and Larese (1973) attribute the porosity in honeycombed feldspar of Paleozoic sandstones in West Virginia and Ohio to direct dissolution. Their evidence (undissolved calcite cement), however, is not conclusive, as leaching of carbonate may be followed by precipitation of carbonate cement. Silicate minerals often are replaced by non-silicate minerals, principally carbonates. These replacive minerals, in turn, frequently are dissolved (Figs. 5E, F) and the resulting secondary porosity may be mistakenly attributed to direct leaching of the silicate minerals.

The dissolution of other non-silicate minerals including, for example, phosphate and oxide minerals contributes very little secondary porosity in sandstones.

Causes.—The conditions which bring about the generation of secondary porosity in sandstones are the result of chemical, physicochemical, physical, biochemical, and biophysical processes.

Chemical processes such as dilution of the interstitial water or changes in the ratios of ion species may cause void-creating dissolution at constant temperatures and pressures. The formation of carbonic acid in the interstitial water is a well known example. It causes lowering of the pH and dissolution of carbonate minerals. In the telogenetic regime carbonic acid forms mainly from atmospheric carbon dioxide. During mesodiagenesis most of the carbon dioxide that forms carbonic acid originates from organic matter undergoing thermomaturation (Fig. 7). Tissot et al. (1974) demonstrated that this release of mesogenetic carbon dioxide occurs over a wide range of temperature/time exposures. Savkevic (1971) suggested that both the interlayer water which is released from montmorillonite during its mesogenetic transformation to illite and CO_2 originating from the oxidation of organic matter could cause subsurface brines to become so undersaturated in respect to carbonate minerals that the latter are dissolved. Parker (1974) concluded that carbon dioxide originating from a plutonic intrusion was responsible for the mesogenetic leaching of carbonate which created the secondary sandstone porosity in the Smackover sandstones of the Thomasville Field, Mississippi. Hydrogen sulfide may also cause mesogenetic dissolution of carbonate (Parker, *ibid.*).

Physiochemical processes can cause dissolution and shrinkage because changes in temperature and pressure affect the solubility of minerals if the salinity and ionic ratio of the interstitial waters remain constant. For instance, pore waters that move from zones of hydrostatic pressure into zones of geopressures may become undersaturated with respect to carbonates on account of the pressure increase. This could lead to mesogenetic dissolution of carbonate.

Physical processes may create conditions for fracturing and shrinkage. Changes of hydraulic and mechanical stress and also thermal expansion and contraction are common causes for fracturing. The stresses that cause fracturing in sandstones may originate externally, e.g. overburden, or internally within the sandstones, e.g. thermal contraction.

Biochemical processes occasionally trigger void-creating dissolution. A host of boring organisms employ biochemical dissolution of rock material to construct their domiciles or to make space for their holdfasts or roots. Metabolic dissimilation by algae and other organisms can result in carbonic acid and leaching of carbonate. Similarly, sulfate reducing bacteria are capable of causing

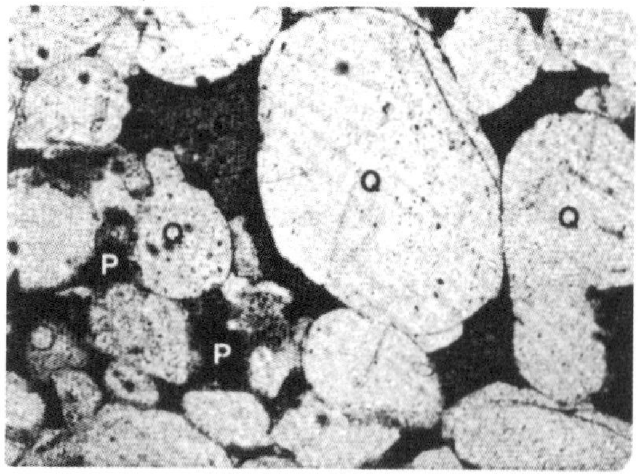

A
.2 MM

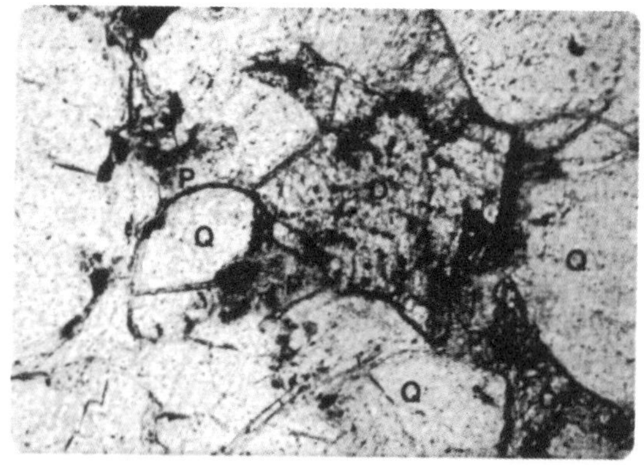

B
.1 MM

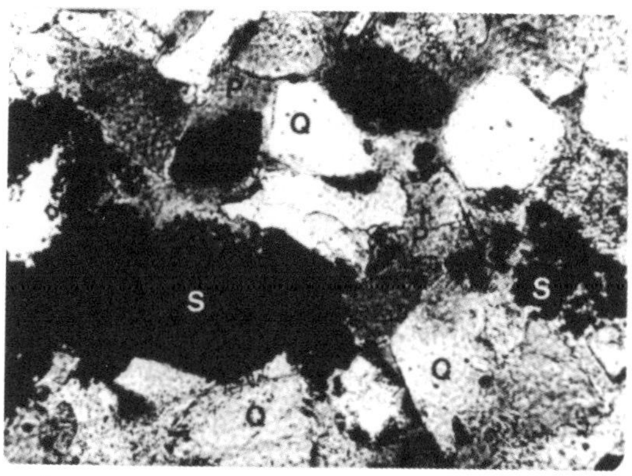

C
.1 MM

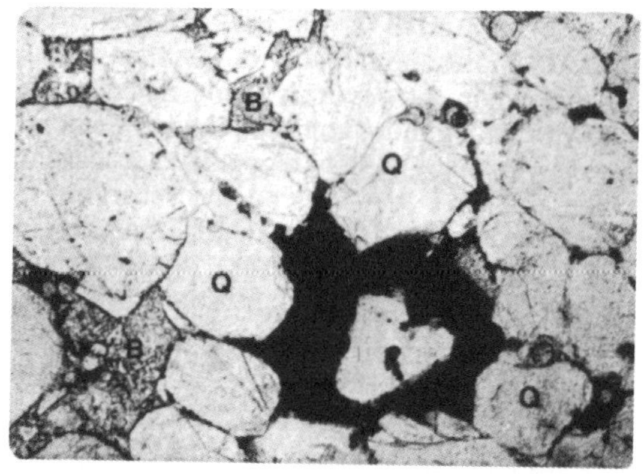

D
.5 MM

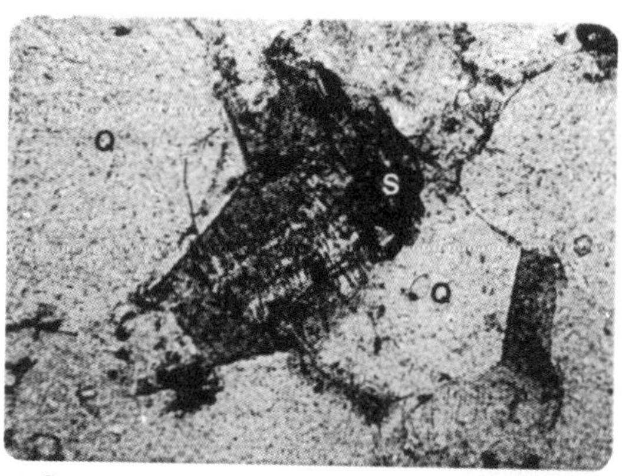

E
.2 MM

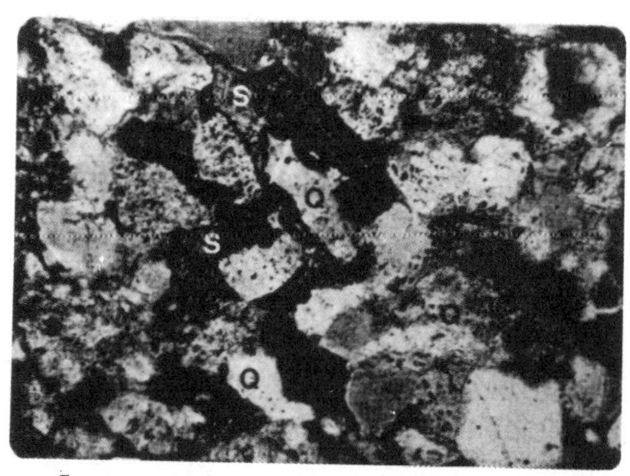

F
.2 MM

the dissolution of sulfate minerals.

Biophysical processes cause very little secondary porosity if one discounts the redistribution of primary porosity by burrowing organisms. Mechanical boring and burrowing can create open voids and the growth of roots may lead to fracturing of sandstones.

Mesogenetic decarbonatization porosity.—It is obvious from the preceding discussion that mesogenetic decarbonatization created the vast majority of secondary porosity in sandstones and is, therefore, of the greatest interest. Decarbonatization is here used as a general term for the dissolution of any textural and mineralogical type of carbonate from sandstones or other rocks. The available information suffices to outline various aspects of the occurrence of mesogenetic decarbonatization porosity.

A selection of important occurrences of mesogenetic decarbonatization porosity is listed in Table 1. In all these examples mesogenetic decarbonatization porosity occurs as the predominant or exclusive type of porosity throughout a considerable part of each basin. The incidence of hydrocarbon accumulations in these sandstones indicates that mesogenetic decarbonatization porosity is a very favorable habitat for the trapping of oil and gas. Many giant fields fall into this category including the Prudhoe Bay Field and the Jurassic fields of the North Sea. They represent a significant percentage of global reserves of natural gas and crude oil. Porosity and permeability of these reservoirs can reach very high values (Table 2). Individual samples may have more than 40 volume percent mesogenetic decarbonatization porosity, and permeability of several darcies. Average porosity for individual reservoir zones ranges from 5 to 30 volume percent.

The distribution of the many known examples of mesogenetic decarbonatization porosity can be correlated with a number of geological parameters (Fig. 8). The porosity is found in sandstones ranging in geological age from Tertiary to Cambrian and may even occur in Proterozoic sandstones. The ratio of decarbonatization porosity to primary porosity increases with increasing geological age given similar depth and temperature of burial, and similar lithology.

Sandstone lithology places no restriction on the occurrence of mesogenetic decarbonatization porosity. It may be present in sandstones of any primary textural and mineralogical composition in any state of diagenetic alteration. Neither paleogeography nor geological setting appear to have any limiting influence. Mesogenetic decarbonatization porosity occurs in every sizable sedimentary basin in which sandstones have been examined using the criteria discussed elsewhere in this volume (Schmidt and McDonald).

Mesogenetic decarbonatization porosity may originate at depths ranging from shallow to 4,000 meters. Greater depths of origin are probable but have not so far been observed. Maximum development of secondary porosity, however, usually occurs within an intermediate depth range, as will be discussed later in this paper. At present, confirmed depths of preservation of mesogenetic decarbonatization porosity range from surface to 6,000 meters. Extrapolation of available data suggests that this porosity type can exist at even greater depths. In sandstones of high mineralogical maturity decarbonatization porosity may possibly extend to a depth of 8,000 meters (Fig. 12).

TEXTURAL STAGES OF MESODIAGENESIS

The improved ability to differentiate between primary and secondary porosity permits the trac-

←

FIG. 5.—Mesogenetic dissolution porosity. *A*, Mesogenetic dissolution of poikilotopic calcite cement created secondary porosity in Cambrian quartz arenite, Northwest Territories, Canada. Polarizer only. P = secondary porosity (blue); Q = quartz grains and mesogenetic syntaxial quartz overgrowth (white); C = mesogenetic calcite cement (red). *B*, Mesogenetic dissolution of dolomite cement created secondary porosity in Lower Cretaceous quartz arenite, Scotian Shelf, Canada. Polarizer only. P = secondary porosity (blue); Q = quartz grains and mesogenetic syntaxial overgrowth; D = dolomite (high relief). *C*, Mesogenetic dissolution of siderite created secondary porosity in Jurassic quartz arenite, Statfjord Field, North Sea. Half crossed nicols. P = secondary porosity (blue); Q = quartz grains; S = siderite (brownish, very high relief). *D*, Mesogenetic dissolution of poikilotopic barite cement created secondary sandstone porosity in Cambrian quartz arenite, Northwest Territories, Canada. Polarizer only. P = secondary porosity (blue); Q = quartz grains and mesogenetic syntaxial quartz overgrowth (white); B = barite (moderately high relief). *E*, Siderite partially replaced feldspar. Subsequent mesogenetic dissolution of replacement siderite created secondary porosity in Lower Cretaceous subfeldspathic arenite, Mackenzie Delta, Canada. Polarizer only. P = secondary porosity (blue); Q = quartz grains and mesogenetic syntaxial quartz overgrowth (white); F = feldspar (honeycombed grain); S = siderite (brownish, very high relief). *F*, Siderite marginally replaced quartz. Subsequent mesogenetic dissolution of replacement siderite and siderite cement created secondary porosity in Triassic quartz arenites, Prudhoe Bay Field, North Slope, Alaska. Half Crossed nicols. P = secondary porosity (blue); Q = quartz grains; S = siderite (very high relief).

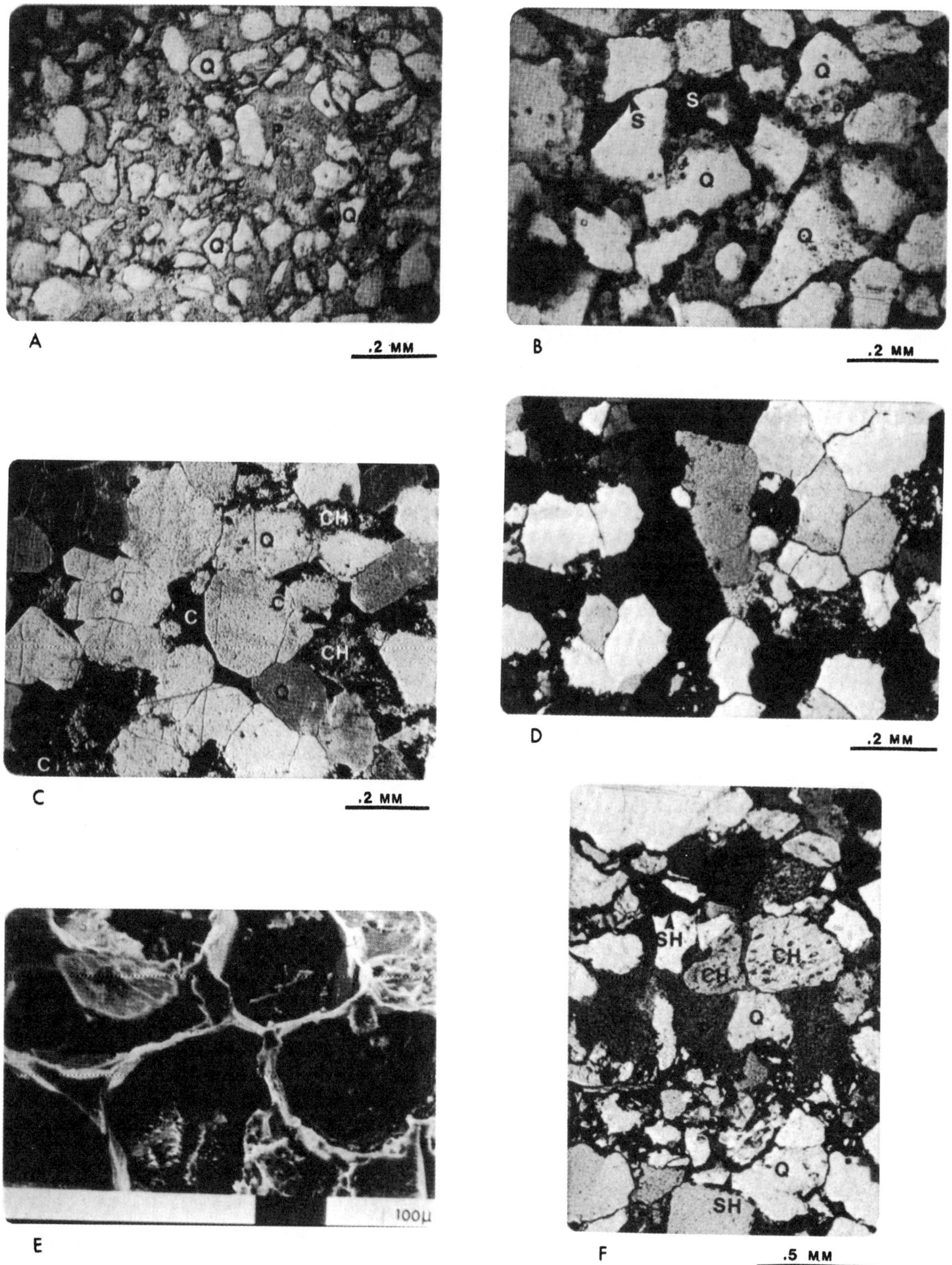
A
.2 MM
Q
P
B
.2 MM
Q
S
C
.2 MM
Q
CH
C
D
.2 MM
E
100μ
F
.5 MM
SH
CH
Q

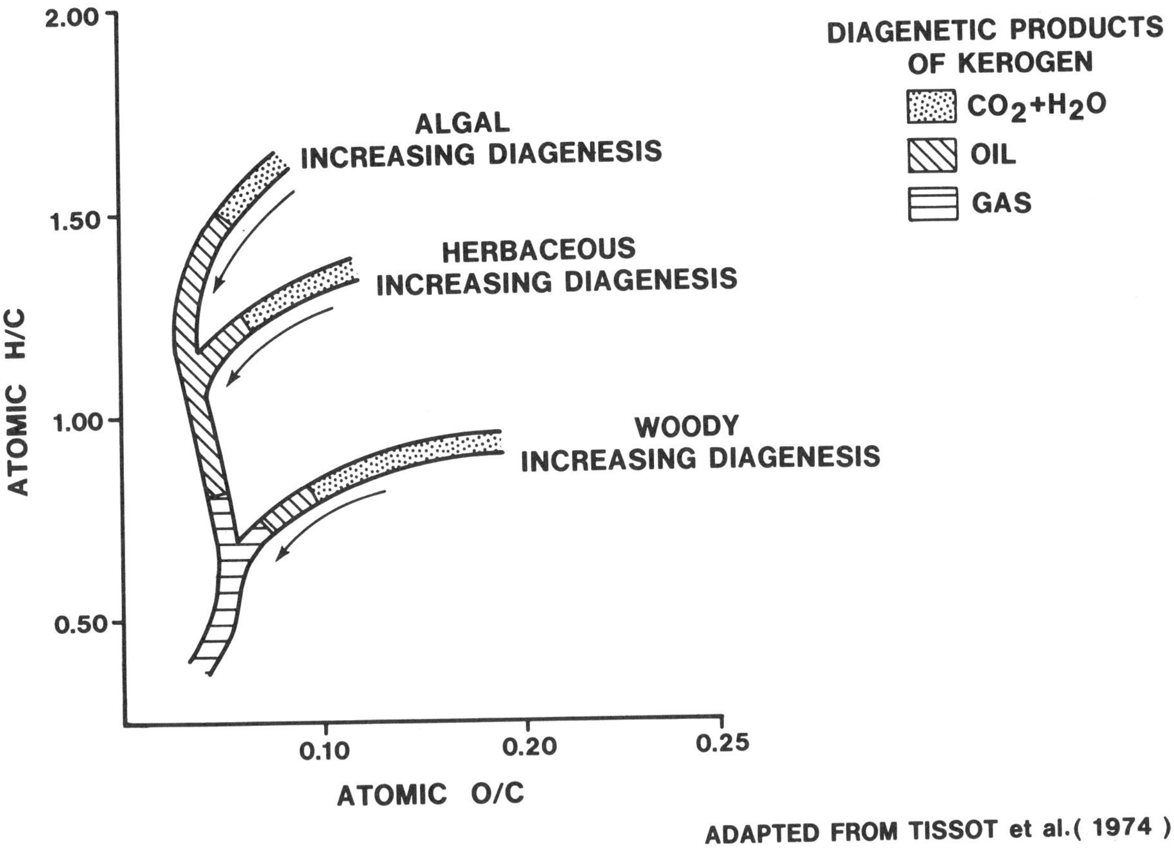

FIG. 7.—Diagenetic evolution of kerogen.

ing of the diagenetic alteration of each of the two porosity classes in response to progressive burial. This provides an excellent means of relating mesogenetic sandstone diagenesis to burial history. We propose four textural stages of sandstone mesodiagenesis that are defined by four petrographically recognizable stages of the burial diagenesis of sandstone porosity (Fig. 9). These are in order of progressive burial: (1) immature stage; (2) semi-mature stage; (3) mature stage;

FIG. 6.—Stages of textural mesodiagenesis. *A*, Immature stage of textural mesodiagenesis, Cretaceous quartz arenite, Scotian Shelf, Canada. Polarizer only. P = primary porosity (blue); Q = quartz grains (white). *B*, Semi-mature stage of textural mesodiagenesis. Incipient chemical compaction of Cretaceous quartz arenite, Scotian Shelf, Canada. Polarizer only. P = primary porosity (blue); Q = quartz grains (white); S = eogenetic siderite cement (brownish, very high relief). *C*, Semi-mature stage of textural mesodiagenesis. Advanced cementation of primary porosity by syntaxial quartz overgrowth followed by cementation of residual primary porosity by mesogenetic poikilotopic calcite, Lower Cretaceous quartz arenite, Mackenzie Delta, Canada. Half crossed nicols. Q = quartz grains and mesogenetic quartz overgrowth; CH = chert grains; C = calcite cement (red). *D*, Mature stage of textural mesodiagenesis. Complete loss of reducible primary porosity by chemical compaction and cementation by syntaxial quartz overgrowth, Lower Cretaceous quartz arenite, Mackenzie Delta, Canada. Crossed nicols. *E*, Mature stage of textural mesodiagenesis. Epoxy cast of lamellar porosity and secondary intergranular porosity of Lower Cretaceous quartz arenite, Mackenzie Delta, Canada. Scanning-electron micrograph. *F*, Mature stage of textural mesodiagenesis. Complete loss of reducible primary porosity by chemical compaction. Subsequent dissolution of sedimentary and diagenetic carbonate created secondary porosity. Grain fracturing resulted from increased stress on grains adjacent to secondary pores. Jurassic quartz arenite, Beryl Field, North Sea. Half crossed nicols. P = secondary porosity (blue); Q = quartz grains; CH = chert grains; SH = stylolitized carbonaceous/argillaceous microlaminae (black).

TABLE 1.—SIGNIFICANT OCCURRENCES OF MESOGENETIC DECARBONATIZATION POROSITY

AREA	GEOLOGICAL AGE OF SANDSTONES	HYDROCARBON RESERVOIR	SOURCE OF INFORMATION
Mackenzie Delta, Canada	Cretaceous, Tertiary	Yes	Authors
Anderson Plain, Canada	Cambrian	Yes	Authors
Alberta, Canada	Devonian	Yes	Alcock & Benteau (1976)
Alberta, Canada	Cretaceous	Yes	Authors
British Columbia Shelf, Canada	Tertiary	No	Authors
Scotian Shelf, Canada	Jurassic, Cretaceous	Yes	Authors
Labrador Shelf, Canada	Cretaceous, Tertiary	Yes	Authors
North Slope, Alaska	Triassic, Jurassic	Yes	Authors
Cook Inlet, Alaska	Jurassic, Cretaceous, Tertiary	Yes	Hayes (this volume)
Gulf Coast, U.S.A.	Cretaceous, Tertiary	Yes	Stanton & McBride (1976)
Wyoming, U.S.A.	Pennsylvanian	Yes	Fox et al. (1975)
Venezuela	Tertiary	Yes	Authors
North Sea	Permian, Triassic, Jurassic	Yes	Authors
Northern Germany	Pennsylvanian to Jurassic	Yes	Authors
South Africa	Pennsylvanian	No	Roswell & DeSwardt (1974)
Ciscaucasus, Russia	Jurassic, Cretaceous	Yes	Proshlyakov (1960)

and, (4) supermature stage.

Description.—The *immature stage* is characterized by mechanical compaction of clean, uncemented sands with primary intergranular porosity (Fig. 6A). The compaction is accomplished by deformation of ductile grains, rotation of grains, and fracturing of grains. It reduces both primary rock volume, and primary porosity and permeability. Chemical compaction as defined below is negligible or absent. Any significant chemical compaction would tend to lithify the rock and make appreciable mechanical compaction impossible. Secondary porosity may exist. It can either be inherited from eodiagenesis or may originate during the immature stage.

TABLE 2.—VALUES OF SECONDARY POROSITY AND PERMEABILITY CREATED BY MESOGENETIC DECARBONATIZATION OF SANDSTONES (BASED ON LABORATORY ANALYSES, MICROSCOPIC ANALYSES, AND LOG ANALYSES)

INDIVIDUAL SAMPLES	
Porosity:	Trace to > 40 vol. %
Permeability:	Nil to several darcies
NET PAY ZONES	
Porosity:	5 vol. % to 30 vol. % (average 18 vol. %)
Permeability:	A few millidarcies to > 1 darcy

The *semi-mature stage* is marked by the onset of pervasive chemical compaction, while, at the same time, mechanical compaction of primary porosity becomes insignificant or ceases entirely. We define chemical compaction as being essentially accomplished through dissolution of sand grains at points and interfaces of contact. In the process rock volume and the percentage of intergranular primary porosity are diminished. The dissolved material may reprecipitate within the sandstone formation as pore cement further reducing the porosity.

The coexistence of reducible primary porosity and pervasive chemical compaction define the range of the semi-mature stage (Figs. 6B, C). Secondary porosity can originate in the semi-mature stage and may coexist with inherited eogenetic and immature-stage secondary porosity. At the end of the semi-mature stage the primary porosity has reached irreducible levels. The "irreducible porosity" consists of narrow gaps between grains or diagenetic crystals a fraction of a micron to several microns wide (Fig. 6D). It is irreducible with respect to chemical compaction and cementation but not to recrystallization. The irreducible porosity in sandstones forms an interconnected system of lamellar pores with extremely low permeability analogous to the irreducible lamellar porosity in carbonate rocks described by Wardlaw (1976). The amount of irreducible porosity decreases with increasing grain size. It com-

GEOLOGICAL ATTRIBUTES	RANGE OF DECARBONATIZATION POROSITY
GEOLOGIC AGE	CAMBRIAN TO TERTIARY
LITHOLOGY	SANDSTONES OF ANY TEXTURAL AND MINERALOGICAL COMPOSITION
TYPE OF SEDIMENTARY BASIN	ANY SIZABLE SEDIMENTARY BASIN
DEPTH OF ORIGIN OF DECARBONATIZATION POROSITY	SHALLOW TO GREATER THAN 3500M(11,500 FT.)
PRESENT DEPTH	SURFACE TO GREATER THAN 6000M(20,000 FT.)

FIG. 8.—Geologic occurrence of significant mesogenetic decarbonatization porosity in sandstones.

monly ranges from about 0.5 percent to 2.5 volume percent.

The *mature stage* is characterized by the absence of primary porosity except for irreducible lamellar pores (Fig. 6E). Secondary porosity can be inherited from preceding stages or may originate during the mature stage (Fig. 6F). The secondary porosity undergoes chemical compaction

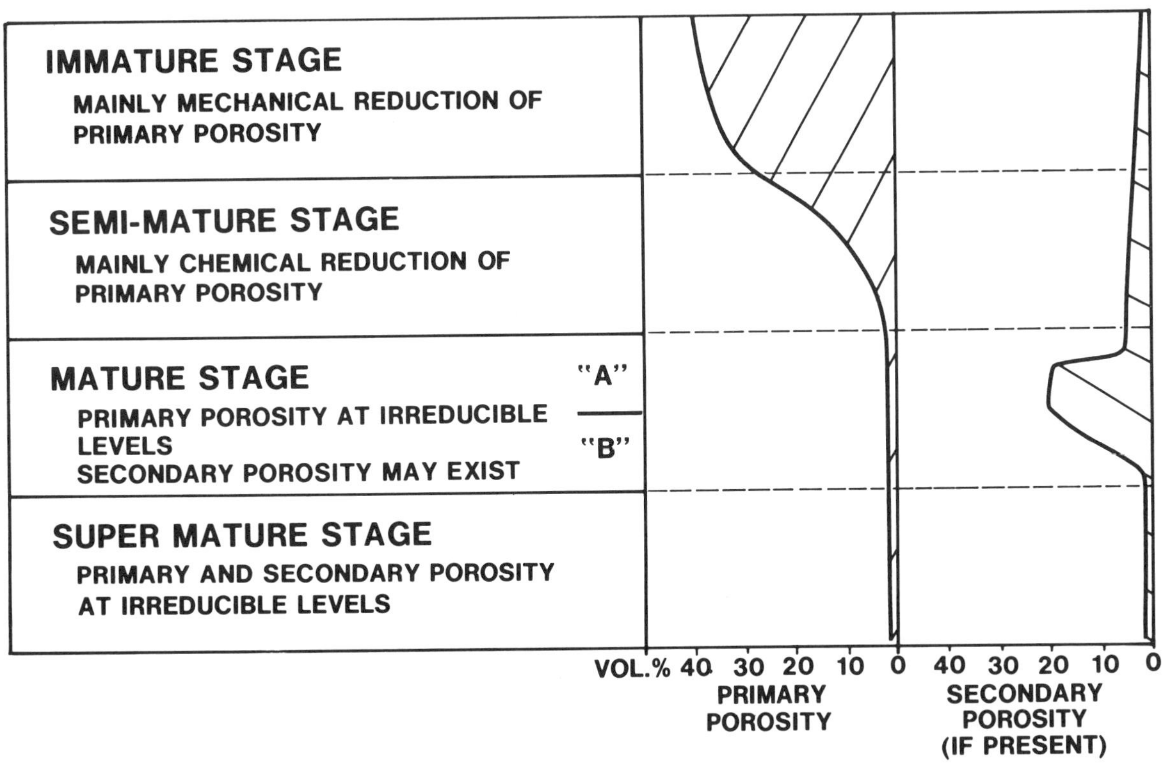

FIG. 9.—Textural stages of mesodiagenesis of sandstone porosity.

and minor mechanical compaction. The mature stage ends when chemical compaction of reducible secondary porosity has reached completion, that is, when it cannot exist any longer for a geologically significant period of time. Two sub-stages can often be identified. *Mature stage "A"* or *early mature stage* is the phase of maximum development of secondary porosity. Chemical compaction of secondary porosity is very subordinate during this interval. *Mature stage "B"* or *late mature stage* is characterized by the gradual destruction of secondary porosity mainly by chemical compaction, and little or no generation of secondary porosity.

The *supermature stage* shows both primary and secondary porosity at irreducible levels. Secondary porosity could conceivably be formed through fracturing and leaching but it would be rapidly eliminated. The advent of metamorphism marks the end of the supermature stage and the conclusion of mesodiagenesis.

In determining the textural stages of mesodiagenesis in sandstones using the above criteria the effects of the following must be discounted: (1) geo-pressures; (2) unusual local chemical conditions; (3) pore filling by hydrocarbons; (4) extrastratal cementation; and (5) unusual local conditions of temperature and pressure such as might be caused by tectonism and vulcanism. The petrographic analysis of a sandstone cannot, by itself, determine whether it presently still resides within the diagenetic zone that affected it last. This question must be answered by geological analysis of the sedimentary basin. It is possible to deduce from the present distribution of regimes of mesodiagenesis the zones of currently active burial diagenesis for specific sandstone lithologies. In similar fashion, one is able to reconstruct ancient regimes of mesodiagenesis.

Influence of sandstone mineralogy.—Figure 10 shows a schematic plot of the depth ranges of textural maturity grades of burial diagenesis versus mineralogical maturity of sandstones. The basic assumptions for the plot are: (1) the subsidence rate of the sedimentary basin is thirty meters per one million years; (2) the sedimentation rate equals the sum of subsidence and compaction; (3) the geothermal gradient is 2.7 degrees Celsius per one hundred meters; and (4) the sandstones are medium grained. Depths of individual zones of textural diagenetic maturity become shallower with decreasing mineralogical maturity, a fact which has been observed by other workers (Galloway, 1974).

Surprisingly, the transition from the semi-mature zone to the mature zone occurs at relatively shallow depths. This implies that in most sedimentary basins effective primary sandstone porosity is restricted to less than half the sedimentary column. The volume of the mature zone in the majority of sedimentary basins exceeds the combined volumes of the immature and semi-mature zones. Thus the volume of sediments which is prospective for secondary porosity is more than double the volume of sediments which is prospective for effective primary porosity.

The large vertical extent of the mature zone compared with that of the semi-mature zone is unexpected. The higher temperature of the mature zone must increase the rate of reaction of chemical compaction if other conditions remain constant. It is obvious that some of these other conditions must have changed thereby hindering chemical compaction. At the present time it is not certain which changes are responsible for the delay in the destruction of secondary porosity in the mature zone. Future investigations will have to solve this problem. It is possible that changes in the chemical environment or changes in the pore geometry and/or changes in the submicroscopic morphology of the interfaces at grain contacts may be the cause.

Influence of temperature and time.—The active stages of burial diagenesis of porosity are schematically plotted in Figure 11 in relationship to residence temperature (residence depth) and residence time. The assumed sandstone lithology is a medium-grained quartz arenite and the chosen geothermal gradient is 2.7 degrees Celsius per one hundred meters. It can be seen that the level of maturity increases with both increasing residence temperature and time. Lengthening of the residence time beyond 400 million years, however, produces very little change.

A similar relationship would be found if the diagram were plotted for other mineralogical classes of sandstone. The position of the zones of textural maturity would shift somewhat towards the upper left (lower residence temperature and shorter residence time) with decreasing mineralogical maturity. These relationships indicate that the texturally mature zone becomes volumetrically more important as the geothermal gradient and the age of the sediments increase.

Porosity range.—Figure 12 shows a schematic plot of the range of primary and secondary porosity versus depth in medium-grained quartz arenites. The assumptions made in constructing the plot are the same as for Figure 10.

Maximum primary porosities decline with increasing depth as a result of mechanical and chemical compaction. At a depth of about 3,000 meters the mature zone is reached and only one to two volume percent of irreducible primary porosity remains.

Maximum secondary porosities exceed 40 vol-

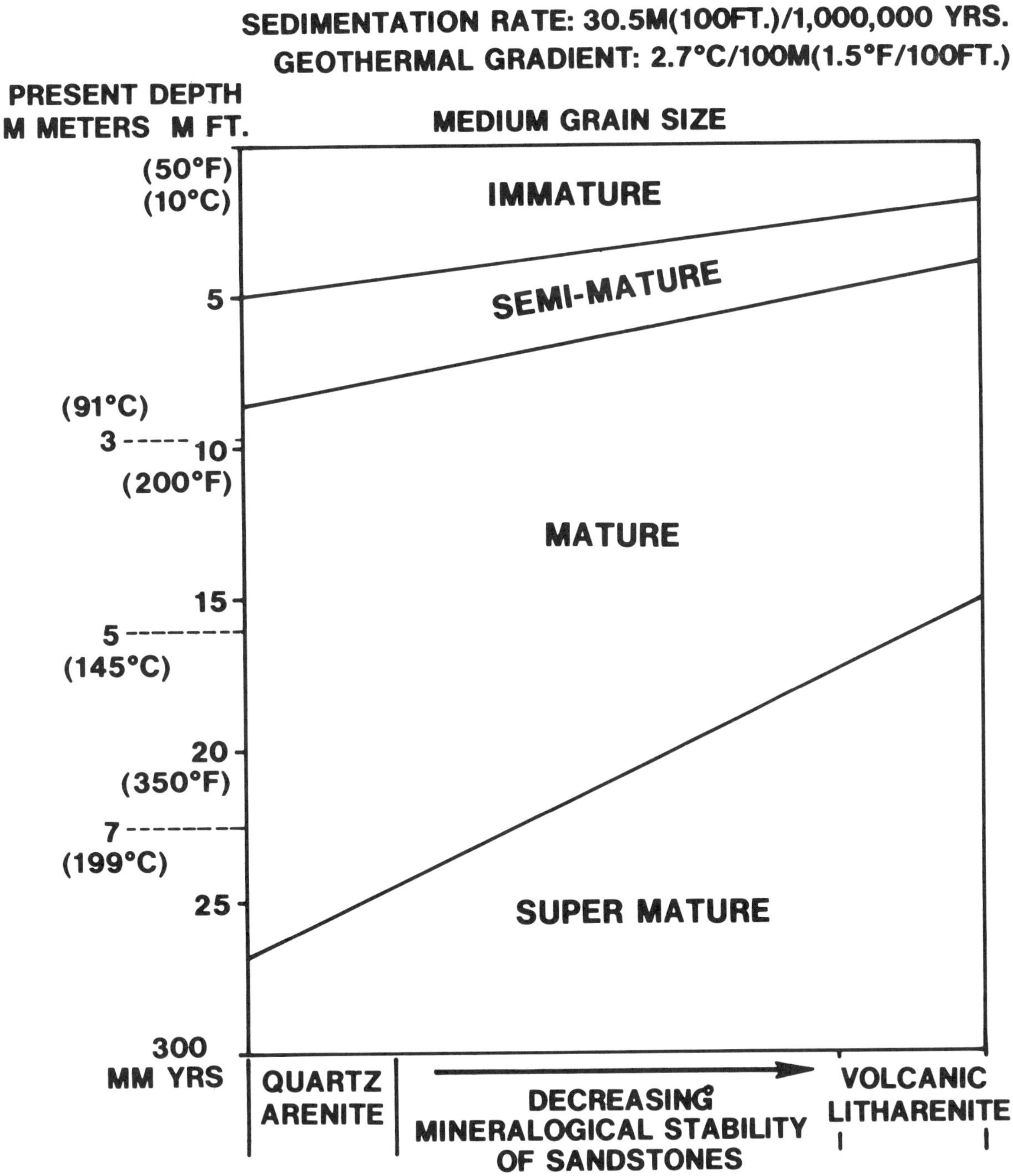

FIG. 10.—Influence of sandstone mineralogy on burial diagenesis of porosity.

ume percent above 4,000 meters, that is, in the immature, semi-mature and upper part of the mature zone. From there they decline steadily with increasing depth in the middle and lower part of the mature zone. At an extrapolated depth of about 8,000 meters secondary porosity should diminish to irreducible levels.

Significantly, secondary porosity can be higher than original primary porosity in the immature, semi-mature, and upper mature zones. In addition, the depth limit for porosity over 10 volume percent is more than twice as deep in the case of secondary porosity compared to that of primary porosity. The mineralogically less mature sandstones ex-

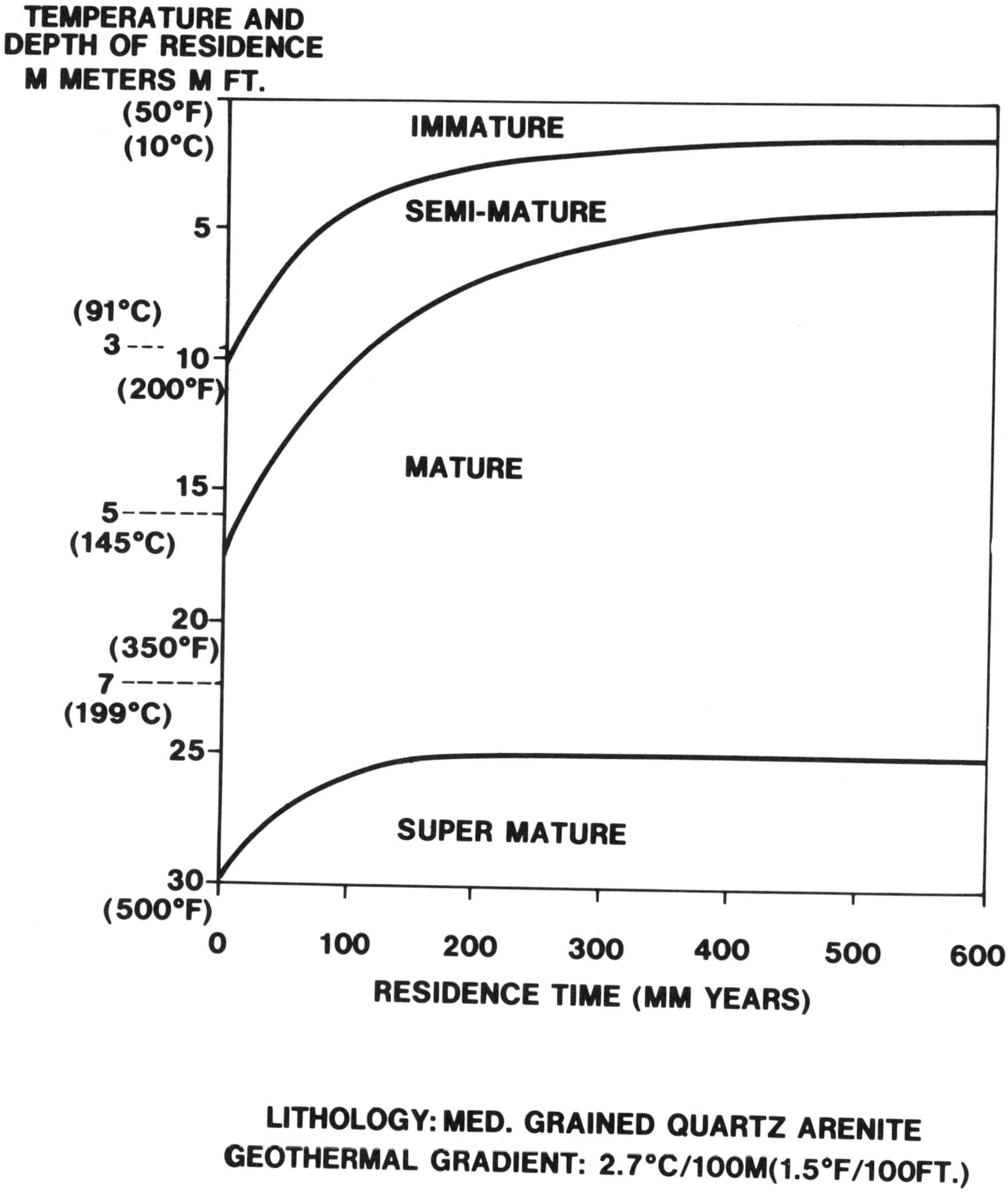

FIG. 11.—Influence of temperature and time on burial diagenesis of porosity.

hibit qualitatively similar relationships between the textural stages of mesodiagenesis and porosity (Galloway, 1974; Hayes, this volume).

Diagenesis of Quartz Arenites

The sandstones of the highest mineralogical stability, the quartz arenites, have been selected to trace in detail the role of secondary porosity during the course of their diagenesis. This choice was made for two reasons: (1) the data base for this sandstone lithology is the most comprehensive; and (2) primary porosity survives much longer in quartz arenites than in sandstones of lower mineralogical maturity. Secondary porosity may be expected to play a relatively less important role in quartz arenites compared to the latter. Thus a baseline for the significance of secondary porosity can be established.

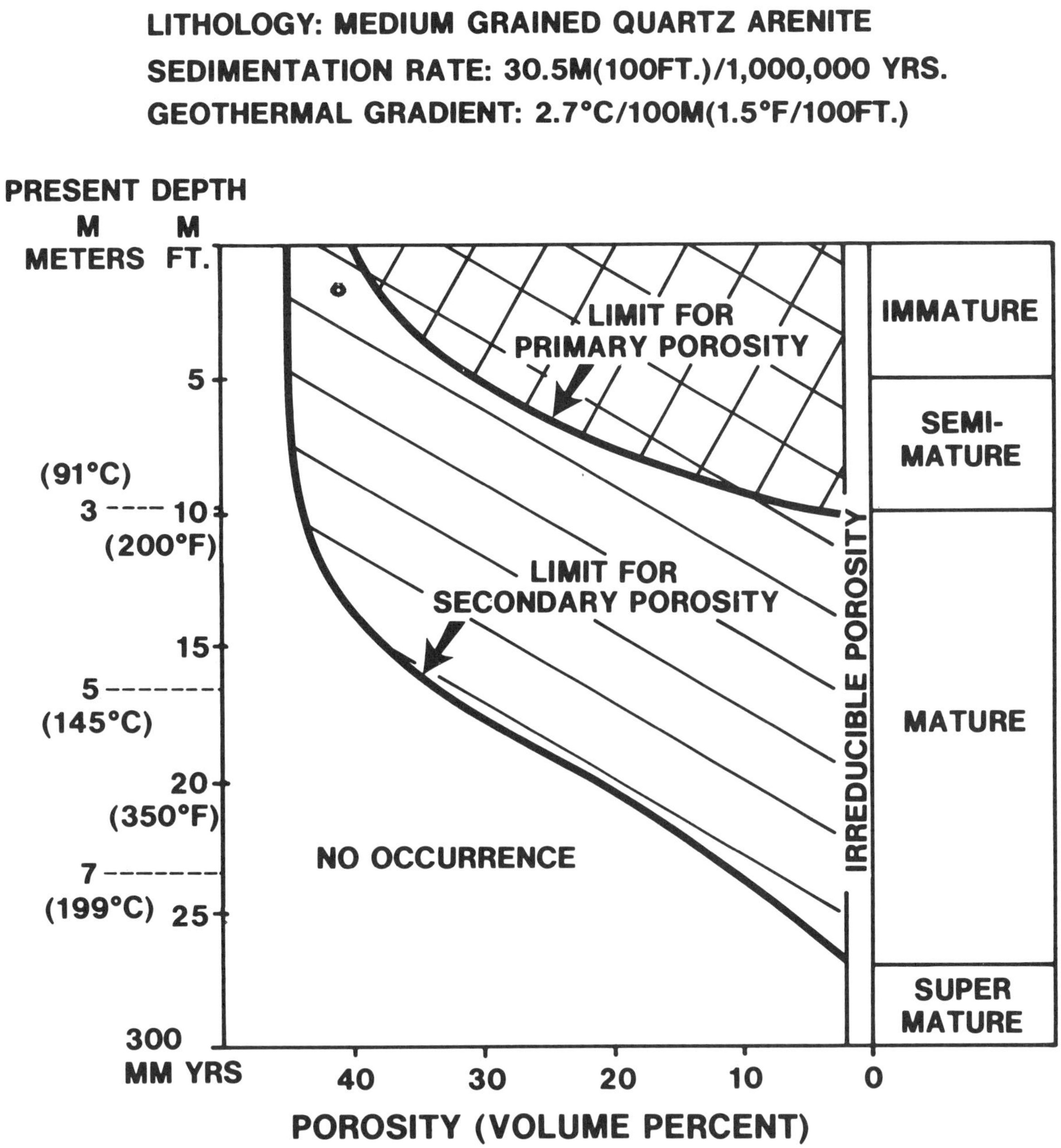

FIG. 12.—Maximum porosity of quartz arenites versus depth of burial.

It is necessary to follow the path of typical diagenetic histories of individual sandstones in order to better understand the vertical distribution of primary and secondary porosity. Four types of diagenetic histories will be discussed: (1) quartz arenites free of carbonate; (2) quartz arenites with sedimentary and eogenetic carbonate; (3) quartz arenites with mesogenetic carbonate; (4) quartz arenites with sedimentary, eogenetic and mesogenetic carbonate.

The diagenesis diagrams (Figs. 15–20) chart the composition of hypothetical quartz arenites from the time of deposition to the advent of metamorphism. These diagrams have been constructed by piecing together petrographical and geological observations of numerous samples.

Quartz arenites free of carbonate.—The loss of primary porosity in quartz arenites is best examined in samples that remain free of carbonate or other readily soluble constituents throughout their diagenesis. The primary intergranular porosity and the rock volume of the quartz arenite shown in Figure 15 are reduced by mechanical compaction during the immature stage. Chemical compaction and reprecipitation of the dissolved silica as cement reduce the intergranular porosity to irreducible lamellar porosity during the semi-mature stage. A considerable reduction in rock volume accompanies this process. Thereafter the sandstone remains unchanged until some recrystallization occurs during the supermature stage (Fig. 13A).

The quartz arenite illustrated in Figure 16 received silica cement from an extrastratal source with the onset of the semi-mature stage. The silica cementation prevented chemical compaction (Fig. 13B). Many coarse grained quartz arenites show this type of diagenesis. Possible sources of the extrastratal silica cement include: (1) chemical compaction of adjacent very fine-grained quartz arenites and quartz siltstone; (2) replacement of quartz grains by carbonate; (3) degradation of feldspars; and (4) clay mineral diagenesis in adjacent shales.

The quartz arenite shown in Figure 17 experienced chemical compaction without reprecipitation of the dissolved silica as cement. Therefore, the rock volume is greatly reduced during the semi-mature stage (Fig. 13C). Of necessity, the dissolved silica migrates away from the sandstone bed. Very fine grained quartz arenites frequently follow this diagenetic path.

The three case histories outlined above demonstrate the early loss of primary porosity. Clearly, mesogenetic decarbonatization porosity cannot develop because of the absence of carbonate throughout the history of the rocks. Such carbonate-free quartz arenites are best suited for determining the first three stages of burial diagenesis.

Quartz arenites with sedimentary and/or eogenetic carbonate.—A large percentage of quartz arenites contain significant amounts of sedimentary and eogenetic carbonate (as also do other sandstone lithologies). The sources of this carbonate are: (1) erosion of carbonate rocks; (2) marine environments; (3) evaporitic environments; (4) deltaic environments; (5) lacustrine and paludal environments; (6) fluviatile environments; and, (7) subaerial environments. The dominant original carbonate minerals are siderite, dolomite, low-magnesium calcite, high magnesium calcite, and aragonite. In most instances the latter two minerals subsequently convert to low-magnesium calcite.

The diagenesis diagram of a fairly common type of quartz arenite is shown in Figure 18. The original sand contains some sedimentary carbonate. Carbonate cement fills part of the intergranular porosity in the eogenetic regime and, in doing so, moderately lithifies the sandstone. Because of this lithification, no mechanical compaction occurs during the immature stage of mesodiagenesis. Chemical compaction takes place during the semi-mature stage, considerably reducing rock volume and porosity. The dissolved silica reprecipitates as intergranular cement within the sandstone bed and further reduces the porosity. At the beginning of the mature stage "A," the sandstone contains only irreducible lamellar porosity (Fig. 13D). However, a considerable amount of porosity is subsequently created through complete decarbonatization (Figs. 13D, E). Minor mechanical and chemical compaction occurs during the remaining mature stage "A." Chemical compaction becomes more active during the mature stage "B" and the secondary porosity declines to an irreducible remnant of lamellar porosity. At this point the supermature stage is reached.

This diagenetic history demonstrates that secondary porosity survives chemical compaction longer during deep burial than primary porosity does at much shallower depth. Decarbonatization can occur at any stage of burial diagenesis and

FIG. 13.—Porosity diagenesis of quartz arenites. *A*, Complete loss of reducible primary porosity by quartz dissolution along grain contacts (chemical compaction) and syntaxial quartz overgrowth. Quartz arenite lacking carbonate. Mature stage. Lower Cretaceous, Mackenzie Delta, Canada. Half crossed nicols. *B*, Complete loss of reducible primary porosity due to cementation by syntaxial quartz overgrowth (chemical compaction). Quartz arenite lacking carbonate. Mature stage. Lower Cretaceous, Mackenzie Delta, Canada. Half crossed nicols. *C*, Complete loss of reducible primary porosity by quartz dissolution at grain contacts (chemical compaction). Quartz arenite, silty/argillaceous, lacking carbonate. Mature stage. Lower Cretaceous, Mackenzie Delta, Canada. Half crossed nicols. Q = quartz grains; M = silty/argillaceous matrix; PY = pyritized material. *D*, Complete loss of reducible primary porosity due to mechanical compaction and dissolution at grain contacts. Quartzose/sideritic/argillaceous/cherty/micaceous litharenite with abundant sedimentary and eogenetic siderite. Mature stage. Jurassic, Statfjord Field, North Sea. Polarizer only. Q = quartz grains (white); S = siderite (black); MI = mica SH/CH = shale fragments and chert grains (brown). *E*, Partial mesogenetic dissolution of sedimentary and eogenetic siderite created secondary porosity. Sideritic/quartzose litharenite. Mature stage. Jurassic, Statfjord Field, North Sea. Polarizer only. P = secondary porosity (blue); Q = quartz grains (white); S = siderite. *F*, Near complete mesogenetic dissolution of sedimentary and eogenetic carbonate created 40 volume percent secondary porosity (core analysis). Same original composition as Figure 13E. Mature stage. Jurassic, Statfjord Field, North Sea. Polarizer only. P = secondary porosity (blue); Q = quartz grains (white); S = siderite (black).

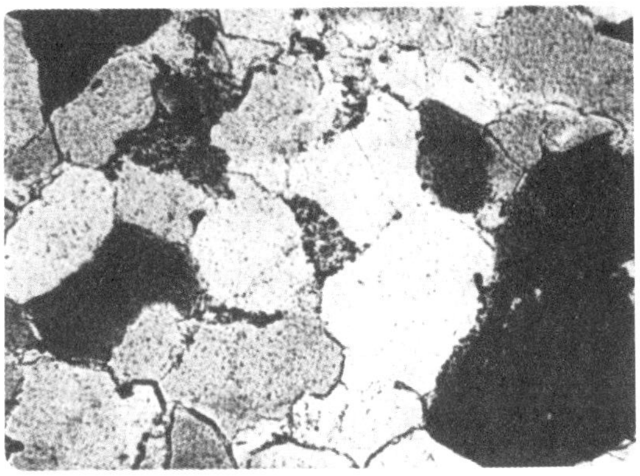

A .2 MM

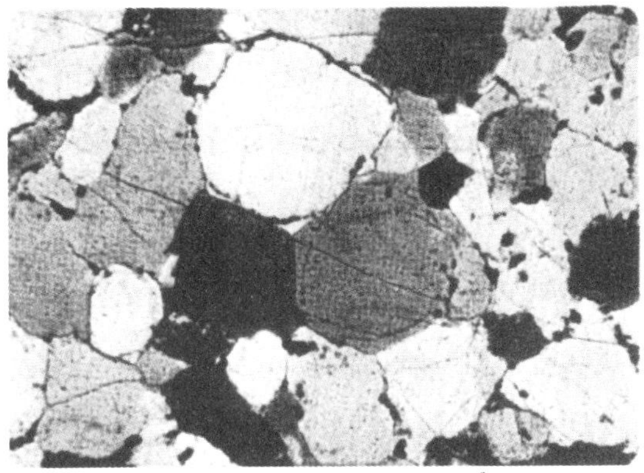

B .3 MM

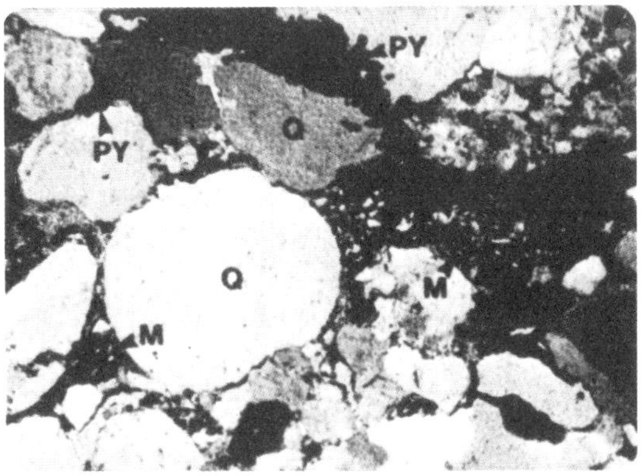

C .2 MM

.2 MM D

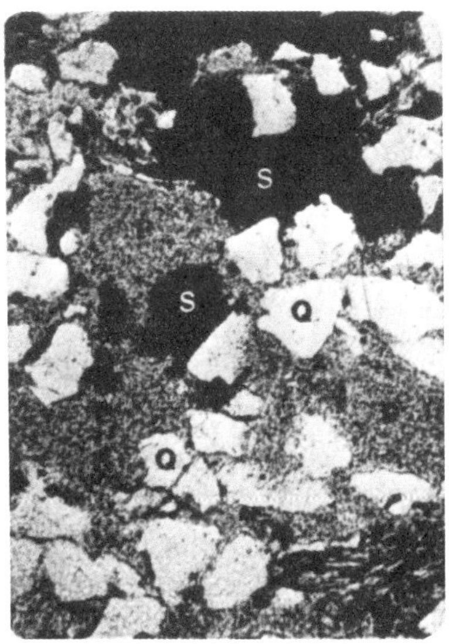

E .2 MM

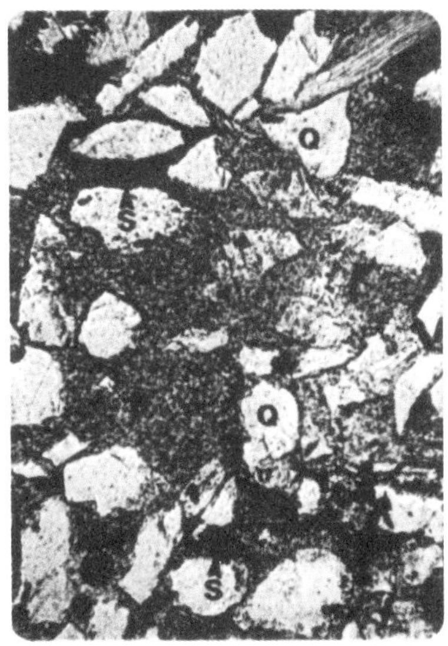

F .2 MM

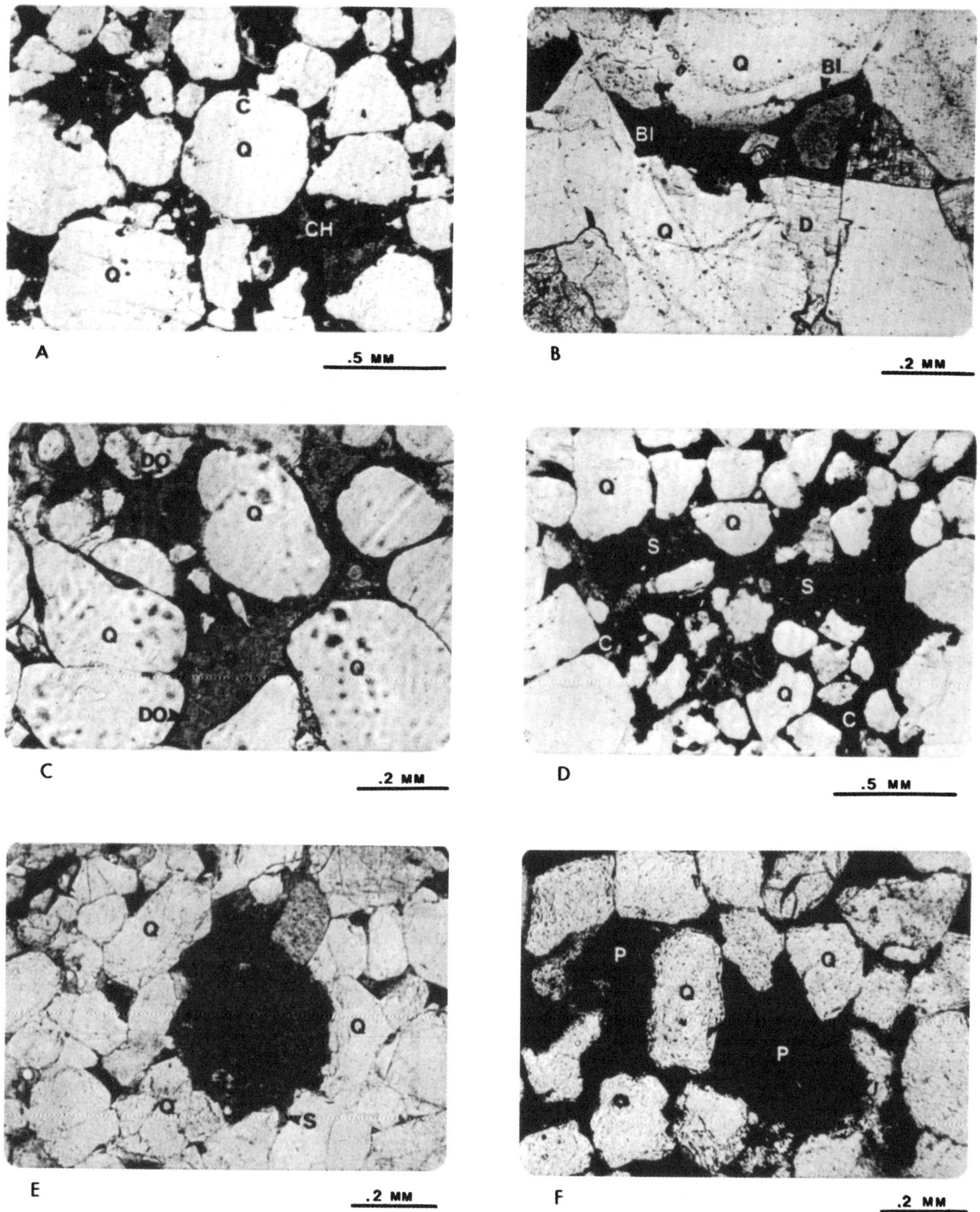

FIG. 14.—Porosity diagenesis of quartz arenites. *A*, Complete loss of reducible primary porosity due to (1) mechanical compaction (fractured chert grain); (2) incipient chemical compaction; and, (3) cementation by mesogenetic poikilotopic calcite. Quartz arenite. Mature zone. Lower Cretaceous, Mackenzie Delta, Canada. Polarizer only. Q = quartz grains (white); CH = chert grain (brown); C = calcite cement (red). *B*, Partial mesogenetic dissolution of mesogenetic dolomite created secondary porosity in Cambrian quartz arenite. Prior to dissolution, mesogenetic dolomite replaced mesogenetic syntaxial quartz overgrowths along the grain boundaries, and completely cemented reducible primary porosity. Migration of crude oil occurred after secondary porosity was formed as documented by bitumen coating secondary pores. Mature zone. Northwest Territories, Canada. Polarizer only. P = secondary porosity (blue); Q = quartz grains and syntaxial quartz overgrowth; D = cementing and replacive dolomite; BI = bitumen. *C*, Complete dissolution of mesogenetic calcite cement created secondary

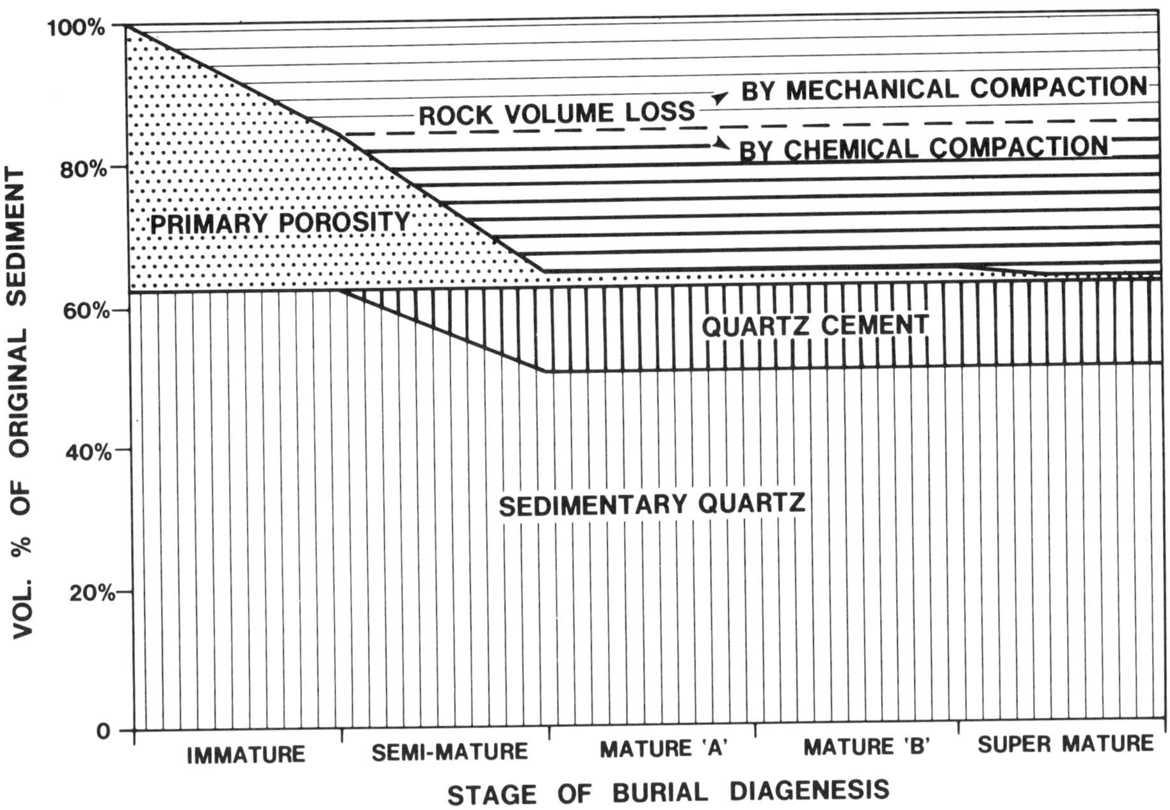

FIG. 15.—Burial diagenesis of quartz arenite with quartz dissolution equal to quartz precipitation.

individual sandstones may be subjected to one or several periods of decarbonatization. However, decarbonatization reaches a pronounced maximum at the beginning of the mature stage "A" in quartz arenites. Decarbonatization is certainly not inevitable; many quartz arenites containing carbonates are not affected by it and remain tight throughout the semi-mature and mature stages.

Quartz arenites with mesogenetic carbonate.— Many quartz arenites that were buried containing no carbonate, later receive mesogenetic carbonate in the form of cement or replacement. The dominant mesogenetic carbonate minerals are calcite, dolomite, and siderite. Mesogenetic carbonatization is frequently succeeded by partial or complete decarbonatization.

Figure 19 illustrates a typical case history of mesogenetic carbonatization and decarbonatiza-

←

porosity in Cambrian quartz arenite (same core as Figure 5A). Cementation by precursor calcite occurred after minor chemical compaction and cementation by syntaxial quartz overgrowth (semi-mature stage). Migration of crude oil occurred after decementation and left secondary pores partially filled with dead oil. Mature zone. Northwest Territories, Canada. Polarizer only. P = secondary porosity (blue); Q = quartz (white); DO = dead oil (yellow); *D*, Complete loss of reducible primary porosity in quartz arenite containing sedimentary and eogenetic siderite and mesogenetic calcite cement. Sequence of porosity loss: (1) mechanical compaction of siderite pellets; (2) chemical compaction and minor mesogenetic cementation by syntaxial quartz overgrowth; and, (3) mesogenetic cementation by poikilotopic calcite. Mature zone. Lower Cretaceous, Mackenzie Delta, Canada. Polarizer only. Q = quartz (white), S = siderite (brown); C = calcite. *E*, Near complete dissolution of sedimentary and eogenetic siderite and mesogenetic calcite created secondary porosity in quartz arenite. Dissolution of carbonate followed after complete loss of reducible primary porosity as described for Figure 14D. Mature zone. Lower Cretaceous, Mackenzie Delta, Canada. Polarizer only. P = secondary porosity (blue); Q = quartz grains and syntaxial quartz overgrowth (white); S = siderite (very high relief). *F*, Complete dissolution of sedimentary, eogenetic and mesogenetic carbonate in quartz arenite. Precursor mesogenetic carbonate marginally replaced quartz grains and syntaxial quartz overgrowth. Dissolution of carbonate followed after complete loss of reducible primary porosity as described for Figure 14D. Mature zone. Lower Cretaceous, Scotian Shelf, Canada. Polarizer only. P = secondary porosity (blue); Q = quartz (white).

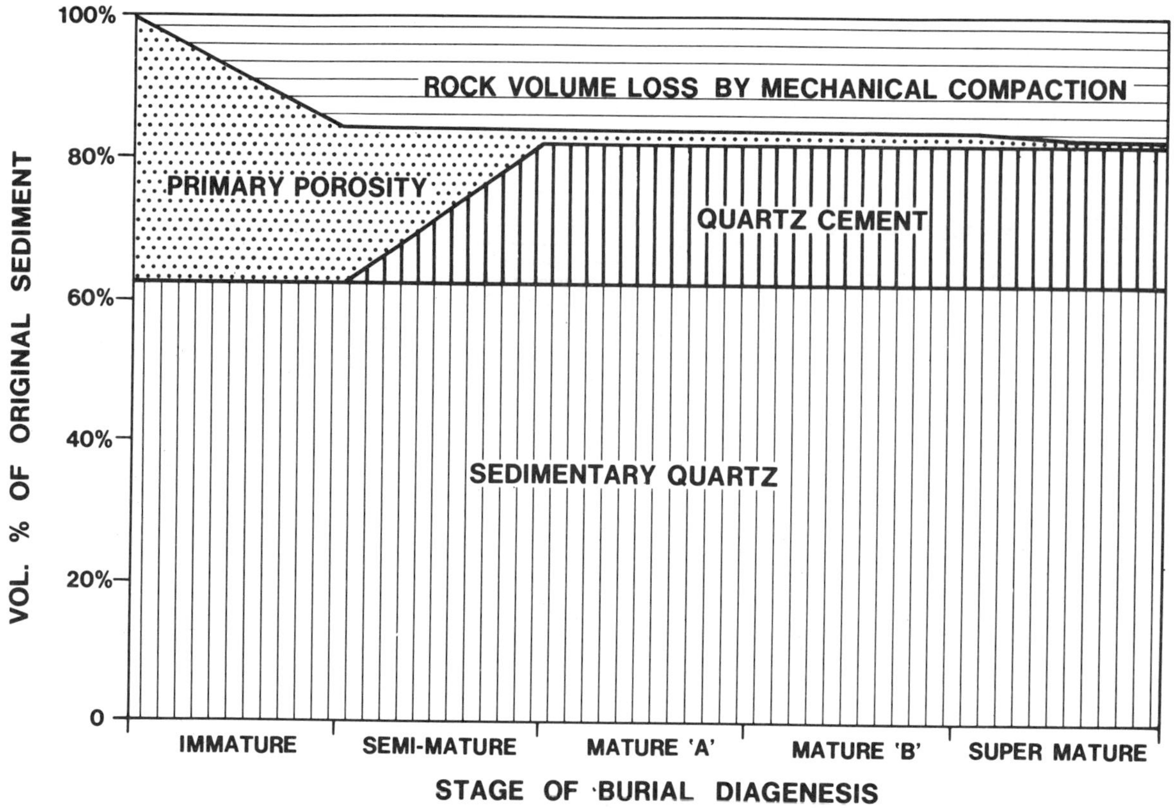

FIG. 16.—Burial diagenesis of a quartz arenite with extrastratal source of quartz cement and without quartz dissolution at grain contacts.

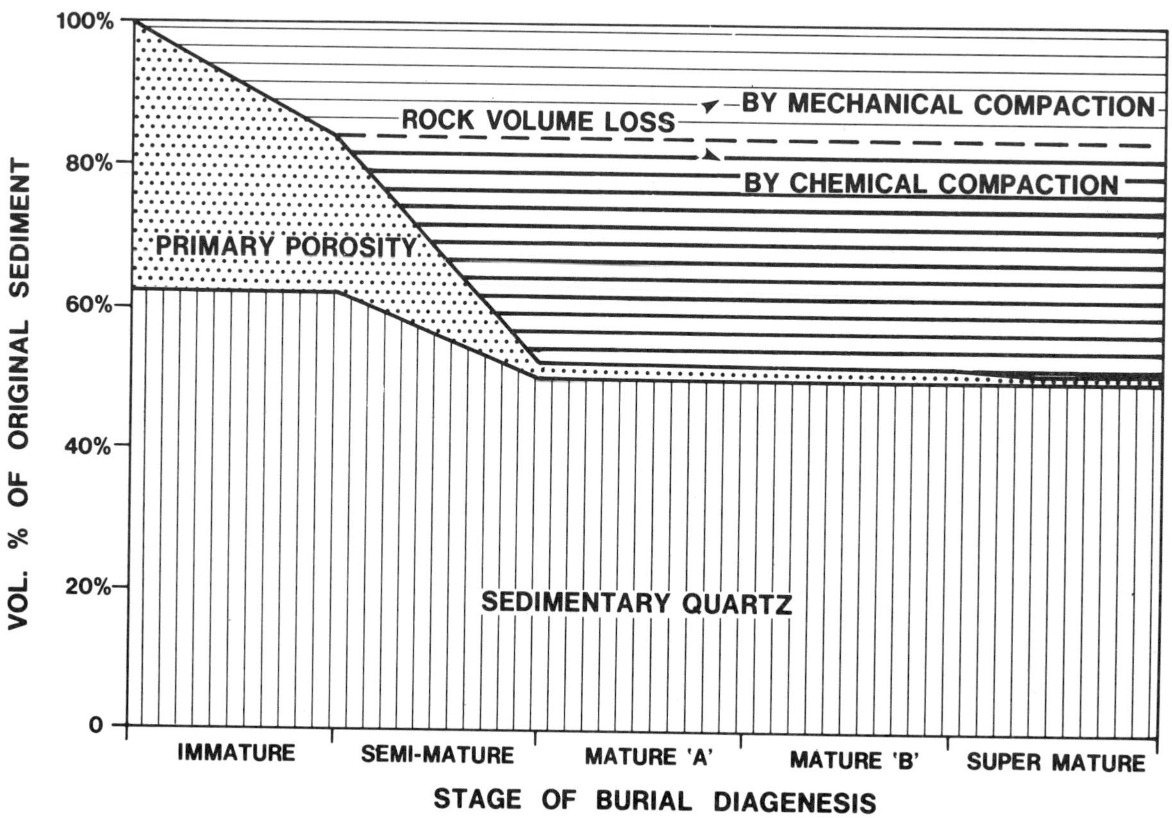

FIG. 17.—Burial diagenesis of quartz arenite with quartz dissolution at grain contacts and without cementation.

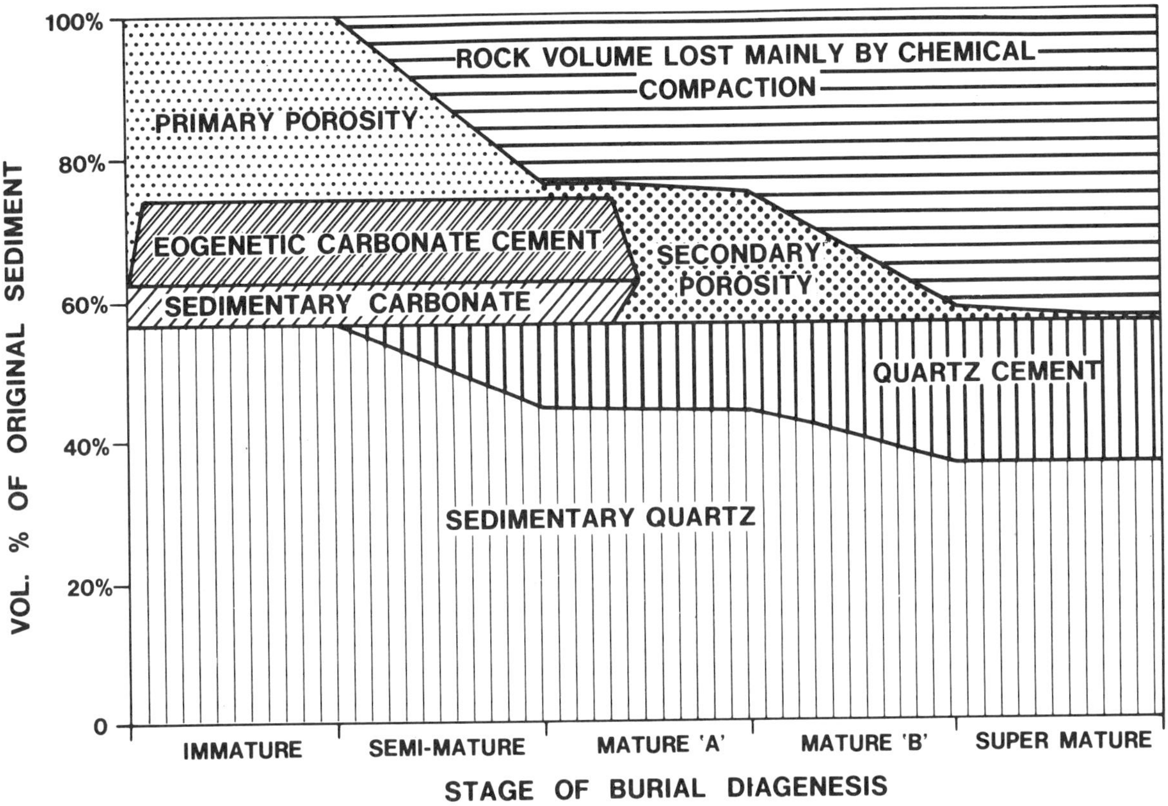

FIG. 18.—Burial diagenesis of a quartz arenite showing eogenetic carbonatization and mesogenetic decarbonatization.

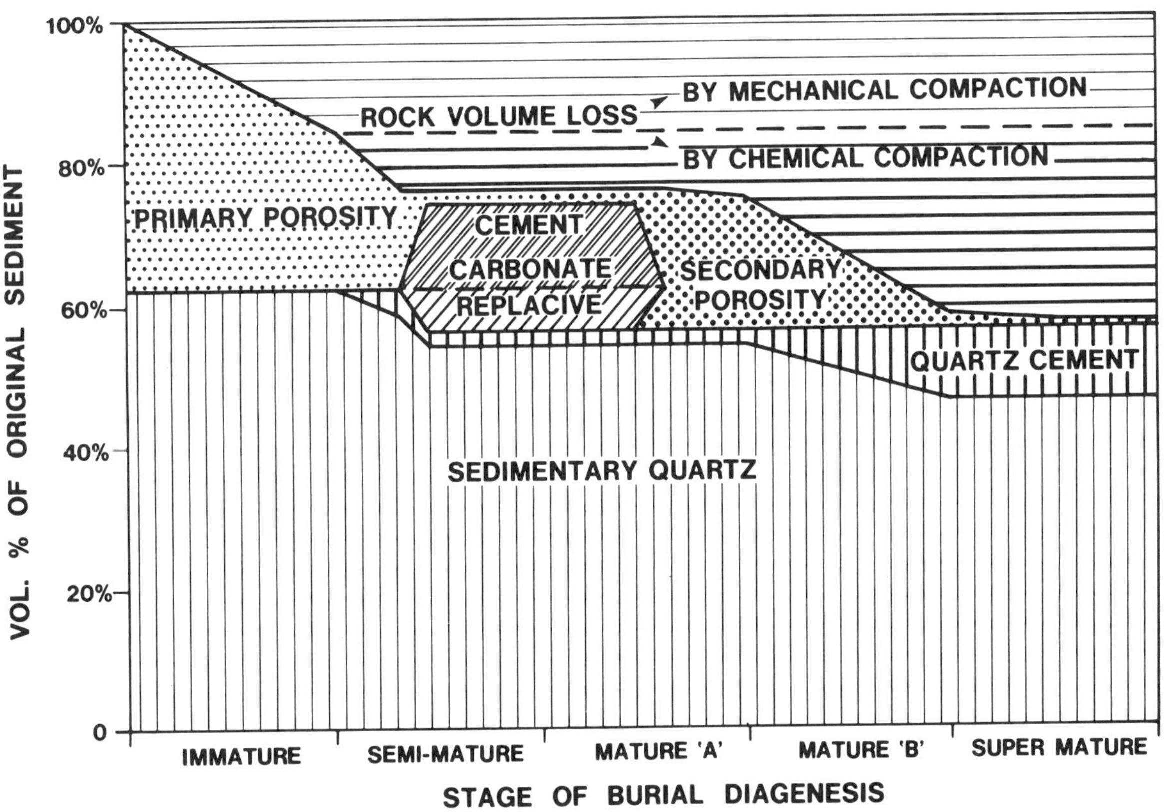

FIG. 19.—Burial diagenesis of a quartz arenite with mesogenetic carbonatization and decarbonatization.

tion. Diagenesis during the immature and early semi-mature stages follows the same path as in Figure 15. Carbonatization during the semi-mature stage reduces the remaining intergranular porosity to an irreducible volume of lamellar porosity and replaces part of the quartz cement and margins of quartz grains. (Fig. 14). The carbonatization also halts chemical compaction. No further change occurs until decarbonatization takes place during the early mature stage (Figs. 14B, C). The volume of the decarbonatization porosity is larger than the volume of primary porosity at the onset of carbonatization because carbonate cement and replacive carbonate have both been completely leached. The subsequent diagenesis follows a similar pattern to that described for Figure 18.

Carbonatization can occur at any stage of mesogenetic sandstone diagenesis. However, in quartz arenites a distinct maximum of carbonatization occurs during the last two-thirds of the semi-mature stage with carbonate cement occupying mainly primary porosity. This zone of maximum carbonatization in many sedimentary basins forms a distinct zone of carbonate enrichment which may be referred to as a "carbonate curtain."

During the process of mesogenetic carbonatization of quartz arenites large volumes of quartz are replaced by carbonate. The silica must go into aqueous solution and is available for reprecipitation. This silica appears to be a major source of the silica overgrowth cement which reduces primary porosity during the semi-mature stage.

Quartz arenites with sedimentary, eogenetic and mesogenetic carbonate.—A fairly common, although complex, case history of diagenesis is shown in Figure 20. The original quartz sand contains a moderate amount of sedimentary carbonate. Eogenetic replacement and cementation add more carbonate and slightly consolidate the sediment. Mechanical compaction during the immature stage is only minor because of the eogenetic carbonatization. Some mesogenetic carbonatization, including both replacement and cementation, occurs after some chemical compaction during the semi-mature stage. Chemical compaction resumes its activity during the remaining part of the semi-mature stage because mesogenetic cementation is incomplete (Fig. 14D). At the same time incomplete decarbonatization takes place creating some secondary porosity. The sandstone enters the mature zone containing considerable amounts of secondary porosity and a

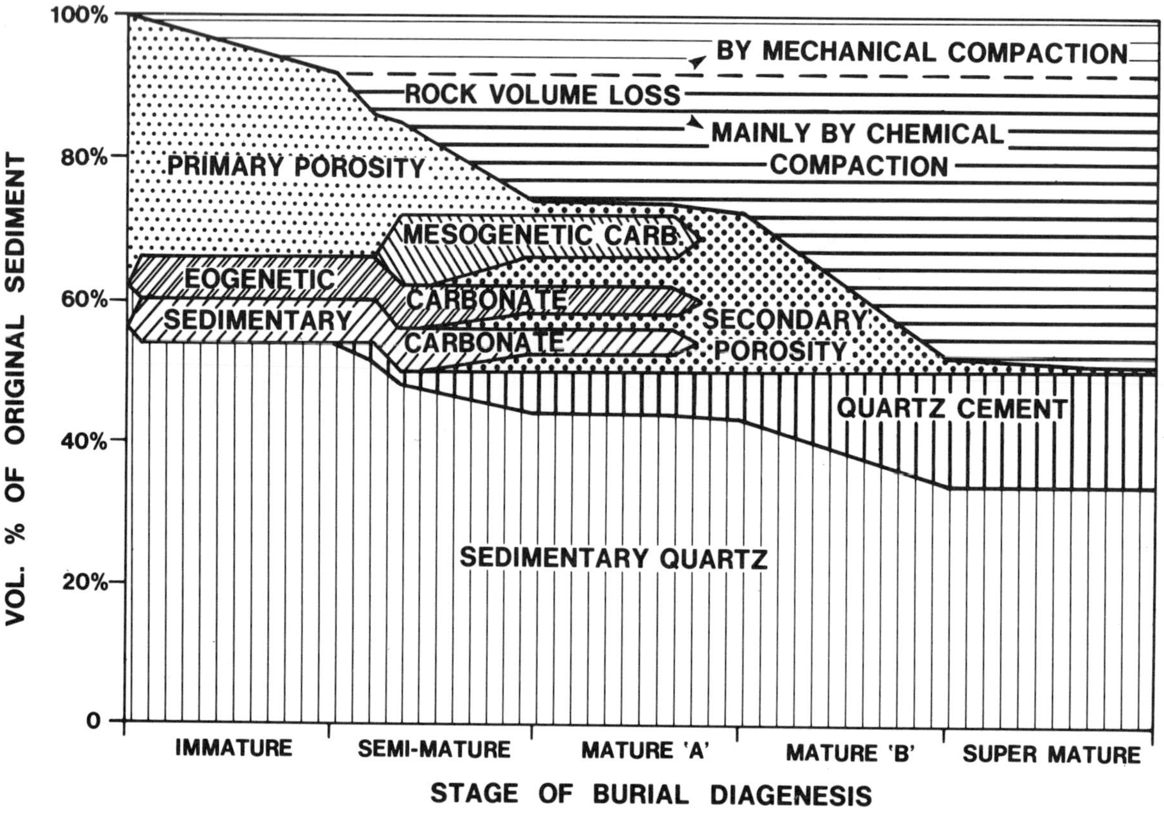

FIG. 20.—Burial diagenesis of a quartz arenite showing eogenetic carbonatization and mesogenetic carbonatization with mesogenetic decarbonatization creating an overlap.

small amount of irreducible primary porosity (Fig. 14E). During mature stage "A" all remaining carbonate constituents are dissolved, greatly increasing the amount of secondary porosity (Fig. 14F). Destruction of the secondary porosity during mature stage "B" follows a similar pattern to that described for Figure 18. The final rock volume in the supermature stage is only half that of the original quartz sand.

The above case history shows that decarbonatization can create exceedingly high porosities in quartz arenites which contain sedimentary, eogenetic, and mesogenetic carbonate including a considerable fraction of replacement carbonate. Multiple pulses of incomplete decarbonatization are very common, but are extremely difficult to recognize microscopically. Individual quartz arenites may also be subjected to several phases of mesogenetic carbonatization, some of which can be readily distinguished under the petrographic microscope. Usually, electron luminescence microscopy and microprobe analysis must be applied to solve the more complex sequences of diagenetic events.

Synthesis.—The principal observations concerning the burial diagenesis of quartz arenites are summarized in Figure 21. The sequence of diagenetic events is mainly controlled by residence temperature and residence time. Vitrinite associated with quartz arenites records the temperature-time exposure by its change in reflectance which can be used as a scale for the thermomaturation of quartz arenites. Thus specific reflectivity values of vitrinite correspond to textural stages of mesodiagenesis.

Mechanical compaction during the immature stage mainly affects primary porosity, reducing it by about one quarter. Mesogenetic carbonatization, decarbonatization, and decarboxylation of organic matter in intercalated shales operate at extremely low levels during the immature stage.

Quartz diagenesis, that is, the chemical compaction of quartz arenites and silica cementation, operates most strongly during the semi-mature stage, mainly reducing primary porosity. Carbonatization reaches its maximum during the last two-thirds of the semi-mature stage, chiefly filling primary porosity but also a considerable volume of secondary porosity. Decarbonatization and decarboxylation are fairly active throughout the semi-mature stage. Moderate generation of liquid hydrocarbons in intercalated shales may take place during the latter part of the semi-mature stage. Loss of porosity predominates over porosity gain though a considerable amount of secondary porosity is formed. All effective primary porosity has been destroyed by the end of the semi-mature stage but an appreciable amount of

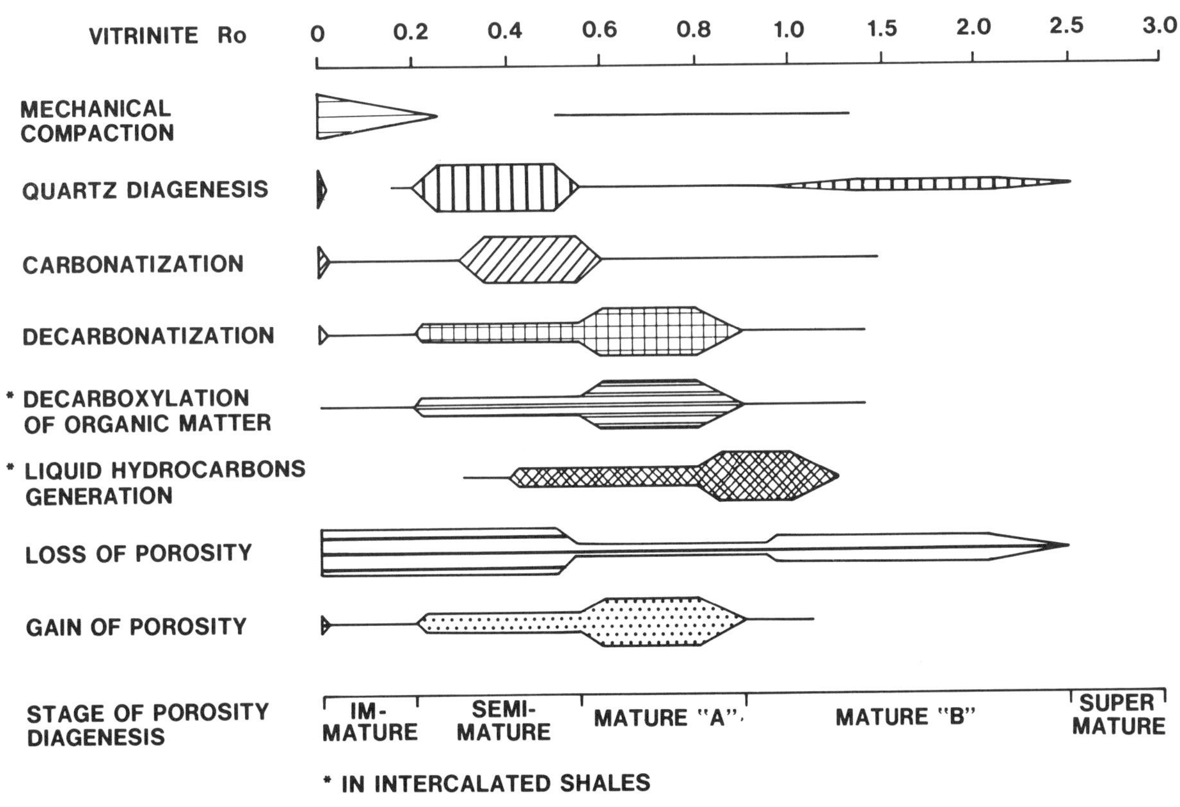

FIG. 21.—Burial diagenesis of quartz arenites.

secondary porosity does exist.

The mature stage "A" is characterized by maximum decarbonatization and decarboxylation of organic matter. Generation of liquid hydrocarbons is, for the most part, moderate but increases to maximum levels towards the end of the mature stage "A." Carbonatization, quartz diagenesis and mechanical compaction operate at low levels. Gain of secondary porosity is at its maximum and results mainly from decarbonatization. It exceeds loss of secondary porosity by a wide margin.

The initial phase of mature stage "B" shows low levels of carbonatization, decarboxylation, decarbonatization, mechanical compaction, and quartz diagenesis, whereas the generation of liquid hydrocarbons continues for a short period, at its maximum level before decreasing. Loss of secondary porosity is minor and gain of secondary porosity is very low.

Quartz diagenesis is the only significant diagenetic activity during the intermediate and final phases of the mature stage "B." It eventually eliminates any reducible secondary porosity. Organic metamorphism generates large volumes of methane during the intermediate phase of the mature stage "B."

The zone of maximum active decarbonatization underlies the zone of maximum active carbonatization. It appears that the former is the source of carbonate for the latter. This means that carbonate must be moving upward in aqueous solution to form the "carbonate curtain." The reasons for the precipitation of carbonate at the zone of carbonatization have not yet been determined satisfactorily. It appears as if the diagenesis of clay minerals in the intercalated shales may influence the chemistry of interstitial waters (change of ionic ratios, alkalinity, pH values) in such a way as to cause carbonate precipitation. Alternatively, formation waters which migrate upwards from a geopressure zone into zones of hydrostatic pressure could precipitate carbonates in the zone of pressure transition.

The occurrence of decarbonatization in most sedimentary basins corresponds very closely with the rate of production of carbon dioxide in intercalated shales or coals resulting from decarboxylation of maturing organic matter. Decarbonatization is most intense in quartz arenites which are closely associated with maturing organic-rich hydrocarbon source rocks. It appears as if generation of carbonic acid in the subsurface is the prime cause of mesogenetic decarbonatization. Other apparently common causes for mesogenetic decarbonatization include: (1) physicochemical gradients at the boundaries of geopressure zones; (2) release of water from clay minerals (Savkevic, 1971); and, (3) generation of hydrogen sulfide.

In many instances mesogenetic decarbonatization takes place in very tight quartz arenites. However, fractures and the irreducible lamellar porosity apparently provide sufficient access for the leaching brines to initiate decarbonatization. As secondary porosity is being created, it continuously improves access for the acid waters. The percentage of secondary porosity which can be generated by complete decarbonatization is limited both by the carbonate content of the sandstone and by the mechanical strength of the non-carbonate framework.

Quartz arenites containing carbonate may escape decarbonatization for one of the following reasons: (1) the subsurface environment does not generate chemical conditions for leaching, e.g. lack of organic matter; (2) the leaching capacity of the migrating formation water has been exhausted previously; and, (3) the leaching formation water has no access to the quartz arenite strata.

In the development of a sedimentary basin intensive decarbonatization will likely occur soon after the oldest quartz arenites have reached the mature zone. These quartz arenites contain mainly sedimentary and eogenetic carbonate. Reprecipitation of this dissolved carbonate in the overlying semi-mature zone may form a thin initial "carbonate curtain." Continued subsidence and sedimentation eventually moves this initial "carbonate curtain" into the mature zone where it may be subjected to strong decarbonatization which, in turn, could lead to a thicker "carbonate curtain" in the quartz arenites which have newly arrived in the semi-mature zone. In this manner both the zone of maximum decarbonatization and maximum carbonatization can move continually into younger sediments. In the process much of the carbonate may be recycled many times over and large volumes of carbonate may accumulate in the moving zones of active carbonatization, provided that sufficient organic matter is present throughout the sedimentary sequence to sustain decarbonatization.

Most liquid hydrocarbons and probably most of the natural gas is generated in the mature zone of burial diagenesis after the peak of development of secondary porosity. This places quartz arenites with secondary porosity in a very favorable situation to trap these hydrocarbons. Indeed, it is reasonable to assume that primary migration emplaced most oil and gas found in secondary sandstone porosity (Figs. 14B, C), whereas secondary migration supplied much of the hydrocarbon encountered in primary porosity.

THE POROSITY DIAGENESIS OF SANDSTONES OF INTERMEDIATE AND LOW MINERALOGICAL MATURITY

Sandstones of lesser mineralogical maturity than quartz arenites exhibit a similar relationship

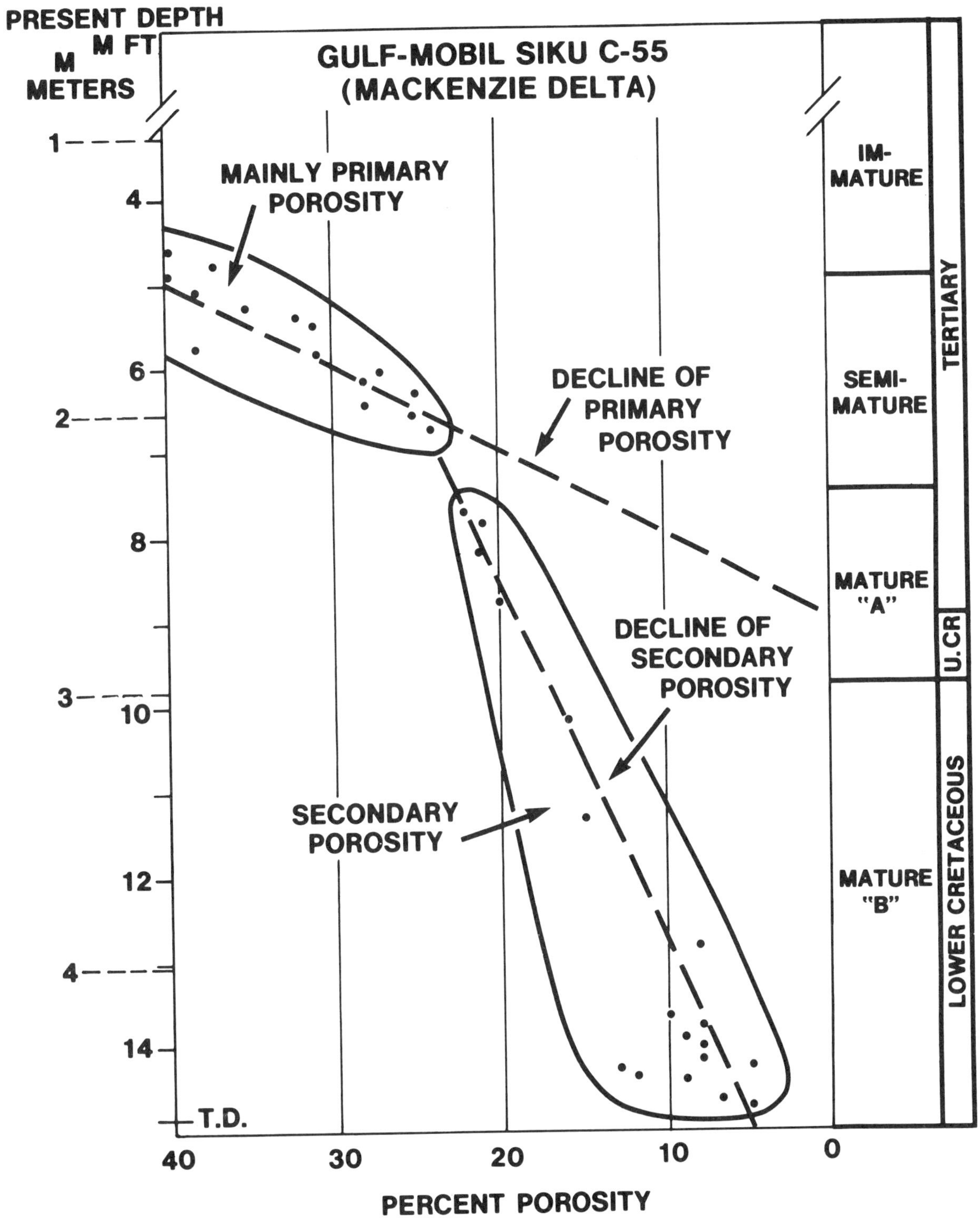

FIG. 22.—Mackenzie Delta: Depth/porosity plot showing decline of primary and secondary porosity (sonic porosities of sandstone intervals >10 ft.).

between the maturation of organic matter and the development of mesogenetic decarbonatization porosity as described above. However, both primary and secondary porosity are reduced by chemical compaction at lower temperatures and at faster rates than observed in quartz arenites. Chemical constituents derived from the dissolution of minerals such as feldspars, micas, amphiboles, and pyroxenes precipitate as mesogenetic cements mainly consisting of clay minerals

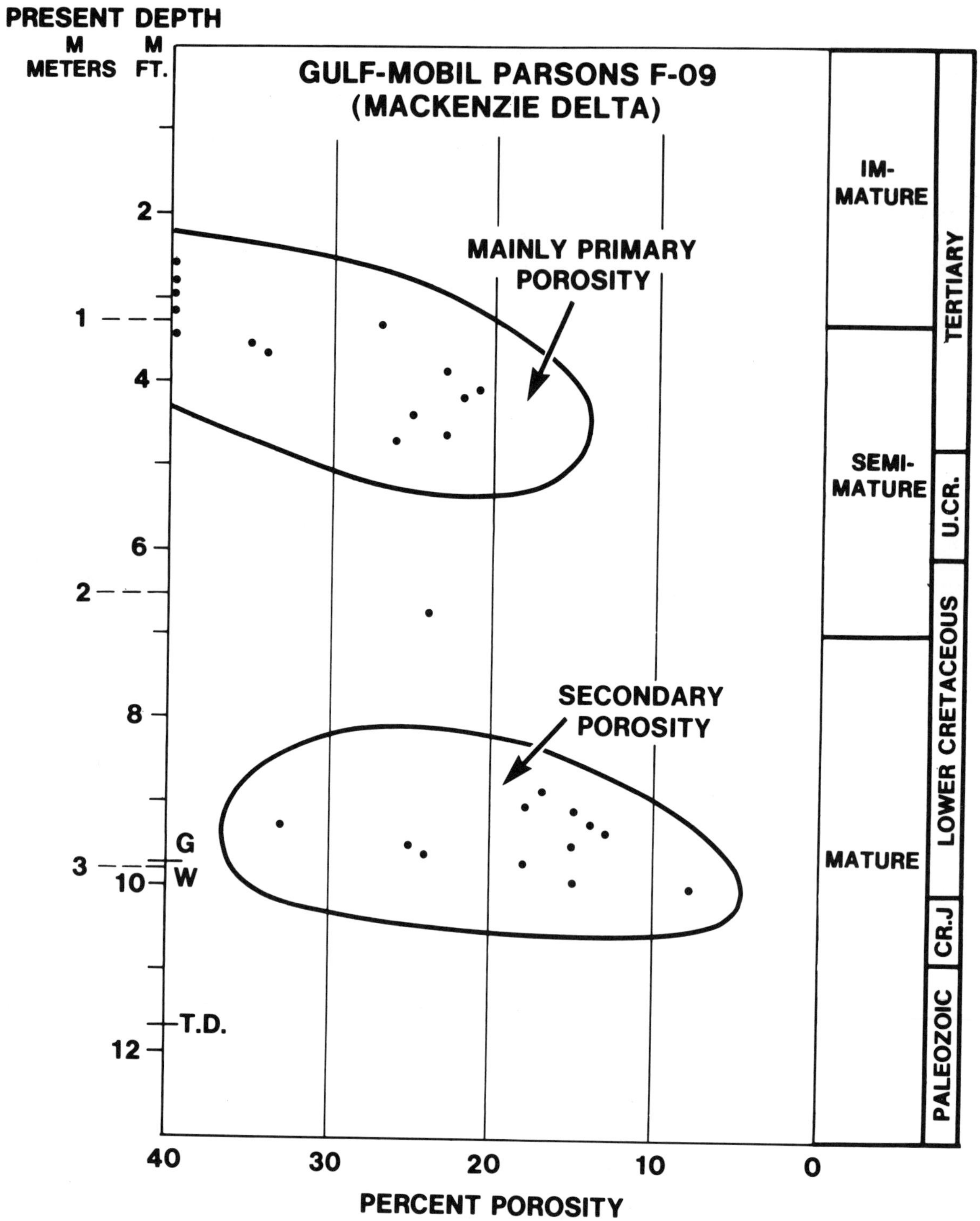

FIG. 23.—Mackenzie Delta: Depth/porosity plot showing wide range of secondary porosity (sonic porosities of sandstone intervals > 10 ft.).

FIG. 24.—Porosity variation in a sandstone zone of mature textural mesodiagenesis, Lower Cretaceous, Mackenzie Delta. *A*, Secondary porosity created by mesogenetic leaching of mainly sedimentary and eogenetic carbonates. Mature stage. Lower Cretaceous quartz arenite, Mackenzie Delta, Canada. Polarizer only. P = secondary porosity (blue); Q = quartz (white). *B*, Secondary porosity created by mesogenetic leaching of sedimentary porosity, eogenetic carbonate and mesogenetic intergranular carbonate cement. Mature stage. Lower Cretaceous quartz

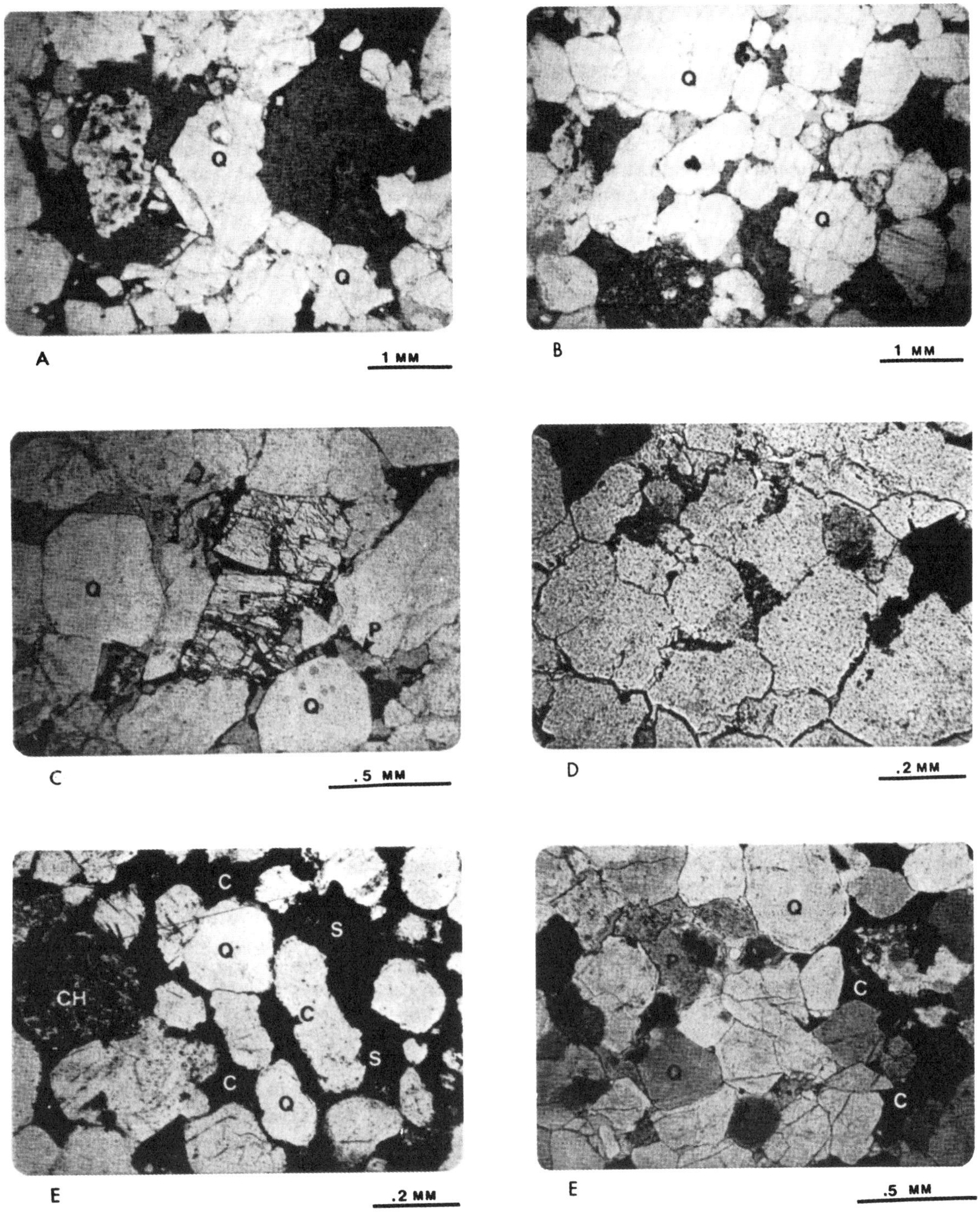

arenite, Mackenzie Delta, Canada. Polarizer only. P = secondary porosity (blue); Q = quartz (white). *C*, Secondary porosity created by mesogenetic leaching of mainly mesogenetic intergranular carbonate cement. Feldspar grains were fractured after leaching of mesogenetic carbonate that partially replaced the feldspar. Mature stage. Lower Cretaceous subfeldspathic arenite. Mackenzie Delta, Canada. Polarizer only. P = secondary porosity (blue); Q = quartz (white); F = feldspar (fractured grains). *D*, Lack of effective porosity due to absence of carbonate when sandstones entered the mature stage. Lower Cretaceous quartz arenite. Mackenzie Delta, Canada. Polarizer only. *E*, Lack of effective porosity due to absence of mesogenetic leaching of sedimentary eogenetic and mesogenetic carbonate. Mature stage. Lower Cretaceous sublithic arenite. Mackenzie Delta, Canada. Polarizer only. Q = quartz grains and mesogenetic syntaxial quartz overgrowth; S = sedimentary siderite pseudomatrix and eogenetic siderite replacing silicate grains along the boundaries and cementing pores (very high relief); CH = chert grain; C = mesogenetic poikilotopic calcite cementing pores and replacing quartz grains and quartz cement (red). *F*, Low effective porosity due to incomplete mesogenetic leaching of mesogenetic calcite. Mature stage. Lower Cretaceous quartz arenite. Mackenzie Delta, Canada. Half crossed nicols. P = secondary porosity (blue); Q = quartz grains and mesogenetic syntaxial quartz overgrowth; C = mesogenetic poikilotopic calcite cement.

and zeolites (Hayes, this volume). Chemical compaction and associated intrastratal precipitation of cements may collectively be termed framework diagenesis, analogous to quartz diagenesis. Mesogenetic carbonatization in these sandstones involves more replacement and less cementation when compared with quartz arenites. Carbonatization tends to culminate during the early phase of mature stage "A" at about the same level of thermomaturation as in quartz arenites.

In general, the development and disappearance of mesogenetic secondary porosity follows a similar pattern in sandstones of intermediate and low mineralogical maturity to that observed in quartz arenites. The detail of the diagenetic histories of sandstones of lesser mineralogical stability is often very complex and much more work remains to be done to establish the basic relationships as clearly as is now possible with quartz arenites. The Cretaceous sandstones of the Labrador Shelf, the Tertiary sandstones of the Mackenzie Delta, and the Mesozoic and Tertiary sandstones of the Cook Inlet are prime candidates for such work. These sandstones of intermediate to low mineralogical maturity contain large hydrocarbon accumulations in reservoirs of secondary sandstone porosity.

EXAMPLES OF POROSITY DISTRIBUTION

Plots of sandstone porosity versus depth are easily constructed using core analyses and/or log interpretations from wells. Petrographic analysis of the sandstones and application of the criteria and concepts developed in this paper allow a much more complete interpretation of such porosity/depth plots than has hitherto been possible.

Mackenzie Delta example.—Figure 22 represents a porosity/depth plot illustrating the rate of decline of primary and secondary sandstone porosity of a well in the Mackenzie Delta which penetrated Tertiary sandstones of intermediate mineralogical maturity and Lower Cretaceous quartz arenites. The Tertiary sandstones above a depth of 2,100 meters have porosity which is predominantly primary and which declines rapidly with depth. Primary porosity has been reduced by both framework diagenesis and by carbonate cementation. Both processes are apparently active at the present time. Below 2100 meters the sandstone porosity is essentially secondary and declines with depth at a much slower rate compared with that of primary porosity. Active maximum decarbonatization occurs between 2,400 and 2,500 meters. The relatively high porosities below 4,000 meters do not necessarily indicate decarbonatization at that depth; rather they represent a zone of exceedingly high decarbonatization with porosities probably exceeding 30 volume percent which subsequently has been reduced to the present levels by quartz diagenesis.

Another well in the Mackenzie Delta with similar stratigraphy to the above well shows a wide range of secondary porosity within the narrow depth interval 2,650 to 3,050 meters (Fig. 23). This zone is in the mature stage "A" of burial diagenesis and the sandstones from 2,650 to 2,900 meters are gas bearing. The secondary porosity is the result of decarbonatization involving the leaching of sedimentary, eogenetic, and mesogenetic carbonate (Figs. 24A–C). High original carbonate content and nearly complete decarbonatization are the attributes of the sandstone intervals with high secondary porosity. Low secondary porosities are shown by sandstone intervals which either originally contained only moderate amounts of carbonate or experienced incomplete decarbonatization (Figs. 24D–F). Accumulation of the gas occurred after decarbonatization during the mature stage "A" of burial diagenesis. The presence of hydrocarbons had little or no effect on the preservation of porosity. However, if the gas bearing sandstones were to be buried deeply enough to enter mature stage "B" of burial diagenesis, then the presence of hydrocarbons would prevent quartz diagenesis and preserve porosity.

Scotian Shelf example.—Porosities of Cretaceous and Upper Jurassic sandstones of a well on the Scotian Shelf are plotted versus depth in Figure 26. The sandstones are principally composed of quartz arenites, feldspathic quartz arenites, and glauconitic quartz arenites, and originally contained considerable amounts of sedimentary and eogenetic carbonate. Intercalated shales vary greatly in their mineralogy and contain very low

FIG. 25.—Sequence of stages of textural mesodiagenesis with depth in Cretaceous sandstones of the Scotian Shelf. *A*, Immature stage. High primary porosity in Cretaceous quartz arenite. Mobil Citnalta I-59, Scotian Shelf, Canada. Polarizer only. P = primary porosity (blue); Q = quartz grains (white). *B*, Semi-mature stage. Primary and secondary porosity in glauconitic quartz arenite. Incomplete mesogenetic cementation by poikilotopic calcite was followed by partial mesogenetic leaching of calcite. Cretaceous. Mobil Citnalta I-59, Scotian Shelf, Canada. Polarizer only. P = primary and secondary porosity (blue); Q = quartz grains and mesogenetic syntaxial quartz overgrowth; C = calcite (red); G = sedimentary and eogenetic glauconite (greenish). *C*, semi-mature stage. Cementation of secondary and primary porosity by mesogenetic syntaxial quartz overgrowth. Cretaceous

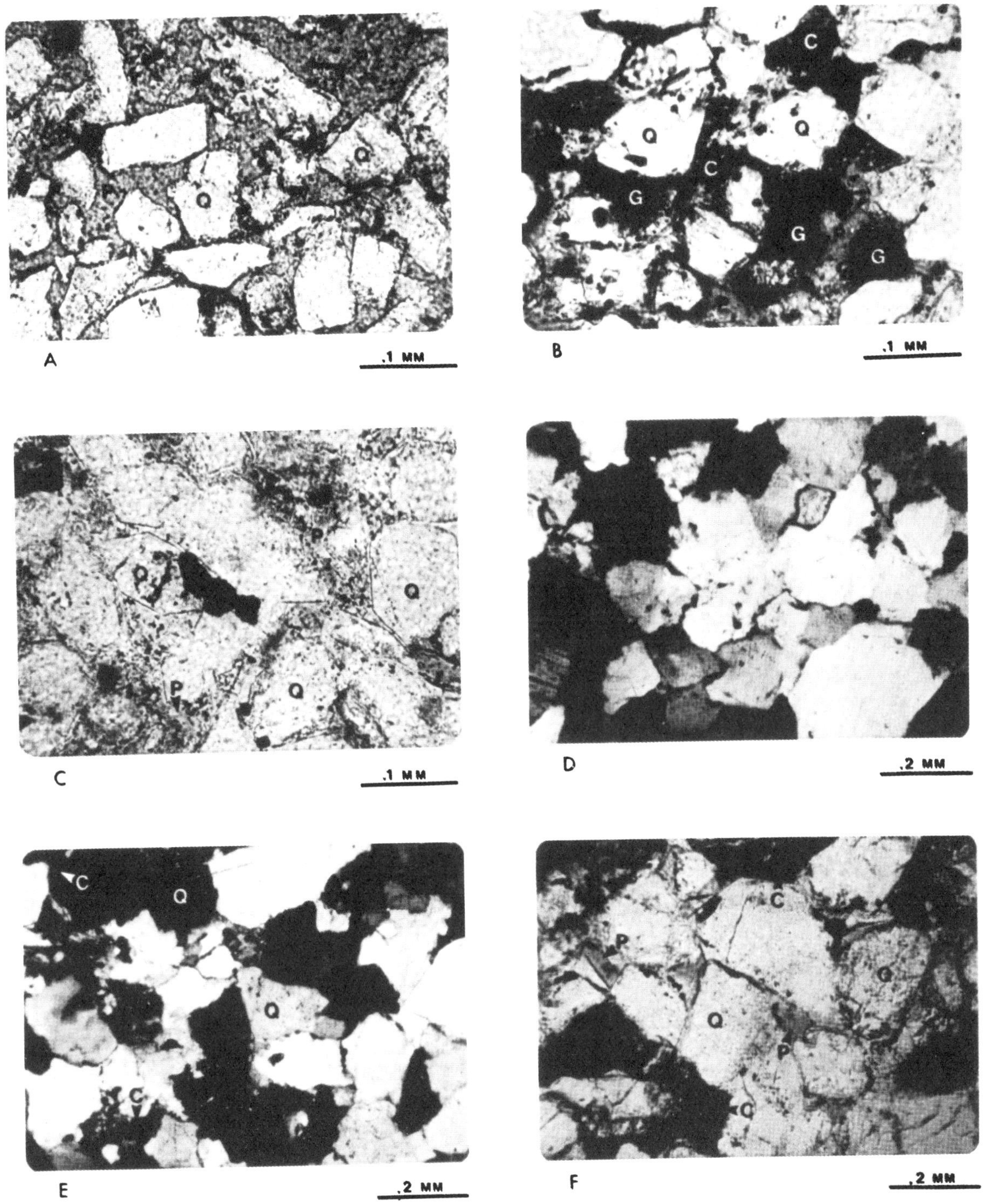

quartz arenite. Mobil Citnalta I-59, Scotian Shelf, Canada. Polarizer only. P = primary and secondary porosity (blue); Q = quartz (white). *D*, Mature stage. Lack of effective porosity due to absence of carbonate when sandstone entered the mature stage. Cretaceous quartz arenite. Mobil Citnalta I-59, Scotian Shelf, Canada. Crossed nicols. *E*, Mature stage. Lack of effective porosity due to absence of mesogenetic leaching of mesogenetic calcite cement. Cretaceous quartz arenite, Mobil Citnalta I-59, Scotian Shelf, Canada. Crossed nicols. Q = quartz grains and mesogenetic syntaxial quartz overgrowth cement; C = mesogenetic poikilotopic calcite cement. *F*, Mature stage. Presence of secondary porosity due to mesogenetic leaching of mesogenetic poikilotopic calcite. Cretaceous subfeldspathic arenite. Mobil Citnalta I-59, Scotian Shelf, Canada. Polarizer only. P = secondary porosity (blue); Q = quartz grains and mesogenetic syntaxial quartz overgrowth cement. C = mesogenetic calcite cement and calcite replacing detrital grains and quartz cement along the boundaries.

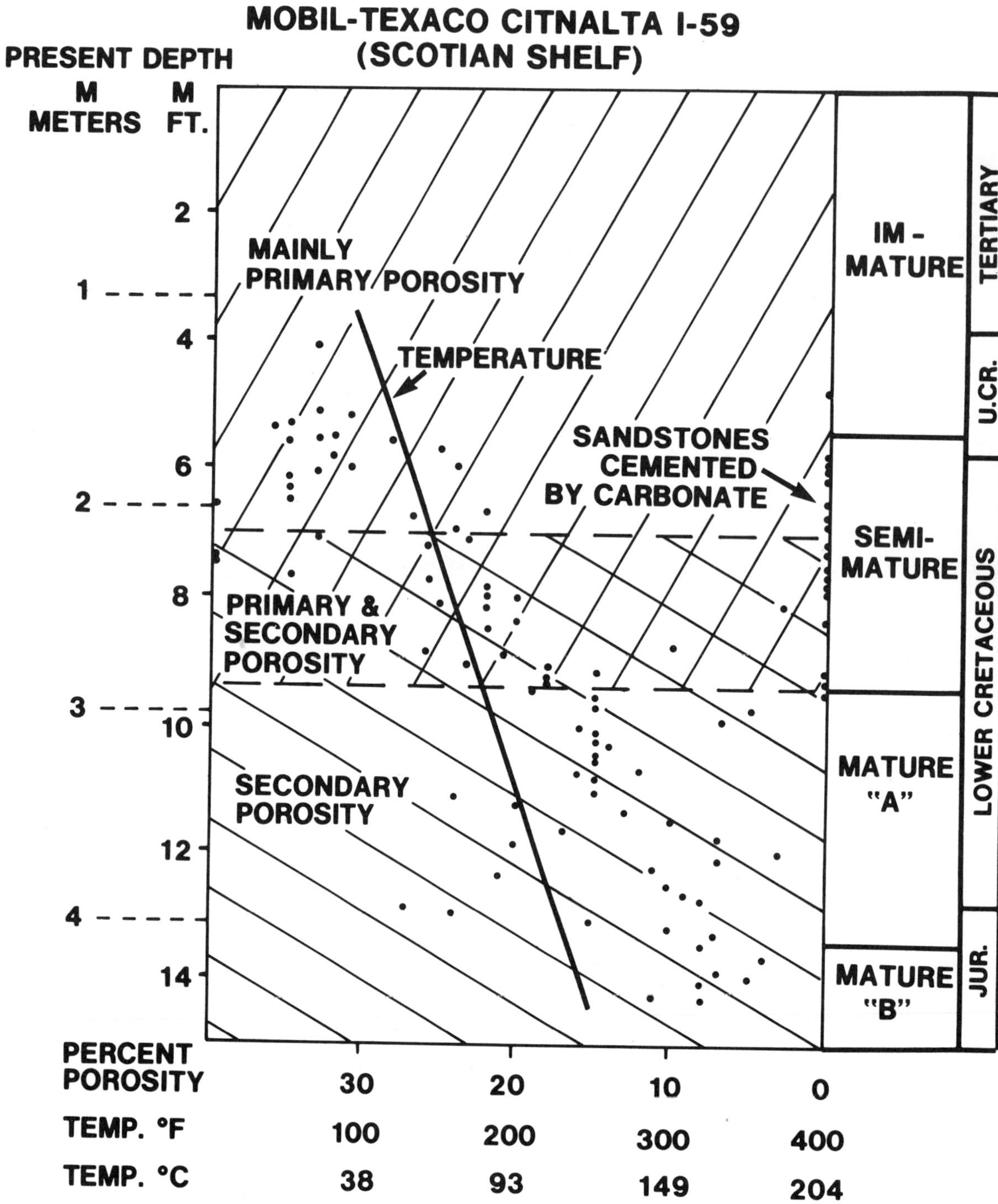

FIG. 26.—Scotian Shelf: Depth/porosity plot showing overlap of primary and secondary porosity (sonic porosities of sandstone intervals >10 ft.).

to moderately high amounts of organic matter of widely varying composition.

Primary and secondary porosity overlap throughout the semi-mature stage of burial diagenesis to form a gradation from very high primary porosities at the top to high secondary porosities at the bottom. Tight sandstones containing abundant mesogenetic carbonate cement are intercalated with the porous sandstones which experienced considerable decarbonatization (Figs. 25A–C). Thus multiple zones of carbonatization alternate with multiple zones of decarbonatization possibly because of the variable content and type of organic matter and lithology of the intercalated

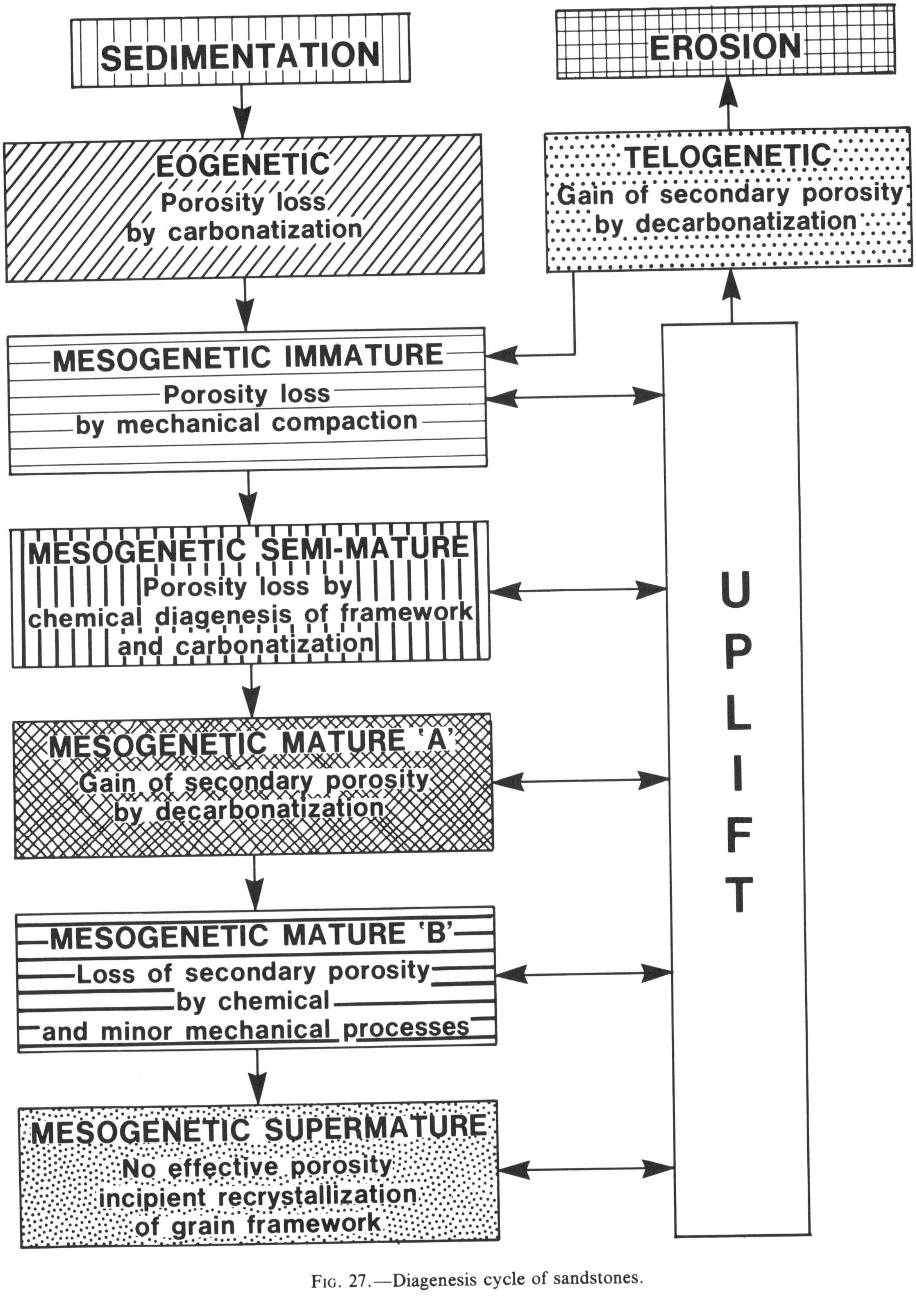

FIG. 27.—Diagenesis cycle of sandstones.

shales. It is likely that a number of horizons of carbonatization and decarbonatization are presently active.

In mature stage "A" of burial diagenesis all sandstones containing carbonate experienced renewed decarbonatization to some degree, whereas carbonatization was negligible (Figs. 25D–F). The sandstones with high secondary porosities are often closely associated with hydrocarbon source rocks, and contain condensate and natural gas. Quartz diagenesis and framework diagenesis of other grains is responsible for the decline of secondary porosity in the mature stage "B" of burial diagenesis.

CONCLUSIONS

The following conclusions concerning the role of secondary porosity in the course of sandstone diagenesis can be drawn from the above discussion.

1. Secondary porosity plays a pre-eminent role in the subsurface mesodiagenesis of sandstones. In addition, a significant amount of secondary porosity may occasionally develop during early diagenesis (eodiagenesis), and frequently develops during erosional diagenesis (telodiagenesis). The origin and destruction of secondary porosity and associated diagenetic processes are key elements in the diagenesis cycle of sandstone (Fig. 27).

2. A large number of textural and genetic types of secondary sandstone porosity exist. Several processes are capable of creating significant volumes of secondary sandstone porosity and these include dissolution of a number of minerals, shrinkage of certain minerals, and fracturing. Dissolution of carbonate minerals creates by far the largest volume of secondary sandstone porosity in all three diagenetic regimes.

3. Decarbonatization during eodiagenesis and telodiagenesis is caused mainly by carbonic acid which is derived from meteoric and biogenic carbon dioxide. Mesogenetic decarbonatization appears to result mainly from leaching by carbonic acid which is derived from the decarboxylation of maturing organic matter in intercalated shales. This decarboxylation occurs over a wide range of temperature-time exposures but most of the decarboxylation occurs after the sandstones have lost their primary porosity.

4. Decarbonatization is the source of dissolved chemical constituents which, for the most part, migrate upwards into overlying sandstones and are precipitated there as mesogenetic cement and replacement. In a subsiding prism of accumulating sediments repeated recycling of carbonate may occur leading to very high concentrations of carbonate in the zone of carbonatization and very high secondary porosities in the underlying zone of decarbonatization.

5. Mesogenetic sandstone decarbonatization is generally controlled by the amount and type of organic matter in intercalated shales and their temperature-time exposure. The texture and composition of the non-carbonate fraction of the sandstones appears to have little influence on decarbonatization. Significantly, decarbonatization can create porosity in tight sandstones because of the presence of irreducible lamellar porosity.

6. Secondary porosity in any sandstone lithology resists reduction by chemical compaction much more than does primary sandstone porosity. However, the rates of reduction of both primary and secondary porosity by chemical compaction are influenced by temperature and mineralogical maturity of the sandstones. Therefore, mapping of the diagenetic maturity of sandstones is only of value if there is reasonable uniformity of mineralogical maturity.

7. The maximum depth of effective secondary porosity created by mesogenetic decarbonatization is more than double that of effective primary porosity in sandstones of high mineralogical stability. It is more than three times the maximum depth of effective primary porosity in sandstones of low mineralogical stability.

8. The maturation of the hydrocarbon source rocks appears to be directly responsible for the generation of most secondary sandstone porosity. The majority of conventional crude oil and natural gas in clastic sequences has been generated in the source rocks after the intercalated sandstones had lost their primary porosity and after they had gained their secondary porosity. For this reason secondary sandstone porosity is a favorable habitat for crude oil and natural gas in clastic sequences.

ACKNOWLEDGEMENTS

Research for this paper was conducted at the geological laboratory of Mobil Oil Canada, Ltd. and since 1976 also at the geological research laboratory of Petro-Canada. The authors would like to thank both companies for their generous support in every phase of the preparation of this paper and for permission to publish. Special thanks are due to Petro-Canada for providing funds for printing of the colored microphotographs. Thanks are also extended to the drafting department of Mobil Oil Canada Ltd. for preparing the excellent illustrations. The authors thankfully appreciate the advice and assistance of the many people that made this paper possible, including: Brian Miller—illustrations and microphotography; David Blair—thin sections; Susan Peters and Sylvia Bates—typing; Jack Thomas, Monte Lerand and Norman Wardlaw—critical reading; John Halsey, Marvin Meloche, Bill Monroe, Hans

Nelson, Henry Nelson, Ron Platt and James Scott—discussion of petrological and geological aspects; Ellis Bray and Trevor Powell—discussion of geochemical aspects, and, last but not least, Nadine McDonald and Sylvia Schmidt are thanked for patience and support as much of this paper was written at the homes of the authors.

REFERENCES

ALCOCK, F. G., AND BENTEAU, 1976, Nipisi Field—A Middle Devonian clastic reservoir, *in* M. M. Lerand (ed.), The sedimentology of selected clastic oil and gas reservoirs in Alberta: Canadian Soc. Petroleum Geology, Spec. Pub., 125 p.

CHOQUETTE, P. W., AND PRAY, L. C., 1970, Geologic nomenclature and classification of porosity in sedimentary carbonates: Am. Assoc. Petroleum Geologists Bull., v. 54, p. 207–250.

FOX, J. E., LAMBERT, P. W., MAST, R. F., NUSS, N. W., AND REIN, R. D., 1975, Porosity variation in the Tensleep and its equivalent, the Weber Sandstone, western Wyoming; a log and petrographic analysis, *in* D. W. Bolyard (ed.), Deep drilling frontiers of the central Rocky Mountains: Denver, Colorado, Rocky Mountain Assoc. Geologists, p. 185–216.

GALLOWAY, W. E., 1974, Deposition and diagenetic alteration of sandstone in northeast Pacific arc-related basins: Implications of greywacke genesis: Geol. Soc. America Bull., v. 85, p. 379–390.

HARTMAN, J. A., 1968, The Norphlet Sandstone, Pelhatchie Field, Rankin County, Mississippi: Gulf Coast Assoc. Geol. Socs. Trans., v. 18, p. 2–11.

HEALD, M. T., AND LARESE, R. E., 1973, The significance of solution of feldspar in porosity development: Jour. Sed. Petrology, v. 43, p. 458–460.

HRABER, S. V., AND POTTER, P. E., 1969, Lower West Baden (Late Mississippian) sandstone body of Owen and Green Counties, Indiana: Am. Assoc. Petroleum Geologists Bull., v. 53, p. 2150–2160.

KRYNINE, P. D., 1941, Petrographic studies of variation in cementing material in the Oriskany sand: Pennsylvania State College Bull. 33, p. 108–116.

LINDQUIST, S., 1976, Leached porosity in overpressured sandstones—Frio Formation (Oligocene), south Texas [abs.]: Am. Assoc. Petroleum Geologists Bull., v. 60, p. 1612.

PARKER, C. A., 1974, Geopressures and secondary porosity in the deep Jurassic of Mississippi: Gulf Coast Assoc. Geol. Socs. Trans., v. 24, p. 69–80.

PROSHLYAKOV, B. K., 1960, Reservoir rocks as a function of their depth and lithology [in Russian]: Geol. Nefti i Gaza, v. 4, no. 12, p. 24–29.

ROWSELL, D. M., AND DE SWARDT, A.M.J., 1974, Secondary leaching porosity in Middle Ecca Sandstones: Geol. Soc. South Africa Trans. and Proc., v. 77, p. 131–140.

SAVKEVIC, S. S., 1971, The possible connection between the leaching-induced porosity of rocks and the primary stage of petroleum generation (migration) [in Russian]: Akad. Nauk SSSR Izv. Ser. Geol., v. 6, p. 70–77.

SCHMIDT, V., 1976, Secondary porosity in the Parsons Lake Sandstone [abs.]: Geol. Assoc. Canada and Mineralog. Assoc. Canada Ann. Meeting Abs., v. 1, p. 50.

——, MCDONALD, D. A., AND PLATT, R. L., 1977, Pore geometry and reservoir aspects of secondary porosity in sandstones [abs.]: Geol. Assoc. Canada and Mineralog. Assoc. Canada Ann. Meeting Abs., v. 2, p. 106.

SEDIMENTATION SEMINAR, 1969, Bethel Sandstone (Mississippian) of western Kentucky and south-central Indiana, a submarine-channel fill: Kentucky Geol. Survey Rept. Inv. 11, ser. X, 24 p.

STANTON, G. D., AND MCBRIDE, E. F., 1976, Factors influencing porosity and permeability of Lower Wilcox (Eocene) Sandstone, Karnes County, Texas [abs.]: Am. Assoc. Petroleum Geologists and Soc. Econ. Paleontologists and Mineralogists Ann. Meeting Abs., v. 1, p. 119.

TISSOT, B., DURAND, B., ESPITALIE, J., AND COMBAY, A., 1974, Influence of nature and diagenesis of organic matter in formation of petroleum: Am. Assoc. Petroleum Geologists Bull., v. 58, p. 499–506.

WARDLAW, N. C., 1976, Pore geometry of carbonate rocks as revealed by pore casts and capillary pressure: Am. Assoc. Petroleum Geologists Bull., v. 60, p. 245–257.

American Association of Petroleum Geologists Memoir 37,
Clastic Diagenesis, edited by D.A. McDonald and R.C.
Surdam, copyright 1984, pp. 3-14.

Hydrogeologic Regimes of Sandstone Diagenesis*

William E. Galloway
Bureau of Economic Geology, The University of Texas
Austin, Texas

ABSTRACT. The subsurface hydrologic system of large, actively filling sedimentary basins includes meteoric, compactional, and thermobaric regimes. Boundaries between the regimes and their contained flow systems evolve as basin filling proceeds. As a result, sands are continuously flushed by a succession of fluids of varying origins and chemistries. Careful examination of the existing hydrologic setting and a reconstruction of generalized hydrologic history and its relationship to observed diagenetic features within a depositional sequence may serve to validate interpretive diagenetic models and may explain or predict paragenetic relationships, regional diagenetic variations within the same depositional episode, and differences in diagenetic products in different episodes in the same or similar basins.

The Frio/Catahoula Formations of the Texas Coastal Plain provide an example that both illustrates the coexistence of hydrologic regimes and relates associations of diagenetic features with existing regimes or with the important mixing zones that occur between regimes. Hydrocarbon geochemistries indicate meteoric flushing to depths approaching 2000 m. The deeper geopressured section is coincident with a thermobaric regime in which clay dewatering recharges the hydrologically restricted portions of the basin fill.

INTRODUCTION

The objective of this short paper is to bring together some general observations about the hydrogeology of sedimentary basins to establish the dynamic framework in which sandstone diagenesis occurs. The Frio Formation (Oligocene) of the northern Gulf Coast Basin has been the subject of both regional and detailed diagenetic studies (see contributions by Loucks et al, Land, and Kaiser in this volume) and of stratigraphic, hydrogeologic, and resource analysis (Galloway, 1977; Bebout et al, 1978; Galloway and Kaiser, 1980; and Galloway et al, 1982). It provides a focus for the discussion.

The subsurface environment must be regarded as a dynamic cauldron in which fluids continuously move. Static conditions rarely, if ever, characterize subsurface fluid flow systems. To further complicate interpretation of diagenetic history, both flow dynamics and fluid chemistries evolve through time. Analysis and description of basin hydrology and extant geochemistry provide a view only of the present diagenetic environment; most diagenetic features may be relicts of earlier fluid regimes. However, such a description is a starting point. Not only does it provide a framework for examining and testing active water–rock interactions, but it also expands our understanding of natural hydrologic systems in large sedimentary basins. The thesis presented here is that a more critical analysis of diagenetic processes within a realistic hydrologic context offers a major step in understanding and predicting diagenetic histories of sandstone reservoirs. Such analysis necessitates a basic understanding of the hydrogeologic regimes and evolutionary pathways of sedimentary basins.

BASIN HYDROLOGIC REGIMES

The hydrologic framework of a large, depositionally active sedimentary basin consists of ground water (the term is used in this paper in its general sense to include all waters of whatever origin that lie in the subsurface environment) moving within several different regimes (Bogomolov et al, 1978; Kissin, 1978) (Fig. 1). Although the boundaries of the regimes and their terminology are sometimes difficult to reconcile, they nonetheless form end members characterized by their relative positions, hydrochemistries, and flow dynamics.

The *meteoric regime* encompasses shallow portions of the basin fill. Water recharges by infiltration of precipitation and surface waters, moves down the topographic gradient in the direction of decreasing gravitational potential energy, and discharges vertically at hydrologic base level. Hydrologic base level is commonly determined by sea (or lake) level. Because there is no head advantage achieved by further movement once the area of base level is attained, there is little potential for additional lateral flow beneath the subaqueous basin center. The shoreline is typically a major discharge area for meteoric aquifers (Fig. 1). On a geologic time scale, meteoric circulation is rapid.

*Publication authorized by the Director, Bureau of Economic Geology, The University of Texas at Austin.

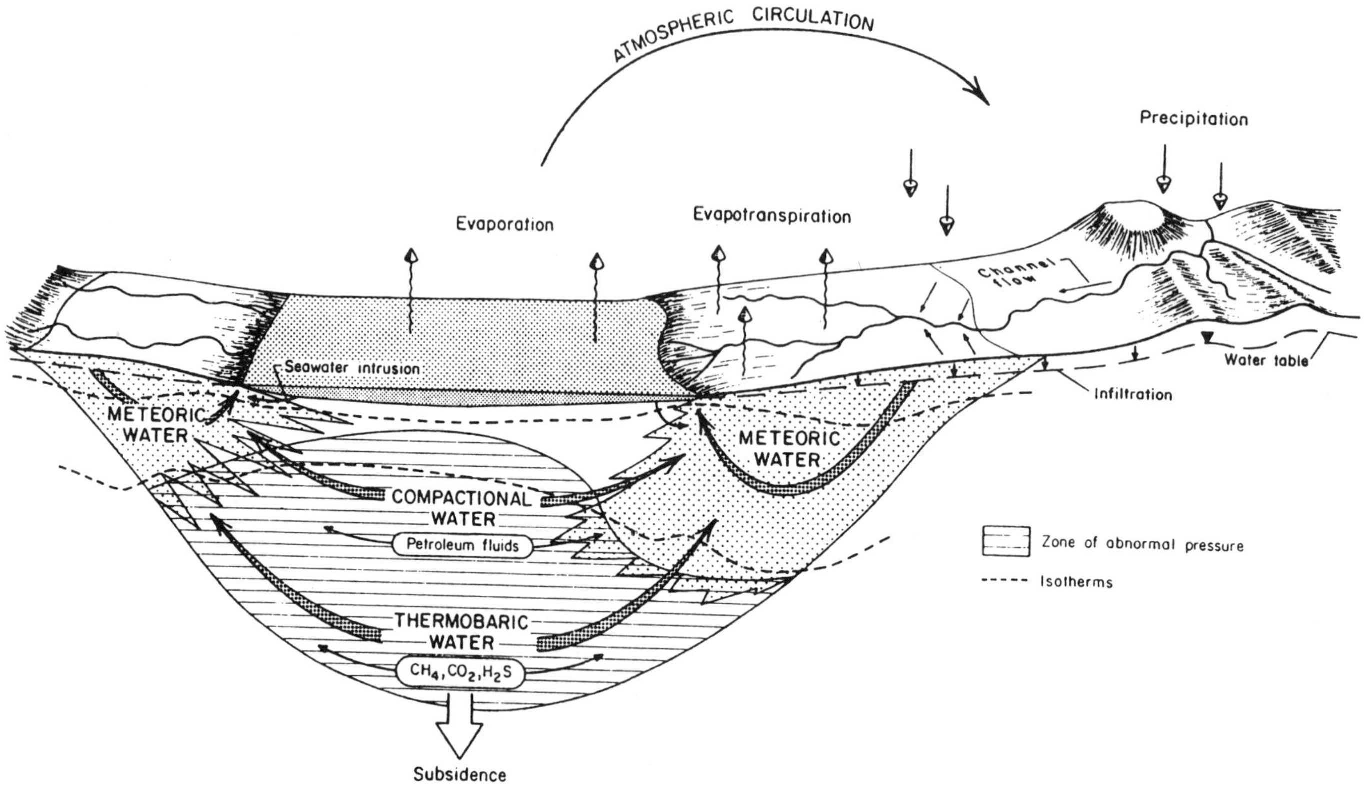

Figure 1—Hydrologic regimes of a large, subsiding depositional basin. From Galloway and Hobday. Copyright © 1983 Springer-Verlag New York. Reprinted with permission.

The *compactional* regime is characterized by upward and outward expulsion of pore waters trapped within the compacting sediment pile (Fig. 1). These waters may include evolved connate water deposited with the sediment and meteoric waters buried below the zone of active meteoric circulation. Lithostatic loading or compressive tectonic stress may generate increased pressure. Pressure head is the driving mechanism for flow. If circulation is impeded by restricted transmissivity of the sediment pile, fluid pressure/depth gradient may exceed that of a free-standing column of water of the same height.

The *thermobaric,* or abyssal, regime lies in the deepest portions of the basin fill where temperature and pressure are greatest. Significant volumes of water may be released by dehydration reactions of clay minerals or of other hydrous mineral phases. Fluids move in response to pressure gradients generated both by phase changes, such as generation of hydrocarbons or release of mineral-bound water, and by ongoing lithostatic loading. However, the extremely low permeability of the compacted sediments commonly restricts water movement over geologically significant time spans. Aquithermal pressuring generated by volume expansion of water may also play a role in raising static pressure, but is rapidly dissipated by even minor amounts of fluid leakage.

Recognition of the thermobaric and compactional regimes as important components of the hydrologic basin necessitates an expanded view of the ground-water cycle as it evolves during the filling of a sedimentary basin. Large volumes of water are diverted from the meteoric and surface flow systems and entombed within the sedimentary pile as pore or mineralogically bound water. Such waters discharge into and mix with the meteoric system slowly, if at all. Unlike the supply of water to the meteoric system, which is continually recharged, the supply of compactional and thermobaric waters, though potentially large, is finite. Processes or products related to fluid flow within these systems must recognize and accommodate this limitation.

EVOLUTION OF HYDROLOGIC BASINS

The compactional and thermobaric regimes persist as long as active basin filling continues. With cessation of subsidence or with tectonic uplift the compaction-driven flow decreases and ultimately ceases. Strata of such a hydrologically "mature" basin are increasingly flushed by meteoric waters recharged along its uplifted margins (Toth, 1980; Cousteau et al, 1975). Regional flow is directed inward toward the topographic floor of the hydrologic basin, and the sediment fill may ultimately be completely flushed by meteoric waters. Well described basins in which ambient flow is dominated by regional meteoric circulation driven by gravitational head include the Paris Basin (Korotchansky and Mitchell, 1972) and the western Canada Mesozoic foreland basin of Alberta (Hitchon, 1969a, 1969b). Long-term

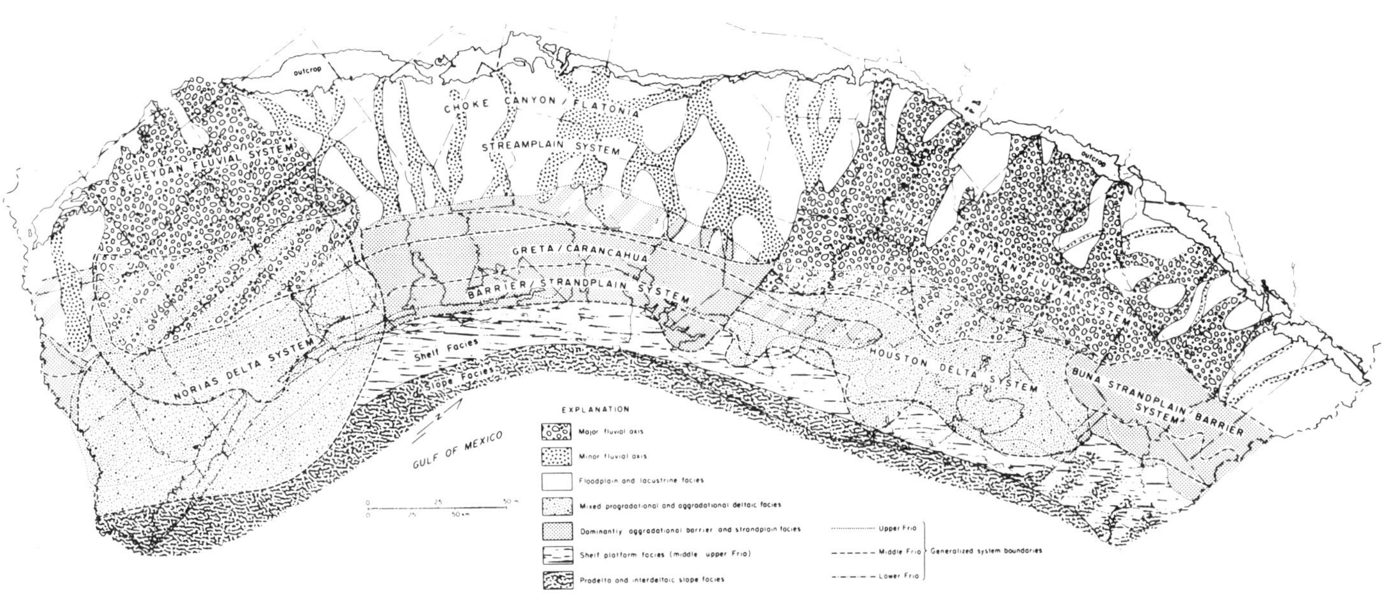

Figure 2—Depositional systems of the Frio (Oligocene) Formation, Texas Gulf Coastal Plain. From Galloway et al (1982).

flushing by regional meteoric circulation replaces residual waters with geochemically evolved meteoric ground water (Clayton et al, 1966).

With long-term tectonic stability and erosional leveling of basin topographic relief, movement of meteoric water slows, and the hydrologic basin may become effectively stagnant. Vertical pressure gradients throughout the basin would be hydrostatic (Cousteau et al, 1975), and only diffusive transfer of soluble constituents could continue. Such extreme stability is, however, rare.

HYDROGEOLOGY OF THE FRIO FORMATION

The Frio Formation, along with its updip equivalent, the Catahoula Formation, constitutes the sedimentary record of one of the major offlap depositional episodes of the Northwestern Shelf of the Gulf of Mexico Basin. Following deposition of a sand and mudstone wedge locally exceeding 3500 m in thickness during Middle to Late Oligocene and Early Miocene time, ongoing basin subsidence has buried downdip portions of the Frio under several thousand meters of younger sediment. Updip, a thin Frio/Catahoula section crops out along a narrow belt across the inner coastal plain of Texas (Fig. 2). Basinward thickening of the section is accomplished largely by a succession of syndepositional growth faults that develop where the Frio wedge prograded beyond the underlying continental platform margin (Fig. 3).

The Frio/Catahoula depositional episode consists of several major depositional systems (Galloway et al, 1982) (Fig. 2). The Oligocene depocenter lay in the Rio Grande Embayment. Here the Gueydan bed-load fluvial system fed the enormous Norias delta system, which prograded the continental margin nearly 100 km gulfward. Along the upper Texas Coastal Plain, two or three major fluvial axes fed downdip into a broad deltaic headland, called the Houston delta system. Between the two deltaic depocenters lay an extensive wave-dominated shore-zone system that includes both well-developed barrier/lagoon and strandplain facies assemblages. Basinward, sand-rich deltaic and shore-zone facies grade into prodelta, shelf, and continental slope mudstones. The deeply buried, growth-faulted sandstones of the delta and shore-zone systems contained world class reserves of hydrocarbons and are still targets for active exploration. Variable reservoir quality of the deep exploration targets has made the Frio the subject of extensive diagenetic study.

Hydrologic Regimes

Hydrologic regimes of the Frio as they exist today are conventionally described in terms of (1) a shallow, updip meteoric regime, (2) an intermediate depth zone characterized by normal or slightly elevated pressure gradients and saline brines, which lies below the base of relatively fresh water and above the top of geopressure, and (3) a deeply buried geopressured regime, which rises up through the Frio section basinward (Fig. 4). Examination of the characteristics of these zones shows that they broadly correspond to the more general hydrologic regimes discussed above, and thus define major environments for sandstone diagenesis. However, as the final section of this paper stresses, boundaries between regimes are both leaky and geologically ephemeral. Any selection of a fixed depth or boundary condition to define the ground-water zones is inherently arbitrary.

Meteoric Regime

Active meteoric circulation, as indicated by extent of low-chlorinity waters, extends to depths of as much as 1.5 km (Fig. 5). In portions of the south Texas Coastal Plain, however, depth of freshwater penetration is limited by shallow faulting of the aquifer or by the sand-poor character of the updip facies

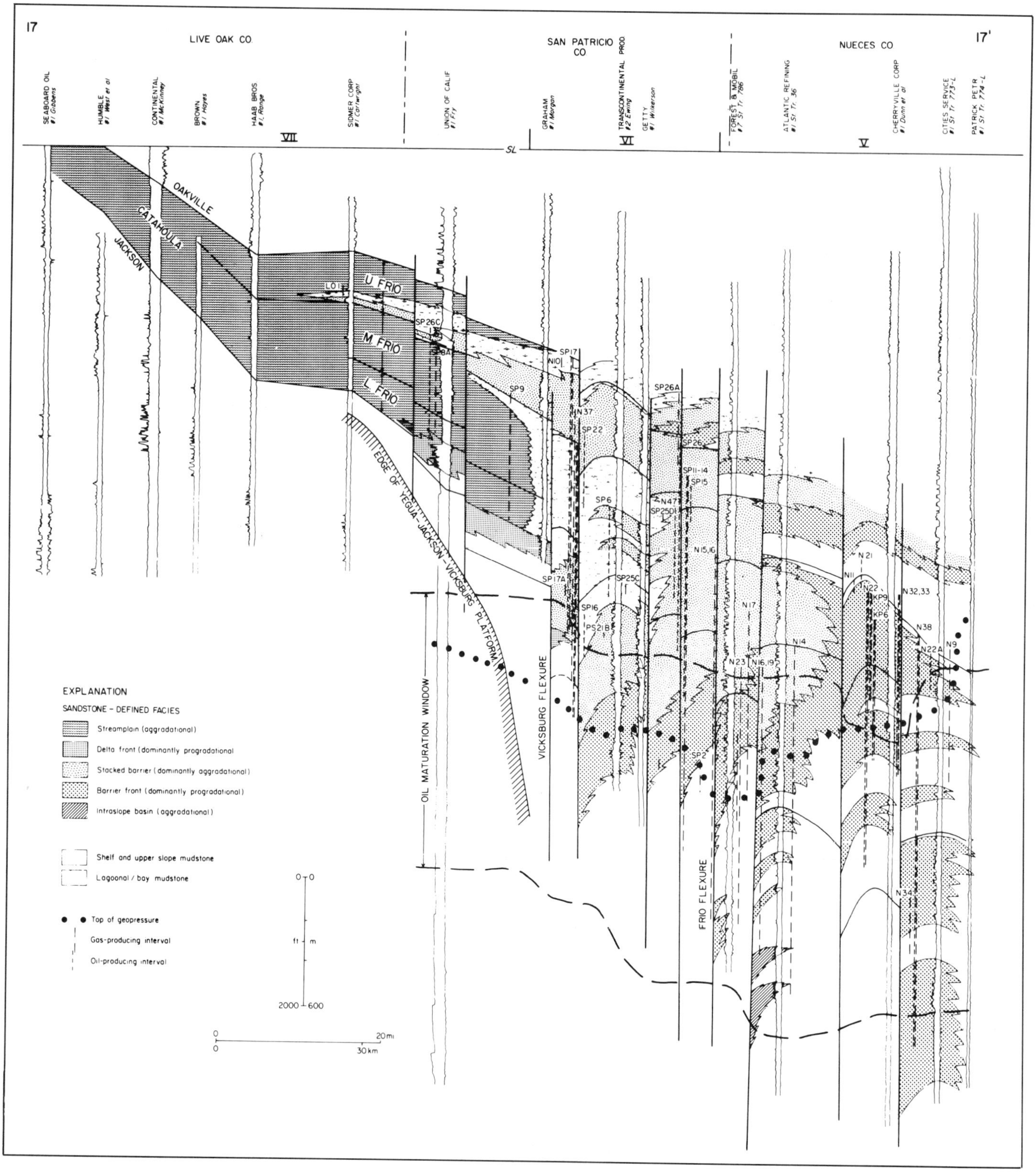

17
17'
LIVE OAK CO.
SAN PATRICIO CO
NUECES CO
SEABOARD OIL #1 Gibbens
HUMBLE #1 West et al
CONTINENTAL #1 McKinney
BROWN #1 Hayes
HAAB BROS #1 Range
SIDMER CORP #1 Cartwright
UNION OF CALIF #1 Fry
GRAHAM #1 Morgan
TRANSCONTINENTAL PROD #2 Ewing
GETTY #1 Wilkerson
FOREST & MOBIL #7 St Tr 786
ATLANTIC REFINING #1 St Tr 36
CHERRYVILLE CORP #1 Dunn et al
CITIES SERVICE #1 St Tr 773-L
PATRICK PETR #1 St Tr 774-L
VII
VI
V
SL
OAKVILLE
CATAHOULA
JACKSON
U. FRIO
M. FRIO
L. FRIO
EDGE OF YEGUA - JACKSON - VICKSBURG PLATFORM
OIL MATURATION WINDOW
VICKSBURG FLEXURE
FRIO FLEXURE
EXPLANATION
SANDSTONE - DEFINED FACIES
Streamplain (aggradational)
Delta front (dominantly progradational)
Stacked barrier (dominantly aggradational)
Barrier front (dominantly progradational)
Intraslope basin (aggradational)
Shelf and upper slope mudstone
Lagoonal / bay mudstone
Top of geopressure
Gas-producing interval
Oil-producing interval
0 0
ft m
2000 600
0 20mi
0 30km

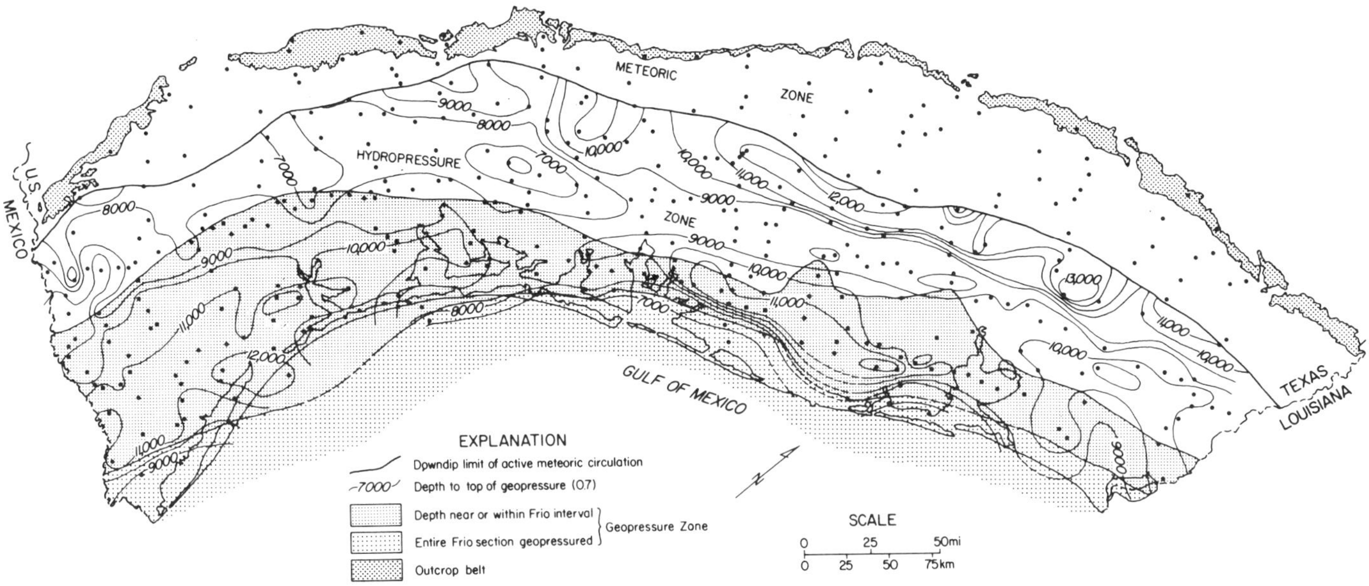

Figure 4—Lateral distribution of hydrologic zones within the Frio Formation. Depth to the "operational" top of geopressure (0.7 psi/ft or 15.8 kPa/m) is contoured in feet.

assemblage. The most active portions of the meteoric system occur primarily within axial facies of the fluvial systems; these shallow deposits have probably experienced meteoric flux throughout their depositional and burial history. Hydrocarbon geochemistry suggests that meteoric influence, as manifested by crude oil alteration by anaerobic bacteria (Fig. 6), has extended to depths of 2 km in the middle and upper coastal plain (Galloway et al, 1982). This deep meteoric penetration must have post-dated migration and accumulation of hydrocarbons. Bacterial destruction of organic acids at temperatures below 70° C (160° F) reported by Carothers and Kharaka (1980) in Miocene deposits supports the interpretation of deep meteoric penetration.

Although total dissolved solids content is characteristically low to moderate, composition of meteoric regime waters varies considerably (Galloway and Kaiser, 1980). Shallow calcium-bicarbonate waters evolve along flow paths into relatively saline sodium-chloride waters.

Diagenetic features of the Catahoula and Frio Formations produced within the meteoric regime include clay cutans, sparry calcite cement, minor amounts of kaolinite, clinoptilolite, opal, chalcedony, and feldspar overgrowths, and locally abundant leached feldspar and rock fragments.

Figure 3—Stratigraphic dip cross section through the Frio Formation and its updip equivalent, the Catahoula Formation, in the middle Coastal Plain of Texas. The downdip progression of facies includes coastal stream plain, lagoon, barrier, and shelf deposits. Growth faults expand the Frio section dramatically where it offlaps the older continental platform. In the expanded section, temperatures are adequate for oil and gas generation, and much of the mud-rich barrier-front and shelf sequence is abnormally pressured. From Galloway et al (1982).

Compactional Regime

Because muddy deltaic and marine sediments are typically deposited with high porosities (commonly exceeding 50%), physical compaction offers a source of water approximately equal in volume to the total volume of sediment. However, compaction and pore-space reduction occur rapidly with earliest burial to only a few hundred meters. Representative volumes of water lost during burial of successive thousand-meter increments of mud that might typify Frio prodelta, shelf, or slope facies are shown graphically in Figure 7. During the first kilometer of burial, the mud releases more than 21,000 cu cm of interstitial water per sq cm of surface area. At a depth of 1 km, average porosity is reduced by physical compaction to slightly more than 20%. Remaining compaction to near zero porosity at depths approaching 10 km releases a comparable volume of water in decreasing proportions with depth (Fig. 7, column B). Potential upward flux across each sq cm (column C) is no greater than the cumulative release of water by underlying sediment volumes. The illustration emphasizes the fact that the volume of available compaction water is finite.

In the Frio Formation, the depth window dominated by the compactional regime is limited by deep penetration of meteoric water (1.4–2 km) and the relatively shallow top of geopressure and associated increase in temperature and pressure gradients and resultant clay dewatering (3–4 km). Waters in the appropriate depth range are typically reducing, slightly acidic to slightly basic sodium-chloride brines containing dissolved carbon dioxide and abundant hydrocarbon gases and liquids.

Mesodiagenetic features described in recent studies that appear to be products of or associated with the compactional regime include ubiquitous quartz overgrowths, albite overgrowths, calcite-pore fill and grain replacements,

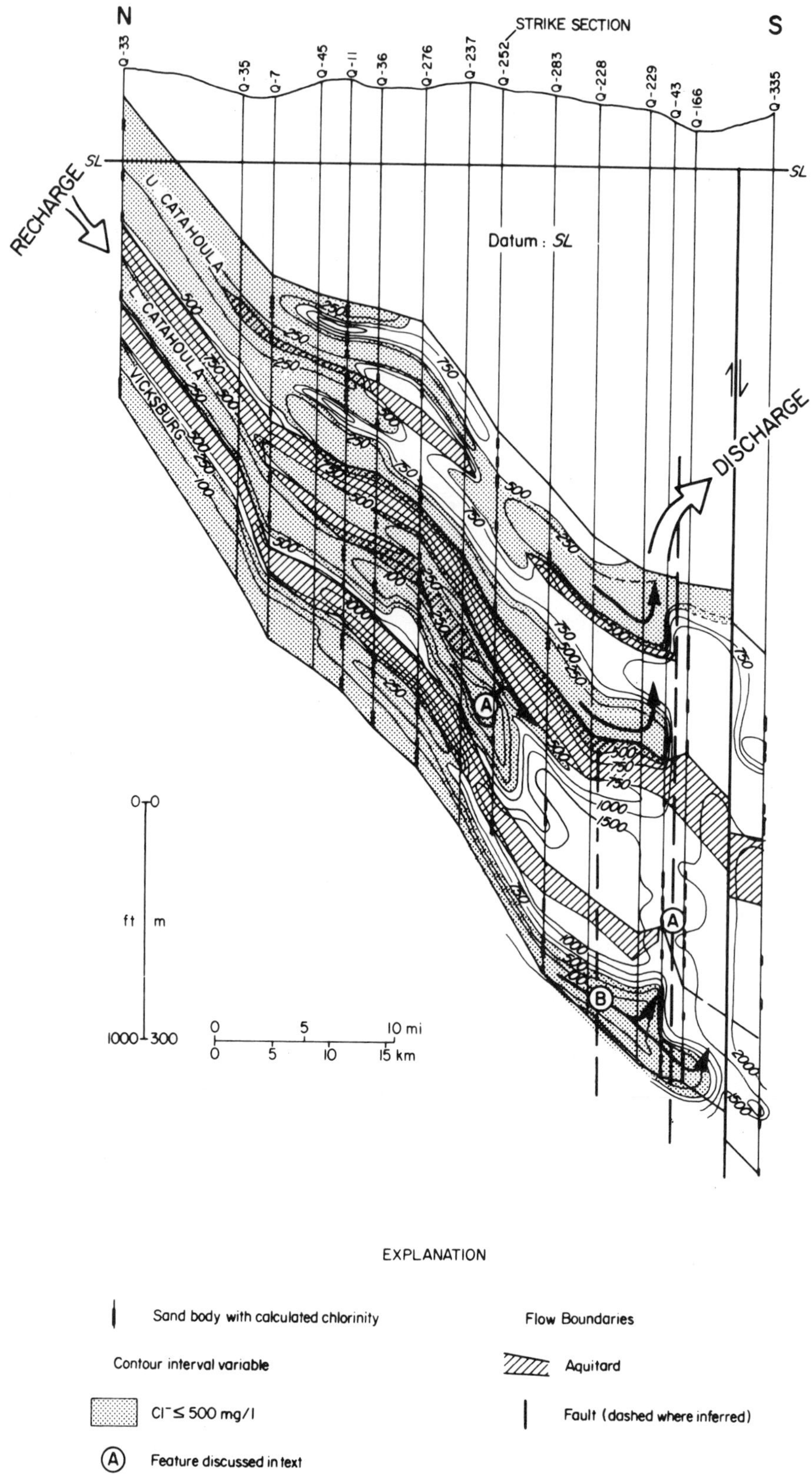

Figure 5—Calculated apparent ground-water chlorinity for a portion of the Chita-Corrigan fluvial system of the Catahoula Formation. In cross section, freshwater tongues are shown to extend into the subsurface to depths in excess of 1 km. Such tongues of relatively fresh water outline areas of most active meteoric circulation. Bunching of salinity contours at depth indicates probable areas of vertical discharge into shallower aquifers. Letters shown in circles point out complexities typical of shallow meteoric flow, such as flow focusing along more permeable channel-fill trends. From Galloway and Kaiser. Copyright © 1980 University of Texas at Austin Bureau of Economic Geology. Reprinted with permission.

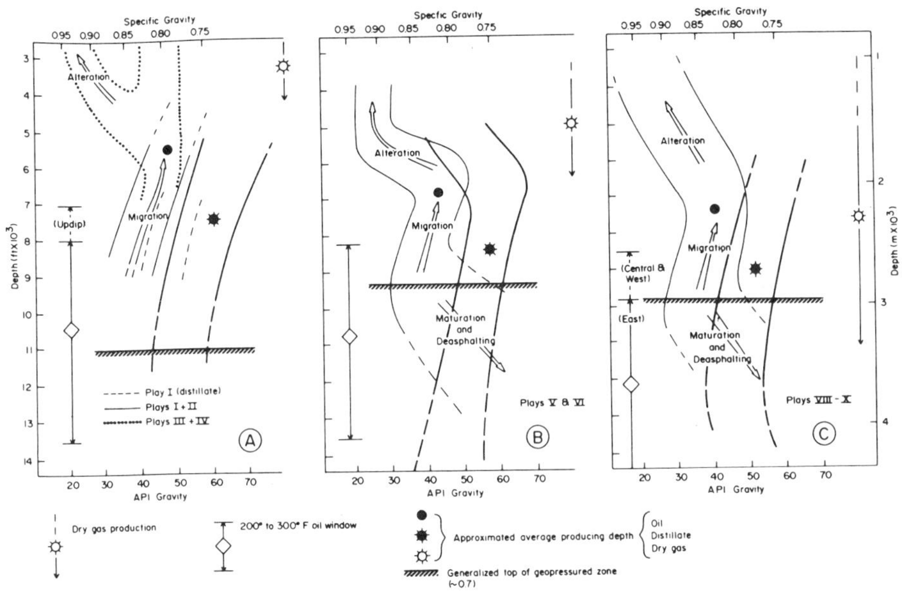

Figure 6—Generalized envelopes around Frio oil and condensate density values plotted against reservoir depth for (**A**) lower, (**B**) middle, and (**C**) upper coastal plain areas. Envelopes include more than 90% of several tens to hundreds of data points and reveal consistent vertical trends in petroleum chemistry. Well symbols show approximated average depth of all producing reservoirs. Average vertical extent of the oil-generation window and the average operational top of geopressure are also shown. Hydrocarbon liquid distributions indicate large-scale upward migration from point of origin to point of entrapment. The increasing specific gravity seen in the shallow parts of curves B and C is interpreted to reflect widespread bacterial alteration of crudes following entrapment. From Galloway et al. Copyright © 1982 University of Texas at Austin Bureau of Economic Geology. Reprinted with permission.

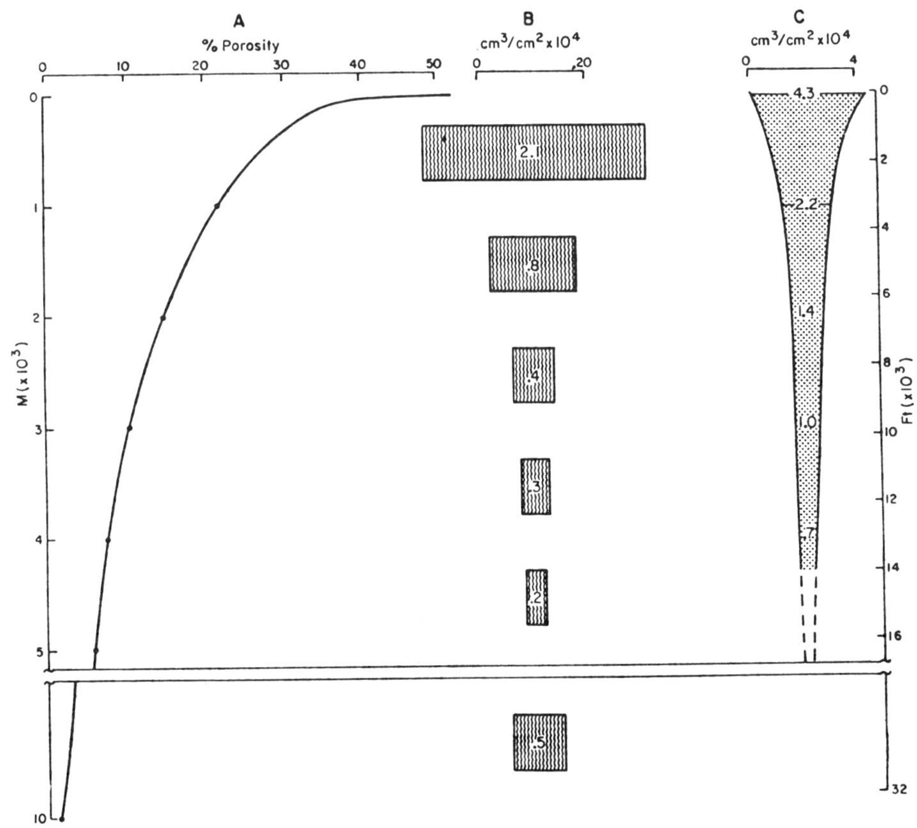

Figure 7—Volumes of expelled pore water generated by burial and compaction of a typical mud. (**A**) Generalized porosity depth profile. (**B**) Comparative volumes of water released from 1000-m-thick columns of sediment (with a surface area of 1 sq cm) for each kilometer of burial. (**C**) Cumulative volume of fluid remaining to be expelled by compaction below successive 1 km depth levels. From Galloway and Hobday. Copyright © 1983 Springer-Verlag New York. Reprinted with permission. Based on calculations provided by K. Bjorlykke.

kaolinite cement and grain replacements, authigenic chlorite, and widespread leached detrital grains and cements and partially to completely albitized plagioclase.

Thermobaric Regime

Deeper portions of the Frio contain waters exhibiting measured pressures considerably above hydrostatic. The top of this geopressured zone ideally is picked on the basis of pressure measurements at the first occurrence of such abnormally high pressures. The depth of the absolute top of geopressure, where gradients based on in-situ measurement exceed 0.465, typically lies 2.5 to 3 km below the surface. An "operational" top of geopressure that is more readily picked indirectly from electrical log measurements is commonly used for regional mapping of the geopressured zone, and is the top of geopressure used in Figures 3, 4, and 6. Fluid pressure gradients at the top of operational geopressure approach 0.7, and the underlying section is substantially overpressured. The top of operational geopressure lies within a thousand meters or so of the top of the first overpressuring.

Near the top of geopressure, temperatures in the Frio Formation are adequate for conversion of smectite layers to illite in mixed-layer clays, with consequent release of interlayer water (Freed, 1981, 1982). Clay dewatering itself may, provided bulk aquifer transmissivity is sufficiently low, be a cause of geopressuring (Burst, 1969). However, the author's observations support those of Magara (1978) and suggest that the mechanism does not fully explain associated petrophysical phenomena such as shale density decrease observed in the overpressure zone. A simpler explanation for young basins characterized by rapid deposition of deltaic and slope systems is that the rate of sediment loading exceeds the potential for fluid discharge under moderate hydraulic gradients (Bredehoeft and Hanshaw, 1968). Within the geopressure zone, decreased thermal conductivity of the sediment column results in an increased thermal gradient that accentuates further clay dewatering. Thus, regardless of the specific cause or causes of overpressuring, the deeper zone of geopressure in the Frio Formation displays the key features that define a distinct ground-water regime.

Actual volume of released water is poorly documented (for a review, see Graf and Anderson, 1981). Nonetheless, clay dewatering offers a mechanism to "recharge" the deeper portion of the compactional regime with a pulse of relatively fresh water. At the same time, higher temperatures associated with the regime lead to thermal alteration of organic materials, producing abundant hydrocarbon liquids and gases (Figs. 3, 6). Analyses of waters of the geopressured zone indicate acidic, highly gas-charged sodium-chloride brines of variable salinity (Galloway, 1982; Kaiser, this volume).

Burial diagenetic features of Frio sandstones commonly ascribed to the geopressure zone include both ferroan and iron-poor carbonate cements, kaolinite and chlorite cements, completely albitized plagioclase feldspars, and thoroughly leached or albitized K-feldspars.

Regime Interactions

The three basin hydrologic regimes can be recognized and described within the Frio Formation today. However, their present distribution reflects only one point in the hydrologic evolution of the northern Gulf Coast Basin. Most, or perhaps all of the diagenetic events recognized in the Frio record early stages in this hydrologic evolution. Distribution and paragenesis of diagenetic features provide indirect evidence of fluxes and interactions between fluids of different regimes. In addition, data such as ambient formation water chemistry, in-situ pressure or temperature trends, and patterns of hydrocarbon distribution and physical properties all provide information on active or recent fluid fluxes.

(1) Pervasive upward flux of water through and out of the geopressured zone into the overlying normally pressured compactional zone is indicated directly by pressure distributions around fault zones, which may act as conduits for cross-stratal discharge (Berg and Habeck, 1982), and by thermal, pressure, and salinity anomalies observed within intermediate-depth portions of the Tertiary section (Harrison, 1980).

(2) Upward fluid migration is indirectly indicated by the vertical displacement of the center of mass of hydrocarbons contained in commercial reservoirs by more than 1 km above their zone of thermal generation (Fig. 6). Upward migration is further suggested by the evidence of chromatographic segregation of hydrocarbon liquids between their point of origin within the thermobaric regime and their point of residence in the compactional regime (Galloway et al, 1982) (Fig. 6).

(3) Indirect evidence for considerable mixing and interaction of waters of the compactional and thermobaric regime includes the pervasive zone of leached porosity development that characterizes intermediate depths (2.2–3.0 km) within the Frio and other Tertiary formations (Loucks et al, 1979).

(4) Intrusion of sulfide-rich thermobaric waters derived from deepest portions of the geopressured zone have played a critical role in the origin of major uranium deposits within the Gueydan fluvial system (Galloway and Kaiser, 1980). These waters, which have discharged all the way upsection into shallow portions of the meteoric regime, have produced tongues of pyritic sandstone within the syndepositionally oxidized shallow aquifers (Fig. 8). Such pyritic zones, which may record multiple pulses of intrusion and sulfidization, are centered around deep-seated growth faults (Galloway and Kaiser, 1980; Galloway, 1982). As in deeper examples of interpreted vertical fluid migration, fault zones appear to have served as principal conduits for discharge from deeper into shallower ground-water regimes. The resultant mixing zone is a focus for diagenetic alteration.

In summary, cursory examination of the Frio/Catahoula Formation as an example of a thick clastic wedge in an actively subsiding basin reveals the continuing presence of each of the three major basin hydrologic regimes. Further, reactions caused by upward flux and mixing of pressure-driven waters from the thermobaric regime continue (Fig. 9), though probably on a reduced scale because the depocenter has shifted basinward. Meteoric waters continue to invade the shallow and

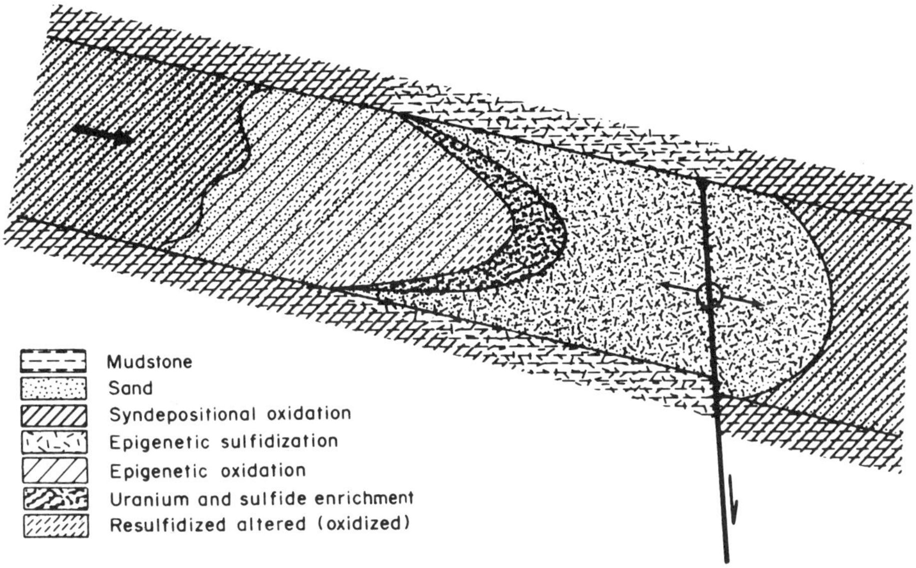

Figure 8—Schematic cross section showing the development of a pyrite-rich tongue within a shallow, oxidized aquifer by intrusion of sulfide-rich thermobaric waters along deep-seated growth faults. Such alteration of the shallow Catahoula sandstones was a prerequisite for uranium mineralization. Multiple pulses of deep-water discharge into the meteoric regime are recorded in many south Texas uranium districts. Arrows indicate the opposed, but temporally alternating counterflows of meteoric and thermobaric waters responsible for the superimposed alteration zones. From Galloway. Copyright © 1982 University of Texas at Austin Bureau of Economic Geology. Reprinted with permission.

intermediate subsurface, expanding their influence into portions of the Frio originally dominated by upward flux of compactional waters and hydrocarbons.

Finally, it should be reemphasized that fluid flow and mixing are not limited to individual depositional wedges. The hydrologic basin encompasses the entire sediment pile. Cross-cutting growth-fault zones and diapirs provide avenues for vertical fluid movement and mixing. The extent of this mixing and the importance of the deep fluids remain topics of speculation and investigation.

In addition, this discussion has focused on description of hydrologic zones and their interactions within the basin fill. Recently, thermally driven convection of waters within aquifers has been proposed as a possible mechanism of fluid mass-transfer within sedimentary basins (Wood and Hewett, 1982). Convection has the attraction of removing some of the perplexing mass-balance limitations implicit in the finite volumes of water ultimately available from compactional and thermal dewatering of the sediment pile. Whether convection can reasonably operate at a scale adequate to explain the requisite vertical displacement of solids proposed in current diagenetic models remains to be tested in the Gulf Coast Tertiary Basin. Factors in a depositional sequence, such as the Frio Formation, that must be accommodated by a convection model include the lateral segmentation of aquifers by growth faults, the vertical pressure differentials and overall upward-directed pressure head, and the lithologic and permeability heterogeneity of the aquifers themselves. Pressures, temperature, and salinity data may provide evidence of the existence and scale of convection cells if examined with this new model in mind.

A FOCUS ON HYDROLOGIC EVOLUTION OF SANDSTONES

From the time of deposition, all but a thin veneer of terrestrial sediment is bathed in subsurface waters. The history of evolving fluxes and hydrochemistries largely determines the diagenetic history of the sediment. Carbonate sedimentologists have recognized this fact and emphasized the hydrologic regime of diagenesis. Its emphasis has been less explicit in most regional models of clastic diagenesis. As indicated in Figure 10, however, the hydrologic pathways experienced by a regional depositional sequence, such as the Frio Formation, may be complex and diverse. Depending on depositional environment, initial waters may be either meteoric or depositional connate. With ongoing deposition, portions of the sediment volume may be swept by semi-confined to confined meteoric waters, which react with the aquifer and evolve chemically (meteoric flush, Fig. 10). With burial, sands contained in mud-rich sequences, particularly in portions of the sediment pile that lie below subaqueous portions of the basin (and therefore seaward of the coastal discharge zone of the meteoric system), are flushed by chemically evolved pore waters squeezed from compactable muds (compaction flush). As the compaction flush wanes, a second pulse of water derived from thermal dehydration of hydrous mineral phases, and most likely characterized by significantly different chemistry, pervades deeply buried portions of the aquifer and discharges up into the overlying section (thermobaric flush). Finally, as subsidence, depositional loading, and thermal alteration run their course, meteoric penetration increasingly dominates the basin hydrology, and sediments may again be flushed by chemically evolved meteoric waters.

As the three hydrologic zones evolve, boundaries between them, which are better described as ill-defined mixing zones, are likely sites of active diagenetic alteration of the aquifer plumbing, the sandstones. Thermally driven convection cells may evolve within the compactional and thermobaric systems, further complicating the dynamics of this already restless system.

The three major hydrogeologic regimes persist within the Frio. Diagenetic products, within both the deep and shallow portions of the Frio/Catahoula section, record the hydrologic evolution of this portion of the Gulf Basin fill. The new data included in several companion contributions in this

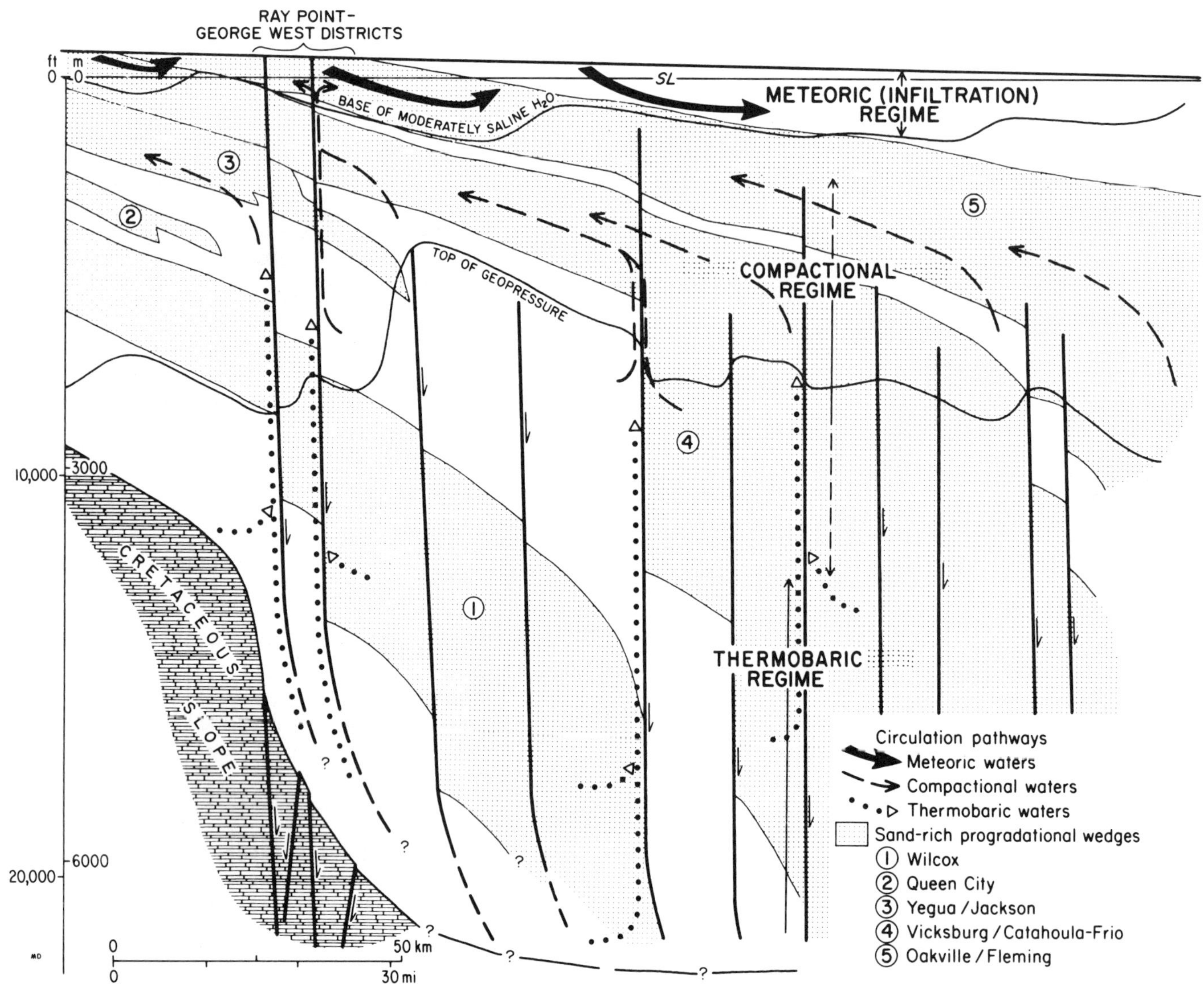

Figure 9—Ground-water regimes and circulation pathways within the offlapping Tertiary fill of the Texas Coastal Plain. Growth fault zones provide conduits for fluid discharge from deeper hydrologic regimes. Major mixing zones lie along the boundary between the geopressured and hydropressured zones and adjacent to shallow growth faults in the meteoric regime. From Galloway. Copyright © 1982 University of Texas at Austin Bureau of Economic Geology. Reprinted with permission.

volume provide a basis for a thorough evaluation of the hydrodynamics of sandstone diagenesis. Study of the modern waters and their hydrochemistries is a starting point for the reconstruction of diagenetic history.

ACKNOWLEDGMENTS

This paper was originally prepared for oral presentation at a symposium on diagenesis at the 1982 International Sedimentological Congress. Comments and discussion by several attendees added to the ideas presented here. The manuscript was reviewed by W. R. Kaiser, G. E. Fogg, and R. A. Morton. S. Doenges edited the draft manuscript. Drafting of figures was under the supervision of D. Scranton and J. Macon.

SELECTED REFERENCES

Bebout, D. G., R. G. Loucks, and A. R. Gregory, 1978, Frio sandstone reservoirs in the deep subsurface along the Texas Gulf Coast: University of Texas, Bureau of Economic Geology Report of Investigations 91, 92 p.

Berg, R. R., and M. F. Habeck, 1982, Abnormal pressures in the lower Vicksburg, McAllen Ranch field, South Texas: Gulf Coast Association of Geological Societies Transactions, v. 32, p. 247–262.

Bogomolov, Y. G., A. V. Kudelsky, and N. N. Lapshin, 1978, Hydrology of large sedimentary basins, *in* Hydrology of great sedimentary basins: Publication of the International Association of Scientific Hydrology, No. 120, p. 117–122.

Bredehoeft, J. D., and B. B. Hanshaw, 1968, On the maintenance of anomalous fluid pressures, 1, Thick sedimentary sequences: Geological Society of America Bulletin, v. 79, p. 1097–1106.

Burst, J. F., 1969, Diagenesis of Gulf Coast clayey sediments and its possible relation to petroleum migration: Bulletin of the

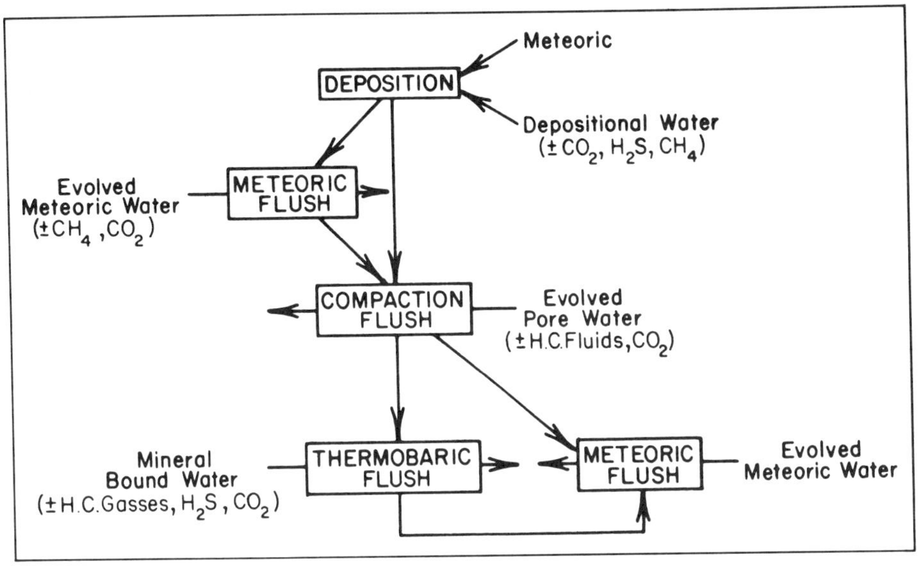

Figure 10—Generalized pathways for hydrologic evolution of a major stratigraphic unit such as the Frio/Catahoula Formation. Through its depositional and subsequent burial history, a volume of sediment may be successively swept by meteoric, compactional, and thermobaric waters. Old, depositionally moribund basins are commonly again permeated by meteoric water. The successive fluxes of these chemically differing waters and evolution of their mixing zones are recorded in the diagenetic history of the sandstone aquifers.

American Association of Petroleum Geologists, v. 53, p. 73–93.
Carothers, W. W., and Y. K. Kharaka, 1980, Stable carbon isotopes of HCO_3 in oil field waters—implications for the origin of CO_2: Geochimica et Cosmochimica Acta, v. 44, p. 323–332.
Clayton, R. N., I. Friedman, D. L. Graf, T. K. Mayeda, W. F. Meents, and N. F. Shimp, 1966, The origin of saline formation waters: Journal of Geophysical Research, v. 71, p. 3869–3882.
Cousteau, H., L. Rumeau, and C. Sourisse, 1975, Classification hydrodynamique des bassins sedimentaires: utilisation combinee avec d' autres methodes pour rationaliser l' exploration des bassins nonproductifs: Proceedings of the 9th World Petroleum Congress, v. 2, p. 105–119.
Freed, R. L., 1981, Shale mineralogy and burial diagenesis of Frio and Vicksburg Formations in twó geopressured wells, McAllen Ranch area, Hidalgo County, Texas: Gulf Coast Association of Geological Societies Transactions, v. 31, p. 289–293.
________, 1982, Clay mineralogy and depositional history of the Frio Formation in two geopressured wells, Brazoria County, Texas: Gulf Coast Association of Geological Societies Transactions, v. 32, p. 459–463.
Galloway, W. E., 1977, Catahoula Formation of the Texas Coastal Plain: depositional systems, composition, structural development, ground-water flow history, and uranium distribution: University of Texas Bureau of Economic Geology Report of Investigations 87, 59 p.
________, 1982, Fluid flow histories and resultant epigenetic zonation of uranium-bearing fluvial aquifer systems, South Texas uranium province: University of Texas Bureau of Economic Geology Report of Investigations 119, 31 p.
Galloway, W. E., and D. K. Hobday, 1983, Terrigenous clastic depositional systems: application to petroleum, coal, and uranium exploration: New York, Springer-Verlag.
Galloway, W. E., D. K. Hobday, and K. Magara, 1982, Frio Formation of the Texas Gulf Coastal Plain—depositional systems, structural framework, and hydrocarbon distribution: Bulletin of the American Association of Petroleum Geologists, v. 66, p. 649–688.
Galloway, W. E., and W. R. Kaiser, 1980, Catahoula Formation of the Texas Coastal Plain: origin, geochemical evolution, and characteristics of uranium deposits: University of Texas Bureau of Economic Geology Report of Investigations 100, 81 p.
Graf, D. L., and D. E. Anderson, 1981, Geochemical inputs for hydrological models of deep-lying sedimentary units: loss of mineral hydration water: Journal of Hydrology, v. 54, p. 297–314.
Harrison, III, F. W., 1980, The role of pressure, temperature, salinity, lithology and structure in hydrocarbon accumulation in Constance Bayou, Deep Lake, and Southeast Little Pecan Lake fields, Cameron Parish, Louisiana: Gulf Coast Association of Geological Societies Transactions, v. 30, p. 113–127.
Hitchon, B., 1969a, Fluid flow in the western Canada sedimentary basin, 1, Effect of topography: Water Resources Research, v. 5, p. 186–195.
________, 1969b, Fluid flow in the western Canada sedimentary basin, 2, Effect of geology: Water Resources Research, v. 5, p. 460–469.
Kissin, I. G., 1978, The principal distinctive features of the hydrodynamic regime of intensive earth crust downwarping areas, *in* Hydrogeology of great sedimentary basins: Publication of the International Association of Scientific Hydrology, no. 120, p. 178–185.
Korotchansky, A. N., and J. Mitchell, 1972, Rejects de dechets dans les nappes profondes: 24th International Geological Congress, Sect. 11, p. 282–295.
Loucks, R. G., M. M. Dodge, and W. E. Galloway, 1979, Importance of secondary leached porosity in lower Tertiary sandstone reservoirs along the Texas Gulf Coast: Gulf Coast Association of Geological Societies Transactions v. 29, p. 1–8.
Magara, K., 1978, Compaction and fluid migration—practical petroleum geology: Amsterdam, Elsevier, 319 p.
Toth, J., 1980, Cross-formational gravity-flow of groundwater: a mechanism of the transport and accumulation of petroleum (the generalized hydraulic theory of petroleum migration): American Association of Petroleum Geologists, Studies in Geology no. 10, p. 121–167.
Wood, J. R., and T. A. Hewett, 1982, Fluid convection and mass transfer in porous sandstones—a theoretical model: Geochimica et Cosmochimica Acta, v. 46, p. 1707–1713.

Reproduced by permission of the Geological Society from the *Journal of the Geological Society*, V. 135, part 1, pp. 107-117.

Possible links between sandstone diagenesis and depth-related geochemical reactions occurring in enclosing mudstones

C. D. Curtis

SUMMARY: Within compacting mudstone sequences several depth zones can be recognised in which distinctive diagenetic reactions take place. Most striking are those involving carbonate precipitation or replacement. The major carbon source is organic matter which is degraded by various microbiological and inorganic reactions to bicarbonate as one product. Other reactions involving silicates and sesquioxides introduce cations to the porewater system and carbonates are precipitated. Their chemical and isotopic composition can be linked with temperature and depth of formation.

Analysis of some of the literature of sandstone diagenesis suggests that carbonate cements in sandstones often originate in mudstones, and that precipitation may take place far from the site of solute generation. Vast volumes of porewater excluded from mudstones during compaction are channelled into and travel great distances within sand units. Early diagenetic reactions drastically modify the permeability of both potential source and reservoir rocks. The component of horizontal flow is greatly increased and this must affect later migration of liquid hydrocarbons.

Sedimentologists have studied post-depositional alteration in sediments from many different viewpoints. This has led to a bewildering array of terminology to describe the various aspects of diagenesis as seen by different workers. Good reviews of the historical development of terms and concepts are included in Larsen & Chilingar (1967).

Diagenetic alteration can be considered to include all processes occurring during sediment burial from environments immediately below the sediment/water interface to environments transitional to metamorphism. Further modifications arise in consequence of tectonism which can bring sediments into contact with deep or shallow meteoric waters. The present paper will be restricted to the burial phase.

There is much evidence that the nature of diagenetic alteration changes with increasing burial depth and with sediment composition. Attempts have been made to collect observations together within a general framework or model of diagenesis within which definite stages or facies can be recognised. Perhaps the most comprehensive version was that of Teodorovich (1946, 1947, 1954—reviewed in Larsen & Chilingar 1967). An important component of this (and other) schemes was a classification of precipitation environments based on pH and Eh. Some aspects of this type of approach, however, particularly with respect to iron mineral assemblages, have been criticised on the basis of thermochemical theory (Curtis & Spears 1968). Another comprehensive discussion of diagenetic 'facies' is to be found in Packham & Crook (1960). Less ambitious approaches to the general problem have been offered by Fairbridge (1967) and, in the specific case of sandstone diagenesis, by Dapples (1962).

The purpose of this paper is to look at the geochemistry of diagenetic alteration from a slightly different viewpoint and also to take some account of the wealth of information that has accumulated in recent years. It is not intended to erect another classification of processes, but to discuss such questions as the source of reactive solute species, the environmental factors which control the amount and rate of alteration, the time and place of a particular cement or replacement within burial history, and the relationships between mineral diagenesis and the evolution and migration of liquid hydrocarbons.

It is acknowledged that much of the raw geochemical evidence mentioned above is derived from marine sediment studies. The conclusions reached therefore apply strictly to marine sediment diagenesis and extrapolation to other environmental situations should be undertaken with care.

An open-system chemical approach to sandstone diagenesis

Solution–mineral equilibria

Since the classic paper of Krumbein & Garrels (1952) demonstrated the potential value of mineral solution equilibrium calculations as an approach to sedimentation and diagenesis, these methods have become established as a basic technique in interpretive sedimentology. R. M. Garrels's text *Mineral Equilibria* has achieved even greater recognition for its lucid account of both the underlying principles of and limitations inherent in the methods of chemical thermodynamics as applied to sedimentary systems.

Mineral stability field diagrams calculated by these authors have been used regularly as a basis for interpretation of ancient mineral assemblages in terms of environments and processes, but these 'secondary' interpretations have often not taken proper account of cautionary remarks in the original texts. Eh/pH stability field diagrams have been used as if these two variables alone completely characterise the aqueous phase in equilibrium with particular solid assemblages. This is not the case; Eh and pH are just two 'activities' (intensity factors) and the stability of solids depends also on the activities of all other dissolved species in the aqueous phase.

All thermochemical calculations are based on equilibrium states. These describe closed systems in which all spontaneous reactions have proceeded to completion and all unstable solids replaced by stable ones. The equilibrium distribution of chemical species extends to the solution phase and its composition must reflect all the solids present. In closed systems similar in composition to sediment–water systems, the bulk of all chemical species (other than pure H_2O) tends to reside in the solid phase. The activities of chemical species in solution will be controlled by the composition and amounts of the solids present rather than the other way round. It follows that it is most unrealistic to think in terms of Eh and pH 'controlling' the precipitation of diagenetic minerals.

Equilibrium calculations are of greatest value in identifying potentially unstable assemblages (which may react) and stable phases which may precipitate from reaction products. It is essential to qualify statements such as these because total internal equilibrium is seldom approached during diagenesis: many potentially unstable phases exist in spite of the calculations.

Eh/pH diagrams based on solution–mineral equilibrium calculations certainly suggest the probable pH/Eh values of the porewaters present at the time of alteration or precipitation. In a sense, however, this really begs the question: asking why a particular diagenetic modification took place is no less sophisticated than (or different from) asking why did the pH and Eh of the porewaters have specific values. In fact, both must be controlled by reactions involving the decomposition of unstable minerals within the sediment system and also by the actual amounts involved. It is these controls which must be identified and assessed.

Open systems and kinetic controls

There are numerous examples of sandstones which are cemented by diagenetic minerals which must have originated somewhere other than in the immediate vicinity. Pore filling by sequential cements of different composition is one example—the cements could not have originated from material in pore space otherwise some would have been included in the earlier cement stages.

Taking the opposite view is instructive: suppose all the cement originated 'somewhere else'. What would then control precipitation and porewater composition? In a simple case, the rate of introduction of bicarbonate, plus the rate of introduction of calcium, minus the rate of precipitation of calcite, minus the rate of removal of 'carbonic acid' (according to equation 1) will leave a slight state of supersaturation with respect to calcite at a slightly decreased pH (due to the dissociation of carbonic acid)

$$Ca^{2+} + 2HCO_3^- \rightarrow CaCO_3 + H_2CO_3 \rightarrow H^+ + HCO_3^- \qquad 1.$$

The pH of the pore fluid and the amount of calcite precipitating clearly reflect the balance between calcium, bicarbonate, hydrogen and hydroxyl ions in the *dynamic, reacting* system. Calcite precipitation and the composition of the pore fluid therefore depend on the rate of introduction of dissolved species which in turn depend on their rate of generation somewhere else and the dynamics of the porewater migration system. The starting point for any enquiry into the controls operating must be to establish what chemical reactions take place in sediments which could generate the relevant dissolved species. Thereafter, the environmental parameters which dictate the rate of solute generation must be assessed.

Supersaturation effects

One other facet of rate controls must be examined before analysing solute generation. This greatly affects the interpretation of petrographic information. It is generally accepted that both rate of crystal nucleation and growth depend (although not exclusively) on degree of supersaturation. Low degrees of supersaturation favour growth on existing crystals rather than nucleation. Coarsening crystal sizes in sequential precipitates indicate reducing supersaturation level and rate of precipitation.

The style of carbonate precipitation also depends on compositional factors. Folk (1974) presented an excellent summary of this dependency and dealt with both degree of supersaturation and composition, particularly the very marked effects of magnesium concentration variation.

Another complication arises from the fact that equilibrium conditions seldom apply during precipitation of carbonates (de Boer 1977). In theory, the ratio of two cations present in solution should bear some fixed numerical relationship to the ratio of the same cations in the precipitate phase (partition constant). For natural or synthetic low temperature calcites, the magnesium content depends on the rate of precipitation as well as the magnesium content of the aqueous solution. Part of the problem stems from surface adsorption phenomena during crystal growth. Almost certainly similar complications arise with other ions (iron, manganese) and other carbonates (dolomite, siderite). All these complexities make equilibrium calculations difficult to apply meaningfully. On the other hand, detailed petrographic/chemical studies of diagenetic assemblages in ancient sediments, combined with experimental work documenting kinetic controls on precipitation, offer the exciting prospect of identifying rate factors operative in ancient diagenetic environments. Having explored some of the links between porewater chemistry and diagenetic assemblages, it now remains to consider the source of the solutes in solution.

Solid reactivity: thermal and surface area controls

Large volumes of porewater are trapped in sediments, especially in muds. The included solutes (even in marine waters) are quite insufficient, however, to make more than a small contribution to diagenetic mineral assemblages. The aqueous phase acts primarily as a medium of reaction and transport.

Detrital solids are the ultimate source of most diagenetic alteration. At the low temperatures of early burial (5 to 30°C), only the most reactive phases (irrespective of true thermodynamic stability) will participate. Probably most significant in this context are the extremely fine-grained and amorphous constituents derived from soils: intimate mixtures of hydrated aluminium and iron oxides together with degraded organic matter. Their presence in sediments is often underestimated simply because they are not revealed by X-ray diffraction. Other very reactive materials include volcanic ash (finely divided glass), biogenic detritus from the depositional environment, and clay minerals (ion-exchange reactions).

Early during diagenesis, therefore, reactions between these constituents are the most likely to generate cement-forming chemical species in aqueous solution. Coarse-grained and less reactive minerals (typical of sands) are much less likely to enter into reaction. Simple pore-filling cementation rather than replacement is to be anticipated.

At the higher temperatures encountered in deeper burial environments, the possibility of coarser material reacting becomes real. Unstable detrital minerals will tend to break down, contribute solutes to the porewater phase and be replaced.

An overall view of sediment composition suggests that muds and silts are the likeliest source of diagenetic cements. Sands consist mostly of mineral detritus which has survived the weathering environment. The clay fraction, on the other hand, represents breakdown products that have formed in an oxygen-rich atmosphere and aqueous solutions of very low ionic strength. These conditions contrast markedly with those of most marine burial environments. In this context it is perhaps significant that several workers have noted that diagenetic alteration of sandstone bodies often shows more intensive development at margins with mudstone units.

The interplay between the various kinetic controls discussed above is diagrammatically represented in Fig. 1. The composition of incoming porewater is dictated by the relative rate of reactions (k_1 to k_5). The degree of supersaturation is that resulting from input reactions less the net precipitation rate, k_6. The composition of the precipitating carbonate phase depends on both the composition of the solution with respect to different solutes and the maintained supersaturation level. Both these factors also affect the fabric of the precipitate. The balance between k_5 and k_6 will dictate pH. The overall style of cementation will depend on the $k_6 : k_7$ ratio. High precipitation rates relative to pore fluid migration rates are likely to lead to localised compositional zoning whereas the reverse situation should lead to widespread precipitation of uniform composition cement.

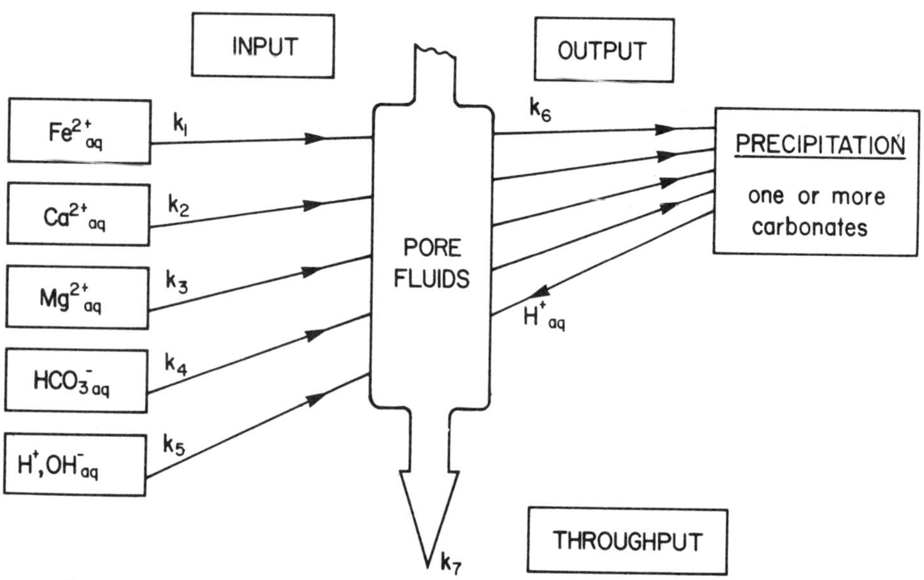

8

FIG. 1.
Rate factors controlling carbonate composition and style of precipitation during diagenesis.

k_1–k_5: rates of generation of different solutes by essentially independent decomposition/dissolution reactions.

k_6: rate of carbonate precipitation.

k_7: rate of porewater migration.

Sources of reactant species: mudstone diagenesis

Distinct depth-related reaction zones: evidence from Recent sediment studies

A much clearer picture of mudstone diagenesis has emerged in recent years. The composition and chemical stability of the terrestrial detritus component can be clearly traced to processes occurring in soil environments (Rateev *et al.* 1969, Gradusov 1972). These are very different from marine depositional environments and soil clays and sesquioxides start to react during the depositional phase. Significant alteration can be detected soon after burial (Russel 1970, Drever 1971*a,b*). These early reactions can also be followed by studying changes in porewater chemistry with depth in Recent sediment columns. Sayles & Manheim (1975) reviewed results from DSDP core studies. Within more rapidly deposited terrigenous clay sediments, sulphate depletion is almost ubiquitous and usually total within a few metres of the sediment–water interface. This is accompanied by bicarbonate and ammonia enrichment, attesting to microbial activity and organic matter degradation. The fate of the reduced sulphur species is well documented (e.g. Berner 1970). Iron sulphides (ultimately pyrite) accumulate and soluble iron is maintained at low levels.

The stable isotope composition of carbon compounds (bicarbonate, methane, etc.) in pore fluids has been established by Presley & Kaplan (1968), Nissenbaum *et al.* (1972), and Claypool *et al.* (1973). This work is of crucial importance because it demonstrates beyond reasonable doubt microbiological degradation of organic matter to bicarbonate. Furthermore, isotopic compositions vary so greatly due to biological fractionations that specific source reactions (sulphate reduction, fermentation) can be identified in certain cases.

Organic matter modification continues to great depth and has been researched diligently for obvious economic reasons. Kerogen modification studies [*in situ*, as in Laplante (1974), or experimental, as in Tissot *et al.* (1974)] suggest four fairly distinct stages. The first, at shallow depths, is dominated by fermentation; the second is characterised by elimination of water and carbon dioxide; the third is the important stage of liquid hydrocarbon evolution, and the last of breakdown to methane and, eventually, graphite. Methane produced at depth can be distinguished from shallow sources by its very different carbon isotopic composition ($\delta^{13}C_{PBB} \sim -50‰$ and $\sim -75‰$ respectively).

Systematic modifications to clay mineral assemblages have also been documented in deeper burial environments. The best known of these is the montmorillonite–mixed-layer–illite progressive transformation. Burst (1969), Perry & Hower (1970, 1972), Schmidt (1973) and Perry (1974) all obtained much useful chemical and structural information. Apart from chemical rearrangements which could well modify porewater composition, the net structural transformation involves dehydration. Water introduced in this way has been linked with anomalously 'fresh' porewaters, over-pressuring and the 'flushing out' of liquid hydrocarbons from source rocks. Some of the links are rather tenuous.

With regard to the overall porewater content of buried sediments, a fairly general pattern has been established, with depositional porosity values around 80 per cent dropping off to about 30 per cent at 1 km, with lower values at depth. Numerous data are available; those in Dickinson (1953), Ham (1966) and Preiss (1968) are representative.

Diagenetic mineral assemblages in ancient sediments

The chemical and isotopic compositions of diagenetic carbonates in ancient sediments (calcite, dolomite, ankerite, siderite) show great variation, implying similar variations in porewater composition during diagenesis. Some examples of note are to be found in Galimov & Girin (1968), Murata *et al.* (1972), Curtis *et al.* (1972), Saas & Kolodny (1972), Pearson (1974) and Curtis *et al.* (1975). The carbon isotope data demonstrate that the source of the carbonate often lies in organic matter degradation reactions and not recrystallization of primary carbonate precipitates derived from ocean reservoir bicarbonate. The very different cation compositions must reflect predominance of particular mineralogical breakdown reactions in specific diagenetic environments.

Diagenetic carbonate cements in mudstones tend to replace porewater such that indications of depth of precipitation should be obtainable from cement/clastic component volume ratios (cementation occurring at different compaction levels). Oertel & Curtis (1972) demonstrated that independent and corroborative information could be obtained from the fabric of included clay minerals. This also changes with compaction from approximately random at deposition to highly orientated at depth. On the basis of information of the kind described in this and the preceding sections, Curtis (1977) proposed that a series of distinctive depth-related zones could be recognised within marine mudstone burial sequences. The chemical reactions occurring in each zone control the style of diagenetic alteration therein.

Depth-related environmental zonation

Table 1 outlines the principal characteristics of six environmental zones spanning sedimentation to metamorphism. It applies to sediment columns consisting of organic-rich clays deposited from marine waters. Zone I is no more than a veneer separating oxygenated bottom waters from the reducing sedi-

TABLE 1: *Diagenetic zones within compacting marine mudstone sequences.*

Zone	Process Description	Zone Base Depth (m)	ΔT^oC	Average Porosity	$\delta^{13}C_{PDB}$ ‰ Carbonate	Carbonates Precipitated	Other Minerals Precipitated	Mineral Modifications
I	Bacterial Oxidation	10^{-2}	0	80	-25	NONE: diffusion of solutes into overlying depositional waters		Partial destruction of organic matter
II	Bacterial Sulphate Reduction	10	0.3	75	-25	Calcite Low Fe, Mn	Pyrite	Rapid destruction of soil colloids Fe, Mn, Al, Si
III	Bacterial Fermentation	10^3	30	30	+15	Ferroan Calcite; Ferroan Dolomite or Siderite	Amorphous Carbonate Apatites	Clay exchange reactions especially $Fe^{2+}/Mg^{2+}/Ca^{2+}$
IV	Decarboxylation	$2.5x10^3$	75	15	-20?	Ferroan Dolomite or Siderite	Kaolinite?	Montmorillonite
V	Liquid Hydrocarbon Generation	$4x10^3$	120	10	Variable: May include carbon from primary or early diagenetic carbonates	Calcite Dolomite Ferroan Dolomite Siderite Depending upon instability of earlier carbonates		Mixed Layer Clays; Illite
VI	Gas Graphite	$> 4x10^3$	120+	<10				Mica Development
	Metamorphism							Chlorite Biotite

ments below. Within this zone, diffusive input of oxygen maintains high oxidation states and supports oxidative bacterial breakdown of organic matter. The soluble reaction products will tend to be lost by back diffusion to the depositional waters and not precipitate as diagenetic cements.

Reducing conditions are rapidly established with depth as molecular oxygen uptake rate exceeds diffusive supply. Sulphate reducing bacteria then thrive as long as diffusive supply of sulphate from depositional waters is maintained. Reduced sulphur species and isotopically light bicarbonate are the principal byproducts of activity. Unstable iron compounds are reduced under these conditions, but iron sulphide precipitation prevents build-up of significant ferrous iron concentrations in solution. Zone II is characterised by the precipitation of diagenetic pyrite and iron poor, isotopically light carbonates. The lower limit of this zone is determined by dissolved sulphate concentrations being reduced to levels insufficient to support bacterial growth.

Zone III is based on fermentative degradation of organic matter. This process is not dependent upon externally derived electron acceptor molecules, but is limited by the availability of suitable organic substrates which, in consequence of activity, will diminish with depth. Another depth-dependent limitation must be temperature. The lower zonal limit is suggested to occur at temperatures some 30°C greater than bottom temperatures, but the justification for choosing this value is not substantial.

Fermentation produces bicarbonate as well as methane. In consequence of the enormous enrichment of ^{12}C in the latter, bicarbonate introduced in Zone III will be isotopically heavy, in marked contrast to the upper zones. Reduction of iron compounds will continue but the absence of sulphide species allows dissolved ferrous iron concentrations to build up. Isotopically heavy, iron-rich carbonates will precipitate. Ion-exchange reactions and further decomposition of soil colloids will introduce calcium and magnesium to porewaters. Phosphate, of organic origin, will tend to precipitate as collophane (amorphous carbonate-apatite) although this process need not be restricted to Zone III: precipitation in Zone II is equally likely.

Bicarbonate will continue to be generated, mostly by decarboxylation, at the increased temperatures of Zone IV. Here, however, its isotopic composition will revert to light values approaching those of the organic substrate. The rate of generation of bicarbonate by this mechanism and also by reduction of iron compounds will tend to increase with depth (temperature) towards some maximum value before dropping again as suitable reactant materials become exhausted. Iron-rich, isotopically-light carbonates will precipitate. Their composition with respect to other cations, however, will depend very much on the composition of the original sediment. The presence of basic volcanic ash

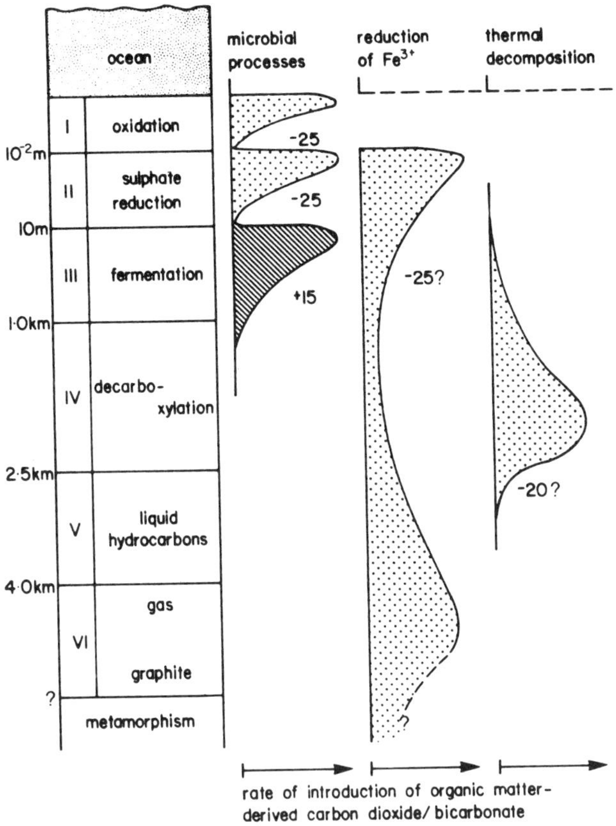

for example, would favour ferroan dolomites in consequence of magnesium availability. Terrestrial sediments poor in primary calcites (which might react at these depths) would tend to favour siderite precipitation.

Zones V and VI are much less likely to be strongly influenced by bicarbonate derived from organic matter. Dissolution, re-precipitation or replacement reactions involving unstable primary carbonates (and early diagenetic carbonates) should be the norm. Their cation composition again will be influenced by increasingly important silicate transformations.

Fig. 2 outlines possible rate trends within the various zones for bicarbonate production and underlines isotopic variation.

Sedimentation-rate facies control

One very important consequence of the zonal model is not accounted for in Table 1. The diagenetic zones must not be likened to metamorphic facies: total internal sediment equilibrium is never even approached.

FIG. 2.
Carbon dioxide production within different diagenetic zones.

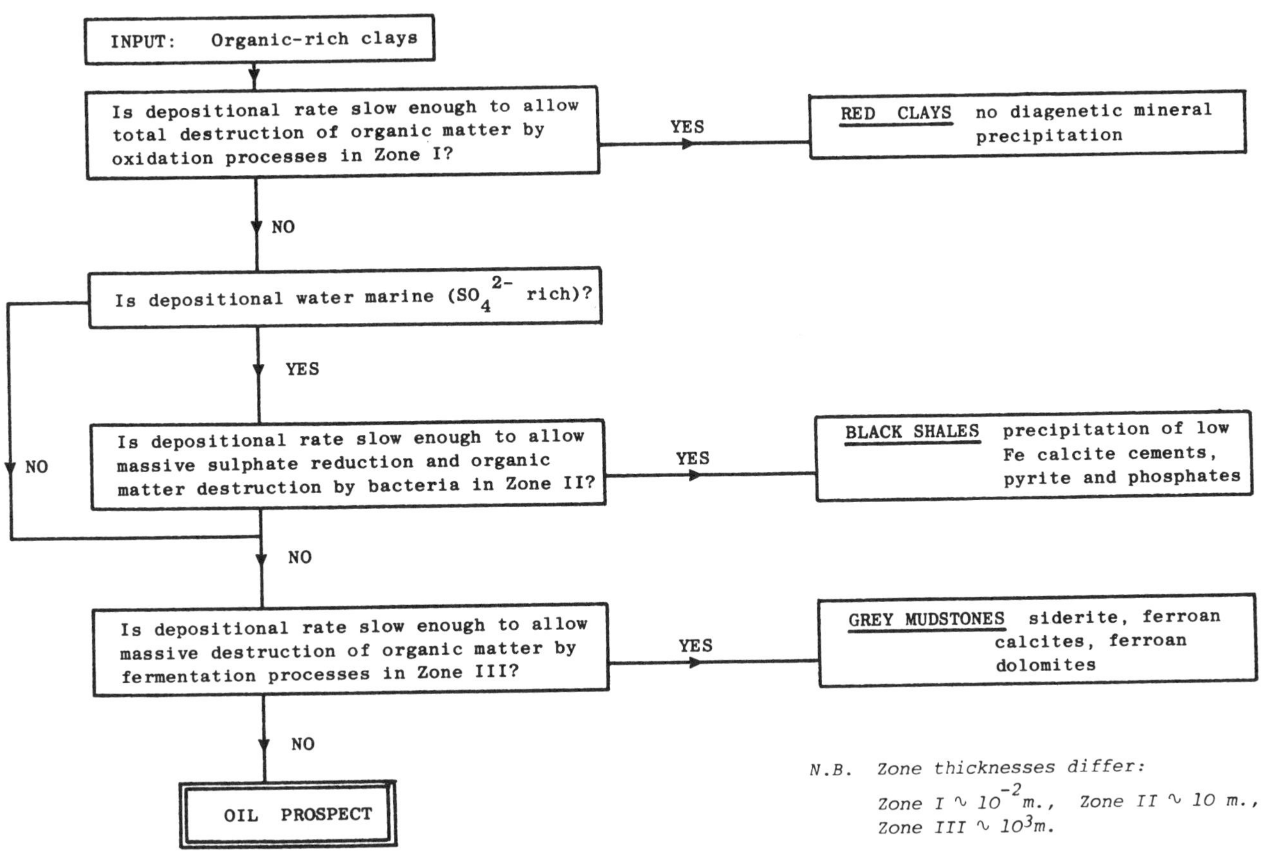

FIG. 3. Sedimentation rate facies control, Zones I to III.

The extent to which sediment is modified by exposure to a particular zone will depend very much on its time of residence therein. The overall effect is simple to anticipate (Fig. 3). Zone I is so thin that only extremely slow depositional rates would allow complete degradation of organic matter to produce red clays. The present-day distribution of marine red clays supports this assertion. Slightly less slow sedimentation would amplify the importance of Zone II (note the very different thicknesses of Zones I, II and III) and produce sediments with the classic black marine shale lithology. Other permutations (including the consequence of non-marine sedimentation) are illustrated (Fig. 3), or can be inferred by applying the same reasoning to deeper zones. This analysis is important in demonstrating that the zonal model is not at variance with numerous observations of the widespread occurrence (throughout the Phanerozoic) of characteristic mudstone lithologies.

Kimmeridge Clay diagenesis: more direct evidence

The zonal model was erected on the basis of fragmentary evidence collected from numerous studies of both ancient and Recent sediments. Irwin *et al.* (1977) reported the results of an investigation into the isotopic composition of carbonates in the Dorset Coast section of the Kimmeridge Clay. This project was set up to test some of the implications of the model within an organic-rich sediment sequence containing much diagenetic carbonate. Both carbon and oxygen isotope ratios were determined.

The 53 samples are plotted in two groups (Fig. 4). The circles are coccolith-rich samples which represent 'primary' carbonate precipitated from surface marine waters. The remaining samples are diagenetic in origin: early calcite concretions (triangles) and ferroan dolomite cementstones (squares).

It is possible to deduce precipitation temperatures for carbonate-water equilibria provided the isotopic composition of the water is known. A reasonable estimate can be made for open ocean-water and values around 26°C are indicated for the coccolith limestones (Irwin *et al.* 1977). This and the carbon isotope composition are in keeping with precipitation from marine reservoir bicarbonate in surface waters.

The diagenetic calcite concretions are very much enriched in ^{12}C and considerably lower precipitation temperatures are indicated. These values are consistent with precipitation at some small distance below the sediment/water interface as a by-product of sulphate reduction in Zone II. It is unlikely that pore-water isotopic composition was different from that in depositional water since the system was open to sulphate diffusion and the sediment contained approximately 80 per cent water.

The remaining ferroan dolomites show a range of $\delta^{13}C$ from +9 to −6 per mil and $\delta^{18}O$ from −1.5 to −7 per mil. Furthermore there is an excellent correlation between these two ratios. If we assume that the change in $\delta^{18}O$ is in part due to increasing temperature of precipitation (this is discussed more fully in Irwin *et al.* 1977), this trend is one of heavy carbon at low temperature (shallow burial) to light carbon at high temperature (deep burial). This very simple interpretation is fully in accord with the zonal model: below Zone II (Sulphate reduction—light carbon), fermentation (heavy carbon) characterises Zone III and this is replaced by inorganic thermally induced processes in Zone IV (light carbon again).

As diagenesis proceeds, of course, the isotopic composition of the enclosed porewater must change. This would be towards lighter oxygen but the magnitude of change is thought unlikely to account entirely for the trend in Fig. 4. What does seem clear is that the carbonates of the Dorset Coast Kimmeridgian formed largely from organic matter. Furthermore the zonal model seems to fit the isotopic data remarkably well and, in this case, the cements probably all pre-date liquid hydrocarbon formation and migration.

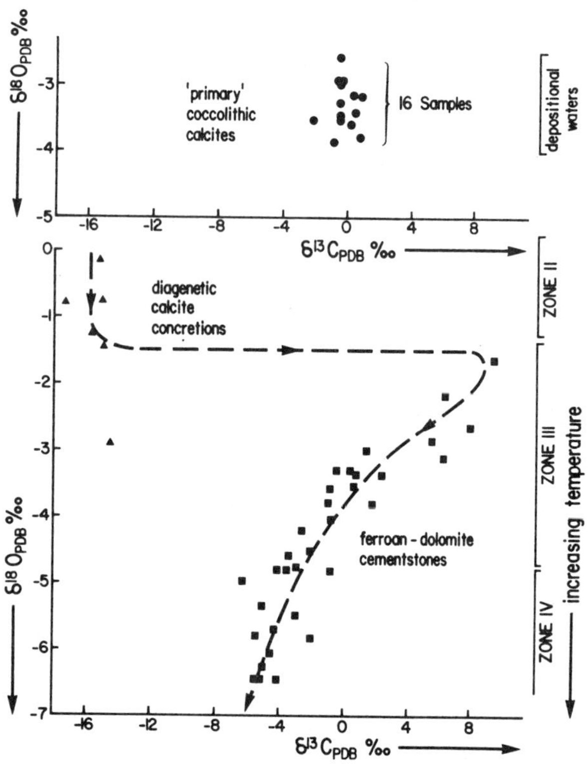

FIG. 4. Stable isotope ratio plots ($\delta^{13}C_{PDB}$ versus $\delta^{18}O_{PDB}$) for Kimmeridge Bay carbonates. Trend line is time of carbonate precipitation relative to time of enclosing detrital sedimentation. For details see text.

Conclusion

The preceding analysis demonstrates conclusively that mudstone sequences are very reactive chemical systems during diagenesis and that, during compaction very large amounts of carbonate cement are generated. Furthermore, very different carbonate minerals (chemical and isotopic composition) precipitate.

A case has also been made for linking specific carbonate cements with distinctive burial depth/temperature zones such that the time, place and mechanism of formation may be predicted. Throughout the time of precipitation, however, porewaters are being continuously expelled. Should the precipitation rate be low relative to the migration rate, solutes generated within mudstones will precipitate, in part at least, at some considerable distance from the site of formation. Their style of precipitation will be controlled by porewater migration pathways.

Porewater migration

Basin analysis

Von Engelhardt (1967) analysed the mode of dewatering of mixed sandstone/mudstone sequences. His primary observations centred on Keuper and Jurassic strata in Germany. At least three stages of diagenetic alteration were documented but significantly, these affected different sandstones in the same sequence to different extents.

From calculations of the total porewater included at deposition and net porosity reduction, von Engelhardt deduced that very large volumes of water had passed through the sandstone units and that these had acted as channels for migration. In general he concluded that porewater flows collect in sandstones and travel great distances in them. On a basinal scale, flow will be conducted towards and up basin margins mainly through sandstone strata. Faulting of any kind obviously will concentrate upward flow to very narrow zones.

The overall picture is very much one of near-horizontal stratiform flow concentrated in low permeability zones within the context of a net upward movement. Within these low permeability units, the distance of flow and the volume of porewater crossing any particular sediment cross-section will be extremely great.

Von Engelhardt concluded that diagenesis should be considered very much from the 'open-system' viewpoint. He suggested that sedimentologists interested in diagenesis should undertake much more work in this area of analysing and documenting porewater exclusion pathways.

Stratiform cementstones

In fine-grained sediment sequences, several authors (Hallam 1967, Raiswell 1971, Curtis *et al.* 1975) have noted that clastic inclusions in concretions or cementstones are both coarser and contain more quartz than the enclosing mudstones. This fact has been linked with preferential migration of porewater (therefore preferential solute supply) along siltier, more permeable horizons. The lateral persistence of very thin diagenetic carbonate horizons of this type is often quite remarkable. 'Bedded' sideritic ironstones of the British Carboniferous fall into this category.

Some of these cementstone horizons form early during burial, long before dewatering is anything like complete (Oertel & Curtis 1972). Their formation will infinitely increase the contrast between vertical (nil) and horizontal permeability. Add this factor to von Engelhardt's analysis, and his conclusions relating to sandstone channelling and horizontal flow are dramatically underlined. In the context of hydrocarbon migration it is interesting to note the establishment of strong stratiform migration pathways within source rocks prior to liquid hydrocarbon formation.

This section has done no more than draw attention to a mere fragment of the available evidence for movement of large volumes of porewaters from mudstones into sandstones and their subsequent migration for great distances therein. Couple this with the evidence for large scale solute generation within mudstones throughout their burial history, and a good case must be made for anticipating close links between mudstone and sandstone diagenesis patterns.

Patterns of diagenetic alteration in sandstones

Dapples's diagenetic stages

Published (and no doubt unpublished) petrographic descriptions of individual sandstone diagenetic histories are legion. Space limitation forbids any kind of detailed survey and only very limited reference to this large volume of information can be made. Dapples (1962, 1967) specifically set out to identify patterns of alteration which appeared to occur generally within sandstones. He concluded that there was widespread evidence for recognising sequential modification reflecting passage through characteristic depth or time related alteration stages. Only two are relevant here in the sense that they cover burial processes. The 'Redoximorphic' stage was suggested to include early reactions often reversible and dominated by oxidation–reduction. Diagenetic minerals of this stage occur mostly as pore-filling cements. The 'Locomorphic' stage was defined to cover reactions occurring at slightly deeper levels (early burial and lithification) and characterised by replacement rather than pore space precipitation. Table 2 is modified from Dapples (1962) and emphasises carbonate mineral diagenesis in terms of appearance without assuming the nature of the chemical reactions involved. Without detailed information as to the precise nature of the carbonate

REDOXOMORPHIC STAGE	CARBONATES PRECIPITATED
Deposition and early burial: generally 'reversible'	Calcite Dolomite Ferrodolomite Siderite
LOCOMORPHIC STAGE	**REPLACEMENT REACTIONS**
Early burial and lithification	Aragonite → Calcite Calcite → Dolomite Quartz, Chert, Feldspars, Clays → Various Carbonates

TABLE 2: *Diagenetic reactions involving carbonates—after Dapples* (1962).

phase compositions, it seems clear that the same wide range of mineral types occurs in sandstones as in mudstones. It is also true that parallels can be drawn between Dapples's stages and the Zones proposed in Table 1. The 'Redoxomorphic' stage includes Zones I to IV and the 'Locomorphic' stage Zones V and VI, although it is difficult and perhaps unwise to seek precise correlation.

Dapples also dwelt on analogies with progressive metamorphism. He recognised the importance of compositional control and suggested four classes corresponding to various oxide 'end-member' assemblages. These would respond in different ways to each burial stage. Dapples acknowledged that interstitial solutions were important as a medium of reaction and also that total equilibrium would not be approached in diagenetic reactions, unlike in metamorphic situations.

In general, however, it seems fair to summarise Dapples's approach as being based largely on information gained from within sandstones and his characteristic reactions 'rearrangements' of chemical species in essentially closed systems. 'Open systems' type reactions were considered by Dapples (1962, p. 927): "As the shales underwent compaction, solutions of high pH and others of high carbonate ion concentration probably were ejected from different shale beds, entered the sandstones, and were reacted with the quartz and clay to cause the replacement relationships observed". The essential difference between Dapples's analysis and that offered here is one of emphasis. It appears likely that much diagenetic alteration in sandstones owes its origins to reactions in mudstones, and these probably take place far away from the site of sandstone modification.

Present state of the art

Some recent detailed petrographic/geochemical work on diagenetic carbonates suggest that Dapples's broad trends are potentially capable of significant subdivision, and thereby probably allow the relevance of the zonal scheme outlined in this paper to sandstone diagenesis to be assessed. Lindholm (1974) studied the geochemistry and fabric of septarian veins in 16 concretions from sediments of differing age and type. He found a surprisingly consistent picture of early Mg-rich, Fe-poor calcites followed by later calcites with the reverse pattern. Ferroan dolomites not uncommonly occur and almost invariably post-date the calcites. Lindholm's results are summarised in Table 3. This kind of approach, much utilised now by carbonate sedimentologists, is clearly capable of identifying definite time sequences for precipitation and evolution of porewater chemistry. Concretions themselves often show changes in cement composition from the centre (presumably early) outwards. Complex precipitation histories are usually revealed by detailed petrographic investigations, however, so that care is needed with this approach. Sideritic concretions, which document long growth periods, often show early Fe- and Mn-rich siderite with much more magnesian siderite (up to 12 mole % $MgCO_3$) in later layers (Curtis *et al.* 1975).

Nash & Pittman (1975) found calcite cements with up to 10 mole % $FeCO_3$ and 39 mole % $MgCO_3$ in Recent sands with ferroan dolomite developing at depth. It is clear that the range of chemical and isotopic composition of carbonate cements is very great indeed. It is equally clear that it has not been comprehensively investigated.

The thesis of this paper is that there is ample evidence linking diagenesis in mudstones and sandstones and that future work must take this into account. Work on detailed chemical and isotopic composition strongly suggests that it should be possible to predict with considerable certainty the time and place of cementation during burial, and following from an understanding of the chemical reactions involved, to predict the environmental controls governing the

Time	Fabric, Mineralogy	mole % $MgCO_3$	mole % $FeCO_3$	Included Organic Matter
Early	Length slow fibrous or bladed calcite	1.6 to 4.4	low	Organic Matter Included
↓	Equant calcite coarsening ↓	0.02 to 1.75	0.5 ↓ Systematic increase 3.5	Clear
Late	Ferroan Dolomite (25% of examples)			

TABLE 3: *Generalised sequence of carbonates precipitated in septarian veins—after Lindholm* (1974).

amount and spatial style of cementation. This should be of considerable value since it has been demonstrated that porosity and permeability of both reservoir and source rocks must be modified dramatically prior to liquid hydrocarbon migration as well as at later stages.

ACKNOWLEDGEMENTS. I would like to thank many colleagues for helpful advice and useful discussions over a considerable time period, during which the ideas in this paper were evolved. Alan Spears, Gerhard Oertel, Max Coleman, Charles Downie, Hilary Irwin, Chari Petrowski, Mike Pearson and Bob Sweeney all helped in this way, but none are responsible for errors of fact or speculation in the paper. Pat Mellor typed the manuscript, and Mike Cooper prepared the diagrams.

This work is being supported by Natural Environment Research Council Grant GR3-3086, which is gratefully acknowledged.

References

BERNER, R. A. 1970. Sedimentary pyrite formation. *Am. J. Sci.* **268,** 1–23.

BURST, J. F. 1969. Diagenesis of Gulf Coast clayey sediments and its possible relation to petroleum migration. *Bull. Am. Ass. Petrol. Geol.* **53,** 73–93.

CLAYPOOL, G., PRESLEY, B. J. & KAPLAN, I. R. 1973. Gas analyses in sediment samples from legs 10, 11, 13, 14, 15, 18 and 19. *Initial Reports of the Deep Sea Drilling Project* **19,** 879–84.

CURTIS, C. D. 1977. Sedimentary geochemistry: environments and processes dominated by involvement of an aqueous phase. *Phil. Trans. R. Soc.* **A286,** 353–71.

——, PEARSON, M. J. & SOMOGYI, V. A. 1975. Mineralogy, chemistry and origin of a concretionary siderite sheet (clay-ironstone band) in the Westphalian of Yorkshire. *Mineralog. Mag.* **40,** 385–93.

——, PETROWSKI, C. & OERTEL, G. 1972. Stable carbon isotope ratios within carbonate concretions: a clue to place and time of formation. *Nature, Lond.* **235,** 98–100.

—— & SPEARS, D. A. 1968. The formation of sedimentary iron minerals. *Econ. Geol.* **63,** 257–70.

DAPPLES, E. C. 1962. Stages in the development of sandstones. *Bull. geol. Soc. Am.* **73,** 913–34.

—— 1967. Diagenesis of sandstones. *In:* LARSEN, G. & CHILINGAR, G. V. (eds) *Diagenesis in sediments: Developments in Sedimentology* **8,** Elsevier, Amsterdam.

DE BOER, R. B. 1977. Stability of Mg-Ca carbonates. *Geochim. cosmochim. Acta* **41,** 265–70.

DICKINSON, G. 1953. Reservoir pressures in Gulf Coast Louisiana. *Bull. Am. Ass. Petrol. Geol.* **37,** 410–32.

DREVER, J. I. 1971*a*. Chemical weathering in a subtropical igneous terrain, Rio Ameca, Mexico. *J. sedim. Petrol.* **41,** 951–61.

—— 1971*b*. Early diagenesis of clay minerals. Rio Ameca Basin, Mexico. *J. sedim. Petrol.* **41,** 982–94.

FAIRBRIDGE, R. W. 1967. Phases of diagenesis and authigenesis. *In:* LARSEN, G. & CHILIGAR, G. V. (eds) *Diagenesis in sediments: Developments in Sedimentology* **8,** Elsevier, Amsterdam.

FOLK, R. L. 1974. The natural history of crystalline calcium carbonate: effect of magnesium content and salinity. *J. sedim. Petrol.* **44,** 40–53.

GALIMOV, E. M. & GIRIN, YU. P. 1968. Variation in the isotopic composition of carbon during the formation of carbonate concretions. *Geokhimiya* **2,** 228–33.

GRADUSOV, B. P. 1972. Razmeshcheniye osnovnykh tipov profiley glinistogo materiala v pochvaka. *Dokl. Akad. Nauk SSSR* **202,** 1164–7.

HALLAM, A. 1967. Siderite- and calcite-bearing concretionary nodules in the Lias of Yorkshire. *Geol. Mag.* **104,** 222–7.
HAM, H. H. 1966. New charts help estimate formation pressures. *Oil Gas J.* **64,** 58–63.
IRWIN, H., COLEMAN, M. & CURTIS, C. D. 1977. Isotopic evidence for several sources of carbonate and distinctive diagenetic processes in organic-rich Kimmeridgian sediments. *Nature, Lond.* **269,** 209–13.
KRUMBEIN, W. C. & GARRELS, R. M. 1952. Origin and classification of chemical sediments in terms of pH and oxidation-reduction potentials. *J. Geol.* **60,** 1–33.
LAPLANTE, R. E. 1974. Hydrocarbon generation in Gulf Coast Tertiary sediments. *Bull. Am. Ass. Petrol. Geol.* **53,** 1281–9.
LARSEN, G. & CHILINGAR, G. V. (eds) 1967. ***Diagenesis in sediments: Developments in Sedimentology*** **8,** Elsevier, Amsterdam.
LINDHOLM, R. C. 1974. Fabric and chemistry of pore filling calcite in septarian veins: models for limestone cementation. *J. sedim. Petrol.* **44,** 428–40.
MURATA, K. J., FRIEDMAN, I. & CREMER, M. 1972. Geochemistry of diagenetic dolomites in Miocene marine formations of California and Oregon. *Prof. Pap. U.S. geol. Surv.* **724-C.**
NASH, A. J. & PITTMAN, E. D. 1975. Ferro-magnesian calcite cement in sandstone. *J. sedim. Petrol.* **45,** 258–65.
NISSENBAUM, A., PRESLEY, B. J. & KAPLAN, I. R. 1972. Early diagenesis in a reducing fjord, Saanich Inlet, British Columbia-I. Chemical and isotopic changes in major components of interstitial water. *Geochim. cosmochim. Acta* **36,** 1007–27.
OERTEL, G. & CURTIS, C. D. 1972. Clay-ironstone concretion preserving fabrics due to progressive compaction. *Bull. geol. Soc. Am.* **83,** 2597–606.
PACKHAM, G. H. & CROOK, K. A. W. 1960. The principle of diagenetic facies and some of its implications. *J. Geol.* **68,** 392–407.
PEARSON, M. J. 1974. Sideritic concretions from the Westphalian of Yorkshire. A chemical investigation of the carbonate phase. *Mineralog. Mag.* **39,** 696–9.
PERRY, E. A. 1974. Diagenesis and the K-Ar dating of shales and clay minerals. *Bull. geol. Soc. Am.* **85,** 827–30.
—— & HOWER, J. 1970. Burial diagenesis in Gulf Coast pelitic sediments. *Clays Clay Miner.* **18,** 165–77.
—— & —— 1972. Late-stage dehydration in deeply buried pelitic sediments. *Bull. Am. Ass. Petrol. Geol.* **56,** 2013–21.
PREISS, K. 1968. In-situ measurement of marine sediment density by gamma-radiation. *Deep-Sea Res.* **15,** 637–41.
PRESLEY, B. J. & KAPLAN, I. R. 1968. Changes in dissolved sulphate, calcium and carbonate from interstitial water of near-shore sediments. *Geochim. cosmochim. Acta* **32,** 1037–48.
RAISWELL, R. 1971. The growth of Cambrian and Liassic concretions. *Sedimentology* **17,** 147–71.
RATEEV, M. A., GORBUNOVA, Z. N., LISITZYN, A. P. & NOSOV, G. L. 1969. The distribution of clay minerals in the ocean. *Sedimentology* **13,** 21–44.
RUSSELL, K. L. 1970. Geochemistry and halmyrolysis of clay minerals, Rio Ameca, Mexico. *Geochim. cosmochim. Acta* **34,** 893–907.
SASS, E. & KOLODNY, Y. 1972. Stable isotopes, chemistry and petrology of carbonate concretions (Mishash Formation, Israel). *Chem. Geol.* **10,** 261–86.
SAYLES, F. L. & MANHEIM, F. T. 1975. Interstitial solutions and diagenesis in deeply buried marine sediments: results from the Deep Sea Drilling Project. *Geochim. cosmochim. Acta* **39,** 103–27.
SCHMIDT, G. W. 1973. Interstitial water composition and geochemistry of deep Gulf Coast shales and sandstones. *Bull. Am. Ass. Petrol. Geol.* **57,** 321–37.
TISSOT, B., DURAND, B., ESPITALIÉ, J. & COMBAZ, A. 1974. Influence of nature and diagenesis of organic matter in formation of petroleum. *Bull. Am. Ass. Petrol. Geol.* **58,** 499–506.
VON ENGELHARDT, W. 1967. Interstitial solutions and diagenesis in sediments. *In:* LARSEN, G. & CHILINGAR, G. V. (eds) ***Diagenesis in sediments: Developments in Sedimentology*** **8,** Elsevier, Amsterdam.

Received and read 9 March 1977; revised typescript received 2 May 1977.

CHARLES D. CURTIS, Department of Geology, The University of Sheffield, Mappin Street, St. George's Square, Sheffield S1 3JD.

THE JOURNAL OF GEOLOGY

March 1979

PLATE-TECTONIC CONTROLS ON DIAGENESIS[1]

RAYMOND SIEVER
Department of Geology, Harvard University, Cambridge, Massachusetts 02138

ABSTRACT

Plate tectonics can be used as a framework for relating diagenesis to geological and geophysical parameters: regional tectonics, including rates of subsidence, uplift and deformation; volcanism and plutonism; heat flow; and transport of formation waters. These parameters affect diagenetic variables: the primary detrital-chemical assemblage; composition of early interstitial waters; rates and depth of burial; length of time buried; geothermal and pressure gradients; and oscillatory vertical movements. The consequences of changes in the values of these variables are illustrated for the diagenesis of the silica polymorphs, Ca- and Mg-carbonates, zeolites, and clay minerals. The unstable nature of biogenic opaline silica components of the primary assemblage leads to successive polymorphic inversions, ultimately to microcrystalline quartz chert. Pressure and temperature control the solubility and rate of inversion of all silica polymorphs. Carbonate systems also are controlled both by the presence of unstable phases (Mg-calcite, aragonite) in the primary assemblage and the effects of pressure and temperature on solubility. In later diagenesis oscillatory movements lead to reversals or replacements of carbonates. Zeolites are derived from the marine and nonmarine alteration of volcaniclastics. Burial, accompanied by pressure and temperature increases, produces the diagenetic zeolite facies. Clay minerals of both terrigenous and marine volcanic derivation are responsive to ion—exchange and silicate reconstitution reactions in early interstitial waters. Pressure and temperature increases lead to the stable chlorite-illite assemblage under deep-burial conditions. Plate-tectonic environments are for simplicity classified as mid-ocean ridge, trailing edge (passive), subduction, continent-continent collision, rift valley, and intraplate (cratonic). Mid-ocean ridges are sites of thin pelagic sediment accumulation affected by high heat flow, hydrothermal activity, and volcanic activity, followed by decreasing heat flow and slow burial as the sediments spread from the ridge. Geothermal gradients are higher in these marine sediments than they are on the continents. Muddy and sandy sediments remain largely unlithified except for layers of chert and limestone. Trailing edges are characterized by lowered heat-flow rates and reduced geothermal gradients, as well as rapid subsidence and burial of terrigenous detrital and chemical sediment after the initial opening phase in which evaporites and heavy metal deposits are formed. The resulting rocks are moderately cemented and lithified by carbonate and quartz and, in deeper zones, have the stable chlorite-illite mineral pair. Volcaniclastics are rare, as are zeolites. Subduction zones, a diverse group of subenvironments, are regions where low heat flow, rapid subsidence, and burial, accompanied by crumpling, operate on turbidites and pelagics, together with large amounts of volcaniclastic debris from nearby island arcs. Zeolites, smectites, and low-temperature-high-pressure mélanges, including ophiolites, are produced. Continent-continent collisions produce enormous quantities of terrigenous detritus deposited in both terrestrial and marine environments, from intermontane basins with subaerial exposure and meteoric water infiltration to submarine fans and plains. Geothermal gradients are low, sedimentation and burial are rapid, and diagenesis is mild, except in sections that have remained deeply buried for 100 m.y. or more. Diagenesis of rift-valley sediment is related to the abundance of volcaniclastics, mixed with alluvial and lacustrine deposits, all buried rapidly in regimes of high heat flow and hydrothermal activity. Intraplate diagenesis is the result of low subsidence and sedimen-

[1] Manuscript received August 4, 1978; revised November 16, 1978.

[JOURNAL OF GEOLOGY, 1979, Vol. 87, p. 127–155]

0022-1376/79/8702-0005$02.28

tation rates in low continental heat-flow regimes. But oscillatory movements are frequent, and complex parageneses of authigenic mineral replacements indicate multiple changes in formation-water composition. A uniformitarian approach to this subject is valid only for the Phanerozoic, for present-day patterns of plate tectonics may not have existed during earlier times when heat flow from the interior was differently distributed.

INTRODUCTION

Much of the literature on diagenesis is concerned with specific mechanisms and geological events directly controlling individual processes. For example, the study of early diagenesis emphasizes the relation between the chemical composition of interstitial waters and mineral reactions in the sediment as controlled by rates of sedimentation and burial. In studies of later diagenesis, the depth of burial and geothermal gradient are discussed as determinants of long-term mineral reactions. In the early diagenesis of carbonate sediments, emphasis is on the nature of the biota and reaction with sea or meteoric waters. Rarely is the larger geologic setting introduced, perhaps because diagenesis has not commonly been a regional geologic study. Yet geologists have attempted to draw the general postdepositional history of a sedimentary rock from local diagenetic processes. It seems in order to examine the larger-scale geological controls on local sets of conditions that ultimately determine the outcome of all diagenetic processes.

The emergence in the past decade of plate-tectonic theory as an overarching theory of the Earth, including the explanation of continental drift, distribution of earthquakes, patterns of volcanism, and facies of sediment on the ocean floors and at continental margins gives a framework for discussing the effects of regional tectonics, volcanism, heat flow, and the paths of migrating solutions on the patterns of post-depositional change in a sediment. We would expect at the outset that some processes are strongly linked to such controls and others weakly coupled. For weakly coupled processes—and perhaps some strongly coupled ones too—one would not necessarily expect to find a specific mechanism operating in only one milieu; rather we would expect much overlap. But an overall pattern should emerge from the noise of all of the individual local diagenetic histories. It is the purpose of this paper to suggest such patterns for future exploration in regional studies of diagenesis. To the extent that regional diagenetic maps correspond to plate tectonic environments they may serve to confirm or modify paleo-plate-tectonic histories based on stratigraphic and structural evidence alone.

I do this by first defining how the geological variables that control diagenesis have their impact on composition and texture, thereby having to review briefly the basic theory of a large and active field of research. The pattern is then illustrated by describing the consequences of changes in the values of these variables for two simple and two complex chemical systems characteristic of large groups of sediments: the silica-water system: SiO_2-H_2O; the carbonate system:

$$CaO-MgO-CO_2-H_2O;$$

the zeolites; a chemical system including common Group I and II elements, alumina, silica, and water; and the clay minerals, containing the same chemical components as the zeolites. Following these preliminary discussions, I propose how all of the major plate-tectonic settings determine the values of the controlling variables. It is beyond the scope of this paper to examine in detail how plate tectonics affect sedimentary compositions and stratigraphic patterns, but some discussion of these relationships is necessary. The literature is so voluminous on all of the subjects covered that individual papers are cited only as examples of a wide variety of contributions, rather than as part of an exhaustive bibliographic review.

DIAGENETIC VARIABLES

Detrital-chemical assemblage.—The raw material on which diagenetic processes operate is the mixture of primary detrital and chemical textural elements of the freshly deposited sediment (fig. 1). The assemblage in a detrital

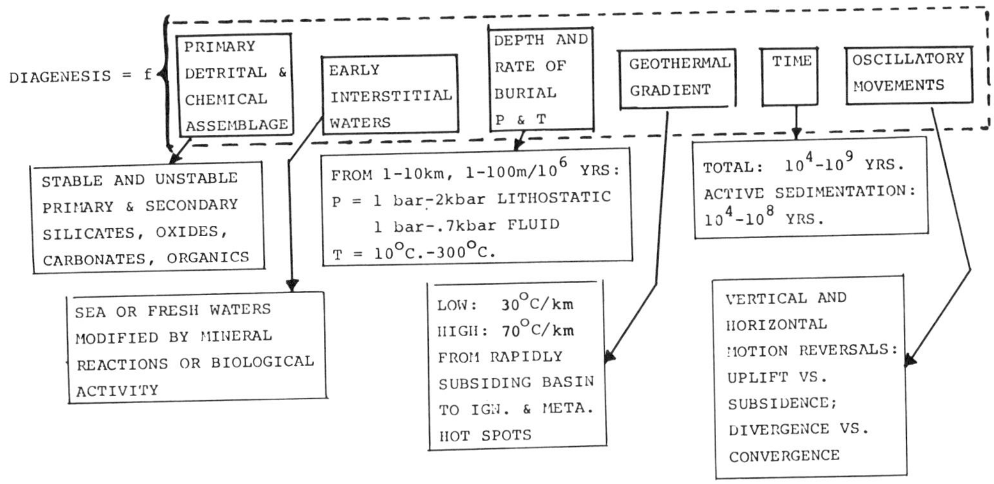

FIG. 1.—Diagenetic variables

sediment typically includes both stable and unstable minerals. The textural elements of chemically precipitated sediments or detrital—chemical mixtures typically include biogenic debris: carbonate, organic matter, and silica. Though there may be some small contribution of nonbiologically produced carbonate and silica, it is doubtful that most common sediments of Phanerozoic age contain any significant proportion of that kind of material. Precambrian sedimentary rocks may include such contributions in large proportion, however.

Early interstitial waters.—Added to this primary mixture is the aqueous component of sediments, the interstitial waters that are trapped during sedimentation. The main origin of such waters is, of course, sea water. For continental deposits it is the fresh water of rivers or lakes and the ground waters with which they connect. Even aeolian deposits, sooner or later, will come into contact with ground waters, which can then participate in diagenetic reactions. Almost as soon as interstitial waters are trapped in the sediment, they may start reacting with the solids. The kind and rate of reaction may be strongly affected by bioturbation.

Burial.—The major operator that affects postdepositional processes is burial. For my purposes I will simply define burial as the net sedimentation rate at the sediment-water or -air interface. Rates of burial vary widely, from a few meters to hundreds of meters per million years. I will consider total depths of sediment burial from 1 to 10 km. Burial is accompanied by increases in lithostatic and hydrostatic pressures. As long as impermeable beds do not seal off a sediment, we may consider the pore waters to be connected to the surface; thus hydrostatic pressures range from 1 bar to about 0.7 kb, depending on depth of burial and the weight of dissolved solids in formation waters. At the same time, depending on sediment density, which does not vary within wide limits, the lithostatic pressure may rise to 2 kb.

Burial is also accompanied by temperature increase along the local crustal geothermal gradient, which may vary from 30°C to 70°C per km, depending on heat flow. The rate of heat flow is strongly linked to the kind of sedimentary basin and its position with respect to plate boundaries and to injection of igneous bodies from the lower crust or mantle. The temperature may rise from a minimum near 0°C at the deep ocean-sediment interface to 300°C and beyond at the bottom of a deeply buried section (Jeffreys 1976: p. 425–427).

The length of time during which diagenetic processes operate on the sediment depends on the rate of subsidence and vertical tectonic movements. The postdepositional history of sediments typically includes burial for a period of time followed by uplift. The sediment may be brought to the zone of meteoric water inflow or even to exposure at the surface. Subsidence may follow again. Some sediments remain deeply buried for only short times. Others remain deep and are sampled only by drill holes. Many sediments have remained buried for hundreds of millions of years at some time in their history. One cannot think of many deeply buried sediments over a billion years old that have never gone through a period of uplift and exposure to the surface—although some may exist. If the sediment is uplifted, it becomes subject to different conditions and "retrograde diagenesis" may result.

A good way to describe burial history is with a sedimentation diagram, by which burial with respect to sea level and the sediment-water or -air interface is plotted against time (Von Bubnoff 1954: p. 31–39). On such a diagram we may plot inferred geothermal gradients and levels of lithostatic and hydrostatic pressure (fig. 2). A graphical representation of the geological history of the sediment, the sedimentation diagram dramatizes the oscillatory movements of the crust through geologic time in response to tectonic forces that express themselves in vertical movement. Not all such vertical movements may be linked to plate tectonics by any means. Much is yet to be learned about vertical movements, particularly in sedimentary basins far from plate boundaries.

MAJOR DIAGENETIC SYSTEMS

The silica-water system.—The system $SiO_2 - H_2O$ is perhaps the simplest one can use to illustrate the effects of the diagenetic variables (table 1). The phases of this system, both stable and unstable, are now fairly well known. So also are the effects of time, temperature, pressure, and chemical variables such as pH (Siever 1972). The components of a primary assemblage that dissolve rapidly to give high concentrations of dissolved silica are, in order of decreasing rate of dissolution, siliceous organisms, volcanic glass and mafic detrital minerals. Studies of dissolution rates of diatoms (Stein 1977) and the compositions of interstitial waters (Siever, Beck, and Berner 1965; Calvert 1974; Manheim and Sayles 1974) indicate that in some places siliceous organisms can contribute enough dissolved silica to saturate interstitial waters with respect to amorphous silica. There is ample evidence that glasses hydrolyze relatively rapidly over geologically short times, liberating large quantities of dissolved silica as they do so (Stefansson 1966; Hurd 1973; Heath 1974). Mafic minerals are only slightly behind the glasses in rates of dissolution, particularly if the interstitial waters are deoxygenated (Siever 1976). Feldspars, particularly the more calcic plagioclases, also may dissolve to give significant quantities of dissolved silica (Wollast 1967; Busenberg and Clemency 1976).

Free silica precipitates from interstitial waters as disordered opal-CT, forming first small nuclei, then radiating fibers and finally lepispheres (Weaver and Wise 1972; Kastner, Keene, and Gieskes 1977; Siever and Stein 1978). Interstitial waters will rarely reach or exceed saturation with respect to amorphous silica. More likely they will become supersaturated only with respect to the less soluble opal-CT, which then starts to nucleate and precipitate. At lower silica concentration levels, the solution may still be supersaturated with respect to quartz, which then precipitates on available detrital quartz grain surfaces.

Increase in hydrostatic pressure affects chemical solubility in this system only slightly, but lithostatic pressure causes pressure solution at points of contact between detrital quartz grains. Pressure solution is a function of time and burial, as the rate of burial carries with it the rate of pressure increase (Weyl 1959).

Temperature has a profound effect; it is perhaps the most important of all the diagenetic variables. Increase in temperature attendant on burial raises both the equilibrium solubility of all the polymorphs of silica and their rate of dissolution—or precipitation—as well. It also increases the rate of polymorphic inversion (Kastner, Keene, and Gieskes 1977). For that reason it is difficult if not impossible to find unstable polymorphs of silica in rocks that have been deeply buried for any length of time—that is, on the order of 3 km for 100 m.y. In regions of high heat flow the solubility and concentration gradients will be high within the sedimentary section, higher yet if there is any hydrothermal activity. Under these circumstances upward movement of formation waters may cause mas-

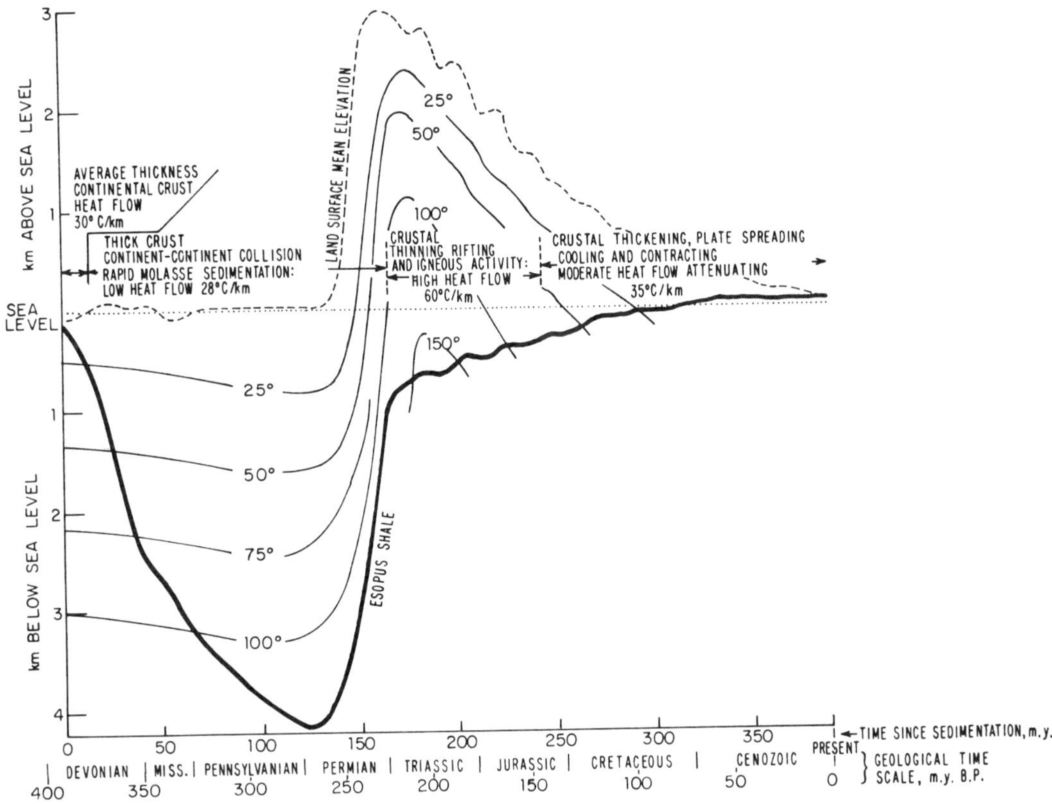

FIG. 2.—Sedimentation diagram of the Esopus Shale of New York, a Devonian formation of the Appalachian Mountain belt, drawn from stratigraphic data in Hepburn (1976). The curve for the shale shows its generalized inferred position with respect to sea level and land surface elevation since the time of sedimentation. Schematic subsurface isotherms are drawn from estimates of crustal thickness and hypothetical heat flow values related to plate movements and igneous activity.

sive transfer of silica from deeper regions, where solubilities are much greater, to shallower emplacement zones. In deeper formation waters rates of quartz nucleation and crystal growth are fast and so metastable opal-CT rapidly inverts to quartz. As a result, concentration levels in deep formation waters tend to be poised somewhat above saturation for quartz but well below that for opal-CT; the precise levels are determined by equilibria between quartz and clay minerals. The times required for quartz precipitation by inversion from unstable silica phases are thus dependent on temperature gradients in the sedimentary pile; average times range from 50–100 m.y. as estimated from secondary quartz in sandstones and limestones and microcrystalline quartz cherts.

Oscillatory vertical movements and what we may term by analogy "retrograde diagenesis" have little effect on the system once quartz is precipitated, unless, as noted above, high heat-flow basins are concerned. In most sedimentary basins heat flow is low enough that once quartz is formed, it remains as a more-or-less inert precipitate: that is, it does not dissolve fast enough to generate precipitation of any of the unstable or metastable silica polymorphs. This is not to say that quartz previously precipitated may not become corroded by replacement with carbonate or clay minerals.

Carbonate systems.—The general pattern followed by carbonates is much the same as for silica, though the situation is complicated by the nonideal activities and ion-pairing of Ca^{++} and Mg^{++} and the multitude of biogenic and abiogenic unstable carbonates in the assemb-

TABLE 1

GEOCHEMICAL AND GEOPHYSICAL PARAMETERS OF DIAGENESIS

Mineral/ Chemical System	Detrital/Chemical Assemblage: Source of Material	Early Pore Waters	Burial Parameters: Pressure	Burial Parameters: Temperature	Geothermal Gradient	Deeper Burial Pore Waters	Time	Oscillatory
SiO_2 Opal Cristobalite Quartz Chert Chalcedony	Volcanic glass, Siliceous organisms, Mafic minerals, High-silica smectites	Amorphous silica: undersat; Opal-CT: supersat or sat	Uniaxial: pressure solution Hydrostatic: slight inc. in solubility	Solubility of all SiO_2 forms inc.; rate of polymorphic inversion inc.	High-T gradients: promotes hydro-thermal transfer of SiO_2	Opal-CT: undersat; Quartz: supersat	To quartz: 40–100 m.y. SiO_2 steadily inc. with time ~ 100–500 m.y.	Most SiO_2 irreversibly pptd.; No reversal of Opal-A → Opal-CT → quartz sequence.
$CaCO_3$—$CaMg(CO_3)_2$ Calcite Aragonite Mg-Calcite Dolomite [Sr as trace constituent]	Biogenic aragonite, calcite, Mg-calcite, oxidizable and refractory organic matter	Calcite: supersat except below $CaCO_3$ compensa-tion depth; refluxing evaporative brines high in Mg/Ca	Uniaxial: pressure solution; Hydrostatic: solubility inc.	Lowers solubility (depending on CO_2) rates of inversion or conversion inc.	High-T gradients promote hydro-thermal transfer of carbonate and leaching of heavy metals	Calcite/ dolomite: sat.	To calc./ dolo.: total carb. steadily dec. with time ~ 200–500 m.y.	Most carbonate eventually lost: Irreversible transforma-tions: Arag/ Mg-calc → calc; rever-sible: calc ↔ dolo.

Zeolites Analcime Phillipsite Clinoptilolite- Heulandite Laumontite	Volcanic glass Ca-plagioclase, Mafic minerals	Amorphous silica: sat or supersat. Alkali/H^+ ratios high	Conversion to prehnite-pumpellyite facies	Conversion to green-schist, (or higher) meta. grades	Low-T, high-P gradients associated with rapid filling of deep troughs	Stable albite-chlorite-mica quartz assemblage: supersat	To ab-Chl-musc-qtz; zeolite gradually lost with time ~ 200 m.y. for complete conversion	Irreversible change to stable assemblage
Clay Minerals Illite Chlorite Kaolinite Smectite	Typical detrital clay mixture; all major groups + quartz, feldspar	Cation balance much like water of environ-ment: sea or fresh water. Ion ex-change, silica sorption	Collapse of smectite by dewatering	Conversion to chlorite/ illite	High-T gradients: approach green-schist facies	Approach equilibrium with chloillite-quartz authigenic feldspar	Ill.-Chl. inc. steadily with time ~ 200 m.y. for complete conversion	Authigenic kaolinite forms as retrograde product in meteoric water

lage. A young carbonate sediment typically includes calcites of varying Mg content, aragonites of varying Sr content, organic matter, and a variable admixture of detrital noncarbonate material.

Early interstitial waters of some carbonate-containing sediments are appreciably affected by rapid dissolution of unstable carbonates, which may result in supersaturation with respect to pure calcite. Oxidation of organic matter may contribute to a temporary rise in CO_2 pressure in the sediment and thus create undersaturation of these solutions with respect to calcite. Boring algae and bioturbating organisms may break down carbonate both physically and chemically. Diagenetic precipitation of magnesian calcites and aragonite may also accompany shallow-water lithification of carbonate sediments (Bricker et al. 1971). Some precipitation may be caused by meteoric water infiltration. In shoreline environments dolomite may form relatively early, and particularly rapidly if Mg^{++}/Ca^{++} ratios are very high, as in briny interstitial waters like those of sabkha flats of the Persian Gulf and elsewhere.

In the deeper waters of the ocean below the calcium carbonate compensation depth (CCD) all carbonate tends to disappear. Here not only bottom waters but slightly buried interstitial waters are undersaturated with respect to both calcite and aragonite. Above the CCD lies an aragonite compensation depth below which the aragonitic remains of pteropods disappear while the calcitic remains of the pelagic foraminifera remain (Berner et al. 1976).

In the carbonate system increased pressure increases solubility in the hydrostatically controlled pore waters and induces pressure solution under lithostatic pressure. Because pressure gradients rarely increase drastically in sedimentary carbonate environments—these normally are not areas of rapid accumulation of thick piles of sediment—the effects of pressure are not overwhelming. On the other hand, they may be important in sedimentary piles in which a component of pelagic carbonate is associated with marine turbidites of continental rises and plains. Even if this material is deposited above the CCD, some of it may later disappear as a result of the pressure effect.

The effects of increased temperature on the carbonate system are more complex because at a given CO_2 pressure, the solubility of carbonate decreases. On the other hand, depending on the total concentration of CO_2-containing species in the interstitial waters ($\sum CO_2$), under certain geological circumstances the pressure effect may dominate and the net solubility of calcite actually increase. There is little doubt that increase in temperature speeds up whatever process is going on, whether dissolution or precipitation. Studies of vertical sections of shales, such as that of Hower et al. (1976), as well as a general comparison of shales of different ages (Garrels and Mackenzie 1971), suggests the diagenetic disappearance of carbonate with age and depth of burial. This would be compatible with the high Ca^{++} content of deep-formation waters and the increased solubility of carbonate along normal geothermal gradients.

Hydrothermal systems may transport in solution large quantities of carbonate that appear as veins and pore fillings. Many sedimentary rocks not obviously affected by hydrothermal activity nevertheless show carbonate precipitates indicative of extensive solution transport by moving formation waters. The wholesale transport of carbonate by solutions such as these suggests that the determining factor may be the $\sum CO_2$ of the solution. This quantity may be controlled primarily by the formations traversed by upward moving acid hydrothermal waters. Thus the temperature-pressure effect on carbonate may be the same as for silica: to cause transport by dissolution and reprecipitation from one part of the sedimentary column to another.

Oscillatory conditions—alternating subsidence and uplift and consequent chemical changes—have a strong effect on carbonate, as inferred from the many reports of reversible replacements of carbonate-silica, carbonate-clay, and calcite-dolomite. Petrographic experience with rocks of various ages suggests that oscillatory conditions lead to the preferential disappearance of calcite in favor of dolomite, and of both carbonates in favor of silica.

Zeolites.—In contrast to the two systems described above, zeolites are rarely present as an original or primary contribution to the sedimentary assemblage. Though zeolites may be transported in rivers and streams from local source rocks, their existence as detrital minerals is ephemeral, for most of these species are relatively soluble in weathering environments. They are typically fine grained or, if coarsely crystalline, subject to abrasion degradation in size. Combined with their chemical reactivity, this makes them susceptible to alteration to clays, feldspars, or other silicates.

The salient fact of zeolite genesis in sediments and volcaniclastic rocks is that zeolites originate primarily by alteration of volcanic glass. They are present in many submarine and lacustrine sediments as by—products of the hydrolysis of volcanic glasses, formed either during early diagenesis or later under conditions of moderate to deep burial (Kastner and Stonecipher, 1978). They also form during the weathering of volcaniclastic sub-aerial deposits.

Both at the interface between the sediment and overlying alkaline lake or sea water, and in shallowly buried interstitial waters, zeolite formation is strongly affected by interstitial water compositions. To look at it another way, compositions of these waters are controlled by the reaction of glass and unstable silicates with solution, normally leading to the liberation of large quantities of dissolved silica, alkali and alkaline earth metals. As a result, Na^{+}/H^{+}, K^{+}/H^{+}, $Ca^{++}/(H^{+})^{2}$, and $Mg^{++}/(H^{+})^{2}$ activity ratios rise together with $Si(OH)_4$ activity, bringing the solution into compositional regions where zeolites can form (see, for example, Garrels and Christ 1965). How transitory concentrations of cations, silica, and alumina as well as pH relate to exactly which zeolite mineral will form is as yet unknown, partly because of the difficulty in establishing equilibrium between various solid phases and solutions at low temperatures.

The effects of temperature and pressure on zeolite diagenesis have been worked out by both sedimentologists and metamorphic petrologists, for zeolite diagenesis grades into zeolite metamorphism—the zeolite facies of Coombs et al. (1959)—illustrating perhaps best of any mineral system the continuum between diagenesis and metamorphism. A series of ascending grades of mineral facies has been worked out for rocks of this type in New Zealand and Australia, tied to increasing depth in sedimentary piles of great thickness. Zeolites at greater depths, about 8–9 km, are gradually converted to a prehnite-pumpellyite facies and ultimately to other greenschist rocks. Critical to a plate-tectonic view of these stages of diagenesis and metamorphism is the comparison of temperature and pressure gradients. The zeolite facies may be characteristic of chemically reactive materials—volcanic glasses and their alteration products—subjected to high pressure gradients accompanied by relatively low geothermal gradients.

Under longer term, deep-burial conditions, zeolites tend to transform to more stable assemblages, in particular to authigenic feldspars and clay minerals or micas. It is unusual to find zeolite in many rocks of early Paleozoic or Precambrian age, even in environments that otherwise appear geologically similar to the zeolite-rich younger alpine belts such as those of New Zealand. What little we know of pore waters of these deeply buried sediments suggests that in early stages, when they are still participating in zeolite equilibria, they are enriched in silica and alkalies, the enrichment gradually being depleted with time.

There is no evidence for reversible changes in the reactions that form and remove zeolites. Once a zeolite has altered to a more stable silicate, it does not reverse. This is probably due to the requirement of a reactive precursor—glass—for zeolite nucleation. Once the glass has gone, the formation of zeolite slows or stops, though there may be transformation from zeolite to smectite and from one zeolite to another.

Clay Minerals.—The clay minerals are perhaps the most complex group of all, for they are both detrital and chemical in nature. The exchangeable cation composition of the clay is, of course, rapidly responsive to any changes of environment, such as the entrance of a river into the ocean or an alkaline or saline lake. Other changes of clay minerals, both structural and compositional, happen more slowly but still fast enough to be detectable over periods of laboratory experimentation. Clays will release or sorb silica from various solutions, irreversibly sorb cations, in particular K^+ and Mg^{++}, and gradually transform from smectite or kaolinite to illite and chlorite (Weaver 1958; Siever and Woodford 1973; Hower et al. 1976). Yet up to a certain point, the basic building blocks of the particular clay mineral that were inherited from its detrital phase can be recognized (Weaver 1958). Most modern sediments contain mixtures of all the common clay mineral groups: kaolinite, illite, chlorite, and smectite. Mixed-layer species are abundant, and there is usually a variable component of X-ray amorphous material that cannot be precisely described.

Ion exchange is probably complete by the time the slowly settling clay particles are deposited at the sediment interface. Further, longer term reactions take place during early diagenesis when interstitial waters may still be much like those of the original environment of sedimentation. The most common authigenic clay minerals formed in early stages are smectites of various compositions, derived from alteration of volcanic glass included in the detrital assemblage. Thus nontronite has been reported from the alteration of greenstones on the Mid-Atlantic Ridge (Melson et al. 1966; Siever and Kastner 1967), and the reports of the Deep Sea Drilling Program include a number of references to authigenic clays (see, for example, Drever 1971; Dunoyer de Segonzac and Hoffert 1973).

The effects of temperature and pressure on clay mineral diagenesis are only now beginning to be understood. In a series of reports, Hower et al. (1976) have detailed the changes that take place in the clay mineralogy of sediments of the U.S. Gulf Coastal Plain as a result of burial and consequent increase in temperature and pressure. Though it has been known for some time that smectites can collapse as a result of increased pressure alone, particularly uniaxial stress such as that induced by the weight of a great thickness of overburden, this effect does not appear to be important in the natural situation. The clays seem to be responsive primarily to temperature increase and less to pressure. Thus the higher the heatflow and geothermal gradients, the more rapid and complete the transformation of clay-mineral assemblages into the stable illite-chlorite pair that is found in so many older shales and sandstones.

Deeper formation waters tend to approach equilibrium with the quartz-feldspar-illite-chlorite assemblage; at the same time they may become enriched in Na^+ and Ca^{++}, as well as chloride, other halides and sulfate.

Oscillatory subsidence and uplift conditions are chiefly reflected by the growth of authigenic kaolinite and dickite, which seem in many occurrences related to infiltration of meteoric, low ionic-strength waters into permeable rocks such as sandstones. This occurs when uplift and erosion bring such rocks into the zone of weathering, even though they may not actually be exposed at the surface. This kaolinite and dickite tend to form well-crystallized books and worms that, once produced, tend to persist. Thus some sandstones showing evidence of such former near-surface conditions favoring kaolinitization are now found deeply buried, containing formation waters that would presumably be in equilibrium with the chlorite-illite assemblage. Conversely, once smectite is converted to illite-chlorite, it does not reappear by reverse alteration as a diagenetic phase.

DIAGENETIC PARAMETERS OF PLATE-TECTONIC ENVIRONMENTS

Variables that govern diagenesis are largely determined by the tectonic framework of (1) the source region of terrigenous detritus, (2) the depositional environment, and (3) the burial environment. They were described by Krynine (1948), Pettijohn (1949) and others from a distinctly petrologic point of view in such terms as quiescent, pre-, syn-, and postorogenic. Their view of mountain building, conditioned by geologists' long experience with the Alps and Appalachians, considered an evolutionary time series in a "cannibalistic" geosyncline. Detrital mineralogy was revealed as the key to provenance and thus ultimately to source-area tectonics. Stille (1936) and Kay (1951) focused on the dynamics of sedimentary basins as parts of larger "mobile belts," geosynclines. Theirs was not necessarily a competing, but a complementary view, emphasizing the relation between tectonic and igneous activity and patterns of sedimentation and stratigraphy of the basins. Dapples, Krumbein, and Sloss (1948) amplified this view by introducing lithofacies analysis in a quantitative way. Yet their view remained a stratigraphic-structural one that emphasized spatial relations among gross units more than an evolutionary time series and more or less by-passed systematic petrology of the rocks concerned. How different tectonic styles made themselves felt in both detrital and chemical components of the sediments was not spelled out.

Plate-tectonic theory allows us to couple geophysical parameters with tectonics and sedimentation in a different and more complete way than before, and thus may provide a better understanding of the ways tectonics influences the course of diagenesis. For each plate-tectonic setting we can specify a more or less distinctive set of geophysical and geochemical variables that control diagenetic histories (table 2).

To simplify the discussion, I shall restrict it to mid-ocean ridges, trailing-plate or passive continental borders, subduction zones, continent-continent collisions, and rift valleys. To this group, all associated with plate borders, I add the continental intraplate regions, which I interpret as essentially cratonic in character. The reader is referred to LePichon et al. (1973) for discussion of these regions and their geophysical characteristics. I recognize that subduction zones in particular may be broken up into a variety of sub-environments, but for the sake of simplicity here will neglect them. The current literature abounds with interpretations of geologic history in plate-tectonic terms, though there is certainly no consensus; Dewey and Bird (1970) and Chase et al. (1975) are two examples of the genre.

Mid-ocean ridge.—The detrital sediments of mid-ocean ridges are a mixture of a well-mixed terrigenous fraction, an oceanic igneous suite—mainly basalt, but including fragments of lower oceanic crust and upper mantle rocks—and a pelagic carbonate component. The amount of terrigenous material is small, as most mid-ocean ridges are some distance from continental source areas and, because of their elevation, are not sites for extensive turbidite sedimentation. There is a very small detrital fraction of oceanic igneous derivation in the thin layer of sediments at the ridge itself, although the composition of that small fraction may be highly diagnostic of its provenance (Siever and Kastner 1967). Since mid-ocean ridges in most places are above the carbonate compensation depth, the dominant sediment type is pelagic carbonate ooze. As the crust created at a mid-ocean ridge spreads away from it, it contracts and sinks (Parsons and Sclater 1977), eventually passing below the CCD and ceasing to accumulate pelagic carbonate.

Sedimentary basins associated with fracture zones of mid-ocean ridges, such as Vema and Romanche of the Mid-Atlantic Ridge, contain the same terrigenous and pelagic components as the central valleys and hills of the ridge itself. Here, however, there may be important—though quantitatively small—detrital components derived from the lower layers of newly created oceanic crust—the dikes, gabbros, cumulates, and even fragments of the surface of the upper mantle—as they become exposed by both transform and vertical movements along the fractures.

The early interstitial waters of mid-ocean ridge sediments are dominated by carbonate and basaltic glass. Limited saturometer measurement on mid-ocean ridge sediments suggest that the activities of bacterial and metazoan infeeders may contribute to under-saturation with respect to calcite (Siever, Beck, and Berner, 1965). Hydrolysis of glass may contribute to high dissolved silica levels, though all the evidence so far shows that levels are between 10^{-3} and 10^{-4} mol/1—that is, poised at a level below saturation with respect to amorphous silica or opal-A at bottom water temperatures. Where regions of high surface productivity of diatoms overlie ridges, abundant diatoms in the sediment may dissolve to raise silica levels of the interstitial water. Heat-flow considerations and the recent direct observation of hot springs by submersible at the Galapagos triple junction (Edmond 1978) indicate that there must be abundant infiltration of sea water and percolation through freshly congealed basalt at the spreading center. These solutions are likely to have some effect on the sediment through which they percolate; but the ridge spreads at a rate of

TABLE 2

CORRELATION OF DIAGENETIC PARAMETERS WITH PLATE TECTONIC ENVIRONMENTS

Parameter	Mid-Ocean Ridge and Oceanic Intraplate	Trailing Plate-Continental Margin	Subduction (Trench)	Continent-Continent Collision	Rift Valley	Intra-Plate (Craton)
Detrital-Chemical Assemblage	Basalt + pelagic carbonate & terrigenous detritus.	Terrigenous detritus (immature + mature) + platform carbonates.	Immature terrigenous detritus + pelagic; volcaniclastics, ophiolites.	Terrigenous detritus: mature early, immature late (unroofing).	Immature alluvium, evaporites, volcanics.	Mature, reworked sedimentary detritus. Platform carbonates.
Early Pore Waters	Carbonate: undersat if below CCD. Silica: undersat; affected by hydrothermal exchange with basalts.	Sea, brackish, or regressive fresh water. Carbonate: sat or supersat; silica: undersat.	Much like sea water. Carbonate: unsat. Amorphous silica: sat or undersat. High alkali/H^+.	Fresh waters, modified by alteration of detritals. Marine turbidites: sea water.	Fresh to saline, alkaline. Amorphous silica: supersat in some alkaline lakes.	Carbonate: sat or supersat, amorphous silica: undersat. Fresh, brackish, or marine water.
Burial Parameters	Little burial: spreading to abyssal plains, where slow burial to 0.5–1 km with normal oceanic gradients P, T.	Normal to low geothermal gradients; moderately rapid burial to 10 km.	Low heat flow: subnormal geothermal gradient; high-P, low-T, rapid burial.	Low geothermal gradient; rapid burial in intermontane basins or in offshore deltas and fans to 10 km.	Normal to high heat flow, rapid moderate burial to 4–7 km.	Normal continental P & T gradients; slow burial rarely exceeds 3–4 km.

Deep Pore Waters	May approach hydrothermal silicate, alkali-rich; abyssal plains: like sea water.	Salt concentration inc. with depth and age to high ionic strength. Quartz: supersat; calcite: sat or undersat; dolomite: supersat. Fresh water infiltration from coast. Salt builup near buried evaporites.	Silica, alkali-rich. May approach hydrothermal if near volcanic belt.	Continental facies: altered meteoric waters, some salt inc. with depth. Marine facies: sea water with some salt inc. with depth.	Some salt concentration; some hydrothermal; may become alkaline.	Salt concentration inc. with depth and age to high ionic strength. Quartz: supersat; calcite: sat or undersat; dolomite: supersat.
Time (of Active Subsidence and Sedimentation)	Spreading rates: 1–6 cm/yr. Abyssal plains: slow burial: 100–200 m.y.	Moderate to great subsidence: 50–200 m.y.	Rapid burial: 10–50 m.y.	Rapid basin fill: 1–30 m.y.	Active subsidence: 5–20 m.y.	Slow subsidence: 100–300 m.y.
Oscillatory Movements	During spreading, some uplift along transforms. Generally irreversible subsidence.	Mostly irreversible burial; some reversals locally and at landward edge.	Irreversible burial up to or more than 20 km followed by orogeny; subduction zone migration.	Many intermittent reversals as orogeny cannibalizes intermontane and marine basins.	Mostly irreversible burial; some reversals at borders.	Many reversals of uplift and subsidence: many unconformalities + mild structural deformations.
Example	Mid-Atlantic Ridge.	North American Atlantic and Gulf Coast Mesozaic and Cenozoic.	Aleutian trench.	Siwalik series of India, Bengal submarine fan.	Triassic of Eastern North America.	Michigan and Eastern interior basins of North America.

Abbreviations:
Sat = Saturated with respect to given mineral
Unsat = Undersaturated with respect to given mineral
Supersat = Supersaturated with respect to given mineral

P = Pressure T = Temperature Inc. = Increases Dec. = Decreases
Pptd. = Precipitated m.y. = Million years

several centimeters per year, while sedimentation rates are only of the order of millimeters or centimeters per 1,000 years (depending on carbonate sedimentation rates). A typical example is as follows: at a half-spreading rate of 2 cm/yr and a gross sedimentation rate of 1 cm/1,000 yrs, in 5 m.y., at 100 km from the ridge, only 50 m of sediment will have accumulated. Furthermore, hydrothermal activity at a ridge seems to be restricted to a very narrow zone, in places only a kilometer wide, right at the spreading line. In that central valley there are rarely more than a few meters of sediment, insufficient to show any profound hydrothermal effects. On the other hand, sediments may reach thicknesses of up to hundreds of meters in fracture zones, where hydrothermal activity may affect the sediment more.

For most of the region, however, burial will be slow until the plate has cooled and contracted sufficiently to have subsided to depths characteristic of abyssal plains. Burial may be moderately rapid if the plate passes under a broad region of high pelagic carbonate productivity before subsiding below the CCD. However slow the average rate of burial, it takes place along more-or-less normal oceanic heat-flow conditions—that is, under conditions of steadily declining heat flow moving away from the ridge. This means that actual geothermal gradients through the sedimentary part of the section will everywhere be higher than those on land, for the heat flow is transmitted through a much thinner crust. Thus we can deal with gradients of the order of 50 or 60°C/km or even higher near a mid-ocean ridge rather than the ordinary continental gradients that average about 30°C/km. Temperatures may be much higher than predicted from continental conditions of burial; at the same time there may be only a modest pressure increase from burial under sediments that retain a relatively high pore—water content and thus have low bulk density.

Once the sediment leaves the central zone of the ridge and is no longer subject to high heat flow and hydrothermal solutions, its interstitial waters relax to a composition much like that of the original trapped sea water, modified by reactions with carbonates, silica, silicates, and oxides. During this period zeolites and smectites will grow authigenically, sometimes associated with opal-CT and microcrystalline quartz cherts if the original sediments were rich in siliceous organisms. Carbonates may become cemented—first to chalks, then to limestones—while chert goes through several stages of formation (Siever and Stein 1978). In general, the volcanic component that engenders zeolites and smectites will be high only if the ridge is close to volcanic belts of subduction zones. Thus sediments of the East Pacific Rise, because of proximity to the Peru–Chile subduction zone, contain much more volcanogenic material than do sediments of the Mid-Atlantic Ridge, far from any important volcaniclastic source other than the point sources of the few volcanic islands of the Azores, Canaries, and others.

Oscillatory movements generally do not significantly affect sediments of the ocean basins. The sediment leaves the ridge, is transported toward the ridge flank, then the abyssal plain, until it finally moves toward a convergence zone; if moving toward subduction it is always subsidising, not subject to vertical uplift. On the other hand, if it moves toward a transform plate boundary, such as the San Andreas Fault between the North American and Pacific plates, it may well be involved in vertical movements and thus be subject to retrograde diagenesis, weathering, and erosion. Sediments of basins associated with large ridge fracture zones, such as Vema or Romanche may also be subject to vertical movements.

Typical diagenetic textures and minerals formed in the mid-ocean-ridge environment (table 3) include baked or cemented foraminiferal oozes formed adjacent to freshly extruded basalt (Bowen et al. 1966) and noted near the bases of many Deep Sea Drilling cores. Smectites, sepiolite, and clinoptilodite-heulandite are formed by alteration of glass and mafic minerals. But most of the muddy and sandy sediments remain unlithified. Much of the section may be cut with a knife, with the noteworthy exception of some hard limestones and cherts. Once the sediments leave the central ridge province, reactions seem to be almost entirely isochemical, although there is some evidence of

TABLE 3

SOME DIAGENETIC MINERALS AND TEXTURES ASSOCIATED WITH PLATE TECTONIC ENVIRONMENTS

Mid-Ocean Ridge and Oceanic Intraplate	Trailing Plate	Subduction	Continent-Continent Collision	Rift Valley	Intraplate (Craton)
Recent sediment may be baked or cemented near basalt flow or sill. Some greenstones and growth of authigenic smectite, sepiolite clinoptilolite-heulandite. Muddy and sandy sediments remain unlithified. Calcareous oozes slowly cemented and lithified. Siliceous oozes transformed to opal-CT, then quartz bedded cherts. Clays almost all detrital, a mixture of all major groups.	Mildly cemented and lithified by carbonate, quartz, and in deeper regions, by authigenic clays. Chlorite-illite formed in deeper zones. Gradual slight metamorphism of organic matter. Some nodular chert, calcareous concretions. Limestones alter to calcite-dolomite from more heterogeneous primary sedimentary mixture.	Well cemented with quartz, clay minerals, zeolites, up to zeolite green-schist, clueshist, or higher metamorphic facies. Zonation from shallow zeolite, to prehnite-pumpellyite to lawsonite to greenschist or glaucophane schist. Graywackes altered to matrix-rich rocks by squashed rock fragments. Bedded chaerts—ophiolites in mélanges.	Continental facies: Mildly cemented by clay matrix from squashed rock fragments, carbonate, some silica, Fe-oxides. Clays largely detrital with authigenic kaolinite additions. Marine facies: turbidite sands cemented by altered clay matrix, recrystallized pelagic carbonate or silica.	Evaporites authigenic zeolites and clays in lake beds and associated sands. Cementation by carbonate, clay, quartz, some authigenic feldspar in arkosic sands. Clays largely detrital, a mixture of all major groups.	Well cemented with quartz and carbonate. Clay minerals alter to chlorite-assemblage over 100–200 m.y. Rarely metamorphosed, little or no zeolite or similar minerals. Chert is nodular, replacement type. Coal and oil only moderately increased in rank and lightness. Limestones may become extensively replaced by dolomite.

chemical exchange with formation waters transporting elements of basement derivation upward. The strongly lithified horizons thus reflect original compositions, such as high amounts of diatoms and/or radiolaria. They may also be zones marking the absence of any significant amount of terrigenous clay that would coat shells and so interfere with cementation. Although clays may be slightly altered, the modest depth of burial of the average oceanic sediment and resulting low temperatures are not sufficient to convert smectite and kaolinite to chlorite-illite. Similarly, although authigenic feldspars can be found (Kastner and Siever 1978), most of the sediments retain the zeolite-clay mineral composition formed during early stages of diagenesis of volcaniclastic material.

All these mild processes in the end produce a generally compacted but unlithified sediment, broken by a few lithified zones of limestones and cherts, which reaches a subduction zone in a easily deformable state. It is not surprising that such sediments may be folded, crumpled, faulted and otherwise deformed in complex ways that resemble those assumed to form from penecontemporaneous slumping. Because such sediments still retain a fairly high water content—as much as 40%—and the pore space in general has not been filled in by authigenic minerals (except for the chert and limestone horizons), the bulk density of these sediments is less than the average figures quoted for older continental sedimentary rocks at similar burial depths. The low density makes it improbable that much of this sediment will be easily entrained along the surface of a downgoing slab and carried into the mantle.

To the extent that the basic components of the ophiolite suite of oceanic rocks are formed at mid-ocean ridges, they should be discussed in this section. However, I defer them to a later section, for they are ultimately emplaced in subduction zones.

Trailing-edge continental margins.—This tectonic environment presents some of the sedimentary rocks most familiar to the geologist used to working with material deposited on and adjacent to the continents. They are deposited on continental shelves, coastal plains, deltas and estuaries, beaches, and in a host of related sedimentary environments. Less familiar in the past have been the sediments of continental rises and slopes, including those of submarine canyons and fans, which are now becoming increasingly better known as many turbidite sections are reassessed (Ricci-Lucchi 1976). The most important characteristic of passive continental margins is their slow subsidence as the continental border moves away from the spreading center, with the plate cooling and contracting as it moves (Sclater and Parsons 1977). I here distinguish early stages in the formation of these tectonic environments from the rift-valley stage that precedes them. The most convenient point of departure is the stage when open, connected seas already separate two rifted continents. At this stage evaporites may form in various barred arms of the sea; some evaporites may actually be recycled via ground water transport from evaporites previously formed on land during earlier stages of rift-valley formation.

The sediment deposited on most trailing edges is dominantly terrigenous—a mixture of immature and mature detritus coming from the rifted edges of the continent and marginal mountain belts that may have formed by an earlier collisional orogeny. Deposition may be largely carbonate if there is little terrigenous detritus, such as will be the case if the paleoslope tilts away from the passive margin. Distribution of the terrigenous sediment depends on the network of major river systems and the pattern of waves, longshore currents, and tides. Where much material is trapped near shore, areas farther offshore may become the scene of undiluted shallow-water carbonate platform deposition: in some cases these carbonates are contemporaneous with the later-stage of evaporites. Our experience with the hot brines of the Red Sea (Degens and Ross 1969) and more recent manned submersible explorations have shown that heavy metal deposits are localized near rift-valley hot springs on the ocean floors. These deposits, along with the evaporites, are at the bottom of the sedimentary pile, overlain by carbonates and terrigenous detritus as the continent recedes from the plate boundary. Volcanic materials are absent from this tectonically quiescent environment.

Early interstitial waters are dominated either by the abundant terrigenous material or carbonate. In early stages the trapped water may be hypersaline if associated with evaporites, brackish if associated with regressive fresh water environments. Waters will tend to be saturated or supersaturated with respect to calcite while undersaturated with respect to amorphous silica and perhaps opal-CT. Low silica concentrations result from the absence of volcaniclastic material and because of the normally low to moderate population of siliceous organisms near shore. These are the interstitial waters that are characteristic of many of the continental shelf sediments of today. In areas dominated by terrigenous material, the sediment experiences only minor changes, but in areas of carbonate deposition there may be extensive diagenetic alteration through reworking by organisms and abiogenic processes such as cementation.

Because heat flow after early stages progressively decreases as the plate contracts and moves away from the plate boundary, the geothermal regime is constantly relaxing. It moves from the high values near the mid-ocean ridge, 60–70°C/km, to the low side of the continental average, 25–30°C, as the continental margin sedimentary pile builds up. At the same time, burial may be rapid, both because of subsidence due to plate contraction and because of abundant supply and rapid sedimentation rate of terrigenous material. The weight of freshly deposited sediment adds its load to the weight on the crust and furthers subsidence. Thus the lower parts of the pile may undergo relatively rapid burial diagenesis, reaching temperatures of 150–200°C at depths of 5 or more kilometers within only a few tens of millions of years. Combining the immature terrigenous sediment at the base of the sequence with the evaporites, carbonates, and organic matter that are frequently associated with early stages of rifting makes it evident that a host of diagenetic reactions may result. They include transformation of the organic fraction into oil and gas, transfer of heavy metals by warm acid solutions, leaching of evaporites and formation of saline pore waters, dolomitization and growth of authigenic minerals, all perhaps accompanied by diapirism of salt beds and other mechanical disturbances.

Above the basal parts of the sequence, the deeper formation waters may become more normal, as the waters of sedimentary basins commonly do by reaction with the surrounding rocks (Von Engelhardt 1977: p. 208–239). Along shoreline regions, in the highest parts of the sequence, meteoric water inflow, under hydraulic head from the adjacent coastal plain, will move downdip into continental shelf formations. This invasion is likely to cause alteration of the clay mineralogy, and cementation by silica and/or carbonate, depending on the rocks through which the waters have passed. Intermittent transgressions and regressions during accumulation of the sedimentary pile make much of it susceptible to such invasion. The sediments built up on the continental rises are less prone to such infiltration by meteoric water. Theirs is a pattern of diagenesis much like that of normal burial in oceanic basins described earlier.

The lifetime of a trailing edge varies with the vagaries of plate motions. The evidence so far indicates that uninterrupted general subsidence is likely to continue over a time interval of 50 to 100 million years. But this period may be part of a longer time span, perhaps many hundreds of millions of years of trailing-edge dominance broken by collisional interruptions of shorter duration. One may cite as an example the history of the edges of the Atlantic and proto-Atlantic Oceans. At least since the beginning of the Cambrian, and perhaps for several hundred million years before that, these continental borders have been dominantly subsiding passive edges interrupted by periods of convergence, both of oceanic and continental plate margins. The well-preserved sedimentary record, however, is deposited inboard, derived almost entirely from the convergence intervals, for the record of older trailing edge intervals is largely transformed by subsequent orogeny, plutonism, and metamorphism.

Trailing plate continental margins generally show irreversible burial, but there is abundant evidence both from studies of coastal plain formations and from continuous seismic profiles of the continental shelves that there have been many local uplifts and deformation episodes leading to unconformities, mild folding, and faulting. These movements have their greatest effect on the sediments at or near the shoreline, where small movements are reflected in large lateral movements of the strand line and migration paths of meteoric waters into coastal-plain formations.

The best-known examples of this tectonic setting are the continental border accumulations along the margins of the Atlantic Ocean: the Mesozoic and Tertiary formations of the North American Gulf Coastal Plain, and to a lesser extent the Atlantic Coastal Plain and shelf. The sedimentary rocks are midly cemented and lithified by carbonate and quartz, and in deeper regions authigenic clay minerals such as chlorite and illite have formed. But even in deeply buried regions the rocks are easily distinguishable from the harder, more lithified, and more completely altered Paleozoic sedimentary rocks beneath them. Organic matter is converted to petroleum fluids by the higher temperatures deeper in the pile and salt diapirism has formed salt domes. Nodular cherts, calcareous concretions, and other penecontemporaneous diagenetic segregations are common along nearshore zones. Limestones are entirely converted to calcite; dolomitization is common only near the base of the sedimentary pile in association with evaporites or along shorelines with supratidal flats and warm temperatures. Paleolatitudes play an important part in these reconstructions, for the predicted occurrence of evaporites, dolomite, and organic matter is linked to warm climates near the equator during polar ice-cap periods (Miocene-Recent, Late Paleozoic, Early Paleozoic, Late Precambrian). These sediment types may be much more widespread in nonice-cap periods, such as the Middle and Upper Mesozoic or Middle Paleozoic.

Subduction zones.—Plate convergence, where one plate is consumed by subduction, produces perhaps the most varied and complex mixture of sediment types and diagenetic regimes. In addition to the normal pelagic deep-sea sediments there are extensive igneous contributions, of both submarine and subaerial volcaniclastic debris and eroded plutonic rocks emplaced to the rear of the island arc. The terrigenous contributions may vary greatly, depending on the position of the subduction zone with respect to any major continental land mass. Indeed, there are so many possible subdivisions of subduction zones, from fore-arc to back-arc basins and marginal seas, that there is a strong limitation on any generalizations for the entire complex of environments. It is possible to focus on the significant proportions of volcaniclastic materials in the primary assemblages of most of the subenvironments. The island arcs are also the source of much reworked older volcaniclastics—weathered basalts and andesites—some of which may contain zeolites and smectites. The following discussion concerns the sediments of the main subduction zone including the materials of the trench and adjacent environments.

The sediments of the subduction zones include blocks and fragments of the ophiolite suite: remnants of siliceous and carbonate pelagic sediment, basaltic volcanics, and fragments of lower oceanic crust and upper mantle inherited from mid-ocean ridge and fracture zones. The deformation process associated with the peeling off of the less dense layers from the top of the subducting plate mixes the ophiolites with the terrigenous sediment derived from the arc and hinterland in a mélange such as those in the Dunnage Formation of Newfoundland (Kay 1951) and the Franciscan of California. The evidence from the olistoliths and exotic blocks found in these rocks suggests that the finest grained mudstones were only partly lithified at the time of their incorporation into mélange. Perhaps only chert and basalt, in addition to the fragments of lower oceanic crust and mantle, were hard rocks before deformation. These would form occasional brittle, lithified strata within a largely plastic, unlithified section. The flow and deformational structures in mélanges, including those of the exotic blocks, are known from seismic profiles to take place under relatively shallow depths of burial in modern trenches; this suggests that flowage did not occur under intense metamorphism.

For the following discussion I take as a typical example the accumulation of a dominantly detrital sediment with a large volcaniclastic component. (Though there may be reef and platform carbonates associated with subduction zones, their diagenesis is probably not grossly different in kind from such rocks in other tectonic settings.) The early diagenesis of such rocks, many of which may be turbidites in deeper basins (either fore- or back-arc), is dominated by interstitial waters that may become fairly alkaline and silica-rich if extensive glass hydrolysis takes place. Zeolites and smectite will be abundant. To the extent that the sediments are turbidites containing a fine-grained upper Bouma interval thick enough to inhibit upward diffusion and connection with the overlying sea water, the changes will be isochemical, and interstitial water compositions may depart from that of average sea water. Proximal turbidites, which are almost entirely sandy, will allow freer movement of pore water, so that interstitial water compositions will be much like that of sea water. Early diagenesis of these turbidites will in any case be interrupted by the next turbid flows eroding fairly deeply into the unit below. During that erosion, the pressure head of the flow above is likely to greatly speed mixing of sea water with pore waters of the underlying sediment by pumping. Thus early diagenesis of these sandy turbidites cannot be considered to start until burial by at least one or possibly two Bouma sequences.

Evidence of mild early cementation comes from the clasts of underlying turbidite units incorporated within the lowest Bouma unit of the overlying turbidite. Commonly these clasts are of the fine intervals and much less commonly of the sandy beds. The generally soft-sediment deformation of the shale pebbles shows that they could form pebbles by virtue of compaction and slight cementation. In contrast, the sandy parts of the sequence even though possibly slightly cemented at grain contacts, could not cohere sufficiently to form pebbles that would last long distances in turbulent transport. Ball-and-pillow structures are also evidence for the lack of cementation in the sandy units. Yet there are places where the turbidity current has bitten deeply into previously deposited sequences and incorporated some sandstone pebbles into the lowest Bouma unit. This suggests that a modest amount of burial, where there is sufficiently labile material to form cement, is sufficient to start the lithification process.

But diagenesis in these sediments occurs mainly after deeper burial. In the main subduction zone, the low-density sediment is deformed early in its postdepositional history, as it is scraped off the top of the descending slab. As yet we know about the contorted sediments of the subduction zone mainly from seismic profiles, but it is clear the rate of accumulation of the sedimentary pile is great. To normal pelagic sedimentation rates, augmented by rapid volcaniclastic and turbidite accumulation, one must add the doubling or tripling of sedimentary thicknesses by crumpling on the hanging wall of the subduction zone.

In this environment, heat flow in the sedimentary section is lower than normal, since this is the cool descending edge of the plate. Ascending magmas from deeper parts of the subducted slab are too distant to affect geothermal gradients. Low heat flow coupled to rapid thickness accumulation leads to subnormal geothermal gradients. Blueschist metamorphic belts, where rocks contain high-pressure-low-temperature mineral facies, are inferred to form in deeper parts of subduction zones of low geothermal gradient (Ernst 1970). At premetamorphic depths, a complex succession of smectite, zeolite, and other silicates is characteristic, as described below.

While pore waters may be cooler for a given depth than in higher heat-flow regimes, at great depths they may become almost hydrothermal, or at least premetamorphic. They would be expected to be of high ionic strength, somewhat acid, silica- and alkali-rich as a consequence of the interaction of hot-water solutions with the various silicates of the primary assemblage. Given enough time, under these conditions the typical volcaniclastic-terrigenous turbidite turns into a graywacke whose matrix is largely a high-grade diagenetic product formed by the conversion of smectite, kaolinite, glassy material, and unstable plagioclase feldspars and mafic minerals into a chlorite-illite assemblage. Mechanically and chemically unstable rock fragments are squashed into matrix (Pettijohn,

Potter, and Siever 1972). If sufficient organic matter is still included in the deeply buried assemblage, it may act as a reductant; then iron and manganese carbonates and ferrous silicates may form in addition to chlorites. Depending on depth and length of burial and subsequent uplift in relation to geothermal gradient, under higher stages of diagenesis, the rock approaches either a blueschist or a greenschist, with growth of albite, better crystallized chlorites, perhaps biotite and garnet.

It is difficult to estimate an average time interval during which sedimentary accumulations form in subduction zones. Instantaneous plate tectonics describes only the kinematics of current plate motions, and deductions of past plate motions from magnetic anomalies do not take us far enough back in time to get a meaningful average. Information from the current oceans indicates that subduction zones may last from 10 to 50 m.y. before being appreciably transformed or altered by changing plate motions or continent-continent collisions.

Subduction zones are not generally characterized by oscillatory vertical movements, for once the material begins to be subducted, it continues until it is immobilized by an orogeny associated with changes in subduction-zone regime. To some extent, limited vertical movements are associated with deformation and crumpling of the mélange, but these movements do not transport the sediment into new zones of diagenesis: for example, from deeply buried to the zone of weathering and back.

There is another aspect of this question, however. It is possible to infer lateral migration of subduction, from terrains assumed to be ancient subduction zones, in some places toward the arc and hinterland, in others toward the subducting oceanic plate and active trench (fig. 3). Migration toward the hinterland implies that the sediments of the former subduction zone continue to be subducted; they are after all still attached to the downward-moving oceanic plate. Migration in the opposite direction means that the formerly subducting sedimentary mass in front of the arc is no longer on the subducting plate, but rather on the stationary plate. In this new location it becomes more likely to be affected by igneous activity, both extrusive and intrusive. There may be appreciable contact metamorphism near igneous channels and a general rise in heat flow and geothermal gradient. This may be the origin of the metamorphosed graywacke sections that are invaded by dikes and sills of andesitic magmas, all of which may later become regionally metamorphosed. The original turbidites of the trench and fore-arc basins thus become transformed into great metagraywacke sections. It may not be primarily collisional orogenies that induce the high-grade diagenesis and low-grade metamorphism of many typical graywackes, but rather the inexorable progression of events at a subduction zone that migrates seaward.

One can pick two different sets of examples of subduction-zone sediments, those currently in trenches or associated basins, and those now uplifted and available for study but inferred to be of subduction-zone origin. The former are known mainly from geophysical exploration, and we can say little of their composition. Recently uplifted trench and fore-arc basin rocks in active subduction zones such as those of Indonesia have not been adequately described petrographically. Older rocks of subduction-zone origin may be exemplified by parts of the Franciscan formation of California and rocks of the Alps of New Zealand. The rocks are tough and well cemented; rock fragments are squashed and the clay matrix is well-crystallized chlorite-illite or smectite-zeolite. In shallower or younger zones free silica is opal-CT; in deeper and older zones free silica has inverted to microcrystalline quartz. The now classic zones of the zeolite facies as explored in New Zealand and Australia show the diagenetic facies expected in younger rocks (Boles and Coombs 1977). Shallow burial leads to the formation of clinoptilolite, analcime, and wairakite. As the sediment is buried more deeply to depths of approximately 6–7 km and temperatures of about 200–250°C, analcime gives way to albite and clinoptilolite to laumontite. At somewhat higher temperatures and greater depths, the prehnite-pumpellyite facies is formed. Coincident with these mineralogical changes are textural changes. There is additional loss of porosity, not so much through

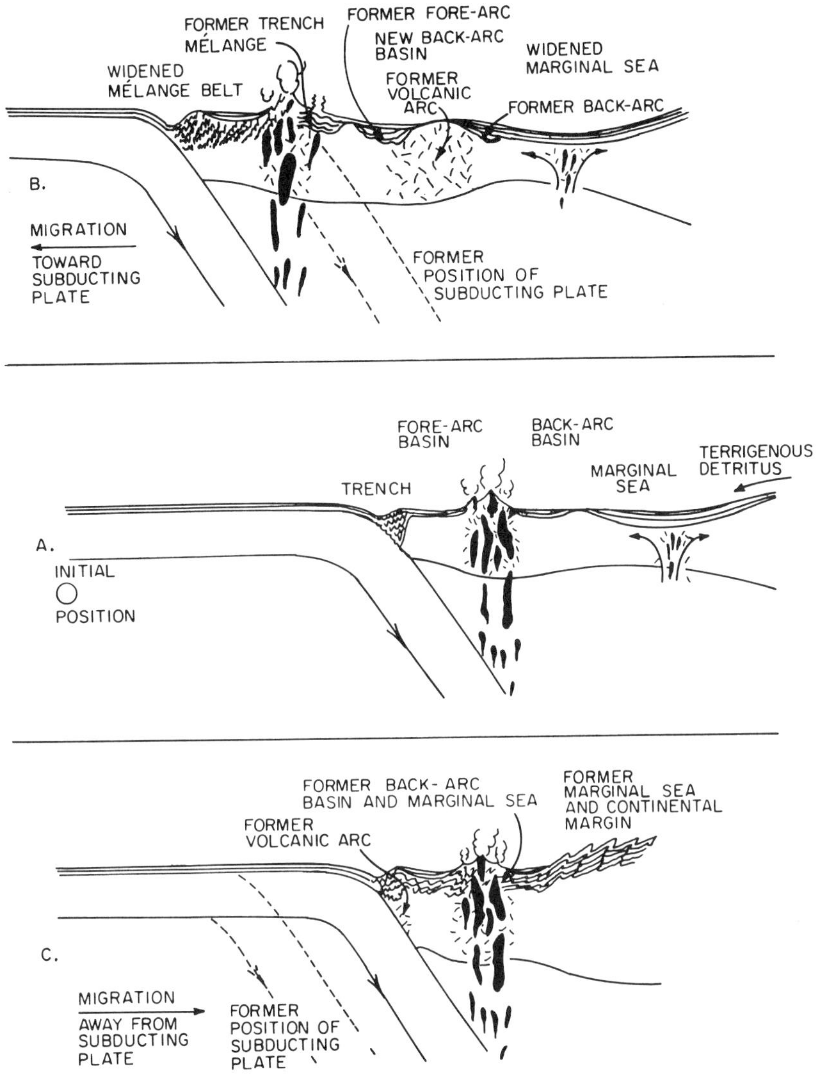

FIG. 3.—Schematic drawing of the effects of lateral migration of a subduction zone. A, Initial complex of sedimentary basins associated with a subduction zone at the convergence of an oceanic and continental plate with a marginal sea between. B, After migration of subduction toward the subducting plate, former arc and associated basins are spread out and overlain by new facies while crust is thickened over a wider area than formerly. C, After migration of subduction away from subducting plate, former arc and associated basins are compressed and deformed against continental margin, while marginal sea disappears and crust is thickened to continental proportions.

physical compaction in deep burial stages, but through precipitation of authigenic minerals in pore spaces. The rocks may become slaty, although much of the typical fracture and flow cleavage is induced by later metamorphism and orogenic movements. Depending on the distribution of pelagic organisms in the vicinity, carbonates may be present; they are generally absent, however, although carbonate turbidites such as those of the Apennines of Italy are common in younger sections. Much carbonate is probably lost from these sections through diagenetic reactions in which alkalies and silica are used up and H^+ is produced, thereby liberating CO_2 and increasing the solubility of the solid carbonate. Over a long period of time these terrains may liberate a great deal of free CO_2. As the carbonate is lost, the Group II elements are taken up in forming calcium-bearing zeolites and metamorphic minerals.

Continent-continent collision.—The collision of two continents at a plate convergence has raised some of the most perplexing problems for those who would translate instantaneous plate tectonics into mountain-building episodes. The Himalayas, frequently given as the archetype of this kind of collision, are by no means satisfactorily explained in many aspects. Nevertheless we may say a good deal about what happens to sediments formed in this setting even though we may be unsure of the general mechanics or even of the precise kinematics. If turbidites of subduction zones can be seen as a modern version of flysch, the deposits shed from collisional orogenies can be correlated with molasse.

The raw material of the sediment is almost entirely terrigenous detrital, shed from high mountain belts. In early stages the detritus may be relatively mature, derived from reworked sediments formerly deposited at continental margins. Later stages of continuing collision and uplift may expose deeper sections of the continental convergence, including regionally metamorphosed rocks and associated plutonic rocks. Volcaniclastics are rare, and chemical sedimentation is swamped by enormous volumes of detritus. Much of the sedimentation is terrestrial, typically the rapid filling of intermontane basins and great downwarped river basins or coastal plains below the mountain front. To the extent that the detritus falls into an integrated drainage system leading to the sea, the debris of orogeny may form large deltaic accumulations and beyond that, great submarine fans and abyssal plains. One can easily see that in these respects the setting is much like that described for trailing edges. The two could be distinguished mainly by the presence of the former rifted margin sediments, such as evaporite and heavy-metal deposits, at the continental margins of trailing edges. In contrast, the base of the continental collision sediment pile, if not too severely deformed structurally during collision, could be deciphered as open sea-floor pelagic sediments gradually making a transition to distal turbidites of abyssal plains.

The early interstitial waters of such sediments would be much like those of the trailing edge; dominated by terrigenous material. But they would be less likely to contain carbonate, other than calclithites—the rapidly eroded and deposited debris of pre-existing limestones. The absence of carbonate is largely the result of swamping by detritus. No doubt there would be terrains in which fresh-water carbonates and other lacustrine deposits would form in intermontane basins, but they would be volumetrically unimportant.

More than in other tectonic environments, early diagenesis at continental plate convergences might take place in alluvial or eolian environments typical of uplifted terrains. The pores of the sediment would be filled with meteoric waters, intermittently if above the water table, permanently if below. This would continue the weathering process begun at the original exposure: the feldspars and mafics would continue their slow breakdown; the clay minerals would alter in accord with the climatic regime; carbonate, organic matter, and siliceous and phosphatic organic remains would disappear fairly rapidly in this oxidizing, slightly acid environment, at least in the dominant sands, gravels, and silts. Such chemical components might be preserved in the muds of floodplain environments or lakes. One would expect relatively little lithification and change of the sediment's gross aspect. Early diagenesis of the turbidites would be as expected for those associated with other tectonic settings, though high supply of detritus would cause that fraction to overwhelm pelagic components and make the preservation of carbonate and organic matter less likely.

The debris of continental collision is buried rapidly, whether in intermontane basins, along compound alluvial plains in front of mountain fronts such as the Ganges-Brahmaputra, or in deltas and submarine fans. All of these environments have in common a low heat flow, although there is much variation. Assuming continental crust—which may be thickened in these regions—we would expect geothermal gradients to be as low as 20–25°C/km. While not leading to the high-pressure-low-temperature metamorphic facies of the blueschists of subduction zones, diagenetic processes driven by rapid rises of temperature associated with burial are retarded. Those processes that are primarily the result of pressure, such as pressure solution, might be relatively accelerated compared to other changes. Pressure solution is a phenomenon that is largely restricted to sandstones with relatively low clay or argillaceous rock-fragment content. But most of the sandstones of the continental collision regime tend to have sufficient clay material to prevent large-scale pressure solution. Clay minerals might alter under thermal regimes similar to those of the zeolite facies, but the lack of any significant amount of volcaniclastics would limit the raw material that could be easily transformed into smectite and zeolite. So these sediments would not show great changes even though deeply buried for relatively short times.

Thus formation waters, even the deeper ones, would not tend to have the composition of solutions equilibrated with extensively altered rocks. In uppermost parts of the section they would have the composition of slightly altered meteoric waters, only briny if connected with saline lake deposits. Reactions of immature detritus with formation waters in sediments deeply buried over 10 m.y. or more would gradually build up the dissolved solids content of formation waters.

Sedimentary basins associated with continental collision fill extraordinarily rapidly: some reach a thickness of 10 km in as little as a few million years, and many in less than 30 m.y. As noted above, the shortness of time limits the extent of diagenetic reactions which at best have slow kinetics.

A most important characteristic of the products of collisional orogenies is the reversibility of uplift and oscillatory epeirogenic movements. The very nature of continent-continent collision, taking place as it does between two continents of generally uneven geometry, means that the first collision leaves large areas still separated. The open seas between are gradually encompassed by the collision as the two plates continue to converge, grinding small continental fragments against each other. Thus a complex history of multiple continuing orogenies is produced. An intermontane basin might be rapidly filled and then uplifted in later stages, causing the sedimentary fill to be largely eroded. We know from the Himalayas—and the prevalence of angular unconformities in many orogenic belts believed to be of this origin—that reversals of subsidence are common. As a result these sediments are particularly prone to move repeatedly into the zone of weathering by meteoric waters. We should expect kaolinitized clays to be common and carbonate cements to gradually disappear in favor of iron oxide or silica.

An unexpected dividend of the general lack of extensive diagenesis in orogenic belts of this type is the relative ease with which sedimentary petrologists have been able to reconstruct source areas and rock types. After many years of subsidence and uplift, the lithic sandstones, polymict gravels, and heterogeneous silts and shales are still good reflectors of their origin. Of course, the longer they stay deeply buried as geothermal gradients relax to normal continental levels, the more likely that they will be generally affected by burial diagenesis. The sediments of Precambrian and Paleozoic basins are likely to reflect this long history.

Rift valley.—As used here, the rift valley setting includes regions such as the classic rift valleys of East Africa and ancient examples such as Triassic basins of eastern North America. Because I do not discuss separately the sediments of transform faults on land, such as the San Andreas of California, I include them here. The hallmark of the rift valley is the presence of the fault basin and the association with volcaniclastics (though this may be less true of transform belts). The rift valleys generate a variety of lacustrine deposits, alluvial plains, and fault-line alluvial fans. The source materials may be a variety of pre-existing sedimentary, metamorphic, and igneous—especially volcanic—rocks, depending on the particular site of rifting. From this mixture of rock types, subject to rapid erosion by virtue of the topography, comes the gravelly detritus of the fans, the finer sands and silts of the alluvial plains, and the silts and muds of the lakes. Depending on the climate, which is itself partly a function of topography, continental position, and paleolatitude, the lakes may produce evaporites, zeolites, alkali silicates, and borates. In higher latitudes (or altitudes) with cooler and more humid climates, fresh-water lakes may produce carbonate or diatom-rich sediment in addition to terrigenous detritus.

The early diagenetic alteration of these kinds of sediments may range from extreme, in the case of the alkaline lake sediments, to mild in the case of the alluvial fan or plain deposits, particularly those containing little volcaniclastics. The combination of evaporation and the hydrolysis of volcanic-glass and mafic minerals lead to a sedimentary assemblage dominated by alkali carbonates, borates, silicates, and organic matter. Few of the minerals found in these deposits are unaltered original precipitates, even though they may form almost immediately after sedimentation. Away from the lakes, meteoric waters may extensively alter the subaerial terrestrial clastics. The arkosic materials from some of these regions continue to weather, losing their more unstable feldspars—those with a high anorthite component. As these feldspars disappear, the Ca^{++} may combine with HCO_3^- to form the common carbonate cement of arkoses. The dominant clay mineral is kaolinite, both detrital and diagenetic. Smectites may also form, abundantly if there is an appreciable volcaniclastic component.

Burial brings the sediment into a high heat-flow regime, perhaps accompanied by hydrothermal solutions. The rate of sedimentary accumulation in rift valleys may be rapid, deep burial to 4–7 km in less than 10 m.y. is not rare. Because of this rapidity of burial, geothermal gradients may not reflect the high heat flow until a long time elapses, during which the sediment gradually is heated. Such conditions force the sediment into a regime where the original constituents and their strong tendency to alter diagenetically cause lithification to proceed relatively rapidly. If the section contains salt or other evaporites and alkaline-lake deposits, large parts of it can become first zeolitized and later feldspathized. Conditions in the middle of thick fanglomerates or other alluvial sequences may not induce extreme change, for once the feldspars have been partially kaolinized, rock alteration slows down. As formation waters slowly become more saline, the sediment tends to approach equilibrium. In these kinds of rocks, hydrothermal activity may form authigenic feldspars (Heald 1956). From these considerations one would expect the Triassic rift-valley sediments of eastern North America to have reached their present aspect relatively early in their history, that is, within 10–20 m.y.

The period of active subsidence in these valleys is of the order of 5–20 m.y., after which they may be caught up in plate separation. Then they become part of the basement of the trailing edge of continental plates, as shown by seismic profiles of the Atlantic continental shelf of North America (Uchupi et al. 1977). As the plates spread farther apart, the original rift valleys become more deeply buried under continental border sediment. The rifts that are farthest inboard on the continental edge tend to be covered only by major transgressions and are never buried deeply beyond their original thickness. As a result of minor general uplift during stages of continent separation, the former rift valleys become exhumed. Thus their present state is not unlike their condition at the time when rifting became continental separation.

The oscillatory subsidence and uplift, then, affect only the parts of the rift-valley zone farthest from the line of faults that becomes an oceanic spreading center. The parts that are buried deeply under the continental shelves tend to irreversible burial, as discussed in the section on trailing edges. The volcanism associated with this plate-tectonic environment ceases when it changes to the trailing-edge setting. As a special case, the region may ride over a hot spot and become involved in plateau basalt formation.

One would expect from the preceding considerations that once the sediments of rift valleys become basement under continental shelves, they become further lithified and altered as they "soak" at high temperatures and pressures while at the same time subject to invasion by formation waters from overlying marine sediments. We might speculate that when we can sample these formations—now seen only through the eyes of geophysical instruments—they would truly be "hard" basement rocks, but without deformation, high grades of regional metamorphism, or contact metamorphic regions.

Intraplate.—We are faced with the odd paradox that most studies of diagenesis have concerned sediments deposited on cratonic platforms, far from the mobile belts associated with plate boundaries. Perhaps that is why until recently much attention was paid to the details of quartz and carbonate cementation, long-term changes in clay-mineral composition, and formation waters of cratonic basins, but relatively little to zeolitization, quasi-hydrothermal activity, deposits of heavy metals in sediments, and the alteration of evaporites and alkaline-lake sediments.

The chief characteristic of the intraplate environment is quiescence. Subsidence is moderate in the cratonic basins and slight in arch areas between such basins. Detritus is dominantly mature for much of the history of basins; any immature materials are from far-distant orogenic belts, as, for example, the Late Paleozoic of the central North American craton. Quartz arenites grade into mature shales, frequently calcareous, and a wide variety of platform limestones and dolomites deposited in shoreline and reef environments. Marine sedimentation prevails over much of the lifetime of cratonic basins, interrupted by brief periods of regression, erosion, and continental sedimentation.

When the history of cratonic basins includes periods of swamping with detritus from a plate-boundary orogeny some distance away, the quartz arenites give way to lithic arenites and shales of diverse composition, usually including much kaolinite and smectite; carbonates become less abundant. Deltaic coastal-plain swamps may appear and coal beds form when the region is poised near sea level. At times of widespread transgression by deeper seas and limited sediment supply, black organic-rich shales may be deposited.

It might be easier to describe the sediments that are missing from this environment: feldspathic and volcanic graywackes, arkoses, volcaniclastic sediments, and widespread pelagic limestones. Turbidites are absent, for cratonic basins are the environments of normal, low-density currents. Above all, only moderate sediment thicknesses accumulate over long periods of time. Over time periods of more than 100 m.y. cratonic basins may receive up to 3–4 km of sediment—rarely much more, frequently a good deal less.

Early interstitial waters reflect the stable character of the detritus or the typically carbonate biogenic component. Solutions are saturated or slightly supersaturated with respect to calcite, the result of an abundant supply of biogenic magnesian calcites and aragonites. The absence of volcaniclastics, unstable silicates, and pelagic silica-secreters lead to low dissolved silica concentrations, except in local clumpings of sponge spicules. Here replacement chert nodules begin to form diagenetically as local concentration gradients are set up and silica diffuses in and carbonate out. Local carbonate concentration gradients in muddy sediments result in carbonate concretions in shales and sandstones and coal balls in coal beds. Other concretions may be ferroan-dolomitic, sideritic, phosphatic, or pyritic. Pebbles of such concretions in clastic sediments overlying minor unconformities are evidence of early lithification before any significant depth of burial. Frequent transgressions and regressions produce repeated alternations of marine, brackish, and fresh waters migrating into formations.

Because rates of subsidence are low, even in large cratonic basins, sources of detritus are distant, and supply of detritus is limited, sediment burial is slow. Heat-flow levels are the lowest of any on the continent, since these basins are far from large volumes of young igneous rocks containing high proportions of uranium and other radioactive materials (Birch et al. 1968; Roy et al. 1968). Geothermal gradients cluster around 25–30°C/km. Because sedimentation rates are low, even relatively impermeable mudstones are likely to transmit fluids sufficiently to prevent abnormal overpressuring of fluids in permeable formations. Salt concentrations of formation waters increase with depth and age; in many basins these waters reach high ionic strengths. Such solutions are typically poised at equilibrium with respect to quartz and stable clay minerals. The waters are depleted in magnesium and enriched in calcium.

These generalizations must be treated with caution, for most well-known cratonic basins are Paleozoic or early Mesozoic: thus we may be looking more at the effects of long periods of moderate burial than at the rapid effects of deep burial. If we were to have looked at some of these basinal sediments during the late Paleozoic or early Mesozoic, close to the time of their deposition, they might have appeared far different: less lithified and containing far fewer authigenic and diagenetically altered minerals than today. One must also take into account the feather edges of sedimentary sections on the borders of such basins, places where the total thickness of sediment may never have exceeded 1 km. Yet in these regions the mineralogical and textural properties of the rocks seem grossly similar to their more deeply buried contemporaries in the central parts of the basin. Regional comparisons of the diagenetic grade of such sediments will be needed to sort out the effects of time, burial, and changes of formation water.

Over the history of cratonic sediment accumulations, oscillatory vertical movements are the rule rather than the exception. Sloss (1963) was the first to point out the importance of unconformities in this milieu. Repeated local and regional vertical movements are also accompanied by mild episodes of deformation. Active and growing structures are typically reflected in both lithology and thickness of sedimentation patterns. How these movements are related to reversible mineralogical replacements, such as of chert and carbonate described by Walker (1962), remains to be worked out.

CONCLUSIONS

What I have summarized here is based on what we know of the dynamics of sedimentation and diagenesis in modern and ancient basins associated with various tectonic settings. In the modern ones it is safe to infer plate-tectonic regimes, for geophysical information is readily available. In basins of Mesozoic and Cenozoic age we are likewise well grounded, for the plate-tectonic history of the last 200 m.y. can be read in the sediments and magnetic anomalies of the ocean floors. Pre-Mesozoic reconstructions are a different matter, for one must avoid circular arguments in the absence of either geophysical or ocean-floor information. To avoid this circularity, I have used stratigraphic and petrologic correspondences to plate-tectonic environments without assuming diagenetic mineralogy and texture to be in any way deterministic for tectonic inferences.

I have also assumed that plate tectonics operated throughout the Phanerozoic, as evidenced by uniformitarian patterns of sedimentation, volcanism and plutonism, orogeny, and metamorphism. One cannot be so sure of the Precambrian; perhaps only the latest part of that era—the time interval from about 1.2–0.57 b.y.—is similar to the Phanerozoic. Older sequences may reflect different patterns of heat flow, reflecting different convective transfer of heat from the interior and thus different patterns of development of continents, their borderlands, and the oceans. This would have had a great impress not only on patterns of sedimentation but also on diagenetic histories of the sediments. To the extent that any of the generalizations given here are true of plate-tectonic settings, they may not be at all applicable to these earlier times.

Another nonuniformitarian aspect of the association of diagenetic histories with tectonic environments is presented by organic macroevolution. Biogenic contributions to sediment were very different before and after the late Mesozoic, when pelagic foraminifera, diatoms and silicoflagellates, and angiosperms first appeared in Earth's surface environments. The carbonate, silica, and organic contributions to sediment were strongly affected; to the extent that the primary assemblage strongly affects diagenesis, this effect must be considered. Were there few, if any, truly pelagic limestones in the Paleozoic? Did radiolarian cherts form in the same ways as diatomaceous cherts have since the Cretaceous? Looking further back in time, we can ask the almost traditional question, whether the appearance of vascular plants in the Silurian affected the nature of weathering and erosion. Before that came the appearance of metazoans at the end of the Precambrian.

Whatever the answers to these and other subsidiary questions, the answer to the main question has been indicated. Tectonics plays a prime role in determining the mineralogy and texture of a sediment, not only in primary assemblages and stratigraphic patterns, but in their diagenetic histories as well. The connection of diagenesis and tectonics has been made before. The definition of tectonic settings by reference to plate tectonics promises to be a better road map for future studies of regional aspects of diagenesis than anything we have had heretofore.

ACKNOWLEDGMENTS.—Much of the background research relating to this work was supported by National Science Foundation Grants EAR76-03803 and EAR76-82862 and the Committee on Experimental Geology and Geophysics of Harvard University. Much of it was written at Imperial College, London, and I am indebted to my colleagues there, in particular Douglas Shearman, for many stimulating conversations. Jutta Binstock critically read the manuscript. Perhaps most of all, I am indebted to the many students who have worked on important parts of the diagenetic story with me over the past twenty years.

REFERENCES CITED

BERNER, R. A.; BERNER, E. K.; and KEIR, R. S., 1976, Aragonite Dissolution on the Bermuda pedestal: It's Depth and Geochemical Significance: Earth and Planetary Sci. Letters, v. 30, p. 160–178.

BIRCH, F.; ROY, R. F.; and DECKER, E. R., 1968, Heat flow and thermal history in New England and New York: Studies of Appalachian Geology: Northern and Maritimes, J. B. Wiley and Sons, p. 437–451.

BOLES, J. R., and COOMBS, D. S., 1977, Zeolite facies alteration of sandstones in the Southland Syncline, New Zealand: Amer. Jour. Sci., v. 277, p. 982–1012.

BOSTROM, K., and PETERSON, M. N. A., 1966, Precipitates from hydrothermal exhalations on the East Pacific Rise: Econ. Geol., v. 61, p. 1258–1265.

———; FARQUHARSON, B.; and EYL, W., 1971, Submarine hot springs as a source of active ridge sediments: Chem. Geol., v. 10, p. 189–203.

BOWEN, V. T.; CIFELLI, R.; and SIEVER, R., 1966, Cemented foraminiferal oozes from the Mid-Atlantic Ridge: Nature, v. 209, p. 32–34.

BRICKER, O., (ed.), 1971, Carbonate Cements: Johns Hopkins Studies in Geology No. 19, Johns Hopkins Press, Baltimore.

BURST, J. F., 1969, Diagenesis of Gulf Coast clayey sediments and its possible relation to petroleum migration: Bull. Amer. Assoc. Petroleum Geol., v. 53, p. 73–93.

BUSENBERG, E., and CLEMENCY, C., 1976, The dissolution kinetics of feldspars at 25°C and 1 atm CO_2 partial pressure: Geochim. Cosmochim. Acta, v. 31, p. 41–49.

CALVERT, S. E., 1974, Deposition and diagenesis of silica in marine sediments, *in* HSÜ, K. J., and JENKYNS, H. C., eds., Pelagic Sediments: On Land and under the Sea: Int. Assoc. of Sediment, Spec. Pub., No. 1: p. 273–299.

CHASE, C. G.; HERON, E. M.; and NORMARK, W. R., 1975, Plate Tectonics: commotion in the ocean and continental consequences: Annual Rev. Earth and Planetary Sci., v. 3, p. 271–292.

COOMBS, D. S.; ELLIS, A. J.; FYFE, W. S.; and TAYLOR, A. M., 1959, The zeolite facies: Geochim. Cosmochim Acta, v. 17, p. 53–107.

DAPPLES, E. C.; KRUMBEIN, W. C.; and SLOSS, L. L., 1948, Tectonic control of lithologic associations: Bull. Amer. Assn. Petroleum Geol., v. 32, p. 1924–1947.

DEGENS, E., and ROSS, D., 1969, Hot brines and recent heavy metal deposits in the Red Sea: Springer-Verlag, N.Y., Inc. 600 p.

DEWEY, J. F., and BIRD, J. M., 1970, Plate tectonics and geosynclines: Tectonophysics, v. 10, p. 625–638.

DREVER, J. I., 1971, Chemical and Mineralogical Studies, Site 66, *in* WINTERER, E. L. et al., Initial Reports of the Deep Sea Drilling Program, v. VII, Washington (U.S. Government Printing Office), p. 965–976.

DUNOYER DE SEGONZAC, G., and HOFFERT, M., 1973, A preliminary investigation of the clay minerals in the western Alboran basin, Site 121, *in* RYAN, W. B. F., HSÜ, K. J. et al., Initial Reports of the Deep Sea Drilling Program, v. XIII, Washington (U.S. Government Printing Office), p. 670–672.

EDMOND, J. M., 1978, Chemistry of the hot springs on the Galapagos Ridge axis, *in* 2nd Ewing Mem. Symp. "Implications of Deep Drilling Results in the Atlantic Ocean," p. 13.

ERNST, W. G., 1970, Tectonic contact between Franciscan mélange and the Great Valley sequence-crustal expression of a late Mesozoic Benioff zone: Jour. Geophys. Res., v. 75, p. 886–901.

GARRELS, R. M., and CHRIST, C. L., 1965, Solutions, minerals, and equilibria: New York, Harper & Row, xii, 450 p.

GARRELS, R. M., and MACKENZIE, F. T., 1971, Evolution of sedimentary rocks: New York, W. W. Norton, 397 p.

GIESKES, J. M., 1975, Chemistry of interstitial waters of marine sediments: Annual Rev. Earth and Planetary Sci., v. 3, p. 433–454.

HATHAWAY, J. C., and SACHS, P. L., 1965, Sepiolite

and clinoptilolite from the Mid-Atlantic Ridge: Amer. Mineral., v. 50, p. 852–867.

HEALD, M. T., 1956, Cementation of Triassic arkoses in Connecticut and Massachusetts: Bull. Geol. Soc. Amer., v. 67, p. 1133–1154.

HEATH, G. R., 1974, Dissolved silica and deep-sea sediments, *in* HAY, W. W., ed., Studies in Paleoceanography: Soc. Econ. Paleo. Min. Spec. Pub. 20, p. 77–93.

HEPBURN, J. REHMER, 1976, Petrology of the Esopus Shale (Lower Devonian), New York and adjacent states, Ph.D. thesis, Harvard Univ., 288 p.

HOLLAND, H. D., 1972, The history of sea water: Geochim. Cosmochim. Acta, v. 36, p. 637–651.

HOWER, J.; ESLINGER, E.; HOWER, M. E.; and PERRY, E. A., 1976, Mechanism of burial metamorphism of argillaceous sediments: 1. Mineralogical and chemical evidence: Bull. Geol. Soc. Amer., v. 87, p. 725–737.

HUMPHRIS, S. E., and THOMPSON, G., 1978, Hydrothermal alteration of oceanic basalts: Geochim. Cosmochim. Acta, v. 42, p. 107–125.

HURD, D. C., 1972, Factors affecting solution rate of biogenic opal in seawater: Earth and Planetary Sci. Letters, v. 15, p. 411–417.

JEFFREYS, H. J., 1976, The Earth, 6th Edition: Cambridge, Cambridge Univ. Press, 574 p.

KASTNER, M.; KEENE, J. B.; and GIESKES, J. M., 1977, Diagenesis of siliceous oozes—I. Chemical controls on the rate of opal-A to opal-CT transformation—an experimental study: Geochim. Cosmochim. Acta, v. 41, p. 1041–1059.

———, and SIEVER, R., 1978, Low temperature feldspars in sedimentary rocks: Amer. Jour. Sci. (in press).

———, and STONECIPHER, S. A., 1978, Zeolites in pelagic sediments of the Atlantic, Pacific, and Indian Oceans, *in* SAND, L. B., and MUMPTON, F. A., eds., Natural Zeolites, Occurrences, Properties, Use: Pergamon Press (in press).

KAY, M., 1951, North American Geosynclines: Geol. Soc. Amer. Memoir 48, 143 p.

KRYNINE, P. D., 1948, The megascopic study and field classification of sedimentary rocks: Jour. Geology, v. 56, p. 130–165.

LEPICHON, S.; FRANCHETEAU, J.; and BONNIN, J., 1973, Plate Tectonics: Amsterdam, Elsevier, 300 p.

MANHEIM, F. T., and SAYLES, F. L., 1974, Interstitial waters of deep-sea sediments, *in* GOLDBERG, E. D., ed., The Sea, v. 5, New York, Wiley-Interscience, p. 527–568.

MELSON, W.; BOWEN, V. T.; VAN ANDEL, TJ. H.; and SIEVER, R., 1966, Greenstones from the central valley of the Mid-Atlantic Ridge: Nature, v. 209, p. 604–605.

MOTTL, M. J., 1976, Chemical exchange between sea water and basalt during hydrothermal alteration of the ocean crust: Ph.D. thesis, Harvard University, 188 p.

PARSONS, B., and SCLATER, J. G., 1977, An analysis of the variation of ocean floor heat flow and bathymetry with age: Jour. Geophys. Res., v. 82, p. 803–827.

PETTIJOHN, F. J., 1949, Sedimentary Rocks (1st Ed.): Harper, New York.

———; POTTER, P. E.; and SIEVER, R., 1972, Sand and Sandstone: Springer-Verlag, Heidelberg, 618 p.

RICCI-LUCCHI, F., 1976, Depositional cycles in two turbidite formations of the northern Apennines (Italy): Jour. Sed. Petrology, v. 45, p. 3–43.

ROY, R. F.; DECKER, E. R.; BLACKWELL, D. D.; and BIRCH, F., 1968, Heat flow in the United States: Jour. Geophys. Res., v. 73, p. 5207–5221.

SIEVER, R.; BECK, K. C.; and BERNER, R. A., 1965, Composition of interstitial waters of modern sediments: Jour. Geology, v. 73, p. 39–73.

———, and KASTNER, M., 1967, Mineralogy and petrology of some Mid-Atlantic Ridge sediments: Jour. Marine Res., v. 25, p. 263–278.

——— 1968, Sedimentological consequences of a steady-state ocean-atmosphere: Sedimentology, v. 11, p. 5–29.

——— 1972, Silicon: Handbook of Geochemistry, WEDEPOHL, K. H. ed.: Heidelberg, Springer-Verlag, Sections 14—D.G.L.

———, and WOODFORD, N., 1973, Sorption of silica by clay minerals: Geochim. Cosmochim. Acta, v. 37, p. 1851–1880.

——— 1976, Rates of dissolution of mafic minerals: Ferrous silicate weathering kinetics: dissolution vs. oxidation (abs.), p. 1106, Abstracts with Programs, v. 8, no. 6, Nov. 8–11, 1976. 1976 Annual Meetings, GSA, Denver, Colorado.

———, and STEIN, C. L., 1978, Models for chert formation: Ms. in prep.

SLOSS, L. L., 1963, Sequences in the cratonic interior of North America: Bull. Geol. Soc. Amer., v. 74, p. 93–114.

STEFANSSON, U., 1966, Influence of the Surtsey eruption on the nutrient content of the surrounding seawater: Jour. Marine Res., v. 24, p. 241–268.

STEIN, C., 1977, Dissolution of diatoms and diagenesis in siliceous sediments: Ph.D. Thesis, Harvard University, 150 p.

STILLE, H., 1936, Present tectonic state of the Earth: Bull. Amer. Assn. Petroleum Geol., v. 20, p. 848–880.

UCHUPI, E.; BALLARD, R. D.; and ELLIS, J. P., 1977, Continental slope and upper rise of western Nova Scotia and Georges Bank: Bull. Amer. Assoc. Petroleum Geol., v. 61, p. 1483–1492.

VAN ANDEL, TJ. H.; THIEDE, J.; SCLATER, J. G.; and HAY, W. W., Depositional history of the South Atlantic Ocean during the last 125 million years: Jour. Geology, v. 85, p. 651–698.

VON BUBNOFF, S., 1954, Fundamentals of Geology, 3rd ed., translated by W. T. Harry, 1963: Edinburgh, Oliver and Lloyd, Ltd., 287 p.

VON ENGELHARDT, W., 1977, The origin of sediments and sedimentary rocks: Wiley, New York, 359 p.

WALKER, T. R., 1962, Reversible nature of chert-carbonate replacement in sedimentary rocks: Bull. Geol. Soc., Amer., v. 73, p. 237–242.

WATSON, J., 1978, Ore deposition through geological time: Proc. Royal Soc. (in press).

WEAVER, C. E., 1958, The effects and geological significance of potassium "fixation" by expandable clay minerals derived from muscovite, chlorite, and volcanic material: Amer. Mineralogists, v. 43, p. 839–861.

WEAVER, F. M., and WISE, S. W., 1972, Ultramorphology of deep-sea cristobalitic chert: Nature Phys. Sci., v. 237, p. 56–57.

WEYL, P. K., 1959, Pressure solution and the force of crystallization—a phenomenological theory: Jour. Geophys. Res., v. 64, p. 2001–2025.

WOLLAST, R., 1967, Kinetics of the alteration of K-feldspar in buffered solutions at low temperature: Geochim. Cosmochim. Acta, v. 31, p. 635–648.